Irrigation and Drainage Engineering

Peter Waller · Muluneh Yitayew

Irrigation and Drainage Engineering

Springer

Peter Waller
Agricultural and Biosystems Engineering
University of Arizona
Tucson, Arizona, USA

Muluneh Yitayew
Agricultural and Biosystems Engineering
University of Arizona
Tucson, Arizona, USA

Additional material to this book can be downloaded from http://extras.springer.com

ISBN 978-3-319-05698-2   ISBN 978-3-319-05699-9 (eBook)
DOI 10.1007/978-3-319-05699-9

Library of Congress Control Number: 2015955638

Springer Cham Heidelberg New York Dordrecht London

Printed on acid-free paper

Springer International Publishing AG Switzerland is part of Springer Science+Business Media (www.springer.com)

Preface

This book was written for three semester-long classes: Soil and Water Resources Engineering, Irrigation Systems Design, and Agricultural Drainage and Effluent Treatment. Almost all the chapters are accompanied with Excel/VBA programs that enable students to understand as well as apply the principles as practicing engineers. Homework problems focus on both the Excel/VBA programs and manual calculations.

The first section presents the fundamental principles of irrigation and drainage engineering: economics, soils, evapotranspiration, lateral design, mainline design, pumps, groundwater and wells, and open channel flow. We started with economics because that is basis for analysis of any agricultural enterprise. Soils and evapotranspiration are the fundamental design variables in irrigation system design. The four fundamental components of irrigation systems are covered in the last five chapters.

The second section describes the design process for the following irrigation systems: center pivot, turf sprinkler, wheel-line sprinkler, orchard sprinkler, landscape drip and bubbler, agricultural subsurface drip, surface, greenhouse hydroponic, and low-head gravity bubbler. The Excel/VBA programs associated with this section enable students to optimize systems with respect to crop growth, water savings, energy use economics, and environmental impact. The programs help students bridge the gap between basic principles and complex designs.

The last part of this book focuses on the use of wetlands and agriculture for waste treatment, subsurface drainage, and the *WINDS* model (Water-use, Irrigation, Nitrogen, Drainage, and Salinity). The theoretical development in these chapters is focused on the algorithms in the WINDS model, which models evapotranspiration and simulates salinity, nitrogen, and water transport in soil layers. Water and solutes are routed with the tipping bucket approach after large storms and the Richards equation (energy differences) between infiltration events. The nitrogen model simulates fertilization, mineralization, denitrification, and plant uptake. The drainage algorithm uses the Kirkham streamtube approach as well as the interaction between the water table and the vadose zone. Traditional drainage algorithms used by the USBR and DRAINMOD are also presented with an emphasis on the economic evaluation of alternative drainage systems designs.

Agricultural food production and water demand are expected to be critical to the survival of a growing global population. Our hope is that this book will help students and practicing engineers develop food production systems with limited water resources and limit the environmental impacts of those systems. In this way, we can develop sustainable food production systems to meet the needs of the growing population. Agricultural engineers with problem solving and water management expertise will play an important role in the coming decades.

Tucson, AZ, USA Peter Waller
Tucson, AZ, USA Muluneh Yitayew

Acknowledgements

We would like to acknowledge the following individuals and organizations for their direct or indirect contributions through their knowledge, research, and development of materials used in this book: Donald Slack, Doug Hunsaker, Armando Barreto Munoz, Armando Uribe Chavez, Eduardo Bautista, Julio Haberland, Curt Reynolds, Stephen Poe, Paul Colaizzi, Alan Mathias, Robert Evans, W-128 Western Regional Research Committee on Microirrigation, Natural Resources Conservation Service, Agricultural Research Service.

We will like to also thank our students who took our courses throughout the years for their feedback and the Agricultural and Biosystems Engineering Department at the University of Arizona for encouraging us to develop and finish this book.

Contents

1 Introduction

Irrigation engineering is the analysis and design of systems that optimally supply the right amount of water to the soil at the right time to meet the needs of the plant system. The enterprise may be growing plants for food, landscape irrigation, or other purposes. Subsurface drainage engineering is the analysis and design of systems that remove water or salts from the soil in order to maintain as close to an optimal plant growth environment as possible.

System design and selection involves three steps. The first step is to characterize crop water requirements, hydrology, and soil characteristics. The second step is the hydraulic design of potential alternatives. The third step is an economic and environmental analysis. To adequately evaluate alternatives, it is important to look at all relevant inputs to the system (water, energy, labor) and at the effect of the system on the environment. Two consistent themes in this book are engineering economic analysis and assessment of environmental impact. Engineering economic analysis sums capital and annual costs in order to calculate the total present value cost of the system. Environmental assessment includes the estimation of such effects as chemical leaching, chemical runoff, soil erosion, and water resource depletion.

Design of irrigation and subsurface drainage systems involves the application of engineering, biology, and soil science in both synthesis and analysis methods of problem solving to assemble components that will fit together for a specific location and crop production system. Thus, aside from the knowledge of hydraulics and other engineering concepts, irrigation and drainage engineering require an understanding of the soil and water environment, soil-water-plant relationships, and hydrology. The feasibility of irrigation and subsurface drainage systems must also be examined from the social point of view. This is particularly true in developing countries where irrigation and drainage projects are undertaken to change the livelihood of the local communities and the region at large.

This book covers irrigation and subsurface drainage fundamentals, systems, and impacts. Chapters 2, 3, 4, 5, 6, 7, 8, 9, 10 and 11 review basic principles and system components. Chapters 12, 13, 14, 15, 16, 17, 18, 19, 20, 21 and 22 focus on design of sprinkler, drip, surface, and bubbler irrigation systems. Chapters 23, 24, 25, 26, 27, 28, 29, 30 and 31 focus on wastewater, soils, nutrients, and subsurface drainage systems (Chaps. 23, 24, 25, 26, 27, 28, 29, 30 and 31). One of the strengths of this book is that almost all of the chapters have associated VBA/Excel programs. The programs enable students to evaluate realistic scenarios in homework problems and to more readily translate the knowledge and techniques in this book to the real world.

History of Irrigation and Subsurface Drainage

Irrigation and civilization have gone hand in hand. Beginning 8000 years ago, the Ubaid and then the Sumerians in Ancient Mesopotamia invented civilization and intensive irrigated agriculture at the same time. Regional canals delivered water from the Tigris and Euphrates rivers to large-scale irrigation systems and farms. Cities had between 50,000 and 80,000 people, with almost 90 % of the people living in the cities. Almost half of the population worked on the large irrigated farms (Fig. 1.1) surrounding each city. With modern agricultural methods, only 1 % of the population can now feed the rest of the population.

The importance of regional canals in Mesopotamia is evident in a statement by the famous Babylonian king Hammurabi (1750 BC):

> When Anu and Bel (gods) gave me the land of Sumer and Akkad to rule,... I dug out the Hammurabi-canal named Nuhus-nisi. Both the banks thereof I changed to fields for cultivation, and I garnered piles of grain, and I procured unfailing water for the land. As for the land of Sumer and Akkad, I collected the scattered peoples thereof, and I procured food and drink for them. In abundance and plenty I pastured them, and I caused them to dwell in peaceful habitation.

P. Waller, M. Yitayew, *Irrigation and Drainage Engineering*, DOI 10.1007/978-3-319-05699-9_1

Fig. 1.1 Sumerian city with canal and surrounding farms (Credit Mariah Dunn, University of Arizona)

Not only were the canal systems remarkable, but the management of irrigation at the farm scale was also technologically advanced, as described in the following selected excerpts from the *Sumerian Farmer's Almanac* (1700 BC, translated from cuneiform by Samuel Kramer):

> In days of yore a farmer instructed his son: When you are about to take hold of your field (for cultivation), keep a sharp eye on the opening of the dikes, ditches, and mounds, (so that), when you flood the field the water will not rise too high in it. When you have emptied it of water, watch the field's water-soaked ground that it stay virile ground for you... plow diagonal furrows where you have plowed straight furrows, (And) Plow straight furrows where you have plowed diagonal furrows. Let your straight furrows make your borders into *tulu*-borders; let the *lu*-furrows make straight your borders. Plow *ab*-furrows where.... (Then) Let all its clods be removed; all its high spots be made into furrows; (and) all its depressions be made into low furrows... When the barley has filled the narrow bottom of the furrow, water the top seed. When the barley stands up high as (the straw of) a mat in the middle of a boat, water it (a second time). Water (a third time)-its royal barley. If the watered barley has turned red, what you say is: "It is sick with the *samana*-disease." But if it has succeeded in producing kernel-rich barley, water it (a fourth time); (and) It will yield you an extra measure of barley in every ten (+10 %)...

We can see the importance that the Sumerians placed on straight furrows, proper maintenance of dikes and ditches, leveling, borders and basins, crop monitoring, and irrigation scheduling. Nothing has changed. Irrigation engineers and managers must still pay careful attention to the design and maintenance of their irrigation systems. Farmers must manage water effectively. As stated in the *Almanac*, yield might be increased by 10 % with optimal irrigation management.

The Sumerian culture vanished from Sumer in 2350 BC. This took place at the same time that the soil became salinized. The Sumerians returned to Sumer in 2055 (short

Fig. 1.2 Diversion structure (Dujiang Weir) for Dujiangyan irrigation system. "Dujiang Weir." (Credit: Huowax at zh.wikipedia)

chronology) and formed the Third Dynasty of Ur for approximately 100 years. Evidence of the salinization is found in cuneiform agricultural records, which report that grain yields in 2400 BC were 2347 L/ha, but production dropped to 1460 L/ha in 2100 BC. Yields then dropped even further to 900 L/ha in 1700 BC. Another indication of salinization is that 80 % of the grain percentage was barley in 2400 BC, but it rose to 99 % barley in 2050 BC.[1] Barley is a salt tolerant crop and is often the first crop used to reclaim saline desert soils for agriculture. A higher percentage of barley in the crop mix indicates soil salinization.

The Egyptians learned about intensive agriculture from the Sumerians. Their irrigation was dependent on the annual flooding of the Nile, which had a different and more predictable annual flood pattern than the Tigris and Euphrates. The water in the Nile covered the Nile floodplain with 1.5 m of water in September and remained for approximately 1 month. Once the fields were dry in November, the farmers planted a winter wheat crop and then harvested the crop in April and May. If necessary, they built basins to impound the water until fields were fully saturated and then allowed the water to drain away. Because the water table was 3–4 m below the ground surface just before the floods, the floodwaters pushed the salts out of the root zone each year.

The Indus River annual flow is three times larger than the Nile River and 10 times larger than the Colorado River. The Indus Valley Civilization (3300–1300 BC) was the largest in the ancient world, with up to five million people. The civilization had small villages with irrigated farms, which produced barley, wheat and other crops. They built extensive irrigation canals and other hydraulic systems such as sewage networks. Much of the civilization collapsed in 1800 BC, possibly due to drought caused by a decrease in monsoon activity in the region. The current Indus River Irrigation System irrigates 16.2 million hectares, which makes it the largest irrigation system in the world.

The Yellow River in China is another cradle of civilization and water management. Hydraulic engineer Zheng-Guo developed a successful irrigation canal network that was fed by a tributary of the Yellow River. The irrigation system enabled the development of 200,000 irrigated acres in Qin province in 246 BC. The system had a main canal and many lateral distribution canals. At the same time, the Qin Dynasty developed the Dujiangyan irrigation system, which is still in use and currently irrigates 5,300 square kilometers. The diversion structure for the Dujiangyan is shown in Fig. 1.2. The plentiful food supplies from irrigated farms and resultant population growth enabled the Qin Dynasty to unify China under its leadership.

Although not well-known outside the Middle East, if you asked a Middle Easterner about major water resource projects in history, then they would immediately think of the Wadi Marib Dam in South Yemen, which is at the southern end of the Arabian Peninsula. It was a triangular structure constructed of packed earth. It was originally 4 m tall and 580 m long in 1700 BC. The height was increased to 7 m and rock was placed on the upstream side in 500 BC, and a sluiceway was constructed for overflow on the side. The Sabean civilization (Kingdom of Sheba) thrived there for over a millennium. The region was taken over by the Himyarites in 115 BC, and they increased the height of the

[1] Altaweel and Watanabe (2012).

Fig. 1.3 Artist's rendition of Hohokam irrigation system (Credit USBR. Artist Peter Hurd)

dam to 14 m and added five spillway channels, two sluices, and a 2 mile channel to a holding tank by 325 AD. At this time, the dam irrigated 100 km^2 in the Marib Plain. The dam was breached and repaired several times after 500 AD, but it was left unrepaired in 570 AD. As a result, 50,000 people were forced to migrate from the region. A new triangular dam has now been reconstructed at a different location in the Wadi Marib.

Native Americans constructed irrigation systems in the southwest United States. One of the most famous was the Hohokam (Fig. 1.3) irrigation network in the Phoenix, Arizona area. The Hohokam civilization lasted hundreds of years, but it was destroyed by an extended drought in thirteenth century AD.

The prehistoric Owens Valley Paiute in California had the most extensive Native American irrigation system. They had a main canal that was several miles long and many small branch canals. Each year they built a new diversion dam in the Owens River with boulders, sticks, and mud in order to divert water into the canal. The Owens River became the focus of one of the most famous battles in water history as Director Mulholland of the Metropolitan Water District in Los Angeles bought out the water rights of Owens Valley farmers (not the original Native American farmers) and ended agriculture in the Owens Valley. This battle between Los Angeles and farmers is described in the book *Water and Power*.

In 1933, a citrus grower in Los Angeles, California named Orton Engelhart invented and patented the impact sprinkler: The family that formed Rain Bird purchased the patent, and the rest is history. Drip irrigation was invented in Israel in the 1950s and 1960s. After these major developments in pressurized irrigation, irrigation companies have continued to refine these basic irrigation technologies in order to reduce cost and improve reliability, uniformity, and efficiency.

Irrigation expanded dramatically in the last century. As the world's population increased, improved food production methods were needed to prevent mass starvation. Along with improved plant varieties and fertilization, irrigation was one of these three reasons for a dramatic increase in world food production. Seventy percent of the world's food is now produced by irrigated agriculture. Many of the rivers in the arid and semiarid western United States (Fig. 1.4) were dammed and used for irrigation. Another example of the impact of irrigation was the Green Revolution in India in the early 1970s. Prior to the Green Revolution, large-scale famines had followed droughts in India. The introduction of dams and irrigation made double-cropping possible. Fed by new dams and irrigation districts, irrigated farms were able to supply the country with adequate food. A recent survey by the International Committee on Irrigation and Drainage found that there are almost 300 million ha of irrigated acreage in the world.

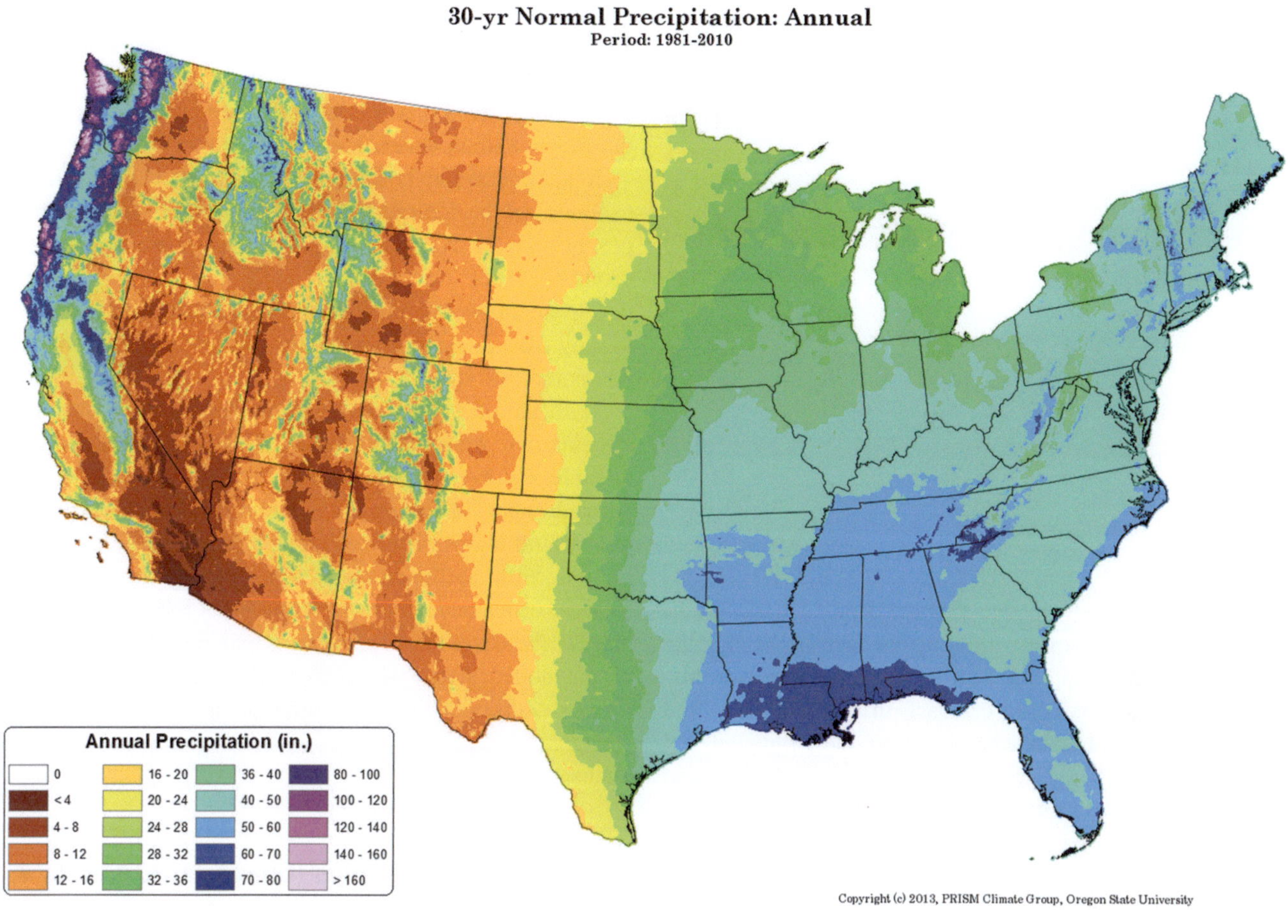

Fig. 1.4 Rainfall patterns in the United States (Credit http://www.prism.oregonstate.edu/normals/ *Work was performed under NRCS contract*)

On the other side, subsurface drainage has been used extensively to drain wetlands and build farms. In the past, "swamps" or "badlands" were considered useless. For example, the goal of the US government was to drain the swamps and convert the land to farms. The government drained wetlands, built levees along rivers, and encouraged the drainage of "swamps" with the Swamp Acts of 1849, 1850, and 1860: swamps were given to any states or private individuals who were willing to drain them. In the 1970s, scientists realized that the loss of wetlands had drastic consequences for the environment. Now, the US government has reversed course. It changed its policy from draining every swamp to protecting even the smallest puddle on a farm. The government is backpedaling as fast as it can in order to preserve, protect, rebuild, and replace wetlands, which are the base of the food web. Wetlands also contribute to the global carbon, nitrogen, water, and sulfur cycles. Wetlands are the breeding ground for 80 % of America's birds, and they are important rest stops for migratory birds. Two-thirds of all birds in the United States reproduce in prairie pothole wetlands. Almost all commercially harvested fish and shellfish are in some way dependent on wetlands. Wetlands also prevent floods by storing flood water during storms. They remove pollutants and sediments from water before they enter rivers. Because of their ecological benefit, areas such as the prairie pothole wetland shown in Fig. 1.5 are now protected and cannot be drained.

It appears that global warming, with its associated droughts and floods, is throwing off previous hydrologic models and design strategies. Rainfed agriculture in some regions is increasingly subject to water stress. Irrigation water allotments, which depended on calculated river flows or groundwater recharge rates may need to change. For example, the Colorado River irrigates 3.5 million acres in the southwest US. These farms produce enough food for 25 million people. The Colorado River was already over allocated because hydrologic studies that were the basis of the legal division of water between western states and Mexico were conducted during a period of higher than normal precipitation. Although some states have yet to require their full allocation of Colorado River water, this period of underutilization by some states is ending with

Fig. 1.5 Prairie pothole wetland on Montana farm (Credit EPA, photo by Dr. Richard Hauer)

population growth and increased water demand in the Southwest. The current severe droughts in California and the Colorado River basin have depleted reservoir levels to historic lows. Water is being cut back in many regions, and farms water allocations are being cut off. Only time will tell whether these droughts are a sign of climate change and whether permanent changes are needed.

One of the most important concepts in water resources development is sustainability. The basic premise of sustainability is that natural resources should not be depleted over time but should continue to provide the same benefit in perpetuity. Sustainability has been defined by law in many states as a 100-year supply of water. With respect to groundwater resources, the rate of extraction from an aquifer should not exceed the rate of aquifer recharge.

If not regulated, water resources naturally become stretched beyond their capability to supply demand. As people begin to develop irrigated farms in a region, well drilling is generally not regulated; thus, the rate of pumping soon exceeds the rate of recharge in a successful farming region. As the groundwater table recedes to hundreds of feet below the ground surface, the cost of pumping increases, and the aquifer yield decreases until farmers eventually start going out of business. As the situation becomes dire, one possible option is that the government might build a canal and deliver water to the region. Another strategy is to take control of the groundwater and not allow excessive pumping. This strategy was employed in Arizona. In 1980, government, agriculture, municipal, mining, and business leaders joined together and developed the Groundwater Management Act, which limited water use by all the entities in the state. Under this plan, no construction of new farms in affected regions is allowed, and water conservation measures are rewarded. Another example of this type of aquifer degradation is the giant Ogallala aquifer, which covers eight states in the High Plains region of the United States. This southern part of the aquifer has been over-pumped for irrigation; and the water table is now so far below the surface that it is no longer economical for many farmers to extract water.

In some cases, over-pumping of aquifers is planned. For example, in the Central Valley of California, planners purposely allowed over-pumping in order to develop the local economy. They calculated correctly that the region, after development, would then have adequate monetary resources to pay for a canal system to deliver water from the Sacramento River.

With the onset of global warming, even water resource systems that were considered sustainable may not be sustainable. Climate models indicate that global warming will cause droughts in some parts of the world. Regions with rain fed agriculture may need to construct irrigation systems, and regions with inefficient irrigation systems may need to increase efficiency. Drainage systems that formerly removed most of the water from spring rains may need to be modified or abandoned. Areas with increased precipitation may need to add additional drainage infrastructure. These types of changes might provide many jobs for irrigation and drainage engineers.

Foundational Principles (Chaps. 2, 3, 4, 5, 6, 7, 8, 9, 10 and 11)

As with the Sumerians, today's engineers and water managers must pay careful attention to their soils, crops, and irrigation technologies in order to maximize production. The next ten chapters cover the fundamental principles and skills required by irrigation and drainage engineers. Knowledge of soils, crop water demand, salinity and water stress, irrigation design principles, pipes, pumps, groundwater, and channels should all be a part of the irrigation and drainage engineer's toolbox.

Decisions are ultimately based on economics. What type of irrigation system and management style will result in the greatest profit for the farmer? Chapter 2 provides a general introduction to irrigation economics, which enables the evaluation of irrigation alternatives based on capital cost and annual costs and benefits. Topics include crop water production functions, which quantify the relationship between depth of water applied and crop yield, and equations for calculating water, energy, and environmental costs.

The next four chapters describe soil physics, crop stress, and crop water demand. Chapter 3 reviews soil water holding capacity, infiltration rate, root zone depth, soil moisture measurement devices, and the relationship between water content and matric potential. Chapter 4 shows how to calculate yield reduction in response to salinity and water stress.

Chapter 5 describes the theory and use of the Penman-Montieth evapotranspiration model. It also includes a program that makes these calculations and automatically downloads weather information from weather stations. Chapter 6 shows how to convert reference evapotranspiration to the evapotranspiration rate for any crop. The crop coefficient is the ratio between crop evapotranspiration and reference evapotranspiration. The dual crop coefficient approach from FAO 56 separates soil evaporation and plant transpiration. The heat unit method calculates the crop coefficient as a function of time and temperature since planting rather than just time since planting

The final five chapters of this section cover the nuts and bolts of irrigation: pipes, pumps, groundwater and wells, and channels. Chapter 7 shows how to design a sprinkler irrigation lateral. Hydraulic equations are used to determine the change in pressure along the pipeline, and statistics are used to generate an application rate distribution along the pipeline. Monte Carlo analysis is used to evaluate the effects of pipe diameter, water cost, energy cost, water application distribution, and crop yield on overall profit. Chapter 8 describes pipe friction and minor losses, energy diagrams, pipe pressure rating, and pipe transients (water hammer). Chapter 9 describes types of pumps, affinity laws, use of pump curves, and the intersection point of the pump curve and the system curve. Chapter 10 describes aquifer types, characteristics, parameters, equations, and classifications. Although irrigation engineers are generally not experts in groundwater hydrology, an introduction to groundwater and wells is helpful as engineers interact with groundwater and well professionals. Both transient and steady state analysis methods are presented. Chapter 11 familiarizes students with the fundamentals of canal and ditch design. Topics include diversion structures, delivery structures, flow measurement structures, conveyance efficiency, Manning's equation, surface roughness, channel sections, uniform flow, varied flow, and Froude number. All of the chapters include VBA/Excel programs that perform the calculations described in the chapters. Homeworks include manual calculations and use of the VBA/Excel programs.

Irrigation Systems (Chaps. 12, 13, 14, 15, 16, 17, 18, 19, 20, 21 and 22)

Contrary to popular belief, there is no such thing as a "best" irrigation system. The selection of an irrigation system is based on soil, crop, economics, water quality, and management considerations. The Natural Resources Conservation Service (NRCS) *National Engineering Handbook* (NEH) describes the four major irrigation methods: surface, sprinkler, micro, and subirrigation.

1. ***Surface***—*Water is applied by gravity across the soil surface by flooding or small channels (i.e., basins, borders, paddies, furrows, rills, corrugations)*
2. ***Sprinkler***—*Water is applied at the point of use by a system of nozzles (impact and gear driven sprinkler or spray heads) with water delivered to the sprinkler heads by surface and buried pipelines, or by both. Sprinkler irrigation laterals are classed as fixed set, periodic move, or continuous or self move. Sprinkler irrigation systems include solid set, handmove laterals, sideroll (wheel) laterals, center pivot, linear move (lateral move), and stationary and traveling gun types. Low Energy Precision Application (LEPA) and Low Pressure In Canopy (LPIC) systems are included with sprinkler systems because they use center pivots and linear move irrigation systems.*
3. ***Micro***—*Water is applied to the point of use through low pressure, low volume discharge devices (i.e., drip emitters, line source emitters,, micro spray and sprinkler heads, gravity and low pressure bubblers) supplied by small diameter surface or buried pipelines.*
4. ***Subirrigation***—*Water is made available to the crop root system by upward capillary flow through the soil profile from a controlled water table. Each irrigation method and irrigation system has specific site applicability, capability, and limitations.*

The relative frequencies of irrigation system types (1997) in the United States are shown in Fig. 1.6. The United States as a whole and the states with largest amount of irrigated land area have a relatively even mix of pressurized and surface irrigation systems. The states with a small irrigated land area tend to exclusively use pressurized irrigation. In 1997, the only state that was almost entirely surface irrigated was Arizona; however, center pivots and drip irrigation have become more popular in recent years in Arizona. In spite of the increasing popularity of pressurized irrigation, the hundreds of thousands of irrigated acres that have been developed by Native American communities in Arizona in recent years are almost entirely surface irrigated. The selection of surface irrigation was driven partially by the fact that the Native American communities have surface irrigated their fields for hundreds, possibly thousands, of years. The other factor was that they installed highly efficient level basin systems that were paid for by the government. With their large flow rates, the water moves quickly across the field, and all parts of the field receive approximately the same depth of water. This type of decision that is driven by past experience and comfort with a certain technology is common among farmers. Many farmers are reluctant to change to new technologies. In many cases they are wise not to change because may be unknown pitfalls associated with new technologies. On the other hand, if other farms adopt technology that enables them to produce higher quality and lower cost food, then the farmers that did not adopt the new technology may be left behind.

One of the primary drivers in irrigation system selection is crop type. For example, vegetable crops cannot be

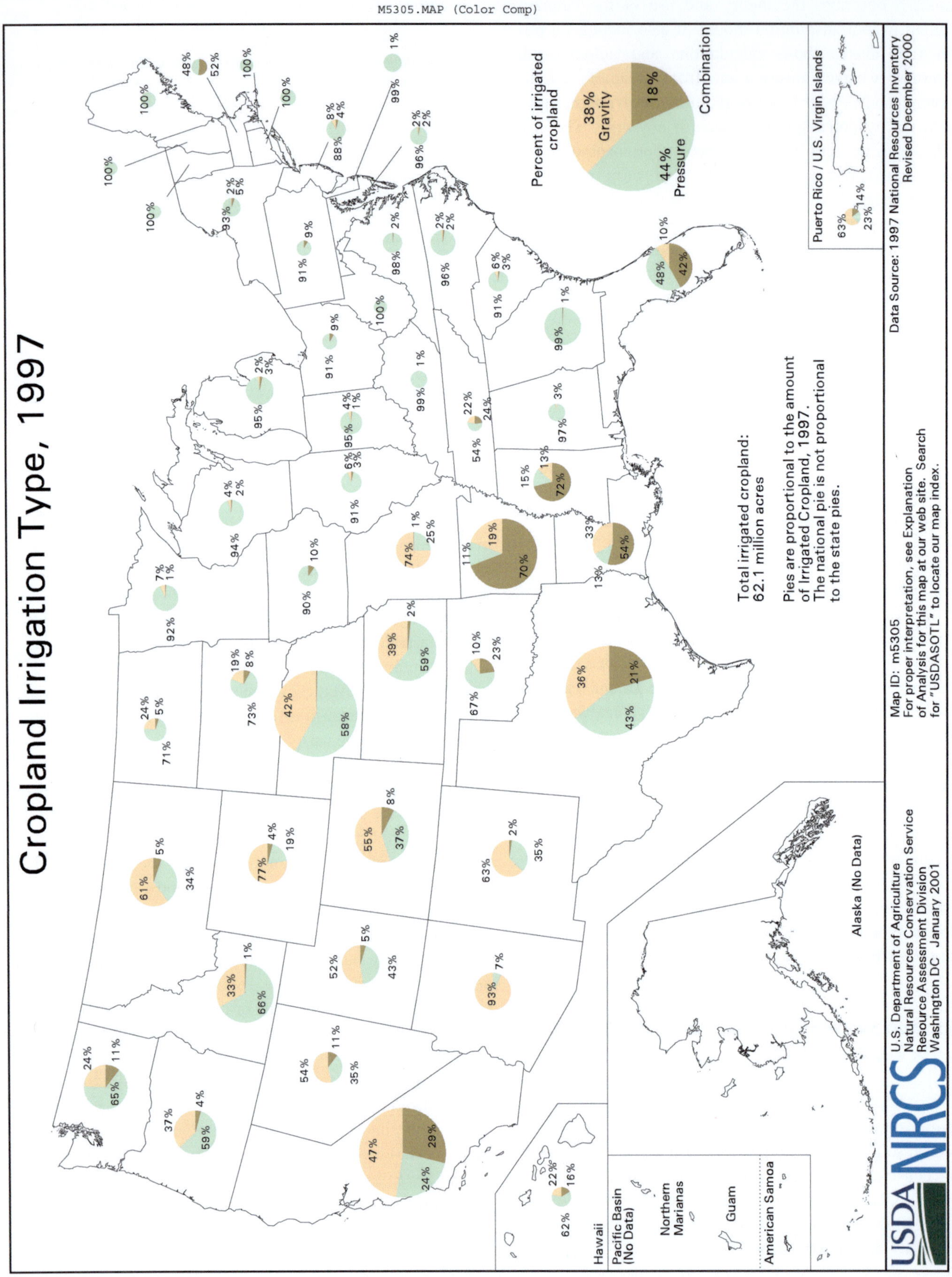

Fig. 1.6 Distribution of irrigation systems in the United States (Credit NRCS NEH)

Table 1.1 Acceptable irrigation systems for crop categories (Credit NRCS NEH Part 652 Ch 3)

Irrigation system	Crop category 1	2	3	4
Surface				
Basins, borders		x	x	x
Furrows, corrugations	x	x		x
Contour levee – rice		x	x	
Sprinkler				
Side (wheel) roll lateral	x	x		
Hand move lateral	x	x		x
Fixed (solid) set		x		x
Center pivot, linear move	x	x		
Big guns – traveling, stationary	x	x		
Micro				
Point source				x
Line source	x			x
Basin bubbler				x
Mini sprinklers & spray heads				x

flooded. The *NRCS NEH* (652- Ch. 3) divided crops into four general categories and defined the acceptable irrigation methods for each crop category in Table 1.1:

Category 1. *Row or bedded crops: sugar beets, sugarcane, potatoes, pineapple, cotton, soybeans, corn, sorghum, milo, vegetables, vegetable and flower seed, melons, tomatoes, and strawberries.*

Category 2. *Close-growing crops (sown, drilled, or sodded): small grain, alfalfa, pasture, and turf.*

Category 3. *Water flooded crops: rice and taro.*

Category 4. *Permanent crops: orchards of fruit and nuts, citrus groves, grapes, cane berries, blueberries, cranberries, bananas and papaya plantations, hops, and trees and shrubs for windbreaks, wildlife, landscape, and ornamentals.*

The crop, soil, water, and climate parameters that should be considered in irrigation selection are listed in Table 1.2. In practice, each agricultural region settles on one or two primary irrigation methods that have proven successful. As an irrigation engineer, I developed a guiding principle: never be the first person to install a new irrigation product in a region, at least don't commit an entire farm to a new irrigation technology, no matter what the product representative claims. The potential pitfalls are unknown because each region is unique, and if the entire farm is committed to an irrigation system that does not work, then there are severe economic consequences. Try out new technologies on small plots and work out the bugs, if possible, on a small scale. In this way, an unforeseen technology failure will not bankrupt the farm. Most farmers have the same level of caution about new products.

Table 1.2 Site considerations in selecting an irrigation method (Credit NRCS NEH. 652 Ch 5)

Crop	Soil	Water	Climate
Crops grown	AWC	Quality	Wind
and rotation	Infiltration rate	Salts, toxic	Rainfall
Water	Depth	elements	Frost
requirement	To water	Sediment	conditions
Height	table	Organic	Humidity
Cultural	To	materials	Temperature
practices	impervious	Fish, aquatic	extremes
Pests	layer	creatures	Rainfall
Tolerance to	Drainage	Quantity	Evaporation
spray	Surface	Reliability	from:
Toxicity	Subsurface	Source	Plant
limitations	Condition	Stream	leaves
Allowable	Uniformity	Reservoir	and stems
MAD level	Stoniness	Well	Soil
Climate control	Slope	Delivery	surface
Frost control	Surface texture	point	Solar
Cooling	Profile textures	Delivery	radiation
Diseases and	Structure	schedule	
Control	Fertility	Freqency	
Crop quality	Temporal	Duration	
Planned yield	properties	Rate	

The ways in which the parameters in Table 1.2 influence irrigation system selection are described in the NRCS NEH (Part 652. Ch. 6):

- ***Surface systems***. *High sediment laden irrigation water generally reduces intake rates, which on coarse textured soils may increase advance rates thereby improving distribution uniformity for the field. On medium and fine textured soils, a reduced intake rate may be undesirable.*
- ***Graded furrow systems***. *On furrow slopes greater than 1 percent and on highly erodible soils, erosion rates can be severe unless protective measures are provided.*
- ***Level basin and graded border systems***. *Larger heads of water are required to meet minimum flow depth requirements in a level basin or border (typically 5 to 7 cubic feet per second) and maintain reasonable field sizes. High uniformity can be attained with level basins on medium and low intake rate soils.*
- ***Sprinkler***. *Low pressure continuous/self move center pivot and linear systems — requires intense water, soil, and plant management for low intake soils, and at least a moderate amount of management on low to medium intake soils.*
- ***Micro***. *Water quality must be high except for basin bubbler systems, which use plastic tubing of 3/8 inch diameter and larger. Chemicals must be used to prevent algae growth in most systems.*

Irrigation efficiency has been defined in many ways. In general, efficiency is the amount used divided by the amount applied; however, irrigation efficiency can be thought of in different ways. Conveyance efficiency is the amount of

Table 1.3 Potential efficiencies of agricultural irrigation systems (Solomon 1988)

Irrigation system	Potential on-farm efficiency
Gravity (Surface)	
Level basin	80–90 %
Furrow	65–75 %
Border	70–85 %
Sprinkler irrigation	
Hand move or portable	65–75 %
Side roll wheel line	65–75 %
Traveling big gun	60–70 %
Center pivot	75–90 %
Linear move	75–90 %
Solid set or permanent	70–80 %
LEPA	80–95 %
Microirrigation	80–90 %

water reaching a field divided by the amount diverted from the irrigation water source. Application efficiency is the amount of water stored in the root zone divided by the amount of water applied to the field. Storage efficiency is the amount of water stored in the root zone during a single irrigation divided by the total water-holding capacity of the root zone. Seasonal irrigation efficiency is the water volume beneficially used by the crop (including leaching) divided by the seasonal amount of water applied.

In general, pressurized irrigation systems (sprinkler and micro) are assumed to have higher application efficiencies than surface irrigation systems. However, loss of water by evaporation or wind, poorly maintained sprinklers or emitters, and poor uniformity due to spacing or pipe hydraulics decrease application efficiency. Surface irrigation system application efficiency is decreased by variation in soil permeability and by variation in ponding time across the field. Runoff also decreases efficiency for surface irrigation systems; however, if a water recycle and reuse system is in place, then runoff generally does not result in a loss of efficiency at the farm scale. At the irrigation district scale, even major losses to leaching and runoff, are generally recycled in the irrigation district as a whole. Thus, improving irrigation efficiency does not necessarily result in an improvement in water use efficiency at the irrigation district scale. Potential application efficiencies for different irrigation systems are shown in Table 1.3.

Sprinkler Irrigation (Chaps. 12, 13 and 14)

These chapters describe five types of sprinkler systems: center pivot, turf, wheel-line and hand-line, orchard, and microsprinkler irrigation. Sprinkler irrigation systems have two major advantages over surface irrigation systems. First, the field does not need to be leveled or graded. In fact, steep slopes can increase uniformity if pipe sizes are selected such that pipe friction loss equals elevation gain (pressure remains constant). Second, if there is no runoff or ponding, then variation in soil properties does not influence sprinkler application uniformity. One of the disadvantages of sprinkler irrigation, especially in arid regions, is evaporation and wind drift before the droplet reaches the soil.

Although 94 % of irrigation in the world is surface irrigation, sprinkler irrigation accounts for 46 % of irrigated acreage in the United States. The major reason for this large percentage is the popularity of center pivots. 75 % of sprinkler-irrigated land in the United States is under center pivot irrigation. Center pivot irrigation began in Nebraska and now waters 4.6 million out of 7 million acres of irrigated agriculture in Nebraska. Considering capital, labor, energy, and water cost, center pivots are generally the most inexpensive way to water row crops and pasture. The major design constraint is that the application rate at the outer end of the pivot cannot exceed the sum of soil infiltration rate and the surface water storage. Otherwise, runoff occurs and application uniformity decreases. Center pivots consist of a steel pipe and truss system that is connected to the water source at the pivot point. They generally use electric power to drive tractor tires that move the structure through the field. Most center pivots water a quarter section (50 ha or 160 acres). Center pivots rotate in a circle around a central pivot point (Fig. 1.7); however, linear move irrigation systems (Fig. 1.8) travel along a straight line and are guided by cables at each end. Chapter 12 describes the sprinkler and pipe calculations for center pivot irrigation. It also includes a VBA/Excel program that calculates sprinkler orifice sizes, pipe diameter, and application uniformity for center pivots.

Turf irrigation has become an important sector of the irrigation industry. Turf irrigation systems (Fig. 1.9) water vegetation that is not sold for profit. Thus, the criteria for turf irrigation management are different than for agricultural irrigation management. For example, a management goal might be that there are no brown spots on a golf course. In this case, the location with the minimum application rate governs the application time of the sprinkler zone. With this in mind, the turf sprinkler industry recommends the scheduling coefficient (average depth applied over minimum depth applied) to evaluate application efficiency. Chapter 13 describes the methods to calculate sprinkler spacing, sprinkler uniformity, pipe design, and the scheduling coefficient. The *Sprinkler Uniformity* program is used to calculate uniformity as a function of sprinkler spacing and sprinkler characteristics.

Chapter 14 describes four types of agricultural sprinkler systems: wheel-lines, hand-lines, undertree sprinklers, and microsprinklers. Hand-lines and wheel-lines (Fig. 1.10) are generally used to water row crops and pasture. Hand lines are small (2–3″ diameter, 50–75 mm) aluminum sprinkler

Fig. 1.7 Center pivot irrigation systems (Credit NRCS)

Fig. 1.8 A canal-sourced linear-move irrigation system in use on a field in Colorado (Credit: GE Cardon, Bugwood.org)

Fig. 1.9 Turf irrigation system (Credit NRCS)

Fig. 1.10 Hand-line and wheel-line (Credit NRCS)

pipes. For crop germination (left side, Fig. 1.10), the hand-line system waters the entire field at one time. For irrigation of established row crops or pasture, single lines of sprinklers are moved once or twice per day and rotated through the field. Similarly, wheel-lines are moved once or twice per day; however, instead of moving the pipes by hand, a small motor rotates the pipeline and wheels.

The *Sprinkler Uniformity* program introduced in Chap. 13 is used in Chap. 14 to develop application distributions for wheel-line and hand-line systems. The calculated spatial distribution statistics are combined with hydraulic calculations of sprinkler flow rate along the pipeline to calculate an overall water application distribution for the entire pipeline.

Undertree orchard sprinklers have a low spray angle so that the stream does not hit the canopy. The primary orchard sprinkler design parameters are application rate (must be less than infiltration rate), number of zones, sprinkler spacing, sprinkler flow rate, and pump selection. Orchard sprinkler spacing is constrained by the fact that trees have a fixed spacing, and sprinkler spacing must be a multiple of tree spacing.

Landscape Irrigation (Chaps. 15 and 16)

Landscape irrigation systems water trees and shrubs in urban landscapes. Two of the primary methods are drip irrigation and bubbler irrigation. The unreliability of single barbed emitters inserted into polyethylene pipe (the conventional method) has led to adoption of more reliable technologies such as in-line drip systems (Fig. 1.11), multiport emitters mounted on PVC pipe, and bubbler systems. Chapter 15 also describes the control zone components: solenoid valves, controllers, valve boxes, pressure regulators, and backflow preventers.

In addition to problem of unreliability, a second problem with most landscape irrigation systems is that homeowners and landscape professionals generally cannot adjust the watering schedule to match the plant evapotranspiration requirements. In addition, they rarely adjust the number of emitters per plant as plants grow. Chapter 16 presents calculation procedures for irrigation schedules and number of emitters per plant. In addition, the *Landscape* program calculates the number of emitters or bubblers required for each plant in an irrigation zone as well as the watering schedule. With the many types and sizes of plants in an irrigation zone, this program is helpful for achieving uniformity in a landscape zone. The program uses the landscape coefficient method to calculate the water requirement of each plant.

Fig. 1.11 Line source in-line drip irrigation tube for landscape irrigation with the emission point near the middle of the picture

Agricultural Drip Irrigation (Chaps. 17, 18 and 19)

Drip irrigation does not represent a large fraction of irrigation systems in the world; however, many new irrigation systems are drip systems. Thus, many irrigation companies focus on drip irrigation system design. One of the most popular applications of drip irrigation is the irrigation of vineyards, where the drip tube is hung from the trellis (Fig. 1.12).

With relatively thin polyethylene laterals and small orifices (Fig. 1.13), drip systems are more fragile than sprinkler, bubbler, or flood systems; thus, drip irrigation systems require a high level of management expertise. Systems may require sand filtration of particulates, chlorination to prevent bacterial growth, acid injection to prevent calcium carbonate

Fig. 1.12 Vineyard drip irrigation system on trellis (Credit NRCS)

Fig. 1.13 Drip emitter and polyethylene tubing hanging from trellis wire (Credit NRCS)

deposition, and regular flushing to remove sediment from drip laterals. Although drip irrigation generally requires a higher level of management expertise, it is growing in popularity for several reasons. One reason is that it maintains a relatively constant level of moisture in the root zone, which increases production of many crops. A second reason for the increasing popularity of drip irrigation is that the potential efficiency (90 %) is greater than for other irrigation systems. Evaporation from the soil surface is dramatically reduced or may be zero for subsurface drip irrigation. Drip irrigation systems are also effective chemigation systems because they apply water directly to the plant root zone.

Subsurface drip irrigation is the most widely used drip irrigation method. In order to protect the drip tubing from cultivation practices, laterals are buried approximately 15 cm below the ground surface (Fig. 1.14). Some of these systems have lasted 15 or more years without replacement of the inline tubing. Reliability depends on management expertise, wall thickness, and soil type. Extremely thin wall tubing might be plowed into the soil each year. Tillage implements and GPS systems have been designed that maintain the bed over the tubing during cultivation practices so that the tubing is not disturbed and remains in the same position in the bed.

Chapters 17 and 18 describe how integrate the components and options in subsurface irrigation in a final design: emitter flow rate, lateral tubing length, lateral and

Fig. 1.14 Subsurface drip irrigation (Courtesy of Paul Colaizzi. USDA-ARS)

emitter spacing, tubing diameter, submain and mainline diameters, flushing, the pump station, and sand filters. The design is analyzed with an economic analysis that includes pipe, tubing degradation, energy, environmental, and water costs, as well as the effect of spatial variation of water application on yield and water cost. The complex analyses are performed in a VBA/Excel program.

Chapter 19 describes chemigation injection systems. It describes calculation procedures for rates of acid injection, chlorination, fertigation, and pestigation. Acid injection rates are calculated as a function of the water alkalinity, acid strength, and the desired endpoint pH.

Surface Irrigation (Chap. 20)

Three of the most popular types of surface irrigation systems are sloping border, graded furrow (Fig. 1.15), and level basin – with furrow irrigation representing half of all surface irrigation systems. Other types include contour level furrows and corrugation irrigation. Many farms now use near-level furrows and borders as the best strategy to achieve high efficiency and uniformity. The four primary design parameters in surface irrigation systems are slope and length of the field (difficult to change), flow rate, and cut-off time. The optimization of these parameters can be performed with the USDA-ARS WINSRFR surface irrigation model. This chapter describes how the advance phase can be modeled with the two-point volume balance method. The infiltration phase is modeled with a numerical model of flow in a furrow that includes decrease in flow rate vs. distance and spatial variation of soil properties. Chapter 20 focuses on the use of the WINSRFR model in the design and analysis of furrow irrigation systems; however, the WINSRFR model has many other "Worlds" for the design of other systems and

Fig. 1.15 Furrow irrigation system in Arizona

applications. The model can also evaluate the effects of such things as leveling, which would have been of interest to the Sumerians.

Hydroponic Drip Irrigation Systems (Chap. 21)

Hydroponic drip systems are used in controlled environment agriculture (greenhouses). The operation and management of these systems is unique. Water is applied to a soilless growing media such as coconut husks, rock wool, sand, and/or perlite. These soilless media give growers the ability "steer" the crop toward desired production characteristics by rapidly varying nutrient and salinity concentrations in the media. The chapter describes the calculation methods for setting up a nutrient recipe for hydroponic fertigation. It also includes a VBA/Excel program for this task.

Low-Head Bubbler Irrigation (Chap. 22)

A system that is gaining in popularity in greenhouse and conventional agriculture is low-head bubbler irrigation. These systems have large openings so they never plug. They have extremely low pressure requirements so power costs are low. They require no specialized components such as emitters or sprinklers so capital costs are low. In these systems, tubing at each plant is cut to a precise length and installed to a precise elevation in order to provide the design flow rate to the plant.

Below the Surface of the Soil (Chaps. 23, 24, 25, 26, 27, 28, 29, 30 and 31)

The last chapters in this book focus on the fate of water, salinity, and nutrients in the soil. These chapters are augmented with the *WINDS* model (Water-use, Irrigation, Nitrogen, Drainage, and Salinity), which models the fate of water and solutes in the soil profile. The chapters describe the algorithms in the *WINDS* model and then use the model to visualize the processes in the soil. The *WINDS* model is also designed to be used in real-world applications. With quick set up and run time for design, evaluation, or management, it can be deployed in a few hours. The *WINDS* model can simulate up to 12 soil layers and over 1,000 field locations;

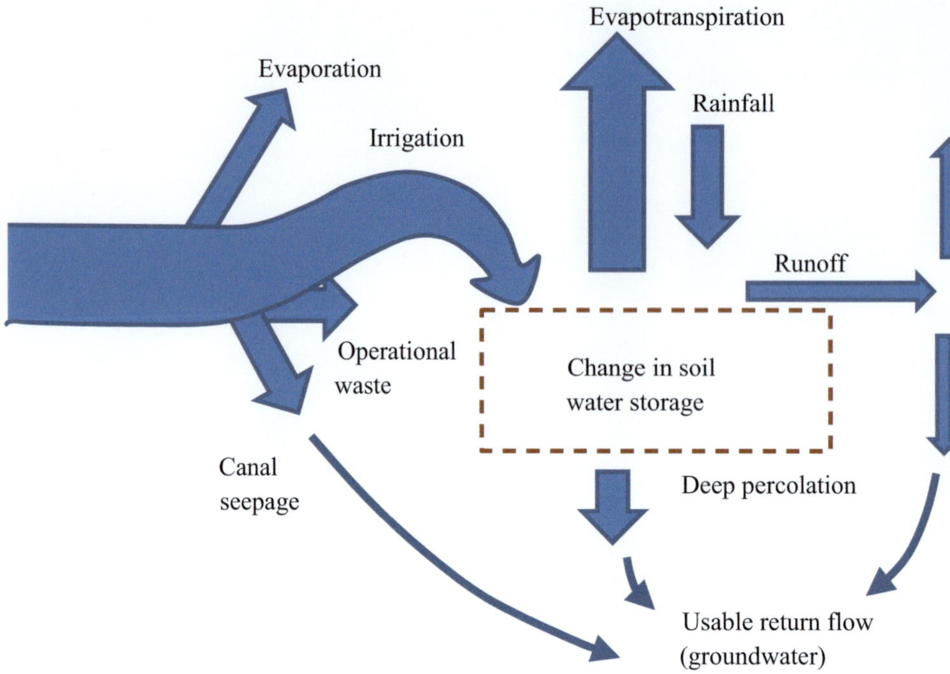

Fig. 1.16 Farm scale irrigation water balance (After NRCS – NEH)

however, the vertical processes in the soil profile are the focus of these chapters. Use of a model that tracks soil profile water and nutrient content enables better estimates of plant requirements, leading to lower nutrient waste and environmental contamination. Two of the important sources of nutrients are manure and wastewater. The first two chapters in this section describe wastewater pathogens and pollution and the application of waste to soils.

Contaminants and Waste Application to Soils (Chaps. 23 and 24)

Irrigation systems and soils are increasingly used as a method of waste disposal. Waste contains beneficial nutrients; however, there are also disease carrying pathogens. Chapter 23 provides a brief introduction to pathogens, disinfection, wastewater characteristics, and waste treatment. Chapter 24 focuses on waste and wastewater application. The primary focus in Chap. 24 is the utilization of nutrients in livestock waste for agriculture. Treatment steps include pond and wetlands treatment, and waste application to fields. Degradation rates, salinity, nutrient concentration, volatilization, percent solids, pond retention time, climate, and other factors must be considered in the calculation of the depth of wastewater that should be applied by irrigation systems.

The *WINDS* Model (Chaps. 25, 26, 27, 28 and 29)

The soil is the control volume in the *WINDS* model, but it is part of a larger system. The interactions of the larger system with the soil profile are shown in Fig. 1.16. Even after touching the field surface, irrigation and rainfall are partitioned between infiltration and runoff. Water that enters the soil may be lost to deep seepage; however, just because water and nutrients are lost to seepage or runoff, does not necessarily mean that they are lost from the farm or irrigation district. The next field may be able to utilize the runoff or the pumped water from the aquifer (Fig. 1.16).

Chapter 25 focuses on nitrogen processes and salinity fate in soils. The nitrogen processes include fertilization, mineralization, denitrification, and plant uptake.

Chapter 26 presents the tipping bucket approach to modeling water, salt, and nitrogen movement in soils during storms or irrigation events. In the tipping bucket model, a mass balance is set up for each soil layer, and water, salt, and nitrogen are routed to the next layer in the soil profile if there is water movement between layers. The foundation of the *WINDS* model is the mass balance: change in storage = input − output + reaction. The water balance begins with the partition between infiltration and runoff. In the WINDS model, the infiltration depth is either specified in a worksheet or it is calculated with the Green-Ampt equation or NRCS curve numbers, based on storm intensity and duration.

Chapter 27 introduces basic concepts associated with the conservation of energy equation. It then describes the relationships between soil energy (gravitational + matric), hydraulic conductivity, and water flux. Energy diagrams are used to show the relationship between matric potential energy, elevation energy, and total energy in soils. The Van Genuchten and Brooks-Corey relationships between matric potential, hydraulic conductivity, and water content are also described.

Chapter 28 combines the conservation of energy and conservation of mass equations in order to model water flux in response to energy gradients. The *WINDS* model is used to demonstrate the Richards equation and the van Genuchten soil parameters. The second part of the chapter focuses on modeling water content in the soil in the presence of a water table. Drainage algorithms calculate the fluctuation of the water table in the presence of periodic infiltration events and evapotranspiration. The *WINDS* drainage algorithm was derived by integrating the van Genuchten equation. Lower soil layers remain in hydraulic equilibrium with the water table and upper layers are disconnected from the water table. A decision algorithm determines which layers are in equilibrium with the water table.

Chapter 29 demonstrates the use of the *WINDS* model in agriculture. It describes the algorithms for crop growth, ET, irrigation, rainfall, water content, and leaching. The chapter focuses on two case studies, an irrigated cotton crop in an arid region, and rainfed agriculture in a tropical climate. The examples demonstrate the relationship between transpiration, evaporation, crop maturity and crop stress. They show how weather station data is used in the WINDS model. Algorithms for modeling partially wetted soils, crop stress, Green-Ampt infiltration, remote sensing and crop coefficients, evapotranspiration in the presence of water or salinity stress, salinity stress, and nitrogen stress are also presented.

Subsurface Drainage Systems (Chaps. 30 and 31)

Chapter 30 reviews engineering and environmental considerations in subsurface drainage design and installation. Engineering equations for drain diameter, particle size fractions in drain envelopes, and auger hole measurement of lateral soil horizontal conductivity are also presented. Chapter 31 focuses on hydraulic and hydrologic models of the water table in the presence of subsurface drains. The traditional drainage model is the Hooghoudt equation for steady-state calculation of water table elevation. The Bower and van Schilfgarde equation and the Bureau of Reclamation equations enable simulation of water table elevation over time. An example shows how to use engineering economic analysis in order to select drain spacing and elevation. Kirkham developed a two-dimensional solution of flow to parallel drains and then modified it for transient and spatial analysis of drained fields. Kirkham's method is implemented in the *WINDS* model, and the chapter concludes with a *WINDS* drainage simulation.

The Value of Water

In today's world, water is becoming scarce in some regions. Who is going to have access to the water, and what are they going to pay for it? Irrigation and drainage engineers will be at the center of some of these controversies. At the very least, engineers will provide the data and models that policy makers will use to answer these questions. As such, engineers should have some familiarity with the value of water and the ways that it can be distributed. One of the important concepts that has arisen in recent years is the concept of virtual water.

The value of water depends on supply, demand, and cost of water resources development. In water stressed regions, the value of water can be greater than the value of food produced by the water. In these regions, one strategy of "water resources development" that is gaining in popularity is the concept of virtual water. A community can import water in the form of food – virtual water. The virtual water concept postulates that if a region imports food, then it is actually importing the amount of water required to grow that food. For example, if water for wheat requires 1,000 m^3/metric ton to produce (Table 1.4), then importing a metric ton of wheat is the same as importing 1,000 m^3 of water.

The United States and Canada, with abundant rainfall in North America, are the largest exporters of water in the form of food commodities (Table 1.5). The largest virtual water

Table 1.4 Water volume for commodities

Wheat	1,000 m^3/metric ton
Citrus	1,000 m^3/metric ton
Rice	2,000 m^3/metric ton
Fresh beef	15,000 m^3/metric ton
Poultry	6,000 m^3/metric ton

Table 1.5 Virtual water exporters

Country	10^9 m^3/yr	Per capita – m^3/yr
USA	758	2,600
Canada	272	8,400
Thailand	233	3,600
Argentina	226	5,800
India	161	150
Australia	146	7,300
Viet Nam	90	1,000
France	88	1,500

importers are, in order from first to last: Sri Lanka, Japan, Netherlands, Korea, China, Indonesia, Spain, Egypt, Germany, and Italy.

Water, on a per volume basis, is much less valuable to agriculture then to municipalities. The value of water in agriculture compared to the municipal value can be compared by looking at the yearly volume required per person for drinking, household management, and food production per year.

1. Drinking water requirement is 1 m^3/person/year.
2. Domestic and municipal uses require 100 m^3/person/year.
3. Food production requires about 1,000 m^3/person/year.

Typically, a large proportion of the most easily developed water resources are allocated to agriculture. As cities grow, they require more water, and they either try to develop more expensive water resources or they buy the water that was originally allocated to agricultural entities.

Within a given region, the value of water can be determined in two ways. It can be established by the actual cost of water delivery (infrastructure and energy costs), or it can be established by the opportunity cost, which is the highest price that someone is willing to pay for the water. For example, if water costs $30/ha-cm to deliver, and a golf course is willing to pay $300/ha-cm for water, and a farmer is only willing to pay $3/ha-cm, then the opportunity cost of the water (value) is $300/ ha-cm, while the actual cost of the water is $30/ha-cm. However, the farmer may continue to receive water at $3/ha-cm because it is his/her legal right. Because of the demand for urban water, some regions have allowed farmers to sell their water to cities rather than continue to grow crops.

The question of whether to allocate water based on opportunity cost or to restrict water to agriculture is often a complex political decision that must consider national food security, maintenance of the rural community, water law, and other considerations. This question is beyond the control of the irrigation engineer. However, one thing is certain. As water resources are stretched between competing demands, the value of the water increases, and the value of efficient water management increases with it. Thus, the value of irrigation engineering increases with the value of water.

Questions

1. List the components of a farm water balance.
2. Describe the major four irrigation methods (NRCS description).
3. Describe the different types of irrigation efficiency
4. In general, pressurized irrigation systems are considered to be more efficient than surface irrigation systems. What are some factors that may decrease the irrigation efficiency of some sprinkler systems to below that of some surface irrigation systems?
5. How do irrigation systems fail?
6. Summarize the virtual water concept.
7. What increases the value of the irrigation engineering?

References

Altaweel M, Watanabe CE (2012) Assessing the resilience of irrigation agriculture: applying a social-ecological model for understanding the mitigation of salinization. J Archeol Sci 39:1160–1171

Natural Resource Conservation Service. Engineering Field Handbook (1997) Part 652 Irrigation Guide. Chapter 5 (652.0501) Methods and systems to apply irrigation water. http://www.scribd.com/doc/12851870/National-Irrigation-Guide

Solomon K (1988) Irrigation water systems and water application efficiencies. Center for Irrigation Technology, Irrigation Notes. California State University, Fresno. http://cwi.csufresno.edu/wateright/880104.asp

Irrigation Economics 2

Economics is the fundamental decision criterion in irrigation and drainage engineering. This chapter helps answer the following questions. Does the improvement in crop income and/or reduction in labor, energy, and water costs justify the capital expense of a new irrigation system? Which alternative irrigation system provides the highest profit and least environmental contamination? What is the optimal rate of water application? Engineering economic analysis uses the project life and expected rate of return to compare the expected present and future costs and benefits of proposed irrigation systems. If the system is profitable at the required rate of return, then the decision is made to invest in the system. Crop water production functions (CWPF) that have been developed in experiments show the relationship between seasonal depth of water applied and crop yield. Combined with water and energy cost information, they enable the calculation of the optimal depth of water application.

Crop Water Production Function

There is a direct relationship between total depth of irrigation water applied during the growing season and crop yield: if water uptake by the plant is limited, then transpiration and other plant processes such as photosynthesis may be limited. The relationships between yield and water application depth (CWPF) for individual crops have been developed in experiments with varying depths of applied water per season in different irrigation treatments. A cotton CWPF (dashed red line) is shown in Fig. 2.1.

In-class Exercise 2.1 Why does increasing the depth of irrigation water applied per season reach a point of diminishing returns?

Why is there a nearly linear relationship between evapotranspiration (ET) and yield and a nonlinear relationship between applied water and yield?

The yield vs. applied water curve (Cuenca 1989), in Fig. 2.1 was based on an experiment conducted by Grimes and El-Zik (1990). They derived the following equation:

$$Y_a = -3954 + 1067(AW_{CWPF})^{0.5} - 54.14(AW_{CWPF}) \quad (2.1)$$

where

AW_{CWPF} = depth of applied water used in calculation of yield, cm.
Y_a = actual yield per ha, kg/ha.

Irrigation (AW) is often supplemented by precipitation. Thus, precipitation should be added to Eq. 2.1.

$$Y_a = -3954 + 1067(AW_{CWPF} + P)^{0.5} - 54.14(AW_{CWPF} + P) \quad (2.2)$$

where

P = precipitation, cm.

Many irrigation systems waste water to surface runoff, nonuniformity, or leaching. If this is the case, then not all of the applied water is utilized for crop growth. The amount used divided by amount applied is the efficiency. The difference in efficiency between the experimental irrigation system used to develop the crop water production function and the field irrigation system should be accounted for. If the experimental (CWPF) irrigation system efficiency is 90 % (a reasonable assumption), then the relationship between gross depth of irrigation water (AW) applied and the depth referred to Eq. 2.1 (AW_{CWPF}) is calculated with the following equation:

$$AW = \frac{AW_{CWPF}}{(100\ \% - (90\ \% - Eff))/100} \quad (2.3)$$

P. Waller, M. Yitayew, *Irrigation and Drainage Engineering*, DOI 10.1007/978-3-319-05699-9_2

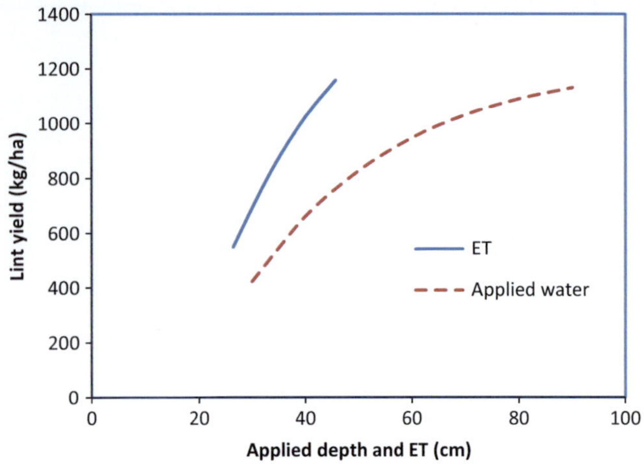

Fig. 2.1 Yield versus applied water (AW_{CWPF}) for cotton (After Grimes and El-Zik 1990)

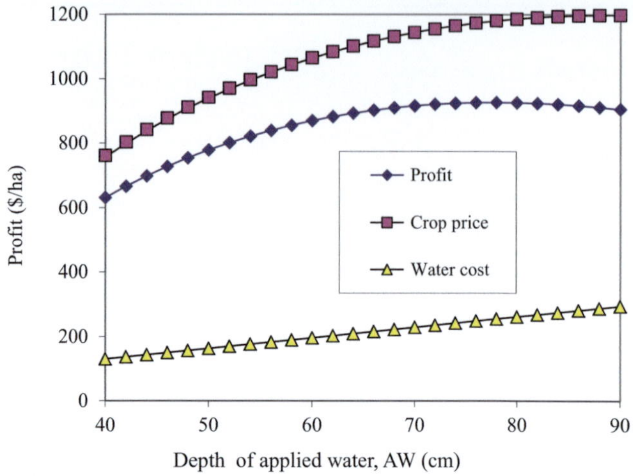

Fig. 2.2 Value of yield, water, and profit per ha for drip irrigated cotton (Example 2.1)

where

Eff = efficiency of field irrigation system (excluding prerrigation), %
AW = Actual gross depth of applied water used to calculate water cost. .

Preirrigation may be needed to germinate the crop, and most of this water is often wasted as leachate or runoff. For example, germination of cotton in a field with a sloping furrow irrigation system often requires 2 ft depth (60 cm depth) of preirrigation water in order to wet the top of the planting bed. Most of this water is wasted as runoff. Preirrigation can be added to Eq. 2.3.

$$AW = \frac{AW_{CWPF}}{(100\ \% - (90\ \% - Eff))/100} + preirrigation \quad (2.4)$$

Water cost is normally (but not always) calculated on a volume basis (m^3, ac-ft, ha-cm). Because it takes 100 m^3 to cover a 1 ha area to a depth of 1 cm, the cost in \$/ ha-cm is equal to 100 * \$/m^3. Thus, if the unit of AW is cm, then the cost of water in \$/ha is

$$WC = \$/m^3 * 100 * AW \quad (2.5)$$

where

WC = water cost, \$/ha.

If only water cost is considered, then profit is calculated as follows:

$$Pr = Y_a * (\$/Y_a) - WC \quad (2.6)$$

where

Pr = benefit or profit, \$/ha.

The maximum profit can be found by setting the derivative d(Pr)/d(AW) equal to zero. Alternatively, a spreadsheet can be used to find the maximum profit.

Example 2.1 Find the depth of applied water (AW) that results in maximum profit with a drip irrigation system for the cotton CWPF by Grimes and El-Zik. Assume that 7.5 cm of precipitation infiltrates during the growing season. The cost of water is \$0.0327/m^3. The selling price of cotton is \$0.92/kg. Assume that preirrigation depth is 0 cm and AW = AW_{CWPF}. The drip irrigation system efficiency is 90 %.

$$Y_a = (-3954 + 1067(AW + 7.5cm)^{0.5} - 54.14(AW + 7.5cm))$$

$$Pr = \$0.92/kg*(-3954 + 1067(AW + 7.5cm)^{0.5} - 54.14(AW + 7.5cm)) - \$3.27*AW$$

The maximum profit (Fig. 2.2) is \$925/ha, which is found at an applied depth of water (AW_{CWPF}) equal to 79 cm (Fig. 2.2).

Example 2.2 Repeat Example 2.1, but use the surface irrigation with 60 cm preirrigation and 60 % efficiency.

$$WC = 3.27*(AW_{CWPF}/(1 - (0.9 - 0.6)) + 60)$$

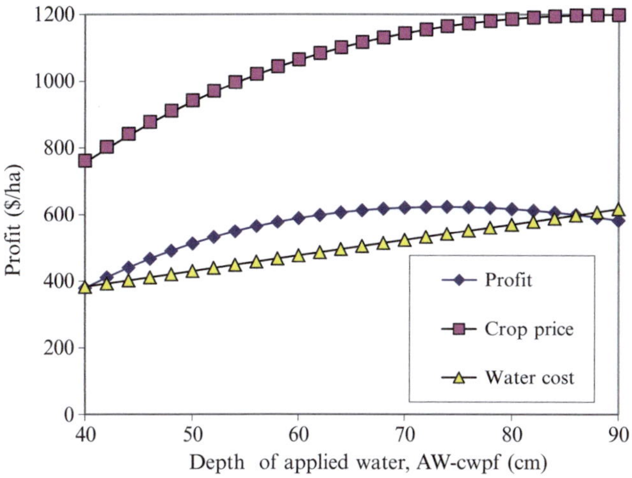

Fig. 2.3 Value of yield, water, and profit for surface irrigated cotton (Example 2.2)

$$\Pr = \$0.92/\text{kg}^*(-3954 + 1067(\text{AW}_{\text{CWPF}} + 7.5)^{0.5} - 54.14(\text{AW}_{\text{CWPF}} + 7.5)) - \text{WC}$$

Maximum profit ($623) is found at $\text{AW}_{\text{CWPF}} = 74$ cm (Fig. 2.3). The gross depth of water that is applied by the irrigation system is

$$\begin{aligned}\text{Depth applied} &= \text{AW} \\ &= \text{AW}_{\text{CWPF}}/(1-(0.9-0.6)) + 60 \\ &= 74/(1-(0.9-0.6)) + 60 = 166 \text{ cm}\end{aligned}$$

The annual benefit is decreased by approximately $300/ha from the drip irrigation system analyzed in Example 2.1. If the capital cost of a drip system is $2,500 per ha, then the annualized cost of the system installation, assuming a 20 year project life and 8 % required rate, is approximately $300/ha. Thus, all other things being equal, the cost of installing a drip system would be paid for by water savings. In addition, if there is a yield increase with drip irrigation, then the drip system would be a worthwhile investment.

Environmental Cost

Although often ignored, the environmental cost of an irrigation strategy should be considered during the design and optimization process. Erosion from sloping furrow irrigation systems is used here as an example. Eroded sediment causes sedimentation of channels, and it costs cities money to remove it from drinking water. In addition, the value of farmland is degraded if topsoil is eroded.

The Twin Falls tract in Idaho, with 85,000 ha, diverts water from the Snake River into its irrigation canal system. The practice of irrigation in the Twin Falls tract adds over 40,000 metric tons of sediment to the Snake River (Brown et al., 1974) or approximately 500 kg/ha. Many of the farms in this area have lost nearly all topsoil from the upper end of fields. As more water is applied to a field, the amount of seasonal erosion increases. Thus, the relationship between sediment load and irrigation amount can be added into the economic analysis. Figure 2.4 shows a relationship between erosion and yearly applied depth of water to a surface irrigated field.

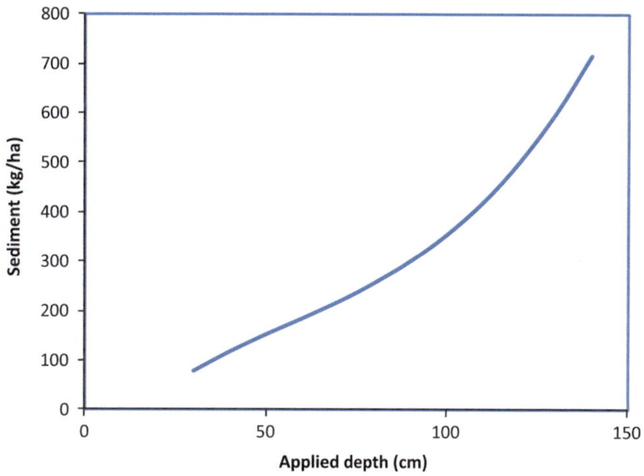

Fig. 2.4 Sediment yield versus applied water depth (cm)

The line in Fig. 2.4 was generated with the following equation:

$$\text{Sediment(kg/ha)} = (4.62{*}10^{-5}i^3 - 7.84{*}10^{-3}i^2 + 0.77\,i - 9.44){*}10^3 \quad (2.7)$$

where

i = seasonal gross depth of water applied by irrigation system, cm.

The environmental cost (EnvC) on a per hectare basis is

$$\text{EnvC}(\$/\text{ha}) = \$/\text{kg} * \text{kg}/\text{ha} \quad (2.8)$$

Equation 2.6 can be modified to include the environmental cost.

$$\text{PR} = \text{Y}_a{*}(\$/\text{Y}_a) - \text{WC} - \text{EnvC} \quad (2.9)$$

Example 2.3 Repeat Example 2.2, but include cost of erosion. The cost of erosion (land degradation and sediment removal from river water) is $0.2/(kg/ha).

The actual depth of applied water is used to calculate the amount of erosion. For example, the irrigation depth at $\text{AW}_{\text{CWPF}} = 64$ cm is

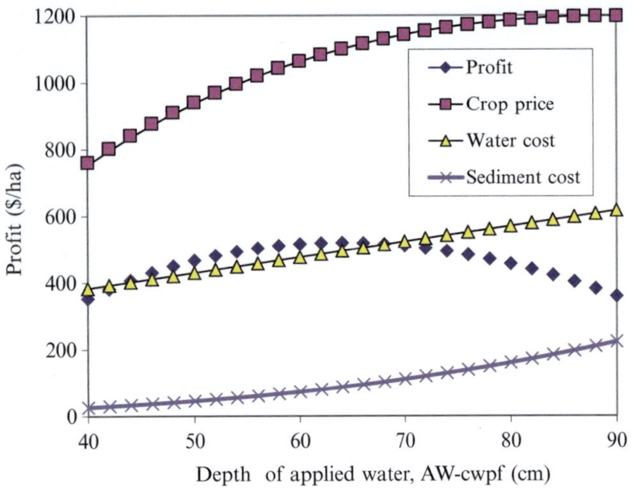

Fig. 2.5 Value of crop, water, profit, and erosion versus AW_{CWPF}

$$
\begin{aligned}
i &= AW_{CWPF}/(1-(0.9-0.6))+60 \\
&= 64/(1-(0.9-0.6))+60 = 151 \text{ cm}
\end{aligned}
$$

$$
\begin{aligned}
\text{Sediment} &= (4.62{*}10^{-5}{*}(151)^3 - 0.00784{*}(151)^2 \\
&\quad + 0.477{*}(151) - 9.4449){*}10 \\
&= 429 \text{ kg/ha}
\end{aligned}
$$

$$
\text{EnvC} = 429 \text{ kg/ha}{*}\$0.2/\text{kg} = \$86/\text{ha}
$$

Using Eq. 2.9, maximum profit ($519/ha) is found at AW = 64 cm with i = 151 cm. Figure 2.5 shows the results of the spreadsheet analysis, including environmental cost.

This example shows that the optimal economic depth of irrigation is generally reduced when environmental cost of irrigation is considered.

In-class Exercise 2.2 Should government charge farmers for sediments from surface irrigation systems that are discharged to streams?

Should farmers pay for the cost of dredging and for sediment removal from public drinking water supplies?

Should the government pay farmers to convert to pressurized irrigation if a farming region has erosive soils?

At the government policy level, should economic analysis of irrigation systems include all environmental degradation due to irrigation?

Transferring a CWPF from One Climatic Zone to Another

Linearized versions of crop water production functions are published by organizations such as the Food and Agriculture Organization (FAO) and used around the world. The cotton

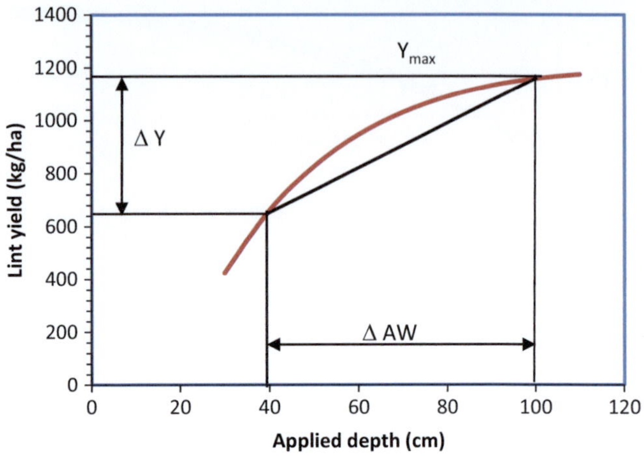

Fig. 2.6 Linearized crop water production curve

CWPF could be linearized (Fig. 2.6), but it obviously loses some accuracy in the process.

With a linear function, a relationship is developed between percent reduction in yield and percent reduction in required depth of water applied %ΔY/%ΔAW. This approach makes the crop water production function more general and applicable to another region with a different water requirement and a different maximum yield.

$$
Y_a = Y_{\max}\left(1 - \frac{\%\Delta Y}{\%\Delta AW}\,\frac{AW_{req} - AW_{CWPF}}{AW_{req}}\right) \tag{2.10}
$$

where

%ΔAW = percent (fraction) change in applied water, dimensionless
%ΔY = percent (fraction) change in yield, dimensionless
Y_{max} = yield with no stress, kg/ha
AW_{req} = applied water depth with no yield reduction, cm.

Example 2.4 Calculate cotton yield. The required depth of applied water in a region is 100 cm, actual applied water depth is 80 cm, and the maximum yield is 1180 kg/ha.

From Fig. 2.6

$$
\frac{\%\ \Delta Y}{\%\ \Delta AW} = \frac{(1161-641)/1161}{(100-40)/100} = 0.746
$$

Equation 2.9

$$
\begin{aligned}
Y_a &= Y_{\max}\left(1 - \frac{\%\Delta Y}{\%\Delta AW}\,\frac{AW_{req} - AW_{CWPF}}{AW_{req}}\right) \\
&= 1,180\ \left(1 - 0.75\frac{100-80}{100}\right) = 1,000\ kg/ha
\end{aligned}
$$

The term %ΔY/%ΔAW is similar to K_y, the yield response factor used by the FAO. Based on Fig. 2.6 and previous discussions, it is easy to appreciate that an understanding of the experimental basis of K_y (and simplifications such as linearization) would help to avoid inaccuracies associated with misapplication of these coefficients. Equation 2.10 can be rearranged and set in the form used by the FAO.

$$1 - \left(\frac{Y_a}{Y_{\max}}\right) = K_y\left(1 - \frac{AW_{CWPF}}{AW_{req}}\right) \tag{2.11}$$

Combined Estimation of Crop Yield

Crop yield is a function of all stresses, included salinity, water, nitrogen, pest, and other stresses. Research has shown that the yield reduction due to combined stress factors (water, pests, and nutrients) can be calculated as the product of the fractions of yield due to the individual stress factors.

Example 2.5 There is 10 % yield loss due to pests, 10 % yield loss due to salinity and the depth of applied water to sugar beets is 80 % of that required. Maximum sugar beet yield is 40 t/ha. What is the yield?

Percent reduction due to water stress from example 2 $- 4 : 37/40{*}100\ \% = 93\ \%$

Expected yield $= 40\ \text{t/ha}{*}(0.93{*}0.9{*}0.9) = 30\ \text{t/ha}$.

Irrigation Uniformity and Economics

Infiltration is generally higher at the furrow inlet than at the end of the furrow (Fig. 2.7): the furrow has an infiltrated depth that is 20 % greater at the beginning than the end; thus, if the end of the furrow is provided the optimal depth of water, then the upper end of the furrow receives excessive water.

Growers do not necessarily apply the target or optimal application depth to the minimum infiltration point at the end of the field. Instead, they often apply less than the target depth at the end of the field in order to avoid excessive overwatering of the upper end of the field. The *Variable water application* worksheet (Fig. 2.8) enables the calculation of yield at each furrow position as a function of the varying infiltration depths along the furrow. In the worksheet, the user inputs data in the white cells and the yellow cells are calculated.

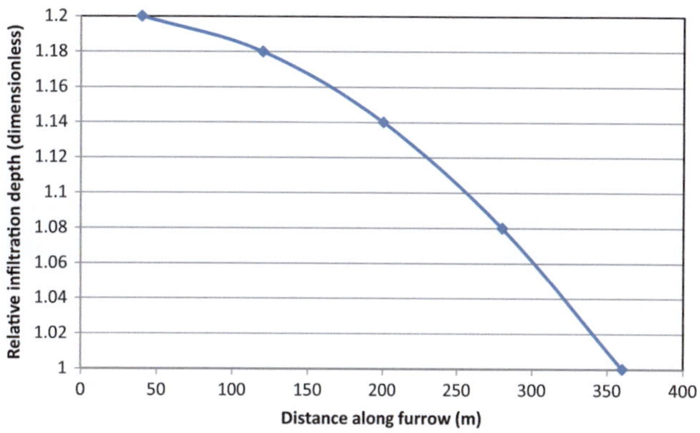

Fig. 2.7 Relative infiltration depths along 400 m length furrow divided into five sections

The target depth is in cell F1. The fraction of the target depth applied at the end of the field is specified in cell F2. The average AW_{CWPF} in column D is multiplied by 1 + runoff fraction (cell 34) in order to find the AW (gross depth of water) applied to the field. The runoff depth is also used to calculate the cost of erosion. The final profit (benefit – cost) in cell C5 is the average yield benefit minus the sum of water cost and sediment cost. The fraction of target depth at end of the furrow (cell F2) can be adjusted until the profit (cell C5) is maximized.

Engineering Economic Analysis

Engineering economic combines present (capital investment) and future (water cost, crop yield, energy cost, and labor) benefits and costs into one number. Conversion to a pressurized system may result in lower water costs and greater crop yield but have higher energy and capital costs; engineering economic analysis shows whether conversion to the pressurized system will ultimately result in greater profit for the farmer. It does this by converting all future costs and benefits to the present value based on the time value of money.

The required rate of return is the interest rate that a company expects to earn on investments. For example, if someone can invest retirement money in a retirement account at a guaranteed fixed return of 6 %, then it would be unwise for the person to invest retirement money elsewhere for an interest rate less than 6 %. In the jargon of engineering economic analysis, the person's required rate of return is 6 %. In engineering economic analysis, if a proposed project returns a profit at the required rate of return, even if the profit is only one dollar, then the decision is made to invest in the project.

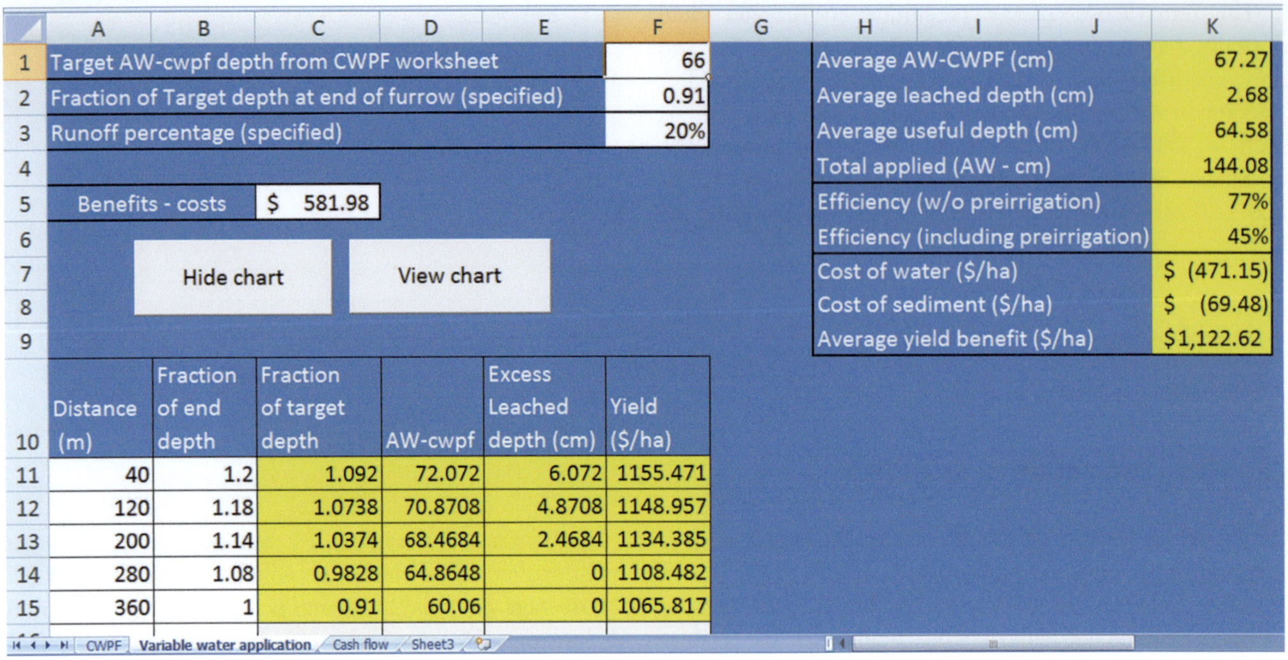

	A–E	F
1	Target AW-cwpf depth from CWPF worksheet	66
2	Fraction of Target depth at end of furrow (specified)	0.91
3	Runoff percentage (specified)	20%
5	Benefits - costs	$ 581.98

Hide chart | View chart

H–J	K
Average AW-CWPF (cm)	67.27
Average leached depth (cm)	2.68
Average useful depth (cm)	64.58
Total applied (AW - cm)	144.08
Efficiency (w/o preirrigation)	77%
Efficiency (including preirrigation)	45%
Cost of water ($/ha)	$ (471.15)
Cost of sediment ($/ha)	$ (69.48)
Average yield benefit ($/ha)	$1,122.62

	Distance (m)	Fraction of end depth	Fraction of target depth	AW-cwpf	Excess Leached depth (cm)	Yield ($/ha)
11	40	1.2	1.092	72.072	6.072	1155.471
12	120	1.18	1.0738	70.8708	4.8708	1148.957
13	200	1.14	1.0374	68.4684	2.4684	1134.385
14	280	1.08	0.9828	64.8648	0	1108.482
15	360	1	0.91	60.06	0	1065.817

CWPF | Variable water application | Cash flow | Sheet3

Fig. 2.8 *Variable water application* worksheet

The future value of money is calculated with the following formula:

$$F = P(1 + i)^n \tag{2.12}$$

where

F = future value
P = present value
i = interest rate
n = number of years.

Example 2.6 If $1.00 is placed into an account for 5 years at an interest rate of 6 %, then the value in 5 years is

$$F = (\$1.00)*(1 + 0.06)^5 = \$1.34$$

On the other hand, money that is received in 5 years is not worth as much as it is today and should be discounted. The present value of money received in the future is

$$P = \frac{F}{(1 + i)^n} \tag{2.13}$$

Example 2.7 The expected annual income from a project is $500/yr for 5 years, and the project requires a $2,000 investment. Determine whether or not to invest in the project at two required rates of return: 6 % and 8 %.

The net present value at a 6 % required rate of return:

$$\begin{aligned} P &= \frac{500}{(1+0.06)^1} + \frac{500}{(1+0.06)^2} + \frac{500}{(1+0.06)^3} + \frac{500}{(1+0.06)^4} + \frac{500}{(1+0.06)^5} - 2,000 \\ &= 472 + 445 + 419 + 396 + 373 - 2,000 = \$106 \end{aligned}$$

Because the present value of the investment is greater than zero ($106) at a 6 % required rate of return, the decision is made to invest in the project.

The net present value at an 8 % required rate of return:

$$P = \frac{500}{(1+0.08)^1} + \frac{500}{(1+0.08)^2} + \frac{500}{(1+0.08)^3} + \frac{500}{(1+0.08)^4} + \frac{500}{(1+0.08)^5} - 2,000$$
$$= 463 + 429 + 397 + 367 + 340 - 2,000 = -\$4$$

Because the present value of the investment is less than zero (−$4) at the 8 % required rate of return, the decision would be not to invest in the project if the ROR was 8 %.

Engineering economic analysis can also account for rate of growth, inflation, depreciation, and other factors. Many of these functions are available in Excel: go to INSERT – FUNCTION – FINANCIAL to access financial functions. The PV calculator (present value from series of future payments) is shown in Fig. 2.9. This calculator will convert an annual series to payments to the present value. For example, the PV calculator in Fig. 2.9 calculates the present value of the 5 payments of $500 at an 8 % interest rate as in Example 2.7. The length of the project in Example 2.7 is 5 years. Engineering economic analysis always includes the lifespan of the project. A typical project lifespan is 20 years.

The following parameters are often incorporated into engineering economic analysis for irrigation systems:

Component replacement (–)
Crop yield or improvement in crop yield with the new irrigation system (+)
Energy cost (–)
Capital investment (–)
Water cost (–)
Labor cost (–)
Environmental cost (–)
Maintenance (–)
Salvage value (+)

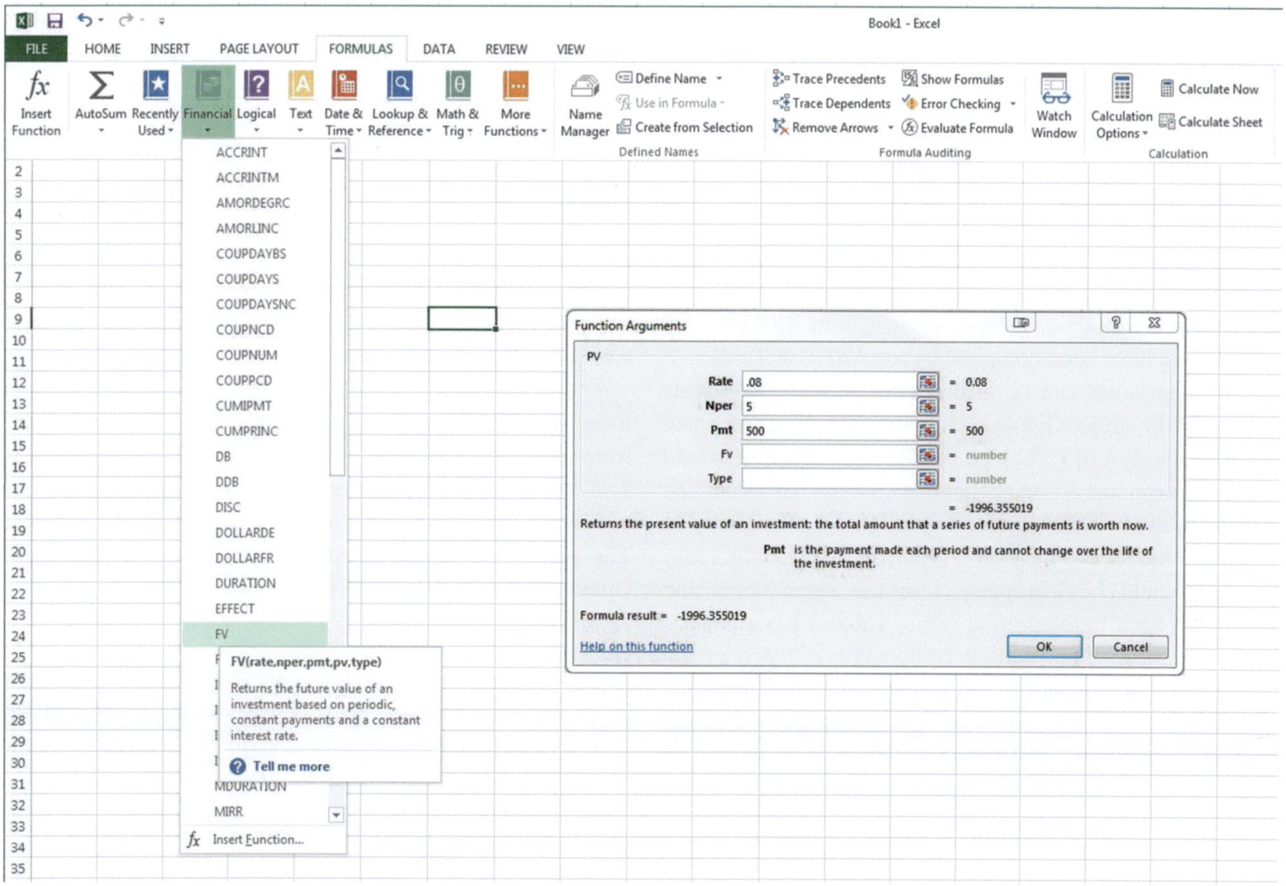

Fig. 2.9 Present value calculator (PV) in excel

Table 2.1 Typical life and annual maintenance cost percentage for irrigation system components (Credit: NRCS NEH, part 652, irrigation guide, Chap 5)

System and components	Life (yr)	Annual maint. (% of cost)	System and components	Life (yr)	Annual maint. (% of cost)
Sprinkler systems	10–15	2–6	**Surface & subsurface systems**	15	5
Hand move	15 +	2			
Side or wheel roll	15 +	2	Related components		
End tow	10 +	3	Pipelines		
Side move w/drag lines	15 +	4	Burled thermoplastic	25 +	1
Stationary gun type	15 +	2	Buried steel	25	1
Center pivot-standard	15 +	5	Surface aluminum	20 +	2
Linear move	15 +	6	Surface thermoplastic	5 +	4
Cable tow	10 +	6	Hurled nonreinforced concrete	25 +	1
Hose pull	15 +	6	Buried galv. steel	25 +	1
Traveling gun type	10 +	6	Buried corrugated metal	25 +	1
Fixed or solid set			Buried reinforced PMP	25 +	1
Permanent	20 +	1	Gated pipe, rigid, surface	10 +	2
Portable	15 +	2	Surge valves	10 +	6
Sprinkler gear driven, impact & spray heads	5–10	6	Pumps		
Valves	10–25	3	pump only	15 +	3
			w/electric motors	10 +	3
Micro systems [1/]	1–20	2–10	w/internal combustion engine	10 +	6
Drip	5–10	3			
Spray	5–10	3	Wells	25 +	1
Bubbler	15 +	2	Linings		
Send-rigid, hurled	10–20	2	Nonreinforced concrete	15 +	5
Semi-rigid, surface	10	2	Flexible membrane	10	5
Flexible, thin wall, buried	10	2	Reinforced concrete	20 +	1
Flexible, thin wall, surface	1–5	10			
Emitters & heads	5–10	6	Land grading, leveling[2/]		
Filters, injectors, valves	10 +	7	Reservoirs[3/]		

Expected annual maintenance costs and component replacement periods for different irrigation systems and components have been compiled by the *NRCS* in Table 2.1. Replacement costs can be incorporated into the economic analysis if the project lifespan is longer than the component life. The lifespan of PVC pipe is 25+ years. Thus, it would not be replaced in a 20 year economic analysis. However, the lifespan of thin wall subsurface drip irrigation tubing may be 2 years. Thus, the expense of replacing the tubing every 2 years would be incorporated in the spreadsheet and discounted back to the present value. Table 2.1 is a general guide. The lifespan of irrigation components is dependent on management, maintenance, soil quality, and water quality.

The value of the crop in future years is a function of many economic variables. Inflation of either crop prices or expenses can have a dramatic effect on profit. If expenses increase and crop prices do not, then farming can become less profitable, as shown in Example 2.8.

Example 2.8 Determine the present value of the income from alfalfa 6 years from now at a required rate of return of 7 %. (1) Assume no inflation. (2) Assume inflation of costs of 5 % per year and no inflation in the selling price of alfalfa. The present value of alfalfa production is $631/acre-yr. Operating expenses are $321/acre-yr.

(1) No inflation.

$$\text{Present value of year 6 profit (without inflation)}\\ (\$631 - \$321)*\left(1/(1+0.07)^6\right) = \$207.$$

(2) Costs inflate at a rate of 5 % per year.

Costs in year 6 :	$\$321(1.05)^6 = \$430/\text{acre}.$
Profit in year 6	$\$631 - \$430 = \$201/\text{acre}.$
Present value of year 6 profit(with inflation)	$\$201*\left(1/(1+0.07)^6\right) = \$134/\text{acre}.$

Crops are generally rotated in order to prevent crop diseases and soil degradation. Previously unfarmed soils or soils that have undergone major land leveling often require development before they can support crops with higher cash value. If soil in newly developed farms have a high initial salt content, as is the case with desert lands, salts must be leached from the soil during the first year of production. Salt tolerant crops such as sorghum, wheat, or barley, can be grown in the first seasons. After 1 or 2 years of grain crops, many farmers plant alfalfa for 3 years. Finally, the field is ready for higher value crops such as vegetables. For tree crops, the economic analysis must include the fact that fruit production does not begin for several years. It should also include an increasing rate of water application as the trees mature. A crop rotation or crop development plan must be specified in order to estimate annual income from a proposed farm or irrigation system.

Energy cost (pumping cost) is a function of pressure requirement, volume of applied water, pump efficiency, friction loss in pipes and valves, and the cost of energy.

Energy (joules) can be calculated based on change in the potential energy of the water.

$$E = mgh/Eff_p = V\rho gh/Eff_p = iA\rho gh/Eff_p \quad (2.14)$$

where

E = energy, J
m = mass of water, kg
g = gravity, m/sec^2
h = pump pressure, m
Eff_p = efficiency of the pump, dimensionless
V = gross volume applied to the field
ρ = density, kg/m^3
i = depth applied, m
A = field area, m^2.

Equation 2.14 can be rewritten in terms of irrigation parameters.

$$E(Joules/ha) = \frac{98{,}000\ (i_{mm})(h)}{Eff_p} \quad (2.15)$$

where

i_{mm} = depth applied per year, mm.

Energy is generally sold in units of kW-hr.

$$E(kW - hr/ha) = \frac{0.0272\ (i_{mm})(h)}{Eff} \quad (2.16)$$

Energy can be also be calculated as the product of pump power and time of pumping. Energy per ha = (pump power) (time)/(field area).

$$\text{Energy(kW - hr/ha)} = P\ t/A \quad (2.17)$$

where

A = Area, ha.
P = Pump power, kW
t = Seasonal pumping time, hr.

Example 2.9 Calculate the cost of energy to pump water for surface irrigation of cotton for 200 ha of land with an irrigation system that requires 3 m pressure head, and a pump that operates at 70 % efficiency. The cost of energy is $0.10/kW-hr. Assume that the surface irrigation system water application efficiency is 65 %. Cotton requires 91 cm of water per year.

The irrigation system must apply 910 mm/ 0.65 = 1,400 mm

$$E = \frac{0.0272(1{,}400\ mm)(3\ m)}{0.70} = 163\ kW - hr/ha$$

$$(\$/ha) = \left(\frac{163\ kW - hr}{ha}\right)\left(\frac{\$0.10}{kW - hr}\right) = \$16.30/ha$$

Total annual cost for 200 ha = $16.30/ ha * 200 ha = $3,260

When determining whether an existing surface irrigation system should be replaced with a new pressurized irrigation system, the economic benefit of the new irrigation system can be estimated based on the change in costs and benefits rather than the total costs and benefits. This concept is demonstrated in the following example.

Example 2.10 A farmer is considering converting his sloping furrow surface irrigation system to a new center pivot irrigation system for alfalfa. The installation cost for the new system is $2,000/ha. Should the farmer install center pivots?

Given

Energy

The center pivot irrigation system will require 20 m pressure head. Alfalfa requires 150 cm/yr with the center pivot system. The cost of energy is $0.10/kW-hr. Pumping efficiency is 80 %. No pumping is required for the surface irrigation system.

Water

Annual cost drops from $250/ha with the surface system to $200/ha with the pivot system.

Labor

One irrigator who makes $25,000 per year is usually required for every 100 acres (40 ha) of sloping furrow surface irrigation. This works out to approximately $625/ha. Assume that yearly labor cost for a pivot irrigation system is $200/ha (this estimate is probably high).

Environmental

Environmental cost drops from $100/ha for surface irrigation due to surface runoff of sediment to $0/ha with pressurized irrigation.

Economics

The required rate of return is 8 %. The length of the project is 20 years.

Maintenance

Maintenance cost of the pivot and surface irrigation systems is the same.

Yield

The yield value increases from $630/ha to $760/ha with the pressurized irrigation system.

Solution

Add all the annual costs and benefits. Convert this number back to a present value and compare to the capital cost.

Calculate energy costs with the new center pivot system.

$$E = \frac{0.0272(1,500\ mm)(20\ m)}{0.80} = 1,020\ kW - hr/ha$$

$$(\$/ha) = \left(\frac{1,020\ kW - hr}{ha}\right)\left(\frac{\$0.10}{kW - hr}\right) = \$102/ha$$

Annual costs and benefits

Input a decrease in costs or increase in benefits as a+.

Input an increase in costs or a decrease in benefits as a–

	Current surface system	Pivot system	Change
Yield Income	$630/ha	$750/ha	+$120/ha
Labor costs	$625/ha	$200/ha	+$425/ha
Water cost	$250/ha	$200/ha	+$50/ha
Environment	$100/ha	$0/ha	+$100/ha
Energy	$0/ha	$102/ha	–$102/ha
Change			+$593/ha

Annual Benefits – Costs = +$593/ha

Convert annual benefits to present value (20 years at 8 %) = $5,822/ha

The benefits of the irrigation system – cost = $5,822 – $2,000 = $3,822/ha

The decision is made to convert to center pivots because the present value is greater than zero.

Example 2.11 A farmer in a developing nation is considering converting his sloping furrow surface irrigation system to a new center pivot irrigation system for alfalfa. The installation cost for the new system is $2,000/ha. Should the farmer install center pivots?

Given

All parameters are the same as Example 2.10 except labor.

Labor

One irrigator who makes $2,000 per year is required for every 40 ha of sloping furrow surface irrigation. Yearly labor costs for a pivot irrigation system are $35/ha.

Solution

Annual costs and benefits

Input a decrease in costs or increase in benefits as a+.

Input an increase in costs or a decrease in benefits as a–

	Current surface system	Pivot system	Change
Yield income	$630/ha	$750/ha	+$120/ha
Labor costs	$50/ha	$35/ha	+$15/ha
Water cost	$250/ha	$200/ha	+$50/ha
Environment	$100/ha	$0/ha	+$40/ha
Energy	$0/ha	$102/ha	–$102/ha
Change			+$123/ha

Annual Benefits – Costs = +$123/ha
Convert Annual Benefits to present value (20 years at 8 %) = $1,207/ha
The benefits of the irrigation system – cost = $1,207–$2,000 = –$793/ha

The farmer should not convert to center pivots because the net present value is less than zero. In a country with lower labor costs, the decision is reversed from the previous example.

Cash Flow Diagrams

Cash flow diagrams show the expected crop income, energy costs, water costs, replacement costs, etc... each year for the life of the project. Inflation and other changes in prices are incorporated into the spreadsheet. It is especially important in agriculture with the development of new lands or planting of trees to accurately calculate the expected yield and costs each year.

A consideration in many economic analyses is risk; however, the analyses in this section only use the expected average yield.

Total benefits minus costs are calculated each year for the cash flow diagram. The total for each year is then discounted to the present value based on the required rate of return. This process is demonstrated in the following example.

Example 2.12 Redo Example 2.10, but use the crop rotation for development of a new farm that is shown in Table 2.2. Also include the following inflation rates and price changes.

Given
It is expected that the selling price of alfalfa will inflate by 7 % each year and that the selling price of cotton, wheat, and sorghum will inflate by 5 % each year. The present annual benefit of growing alfalfa, wheat/sorghum, and wheat/cotton, neglecting water, labor, and energy costs, is $1,000/ha, $500/ha, $700/ha, respectively.

The present annual cost of energy is $100/ha; and the cost of energy is expected to inflate by 8 % per year for all crop rotations. The present annual cost of water is $200/ha; and the cost of water is expected to inflate by 4 % per year for all crop rotations. The present annual cost of irrigation labor and maintenance is $200/ha, and is expected to inflate by 5 % per year for all crop rotations.

There is no environmental cost.

The capital cost of purchasing the land ($500/ha) and developing the farm ($2,500 / ha) is $3,000/ha. In addition, $300 must be added for repairs and maintenance each year.

Table 2.2 Crop rotation

Year	Winter	Summer	Year	Winter	Summer
1	Wheat	Sorghum	11	Wheat	Cotton
2	Alfalfa	Alfalfa	12	Wheat	Cotton
3	Alfalfa	Alfalfa	13	Wheat	Cotton
4	Alfalfa	Alfalfa	14	Alfalfa	Alfalfa
5	Wheat	Cotton	15	Alfalfa	Alfalfa
6	Wheat	Cotton	16	Alfalfa	Alfalfa
7	Wheat	Cotton	17	Wheat	Cotton
8	Alfalfa	Alfalfa	18	Wheat	Cotton
9	Alfalfa	Alfalfa	19	Wheat	Cotton
10	Alfalfa	Alfalfa	20	Alfalfa	Alfalfa

The crop rotation and expected annual benefit at current prices are shown in Table 2.2.

Solution
A cash flow diagram for the 20 year project is constructed in the *Cash flow* worksheet (Figs. 2.10 and 2.11). The future prices of the crops are based on their own inflation rates in cells C2:C8. The profit each year is summed and converted to the present value at the required rate of return.

The sum of present value annual costs and benefits at an 8 % required rate of return is $326/ha. Thus, the project has a profit greater than the required rate of return and is recommended.

Customer Relations and Sales

From the point of view of many irrigation companies, the most important economic concern is selling irrigation systems or selling the design service for irrigation systems. In order to do this, one must gain the confidence of the farmer. One of the first things a farmer wants to know when considering a new system is the water and power requirement. The best way to lose the job is to stand there with your mouth open and not have an answer to these questions. A few simple equations for estimating water and power requirements should be memorized.

Irrigation system water requirements are based on the worst case – the part of the year when the plants require the most water. The rationale is that if the irrigation system supplies enough water during the time of the year with maximum evapotranspiration, then it will provide enough water during the rest of the year. It is up to the farmer to schedule the irrigation frequency during the rest of the year

	Present Costs ($/ha)	Inflation rate / yr
Water cost ($/ha)	-200	0.04
Energy cost ($/ha)	-100	0.08
Labor cost ($/ha)	-200	0.05
Required rate of return		0.08
Wheat/sorghum	400	0.05
Alfalfa	1000	0.07
Wheat/cotton	700	0.05

Net present value $/ha	$ 326.04

Hide chart | View chart

Year	Winter	Summer	Present benefit	Inflation rate	Capital or salvage cost	Crop value	Water cost	Energy cost	Labor cost	Sum of annual	Present value
					Annual costs and benefits ($/ha)						
0	-	-		-	-3000	0	0	0	0	$(3,000.00)	$(3,000.00)
1	Wheat	Sorghum	400	0.05	-300	420	-208	-108	-210	$ (406.00)	$ (375.93)
2	Alfalfa	Alfalfa	1000	0.07	-300	1145	-216	-117	-221	$ 291.44	$ 249.86
3	Alfalfa	Alfalfa	1000	0.07	-300	1225	-225	-126	-232	$ 342.57	$ 271.95
4	Alfalfa	Alfalfa	1000	0.07	-300	1311	-234	-136	-243	$ 397.67	$ 292.30
5	Wheat	Cotton	700	0.05	-300	893	-243	-147	-255	$ (52.12)	$ (35.47)
6	Wheat	Cotton	700	0.05	-300	938	-253	-159	-268	$ (41.70)	$ (26.28)
7	Wheat	Cotton	700	0.05	-300	985	-263	-171	-281	$ (31.02)	$ (18.10)
8	Alfalfa	Alfalfa	1000	0.07	-300	1718	-274	-185	-295	$ 663.89	$ 358.68
9	Alfalfa	Alfalfa	1000	0.07	-300	1838	-285	-200	-310	$ 743.63	$ 372.00
10	Alfalfa	Alfalfa	1000	0.07	-300	1967	-296	-216	-326	$ 829.43	$ 384.19
11	Wheat	Cotton	700	0.05	-300	1197	-308	-233	-342	$ 14.11	$ 6.05
12	Wheat	Cotton	700	0.05	-300	1257	-320	-252	-359	$ 25.90	$ 10.29
13	Wheat	Cotton	700	0.05	-300	1320	-333	-272	-377	$ 37.85	$ 13.92
14	Alfalfa	Alfalfa	1000	0.07	-300	2579	-346	-294	-396	$ 1,242.49	$ 423.02
15	Alfalfa	Alfalfa	1000	0.07	-300	2759	-360	-317	-416	$ 1,365.84	$ 430.57
16	Alfalfa	Alfalfa	1000	0.07	-300	2952	-375	-343	-437	$ 1,498.40	$ 437.37
17	Wheat	Cotton	700	0.05	-300	1604	-390	-370	-458	$ 86.43	$ 23.36
18	Wheat	Cotton	700	0.05	-300	1685	-405	-400	-481	$ 98.54	$ 24.66
19	Wheat	Cotton	700	0.05	-300	1769	-421	-432	-505	$ 110.54	$ 25.61
20	Alfalfa	Alfalfa	1000	0.07	-300	3870	-438	-466	-531	$ 2,134.70	$ 458.00
											$ 326.04

Fig. 2.10 Annual crop value, water, energy, and labor from *Cash flow* worksheet

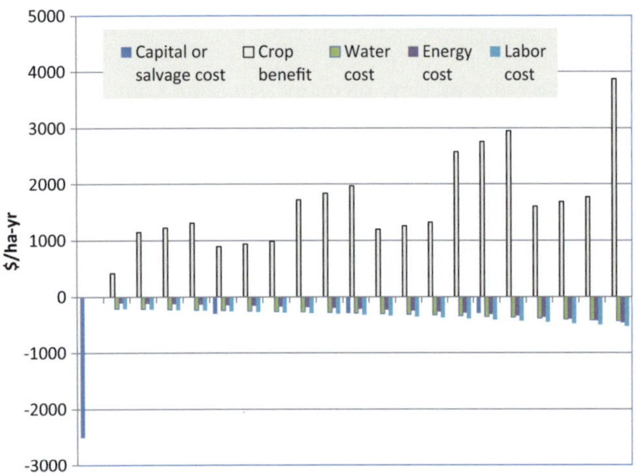

Fig. 2.11 Cash flow diagram

such that the water application matches the rate of evapotranspiration.

The irrigation system must supply a flow rate that is equal to the product of maximum crop evapotranspiration (ET_{c-max} mm/day) and farm area (A) divided by the efficiency of the system.

$$Q = ET_{c\text{-}max} * A/Eff \tag{2.18}$$

Rules of thumb for estimating flow rate requirements can be developed in each region. For example, if the maximum expected evapotranspiration rate is 13 mm/day in a region, then the water requirement per acre or per hectare is calculated as follows:

$$ET_{c-\max} = \frac{13\ mm}{day}\left(\frac{10,000\ m^2}{ha}\right)\left(\frac{1\ m}{1,000\ mm}\right) = \frac{130\ m^3}{day\ ha}$$

$$ET_{c-\max} = \frac{130\ m^3}{day}\left(\frac{1\ day}{24\ hr}\right)\left(\frac{1\ hr}{60\ \min}\right)\left(\frac{1,000\ L}{1\ m^3}\right) = \frac{90\ L/\min}{ha}$$

$$ET_{c-\max} = \frac{90\ L/\min}{ha}\left(\frac{1\ gal}{3.785\ L}\right)\left(\frac{0.4047\ ha}{1\ acre}\right) = \frac{9.65\ gpm}{ac}$$

In-class Exercise 2.3 Develop a rule of thumb for pump flow rate requirement for a region that has a max ET = 10 mm/d. Assume system water application efficiency is 80 % and 10 % down time?

What pump flow capacity will be required for a 10 ha farm if ET = 10 mm/day?

The other number that a farmer wants to know is the required pump horsepower. The power is a function of flow rate, pressure, and efficiency.

$$P = Q\rho gh/Eff \quad (2.19)$$

where

P = power, W
Q = flow rate, m^3/sec
ρ = density of water, 1,000 kg/m^3
g = gravity, 9.8 m/sec^2
h = pressure head, m
Eff = efficiency, fraction.

Equation 2.20 can be simplified for metric and United States units. The Power is calculated for typical irrigation units as follows:

Metric:

$$\begin{aligned}\text{Power(kW)} &= (\text{Q}(\text{m}^3/\sec)*\text{H})/(0.102*\text{Eff})\\ \text{Power(mhp)} &= (\text{Q(L/min)}*\text{H})/(4634*\text{Eff})\end{aligned} \quad (2.20)$$

where

Q = flow rate, m^3/sec (upper) or L/min (lower)
H = discharge pressure, m
Eff = pump energy efficiency, fraction
mhp = metric horsepower.

U.S.

$$HP = \frac{gpm*ft}{3,960*Eff} \quad (2.21)$$

where

HP = horsepower, U.S. units
ft = ft of head
gpm = flow rate in gpm
Eff = efficiency of the pump, fraction.

Energy is generally sold in metric units: kW-hr. To convert from HP to kW, multiply by 0.746: kW = HP * 0.746. Metric horsepower (mhp) is almost the same as U.S. HP. HP = mhp * 1.014.

In-class Exercise 2.4 Students act out the following play about a typical meeting between the irrigation engineer and the client (this may be the only irrigation engineering play in history).

Josephine, the irrigation sales engineer from Irrigation Systems Design Company, drives up to McGillycutty Wellington Growing and Harvesting Farms (MWGH farms or My Way or Get Back to the Highway Farms) in her Chevrolet S10 pickup truck.

Mr. McGillycutty: "So Josephine, come on in and have a cup of coffee."
Josephine: "Don't mind if I do, Mr. McGillycutty."
Mr. McGillycutty: "Josephine, what size pump do I need if I want to convert my section of land (640 acres) to drip?"
Josephine makes the following silent mental calculation while she sips a cup of coffee.

$$640*10 = 6,400 \text{ gpm}.$$

Josephine: I can see that you are growing cotton and alfalfa, so Mr. McGillycutty, you would need somewhere between 6,000 and 7,000 gpm.
Mr. McGillycutty: "How much is that in cubic feet per second?"
Josephine: "Mr. McGillycutty, 450 gpm equals about 1 cfs, so that would be approximately 8 cfs."
Mr. McGillycutty: "No problem, we use quite a bit more than that for our current flood system. I am worried about bringing electrical power out for the pump. What horsepower pump would that require?"
Josephine: "If you want to irrigate 640 acres with drip irrigation, then you would probably need to generate about 30 psi pressure. Large pumps probably have an 80 % electrical efficiency." She then puts her coffee down, pulls out her calculator, and makes the following calculation:

$$30 \text{ PSI}*2.31 = 70 \text{ ft}$$

$$\left(\frac{6,400\ gpm*70ft}{3,960*0.8}\right) = 150\ HP$$

Josephine puts her cup of coffee down and says; "Mr. McGillycutty, you need 150 HP."
Mr. McGillycutty: "What is that in kilowatts?"
Josephine: "That would be about 110 kW."
Mr. McGillycutty: "That sounds reasonable. How soon can you design the system?"
Josephine: "I will get back to you in a few weeks."

During the process of design, it is quite typical for growers to make changes. In fact, many growers will make three or more major changes during the design process. This causes extra work.

There are many other practices that should be followed by the engineer in order to maintain good relationships with customers. Many of these are detailed in the *Turf and Landscape Best Management Practices* by the Irrigation Association.

Questions

1. Why does increasing the depth of irrigation water applied per season reach a point of diminishing returns?
2. Why is there a nearly linear relationship between evapotranspiration (ET) and yield and a nonlinear relationship between applied water and yield?
3. How are the crop water production function and engineering economic analysis used to assess or manage irrigation systems.
4. Find the profit with the following parameters for the Grimes and El-Zik CWPF for drip irrigated cotton: 15 cm depth of precipitation, cost of water is $0.02/m^3$. The selling price of cotton is $0.90/kg. Depth of applied water, $AW_{CWPF} = 62$ cm. Show your work.
5. For the parameters in question 4, calculate the optimal depth of water application, AW_{CWPF}. Do this problem in Excel and turn in the graph that shows the yield, cost of water, and profit vs. AW_{CWPF}.
6. For the parameters in question 4, calculate the profit for a surface irrigation system with 70 % efficiency at $AW_{CWPF} = 70$ cm. Preirrigation is 45 cm. The water from preirrigation provides no benefit for crop growth in this case. Do not consider erosion. Show your work.
7. For the parameters in question 6, find the optimal depth of water application, AW_{CWPF}. Do this problem in Excel, and turn in the graph that shows the yield, cost of water, and profit vs. AW_{CWPF}.
8. For the parameters in question 6, calculate the profit per ha, but include erosion with erosion calculated by Eq. 2.7 and the cost of erosion equal to $0.004/kg.
9. For the parameters in question 8, find the optimal depth of water application, AW_{CWPF}. Do this problem in Excel, and turn in the graph that shows the yield, cost of water, cost of erosion, and profit vs. AW_{CWPF}.
10. Use the *Variable water application* worksheet. Keep all other parameters the same, but double the cost of water. Find the optimal depth of water application.
11. Use the *Variable water application* worksheet but change the number of furrow sections to 6 with the following multipliers of end furrow application. 1.5, 1.45, 1.33, 1.25, 1.13, and 1. Keep all other parameters the same as the original version. Find the optimal depth of water application.
12. Should government charge farmers with surface irrigation systems that discharge sediment to streams for the cost of dredging and for sediment removal from public drinking water supplies? Why or why not?
13. Should the government pay farmers to convert to pressurized irrigation if a farming region has erosive soils?
14. At the government policy level, should economic analysis of irrigation systems include all environmental degradation due to irrigation?
15. Calculate cotton yield if the required depth of applied water in a region is 120 cm, actual applied water depth is 80 cm, and the maximum yield is 1200 t/ha. Look at Fig. 2.6. Is your answer realistic?
16. There is 10 % yield loss due to pest stress in addition to the loss due to water stress calculated in question 15. What is the expected yield?
17. The expected annual income from a project is $600/yr for 5 years, and the project requires a $2,000 investment. Determine whether or not to invest in the project at a 9 % required rate of return.
18. Determine the present value of the income from cotton 4 years from now for full water application and a required rate of return of 7 %. (1) Assume no inflation. (2) Assume inflation of 5 % per year in costs and no inflation in the selling price of cotton. Current yearly income is $854.25, and current yearly expenses are $721.16
19. Calculate the cost of energy to pump water for sprinkler irrigation alfalfa for 100 ha of land with an irrigation system that requires a 30 m pressure head, and a pump that operates at 80 % efficiency. The cost of energy is $0.10/kW-hr. Assume that the sprinkler irrigation system water application efficiency is 75 %. Assume that alfalfa requires 5 m of water per year
20. Develop a rule of thumb for pump flow rate requirement for an area that has a max ET = 10 mm/d in terms of gpm/acre? The expected irrigation efficiency is 70 %, and the expected downtime is 10 %. What pump flow capacity will be required for a 10 acre farm? If the farm is divided into five irrigation zones, then what pump size is required?
21. For the flow rate in question 20, what motor horsepower is required and wattage is required if the pump operates at 100 ft pressure and is 85 % efficient?
22. Using the *Cash flow diagram* worksheet, redo the analysis for alfalfa planted every year.

References

Brown M, Carter D, Bondurant J (1974) Sediment in irrigation and drainage waters and sediment inputs and outputs for two large tracts in Southern Idaho. J Environ Qual 3(4):347–351
Cuenca RH (1989) Irrigation system design. Prentice Hall, Englewood Cliffs
Grimes DW, El-Zik KM (1990) Cotton. In: Stewart BA, Nielsen DR (eds) Irrigation of agricultural crops. American Society of Agronomy, Madison, pp 741–773, Monograph # 30
Irrigation Association. Turf and landscape best management practices. http://www.irrigation.org/landscapebmps/

Soil Physics

3

For some, it is difficult to appreciate the discipline of soil science (Fig. 3.1). It sounds boring and like the word that soil scientists hate to hear: dirt. However, soil is a living ecosystem with nutrients, water, air, and microorganisms. It is the anchor for the roots and allows the plant to stand upright. The earth did not always have soil. There was only rock. In combination with chemical and physical weathering processes, plant roots built the soil. Early plants extended roots into rocks, and thus split apart the rocks into soil particles. Soils form slowly, which is why soil conservation is such an important concept.

Understanding the soil is the first key to successful irrigation and drainage system design. Infiltration rate is controlled by soil texture and structure. The water holding capacity and root depth determine the length of time between irrigation events. Matric potential becomes more negative as the soil dries. The matric potential level determines the ease with which plants remove water from the soil. When the soil dries beyond the management allowable depletion (MAD), the matric potential becomes too negative, and the plant uptake lags behind the crop water demand. One way to prevent excessive soil moisture depletion is to monitor the soil with soil moisture sensors. The soil moisture depletion can also be calculated based on the field capacity, permanent wilting point, and rate of evapotranspiration. The following soil parameters are reviewed in this chapter: soil layering, soil texture and structure, root zone depth, soil water holding capacity, and infiltration rate.

Soil Development and Layering

Layers (soil horizons) form in the soil with different structures and textures. Understanding the response of soil layers to irrigation and agricultural practices is essential to maintaining a productive soil environment. In saturated soils, tractors and overlying soil can cause the formation of plow layers, also known as hardpan. Plow layers restrict infiltration and root growth. Even sand layers can restrict infiltration. If a fine-textured soil layer is underlain by a sand layer, then water may remain perched above the sand because the capillary suction of the fine soil prevents the downward movement of water into the sand, which has very little capillary suction. Soil layering is also an important consideration for subsurface drainage systems design. If possible, drains should be placed in layers with high hydraulic conductivity.

Soil Texture

Soil particles are classified based on particle diameter. There are three major categories of soil particle size: sand, silt, and clay. The USDA particle size classification scheme is in Table 3.1.

Coarse sand has approximately the same diameter as pencil lead (Fig. 3.2) while the largest clay particle diameters are three orders of magnitude smaller. Based on the percentages of sand, silt, and clay, soils are classified by "textural class names" such as sandy loam, clay, and silt loam.

The USDA soil textural triangle is used to classify soil textures (Fig. 3.3).

In-class Exercise 3.1 Find the textural class name of soil with 40 % clay, 20 % sand, and 40 % silt?

Soil textures can also be classified (NRCS-NEH Part 652-2) based on the way that the soil responds in the hand (the feel and appearance method).

Sand—Sand is loose and single-grained. The individual grains can be readily seen and felt. Squeezed in the hand when dry, sand falls apart when pressure is released. Squeezed when moist, it forms a cast, but crumbles when touched.

Sandy loam—A sandy loam is soil containing a high percentage of sand, but having enough silt and clay to make it somewhat coherent. The individual sand grains can be readily seen and felt. Squeezed when dry, a sandy loam forms a cast that falls apart readily. If squeezed when moist, a cast can be formed that bears careful handling without breaking.

P. Waller, M. Yitayew, *Irrigation and Drainage Engineering*, DOI 10.1007/978-3-319-05699-9_3

Fig. 3.1 Soil scientist describing soil layers and roots (Credit Jeff Vanuga, NRCS)

Table 3.1 Particle size classification by USDA

Material	Diameter
Stones	>250 mm
Cobbles	80–250 mm
Course gravel	12.5–80 mm
Fine gravel	2.0–12.5 mm
Very coarse sand	1.0–2.0 mm
Coarse sand	0.5–1.0 mm
Medium sand	0.25–0.5 mm
Fine sand	0.1–0.25 mm
Very fine sand	0.05–0.1 mm
Silt	0.002–0.05 mm
Clay	Less than 0.002 mm

Loam—A loam is soil having a relatively even mixture of different grades of sand, silt, and clay. It is friable with a somewhat gritty feel, but is fairly smooth and slightly plastic. Squeezed when dry, it forms a cast that bears careful handling, and the cast formed by squeezing the moist soil can be handled freely without breaking.

Silt loam—A silt loam is soil having a moderate amount of fine sand with a small amount of clay. Over half of the particles are silt size particles. When dry, a silt loam appears cloddy, but the lumps can be readily broken. When pulverized, it feels soft and floury. When wet, the soil runs together readily and puddles. Either dry or moist, silt loam forms a cast that can be handled freely without breaking. When moist and squeezed between thumb and finger, it does not ribbon, but has a broken appearance.

Clay loam—A clay loam is moderately fine-textured soil that generally breaks into clods or lumps that are hard when dry. When the moist soil is pinched between the thumb and finger, it forms a thin ribbon that breaks readily, barely sustaining its own weight. The moist soil is plastic and forms a cast that bears much handling. When kneaded in the hand, clay loam does not crumble readily, but works into a heavy compact mass.

Clay—A clay is fine-textured soil that usually forms very hard lumps or clods when dry and is very sticky and plastic when wet. When moist soil is pinched between thumb and finger, it forms a long flexible ribbon. Some clays are very high in colloids are friable and lack plasticity.

Three laboratory methods are commonly used to measure soil particle size distribution: laser particle size analysis, hydrometer and sieve. The state of the art is the laser system, which can measure hundreds of samples per day. The traditional method for measuring small particle size is the hydrometer, which measures the rate that particles sink in quiescent water. This method is based on Stoke's law of falling bodies. It requires manual observation and can only measure one sample per day. A series of standard sieves can be used to measure larger particles with diameters greater than 2 microns (coarse sand and larger).

Once the soil texture is known, the Soil Water Characteristics Calculator (URL in References) calculates soil hydraulic properties. The Calculator also includes compaction, which incorporates the soil structure concepts described below.

Soil Structure

Soil structure has a strong influence on soil properties. Soil structure is defined by the NRCS as follows:

> Soil structure is the arrangement and organization of soil particles into natural units of aggregation. These units are separated from one another by weakness planes that persist through cycles of wetting and drying and cycles of freezing and thawing. Structure influences air and water movement, root development, and nutrient supply.

Soil structure strongly influences infiltration rate. Single grain soils are sands with no cementing agents attaching particles to each other. Granular soils are aggregates of small particles. Both of these soils have rapid infiltration rates. Blocky and prismatic soils are larger aggregates and have moderate infiltration rates. Although platy structures are aggregates, they have poor vertical infiltration. Massive

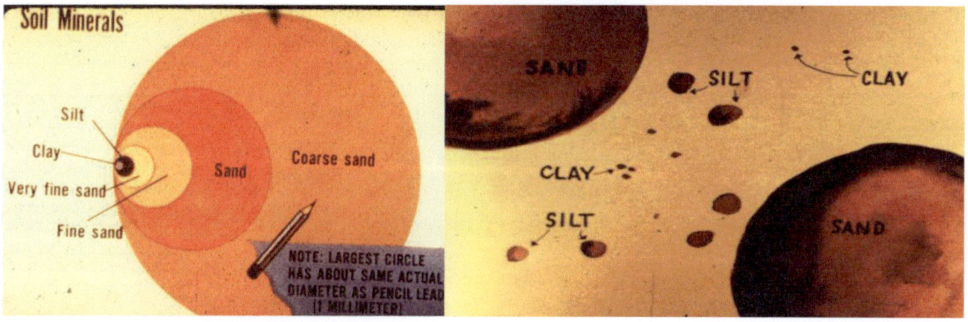

Fig. 3.2 Soil particle sizes (Courtesy of Don Post. The University of Arizona)

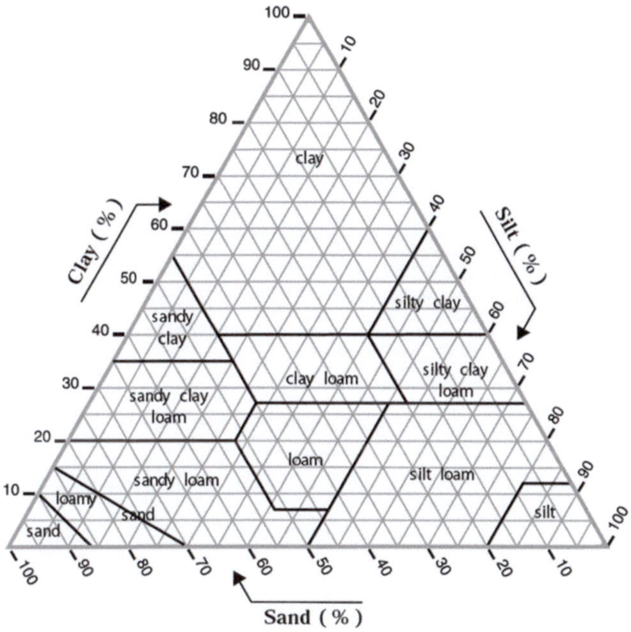

Fig. 3.3 USDA soil textural triangle (Credit NRCS, National Agronomy Manual. Part 504)

soils are generally grossly swelling soils (montmorrilonite clay). These soils have extremely low infiltration rates.

Cultural practices can either improve (increase porosity and hydraulic conductivity) or degrade soil texture (increase density). For example, driving a tractor in a wet field can decrease soil structure and form hardened layers with low infiltration rates.

Root Zone

The depths to which roots of various mature crops will extract available water are shown in Table 3.2. The reported depths in Table 3.2 are 80 % of actual root zone depths because plants extract most of their water from the upper portion of the root zone.

A rule of thumb that is applied to root water uptake is that 40 % of plant water uptake is from the upper 25 % of the root zone, 30 % from the next, 20 % from the next, and 10 % from the lowest quarter of the root zone (Fig. 3.4).

Soil-Water Relationships

The two most important parameters that describe the status of water in soil are water content and total water potential. Water content refers to the fraction of the soil that is occupied by water and can be measured by mass or volume. Total water potential refers to the energy of the water in soil and generally includes matric potential (also called capillary potential or moisture potential), gravitational potential due to elevation, and osmotic potential due to salinity.

Water content can be calculated gravimetrically or volumetrically. Gravimetric soil water content is the mass of water divided by the mass of dry soil. It can be measured by weighing a mass of wet soil, drying the soil for 24 hours at 105 °C, and then reweighing the sample.

$$\theta_{grav} = \frac{m_{water}}{m_{dry\ soil}} = \frac{m_{wet\ soil} - m_{dry\ soil}}{m_{dry\ soil}} \quad (3.1)$$

where

θ_{grav} = gravimetric water content, gm/gm
m_{water} = mass of water, gm.
$m_{dry\ soil}$ = mass of soil after drying, gm.
$m_{wet\ soil}$ = mass of soil before drying, gm.

Example 3.1 A soil sample is collected just before irrigation, and another soil sample is collected two days after irrigation. The soil sample collected before irrigation weighs 1.73 kg, and the soil collected just after irrigation weights 1.94 kg. Both soils are placed in an oven and dried at 100 °C

Table 3.2 Depth to which roots of mature crops will extract available water from a deep uniform, well-drained soil under average unrestricted conditions (depths shown are 80 % of the entire root zone depth). (Credit NRCS NAM 504-4)

Crop	Depth (m)	Crop	Depth (m)
Alfalfa	1.5	Milo	0.6–1.2
Asparagus	1.5	Mustard	0.6
Bananas	1.5	Onions	0.3–0.6
Beans, dry	0.6–0.9	Parsnips	0.6–0.9
Beans, green	0.6–0.9	Peanuts	0.6–0.9
Beets, table	0.6–0.9	Peas	0.6–0.9
Broccoli	0.6	Peppers	0.3–0.6
Berries, blue	1.2–1.5	Potatoes, Irish	0.6–0.9
Berries, cane	1.2–1.5	Potatoes, sweet	0.6–0.9
Brussels sprouts	0.6	Pumpkins	0.9–1.2
Cabbage	0.6	Radishes	0.3
Cantaloupes	0.9	Safflower	1.2
Carrots	0.6	Sorghum	1.2
Cauliflower	0.6	Spinach	0.3–0.6
Celery	0.3–0.6	Squash	0.9–1.2
Chard	0.3–0.6	Strawberries	0.3–0.6
Clover, Ladino	0.6–0.9	Sudan grass	0.9–1.2
Cranberries	0.3	Sugar beets	1.2–1.5
Corn, sweet	0.6–0.9	Sugarcane	1.2–1.5
Corn, grain	0.9–1.2	Sunflower	1.2–1.5
Corn, seed	0.9–1.2	Tobacco	0.9–1.2
Corn, silage	0.9–1.2	Tomato	0.9
Cotton	1.2–1.5	Turnips	0.6– 0.9
Cucumber	0.3–0.6	Watermelon	0.9– 1.2
Eggplant	0.6	Wheat	1.2
Garlic	0.3–0.6	Trees	
Grains and flax	0.9–1.2	Fruit	1.2–1.5
Grapes	1.5	Citrus	0.9–1.2
Grass pasture/hay	0.6–1.2	Nut	1.2–1.5
Grass seed	0.9–1.2		
Lettuce	0.3–0.6		
Melon	0.6–0.9		

for 24 hours. After drying, the soil collected just before irrigation weighs 1.49 kg and the soil collected just after irrigation weighs 1.52 kg. What are the volumetric water contents just before and just after irrigation?

Solution:

Gravimetric water content just before irrigation

$$\theta_{grav} = \frac{1.79\ kg - 1.49\ kg}{1.49\ kg} = 0.20 = 20\%$$

Gravimetric water content just after irrigation

$$\theta_{grav} = \frac{1.94\ kg - 1.52\ kg}{1.52\ kg} = 0.27 = 27\%$$

Water content collected two days after irrigation is close to field capacity because the soil has drained, but there has been little evaporation after drainage.

Gravimetric water content soil samples can be collected with a shovel or auger because the measurement is not dependent on the volume of soil. Volumetric water content is defined as the percent of total soil volume that is occupied by water and is more difficult to measure than gravimetric water content because it is based on soil volume. Gravimetric water content is converted to volumetric water content by multiplying by the soil bulk density.

$$\theta_v = \theta_{grav}\rho_b = \theta_{grav}\frac{\rho_b}{1} = \theta_{grav}\frac{\rho_b}{\rho_w} = \theta_{grav}\ \rho_b \tag{3.2}$$

where

θ_v = volumetric water content, cm^3/cm^3
ρ_b = soil bulk density, gm/cm^3.
ρ_w = density of water, 1 gm/cm^3.

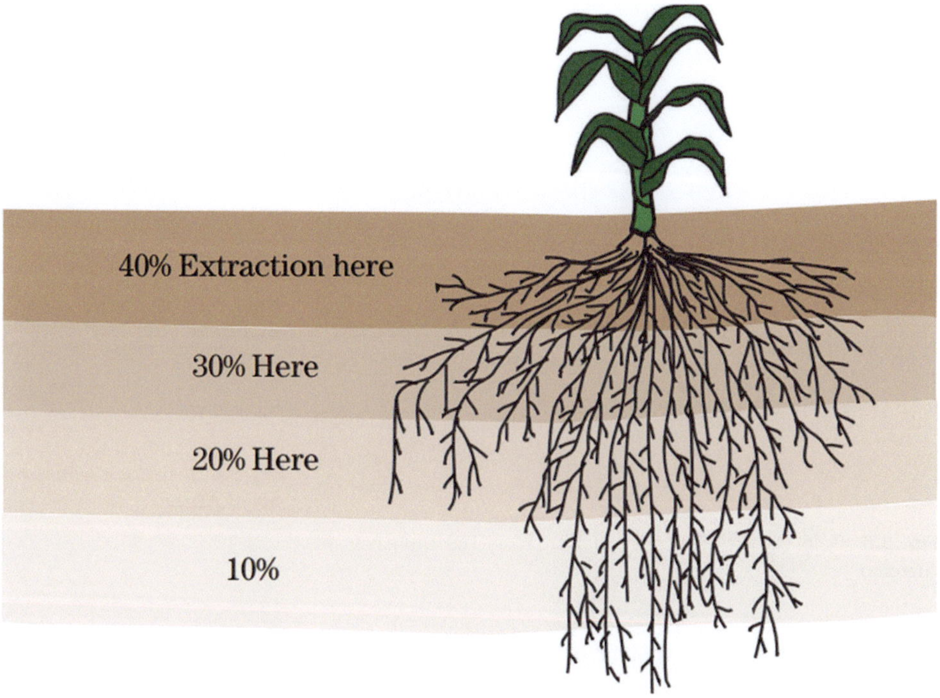

Fig. 3.4 Typical water uptake pattern (Credit: NRCS)

The soil bulk density (ρ_b) is the mass of dry soil divided by the volume of soil. Specialized soil sampling tools are used for collection of undisturbed soils for measurement of bulk density – the soil sample must be collected in a cylinder of known size, and the soil sample within the cylinder must be undisturbed. One such device is a bucket auger (cylinder), with slots at the top and bottom of the cylinder. Knives are inserted into the slots, and soil is removed above and below the sample. Then the knives are removed and the soil is pushed out of the auger into a drying can, which is then dried in the oven. A second type of bulk density sampling device is a small metal cylinder that is inserted into the soil and removed with the soil sample. Excess soil is then scraped off above and below the cylinder.

Soil bulk density can range from 1 to 2 g/cm^3. Typical clay, loam, and sandy soils have bulk densities equal to 1.3-, 1.5-, and 1.8-g/cm^3, respectively. Organic matter or fossilized shells can cause the bulk density to even be less than 1 g/cm^3, the density of water.

Porosity is the fraction of voids in a soil (air + water). It can be calculated based on the bulk density of a soil and the average particle density of soil particles. Average particle density of mineral soils is 2.65 g/m^3.

$$\phi = 100 - \left(\frac{\rho_b}{\rho_p}\right)(100) \tag{3.3}$$

where

ϕ = porosity (fraction of voids), percent
ρ_p = particle density, 2.65 g/cm^3 for mineral soils.

Example 3.2 The soils described in Example 3.1 have bulk density 1.3 g/cm^3. Find the volumetric water contents and porosity of the two samples.

Solution:
Volumetric water content just before irrigation

$$\theta_v = \theta_{grav}\rho_b = (0.20\ g/g)(1.3\ g/cm^3) = 0.26 = 26\%$$

Volumetric water content just after irrigation

$$\theta_v = \theta_{grav}\rho_b = (0.27\ g/g)(1.3\ g/cm^3) = 0.36 = 36\%$$

The bulk density is 1.3 g/cm^3 and soil particle density is 2.65 g/cm^3.

$$\phi = 100 - \left(\frac{1.3}{2.65}\right)(100) = 51\%$$

This means that 51 % of the soil volume is filled with air or water.

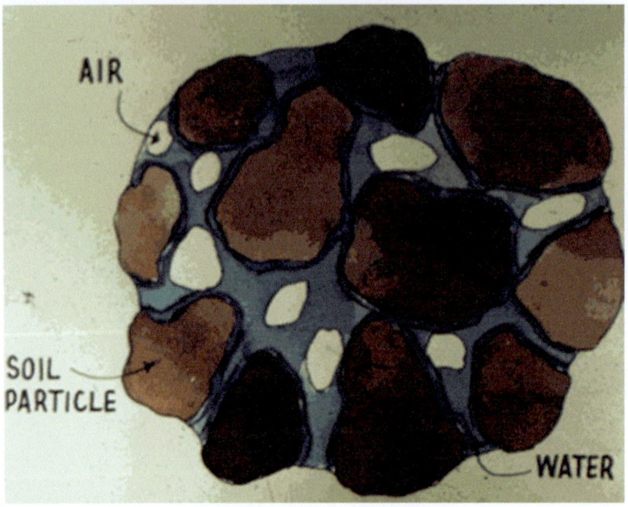

Fig. 3.5 Field capacity (Courtesy of Don Post. The University of Arizona)

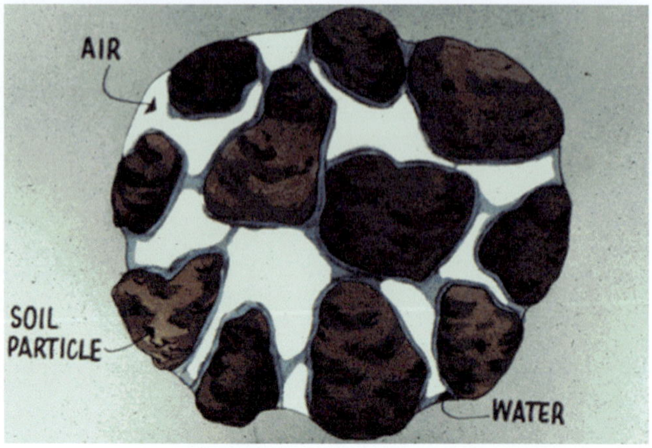

Fig. 3.6 Permanent wilting point (Courtesy of Don Post. The University of Arizona)

Three primary water content percentages are generally considered in irrigation and drainage system designs: saturation (θ_{sat}), field capacity (θ_{FC}), and permanent wilting point (θ_{PWP}).

Saturated soils have nearly all pore space filled by liquid, except for a small percentage of pores with trapped air: thus, saturated water content is approximately equal to porosity. High water content is not desirable because it restricts oxygen diffusion from the atmosphere into the soil because most of the soil pores are occupied by water: oxygen diffusion is approximately 1,000 times greater in a gas than it is in water. This lack of oxygen restricts oxygen uptake by plant roots during respiration.

Field capacity (Fig. 3.5) is the amount of water left in the soil after free drainage by gravity (larger soil pores are drained by gravity). This takes place quickly in coarse textured soils (a few hours), less quickly in medium textured soils (24 hr) and slowly (several days) in fine textured soils.

Field capacity can be estimated in the laboratory by applying a vacuum. A suction of – 0.01 MPa (−0.1 atm) is applied to sandy soils and a suction of – 0.03 MPa (−0.3 atm) is applied to clay soils. Then, the soil is weighed, dried, and then reweighed in order to calculate gravimetric water content at field capacity. Volumetric water content at field capacity can be calculated if porosity and bulk density of the soil are known.

The wilting point (Fig. 3.6) is the water content at which a crop cannot remove water from the soil quickly enough to prevent wilting and tissue damage. This point varies for different crops and varies with evapotranspiration demand because the process of water moving to the root is time dependent. Permanent wilting point (PWP) is generally defined as a soil water potential equal to −1.5 MPa (−15 atm). In the laboratory, permanent wilting point is measured by placing the soil in a cylinder on a porous plate, applying a positive pressure of 1.5 MPa (15 atm), and pushing the water from the soil through the porous plate.

Theoretical field capacities and permanent wilting points for various soils can be calculated with the Soil Water Characteristics Calculator.

The total energy of water in soils is generally negative due to capillary suction. It becomes more negative when the soils have high salinity because the salts hold the water by osmotic potential. Thus, saline soils have less water available to the plant than nonsaline soils.

Chemical (osmotic), elevation (gravitation), and matric (hydraulic) potential energy are all components of the total energy of water in soils, H. Thermal energy can also influence the direction of water flow, although it is not included in soil water calculations in this text. Water velocity is extremely slow in soils so kinetic energy is not a significant component of water energy in soils. Thus, the total potential of water in soils is expressed as follows:

$$H = \frac{P}{\rho g} + z + \psi_s = \psi_p + z + \psi_s \qquad (3.4)$$

where

ψ_s = osmotic potential, m
ψ_p = capillary or hydraulic (pressure) potential (also equal to h_c in Eq. 3.4), m
z = elevation, m
P = pressure, N/m^2
ρ = density of water, kg/m^3
g = gravity, 9.8 m/sec^2

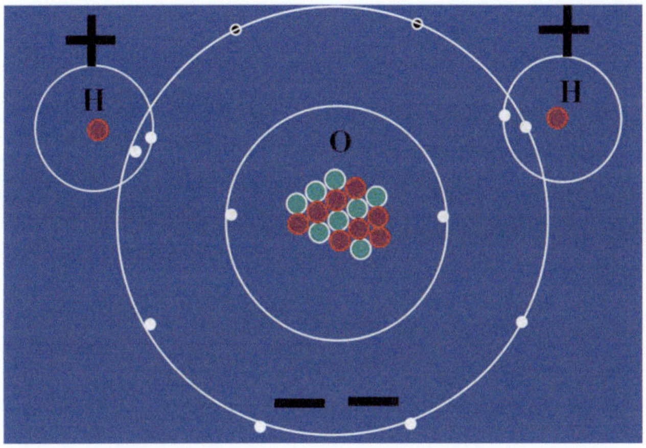

Fig. 3.7 Water molecule

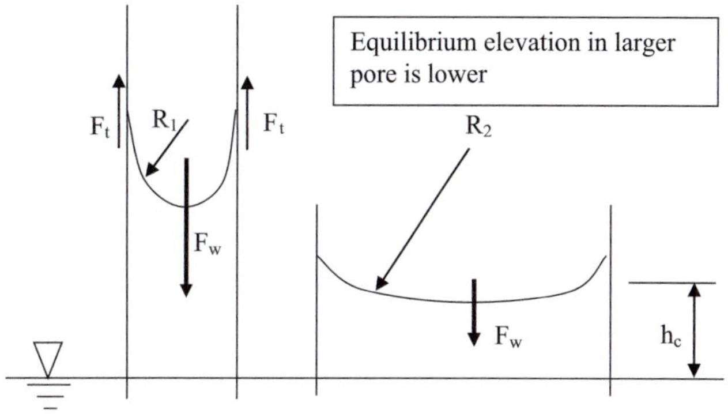

Fig. 3.8 Capillary force (matric potential)

Water molecules are attracted to each other because the negative side of one water molecule (Fig. 3.7) is attracted to the positive side of another. This force of attraction due to polarity of the water molecule has several names: hydrogen bonds, Van der Waal's force, or electrostatic attraction. In the liquid state, water demonstrates cohesion up to a suction of approximately 30 MPa or 30,000 kPa (300 atmospheres). Water cohesion is an important water property in soils and plants since water must flow from the soil to the plant root and then upward in the plant at extremely negative water potentials. Cohesion also causes water to form a meniscus under negative pressure due to water surface tension. If the radius of the meniscus in a pore is small, then the force, $-F_\tau$, is large, and it is very difficult to remove water from the pore; thus, the energy of the water can be much more negative in small pores.

The energy in hydrogen bonds between water molecules is approximately 20 kJ/mole of water;

$$\left(\frac{20\ kJ}{mole}\right)\left(\frac{mole}{18\ g}\right)\left(\frac{1,000\ g}{kg}\right) = 1,100\ kJ/kg$$

The energy required to evaporate water, the latent heat of vaporization, is 2,450 kJ/kg. Thus, significant part of the energy used to evaporate water is used to break hydrogen bonds.

The downward force on the meniscus in Fig. 3.8 is gravity.

The equation that determines the negative hydraulic pressure (capillary potential) exerted by a meniscus is based on the law of conservation of momentum: summation of forces equals zero.

$$F_\tau = -F_w \quad 2\pi R\tau = -\rho g h_c \pi R^2 \quad \psi_p = h_c = -\frac{2\tau}{\rho g R} \qquad (3.5)$$

where

F_τ = upward force of clay surface on water, N
F_w = downward force of water column, N
h_c = matric potential, m
τ = surface tension of water, 0.073 N/m = 73 dyne/cm
R = pore radius, m.

Matric (capillary) potential energy is commonly expressed in atmospheres (also called bars). Air density at sea level is 1.23 kg/m^3 or 101.3 kPa (1 atmosphere). Units of pressure (kPa) are converted to units of length by dividing by the specific weight (ρg) of water.

$$\frac{P}{\rho g} = \frac{1.013*10^5\ Pa}{1,000\ kg/m^3*9.801\ m/\sec^2} = 10.34\ m \qquad (3.6)$$

Example 3.3 Calculate the matric potential at which pores begin to drain in clay soils (R = 1 *10^{-6} m) and fine sand soils (R = 50 microns).

Clay
$$h_c = -\frac{2\tau}{\rho g R} = \frac{-2*0.073}{1,000*9.8*10^{-6}} = -15\ m*\left(\frac{atm}{10.34\ m}\right) = -1.45\ atm$$

Sand
$$h_c = \frac{-2*0.073}{1,000*9.8*0.00005} = -0.3\ m*\left(\frac{atm}{10.34\ m}\right) = -0.03\ atm$$

Water is strongly attracted to negatively charged clay surfaces. This attraction to other materials is called adhesion. The closer water molecules are to clay surfaces, the more negative is their potential energy and the more difficult it is to remove them from the soil.

The energy of water in air is even more negative than energy of water in soils. Thus, water flows from the relatively higher energy in the soil to the lower energy in the atmosphere. The plant acts as a passive conduit for the water as it flows from the soil to the atmosphere. The plant maintains positive hydraulic pressure during this process by having negative osmotic potential due to high sugar concentration in cells.

The relationship between matric potential (moisture potential) and water content for typical soils is shown in Fig. 3.9. The moisture potential axis has a log scale: moisture potential drops by 2 orders of magnitude in Fig. 3.9 between saturation and field capacity and then between field capacity and permanent wilting point.

In general, soil water content should remain above the 30 % depletion line (MAD = 30 %) for drought sensitive crops and above 50 % depletion for drought tolerant crops in order to avoid yield reduction.

The water content – moisture potential relationship varies dramatically between soils. Soils with medium texture have the most water availability. Clay soils have a large fraction of small, flat pores, and much of the water in clays is unavailable to the plant because plants cannot develop a negative enough osmotic potential to extract water from small pores. One the other hand, large pores in sand drain quickly after rainfall before plants can utilize the water.

The difference between field capacity and permanent wilting point is called available water capacity. AWC is the fraction of soil volume between field capacity and permanent wilting point.

$$AWC = \theta_{FC} - \theta_{PWP} \quad (3.7)$$

where

AWC = available water capacity, fraction or percent.

The soil water holding capacity (SWHC) is the depth of water in the soil available for plant growth. SWHC is also known as total available water (TAW)

$$SWHC = TAW = (AWC)(z) \quad (3.8)$$

where

SWHC or TAW = soil water holding capacity or total available water, m

z = root zone depth, m

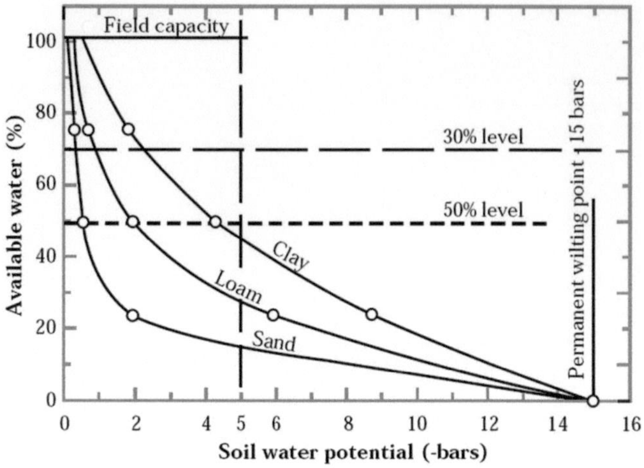

Fig. 3.9 Depletion percentages for different soils versus soil water potential (Credit NRCS)

Example 3.4 The root depth of an orange tree is 1.5 m. The soil has 56 % solid particles, and 44 % voids. After drainage, the 50 % of the void volume contains water and 50 % holds air. At the wilting point, the void volume is 25 % water. What is the soil water holding capacity (SWHC).

Solution:

Field capacity = (0.44) (0.5) = 0.22 = 22%
Permanent wilting point = (0.44) (0.25) = 0.11 = 11%
$AWC = \theta_{FC} - \theta_{PWP} = 0.22 - 0.11 = 0.11$
$TAW = SWHC = (AWC)(z) = 0.11(1.5\text{ m})$
$= 0.165\text{ m} = 16.5\text{ cm}.$

Note: Only about half of the SWHC or TAW (total available water) is available to the plant so only about 8 cm of water would be available between irrigation events. This means that if the evapotranspiration rate was 1 cm/day, then the field would need to be irrigated every 8 days.

Depletion is the depth of water removed from the soil (after the soil reaches field capacity by gravity drainage) by evapotranspiration. Percent depletion is percent of the TAW that is removed by evapotranspiration. Percent depletion is 0 % at field capacity and 100 % at permanent wilting point. Depletion and percent depletion are calculated as follows.

$$\text{Depletion} = D_r = (\theta_{FC} - \theta)z$$
$$\%Dep = \frac{D_r}{TAW} * 100 = \frac{(\theta_{FC} - \theta)}{(\theta_{FC} - \theta_{PWP})}(100) \quad (3.9)$$

Example 3.5 The soil described in Example 3.2 has a permanent wilting point equal to 20 % (volumetric water content). Calculate the percent depletion in the soil sample in Example 3.2 taken just before irrigation. If the root zone is 1 m deep, then what is the depth of depletion?

Solution:

Find the percent depletion

$$\%Dep = \frac{(\theta_{FC} - \theta)}{(\theta_{FC} - \theta_{PWP})}(100) = \frac{(0.36 - 0.26)}{(0.36 - 0.2)}(100)$$
$$= 63\%$$

Find the total available water in order to find the depth of depletion.

$$AWC = \theta_{FC} - \theta_{PWP} = 0.36 - 0.2 = 0.16$$
$$TAW = SWHC = (AWC)(z) = 0.16\ (1\ \text{m})$$
$$= 0.16\ \text{m} = 16\ \text{cm}$$

Find the depth of depletion

$$\frac{\%Dep}{100}TAW = D_r = \frac{63}{100}(16\ cm) = 10\ cm$$

The percent depletion with no yield reduction for specific crops is called the management allowed depletion (MAD). The MAD value for many crops changes with growth stage (Table 3.3) and is based on the crop sensitivity to water stress at that stage. The MAD value for a drought tolerant crop is approximately 0.5 and the MAD value for a drought sensitive crop is approximately 0.2.

The readily available water (RAW) is the depth of water available to the plant between irrigation events.

$$RAW = (TAW)(MAD) \quad (3.10)$$

where
MAD = management allowed depletion, fraction
RAW = readily available water, m.

Example 3.6 Use the soil described in Example 3.5. A crop has an MAD value of 0.4 and a root zone depth of 1 m. What is the RAW? The depth of evapotranspiration is 1 cm/day. How often would the field need to be irrigated?

Solution:

$$RAW = (TAW)(MAD) = (16\ \text{cm})(0.43) = 7\ \text{cm}.$$

Thus, if the evapotranspiration rate was 1 cm/day, then the field would need to be irrigated once per week.

In order to find the preferred volumetric water content at which the next irrigation should take place, subtract AWC (MAD) from the field capacity.

Example 3.7 Use the soil described in Example 3.5. Find the preferred water content at which irrigation should be take place for a plant with 40 % MAD.

Solution:

$$\text{Water content at next irrigation} = FC - (AWC)(MAD)$$
$$= FC - (FC - PWP)(MAD)$$
$$= (0.36) - (0.36 - 0.20)(0.40)$$
$$= 0.30 = 30\%$$

Note: 30 % is a higher water content than the soil water content taken just before irrigation in Example 3.2. Thus, a plant with a 0.4 MAD would have been overly stressed before irrigation.

With infrequent irrigation, such as with orchard sprinkler systems or surface irrigation systems, irrigation should replace the RAW. Frequent irrigation with drip or center pivots keeps the soil moist and does not approach the MAD.

Table 3.3 Recommended Management Allowed Depletion (MAD) values for loamy soils (Credit NRCS)

	Establishment	Vegetative	Flowering	Ripening
Alfalfa hay	50	50	50	50
Alfalfa seed	50	60	50	80
Beans, green	40	40	40	40
Beans, dry	40	40	40	40
Citrus	50	50	50	50
Corn, grain	50	50	50	50
Corn, seed	50	50	50	50
Corn, sweet	50	40	40	40
Cotton	50	50	50	50
Cranberries	40	50	40	40
Garlic	30	30	30	30
Grains, small	50	50	40	60
Grapes	40	40	40	50
Grass pasture/hay	40	50	50	50
Grass seed	50	50	50	50
Lettuce	40	50	40	20
Milo	50	50	50	50
Mint	40	40	40	50
Nursery stock	50	50	50	50
Onions	40	30	30	30
Orchard, fruit	50	50	50	50
Peas	50	50	50	50
Peanuts	40	50	50	50
Potatoes	35	35	35	50[4/]
Safflower	50	50	50	50
Sorghum, grain	50	50	50	50
Spinach	25	25	25	25
Sugar beets	50	50	50	50
Sunflower	50	50	50	50
Vegetables				
30–60 cm root depth	35	30	30	35
90–120 cm root depth	35	40	40	40

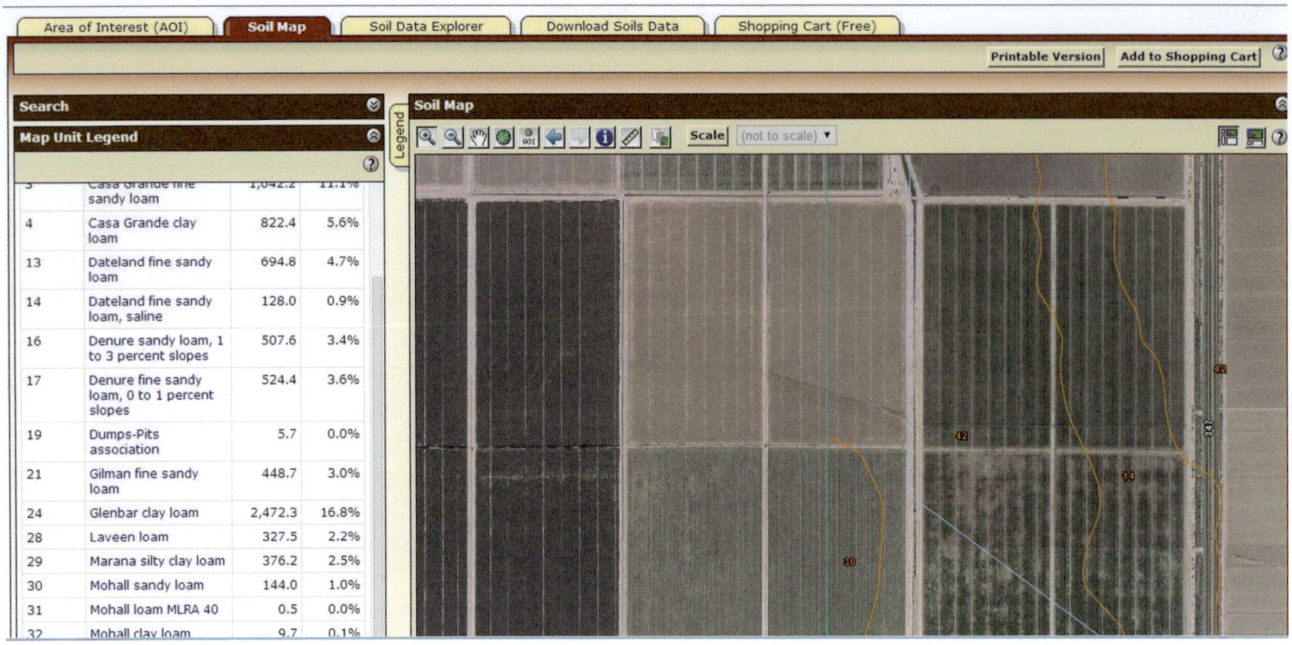

Fig. 3.10 Web soil survey map

Recommended MAD values from NRCS for loamy soils are shown in Table 3.3. A document called FAO56 (Allen et al. 1998) is a standard reference for evapotranspiration and plant stress and it presents a slightly different concept for MAD estimation: the crop stress threshold, p, is similar to the MAD but not the same. FAO 56 lists p values in its Table 22 and adjusts p values based on ETc as follows: $p = p_{Table\ 22} + 0.04\,(5 - ET_c)$. Thus, the p value decreases for high evapotranspiration. This is because plant water uptake is a time dependent process and if water is needed more quickly, then the soil water content should be higher. Soil texture also affects p. FAO56 recommends increasing p by 5–10 % in a coarse (sandy) textured soil and decreasing p by 5–10 % in a fine textured soil.

Many factors can change available water capacity. The available water capacity (AWC) is obviously reduced if the soil has rocks. Organic matter can decrease bulk density and increase porosity by up to 10 %. Good soil structure (granular, blocky, or prismatic) can increase AWC by 10 % whereas poor soil structure (massive, platy, or single grain) can decrease AWC by 10 %. Compaction can decrease AWC by 20 %. Restrictive layers can increase the AWC in the layers above by 10 % by restricting downward flow of water by gravity. Soils increase in density with depth because of overburden pressure, decreased structure, and decreased organic matter. Thus, AWC decreases by approximately 5 % per 30 cm depth in the soil. Vegetation can increase AWC because roots improve soil structure and condition.

Characterization of Soil Layers

In layered soils, the SWHC for the soil profile is the sum of the products of the AWCs and the layer thicknesses

$$SWHC = \sum_{i=1}^{n} SWHC_i = \sum_{i=1}^{n} \Delta z_i * AWC_i \qquad (3.11)$$

where

Δz = thickness of soil layer, cm
i = soil layer number
n = number of layers.

In preparation for an irrigation or drainage design, soils analysis optimally includes drilling soil cores and digging soil pits in order to characterize soil. A less preferable, but less expensive, alternative is to use published soil map descriptions; for example, Web Soil Survey (websoilsurvey.sc.egov.usda.gov) is hosted by the United States Department of Agriculture (USDA). After defining an area of interest (AOI), you can click on the Soil Map tab to find the soils in the AOI (Fig. 3.10). Each of the numbered soil types in the Map Unit Legend on the left corresponds to a number in the soil map.

The soils listed in the Map Unit legend can be found at the Soil Series Description web site (https://soilseries.sc.egov.usda.gov/osdname.asp). For example, one of listed soils is

Mohall sandy loam (#30). Just type Mohall into the textbox at the Soil Series Description web site, and the following description appears. The sandy loam should not be typed into the textbox because it just refers to the upper layer of soil.

Mohall Series

The Mohall series consists of very deep, well-drained soils formed in fan and stream alluvium from mixed sources. Mohall soils are on fan terraces, stream terraces, and relict basin floors and have slopes of 0–8 %. The mean annual precipitation is about 6 inches and the mean annual air temperature is about 73°F.

TAXONOMIC CLASS: Fine-loamy, mixed, superactive, hyperthermic Typic Calciargids

TYPICAL PEDON: Mohall coarse sandy loam - rangeland. (Colors are for dry soil unless otherwise noted.)

A--0–4 inches; reddish yellow (7.5 YR 6/6) coarse sandy loam, brown (7.5 YR 4/4) moist; weak medium platy structure parting to weak fine granular; slightly hard, very friable, slightly sticky and slightly plastic; few fine roots; many very fine and fine irregular and common fine tubular pores; moderately alkaline (pH 7.9); abrupt wavy boundary. (2–7 inches thick)

BA--4–10 inches; strong brown (7.5 YR 5/6) coarse sandy loam, brown (7.5 YR 4/4) moist, weak coarse angular blocky structure parting to weak medium subangular blocky; slightly hard, very friable, slightly sticky and slightly plastic; common fine and very fine roots; common very fine and few fine tubular pores; few faint clay films on faces of peds and lining pores; moderately alkaline (pH 7.9); clear wavy boundary. (2–8 inches thick)

Bt--10–19 inches; brown (7.5 YR 5/4) sandy clay loam, reddish brown (5 YR 4/4) moist; moderate medium and coarse prismatic structure parting to moderate medium and coarse angular blocky; very hard, friable, very sticky and very plastic; few fine roots; few fine and very fine tubular pores; many faint clay films on faces of peds and lining pores; moderately alkaline (pH 8.0); clear wavy boundary. (5–10 inches thick)

Btk1--19–27 inches; brown (7.5 YR 5/4) clay loam, reddish brown (5 YR 4/4) moist; weak medium prismatic structure parting to moderate medium subangular blocky; very hard, friable, very sticky and very plastic; few very fine roots; common faint clay films on faces of peds; strongly effervescent as common fine calcium carbonate filaments and few small concretions; moderately alkaline (pH 8.2); clear wavy boundary. (6–12 inches thick)

Btk3--27–37 inches; strong brown (7.5 YR 5/6) loam, brown (7.5 YR 4/4) moist; weak fine and medium subangular blocky structure; slightly hard, very friable, moderately sticky and moderately plastic; few very fine roots; common very fine and fine tubular pores; few faint clay films on faces of peds and lining pores; strongly effervescent as common medium distinct calcium carbonate filaments and masses; moderately alkaline (pH 8.2); clear wavy boundary. (5–15 inches thick)

2Bk--37–54 inches; pinkish white (7.5 YR 8/2) gravelly sandy loam, pinkish gray (7.5 YR 7/2) moist; massive; hard, friable, slightly sticky and slightly plastic; common very fine and fine tubular pores; 20 % gravel; violently effervescent as common soft calcium carbonate masses; moderately alkaline (pH 8.0); gradual wavy boundary. (12–24 inches thick)

2C1--54–76 inches; pinkish white (7.5 YR 8/2) gravelly sandy loam, pink (7.5 YR 7/4) moist; massive; very hard, friable, nonsticky and nonplastic; common fine tubular pores; 25 % gravel; violently effervescent; moderately alkaline (pH 8.2); gradual wavy boundary. (20–30 inches thick)

2C3--76–98 inches; pinkish gray (7.5 YR 7/2) gravelly loamy coarse sand, light brown (7.5 YR 6/4) moist; massive; soft, loose; slightly effervescent; moderately alkaline (pH 8.2).

Example 3.8 For a Mohall sandy loam soil profile, calculate the available water content (AWC) with the Soil Water Characteristics Calculator for each horizon down to 130 cm. Then, divide the soil into 20 cm increments, with a 30 cm increment at the top of the soil profile, such as might be necessary to correspond with moisture probe measurements. Calculate SWHC and RAW for the 20 cm layers, and then sum the SWHCs and calculate the RAW. Finally, calculate the number of days between irrigation events. In Table 3.3, p is 0.5; however, ET is high (1 cm/day) so MAD is reduced from 45 % to 50 %. The farm is in the desert and soil salinity is 3 dS/m

Estimation of parameters in soil layers.

The upper A horizon is defined by the name (Mohall sandy loam), and is not necessarily the same as the upper layer in the soil description. However, in this case, the upper layer in the Mohall description (course sandy loam) is the same. Use the Soil Water Characteristics Calculator (Fig. 3.11) to define the AWC. Because it is probably a coarse sandy loam, select the upper part of the Sandy loam (SaL) area, with 72 % sand and 10 % clay. At this soil texture, the FC is 13.8 % and the PWP is 6.8 %. Thus, the AWC is 8 % (Table 3.4). Because it is the surface layer, assume that the compaction is between loose and normal. There is no gravel in the soil description for the upper layer so that is zero. It is a desert soil, so assume that the organic matter fraction is just above zero.

The BA horizon is also a coarse sandy loam and has a blocky structure. Change to normal density and leave other

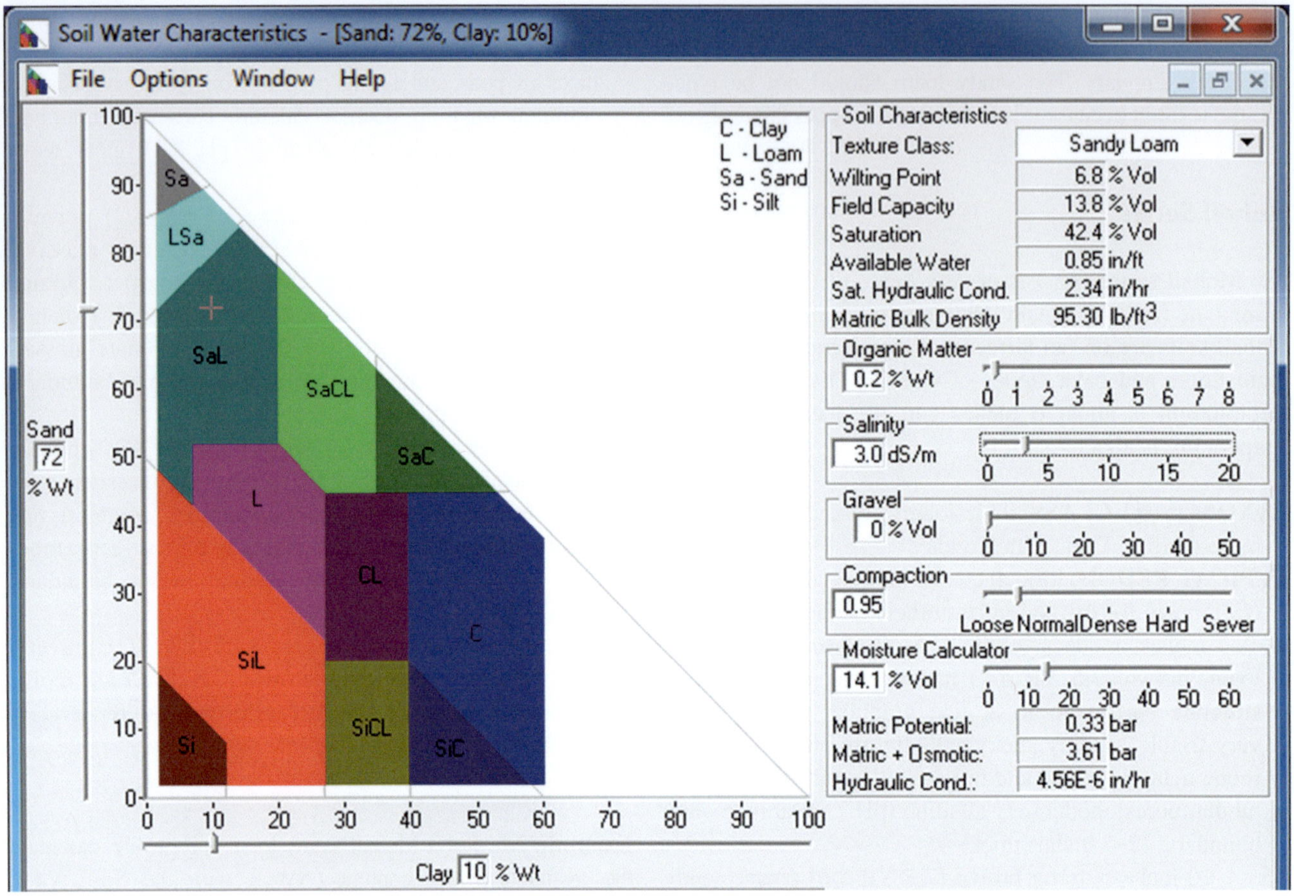

Fig. 3.11 Soil water characteristics calculator estimation of soil properties of upper layer of Mohall sandy loam soil

Table 3.4 Mohall sandy loam Available Water Capacity (AWC)

Horizon	Depth	Texture	AWC (%)
A	0–4 in (0–10 cm)	Coarse sandy loam	8.0
BA	4–10 in (10–25 cm)	Coarse sandy loam	6.6
Bt	10–19 in (25–48 cm)	Sandy clay loam	8.5
Btk1	19–27 in (48–58 cm)	Clay loam	12
Btk2	27–37 in (58–83 cm)	Loam	11
2Bk	37–54 in (83–137 cm)	Gravelly sandy loam	5.7

parameters the same in the HPC → PWP = 6.7%, FC = 13.3 %, and AWC = 6.6 %.

The Bt horizon is a sandy clay loam (Sand = 59 % and Clay = 28 %). The Bt horizon has a prismatic and blocky structure but it is 25–48 cm so increase the compaction to between normal and dense: SWCC → PWP = 17.2%, FC = 25.7 %, and AWC = 8.5 %.

The Btk1 horizon is a clay loam. The structure is blocky and prismatic so the AWC is at the upper, but the average depth is 60 cm so increase the compaction to dense: SWCC → PWP = 20.3%, FC = 32.3 %, and AWC = 12 %.

The Btk2 horizon is a loam (Sand = 42 % and Clay = 17 %). The soil has a subangular blocky structure, but the depth is 60 cm so characterize the compaction as dense: SWCC → PWP = 11.1%, FC = 22.2 %, and AWC = 11 %.

2Bk layer is a gravelly sandy loam (66 % sand and 10 % clay) with 20 % gravel; thus, move the gravel percentage to 20 %. It is a massive structure (no structure) so move the compaction to hard: SWCC → PWP = 6.4%, FC = 12.1 %, and AWC = 5.7 %.

Adjust Table 3.4 for measurement depths, and calculate SWHC for each layer (Table 3.5).

Table 3.5 Water holding capacity corresponding to neutron probe depths

Depth increment	Texture	AWC (%)	Depth of layer (cm)	SWHC/layer (cm)
0–30 cm (grav)	Sandy loam	7	30	2.1
30–50 cm	Sandy clay loam	8.5	20	1.7
50–70 cm	Clay loam and loam	12	20	2.4
70–90 cm	Loam and GSL	9	20	1.8
90–110 cm	Gravelly sandy loam	5.7	20	1.4
110–130 cm	Gravelly sandy loam	5.7	20	1.4

Table 3.6 Saturated conductivity (steady state infiltration rates) for different soil textures

Texture		Sand	Sandy loam	Loam	Clay loam	Silty clay	Clay
Average infiltration rate	(mm/hr)	50	25	13	8	2.5	0.5
Range	(mm/hr)	(25–250)	(13–76)	(8–20)	(2.5–15)	(0.5–5)	(0.1–1)

The SWHC or TAW for the entire root zone is 10.8 cm.

The RAW for the entire root zone is AWC(MAD = 10.8 cm (0.45) = 4.9 cm.

Thus, for 1 cm/day evaporation, there are approximately 5 days between irrigation events.

Infiltration

Infiltration (i) is the process of water from precipitation or irrigation entering the soil. The rate of infiltration (di/dt) can be described as the rate of change in ponded depth of water on the soil surface, which is called the Darcy velocity. If the rate of precipitation is less than the maximum infiltration rate of the soil, then water does not pond on the surface. The infiltration rate is initially high and then decreases over time to nearly steady state. Typical steady-state infiltration rates are given in Table 3.6. Notice the broad range of infiltration rates within each textural classification. In fact, the sand infiltration rate varies by an order to magnitude. This is why it is important to measure infiltration rates on the farm rather than rely on classifications.

An empirical model of infiltration over time can be developed by curve fitting infiltration data. One popular infiltration equation is the Kostiakov equation.

$$i = kt^a \tag{3.12}$$

where

i = depth of infiltration, mm
t = time, hr
k, a = constants.

Because the Kostiakov equation is an exponential equation, the coefficients can be calculated by taking the natural log of both sides of Eq. 3.13 and inserting two measured infiltration depths and times: two equations and two unknowns. The logarithm of infiltration rates is generally linear and fits the equation for the slope of a line in Eq. 3.14.

Table 3.7 Infiltrometer data

Time (min)	Depth of water (cm)	Infiltrated depth (cm)
0	15	0
10	13.9	1.1
20	13.1	1.9
40	11.8	3.2
60	10.6	4.4

$$\ln(i) = a\ln(t) + \ln(k) \; (y = mx + b) \tag{3.13}$$

$$y = mx + b \tag{3.14}$$

Infiltration rate can be measured with infiltrometers in which water is ponded above the soil surface, or disk permeameters, which create a negative potential above a porous plate placed on the soil surface. There are many types of ponded infiltrometers (single ring, double ring, furrow, etc.). After water is ponded in an infiltrometer, the depth of water in the cylinder vs. time is recorded. Double ring infiltrometers (Fig. 20.8) have two concentric cylinders, and it is assumed that infiltration in the inner cylinder, where depth is measured, is vertical. Automated infiltrometers add water to the infiltrometer with a float valve and record the volume of water addition to the infiltrometer over time. An entire furrow or a furrow section can be used as an infiltrometer: a volume balance (in – out = infiltration) can be used to estimate the rate of infiltration into a furrow section if the depth is constant.

Example 3.9 Calculate k and a in the Kostiakov equation for the ponded infiltrometer data in Table 3.7.

Take the natural log of time and infiltrated depth (Table 3.8). For example, the natural logs of time and infiltrated depth after 10 minutes are calculated as follows.

Table 3.8 Natural logarithm of data

Time (min)	ln (time)	ln (Infiltrated depth)
0		
10	2.302	0.0953
20	2.995	0.642
40	3.689	1.16
60	4.094	1.48

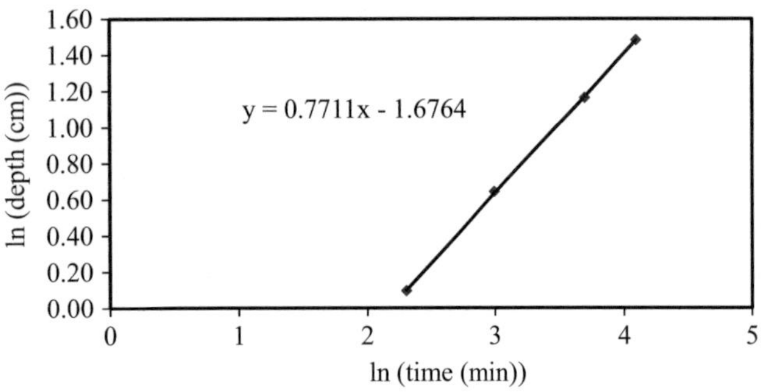

Fig. 3.12 Regression of logarithm data

$$\text{Infiltrated depth} = 15\text{ cm} - 13.9\text{ cm} = 1.1\text{ cm}.$$
$$\ln(i) = \ln(1.1\text{ cm}) = 0.095$$
$$\ln(t) = \ln(10\text{ minutes}) = 2.3$$

The line formed by the natural log data has slope 0.77 (Fig 3.12).

The slope (0.7711) is "a" in the Kostiakov equation. ln (i) = a ln (t) + ln (k).

Now solve for k by using the data from the 20-minutes infiltration point.

$$k = \frac{i}{t^a} = \frac{1.9}{20^{0.7711}} = 0.189$$

The short time represented by this infiltration example is generally not representative of the entire irrigation event, which may take 12–24 hours.

Instantaneous infiltration rate is the rate of water infiltration at any moment in time. The instantaneous Kostiakov infiltration rate is found by taking the derivative of Eq. 3.12 with respect to time.

$$di/dt = ak(t)^{(a-1)} \tag{3.15}$$

where

t = time, minutes
di/dt = infiltration rate, cm/min.

Soil surface storage can be added to the infiltration depth. If there are surface depressions, then water that does not infiltrate into the soil during a storm may be stored on the soil surface until it infiltrates. Typical soil surface storage values are 1.2, 0.8, 0.2, and 0 cm for soils with slopes of 0–1, 1–3, 3–5, and >5 %, respectively (*NRCS NEH* values); however, these value depend on tillage method.

Table 3.9 NRCS intake family coefficients for equation 3–16 (c is 0.6985 for cm and 0.275 for inches for all intake families) for time, t, in minutes (After Cuenca 1989)

Intake family	a (cm)	a (in)	b
0.05	0.0533	0.0210	0.618
0.1	0.0620	0.0244	0.661
0.15	0.0701	0.0276	0.683
0.2	0.0771	0.0304	0.699
0.25	0.0853	0.0336	0.711
0.3	0.0925	0.0364	0.720
0.35	0.0996	0.0392	0.729
0.4	0.1064	0.0419	0.736
0.45	0.1130	0.0445	0.742
0.5	0.1196	0.0471	0.748
0.6	0.1321	0.0520	0.757
0.7	0.1443	0.0568	0.766
0.8	0.1560	0.0614	0.773
0.9	0.1674	0.0659	0.779
1.0	0.1786	0.0703	0.785
1.5	0.2283	0.0899	0.799
2	0.2753	0.1084	0.808
3	0.3650	0.1437	0.816
4	0.4445	0.1750	0.823

Soils have been classified by the Natural Resource Conservation Service (NRCS) by intake families. The NRCS equation for infiltration is a modified form of the Kostiakov equation.

$$i = a(t)^b + c \tag{3.16}$$

where

i = cumulative infiltration depth (cm)
a, b, c = constants
t = time after initiation of infiltration (min)

The c value was not included in the Kostiakov equation. This term represents initial infiltration into cracks and worm holes. This type of infiltration takes no time so it is not a function of t. It is assumed by the NRCS that c is the same for all intake families. However, c can be much larger for some soils. For example, most of the infiltration into swelling clay soils is into cracks. Once the cracks close within a few minutes, infiltration effectively ceases.

Intake family coefficients are listed in Tables 3.9 and 3.10. Although it is preferable to measure the infiltration rates on the farm with an infiltrometer or by observing the performance of a surface irrigation event, the intake families can be used for design. The NRCS intake families are groups of soils with similar intake rates and they are classified by saturated intake rate (inches/hour). NRCS soil maps list intake families for each soil, but these are rough guesses

Table 3.10 Soil intake families by surface texture (Credit NRCS)

	Intake family		
Soil texture	Sprinkler	Furrow	Border and basin
Clay, silty clay	0.1–0.2	0.1–0.5	0.1–0.3
Sandy clay, silty clay loam	0.1–0.4	0.2–0.8	0.25–0.75
Clay loam, sandy clay loam	0.1–0.5	0.2–1.0	0.3–1.0
Silty clay, loam	0.5–0.7	0.3–1.2	0.5–1.5
Very fine sandy loam, fine sandy loam	0.3–1.0	0.4–1.9	1.0–3.0
Sandy loam, loamy very fine sand	0.3–1.25	0.5–2.4	1.5–4.0
Loamy fine sand, loamy sand	0.4–1.5	0.6–3.0	2.0–4.0
Fine sand, sand	0.5+	1.0+	3.0+
Coarse sand	1.0+	4.0+	4.0+

rather than measured rates. Another method of estimating infiltration rate is to observe the rate of advance and recession during surface irrigation events (Chap. 20). For further information, I recommend the following papers (listed in references) that pertain to infiltration estimation. Clemmens and Bautista (2007) described some of the challenges and alternatives for estimating infiltration rates. Walker et al. (2006) described the development of modified NRCS intake families.

The infiltration behavior of soils can vary with irrigation method (Table 3.10).

The infiltration rate is found by taking the derivative of Eq. 3.16 with respect to time. Infiltration rate in cm/hr is found by multiplying Eq. 3.17 by 60.

$$di/dt = ab(t)^{(b-1)} \tag{3.17}$$

where

t = time, minutes
di/dt = infiltration rate, cm/min.

Example 3.10 Calculate instantaneous infiltration rate (cm/hr) and cumulative infiltration (cm) vs. time, and plot both curves for the first 60 minutes after infiltration begins for intake family 4 – sandy soil.

The depth of infiltration after 60 minutes is $i = 0.4445\,(60)^{0.823} + 0.6985 = 13.6$ cm

Infiltration rate after 60 minutes is

$$\begin{aligned} di/dt\ (cm/hr) &= 60{*}ab(t)^{b-1} \\ &= 60{*}0.4445{*}0.823(60)^{(-0.177)} \\ &= 10.6\ cm/hr = 106\ mm/hr. \end{aligned}$$

The infiltration rate in Fig. 3.13 is approaching 10 cm/hr (100 mm/hr). This is within the range listed for sand in Table 3.6 (25–250 mm/hr). It is also equal to 4.0 inches per hour, which is the same as the intake family name. The names of intake families are based on their steady state infiltration rate, in/hr.

Saturated hydraulic conductivities are also estimated in the Soil Water Characteristics Calculator. For example, a sandy loam soil has a saturated hydraulic conductivity of 2.34 in/hr, which would be in between and intake family 2 and intake family 3 soil; Another method of estimating infiltration rate is to observe the rate of advance and recession during surface irrigation events (Chap. 20).

Soil Moisture Sensors

Management of systems is improved by feedback. Many sensors have been developed that detect plant and soil moisture status, which enable the grower to more accurately assess plant water requirements and irrigation scheduling. Most are in-situ (in contact with the soil) sensors. The advantage of in-situ soil moisture sensors is that they measure soil moisture change before the crop begins to experience stress. The disadvantage is that they must be physically placed in the soil, monitored, and possibly removed if they are in the way of cultivation operations. A second disadvantage is that many provide an inaccurate estimate of soil moisture status because of small sample volume, poor design, or poor installation.

Tensiometers have a porous ceramic cup in content with the soil, and they directly measure soil water potential. The normal operating range is −0 to −80 kPa (−0 to −0.8 atm or bar). They are suitable for drip irrigated crops where the water potential does not become extremely negative. They are sometimes wired to drip irrigation controllers in order to automatically trigger an irrigation.

Electrical resistance is inversely related to soil moisture content. The most common electrical resistance devices are gypsum blocks with electrodes inside the blocks. The effective range of gypsum blocks is −150 to −600 kPa (−1.5 to −6 bar). The disadvantages of gypsum blocks include slow response to changes in soil water content, sensitivity to soil salinity, chemical degradation, possible damage by field equipment, and insensitivity to soil moisture potentials less negative than −1 bar.

Heat dissipation devices measure matric potential in a porous material that is buried in the soil and is in hydraulic equilibrium with the soil. The porous medium contains a heating element and temperature sensor in the center. The temperature in the porous medium is measured, then the heating element gives a small heat pulse, and the temperature is measured again. The rise in temperature depends on the amount of water present in the porous material (less water means a larger rise in temperature). These sensors are more durable than gypsum blocks. They are insensitive to soil temperature and

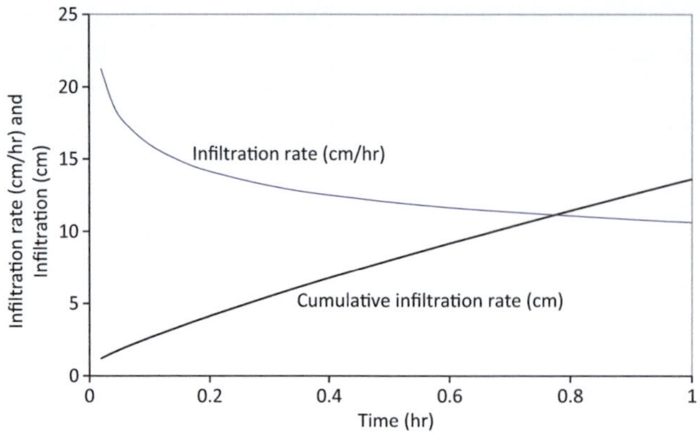

Fig. 3.13 Cumulative infiltration and infiltration rate for intake family 4

Fig. 3.14 Neutron probe and access tube (Credit Paul Colaizzi, USDA-ARS)

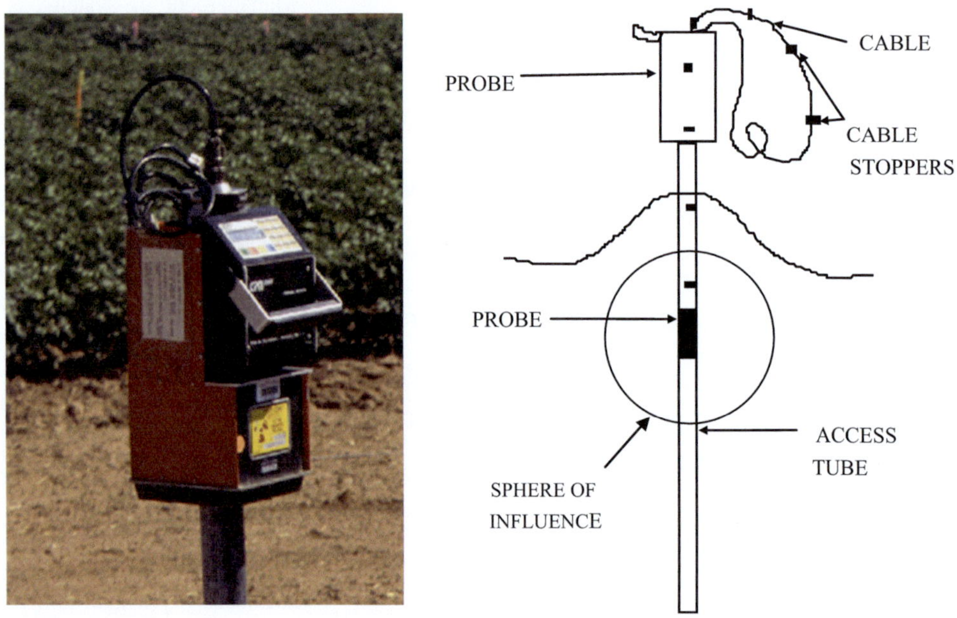

salinity, and seldom, if ever, require recalibration. They perform best at matric potentials less negative than −3 bar.

In a study by Evett et al. (2009), the neutron probe was recommended as the only practical, consistent, and accurate soil moisture measurement device for irrigation management. Electromagnetic sensors were found to be consistently inaccurate. The neutron probe (Fig. 3.14) consists of a sealed neutron source (Americium, Plutonium, or Radium mixed with Beryllium). The probe is lowered into an access tube (typically a 2-inch aluminum pipe). Neutrons emitted from the probe scatter into the surrounding soil, collide with hydrogen atoms in water molecules, and bounce back toward the probe. Since hydrogen atoms have the same mass as a neutron, a neutron will lose kinetic energy after colliding with hydrogen. Neutrons are then slowed down, and a detector tube (boron tri-fluoride) can count them over a specified time interval (15–30 seconds). The detector is not sensitive to fast neutrons bouncing back from large soil molecules.

There are several advantages to neutron probes. They are accurate over a large range of soil moisture, they can measure soil moisture at lower depths (1–3 m), and they are practically maintenance free. They also measure a much larger soil volume than other sensors so they provide a better estimate of the overall field moisture content. The neutron probe access tubes, when placed in center of beds, seldom interfere with field equipment. There are also several disadvantages to neutron probes. In field locations that require tillage, neutron probe tubes must be installed with a Giddings soil corer before and removed after each growing season. The access tubes can be buried below the soil surface during tillage if they have extensions at the soil surface that can be removed. RTK GPS (accurate to a cm) makes it easier to bury and find the access tubes after tillage. Neutron probes are time and labor intensive. It takes approximately 6 minutes to read 10 readings from one tube. Finally, the neutron scattering "sphere of influence" requires the probe to be 40–60 cm below the surface before readings are accurate, but most of the soil water content change occurs above 60 cm when sprinklers and drip systems are used. Last, but not least, neutron probes are radioactive, requiring substantial safety training and bureaucracy. However, the level of radiation is not dangerous if used correctly.

Neutron probes are calibrated by measuring the gravimetric water content at each depth from the soil removed from a new access hole. The calibration is made by setting a regression line through the neutron probe count number vs. soil water content measurements (measured by drying soil in oven). Neutron counts are then related to volumetric soil moisture by Eq. 3.18.

$$\theta_v = m\frac{x_n}{x_s} + b \qquad (3.18)$$

where

θv = volumetric soil moisture
m = the slope
x_n = field count
x_s = average of all standard counts
b = the intercept.

Because the neutron source radioactivity decreases over time, a series of standard counts must be taken with the neutron probe inside the neutron probe housing before each day's measurements.

Field capacity can be estimated with a neutron probe by collecting measurements a few days after irrigation (after the field has drained). The lower limit of soil moisture, permanent wilting point can be estimated based on soil texture.

There are many types of capacitance measurement devices. Time domain reflectometers (TDR) are the most popular. In the past they have been very expensive; however, new models are less expensive and provide more consistent data. The original TDR probes consist of a buried probe, cable, and the TDR instrument which sends a single electromagnetic pulse down the cable to the probe. The pulse is reflected back to the instrument as a wave. The shape of the wave indicates the water content and salinity.

Example 3.11 Data in Table 3.11 was collected three days after irrigation. Field capacity for 40 cm depth is calculated below and field capacities for other depths are shown in Table 3.11.

$$\theta_v = 0.1823\frac{22043}{13308.1} - 0.03429 = 0.268 = 26.8\%$$

Questions

1. What is the textural class name of a soil that has 40 % clay, 20 % sand, and 40 % silt?
2. What is the textural class name of a soil that has 35 % clay, 15 % sand, and 50 % silt?
3. Download the Soil Water Characteristics calculator from the website listed in the References, and calculate the field capacity and permanent wilting point for the soil described in question 2. Use the field capacity and permanent wilting point values to calculate AWC.
4. Use the Soil Water Characteristics Calculator to determine whether Field Capacity or Permanent Wilting Point changes more with soil compaction and explain why. What is the percent change from Loose to Hard.
5. Use the Soil Water Characteristics Calculator to evaluate changes in Field Capacity or Permanent Wilting Point from zero to 8 % organic matter. Make sure your salinity is below 5 dS/m. Is the change greater for a sandy loam or a clay? Does the change increase the AWC?
6. A soil sample is removed from the field and weighed (130 g). The soil is then dried and the weight is 100 g. What is the gravimetric water content?
7. If gravimetric water content θ_{grav} is 30 % and bulk density ρ_b is 1.30 g/cm^3, then what is the volumetric water content? What is the porosity?
8. What is the FC, PWP, and AWC of a sandy loam? Use the Soil Water Characteristics Calculator (0 % and 4 % organic matter, salinity = 3 dS/m, gravel = 0 %, and compaction is normal. Discuss the impact of properties other than soil texture on hydraulic properties.
9. What is the depth of readily available water (RAW) for sandy loam (4 % organic matter) if the effective root zone depth is 1.5 m and MAD = 0.4?
10. Define MAD and answer the following questions. What is meant by 40 % MAD? Does 40 % MAD have a water content closer to PWP or FC?
11. What is the percent depletion if measured water content is 19 %, field capacity is 25 % and permanent wilting point is 10 %? If the MAD is 50 %, at what water content must the next irrigation take place? If the root zone depth is 1.5 m, then what depth of available water remains for plant use before the next irrigation? If evapotranspiration rate is 1 cm/day, then what is the maximum length of time before the next irrigation?
12. How much irrigation water (ft) should be applied in the next irrigation if porosity is 50 %, field capacity is 27 %, and permanent wilting point is 12 % in all layers? Measured soil water content in the upper 4 ft of soil (root zone) is as follows: 0–1 ft = 21 %, 1–2 ft = 22 %, 2–3 ft = 17 %, and 3–4 ft = 22 %. Assume that the irrigation efficiency is 100 %. Redo assuming that the irrigation efficiency is 80 %.
13. What is the depth of available water in the root zone if the readily available water in the root zone 1 week ago was 10 cm and the rate of evapotranspiration was 1 cm/day? During this time, a storm added 2 cm water to the soil. When should the next irrigation take place?
14. Use the Web Soil Survey (WSS) to find a soil at the agricultural field station in your area (or a location

Table 3.11 Neutron probe readings taken three days after irrigation

DAYS AFTER IRRIGATION		3	
Calibration Slope = m		0.1823	
Calibration Intercept = b		−0.03429	
Average of standard counts =		13308.1	
Standard counts	Depth	Field counts	Moisture content
13,340	40 cm	22,043	26.8 %
13,153	60 cm	12,354	13.5 %
13,160	80 cm	13,012	14.4 %
13,376	100 cm	11,243	12.0 %
13,330	120 cm	10,142	10.5 %
13,300	140 cm	10,876	11.5 %
13,015	160 cm	10,564	11.0 %
13,394	180 cm	10,743	11.3 %
13,528	200 cm	9,876	10.1 %

specified by the instructor) and repeat Example 3.8 for your soil. First, go to the WSS URL listed in the References and click "Start Web Soil Survey" in the upper right corner. There are four tabs at the top of WSS. Find your location under the "Area of Interest (AOI)" tab. You can make the scaling process faster by using your mouse and outlining the location you are interested in. Then outline one field at the research station with the red area of interest rectangle tool (Second button from right along top). After outlining the area of interest, press the Soil Map tab at the top. The soils in the AOI are listed on the left. Click on the soil name for a short description of that soil. For a more extensive description, go to the Natural Resources Conservation Service Soil Series Descriptions at the URL listed in the References. Type in only the name of the series, but not the texture. Define the field capacity and permanent wilting with the Soil Water Characteristics Calculator Assume an appropriate crop in your area. Then calculate the RAW of the soil based on the MAD (Table 3.3) and root zone depth of the crop. Find the root zone depth in Table 3.2.

15. Estimate the long-term ponded steady-state infiltration rate for a sandy clay loam with the Soil Water Characteristics Calculator.
16. For the following infiltration data, determine the SCS intake family as shown in Example 3.3.

Time (min)	Infiltrated depth (cm)
0	0.6985
5	1.33
10	1.79
50	4.55
100	7.33
150	9.82
200	12.13

17. Calculate the depth of infiltration and infiltration rate over time and plot the two curves for an intake family 1.0 soil. Plot your infiltration rate curves in terms of cm/hr and in/hr and calculate out to 1,000 minutes. At what time is the intake rate equal to 1.0 in/hr? Is this the steady state intake rate?
18. Calculate the moisture contents in Table 3.11 in the upper four layers if the calibration slope is changed to 0.2. If permanent wilting point is 11 % in the upper four layers, calculate the total available water in the upper four layers assuming that the neutron measurements were taken a few days after irrigation. Calculate the readily available water in the upper four layers if MAD is 0.45.

References and Resources

Allen R, Pereira L, Raes D, Smith M (1998) Crop evapotranspiration – guidelines for computing crop water requirements – FAO irrigation and drainage paper 56. Food and Agriculture Organization. http://www.fao.org/docrep/x0490e/x0490e00.htm

Clemmens A, Bautista E (2007) Toward physically based estimation of surface irrigation infiltration. J Irrig Drain Eng/ASCE. Sept/Oct 2007: 588–596. http://naldc.nal.usda.gov/naldc/download.xhtml?id=42346&content=PDF

Cuenca RH (1989) Irrigation system design. Prentice Hall, Englewood Cliffs.

Evett S, Schwartz R, Tolk J, Howell T (2009) Soil profile water content determination: spatiotemporal variability of electromagnetic and neutron probe sensors in access tubes. Vadose Zone J 8(4):926–941

Ley T, Stevens R, Topielec R, Neibling W (1994) Soil water monitoring and measurement. A Pacific Northwest Publication # 475. Washington State University Extension. http://cru.cahe.wsu.edu/CEPublications/pnw0475/pnw0475.html

Natural Resources Conservation Service, Section 15 Irrigation. Chapter 1 Soil plant water relationships. In: National engineering handbook. http://www.info.usda.gov/CED/ftp/CED/neh15-01.pdf

Natural Resources Conservation Service (1997) Part 652. Engineering field handbook. irrigation guide. Chapter 2. Physical soil characteristics. http://directives.sc.egov.usda.gov/OpenNonWebContent.aspx?content=17551.wba

Natural Resources Conservation Service Soils Website with links to publications and other sites. http://soils.usda.gov/

Natural Resources Conservation Service Soil Series Descriptions http://www.nrcs.usda.gov/wps/portal/nrcs/detailfull/national/home/?cid=nrcs142p2_053587

Natural Resources Conservation Service Web Soil Survey http://websoilsurvey.nrcs.usda.gov/app/

Soil Water Characteristics Calculator: download from http://hydrolab.arsusda.gov/soilwater/Index.htm

Walker W, Prestwich C, Spofford T (2006) Development of the revised USDA-NRCS intake families for surface irrigation. Agric Water Manag 85:157–164

Water and Salinity Stress

4

Continually applying salt-laden irrigation water to soils can lead to soil salinization because plants leave most of the salts behind as they uptake water. Osmotic potential energy in saline soils is negative and resists the movement of water toward plant roots. According to the FAO, approximately 3 ha per minute are lost to soil salinization in the world, and 80 million ha have already been lost to soil salinization. Salinity management practices such as leaching water below the root zone are needed to prevent salt accumulation in the root zone. Leaching takes place when irrigation is increased beyond the evapotranspiration requirement. On the other hand, excess water in the soil restricts the movement of oxygen into the soil. Plants require oxygen for the roots as the conduct respiration at night. Insufficient water reduces crop evapotranspiration, and there is generally a linear relationship between percent depletion beyond the management allowed depletion and yield reduction. Crop yield reduction due to water or salt stresses can be quantified by plant stress coefficients, which reflect the plant sensitivity to stress. Water and salinity stress coefficients as well as methods to measure and control salinity are described in this chapter.

Salinity Units

Salts, in the solid phase, are composed of a metal and a ligind joined together by an ionic bond: for example, sodium (Na^+) and chloride (Cl^-). Because of differences in electronegativity, the metal loses one or more electrons and becomes a positively charged ion (cation), and the ligind gains one or more electrons and becomes a negatively charged ion (anion) when salts are dissolved. Common cations and anions in irrigation water are listed in Table 4.1. The equivalent mass of an ion is the molecular mass divided by the valence: the mass of substance required to form one mole of charge when dissolved in water.

Because ions carry electric charge, electrical conductivity of water increases with salinity. Thus, salinity can be measured with an electrical conductivity meter. Common units are decisiemens/meter (dS/m) and mmhos/cm, which equal each other. The conductivity measurement, dS/m, is multiplied by 640 in order to obtain ppm or mg/L of salts in solution. Depending on the proportion of different salts in water, the constant 640 varies from region to region.

$$C_{iw} = EC_{iw}*640 \tag{4.1}$$

where

C_{iw} = total salinity of water, mg/L
EC_{iw} = electrical conductivity of the irrigation water, dS/m.

Saturated paste extract salinity, EC_e, is the standard measurement for soil salinity. In this test, a portion of soil is ground to a fine powder. Water is then added to the soil until it is saturated and the surface glistens. After allowing the soil to equilibrate for 1 day in a Buchner funnel, a vacuum is applied to the funnel, and soil water is extracted. Then, the electrical conductivity of the water extract is measured. Instructions on conducting a saturated paste extract test can be found in Methods of Soil Analysis, Part 1 – Physical Methods, published by the Soil Science Society of America. It is important to follow these instructions precisely in order to obtain a standardized measurement of saturated paste extract salinity.

Several unit systems are used to express salt concentration in water. The selection of the unit system generally depends on the purpose of the measurement. For example, the total mass of salts in water is expressed by mass (mg/L) whereas the quantity of carbonate that must be neutralized by acid is expressed in meq/L (moles of charge). The valence is denoted by z.

P. Waller, M. Yitayew, *Irrigation and Drainage Engineering*, DOI 10.1007/978-3-319-05699-9_4

Table 4.1 Molecular mass and valence of common ions in water

	Molecular mass (g/mole)	Valence	Equiv mass (g/equiv)
Cations			
Ca^{2+}, calcium	40.1	2	20.05
Na^{+}, sodium	23	1	23
K^{+}, potassium	39.1	1	39.1
Mg^{2+}, magnesium	24.3	2	12.15
NH_4^{+}, ammonium	14 + 4 = 18.0	1	18.0
Anions			
Cl^{-}, chloride	35.5	1	35.5
SO_4^{2-}, sulfate	32 + 4 * 16 = 96	2	48
NO_3^{-}, nitrate	14 + 3 * 16 = 62	1	62
CO_3^{2-}, carbonate	12 + 3 * 16 = 60	2	30
PO_4^{2-}, phosphate	31.0 + 4 * 16 = 95	2	47.5
HCO_3^{-}, bicarbonate	1 + 12 + 3 * 16 = 61	1	61

Molarity, mol/L = moles of solute (dissolved substance)/volume of solution
Concentration, mg/L = mass of solute/volume of solution
Equivalents/L, eq/L = (moles/L) (z)
Normality, meq/L = (moles/L) (z) (1,000)
Parts/million, ppm = g of solute/10^6 g solution (same as conc. (mg/L))

Example 4.1 Calculate the mass of calcium chloride, $CaCl_2$, dissolved in 3 liters of water, which results in chloride molarity equal to 0.15. Calculate the concentration of chloride ions.

Molarity is the number of moles of substance (calcium or chloride ions in solution) in one L of water. If the molarity of chloride ions is 0.15, then the number of moles of chloride in the water is calculated as follows: $0.15\ \text{moles/L} * 3\ \text{L} = 0.45\ \text{moles}\ Cl^{-}$.

Two moles of chloride ions are produced for every mole of calcium chloride dissolved. Thus, 0.45/2 = 0.225 moles of calcium chloride are required.

The molecular weight of calcium chloride, $CaCl_2$, is calculated as follows:

$$\begin{aligned} &\text{Ca} - 40.1\ \text{g/mole} \\ &\text{Cl} - 35.5\ \text{g/mole} \\ &\text{CaCl}_2 - 40.1 + 2\,(35.5) = 111.1\ \text{g/mole} \end{aligned}$$

Calculate the mass of $CaCl_2$ required. 0.225 moles (111.1 g/mole) = 27.8 g $CaCl_2$

Calculate the concentration of chloride in solution.

$$\left(\frac{0.15\ moles\ Cl^{-}}{Liter}\right)\left(\frac{35.5\ g}{mole}\right)\left(\frac{1,000\ mg}{g}\right)$$
$$= 5,325\ mg/L$$

In-class Exercise 4.1 Calculate the mass of calcium carbonate, $CaCO_3$, dissolved in 4 L of distilled water that results in carbonate molarity equal to 0.0015. Calculate the concentration of carbonate ions in the water.

Osmotic Potential Energy in Soils

Osmotic potential is caused by the fact that water with a higher concentration of salts has a lower energy (more negative) than water with a lower concentration of salts. Thus, water is driven toward the higher salinity region, or it is impeded from flowing away from the high salinity region in the case of plant water uptake. Thus, if salinity is high in the soil, then the plant has difficulty pulling water from the soil.

The osmotic potential energy in saturated soils can be calculated with the following equation:

$$\psi_s = -3.6 * EC_e \tag{4.2}$$

where

EC_e = electrical conductivity of the saturated paste extract, dS/m
ψ_s = Osmotic potential, m.

As the soil dries, the salinity increases. The osmotic potential as a function of water content is

$$\psi_s = -3.6 * EC_e * \frac{\theta_{sat}}{\theta} \tag{4.3}$$

where

θ_{sat} = saturated water content (by volume), ml/ml
θ = actual water content (by volume), ml/ml

Example 4.2 Calculate osmotic potential in the soil at 25 % water content if EC_e = 1 dS/m, and saturated water content is 50 %.

$$\psi_s = -3.6 * EC_e * \frac{\theta_s}{\theta} = -3.6 * 1 * 2 = -7.2\ m$$
$$= -0.72\ atmospheres$$

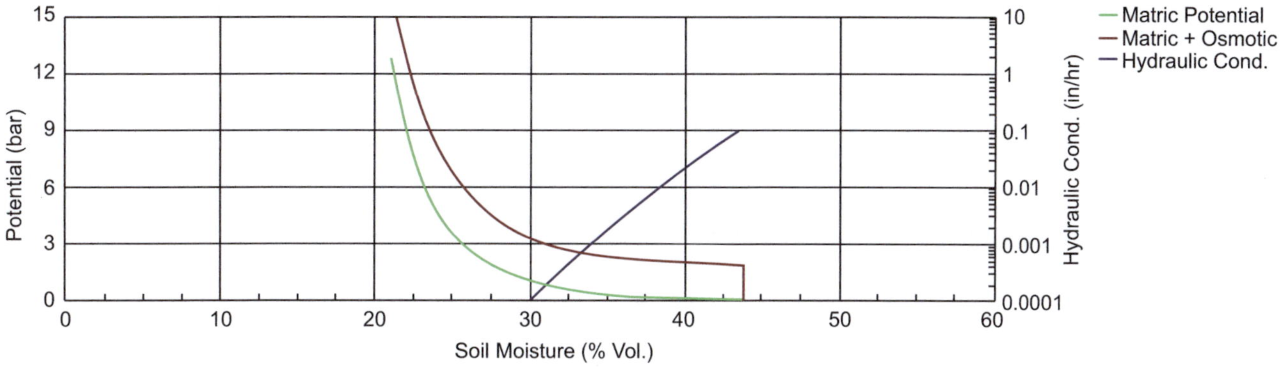

Fig. 4.1 Effect of salinity on soil-water retention in Soil Water Characteristics Calculator

The effect of salinity in a soil can be observed in the graph at the bottom of the Soil Water Characteristics Calculator. Figure 4.1 was made in the Soil Water Characteristics Calculator for a clay loam soil and a saturated paste extract salinity of 5 dS/m. The red line represents matric + osmotic potential, and the green line represents matric potential alone. At 3 bar osmotic potential, the saline soil has 31 % water content and the nonsaline soil has 25 % water content (Fig. 4.1). If a plant can only remove water to 3 bar potential, there would be 6 % less water available to the plant in the saline soil. This difference represents the deceased availability of water for the plant. The overall effect of salinity is that it increases the permanent wilting point water content but leaves the field capacity the same. A change of 5 dS/m only decreases the water content of the permanent wilting point (15 bar) by 1 %, but there is a much greater difference at less negative matric potentials. Thus, the effect of salinity is most detrimental to salt sensitive and drought sensitive plants, which extract less water from the soil between irrigations.

Plants develop extremely negative osmotic potential energy by maintaining high solute (sugars and salts) concentration within cells. This creates an energy gradient that draws water from the soil. The osmotic potential within the plant must be significantly lower (more negative) than the combined matric and osmotic potential in the soil in order to draw water into the plant at the potential evapotranspiration rate.

Water moves through the plant from soil to atmosphere because the energy of water in the atmosphere is lower than the energy of water in the plant or the soil. The energy of water in the atmosphere is a function of the relative humidity. The energy of water is zero in the atmosphere at 100 % relative humidity, but it decreases to negative 1,000 atmospheres at 50 % relative humidity. The plant water energy, due to high sugar and salt concentration, ranges from −10 to −20 atm at midday, while the soil water energy ranges from 0 to −15 atm. Because the partial pressure of water in the atmosphere is less than that of the plant, which is less than that of the soil, water moves from the soil, through the plant, and to the atmosphere. Thus, the plant is a passive conduit that allows water to pass through based on the overall energy gradient between the atmosphere and the soil. The water column does not separate in the plant at negative energy potentials because of water's cohesive properties.

Water transpires from the roots, through the plant, and to the leaf surface through long, hollow, dead cells called xylem. Plants form sugars in the leaves through photosynthesis, and sugars are carried downward from the leaves to the roots in long, hollow, live cells called phloem. Xylem and phloem are like veins; thus, plants are referred to as vascular systems.

Plant cells have a selective semipermeable membrane lying just inside the cell wall. Some molecules, such as water, passively diffuse through the membrane. Other molecules are actively transported through the membrane wall by proteins that are embedded in the membrane. Thus, if the cell would like some extra sugar for dinner (for use in respiration), then it would tell the protein to pass some sugar through the membrane.

Sugars are transported from source leaves through the phloem to other locations within the plant. Phloem tissue expands as the quantities of translocated sugars increase. Unlike transpiration's one-way upward flow of water sap in the xylem, sugars in phloem sap can be transported in any direction needed so long as there is a source of sugar and a sink able to use or store the sugar. The source and sink may be reversed depending on the season, or the plant's needs. Sugar stored in roots may be mobilized to become a source of food in the early spring when the buds of trees, the sink, need energy for growth and development of the photosynthetic apparatus. Phloem sap is mainly water and sucrose; but other sugars, hormones and amino acids are also transported.

Because sugars are concentrated in the cell, the osmotic potential (chemical potential) of water in the cell is negative. The osmotic potential (Gibb's free energy) of cell water is

$$G(J/mole) = RT\ \ln(C) \tag{4.4}$$

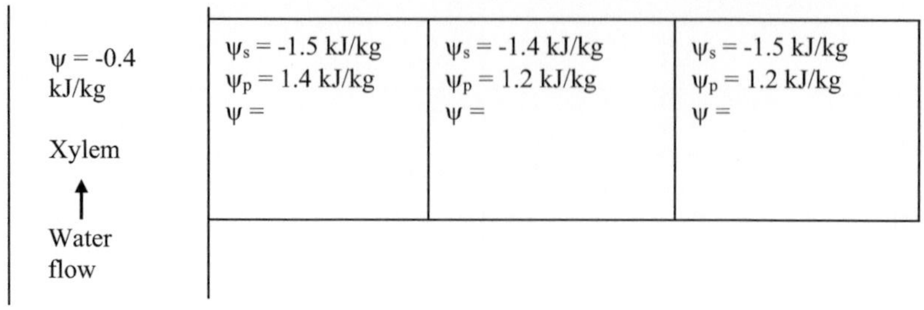

Fig. 4.2 Osmotic potential and turgor pressure in xylem and adjacent plant cells

where

C = mole fraction of water in the cell
T = temperature, °K
R = Avogadro's constant, 8.314 (kPa * L)/(mole * K)
G = Gibbs free energy, J/mole.

C (mole fraction) is found by dividing the number of moles/L of water by the total number of moles/L in the cell solution (mole fraction).

Example 4.3 Calculate the Gibb's free energy of water in a cell if the concentration of sucrose in the cell is 0.5 $\text{mole}_{\text{sucrose}}$/L water at a temperature of 27 °C. Express your answer in J/mole and J/kg (kPa).

$$mole/L\ water = \left(\frac{1{,}000\ g}{L}\right)\left(\frac{mole}{18\ g}\right)$$
$$= 55.56\ mole/L\ water$$

$$\text{mole fraction of water} = (55.56)/(55.56 + 0.5) = 0.991$$

$$\text{G} = \text{RT}\ \ln(\text{C}) = 8.314{*}300\ \ln(0.991) = -23\ \text{J/mole}$$

Multiply by 55.56 kg/L to obtain the answer in J/kg (kPa).

$$\text{G} = \text{RT}\ln(\text{C}) = 55.56{*}8.314{*}300\ \ln(0.991)$$
$$= -1{,}250\ \text{J/kg(kPa)} = -12.5\ \text{atm}$$

The total energy in the cell is the sum of the turgor pressure and osmotic potential

$$\psi = \psi_s + \psi_p \tag{4.5}$$

where

ψ_s = osmotic potential, J/kg
ψ_p = hydraulic pressure, J/kg
ψ = total Gibb's free energy of water in cell, J/kg.

Cell osmotic potential remains relatively constant over the cycle of a 24-hour day; however, the turgor pressure (matric potential or hydraulic pressure, ψ_p) varies during the day and night. When the turgor pressure is just slightly higher than zero during the day, as the plant is transpiring water, the total potential energy in plant cells is close to the osmotic potential, which is between −1 and −2 kJ/kg (−10 to −20 atm). If this total potential is more negative than the soil water potential (turgor + osmotic) , then the plant will remove water from the soil. At night, the cell turgor potential increases to between 1 and 2 kJ/kg (10–20 atm); thus, at night the total potential energy in the cell is close to zero. Thus, the plant does not remove water from the soil at night.

The hydraulic (turgor) pressure in plant cells enables the plant to maintain its shape. If water is limited, and the turgor pressure drops below zero gauge pressure, then the plant wilts. Wilting is more likely to occur when soil has high salinity because the soil osmotic potential and the total soil water potential are more negative, making it more difficult for the plant to draw water from the soil. Even if wilting does not occur, higher salinity in the soil can decrease the rate at which water enters the plant; thus, slowing water uptake → photosynthesis → plant growth.

In-class Exercise 4.2 Fill in the missing total potential values and show direction of water flow. Fill in values for total potential. Are the total potentials in Fig. 4.2 more likely to occur in the day or the night?

If the soil has a low (very negative) energy potential, whether due to osmotic or capillary potential, then it takes longer for the plant to draw water from the soil because the slope of the energy gradient is less. The plant's physiological response to water stress and salinity stress is not the same. Plants compensate for low soil osmotic potential by absorbing ions from the soil solution and manufacturing organic osmolytes within plant cells (FAO 56) in order to lower the energy potential of water in the plant. Synthesis of organic osmolytes requires expenditure of metabolic energy, so this process reduces the amount of energy directed toward crop growth and yield (FAO 56). Partial stomatal closure may also occur under high salinity conditions and thus decrease ET_c (FAO 56).

Salinity Stress

The classic book on salinity and plant stress is

Due to the decreased water uptake in the presence of soil salinity, yield decreases as salinity increases. Yield-salinity curves have been generated for many crops. Figure 4.3

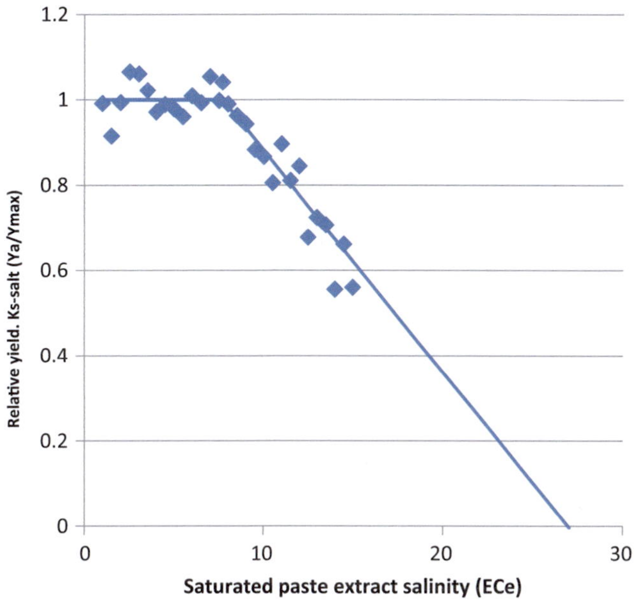

Fig. 4.3 Response of cotton to increasing saturated paste extract salinity

shows the response of cotton to soil salinity. When EC_e exceeds 7.7, the yield begins to decrease. Thus, EC_e is the salinity threshold for cotton. If saturated paste extract salinity is less than 7.7, then there is no yield decrease. Figure 4.3 shows an abrupt change in slope at the threshold with a linear relationship between soil salinity and yield above the threshold. Although recent work by Van Genuchten and others has shown that a curved line near the threshold has a closer fit to most data, this text uses the linear model in order to use standard FAO methods with linear slopes of the yield – salinity line for calculation of yield decrease as a function of salinity.

Crops that are classified as sensitive, moderately sensitive, moderately tolerant, and tolerant to salinity have no reduction in yield at $EC_e = 1.0$-, 3.0-, 6.0-, and 9.5-dS/m, respectively (Fig. 4.4). Vegetable crops are generally sensitive to moderately sensitive to salinity. Field crops such as cotton, wheat, and barley tend to be less sensitive to salinity. Increasing salinity is not always negative. For example, tomatoes are sometimes grown at higher salinity in greenhouse hydroponic systems to increase sugar concentration in the fruit.

The NRCS classifies irrigation salinity levels as no restriction on use ($EC_{iw} < 0.7$), slight to moderate restriction ($0.7 < EC_{iw} < 3.0$), and severe restriction ($3.0 < EC_{iw}$). Crops that are classified as sensitive, moderately sensitive, moderately tolerant, and tolerant to salinity have no reduction in yield at 1.0-, 3.0-, 6.0-, and 9.4-dS/m, respectively. Vegetable crops are generally classified as sensitive to moderately sensitive to salinity. Field crops (grains and fibers) tend to be less sensitive to salinity.

Table 4.2 lists b values, the slope of the line, and $EC_{e\text{-}t}$ values, the threshold salinity, for different crops. In Eq. 4.6, b is divided by the crop sensitivity to water stress because the water stress and salinity stress equations are used together to calculate yield decrease due to both factors (Eqs. 4.9, 4.10, and 4.11). The salinity stress coefficient is equal to 1.0 if $EC_e < EC_{e\text{-}t}$, otherwise

$$K_{s-salt} = 1 - \frac{b}{100{*}K_y}(EC_e - EC_{e-t}) \qquad (4.6)$$

where

b = slope of EC_e/yield line, %/(dS/m)
$EC_{e\text{-}t}$ = threshold saturated paste extract EC_e with no yield decrease, dS/m
$K_{s\text{-}salt}$ = salinity stress coefficient, varies from 0 to 1.

The data in Table 4.2 was acquired in crop yield – salinity stress experiments that were generally conducted with the soil near field capacity (FAO 56). Thus, the data is valid for soils with water content near field capacity. It may be possible to adjust EC in Eq. 4.6 based on water content (EC increases when water content decreases); however, FAO 56 expresses doubt that this procedure will provide accurate results.

Water Stress Coefficient

As with salinity stress, the water stress coefficient is generally calculated under the assumption that yield decreases linearly with increased percent water depletion. The threshold percent depletion, θ_t, is the water content at which yield starts to decrease. Thus, $K_{s\text{-water}}$ decreases linearly from 1 at θ_t to 0 at θ_{pwp} (Fig. 4.5), and $K_s = 1$ at depletion levels less than the threshold depletion. The equation can be rearranged and written in terms of threshold water content and actual water content. Threshold water content should be found with p (p is the point at which yield decreases in contrast to MAD, which is the management water depletion, which may be equal to p or less than p). The key point is that crop stress is initiated at p.

$$K_{s-water} = \frac{\theta - \theta_{pwp}}{\theta_t - \theta_{pwp}} \qquad (4.7)$$

$$\theta_t = \theta_{FC} - (p/100)(\theta_{FC} - \theta_{PWP}) \qquad (4.8)$$

The combined salt and water stress coefficient is the product of the two.

Fig. 4.4 Salinity tolerance and yield reduction of various crops in response to salinity

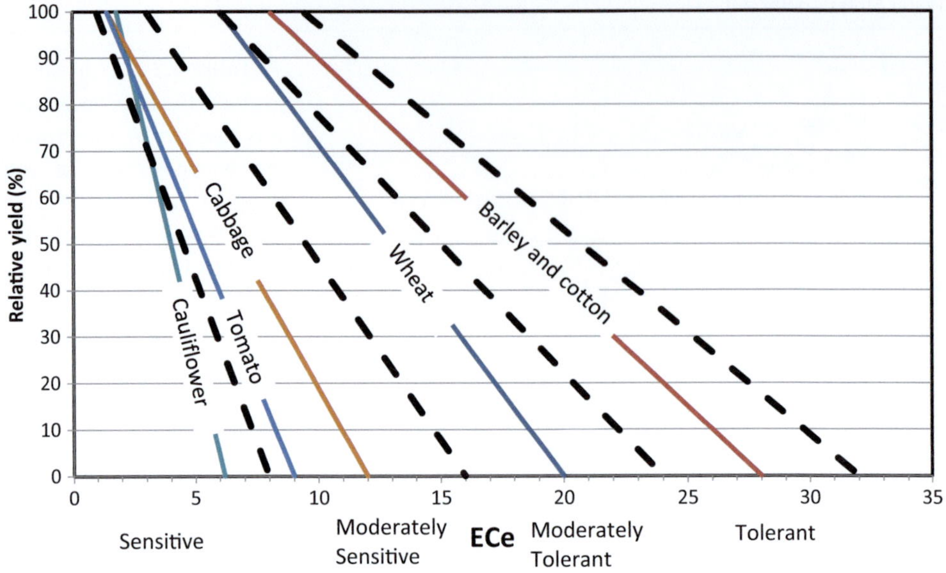

Table 4.2 Threshold EC_e and b values (from FAO 56)

Crop[1]	$EC_{e\ threshold}$[2] ($dS\ m^{-1}$)[3]	b[4] (%/$dS\ m^{-1}$)	Rating[5]
a. Small vegetables			
Broccoli	2.8	9.2	MS
Brussels sprouts	1.8	9.7	MS
Cabbage	1.0–1.8	9.8–14.0	MS
Carrots	1.0	14.0	S
Cauliflower	1.8	6.2	MS
Celery	1.8–2.5	6.2–13.0	MS
Lettuce	1.3–1.7	12.0	MS
Onions	1.2	16.0	S
Spinach	2.0–3.2	7.7–16.0	MS
Radishes	1.2–2.0	7.6–13.0	MS
b. Vegetables – Solanum family *(Solanaceae)*			
Eggplant	–	–	MS
Peppers	1.5–1.7	12.0–14.0	MS
Tomato	0.9–2.5	9.0	MS
c. Vegetables cucumber family *(Cucurbitaceae)*			
Cucumber	1.1–2.5	7.0–13.0	MS
Melons		–	MS
Pumpkin, winter squash	1:2	13.0	MS
Squash, Zucchini	4.7	10.0	MT
Squash (scallop)	3.2	16.0	MS
Watermelon	–	–	MS
d. Roots and tubers			
Beets, red	4.0	9.0	MT
Parsnip	–	–	S
Potato	1.7	12.0	MS
Sweet potato	1.5–2.5	10.0	MS
Turnip	0.9	9.0	MS
Sugar beet	7.0	5.9	T
e. Legumes *(Leguminosae)*			
Beans	1.0	19.0	S
Broadbean (faba bean)	1.5–1.6	9.6	MS
Cowpea	4.9	12.0	MT
Groundnut (peanut)	3.2	29.0	MS
Peas	1.5	14.0	S
Soybeans	5.0	20.0	MT

(continued)

Table 4.2 (continued)

Crop[1]	$EC_{e\ threshold}$[2] ($dS\ m^{-1}$)[3]	b[4] (%/$dS\ m^{-1}$)	Rating[5]
f. Perennial vegetables (with winter dormancy and initially bare or mulched soil)			
Artichokes	–	–	MT
Asparagus	4.1	2.0	T
Mint	–	–	–
Strawberries	1.0–1.5	11.0–33.0	S
g. Fibre crops			
Cotton	7.7	5.2	T
Flax	1.7	12.0	MS
h. Oil crops			
Casterbean	–	–	MS
Safflower	–	–	MT
Sunflower	–	–	MS
i. Cereals			
Barley	8.0	5.0	T
Oats	–	–	MT
Maize	1.7	12.0	MS
Maize, sweet (sweet corn)	1.7	12.0	MS
Millet	–	–	MS
Sorghum	6.8	16.0	MT
Rice [6]	3.0	12.0	S
Wheat *(Triticum aestivum)*	6.0	7.1	MT
Wheat, semidwarf *(T. aestivum)*	8.6	3.0	T
Wheat, durum *(Triticum turgidum)*	5.7–5.9	3.8–5.5	T
j. Forages			
Alfalfa	2.0	7.3	MS
Barley (forage)	6.0	7.1	MT
Bermuda	6.9	6.4	T
Clover, Berseem	1.5	5.7	MS
Clover (alsike, ladino, red, strawberry)	1.5	12.0	MS
Cowpea (forage)	2.5	11.0	MS
Fescue	3.9	5.3–6.2	MT
Foxtail	1.5	9.6	MS
Hardinggrass	4.6	7.6	MT
Lovegrass	2.0	8.4	MS
Maize (forage)	1.8	7.4	MS
Orchardgrass	1.5	6.2	MS
Rye-grass (perennial)	5.6	7.6	MT
Sesbania	2.3	7.0	MS
Sphaerophysa	2.2	7.0	MS
Sudangrass	2.8	4.3	MT
Trefoil, narrowleaf birdsfoot	5.0	10.0	MT
Trefoil, big	2.3	19.0	MS
Vetch, common	3.0	11.0	MS
Wheatgrass, tall	7.5	4.2	T
Wheatgrass, fairway crested	7.5	6.9	T
Wheatgrass, standard crested	3.5	4.0	MT
Wildrye, beardless	2.7	6.0	MT
k. Sugar cane	1.7	5.9	MS
l. Tropical fruits and trees			
Banana	–	–	MS
Coffee	–	–	–
Date palms	4.0	3.6	T
Palm trees	–	–	T
Pineapple (multi-year crop)	–	–	MT
Tea	–	–	–

(continued)

Table 4.2 (continued)

Crop[1]	$EC_{e\ threshold}$[2] $(dS\ m^{-1})$[3]	b[4] $(\%/dS\ m^{-1})$	Rating[5]
m. Grapes and berries			
Blackberry	1.5	22.0	S
Boysenberry	1.5	22.0	S
Grapes	1.5	9.6	MS
Hops	–	–	–
n. Fruit trees			
Almonds	1.5	19.0	S
Avocado	–	–	S
Citrus (grapefruit)	1.8	16.0	S
Citrus (orange)	1.7	16.0	S
Citrus (lemon)	–	–	S
Citrus (lime)	–	–	S
Citrus (pummelo)	–	–	S
Citrus (tangerine)	–	’	S
Conifer trees	–	–	MS/MT
Apples	–	–	S
Peaches	1.7	21.0	S
Cherries	–	–	S
Pear	–	–	S
Apricot	1.6	24.0	S
Plum, prune	1.5	18.0	S
Pomegranate	–	–	MT
Olives	–	–	MT

S sensitive, *MS* moderately sensitive, *MT* moderately tolerant, *T* tolerant
[1]The data serve only as a guideline - Tolerance vary depending upon climate, soil conditions and cultural practices. Crops are often less tolerant during germination and seedling stage.
[2]ECe, threshold means average root zone salinity at which yield starts to decline
[3]Root zone salinity is measured by electrical conductivity of the saturation extract of the soil, reported in deciSiemens per metre (dS m-1) at 25°C
[4]b is the percentage reduction in crop yield per 1 dS/m increase in ECe beyond ECe threshold
[5]Ratings are: T = Tolerant, MT = Moderately Tolerant, MS = Moderately Sensitive and S = Sensitive
[6]Because paddy rice is grown under flooded conditions, values refer to the electrical conductivity of the soil water while the plants are submerged

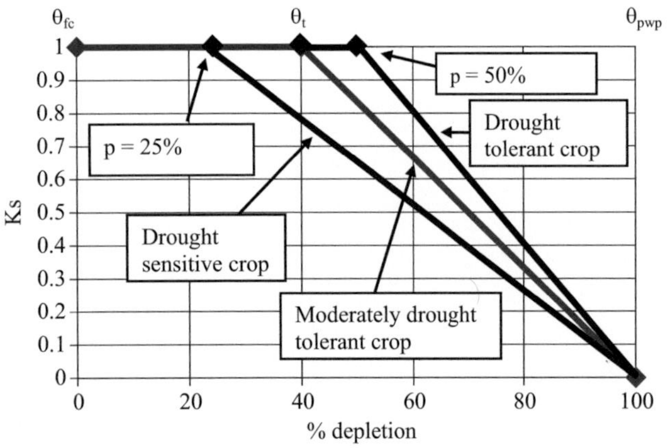

Fig. 4.5 Ks versus water content curve (after FAO 56)

$$K_s = (K_{s-salt})\ (K_{s-water}) \tag{4.9}$$

where

K_s = combined salt and water stress coefficient.

The relationship between yield decrease and combined stress is

$$1 - \left(\frac{Y_a}{Y_{max}}\right) = K_y(1 - K_s) \tag{4.10}$$

where

K_y = crop sensitivity to water stress.
Y_a = actual yield, kg/ha
Y_{max} = maximum potential yield, kg/ha.

Equation 4.10 can be rearranged to solve for actual yield

$$Y_a = (1 - K_y(1 - K_s))Y_{max} \tag{4.11}$$

Crop sensitivity to water stress, K_y, for different crops is tabulated in Table 4.3.

Table 4.3 Crop sensitivity to water stress, K_y (FAO 56)

Crop	K_y	Crop	K_y
Alfalfa	1.1	Potato	1.1
Banana	1.2–1.35	Safflower	0.8
Beans	1.15	Sorghum	0.9
Cabbage	0.95	Soybean	0.85
Citrus	1.1–1.3	Spring Wheat	1.15
Cotton	0.85	Sugarbeet	1.0
Grape	0.85	Sugarcane	1.2
Groundnet	0.70	Sunflower	0.95
Maize	1.25	Tomato	1.05
Onion	1.1	Watermelon	1.1
Peas	1.15	Winter wheat	1.05
Pepper	1.1		

Example 4.4 Calculate actual yield for cotton for a growing season if average salinity during the growing season is 10.4 dS/m, and average water content is 14 %. $\theta_{fc} = 20$ %, $\theta_{pwp} = 10$ %. $K_y = 0.85$. Max yield = 1,285 kg/ha. MAD = 50 % (set equal to p from FAO56, but don't worry about this). Threshhold EC_{et} is 7.7 dS/m and b is 5.2.

$$\theta_t = \theta_{FC} - (MAD/100)(\theta_{FC} - (\theta_{PWP})$$
$$\theta_t = 20 - (0.5)(20 - 10) = 15\%$$

$$K_{s-water} = \frac{\theta - \theta_{pwp}}{\theta_t - \theta_{pwp}} = \frac{0.14 - 0.10}{0.15 - 0.10} = 0.8$$

Calculate K_{s-salt}

$$K_{s-salt} = 1 - \frac{b}{100*K_y}(EC_e - EC_{e-t})$$
$$= 1 - \frac{5.2}{100*0.85}(10.4 - 7.7) = 0.83$$

Calculate total K_s

$$K_s = K_{s-water}K_{s-salt} = 0.80*0.83 = 0.67$$

Calculate yield

$$Y_a = (1 - K_y(1 - K_s))Y_m = (1 - 0.85(1 - 0.67))1,285 = 924\ kg/ha$$

Example 4.5 Calculate the yield reduction for sugar beets. Assume 80 cm is applied and 100 cm is required. $Y_{max} = 40$ tons/ha.

From Table 4.3, $K_y = 1.0$. The yield based on the FAO K_y slope is calculated as follows:

$$Y_a = \left(1 - K_y\left(1 - \frac{ET_{c-adj}}{ET_c}\right)\right)Y_m$$
$$= \left(1 - 1.0\left(1 - \frac{80}{100}\right)\right)40 = 32\ t/ha$$

Modeling Soil Salinity

The basis for the salinity model is the conservation of mass equation. Salts may change from liquid to solid phase or be removed from the soil by plant uptake. Although different salts are taken up by plants in different proportions, have different absorption potentials, and different solubilities, a simple model of root zone salinity is often used, which lumps all salts into one salinity term and ignores precipitation to the solid phase, adsorption, plant uptake, and dissolution. Hillel (1998) presented the following equation with all of the possible terms for overall salinity within the liquid phase in the root zone.

$$[\rho_w(V_rC_r + V_iC_i + V_gC_g) + M_s + M_a] - (M_p + M_c + \rho_w V_d C_d) = \Delta M_{sw} \tag{4.12}$$

where

ρ_w = density of water, 1,000 mg/L
V_r = volume of rainwater entering the soil, L
C_r = salinity concentration in rainwater, mg/L
V_i = volume of irrigation water entering the soil, L
C_i = salinity concentration in irrigation water, mg/L
V_g = volume of ground water entering the soil, L
C_g = salinity concentration in ground water, mg/L
M_s = mass of salt dissolved from the soil, mg
M_a = mass of salt from agricultural inputs, mg
M_p = mass of salt precipitated (turned to solid), mg
M_c = mass of salt removed by the crop, mg
V_d = volume of water removed from the soil by drainage, L
C_d = concentration of salts in drainage water
ΔM_{sw} = change in mass of salt in the soil's liquid phase.

Crops only remove a negligible amount of salt (Hillel 1998). For a steady state model, assuming constant salinity concentration during the season and the year, and assuming that in-situ precipitation (liquid to solid) and dissolution (solid to liquid) of salt is negligible in the case of a steady-state model, then the salt balance becomes (Hillel 1998)

$$V_iC_i + V_rC_r = (V_d - V_g)C_d \tag{4.13}$$

If precipitation and irrigation water salinity are averaged so that C_{in} is the average weighted salinity of the two water sources, and if water movement upward into the root zone from a water table is negligible, then Eq. 4.13 becomes

$$V_{in}C_{in} = V_{out}C_{out} \tag{4.14}$$

Water volume can be replaced by water depth.

$$d_{in}C_{in} = d_{out}C_{out} \tag{4.15}$$

Leaching Fraction

The strategy for keeping soil salinity within an acceptable range is leaching the soil: applying extra water and leaching salts below the root zone. Although the steady state leaching fraction equation described in this section are inferior to transient analysis (Letey et al. 2011), it is the most common method of calculating the leaching fraction. The *WINDS* model is used to compare transient analysis of salinity leaching to the steady state equations in Chapter 26. The leaching fraction, LF, is the leached depth or seepage depth divided by the applied depth where i in Eq. 4.16 includes precipitation as well as irrigation if C_{in} is the average salinity of both.

$$LF = \frac{d_{out}}{d_{in}} = \frac{d_{seepage}}{i} \tag{4.16}$$

where

LF = leaching fraction, dimensionless.

Substitute Eq. 4.16 into Eq. 4.15 in order to solve for the leachate concentration

$$d_{in}C_{in} = LF{*}i\,C_{out}$$

$$C_{out} = \frac{C_{in}}{LF} \tag{4.17}$$

Because EC is proportional to the concentration, C, the leachate salinity for uniform water application can be written as (Ayers and Westcott 1985).

$$EC_{dw} = \frac{EC_{iw}}{LF} \tag{4.18}$$

where

EC_{dw} = electrical conductivity of drainage water (leachate or seepage), dS/m.
EC_{iw} = electrical conductivity of irrigation water, dS/m.

Depth of drainage water is equal to the irrigation water minus d_{ET}. Substitute into Eq. 4.16

$$d_{out} = i - d_{ET} \quad LF = \frac{i - d_{ET}}{i} \tag{4.19}$$

The maximum allowable EC_e can be calculated based on crop sensitivity to salinity stress; and the required leaching fraction can be calculated based on the maximum allowable EC_e. Equation 4.20 an empirical equation based on field experiments conducted with low frequency irrigation systems. It has been a standard method for calculation of leaching fraction, although updated methods are often recommended in it place. Nevertheless, the equation is reliable within normal ranges of salinity and leaching fraction.

$$LF = \frac{EC_{iw}}{5{*}EC_e - EC_{iw}} \tag{4.20}$$

Example 4.6 Irrigation water salinity (EC_{iw}) = 1 dS/m. Applied water depth (d_{in}) = 1176 mm/season. Crop water demand (ET_c) = 1,000 mm/season. Assume that plants extract 40 %, 30 %, 20 %, and 10 % of their water from the upper quarter, 2nd quarter, 3rd quarter, and lowest quarter of the root zone, respectively. First, determine the leachate salinity treating the root zone as a single layer. Next, determine the seepage salinity from each of 4 layers and the average salinity for the 4 layers (from Ayres and Westcott 1985).

Treating the entire root zone as a single layer, calculate seepage salinity.

$$LF = \frac{i - ET}{i} = \frac{1,176 - 1,000}{1,176} = 0.15$$

$$EC_{dw} = \frac{EC_{iw}}{LF} = \frac{1}{0.15} = 6.7 \ dS/m$$

Use the same equations to determine soil salinity at the bottom of each of the four quarters of the root zone.

$$LF_1 = \frac{1,176 - 0.4{*}ET}{1,176} = \frac{1,176 - 0.4{*}1,000}{1,176} = 0.66$$

$$EC_1 = EC_{iw}/LF_1 = 1/0.66 = 1.5 \ dS/m$$

$$LF_2 = \frac{776 - 0.3{*}ET}{776} = \frac{776 - 0.3{*}1,000}{776} = 0.61$$

$$EC_2 = EC_1/LF_2 = 1.5/0.61 = 2.5 \ dS/m$$

$$LF_3 = \frac{476 - 0.2{*}ET}{476} = \frac{476 - 0.2{*}1,000}{476} = 0.58$$

$$EC_3 = EC_2/LF_3 = 2.5/0.58 = 4.3 \ dS/m$$

$$LF_4 = \frac{276 - 0.1{*}ET}{276} = \frac{276 - 0.1{*}1,000}{276} = 0.64$$

$$EC_4 = EC_3/LF_4 = 4.3/0.64 = 6.7 \ dS/m$$

The calculated seepage salinities, treating the soil as a whole and in layers, agree: 6.7 dS/m.

The average soil salinity is the average of the irrigation water salinity and the salinities at the bottom of the 4 layers.

$$EC_{ave} = (1/2 + 1.5 + 2.5 + 4.3 + 6.7/2)/4 = 3.0 \ dS/m.$$

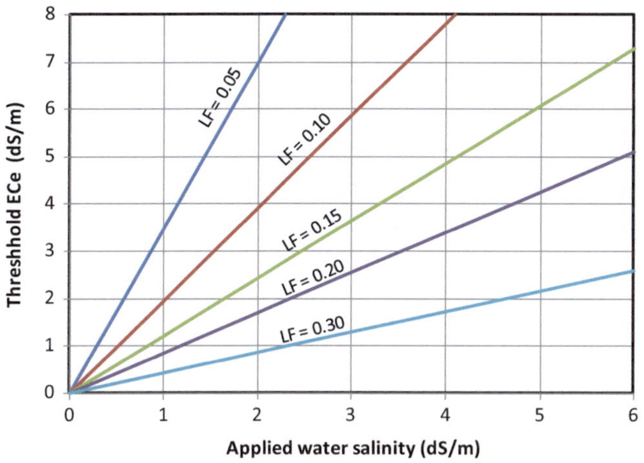

Fig. 4.6 Recommended leaching fractions from Hoffman and Ven Genuchten equation where EC_a is salinity of irrigation water and C_v is saturated paste extract EC

Soil water salinity is greater than the saturated paste extract salinity, because, during the saturated paste extract test, distilled water is added to the soil until saturation is reached. If the water content in the soil is at field capacity, and the field capacity is approximately 50 % of saturated water content (porosity), then, the average EC_e in the soil, based on Eq. 4.3, is

$$EC_{e\ (sat)} = EC_{ave}*(\theta_{FC}/\theta_{sat}) = 3.0*(0.5) = 1.5\ \text{dS/m}$$

Rhoades (1974) recommended the following leaching fraction equation for high frequency sprinkler or trickle irrigation; however, the need for this equation has been questioned by some. It results in higher recommended leaching fractions for high frequency irrigation.

$$LF = EC_{iw}/(2\ EC_e) \quad (4.21)$$

Hoffman and Van Genuchten (1983) developed the following theoretical equation for LF. In most cases, Eq. 4.22 yields similar results to Eq. 4.20, but because it is based on theory, it has a wider range of salinity parameters for which it is accurate. Hoffman and Van Genuchten developed Fig. 4.6 based on Eq. 4.22.

$$\frac{EC_e}{EC_{iw}} = \left(\frac{1}{LF} + \frac{\delta}{Z*LF}\ln\left(LF + (1 - LF)e^{-z/\delta}\right)\right) \quad (4.22)$$

where

Z = root zone depth
δ = empirical constant set to 0.2 Z.

One of the key factors in leaching fraction calculations is the leaching efficiency. In some soils, there are mobile and immobile zones. There may be very little mixing of irrigation water and the majority of soil water in immobile zone, with most of the infiltrated water leaching through the soil without mixing with the immobile water. For example, the leaching efficiency may be only 20 % in a clay soil with most soil water between clay particles, so the soil water salinity after leaching is much higher than calculated with the assumption that leaching efficiency is 100 %.

Example 4.7 Calculate the leaching fraction with no reduction in yield for cotton and corn if irrigation water salinity is 960 ppm. Compare results from Eqs. 4.20, 4.21, and Fig. 4.6 (Eq. 4.22).

$$\text{Irrigation water salinity} = 960\ \text{ppm} = 960/640\ \text{dS/m} = 1.5\ \text{dS/m}$$

Cotton

Max. soil salinity in the saturated paste extract (Table 4.2) for cotton with no yield reduction = 7.7 dS/m.

Equation 4.20 $LF = \frac{EC_{iw}}{5(EC_e) - EC_{iw}} = \frac{1.5}{5(7.7) - 1.5} = 0.04$

Equation 4.21 LF = $EC_{iw}/(2*EC_e) = 1.5/(2*7.7) = 0.10$

From Fig. 4.6, with ECa = 1.5 and Cv = 7.7, LF = 0.04

Corn

Max. soil salinity in the saturated paste extract (Table 4.2) for corn with no yield reduction = 1.7 dS/m.

Equation 4.20 $LF = \frac{EC_{iw}}{5(EC_e) - EC_{iw}} = \frac{1.5}{5(1.7) - 1.5} = 0.21$

Equation 4.21 LF = $EC_{iw}/(2*EC_e) = 1.5/(2*1.7) = 0.44$

From Fig. 4.6, with ECa = 1.5 and Cv = 1.7, LF = 0.16

As expected Eq. 4.20 and Fig. 4.6 had similar results, and Eq. 4.21 estimated a higher leaching fraction.

With drip irrigation systems, salts may be concentrated at the edge of the wetted zone, leaving a ring of salts around the emitters on the soil surface. Similarly, salts may be pushed to the midpoint between furrows and concentrate at the centers of beds. If this is the case, then it may be beneficial to plant crops at an offset position on the bed.

If water is less expensive in winter, then growers can leach the soil during the winter, and use less water in summer. However, transient changes in soil salinity concentration can occur quickly during the growing season. In fact, soils can become salinized within one growing season if there is no leaching and high salinity in irrigation water. Plants are most sensitive to salinity during germination, so it is most important to control salinity at the beginning of the season.

Irrigation Application Depth and Leaching Fraction

If the goal is to maintain salinity within an acceptable range during the entire growing season, then the depth of irrigation water that should be applied during any one

Table 4.4 Maximum acceptable salinity in irrigation water as a function of soil type, rainfall per year, and crop sensitivity to salinity (After Rao et al. 1994)

Soil texture (percent clay)	Crop tolerance	Annual rainfall <350 mm	350–500	550–750
Fine (>30 %)	Sensitive	1	1	1.5
	Semi-tolerant	1.5	2	3
	Tolerant	2	3	4.5
Moderately fine (20–30 %)	Sensitive	1.5	2	2.5
	Semi-tolerant	2	3	4.5
	Tolerant	4	6	8
Moderately coarse (10–20 %)	Sensitive	2	2.5	3
	Semi-tolerant	4	6	8
	Tolerant	6	8	10
Coarse (<10 %)	Sensitive	–	3	3
	Semi-tolerant	6	7.5	9
	Tolerant	8	10	12.5

irrigation event is the RAW divided by the irrigation efficiency and (1-LF).

$$IR = \frac{100}{IE(1-LF)}RAW \quad (4.23)$$

where

IR = irrigation requirement during single irrigation event, cm
IE = irrigation efficiency, percent.

Example 4.8 Calculate the depth of irrigation water required (average for the field), IR, for melons based on Eq. 4.23. The MAD is 45 %, the irrigation system efficiency is 70 %, the irrigation water EC_{iw} is 1.09 dS/m, and the TAW is 24 cm.

Max. soil salinity in the saturated paste extract (Table 4.2) for melons with no yield reduction = 2.2 dS/m.

$$LF = \frac{EC_{iw}}{5(EC_e) - EC_{iw}} = \frac{1.09}{5(2.2) - 1.09} = 0.11$$

$$IR = \frac{100}{IE(1-LF)}RAW = \frac{100}{70(1-0.11)}(0.45*24\ cm)$$
$$= 17\ cm$$

Project Planning and Salinity

Irrigation water salinity is the primary factor in determining the susceptibility of a region to salinization. Irrigation water with a salinity lower than 450 mg/L ($EC_{iw} = 0.7$) does not present a hazard for irrigation salinity. Irrigation water with salinity in excess of 2,000 mg/L ($EC_{iw} = 3$) presents a hazard for many crops. However, rainfall, soil type, and crop sensitivity to salinity are critical factors in determining the salinity hazard from irrigation water. Rao et al. (1994) developed Table 4.4 that specifies the maximum acceptable salinity of irrigation water as a function of these three factors.

Sodicity

Excess sodium reduces water availability because it leads to breakdown of clay particle structure (dispersion), and these particles can clog the soil and reduce infiltration rate to nearly zero. The reason that sodium causes clay particle dispersion is that sodium molecules, with a valence of 1 (Na^+), are loosely attracted to negatively charged clay layers, and they maintain a hydration shell of approximately ten water molecules. The other major cation in water, calcium, is attracted much more strongly to the clay particles because they have a charge of +2. As a result, they are positioned very close to the clay particle surface, and they have a much smaller shell of water hydration (Fig. 4.7). If too many sodium molecules,with their large shell of hydration, are in the soil water solution, then they will be the primary ion in the interlayer between clay particles, and the shells of hydration will force the clay layers apart and break down the soil structure.

Equation 4.24 calculates the sodium adsorption ratio (SAR), which can be used to determine the sodium hazard associated with irrigation water.

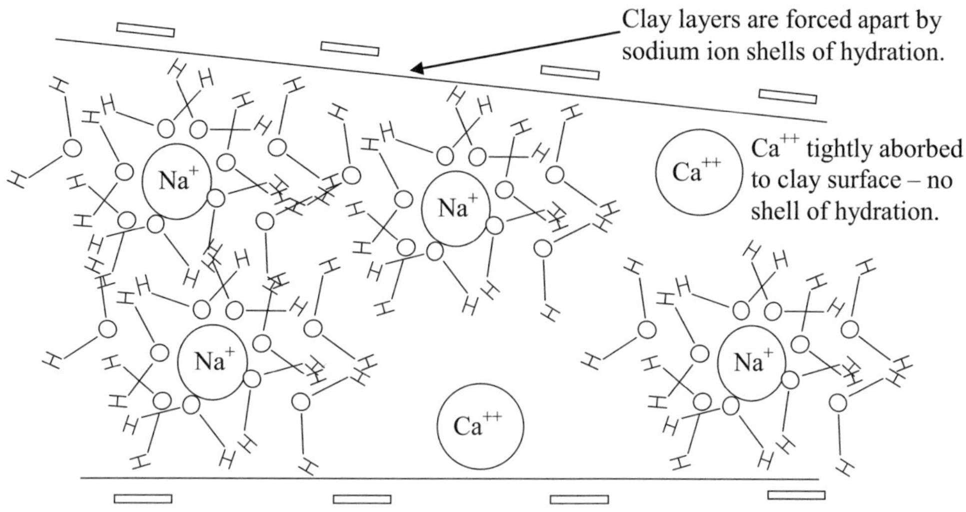

Fig. 4.7 Sodium hydration shell and calcium ions between clay layers

Table 4.5 Sodicity hazard to soils as a function of irrigation water sodicity and salinity

	EC_{iw}		
SAR	None	Slight to moderate	Severe
0–3	>0.7	0.7–0.2	<0.2
3–6	>1.2	1.2–0.3	<0.3
6–12	>1.9	1.9–0.5	<0.5
12–20	>2.9	2.9–1.3	<1.3
20–40	>5.0	5.0–2.9	<2.9

$$SAR = \frac{|Na^+|}{\sqrt{\frac{|Ca^{++}|+|Mg^{++}|}{2}}} \quad (4.24)$$

where

SAR = sodium absorption ratio, dimensionless
Na^{++} = sodium normality, meq/L
Ca^{++} = calcium normality, meq/L
Mg^{++} = magnesium normality, meq/L.

The hazard level of sodium is a function of both the SAR and the overall salinity (Table 4.5). Higher salinity in the soil water decreases the osmotic potential (more negative) of water in the soil water solution, and, as such, decreases the amount of water in the hydration shells around the sodium ions in the interlayer between clay particles. Thus, it may be very detrimental to irrigate with low salinity water in a field that was previously irrigated with high salinity and sodicity water.

Example 4.9 Irrigation water has 460 mg/L sodium (Na^+), 40.1 mg/L calcium (Ca^{++}), and 24.3 mg/L magnesium (Mg^{++}). If irrigation water salinity is 1,280 ppm, then what level of hazard is presented by sodicity?

Equivalent masses are given in Table 4.1. Calculate meq/L for each cation.

$$460 \text{ mg/L } Na^+/23 \text{ mg/meq} = 20 \text{ meq/L } Na^+$$
$$40.1 \text{ mg/L } Ca^{++}/20.05 \text{ mg/meq} = 2 \text{ meq/L } Ca^{++}$$
$$24.3 \text{ mg/L } Mg^{++}/12.15 \text{ mg/meq} = 2 \text{ meq/L } Mg^{++}$$

$$SAR = \frac{|Na^+|}{\sqrt{\frac{|Ca^{++}|+|Mg^{++}|}{2}}} = \frac{20}{\sqrt{\frac{2+2}{2}}} = \frac{20}{\sqrt{2}} = 7$$

The EC_{iw} of the irrigation water is 1,280 ppm/640 dS/m/ppm = 2. From Table 4.5, there is no possible hazard due to sodicity from this water.

Specific Ion Toxicity

Specific ion toxicity is also generally not quantified by crop stress coefficients; however, it can also lead to yield loss when certain salts (ions) become toxic at high concentration in soils. Table 4.6 lists specific ion toxicity levels of sodium, chloride, and boron. Certain crops are more susceptible to specific ions than others. Not even the most sensitive crops are susceptible to specific ion toxicities at levels less than those list in the "none" column. Nearly all crops are affected at the severe level.

Table 4.6 Specific ion toxicity (Credit NRCS)

			Degree of restriction on use		
Element	Irrigation type	Units	None	Slight to moderate	Severe
Sodium	Sprinkler irrigation	meq/L	<3	3–9	>9
Sodium	Surface irrigation	meq/L	<3	>3	
Chloride	Sprinkler irrigation	meq/L	<4	4–10	>10
Chloride	Surface irrigation	meq/L	<3	>3	
Boron		meq/L	<0.7	0.7–3.0	>3.0

Questions

1. What is the molecular mass (g/mole) of calcium carbonate ($CaCO_3$)?
2. What is the molarity and ppm of Na^+ and Cl^- if 0.02 g NaCL is dissolved in 4 L of water?
3. What is the concentration of salts (mg/L) in water with $EC_{iw} = 2.4$ dS/m?
4. What is the soil salinity (mg/L) at saturation if the saturated paste extract EC_e is 4 dS/m?
5. How many moles of sodium chloride "NaCl" are required in 4 L of water to develop a solution that has 1,000 mg/L sodium? What is the concentration of chloride in the water?
6. How many grams of NO_3 are dissolved in 4 L of water with a nitrate concentration of 20 mg/L?
7. Calculate the mass of ammonium nitrate NH_4NO_3 dissolved in 200 L water to obtain a nitrate concentration of 30 mg/L in water.
8. If saturated paste extract EC_e is 2 dS/m, then what is the osmotic potential ψ_s of the water if the water content θ is 15 % and the saturated water content θ_s is 45 %?
9. Explain Fig. 4.1.
10. Explain the function of the xylem and the phloem in the plant.
11. Calculate the Gibb's free energy of water "G" in a cell if the concentration of sucrose in the cell is 0.3 mole sucrose/L water at a temperature of 20 °C. Express your answer in J/mole and J/kg (a.k.a. kPa).
12. Calculate yield for cotton for a growing season if average values of EC_e and water content during the growing season are 12 dS/m and 13.5 %, respectively. $\theta_{fc} = 0.2$, $\theta_{pwp} = 0.1$. $K_y = 0.85$. Max yield = 1,000 kg/ha. MAD = 0.55. Threshold EC_e is 7.7 dS/m and b is 5.2. $K_y = 0.85$ for cotton
13. Why does high sodium ruin some soils? What types of soils are most vulnerable?
14. Irrigation water has 230 mg/L sodium (Na+), 60.15 mg/L calcium (Ca++), and 24.3 mg/L magnesium (Mg++). If irrigation water salinity is 1,000 ppm, then what level of hazard is presented by sodicity?
15. Fill in the missing total potential values and show direction of water flow. Are these total potentials more likely to occur in the day or the night?

$\psi = -1.1$ kJ/kg Xylem ↑ Water flow	$\psi_s = -1.5$ kJ/kg $\psi_p = 0.5$ kJ/kg $\psi =$	$\psi_s = -1.4$ kJ/kg $\psi_p = 0.6$ kJ/kg $\psi =$	$\psi_s = -1.5$ kJ/kg $\psi_p = 0.7$ kJ/kg $\psi =$

16. Determine the leachate salinity. Irrigation water salinity (EC_{iw}) = 2 dS/m. Applied water depth (d_{in}) = 1,250 mm/season. There is no precipitation during the growing season. Crop water demand (ET_c) = 1,000 mm/season. Average soil moisture content is the same at the beginning and end of the growing season.
17. Irrigation water salinity (EC_{iw}) = 2 dS/m. Applied water depth (d_{in}) = 1300 mm/season. Crop water

demand (ET_c) = 900 mm/season and assume that plants extract 40 %, 30 %, 20 %, and 10 % of their water from the upper quarter, next quarter, next, and lowest quarter of the root zone, respectively. Calculate the salinity at the bottom of the root zone by treating the root zone as a whole and calculate salinity at the bottom of each of the 4 layers. Calculate the average salinity in the root zone by assuming that the field capacity is half of the saturated water content (divide average salinity in half for EC_e). Then calculate the leaching fraction that would be required to have this average salinity in the root zone with Eq. 4.20 and compare with the leaching fraction in this problem.

18. Calculate the depth of irrigation water required (average for the field), IR, for cotton based on Eq. 4.23. The MAD is 50 %, the irrigation system efficiency is 60 %, the irrigation water EC_{iw} is 2 dS/m, and the TAW is 20 cm.

References and Resources

Allen R (1997) Crop evapotranspiration. In: "Guidelines for computing crop water requirements" – FAO Irrigation and drainage paper 56. http://www.fao.org/docrep/X0490E/X0490E00.htm

Ayers RS, Westcott DW (1985) Water quality for agriculture. In: FAO irrigation and drainage paper 29. http://www.fao.org/DOCREP/003/T0234E/T0234E00.htm#TOC

Gupta R (1994) Ground water quality for Irrigation. In: Rao DLN, Singh NT, Gupta RK, Tyagi NK (eds) Salinity management for sustainable agriculture. Central Soil Salinity Research Unit, Karnal, p 94 (Table 3). http://krishikosh.egranth.ac.in/HD/common/split_document.jsp?doc=2.pdf1/2046581/1/

Hillel D (1998) Environmental soil physics. Elsevier, Amsterdam/Boston, p 771

Hoffman GJ, Van Genuchten MT (1983) Soil properties and efficient water use: water management for salinity control. In: Taylor HM et al (eds) Limitations to efficient water use in crop production. American Society of Agronomy, Madison, pp 73–85

Kimberly Idaho Research and Extension Center FAO 56 Website (includes spreadsheets by Richard Allen). http://www.kimberly.uidaho.edu/water/fao56/index.html

Letey J, Hoffman G, Hopmans J, Grattan S, Suarez D, Corwin D et al (2011) Evaluation of soil salinity leaching requirement guidelines. Agric Water Manag 98:502–506

Rhoades J (1974) Drainage for salinity control. In: van Schilfegarde J (ed) Drainage for agriculture, vol 17. American Agronomy Society Monograph, Madison, pp 433–462

United States Salinity Laboratory. Handbook #60 *Diagnosis and Improvement of Saline and Alkali Soils*. http://www.ars.usda.gov/Services/docs.htm?docid=10155

5 Evapotranspiration

The crop water requirement determines the irrigation system flow rate. Evapotranspiration (ET) is the combined process of evaporation from the soil and transpiration from the plant (Fig. 5.1). Transpiration is the process by which water passes from the soil into the atmosphere through the tissues of the plant (Fig. 5.1). Without quantifying evapotranspiration, design and management of irrigation systems is guesswork. Quantification of evapotranspiration through crop experiments and theoretical physics has led to increased irrigation efficiency. In addition, governments have invested heavily in the establishment and maintenance of agricultural weather station networks. Most weather station networks use the Penman-Montieth model to calculate reference evapotranspiration. The model is a surface energy balance with sensible heat flux, latent heat of vaporization, net radiation, and soil heat flux. In addition to theoretical development and demonstration of the calculation procedure, the chapter includes Excel worksheets that calculate the Penman-Montieth hourly and daily ET. The chapter ends with a program that automatically acquires and downloads weather station data for use in evapotranspiration and crop models.

For an open pond, there is a free transfer of liquid water between the water surface and the atmosphere (Fig. 5.2). The rate of change in liquid water volume, evaporation, is the difference between the rate that molecules escape the liquid surface and the rate that molecules are entrapped by the surface. The rate that molecules are entrapped by the water surface increases with relative humidity; thus, evaporation rate is higher in dry air. The dew point is the temperature at which the rate of molecules escaping from and molecules trapped at the water surface is equal.

Energy is required for evaporation from a leaf or water surface. This energy, called the latent heat of vaporization, changes with temperature and is calculated as follows:

$$\lambda_v = 2,503 - 2.39\ T \tag{5.1}$$

where

λ_v = latent heat of vaporization, kJ/kg
T = temperature, °C.

Mass and Energy Transfer from Canopy to Atmosphere

Water vapor movement from the plant canopy to the atmosphere is calculated based on the vapor pressure gradient (Gibbs free energy difference) between the plant canopy and the atmosphere. As long as there is sufficient energy for vaporization, the maximum rate of vapor transfer into the atmosphere from the plant canopy based on the vapor pressure gradient is calculated as follows:

$$ET = 0.622\rho_a h_v \frac{(e_c - e_a)}{P} \tag{5.2}$$

where

e_c = canopy vapor pressure, kPa
e_s = saturated vapor pressure; partial pressure of water at 100 % humidity, kPa
e_a = partial pressure of water in atmosphere, kPa.
h_v = water vapor transfer conductance $=1/(r_{av} + r_s)$, m s^{-1}
r_{av} = resistance to momentum transfer in atmosphere above canopy, s m^{-1}
r_s = bulk surface resistance, s m^{-1}
0.622 = ratio of mass of water vapor to mass of dry air, dimensionless
ET = evapotranspiration, kg m^{-2} s^{-1} = mm/s
ρ_a = density of air, kg m^{-3}, $\frac{e_a}{R_aT} + \frac{e_d}{R_dT} \sim 1.23 - 0.000112{*}z$
R_a = specific gas constant for water vapor, 0.4615 kJ kg^{-1} K^{-1}
R_d = specific gas constant for dry air, 0.287 kJ kg^{-1} K^{-1}
e_d = partial pressure of dry air, kP
z = elevation above mean sea level, m.

The bulk surface resistance in Eq. 5.2, r_s, is the resistance of water vapor transfer from the soil and foliage to the surface boundary layer (layer of atmosphere directly above

P. Waller, M. Yitayew, *Irrigation and Drainage Engineering*, DOI 10.1007/978-3-319-05699-9_5

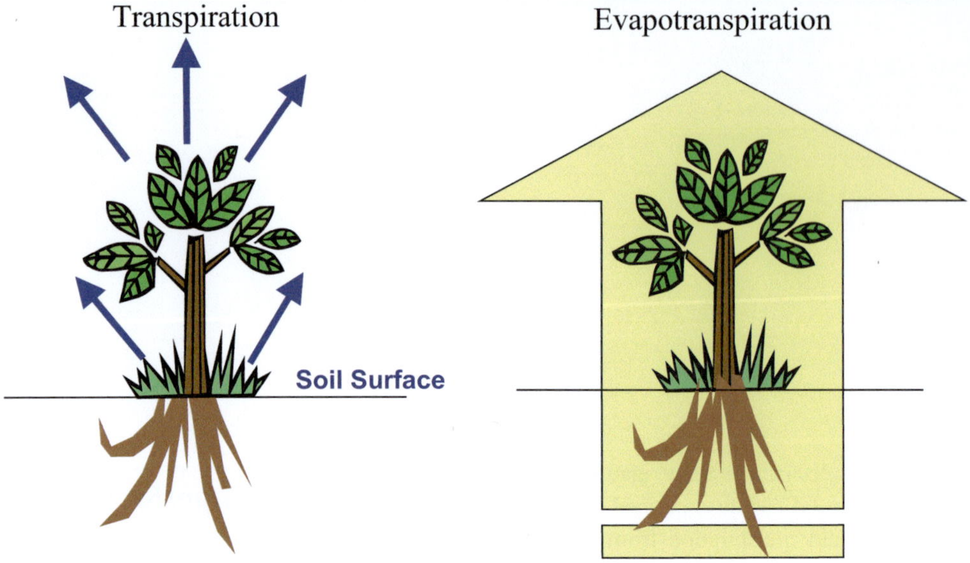

Fig. 5.1 Transpiration and evapotranspiration (Courtesy of Don Slack, The University of Arizona)

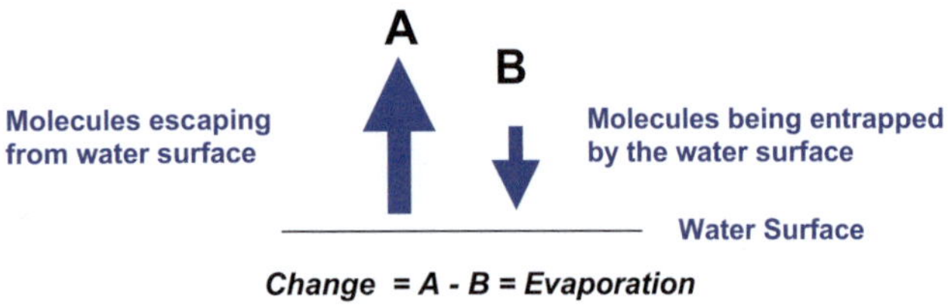

Fig. 5.2 Evaporation (Courtesy of Don Slack, The University of Arizona)

the canopy). It includes (1) resistance to movement of water through soil and through the plant, (2) resistance of stomata in the multiple layers of leaves, and (3) resistance to vapor movement vertically upward through the canopy into the boundary layer of air above the canopy (Fig. 5.3).

The bulk surface resistance is the sum of the three resistances in Fig. 5.3.

$$r_s = r_{soil/plant} + r_{stomate} + r_{vapor\ movement\ within\ canopy} \quad (5.3)$$

The total resistance to water vapor transfer is the sum of bulk surface resistance, r_s, and aerodynamic resistance, r_{av}, to water vapor transfer. The aerodynamic resistance to water vapor transfer for well-watered turf (reference crop for evapotranspiration calculations) can be calculated as $208/U_2$, where U_2 is the wind speed at 2 m elevation in units of m/sec.

Turf and alfalfa have been selected as ET reference crops in order to standardize the effects of weather on ET. Reference evapotranspiration (ET_0) is the rate of evapotranspiration (mm/day) from well-watered 10 cm tall turf grass or 30 cm tall alfalfa that is free of disease, insect infestations or other factors which may stress or limit its growth and transpiration rate. As a result, it is desirable to place weather stations in turf or alfalfa areas.

Air pressure in Eq. 5.2 is a function of elevation.

$$P = 101.3\ \left(\frac{293-0.0065z}{293}\right)^{5.26} \sim 101.3 - 0.01055\ z \quad (5.4)$$

where

z = elevation above mean sea level, m.
P = atmospheric pressure, kPa

The canopy vapor pressure, e_c, is assumed to be the saturated vapor pressure, e_s.

Example 5.1 Calculate the total resistance to water vapor transfer and the water vapor transfer conductance for a well-watered turf crop. Wind speed at 2 m elevation is 3 m/sec. Bulk surface resistance is 70 s/m. Calculate the maximum rate of water vapor transfer due to the vapor pressure gradient if the relative humidity is 30 %, temperature is 30 °C, and the elevation is sea level. Use the following resistances for turf.

$$r_s = 70\ s/m$$
$$r_{av} = 208/U_2 = 208/3 = 69\ s/m$$
$$r_{total} = r_s + r_{av} = 70 + 69 = 139\ s/m$$
$$h_v = \frac{1}{r_{total}} = \frac{1}{139} = 0.0072\ m/s$$

The canopy vapor pressure, e_c, is the saturation vapor pressure.

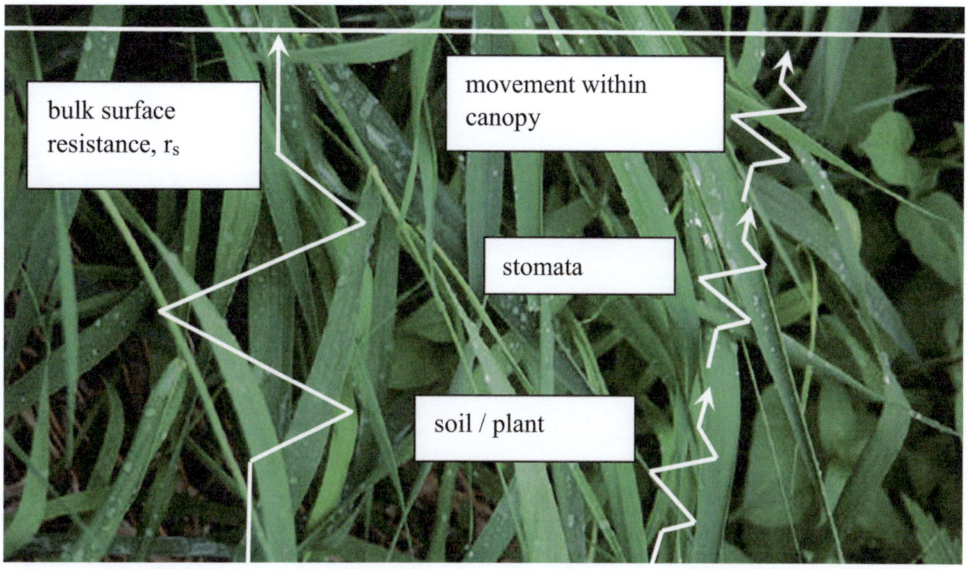

Fig. 5.3 Components of the bulk surface resistance, r_s

$$e_c = e_s = 0.6108\ \exp\left(\frac{17.27(30)}{30 + 237.3}\right) = 4.24\ kPa$$

The atmospheric vapor pressure is the product of saturation vapor pressure and relative humidity.

$$e_a = e_s{*}(RH)/100\% = 4.24\ kPa{*}0.3 = 1.27\ kPa$$

Calculate the ET based on the vapor pressure gradient.

$$\begin{aligned} ET_0 &= 0.622\rho_a h_v \frac{(e_c - e_a)0.622}{P} \\ &= 0.622{*}1.23{*}0.0072\frac{(4.24 - 1.27)}{101.3} \\ &= 0.00016\ kg/(m^2 - \sec) = 0.00016\ mm/\sec \\ &= 0.58\ mm/hr \end{aligned}$$

In the previous equations, we looked at the transfer of water vapor. Sensible heat flux, H, is the transfer of energy in response to a temperature gradient between the plant canopy and the atmosphere. The sensible heat flux equation (Fourier's heat transfer law) is similar to the vapor pressure gradient equation (Eq. 5.2), but the difference is between canopy and atmospheric temperature rather than canopy and atmospheric vapor pressure.

$$\begin{aligned} H = k\ dT/dz\ -> \ &\rho_a c_p h_a(T_c - T_a) \\ &= (\rho_a c_p/r_a)(T_c - T_a) \end{aligned} \tag{5.5}$$

where

h_a = aerodynamic conductance of sensible heat flux = $1/r_a$, m s^{-1}
r_a = aerodynamic resistance to sensible heat flux, s m^{-1}
ρ_a = density of air, kg m^{-3}
c_p = heat capacity of air, 1,013 J kg^{-1} $°C^{-1}$
T_c = surface (crop or soil) temperature, °C
T_a = atmospheric temperature, °C
z = vertical elevation, m.
k = conductivity, m/s.

The aerodynamic resistance, r_a, to energy transfer is proportional to the aerodynamic resistance to vapor transfer, r_{av}. In agricultural crops, both resistances are calculated based on four parameters: measured wind speed (U), the elevation at which wind speed is measured (z), elevation in the canopy at which wind speed is zero (d), and the roughness length (z_o).

The wind speed vs. elevation profile above a canopy can be modeled with a logarithmic equation. The logarithmic profile intersects the ground surface if there is no vegetation; however, the profile changes with a crop canopy (Fig. 5.4); the elevation, $d + z_0$, is the height in the crop canopy at which the logarithmic profile of the wind speed intersects the z-axis.

The wind speed profile changes with atmospheric stability conditions (Fig. 5.5). Atmospheric stability refers to the buoyancy of air and the tendency for vertically oriented eddies to form. Buoyancy is caused by temperature decrease with elevation that leads to increased air density with elevation. If the air at the ground surface is hotter (less dense) than the air above it, then vertical eddies are formed; and these eddies promote heat transfer from the canopy to the atmosphere. There vertical eddies tend to transfer wind energy to the ground surface. For neutral conditions, the surface temperature is nearly the same as the atmosphere, and the effects of buoyancy on eddies is negligible. The wind speed is slower near the ground surface with stable atmospheric conditions with more of a laminar flow. The difference is similar to the difference in velocity profiles between laminar and turbulent flow in a pipe.

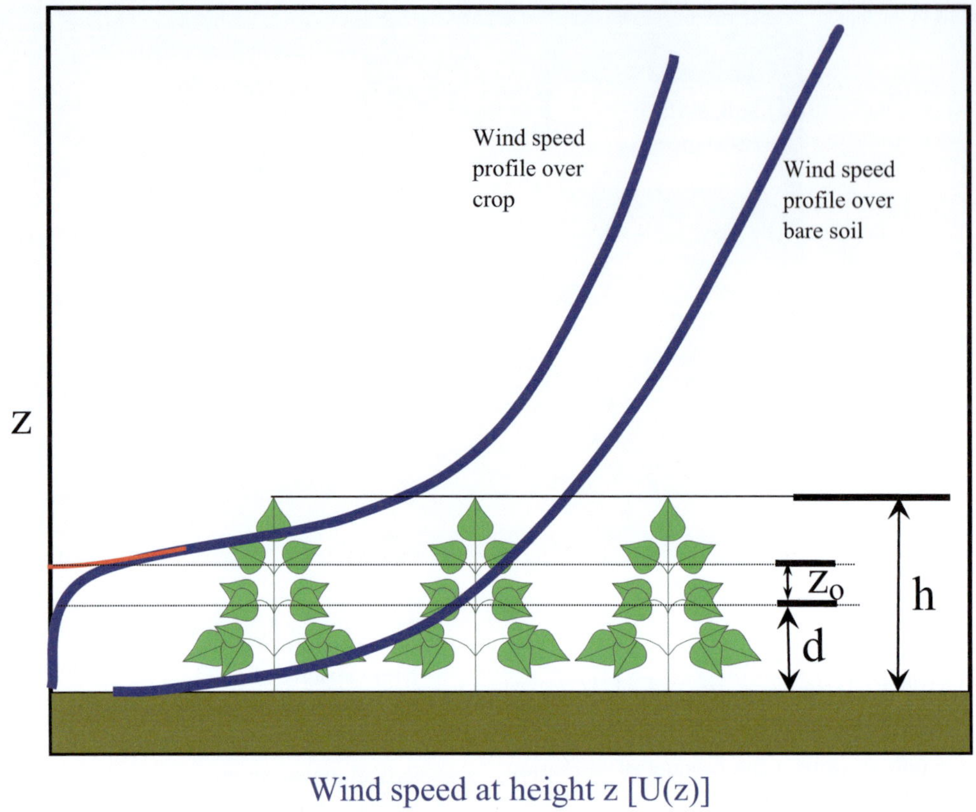

Fig. 5.4 Wind speed profile in and over canopy and over bare soil (Courtesy of Don Slack, The University of Arizona)

Eddies also increase the energy transfer (decrease the resistance) from the canopy to the atmosphere. The least energy transfer occurs if the wind profile is laminar (stable). Values of aerodynamic resistances at different atmospheric stabilities are shown in Fig. 5.6. The lowest resistance to energy and vapor transfer is observed when surface/canopy temperatures are higher than air temperature (Ts − Ta = +2 °C): a buoyant condition with intensive mixing. There is no decrease in resistance as wind speed increases because maximum mixing is already occurring. The maximum resistance to energy and water vapor transfer occurs when wind speed is low and air temperature is higher than surface/canopy temperature. There are very few eddies, and energy and vapor must transfer by molecular diffusion. As the wind speed increases, all conditions approach nearly the same resistance, since high wind generates turbulent eddies and promotes mixing just as buoyant conditions promote mixing.

Radiation

Radiation from the Sun is the primary source of energy for evapotranspiration. Net radiation, R_n, is the amount the incoming short wave solar radiation (R_S) minus the reflected short wave radiation minus the long wave infrared radiation from the soil surface and crop – all are forms of electromagnetic radiation. The albedo, α, is the ratio of reflected radiation to incoming radiation. Albedos for several surfaces are listed in Table 5.1. The albedo can vary significantly with solar zenith angle (0° when sun is directly overhead). Water has a large variation with solar zenith angle. The albedo of water with the sun directly overhead (zero solar zenith angle) is 0.1 (nearly all energy absorbed by water), while the albedo of water with the sun at a 50° angle is 0.975 (nearly all solar energy reflected back to space).

Equation 5.6 calculates net radiation over well-watered turf during daylight hours based on hourly weather station solar radiation measurements. The number (−0.3) represents the long wave (infrared) radiation from the crop. Albedo is 0.23 for turf.

$$\begin{aligned} R_n &= 277.8*(-0.3 + (1 - \alpha)*R_S) \\ &= 277.8*(-0.3 + (1 - 0.23)*R_S) \end{aligned} \tag{5.6}$$

where

α = albedo
R_n = Net radiation intensity over turf, W/m^2
R_S = Hourly solar radiation, MJ m^{-2} hr^{-1}.

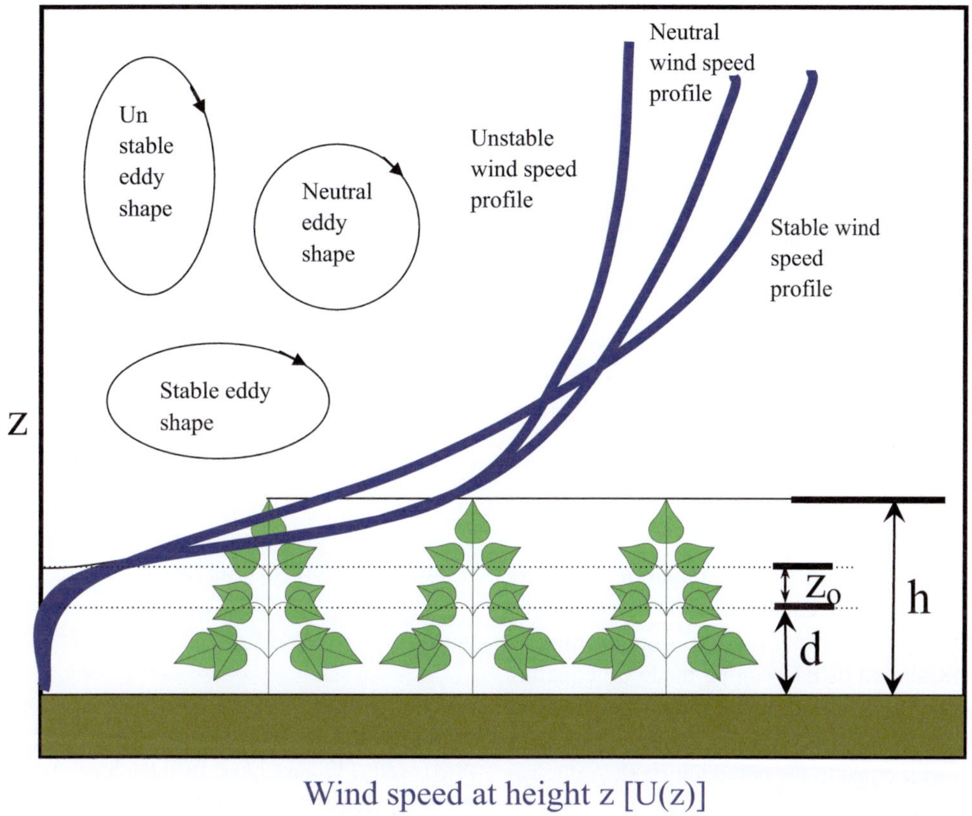

Fig. 5.5 Wind speed profiles under three different atmospheric stability regimes (Courtesy of Don Slack, The University of Arizona)

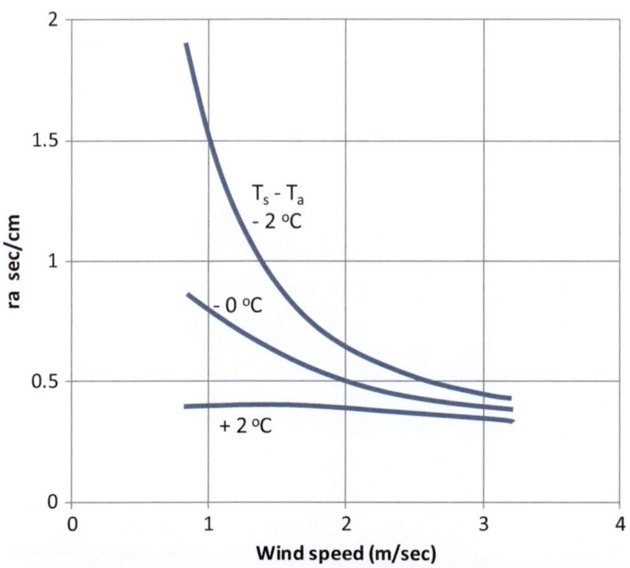

Fig. 5.6 Aerodynamic resistances at different atmospheric stabilities for a 30 cm-tall crop (d = 20 cm, z_0 = 4.0 cm at u = 1.0 m/sec and 2.0 cm at u = 3.0 m/sec; u measured at 150 cm above the canopy (After Szeicz and Long (1969))

Table 5.1 Albedo for natural surfaces (Credit NRCS)

Surface	Albedo
Fresh snow	0.80–0.95
Old snow	0.42–0.7
Dry sandy soils	0.25–0.45
Dry clay soils	0.20–0.35
Peat soils	0.05–0.15
Most field crops	0.2–0.3
Turf	0.23
Forests, deciduous	0.15–0.2
Forests, coniferous	0.1–0.15
Forests, deciduous with snow on ground	0.2

Some weather stations report hourly radiation in units of Langleys, which is the number of calories of energy from the Sun per cm^2 of soil surface area.

Example 5.2 Calculate total solar radiation intensity (power) and net radiation intensity (power) over turf if hourly radiation energy input (R_S) is 43.1 Langleys (cal/cm^2).

First convert Langleys to MJ m^{-2} hr^{-1}

$$43.1\ \text{cal/cm}^2\text{/hr}\ (4.18\ \text{Joules/calorie}) = 1.80{*}10^2\text{J cm}^{-2}\text{hr}^{-1} = 1.80{*}10^6\text{J m}^{-2}\text{hr}^{-1} = 1.8\text{MJ m}^{-2}\text{hr}^{-1}$$

Calculate total solar radiation intensity

$$R_S(\text{power}) = \text{energy/time} = 1.80{*}10^6\text{J/m}^2/3,600\ \text{seconds} = 500\ \text{W/m}^2$$

Calculate net radiation intensity over turf (turf albedo = 0.23).

$$R_n = 277.8{*}(-0.3 + 0.77{*}1.8) = 300\ \text{W/m}^2.$$

Latent heat of vaporization and net radiation generally dominate the energy balance equation. In fact, for a sunny day in an arid climate (water vapor transfer is not limiting), a reasonable approximation of ET_0 can be made by equating the latent heat of vaporization to net radiation. Over a 1.0 m^2 area, 1.0 mm depth of water is equal to 1.0 kg of water. In this case, ET_0 (mm) is equal to the net radiation (kJ/m^2) over the latent heat of vaporization.

$$(1\ mm)(1\ m^2)\left(\frac{m}{1,000\ mm}\right)\left(\frac{1,000\ kg}{m^3}\right) = 1\ kg$$

$$ET_0(mm) = \frac{R_n}{\lambda}\left(\frac{\frac{\text{kJ}}{\text{m}^2}}{\frac{\text{kJ}}{\text{m}^2\ \text{mm}}}\right) \quad (5.7)$$

Example 5.3 Calculate the depth of water transpired from turf in 1 hour if all net solar radiation is converted to the latent heat of vaporization. Cumulative radiation energy (R_s) over 1 hour is 43.1 Langleys. Temperature is 30 °C.

As calculated in Example 5.2, R_n = 300 W/m^2. Net radiation over 1 hour is

$$R_n = \frac{300\ W}{m^2} = \left(\frac{300\ J}{\sec - m^2}\right)\left(\frac{3,600\ \sec}{hr}\right)\left(\frac{kJ}{1,000\ J}\right) = 1,080\ kJ/m^2/hr$$

$$\lambda = 2,503 - 2.39\ T = 2,503 - 2.39(30) = 2,431\ \text{kJ/kg} = 2,431\ \text{kJ/(mm-m}^2).$$

$$ET_0 = R_n/\lambda = (1,080\ \text{kJ/m}^2)/(2,431\ \text{kJ/(mm-m}^2)) = 0.44\ \text{mm/hr}$$

The Penman Model of Evapotranspiration

Evapotranspiration can be calculated by setting up an energy balance at the Earth's surface (Fig. 5.7) and solving for the radiation energy that is converted to the latent heat of vaporization.

There are four terms in the energy balance equation for the Earth's surface, which can be written in terms of energy or power (energy per unit time).

$$R_n + G + H + LE = 0 \quad (5.8)$$

where

R_n = net radiation, W/m^2 or J/m^2
G = soil heat flux, W/m^2 or J/m^2
H = sensible heat flux (energy exchange between surface and air), W/m^2 or J/m^2
LE = energy lost to latent heat of vaporization,W/m^2 or J/m^2.

Penman (1948) developed an evapotanspiration model based on the terms in Eq. 5.8 for agricultural crops and open water. Others have further developed and refined this model. Penman rearranged Eq. 5.8 and solved for the energy used to vaporize water (evapotranspiration), LE.

$$LE = R_n - G - H = R_n - G - \rho_a c_p h_a (T_c - T_a) \quad (5.9)$$

The soil heat flux, G, is calculated with Fourier's law (heat transfer gradient equation). For models of daily ET, soil heat flux is generally dropped from the Penman equation because the sum of soil heat flux over 24 hours is generally zero.

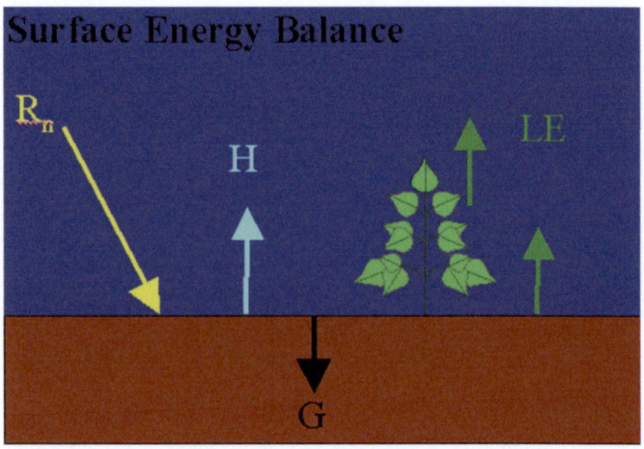

Fig. 5.7 Surface energy balance (Courtesy of Don Slack, The University of Arizona)

$$G = k\ dT/dz \sim k(T_s - T_z)/z \tag{5.10}$$

where

k = thermal conductivity – varies with soil type and water content, W/ °C-m
T_z = soil temperature at depth z below the soil surface, °C.
T_s = soil temperature at soil surface, °C.
z = depth in soil, m.
G = soil heat flux, W/m^2.

The energy contributed to evapotranspiration, LE, is equal the product of depth evaporated and latent heat of vaporization as shown in Eq. 5.9.

$$\lambda ET_0 = R_n - G - \rho_a c_p h_a (T_c - T_a) \tag{5.11}$$

Because it was difficult to measure canopy temperature at the time the Penman equation was derived, Penman modified Eq. 5.5 to use the difference in vapor pressure between the canopy and the atmosphere rather than the temperature difference. He based this conversion on the fact that there is a direct relationship between saturated vapor pressure and air temperature (Eq. 5.12). The slope of the saturated vapor pressure-temperature curve de_s/dT is designated as Δ in the Penman equation.

$$\frac{de_s}{dT} = \Delta = \frac{(e_{sc} - e_{sa})}{(T_c - T_a)} \tag{5.12}$$

$$\Delta = 2,503 \frac{\exp\left(\frac{17.27T}{T+237.3}\right)}{(T+237.3)^2} \tag{5.13}$$

$$H = \rho_a c_p h_a (T_c - T_a) = \frac{\rho_a c_p h_a}{\Delta}[(e_{sc} - e_{sa})] \tag{5.14}$$

where

e_{sc} = saturated vapor pressure at canopy temperature, kPa
e_{sa} = saturated vapor pressure at air temperature, kPa
Δ = slope of the saturated vapor pressure-temperature curve, kPa/°C
T = temperature, °C.

Substitute Eq. 5.14 for sensible heat flux.

$$\lambda ET_0 = R_n - G - \frac{\rho_a c_p h_a}{\Delta}[(e_{sc} - e_{sa})] \tag{5.15}$$

In order to make a substitution later based on Monteith's modification of the Penman equation, Eq. 5.15 is initially made more complex as follows.

$$\lambda ET_0 = R_n - G - \frac{\rho_a c_p h_a}{\Delta}[(e_{sc} - e_c) + (e_c - e_a) + (e_a - e_{sa})] \tag{5.16}$$

Equation 5.2 can be modified with the psychrometric constant and rearranged and then substituted for $e_c - e_a$ in Eq. 5.16. The psychrometric constant is the relationship between temperature and partial pressure of water in the atmosphere and can be used to calculate the actual vapor pressure from dry bulb and wet bulb temperature.

The psychrometric constant is not really "constant", and with units of (kPa $°C^{-1}$) can be calculated as

$$\gamma = (c_p P)/(0.622\lambda) = 0.000665\ P \tag{5.17}$$

where

γ = psychrometric constant, kPa $°C^{-1}$
P = total atmospheric pressure, kPa
λ = latent heat of vaporization (different units than above), J/kg.

$$\lambda ET_0 = \rho_a c_p h_v \frac{(e_c - e_a)0.622\lambda}{c_p P} = \rho_a c_p h_v \frac{(e_c - e_a)}{\gamma}$$

$$\frac{\gamma \lambda ET_0}{\rho_a c_p h_v} = (e_c - e_a) \tag{5.18}$$

Substitute Eq. 5.18 into Eq. 5.16 and solve for λET_0. Equation 5.19 is the Penman equation.

$$\lambda ET_0 = R_n - G - \frac{\rho_a c_p h_a}{\Delta}\left[(e_{sc} - e_c) + \frac{\gamma \lambda ET_0}{\rho_a c_p h_v} + (e_a - e_{sa})\right]$$

$$\lambda ET_0 = \frac{(R_n - G)\Delta + \rho_a c_p h_a[(e_{sa} - e_a) - (e_{sc} - e_c)]}{\left(\Delta + \gamma \frac{h_a}{h_v}\right)} \tag{5.19}$$

The Penman equation was modified by Monteith. The Monteith modification assumes that the bulk surface resistance, r_s, is 0; and, as a result, $h_a \sim h_v$ or $h_a/h_v = 1$. This assumption is made because under reference evapotranspiration conditions, the plant is theoretically not water stressed (no resistance to water movement through the soil to the top of the plant canopy). However, this assumption of no bulk surface resistance is probably erroneous. Nevertheless, we will move on with Montieth's assumption. If water is non-limiting (definition of reference crop: no water stress), then it can also be assumed that the canopy atmosphere is saturated so that $e_c = e_{sc}$ and

$$\lambda ET_0 = \frac{\Delta}{(\Delta+\gamma)}(R_n - G) + \frac{\rho_a c_p h_a (e_{sa} - e_a)}{(\Delta+\gamma)} \tag{5.20}$$

Remove the heat capacity of air from Eq. 5.20 with the following substitution.

$$c_p = 0.622\ \lambda\gamma/P \tag{5.21}$$

$$\lambda ET_0 = \frac{\Delta}{(\Delta+\gamma)}(R_n - G) + \frac{0.622\lambda h_a \gamma (e_{sa} - e_a)}{P(\Delta+\gamma)} \tag{5.22}$$

The term 0.622 λ/P is a constant. Monteith rolled these parameters and the aerodynamic conductance, h_a, into a somewhat empirical wind transfer function, f(u). The original Penman-Monteith equation is divided into a radiation term (left) and an aerodynamic term (right).

$$\lambda ET_0 = \frac{\Delta}{(\Delta+\gamma)}(R_n - G) + \left(\frac{\gamma}{\Delta+\gamma}\right) f(u)(e_{sa} - e_a) \tag{5.23}$$

Equation 5.23 the general form of the Penman-Monteith equation. Over the years, researchers in different regions developed their own variation of the Penman-Monteith equation in order to fine tune the equation for the specific climate in their region. However, this resulted in different (30–40 % variation) estimations of reference ET, even with the same weather data. In order to solve this problem, the United Nations Food and Agriculture Organization (FAO) and the American Society of Civil Engineers Task Committee (ASCE TC) selected a standardized reference ET calculation procedure that was developed by Allen et al. (1998 and 2005) (ASCE, 2005) and is the reference evapotranspiration calculation for the FAO 56 crop evapotranspiration calculation.

$$ET_{sz} = \frac{0.408\ \Delta\ (R_n - G) + \gamma\frac{C_n}{T+273}U_2(e_s - e_a)}{\Delta + \gamma(1 + C_d U_2)} \tag{5.24}$$

where

ETsz = standardized reference crop evapotranspiration, mm d^{-1} or mm h^{-1}
Δ = slope of the saturation vapor pressure temperature curve, kPa °C^{-1}
Rn = calculated net radiation at the crop surface, MJ m^{-2} d^{-1} or MJ m^{-2} h^{-1}
G = soil heat flux density at the soil surface, MJ m^{-2} d^{-1} or MJ m^{-2} h^{-1}
γ = psychrometric constant, kPa °C^{-1}
C_n = numerator constant that changes with reference type and time step
T = mean daily air temperature measured at 1.5–2.5 m above ground, °C
U_2 = mean daily wind speed wind speed measured at 2 m above ground, m s^{-1}
e_s = saturation vapor pressure measured at 1.5–2.5 m above ground, kPa
e_a = mean actual vapor pressure measured at 1.5–2.5 m above ground, kPa
C_d = denominator constant that changes with reference type and time step.

Reference ET Calculation with Weather Station Data

Most agricultural regions have a network of agricultural weather stations operated by the state agricultural college. These agricultural weather stations report hourly and daily cumulative radiation, vapor pressure, wind speed, and evapotranspiration parameters. The Arizona, Meteorological Network (AZMET) weather station locations are shown in Fig. 5.8. A typical weather station is shown in Fig. 5.9. The Arizona stations use a data logger, which samples weather parameters every 10 seconds. At the end of each hour, the data logger averages or sums, all weather parameters. Maintenance is critical for successful weather station operation. Once a year, all solar sensors and wind speed sensors are replaced with recalibrated sensors. Humidity sensor chips are replaced twice a year. It is more accurate to have professionally maintained weather stations distributed in a region than for individual farmers to try to maintain their own weather stations, even considering local climatic variability (Table 5.2).

Weather stations should be placed over an extended turf or alfalfa field (Fig. 5.9) in order to avoid disturbances in wind and other parameters. If a weather station is placed over a bare soil, then humidity and temperature measurements will not be the same as they would be over transpiring vegetation.

The following description of the ASCE and FAO calculation can be found in online publications by Dr. Paul Brown, Dr. Richard Snyder (2007), and Dr. Richard Allen. Links to these publications are in the references section at the end of this chapter. The three authors all served on the ASCE committee that selected the standardized equation.

The standardized Eq. 5.30 has been calibrated to the two types of reference surfaces (turf and alfalfa) and to two time scales (hourly and daily). Thus, there are a total of four equations. Alfalfa (tall crop, 0.5 m) has a higher reference evapotranspiration value than turf (short crop, 0.12 m) and is

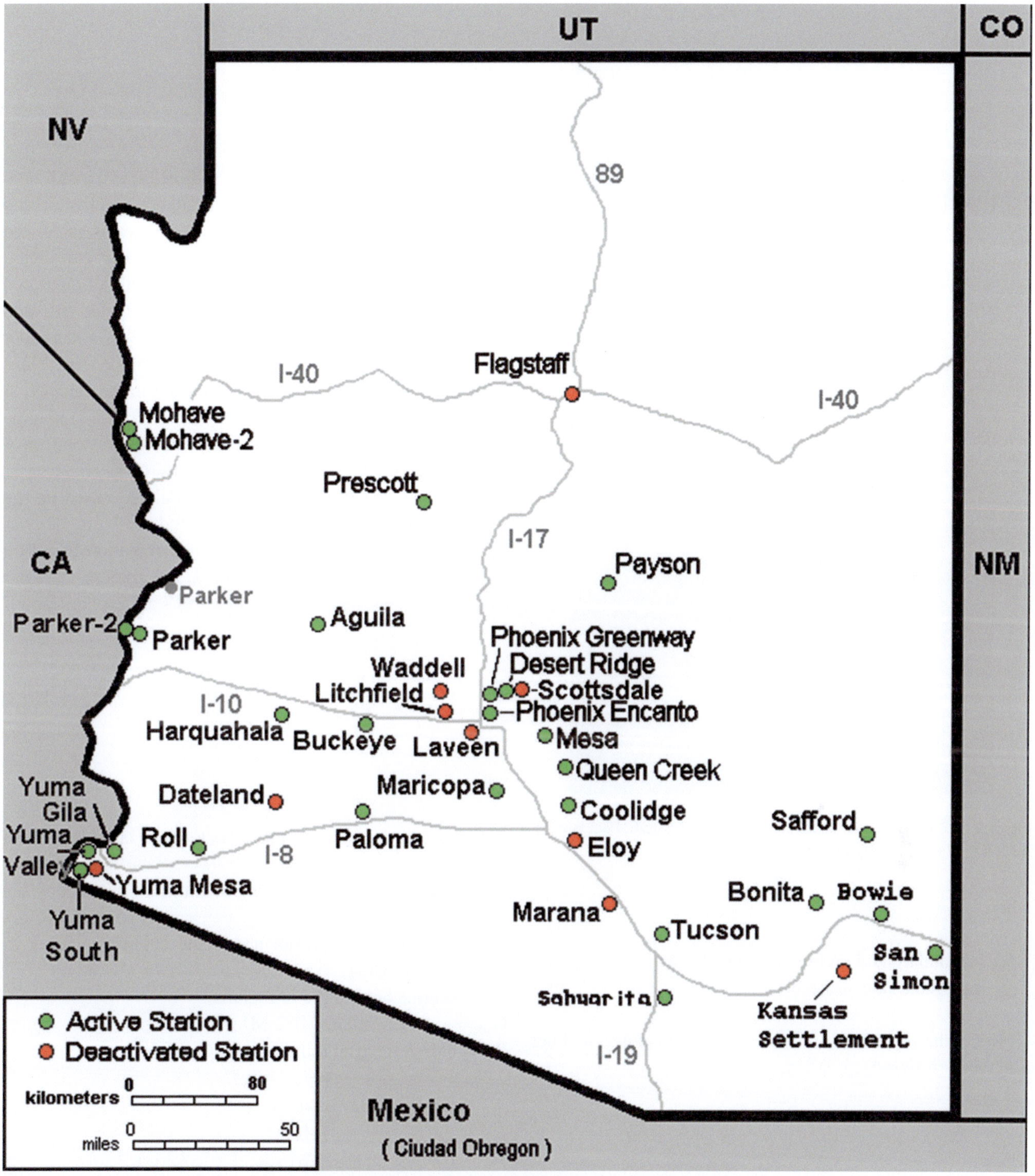

Fig. 5.8 AZMET meteorological network (Credit Paul Brown, AZMET, The University of Arizona)

designated as ET_r while turf is designated as ET_0. Although earlier research had shown that hourly calculations were more accurate, the ASCE committee found that the standardized equations with hourly and daily time steps only varied by 2 %.

For daily time step calculations of turf-based ET_0, Eq. 5.23 written as follows:

$$ET_{0s} = \frac{0.408\ \Delta\ (R_n - G) + \gamma \frac{900}{T+273} U_2 (e_s - e_a)}{\Delta + \gamma(1 + 0.34\ U_2)} \tag{5.25}$$

Fig. 5.9 AZMET weather station in Phoenix, AZ (Credit Paul Brown, AZMET, The University of Arizona)

where

ET_{0s} = standardized reference ET from a short crop (s represents standardized).

Net radiation as the sum of net shortwave (solar) and net longwave (infrared) radiation.

$$R_n = R_{nS} - R_{nL} \tag{5.26}$$

where

R_{nS} = net short wave radiation, MJ m^{-2} day^{-1}
R_{nL} = net long wave radiation (positive outward), MJ m^{-2} day^{-1}.

The next shortwave radiation is the measured daily shortwave radiation minus reflected shortwave radiation, and is calculated based on the albedo.

$$R_{nS} = R_S(1 - \alpha) \tag{5.27}$$

where

R_{nS} = net short wave radiation.

Short wave radiation is measured directly with a radiometer on weather stations. These sensors are not sensitive to long wave (infrared) radiation. Long wave radiation is calculated based on estimated cloud cover and clear sky emissivity. Cloud cover can be estimated based on the ratio of measured short wave radiation to clear sky radiation (solar radiation with no clouds). Clear sky emissivity varies with water vapor pressure because water vapor stores and emits energy; thus, humidity (vapor pressure) is also included in the calculation of net long wave radiation. Clear sky emissivity and radiation from clouds are represented in Fig. 5.10 by the "long-wave radiation from clouds and greenhouse gases."

The clear sky solar radiation is calculated based on the extraterrestrial radiation at a given latitude and time of year. This value is then converted to ground radiation based on elevation. Extraterrestrial radiation is the amount of radiation that reaches the outside of the earth's atmosphere, and is calculated in Eq. 5.28.

$$R_a = \left(\frac{24*60}{\pi}\right) G_{sc} d_r [\omega_s \sin\delta \sin\varphi + \cos\varphi \cos\delta \sin\omega_s] \tag{5.28}$$

where

ϕ = latitude in radians converted from latitude (L) in degrees
d_r = correction for eccentricity of Earth's orbit around the sun on DOY
δ = declination of the sun above the celestial equator in radians on DOY
ω_s = sunrise hour angle in radians
G_{sc} = solar constant, 0.082 MJ m^{-2} min^{-1}
R_a = extraterrestrial radiation, MJ/(m^2-day).

$$\phi = \frac{\pi * Latitude}{180}$$

$$d_r = 1 + 0.033 \cos\left(\frac{2\pi}{365} DOY\right)$$

$$\delta = 0.409 \sin\left(\frac{2\pi}{365} DOY - 1.39\right)$$

$$\omega_s = \cos^{-1}[-\tan\phi \tan\delta]$$

Clear sky solar radiation (R_{so}) is then adjusted based on elevation.

Table 5.2 AZMET weather station sensors (Credit Paul Brown, AZMET, The University of Arizona)

Measurement	Instrument model	Sensor type	Siting	Accuracy
Air Temp.	Vaisala	Thermistor	1.5 m	+/−0.4 °C
1992-Present	HMP35C, HMP45C	Fenwall	Above	
		(UUT51J1)	Surface	
Rel. Humidity	Vaisala	Vaisala	1.5 m	+/−2 % RH
1992-Present	HMP35C, HMP45C	Capacitive	Above	
			Surface	
Soil Temp.	Thermocouple	Type 1	5 and 10 cm	+/−0.2 °C
		Copper-	Depth	
		constantan	Bare surface	
Solar Radiat.	LiCor LI200S	Silcon cell	2.5 m	+/−5 %
	Pyranometer	Pyranometer	Above	
			Surface	
Precipitation	Sierra Misco	Tipping	Gage top	+/−1 mm
	RG2501 Rain	Bucket with	At ~43 cm	
	Gage	Magnetic	Above	
		ReedSwitch	surface	
Wind Speed	Met one model	Anemometer	3.0 m	+/−1.5 %
	014A wind speed	Using reed	Above	
	Sensor	Switch	Surface	
Wind direct.	Met one model	Wind vane	3.0 m	+/−5°
	024A wind dir.	Attached to	Above	
	Sensor	Potentiometer.	Surface	

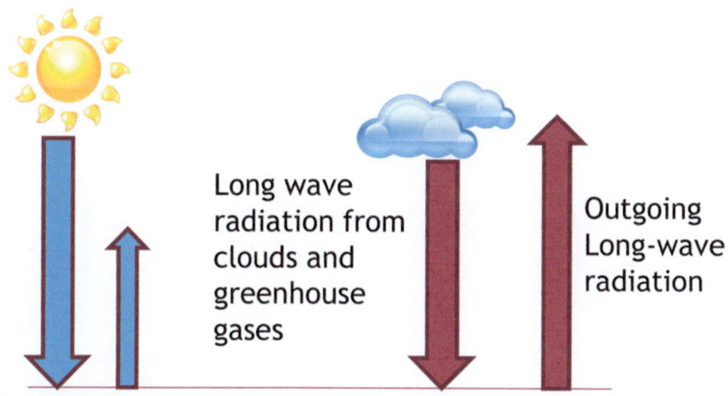

Fig. 5.10 Net radiation components

$$R_{SO} = (0.75 + 2*10^{-5}z)R_a \quad (5.29)$$

where

z = elevation above sea level, m
R_{SO} = clear sky solar radiation, MJ m^{-2} day^{-1}

Net daily long wave radiation is then calculated as follows with the first term representing black body radiation at the average temperature of the Earth's surface, the second term representing clear sky emissivity (global warming gases) and the third term representing cloud cover:

$$R_{nL} = \sigma\left(\frac{T_{\min}^4 + T_{\max}^4}{2}\right)(0.34 - 0.14\sqrt{e_a})\left[1.35\left(\frac{R_S}{R_{SO}}\right) - 0.35\right] \quad (5.30)$$

where

σ = Stefan-Boltzman constant, 4.901 × 10^{-9} MJ/(K^4 m^2 d)
T_{min} = daily minimum temperature, °K
T_{max} = daily maximum temperature, °K
e_a = mean vapor pressure measured 1.5 m above ground, kPa
R_s = measured short wave solar radiation, MJ m^{-2} day^{-1}
R_{nL} = net long wave infrared radiation, MJ m^{-2} day^{-1}.

The ratio R_S/R_{SO} cannot be less than 0.3 or greater than 1.0. For values less than 0.3, the ratio is set to 0.3 and for values greater than 1.0, the ratio is set to 1.0.

Mean air temperature, T, is the mean of the maximum and minimum temperatures recorded by the datalogger.

$$T = \left(\frac{T_{\min} + T_{\max}}{2}\right) \quad (5.31)$$

In Eq. 5.25, saturated vapor pressure is the average of the saturated vapor pressures at the maximum and minimum temperatures.

$$e_s = \left(\frac{e_{s-T\min} + e_{s-T\max}}{2}\right) \quad (5.32)$$

Average actual vapor pressure in Eqs. 5.25 and 5.30 can be calculated based on max and min temperatures and associated relative humidities. However, some biometeorologists prefer to calculate e_a based on an average of the 24 hourly averages.

$$e_a = \left(\frac{\frac{RH_{T-\min}}{100} e_s(T_{\min}) + \frac{RH_{T-\max}}{100} e_s(T_{\max})}{2} \right) \quad (5.33)$$

Average actual vapor pressure can also be calculated at the mean daily dewpoint, T_d (°C).

$$e_a = 0.6108 \ \exp\left(\frac{17.27(T_d)}{T_d + 237.3} \right) \quad (5.34)$$

Wind speed is another component of Eq. 5.25. If wind speed is measured at any elevation other than 2 m, then it is converted to 2 m wind speed with the following equation.

$$U_2 = U_Z \frac{4.87}{\ln(67.8z_w - 5.42)} \quad (5.35)$$

where

U_z = wind speed measurement at elevation z_w, m/s
z_w = elevation of wind speed measurement, m.

Example 5.4 Calculate daily evapotranspiration in Fort Collins, Colorado on DOY 195 (July 15). The latitude of Fort Collins is 40.5°N. Elevation is 1,524 m. Maximum temperature is 30 °C and minimum temperature is 20 °C. Relative humidity at maximum temperature is 40 % and at minimum temperature is 60 %. Measured solar radiation is 25 MJ m^{-2} day^{-1}. Average wind speed at 3 m elevation is 3.2 m/sec.

$$\varphi = \frac{\pi * Latitude}{180} = \frac{\pi * 40.5}{180} = 0.706$$

$$d_r = 1 + 0.033 \cos\left(\frac{2\pi}{365} DOY \right)$$
$$= 1 + 0.033 \cos\left(\frac{2\pi}{365} 195 \right) = 0.968$$

$$\delta = 0.409 \sin\left(\frac{2\pi}{365} DOY - 1.39 \right) = 0.377$$

$$\omega_s = \cos^{-1}[-\tan\varphi \tan\delta]$$
$$= \cos^{-1}[-\tan(0.706)\tan(0.377)] = 1.915$$

$$R_a = \left(\frac{24*60}{\pi} \right) G_{sc} d_r [\omega_s \sin\delta \sin\phi + \cos\phi \cos\delta \sin\omega_s]$$
$$= \left(\frac{24*60}{\pi} \right) *0.082*0.97*[1.915* \sin(0.377)\sin(0.706)$$
$$+ \cos(0.706)\cos(0.377)\sin(1.92)]$$
$$= 41 \ MJ \ m^{-2} day^{-1}$$

Clear sky solar radiation at 1,524 m elevation is

$$R_{SO} = (0.75 + 2*10^{-5} z) R_a = (0.75 + 2*10^{-5}*1,524) 41$$
$$= 31.9 \mathrm{MJ} \ \mathrm{m}^{-2} \mathrm{day}^{-1}$$

Net solar radiation based on a radiometer and albedo is

$$R_{nSO} = R_S*(1 - \alpha) = 25*(1 - 0.23) = 19.25 \ \mathrm{MJ} \ \mathrm{m}^{-2} \mathrm{day}^{-1}$$

Calculate saturated vapor pressure at maximum and minimum temperature. First calculate saturated vapor pressure at the two temperatures and average saturated vapor pressure.

$$e_s = 0.6108 \ \exp\left(\frac{17.27(30)}{30 + 237.3} \right) = 4.24 \ kPa$$
$$e_s = 0.6108 \ \exp\left(\frac{17.27(20)}{20 + 237.3} \right) = 2.34 \ kPa$$
$$e_{s\text{-}ave} = (4.24 + 2.34)/2 = 3.29 \ \mathrm{kPa}$$

Calculate average daily vapor pressure, e_a

$$e_a = \left(\frac{\frac{RH_{T-\min}}{100} e_s(T_{\min}) + \frac{RH_{T-\max}}{100} e_s(T_{\max})}{2} \right)$$
$$= \left(\frac{0.6*2.34 + 0.4*4.24}{2} \right) = 1.55 \ kPa$$

Calculate net long wave radiation

$$R_{nL} = \sigma \left(\frac{T_{\min}^4 + T_{\max}^4}{2} \right) (0.34 - 0.14\sqrt{e_a}) \left[1.35 \left(\frac{R_S}{R_{SO}} \right) - 0.35 \right]$$

$$R_{nL} = 4.901*10^{-9} \left(\frac{293^4 + 303^4}{2} \right) (0.34 - 0.14\sqrt{1.55}) \left[1.35 \left(\frac{25}{32.4} \right) - 0.35 \right] = 4.55 \ \mathrm{MJ} \ \mathrm{m}^{-2} \mathrm{day}^{-1}$$

Calculate total net radiation

$$R_n = R_{nS} - R_{nL} = 19.25 - 4.55 = 14.7 \mathrm{MJ} \ \mathrm{m}^{-2} \mathrm{day}^{-1}$$

	A	B	C	D	E	F	G	H	I	J	K	L
1	Max T	30	C			Latitude	40.5	degrees		Elevation	1524	m
2	Min T	20	C			DOY	195	days		Measured Rs	25	MJ/m2/d
3	Tave	25.00	C			Phi	0.71			solar radiation		
4	es Max T	4.24	kPa			dr	0.97			Wind speed	3.2	m/s
5	es Min T	2.34	KPa			delta	0.38			Wind z	3	m
6	es-ave	3.29	kPa			w	1.92			Wind, 2 m	2.95	
7	Rhmax	0.6	fraction			Ra	40.89	MJ/m2/d				
8	Rhmin	0.4	fraction			Rs/Rso	0.783					
9	ea RH max	1.40										
10	ea RH min	1.70				Calculation of daily ET with standardized Penman equation						
11	ea-ave	1.55	kPa			Input data required in white cells						
12	es-ea	1.74										
13	Rso	31.92	MJ/m2/d									
14	Rnl	4.55	MJ/m2/hr									
15	Rns	19.25	MJ/m2/d									
16	Rn	14.70	MJ/m2/d									
17	P	84.54	kPa									
18	DELTA	0.19										
19	gamma	0.06										
20	U2	2.95	m/s									
21	G	0	MJ/m2/d									
22	R(f) rad. Fn.	1.132										
23	w(f) wind Fn.	0.871										
24	denominator	0.301										
25	ETO	6.649	mm/day									
26												

Daily ETo Example 5-9 / Hourly ETo Example 5-10 / Hourly ETo Tucson / Daily ETo Tucson / Active_year_w

Fig. 5.11 *Daily ETo* worksheet in *Chapter 5 ET calculator*

Calculate wind speed at 2 m, U_2.

$$U_2 = U_z \frac{4.87}{\ln(67.8z - 5.42)} = 3.2 \frac{4.87}{\ln(67.8*3 - 5.42)} = 2.95\ m/\sec$$

Calculate P, γ, Δ, and average temperature, T

$$P = 101.3\left(\frac{293 - 0.0065*1,524}{293}\right)^{5.26} = 84.5\ kPa$$

$$\gamma = 0.000665\ \mathrm{P} = 0.000665*84.5$$
$$= 0.056(\text{psychrometric constant})$$

$$T = \left(\frac{T_{\min} + T_{\max}}{2}\right) = \left(\frac{20 + 30}{2}\right) = 25$$

$$\Delta = 2,504\,\frac{\exp\left(\frac{17.27T}{T + 237.3}\right)^{5.26}}{(T + 237.3)^2} = 2,504\,\frac{\exp\left(\frac{17.27(25)}{25 + 237.3}\right)^{5.26}}{(25 + 237.3)^2}$$
$$= 0.189(\text{slope of } e_s/\mathrm{T})$$

Finally calculate ET (mm/day). Assume that G is zero for a 24-hr period because soil temperature does not change significantly from 1 day to the next.

$$ET_{OS} = \frac{0.408\,\Delta\,(R_n - G) + \gamma \frac{900}{T + 273} U_2(e_s - e_a)}{\Delta + \gamma(1 + 0.34\,U_2)}$$

$$ET_{OS} = \frac{0.408*0.189\,(14.9) + 0.056\frac{900}{25 + 273}*2.95(3.29 - 1.55)}{0.189 + 0.056(1 + 0.34*2.95)}$$
$$= 6.65\,mm$$

The same calculations are made in the Daily ETo worksheet (Fig. 5.11)

The next section describes hourly ET calculation with the ASCE standardized equations. One would think that calculations of daily ET by summing hourly calculations would be more accurate than calculations based on daily averages; however, this is not necessarily the case. Thus, many weather station managers choose to make daily calculations of ET based on Eq. 5.25 rather than sum the hourly calculations, even though hourly data are available.

There are two models for hourly calculation of ET, one for day and one for night. The night equation is selected when the solar radiation drops below a threshold value.

$$ET_{OS} = \frac{0.408\Delta(R_n - G) + \gamma\frac{37}{T+273}U_2(e_s - e_a)}{\Delta + \gamma(1 + 0.24\ U_2)} \quad \text{Day} \tag{5.36}$$

$$ET_{OS} = \frac{0.408\Delta(R_n - G) + \gamma\frac{37}{T+273}U_2(e_s - e_a)}{\Delta + \gamma(1 + 0.96\ U_2)} \quad \text{Night} \tag{5.37}$$

For hourly calculation of ET, the clear sky solar radiation as a function of time of day is calculated with the following equations:

$$b = \frac{2\pi(DOY - 81)}{364} \tag{5.38}$$

$$Sc = 0.1645 * \sin(2b) - 0.1255\cos(b) - 0.025 * \sin(b) \tag{5.39}$$

$$w = \frac{\pi}{12}\left((t - 0.5) + \frac{(L_c - L_m)}{15} - 12 + Sc\right) \tag{5.40}$$

$$w_2 = w + \left(\frac{1}{2}\right)\left(\frac{\pi}{12}\right) \quad w_1 = w - \left(\frac{1}{2}\right)\left(\frac{\pi}{12}\right) \tag{5.41}$$

$$\sin(\theta) = (w_2 - w_1)\sin(\phi)\sin(\delta) + \cos(\phi)\cos(\delta)(\sin(w_2) - \sin(w_1)) \tag{5.42}$$

where

Sc = hourly time correction for Earth's wobble in Earth's rotation
L_c = station longitude in degrees
L_m = longitude of the local time meridian
t = local standard ending time, hr.
w_1 = hour angle ½ hr before w in radians
w_2 = hour angle ½ hr after w in radians
w = hour angle in radians
θ = solar altitude angle in radians
ϕ, δ, d_r = calculated in Eqs. 5.25, 5.26 and 5.27.

$$R_a = \left(\frac{12}{\pi}\right)60\ G_{sc}d_r\sin\theta \tag{5.43}$$

For hourly calculations of ET, soil heat flux is significant because energy enters the soil during the day and is released from the soil at night. It is estimated as follows for a short (turf) reference crop (ET_{os}):

$$G = 0.1\ R_n \quad \text{Day} \tag{5.44}$$

$$G = 0.5\ R_n \quad \text{Night} \tag{5.45}$$

Example 5.5 Calculate hourly ET over turf in Fort Collins, Colorado, if solar radiation is 3.47 MJ m^{-2} hr^{-1}. Average temperature is 30 °C, and wind speed is 1.5 m/s at 2 m elevation. Elevation is 1,524 m, vapor pressure deficit is 2.0 kPa. Longitude of Fort Collins is 105.07°. Fort Collins is in the Mountain Time Zone. The longitude of the Mountain Time meridian is 105°. The DOY is 195. The period is from 12 to 1 p.m.

Saturated vapor pressure at 30 °C is 4.24 kPa. Vapor pressure deficit (vpd) is equal to saturated vapor pressure minus actual vapor pressure, thus

$$e_a = e_s - v\,pd = 4.24 - 2.0 = 2.24\ \text{kPa}$$

Calculate solar radiation at given time

$$b = \frac{2\pi(DOY - 81)}{364} = \frac{2\pi(195 - 81)}{364} = 1.97$$

$$Sc = 0.1645 * \sin(2 * 1.97) - 0.1255\cos(1.97) - 0.025 * \sin(1.97) = -0.092$$

$$w = \frac{\pi}{12}\left((t - 0.5) + \frac{(L_c - L_m)}{15} - 12 + Sc\right) = \frac{\pi}{12}\left((13 - 0.5) + \frac{(0.07)}{15} - 12 - 0.092\right) = 0.108$$

$$w_1 = w - \left(\frac{1}{2}\right)\left(\frac{\pi}{12}\right) = 0.108 - \left(\frac{1}{2}\right)\left(\frac{\pi}{12}\right) = -0.0228$$

$$w_2 = w + \left(\frac{1}{2}\right)\left(\frac{\pi}{12}\right) = 0.108 + \left(\frac{1}{2}\right)\left(\frac{\pi}{12}\right) = 0.239$$

$$\sin(\theta) = (w_2 - w_1)\sin(\phi)\sin(\delta) + \cos(\phi)\cos(\delta)(\sin(w_2) - \sin(w_1))$$

$$\sin(\theta) = (0.239 + 0.0228)\sin(0.707)\sin(0.378) + \cos(0.707)\cos(0.378)(\sin(0.239) - \sin(-0.0228))\ \sin(\theta) = 0.246$$

Calculate extraterrestrial radiation during the 1 hour period.

$$R_a = \left(\frac{12}{\pi}\right)60\ G_{sc}d_r\sin\theta = \left(\frac{12}{\pi}\right)60\ (0.082)(0.97)(0.246) = 4.48\ MJ\ m^{-2}\ hr^{-1}$$

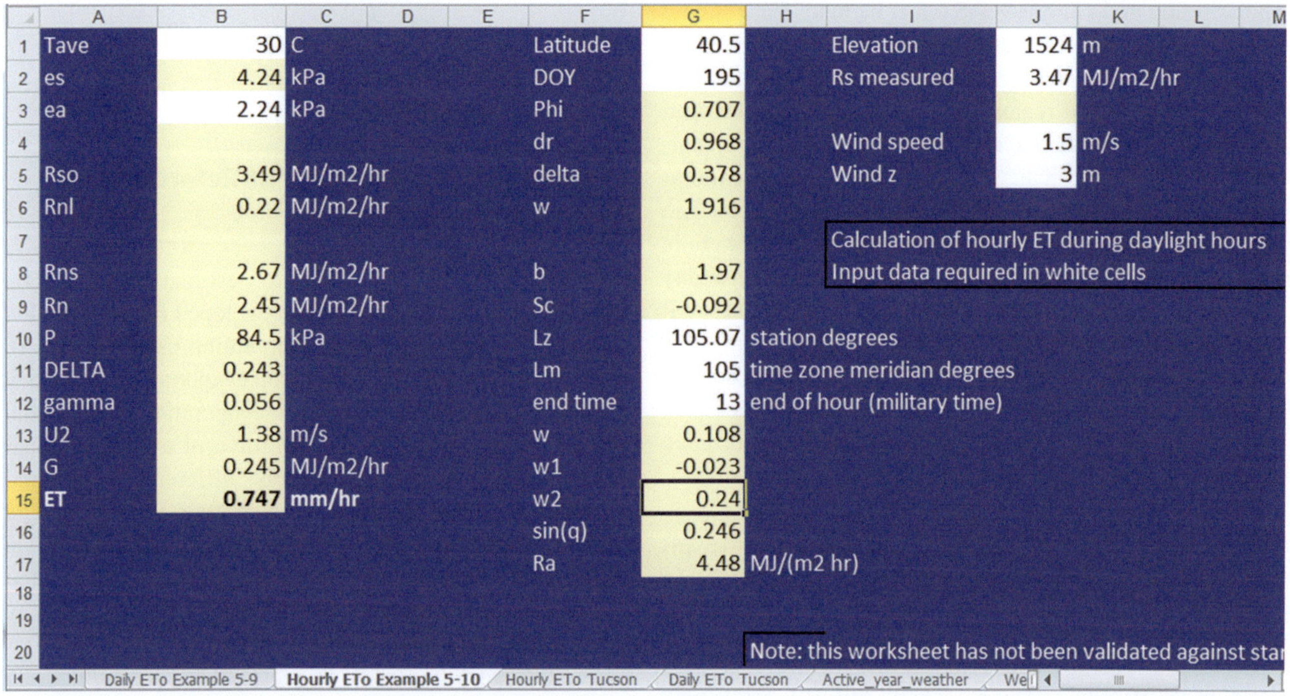

	A	B	C	D	E	F	G	H	I	J	K	L	M
1	Tave	30	C			Latitude	40.5		Elevation	1524	m		
2	es	4.24	kPa			DOY	195		Rs measured	3.47	MJ/m2/hr		
3	ea	2.24	kPa			Phi	0.707						
4						dr	0.968		Wind speed	1.5	m/s		
5	Rso	3.49	MJ/m2/hr			delta	0.378		Wind z	3	m		
6	Rnl	0.22	MJ/m2/hr			w	1.916						
7									Calculation of hourly ET during daylight hours				
8	Rns	2.67	MJ/m2/hr			b	1.97		Input data required in white cells				
9	Rn	2.45	MJ/m2/hr			Sc	-0.092						
10	P	84.5	kPa			Lz	105.07	station degrees					
11	DELTA	0.243				Lm	105	time zone meridian degrees					
12	gamma	0.056				end time	13	end of hour (military time)					
13	U2	1.38	m/s			w	0.108						
14	G	0.245	MJ/m2/hr			w1	-0.023						
15	**ET**	**0.747**	**mm/hr**			w2	0.24						
16						sin(q)	0.246						
17						Ra	4.48	MJ/(m2 hr)					
18													
19													
20								Note: this worksheet has not been validated against sta					

Daily ETo Example 5-9 | **Hourly ETo Example 5-10** | Hourly ETo Tucson | Daily ETo Tucson | Active_year_weather | We

Fig. 5.12 *Hourly ETo* worksheet in *Chapter 5 ET calculator*

Clear sky solar radiation is

$$\begin{aligned} R_{SO} &= \left(0.75 + 2{*}10^{-5}z\right)R_a \\ &= \left(0.75 + 2{*}10^{-5}{*}1{,}524\right)0.0746 \\ &= 3.49\ \text{MJ}\ \text{m}^{-2}\text{hr}^{-1} \end{aligned}$$

Measured solar radiation that is absorbed by the surface is

$$R_{nS} = R_S{*}(1-\alpha) = 3.47{*}(1-0.23) = 2.67\ \text{MJ m}^{-2}\text{hr}^{-1}$$

Calculate Δ at 30 °C.

$$\Delta = 2{,}504\ \frac{\exp\left(\frac{17.27T}{T+237.3}\right)^{5.26}}{(T+237.3)^2} = 2{,}504\ \frac{\exp\left(\frac{17.27(30)}{25+237.3}\right)^{5.26}}{(30+237.3)^2}$$
$$= 0.243$$

Net long wave radiation is calculated by dividing Eq. 5.30 by 24 and using average air temperature to calculate black body radiation.

$$R_{nL} = \frac{\sigma T^4}{24}\left(0.34 - 0.14\sqrt{e_a}\right)\left[1.35\left(\frac{R_S}{R_{SO}}\right) - 0.35\right]$$
$$\begin{aligned} R_{nL} &= \frac{\sigma(273+30)^4}{24}\left(0.34 - 0.14\sqrt{2.24}\right)\left[1.35\left(\frac{3.47}{3.49}\right) - 0.35\right] \\ &= 0.22\ MJ\ m^{-2}\ hr^{-1} \end{aligned}$$

Calculate net radiation

$$R_n = R_{nS} - R_{nL} = 2.67 - 0.22 = 2.45\ \text{MJ}\ \text{m}^{-2}\text{hr}^{-1}$$

Calculation soil heat flux, G for daytime conditions in turf

$$G = 0.1\ R_n = 0.1{*}2.45 = 0.245\ \text{MJ}\ \text{m}^{-2}\text{hr}^{-1}$$

Calculate reference evapotranspiration for daytime conditions in turf

$$ET_{OS} = \frac{0.408\Delta(R_n - G) + \gamma\frac{37}{T+273}U_2(e_s - e_a)}{\Delta + \gamma(1 + 0.24\ U_2)}$$
$$\begin{aligned} ET_{OS} &= \frac{0.408{*}0.243(2.45 - 0.245) + 0.056\frac{37}{30+273}1.5(2.0)}{0.243 + 0.056(1 + 0.24{*}1.5)} \\ &= 0.75\ mm/hr \end{aligned}$$

The same calculations are made in the Hourly ETo worksheet (Fig. 5.12).

Hargreaves-Samani Reference ET Calculation Method

If weather station data is insufficient for the Penman equation, then other methods of estimating ET_0 based on latitude, daily solar radiation, daily high and low temperature or other parameters are available. These include the Blaney-Criddle,

Thornewaite, and Hargreaves-Samani method. Care should be taken to select the model that is most appropriate for the given climate and data set. Some of these models were developed for humid regions while others were developed under arid conditions. One model that is gaining in popularity is the Hargreaves-Samani model. Data used for calibration of the Hargreaves-Samani equation was collected in Davis, California; and the model seems to work well in most arid or semiarid regions and tends to be highly correlated with Penman-Montieth estimates. It overestimates ET in humid climates and underestimates ET in high wind (FAO 56); however, Turk developed a modification for humid climates. If data is limited to mean air temperature and daily solar radiation, then the original Hargreaves formula may be used.

$$ET_0 = 1,350\frac{R_s}{\lambda}(T_m + 17.8) \tag{5.46}$$

where

R_s = measured solar radiation (at earth surface), MJ/m^2/day
ET_0 = reference ET, mm/day
λ = latent heat of vaporization, J/kg
T_m = mean daily temperature, °C.

If data includes average, maximum, and minimum daily temperatures, then the Hargreaves-Samani modification can be used to estimate ET_0 (Hargreaves and Samani 1985). The general form of the Hargreaves-Samani equation is

$$ET_0 = 0.00094\left(R_a[T_m + 17.8]\sqrt{T_x - T_n}\right) \tag{5.47}$$

where

R_a = extraterrestrial solar radiation, MJ/m^2/day
T_x = maximum daily air temperature, °C
T_n = minimum daily air temperature, °C
T_m = mean daily temperature, °C.

Example 5.6 Calculate the daily ET in Fort Collins, Colorado on DOY 195 (July 15) with the Hargreaves-Samani method. Max and min temperatures are 30 °C and 20 °C, respectively

$$\begin{aligned} ET_0 &= 0.00094\left(R_a[T_m + 17.8]\sqrt{T_x - T_n}\right) \\ &= 0.00094\left(41*[25 + 17.8]\sqrt{30 - 20}\right) = 5.2\ mm \end{aligned}$$

The Turk modification of the Hargreaves equation was developed for regions with humidity greater than 50 %.

$$ET_0 = 0.013\frac{T_m}{T_m + 15}(23.9\ R_a + 50) \tag{5.48}$$

Evaporation Pan Estimate of Reference Evapotranspiration

Evaporation pans are used to estimate daily ET_0. Standardized methods have been developed for construction and use of Standard Class A evaporation pans (Fig. 5.13). Although evaporation pans and crop evapotranspiration may sometimes have similar rates of evaporation, the diurnal characteristics and major driving force of evaporation and evapotranspiration are completely different. Lakes and evaporation pans continue to evaporate water at night while plants close their stomata and stop transpiring at night. The fundamental equations governing evaporation and evapotranspiration are also different. The major energy input in the Penman-Monteith equation of evapotranspiration is solar radiation while this term is completely missing from the equation for open pond evaporation. This is probably because most solar radiation passes through the lake surface and is converted to heat in the bulk water body. Conversely, in plants, most solar radiation is directly converted to plant transpiration. The major driving force in lake evaporation is the Gibbs free energy difference (vapor pressure difference) between the water surface and the atmosphere. For this reason, this author tends to lean away from recommending evaporation pans as an estimate of crop evapotranspiration and recommends the Penman-Montieth calculation with weather station data. Nevertheless, if no other information is available, the daily evaporation from a lake can be approximately estimated with the weather station Penman-Montieth daily ET_0. Likewise, evaporation pans can be used as an estimate of crop evapotranspiration.

In general, reference evapotranspiration, ET_0, is estimated as 0.7 * ET_{pan} in winter and 0.6 * ET_{pan} in summer. K_p, the pan coefficient, is the ratio of daily reference evapotranspiration to pan evaporation. Thus, the pan coefficient, K_p, is 0.7 in winter and K_p = 0.6 in summer.

Pan evaporation is influenced by humidity, wind speed, and surrounding vegetation. If there is no surrounding vegetation, then air humidity is lower and ground temperature is higher; thus, pan evaporation increases. As a result, the pan coefficient can be 15 % higher in vegetated fields than in bare soil. Higher wind results in a greater increase in pan evaporation than evapotranspiration; as a result, K_p is lower in higher wind. Likewise, lower mean relative humidity results in greater increase in pan evaporation than reference ET_0, so the pan coefficient is lower in low humidity

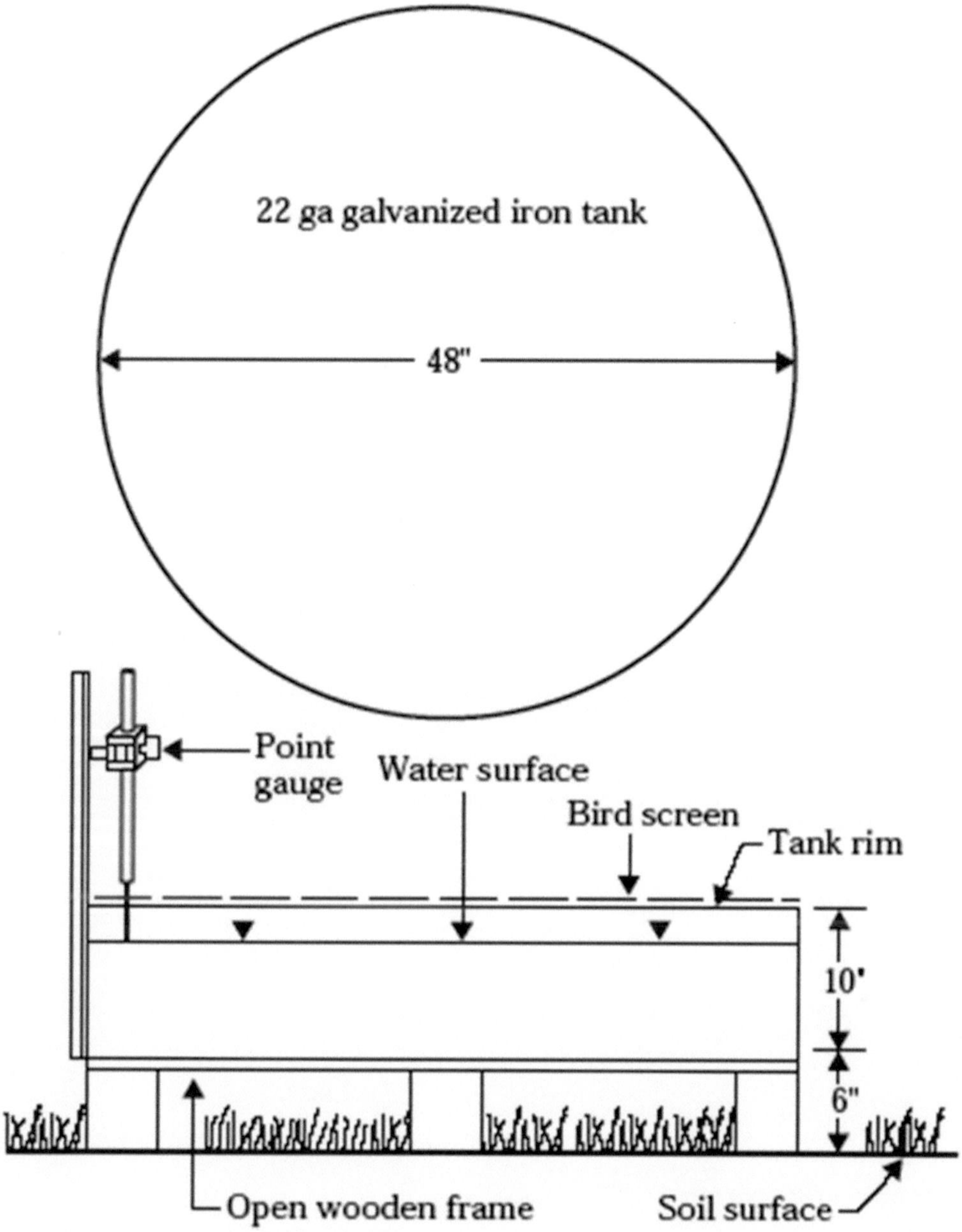

Fig. 5.13 Schematic diagram of Class A evaporation pan (Credit NRCS NEH. Section 15, Chapter 2)

conditions. In general, these differences are due to the fact that lake evaporation is a direct function of water and air temperature, wind speed and humidity, and evapotranspiration is primarily driven by solar radiation. Tables are available for adjustment of the pan coefficient.

Acquiring Weather Data

The Chapter 5 ET Calculator automatically acquires data from the AZMET weather network (see Fig. 5.14).

The Weather Acquisition form acquires data from the current year, or any series of previous years from any of the AZMET weather stations. It can also calculate historical averages from the imported series. The VBA program is unlocked and can be easily modified by an experienced VBA programmer to acquire weather data from other weather networks. The Weather Form is triggered from the *Active year weather* worksheet (Fig. 5.15). The data represented by Fig. 5.15 is shown in Fig. 5.16.

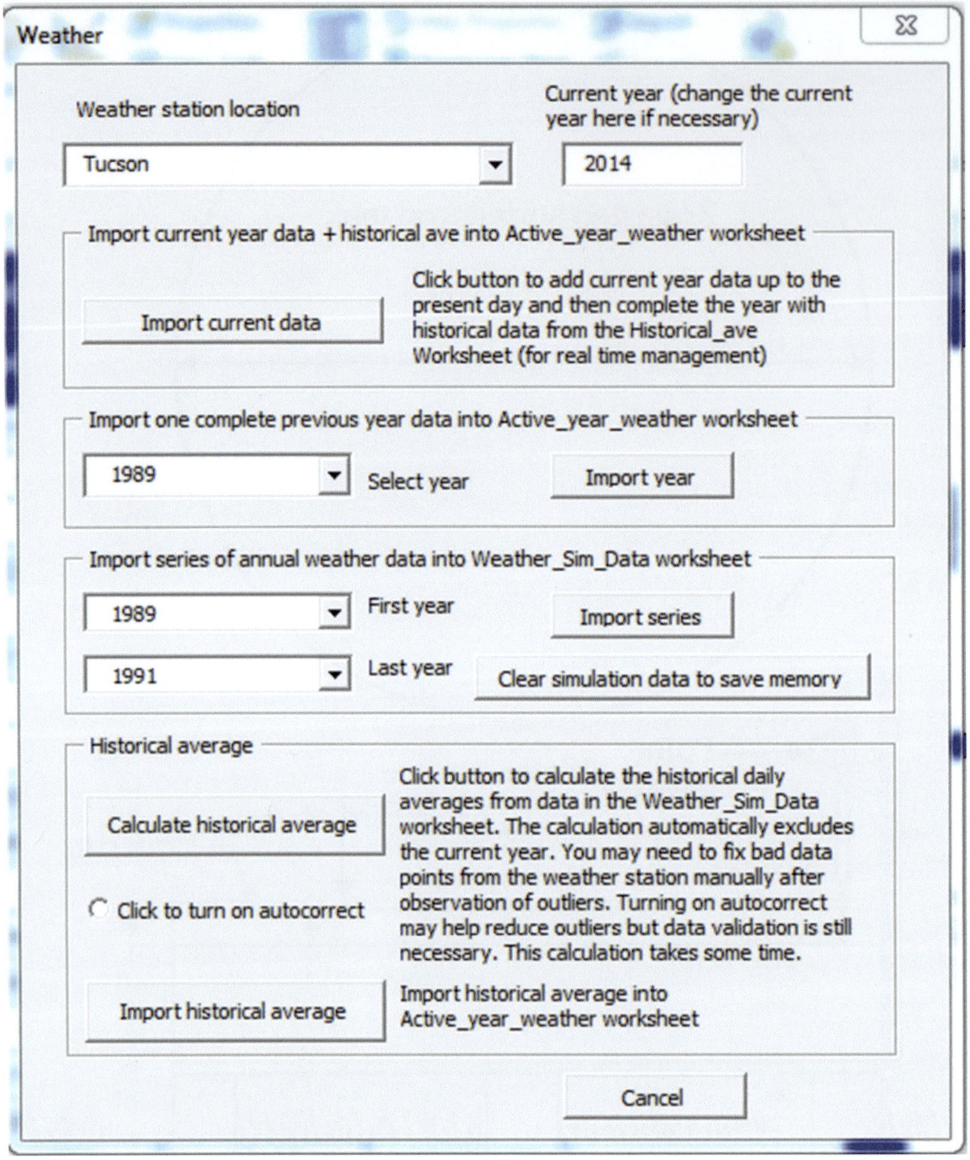

Fig. 5.14 Userform to acquire weather data from AZMET network

Fourier Series Simulation of Weather Data

Annual weather parameters, such as ET or temperature (Fig. 5.16), approximate a sin curve and are suitable for simulation with Fourier series.

$$\begin{aligned} W = W_0 + W_1{*}\cos(P) + W_2{*}\sin(P) \\ + W_3{*}\cos(2P) + W_4{*}\sin(2P) \\ + W_5{*}\cos(3P) + W_6{*}\sin(3P) \end{aligned} \tag{5.49}$$

where

W = coefficients
P = 2π * (day of year)/365.

The value, P, ranges from zero to 2π during the year, and is used to represent the 365 day annual weather cycle.

The Fourier coefficients, $W_0 - W_6$, in Eq. 5.49 can be calculated with the Regression Function in the *Fourier T and ETo* worksheet (Fig. 5.16). The Regression Function is found in Data → Data Analysis. One of the dependent variables in columns J:K is selected as the "Input Y Range" in the multilinear regression. The six independent variables (Input X range) are the sin and cos variables in columns Q:V.

Alternatively, the *Make Fourier Series* button triggers a subroutine that runs Regression for each of the dependent variables and automatically places the coefficients in cells B2:G8 (Fig. 5.16). The subroutine first copies the environmental parameters from the *Active year weather* worksheet and pastes them into columns J:O. It then runs the regression

	A	B	C	D	E	F	G
1	Run Weather form				Tucson	Hist_ave	1992
2	DOY	RHmin	WS	ET0 (mm)	Rain (mm)	Tmax ©	Tmin ©
3	1	29.907	1.4	1.827477	0.016666667	19.86	2.01333333
4	2	27.673	1.6067	2.018851	0.508	19.99333333	2.38
5	3	28.14	1.92	2.101266	1.382	18.94666667	3.06666667
6	4	31.56	1.5867	1.847711	1.024666667	18.8	1.71333333
7	5	28.633	1.76	1.972004	0.83	19.02666667	2.16
8	6	32.173	1.96	2.059613	2.814666667	17.72666667	1.60666667
9	7	28.533	1.9067	2.143521	2.1	19.28	1.56
10	8	26.173	1.54	2.023296	1.708	20.28	1.90666667
11	9	25.527	1.36	1.93834	0.608666667	20.24666667	1.27333333
12	10	24.66	0.45	2.336027	1.068666667	20.1	2.30666667
13	11	23.9	1.8267	2.250916	0.302	20.32	3.24
14	12	28.9	1.8333	2.039189	1.043333333	18.70666667	1.7
15	13	30.1	1.6133	2.002887	1.567333333	19.17333333	1.42666667
16	14	24.953	1.4667	2.042373	0.136	20.74666667	2.19333333
17	15	23.373	1.5667	2.170453	0.016666667	20.2	2.01333333
18	16	25.66	1.9533	2.302706	0.589333333	19.19333333	2.58
19	17	27.933	1.7533	2.166377	0.796	19.18	1.39333333

Active_year_weather | Weather_stations | Weather_Sim_Data | Historical_ave | New_AZMET_format

Fig. 5.15 *Active year weather* worksheet

for each parameter. Coefficients for each of the six parameters are placed in columns B:G and simulated values are placed in columns X:AC.

For example, the Fourier series equation for Tmax in column F in Fig. 5.17 is as follows:

$$W = W_0 + W_1{}^* \cos(P) + W_2{}^* \sin(P) + W_3{}^* \cos(2P) + W_4{}^* \sin(2P) + W_5{}^* \cos(3P) + W_6{}^* \sin(3P)$$

$$T = 28.62 - 9.35\cos(P) - 2.93\sin(P) - 0.376\cos(2P) + 0.140\sin(2P) - 0.177\cos(3P) - 0.976\sin(3P)$$

Example 5.7 Calculate maximum temperature for day of year (DOY) 100, based on Fourier series coefficients in Fig. 5.17.

Calculate P on DOY 100

$$P = 2\pi{}^*100/365 = 1.72$$

Tmax = 28.62 − 9.35cos(1.72) − 2.93sin(1.72) − 0.376cos(2*1.72) + 0.140sin(2*1.72) − 0.177cos(3*1.72) − 0.976sin(3*1.72) = 26.5 °C (cell N102)

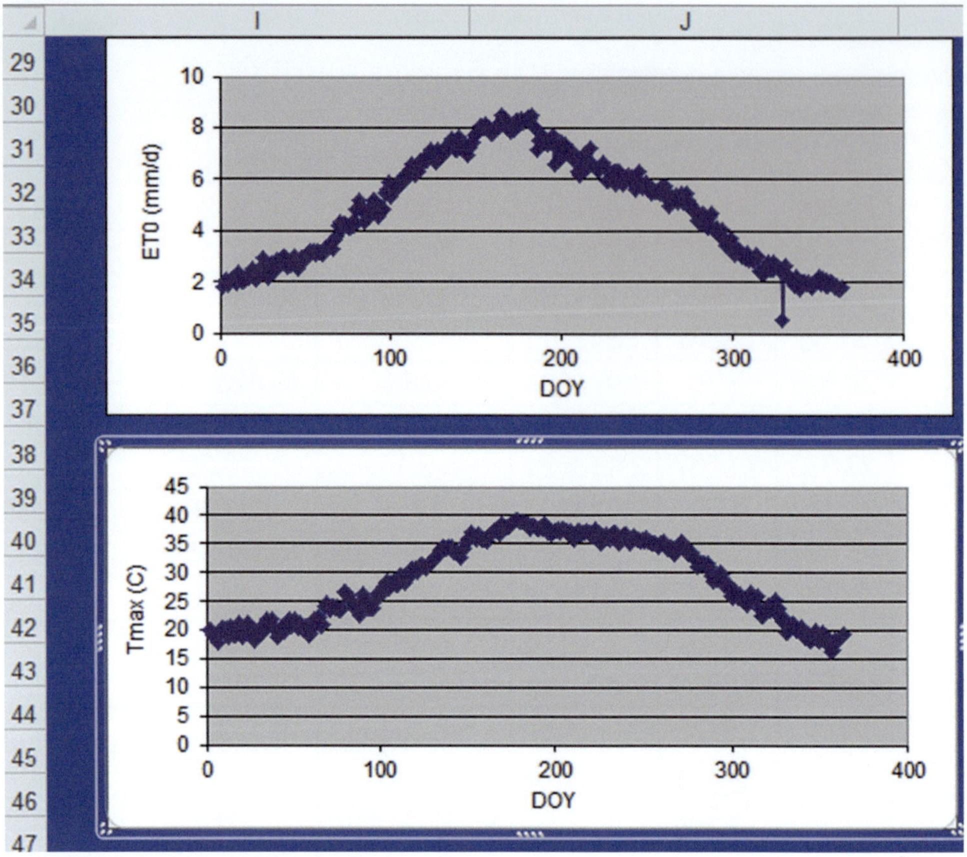

Fig. 5.16 Graphs in *Active year weather* worksheet

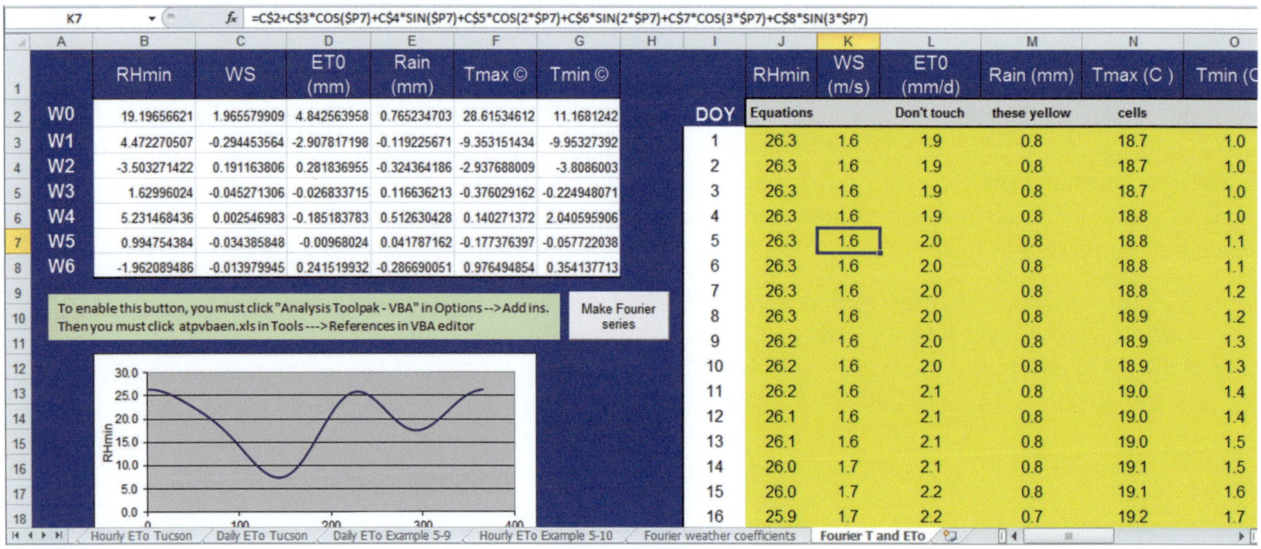

	RHmin	WS	ET0 (mm)	Rain (mm)	Tmax ©	Tmin ©
W0	19.19656621	1.965579909	4.842563958	0.765234703	28.61534612	11.1681242
W1	4.472270507	-0.294453564	-2.907817198	-0.119225671	-9.353151434	-9.95327392
W2	-3.503271422	0.191163806	0.281836955	-0.324364186	-2.937688009	-3.8086003
W3	1.62996024	-0.045271306	-0.026833715	0.116636213	-0.376029162	-0.224948071
W4	5.231468436	0.002546983	-0.185183783	0.512630428	0.140271372	2.040595906
W5	0.994754384	-0.034385848	-0.00968024	0.041787162	-0.177376397	-0.057722038
W6	-1.962089486	-0.013979945	0.241519932	-0.286690051	0.976494854	0.354137713

	RHmin	WS (m/s)	ET0 (mm/d)	Rain (mm)	Tmax (C)	Tmin (C
DOY	Equations		Don't touch	these yellow	cells	
1	26.3	1.6	1.9	0.8	18.7	1.0
2	26.3	1.6	1.9	0.8	18.7	1.0
3	26.3	1.6	1.9	0.8	18.7	1.0
4	26.3	1.6	1.9	0.8	18.8	1.0
5	26.3	1.6	2.0	0.8	18.8	1.1
6	26.3	1.6	2.0	0.8	18.8	1.1
7	26.3	1.6	2.0	0.8	18.8	1.2
8	26.3	1.6	2.0	0.8	18.9	1.2
9	26.2	1.6	2.0	0.8	18.9	1.3
10	26.2	1.6	2.0	0.8	18.9	1.3
11	26.2	1.6	2.1	0.8	19.0	1.4
12	26.1	1.6	2.1	0.8	19.0	1.4
13	26.1	1.6	2.1	0.8	19.0	1.5
14	26.0	1.7	2.1	0.8	19.1	1.5
15	26.0	1.7	2.2	0.8	19.1	1.6
16	25.9	1.7	2.2	0.7	19.2	1.7

Fig. 5.17 Fourier series calculation and graphs in *Fourier T and ETo* worksheet

Questions

1. Atmospheric pressure is 100 kPa and the atmosphere is holding 2 % water. What is the partial pressure of water in the atmosphere?
2. Why is relative humidity an important factor in evapotranspiration?
3. Calculate the relative humidity if the partial pressure, e_a, of water in the air is 1.1 kPa and T = 40 °C. See Example 5.1 for saturated vapor pressure equation.
4. The density of water in the atmosphere at sea level is 0.018 kg/m^3, atmospheric pressure is 101 kPa, and temperature is 40 °C. What is the relative humidity?
5. Calculate the total resistance to water vapor transfer and the water vapor transfer conductance for a well-watered turf crop. The wind speed at 2 m elevation is 2 m/sec and the bulk surface resistance is 70 s/m. Calculate the maximum depth of water vapor transfer during 1 hour if the relative humidity is 60 %, temperature is 30 °C, and elevation is sea level.
6. Write a *sentence* describing each of the four terms in the crop evapotranspiration energy balance equation.
7. Calculate cumulative net radiation (MJ/m^2) over turf for 2 hour if average hourly radiation energy input (R_S) is 50 Langleys (cal/cm^2).
8. The plant canopy of a transpiring plant is generally cooler than the atmosphere above it. Why?
9. How would the height of the crop influence the 2 m wind speed measurement? Would a calculation of aerodynamic resistance based on 2 m wind speed without accounting for roughness elevation be in error? Why isn't this a problem for reference evapotranspiration calculations?
10. Explain why there is a difference between the stable wind speed and the unstable wind speed profile.
11. Explain why all of the curves in Fig. 5.6 have the same aerodynamic resistance at 3 m/sec wind speed. Why does the −2 °C curve have a higher aerodynamic resistance than the +2 °C curve?
12. Conduct dimensional Eqs. 5.15, 5.17, 5.19, 5.20, and 5.22 and make sure that the units in the equations are consistent.
13. In Eq. 5.19 the radiation term on the left and the aerodynamic term on the right are independent. Can Eq. 5.20 be broken down in the same way?
14. Compare the results of Examples 5.4 and 5.6. Evaluate the percent difference and discuss the reason for the difference. Does the Hargreaves-Samani equation in any way include humidity?
15. Use data from April 10, 2009, Tucson weather station data (http://ag.arizona.edu/azmet/data/0109eh.txt) to make hand calculations of daily ET with the equations. Then use the *Chapter 5 ET calculator* program and compare with your results. The latitude of Tucson is 33.95°N. Elevation is 655 m. Maximum temperature is 28 °C and minimum temperature is 8 °C. Relative humidity minimum is 7 % and relative humidity maximum is 57 %. Measured solar radiation is 27 MJ m^{-2} day^{-1}. Average wind speed at 3 m elevation is 2.8 m/sec. Check your answer against the AZMET calculated ET.
16. Use hand calculations to calculate hourly ET over turf in Tucson, Arizona on DOY 100. Solar radiation is 2.84 MJ m^{-2} hr^{-1}. Average temperature is 27 °C, and wind speed is 4.2 m/s at 3 m elevation. Elevation is 655 m, relative humidity is 9 %. Longitude of Tucson is 113.18^0. Tucson is in the Mountain Time Zone (Lm = 105). The longitude of the Mountain Time meridian is 105^0. The period is from 2 to 3 p.m. Check your results against the spreadsheet calculations and against the AZMET calculated ET.
17. Use average *monthly* values of ET_0, maximum temperature, and minimum temperature from a weather series in your area to calculate Fourier series coefficients for maximum and minimum temperature and ET_0.

References

Allen RL, Pereira DR, Smith M (1998) *Crop evapotranspiration – guidelines for computing crop water requirements* – FAO Irrigation and drainage paper 56. FAO – Food and Agriculture Organization of the U. N http://www.fao.org/docrep/X0490E/x0490e00.htm

Allen R, Walter I, Elliott R, Howell T, Itenfisu D, Jensen M, Snyder R (2005) The ASCE standardized reference evapotranspiration equation. American Society of Civil Engineereing, Reston, p 192

Brown P Standardized reference evapotranspiration, University of Arizona Extension Publication 1324. http://ag.arizona.edu/pubs/water/az1324.pdf

Hargreaves GH, Samani ZA (1985) Reference crop evapotranspiration from temperature. Appl Eng Agric 1(2):96–99, NRCS National Engineering Handbook. Irrigation water requirements

Martin D, Gilley J (1993) NRCS NEH national engineering handbook, Part 623, Irrigation Water Requirements, Section 15, Ch. 2., 1993. http://www.info.usda.gov/CED/ftp/CED/neh15-02.pdf

Penman HL (1948) Natural evaporation from open water, bare soil and grass. Proc Roy Soc Lond A193:120–146

Snyder R (2007) Biometeorology page at UC Davis. This page contains many excel programs for calculating evapotranspiration. http://biomet.ucdavis.edu

Szeicz G, Long IF (1969) Surface resistance of crop canopies. Water Resour Res 5:622–633

Crop Evapotranspiration

6

Crop evapotranspiration is a function of both weather (reference ET) and the growth stage of the plant. The ratio between reference ET and crop ET is the crop coefficient, which changes during the season with crop physiological changes. At the time of planting, most ET is evaporation from the bare soil surface, but as the canopy matures ET is predominantly transpiration from the crop (Fig. 6.1). The single crop coefficient combines evaporation and transpiration. A theoretically more accurate approach divides the evaporation and transpiration into two separate coefficients (dual crop coefficient). Both methods are extensively described in the Food and Agriculture Organization Irrigation and Drainage Paper No. 56 (FAO 56). Crop coefficients are normally calculated based on days after planting, but they can also be calculated based on heat units after planting. Heat units are the integrated temperature and time after planting. This chapter also includes a Fourier series approach to modeling crop coefficients and weather. Finally, the chapter includes a brief introduction to the use of remote sensing for irrigation scheduling.

Crop Coefficient

Evapotranspiration can be calculated as a function of two primary parameters: (1) weather and (2) crop physiology and maturity. The weather parameters are used to calculate reference ET_0 as described in Chap. 5. The crop physiological parameters are quantified in the crop coefficient K_c. The traditional single crop coefficient is defined as the ratio of total (evaporation + transpiration) crop evapotranspiration, ET_c, to ET_0. K_c is crop-specific and ranges from zero to over unity, depending on the crop growth stage. Crop ET_c at any time during the growing season is calculated as the product of ET_0 and the crop coefficient.

$$ET_c = (ET_0)(K_c) \tag{6.1}$$

where

ET_c = crop evapotranspiration, mm/day
K_c = single crop coefficient, dimensionless.

Linearized crop coefficient curves (Fig. 6.3) have been developed for nearly all crops by measuring crop water use with lysimeters (Fig. 6.2) and dividing the daily crop water use by daily ET_0 during the growing season.

In-class Exercise 6.1 If reference ET_0 is 10 mm/day for 1 week, and the crop coefficient is 0.5, then what is the depth of crop evapotranspiration during 1 week?

If reference ET_0 is calculated as 7 mm/day, pan evaporation is 9 mm/day, and measured crop evapotranspiration is 5 mm/day, as measured by a lysimeter, then what are the pan and reference ET_0 crop coefficients?

Linearized crop coefficient curves (Fig. 6.3) generally have four stages: the initial stage from the beginning of the season to 10 % canopy development, the development stage from 10 % canopy cover to full canopy cover, the mid-season stage from full canopy cover to the start of maturity, and the late season stage from maturity to harvest. Some crops also require an end season stage, as shown in Fig. 6.6.

Full canopy cover, which corresponds with the beginning of the midseason stage, is generally defined for row crops as the time when leaves from adjoining rows are interlocked. It may be defined based on crop height for grasses and forage crops. Alternatively, a leaf area index (LAI) value of 3.0 can be defined as full canopy cover; the LAI is the ratio of leaf surface area to soil surface area. Lengths of stages and single crop coefficient values for many crops are listed in FAO 56 Tables 11 and 12, respectively, for different crops and climatic regions.

The pattern of crop water use (crop coefficient) during the early growing season is influenced by the irrigation method. Irrigation effects are greatest when the entire soil surface is wetted and the crop canopy is not full during the initial and

P. Waller, M. Yitayew, *Irrigation and Drainage Engineering*, DOI 10.1007/978-3-319-05699-9_6

development phases. Evaporation spikes due to a wet soil surface after irrigation or rainfall during the early season can be as great as midseason ET (Fig. 6.4). For this reason, the FAO56 single crop coefficient during the early season is calibrated based on depth applied during each irrigation. The midseason crop coefficient is adjusted based on wind speed and relative humidity, as described in the FAO 56 *Single Crop Coefficient* chapter (Allen et al. 1998). Stress events with insufficient water during midseason can reduce ET (Fig. 6.4). Reduction in ET due to water stress can be calculated based on the soil moisture content and the crop sensitivity to water stress as described in Chaps. 2 and 4.

Normal crop evapotranspiration and the excess evaporation caused by irrigation can be separated into two terms: the basal crop coefficient and the soil evaporation coefficient. The sum of the two is called the average crop coefficient (Fig. 6.5). The basal crop coefficient is lower than the single crop coefficient during the initial and development phases. The single crop coefficient is similar to the average crop coefficient as shown in Fig. 6.5.

$$K_a = K_s K_{cb} + K_e \tag{6.2}$$

$$ET_c = (ET_0)(K_a) \tag{6.3}$$

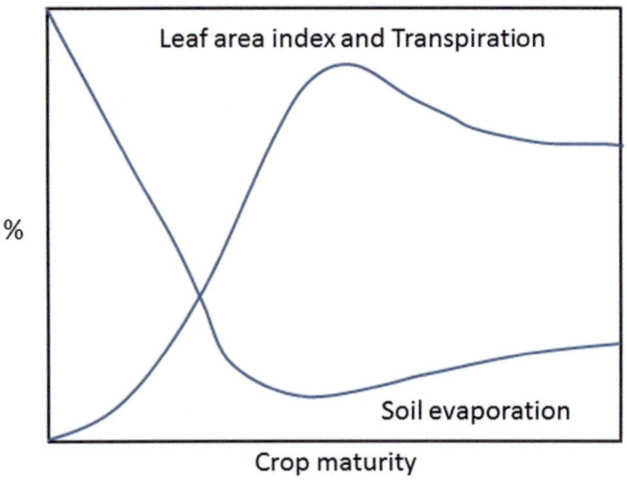

Fig. 6.1 Relative percent of evaporation and transpiration vs. crop maturity (After FAO 56)

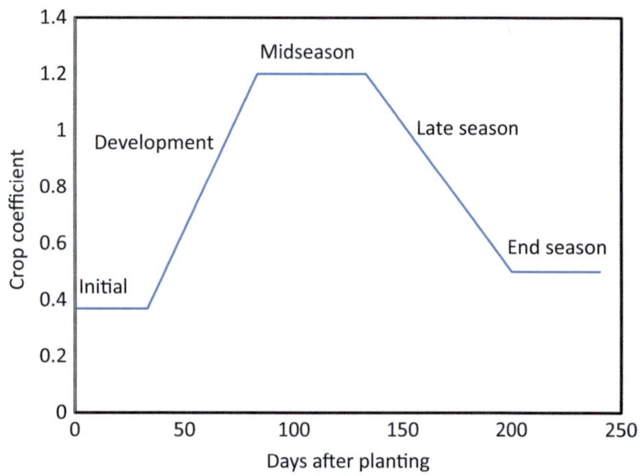

Fig. 6.3 Single crop coefficient curve, K_c, for cotton ET in central Arizona (Values from Table 11 in FAO 56)

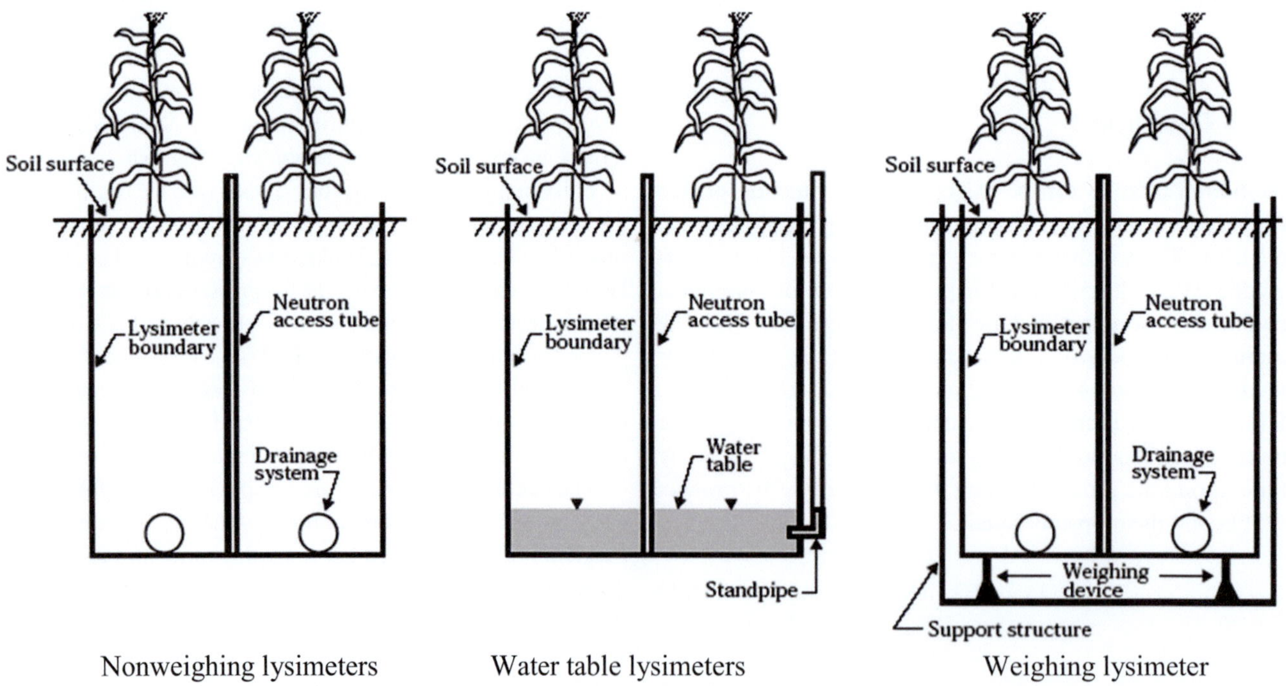

Fig. 6.2 Lysimeters for measurement of crop coefficient (Credit NRCS NEH Sec 15, Ch 2)

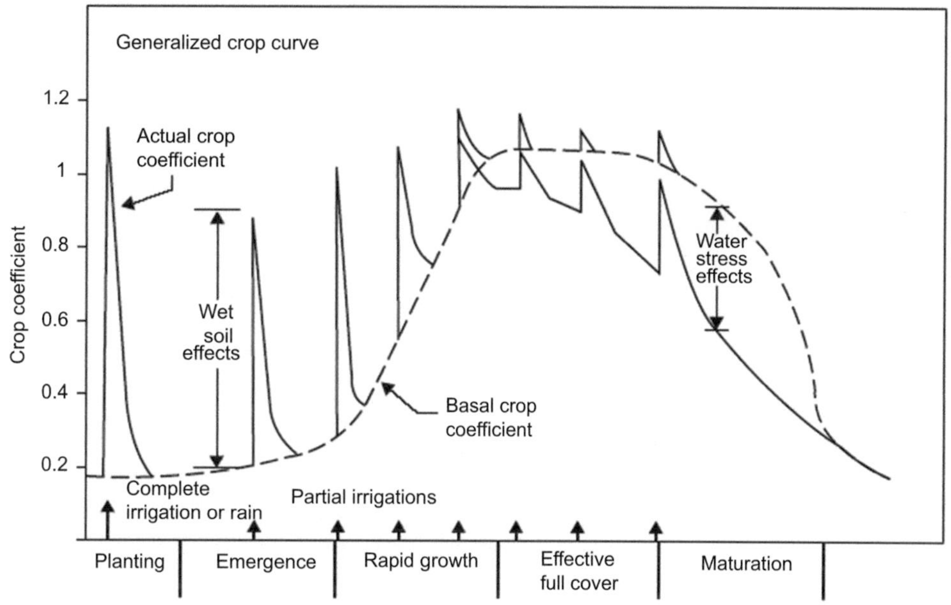

Fig. 6.4 Crop evapotranspiration with wetting and stress events (Credit NRCS NEH, Sec 15–2)

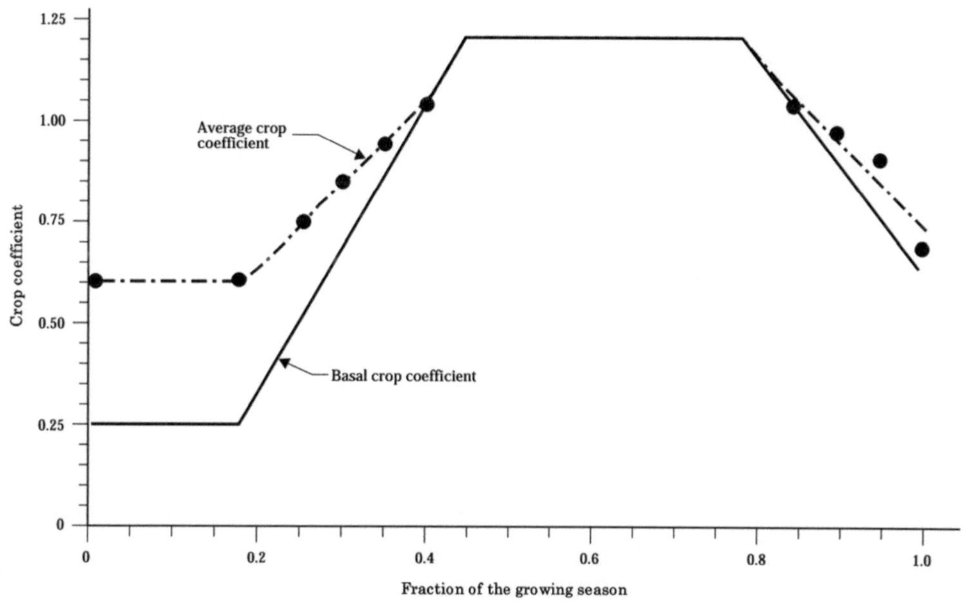

Fig. 6.5 Average and basal crop coefficients (Credit NRCS)

where

K_s = stress coefficient due to dry soil and/or high salinity, dimensionless
K_a = average crop coefficient, dimensionless
K_e = soil evaporation coefficient, dimensionless
K_{cb} = basal crop coefficient, dimensionless.

The method with evaporation and basal crop coefficients is called the dual crop coefficient method. Although the single crop coefficient can be adjusted based on the frequency of irrigation events (see FAO 56), the dual crop coefficient approach more naturally integrates the effects of frequency of irrigation, stress, and fraction of soil wetting into the ET_c calculation because it is separates the evaporation and transpiration (basal) terms. Accounting for irrigation effects on ET enables the agronomic evaluation of different types of irrigation systems and management strategies (*WINDS* model in Chap. 29) based on fraction of soil surface wetting and irrigation frequency. For this reason, the dual crop coefficient method is emphasized in this chapter and is the ET calculation method in the *WINDS* model.

The basal crop coefficient is defined as the ratio of the crop evapotranspiration to the reference evapotranspiration when the soil surface is dry, with transpiration occurring at the maximum potential rate (FAO 56 definition). Basal crop coefficients are listed in FAO 56 Table 17. These values represent K_{cb} at 45 % relative humidity and moderate wind speed (2 m/sec) conditions. The K_{cb} from Table 17 is adjusted with the following equation based on average weather conditions during the mid and late season growth stages:

$$K_{cb} = K_{cb(table\ 17)} + [0.04(u_2 - 2) - 0.004(RH_{min} - 45)]\left(\frac{h}{3}\right)^{0.3} \tag{6.4}$$

where

K_{cb} (Table 17) = value for $K_{cb\ mid}$ or $K_{cb\ end}$ taken from Table 17

U_2 = the mean value for daily wind speed at 2 m height over grass during the mid or late season stage [m s^{-1}] for 1 m s^{-1} $\leq$ u2 $\leq$ 6 m s^{-1}

RH_{min} = the mean value for daily minimum relative humidity during the mid- or late season growth stage [%] for 20 % $\leq RH_{min} \leq$ 80 %

h = the mean plant height during the mid or late season stage [m] (from Table 12).

Example 6.1 Adjust K_{cb} for cotton grown in central Arizona. Average minimum relative humidities during midseason and late season growth stages are 20 % and 25 %, respectively. Average wind speed at 2 m elevation is 2 m/sec during mid and late season growth stages.

From Table 2;17, $K_{cb\ mid}$ = 1.10 – 1.15 $K_{cb\ end}$ = 0.50 – 0.40

From Table 12, h = 1.2 – 1.5

$$K_{cb\ mid} = 1.125 + [0.04(2-2) - 0.004(20-45)]\left(\tfrac{1.2}{3}\right)^{0.3} = 1.20$$
$$K_{cb\ end} = 0.45 + [0.04(2-2) - 0.004(25-45)]\left(\tfrac{1.2}{3}\right)^{0.3} = 0.51$$

$K_{cb\ ini}$ is unchanged (0.15).

The basal crop coefficient curve for cotton in central Arizona is shown in Fig. 6.6. An end season phase has been added by Doug Hunsaker – USDA-ARS ALARC.

The late season and end season K_{cb} decreases due to loss of foliage or drying at the end of the season. Other crops may be harvested before any reduction in growth rate and thus do not have a late season decrease in K_{cb}. Daily K_{cb} values are found using linear interpolation between the inflection points shown in Fig. 6.6.

Rooting depth determines the soil water holding capacity. Rooting depth is calculated with a linear model. The roots are given an initial constant depth during the initial stage because roots are allowed to grow down toward the water during the initial stage, and the model should not be triggered to irrigate by with an unrealistically small RAW. The roots then grow from the initial depth to their full depth during the development stage. Root growth is modeled during the development stage with a linear root growth model (Eq. 6.5 and Fig. 6.7).

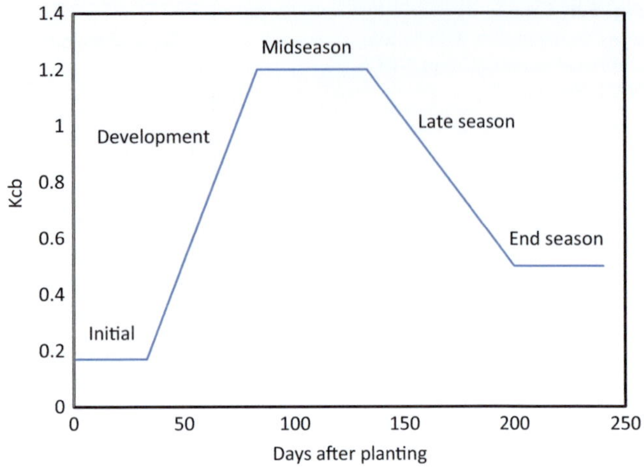

Fig. 6.6 Basal crop coefficient curve for cotton in Arizona

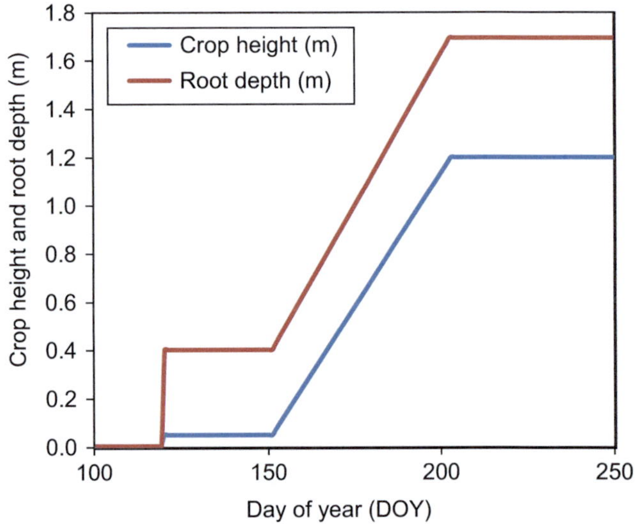

Fig. 6.7 Model of rooting depth and crop height versus time for cotton in Arizona

$$Z_j = Z_{ini} + (Z_{max} - Z_{ini})(j - Dev - Plant)/(Mid - Dev) \tag{6.5}$$

where

Z_{ini} = initial root depth, m

Z_j = root depth on day j, m

Z_{max} = final root depth, m

Dev = Number of days after planting date for beginning of development stage

Mid = Number of days after planting date for beginning of midseason stage

Plant = Day of year of planting date

j = Day of year.

Crop height and rooting depth have similar early patterns in crop ET models, as shown in the following two examples.

Example 6.2 The initial depth for cotton roots is 0.4 m and the final depth is 1.7 m. Planting date is DOY 120. The development stage begins 31 days after planting and the midseason stage begins 83 days after planting. With Eq. 6.5, calculate the rooting depth on DOY 165, and plot the root growth vs. days after planting.

$$Z_j = Z_{ini} + (Z_{max} - Z_{ini})(j - \text{Dev} - \text{Plant})/(\text{Mid} - \text{Dev})$$
$$Z_{165} = 0.4 + (1.7 - 0.4)(165 - 31 - 120)/(83 - 31) = 0.75 \text{ m}$$

Rooting depth vs. time is shown in Fig. 6.7.

Example 6.3 The initial crop height is 0.05 m and the final crop height is 1.2 m. Plot the crop height vs. days after planting. Calculate the crop height on DOY 165.

The height on DOY 165 is calculated as follows:

$$H_j = H_{ini} + (H_{max} - H_{ini})(j - \text{Dev} - \text{Plant})/(\text{Mid} - \text{Dev})$$
$$H_{165} = 0.05 + (1.2 - 0.05)(165 - 31 - 120)/(83 - 31) = 0.36 \text{ m}$$

The crop height is also shown in Fig. 6.7.

The soil evaporation coefficient is a function of crop height and is calculated as a function of soil type, fraction of wetted soil, crop height, humidity, and wind speed. The maximum allowable evapotranspiration, from both evaporation and transpiration, K_{c-max}, is constrained by the energy available and the vapor pressure gradient. The basal crop coefficient K_{cb} plus the evaporation coefficient K_e cannot exceed K_{c-max}.

$$K_e + K_{cb} \leq K_{c-max} \tag{6.6}$$

K_{c-max} ranges from 1.05 to 1.3 for a grass reference crop, and is calculated as the maximum of Eqs. 6.7 and 6.8.

$$K_{c-max} = 1.2 + [0.04(u_2 - 2) - 0.004(RH_{min} - 45)]\left(\frac{h}{3}\right)^{0.3} \tag{6.7}$$

$$K_{c-max} = K_{cb} + 0.05 \tag{6.8}$$

In the FAO 56 dual crop coefficient model, evaporation takes place from the surface layer. Evaporation is high when this layer has high water content and the entire surface is wetted. It decreases to zero when water content in the surface layer is half of the permanent wilting point. The soil evaporation coefficient is a function of the fraction of soil surface that is wetted by irrigation or precipitation (100 % for precipitation) and the depletion of water from the surface layer. The evaporation reduction coefficient, K_r, quantifies the reduction in evaporation rate as a function of moisture content. It is calculated in two stages: the energy limiting stage and a falling rate stage. During the energy limiting stage with higher water content, evaporation is only limited by atmospheric conditions and solar energy, so $K_r = 1.0$. The depth of water that evaporates during stage 1 is called the readily evaporable water, REW. The total evaporable water, TEW, is the range between field capacity and half of permanent wilting point. When the depletion is greater than REW, K_r is calculated as a linear function that is similar to the water and salinity stress coefficient calculations in Chap. 4 and as shown in Fig. 6.8 if D_r exceeds REW.

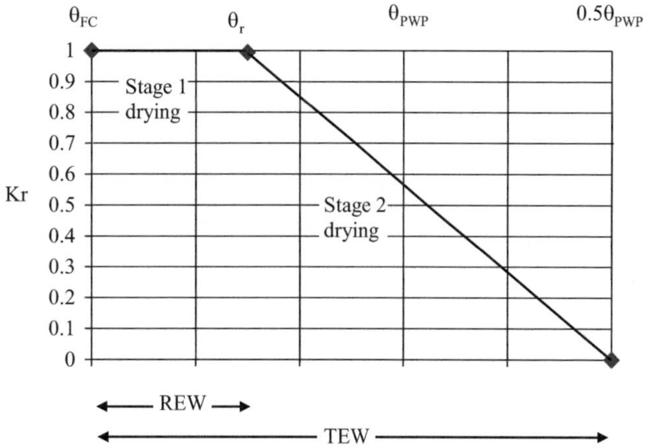

Fig. 6.8 Evaporation reduction coefficient

$$K_r = \frac{TEW - D_r}{TEW - REW} \tag{6.9}$$

where

K_r = dimensionless evaporation reduction coefficient dependent on the cumulative depth of water depleted (evaporated) from the topsoil
TEW = total evaporable water when surface has been wetted and then drained to field capacity, mm
REW = depth of water evaporated during stage 1, mm
D_r = depletion below field capacity, mm
θ_r = water content at threshold between stage 1 and stage 2 evaporation.

$$\text{TEW} = 1000\ (\theta_{FC} - 0.5\theta_{PWP})Z_e \tag{6.10}$$

where

Z_e = depth of surface layer, mm.

The depth of the surface layer ranges from 0.1 to 0.15 m with the deeper value for finer textured soils because of capillary suction. Values of TEW and REW for different soil textures are reported in FAO 56 Table 19. An accurate estimation of REW is not as critical as an accurate estimation of TEW, because TEW quantifies the total depth evaporated after each irrigation or precipitation event while shifting REW only changes the rate of evaporation.

Example 6.4 Field capacity is 23.5 % and permanent wilting point is 11 %. The depth of the surface layer is 0.11 m. REW = 9 mm. Calculate TEW. Calculate K_r for the surface layer depletion equal to 12 mm.

$$\begin{aligned}\text{TEW} &= 1{,}000(\theta_{FC} - 0.5\theta_{PWP})Z_e \\ &= 1{,}000(0.235 - (0.5)(0.11))(0.11) = 20 \text{ mm}\end{aligned}$$

$$K_r = \frac{TEW - D_r}{TEW - REW} = \frac{20 - 12}{20 - 9} = \frac{8}{11} = 0.73$$

The evaporation coefficient, K_e, is a function of the soil moisture status as quantified by the evaporation reduction coefficient, K_r, and crop height and shading, which is quantified with the basal crop coefficient. The evaporation must also be less than the product of $K_{c\text{-max}}$ and the fraction of exposed and wetted soil (FAO 56).

$$K_e = K_r(K_{c-\max} - K_{cb}) \le f_{ew}K_{c-\max} \tag{6.11}$$

where

f_{ew} = fraction of the soil that is both exposed to the sun and wetted; i.e., the fraction of soil surface from which most evaporation occurs.

The fraction of the soil that is both exposed and wetted may be dependent on one or the other, or a combination of the two; f_{ew} is the minimum of the exposed and/or wetted area (FAO 56).

$$f_{ew} = \min(f_w, 1 - f_c) \tag{6.12}$$

where

f_w = fraction of the soil that is wetted
f_c = fraction of the soil that is covered by shade.

The fraction exposed and wetted depends on where the water is applied. For example, if a drip irrigation system applies water directly under the crop, then very little of the soil is both exposed and wetted (FAO 56). In this case, a refined approach is needed to determine the fraction of the drip wetted area that is exposed to sunlight. In the case of flood or furrow irrigation, where the water is applied over the entire field or just between rows, the minimum of $(1 - f_c)$ or f_w can be used as the exposed and wetted fraction, as described in Eq. 6.12. The fraction of shaded area can be measured or it can be calculated based on crop height as follows:

$$f_c = \left(\frac{K_{cb} - K_{c-\min}}{K_{c-\max} - K_{c-\min}}\right)^{1+0.5h} \tag{6.13}$$

where

$K_{c-\max}$ = maximum value of K_c immediately following wetting
$K_{c-\min}$ = minimum value of K_c for a dry bare soil (0.15–0.2)
H = average plant height, m
K_{cb} = the value of the basal crop coefficient on that day.

The dual crop coefficient model is demonstrated with the WINDS model in Chap. 29.

Heat Unit Crop Coefficient

Weather varies between regions and between years. Thus, the crop maturity at a certain time after planting will not be the same every year and in every location. The heat unit (HU) approach accounts for varying growth rate and provides a method of transferring crop coefficient curves between different regions and different years. The rationale is that crop growth stage is more closely correlated with the amount of heat since planting than the number of days since planting.

The HU approach calculates the crop coefficient, Kc, based on accumulated energy (growing degree days) during the growing season rather than on number of days since planting. The growing degree-day is the product of time and temperature. The GDD for each day is calculated with the maximum and minimum temperature in Eq. 6.14. The lower threshold temperature (THR) is the temperature below which crop growth does not occur. In this chapter, evapotranspiration from four crops are simulated with the heat unit approach: alfalfa, corn, cotton, and melons. Threshold values for each of these crops are in Table 6.1.

$$GDD_i = (t_{max} + t_{min})/2 - THR \tag{6.14}$$

where

THR = lower threshold temperature at which plant growth stops, °C
t_{max} = maximum temperature during the day, °C
t_{min} = minimum temperature during the night, °C
GDD_i = Growing degree days on day i, day -°C.
If $(t_{max} + t_{min})/2$ is less than THR, then no crop growth occurs and GDD is zero.

Table 6.1 Crop Kc parameters

Crop	Alfalfa	Corn	Watermelon
THR (°C) (lower threshold)	5	10	10
UPP (°C) (upper threshold)	25	30	30
C_0 (half period length coefficient)	4550	1800	2600
C_1	0.811	1.05	0.942
C_2	2.74E-03	0.0675	−0.260
C_3	0.852	0.0109	0.178
C_4	−1.28E-02	0.00348	−0.103
C_5	1.56E-02	−0.00854	0.124
C_6	0.743	0	−0.0301
GDD_{total}	550	883	1500
Max days (from time of planting)	365	145	135

Just as crop growth does not occur below a certain minimum temperature (THR), crop growth also shuts down above a certain maximum temperature, UPP. When the average ambient temperature $(t_{max} + t_{min})/2$ exceeds UPP, then GDD is the difference between UPP and THR (Eq. 6.5). A more complicated method than Eq. 6.5 was described by Snyder (1985), but the results do not change significantly for the crops modeled in this text, so Eq. 6.5 is used.

$$GDD_i = UPP - THR \tag{6.15}$$

GDD is set equal to zero at the time of planting. The cumulative growing degree days at any time during the growing season is the sum of growing degree days up to that point in the growing season.

$$GDD_{cum} = \sum_{i=1}^{n} GDD_i \tag{6.16}$$

where

n = length of time after beginning of growing season, days
GDD_{cum} = cumulative growing degree days, day -°C

Example 6.5 On the first day after planting corn, t_{max} and t_{min} are 32 °C and 8 °C, respectively. On the second day after planting, t_{max} and t_{min} are 42 °C and 22 °C, respectively. Find GDD_1, GDD_2, and the cumulative growing degree days after day 2, GDD_{cum}. From Table 6.1, UPP and THR are 30 °C and 10 °C, respectively.

The accumulation of growing degree days on 1st day: $GDD_1 = (32 + 8)/2 - 10 = 10$ day-°C

On 2nd day, $(t_{max} + t_{min})/2 >$ UPP: $GDD_2 = UPP - THR = 30 - 10 = 20$ day-°C

Cumulative growing degree days after 2nd day: $GDD_{cum} = GDD_1 + GDD_2 = 10 + 20 = 30$ day-°C

Paul Brown, at the University of Arizona, developed the *Arizona Heat Unit Program* (Fig. 6.9). It uses a sin curve based approach rather than the simple approach in Eq. 6.5. The calculation method is summarized in a document at http://ag.arizona.edu/azmet/AHUP-Manual-2012.pdf. Some heat unit equations do not impose an upper temperature limit; In this case, the value 200 (no limit) is entered into cell B4.

The cotton crop coefficient curve, as a function of heat units, is shown in Fig. 6.10.

Fourier Series Heat Unit Crop Coefficient

At the University of Arizona, Don Slack and Edward Martin (2001) developed Fourier sin series models of crop coefficient curves (K_c vs. heat units) in the AZSCHED model. The crop coefficient constants, C_1 to C_6, are calculated by using the Excel Regression function with data from crop experiments. The Fourier series for the crop coefficient, K_c is a sine series because values are always positive and do not oscillate about a mean as with yearly weather data. Unlike the weather parameters, the values are based on cumulative growing degree days (GDD) rather than time. Alfalfa, corn, and watermelon values for C_0 and C_1 to C_6 are shown in Table 6.1.

$$K_c = C_1 \sin(P) + C_2 \sin(2P) + C_3 \sin(3P) + C_4 \sin(4P) + C_5 \sin(5P) + C_6 \sin(6P) \tag{6.17}$$

where

C_1 to C_6 = constants for a specific crop
$P = \pi\, (GDD_{cum})/\, C_0$
C_0 = Half period length coefficient, days -°C.

The half period length coefficient, C_0, is a curve fitting parameter: the time at which the crop coefficient curve returns to zero. It is not the same as the end of the growing season.

The crop is harvested when the required heat units (GDD_{total}) for the crop growth cycle are exceeded by GDD_{cum}. There is the possibility, especially in areas with early winters, that the number of heat units during a season is not sufficient to complete the crop growth cycle during a cold year. In this case, the growing season ends when days since planting exceeds the maximum length of the growing season.

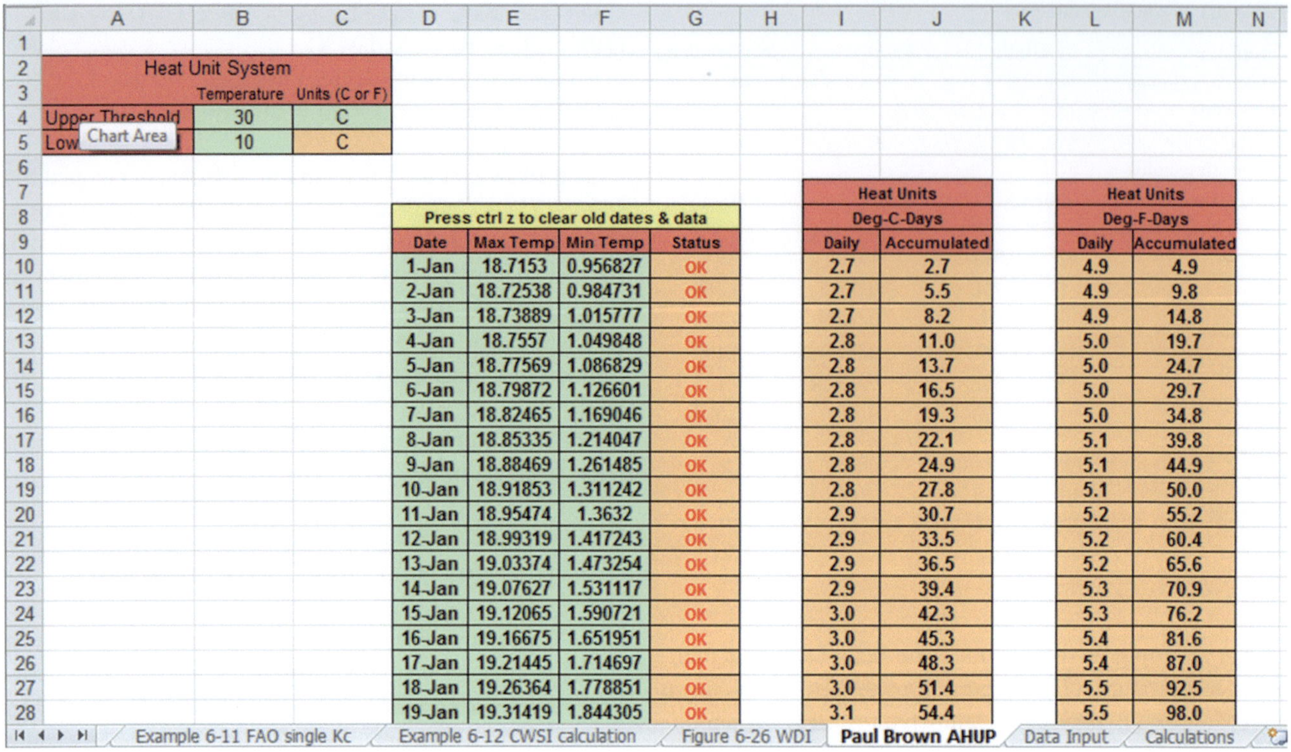

Heat Unit System

	Temperature	Units (C or F)
Upper Threshold	30	C
Low[...]	10	C

Press ctrl z to clear old dates & data

Date	Max Temp	Min Temp	Status	Heat Units Deg-C-Days Daily	Heat Units Deg-C-Days Accumulated	Heat Units Deg-F-Days Daily	Heat Units Deg-F-Days Accumulated
1-Jan	18.7153	0.956827	OK	2.7	2.7	4.9	4.9
2-Jan	18.72538	0.984731	OK	2.7	5.5	4.9	9.8
3-Jan	18.73889	1.015777	OK	2.7	8.2	4.9	14.8
4-Jan	18.7557	1.049848	OK	2.8	11.0	5.0	19.7
5-Jan	18.77569	1.086829	OK	2.8	13.7	5.0	24.7
6-Jan	18.79872	1.126601	OK	2.8	16.5	5.0	29.7
7-Jan	18.82465	1.169046	OK	2.8	19.3	5.0	34.8
8-Jan	18.85335	1.214047	OK	2.8	22.1	5.1	39.8
9-Jan	18.88469	1.261485	OK	2.8	24.9	5.1	44.9
10-Jan	18.91853	1.311242	OK	2.8	27.8	5.1	50.0
11-Jan	18.95474	1.3632	OK	2.9	30.7	5.2	55.2
12-Jan	18.99319	1.417243	OK	2.9	33.5	5.2	60.4
13-Jan	19.03374	1.473254	OK	2.9	36.5	5.2	65.6
14-Jan	19.07627	1.531117	OK	2.9	39.4	5.3	70.9
15-Jan	19.12065	1.590721	OK	3.0	42.3	5.3	76.2
16-Jan	19.16675	1.651951	OK	3.0	45.3	5.4	81.6
17-Jan	19.21445	1.714697	OK	3.0	48.3	5.4	87.0
18-Jan	19.26364	1.778851	OK	3.0	51.4	5.5	92.5
19-Jan	19.31419	1.844305	OK	3.1	54.4	5.5	98.0

Fig. 6.9 Arizona Heat Unit Program developed by Paul Brown

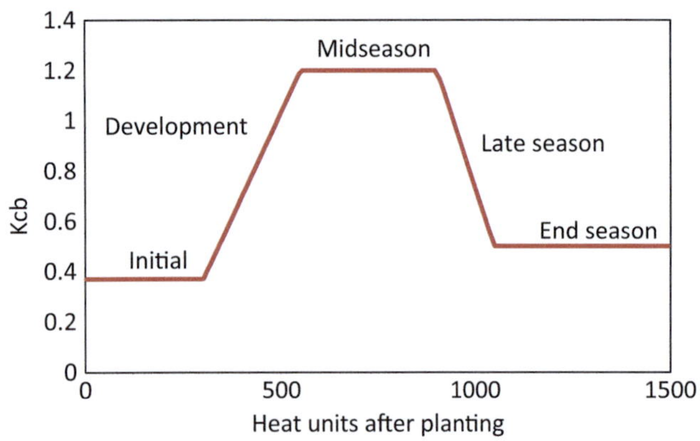

Fig. 6.10 Cotton crop coefficient curve based on heat units

Table 6.2 Root depth parameters

Crop	Alfalfa	Corn	Watermelon
Z_{ini} (m)	1.8	0.1	0.15
Z_{max} (m)	1.8	0.8	1
GDD_{root}	550	883	1500

Rooting depth can be modeled as a function of growing degree days. Roots are fully extended at the time of GDD_{root} in Table 6.2. Because alfalfa remains in the field for several years, alfalfa root depth is normally the maximum root depth for the entire season. For other crops, Eq. 6.18 calculates root depth until GDD_{cum} exceeds GDD_{root}. Then, root depth equals Z_{max} during the remainder of the growing season.

$$Z_i = Z_{ini} + (Z_{max} - Z_{ini})(GDD_{cum}/GDD_{root}) \tag{6.18}$$

where

Z_{ini} = initial root depth, m
Z_i = root depth at period i, m
Z_{max} = final root depth, m
GDD_{root} = GDD_{cum} when roots stop growing, day -°C.

Irrigation Scheduling

The approach to irrigation scheduling is dependent on the irrigation system. High frequency drip and center pivot irrigation systems generally irrigate all zones every day or nearly every day. The irrigation depth corresponds with the daily evapotranspiration. In contrast, orchard, wheel line, and surface irrigation systems apply a deep irrigation to one zone on one day and then apply a deep irrigation to another zone on the next day. After irrigating all zones, the first zone is irrigated again and the cycle continues. For these systems, irrigation takes place when soil water moisture

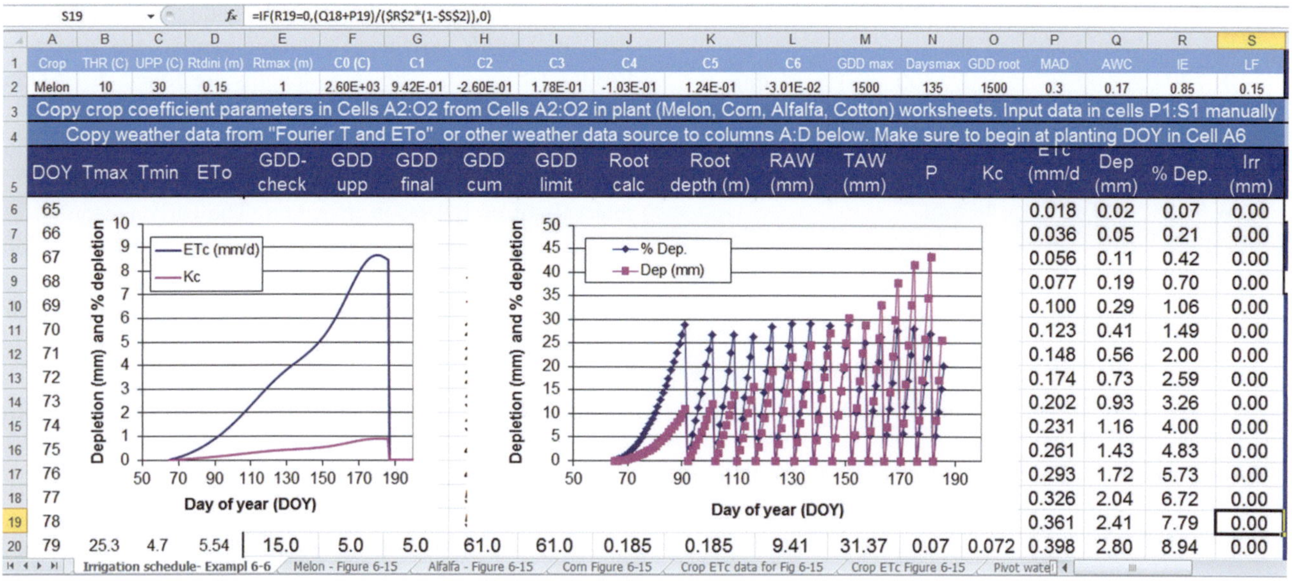

Fig. 6.11 Melon irrigation schedule in *Irrigation schedule* worksheet

content nears the management allowable depletion (MAD). This section shows how to develop an irrigation schedule within a spreadsheet with the Fourier series heat unit approach and the FAO56 linearized single crop coefficient method. Chapter 29 uses the FAO56 dual crop coefficient method within the *WINDS* model, with calculations in VBA.

The *Irrigation schedule* worksheet (Fig. 6.11) calculates GDD and root depth, and thus calculates ET and remaining available water, respectively. RAW and TAW are calculated in columns L and M based on the AWC and MAD, and root depth (column K). The crop coefficient, K_c, is calculated in column O based on the GDD in column I. K_c (column O) and ET_0 (column D) are then multiplied to calculate ET_c in column P. Depletion and percent depletion are then calculated in columns Q and R. Irrigations are specified when the percent depletion reaches the MAD value specified in cell P2. Irrigation is calculated as the previous depletion + present ET_c and is divided by irrigation efficiency and (1-LF) to find gross depth of irrigation.

Example 6.6 Calculate K_c for melons when $GDD_{cum} = 145$. ET_0 on the corresponding day is 6.98 mm/day. $P = \pi\,(GDD_{cum})/\,C_0 = \pi\,(145)/2{,}600 = 0.175$

$$K_c = C_1 \sin(P) + C_2 \sin(2P) + C_3 \sin(3P) + C_4 \sin(4P) + C_5 \sin(5P) + C_6 \sin(6P)$$

$$K_c = 0.942 \sin(P) - 0.260\,\sin(2P) + 0.178\,\sin(3P) - 0.103 \sin(4P) + 0.124\,\sin(5P) - 0.0301 \sin(6P)$$

$$K_c = 0.942(0.174) - 0.260(0.343) + 0.178(0.501) - 0.103(0.644) + 0.124(0.768) - 0.0301(0.867) = 0.167$$

The crop evapotranspiration (ETc) curves for alfalfa,

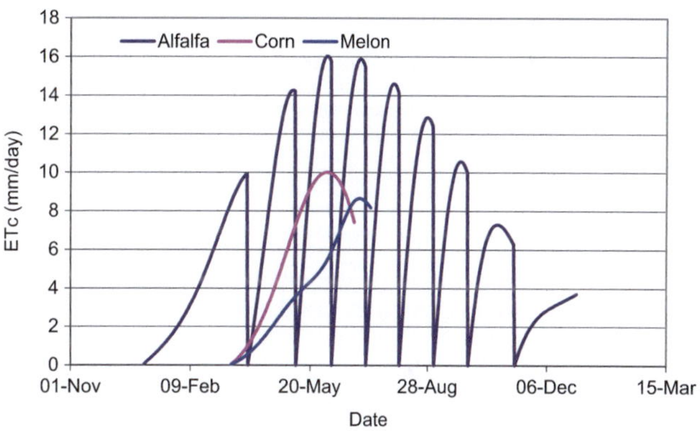

Fig. 6.12 Fourier series ET_c for three crops

corn, and watermelon (Fig. 6.12) were implemented with the Fourier temperature and ET_0 models (Eq. 5.49) and the Fourier crop coefficient model (Eq. 6.7). In order to construct the crop evapotranspiration curve for alfalfa (Fig. 6.15), which is harvested repeatedly during the year, GDD_{cum} is set back to zero every time that it exceeds GDD_{total} (harvest is triggered). Irrigation of sweet corn and watermelon stops near the seasonal peak of evapotranspiration.

Example 6.7 Calculate the rooting depth for cotton when GDD = 200. $Z_{ini} = 0.15$, $Z_{max} = 1.22$, $GDD_{root} = 1544$.

$$Z_{200} = 0.1 + (0.8 - 0.1)(200/883) = 0.26\ \text{m}$$

Example 6.8 Calculate the TAW if the root depth is 58 cm, and the AWC and layer depths are as shown in Table 3.6.

What is the maximum allowable depletion if MAD = 0.5?

$Layer_i$	AWC_i	TAW_i
0 – 30 cm	7 %	0.07 (300 mm) = 21 mm
30 – 50 cm	8.5 %	0.085 (200 mm) = 17 mm
50 – 58 cm	12 %	0.12 (80 mm) = 10 mm

$$TAW = 21 + 17 + 10 = 48 \text{ mm}$$

$$RAW = MAD(TAW) = 48(0.5) = 24 \text{ mm}$$

Maximum allowable depletion = 24 mm

Keeping track of the depth of depletion is like keeping track of a checkbook. The depth of depletion is calculated by adding the depth of depletion from the previous day to the evapotranspiration of the present day. When depletion is equal to the depth of readily available water, RAW, then irrigation takes place, and depletion is set back to zero.

Example 6.9 ET_c is 6 mm/day for the soil parameters in Example 6.8. Calculate the depletion and % depletion 4 days after irrigation and determine when the next irrigation should take place. Determine the normal irrigation frequency and gross depth applied to the field during each irrigation if the irrigation efficiency is 80 % and the required leaching fraction is 0.2.

$$\begin{aligned}&\text{Depletion} = 4(6 \text{ mm/day}) = 24 \text{ mm.}\\&\%\text{depletion} = 24/80 = 30\%\\&\text{Maximum depletion} = 40 \text{ mm or } 50\%.\end{aligned}$$

From this point forward, there is 16 mm available for evapotranspiration. If the ET_c rate is 6 mm/day, irrigation should take place within 2 days in order to avoid exceeding the maximum depletion of 40 mm.

The maximum depletion is 40 mm; thus, water replacement should be 36 mm every 6 days.

$$\begin{aligned}IR &= 100 \text{ RAW}/((IE)(1 - LF))\\&= 100(36 \text{ mm})/(80(1 - 0.2)) = 56 \text{ mm.}\end{aligned}$$

The *Irrigation schedule* worksheet uses an If statement to set the depletion back to zero when MAD is reached (see highlighted cell formula in Fig. 6.11). The required irrigation depth is the sum of previous depletion and present ET_c. The required depth of irrigation is calculated as the net depth required divided by 1 – LF (cell S2) and irrigation efficiency (cell R2) as shown in the equation for cell S33.

Example 6.10 demonstrates how to use the FAO56 linearized single crop coefficient curves. As with the basal crop coefficient curve, there are adjustments for humidity and other conditions, but there are also adjustments for the initial irrigation management practices, which govern the wetness of the soil surface and thus evaporation. These adjustments are accounted for in the Fig. 6.13 worksheet; however, care must be taken to understand the procedures in FAO56. Details on the calculation method are found in FAO56 Chapter 6. The stage lengths are found in FAO56 Table 6.11. The K_c values are found in FAO56 Table 6.12. The early season $K_{c\text{-}ini}$ may require interpolation between FAO56 Figs. 29 and 30, depending on the depths and intervals of early season irrigation. The early season $K_{c\text{-}ini}$ is also a function of reference ET. The midseason crop coefficient is adjusted in the same way as for the dual crop coefficient method since the conditions of evaporation for full canopy cover are the same.

Example 6.10 Calculate the irrigation schedule for corn in Arizona with the FAO56 single crop coefficient. The crop is planted on January 1. Irrigation efficiency is 80 % and LF is zero.

Calculations are made in the *Example* 6.10 worksheet. Stage lengths, reference midseason and endseason crop coefficients (Fig. 6.13), rooting depths, and crop heights (cells T3:X6), are found in Tables 12 and 13 in FAO 56. The initial crop coefficient (cell X24 and V3) is calculated based on the average depth applied during the initial phase (cell X9). The midseason crop coefficient (cell X18) is adjusted based on the average wind speed, crop height, and relative humidity.

The root depth and Kc vs. day of year are shown in Fig. 6.14, and the irrigation schedule is shown in Fig. 6.15.

High Frequency Irrigation Systems

Drip and center pivot systems generally replace the daily ET_c (mm/day). The system should be designed for the worst case, which is peak ET_c. The required gross application rate is the peak ET_c divided by the efficiencies associated with evaporation loss from sprinklers (L_e), water lost due to compensating for lack of application uniformity by overapplying water (L_d), and the required leaching fraction, LF.

$$\begin{aligned}i_g &= \frac{ET_c}{(1 - L_e)(1 - L_d)(1 - LF)}\\&= \frac{i_n}{(1 - L_e)(1 - L_d)(1 - LF)}\end{aligned} \tag{6.19}$$

where

i_g = gross depth of application per day, mm/day
i_n = net depth of application per day equal to ET_c, mm/day
L_e = fraction of water lost to evaporation, dimensionless
L_d = fraction of water lost due to overapplication to compensate for low uniformity, dimensionless

S	T	U	V	W	X
Maize (sweet) in California desert			Plant at end of October (Day 300)		
	Stage lengths and Kc from Tables 11 and 12 in FAO 56				
	stage length	Days after planting	Kc	Root (m)	Crop height
Initial	20	20	0.47	0.40	0.10
Dev	40	60			
Mid	70	130	1.15	0.80	1.50
End	10	140	1.05		

Adjust Kc ini based on the following values with Table 29 and 30	
Average depth applied during initial phase (mm)	20.00
Average number of days between irrigations	10.00
Maximum gross depth applied per day or irrigation (mm)	100.00
Typical midseason minimum relative humidity %	15.00
Typical midseason average daily wind speed (m/s)	3.00
Adjusted midseason crop coefficient	1.28
Average early season reference ET (mm/day)	3.90
Table 29 crop coefficient based on frequency	0.40
Table 30 crop coefficient based on frequency	0.50
Calculated Kc ini based on table 29 and table 30	0.47

Fig. 6.13 Crop coefficient calculation cells for corn with FAO56 method

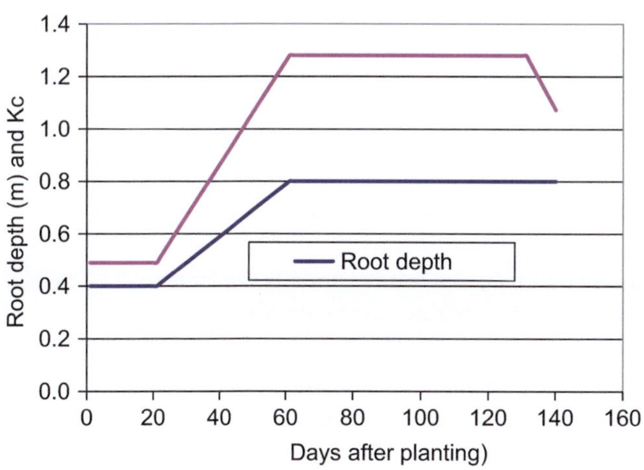

Fig. 6.14 Root depth and Kc for corn with FAO56 method

LF = fraction of water used to leach salts below the root zone, dimensionless.

It is sometimes possible to utilize soil water storage to carry the crop through the peak ET interval, which makes it possible to design for less than the peak ET_c, thus reducing system flow rate and cost. This is the case for deep rooted orchards in the deep loamy soils of eastern Washington, which have enough soil water holding capacity to carry the trees through short intervals (week to 2 weeks) of high temperature.

This strategy can also be used for center pivot irrigation of alfalfa and other hay crops, which have high evaporation before cutting and low evapotranspiration just after cutting. Soil water storage capacity can carry the crop through the

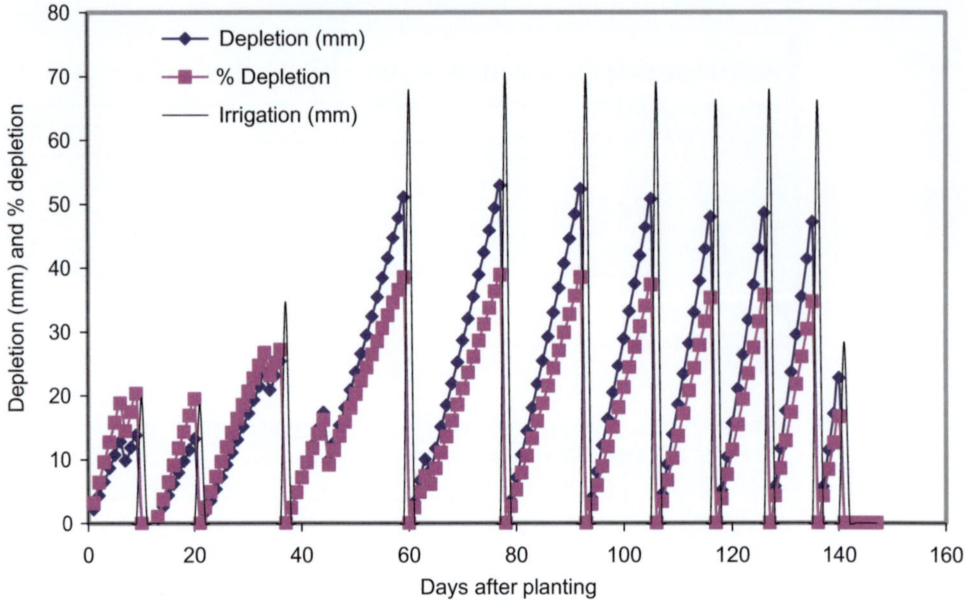

Fig. 6.15 Depletion, percent depletion, and irrigation depth for corn with FAO56 method

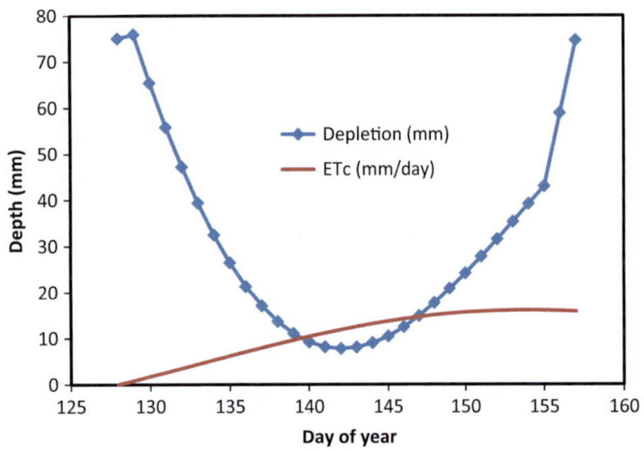

Fig. 6.16 Depletion and ETc for high frequency (daily) pivot irrigation during one alfalfa cutting cycle

high evaporation periods until irrigation refills the soil during low evaporation periods.

Example 6.11 Calculate the required irrigation rate for irrigation of alfalfa during peak ET_0 (midsummer) and check to see that MAD is not exceeded with the *High frequency* worksheet. The ET_c is shown in Fig. 6.12. A center pivot system is turned off for 2 days before and 2 days after cutting in order to dry the field for harvesting processes. The TAW = 150 mm and MAD = 0.5 so RAW = 75 mm. The farmer does not apply extra water during peak ET to leach salts so LF = 0; however, 11 % of applied water is lost to evaporation and application efficiency is 90 %.

The maximum ET_c for alfalfa (Fig. 6.12) occurs on day of year 177, 16 mm/day. The maximum period of evapotranspiration occurs between DOY 153 (June 1) and DOY 179 (June 28), and the total evapotranspiration during this period is 300 mm. The period between cuttings is 27 days long, but the first 2 days and the last 2 days of the cycle are not available for irrigation, so the required 300 mm is applied in 23 days. The depletion does not exceed 75 mm at the end of the cycle (Fig. 6.16) so this scheme is acceptable. The net irrigation rate, $i_{n,}$ is the total ET_c during the cycle divided by the number of days of application: i_n = 300 mm/ 23 days = 13 mm/day. Thus, the required net application rate is 3 mm/day less than the peak ET_c.

Calculate the gross application rate based on the net application rate, evaporation, and efficiency.

$$i_g = \frac{i_n}{(1 - L_e)(1 - L_d)(1 - LF)}$$

$$= \frac{13\ mm/day}{(1 - 0.11)(1 - 0.10)(1 - 0)} = 16.2\ mm/day$$

Remote Sensing of Plant Stress

In a dry climate, the leaf is normally cooler than the air when the plant is transpiring due to evaporative cooling. Transpiration is reduced if the plant is under stress. Thus, leaf temperature is an indication of water stress because the temperature of the leaf increases if transpiration rate decreases. It is not transpiring, then the leaf is warmer than the air because of absorbed solar radiation.

Infrared thermometers (hand-held or equipment-mounted) or thermal scanners (satellite or aircraft) remotely can measure a crop's surface temperature. They calculate temperature by the Stefan-Boltzmann blackbody law:

$$R = M\sigma T_s^4 \tag{6.20}$$

where

R = the radiation emitted by the surface, W m^{-2}
M = the emissivity of the surface, assume 1.0 for plants
σ = the Stefan-Boltzmann constant 5.674×10^{-8} W m^{-2} K^{-4}
T_s = surface temperature, K.

The Crop Water Stress Index (CWSI) quantifies the level of water stress in the field based on the difference between crop temperature and air temperature.

$$\text{CWSI} = (dT - dT_L)/(dT_u - dT_L) \tag{6.21}$$

where

dT = the measured difference between crop canopy and air temperature
dT_u = upper limit of canopy minus air temperature (non-transpiring crop)
dT_L = lower limit of canopy minus air temperature (well-watered crop).

The lower limit of canopy minus air temperature is a vapor pressure deficit (VPD which equals $e_s - e_a$). The upper limit (maximum canopy temperature with no evapotranspiration) is not related to VPD (Fig. 6.17). Slope and intercept values of the lower limit (full evapotranspiration) for various crops are listed in Table 6.3.

$$dT_L = \text{Intercept} + \text{Slope}(\text{VPD}) \tag{6.22}$$

The upper limit is solely a function of air temperatura

$$dT_u = \text{Intercept} + \text{Slope}\ (e_s\{T_a\} - e_s\{T_a + \text{Intercept}\}) \tag{6.23}$$

Example 6.12 Calculate the CWSI for soybeans. Air temperature is 30 °C, canopy temperature is 29 °C, and vapor pressure is 1.06 kPa.

From Table 6.3, Intercept = 1.44 and slope = −1.34

$$e_{s-30} = \exp\left(\frac{16.78(30) - 116.9}{30 + 237.3}\right) = 4.25\ kPa$$

$$\text{VPD} = e_s - e_a = 4.25 - 1.06 = 3.19\ \text{kPa}(\text{RH} = 25\%)$$

$$dT_L = \text{Intercept} + \text{Slope}(\text{VPD}) = 1.44 - 1.34(3.19) = -2.83$$

$$T_a + \text{Intercept} = 30 + 1.44 = 31.44$$

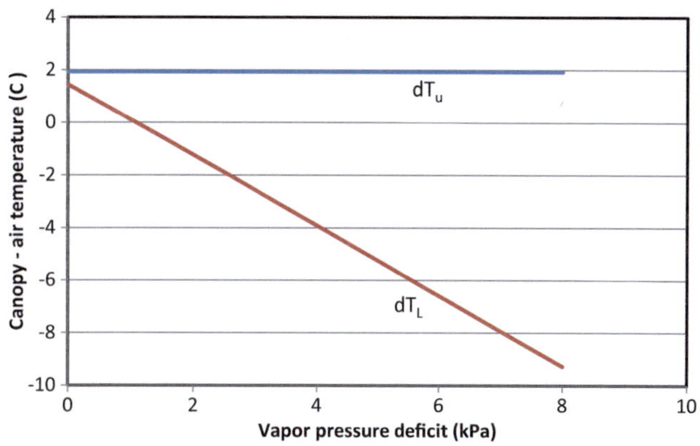

Fig. 6.17 Baseline values for difference between canopy and air temperature for soybeans (Courtesy of Ed Barnes, USDA-ARS Water Conservation Laboratory)

Table 6.3 Baseline parameters for CWSI for various crops – sunlit conditions (From Idso 1982) (Courtesy of Ed Barnes, USDA-ARS Water Conservation Laboratory)

Crop	Intercept	Slope
Alfalfa,	.51	−1.92
Barley (pre-heading)	2.01	−2.25
Barley (post-heading)	1.72	−1.23
Bean	2.91	−2.35
Beet	5.16	−2.30
Corn (no tassels)	3.11	−1.97
Cotton	1.49	−2.09
Potato	1.17	−1.83
Soybean	1.44	−1.34
Tomato	2.86	−1.96
Wheat (post-heading)	2.88	−2.11

$$e_{s-31.44} = \exp\left(\frac{16.78(31.44) - 116.9}{31.44 + 237.3}\right) = 4.61\ kPa$$

$$dT_u = \text{Intercept} + \text{Slope}(e_s\{T_a\} - e_s\{T_a + \text{Intercept}\})$$
$$= 1.44 - 1.34(4.25 - 4.61) = 1.92$$

$$dT = \text{canopy} - \text{air temperature} = 29 - 30 = -1^0\text{C}$$

$$\text{CWSI} = (dT - dT_L)/(dT_u - dT_L)$$
$$= (-1 - (-2.83))/(1.92 - (-2.83)) = 0.38$$

Irrigation is typically triggered by CWSI values above 0.24. Thus, this soybean crop needs water.

The CWSI, as discussed above, applies to full canopy cover. However, fields with incomplete canopy cover have higher temperatures because the soil is hotter than the crop. Percent canopy cover can be determined by another remotely sensed index called the normalized difference vegetation index (NDVI). NDVI radiometers include narrow band red (660 nm) and near infrared-NIR (800 nm) sensors,

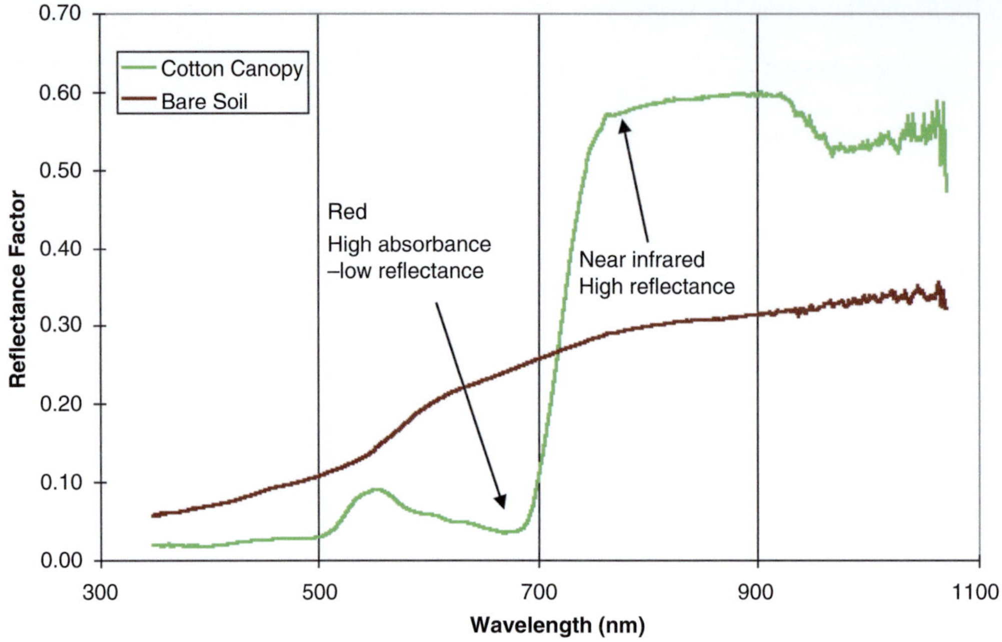

Fig. 6.18 Typical crop canopy and soil reflectance (Courtesy of Ed Barnes. USDA-ARS Water Conservation Laboratory)

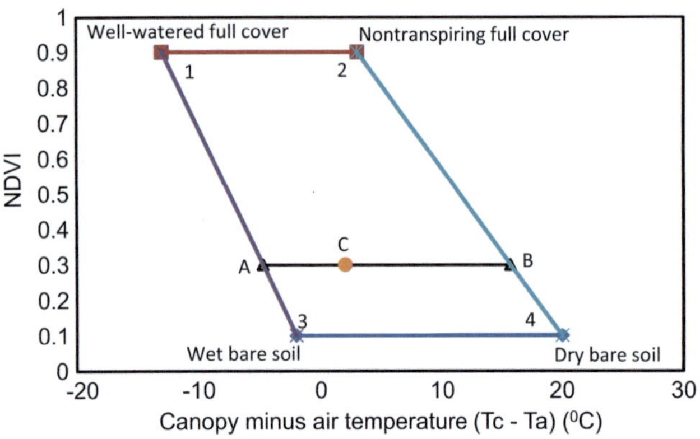

Fig. 6.19 Water deficit index trapezoid for cotton (Courtesy of Ed Barnes. USDA-ARS Water Conservation Laboratory)

and the index compares the ratio of the red to NIR radiation. Plants have low red reflectance (Fig. 6.18) because red light is absorbed by chlorophyll. However, a plant reflects most of the NIR. Conversely, bare soil has very little difference between reflectance of red and near infrared radiation. The ratio of red to NIR reflectance is "normalized" from 0 to 1 with the NDVI index. Theoretically, an NDVI near 1 represents full canopy cover and an NDVI near 0 is soil.

$$NDVI = \frac{\rho_{NIR} - \rho_{RED}}{\rho_{NIR} + \rho_{RED}} \qquad (6.24)$$

where

ρ_{NIR} = NIR reflectance, dimensionless (reflected NIR radiation/NIR insolation)

ρ_{RED} = Red reflectance, dimensionless (reflected red radiation/red insolation).

The water deficit index (WDI) utilizes the NDVI and CWSI within a two-dimensional trapezoid (Fig. 6.19). Points 1 and 2 on the trapezoid are the points that correspond to the upper and lower limits for the CWSI for full canopy cover. The corresponding NDVI for full canopy cover is typically 0.9. Points 3 and 4 are the temperatures of a wet and dry bare soil surface, respectively, and are measured in the field. The NDVI for bare soil is typically 0.1. The WDI is calculated as follows with dT_{L13} taken from line 1–3 at the given vegetation index, and dT_{L24} taken from line 2–4 (Fig. 6.19).

$$WDI = (dT - dT_{L13})/(dT_{L24} - dT_{L13}) \qquad (6.25)$$

The *WDI* worksheet includes a calculation for WDI (Fig. 6.19). The line at NDVI = 0.3 represents cotton in Arizona in June before full canopy development. The lower line represents bare soil at NDVI = 0.1. Bare dry soil at midday is typically 20 °C hotter than atmospheric temperature and bare wet soil may be 2 °C cooler than atmospheric temperature. The upper line represents the full canopy at NDVI = 0.9. Based on the theoretical CWSI, the canopy temperature is 13 °C cooler than air temperature with full transpiration and 3 °C warmer than air temperature with no transpiration. The endpoints of the NDVI = 0.3 line are found by interpolation from the upper and lower corners of the WDI trapezoid. These parameters were given by Doug Hunsaker at the USDA-ARS Arid Land Agricultural Research Center. In this case, the WDI is 0.33, which is 1/3 the distance from left to right on the NDVI = 0.3 line.

Research indicates that a single daily measurement of CWSI or WDI is not an effective method to schedule irrigation. By the time the crop shows stress, it should have already been irrigated; however, other similar methods that take more measurements per day or use other creative techniques appear to be promising.

Questions

1. Reference ET_0 is 10 mm/day for 1 week, and the crop coefficient is 0.5. What is the depth of water required during 1 week?
2. Reference ET_0 is 7 mm/day, pan evaporation is 8 mm/day, and measured crop evapotranspiration is 5 mm, as measured by a lysimeter. What are the pan and reference ET_0 crop coefficients?
3. Explain the difference between the single crop coefficient and the dual crop coefficient.
4. Explain the spikes in transpiration rate in Fig. 6.4 and the average crop coefficient in Fig. 6.5.
5. Find the Kcb values for winter wheat (dual crop coefficient) in FAO56 Table 17, and adjust $K_{cb\text{-}mid}$ for average minimum relative humidity during mid-season and late season growth stages equal to 20 % and 30 %, respectively. Average wind speed at 2 m elevation is 2 m/sec during mid and late season growth stages. There is less than 10 % ground cover during the initial phase. There is high grain moisture at harvest. The crop is grown in California. Plot the linearized crop coefficient curve for the dual component model.
6. Repeat question 5, but plot the single crop coefficient curve (FAO Table 12) for winter wheat. Average depth of irrigation during the initial phase is 50 mm. Average number of days between irrigations during the initial period is 20 days. Minimum relative humidity is 20 % and average wind speed is 2 m/sec. The crop is machine harvested (late season K_c). During the early season, the crop is irrigated every 10 days, irrigation depths are greater than 40 mm, and the ET_0 is 7 mm/day. Also plug values into the Example 6.10 worksheet in order to solve the problem.
7. Calculate the single day crop basal transpiration for winter wheat 70 days after planting. Reference ET_0 is 7 mm/day and the crop stress coefficient is 0.8. Relative humidity is 20 %, and wind speed is 3 m/sec at 3 m elevation.
8. Plot the winter wheat root growth curve as defined in FAO56. Initial depth is 0.4 m and the final depth is 0.8 m.
9. What is the reason that soil evaporation + basal transpiration cannot exceed $K_{c\text{-}max}$?
10. Calculate K_{c-max} if relative humidity is 20 %, crop height is 0.5 m, and wind speed is 2 m/sec.
11. Explain the difference between REW and TEW.
12. Field capacity is 20 % and permanent wilting point is 10 %. The depth of the surface layer is 0.11 m. REW = 7 mm. Calculate TEW. Calculate K_r for the surface layer depletion equal to 4 mm.
13. Alfalfa has a low ET just after cutting and high ET just before cutting. For arid conditions with moderate wind, calculate alfalfa evapotranspiration just before and just after cutting if reference ET_0 is 10 mm/day. Use FAO 56 Table 12.
14. During the first week after planting watermelons, t_{max} and t_{min} are 32 °C and 8 °C, respectively. Calculate the number of growing degree days accumulated after 1 week.
15. Calculate heat unit Fourier series K_c and ET_c curves for watermelon and corn for (Instructor selects year and location) weather data. You can download this data with the *Chapter 5 ET Calculator – Active Year Weather* worksheet from a city in Arizona or use weather from your home state. Click the *Run Weather Form* button. Calculate the Fourier W coefficients for the Tmax, Tmin and ETo curves in the *Chapter 5 ET Calculator – Fourier T and ETo* worksheet. Then, insert your W coefficients for Tmax, Tmin and ETo into the *Chapter 6 Crop ET and scheduling – Fourier T and ETo* worksheet. Then, copy your Tmax, Tmin and ETo values into the *Chapter 6 Crop ET and scheduling – ETo and temp – ch 5* worksheet. Soil is heavy-textured.
16. As in Example 6.6, calculate corn K_c and ET_c 50 days after planting by hand (with a calculator) and show that calculated K_c and ET_c in the spreadsheet agree with the equations.
17. According to the Fourier series GDD equations, calculate the rooting depth for melons when GDD = 500 and 1,000. AWC is 10 %, MAD is 0.5, and ET_c is 7 mm/day at the first rooting depth and 9 mm/day at the second rooting depth. What is the required frequency of irrigation at the two rooting depths.
18. For the 2008 Fourier series Tucson weather data, and the Fourier series crop coefficient for alfalfa, input the cutting cycle with peak ET into the High frequency alfalfa Irr. Worksheet. Then determine the required depth of irrigation per day with a center pivot irrigation system in order to keep up with crop water needs. Check to make sure that the percent depletion does not exceed 50 %. Use TAW = 140 mm. Assume a leaching fraction of 10 % and irrigation efficiency of 90 % in order to calculate the required gross application rate (divide net application rate in Worksheet by efficiency and (1 – LF)).

19. Calculate the CWSI for potatoes if the air temperature is 25 °C, canopy temperature is 24 °C, and actual vapor pressure is 1.32 kPa. Determine whether irrigation is needed.

References

Allen RL, Pereira DR, Smith M (1998) Crop evapotranspiration guidelines for computing crop water requirements – FAO Irrigation and drainage paper 56. FAO – Food and Agriculture Organization of the U. N http://www.fao.org/docrep/X0490E/x0490e00.htm

Idso SB (1982) Non-water-stressed baselines: a key to measuring and interpreting plant water stress. Agric Meteorol 27:59–70

Martin E, Slack D (2001) Arizona Irrigation Scheduling System (AZSCHED). The University of Arizona. http://cals.arizona.edu/cgi-bin/formhandler/formhandler.cgi

Snyder RL (1985) Hand calculating degree days. Agric For Meteorol 35:353–358

Irrigation Lateral Design

7

Water is conveyed to drip emitters and sprinklers through multioutlet pipelines called laterals. This chapter begins with a basic introduction to hydraulics and statistics. These introductory materials may be unnecessary for many students. Then these principles are used to simulate the distribution of water applied by an irrigation lateral. Hydraulics is used to determine the change in pressure and resultant application variability along the lateral. Normally, the design objective is high application uniformity, which is accomplished by keeping pressure variation along the lateral within an acceptable range. Using larger pipe reduces pressure loss due to friction, but it also increases capital cost. Irrigation system application uniformity is not only a function of hydraulics. Other causes of nonuniformity include emitter manufacturing variation and application rate nonuniformity. The manufacturer's coefficient of variation or expected field application variability is used to add a statistical distribution to the hydraulic application rate distribution along the lateral. Monte Carlo analysis utilizes the random number generator in Excel to evaluate the effect of the coefficient of variation on application distribution and profit.

Hydraulics

Pressure is a form of energy. Pressure is the amount of force per unit area exerted on the walls of a vessel or pipe. Other forms of energy that may be important in pressurized irrigation systems design are elevation and kinetic energy.

In-class Exercise 7.1 A one cubic foot container of water contains 62.4 pounds of water.

What is the pressure at the bottom of the container in pounds per square foot?

What is the pressure at the bottom of the container in pounds per square inch (PSI)?

What is the pressure in the container at a depth 0.5 ft below the top of the container (PSI)?

The weight of water is equal to volume * density * gravity. The density of water is 1,000 kg/m^3. Gravity is 9.81 m/sec^2. Weight is the same as force and can be measured in units of Newtons (N). In metric units, 10 m head is approximately equal to one atmosphere and to 100 kPa (10^5 Pa).

$$P = z\rho g = 10.3\ \text{m}\left(1{,}000\ \text{kg/m}^3\right)\left(9.80\ \text{m/sec}^2\right) = 1.013*10^5\text{Pa} = 101.3\ \text{kPa} = 1\ \text{atm} \tag{7.1}$$

where

z = depth of water, m,
ρ = density, kg/m^3,
g = acceleration due to gravity, 9.8 m/sec^2,
P = pressure, N/m^2.

Pressure in metric units is measured in Pascals (Pa), which is equal to N/m^2.

In-class Exercise 7.2 A cubic meter of water contains 1,000 kg of water.

What is the mass of the water in the container (kg)?
What is the weight of the water in the container (N)?
What is the pressure at the bottom of the container in Pa?

Water resources engineers often refer to the energy of water as head and use units of length. For example, atmospheric pressure can be expressed as 34 ft or 14.7 PSI or 1 atm. The relationship between pressure and length in United States units is 2.31 ft = 1 PSI. The relationship between metric and United States pressure units is 6.9 kPa = 1 PSI (kPa is thousands of Pa).

In-class Exercise 7.3 Calculate the pressure in units of feet at the bottom of the container from Exercise 7.1. Use pressure calculated in Exercise 7.1 and convert to ft with the unit conversion 2.31 ft = 1 PSI.

P. Waller, M. Yitayew, *Irrigation and Drainage Engineering*, DOI 10.1007/978-3-319-05699-9_7

Calculate the pressure in units of feet 0.5 ft from the top of the container. Use pressure calculated in Exercise 7.1 and convert to ft with the unit conversion 2.31 ft = 1 PSI.

Calculate the pressure in units of m at the bottom of the container from Exercise 7.2. Use pressure calculated in Exercise 7.2 and convert to m with the unit conversion 10.3 m = 101 kPa.

In-class Exercise 7.4 Calculate the gauge pressure and absolute pressure (PSI and ft of head) at the bottom of the swimming pool (at sea level) that is 9 ft deep.

There are two types of pressure in pipe systems: static pressure and dynamic pressure. Static pressure is the pressure in a pipe system when water is not flowing. Calculating static pressure is like calculating the pressure at the bottom of a swimming pool; it is just based on the depth below the surface.

In-class Exercise 7.5 Write down the static pressure in PSI and ft of head at each of the fittings in Fig. 7.1.

	PSI	FT
A		
B		
C		
D		
E		
F		

Dynamic pressure is the pressure that is measured when water is flowing in the pipe. It is different from static pressure because energy is lost due to turbulence in the pipe. Water molecules sliding past each other in the turbulent flow and past the pipe walls create friction, which gives off heat energy. The energy is then lost through the walls of the pipe. Because pressure is the form of energy in the pipe, pressure drops due to pipe friction.

Energy loss is primarily a function of pipe diameter and flow rate. Equations and tables have been developed to calculate the amount of energy loss in a pipe due to friction as a function of pipe diameter and flow rate. Friction loss expressed as feet of pressure loss per hundred feet of pipe are tabulated in Tables 7.1, 7.2, and 7.3 for Class 160, Class 200, and Schedule 40 pipe, respectively. The friction loss at corresponding positions in the tables are different because the wall thickness and inside diameters vary with different classes of pipe.

Example 7.1 Calculate the pressure loss in 200 ft of 1½ inch Class 160 pipe at a flow rate of 25 GPM.

The friction loss is 2.442 ft/100 ft.

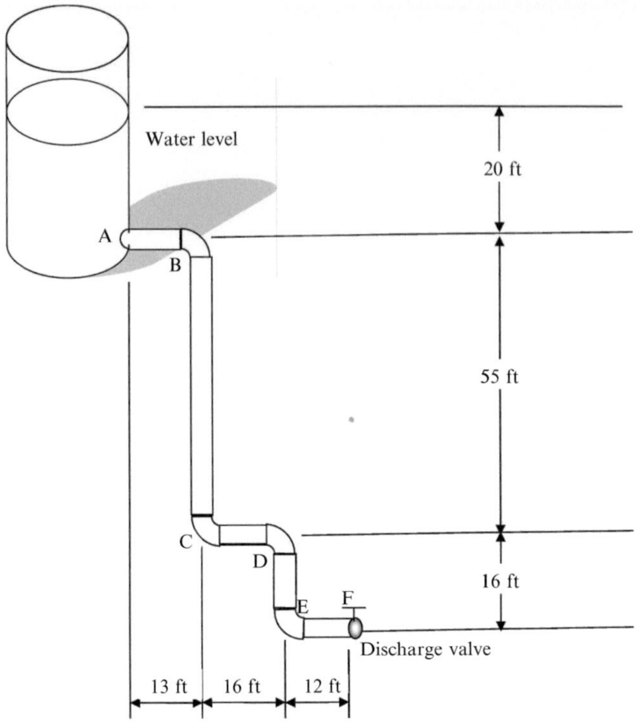

Fig. 7.1 Pipe layout for examples and in-class exercises

Because the pipe is 200 ft long, the total friction loss in the entire length of pipe is

$$(2.442 \text{ ft}/100 \text{ ft})(200 \text{ ft}) = (2.442 \text{ ft})(2) = 4.884 \text{ ft}.$$

Example 7.2 Calculate the pressure loss in 50 ft of 1½ inch Schedule 40 pipe at a flow rate of 25 GPM.

The friction loss is 3.70 ft/100 ft. Because the pipe is 50 ft long, the total friction loss in the entire length of pipe is

$$3.706 \text{ ft}/100 \text{ ft}*50 \text{ ft} = 3.706 \text{ ft}*0.5 = 1.85 \text{ ft}.$$

Example 7.3 Calculate the water velocity in the pipe in Example 7.2 if the ID is 35.1 mm.

$$25 \text{ GPM} = 1.58 \text{ L/sec} = 0.00158 \text{ m}^3/\text{sec}\ \text{v} = \text{Q}/\text{A}$$
$$= 0.00158/\left(\pi(0.0351)^2/4\right) = 1.63 \text{ m/sec}$$

In addition to friction loss, a criterion for pipe diameter selection is that water velocity in the pipe cannot be greater than 5 ft/sec (1.5 m/sec). The primary reason for this is that higher velocities lead to unacceptable pressure fluctuations when valves are closed and cause water hammer. An indication that water velocity would be greater than 5 ft/sec in the

Table 7.1 Friction loss in PVC Class 160 pipe (ft/100 ft). C = 150. 73.4 °C

Flow rate	Nominal pipe diameter (inches)						
GPM	1	1 1/4	1 1/2	2	2 1/2	3	4
1	0.0408	0.0122	0.0063	0.0021	0.0008	0.0003	0.0001
2	0.1473	0.0439	0.0227	0.0077	0.003	0.0012	0.0003
3	0.312	0.0931	0.0481	0.0162	0.0064	0.0025	0.0007
4	0.5316	0.1585	0.082	0.0276	0.0109	0.0042	0.0012
5	0.8036	0.2397	0.124	0.0418	0.0165	0.0063	0.0019
6	1.1264	0.3359	0.1738	0.0585	0.0231	0.0089	0.0026
7	1.4985	0.4469	0.2312	0.0779	0.0307	0.0118	0.0035
8	1.9189	0.5722	0.296	0.0997	0.0393	0.0151	0.0044
9	2.3866	0.7117	0.3682	0.1241	0.0489	0.0188	0.0055
10	2.9008	0.8651	0.4475	0.1508	0.0594	0.0229	0.0067
11	3.4607	1.0321	0.5339	0.1799	0.0709	0.0273	0.008
12	4.0658	1.2125	0.6273	0.2113	0.0833	0.0321	0.0094
13	4.7154	1.4062	0.7275	0.2451	0.0966	0.0372	0.0109
14	5.4091	1.6131	0.8345	0.2812	0.1108	0.0427	0.0125
15	6.1463	1.8329	0.9482	0.3195	0.1259	0.0485	0.0142
16	6.9265	2.0656	1.0686	0.36	0.1419	0.0546	0.016
17	7.7495	2.311	1.1956	0.4028	0.1588	0.0611	0.0179
18		2.5691	1.329	0.4478	0.1765	0.0679	0.0199
19		2.8396	1.469	0.4949	0.1951	0.0751	0.022
20		3.1226	1.6154	0.5443	0.2145	0.0826	0.0242
21		3.4178	1.7681	0.5957	0.2348	0.0904	0.0265
22		3.7253	1.9272	0.6493	0.2559	0.0985	0.0289
23		4.045	2.0926	0.705	0.2779	0.107	0.0314
24		4.3767	2.2642	0.7628	0.3007	0.1157	0.034
25		4.7203	2.442	0.8227	0.3243	0.1248	0.0367

Table 7.2 Friction loss in PVC Class 200 pipe (ft/100 ft). C = 150. 73.4°C

Flow rate	Nominal pipe diameter (inches)						
GPM	1	1 1/4	1 1/2	2	2 1/2	3	4
1	0.0403	0.0134	0.0069	0.0023	0.0009	0.0004	0.0001
2	0.1455	0.0484	0.025	0.0084	0.0033	0.0013	0.0004
3	0.3083	0.1025	0.053	0.0179	0.0071	0.0027	0.0008
4	0.5252	0.1746	0.0902	0.0305	0.012	0.0046	0.0014
5	0.7939	0.2639	0.1364	0.0461	0.0182	0.007	0.0021
6	1.1127	0.3699	0.1912	0.0646	0.0255	0.0098	0.0029
7	1.4803	0.4921	0.2543	0.086	0.0339	0.013	0.0038
8	1.8956	0.6301	0.3257	0.1101	0.0434	0.0167	0.0049
9	2.3576	0.7837	0.405	0.1369	0.054	0.0207	0.0061
10	2.8656	0.9525	0.4923	0.1664	0.0657	0.0252	0.0074
11	3.4187	1.1364	0.5873	0.1985	0.0784	0.0301	0.0088
12	4.0165	1.3351	0.69	0.2333	0.0921	0.0353	0.0104
13	4.6582	1.5484	0.8002	0.2705	0.1068	0.041	0.012
14	5.3434	1.7762	0.918	0.3103	0.1225	0.047	0.0138
15	6.0717	2.0182	1.0431	0.3526	0.1392	0.0534	0.0157
16	6.8425	2.2745	1.1755	0.3974	0.1568	0.0602	0.0177
17	7.6554	2.5447	1.3151	0.4446	0.1755	0.0674	0.0198
18		2.8288	1.462	0.4942	0.1951	0.0749	0.022
19		3.1267	1.6159	0.5463	0.2156	0.0828	0.0243
20		3.4383	1.777	0.6007	0.2371	0.091	0.0267
21		3.7634	1.945	0.6575	0.2595	0.0996	0.0292
22		4.102	2.12	0.7167	0.2828	0.1086	0.0319
23		4.4539	2.3019	0.7782	0.3071	0.1179	0.0346
24		4.8192	2.4906	0.842	0.3323	0.1276	0.0374
25		5.1976	2.6862	0.9081	0.3584	0.1376	0.0404

Table 7.3 Friction loss in PVC Schedule 40 pipe (ft/100 ft). C = 150. 73.4 °C

Flow rate	Nominal pipe diameter (inches)						
GPM	1	1 1/4	1 1/2	2	2 1/2	3	4
1	0.077	0.0202	0.0096	0.0028	0.0012	0.0004	0.0001
2	0.2778	0.0731	0.0345	0.0102	0.0043	0.0015	0.0004
3	0.5886	0.1548	0.0731	0.0216	0.0091	0.0032	0.0008
4	1.0028	0.2637	0.1245	0.0369	0.0155	0.0054	0.0014
5	1.5159	0.3987	0.1882	0.0557	0.0234	0.0081	0.0022
6	2.1248	0.5588	0.2637	0.0781	0.0329	0.0114	0.003
7	2.8267	0.7434	0.3509	0.1039	0.0437	0.0152	0.004
8	3.6198	0.9519	0.4493	0.1331	0.056	0.0194	0.0052
9	4.502	1.1839	0.5588	0.1655	0.0696	0.0242	0.0064
10	5.472	1.439	0.6792	0.2011	0.0846	0.0294	0.0078
11	6.5282	1.7168	0.8103	0.24	0.101	0.0351	0.0093
12	7.6696	2.017	0.952	0.2819	0.1186	0.0412	0.011
13	8.895	2.3392	1.1041	0.327	0.1376	0.0478	0.0127
14		2.6833	1.2665	0.3751	0.1578	0.0548	0.0146
15		3.049	1.4391	0.4262	0.1793	0.0623	0.0166
16		3.4361	1.6218	0.4803	0.2021	0.0702	0.0187
17		3.8443	1.8145	0.5374	0.2261	0.0785	0.0209
18		4.2736	2.0171	0.5973	0.2514	0.0873	0.0232
19		4.7236	2.2296	0.6603	0.2779	0.0965	0.0257
20		5.1943	2.4517	0.726	0.3055	0.1061	0.0282
21		5.6855	2.6836	0.7947	0.3344	0.1161	0.0309
22		6.197	2.925	0.8662	0.3645	0.1266	0.0337
23		6.7287	3.176	0.9405	0.3958	0.1374	0.0366
24			3.4364	1.0176	0.4283	0.1487	0.0396
25			3.7062	1.0976	0.4619	0.1604	0.0427

1¼ inch pipe at 25 GPM is that there is no listed friction loss at this flow rate in Table 7.3. In general, PVC friction loss tables do not list friction loss values at flow velocities that are in excess of 5 ft/sec.

Dynamic pressure can be calculated from one point to the next in a pipe network as follows:

$$\text{Pressure}_2 = \text{Pressure}_1 + \text{elevation gain} - \text{friction loss}.$$

Example 7.4 Calculate the pressure at points, A, B and C in Fig. 7.1 if the flow rate in the pipe is 25 GPM and the pipe is 1½ inch class 160 pipe.

From Example 7.1, the friction loss in the pipe is 2.44 ft/100 ft.

The pressure at A is 20 ft since there is no pressure loss in the pipe.

The length of pipe from A to B is 13 ft. Thus, the friction loss in that pipe section is

$$(13 \text{ ft})(2.44 \text{ ft}/100 \text{ ft}) = 0.32 \text{ ft}.$$

There is no elevation gain between A and B; thus, the pressure at B is calculated as follows:

$$\text{Pressure}_\text{B} = \text{Pressure}_\text{A} + \text{elevation gain} - \text{friction loss}.$$
$$\text{Pressure}_\text{B} = 20 + 0 - 0.32 = 19.68 \text{ ft}$$

The length of pipe between B and C is 55 ft. Thus, the friction loss between B and C is

$$(55 \text{ ft})(2.44 \text{ ft}/100 \text{ ft}) = 1.34 \text{ ft}.$$

The elevation gain is 55 ft; thus, the pressure at C is calculated as follows:

$$\text{Pressure}_\text{C} = \text{Pressure}_\text{B} + \text{elevation gain} - \text{friction loss}.$$
$$\text{Pressure}_\text{C} = 19.68 + 55 - 1.34 = 73.34 \text{ ft}$$

In-class Exercise 7.6 If the pipe in Fig. 7.1 is 2 inch Class 160 pipe and the flow rate is 22 GPM, calculate the pressure at all locations in the pipe. Report the pressure in ft.

A
B
C
D
E
F

In-class Exercise 7.7 What is the total friction loss in the pipe system in Fig. 7.1 if the flow rate is 10 GPM and the pipe is 1 inch Class 200?

What is the pressure at the discharge valve?

Statistics

Monte Carlo analysis can be used to analyze spatially varying irrigation. Statistical distributions are generated using the expected coefficients of variation for irrigation emitters and/or soil properties. Variation in water application due to pressure variation along laterals or ponding time variation in furrows can be incorporated into the generation of statistical distributions. This variation is then used to calculate an expected salinity and yield distribution and is used to find the optimal depth of seasonal applied water. Spatial variation of water application has been evaluated in many research studies. For example, Heermann et al. (1992) found that a normal distribution was a good representation of center pivot application rates. A normal random variable having mean μ and a variance σ^2 can be described by the following normal probability density function where f(x) is the frequency of occurrences at any x:

$$f(x) = \frac{1}{\sqrt{2\pi\sigma}}\exp\left[-\frac{1}{2}\left(\frac{x-\mu}{\sigma}\right)^2\right] \quad for \quad -\infty < x < \infty \tag{7.2}$$

where

x = random variable,
σ = standard deviation,
μ = mean.

The normal distribution probability density function (pdf) is bell-shaped and symmetric with respect to the mean, μ, (Fig. 7.2). This is not necessarily the correct distribution of application depths in an irrigation scenario, but it is the assumed distribution in this chapter. The standardized Z distribution (Fig. 7.2) is the normal distribution with mean zero and unit (1) variance (σ^2): the area under the standardized normal Z distribution curve is one. The probability that a value falls within a range is the area under the curve within that range. For example, the area between −1 and 1 standard deviation in Fig. 7.2 is 0.68: thus, there is a 68 % chance that the Z random variable falls between −1 and 1.

Excel can be used to generate a Z distribution. First, the random number generator (+RAND) generates a uniform distribution between 0 and 1: randomly generated numbers that are evenly distributed between 0 and 1. Second, the function NORMSINV() converts the 0–1 uniform distribution generated by RAND to a standardized normal Z distribution. The combined function can be entered as follows: NORMSINV(+RAND()). Filling a column with this function generates a normally distributed set of data with mean 0 and standard deviation equal to 1.

Any normally distributed random variable, X, can be converted to the Z distribution by subtracting the mean of that distribution and dividing by the standard deviation.

$$\frac{(X-\mu)}{\sigma} = Z \tag{7.3}$$

Conversely, if mean and standard deviation are known, then an expected distribution can be generated by rearranging Eq. 7.3. If depth of infiltration, i, replaces X in Eq. 7.3, then the Z values can be converted to an infiltration distribution as follows:

$$i = \bar{i} + Z\sigma \tag{7.4}$$

where

i = water application depth in any section of the field, mm,
$\bar{i}$ = average water application depth in the field, mm,
Z = randomly generated Z values, dimensionless.

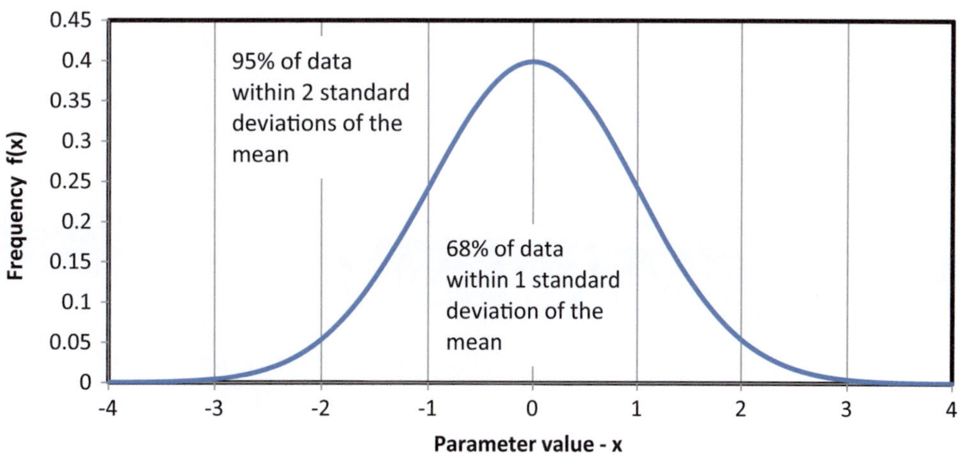

Fig. 7.2 Standardized normal Z distribution

Many products used in engineering designs are characterized by the factory with the manufacturer's coefficient of variation. The coefficient of variation of a product is a function of the level of precision of the manufacturing process. Precision refers to the repeatability of a manufacturing process or of an instrument's ability to measure a parameter. The coefficient of variation along with the physics of a process can be used to simulate the performance of the product in the field or in an industrial application. The coefficient of variation (CV) or standard deviation can be used to simulate the application variability in the field. The coefficient of variation is the standard deviation divided by the mean. The standard deviation (σ) of irrigation application depths over the field is equal to the product of the mean and coefficient of variation.

$$\begin{aligned} CV &= \sigma/\mu \\ \sigma &= \mathrm{CV}^{*}\bar{i} \end{aligned} \tag{7.5}$$

Substitute Eq. 7.5 into Eq. 7.4 to calculate irrigation depths as a function of CV and Z.

$$i = \bar{i} + Z^{*}CV^{*}\bar{i} \tag{7.6}$$

The CV is a convenient statistical parameter to represent irrigation uniformity because it gives the relative difference in application across the field. Heermann et al. (1992) found that coefficients of variation (CV) for center pivot applications were 0.2, 0.24, and 0.30 for wind speeds <2.2 m/sec, 2.2–4.5 m/sec, and >4.5 m/sec, respectively. The expected distribution for an average application depth of 2 cm and CV = 0.25 (standard deviation = 0.5) is shown in Fig. 7.3.

A statistical distribution can be generated with the random number generator in Excel. A uniform (0, 1) distribution is generated by filling a column with the RAND() function (Fig. 7.4). The uniform (0, 1) distribution is a uniform set of numbers generated between the limits of 0 and 1.

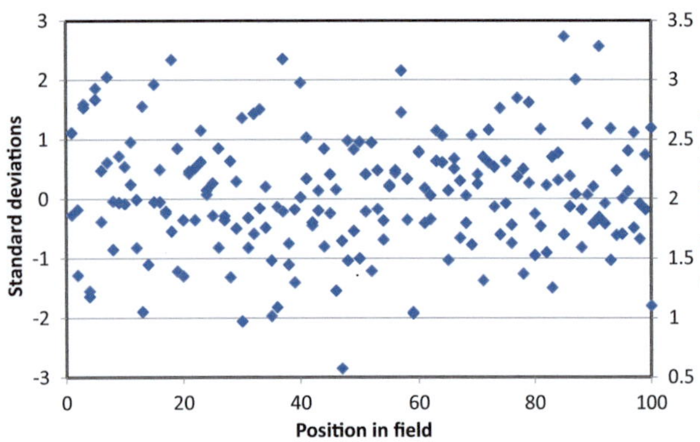

Fig. 7.3 Distribution of application depths with average application depth equal to 2 cm and CV = 0.25

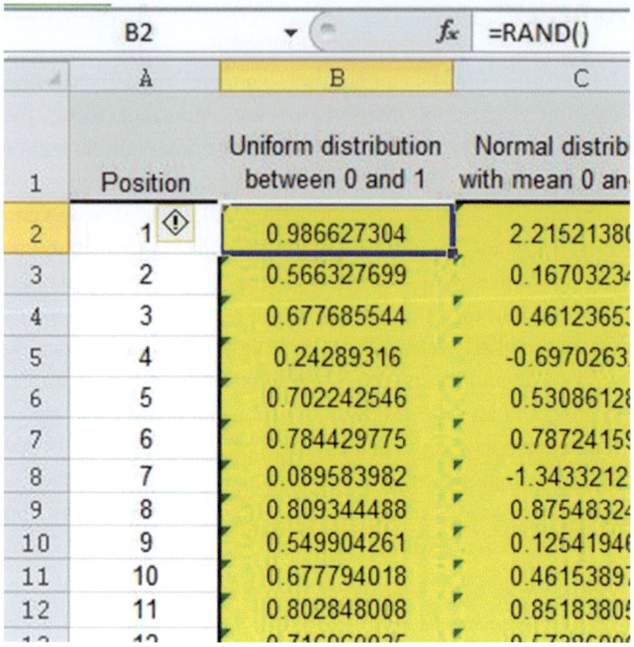

B2		=RAND()
	A	B	C

	A	B	C
1	Position	Uniform distribution between 0 and 1	Normal distrib with mean 0 an
2	1	0.986627304	2.2152138
3	2	0.566327699	0.1670323
4	3	0.677685544	0.4612365
5	4	0.24289316	-0.6970263
6	5	0.702242546	0.5308612
7	6	0.784429775	0.7872415
8	7	0.089583982	-1.3433212
9	8	0.809344488	0.8754832
10	9	0.549904261	0.1254194
11	10	0.677794018	0.4615389
12	11	0.802848008	0.8518380

Fig. 7.4 RAND() function in an excel worksheet

The uniform (0, 1) distribution is converted to a normal distribution (0, 1) with the NORMSINV() function as shown in Fig. 7.5. The normal (0, 1) distribution has mean 0 and standard deviation 1.

Alternatively, the RAND() function can be entered directly into the NORMSINV() function so that column B is not needed: NORMSINV(RAND()). A normal distribution with any mean and standard deviation can be generated with the NORMINV function. In Fig. 7.6, the mean and standard deviation are from named cells G2:G3 (Fig. 7.11). Cells are named by placing the cursor in the cell and then clicking the Name Box in the upper left corner of the worksheet. The following equation is entered in column D:

$$= \mathrm{NORMINV(RAND(),\ mean,\ sd)} \tag{7.7}$$

The standard deviation is equal to the product of the mean and CV. Thus, Eq. 7.7 can be modified to incorporate the coefficient of variation (Fig. 7.7) where CVRAND is the name of cell G4.

$$= \mathrm{NORMINV(Rand(),\ mean,\ mean^{*}CVRAND)} \tag{7.8}$$

Optimization of Irrigation Design and Management

Ideally, all portions of a field receive the optimal depth of water such that profit (yield – water cost) is maximized, environmental contamination is minimized (runoff and

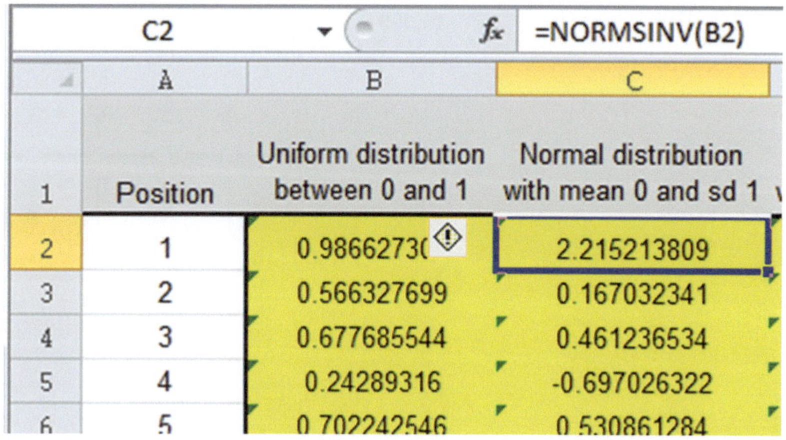

Fig. 7.5 Converting uniform (0, 1) distribution to normal (0, 1) distribution with NORMSINV()

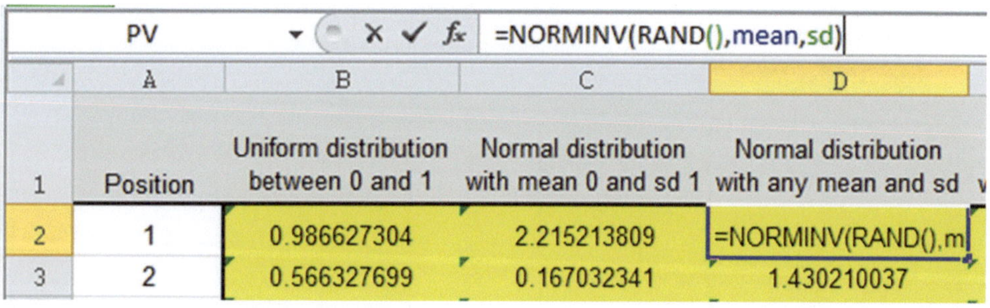

Fig. 7.6 NORMINV function to calculate normal distribution with a mean and standard deviation

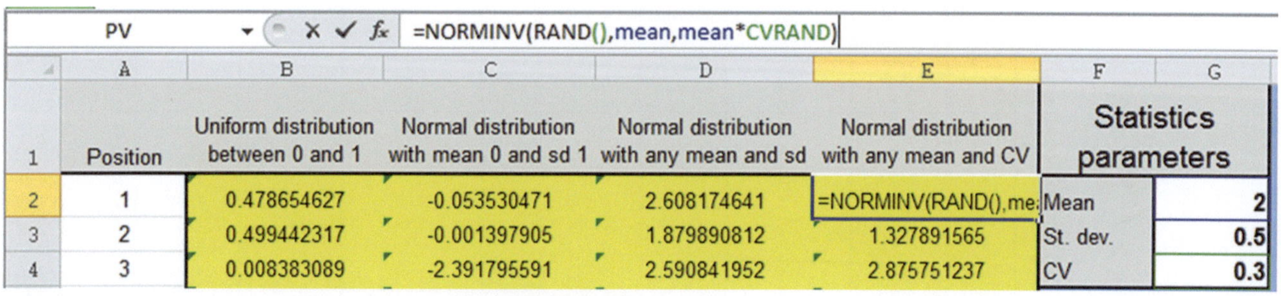

Fig. 7.7 NORMINV function with CV in *various examples* worksheet

chemical leaching), and salinity is maintained below the threshold value in all parts of the field.

Maximizing profit and minimizing environmental contamination may be competing objectives in fields with low uniformity. If uniformity is low, then excess water is often applied to the field in order to ensure that all sections receive the required depth of water; however, this practice increases water waste, and possibly causes aquifer and surface water contamination by agrochemicals.

Rehabilitating an old system or putting in a new irrigation system can increase uniformity, decrease water waste, and increase yield; however, new equipment requires capital investment. In order to determine whether a new system is a good investment, the benefit from increased yield and reduced environmental contamination and water use can be compared to the cost of rehabilitating or replacing the irrigation system. For a realistic assessment, the analysis should include spatial and temporal variation of water application and soil water content.

Simulation of Hydraulic and Statistical Variation

A typical sprinkler lateral (Fig. 7.8) has evenly spaced sprinklers. Flow rate in each pipe section is the sum of all downstream sprinklers. For example, if a sprinkler lateral had 15 sprinklers with 1,000 L/hr flow rate, then the flow rate into the lateral would be 15,000 L/hr. The flow rate in the second to last pipe section would be 2,000 L/hr (Fig. 7.8).

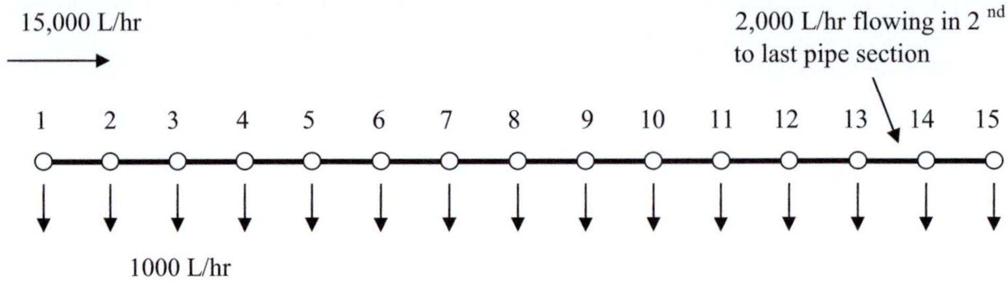

Fig. 7.8 Sprinkler lateral schematic

F7 =3163*(dx)*(E7/150)^1.852/D7^4.87

	A	B	C	D	E	F	G	H	I	J	K
1	K	200		CV	0.05	CV Opt	Profit	$ 121.60		Price ($/kg)	$ 0.20
2	x	0.58		Minimum depth in field	70	Depth Opt					Price_Opt
3	End pressure	16.2	m	Energy cost	$ 57.77	per ha-yr	CWPF		a x^0	b x^0.5	c x^1
4	Spacing bet. Sprinklers	15	m	Water cost	$ 24.65	per ha-yr	Coefficients		-3954	1067	-54.14
5				Pipe cost	$ 47.78	per ha-yr		Average (cm)	82.18	Average	$ 251.80
6	Sprinkler number	Press. (m)	Sp. Flow (LPH)	Diameter (mm)	Pipe flow (LPH)	Pressure loss (m)	Rand	Relative flow rate	Depth (cm)	Yield (kg/ha)	Value ($/ha)
7	30	16.20	1006	25	1006	0.251	1098	1.09	76.4	1236	$ 247.22
8	29	16.45	1015	25	2021	1.036	1047	1.04	72.9	1209	$ 241.80
9	28	17.49	1051	50	3072	0.077	1125	1.12	78.3	1249	$ 249.71
10	27	17.56	1054	50	4126	0.133	1064	1.06	74.1	1219	$ 243.78
11	26	17.70	1059	50	5185	0.203	957	0.95	66.6	1148	$ 229.66
12	25	17.90	1066	50	6251	0.287	1060	1.05	73.8	1217	$ 243.33
13	24	18.19	1076	50	7327	0.385	1052	1.05	73.2	1212	$ 242.37
14	23	18.57	1089	50	8415	0.498	1065	1.06	74.1	1219	$ 243.84
15	22	19.07	1106	50	9521	0.625	1025	1.02	71.4	1196	$ 239.21
16	21	19.69	1127	50	10648	0.769	1159	1.15	80.6	1262	$ 252.36
17	20	20.46	1152	50	11799	0.930	1064	1.06	74.0	1219	$ 243.72
18	19	21.39	1182	75	12981	0.154	1104	1.10	76.8	1239	$ 247.81
19	18	21.55	1187	75	14168	0.181	1236	1.23	86.0	1285	$ 256.99
20	17	21.73	1193	75	15361	0.210	1295	1.29	90.1	1296	$ 259.21
21	16	21.94	1199	75	16560	0.242	1132	1.13	78.8	1251	$ 250.28
22	15	22.18	1207	75	17767	0.276	1210	1.20	84.2	1278	$ 255.65
23	14	22.46	1216	75	18983	0.312	1178	1.17	82.0	1268	$ 253.69
24	13	22.77	1225	75	20208	0.350	1190	1.18	82.8	1272	$ 254.48
25	12	23.12	1236	75	21445	0.390	1351	1.34	94.0	1302	$ 260.36
26	11	23.51	1248	75	22693	0.434	1169	1.16	81.3	1265	$ 253.09
27	10	23.94	1262	75	23955	0.479	1204	1.20	83.8	1277	$ 255.33
28	9	24.42	1276	100	25231	0.130	1352	1.34	94.1	1302	$ 260.38
29	8	24.55	1280	100	26511	0.142	1279	1.27	89.0	1294	$ 258.72
30	7	24.69	1285	100	27796	0.156	1259	1.25	87.6	1290	$ 257.97

Irrigation_simulation with VBA / small pipes example figure 7-14 / Standard deviation figure 7-15 / Irrigation_with slope

Fig. 7.9 Sprinkler flow and yield

Each row in the *Irrigation simulation* worksheet (Fig. 7.9) represents a pipe section. The spreadsheet starts with the last sprinkler on the lateral (Row 7, Sprinkler 30). This may seem odd, but the rationale is that the design generally begins with the criterion that the last sprinkler has sufficient pressure. In addition, the following procedure only works if you begin with the last sprinkler. The pressure (cell B7) at the last sprinkler (Row 7) is read from cell B3. The flow rate of the last sprinkler is calculated in column C based on the pressure. The pipe diameter in each pipe section is input in column D. The pressure loss in the last pipe section is calculated in column F based on the pipe diameter (column D) and the pipe flow rate (column E). The pressure of the next to last sprinkler (Row 8) is the sum of the last sprinkler pressure (cell B7) and the pressure loss in the last pipe section (cell F7); The pipe flow rate (column E) is the sum of all downstream sprinklers. This procedure continues until the first sprinkler.

The statistics begin in column G. Normally distributed flow rates are calculated in column G based on the deterministic flow rate in column C. A normal distribution may not be realistic for sprinklers, but application distributions are reviewed in Chaps. 13 and 14. Crop yield is a function of the depth of water applied during the growing season (Chap. 2). Thus, sprinkler flow rates must be converted to depth applied. The way that this is done is that a relative flow rate is calculated in column H: the flow rate at each position (column G) is divided by the minimum flow rate in column C (cell C7) in order to obtain the relative flow rate. Column I is the expected depth applied at each position, calculated as the product of the minimum depth applied (cell E2) and the relative flow rate at each position (column H). The yield is then calculated as a function of depth applied (column I) with a polynomial equation (coefficients in cells I4:K4). The value of the yield (column K) is the product of the value of the crop (cell K1) and the yield (column J).

Because of spatially variable application rate over the field, farmers cannot give each plant the optimal depth of water. Instead, some parts of the field are overirrigated and other parts are underirrigated. Farmers must generally select a minimum amount to apply to the field during the growing season. This depth is specified in cell E2.

The Hazen-Williams equation calculates the friction loss between sprinklers in column F.

$$h_f = 3163\ L\frac{\left(\frac{Q}{150}\right)^{1.85}}{D^{4.87}} \tag{7.9}$$

where

h_f = pressure loss, m.
L = length of the pipe, m,
Q = pipe flow rate, L/hr,
D = pipe diameter, mm.

For example, the pressure loss in the last pipe section (cell F7), Q = 1006 L/hr (cell E7), D = 25 mm (cell D7), L = 15 m (cell B4) is calculated as follows:

$$h_f = 3163\ L\frac{\left(\frac{Q}{150}\right)^{1.85}}{D^{4.87}} = 3163\ (15)\frac{\left(\frac{1006}{150}\right)^{1.85}}{25^{4.87}} = 0.25\ m$$

The relationship between pipe pressure and sprinkler flow rate is

$$\mathrm{Q} = \mathrm{K\ H^x} \tag{7.10}$$

where

Q = outlet flow rate, L/hr
K = constant
H = pipe pressure, m,

The flow rate at each sprinkler is calculated in column C. For example, the sprinkler flow rate at the last sprinkler (cell C7), with K = 200 (cell B1), x = 0.58 (cell B2), and H = 16.2 (cell B7), is calculated as follows:

$$\mathrm{Q} = \mathrm{K\ H^x} = 200(16.2)^{0.58} = 1006\ \mathrm{LPH}$$

The statistical calculation in column G uses the NORMINV function. For example, the formula in cell G7 is NORMINV (RAND(),C7,C7*CV). In Fig. 7.9, the CV is 0.05, and the flow rate in cell C7 is 1006:

$$\begin{aligned}\text{Cell G7} &= \text{NORMINV(RAND(), mean, mean*CV)}\\ &= \text{NORMINV(RAND(), 1006, 1006*0.05)}\end{aligned}$$

The flow rate in column C is the solid red line in Fig. 7.10 and the statistically varying flow rate in column G is represented by the blue squares. The reason that the red line is not smooth is that pipe size changes intermittently along the lateral.

The relative flow rate at each field position is calculated in column H by dividing the flow rate at each field position in column G by the minimum hydraulic flow rate in column C. For example, the relative flow rate in cell H7 is calculated as follows:

$$= \text{G7/MIN(\$C\$7:\$C\$36)} = 1040/1006 = 1.034$$

The spatially varying flow rates are then converted to spatially varying depths of application in column I. Thus, in this case, there is a normally distributed variation in depths applied.

All yields and costs are calculated on a per hectare basis. A hectare (ha) is approximately 2.2 acres. In this example, the yield is calculated in column J with the CWPF in Chap. 2 for cotton (Fig. 7.11).

$$\begin{aligned}\text{Yield(kg/ha)} &= \mathrm{a} + \mathrm{bx}^{0.5} + \mathrm{cx}\\ &= -3954 + 1067\ \text{depth}^{0.5} - 54.14\ \text{depth}\end{aligned}$$

For example, the yield in cell J7 is

$$\begin{aligned}&= -3954 + 1067\text{*I7\^{}0.5} - 54.14\text{*I7}\\ &= -3954 + 1067\text{*72.4\^{}0.5} - 54.14\text{*72.4} = 1205\ \text{kg/ha}\end{aligned}$$

Finally the value of the yield at each position is calculated in column K based on the selling price of the crop (cell K1). For example, in Row 7, the yield value is calculated as follows:

$$\begin{aligned}\text{Cell K7} &= \text{J7*\$K\$1} = 1205\ \text{kg/ha*\$0.2/kg}\\ &= \text{\$241.02/ha}\end{aligned}$$

Next, the average yield value (cell K5) per ha is found:

$$= \text{AVERAGE(K7 : K35)} = \text{\$252.16}$$

Optimization is often based on profit. In this case, the annual costs are pumping energy cost, water cost, and the annualized capital cost of pipe. The pumping energy is calculated based on the pressure input to the system and the total volume of water applied (Chap. 2).

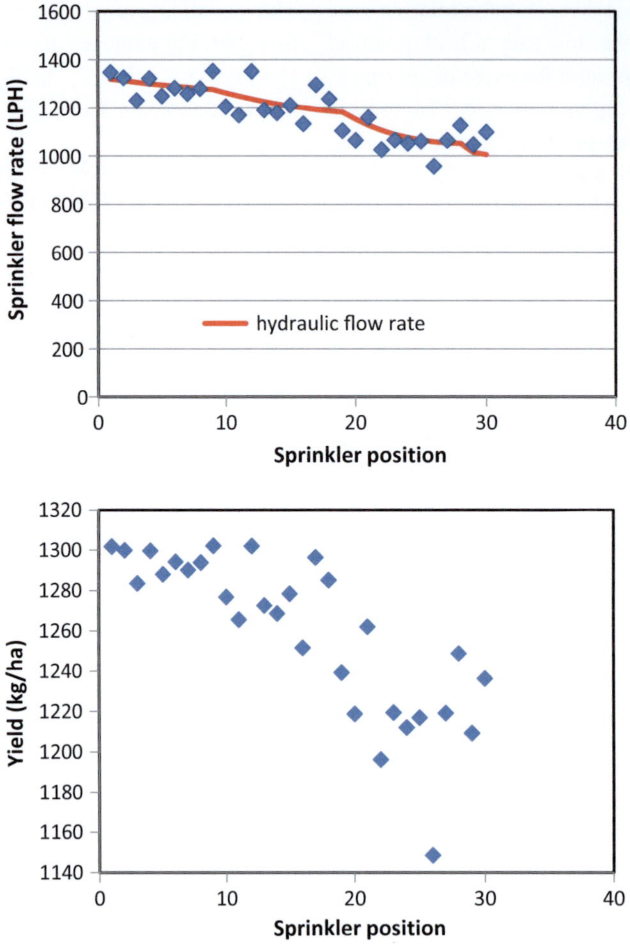

Fig. 7.10 Flow rate and yield at each sprinkler position

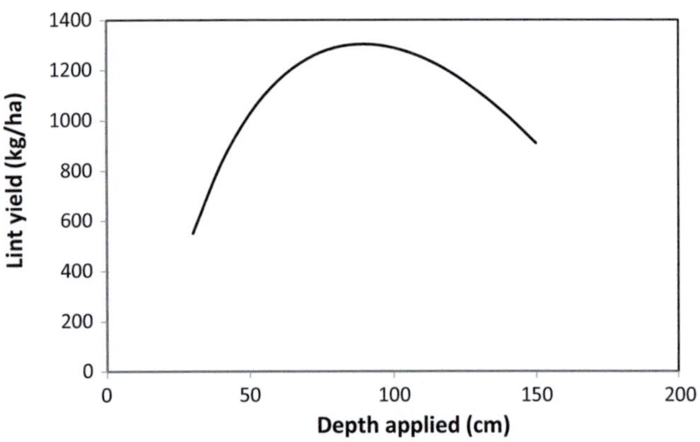

Fig. 7.11 Cotton yield versus depth of water applied per growing season (From Chap. 2)

$$\$/\text{ha} = 0.0272(\text{input pressure, m})(\text{average mm applied})$$
$$(\$/\text{kW} - \text{hr})/(\text{pump efficiency})$$

The average depth applied is 82.09 cm (cell I5) = 821 mm.

The input pressure is the pressure at the first sprinkler (cell B36) = 25.8 m

The cost of energy is \$0.10/kW-hr

The annual cost of energy (cell E3) is calculated as follows:

$$\$/\text{ha} = 0.0272(25.8\ \text{m})(821\ \text{mm})(\$0.10/\text{kW} - \text{hr})$$
$$= \$57.71/\text{ha}(\text{cell E3})$$

The cost of water is calculated based on the average depth of water applied (cell I5). The cost of water is \$30/ha-m for this example. Thus, the annual cost of water (cell E4) per ha is calculated as follows:

$$\$/\text{ha} = 82.1/100(\text{cm/m})^{*}\$30 = \text{I5}/100^{*}30 = \$24.63/\text{ha}$$

Finally, the cost of pipe is included. In this case, the diameter of the pipe (column D) is divided by 1.5 in order to obtain the annualized cost of pipe per ha. This is a simplistic relationship that is used for the purpose of this example. Thus, the annualized cost of pipe is calculated in cell E5 as follows:

$$= \text{AVERAGE(D7:D36)}/1.5 = \$47.78/\text{ha}$$

The profit (cell E1) is the Average yield value – Energy cost – Water cost – Pipe cost.

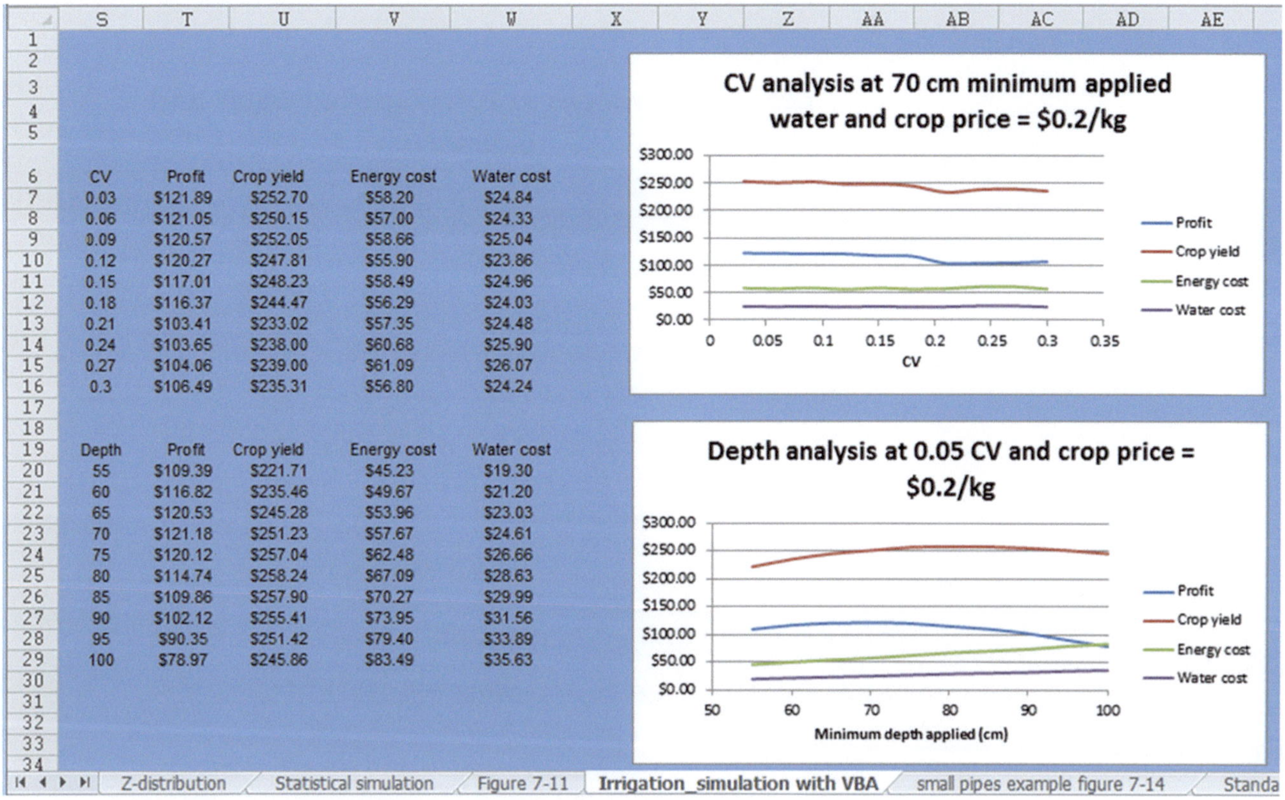

CV	Profit	Crop yield	Energy cost	Water cost
0.03	$121.89	$252.70	$58.20	$24.84
0.06	$121.05	$250.15	$57.00	$24.33
0.09	$120.57	$252.05	$58.66	$25.04
0.12	$120.27	$247.81	$55.90	$23.86
0.15	$117.01	$248.23	$58.49	$24.96
0.18	$116.37	$244.47	$56.29	$24.03
0.21	$103.41	$233.02	$57.35	$24.48
0.24	$103.65	$238.00	$60.68	$25.90
0.27	$104.06	$239.00	$61.09	$26.07
0.3	$106.49	$235.31	$56.80	$24.24

Depth	Profit	Crop yield	Energy cost	Water cost
55	$109.39	$221.71	$45.23	$19.30
60	$116.82	$235.46	$49.67	$21.20
65	$120.53	$245.28	$53.96	$23.03
70	$121.18	$251.23	$57.67	$24.61
75	$120.12	$257.04	$62.48	$26.66
80	$114.74	$258.24	$67.09	$28.63
85	$109.86	$257.90	$70.27	$29.99
90	$102.12	$255.41	$73.95	$31.56
95	$90.35	$251.42	$79.40	$33.89
100	$78.97	$245.86	$83.49	$35.63

Fig. 7.12 VBA simulation results for profit versus coefficient of variation and depth applied

= Yield value − Energy cost − Water cost − Pipe cost
= K5 − E3 − E4 − E5
= $252.16 − $57.71 − $24.63 − $47.78 = $122.05

It would be laborious to compare all variations by changing values in cells one at a time. However, you can automate this type of analysis in Excel worksheets (Fig. 7.12) with Visual Basic for Applications, which is the programming language in Excel. You need to add the Developer tab to your menu bar in Options and then click Visual Basic in order to access the programming environment. For example CV and depth were varied in Fig. 7.12.

Buttons in cells F1, F2, and K2 trigger the depth, CV, and crop price VBA programs. The VBA programs that are triggered by the buttons are below.

```
Sub Optimization_CV()
    cv_int = 0.02
    min_cv = 0.03
    For i = 0 To 9
      CV = min_cv + i * cv_int
      Range("E1").Value = CV
      Profit = Range("H1").Value
      Energy_cost = Range("E3").Value
      Water_cost = Range("E4").Value
      Crop_yield = Range("K5").Value
      Cells(i + 6, 19).Value = CV
      Cells(i + 6, 20).Value = Profit
      Cells(i + 6, 21).Value = Crop_yield
      Cells(i + 6, 22).Value = Energy_cost
      Cells(i + 6, 23).Value = Water_cost
    Next i
End Sub
Sub Optimization_depth()
    depth_int = 5
    min_depth = 50
    For i = 1 To 10
      Depth = min_depth + i * depth_int
      Range("E2").Value = Depth
      Profit = Range("H1").Value
      Energy_cost = Range("E3").Value
      Water_cost = Range("E4").Value
      Crop_yield = Range("K5").Value
      Cells(i + 19, 19).Value = Depth
      Cells(i + 19, 20).Value = Profit
      Cells(i + 19, 21).Value = Crop_yield
      Cells(i + 19, 22).Value = Energy_cost
      Cells(i + 19, 23).Value = Water_cost
    Next i
End Sub
```

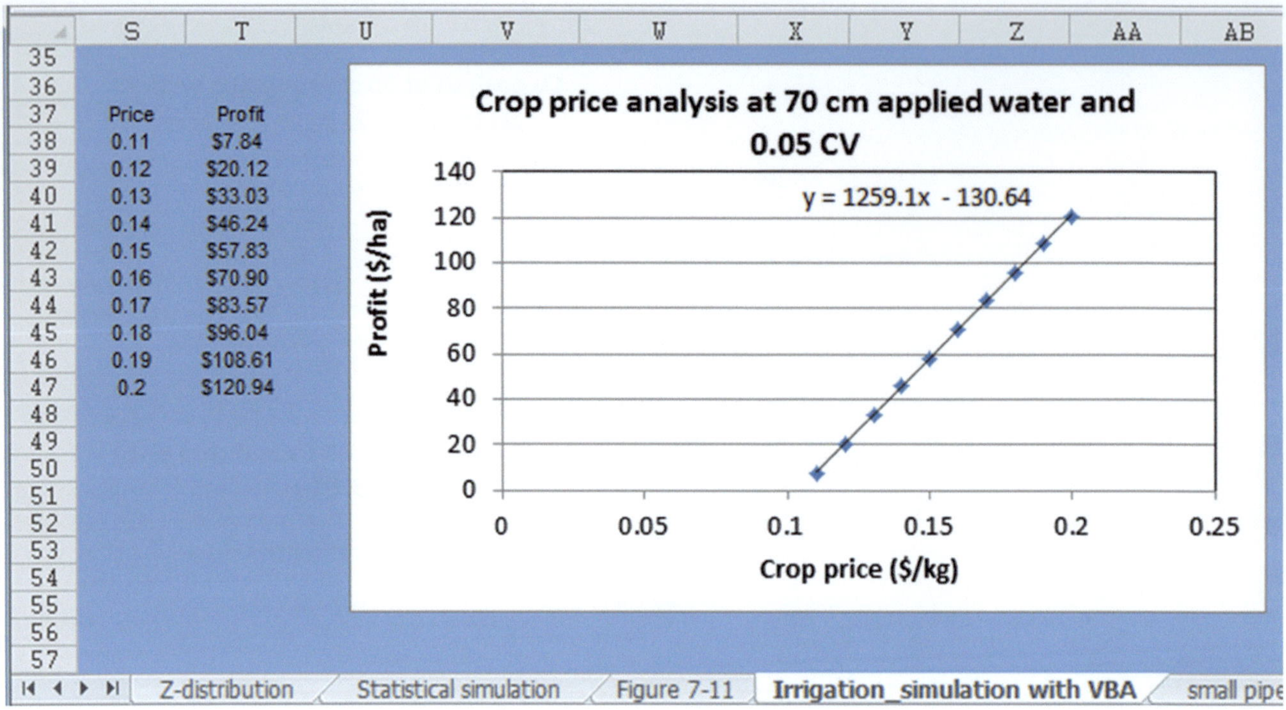

Fig. 7.13 Profit analysis in *irrigation simulation with VBA* worksheet

The programs change values in cells E1, E2, and K2 and then record the CV, yield value, energy, water, and profit in columns S:W. Optimal CV is obviously the lowest CV and optimal depth is between 65 cm and 70 cm.

Alternatively, CV and depth could be fixed is between 65 and 70, and the crop price could be changed in order to find the breakeven price (Fig. 7.13).

```
Sub Break_even_price()
    price_int = 0.01
    min_price = 0.1
    For i = 1 To 10
      Price = min_price + i * price_int
      Range("K1").Value = Price
      Profit = Range("H1").Value
      Cells(i + 37, 19).Value = Price
      Cells(i + 37, 20).Value = Profit
    Next i
  End Sub
```

The effect of reducing pipe size on energy cost is seen in Fig. 7.14. In this case, energy cost increased to $221, and there was no profit (cell H1). The sprinkler flow rate at the beginning of the lateral was twice as high as the flow at the end due to pressure variation.

Figure 7.15 shows that the standard deviation could be used instead of the coefficient of variation in cell E2. In this case, the equation in cell G7: is = NORMINV(RAND(),C7, SD), with the name of cell E1 changed to SD.

Simulation of Slope

The previous analyses were conducted for laterals in a level field; however, many laterals are laid on an uneven slope. In order to account for slope, elevation (column D) was added to the *Irrigation with slope* worksheet (Fig. 7.16). The change in elevation between sprinklers is automatically calculated as the product of slope (cell B5) and the distance between sprinklers (cell B4). The elevation of the last sprinkler is set equal to cell D4. Alternatively, elevations at each sprinkler could be manually entered into column D. The calculation of pressure at each sprinkler in column B includes the sum of elevation change and friction loss. In the case of Fig. 7.16, the slope is downhill from the inlet so the amount of pressure change from one end of the lateral to the other is decreased, in comparison to the level field in Fig. 7.15.

Simulation of Environmental Parameters

With a properly designed sprinkler system, there is no runoff; thus, the major concern is leaching. Leaching carries agricultural chemicals out of the root zone and into the groundwater. Many aquifers in sprinkler irrigated areas are

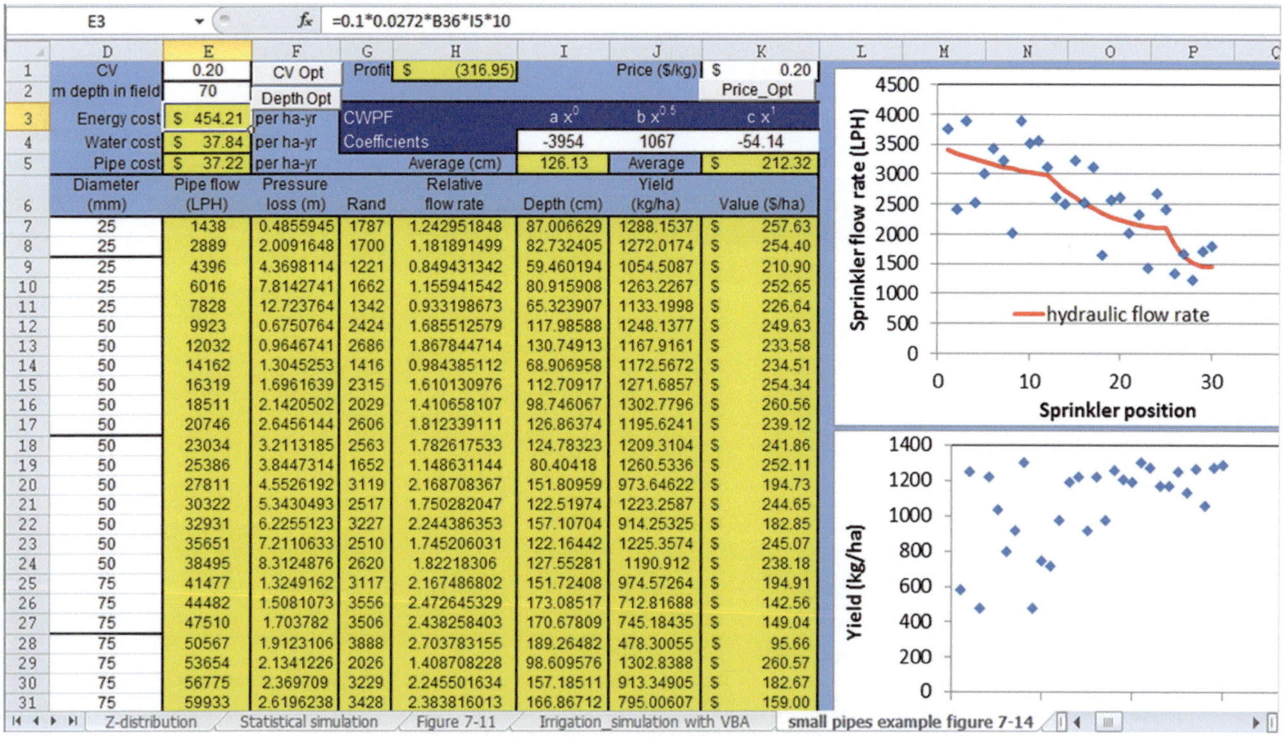

	D	E	F	G	H	I	J	K
1	CV	0.20	CV Opt	Profit	$ (316.95)		Price ($/kg)	$ 0.20
2	m depth in field	70	Depth Opt					Price_Opt
3	Energy cost	$ 454.21	per ha-yr	CWPF		a x^0	b x^0.5	c x^1
4	Water cost	$ 37.84	per ha-yr	Coefficients		-3954	1067	-54.14
5	Pipe cost	$ 37.22	per ha-yr		Average (cm)	126.13	Average	$ 212.32
6	Diameter (mm)	Pipe flow (LPH)	Pressure loss (m)	Rand	Relative flow rate	Depth (cm)	Yield (kg/ha)	Value ($/ha)
7	25	1438	0.4855945	1787	1.242951848	87.006629	1288.1537	$ 257.63
8	25	2889	2.0091648	1700	1.181891499	82.732405	1272.0174	$ 254.40
9	25	4396	4.3698114	1221	0.849431342	59.460194	1054.5087	$ 210.90
10	25	6016	7.8142741	1662	1.155941542	80.915908	1263.2267	$ 252.65
11	25	7828	12.723764	1342	0.933198673	65.323907	1133.1998	$ 226.64
12	50	9923	0.6750764	2424	1.685512579	117.98588	1248.1377	$ 249.63
13	50	12032	0.9646741	2686	1.867844714	130.74913	1167.9161	$ 233.58
14	50	14162	1.3045253	1416	0.984385112	68.906958	1172.5672	$ 234.51
15	50	16319	1.6961639	2315	1.610130976	112.70917	1271.6857	$ 254.34
16	50	18511	2.1420502	2029	1.410658107	98.746067	1302.7796	$ 260.56
17	50	20746	2.6456144	2606	1.812339111	126.86374	1195.6241	$ 239.12
18	50	23034	3.2113185	2563	1.782617533	124.78323	1209.3104	$ 241.86
19	50	25386	3.8447314	1652	1.148631144	80.40418	1260.5336	$ 252.11
20	50	27811	4.5526192	3119	2.168708367	151.80959	973.64622	$ 194.73
21	50	30322	5.3430493	2517	1.750282047	122.51974	1223.2587	$ 244.65
22	50	32931	6.2255123	3227	2.244386353	157.10704	914.25325	$ 182.85
23	50	35651	7.2110633	2510	1.745206031	122.16442	1225.3574	$ 245.07
24	50	38495	8.3124876	2620	1.82218306	127.55281	1190.912	$ 238.18
25	75	41477	1.3249162	3117	2.167486802	151.72408	974.57264	$ 194.91
26	75	44482	1.5081073	3556	2.472645329	173.08517	712.81688	$ 142.56
27	75	47510	1.703782	3506	2.438258403	170.67809	745.18435	$ 149.04
28	75	50567	1.9123106	3888	2.703783155	189.26482	478.30055	$ 95.66
29	75	53654	2.1341226	2026	1.408708228	98.609576	1302.8388	$ 260.57
30	75	56775	2.369709	3229	2.245501634	157.18511	913.34905	$ 182.67
31	75	59933	2.6196238	3428	2.383816013	166.86712	795.00607	$ 159.00

Fig. 7.14 The effect of decreasing pipe sizes

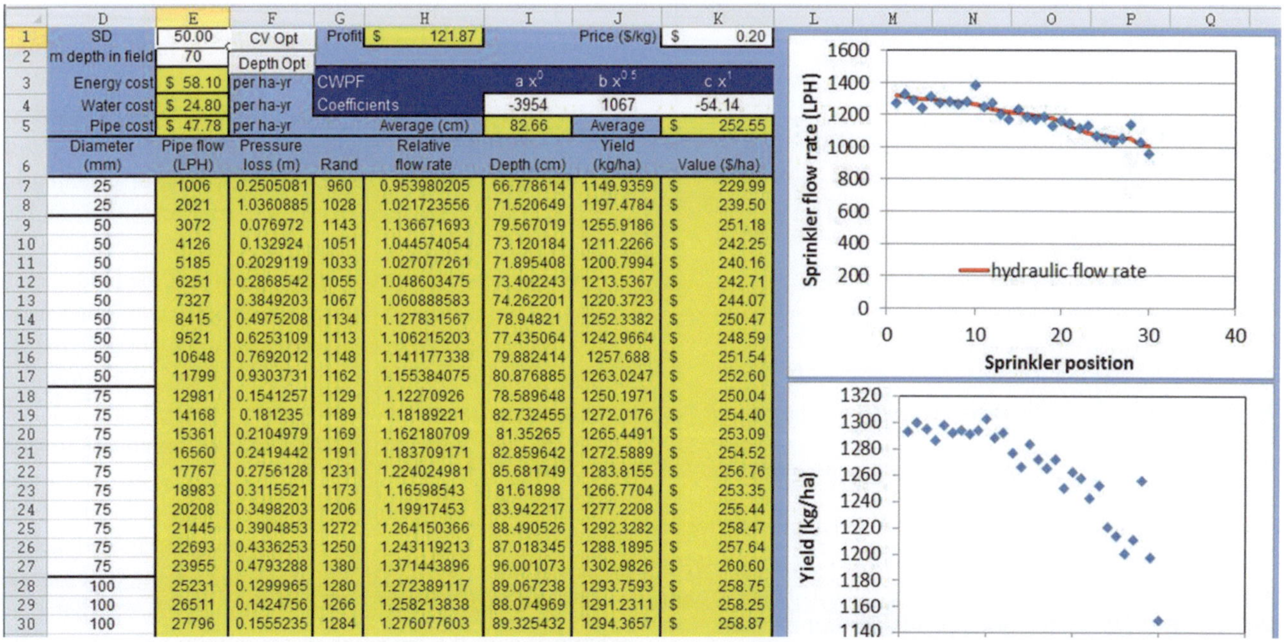

	D	E	F	G	H	I	J	K
1	SD	50.00	CV Opt	Profit	$ 121.87		Price ($/kg)	$ 0.20
2	m depth in field	70	Depth Opt					
3	Energy cost	$ 58.10	per ha-yr	CWPF		a x^0	b x^0.5	c x^1
4	Water cost	$ 24.80	per ha-yr	Coefficients		-3954	1067	-54.14
5	Pipe cost	$ 47.78	per ha-yr		Average (cm)	82.66	Average	$ 252.55
6	Diameter (mm)	Pipe flow (LPH)	Pressure loss (m)	Rand	Relative flow rate	Depth (cm)	Yield (kg/ha)	Value ($/ha)
7	25	1006	0.2505081	960	0.953980205	66.778614	1149.9359	$ 229.99
8	25	2021	1.0360885	1028	1.021723556	71.520649	1197.4784	$ 239.50
9	50	3072	0.076972	1143	1.136671693	79.567019	1255.9186	$ 251.18
10	50	4126	0.132924	1051	1.044574054	73.120184	1211.2266	$ 242.25
11	50	5185	0.2029119	1033	1.027077261	71.895408	1200.7994	$ 240.16
12	50	6251	0.2868542	1055	1.048603475	73.402243	1213.5367	$ 242.71
13	50	7327	0.3849203	1067	1.060888583	74.262201	1220.3723	$ 244.07
14	50	8415	0.4975208	1134	1.127831567	78.94821	1252.3382	$ 250.47
15	50	9521	0.6253109	1113	1.106215203	77.435064	1242.9664	$ 248.59
16	50	10648	0.7692012	1148	1.141177338	79.882414	1257.688	$ 251.54
17	50	11799	0.9303731	1162	1.155384075	80.876885	1263.0247	$ 252.60
18	75	12981	0.1541257	1129	1.12270926	78.589648	1250.1971	$ 250.04
19	75	14168	0.181235	1189	1.18189221	82.732455	1272.0176	$ 254.40
20	75	15361	0.2104979	1169	1.162180709	81.35265	1265.4491	$ 253.09
21	75	16560	0.2419442	1191	1.183709171	82.859642	1272.5889	$ 254.52
22	75	17767	0.2756128	1231	1.224024981	85.681749	1283.8155	$ 256.76
23	75	18983	0.3115521	1173	1.16598543	81.61898	1266.7704	$ 253.35
24	75	20208	0.3498203	1206	1.19917453	83.942217	1277.2208	$ 255.44
25	75	21445	0.3904853	1272	1.264150366	88.490526	1292.3282	$ 258.47
26	75	22693	0.4336253	1250	1.243119213	87.018345	1288.1895	$ 257.64
27	75	23955	0.4793288	1380	1.371443489	96.001073	1302.9826	$ 260.60
28	100	25231	0.1299965	1280	1.272389117	89.067238	1293.7593	$ 258.75
29	100	26511	0.1424756	1266	1.258213838	88.074969	1291.2311	$ 258.25
30	100	27796	0.1555235	1284	1.276077603	89.325432	1294.3657	$ 258.87

Fig. 7.15 Using standard deviation instead of coefficient of variation

so high in nitrate that they are unsuitable for potable water systems. In some location, farmers don't even need to add nitrate fertilizer because of the excessive nitrate in the groundwater. This section applies an environmental cost to leached water. Although this may not be a current cost of farming, regulations are tightening.

A simplistic relationship for leached depth vs. applied water was set up for this example (Fig. 7.17). The function

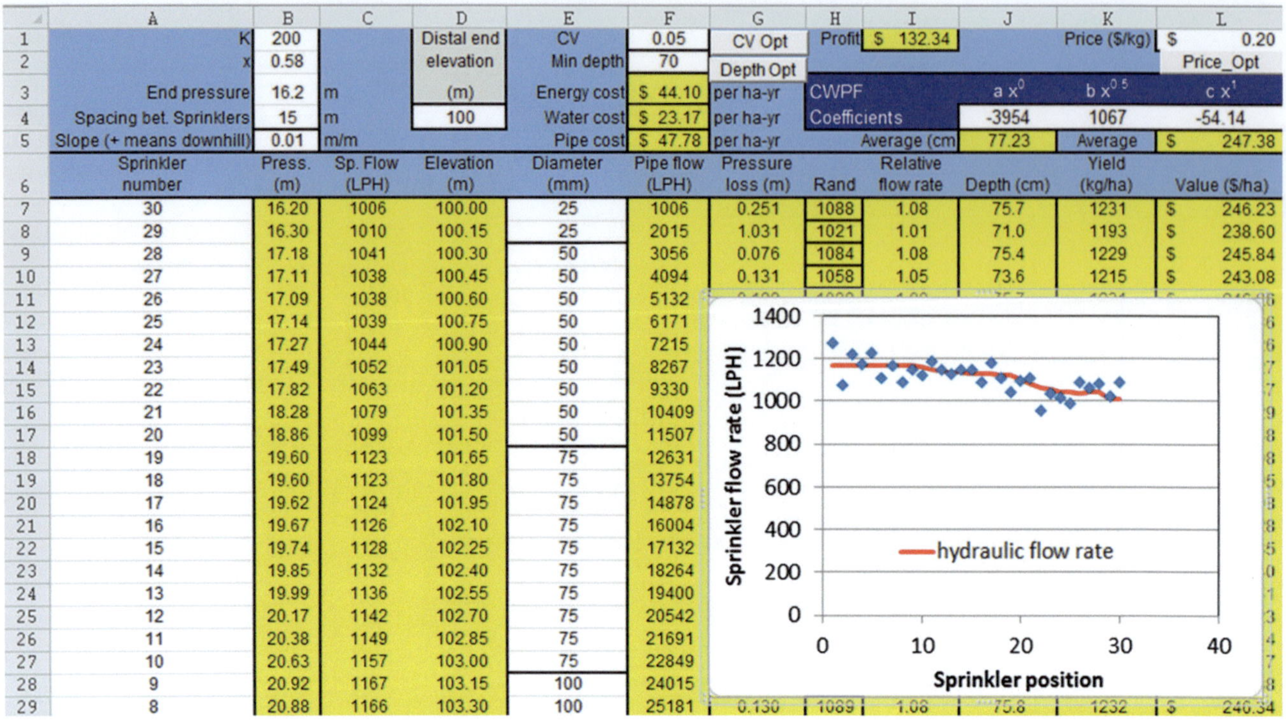

	A	B	C	D	E	F	G	H	I	J	K	L
1	K	200		Distal end	CV	0.05	CV Opt	Profit	$ 132.34		Price ($/kg)	$ 0.20
2	x	0.58		elevation	Min depth	70	Depth Opt					Price_Opt
3	End pressure	16.2	m	(m)	Energy cost	$ 44.10	per ha-yr	CWPF		a x^0	b x^0.5	c x^1
4	Spacing bet. Sprinklers	15	m	100	Water cost	$ 23.17	per ha-yr	Coefficients		-3954	1067	-54.14
5	Slope (+ means downhill)	0.01	m/m		Pipe cost	$ 47.78	per ha-yr		Average (cm	77.23	Average	$ 247.38
6	Sprinkler number	Press. (m)	Sp. Flow (LPH)	Elevation (m)	Diameter (mm)	Pipe flow (LPH)	Pressure loss (m)	Rand	Relative flow rate	Depth (cm)	Yield (kg/ha)	Value ($/ha)
7	30	16.20	1006	100.00	25	1006	0.251	1088	1.08	75.7	1231	$ 246.23
8	29	16.30	1010	100.15	25	2015	1.031	1021	1.01	71.0	1193	$ 238.60
9	28	17.18	1041	100.30	50	3056	0.076	1084	1.08	75.4	1229	$ 245.84
10	27	17.11	1038	100.45	50	4094	0.131	1058	1.05	73.6	1215	$ 243.08
11	26	17.09	1038	100.60	50	5132						
12	25	17.14	1039	100.75	50	6171						
13	24	17.27	1044	100.90	50	7215						
14	23	17.49	1052	101.05	50	8267						
15	22	17.82	1063	101.20	50	9330						
16	21	18.28	1079	101.35	50	10409						
17	20	18.86	1099	101.50	50	11507						
18	19	19.60	1123	101.65	75	12631						
19	18	19.60	1123	101.80	75	13754						
20	17	19.62	1124	101.95	75	14878						
21	16	19.67	1126	102.10	75	16004						
22	15	19.74	1128	102.25	75	17132						
23	14	19.85	1132	102.40	75	18264						
24	13	19.99	1136	102.55	75	19400						
25	12	20.17	1142	102.70	75	20542						
26	11	20.38	1149	102.85	75	21691						
27	10	20.63	1157	103.00	75	22849						
28	9	20.92	1167	103.15	100	24015						
29	8	20.88	1166	103.30	100	25181	0.130	1089	1.08	75.8	1232	$ 246.34

Fig. 7.16 Simulation of sprinkler lateral in sloping field

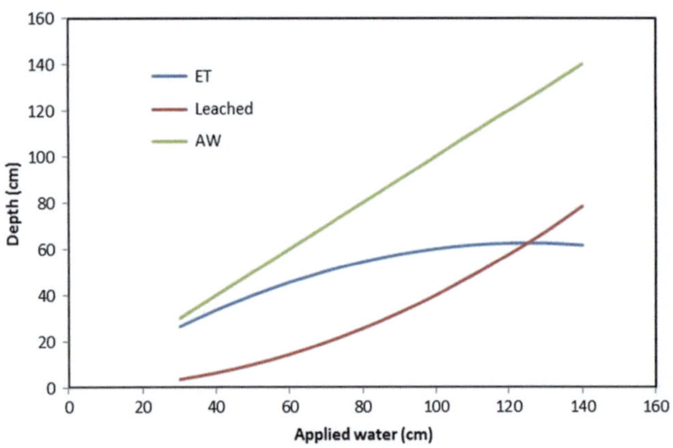

Fig. 7.17 Hypothetical relationship between leached depth and seasonal applied depth

is based on the relationship between ET and applied water in Chap. 2. In reality, leached depth is a function of soil type, crop water uptake, frequency, timing, and depth of irrigation. The *WINDS* model, described in the last chapters, can give a more realistic assessment of leached depth.

$$\text{Leached depth(cm)} = \text{AW}(1 - (1 - 4\ \text{AW}/1000)) \quad (7.11)$$

where

AW = depth of applied water (cm)

Two columns were added, M and N, for the leaching analysis (Fig. 7.18).

The depth of leaching in column M is calculated based on Eq. 7.11 and the depth applied in column J. The *leach cost evaluation* button in cell M3 varies the cost of leaching in cell N1 in order to develop the profit vs. cost of leaching graph (Fig. 7.18).

Overapplication of water in some parts of the field causes leaching, but at the same time underirrigation in other parts of the field causes an increase in salinity and possible reduction in yield due to salinity. The soil salinity variation in a field can be calculated based on the applied and leached depths at each position. A leaching fraction equation such as the Rhoades equation can be rearranged (Eq. 7.12) and used to calculate salinity at each position in the field (Fig. 7.19). The leaching fraction in column O is the leached depth (column M) divided by the applied depth (column J). Then, the saturated paste extract salinities are calculated in column P with the irrigation water salinity value in cell P1 and the leaching fraction in column O.

$$EC_e = \frac{\frac{EC_{iw}}{LF} + EC_{iw}}{5} \quad (7.12)$$

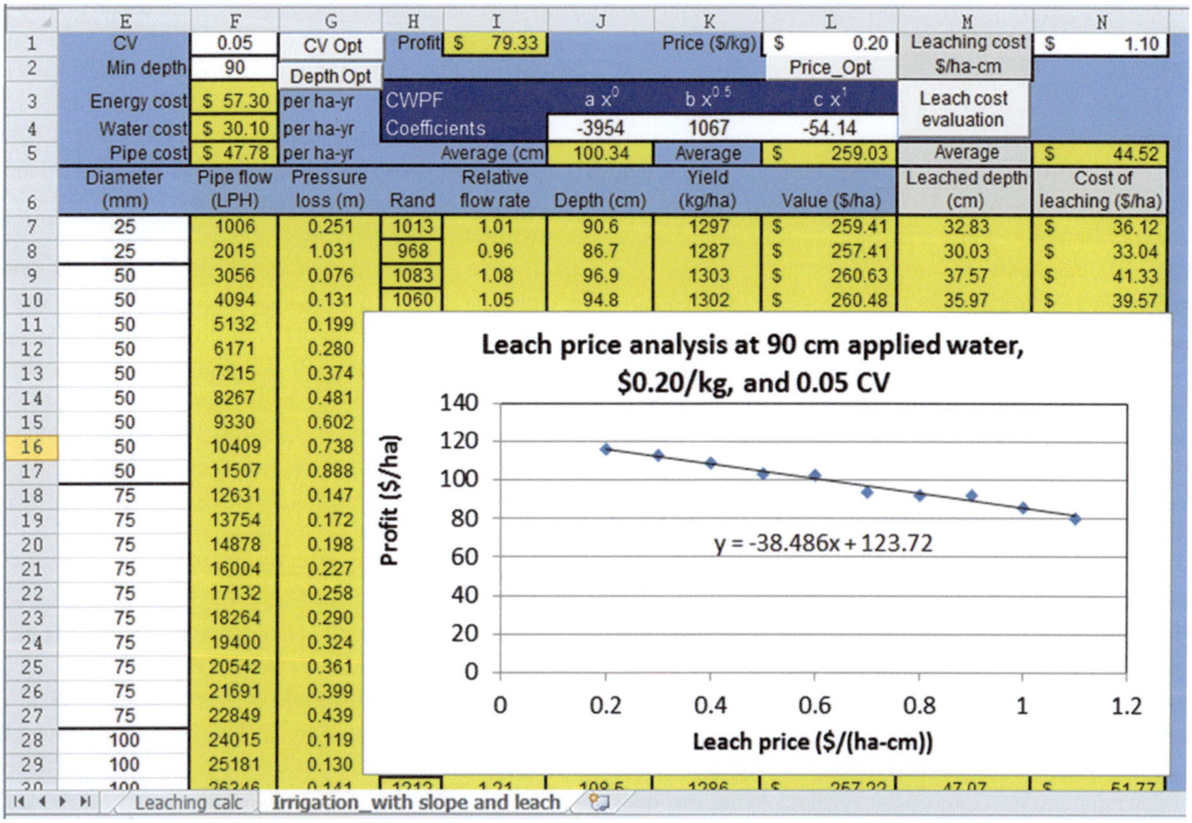

Fig. 7.18 Effect of leaching penalty on environmental cost and overall profit

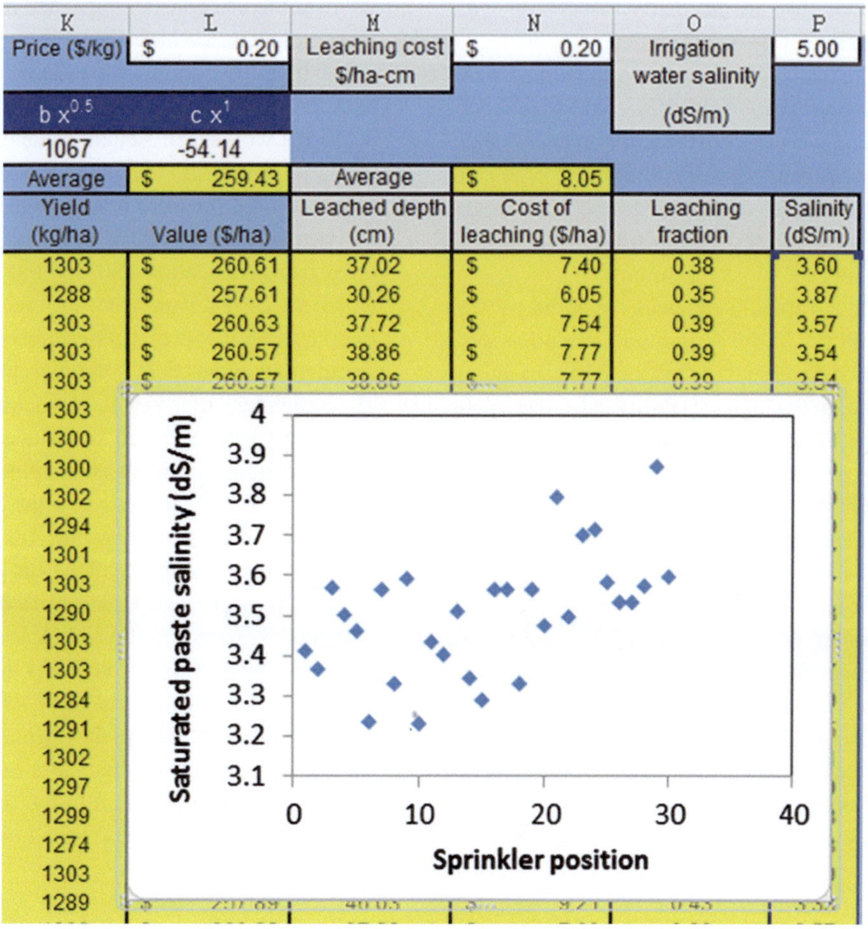

Fig. 7.19 *Salinity analysis* worksheet

Questions

1. One cubic foot of water weighs 62.4 pounds. Assume you have a 1 ft × 1 ft × 1 ft container filled with water.
2. What is the pressure (in feet of head) when the pressure is 2 psi? Remember that 1 psi = 2.31 ft of head.
3. Calculate the pressure in units of feet at the bottom of a one cubic foot container.
4. Calculate the pressure (psi and ft of head) at the bottom of the swimming pool (at sea level) that is 9 ft deep.
5. Using the table below, write down the static pressure in psi and ft of head at each of the fittings found in Fig. 7.1:
6. What is the minimum acceptable inside pipe diameter for a Class 200 PVC pipe with a flow rate 90 gpm?
7. Using Fig. 7.1, calculate the dynamic pressure (in ft of head) at each point of the fittings if the flow rate is 30 gpm through a 2″ Schedule 40 PVC pipe.
8. Describe the relationships between the Z distribution, standard deviation, coefficient of variation, and mean, and describe how you can use those concepts to generate a normal distribution with mean equal to average sprinkler flow rate or application. Ignore pipe hydraulics, and use an equation to convert from Z values to the sprinkler distribution.
9. Generate 50 values that are normally distributed with mean 5 and standard deviation 1.
10. Generate 50 values that are normally distributed with mean 50 and coefficient of variation 5 %.
11. Changing no more than 5 pipe sizes, adjust the pipe sizes in the *Irrigation Simulation with VBA* worksheet (column D) such that the energy cost is equal to or less than pipe cost. State which pipe lengths were changed. Describe how the process of selecting pipe sizes might be automated in a computer program so that total annual energy cost + pipe cost was a minimum.
12. Observe the spatial variation of yield for sprinkler CV (cell E1) 0.05 and 0.5 in the *Irrigation Simulation with VBA* worksheet and observe the effect on profit, rounding to the nearest $/ha. In the high variability case, you might get a flow rate that is negative. Put an IF statement in the spreadsheet that prevents negative flow rates. In each case, observe whether the major cause of application variability and yield variability is due to pressure variation in the pipeline or nonuniformity of application.
13. Using the *Standard deviation* Fig. 7.15 worksheet, increase the standard deviation to 200 LPH, and report the profit. If the standard deviation is 200, then what is the coefficient of variation at the first sprinkler (column C). How does using the standard deviation rather than the coefficient of variation change the distribution along the pipeline. In your estimation, which is more representative of variation along a pipeline and why.
14. Using the *Standard deviation* Fig. 7.15 worksheet, decrease the standard deviation to 20 LPH, and increase the pipe sizes so that energy cost decreases to $37/hr and pipe cost increases to $61. Report the profit. Does the overall profit change significantly from question 7.12, with low CV? Why or why not.
15. Increase the number of sprinklers in the *Irrigation Simulation with VBA* worksheet to 40 by changing the value in cell A7 and copying row 36 downward. Keep end pressure the same (cell B3) and redo the pipe sizes so that the inlet pressure is no more than 20 m (no more than 20 % pressure variation). Try to vary pipe sizes such that flow rate decreases linearly along the pipeline. The next largest pipe size above 100 is 150. Report the number of sections with each pipe size.
16. Adjust the depth in cell E2 in the *Irrigation Simulation with VBA* worksheet, with CV = 0.05, to the nearest integer value such that profit is maximized. The value is between 65 and 70 cm. Also, make a graph of profit vs. depth applied at 5 cm intervals between 40 and 80 cm. Click the Depth Opt. button and make a graph for profit vs. depth applied (Columns S and T). Explain the cause of the curve.
17. Using the *Irrigation_simulation with VBA* worksheet, observe the flow rate vs. distance plot with all pipe diameters equal to 75 mm. Explain the difference from the original graph with varying pipe size. Why wouldn't you want to use constant pipe size if friction loss is less?
18. Insert actual inside pipe diameters for schedule 40 pipe in rather than the nominal pipe diameters column D rather than the nominal diameters in the *Irrigation_simulation with VBA* worksheet. Actual inside pipe diameters can be found in Table 8.1. Compare pressures in cell B36 and calculate total pressure loss in both cases, and calculate the percent difference in pressure loss by subtracting them from each other and dividing by the mean. State whether this is a significant difference and whether it is important to use actual pipe diameters in pressure loss calculations rather than nominal pipe diameters.
19. The roughness of pipes is characterized by the Hazen-Williams C value, which is included in the equation in column F in the spreadsheet. As pipes age, the roughness can increase, which decreases the C value. Using the actual pipe sizes from question 18, for the equations in column F, decrease the C value from 140 to 100 and report the total pressure loss. Compare to the pressure losses in question 18. What does this example say about the need to estimate long term pipe roughness in sprinkler systems.
20. Beginning with the spreadsheet with actual pipe sizes from question 18, change the Sprinkler K to 100 and x to 0.5 and input new pipe diameters such that pressure loss from one end of the lateral to the other is 20 %. Calculate the percent difference based on the average of the

inlet and distal end pressure. Report the decrease in pipe cost and copy the pipe sizes into the answer.

21. Write a VBA program to optimize energy cost and pipe sizing in the Irrigation Simulation worksheet.
22. With the *Irrigation Simulation* worksheet, write a VBA program to calculate lateral end pressure if lateral input pressure is known. This will probably require an iterative procedure.
23. Increase the slope to 3 % in the *Irrigation with slope* worksheet. Using the nominal pipe sizes, adjust the pipe sizes so that the difference between maximum and minimum pressure in the pipeline is no more than 2 m. Report on the number of each pipe length.
24. Make an algorithm in the Irrigation Simulation worksheet that calculates the percent difference between the maximum and minimum pressure in the lateral.
25. Write a VBA program that evaluates and changes pipe sizes to column D in the *Irrigation Simulation* worksheet such that maximum pressure difference in the lateral is no more than 20 % and pipe cost is minimized.
26. Using the *Leaching analysis* worksheet, run the "leach cost evaluation" for CV values of 5 %, 25 %, and 50 %, all at 80 cm applied depth. Copy the graphs of profit vs. leach price and evaluate. The current title of the graph is "Leach price analysis at 90 cm applied water." Make sure that you paste pictures of the figures into the Word document. Otherwise, they will be automatically updated when you rerun the simulation. You might need to rerun the simulation a few times if you get NAN for the 50 % CV. Determine whether the decrease in profit is primarily due to lost yield or increased leaching cost.
27. Using the Salinity analysis worksheet, copy the plots of saturated paste extract salinity vs. distance for CV values of 5 %, 25 %, and 50 %. Let irrigation water salinity equal 6 dS/m. State whether variability is primarily due to hydraulics or spatial variation of application in each case
28. Write a VBA program that evaluates and changes pipe sizes in column D in the *Irrigation Simulation* worksheet, such that profit is maximized.

Reference

Heermann DF, Duke HR, Serafim AM, Dawson LJ (1992) Distribution functions to represent center-pivot water distribution. Trans ASAE

8 Mainline Design

Mainlines deliver water from the water source to irrigation zones or from point A to point B. Zones are activated by solenoid values that supply water to the submains, which then supply water to the laterals (Fig. 8.1). Buried mainlines are generally constructed with PVC pipe, and surface mainlines are generally constructed with aluminum or polyethylene pipe. As with laterals, mainline pipe friction loss is generally minimized; however, instead of the goal of high uniformity, the goal is to save energy. Pipe friction loss is calculated with the Hazen-Williams or Darcy-Weisbach equation and with equations for minor losses in fittings and valves. Pipe selection is based on pipe cost, energy cost (friction loss), and pipe pressure rating. Because mainlines are generally blocked at the end by valves, pipe transients such as water hammer are of concern. Sudden changes in velocity can cause pressure surges that might burst the pipe. Pressure relief valves dissipate pressure surges, and air vents allow air to escape as the pipe is filled. Proper installation techniques and adequate pipe structural strength prevent pipe structural failure.

Pipe Size Classifications

There are three PVC pipe dimensional classification systems: Class (Standard dimension ratios), Schedule (based on steel pipe dimensions), and PIP (Plastic Irrigation Pipe). Class and Schedule pipes are manufactured with outside diameters corresponding to the IPS (Iron Pipe Size) system while PIP pipe has different outside diameters; thus, PIP pipe and IPS pipe are not mixed within the same irrigation system because PVC fittings glue to the outside of pipes and PIP and IPS fittings have different inside diameters. Within each PVC pipe classification system, pipes are classified based on wall thickness. While all pipes of a given size have the same outside diameter, they have different wall thicknesses and inside diameters. The Class system refers to the pressure rating of PVC pipe (maximum allowable pressure for the pipe). Pressure ratings are calculated based on the ratio of wall thickness to pipe diameter (dimension ratio). The Class system standard dimension ratios (SDR) are: 41, 32.5, 26, and 21 for Class 100, 125, 160, and 200, respectively (Class 200 pipe has a pressure rating of 200 psi). Dimension ratios are ratio of wall thickness to pipe diameter, Schedule 40 and Schedule 80, which are based on iron pipe size specifications. Pipe class and schedule sizes in the United States and metric units for pipe sizes less than 12″ (300 mm) are listed in Tables 8.1 and 8.2, respectively.

Large diameter pipe is generally sold as PIP pipe (gasketed pipe) or Schedule 40 pipe (glued fittings). Large diameter Schedule 40 pipe diameters and pressure ratings are listed in Table 8.3.

In-class Exercise 8.1 Determine whether Schedule 40 25 mm (1 in.) pipe has a thicker wall than Class 200 25 mm (1 in.) pipe.

Determine the inside diameter (m) of Class 200 25 mm (1 in.) pipe.

Pipeline Energy Calculations

Pressure and elevation change (energy gradients) propel water through pipelines while water molecules moving past each other cause friction, which dissipates energy and resists flow. In turbulent flow, water molecules move past each other in three scales of eddies; in laminar flow, water molecules in the layered flow profile move past each other in concentric layers. Although friction energy loss is dissipated as heat in irrigation pipelines, the heat does not increase the pipe temperature to the point where it is noticeable to human touch.

In irrigation pipelines, the energy of water includes elevation potential energy, kinetic energy and pressure energy, and is calculated with the Bernoulli equation.

P. Waller, M. Yitayew, *Irrigation and Drainage Engineering*, DOI 10.1007/978-3-319-05699-9_8

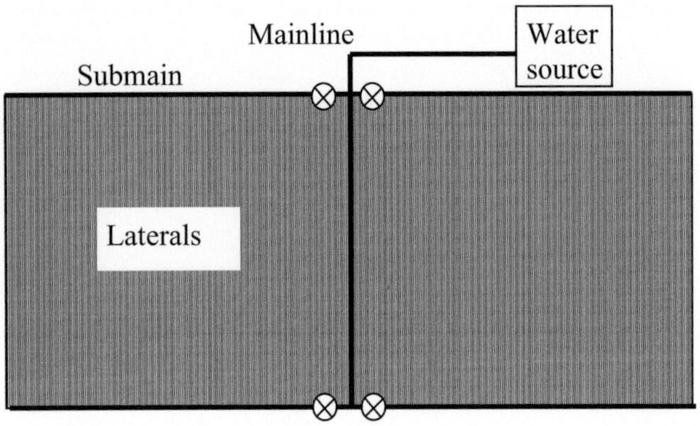

Fig. 8.1 Irrigation pipe network

$$H = \frac{v^2}{2g} + \frac{P}{\rho g} + z \quad (8.1)$$

where

z = elevation of pipeline above datum, m
P = pressure, Pa
v = velocity, m/sec
H = head, m.

Energy of water in pipelines is typically divided into hydraulic head (elevation + pressure) and velocity head: total energy of the water is the sum of hydraulic head and velocity head. Hydraulic head is the elevation that water rises in a vertical pipe (manometer) attached to an irrigation pipeline and is also called piezometric head. Velocity head is generally negligible in comparison to hydraulic head in sprinkler and drip irrigation pipelines but may be significant in low pressure (surface irrigation) pipelines or channels.

In-class Exercise 8.2 Calculate the hydraulic head and total energy of water flowing in a pipe that is 5 m above the datum. Pressure is 35 kPa (~3.5 m) and water velocity is 1.5 m/sec. Draw the pipe with a vertical manometer extending upward and show the pipe elevation, piezometric head elevation (water level in manometer) and total energy elevation.

For friction loss calculations, irrigation pipelines are assumed to operate at steady state; thus, there is no change in internal energy within the pipeline over time. The law of conservation of energy states that energy out (point 2 downstream) = energy in (point 1 upstream) − energy lost + energy added.

$$H_2 = H_1 - h_f - h_m + H_p$$
$$\frac{v_2{}^2}{2g} + \frac{P_2}{\rho g} + z_2 = \frac{v_1{}^2}{2g} + \frac{P_1}{\rho g} + z_1 - h_f - h_m + H_p \quad (8.2)$$

where

h_f = energy lost to pipe friction loss between points 1 and 2, m
h_m = minor losses (energy lost in fittings and valves), m
H_p = energy added by pump, m.

Pipe friction loss in straight pipes, h_f, can be calculated with either of two equations, the Hazen-Williams equation or the Darcy-Weisbach equation. The Hazen-Williams equation is simpler than the Darcy-Weisbach equation; however, the Hazen-Williams does not differentiate between laminar and turbulent flow or account for change in viscosity due to temperature. The Hazen-Williams equation is valid within the normal ranges of flow velocity and temperature found in irrigation pipelines. Because of its simplicity, the Hazen-Williams equation is generally used to calculate friction loss in larger pipelines.

$$h_f = k_1 L \frac{\left(\frac{Q}{C}\right)^{1.85}}{D^{4.87}} \quad (8.3)$$

where

k_1 = conversion factor (Table 8.4)
L = length of the pipe, m of ft
Q = pipe flow rate, various units
C = roughness coefficient
D = diameter, mm.

The Hazen-Williams C value assumes viscosity = $1 * 10^{-6}$ m^2/s and so is only valid between 40 and 75°F (4 °C and 24 °C). The design Hazen-Williams C value ranges from 140 to 150 for smooth pipes constructed from polyvinyl chloride (PVC), polyethylene (drip tubing), and epoxy coated steel. However, C is lower for rough pipes because surface roughness increases turbulence. Iron pipe degrades over time, and the C value decreases from 130 for new iron pipe to 80 for 40-year-old iron pipe. The C value is approximately 130 for new galvanized steel pipe and new concrete pipe. Aluminum pipe has a range of reported C values depending on the type of couplings between pipes (from 130 to 90) and the application. Conventional 3″ (75 mm) aluminum hand lines with 30 ft pipe lengths joined by aluminum couplers have a C value of 130 (*NRCS NEH, Sec. 15, Ch. 11*).

The Darcy-Weisbach equation has different forms for laminar and turbulent flow.

$$h_f = f \frac{L}{D} \frac{v^2}{2g} \quad (8.4a)$$

Table 8.1 Nominal diameters, wall thickness, and inside diameters for PVC irrigation pipe (United States units: inches); the Schedule 40 dimensions are Schedule 40 steel pipe dimensions

	Spec	¾″	1″	1¼″	1½″	2″	2½″	3″	4″	6″	8″	10″	12″
Type	OD (in)	1.050	1.315	1.660	1.90	2.375	2.875	3.500	4.500	6.625	8.625	10.75	12.75
Cl 100	Wall (in)							0.085	0.110	0.162	0.210	0.262	0.311
Cl 125				0.051	0.058	0.073	0.088	0.108	0.138	0.204	0.265	0.331	0.392
Cl 160				0.064	0.073	0.091	0.110	0.135	0.173	0.255	0.332	0.413	0.490
Cl 200		0.060	0.063	0.079	0.090	0.113	0.137	0.167	0.214	0.316	0.410	0.511	0.606
SC 40		0.113	0.133	0.140	0.145	0.154	0.203	0.216	0.237	0.280	0.322	0.365	0.406
Cl 100	ID (in)							3.33	4.28	6.301	8.205	10.226	12.128
Cl 125				1.56	1.78	2.23	2.70	3.284	4.224	6.217	8.095	10.088	11.966
Cl 160				1.53	1.75	2.193	2.655	3.23	4.154	6.115	7.961	9.924	11.77
Cl 200		0.93	1.189	1.50	1.72	2.149	2.601	3.166	4.072	5.993	7.805	9.728	11.538
SC 40		0.824	1.049	1.38	1.61	2.067	2.469	3.068	4.026	6.065	7.981	10.02	11.938
SC 40	PSI	480	450	370	330	280	300	260	220	180	160	140	130

Table 8.2 Nominal diameters, wall thickness, and inside diameters for PVC pipe (metric: mm); Schedule 40 dimensions are Schedule 40 steel pipe dimensions

	Spec	18	25	31	37	50	62	75	100	150	200	250	300
kPa rating	OD (mm)	26.7	33.4	42.2	48.3	60.3	73.0	88.9	114.3	168.3	219.1	273.1	323.9
689	Wall (mm)							2.16	2.79	4.11	5.33	6.65	7.90
862				1.30	1.49	1.86	2.25	2.74	3.51	5.18	6.73	8.41	9.96
1103				1.63	1.85	2.31	2.79	3.43	4.39	6.48	8.43	10.49	12.45
1379		1.52	1.60	2.01	2.29	2.87	3.48	4.24	5.44	8.03	10.41	12.98	15.39
SCH40		2.87	3.38	3.56	3.68	3.91	5.16	5.49	6.02	7.11	8.18	9.27	10.31
689	ID (mm)							84.6	108.7	160.0	208.4	259.7	308.1
862				39.6	45.3	56.6	68.5	83.4	107.3	157.9	205.6	256.2	303.9
1103				38.9	44.6	55.7	67.4	82.0	105.5	155.3	202.2	252.1	299.0
1379		23.6	30.2	38.2	43.7	54.6	66.1	80.4	103.4	152.2	198.2	247.1	293.1
SCH40		20.9	26.6	35.1	40.9	52.5	62.7	77.9	102.3	154.1	202.7	254.5	303.2
SCH40	kPa	3360	3150	2590	2310	1960	2100	1820	1540	1260	1120	980	900

Table 8.3 Larger diameter PVC and steel Schedule 40 pipe dimensions

United States units						Metric units				
Nominal diameter (in)	OD (in)	Wall thickness (in)	ID (in)	Rating (PSI)	DR	Nominal diameter (mm)	OD (mm)	Wall thickness (mm)	ID (mm)	rating (kPa)
14	14	0.437	13.126	130	32	350	356	11.1	333	900
15	15	0.469	14.062	130	32	375	381	11.9	357	900
16	16	0.5	15	130	32	400	406	12.7	381	900
18	18	0.562	16.876	130	32	450	457	14.3	429	900
20	20	0.593	18.814	120	34	500	508	15.1	478	830
21	21	0.617	19.766	120	34	525	533	15.7	502	830
24	24	0.687	22.626	120	35	600	610	17.4	575	830

Table 8.4 Conversion constants for the Hazen-Williams equation given different combinations of units (From Cuenca 1989)

h_f	length	Q	D	k_1
m	m	L/sec	mm	$1.22 * 10^{10}$
m	m	L/hr	mm	3163
m	m	m^3/day	mm	$3.162 * 10^6$
ft	ft	ft^3/sec	ft	4.73
ft	ft	GPM	in	10.46

Table 8.5 Conversion constants for the Darcy-Weisbach equation given different combinations of units (From Cuenca 1989)

h_f	Length	Q	D	k_2
m	m	L/sec	mm	$8.26 * 10^7$
m	m	L/hr	mm	6.376
m	m	m^3/day	mm	$1.107 * 10^4$
ft	ft	ft^3/sec	ft	0.0252
ft	ft	gpm	in	0.0311

where

L = pipe length, m
D = lateral inside diameter, m
f = Darcy-Weisbach friction factor
h_f = head loss, m
v = velocity, m/sec.

Darcy-Weisbach friction factor can be modified in order to make it more convenient for use in pipe flow calculations. (Table 8.5)

$$h_f = k_2 f L \frac{Q^2}{D^5} \quad (8.4b)$$

where

k_2 = conversion factor (Table 8.8)
L = lateral length, m
Q = pipe flow rate, L/h or other flow units
D = lateral inside diameter, mm or other flow units
f = Darcy-Weisbach friction factor
h_f = head loss, m.

The Blasius friction factor is appropriate for smooth pipe (such as PVC) with turbulent flow (Re > 2,000).

$$f = \frac{0.316}{Re^{1/4}} \quad (8.5)$$

$$Re = \frac{vD}{\nu} \quad (8.6)$$

where

v = velocity in the pipeline, m/sec
D = pipe inside diameter, m
ν = kinematic viscosity of water, $1 * 10^{-6}$ m^2/sec at standard temperature
Re = Reynold's number, dimensionless.

For laminar flow (Reynolds number < 2,000) with any pipe roughness, the friction factor is.

$$f = 64/Re \quad (8.7)$$

In-class Exercise 8.3 What ratio does the Reynold's number represent, and why are higher Reynold's numbers associated with turbulent flow?

In the transitional flow regime between 2,000 < Re < 4,000, there is a discontinuity between Eqs. 8.5 and 8.7. In general, the choice of equations in this range is not critical because friction loss is insignificant at low velocities.

Water velocity is the flow rate over the pipe cross-sectional area, A

$$v = Q/A = Q/\left(\pi D^2/4\right) \quad (8.8)$$

where

Q = flow rate, m^3/sec
A = pipe cross-sectional area, m^2.

Table 8.6 Equivalent lengths of valves and steel fittings

Type of fitting	Equiv. length/diameter
Globe valves, fully open (solenoid valve)	340
Gate valves, fully open	13
Gate valves, ¾ open	35
Gate valves, ½ open	160
Swing check valves, fully open	135
In-line check valves, fully open	150
90° standard elbow	20–30
45° standard elbow	16
90° street elbow (larger radius)	50
Standard tee, flow through run	20
Standard tee, flow through branch	60

Minor losses in fittings, valves, and transitions are a function of velocity squared.

$$h_m = K_m \frac{v^2}{2g} \tag{8.9}$$

where

h_m = minor energy losses in fittings, valves, and transitions, m
v = velocity, m/sec
K_m = minor loss coefficient.

Minor loss coefficients for most valves and fittings can be found online by typing "minor loss coefficients" into Google. Many valves used in irrigation systems, such as solenoid valves, backflow prevention valves, and pressure regulators, have geometries that are specific to each manufacturer. Manufacturer's catalogs should be consulted to find minor loss coefficients or graphs of pressure loss vs. flow rate for these valves.

In some cases, it is convenient to set up Eq. 8.2 with minor losses written in terms of the equivalent length of pipe rather than as a function of velocity squared. The equivalent length is a function of the pipe diameter. Table 8.6 lists losses in fittings in terms of equivalent length.

Minor loss coefficients for long-radius butt-welded or flanged 90° elbows are listed in Table 8.7, which are a function of the radius of the angle divided by the inside diameter of the pipe. L/D represents the equivalent length to pipe diameter ratio.

Minor loss coefficients for mitered bends (cut and welded pipes) are listed in Table 8.8.

Example 8.1 Calculate equivalent length and length to diameter ratio (L/D) of a square edged inlet and pipe discharge to a pond if water velocity in the pipe is 1.0 m/sec and pipe diameter is 0.1 m. Assume that the Darcy-Weisbach friction factor is calculated with the Blasius equation for smooth pipe. Use the minor loss coefficient from *Chapter 8 minor loss coefficients worksheet*: $K_m = 0.5$. Note: the purpose of this example is to show that there is a relationship between equivalent lengths and minor loss coefficients (one would never actually do this in real life).

The equivalent length is found by setting the minor loss equation equal to the friction loss in the straight pipe equation and solving for L/D.

$$\frac{fL}{D}\frac{v^2}{2g} = K_{ent}\frac{v^2}{2g} \rightarrow fL = DK_{ent} \rightarrow L = \frac{K_{ent}D}{f} = \frac{0.5D}{f}$$

$$f = \frac{0.316}{\mathrm{Re}^{1/4}} = \frac{0.316}{\mathrm{Re}^{1/4}} = \frac{0.316(\nu^{0.25})}{(vD)^{0.25}} = \frac{0.316\left((1*10^{-6})^{0.25}\right)}{(vD)^{0.25}}$$

$$= 0.01(vD)^{-0.25}$$

$$L = \frac{0.5D}{\left(^{0.01}/_{(vD)^{0.25}}\right)} = 50v^{0.25}D^{1.25} = (50)(1^{0.25})(0.1^{1.25}) = 2.8\ m$$

$$\frac{L}{D} = \frac{0.5}{\left(^{0.01}/_{(vD)^{0.25}}\right)} = 50(vD)^{0.25} = 50(1.0*0.1)^{0.25} = 28$$

PVC fittings have a particular geometry, and the minor loss coefficients are listed in Table 8.9.

For gravity flow systems, all terms in the Bernoulli equation may be relevant. The goal in setting up these problems is to set up the Bernoulli equation (Eq. 8.2) so that as many terms as possible are known. This is generally accomplished by selecting the open water surface or open discharge at either end of the pipeline as points 1 and 2. The reason is that velocity and pressure at the pond surfaces are zero, thus eliminating those terms.

Example 8.2 There is a 6 m elevation difference between two reservoir surfaces (Fig. 8.2). They are connected by 2″ Class 125 pipe that is 200 m long. The inlet is square-edged and flush with the wall. There is no screen or valve at the inlet. The inlet and outlet pipes are 1.0 m below the water surfaces. Calculate the flow rate and plot the elevation, hydraulic head, and total energy lines over the length of the pipeline and at the inlet and outlet. Set the datum at the elevation of the lower reservoir water surface. Let C = 150.

Set the control volume limits at the water surfaces as shown in Fig. 8.3. The minor loss coefficient for a square-edged inlet is 0.5, and for discharge to a reservoir is 1.0. The inside diameter of 2″ Class 125 pipe is 56.6 mm (Table 8.5). Substitute terms into the Bernoulli Eq. (8.2).

Table 8.7 Minor loss coefficients for long-radius butt-welded or flanged 90° elbows

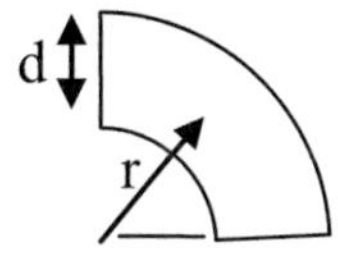

ID (in) (mm)	L/D	½ 12	¾ 18	1 25	1¼ 31	1½ 37	2 50	3 75	4 100	6 150	8 200	12 300	18 450
r/d = 1	20	0.54	0.5	0.46	0.44	0.42	0.38	0.36	0.34	0.3	0.28	0.26	0.24
r/d = 2	12	0.32	0.3	0.28	0.26	0.25	0.23	0.22	0.2	0.18	0.17	0.16	0.14
r/d = 3	12	0.32	0.3	0.28	0.26	0.25	0.23	0.22	0.2	0.18	0.17	0.16	0.14
r/d = 4	14	0.38	0.35	0.32	0.31	0.29	0.27	0.25	0.24	0.21	0.2	0.18	0.17
r/d = 6	17	0.46	0.43	0.39	0.37	0.36	0.32	0.31	0.29	0.26	0.24	0.22	0.2
r/d = 8	24	0.65	0.6	0.55	0.53	0.5	0.46	0.43	0.41	0.36	0.34	0.31	0.29
r/d = 10	30	0.81	0.75	0.69	0.66	0.63	0.57	0.54	0.51	0.45	0.42	0.39	0.36
r/d = 12	34	0.92	0.85	0.78	0.75	0.71	0.65	0.61	0.58	0.51	0.48	0.44	0.41
r/d = 14	38	1.03	0.95	0.87	0.84	0.8	0.72	0.68	0.65	0.57	0.53	0.49	0.46
r/d = 16	42	1.13	1.05	0.97	0.92	0.88	0.8	0.76	0.71	0.63	0.59	0.55	0.5
r/d = 18	45	1.24	1.15	1.06	1.01	0.97	0.87	0.83	0.78	0.69	0.64	0.6	0.55

Table 8.8 Minor loss coefficients for mitered bends

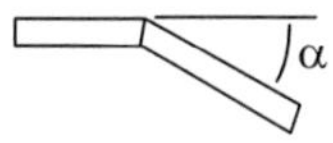

ID (in) (mm)	L/D	½ 12	¾ 18	1 25	1¼ 31	1½ 37	2 50	3 75	4 100	6 150	8 200	12 300	18 450
α = 0°	2	0.05	0.05	0.05	0.04	0.04	0.04	0.04	0.03	0.03	0.03	0.03	0.02
α = 15°	4	0.11	0.1	0.09	0.09	0.08	0.08	0.07	0.07	0.06	0.06	0.05	0.05
α = 30°	8	0.22	0.2	0.18	0.18	0.17	0.15	0.14	0.14	0.12	0.11	0.1	0.1
α = 45°	15	0.41	0.38	0.35	0.33	0.32	0.29	0.27	0.26	0.23	0.21	0.2	0.18
α = 60°	25	0.68	0.63	0.58	0.55	0.53	0.48	0.45	0.43	0.38	0.35	0.33	0.3
α = 75°	40	1.09	1	0.92	0.88	0.84	0.76	0.72	0.68	0.6	0.56	0.52	0.48
α = 90°	60	1.62	1.5	1.38	1.32	1.26	1.14	1.08	1.02	0.9	0.84	0.78	0.72

Table 8.9 Equivalent lengths of PVC fittings (From Harvel Plastics PVC and CPVC design manual. Courtesy of Spears Manufacturing)

Equivalent length of pipe (ft)											
Size (in.)	½"	¾"	1"	1¼"	1½"	2"	2½"	3"	4"	6"	8"
Tee Run	1	1.4	1.7	2.3	2.7	4	4.9	6.1	7.9	12.3	14
Tee Branch	3.8	4.9	6	7.3	8.4	12	14.7	16.4	22	32.7	49
90° Ell	1.5	2	2.5	3.8	4	5.7	6.9	7.9	11.4	16.7	21
45° Ell	0.8	1.1	1.4	1.8	2.1	2.6	3.1	4	5.1	8	10.6
Equivalent length of pipe (m)											
Size (mm)	12	18	25	32	28	50	62	75	100	150	200
Tee Run	0.3	0.4	0.5	0.7	0.8	1.2	1.5	1.9	2.4	3.7	4.3
Tee Branch	1.2	1.5	1.8	2.2	2.6	3.7	4.5	5.0	6.7	10.0	14.9
90° Ell	0.5	0.6	0.8	1.2	1.2	1.7	2.1	2.4	3.5	5.1	6.4
45° Ell	0.2	0.3	0.4	0.5	0.6	0.8	0.9	1.2	1.6	2.4	3.2

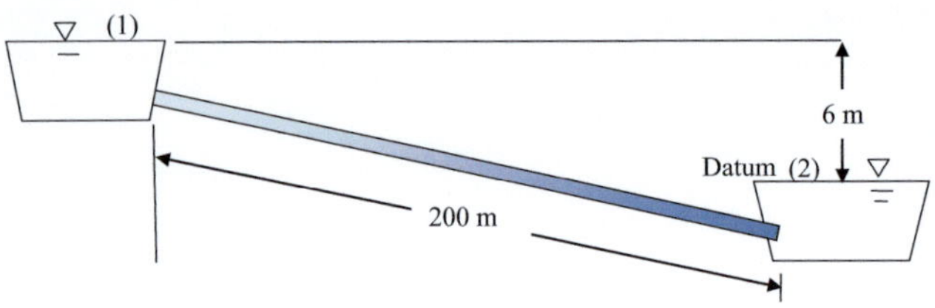

Fig. 8.2 Flow between two reservoirs

$$\frac{v_2^2}{2g}+\frac{P_2}{\rho g}+z_2=\frac{v_1^2}{2g}+\frac{P_1}{\rho g}+z_1-h_f-h_m+H_p\frac{0^2}{2g}+\frac{0}{\rho g}+0$$

$$=\frac{0^2}{2g}+\frac{0}{\rho g}+6-h_f-h_m+0\rightarrow h_f+h_m=6$$

$$k_1L\frac{\left(\frac{Q}{C}\right)^{1.85}}{D^{4.87}}+K_{entrance}\frac{v^2}{2g}+K_{exit}\frac{v^2}{2g}=6$$

Calculate the Hazen-Williams friction loss in terms of velocity because the minor loss term also uses velocity. The entrance minor loss coefficient for a square-edged entrance is 0.5, and the discharge coefficient is 1.0. Convert to appropriate units as follows.

$$(1.22*10^{10})(200)\frac{\left(\frac{1000\ vA}{C}\right)^{1.85}}{D^{4.87}}+(0.5+1)\frac{v^2}{2g}=6\ m$$

$$(1.22*10^{10})(200)\frac{\frac{\left(1000\ v\left(\frac{\pi*0.0566^2}{4}\right)\right)^{1.85}}{150^{1.85}}}{56.6^{4.87}}+(1.5)\frac{v^2}{(2)(9.81)}=6\ m$$

The solution of the equation is v = 1.286 m/sec. Calculate the flow rate Q, L/hr.

$$Q=(1{,}000\ L/m^3)(3{,}600\ \sec/hr)(1.286\ \text{m}/\sec)\left(\frac{0.0566^2\pi}{4}m^2\right)=11{,}650\ L/hr$$

The Pipe flow calc worksheet in the *Chapter 8 Pipes* program performs the calculations in this example (Fig. 8.3). The sum of minor loss coefficients is input in cell B2. The elevation difference between the two reservoir surfaces is input in cell B3. The kinematic viscosity is input in cell B4. The Hazen-Williams C value is input in cell B5 and the surface roughness used to calculate the Darcy-Weisbach f is input in cell F5: the roughness is specified as zero in this example since the pipe is smooth. The pipe length and inside diameter are input in cells B6:B7. An initial guess for velocity is in cell B10. The program then calculates a velocity in cell B12. The initial solution in cell B12 is generally the correct solution, but the user might want to input this value in B10 to see if the program converges to a different value. The Darcy-Weisbach solution in column F uses the Hazen-Williams solution as the first guess. The velocity calculated with the Hazen-Williams equation, 1.286 m/sec, is approximately the same as that calculated with the Darcy-Weisbach equation, 1.30 m/sec (cell F14 in Fig. 8.3).

The lower section of the *Pipe flow calc* worksheet (Fig. 8.4) illustrates the elevation, hydraulic and velocity head just outside the pipeline and just inside the pipeline.

The datum is the lower reservoir water surface. Thus, the elevation of the upper pipe inlet is 5 m since it is 1 m below the upper reservoir surface (cell B24). The pressure head at the inlet is 1 m since that is the depth of the pipe in the reservoir (cell B25). The hydraulic head, 6 m, is the sum of elevation and pressure head (cell B26). The velocity head in the upper reservoir is zero since the water is not moving in the reservoir (cell B27). The energy difference between the reservoir and the inside of the pipe is equal to the entrance loss.

$$K_{entrance}\frac{v^2}{2g}=0.5\frac{1.30^2}{(2)(9.81)}=0.043\ m$$

Thus, the total energy inside the pipeline is 6 m – 0.043 m = 5.957 m (cell C28). The velocity head inside the pipeline is $1.303^2/(2)(9.81)=0.087$ m (cell C27). The hydraulic head is the total energy minus the velocity head: 5.957 – 0.087 = 5.87 m (cell C26). The pressure head inside the pipeline is the difference between the hydraulic head and elevation head, 0.87 m (cell C26). Energy is lost due to pipe friction between the upper and lower reservoir. The energy lines from the upper to the lower reservoir are shown in Fig. 8.5.

In Fig. 8.5, the total energy and the hydraulic head are almost indistinguishable; this illustrates the insignificance of velocity head in many irrigation calculations, and it is generally ignored. The energy lines from inside the discharge outlet to the lower reservoir are shown in Fig. 8.6. The energy loss from the pipeline to the lower reservoir is equal to the velocity head, which is dissipated as water enters the lower reservoir. The elevation is –1.0 because the pipe enters the reservoir 1.0 m below the datum (water surface of lower reservoir).

The following example also defines the control volume limits as the water surfaces, but uses equivalent lengths rather than minor losses to solve the problem.

Example 8.3 A surface irrigation system requires a maximum flow rate of 20 L/sec and is supplied by a distribution box (36″ stacked concrete rings). The distance from the

	A	B	C
1	Hazen-Williams calculation of velocity		
2	Minor + vel. losses Km	1.5	
3	Pot + Press. Energy diff	6	m
4	Pipe parameters		
5	C	150	
6	Pipe length	200	m
7	Inside pipe diameter	0.0566	m
8	Cross-sectional area	0.0025	m2
9	Adjust cell C10 until cell C13 = cell C3		
10	Initial main velocity guess	1.2858	m/sec
11	Initial energy diff	6.00	m
12	Adjusted velocity	1.2861	m/sec
13	Calculated energy diff	6.000	m
14	Flow rate	3.24	LPS
15	Velocity head	0.0843	m
16			
17			
18	Input data in white cells		

	E	F	G
1	Darcy-Weisbach calculation of velocity (main pipeline)		
2	Minor + vel. losses Km	1.5	
3	Pot + Press. Energy diff	6	m
4	Kinematic viscosity	1.00E-06	m2/sec
5	Surface roughness K	0.00000	mm K/D = 0.0000000
6	Pipe length	200	m
7	Inside pipe diameter	0.0566	m
8	Cross-sectional area	0.0025	m2
9	Laminar flow?	FALSE	
10	Velocity from Hazen-W	1.286	m/sec
11	Reynolds number	72793	
12	f factor	0.0192	
13	Initial energy diff	5.8574	m
14	Adjusted velocity	1.303	m/sec
15	Adjusted f factor	0.0192	
16	Adjusted flow rate	3.28	LPS
17	Calculated energy diff	5.99	m
18	Velocity head	0.0865	m

Fig. 8.3 Upper section of *Pipe flow calc* worksheet in *Chapter 8 Pipes* program

	A	B	C	D	E	F	G
20	Calculation below for pipe between two reservoirs with free flow (graphs are below)						
21		Upper reservoir			Lower reservoir		
22		Outside pipe	Inside pipe		Inside pipe	Outside pipe	
23	Distance	0.000	0.000		200.000	200.000	
24	Elevation head	5.000	5.000	Eneryg	-1.000	-1.000	
25	Pressure head (m)	1.000	0.870		1.000	1.000	
26	Hydraulic head	6.000	5.870		0.000	0.000	
27	Velocity head	0.000	0.087		0.087	0.000	
28	Total energy	6.000	5.957		0.087	0.000	
29							
30	This section is for plotting	Distance	Elevation	Pressure head (m)	Hydraulic head	Velocity head	Total energy
31	Inside upper reservoir	0	5	1.000	6.000	0.000	6.000
32	Inside pipe	0	5	0.870	5.870	0.087	5.957
33	Inside pipe	200	-1	1.000	0.000	0.087	0.087
34	Inside lower reservoir	200	-1	1.000	0.000	0.000	0.000

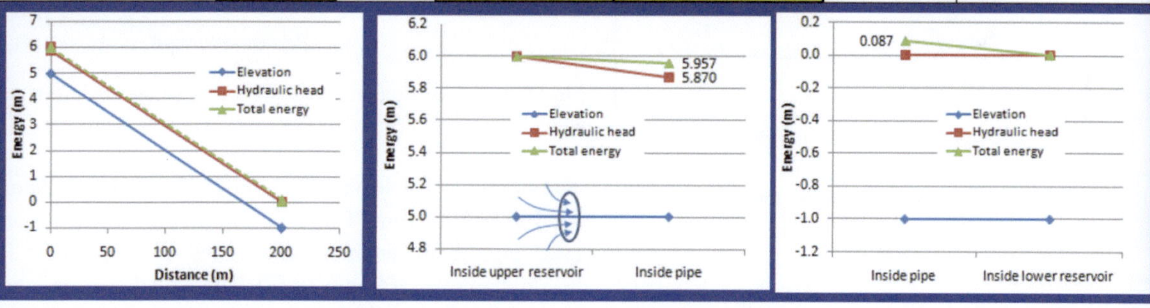

Fig. 8.4 Lower section of *Pipe flow calc* worksheet in *Chapter 8 Pipes* program

reservoir to the distribution box is 300 m, and the working water surface elevation of the distribution box is 1.0 m lower than the water surface elevation of the reservoir (Fig. 8.7). Determine the required diameter of the PVC pipe. There is a sharp-edged entrance to the pipe that is flush with the wall in the reservoir (1). Solve the problem in terms of equivalent length rather than minor loss coefficients. Use Hazen-Williams equation and let C = 145. Use Class 100 pipe.

The L/D ratio for the entrance loss is 28. Because the exit loss K_m is twice the entrance K_m, assume that the exit L/D is 56. The pipe size is unknown, so make an initial guess of 100 mm. Thus the equivalent length of the entrance plus exit losses is 2.8 m + 5.6 m = 8.4 m. Therefore, the length of pipe used in the Hazen-Williams equation is 308.4 m.

By choosing the water surfaces as the control volume boundaries and by using equivalent lengths, Bernoulli's equation reduces to

$$h_f = 1\ m = kL\frac{\left(\frac{Q}{C}\right)^{1.85}}{D^{4.87}}$$

Rearrange the equation and solve for the required pipe diameter, D.

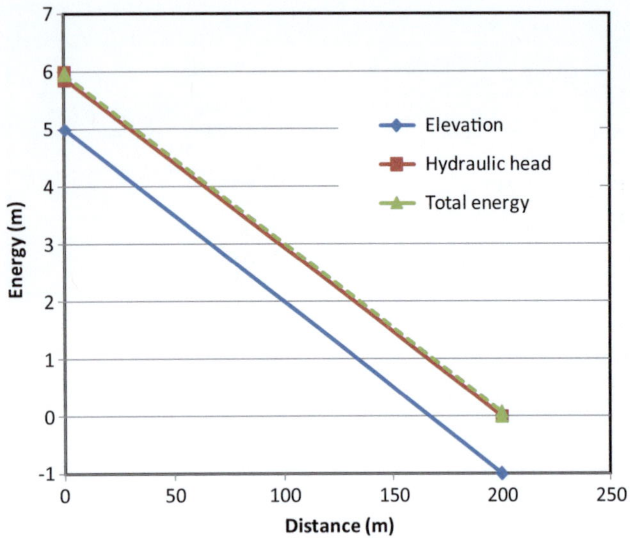

Fig. 8.5 Energy lines between upper and lower reservoirs

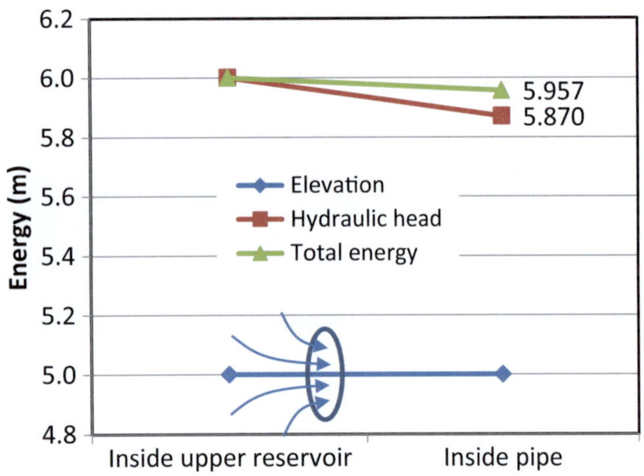

Fig. 8.6 Energy change from inside pipe to lower reservoir

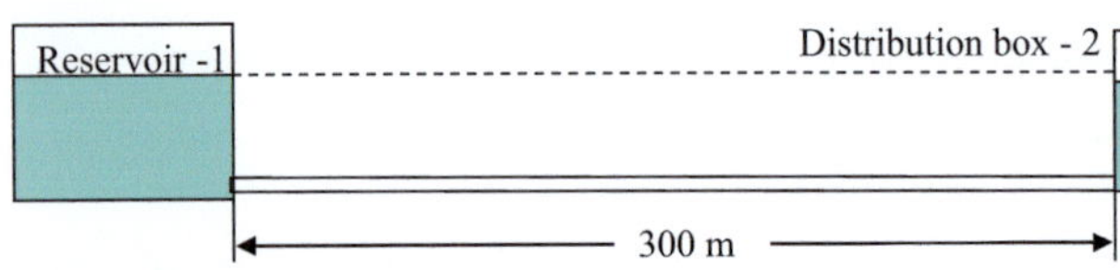

Fig. 8.7 Irrigation distribution box supplied by reservoir

$$D = \left(kL\frac{\left(\frac{Q}{C}\right)^{1.85}}{h_f}\right)^{1/4.87}$$

$$= \left((1.22{*}10^{10})(308.4)\frac{\left(\frac{20}{145}\right)^{1.85}}{1}\right)^{1/4.87} = 180\ mm$$

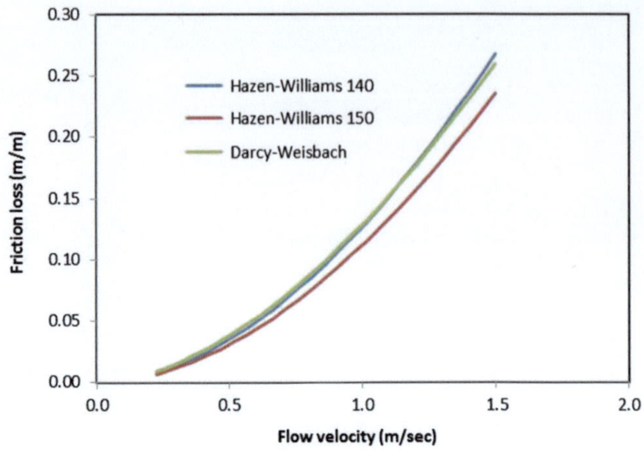

Fig. 8.8 Comparison of friction loss equations (m/m) for 12 mm ID smooth pipe

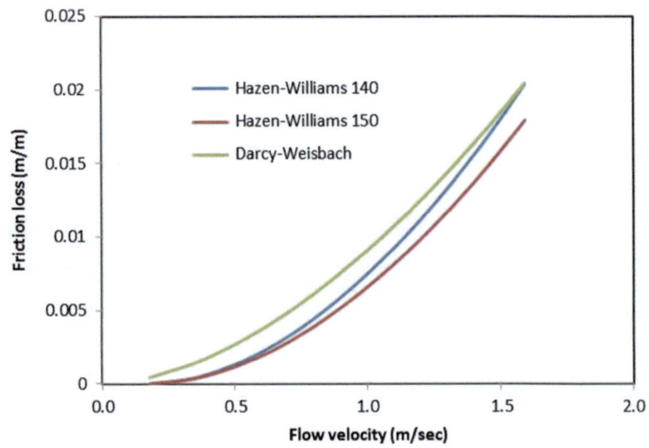

Fig. 8.9 Comparison of friction loss equations for 100 mm ID smooth pipe

The closest available manufactured PVC pipe inside diameters are 160 and 208 mm (Table 8.2). In problems like this, calculated pipe diameter must be rounded up to the next larger inside diameter, 208 mm (8 in). The calculation can then be refined based on the equivalent length for 208 mm ID pipe. The equivalent length for the 208 mm pipe is 0.208 (28 + 56) = 17.5 m, which results in a total length of 317.5 m. The required pipe diameter with L = 317.5 m is 181 mm instead of 180 mm as calculated previously, so there is no change in the selection of 208 mm ID pipe for the project.

There is often uncertainty over the correct C value to select for smooth pipe. Most commonly, the Hazen-Williams C for smooth PVC is assumed to be 140; however, 150 is also recommended in some cases. Hazen-Williams C = 140 generally results in a close agreement with the Darcy-Weisbach equation (Figs. 8.8 and 8.9). Figure 8.8 shows that the Darcy-Weisbach equation and the Hazen-Williams equation with C = 140 have almost the same

Table 8.10 Temperature-pressure derating for PVC fittings and pipes (Courtesy of Spears Manufacturing Company)

Temp (°F)	73	80	90	100	110	120	130	140	150
Temp (°C)	23	27	32	38	43	49	54	60	66
% PR	100 %	90 %	75 %	62 %	50 %	40 %	30 %	22 %	0 %

friction loss for 12 mm ID tubing, whereas the Hazen-Williams equation with C = 150 has a significantly lower estimate of friction loss. The Hazen-Williams equation has reasonable agreement with the Darcy-Weisbach equation up to 1.5 m/sec velocity, which is generally the maximum allowable velocity for PVC pipe. Engineers often use C = 130 or less in order to account for gradual degradation of pipe or crusting of interior pipe surfaces.

Pressure Rating

The pressure rating is calculated based on the dimension ratio, which is the ratio of the wall thickness to the outside diameter.

$$DR = \frac{OD}{T} \tag{8.10}$$

where

OD = average outside diameter, mm
T = minimum wall thickness, mm.

The pressure rating of PVC pipe is calculated as follows:

$$PR = \frac{2S}{DR - 1} \tag{8.11}$$

where

PR = pressure rating of pipe, kPa
S = hydrostatic design stress, kPa.

The hydrostatic design stress is the maximum tensile stress due to internal hydrostatic pressure that can be applied continuously with a high degree of certainty that failure will not occur (Cuenca 1989). The hydrostatic design stresses for PVC 1120 and polyethylene (PE 3408) are 13.8 and 5.5 MPa, respectively.

PVC pipe manufacturers are required to pressure test pipe samples at 4 times the rated pressure for five minutes. They also perform a burst pressure test of PVC pipes at five times the rated pressure, and they perform semiannual tests of 24 hours at 3.5 times the rated pressure.

Example 8.4 Calculate the pressure rating for 4″ Class 200 PVC 1120 pipe.

The standard dimension ratio (SDR) for Class 200 pipe is 21.

$$PR = \frac{2S}{DR - 1} = \frac{(2)(13.8)}{21 - 1} = \frac{27.6}{20} = 1.38\ Mpa$$
$$= 1{,}380\ kPa$$

This same pressure rating (1,379 kPa) is listed in Table 8.2 for Class 200 pipe.

Pressure ratings for PVC pipe and fittings are calculated at 73 °F (23 °C). If water temperature is hotter, then plastic becomes weaker, and the pressure rating decreases. Pressure ratings are listed as a percentage of the maximum pressure rating in Table 8.10. For example, at 43 °C, 125 PSI pipe would have a pressure rating of 62 PSI.

The weakness in PVC pipe networks is at the fittings. While plastic pipe is symmetrical and pressure is directed outward, leading to uniform stress in the plane of the pipe wall, fittings are not symmetrical and stresses are directed at angles in the fitting. Thus, even though fittings may be thicker than the pipe, they are generally weaker. Keller/Bleisner Engineering prepared Table 8.11, which gives the pressure rating for glued and threaded fittings. Although manufacturing companies may report pressure ratings for PVC fittings, there is no national or professional standard by which these pressure ratings are calculated so they are cannot be relied on.

In-class Exercise 8.4 Discuss the impact of fitting strength (Table 8.14) on design of pipelines and selection of fittings.

Transient Flow, Surge Pressure, and Air

After installation of a long 250 mm diameter (10″) irrigation pipeline with a closed valve at the end, a worker suddenly opened the inlet valve, and a short time later, the pipe blew a hundred feet out of the trench (story from an old irrigation engineer). If moving water suddenly stops, then Newton's second law states that the resultant force is equal to the rate of change of momentum, F = ma. When the worker opened the valve, the water rushed down the pipe at high velocity until it reached the end and suddenly stopped. This resulted in a huge force and resultant pressure surge (water hammer). Pressure waves traveled along the pipeline at extremely high velocity, and blew the entire pipeline out of the trench.

Irrigation system operators should slowly fill a pipe with water, and then, only when the pipe is full, completely open

Table 8.11 Pressure ratings for pipe and fittings (at 73 °F) (Courtesy of Spears Manufacturing Company. Prepared for Spears by Keller/Bleisner Engineering). SR refers to Special Reinforced plastic thread

	Schedule 40						Schedule 80					
Nom. Size (in)	Pipe		Solvent cemented joint and Spears SR threaded joint		Standard threaded joint		Pipe		Solvent cemented joint and Spears SR threaded joint		Standard threaded joint	
	psi	MPa	psi	MPa	psi	MPa	psi	MPa	psi	MPa	psi	MPa
¼	780	5.4	468	3.2	390	2.7	1130	7.8	678	4.7	565	3.9
½	600	4.1	360	2.5	300	2.1	850	5.9	510	3.5	425	2.9
¾	480	3.3	288	2.0	240	1.7	690	4.8	414	2.9	345	2.4
1	450	3.1	270	1.9	225	1.6	630	4.3	378	2.6	315	2.2
1 ¼	370	2.6	222	1.5	185	1.3	520	3.6	312	2.2	260	1.8
1 ½	330	2.3	198	1.4	165	1.1	470	3.2	282	1.9	235	1.6
2	280	1.9	168	1.2	140	1.0	400	2.8	240	1.7	200	1.4
2 ½	300	2.1	180	1.2	150	1.0	420	2.9	252	1.7	210	1.4
3	260	1.8	156	1.1	130	0.9	370	2.6	222	1.5	185	1.3
4	220	1.5	132	0.9	110	0.8	320	2.2	192	1.3	160	1.1
5	190	1.3	114	0.8	95	0.7	290	2.0	174	1.2	145	1.0
6	180	1.2	108	0.7	90	0.6	280	1.9	168	1.2	140	1.0
8	160	1.1	96	0.7	80	0.6	250	1.7	150	1.0	125	0.9
10	140	1.0	84	0.6	70	0.5	230	1.6	138	1.0	115	0.8
12	130	0.9	78	0.5	65	0.4	230	1.6	138	1.0	115	0.8

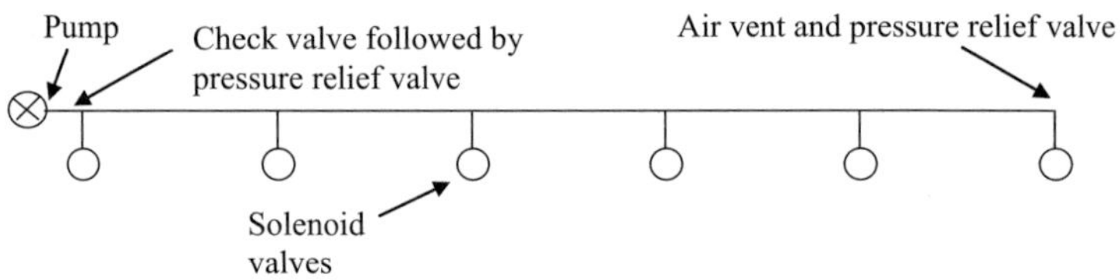

Fig. 8.10 Irrigation main line with solenoids, air vents and pressure relief valves

the valve in order to prevent catastrophes like this. ASAE Standard 376.1 recommends a filling velocity of no more than 0.1 m/sec. This fill velocity would result in a 10 minute filling time for a 300 m length pipe (ASAE 376.1). This principle also applies to centrifugal pump start up. The pump should be started with the discharge valve closed; then the valve should be cracked open, and water should be allowed to slowly fill the pipe before the valve is fully opened. Pumps should not be allowed to run for an extended period against a completely closed valve because pumps are water cooled.

Because valves should always be opened or closed slowly, gate valves or gear operated butterfly valves are preferable on large diameter pipelines because these valves require several turns to open, unlike a ball valve or lever operated butterfly valve. Large pipeline designs include calculation of the minimum allowable valve closure or opening time, and valves are generally designed so that it is impossible for a worker to open or close a valve too quickly.

Although the pipe blowout was caused by suddenly opening an inlet valve, most pressure surges in pipelines are caused by suddenly closing a valve at the end of the pipeline. These surges are accompanied by a banging sound called water hammer. The banging sound is caused by pressure waves expanding and contracting the pipe as they move back and forth in the pipeline. Even if they do not burst the pipe, repeated incidences of pressure surge may eventually lead to pipe or joint failure. Irrigation systems typically have several electronic solenoid valves along the mainline that activate irrigation zones at a signal from of the irrigation controller (Fig. 8.10). These valves typically take a few seconds to close although they can suddenly finish closing during the last phase of valve closure and cause water hammer.

Pressure relief valves allow water to escape if the pipe pressure surges beyond a given threshold; they are typically placed just before the last valve on the mainline (Fig. 8.10), and just after a check valve at the pump. The check valve helps prevent water hammer the next time the system is turned on because it prevents the mainline from draining when the system is shut down. Pressure relief valves should be set to release water at no more than 5 PSI (35 kPa) greater pressure than the design operating pressure of the system (ASAE S376.1). Air vents are also placed on irrigation mainlines; they allow air to escape from the pipe as the pipe is filling, or they allow air to replace the water in the pipe when the pipe is draining. They have a small buoyant ball that is pressed against the upper vent hole when the pipe has water, but the ball drops down and allows air to escape when the pipe has air.

Although pressure relief valves and air vents should always be installed on large agricultural irrigation pipelines, the irrigation system should also be designed with the criterion that the operating pressure plus possible surge pressure (as if the valve were not there) should be no greater than 150 % of the rated pipe pressure (Harvel Plastics Engineering Guide). This number, 150 %, is based on an expected number of surges at which pipe failure will occur at a given surge pressure percentage over the rated pressure. The expected surge pressure in the event of rapid valve closure or other causes is a function of design water velocity. The rule of thumb is that water velocity in closed end PVC pipes should be kept below 1.5 m/sec (5 ft/sec) in order to prevent excessive surge forces. The 1.5 m/sec rule does not apply to pipes that have an open discharge.

The point in the irrigation system that has the highest flow velocity is most susceptible to water hammer. If water hammer is noticed (banging sound) in a landscape irrigation pipe system, then troubleshooting the water hammer problem should be conducted as follows (Stryker 2015): (1) check for under-designed pipe diameters (>1.5 m/sec velocity), (2) check for more than one zone turning on at a time resulting in double the design flow rate, (3) check for valves closing too quickly, which may be solved by installing valves that close slowly, (4) check for lack of pressure relief valves and air vents at the end of mainlines, (5) and check for pressure surges caused by abrupt changes in direction such as at 90° elbows..

In order to prevent water hammer and sudden changes in velocity, the maximum recommended pipe velocity in PVC pipe is 1.5 m/sec (5 ft/sec). The minimum acceptable pipe diameter based on the 1.5 m/sec rule is

$$D = \sqrt{\left(\frac{4Q}{v_{max}\pi}\right)} \tag{8.12}$$

where

Q = flow rate, m^3/sec
v_{max} = maximum allowable velocity, 1.5 m/sec for PVC pipe
D = pipe diameter, m.

Example 8.5 Find the minimum required PVC pipe inside diameter for flow rate 95.3 L/sec.

$$D = \sqrt{\left(\frac{4Q}{v_{max}\pi}\right)} = \sqrt{\left(\frac{(4)(0.0953\ m^3/\sec)}{(1.5\ m/\sec)\pi}\right)} = 0.284\ m$$
$$= 28.4\ cm$$

The expected surge pressure, the increase in pressure over the operating pressure, is proportional to the product of the design water velocity and the pressure wave velocity.

$$\Delta H = \Delta v \frac{a}{g} \tag{8.13}$$

where

Δv = change in water velocity (design water velocity), m/sec
a = pressure wave velocity, m/sec
g = gravity, 9.8 m/sec^2
ΔH = pressure surge, m.

The pressure wave velocity is a function of pipe material and the water properties.

$$a = \frac{\left[\frac{K}{\rho}\right]^{0.5}}{\left[1 + \left(\frac{K}{E}\right)\left(\frac{D}{t}\right)C_1\right]^{0.5}} \tag{8.14}$$

where

K = bulk modulus of elasticity of water, $2.2 * 10^9$ Pa at 25 °C
ρ = density of water, 1,000 kg/m^3
D = inside pipe diameter, m
t = pipe wall thickness, m
C_1 = pipe support coefficient
E = pipe modulus of elasticity (Table 8.15), Pa.

The pipe support coefficient, C_1, is a function of how the pipe is anchored (Cuenca 1989) by thrust blocks or other constraints.

$C_1 = 1.25\mu$, pipes anchored at one end,
$C_1 = 1\mu^2$, pipes anchored at both ends,
$C_1 = 1.0$, pipes with expansion joints.

where

μ = Poisson ratio (Table 8.12).

Typically, irrigation pipelines are buried and constrained at both ends so $C_1 = 1 - \mu^2$. However, gasketed – bell

Table 8.12 Bulk modulus of elasticity and Poisson ratio for pipe materials (After Cuenca 1989)

Material	Modulus of elasticity (MPa) (E)	Poisson's ratio (μ)
Asbestos-cement	$2.07 * 10^4$	0.2
Cast iron	$1.03 * 10^5$	0.29
Ductile iron	$1.65 * 10^5$	0.29
Polyvinyl chloride	$2.76 * 10^3$	0.46
Polyethylene	$6.89 * 10^2$	0.40
Steel	$2.06 * 10^5$	0.30

ended pipe is not glued but slipped together with lubricated gaskets. The gasketed joints serve as expansion joints so C_l is 1.0 for gasketed bell-ended pipe.

Example 8.6 Calculate the surge pressure in a 21.9 cm OD (8.625″) 6 gage steel pipe buried in the ground and carrying a design flow velocity of 2.3 m/sec. The pipe is anchored at both ends.

$$C_l = 1\mu^2 = 10.3^2 = 0.91$$

Calculate the velocity of pressure wave: 6 gage (0.2″ thick walls) tubing has an inside diameter of 8.625 – (0.2) (2) = 8.225″.

$$a = \frac{\left[\frac{K}{\rho}\right]^{0.5}}{\left[1 + \left(\frac{K}{E}\right)\left(\frac{D}{t}\right)C_1\right]^{0.5}} = \frac{\left[\frac{2.2*10^9}{1{,}000}\right]^{0.5}}{\left[1 + \left(\frac{2.2*10^9}{2.06*10^{11}}\right)\left(\frac{8.225}{0.2}\right)0.91\right]^{0.5}}$$
$$= 1{,}253\ m/\sec$$

Calculate the magnitude of pressure wave.

$$\Delta H = \Delta v \frac{a}{g} = (2.3)\left(\frac{1{,}253}{9.8}\right) = 294\ m$$

Because the water bulk modulus of elasticity is a constant (with slight changes due to temperature), Eqs. 8.13 and 8.14 can be simplified. For each standard dimension ratio and pipe material, the surge pressure is directly proportional to water velocity. The surge pressure coefficients for PVC pipe at SDR 21, 26, 32.5, and 41 are 16.1-, 14.4-, 12.9-, and 11.4- (psi/(ft/sec)) or 364-, 326-, 292-, and 258- (kPa/(m/sec)).

$$\Delta H = C_{sp} v \tag{8.15}$$

where

v = design velocity, m/sec
ΔH = surge pressure, m
C_{sp} = surge pressure coefficient, sec.

If the total pressure in the pipe under surge (surge pressure plus design operating pressure) is greater than 1.5 * rated pressure (left column of Table 8.5 or calculated with Equation 8.11), then the pipe diameter should be increased in order to increase flow velocity, or the wall thickness should be increased to increase the pressure rating.

Example 8.7 A pipe must carry a flow rate of 286 L/sec at a design pressure of 530 kPa. Calculate the minimum acceptable pipe diameter according to the 1.5 m/sec rule. Use Schedule 40 PVC pipe. Fittings have a pressure rating that is 80 % of pipe pressure rating, and water temperature is 27 °C. The pipe support coefficient, C_1, is 1.0. Select a pipe size based on surge pressure.

As a first guess, find the minimum acceptable diameter based on the 1.5 m/sec rule.

$$D = \sqrt{\left(\frac{4Q}{v_{max}\pi}\right)} = \sqrt{\left(\frac{(4)(0.286\ m^3/\sec)}{(1.5\ m/\sec)\pi}\right)} = 0.497\ m$$
$$= 497\ mm$$

The next larger pipe size is 21-inch (502 mm ID) SCH 40 pipe (Table 8.3).

$$v = \frac{Q}{A} = \frac{0.286\ m^3/\sec}{(0.502\ m/2)^2\pi} = 1.44\ m/\sec$$

The rated pressure of 21-inch pipe is 830 kPa; however, this pressure must be reduced because of fittings and high temperature. At 27 °C, the plastic has a pressure rating that is 90 % of the reported pressure rating. Fittings have a pressure rating that is 87 % of the pipe pressure rating.

$$\text{Maximum design pressure} = (830)(0.9)(0.87) = 650\ \text{kPa}$$

The maximum allowable design + surge pressure is equal to 150 % of the rated pressure.

$$\text{Maximum surge + design pressure} = 650(1.5) = 975\ \text{kPa}$$

The dimension ratio for 21-inch SCH 40 pipe is 34 (Table 8.3). The surge pressure coefficient can be found by interpolation between surge pressure coefficients for Class rated pipe.

$$C_{sp} = 258 + (292 - 258)\frac{41 - 34}{41 - 32.5} = 286\ kPa/(m/\sec)$$

$$\Delta H = C_{sp}\ v = (286\ \text{kPa}/(\text{m}/\sec))(1.44\ \text{m}/\sec)$$
$$= 412\ \text{kPa}$$

The design pressure + surge pressure (530 kPa + 412 kPa = 942 kPa) does not exceed the maximum allowable design + surge pressure of the pipe (975 kPa).

Thus, 21-inch pipe is acceptable. Check the next smaller pipe diameter, 20-inch.

$$v = \frac{Q}{A} = \frac{0.286\ m^3/\sec}{(0.478\ m\ /2)^2\pi} = 1.59\ m/\sec$$
$$\Delta H = C_{sp}\ v = (286)(1.59) = 455\ \text{kPa}$$

The design pressure + surge pressure (455 kPa + 530 kPa = 985 kPa) **exceeds** the maximum allowable design + surge pressure of the pipe (975 kPa). Thus, 20-inch pipe is not acceptable for this application, and the 1.5 m/sec rule led to the correct pipe selection in this case. The *Surge* worksheet in the *Chapter 8 Pipes* program performs the calculations shown in this example.

Minimum valve closure time can be calculated based on the length of the pipeline and the pressure wave velocity. The valve should not finish closing before the pressure wave reaches the other end of the pipeline and returns. Thus, minimum valve closure time is calculated as follows:

$$t_c = \frac{2L}{a} \tag{8.16}$$

where

t_c = valve closure time, sec
L = pipeline length, m
a = pressure wave velocity, m/sec.

During operation, air tends to collect at high points in pipelines. Air pockets in pressurized pipelines can increase friction/pressure loss. Even worse, air pockets at high points in low-pressure gravity-flow pipelines (as in Example 8.4) can stop flow completely. For pipelines that travel over high and low points, air vents should be placed at high points in order to remove accumulated air.

Due to high tensile strength, PVC pipes are able to resist breakage under high positive pressure. However, low-pressure gravity flow pipes ("Class 63" and "100 ft head") with thin walls cannot resist collapse under a negative pressure of one atmosphere (zero gauge pressure). If water drains from the lower end of a sloping pipe, and the upper end does not have a vent that allows air to enter the pipe, then a vacuum forms because nothing replaces the drained water. In order to prevent development of a vacuum, all high points on pipelines (closed ends or middle) should have a vacuum relief valve to prevent the formation of a vacuum when the pipe is drained. Vacuum relief valves have a vent at the top of the valve that opens when a vacuum forms. Because air vents and vacuum relief valves are both needed at high points, combination air vent/vacuum relief valves are manufactured.

ASAE Standard S376.1 specifies air vent dimensions (diameter of the threaded connection to the PVC pipe) for low-pressure and high-pressure pipelines (Table 8.13). Air vents and vacuum relief valves on gravity flow pipelines are larger because prevention of vacuum and air locks is more important so air must be allowed to rush into or out of the pipe more quickly.

The *Chapter 8 Pipes* program includes two worksheets that model the process of a suddenly filling pipe. This would be the case if someone suddenly opened a valve and allowed water to flow into an unfilled pipe (this should never be done except in a computer model). The first worksheet has no air vent at the end of the pipe so the pressure builds at the end of the pipe as the pipe fills and the air is compressed (Fig. 8.11).

Table 8.13 Air vent and vacuum relief valve standards (After Cuenca 1989)

High pressure systems			
Pipe diameter		Valve threaded connection diameter	
≤102 mm	(≤4″)	13 mm	(0.5″)
125–200 mm	(5–8″)	25 mm	(1″)
250–500 mm	(10–20″)	51 mm	(2″)
≥525 mm	(≥21″)	0.1 * pipe diameter	
Low pressure systems			
Pipe diameter		Valve threaded connection diameter	
≤150 mm	(≤6″)	51 mm	(2″)
200–250 mm	(8–10″)	76 mm	(3″)
≥300 mm	(≥12″)	102 mm	(4″)

The rate of pipe filling is based on the pressure (force) differential between the advancing front of the water and the pipe inlet, and the resistance to flow of pipe friction (h_f). The model assumes that the pipe is initially filled to a length that is specified in cell B4. The unfilled length of the pipe is specified in cell B5. The rate that water increases or decreases in velocity in the pipe is based on the force differential divided by the mass of water in the pipe. Thus, the pressure and velocity fluctuate over time as the air pressure fluctuates in the closed end of the pipe. Note that this worksheet and the *air vent surge* worksheet are only for demonstration and are not calibrated for use in actual engineering designs.

The *air vent surge* worksheet (Fig. 8.12) assumes that the air is released as the pipe fills but that there is no pressure relief valve placed on the pipe. Thus, the only resistance to pipe filling is pipe friction (h_f). The velocity initially increases to a maximum and then decreases as the length of water in the pipe increases. The surge pressure is then calculated at the velocity of the water when it reaches the end of the pipe with Eqs. 8.13 and 8.14.

Installation

Irrigation system installation should include the following steps:

1. Laying out the locations of the trenches and possibly marking them with paint on the ground.
2. Verifying the locations of existing utilities and irrigation pipes.
3. Trenching.
4. Gluing the mainline pipe outside the trench.
5. Running water through the mainline in order to flush out dirt from the pipe.
6. Installing valves on the mainline.
7. Placing solenoid valve wires into the trench.
8. Dropping the mainline into the trench on top of the solenoid valve wires.
9. Gluing submains and laterals outside the trench.

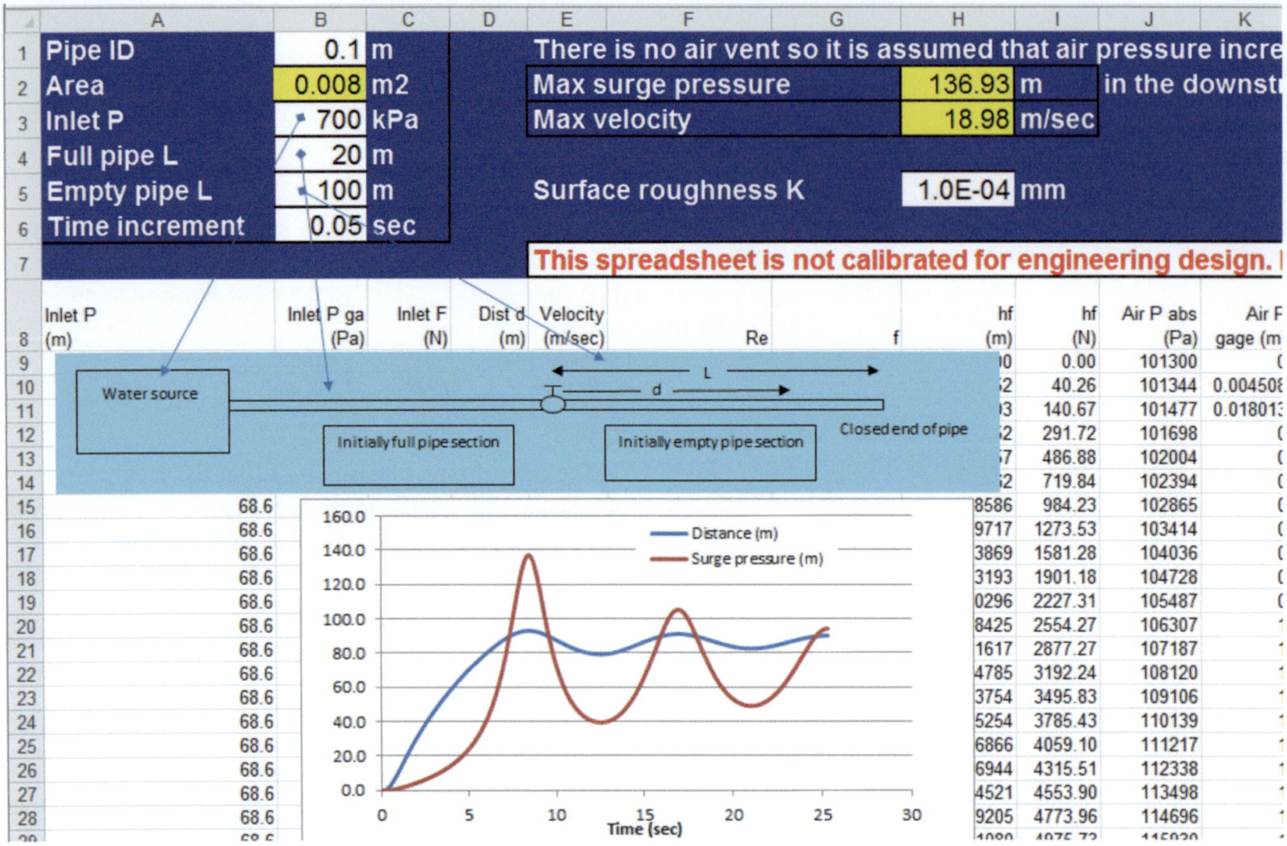

Fig. 8.11 Surge calculator for pipe filling in *No air vent* worksheet

10. Attaching submains and/or laterals to the mainline solenoid valves.
11. Dropping the submains and laterals into trenches.
12. Running water through the submains and/or laterals in order to flush out dirt from the pipes.
13. Attaching emitters, tubing, or sprinklers to the laterals.
14. Pressure checking the entire system for leaks.
15. Backfill the trench.

Trenching, Installation, Pipe Strength, and Corrosion

Trench locations should be carefully laid out and verified with all stakeholders: it is very hard to move a trench. In some installations, it is helpful to mark the locations of trenches with spray paint in order to guide the trencher operator. Some trenchers are guided by lasers or GPS and do not require marked trench locations. Urban installations should be "Blue Staked:" utility companies mark the locations of all underground utilities. Neglecting this step can be hazardous and expensive. For example, trenching through 440 V electrical power lines will kill the tractor operator, and cutting through fiber optic communication lines can result in a fine of hundreds of thousands of dollars. Even marked utility locations may not be correct, especially in the case of old utilities such as 440 V power lines. New installations have foil tape several cm above utility lines; thus, the location of the line can be found and exposed with a shovel before the trencher moves across the line. In agricultural fields, it is often necessary to find existing irrigation pipes in fields. Some specially endowed individuals can walk through a field with L-shaped metal rods in each hand pointing straight ahead; when the person crosses a buried pipe, the metal rods cross over each other. It is thought that lines of magnetic or gravitational force cause the rods to change direction. Others ascribe this talent to witchcraft and don't want to have anything to do with it. Regardless of the reason, it does work.

There are many types of trenchers, backhoes, and pipe pullers. Efficient installation of irrigation or pipe systems requires an adequately sized machine. An inadequately sized trencher is dangerous and can result in the crew and trencher taking an extra week to install the job, which is not profitable. Making bids on excavation and trenching in potentially rocky other otherwise difficult soils should always be preceded by taking extensive soil cores at the site, or the bidders should assume the worst possible conditions (for example, rock) and bid a trencher accordingly. Recommended trench widths and depths are specified in ASAE standard S376.1. In general, the minimum trench width is 30 cm (12 in.) wider than the pipe

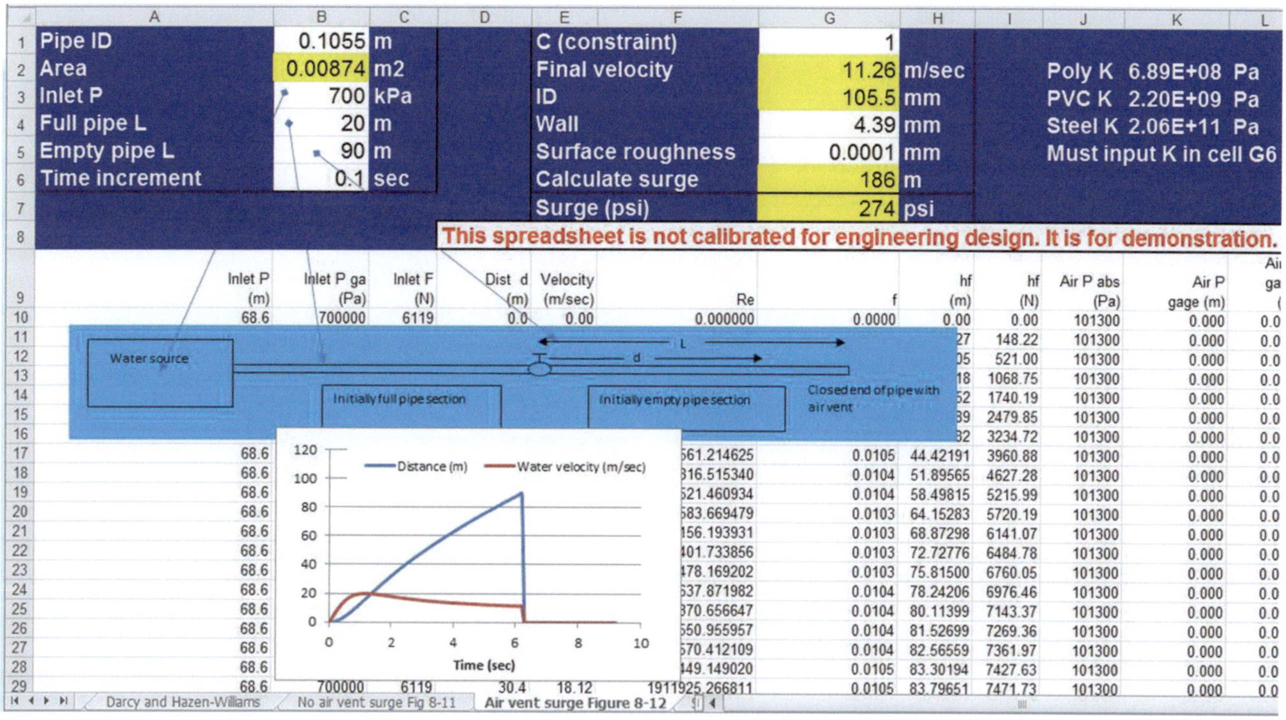

Fig. 8.12 *Air vent surge* worksheet

diameter, and the maximum trench width is 75 cm (30 in.) for pipes less than 37 cm (15 in.).

In the case of low-pressure, gravity-flow pipelines that are installed on nearly level ground, the grade of the pipe and the trench bottom should be carefully surveyed or the pipe should be installed with a laser-guided machine in order to avoid high points that would develop air pockets that stop the flow of water.

Pipe structural strength and stresses due to overload, sharp objects, and bends in pipe must be considered when selecting the correct backfill material. In general, small diameter, pressurized (Class 125 and heavier) irrigation pipes installed in shallow trenches with 18″ (45 cm) of backfill over the pipe are structurally stable. The soil removed from the trench can be placed back over the pipe. However, large rocks should not be placed back in the trench. Larger pipes, lower pressure ratings (thinner wall), and deeper pipes are more susceptible to structural failure. Pipes buried deep in the ground should be "shaded" with sand backfill around the pipe. No rocks should be backfilled over the sand because the rocks will migrate downward in the trench over time.

Pipes can collapse due to an internal vacuum or external load. Pipe resistance to collapse is based on the ratio of pipe wall thickness to diameter. The maximum differentials between pipe internal pressure and external load (critical collapse pressure rating) for SDR 41 (Class 100), SDR 26 (Class 160), and SDR 21 (Class 200) are 17-, 74-, and 126-psi (120-, 510-, and 870-kPa) for all pipe diameters. Schedule 40 pipe, which does not have a constant SDR for different diameters, has a maximum critical collapse pressure rating of 451 psi at 2.5″ diameter and decreases to 25 psi at 24″ diameter. Vehicular traffic or heavy soil loads can damage pipes. Tractors with large tires and uniform load distribution can be driven over lower SDR PVC pipes or pipe trenches without breaking the pipes. However, trucks with small tires, if they are oriented in the same direction can collapse large diameter pipes buried in trenches. The soil overburden pressure is added to the vehicle load in order to find the total pressure on the pipe.

Example 8.8 A 6″ Class 100 pipe is buried in a trench with 18″ cover over the pipe. After a storm, a farm vehicle with small tires is driven over the wet field. A tire is oriented in the same direction as the trench and applies a load of 1,000 lbs over a 1 ft length of the empty pipe. The dry bulk density of the soil is 1.3 g/cm^3, and the soil has a 40 % by volume water content. Determine whether the pipe will collapse.

Calculate the soil density under wet conditions. $1.3\ \text{g/cm}^3$

$$+0.4\ \text{g/cm}^3$$
$$= 1.7\ \text{g/cm}^3.$$

Calculate the soil overburden pressure.

$$1.7\ \text{g/cm}^3 = 0.61\ \text{lb/in}^3$$
$$(0.61\ \text{lb/in}^3)(18\ \text{in}) = 1.1\ \text{psi}$$

Calculate the vehicle load and total load.

$$1{,}000\ \text{lb}/((12\ \text{in})(6\ \text{in})) = 13.9\ \text{psi}$$
$$13.9\ \text{psi} + 1.1\ \text{psi} = 15\ \text{psi}$$

The critical collapse pressure is 17 psi so the pipe would probably not collapse under the load. However, if the load is suddenly placed on the pipe (+ momentum force), then it may collapse.

Because collapse is based on the difference between internal and external pressure, a filled pipe is more resistant to collapse than an empty pipe.

At points along the pipeline with changes in direction, momentum change can exert forces on the pipe that may cause the pipe to come apart. For this reason, concrete is poured into the trench around the pipe at these locations. Thrust blocks give pipe the strength to resist deformation or breakage. Required dimensions for thrust blocks are given in ASAE/ASABE Standard S376.1. Thrust blocks are especially important for gasketed pipe that does not have a fixed connection.

One of the new innovations in trenching is a machine that shanks in pipe or drip tubing without digging a trench. When these machines are used, the pipe is glued together prior to installation.

Steel pipes are subject to corrosion. In order to prevent corrosion, the pipes are often coated with cement or epoxy on the inside and outside of pipes, and they should have cathodic protection to prevent electrons from being transferred from the pipe to the soil. Electrical cells that transfer electrons from pipe to the soil (iron is oxidized) can be microscopic in size or up to several miles long if the pipe is laid in different types of soils. A small investment in cathodic protection can essentially preserve steel pipe indefinitely, whereas ignoring cathodic protection may cause the pipe to degrade within several years.

In cold regions, the required depth of cover over pipes is based on the depth of soil freezing in winters. In cold areas where it is difficult to make deep trenches through bedrock, pipes may be installed at a shallower depth, but they must be blown out with air before winter. One of the concerns with thermoplastic pipeline installation is expansion due to changes in temperature. The coefficient of thermal expansion for PVC pipe is $2.9 * 10^{-5}$ ft/ft/°F (5.2 m/m/°C).

Standard PVC pipe is not resistant to UV degradation. Thus, if the pipe is left in the sun for an extended period (a few months) the pipe turns brown and is degraded because UV light breaks polymer bonds in the plastic. Painting the outside of pipe prevents UV degradation. Although it is standard practice to use steel pipe for all above ground pipe installations, some farmers successfully use PVC above ground in certain applications such as valve stations by painting the PVC pipe with latex or acrylic paint. Unlike PVC pipe, polyethylene pipe is not sensitive to UV degradation and is often used for above ground applications.

Pipe Connections

Connecting pipe sections correctly is a critical step in producing a high quality irrigation system because leaks and pipe failure generally occur at joints. Large diameter polyethylene pipe is fused together by placing the ends together and melting the pipes together. The joint is as strong as the pipe. PVC joint types include glued, threaded, flanged, and gasketed/bell-ended connections. Glued PVC fittings and pipes are connected by softening the outer pipe surfaces with a primer or cement and then fusing the surfaces with cement. Primer is generally used to soften large diameter PVC pipe prior to cement application while a more active cement (blue glue) can be used to both soften and cement small diameter pipes (<2″). There are many solvent cement formulations. Higher viscosity and slower drying cements are used for large diameter pipe. Check the label in order to select the right cement for the given application. It is important to press the pipes completely together during gluing and to apply enough glue such that a small bead appears around the joint after pressing.

Pressure Regulation

On farms with major elevation changes, pressure regulators and other strategies may be needed in order to prevent pressure from increasing beyond the rated pressure of the pipe or irrigation components. For example, a sprinkler irrigated apple orchard in Washington State is constructed on a deep river canyon (700 m deep). The river water is first pumped by high pressure pumps through a steel pipeline to a reservoir at the upper end of the farm (Fig. 8.13). A much lower pressure pump removes the water from the reservoir and pressurizes the irrigation mainline. The irrigation mainline runs back down the hill and feeds each of the irrigation zones. The irrigation mainline can be designed so that pressure loss is equal to elevation gain; however, when irrigation is turned off the pressure greatly exceeds the rated pressure of the pipe. At this pressure, the pipes burst and can create huge holes in the orchard because of the high flow rate from mainlines. In order to prevent such catastrophes, pressure regulators are placed at strategic points along mainline in order to prevent pressure from building beyond the rated pressure of the pipe.

Large pressure regulators are expensive and tend to wear out after 3–4 years. Some farmers control pressure to low elevation zones with gate valves. The gate valves wear out after a few years, but a flanged gate valve is easy to replace and much cheaper than a large pressure regulator. The gate valve, however, would not work in the case of Fig. 8.13.

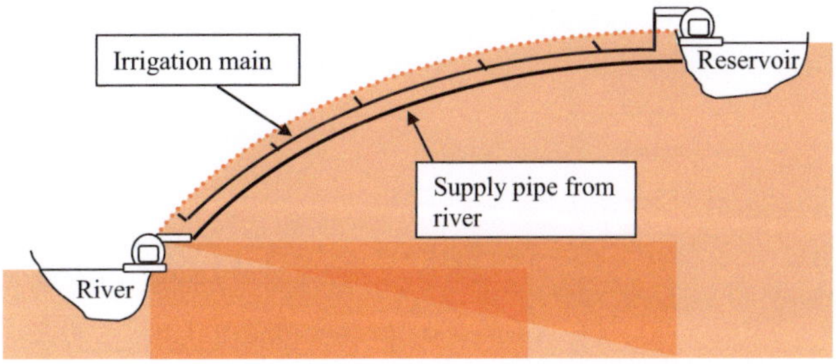

Fig. 8.13 Apple orchard irrigation system on river canyon

Table 8.14 Pipe prices for Example 8.9

Nominal diameter (in)	Nominal diameter (mm)	Wall thickness	Cost ($/m)
3/4	18	Sch 40	0.39
1	25	Sch 40	0.56
1.5	37	Cl 125	0.49
2	50	Cl 125	0.75
3	75	Cl 125	1.64
4	100	Cl 125	2.72
6	150	Cl 125	5.87
8	200	Cl 125	9.71
10	250	Cl 125	18.60
12	300	Cl 125	24.54
15	375	SCH 40	26.51
18	450	SCH 40	34.51
21	525	SCH 40	50.00
24	600	SCH 40	70.00

Economics of Pipe Sizing: Energy Cost Versus Capital Cost

Pipe diameters are selected based on the capital cost of pipe and the present value of the annual cost of energy (pumping) over the life of the project. For the purpose of questions in this book, pipe prices are shown in Table 8.14; however, the price of PVC fluctuates with the price of oil. It also fluctuates during the year with supply and demand, the amount in inventory, and the date at which the inventory tax is calculated.

The determination of best economic diameter is made in the *Surge and Energy Cost worksheets* based on the price of plastic and the price of energy. However, in the next example, prices used in Table 8.14 are used.

Example 8.9 An 800 m pipe supplies 3 center pivots. Each pivot flow rate is 95.3 L/sec. Based on Example 8.10, 502 mm (21 in. pipe is the minimum acceptable diameter to prevent surge pressure damage at 286 L/sec. Thus, select either 502 mm or 575 mm (24 in. pipe.

Table 8.15 Hours of operation, friction loss, power, and annual cost of energy for 502 mm pipe

Flow rate (cms)	Hours of operation	Friction loss (m)	Power (kW)	Energy (kW-hr)	Cost ($/year)
0.286	1,547	2.58	9.0	13,989	$1,399
0.190	294	1.21	2.8	828	$ 83
0.095	2,076	0.34	0.4	822	$ 82
Total	3,917				$1,564

Table 8.16 Hours of operation, friction loss, power, and annual cost of energy for 575 mm pipe

Flow rate (cms)	Hours of operation @ diff. flow rates	Friction loss (m)	Power (kW)	Energy (kW-hr)	Cost ($/year)
0.286	1,547	1.33	4.7	7211	$ 721
0.190	294	0.62	1.4	424	$ 42
0.095	2,076	0.17	0.2	411	$ 41
Total	3,917				$ 804

The financial parameters are 20 yr project life and 8 % ROR. Pumps are variable speed so that pressure and flow can be decreased as needed. It is expected that the system will have the following operation times during the year.

$$\begin{aligned} &3 \text{ pivots operating at one time}(286\ \text{L/sec}): && 1{,}547\ \text{hr} \\ &2 \text{ pivots operating at one time}(191\ \text{L/sec}): && 294\ \text{hr} \\ &1 \text{ pivot operating alone}(96\ \text{L/sec}): && 2{,}076\ \text{hr} \end{aligned}$$

Costs are calculated in Tables 8.15 and 8.16. Example cost calculations are shown for the 502 mm pipe. First, evaluate capital + annual costs for 502 mm pipe. Calculate pipe friction loss in the 502 mm pipe.

$$h_f = (1.22{*}10^{10})(800\ m)\left(\left(\frac{286}{140}\right)^{1.852}/502^{4.87}\right)$$

$$= 2.58\ m$$

Friction loss can be converted to a power requirement for 502 mm pipe follows:

$$\begin{aligned} \text{Power(kW)} &= (\text{Q H})/(.102\ \text{Eff}). \\ &= (0.286\ \text{cms})(2.58\ \text{m})/((.102)(.80)) \\ &= 9.0\ \text{kW} \end{aligned}$$

Present value of energy costs for the 502 mm pipe is calculated based on 20 yr project life and 8 % ROR. Present value ($1,564/year) is $15,355 (PV function).

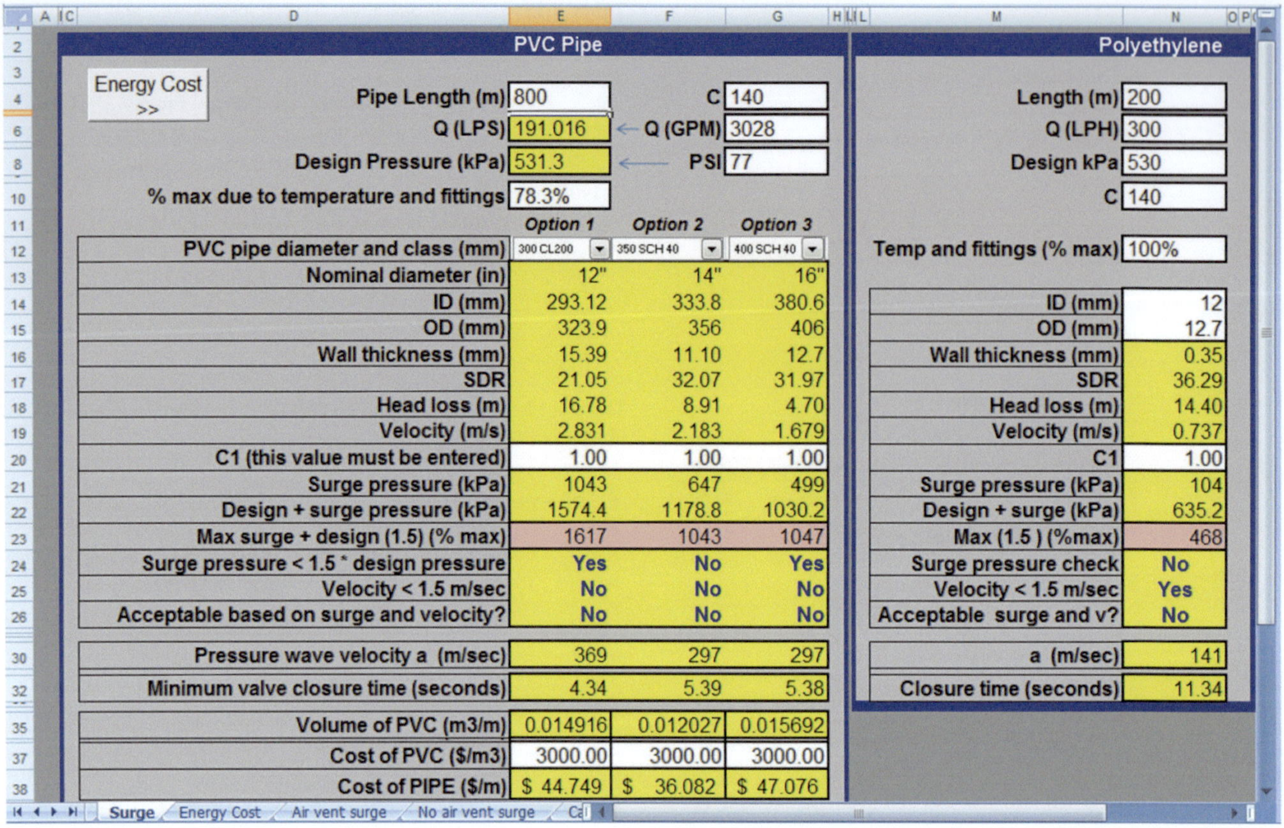

Fig. 8.14 Surge pressure analysis for Example 8.10

$$\text{Capital cost of 502 mm pipe (Table 8.14)} \\ = (\$50/\text{m})(800\ \text{m}) = \$40,000$$

Total cost of the 502 mm pipe is equal to capital + annual costs: $40,000 + $15,355 = $55,355

Next, calculate the total cost (energy + capital) of 575 mm pipe (Table 8.15).

$$\text{Present value}(\$804/\text{year})\text{is } \$7,894$$
$$\text{Capital cost of 575 mm pipe} = (\$70/\text{m})(800\ \text{m}) = \$56,000$$
$$\text{Total cost of the 575 mm pipe is } \$56,000 + \$7,894 = \$63,894$$

Because the total cost of 502 mm pipe, $55,355, is less than $63,894, select 502 mm pipe.

Example 8.10 Change the number of pivots in Example 8.9 to two. Perform surge and economic analysis as in Examples 8.7 and 8.9, respectively. Calculate PVC cost based on volumetric cost of PVC: $3,000/m^3. Financial parameters are 20 years and 8 % rate of return. Cost of energy is $0.1/kW-hr. Pump efficiency = 80 %. Design pressure is 530 kPa. Solve with the *Surge* and *Energy Cost* worksheets. Assume that PVC only has 78 % of maximum strength as in Example 8.7.

$$\text{2 pivots operating at one time}(191\ \text{L/sec}) : 294\ \text{hr}$$
$$\text{1 pivot operating alone}(96\ \text{L/sec}) : 2,076\ \text{hr}$$

Surge pressure is evaluated in Fig. 8.14. Although none are acceptable based on velocity, two out of three are acceptable based on surge pressure is less than design pressure.

300 mm(12 in)Class 200	acceptable
350 mm(14 in)Schedule 40	not acceptable
400 mm(16 in)Schedule 40	acceptable

The next step is economic analysis. Click the *Energy Cost* button on the *Surge* worksheet (Fig. 8.15). First evaluate at the high flow rate for 294 hours.

The present value of energy costs at the low flow rate is shown in Fig. 8.16.

The sum of present value costs are shown in Table 8.17. The pipe diameter with the lowest present value cost for the sum of energy and capital is the 20" pipe.

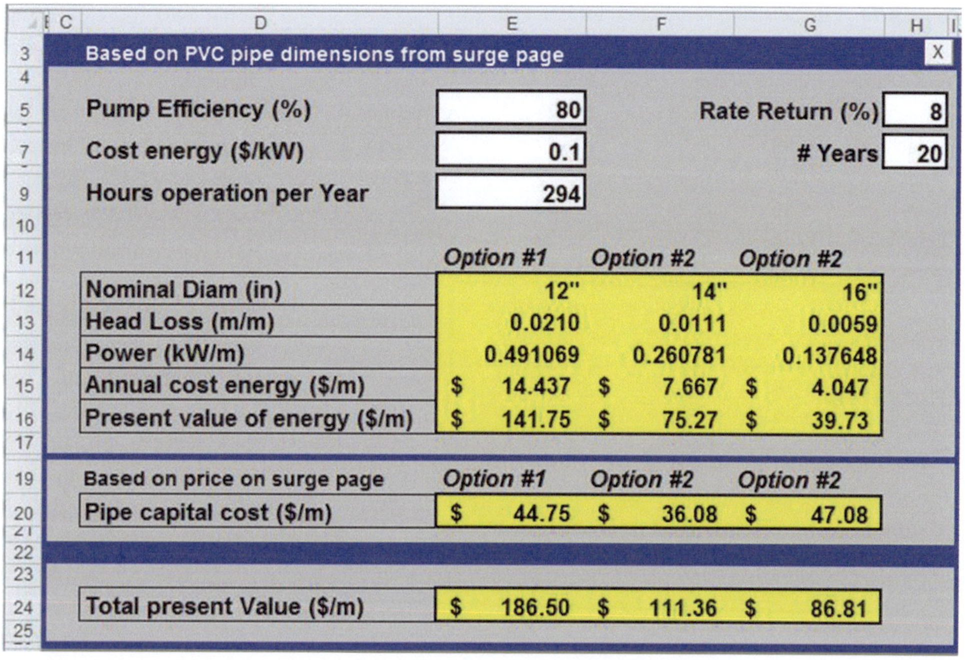

Based on PVC pipe dimensions from surge page

Pump Efficiency (%)	80	Rate Return (%)	8
Cost energy ($/kW)	0.1	# Years	20
Hours operation per Year	294		

	Option #1	Option #2	Option #2
Nominal Diam (in)	12"	14"	16"
Head Loss (m/m)	0.0210	0.0111	0.0059
Power (kW/m)	0.491069	0.260781	0.137648
Annual cost energy ($/m)	$ 14.437	$ 7.667	$ 4.047
Present value of energy ($/m)	$ 141.75	$ 75.27	$ 39.73

Based on price on surge page	Option #1	Option #2	Option #2
Pipe capital cost ($/m)	$ 44.75	$ 36.08	$ 47.08
Total present Value ($/m)	$ 186.50	$ 111.36	$ 86.81

Fig. 8.15 Present value of capital and energy costs at high flow rate for Example 8.10

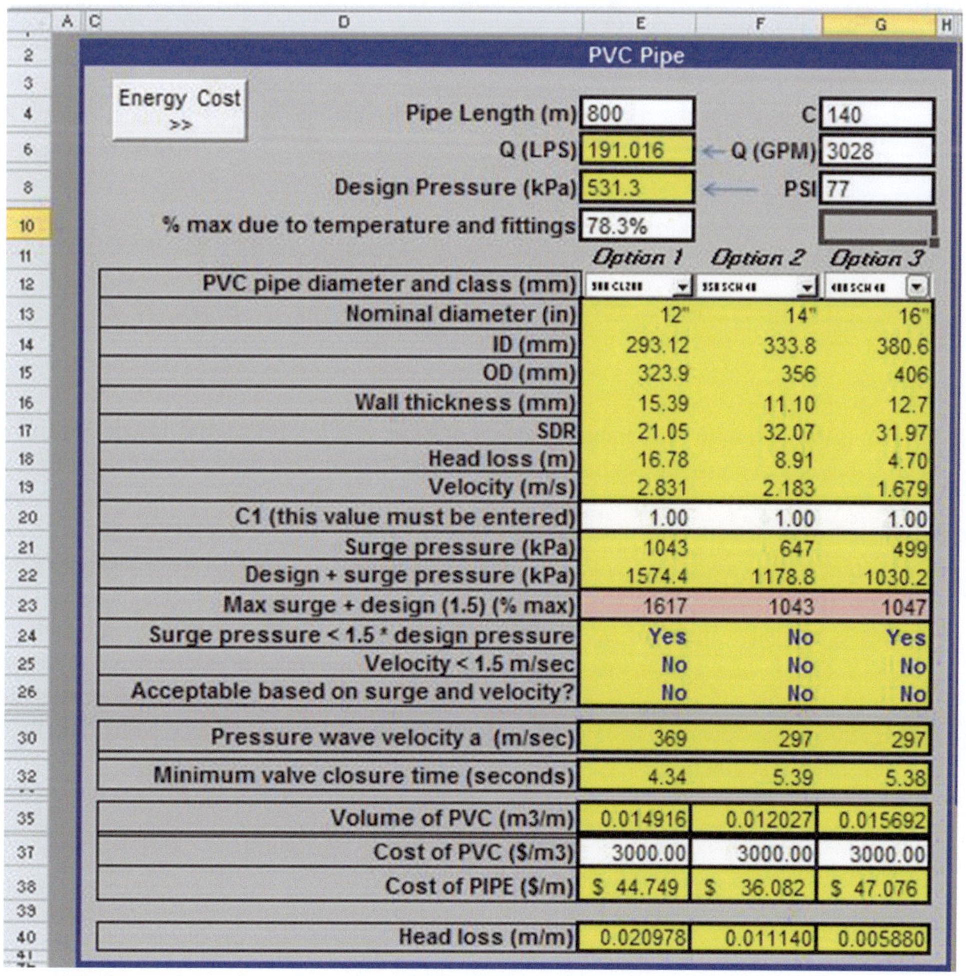

PVC Pipe

Energy Cost >>

Pipe Length (m)	800	C	140
Q (LPS)	191.016	Q (GPM)	3028
Design Pressure (kPa)	531.3	PSI	77
% max due to temperature and fittings	78.3%		

	Option 1	Option 2	Option 3
PVC pipe diameter and class (mm)	[illegible]	[illegible]	[illegible]
Nominal diameter (in)	12"	14"	16"
ID (mm)	293.12	333.8	380.6
OD (mm)	323.9	356	406
Wall thickness (mm)	15.39	11.10	12.7
SDR	21.05	32.07	31.97
Head loss (m)	16.78	8.91	4.70
Velocity (m/s)	2.831	2.183	1.679
C1 (this value must be entered)	1.00	1.00	1.00
Surge pressure (kPa)	1043	647	499
Design + surge pressure (kPa)	1574.4	1178.8	1030.2
Max surge + design (1.5) (% max)	1617	1043	1047
Surge pressure < 1.5 * design pressure	Yes	No	Yes
Velocity < 1.5 m/sec	No	No	No
Acceptable based on surge and velocity?	No	No	No
Pressure wave velocity a (m/sec)	369	297	297
Minimum valve closure time (seconds)	4.34	5.39	5.38
Volume of PVC (m3/m)	0.014916	0.012027	0.015692
Cost of PVC ($/m3)	3000.00	3000.00	3000.00
Cost of PIPE ($/m)	$ 44.749	$ 36.082	$ 47.076
Head loss (m/m)	0.020978	0.011140	0.005880

Fig. 8.16 Present value of capital and energy costs at low flow rate for Example 8.10

Table 8.17 Pipe selection comparison for Example 8.10

Pipe Diam.	Capital	Low flow rate	High flow rate	Total
12 inch	44.75	138.63	141.75	Astronomical
14 inch	36.08	73.62	75.27	Astronomical
16 inch	47.08	38.86	39.73	$126/m
18 inch	59.86	21.84	22.33	$104/m
20 inch	70.15	12.84	13.12	$95/m***
21 inch	76.54	10.13	10.36	$96/m

Questions

1. On the following drawing of a drip irrigation system network, label the type in the well, pump station, submains, mainlines, and laterals.

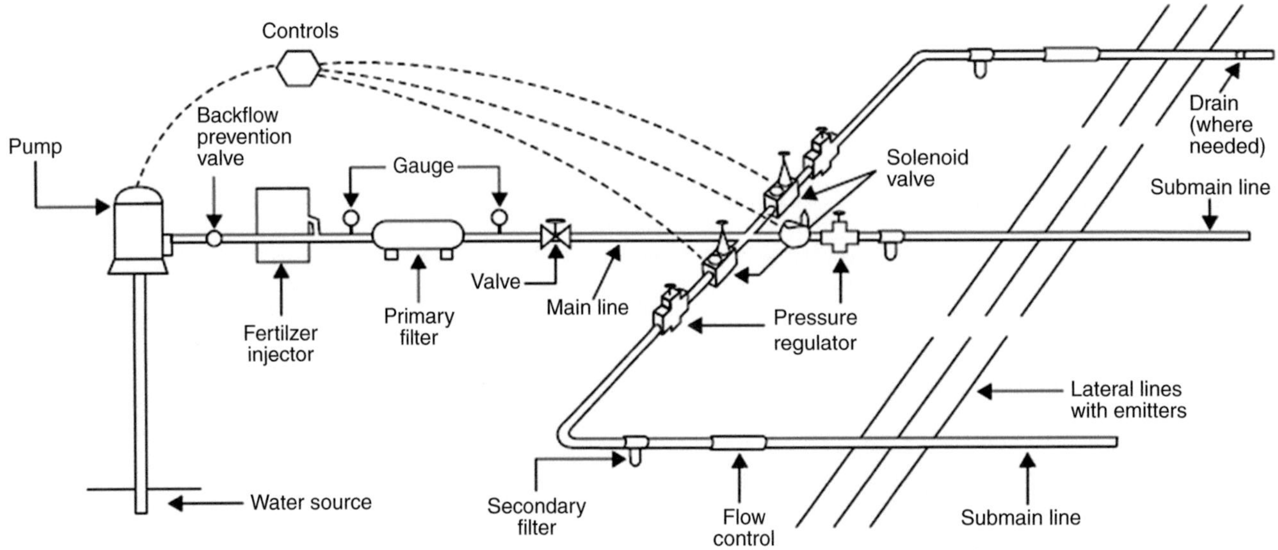

2. What is the hydraulic head and total energy of water in a pipe that is 5 m above the datum with pressure 350 kPa and water velocity 1.5 m/sec?
3. When does nonsteady state flow occur, and what are the possible hazards associated with nonsteady state flow?
4. Maximum allowable flow velocity in PVC irrigation pipes is typically specified as 1.5 m/sec. What is the kinetic energy of water at this velocity? Express your answer in terms of m (length) and kPa (pressure). If the irrigation system operates at 350 kPa, then what percent of the energy is kinetic?
5. What two forces are included in the Reynolds number, and why is turbulent flow observed at higher Reynolds numbers?
6. Calculate the friction loss in 1,000 m of 50 mm nominal diameter class 125 pipe. Calculate for flow velocities of 0.05, 1, and 3 m/sec with the Hazen-Williams and Darcy-Weisbach equations. Use C = 140 and 150 in the Hazen-Williams equation. Show your work.
7. Redo Example 8.2, but the pipe discharges into a pond with a water surface elevation that is 100 meters below the upper pond water surface elevation. The pipe inlet projects into the upper pond. Use the Hazen-Williams equation.
8. For the problem described in Example 8.3, find the flow rate with a known diameter pipe with the Hazen-Williams equation: the nominal diameter is 10 in (250 mm) Class 125 pipe, the pipe length is 300 m, and the Hazen-Williams C value is 150. Recalculate flow rate with the Darcy-Weisbach equation. The inlet pipe projects into the reservoir.
9. Calculate the pressure rating in metric units and convert to psi for 4 inch (100 mm) class 160 PVC (1120) pipe.
10. Ten inch (250 mm) Class 160 bell end pipe has expansion joints. Calculate the velocity and magnitude of the pressure wave if the operating velocity is 2.3 m/sec, and a valve suddenly closes. If the operating pressure is 50 psi (345 kPa), then what is the maximum surge pressure?

11. Flow rate is 10 GPM (37.9 L/min) and the design pressure is 50 psi (345 kPa). Select a pipe class and diameter that does not exceed the maximum allowable surge pressure. Use the surge equations in this case and not just the 1.5 m/s rule. Also perform an economic analysis for the best pipe diameter. Project parameters are \$3,000/m^3 PVC, 20 year, 8 %, 1440 hr/year, \$0.1/kW-hr, pump efficiency = 80 %.
12. Calculate the maximum surge pressure in a long (100 m) drip irrigation tube (12 mm ID polyethylene). Solve the problem in two ways: assume an air vent and no air vent. Inlet pressure is 210 kPa. Conclude by discussing whether pressure relief valves and air vents are needed on long drip irrigation tubes and state the reason for your answer. Also consider sprinkler laterals (PVC) with and without sprinklers along the pipe.
13. For the following parameters, find the best economic pipe size. The price of PVC is \$3,000/m^3. Project parameters are 20 year, 8 %, \$0.1/kW-hr, pump efficiency = 80 %.

References

ASAE/ASABE Standards. American Society of Agricultural and Biological Engineers. Standard S376.1. http://www.asabe.org/publications.aspx

Cuenca RH (1989) Irrigation system design. Prentice Hall, Englewood Cliffs

Stryker J (2014) Landscape sprinkler tutorials website. http://www.irrigationtutorials.com/landscape-sprinkler-system-design-tutorial/

Stryker (2015) Water hammer and air in pipes tutorial website. http://www.irrigationtutorials.com/waterhammer.htm

Pumps

9

Selecting a pump is generally the last step in an irrigation system design. The pump is selected based on the required flow rate and pressure requirements of the irrigation system. The most common pumps used in sprinkler and drip irrigation systems are centrifugal pumps. Pump selection is generally a process of looking through pump catalogs and selecting the pump with highest efficiency at the required flow rate and pressure. The process also includes motor power selection, calculation of the net positive suction head, and possibly trimming the impeller in order to fine-tune the pump to the irrigation system requirements. The affinity laws govern the relationships between impeller diameter, motor frequency (RPM), flow rate, and pressure. The flow rate and pressure relationship for a given impeller diameter is called the pump curve. Pump curves and irrigation system curves can be mathematically combined in order to find the operating pressure and flow rate of the system. There are several possible sources of energy for pumps. The costs of three energy sources (solar, diesel, and electric) are compared in an example. Finally, the chapter covers basic principles of chemigation injection system design. Designing the pump station correctly is an essential final step in the provision of a reliable source of water.

Pump Types

Pumps add pressure energy and/or kinetic energy (velocity). The ratio of the kinetic to pressure energy varies widely between pumps. Axial flow pumps (water passes straight through a propeller or auger) primarily add kinetic energy to water: a propeller increases the velocity of water and adds a small amount of pressure. The water stream does not change direction as it passes through the pump. Propeller pumps are used for low pressure applications, such as pumping water from a canal for surface irrigation. At the other extreme, centrifugal pumps (Fig. 9.1) primarily add pressure energy to water, although it first has the form of kinetic energy within the pump. In centrifugal pumps, water enters into the center of the pump, is thrown by vanes in the perpendicular direction (90° change of direction), and water also exits the pump in a perpendicular direction to the inflow. Centrifugal pumps increase pressure by increasing water velocity in pump vanes (kinetic energy), and then convert kinetic energy to pressure energy as the water velocity slows in the volute (casing). Unlike the axial flow pump, the water exits the centrifugal pump in the perpendicular direction to the inflow direction. A typical application of a centrifugal pump is pressurizing water from a surface water body (pond or canal) for use in a pressurized irrigation system (sprinkler or drip). Centrifugal pumps can be driven by an electric motor (Fig. 9.2) or an engine. Submersible and Line shaft turbines are used to pump water from wells. Line shaft turbines can also be driven by an engine (Fig. 9.3) or motor.

Well pumps are Francis impeller or mixed flow pumps. These pumps throw off water at intermediate angles, between 0° (axial) and 90° (centrifugal). Francis impellers add comparable magnitudes of pressure and kinetic energy to water. The advantage of these pumps is that they can be stacked in wells (Fig. 9.2), one on top of the other. Each of the pumps (called bowls) adds some pressure to the water. Turbine pumps with many bowls can add hundreds of meters of pressure to water and remove water from very deep aquifers. Tens of bowls may be stacked for deep well applications.

Specific speed (ratio of discharge * RPM to pressure) is a measure of the ratio of kinetic to pressure energy. It is calculated to determine the best impeller type and RPM for a given application. Centrifugal, Francis impeller, mixed flow, and propeller flow pumps have specific speeds of 500, 2000, 5,000, and 10,000, respectively. Specific speed is not generally considered by irrigation engineers because the flow and pressure characteristics of available pumps are listed in pump catalogs; however, the number may be useful when developing a new application.

P. Waller, M. Yitayew, *Irrigation and Drainage Engineering*, DOI 10.1007/978-3-319-05699-9_9

$$N_s = 0.2108N\left(\frac{Q^{0.5}}{H^{0.75}}\right) \quad (9.1)$$

where

N_s = specific speed, dimensionless,
N = revolutionary speed of the pump, rev/min,
Q = pump discharge, L/min,
H = discharge pressure, m.

In-class Exercise 9.1 The revolutionary speed of electric pumps is generally slightly less than one of the divisors of 3600; for example, typical pump rpm's are 875, 1750, and 3500. Why are most pumps manufactured with only these revolutionary speeds? Wouldn't it provide more flexibility if the pump could be adjusted to any speed?

Of the four impeller types, Frances impeller and mixed flow impellers, used in deep well turbines, have the highest potential efficiency (in the range of 90 %). Large centrifugal pumps have efficiencies approaching 85 %. The lowest efficiency pumps are axial flow pumps with maximum efficiencies in the lower 80s. Smaller pumps (<30 HP) generally have lower efficiency than large pumps, in the range of 70 %. A caveat to these efficiencies is that pumps are not efficient outside of their design pressure and flow rate. For example, a centrifugal pump is not efficient for low lift applications.

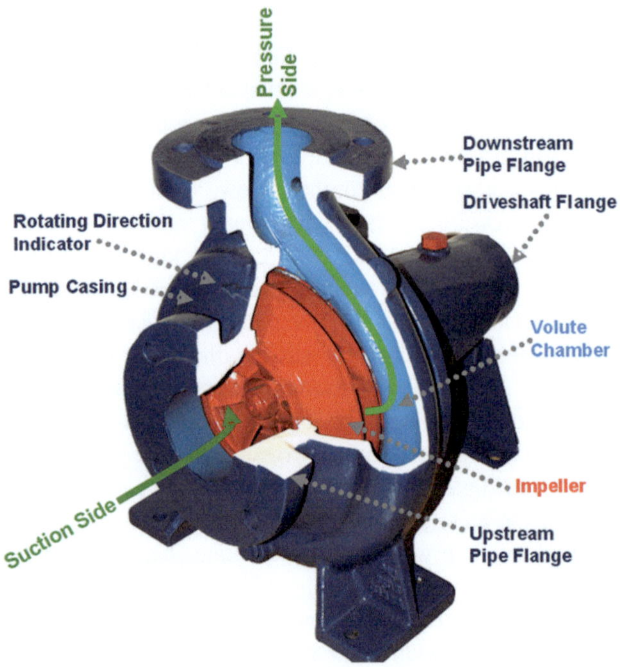

Fig. 9.1 Centrifugal pump (Wikipedia)

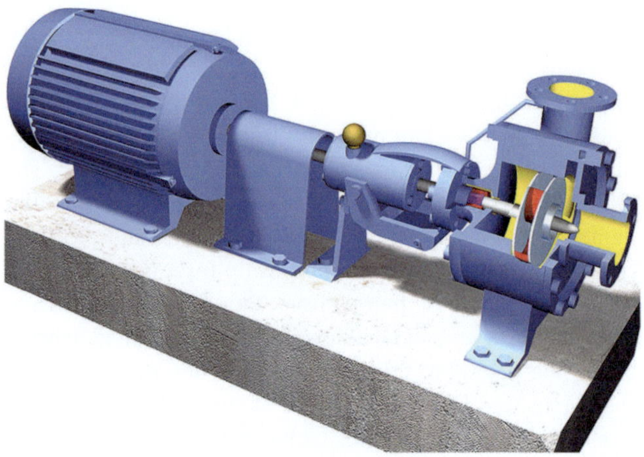

Fig. 9.2 Centrifugal pump and motor (Wikipedia)

Fig. 9.3 Deep well turbine connected to engine (Credit NRCS)

Pump Performance Curves

Pump performance curves are developed by controlling flow with a throttling valve and pressure gauge and evaluating the discharge pressure over a range of flow rates. Pump companies publish pump performance curves, also called pump characteristic curves or head-capacity curves, with this information. The head/capacity curve (Fig. 9.4) is a graph of flow rate vs. pressure. The total dynamic head is the pressure added by the pump at a given flow rate. The pump total dynamic head at zero flow rate is called the maximum shutoff pressure. Many pump curves are available at http://www.wcc.nrcs.usda.gov/ftpref/wntsc/Pump%20 Curves/

In-class Exercise 9.2 Think about the relationships between power, flow rate, and head. Based on those relationships, what would the shape of the head/capacity curve look like if efficiency was constant over a range of flow rates?

Each centrifugal pump model comes with a range of impellers and motors. For example, the impeller diameters for the pump in Fig. 9.4 range from 4.75 in to 5.9375 in. Total dynamic head can be adjusted for centrifugal pumps by adjusting the diameter of the impeller.

In-class Exercise 9.3 The irrigation system requirement is 600 gpm and 160 ft head. Select the impeller that results in this flow rate in Fig. 9.5.

Pumps designed to operate at high pressure should not be operated with no backpressure. If pumps are operated at a higher flow rate than the recommended operating range, then the motor is spinning too fast, which pushes high amperage through the windings of the motor, and the motor overheats. Head-capacity curves show the limit of the allowable operating range by ending the pump curve; for example, the 5.625 impeller in Fig. 9.4 should not be pumped at less than 6 m (15 ft) back pressure.

In Fig. 9.5, a head-capacity curve is plotted for each of the standard impeller diameters for the model B4JPBH pump. For example, the 11 3/8″ diameter impeller has a TDH of 145 ft at 200 gpm. Efficiency curves in Fig. 9.5 show that the 11 3/8″ impeller is most efficient at 740 gpm where the 11 3/8″ curve crosses the 80 % efficiency line.

For an extra fee, the pump company will trim a standard impeller to match the flow and pressure requirements of an irrigation system. The relationship between impeller diameter and flow, head, and power can be calculated with affinity laws. The discharge is directly proportional to the impeller diameter (Fig. 9.2), the flow rate is proportional to the square

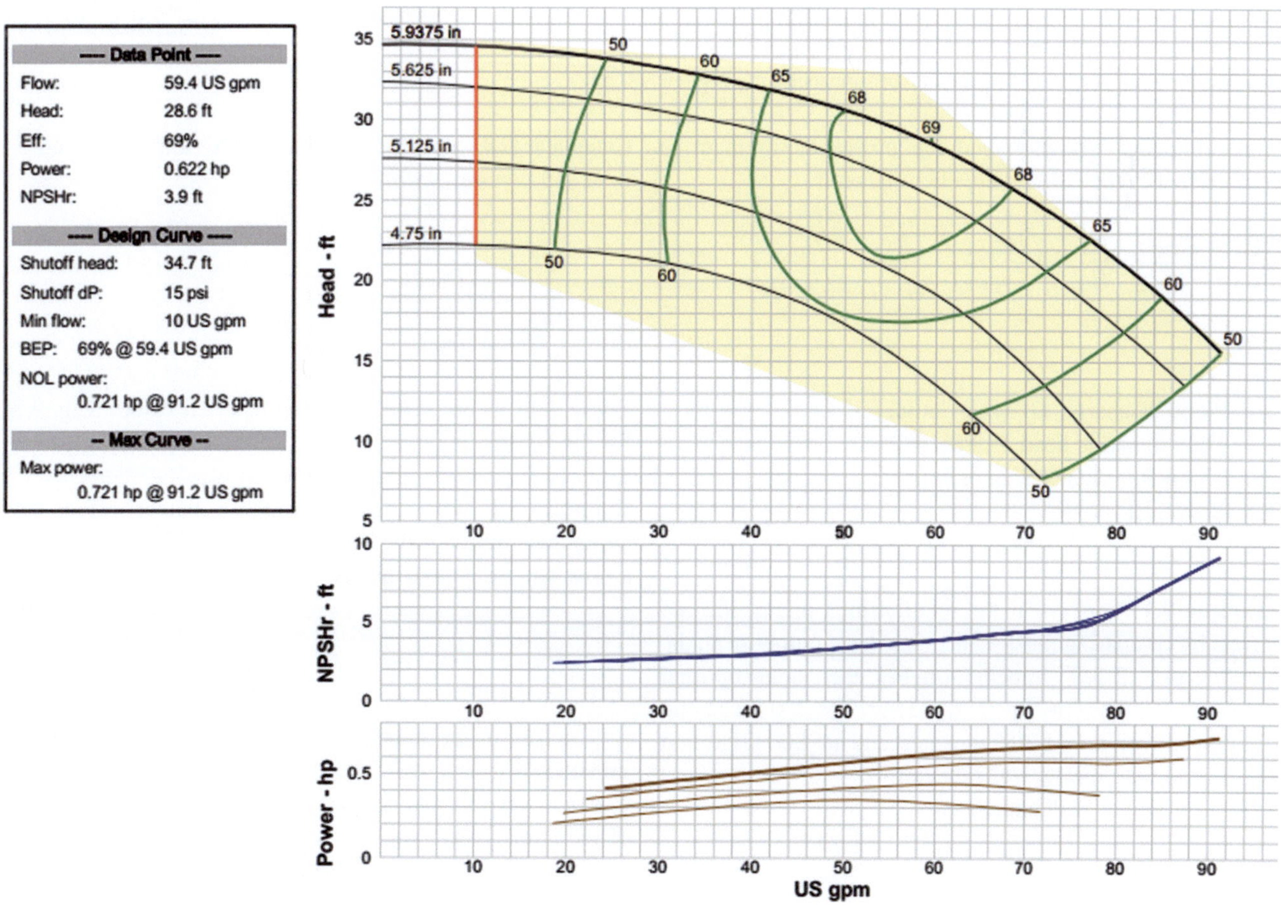

Fig. 9.4 Pump head-capacity curve for Goulds pump (Credit NRCS)

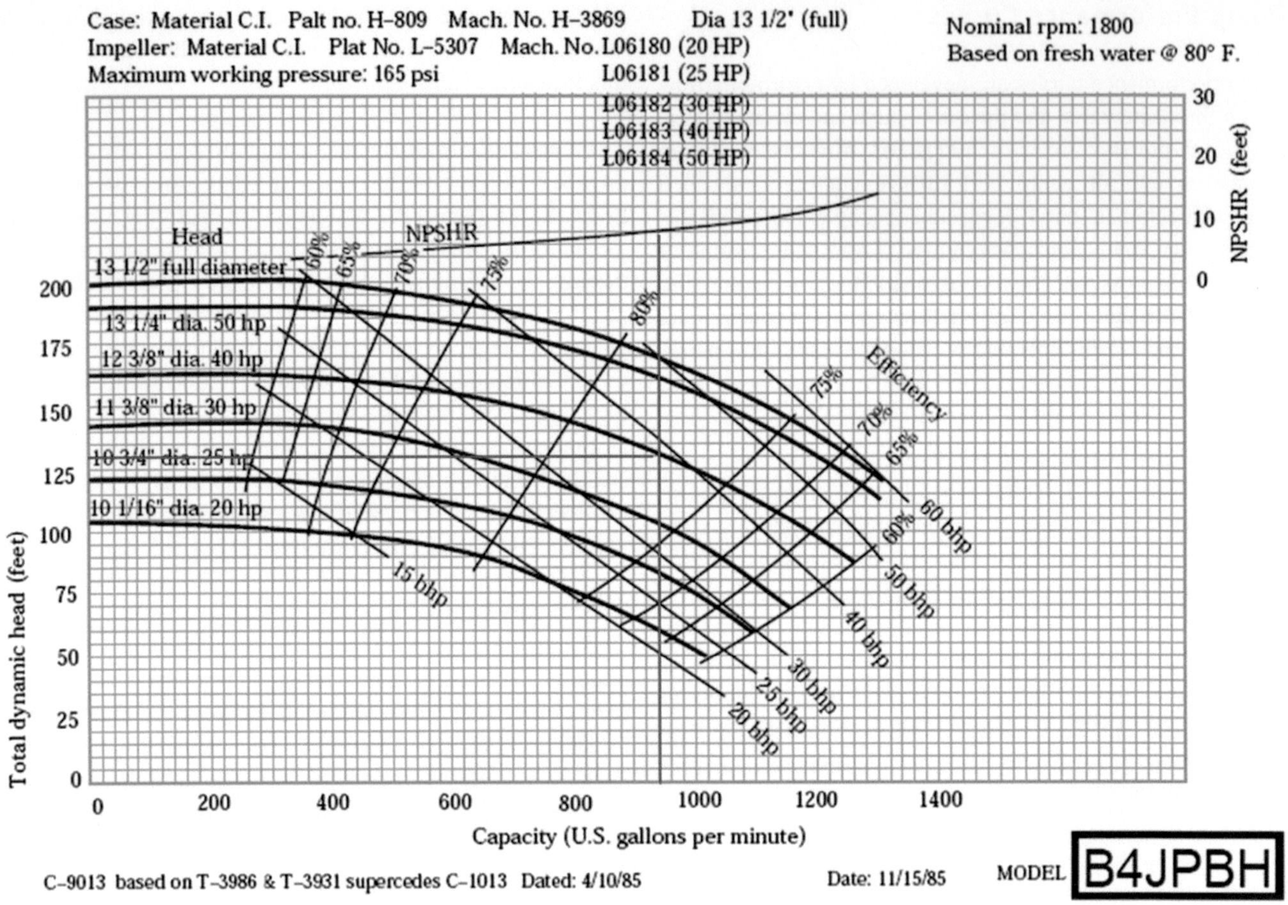

Fig. 9.5 Berkeley pump centrifugal pump performance curves (From NRCS)

of the impeller diameter, and the power requirement is proportional to the cube of impeller diameter.

$$\frac{Q_1}{Q_2} = \frac{D_{\mathrm{Im-1}}}{D_{\mathrm{Im=2}}} \quad \frac{H_1}{H_2} = \left(\frac{D_{\mathrm{Im-1}}}{D_{\mathrm{Im=2}}}\right)^2 \quad \frac{P_1}{P_2} = \left(\frac{D_{\mathrm{Im-1}}}{D_{\mathrm{Im=2}}}\right)^3 \qquad (9.2)$$

where

Q = Pump discharge, GPM or L/min,
H = Total dynamic head, ft, PSI, or m,
D_{im} = Impeller diameter, in or cm,
P = Power, HP or kW.

Example 9.1 Verify that the relationships in Eq. 9.2 are correct by comparing the discharge and head output from the 11 3/8″ and 12 3/8″ impellers in Fig. 9.5 at the optimal efficiency (80 %): flow rate and pressure are 740 GPM and 122 ft head with the 11 3/8″ impeller, and are 800 GPM and 145 ft head with the 12 3/8″ impeller.

Ratio of impeller diameters is

$$12.375/11.375 = 1.08$$

Calculate TDH and flow for the 12 3/8″ impeller based on flow and pressure with the 11 3/8″

$$H_2 = \left(\frac{D_{\mathrm{Im-2}}}{D_{\mathrm{Im=1}}}\right)^2 H_1 = 1.08^2{*}122 = 144\ ft$$

$$Q_2 = \left(\frac{D_{\mathrm{Im-2}}}{D_{\mathrm{Im=1}}}\right) Q_1 = 1.08{*}740 = 800\ gpm$$

These values correspond with the reported values (Fig. 9.5) for the 12 3/8" impeller.

Example 9.2 What trimmed impeller diameter (Fig. 9.5) would be required if the customer needed a TDH of 133 ft at a flow rate of approximately 760 gpm?

$$D_{\mathrm{Im-2}} = D_{\mathrm{Im-1}}\sqrt{\frac{H_2}{H_1}} = 11.375\sqrt{\frac{133}{122}} = 11.876$$

The trim of the impeller refers to the outer radius of the impeller (R in Fig. 9.6). The manufacturer leaves everything else the same in a given pump model but just trims the radius of the already machined impeller. The volute also remains the same. Thus, the distance between the volute casing and

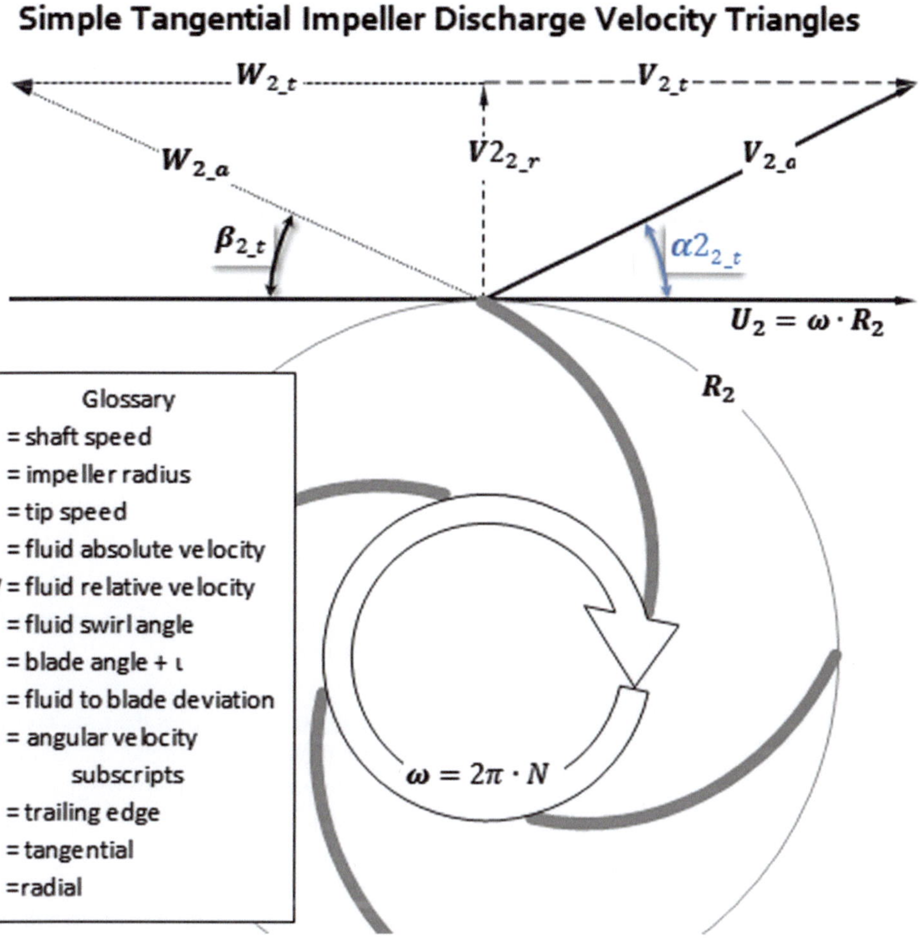

Fig. 9.6 Impeller characteristics (Credit Wikipedia, Koronowski)

the impeller increases when the impeller is trimmed. Because of the shape of the impeller, the discharge velocity angle decreases with the radius, and high efficiency is maintained. Note that kinetic energy is converted to pressure energy in the volute.

Motor brake horsepower (BHP) curves are also plotted in Fig. 9.5. In order to protect the pump motor, the pump must never be operated at a point where the impeller curve (water power) exceeds the BHP curve by even a small amount. This also causes excessive current to travel through the windings in the motor, and this excess current heats and burns up the motor. Thus, the selected motor BHP must always exceed the WHP at the range of flow rates and pressures expected for the pump/irrigation system. This concept is associated with the low backpressure rule because the water horsepower (WHP) curve exceeds the BHP curve at low backpressure. Note that the BHP curves in Fig. 9.5 are straight lines while the pump curves are curved, which causes the BHP requirement to exceed the WHP curve at some point.

Centrifugal pumps should always be started with a closed throttling valve (gate valve), and then the pipeline should be filled slowly in order to avoid overheating the motor. Pumps should not be run against a completely closed valve for more than 30 seconds. Pumps are water cooled, and if no water is moving through the pump, then the pump will overheat.

Example 9.3 Select the appropriate motor (BHP) in Fig. 9.5 if the expected flow rate and head are 800 GPM and 145 ft head, respectively.

The 12 3/8″ impeller head-capacity curve matches the specified requirements. The 40 HP motor exceeds the 12 3/8″ impeller curve (Fig. 9.5) at the operating point. If it is expected that the irrigation system will never exceed 1,000 GPM (see Fig. 9.3), then the 40 HP motor is sufficient. However, if there is some reason that the pump will exceed 1,000 GPM for a process such as flushing the irrigation system, then the 40 HP motor would not be large enough.

According to Fig. 9.5, the pump operates at 80 % efficiency at the operating point. Although the plotted head-capacity curve and BHP curve are normally used to determine the required motor HP, the pump BHP requirement can also be calculated with Eq. 2.21 as follows:

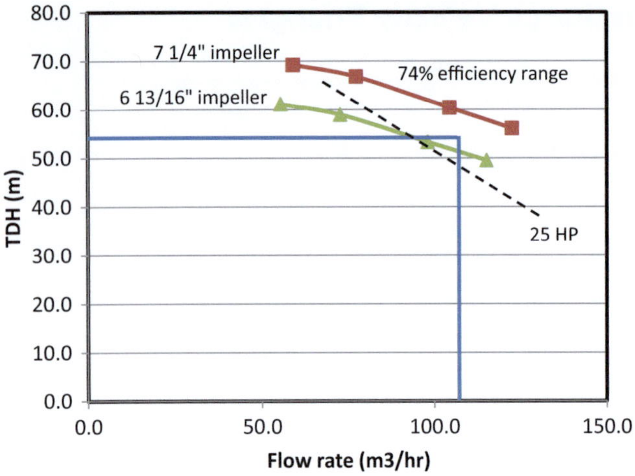

Fig. 9.7 Goulds Model 16BZ with 7¼″ and 6 13/16″impeller curves (units converted to metric) and required operating point for the irrigation system in Example 9.4

$$BHP = HP = \frac{GPM*ft}{3,960*Eff} = \frac{800*145}{3,960*0.8} = 37\ HP$$

Select the 40 HP motor because the power requirements are rounded up to the next motor size.

Example 9.4 An irrigation system design requires 107 m^3/hr, and TDH = 49 m. In addition, there is 2 m head loss in pump fittings and valves and 3 m pressure is added to account for degradation of the pump over time. Thus, the required pump discharge pressure is 49 + 2 + 3 = 54 m. Find the operating point.

The pump is a Goulds Model 16BZ 3 × 4 – 8). Only the 7 ¼″ and 6 13/16″ impeller curves are shown (units converted to metric). The numbers, 3 × 4 – 8, refer to the discharge and inlet diameters, 3″ and 4″, respectively, followed by the maximum impeller diameter (8″).

The blue lines in Fig. 9.7 show the required flow rate and pressure. The 25 HP motor line is the dashed line. The pump curve calculations for this example are in the *Pump curves* worksheet.

The selection of the impeller diameter causes a bit of a dilemma because the design point is just slightly higher than one of the standard impeller diameters, (6 13/16″). Remember that 3 m head was added as a safety factor. One could cut the safety factor, but this should be explained to the farmer.

Do you want to purchase a trimmed impeller with a safety factor or do you want to spend less money on a standard impeller and have no safety factor? The smaller impeller will use less energy and be less expensive, but when the pump degrades over time, there might be inadequate pressure for the design flow rate; thus, sprinkler flow rates will decrease.

If the farmer selects the standard 6 13/16″ impeller, then the pump operates at a slightly higher pressure than 25 HP motor curve. Because it is never acceptable to select a HP curve that is below the pump curve at the operating point, select a 30 HP motor. Selecting the 30 HP motor does not mean that 4 HP will be wasted. The 30 HP motor will only use the fraction of HP required (approximately 26 HP) to run the pump. If the farmer decides to use a trimmed impeller on the 16BZ pump, then affinity laws can be used to select the correct impeller size.

$$D_{\mathrm{Im}-2} = D_{\mathrm{Im}-1}\sqrt{\frac{H_2}{H_1}} = 6.8125\sqrt{\frac{53}{50.5}} = 7''$$

The calculation of the 7″ impeller flow rate and head are shown in the *Pump curves* worksheet.

Finding the Intersection of the System Curve and Pump Curve

A system curve describes the pressure-flow relationship for the irrigation system: higher pressure generally results in higher flow rate. For many irrigation systems, the flow is proportional to the square root of pressure because many irrigation components are orifices or dissipate energy with turbulent flow. A typical irrigation system has a flow rate that is approximately proportional to the square root of pump pressure (x = 0.5).

$$Q_{system}\left(m^3/h\right) = C_{system} TDH^x \tag{9.3}$$

where

TDH = pump pressure, m,
Q_{system} = irrigation system flow rate, m^3/hr,
x = system flow exponent,
C_{system} = system coefficient.

If an equation for the pump head-capacity curve is found by regression from the manufacturer's pump curve, then the head-capacity equation and the system equation can be solved simultaneously in order to find the system operating point.

Example 9.5 The 16BZ pump (Fig. 9.7) with a 7¼″ impeller is used to run a sprinkler system with the following system curve:

$$Q_{system}\left(m^3/h\right) = 14.175\left(H_{system}\right)^{0.531}$$

Find the operating point of the irrigation system. This is the intersection of the pump curve and system curve.

Assume that there is a 4 m pressure loss in pump fittings and filters. This problem is solved with the *System and pump curves* worksheet. Adjust the system curve to account for pump fitting and filter losses ($H_{system} = TDH - 4$ m).

$$Q_{system}\left(m^3/h\right) = 14.175(TDH - 4)^{0.531}$$

The 7¼″ impeller pump-capacity curve and the system curve are plotted in Fig. 9.8 (metric units). The operating point is at the intersection of the two curves. In order to calculate the operating point, a second order polynomial can be calculated for the pump head-capacity curve with the 7¼″ impeller with Excel Trendline (Fig. 9.8). Next substitute the head-capacity equation for TDH in the system equation.

$$TDH = -0.00111Q^2 - 0.00926Q + 73.74$$
$$Q = 14.175\left(-0.00111Q^2 - 0.00926Q + 73.74 - 4\right)^{0.531}$$
After iteration, $Q = 118\,m^3/hr$, $TDH = 58$ m.

If a farm has multiple center pivots, then some or all pivots may run at the same time. Multiple, pumps can be connected to a central manifold. Each of the pumps in the manifold is equal to the flow rate of one center pivot; however, all of the pumps must be sized for the worst case (pivot with highest pressure requirement).

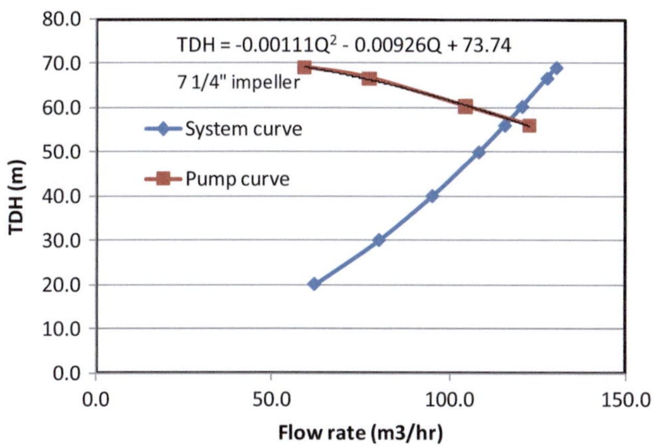

Fig. 9.8 System curve and pump curve for Example 9.5

Example 9.6 Calculate the TDH required for four parallel pumps in a pump manifold that supply the four center pivots shown in Fig. 9.9. A filter is located at the pump station with a 2 m pressure loss. Pump fittings losses are 3 m. Supply 2 m extra pressure as a safety factor. Pivot pipe loss is 2 m. Minimum sprinkler pressure is 12 m and sprinklers are 1 m above the ground surface.

Construct the following table to find the pressure requirement at each pivot inlet. Calculate the pressure required at each pivot inlet by summing all pressure losses and elevation gains (Table 9.1).

All pumps must supply 50 m TDH, the worst case. Note that the system is designed for pivot 2, not for the farthest pivots, 3 and 4, because pivot 2 is the worst case with respect to pressure. Pivots 3 and 4 are down a hill and gain pressure by elevation change. Pressure regulating valves may be needed at pivots 1, 3, and 4 in order to dissipate extra energy. It is possible that a lower pressure pump could be used to supply water when only pivot 1 is running. A second alternative would be to install booster pumps at the three pivots with higher pressure requirements and run the pumps at the main pump station at 30 m TDH.

In-class Exercise 9.4 Redo Example 9.6 except assume that pivot 2 is at 100 m elevation and pivots 3 and 4 are at 120 m elevation. Select the pump operating pressure. If each pivot requires 100 L/sec, then how many pumps should be installed and what should their flow rate be?

Variable Speed Pump Controllers

As shown in the previous examples, if a farm has irrigation zones that operate at different pressures, then using a single-speed pump will result in wasted energy. Unlike pivot farms where all pivots may operate at the same time, orchard and turf (permanent sprinklers) irrigation systems only operate one zone at a time. However, the problem with variable pump requirements is the same. Orchards and turf systems

Fig. 9.9 Maximum pivot elevations (E) and pipeline pressure (h_f) losses

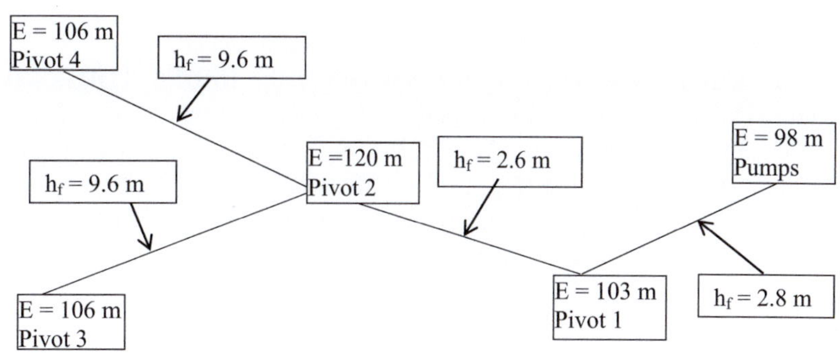

Table 9.1 Pump pressure requirements

Component (All hf in meters)	Pivot 1	Pivot 2	Pivot 3	Pivot 4
Sprinkler pressure (+ elevation) required	13	13	13	13
Pressure loss in pivot pipeline	2	2	2	2
Elevation difference (Max pivot – pump Elev)	5	22	8	8
Cum. pressure loss in mainline prior to pivot	2.8	5.4	15	15
Screen filter	2	2	2	2
Pump fittings losses	3	3	3	3
Safety factor	2	2	2	2
Total pressure requirement (TDH)	29.8	49.4	45	45

are often laid out on hills, which may have zones with dramatically different pressure requirements. If a pump supplies two or more zones with different pressure and flow requirements, then the farmer has four options. First, pressure can be reduced to the lower pressure zone with a pressure regulator. Second, the size of the upper zone can be reduced so that it runs at a different point on the pump curve with more pressure and less flow. Third a booster pump can be installed to increase pressure to the upper zone. Fourth a variable speed pump controller can vary the frequency of AC current that is delivered to the pump motor, which then changes the rotational velocity and discharge characteristics of the pump.

Pump discharge, pressure, and power relationships for variable speed pumps are calculated with the following affinity laws, which are similar to Eq. 9.2 but vary RPM rather than diameter:

$$\frac{Q_1}{Q_2} = \frac{RPM_1}{RPM_2} \quad \frac{H_1}{H_2} = \left(\frac{RPM_1}{RPM_2}\right)^2 \quad \frac{P_1}{P_2} = \left(\frac{RPM_1}{RPM_2}\right)^3 \tag{9.4}$$

where

RPM = rotational speed, revolutions per minute.

Variable speed pump controllers can save significant energy if an irrigation pump is sized for a zone with one flow rate and pressure but must also service a smaller zone. They can increase or decrease flow with constant pressure, increase or decrease pressure with constant flow, or match the pressure and flow rate to the system curve requirements.

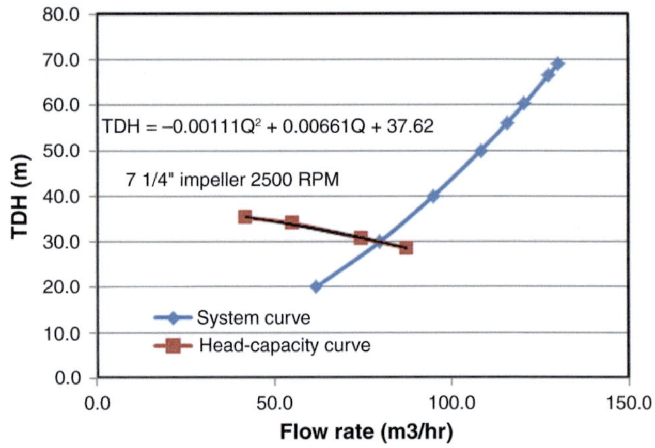

Fig. 9.10 Head-capacity curve for Model 16BZ pump at 2500 RPM for Example 9.7

Example 9.7 A farmer decides to operate his irrigation system at half of the original pressure and use low-pressure sprinkler nozzles (hexagon-shaped orifice). A variable speed pump controller is used to vary the flow rate of the 16BZ pump (Fig. 9.7) with the 7¼″ impeller. The revolutionary speed is lowered from 3500 to 2500 RPM. The system curve is Q_{system} (m^3/h) = 14.175 $(H_{system})^{0.531}$. There is 5 m head loss in the pump fittings and filters. Find the operating point TDH and flow rate. This problem is solved in the *Variable speed pumps* worksheet. Assume that efficiency remains the same at the new RPM, which is probably not a good assumption.

The pump head-capacity curve can be adjusted downward by calculating the adjusted flow rate at four points with Eq. 9.4. For example, at one point on the curve, the flow rate is 260 gpm and 227 ft.

$$Q_2 = \frac{RPM_2}{RPM_1} Q_1 = \frac{2,500}{3,500} 260 = 186 \ gpm$$

$$H_2 = \left(\frac{RPM_2}{RPM_1}\right)^2 H_1 = \left(\frac{2,500}{3,500}\right)^2 227 \ ft = 116 \ ft$$

The adjusted pump curve, its equation, and the system curve (in metric units) are plotted in Fig. 9.10.

$\mathrm{TDH} = -0.00111\mathrm{Q}^2 - 0.00661\mathrm{Q} + 37.62$

$\mathrm{Q} = 14.175\,(-\,0.00111\mathrm{Q}^2 - 0.00661\mathrm{Q} + 37.62 - 5)^{0.531}$

After iteration, Q = 79 m^3/hr, TDH = 30 m.

The power requirement is 1/3 of the original. Power is calculated with Eq. 2.20

$$\text{Power(kW)} = (\text{Q(m}^3/\text{ sec)*H})/(0.102\text{*Eff}) = (118/3600\text{*}58)/(0.102\text{*}0.74) = 25\text{ kW}$$
$$\text{Power(kW)} = (\text{Q(m}^3/\text{ sec)*H})/(0.102\text{*Eff}) = (79/3600\text{*}30)/(0.102\text{*}0.74) = 8.7\text{ kW}$$

The energy savings would not be as great since the lower pressure version would need to operate for a longer time in order to supply the same amount of water.

Centrifugal Pump Installation

Centrifugal pumps should be mounted and leveled on a concrete pad. Whenever possible, install a centrifugal pump at an elevation below the water surface. Otherwise, pumps must be primed. Even though it is desirable to have the pump intake below the water surface, pumps should not be placed in a hole because the pump will overheat.

If the pump must be installed above the static water level, then the pump should be installed as shown in Fig. 9.11. The suction must be placed at a depth in the water that is not going to cause a vortex when water is sucked into the pipe. The rule of thumb is typically that the intake should be 30 cm below the water surface; however, equations are available that calculate minimum elevation below the water surface as a function of flow rate. If there is a vortex, a plastic sheet can be floated on the water surface or a vortex breaker can be placed near the pump suction inlet.

The goals in pump suction design are to avoid excessive turbulence before water enters the pump, which reduces efficiency, and to avoid cavitation, which is caused by low pressure in the suction line, an air vapor leak in the suction line, or an air pocket at a high point in the suction line. The flow velocity in the pump suction is typically 1.5 m/s and up to 6 m/s in the pump inlet. Thus, a larger diameter pipe than the inlet nozzle delivers water from the pond to the inlet nozzle. Excessive turbulence can be avoided by using a

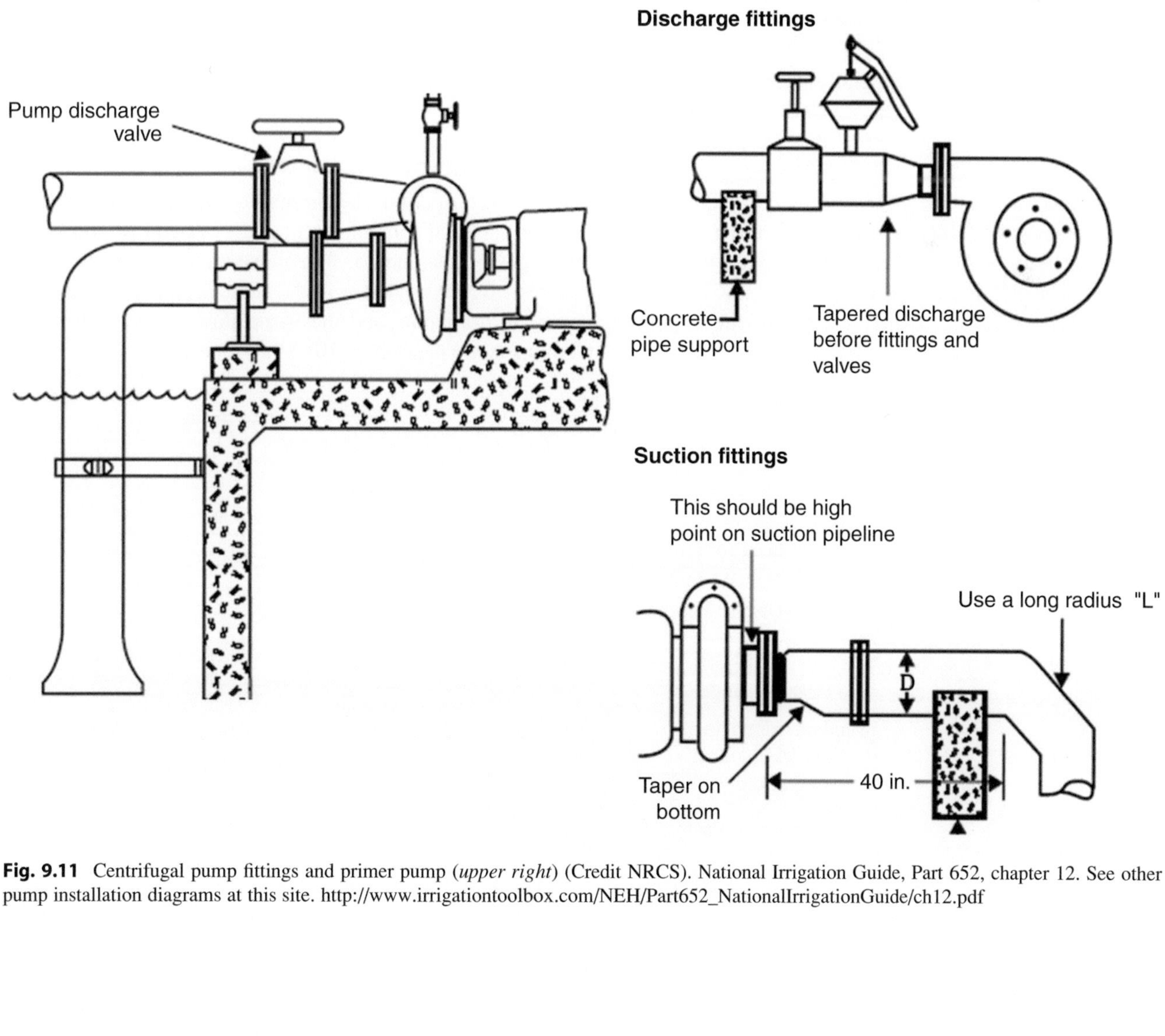

Fig. 9.11 Centrifugal pump fittings and primer pump (*upper right*) (Credit NRCS). National Irrigation Guide, Part 652, chapter 12. See other pump installation diagrams at this site. http://www.irrigationtoolbox.com/NEH/Part652_NationalIrrigationGuide/ch12.pdf

sweep 90 (Long radius) or two mitered 45 degree ells. This is followed by a horizontal pipe leading into the pump that is at least four pipe diameters in length. Pipe diameter is reduced with an eccentric (Fig. 9.11), which is a cone cut at an angle. The purpose of the eccentric is to maintain the top of the pipe at one elevation so that gas pockets are not trapped prior to the reducing cone, which would then be entrained into the flow, and then cause cavitation into the pump. Although use of an eccentric is a standard design practice on horizontal inlet pipes, is recommended by pump manufacturers, and should therefore be followed, some engineers disagree with this practice and think that it causes excess turbulence and does not reduce cavitation. On the discharge side of the pump, an expansion cone (Fig. 9.11) is placed just after the outlet. A sudden expansion right after the pump causes turbulence and loss of energy. A gate valve is placed after the cone.

For pumps that are not self-priming and are located above the water level, a primer pump can be connected to the discharge side of the pump (Fig. 9.11). If the pump is powered by an internal combustion engine, then the exhaust from the engine can be connected to a Venturi valve in order to create a priming pump. The Venturi creates a vacuum and thus sucks water into the pump.

The pump suction inlet generally includes a foot valve and a screen. A foot valve (check valve) can keep water in the pump suction after shut off so that the pump does not need to be primed before each use (Fig. 9.12). If there is debris in the water source, then screens are installed at the pump inlet (Fig. 9.12). For large debris loads, large cylindrical screens with an internal rotating sprinkler to keep the screen free of debris may be needed. Alternatively, water can be screened before or after the pump with a stainless steel screen.

In order to safeguard pumps, pump switches should be installed on the discharge side of pumps. These switches measure the discharge pressure from the pump and shut down the pump if there is no pressure.

When a centrifugal pump is higher than the water source, then the water in the suction pipe will be under negative pressure (less than atmospheric). If the pressure is too negative, then there is a danger that water will enter the gas phase and form bubbles in the water. The pump increases the water velocity inside the pump and thus lowers the water pressure even further. The bubbles implode when they hit the impeller. This causes pitting and eventually destroys the impeller. Each pump has a specified net positive suction head required ($NPSH_R$) which is the minimum absolute pressure (not gage pressure) in the suction pipe at the pump inlet that will not result in bubble formation and cavitation in the pump. The $NPSH_R$ increases with pump flow rate; for example, the $NPSH_R$ curve is shown on the B4JPBH pump curve diagram (Fig. 9.5) with $NPSH_R$ increasing with pump flow rate.

The actual NPSH at the pump inlet is calculated as follows:

NPSH = Atmos. pressure – water vapor pres. – pump elev above water – friction loss in suction

Water vapor pressure at 30 °C is 0.43 m. Atmospheric pressure is 9.9 m.

Example 9.8 Calculate NPSH and determine whether a pump suction with the following design characteristics meets the $NPSH_R$ requirement at the design operating point in Fig. 9.5. The pump is 3 m above the surface of the pond, there is 0.5 m head loss in the suction pipe and foot valve prior to the pump, and the pump flow rate is 1,000 gpm. Atmospheric pressure is 101 kPa (9.9 m).

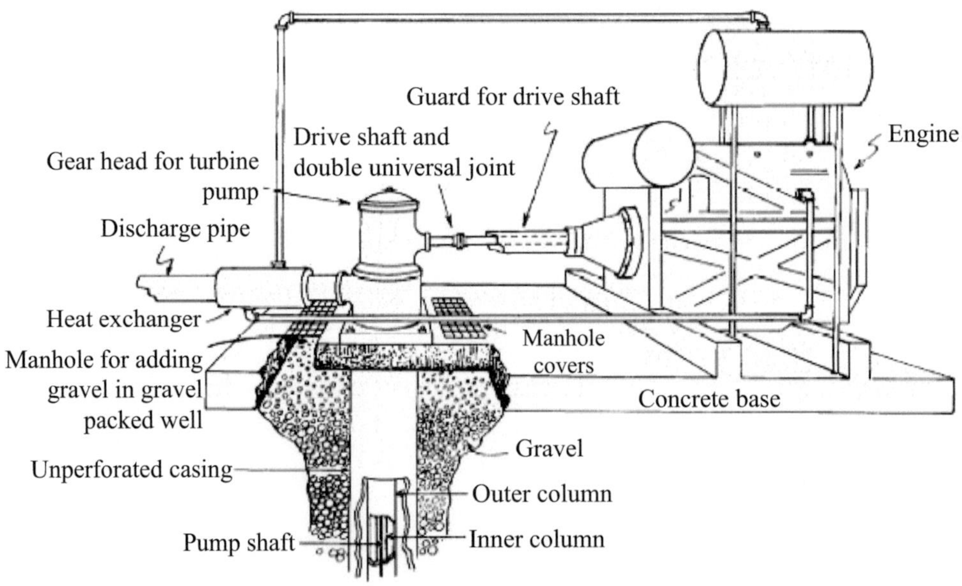

Fig. 9.12 Line shaft turbine pump (Credit NRCS)

The net positive suction head (NPSH) is calculated as follows:

$$\text{NPSH} = \text{Atmospheric pressure} - \text{water vapor pressure} - \text{pumpelevation} - \text{friction loss}$$
$$\text{NPSH} = 9.9\text{ m} - 0.43\text{m} - 3\text{m} - 0.5\text{m} = 6\text{m}.$$

The $NPSH_R$ requirement for 1,000 gpm flow rate in Fig. 9.5 is 8 ft (2.4 m). The design NPSH at 6.0 m is greater than the $NPSH_R$ requirement (2.4 m); thus, the pump installation will not cause cavitation and is acceptable.

Turbine Pumps (Well Pumps)

There are two primary types of irrigation well pumps: submersible pumps and line shaft turbines. Line shaft turbines can be driven by a diesel engine or an electric motor at the ground surface (Fig. 9.12). Alternatively, an electric motor can be mounted directly over the well. The pump hangs from a steel pipe called the outer column (Fig. 9.12). The shaft that delivers torque to the pump is contained inside the inner column. The pump shaft is kept in the center of the inner column by bearings. Line shaft turbines and submersible pumps have a series of bowls with their own impellers, and all impellers are connected to the same shaft.

Turbines can also be used to pump from shallow wells, reservoirs (pump vaults) near the ground surface, and can be used as booster pumps.

The pump shown in Fig. 9.13 is a submersible turbine. The motor hangs below the the pump, and power is delivered to the motor with a power cable in the well. The advantage of "submersibles" is that they do not need a long shaft in the well connecting the pump and motor.

Some deep well pumps are equipped with a "soft start" mechanism, which varies RPM with a variable speed controller. The soft start allows the pump to start slowly and then increases the revolutionary speed of the pump over time. Starting the pump slowly prevents damage to the pump and damage to the irrigation system. The variable speed controller only operates during start up and does not change the revolutionary speed of the pump during normal operation. The reason for this is that turbine pumps in wells use mixed flow impellers, which are generally designed to operate at just one rotational speed; their efficiency decreases if the speed changes. The reason for this is that the water must come off the impeller blades at the same angle as the vanes in the bowl in order to achieve maximum pump efficiency. If the speed changes, then the angle changes.

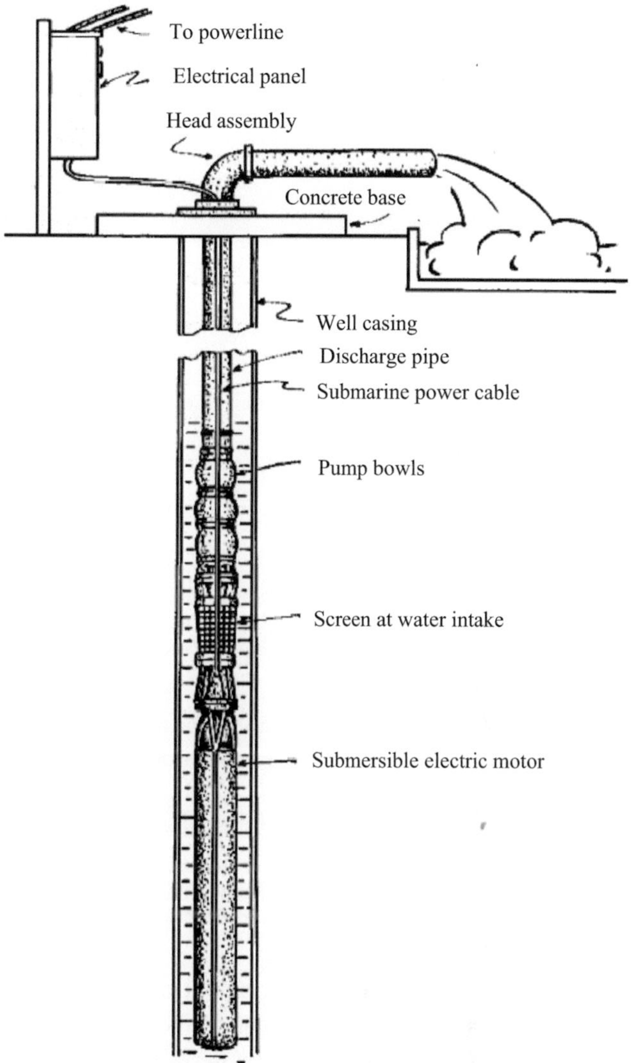

Fig. 9.13 Submersible pump (Credit: Natural Resource Conservation Service, National Engineering Handbook, Sec. 15)

Pump Station Power and Cost

Pump station power can be separated into water horsepower, brake horsepower, engine or motor power, and fuel source power. Energy losses are found in each step of the transfer of energy from the fuel source to the water. In addition, energy losses take place in pump station fittings, pipes, valves, and filters.

The water horsepower (not including efficiency) is the power actually delivered to the water and can be calculated for different units as follows.

$$\begin{aligned} &\text{WHP} = (\text{Qrg}) \\ &\text{HP} = \text{gpm*ft}/3960 \quad \text{(U.S. units horsepower)} \\ &\text{mhp} = \text{L/min*m}/4634 \quad \text{(metric horsepower)} \\ &\text{kW} = \text{HP*0.746} \quad \text{mhp} = \text{HP}/1.01422 \quad \text{kW} = \text{mhp*0.757} \end{aligned} \tag{9.5}$$

Table 9.2 Nebraska pumping plant performance criteria for fossil fuel powered pumps (After Huffman et al. (2013)). Listed values are power delivered to the pump, and not raw energy content of fuel

Energy source	b-kW-hr per unit	w-kW-hr per unit	Unit	Overall efficiency
Diesel	3.282	2.46	Liter	23
Gasoline	2.273	1.71	Liter	17
Liquid propane	1.813	1.36	Liter	18
Natural gas	2.166	1.62	Liter	17

where

WHP = water horsepower, HP.

The brake horsepower is the power delivered by the motor or engine to the pump.

$$BHP = WHP/E_p \tag{9.6}$$

where

BHP = brake horsepower (U.S. units),
E_p = efficiency of pump.

The power required by the pump/motor assembly is the BHP divided by the efficiency of the motor or engine. In the case of an electric pump, the pump and motor are generally sold as a unit and the efficiencies of the pump and motor are not separated. Thus, the HP requirement for electric pumps (Figs. 9.5, 9.6, and 9.7) refers to the power requirement of the motor, and not to the BHP supplied by the motor to the pump (Beard and Hill 2000).

Even though an electric motor is rated at a certain HP, this is not necessarily the power output of the motor during operation. The motor will only draw the amount of power required by the pump; for example, if a 40 HP motor was used to supply the energy for pump in Example 9.4, the motor would only draw 31 HP from the power grid. Thus, it is not necessarily a waste of energy to select a motor that is larger than required (neglecting possible changes in motor efficiency).

The selection between different types of pump power sources depends on the cost of energy, and energy prices vary regionally. For example, electricity may be expensive and natural gas inexpensive in one region so most irrigation systems will use natural gas engines. In another region, natural gas may be unavailable and most pumps will be powered by electricity. Even if the cost of a given form of energy is inexpensive, the cost of energy delivery may be high; for example, a new farm may be miles from the nearest connection to the electric power grid.

Internal combustion engines are inefficient converters of chemical energy in fuel to mechanical pumping energy. The Nebraska Pumping Plant Performance Criteria (Table 9.2) give the overall kW-hr delivered to the pump and associated efficiencies.

Example 9.9 A turbine well pump to be installed at a new farm requires 36 BHP and delivers 27 HP to the water (75 % efficiency). The pump will operate for 1,800 hours per year and deliver water for surface irrigation. Select between a diesel engine or an electric motor as the power source. Assume a 20 yr project life and 8 % ROR. Ignore inflation, but include replacement costs. For both pump systems, the cost of the pump is $3,000.

The cost of diesel delivered to the tank is $2.70/gallon. The diesel engine costs $15,000, and the diesel tank costs $3,000 installed. The tank has a service life of 20 years. Annual maintenance cost of pump and engine is $1,000/yr. The pump will need to be replaced after 16 years, and the diesel engine replaced after 14 years.

The cost of tying into the existing power grid and bringing an electric power line to the pump is $40,000, the cost of electrical energy is $0.09/kW-hr, and the cost of the electric motor is $3,000. The electric motor efficiency is 90 %. Annual maintenance cost is $400/yr. The electric motor will not need to be replaced, and the pump will be replaced after 16 years.

This problem is solved in the *Fuel and Pump costs* worksheet

Solution:
See Table 9.3

The present value of the electric pump is less so choose the electric system.

Pump Station Friction Loss and Head Requirement

Steel pipe used in wells is sold in two classifications: gage and schedule. Gage refers to the thickness of the pipe wall (Table 9.4), which is sometimes reported in mils, where mils are thousandths of an inch. Schedule 40 steel pipe dimensions are the same as PVC (Table 8.2). Dimensions of other Schedules (10, 20, 30, and 80) are available in handbooks or on the Internet. Outside diameters are the same as PVC. Subtract 2x wall thickness from outside diameter to find the inside diameter.

The head requirement for a well pump is the sum of the following (Fig. 9.4): the difference between the discharge elevation and the dynamic (pumping) water level, friction

Table 9.3 Costs of pumping for Example 9.9

Diesel fuel cost	**Diesel fuel costs**
The cost of fuel \$2.70/gallon	\$0.713/liter
Energy density	3.282 b-kW-hr/liter
Power required 36 HP * 0.746	26.9 kW
Energy required 1,800 hours * 26.9	48,300 b-kW-hr
Volume of fuel 48,300 b-kW-hr/3.28 b-kW-hr/L	14,729 L/yr
Annual cost 14,729 L * \$0.713/L	\$10,507/yr
Present value PV(0.08, 20, 10,507)	\$103,158
Diesel pump cost	
Initial installation of engine, pump, and tank	\$21,000
Present value pump year 16 \$3,000 $(1 + 0.08)^{-16}$	\$876
Present value engine year 14 \$15,000 $(1+ 0.08)^{-14}$ =	\$5,106
Annual maintenance cost present value PV(0.08, 20, 1,000)	\$9,818.
Total PV \$103,158 + \$21,000 + \$876 + \$5,106 + \$9,818	**\$139,959**
Electric power cost.	
Cost of electricity	\$0.09/kW-hr
Power required 36 HP * 0.746	26.9 kW
Electric power required 26.9 kW/Eff = 26.9 / 0.9 (40 HP)	29.89 kW
Energy required. 1,800 * 29.89	53,800 kW-hr
Annual cost 53,800 kW-hr * \$0.09/kW-hr	\$4,842
Present value PV(0.08, 20, 4,842)	\$47,462
Electric pump costs	
Initial installation of pump, motor and power line \$6,000 + \$40,000	\$46,000
Present value pump year 16 \$3,000 $(1 + 0.08)^{-16}$	\$876
Annual maintenance cost present value PV(0.08, 20, 400)	\$3,927
Total PV \$46,000 + \$47,462 + \$876 + \$3,927	**\$98,265**

Table 9.4 Wall thickness for steel gage pipe

Gage	Wall thickness (inch)	Wall thickness (mm)
20	0.040	1.02
18	0.052	1.32
16	0.064	1.63
14	0.079	2.01
12	0.109	2.77
10	0.138	3.51
8	0.164	4.17
6	0.203	5.16
3	0.259	6.58

losses in the pipe and fittings, and velocity and/or pressure head at the discharge.

Example 9.10 Calculate the discharge head requirement and HP for the pump shown in Fig. 9.14. Flow rate 64.6 L/sec. Pipe is 8″ (200 mm) 12 gage steel. Use Hazen-Williams C = 120.

Outside diameter of 200 mm pipe is found in Table 8.2: 219.1 mm. The wall thickness of 12 gage pipe is 2.77 mm. Thus, the inside diameter of pipe is 219.1 − 2.77 * 2 = 213.6 mm.

The equivalent length of a 45° standard steel elbow for 200 mm pipe as 16 * 200 mm = 3.2 m The two 45° elbows + length of pipe = 100 m + 6.4 m = 106.4 m.

Pipe friction pressure loss:

$$h_f = 1.22*10^{10}*106.4\ m\left(\left(\frac{64.6}{120}\right)^{1.852}/213.6^{4.87}\right)$$

$$= 1.9\ m$$

The kinetic energy at the discharge:

$$V = \frac{Q}{A} = \frac{0.0646\ m^3/\sec}{(0.2136\ m/2)^2*\pi} = 1.80\ m/\sec \qquad \frac{V^2}{2g} = \frac{1.80^2}{2*9.8}$$

$$= 0.33\ m$$

Energy losses are summed for Pump TDH. The difference between the pumping water level (water level is drawn down during pumping) and the discharge elevation is the elevation gain:

7 m + 18 m = 25 m.

Elevation gain	7 + 18 m
Loss in 45°elbows + pipe	1.9 m
Velocity head lost at discharge	0.33 m
Total dynamic head requirement (TDH)	27.2 m

$$\text{Power} = Q(m^3/\sec)*H)/(0.102*\text{Eff})$$
$$= (64.6/1000*27.2)/(0.102*0.75) = 23\ \text{kW}$$

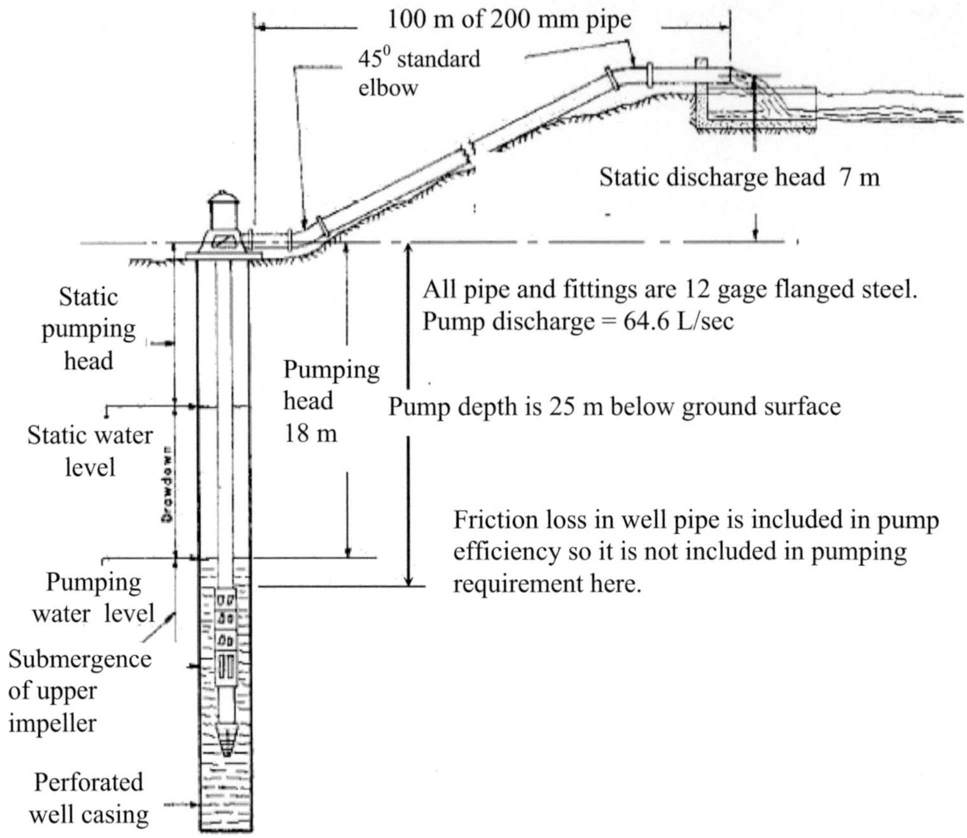

Fig. 9.14 Turbine pump pressure requirement calculation (Credit NRCS (converted to metric))

Fig. 9.15 Concentric cone expansion and reducer: equations for minor loss coefficients

$\theta < 45^0$ $$K = 2.6 \sin\frac{\theta}{2}\left(1-\frac{d_1^2}{d_2^2}\right)^2$$

$\theta > 45^0$ $$K = \left(1-\frac{d_1^2}{d_2^2}\right)^2\sqrt{sin\frac{\theta}{2}}$$

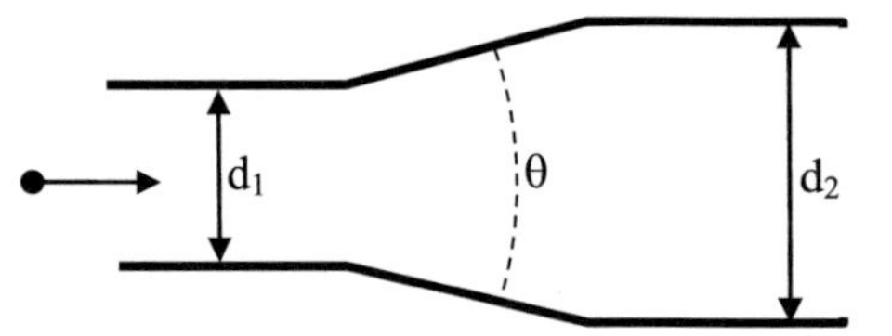

Concentric cone reducer.

$\theta < 45^0$ $$K = 0.8 \sin\frac{\theta}{2}\left(1-\frac{d_1^2}{d_2^2}\right)^2$$

$\theta > 45^0$ $$K = 0.5\left(1-\frac{d_1^2}{d_2^2}\right)^2\sqrt{sin\frac{\theta}{2}}$$

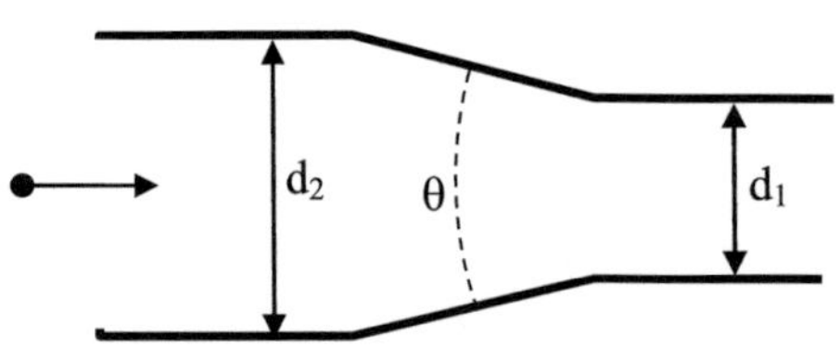

Centrifugal pump stations often use concentric cone expansion fittings and eccentric reducers to reduce friction loss as pipe size changes (Fig. 9.11). The minor loss coefficients are a function of inlet and discharge diameter and cone angle (Fig. 9.15). As with minor loss coefficients for a sudden expansion or sudden contraction, minor loss coefficients in these gradual transitions are multiplied by the higher velocity in the transition; thus, the minor loss coefficient is multiplied by the inlet velocity for an expansion cone, and by the discharge velocity for a reducing cone.

Example 9.11 A pump (Fig. 9.16) sucks water from a canal at a flow rate of 10 L/sec. The suction pipe is 4″. The suction section includes the following: 4″ fittings and pipe: basket strainer, foot valve, flanged steel sweep 90 with r/d = 4, and 2 m of straight pipe. An eccentric reducer is used to reduce pipe diameter from 4″ to the 0.2 m length 2″ diameter inlet nozzle pipe diameter (angle). Use the concentration cone equation to calculate eccentric K with angle θ = 44°. The pump discharge is located 1 m above the water surface of the supply canal. The discharge side of the pump is connected to

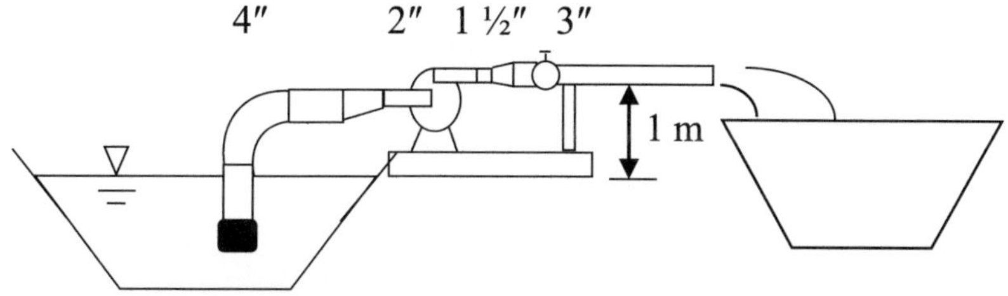

Fig. 9.16 Centrifugal pump station

a 4 cm length of 1½″ steel outlet pipe. A concentric cone with angle θ = 30° transitions from 1 ½″ to the 3″ steel discharge pipe. The 3″ discharge section includes a 3″ threaded gate valve and a 1.5 m length of steel pipe, which then has an open discharge that feeds water to a second canal. All pipe is 6 gage steel. Find the total friction loss in the pump station fittings, and the fitting with greatest head loss. Note: a centrifugal pump is not generally used for such a low head application as the one shown in Fig. 9.16, but this example demonstrates the calculation method.

The wall thickness of 6 gage pipe is 5.16 mm (Table 9.4), and outside diameters are found in Table 8.2. Inside diameters are calculated as follows:

Diam	OD (mm)	Wall thickness (mm)	ID (mm)
4	114.3	5.16	104.0
2	60.3	5.16	50.0
1½	48.3	5.16	38.0
3	88.9	5.16	78.6

The *Centrifugal pump fittings* worksheet (Fig. 9.17) calculates the head loss. The worksheet is divided into four sections because there are four pipe diameters represented in the fittings surrounding a centrifugal pump. The four sections are the suction, inlet, outlet, and discharge sections. There is also a mainline section, but this example does not have a mainline.

Fitting minor losses can be found in Appendix A. In this example, the basket strainer is assigned a K value of 1.3, which corresponds with the point at which the basket strainer K value line crosses the 4 inch line, which is the nominal diameter of the suction pipe. The sweep 90° has a minor loss coefficient of 0.24, which is found in Table 8.10 for a 4 in diameter pipe with r/d = 4.

The eccentric minor loss coefficient is calculated as follows:

$$\theta < 45^\circ \quad K = 0.8\ \sin\left(\frac{44(\pi/180)}{2}\right)\left(1 - \frac{0.05^2}{0.104^2}\right)^2 = 0.18$$

The expansion cone minor loss coefficient is calculated as follows:

$$\theta < 45^\circ \quad K = 2.6\ \sin\left(\frac{30(\pi/180)}{2}\right)\left(1 - \frac{0.038^2}{0.0786^2}\right)^2 = 0.40$$

Use the larger velocity for the expansion cone minor loss calculation:

$$h_m = K\frac{V^2}{2g} = 0.4\frac{8.82^2}{2{*}9.8} = 1.6\ m$$

The discharge pipe losses include the gate valve minor loss (K = 0.09 for screwed gate valve), pipe friction loss, and loss of kinetic energy at the discharge point (K = 1).

There are no mainline losses since there is no mainline pipe. Thus, the length of the mainline pipe is set at zero, and sum of K values for the mainline is also set to zero.

Total head loss in all fittings is 2.58 m. The head loss in the expansion cone, 1.6 m, is greater than all other losses in the pump station combined. If a 10° angle cone expansion joint were used, which allows for a smoother transition and less turbulence, then the friction loss is reduced to 0.53 m or 1/3 of the head loss with a 30° angle; however, a 10° cone expansion is not a normal fitting.

Example 9.12 A pump (Fig. 9.18) sucks water from a canal and discharges to a reservoir 20 m above the canal. All suction, inlet, and outlet parameters are the same as in Example 9.11. The discharge section includes the same length of pipe but also includes a steel threaded 102 mm (4 in. 90°. The length of the mainline pipe is 500 m and the pipe classification is 102 mm (4 in. SCH 40 PVC with C = 140. The mainline pipe also includes two 102 mm (4 in. PVC 90° in the ground just below the pump station. The pipe discharges to the upper reservoir through a submerged open outlet. Calculate the pump TDH required. Compare the percent of TDH due to pump station losses to the mainline friction loss and the elevation gain.

The *Centrifugal pump fittings* worksheet was used to calculate the head requirement. Changes are made to the discharge and mainline sections (Fig. 9.19).

The 90° Elbow loss in the discharge pipe is added here as an equivalent length of pipe rather than a minor loss. The equivalent length to diameter ratio is 20–30 and the diameter is 0.1 m; thus, the equivalent length is 2.5 m. Thus, the equivalent discharge pipe length is 1.5 m + 2.5 m = 4.0 m. Because the 90° elbow was included as an equivalent pipe length, the K value for angle bends in Cell B38 remains as zero. There is also a 75 mm (3 in. to 102 mm (4 in. bushing

	A	B	C	D	E	F
1	Component			Head requirement		Input data in
2	Required pump TDH			3.58	m	white cells
3	Total friction loss			2.58	m	
4	Mainline friction loss			0.00	m	
5	Total pump station friction			2.58	m	
6	Elevation gain			1	m	
7	Flow rate	10	L/sec			
8	Steel pipe C	120		Head loss		
9	Suction pipe total			0.2023	m	
10	Suction pipe length	2	m	0.0370	m	Pipe friction
11	Suction pipe diameter	0.104	m			
12	Basket strainer K	1.3		0.0919	m	Fittings
13	Entrance coefficient K	0		0.0000	m	
14	Foot valve K	0.8		0.0565	m	
15	Sweep 90 K	0.24		0.0170	m	
16	Cross-sectional Area	0.0085	m3			
17	Flow velocity	1.18	m/sec			
18	Inlet pipe total			0.3652	m	
19	Inlet pipe length	0.2	m	0.1309	m	Pipe friction
20	Inlet pipe diameter	0.05	m			
21	Eccentric angle (degrees)	44				
22	Eccentric K	0.18		0.2343	m	Fittings
23	Cross-sectional Area	0.001963	m2			
24	Flow velocity	5.09	m/sec			
25	Outlet pipe total			1.6660	m	
26	Outlet pipe length	0.04	m	0.0996	m	Pipe friction
27	Outlet pipe diameter	0.038	m			
28	Cone angle	30				
29	Cone expansion K	0.40		1.5664	m	Fittings
30	Other fittings K	0.0		0.0000	m	
31	Cross-sectional Area	0.001134	m2			
32	Flow velocity	8.82	m/sec			
33	Discharge pipe total			0.3445	m	
34	Discharge pipe length	1.5	m	0.1085	m	Pipe friction
35	Discharge pipe diameter	0.0786	m			
36	Discharge valve K	0.09		0.0195	m	Fittings
37	Discharge loss K (kinetic)	1		0.2166		
38	Angle bends K (sum)	0		0.0000	m	
39	Cross-sectional Area	0.004852	m2			
40	Flow velocity	2.06	m/sec			
41	Mainline pipe total			0.0000		
42	Mainline pipe length	0	m	0.0000	m	Pipe friction
43	Mainline pipe diameter	0.1023	m			
44	Mainline pipe C	140				
45	Mainline pipe total K	0		0.0000	m	Fittings
46	Cross-sectional Area	0.008219	m2			
47	Flow velocity	1.22	m/sec			

Steel pipe diameter			
6 gage	OD	t	ID (mm)
4	114.3	5.16	104.0
2	60.3	5.16	50.0
1.5	48.3	5.16	38.0
3	88.9	5.16	78.6

PVC mainline diameter			
Sch 40	OD	t	ID (mm)
4	114.3	6.02	102.3

Fig. 9.17 *Centrifugal pump fittings* worksheet for Example 9.11

as the diameter increases from the discharge pipe to the mainline pipe. This is classified as a sudden enlargement in *Chapter 8 minor losses* worksheet. All of the kinetic energy is lost in a sudden enlargement so the K value is 1.0. This should be included in the discharge section since it is the kinetic energy in the 75 mm (3 in. pipe that is lost. As a result, there is a discharge K value equal to 1.0 in the discharge section.

In the mainline, the two 102 mm (4 in. PVC 90° elbows in the ground have an equivalent length of 3.5 m each (Table 8.12). As a result of these fittings and the length of pipe, the equivalent length of mainline pipe is 507 m in cell B42. There is also a loss of kinetic energy at the discharge to the upper canal so the K value for the mainline pipe is 1.

With the additional energy loss and elevation gain, the total dynamic head that must be supplied by the pump is

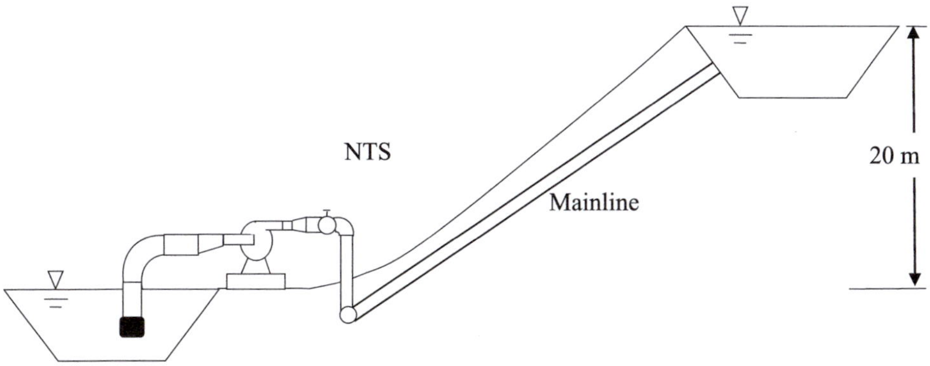

Fig. 9.18 Centrifugal pump lifting water 20 m from one canal to another

33	Discharge pipe total			0.5253	m	
34	Discharge pipe length	4	m	0.2892	m	Pipe friction
35	Discharge pipe diameter	0.0786	m			
36	Discharge valve K	0.09		0.0195	m	Fittings
37	Discharge loss K (kinetic)	1		0.2166	m	
38	Angle bends K (sum)	0		0.0000	m	
39	Cross-sectional Area	0.004852	m2			
40	Flow velocity	2.06	m/sec			
41	Mainline pipe total			7.8538	m	
42	Mainline pipe length	507	m	7.6372	m	Pipe friction
43	Mainline pipe diameter	0.1023	m			
44	Mainline pipe C	140				
45	Mainline pipe total K	1		0.2166	m	Fittings
46	Cross-sectional Area	0.008219	m2			
47	Flow velocity	1.22	m/sec			

Fig. 9.19 *Centrifugal pump fittings* worksheet for Example 9.12

30.61 m. Of that total 10.61 m is friction loss and 20 m is elevation. The pump station friction losses are 2.76 m. Thus, pump station friction losses are 9 % of the total dynamic head requirement.

Chemigation Systems

Many irrigation systems include the capability to inject chemicals, such as fertilizer, into the irrigation water. Chemigation systems include a chemical supply tank, injection system and safety devices (Fig. 9.20).

The three most common chemigation injection devices are piston pumps, diaphragm pumps, and Venturi injectors. Piston pumps and diaphragm pumps are classified as positive displacement pumps, whereas Venturi injectors rely on the Venturi pressure drop principle to draw the chemical from the tank into the irrigation pipeline. Positive-displacement pumps (Fig. 9.20) are recommended where precise control of chemical flow rate is required, because these pump flow rates remain stable over a range of irrigation pipeline pressures and flow rates.

In Venturi injector systems (Fig. 9.21), water is extracted from the main line and pressure is added with a centrifugal pump or by a pressure differential created in the mainline. The water velocity increases within the Venturi, and the pressure decreases below atmospheric pressure in the throat of the Venturi, causing the chemical to be sucked into the injector from the chemical reservoir. Because chemical is sucked into the irrigation system after the centrifugal pump, there is no contact between the chemical and the pump; thus, Venturi injection systems are less susceptible to corrosion than positive displacement pumps; however, Venturi injection flow rate is dependent upon chemical viscosity as well

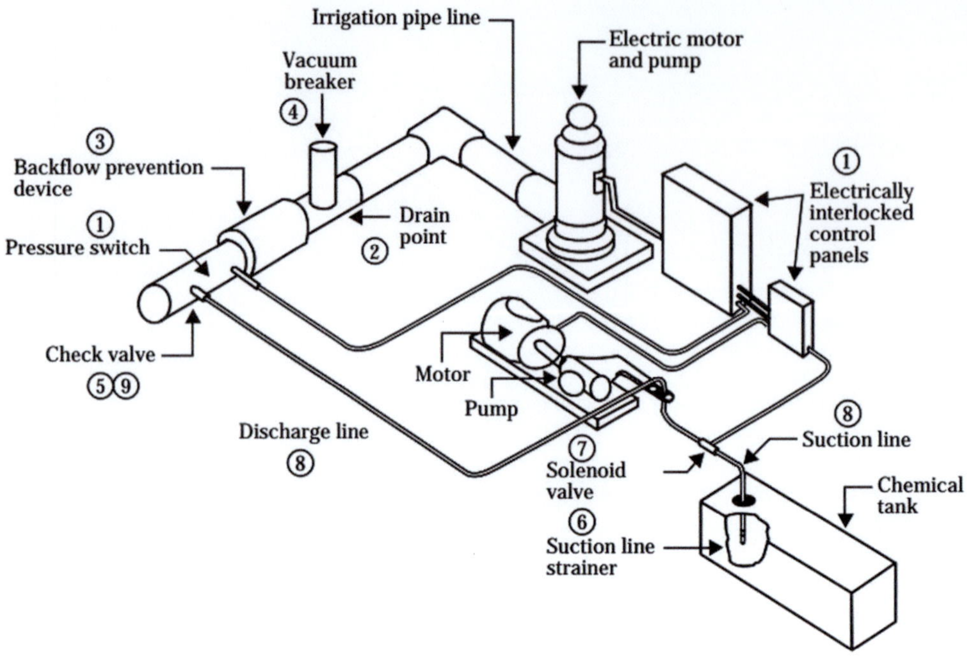

Fig. 9.20 Chemigation injection system (Credit NRCS 1997. NEH, Part 652, chapter 7)

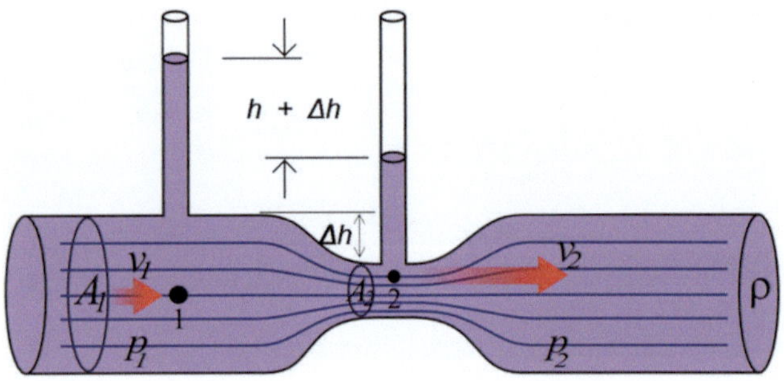

Fig. 9.21 Venturi injector (Wikipedia)

as irrigation pipeline pressure. Thus, the injection rate control is not precise. The change in Venturi injection flow rate can be in the range of 5–10 % for a temperature change of 20 °C for viscous fertilizers

In addition to flow rate range and precision, the primary selection criteria for chemigation systems are durability, accuracy, ease of operation and repair, service life, and susceptibility to corrosion.

It is convenient to place a flow meter in the injection line in order to adjust the injection system to the proper flow rate.

Injection systems should always provide for complete mixing and uniform concentrations before the chemicals reach the field. Chemical will mix naturally with water in an irrigation pipeline due to turbulent eddies. However, a minimum length of pipe is required for complete mixing.

Proportional pumps vary the injection flow rate with water flow rate, providing a constant ratio of chemical to irrigation water. These are especially useful in hydroponic systems, where maintaining an injection ratio is the goal.

In-class Exercise 9.5 Venturi injectors are designed based on the principle that if water velocity increases, then pressure decreases, as shown by the Bernoulli equation. A narrow throat increases the velocity at the suction point. Concentric cones are used to gradually increase flow rate to the throat and decrease flow rate from the throat. Based on what you know about concentric cones, draw a Venturi injector geometry that has minimum head loss.

Questions

1. The revolutionary speed of electric pumps is slightly less than divisors of 3600. Typical pump rpm's are 875, 1750, and 3500. Why are most pumps manufactured with these revolutionary speeds?
2. What would be a typical TDH for a centrifugal pump with flow rate 1,000 m^3/hr based on the typical specific speed for a centrifugal pump? Recalculate for pump flow rates of 100 m^3/hr and 10 m^3/hr. What type of pump would be appropriate for a very high flow rate and very low head?
3. Using the equations for the relationships between power, flow rate, and head, describe the shape of the head/capacity curve if efficiency was constant over a range of flow rates?
4. Verify that the water horsepower generated by the 5.9375 impeller curve in Fig. 9.4 corresponds with the efficiency and brake horsepower (the curve below the head capacity curve). Calculate at the point of highest efficiency.
5. Describe the relationship between efficiency and flow rate in Fig. 9.4.
6. An irrigation system requires 600 gpm and 160 ft head. Select the best impeller for this application on the B4JPBH (Fig. 9.5) pump curve.
7. What is the maximum allowable flow rate of a B4JPBH pump (Fig. 9.5) with a 12 3/8″ impeller and a 40 HP motor? What is the maximum flow rate for the 50 HP motor with the same impeller?
8. An irrigation system requires TDH = 168 ft and Q = 600 gpm. Select an impeller diameter (trimmed if necessary) and select a motor HP with the B4JPBH pump.
9. The 16BZ pump (Fig. 9.7) with a 5¾″ impeller is used to run a sprinkler system. There is a 2 m pressure loss in pump fittings and filters. Find the operating point. Plot the two curves and verify that the calculated point is the correct point. The 5¾″ head-capacity curve and the irrigation system curve are:

$$\mathrm{TDH(m)} = -0.00170\mathrm{Q}^2 + 0.0743\mathrm{Q} + 43.76$$
$$\mathrm{Q_{system}}(\mathrm{m^3/h}) = 14.175\left(\mathrm{H_{system}}\right)^{0.531}$$

10. In Example 9.6, change the elevation of pivot 2 to 100 m elevation and pivots 3 and 4 to 120 m elevation. Select the pump operating pressure. Each pivot requires 100 L/sec. Determine the number of pumps, flow rate, and TDH of the pump station. Discuss options to reduce energy.
11. A variable speed pump controller is used to vary the flow rate of the 16BZ pump with the 5 3/4″ impeller. The revolutionary speed is lowered from 3500 to 3000 RPM. The system curve is Q_{system} (m^3/h) = 14.175 $(H_{system})^{0.531}$. There is 3.5 m head loss in the pump fittings and filters. Find the operating point TDH and flow rate.
12. Imagine that a new technology was developed that enabled farmers to produce biodiesel from crop residue. The biodiesel production unit has a capital equipment cost of $50,000; a labor, maintenance, and energy cost of $0.30/L, and produces 15,000 L of biodiesel per year. Calculate whether this would be a less expensive alternative than the electric pump system in Example 9.9. Use the *Fuel and pump costs* worksheet in Chapter 9 Excel program.
13. Redo Example 9.9 with a solar powered pump. Based on the cost of materials and the service life and replacement cost of solar components, the solar panel array provides electrical energy at a cost of $0.08/kW-hr for the 20 year project life. The solar pump can only be used during daylight; thus a larger pump is required and a reservoir must be constructed for storage. Increased capital cost of hydraulic components is $50,000 and replacement and maintenance costs remain the same as Example 9.9. Recalculate if carbon credits for the system are worth $1,000/yr.
14. A pump sucks water from a canal and discharges to a reservoir 100 m above the canal. Pump station valves and fittings are the same as in Example 9.11 except that the pipe diameters are 6″, 3″, 2.5″, and 4″ instead of 4″, 2″, 1½″, and 3″. Two other changes are that the eccentric angle is 50° and the cone angle is 40°. Flow rate is 20 L/sec. All pump station pipe is 6 gage steel, and the mainline pipe is 4″ SCH 40 and is 500 m long. Assume an open discharge to the upper reservoir. Calculate the pump TDH required. Show calculations for the pressure loss in the eccentric reducer and the concentric cone. Calculate the percent of required TDH due to pump station losses, and the percent of total friction loss that is due to pump station losses. (Use worksheet)

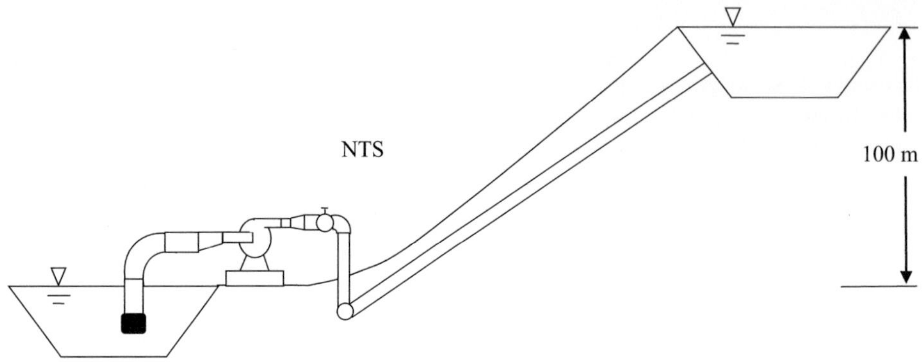

15. Redo question 15, but discard the eccentric, and let the suction pipe be all 75 mm (3 in. pipe. Second, use a bushing on the discharge side (sudden expansion) rather than a cone expansion. Determine which change results in the greatest increase in head loss.
16. Use the 16BZ pump with 5 3/4″ impeller (M) to deliver water to the upper reservoir for the system shown below. Select pipe diameters equal to 6″, 4″, 3″, and 4″ for the four pump station pipe sections. Use 4″ Schedule 40 PVC for the mainline, which is 493 m long. Draw a system curve (develop with *Centrifugal pump fittings* worksheet by inputting different flow rates and corresponding TDH requirement) and pump head-capacity curve based on Fig. 9.6. Find an exponential equation for the system curve and equation for the head-capacity curve, and calculate the point of intersection (operating point) for the system

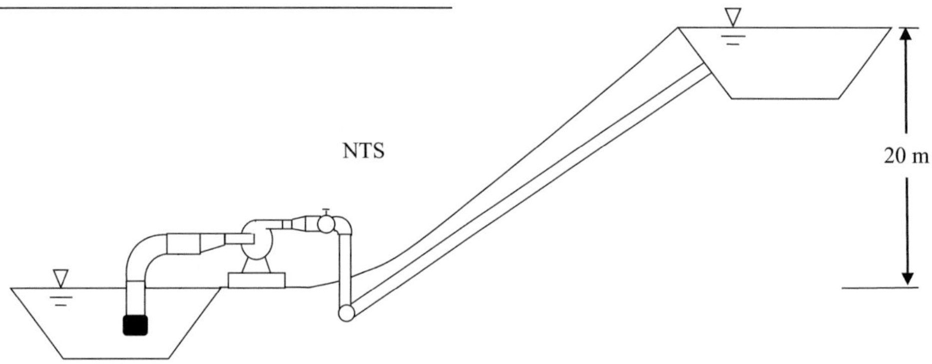

17. Venturi injectors are designed based on the principle that if water velocity increases, then pressure decreases as shown by the Bernoulli equation. A narrow throat increases the velocity at the suction point. Concentric cones are used to gradually increase flow rate to the throat and decrease flow rate from the throat. Based on what you know about concentric cones, draw a Venturi injector geometry that has minimum head loss.
18. Some people recommend creating the pressure differential across a Venturi by restricting mainline flow. It is a much better idea to have a separate centrifugal pump provide the pressure differential, as shownin this example. Mainline flow rate is 200 L/sec, and Venturi flow rate is 0.90 L/sec. Venturi injection time is 1,000 hours per year. The required pressure differential across the Venturi is 283 kPa. The cost of energy is $0.10/kW-hr. Calculate the energy cost per year for providing the required pressure differential across the Venturi by constricting the mainline flow with a valve. Calculate the energy cost of using a centrifugal pump in the bypass line to provide the pressure differential needed by the Venturi.
19. How could a Venturi be used within a pump to lift groundwater up to the surface in a well. (Hint: look up jet pumps).
20. Calculate throat pressure (gage pressure and absolute pressure) and discharge pressure in a Venturi injector that that has a 30 mm internal diameter at both ends and that has a flow rate of 0.9 L/sec. The length of the entire Venturi is 15 cm and the length of the throat is 2 cm; however, assume that the equivalent length of the throat is 10 cm due to flow entering the throat through the suction tube. The upstream pressure is 300 kPa. The

reducer cone angle (inlet side), θ, is 30°, and the expansion cone angle (discharge side) is 15°. Assume that Hazen-Williams C in the throat is 100. The inside diameter of the throat is 7 mm.

21. Redo question 20 but optimize the inlet and discharge angle in order to minimize pressure loss across the Venturi. Keep the same throat dimension and Venturi length. Derive an equation based on the geometry of the system that calculates discharge angle as a function of inlet angle.

References

Beard R, Robert Hill (2000) Maintaining electric motors used for irrigation. Utah State University Cooperative Extension. ENGR/BIE/WM/06 http://extension.usu.edu/files/publications/publication/ENGR_BIE_WM_06.pdf

NRCS (1997) Energy use and conservation. National Engineering Handbook. Part 652. Chapter 12. http://www.irrigationtoolbox.com/NEH/Part652_NationalIrrigationGuide/ch12.pdf

NRCS (2010) Variable Speed Drive (VSD) for irrigation pumping, engineering technical note MT-14. http://www.nrcs.usda.gov/Internet/FSE_DOCUMENTS/nrcs144p2_054026.pdf

Huffman R, Fangmeier D, Elliott W, Workman S (2013) Soil and water conservation engineering. American Society of Agricultural and Biological Engineering

Groundwater

10

An aquifer is an underground body of water perched on top of bedrock or other impermeable layer. Most usable freshwater storage is within aquifers (Table 10.1). The residence time of groundwater can vary from days to millennia (Fig. 10.1). Groundwater's often long period of water exchange (Table 10.1) makes it useful as a stable source of water but also leads to its susceptibility to long term degradation by aquifer contamination. Groundwater supplies approximately 40 % of the water used by irrigated agriculture. Wells are drilled into high conductivity layers in geologic formations. Sand and gravel aquifers are the primary source of groundwater for irrigation. Measurement of the piezometric gradient and aquifer hydraulic conductivity enables hydrologists to calculate aquifer flow velocity and direction. The product of aquifer thickness and hydraulic conductivity is the transmissivity. The Thiem equation calculates aquifer transmissivity in confined aquifers based on steady-state pumping in a pumping well and draw down (decline in water table) in an observation well. The steady-state unconfined aquifer equation adds aquifer thickness to the equation. The Theis equation evaluates transient drawdown in an observation well.

The rate that water is replaced in an aquifer is called the rate of exchange. The average rate of groundwater exchange in the world is 280 years (Table 10.1). Primarily because of the slow rate of exchange, polluted aquifers are much more difficult to reclaim than polluted rivers and lakes. In some desert regions, the water in the aquifer is called fossil groundwater because the water was deposited during a different climatic period. There is no water exchange in these aquifers. Agriculture has polluted some aquifers with unacceptable levels of nutrients and pesticides. Many of these have very slow exchange rates so there is no possibility of remediation in the foreseeable future. Agriculture has also improved the water quality of some aquifers by increasing the rate of exchange. When higher quality water is used for irrigation than is in the underlying aquifer, and drainage is used to remove water from the aquifer and prevent water table buildup, the irrigation water can flush the aquifer and improve water quality.

Aquifers are defined as confined or unconfined depending on whether they have a confining bed (aquitard) above them (Fig. 10.1). The unsaturated zone above an aquifer called the vadose zone (Fig. 10.2). The water table is defined as the elevation at which water pressure is zero gauge pressure. Water above the water table is called the capillary fringe. This water wicks up above the water table because of capillary potential of the small pores in soil. The capillary fringe is thin for coarse textured soils and can be up to 15 m thick in fine textured soils.

Aquifers

There are five major types of groundwater aquifers (Fig. 10.3): unconsolidated sand and gravel aquifers, semi-consolidated sand and gravel aquifers, volcanic aquifers, sandstone and carbonate rock aquifers, and sandstone aquifers.

Most groundwater-irrigated land draws water from unconsolidated sand and gravel aquifers (Fig. 10.3). In the US, the High Plains aquifer is in the central US, the basin and range aquifers are the southwest US, Coastal and Central Valley aquifers are in California, Pacific Northwest basin fill aquifers are in Washington, the Surficial aquifer system is in eastern Florida, Georgia, and South Carolina, and the Mississippi Valley alluvial aquifer and the glacial-deposit (glacial-fill) aquifers are in the northern United States.

There are three types of unconsolidated sand and gravel aquifers: basin fill, blanket sand and gravel, and glacial-deposit (glacial-fill). Basins formed due to uplift of mountains on either side of the basin and are filled with unconsolidated alluvial sediments underlain by bedrock. Finer grained sediments are typically located near the center of basins.

Water yielding sand and gravel layers are typically confined by finer grained sand or clay layers that have low hydraulic conductivity. Geologists can observe the sand

P. Waller, M. Yitayew, *Irrigation and Drainage Engineering*, DOI 10.1007/978-3-319-05699-9_10

Table 10.1 Freshwater of the hydrosphere and its rate of exchange. [Modified from Lvovich (1979), Tables 2 and 10] Ground-Water Hydrology (Credit USGS: Paper 2220)

Parts of the hydrosphere	Volume of freshwater km^3	Mile^3	Share in total volume of freshwater (percent)	Rate of water exchange (yr)	Recharge km^3/year
Ice sheets and glaciers	24,000,000	5,800,000	84.9	8,000	3,000
Ground water	4,000,000	960,000	14.2	280	14,000
Lakes and reservoirs	155,000	37,000	0.549	7	22,000
Soil moisture	83,000	20,000	0.294	1	83,000
Vapors in atmosphere	14,000	3,400	0.049	0.027	52,000
River water	1,200	300	0.004	0.031	39,000
Total	28,253,200	6,820,700	100		

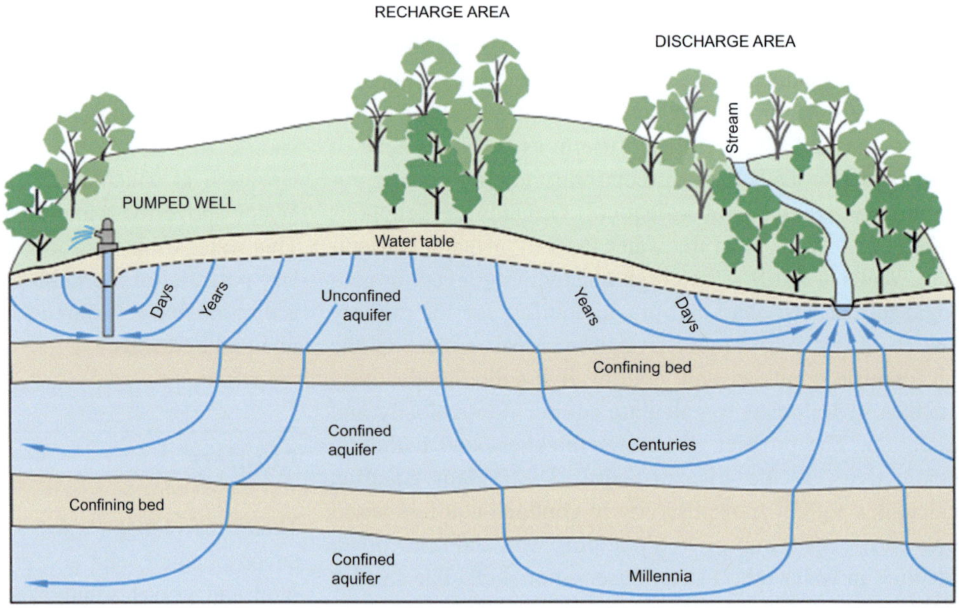

Fig. 10.1 Groundwater residence times (Credit: U.S. Geological Survey Circular 1139)

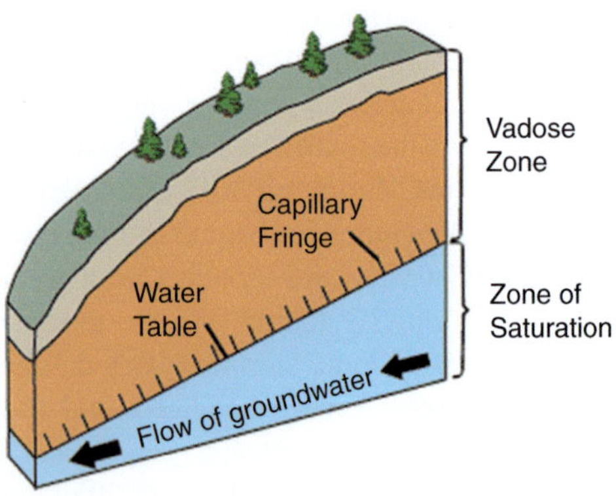

Fig. 10.2 Vadose zone (Credit USGS. http://geology.er.usgs.gov/eespteam/brass/ground/groundintro.htm)

and rock removed during drilling and conduct other electrical resistance and sound tests in order to determine the location of the water yielding formations.

Although basins have a regional elevation of the water table there is generally a complex network of aquifers within sand and gravel layers and perched above impermeable layers. There are often hydraulic connections between aquifers and sometimes between the aquifers and surface water bodies.

In their natural state, many basin aquifer systems supply streams with groundwater that originates from mountain front recharge and surface infiltration (Fig. 10.4).

The geology of the hill slopes and mountains surrounding an aquifer is an important factor in the assessment of a basin's suitability for irrigation. For example, hills on the west side of the Central Valley of California have high levels of selenium. Wells were dug in the valley in order to supply water for irrigation. Excessive selenium in the system and in the drainage effluent caused bird deformities. This example points to the need to conduct an extensive soil and geological survey throughout a hydrologic basin before beginning an irrigation project.

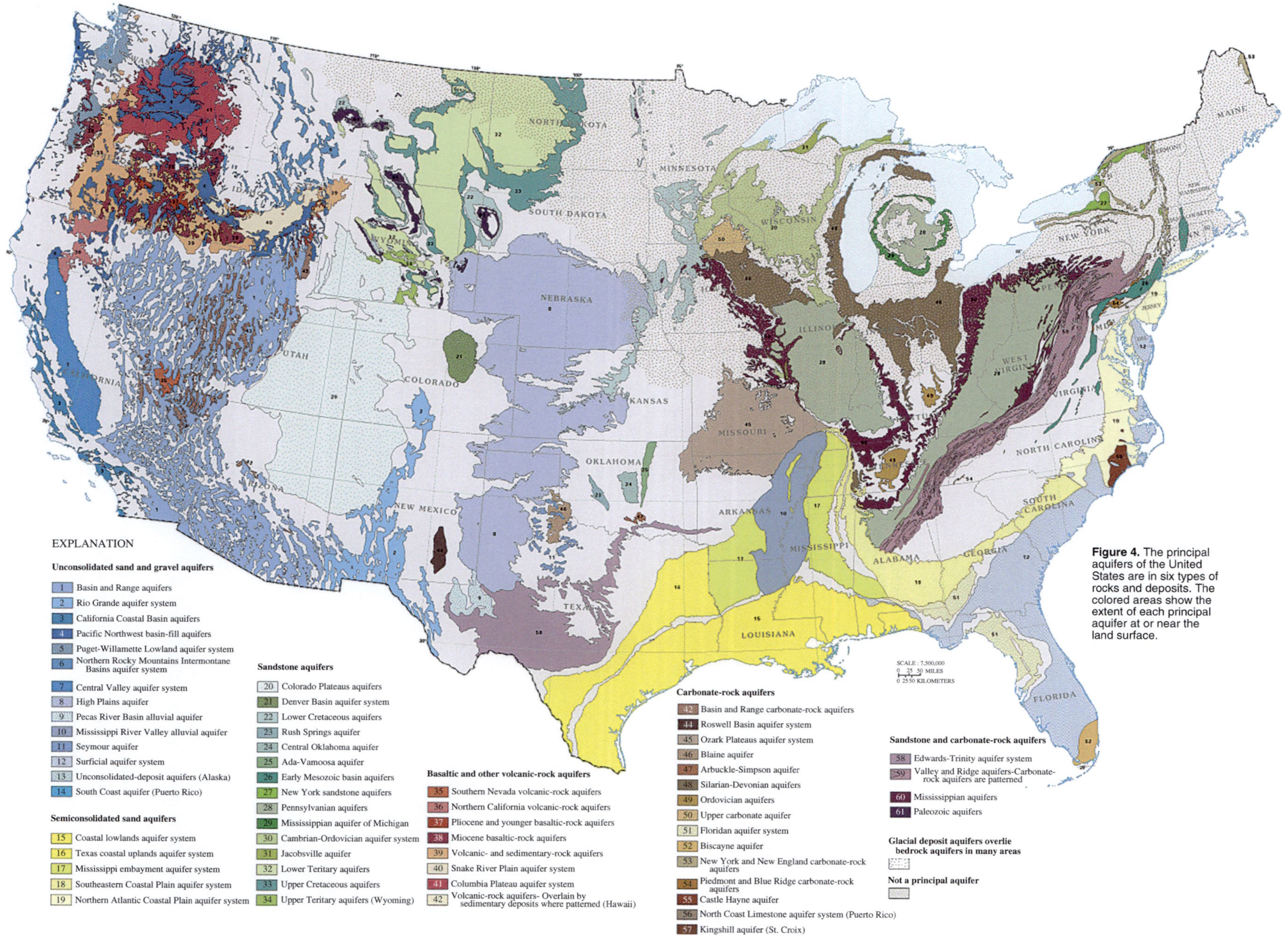

Fig. 10.3 Aquifers in the United States (Credit: USGS)

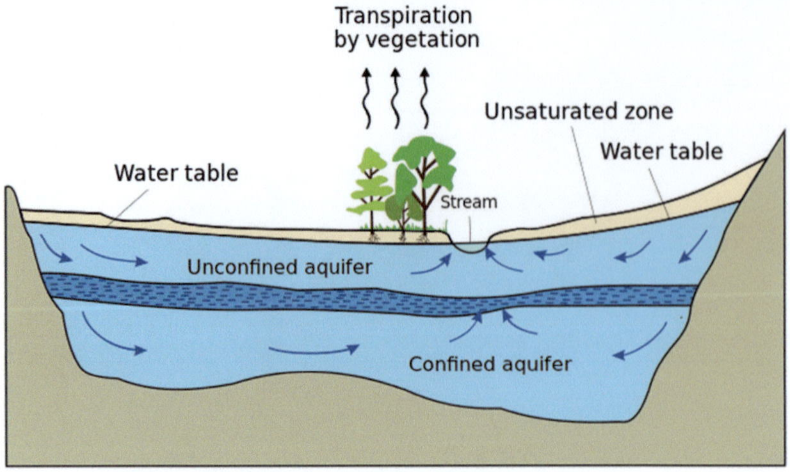

Fig. 10.4 Natural ground water flow in basin (Credit Wikipedia, Hans Hillewaert)

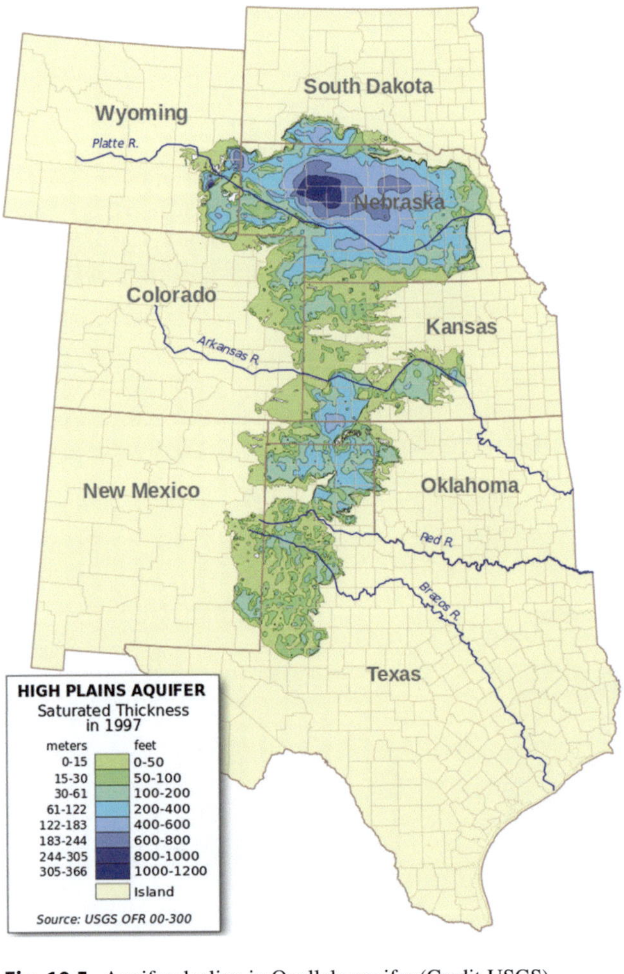

Fig. 10.5 Aquifer decline in Ogallala aquifer (Credit USGS)

The components of the water balance in a basin aquifer system include mountain front recharge, surface infiltration and streamflow infiltration, interbasin flow, and underflow. Streamflow infiltration is generally the largest component and mountain front recharge is the second largest recharge component (USGS, Paper 2220).

Groundwater pumping changes the hydrologic flow patterns in basins. Aquifers in a basin may have naturally provided water to streams (Fig. 10.4); however, wells may drop the water table below the river and dry up the river. In arid regions, the streams and rivers only flow during major storms; thus, the stream ecosystems have been destroyed.

The High Plains Aquifer (contains the Ogallala aquifer) in Nebraska, Colorado, Kansas, Oklahoma, Texas, South Dakota, Wyoming, and New Mexico (Fig. 10.5) is a blanket sand and gravel aquifer. These aquifers are formed from wind-blown sand. They are unconfined by impermeable layers above, and are perched above bedrock. As a result, they are susceptible to pollution by agricultural chemicals. Some regions of the High Plains Aquifer in Nebraska have such high levels of nitrogen due to nitrogen leaching from agricultural fields that farmers do not even have to add nitrogen fertilizer to fields that use pumped groundwater. Although the lack of need for fertilizer is an agricultural benefit, the farmers cannot drink the water because of the high nitrate concentration.

In regions where the rate of pumping has exceeded the recharge rate, the case in the southern part of the Ogallala aquifer, water tables have declined. They can decline by hundreds of feet. In contrast, irrigated agriculture can also raise the water table, which has been the case in the northern sections of the Ogallala aquifer. The groundwater table has fallen so far that it has become uneconomical for farmers to irrigate. The same problem is found in other dry regions. In the Middle East, large sand and gravel aquifers underlie the desert; however, as countries such as Syria and Saudi Arabia try to develop "food security" and become self-sufficient food producers, sand and gravel aquifers that were filled during a different climatic era are becoming depleted (Elhadj 2008). The Punjab area of northern India produces a significant fraction of the wheat and rice requirements of India; however, aquifers are being depleted. Saudi Arabia, US, northern China, and northern India are the top regions in the world with aquifer depletion.

A dramatic increase in recharge due to irrigation can cause water tables to rise. In some irrigated regions where no subsurface drainage has been installed and an unconfined aquifer is near the soil surface, rising water tables have actually reached the soil surface. In fact, approximately 50 million ha of irrigated agriculture in the world has high water tables, limited leaching capability, and, as a result, rising salinity. Installation of subsurface drainage systems is needed in these areas.

The third type of unconsolidated sand and gravel aquifer is the glacial-fill aquifer. These aquifers, laid down by glaciers during the last ice age, contain alternating layers of permeable sand and gravel and impermeable clay and silt layers.

Aquifer Parameters

Groundwater hydrologists focus on the amount of water that will drain from a porous media during pumping rather than the amount of water that a soil will hold after drainage (field capacity). The fraction of water that drains naturally from an aquifer is called the specific yield. Conversely, the specific retention is the amount of water that remains in the rock after gravity drainage (like field capacity). The specific retention (S_r) + specific yield (S_y) is equal to the porosity (n). The specific yield is also called the storativity. The depth of water removed from an aquifer divided by the change in aquifer elevation is defined as the storativity or the specific yield. For example, if 1 m depth of water was removed from an aquifer and the ground water elevation decreased by 4 m, then the storativity or specific yield would be ¼ = 25 %. Values of porosity and specific yield for various rock materials are shown in Table 10.2.

Table 10.2 Approximate characteristics of various natural porous media (After Huffman et al. 2013)

Media	Porosity	Specific yield	Conductivity K (low) m/d	Conductivity K (high) m/d
Gravel	25–40	19	100	100,000
Coarse sand	30–45	22	10	1000
Sand, mixture	20–35	22	5	10
Fine sand	25–50	22	1	50
Silt	35–50	22	0.01	1
Clay	40–70	2	0.001	and lower
Loam Soil	55	40		
Karst Limestone	5–50	18	0.07	1000
Fractured Ign. & met, rock	0–10	9	0.00050	20
Limestone and dolomite	0–20		0.00003	0.2
Sandstone	5–30	6	0.00001	0.2
Shale	0–10		100	0.0001
Unfractured rock	0–5		<0.00000001	0.00002

Values of porosity do not vary dramatically between different unconsolidated materials (Table 10.2); however, conductivity, which determines how fast the water moves to the well, can vary by 10 orders of magnitude (Table 10.2). Only the high conductivity materials are suitable for wells and water supply.

The piezometric (potentiometric) surface is defined as the sum of elevation + pressure. Confined aquifers can have positive pressure if water flows to the aquifer from a high elevation. If the potentiometric surface of a confined aquifer is greater than the ground surface, then a well drilled into the aquifer will be a flowing artesian well (Fig. 10.6). The piezometric surface for an unconfined aquifer is the same as the water table since the water table is defined as the point of zero pressure.

Aquifer Flow

Groundwater hydrologists can determine the rate that water is flowing through an aquifer by measuring the piezometric gradient and aquifer hydraulic conductivity. They measure the hydraulic gradient (slope of water table) and direction of flow by setting up networks of two or more observation wells. Three wells are needed to determine the direction of groundwater flow. The USGS outlines a procedure for determining the groundwater flow direction and gradient using three wells based on the geographic position, distance between wells, and total head in each well.

(a) Identify the well that has the intermediate water level (that is, neither the highest head nor the lowest head).
(b) Calculate the position between the well having the highest head and the well having the lowest head at which the head is the same as that in the intermediate well. This is accomplished by interpolation using the elevation of water in the intermediate well.
(c) Draw a straight line between the intermediate well and the point identified in step b as being between the well having the highest head and that having the lowest head. This line represents a segment of the water-level contour along which the total head is the same as that in the intermediate well.
(d) Draw a line perpendicular to the water-level contour and through either the well with the highest head or the well with the lowest head. This line parallels the direction of ground-water movement.
(e) Divide the difference between the head of the well and that of the contour by the distance between the well and the contour. The answer is the hydraulic gradient.

These steps are included in the *Well triangulation* worksheet; however some of the graphical techniques are replaced by equations.

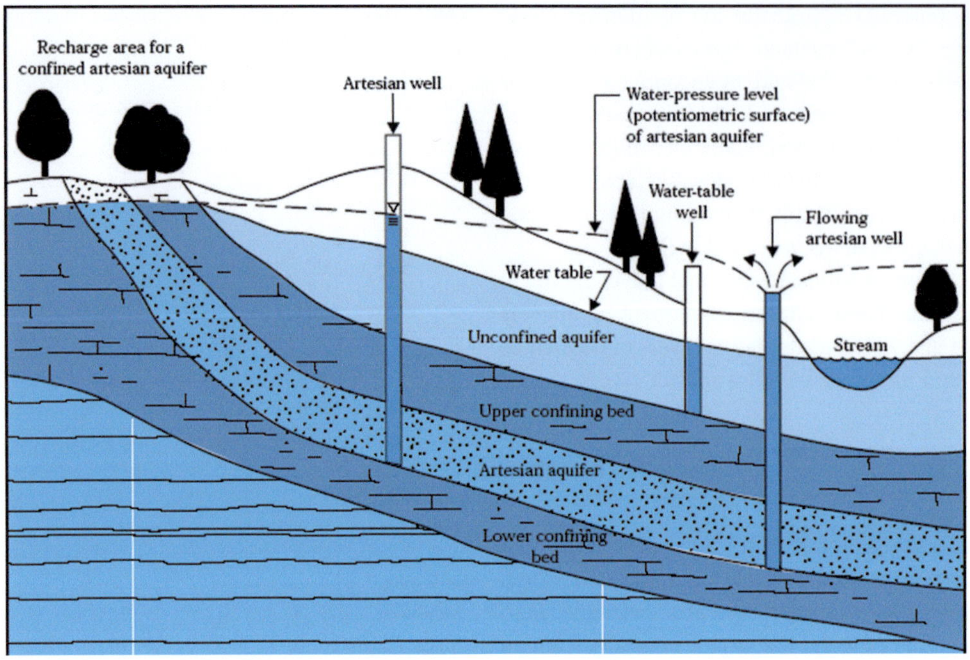

Fig. 10.6 Potentiometric or piezometric surface over confined aquifers (Credit USGS)

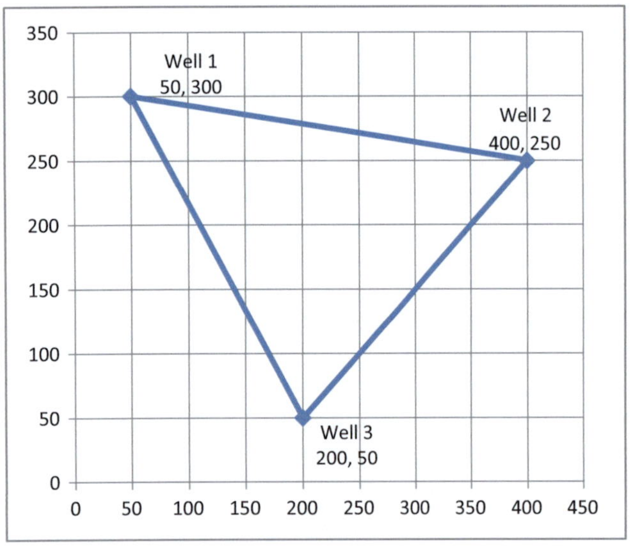

Fig. 10.7 Well coordinates for Example 10.1

Example 10.1 Three wells are located as shown in Fig. 10.7. Use the *Well triangulation* worksheet to find the hydraulic gradient and the direction of flow. The East axis (x-axis) is zero degrees.

Well 1 elevation = 104 m
Well 2elevation = 105 m
Well 3 elevation = 100.8 m

Step a. The first step is to find the well with the intermediate water level (M). This is well 1, and the elevation is 104 m.

Steps b and c. The next step is to find the point at which the contour line through well 1 crosses the line between the high (H) and low (L) elevation wells, wells 2 and 3, respectively. This point (Int) is found with the ratio of the well elevations.

$$\mathrm{Ratio}_{H,L} = (z_H - z_M)/(z_H - z_L) = (105 - 104)/(105 - 100.8) = 0.3125\,\mathrm{m}.$$

$$x_{Int} = x_H - (x_H - x_L)^*\mathrm{Ratio}_{H,L} = 400 - (400 - 200)^*0.3125 = 352\,\mathrm{m}.$$

$$y_{Int} = y_H - (y_H - y_L)^*\mathrm{Ratio}_{H,L} = 250 - (250 - 50)^*0.3125 = 202\,\mathrm{m}.$$

The next step is to find the equation for the line between the intermediate well (M) and the point at which the contour line intersects the line between the high and low wells (x_i, y_i).

$$m = (y_{Int} - y_M)/(x_{Int} - x_M) = (202 - 300)/(352 - 50) = -0.323 \quad \text{(cell H5)}$$

$$y = mx + b \quad b = y_{Int} - mx_{Int}$$

$$b = 202 - (-0.323)^*352 = 316 \quad \text{(cell H6)}$$

$$y = -0.323\ x + 316$$

The slope of the line that is perpendicular to this line, which is the direction of flow, is

$$m_{Flow} = -1/m = -1/-0.323 = 3.44$$

The y-intercept of the flow direction is found with the coordinates of the well with the lowest water table elevation.

$$b_{Flow} = y_L - m_{Flow}x_L = 50 - 200^*3.44 = -638$$

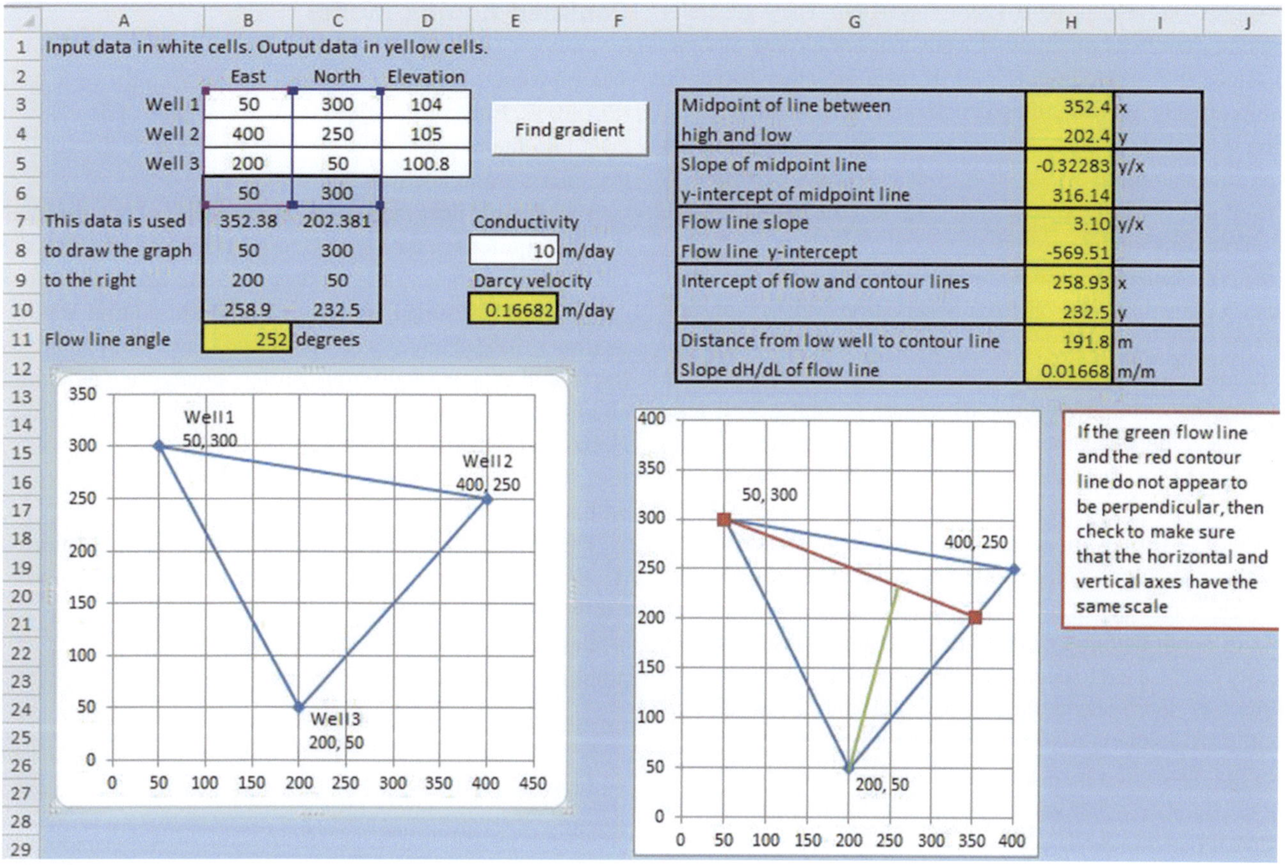

Fig. 10.8 Determination of flow direction based on water table elevations in three wells

The intersection point of the contour line and the flow direction line is found by solving the two equations simultaneously.

$$y = (-m_{Flow}{*}b/m + b_{Flow})/(1 - m_{Flow}/m)$$
$$y = (-3.44{*}316/(-0.323) - 639)/(1 - 3.44/(-0.323)) = 233 \quad \text{(cell H10)}$$
$$x = (y - b)/m = (233 - 316)/(-0.323) = 259 \quad \text{(cell H9)}$$

The distance between the well with the lowest elevation and the contour line is found with the Pythagorean Theorem.

$$\text{Distance} = \left((x_L - x)^2 + (y_L - y)^2\right)^{0.5} = \left((200 - 259)^2 + (50 - 233^2\right)^{0.5} = 192 \quad \text{(cell H11)}$$

The hydraulic gradient, dH/dL, is the elevation difference between the contour line and well with the lowest elevation divided by the distance between the two. The elevation of the contour line is 104 m.

$$dH/dL = (z_M - z_L)/\text{Distance} = (104 - 100.8)/192 = 0.0167\,\text{m/m}. \quad \text{(cell H12)}$$

The Darcy velocity (cell E10) is calculated as the product of conductivity and dH/dL.

The flow line angle in cell B11 (252 degrees) is the direction of water flow. Zero degrees would be east (to the right as in the positive x-direction). In the case of Fig. 10.8, the green line is the flow line, and the flow line angle is 252 degrees from zero (counterclockwise), which agrees with the fact that the direction of flow (direction of the green line) is to the lower left (SW). Thus, one must determine the flow direction based on the angle specified in cell B11.

Steady State Models of Well Flow Rate

Once a well is drilled, the flow rate of a well is tested before the pump is purchased. The flow rate of a well is determined by the conductivity and thickness of an aquifer. Groundwater hydrologists use pumping tests and mathematical models to predict well flow rate. Finite difference and finite element models of groundwater flow (i.e. MODFLOW) are used by hydrologists and groundwater engineers, but this section focuses on analytical models.

Aquifer and well pump tests can be divided into transient and steady-state tests. Steady-state well tests are less complex but also provide less information. They generally require data from observation wells that are located at some distance from the pumping well. If there are other irrigation wells already in the aquifer, then they can be used as observation wells. Steady-state methods require that the pumping test continues until a steady state water surface elevation is observed in both the observation wells and the pumping well.

Transient models are based on the change in water surface elevation over time within the pumping well, and some also require information from observation wells. Transient pump tests can reveal the extent and yield of individual aquifers, and sometimes reveal whether aquifers are connected to other aquifers. The more complex well tests require complex mathematical models and a more in-depth understanding of groundwater hydrology than is presented in this book.

In this text, two simple steady state models/tests for confined and unconfined aquifers are presented that calculate the transmissivity of an aquifer. Transmissivity is the product of depth and conductivity of the aquifer. The transmissivity can be used to determine the relationship between well flow rate and water table drawdown around the well. This will provide the necessary information for sizing the pump, setting the elevation of the pump, and determining the maximum flow rate that can be pumped without exceeding the capacity of the well. The next section presents the steady state test for a confined aquifer, which is followed by the steady state model for an unconfined aquifer. Although flow to a well may include a vertical component, these models ignore the vertical component of flow with the Dupuit-Forchheimer assumption.

Confined Aquifer Model

The piezometric surface declines in the vicinity of a well (Fig. 10.9). If the piezometric surface and cone of depression drop into the aquifer, then the upper portion of the aquifer becomes unsaturated near the well.

The analytic flow model presented in this section assumes that the aquifer remains saturated near the well and that the piezometric surface does not drop into the aquifer. Several assumptions are required for this analytic aquifer model: isotropic (no change with direction) and homogeneous (no spatial change) aquifer properties, steady state pumping, and infinite aquifer extent. It also assumes horizontal flow. Horizontal Dupuit-Forchheimer flow lines are shown in Fig. 10.9. This assumption is valid except for deep aquifers with a large vertical component of flow up to the well.

The elevation of the piezometric surface at two distances (r_1 and r_2) away from the well can be measured with observation wells (Fig. 10.9).

Based on Darcy's law, the horizontal flow of water at any point in the aquifer is

$$v = -K\frac{dH}{dr} \tag{10.1}$$

where

r = radial distance from the well, m,
H = piezometric surface (elevation + pressure), m,
K = hydraulic conductivity, m/day,
v = Darcy velocity, m/day.

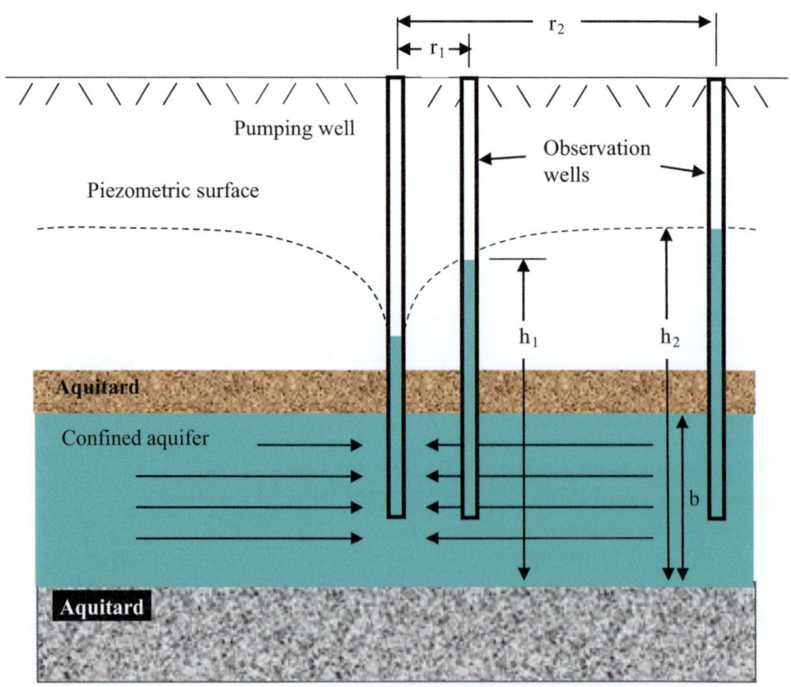

Fig. 10.9 Confined aquifer with Dupuit Forchheimer assumption of horizontal flow

Total flow, Q, equals the product of velocity of flow and cross sectional area of flow.

$$Q = vA \tag{10.2}$$

where

A = cross-sectional area of aquifer perpendicular to flow direction, m^2,
Q = flow rate, m^3/day.

The cross sectional area of flow at any radial distance from the well is the product of the circumference and the depth of the aquifer b.

$$A = b*2\pi r \tag{10.3}$$

Substitute Eqs. 10.2 and 10.3 into Eq. 10.1.

$$-Q = -K\frac{dH}{dr}*2\pi rb \tag{10.4}$$

Q is negative because flow is in the negative r direction (toward the well).

It is conventional in groundwater modeling to combine the depth and conductivity of the aquifer into one term called the transmissivity.

$$T = Kb \tag{10.5}$$

where

T = transmissivity, m^2/day.

$$Q = \frac{dH}{dr}*2\pi rT \tag{10.6}$$

Integrate Eq. 10.6 and solve for elevation of the piezometric surface with distance from the well.

$$\frac{Q}{2\pi T}\int_{r_1}^{r_2}\frac{dr}{r} = \int_{h_1}^{h_2} dH \tag{10.7}$$

$$\frac{Q}{2\pi T}(\ln(r_2) - \ln(r_1)) = h_2 - h_1 \tag{10.8}$$

$$T = \frac{Q\ln(^{r_2}/_{r_1})}{2\pi(h_2 - h_1)} \tag{10.9}$$

Equation 10.9 is called the Thiem equation.

Example 10.2 Find the transmissivity and conductivity in a confined aquifer in which the flow rate to a well is 200 gpm, and observation wells at distances of 100- and 200-m from the pumping well have depths to the water table of 100-m and 95-m, respectively. The upper surface of the aquifer is 150 m below ground and the aquifer is 50 m in depth. The same calculations are also performed in the *Confined aquifer* worksheet.

Calculate flow rate in m^3/day.

$$\left(\frac{200\ gal}{\min}\right)\left(\frac{0.003785\ m^3}{gal}\right)\left(\frac{24*60\ \min}{day}\right) = 1,090\ m^3/day$$

Find the elevations h_1 and h_2. Let the datum = 200 m below the ground surface, which is at the boundary between the aquifer and the lower aquitard.

$$h_1 = 200 - 100 = 100\ \text{m} \quad h_2 = 200 - 95 = 105\ \text{m}$$

$$T = \frac{1,090\ln(^{200}/_{100})}{2\pi(105 - 100)} - 24\ m^2/day \quad K = \frac{T}{b} = \frac{24}{50} = 0.48\ m/day$$

Unconfined Aquifer Model

Unconfined aquifers have no confining aquitard above; the water table is at atmospheric pressure, and the piezometric surface is the water table. Nevertheless, Dupuit-Forchheimer (horizontal) flow is also assumed in this unconfined aquifer model (Fig. 10.10). Transmissivity of an unconfined aquifer is variable because the cross-sectional area of flow decreases near the well as the water table declines (Fig. 10.10).

The depth of the aquifer is equal to h, the water table elevation, at any radial distance from the pumping well. The cross-sectional area is 2πrh, and flow rate is Darcy velocity * cross-sectional flow area.

$$-Q = -K\frac{dH}{dr}*2\pi rh \tag{10.10}$$

$$\frac{Q}{2\pi K}\int_{r_1}^{r_2}\frac{dr}{r} = \int_{h_1}^{h_2} hdH \tag{10.11}$$

$$\frac{Q}{2\pi K}\ln(r_2) - \ln(r_1) = \frac{h_2^2 - h_1^2}{2} \tag{10.12}$$

$$K = \frac{Q\ln(^{r_2}/_{r_1})}{\pi(h_2^2 - h_1^2)} \tag{10.13}$$

Equation 10.13, the Thiem equation for unconfined aquifers, is different from Eq. 10.9 for confined aquifers because the

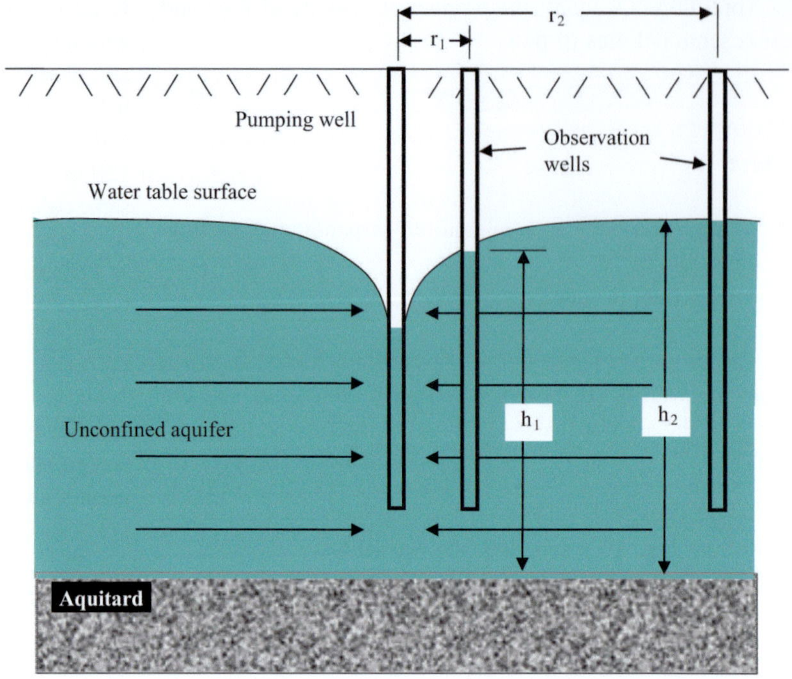

Fig. 10.10 Unconfined aquifer

cross-sectional area of flow changes with distance from the well.

Example 10.3 Find the conductivity and transmissivity of an unconfined aquifer. The flow rate is 200 gpm, and observation wells at distances of 100- and 200-m from the pumping well have depths to the water table of 100-m and 95-m, respectively. Depth to the aquitard is 130 m.

Set the datum equal to the upper surface of the aquitard.

$$h_1 = 130 - 100 = 30 \text{ m}$$
$$h_2 = 130 - 95 = 35 \text{ m}$$

$$K = \frac{1,090 \ln(^{200}/_{100})}{\pi(35^2 - 30^2)} = 0.74\ m/day$$

The product of conductivity and average aquifer thickness is a reasonable estimate of transmissivity.

$$\bar{h} = \left(\frac{35 + 30}{2}\right) = 32.5m\ \ T = K\bar{h} = 0.74*32.5 = 24\,m^2/day$$

These calculations are also performed in the Unconfined aquifer worksheet

Specific Capacity

Once the transmissivity of the aquifer is calculated, the specific capacity of the well can be determined, which is the ratio of the pumping flow rate to the drawdown. The drawdown is the difference between the dynamic water level in the well during pumping and the static water level in the well with no pumping. For a confined aquifer, the specific capacity is calculated as follows by rearranging the Thiem equation:

$$SC = \frac{Q}{(h_e - h_w)} = \frac{2\pi T}{\ln(^{r_e}/_{r_w})} \tag{10.14}$$

where

r_w = well radius
r_e = distance where there is no drawdown, m.

In this case, the radial distances in the Thiem equation are well radius and the distance to a point in the aquifer at which there is no drawdown. If this distance is unknown, then it is typically assumed that $r_e = 1{,}000\ r_w$. If this is the case, ln $(1{,}000\ r_w/r_w) = 6.9$, and Eq. 10.14 reduces as follows:

$$SC = \frac{Q}{(h_e - h_w)} = \frac{2\pi T}{6.9} \approx T \tag{10.15}$$

The specific capacity of an unconfined aquifer is more difficult to calculate and must be adjusted with correction factors.

Example 10.4 Calculate the drawdown in the well that was assessed in Example 10.2 if pumping flow rate is 200 gpm (1,090 m^3/day). A "good well" is generally considered to be able to support a flow rate of 1,000 gpm. Reassess at the "good well" flow rate.

In Example 10.2, T was calculated as 24 m^2/day. Thus, SC is approximately equal to 24 m^2/day.

$$h_e - h_w = \text{drawdown} = Q/SC = 1{,}090/24 = 45 \text{ m}$$

This would be acceptable since the dynamic water level in the well would be approximately 5 m above the aquifer since the static water level is approximately 50 m above the top of the aquifer.

At 1,000 gpm, the drawdown would be

$$h_e - h_w = \text{drawdown} = Q/SC = 5(1{,}090)/24 = 227 \text{ m}$$

This would be unacceptable since the water table would theoretically drop below the aquifer.

Transient Aquifer Models

There are many transient models for confined, unconfined, and leaky confined aquifers). This section presents the original confined aquifer model for a fully penetrating well (water flows into the well for the entire thickness of the aquifer), which was adapted by C.V. Theis from heat transfer equations.

C.V. Theis adapted heat transfer equations to groundwater and developed the Theis equation for transient water table behavior. This equation can be used to calculate the water table at any distance from a well at any time after initiation of pumping. It can also be used to fit observed transient water table behavior in order to calculate the transmissivity and storativity of an aquifer. The traditional method developed by Theis is called the type curve method; he overlaid the observed drawdown vs. time logarithmic curve on a theoretical logarithmic curve in order to calculate storativity and transmissivity. The following equation developed by Theis calculates the drawdown s as a function of flow Q, transmissivity T, and the well function, W(u).

$$(h_e - h_w) = s = \frac{Q}{4\pi T}\int_u^{\infty} \frac{e^{-u}}{u} du \qquad (10.16)$$

$$\begin{aligned}(h_e - h_w) &= s \\ &= \frac{Q}{4\pi T}\left[-0.5772 - \ln u + u - \frac{u^2}{2*2!} + \frac{u^3}{3*3!} - \frac{u^4}{4*4!} + \ldots\right] \\ &= \frac{Q}{4\pi T}W(u) \end{aligned} \qquad (10.17)$$

where W(u) is called the well function and

$$u = \frac{r^2 S}{4Tt} \qquad (10.18)$$

where

S = Storativity or specific yield, m/m.

The equation can be rearranged and s can be plotted as a function of t/r^2 as follows or as a function of t as in Eq. 10.18:

$$s = \frac{Q}{4\pi T}W\left(\frac{r^2 S}{4Tt}\right) \qquad (10.19)$$

The derivation and theory behind the Theis equation are not easy to understand; however, the use of the Theis equation to find T and S is made quite simple in the *Chapter 10 Groundwater* Excel program. The *Theis scrollbars* worksheet allows adjustment of S and T until the Theis theoretical curve (based on Eq. 10.19) matches the shape and magnitude of a drawdown vs. time curve (s vs. t/r^2) curve for a group of wells. Adjusting transmissivity changes the magnitude of the curve, and adjusting storativity changes the shape (slope) of the curve. This scrollbar method for matching the Theis curve was originally developed by Gao, but the reference has been lost, and cannot be found on the Internet. This version is not the original program by Gao, but follows the same procedure. The "expint1" W(u) function is in the Excel VBA editor.

Example 10.5 Calculate the transmissivity and storativity of a confined aquifer with the *Theis scrollbar* worksheet. The following drawdown data was collected from an observation well 20 m from the pumping well. The well flow rate was 5,000 m^3/day.

Time after initiation of pumping	Drawdown s (m)
5 min	2.98
10 min	4.23
20 min	5.54
1 hour	7.67
2 hour	9.04
3 hour	9.84
4 hour	10.41
5 hour	10.86
6 hour	11.22
7 hour	11.53
10 hour	12.23
15 hour	13.04
20 hour	13.61
24 hour	13.97

Convert the data to t/r^2 with units of days/m^2

Time	Time (days)	t/r^2	s
5 min	0.0035	0.0000087	2.98
10 min	0.0069	0.0000174	4.23
20 min	0.0139	0.0000347	5.54

(continued)

Time	Time (days)	t/r^2	s
1 hour	0.0417	0.0001042	7.67
2 hour	0.0833	0.0002083	9.04
3 hour	0.1250	0.0003125	9.84
4 hour	0.1667	0.0004167	10.41
5 hour	0.2083	0.0005208	10.86
6 hour	0.2500	0.0006250	11.22
7 hour	0.2917	0.0007292	11.53
10 hour	0.4167	0.0010417	12.23
15 hour	0.6250	0.0015625	13.04
20 hour	0.8333	0.0020833	13.61
24 hour	1.0000	0.0025000	13.97

Input the right two columns of data into the *Theis scrollbar* worksheet under the Actual data columns (columns C:D) as shown in Fig. 10.11. Note that because time is expressed as t/r^2, drawdown data could have been taken from any location in the aquifer where each drawdown is normalized by its distance, r, from the pumping well.

Adjust the T and S scrollbars until the Actual data and Calculated values curves are aligned. It is best to reduce the range of s on the y-axis scale in order to improve accuracy (Fig. 10.12). rom Fig. 10.27, it is observed that storativity, S, is 0.001 and transmissivity, T, is 193 m^2/day. It should be noted that actual drawdown curves will not generally fit as tightly as in Fig. 10.12.

If storativity and transmissivity are known, then the Theis equation can be used to plot drawdown vs. time. worksheets are available for this purpose *Theis s* vs. *t* worksheet and *Theis s* vs. t/r^2 worksheet.

The Theis equation is only applicable to homogenous aquifers. Numerical groundwater models have been developed to calculate aquifer parameters for nonhomogenous, nonisotropic and irregular aquifers. One popular aquifer model developed by the USGS is called MODFLOW.

After rough estimates of well flow rate with the Thiem or Theis models, or steady state tests, a step drawdown test with a test pump can be used to determine the actual size of the pump that should be installed in the well. The first phase of a step drawdown test is to find the maximum permissible pumping flow rate, which is found by increasing the pumping rate until the steady state drawdown is equal to the maximum permissible drawdown in the well. This maximum flow rate should be maintained for at least 8 hours, and once it is confirmed that the drawdown has reached steady state, the test should be maintained for at least one more hour. However, a 24 hour test is preferable, especially in low permeability aquifers (ASAE EP400.2 T). Next, the drawdown at 80 % of the maximum discharge is determined by lowering the pump flow rate by 20 % and continuing to pump. Once the drawdown level is maintained at a constant level for at least 30 minutes and recorded, then flow is dropped to the 60 %, 40 %, and 20 % of maximum discharge flow rates. These drawdown and discharge rates can be used to select a pump and to find the intersection point of the pump curve and the irrigation system curve.

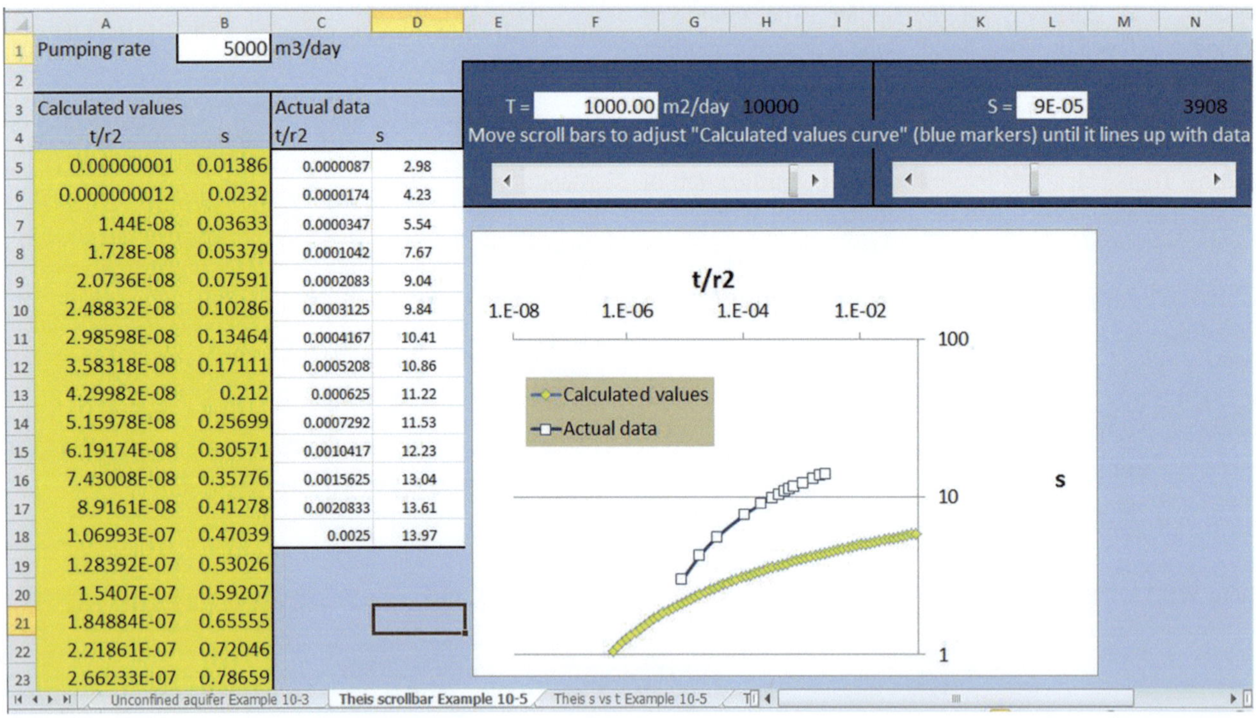

Fig. 10.11 Theis scrollbars worksheet before adjustment of T and S scrollbars

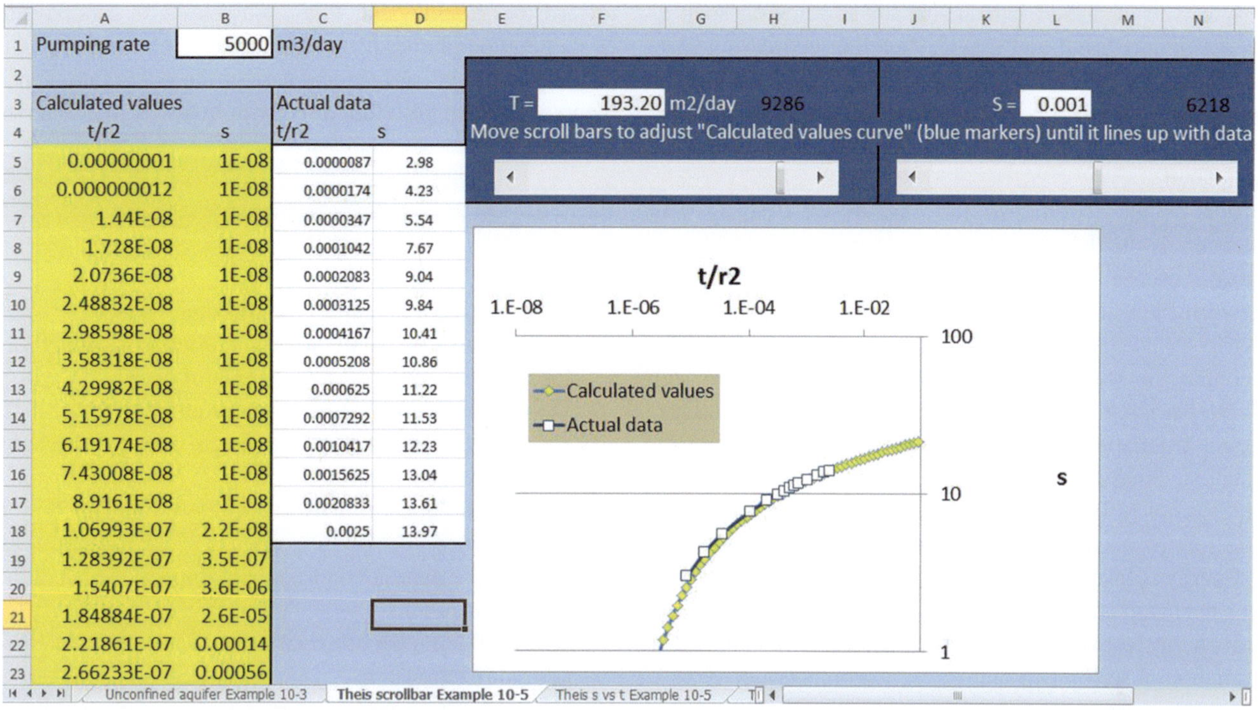

Calculated values t/r2	s	Actual data t/r2	s
0.00000001	1E-08	0.0000087	2.98
0.000000012	1E-08	0.0000174	4.23
1.44E-08	1E-08	0.0000347	5.54
1.728E-08	1E-08	0.0001042	7.67
2.0736E-08	1E-08	0.0002083	9.04
2.48832E-08	1E-08	0.0003125	9.84
2.98598E-08	1E-08	0.0004167	10.41
3.58318E-08	1E-08	0.0005208	10.86
4.29982E-08	1E-08	0.000625	11.22
5.15978E-08	1E-08	0.0007292	11.53
6.19174E-08	1E-08	0.0010417	12.23
7.43008E-08	1E-08	0.0015625	13.04
8.9161E-08	1E-08	0.0020833	13.61
1.06993E-07	2.2E-08	0.0025	13.97
1.28392E-07	3.5E-07		
1.5407E-07	3.6E-06		
1.84884E-07	2.6E-05		
2.21861E-07	0.00014		
2.66233E-07	0.00056		

Fig. 10.12 Aligned actual data and calculated values curves

Wells

This section is intended to give a brief summary of well drilling steps, screens, gravel pack, and casings so that an irrigation engineer can communicate with well designers. It is not at all intended to give the necessary background to pick well casings, gravel packs, screens, pumps, or pipes. These should be designed by a certified well driller or professional with experience in the region in which the well is to be drilled. Selecting the proper casing, screen, etc. is a science in and of itself. Some well equipment manufacturers will design the components of the well if the materials are purchased from them.

The well casing is solid steel pipe except where the well penetrates the aquifer. Slotted steel pipe or wound stainless steel screens are surrounded by a gravel pack in order to facilitate water movement into the well but prevent aquifer materials from entering the well (Fig. 10.13). The well is sealed above and below the aquifer to prevent pollutant leakage into the aquifer by placing a bentonite seal above and below the gravel pack. The bentonite is poured into the well as a dry aggregate at the base of and at the upper end of the water bearing aquifer. As it hydrates between the well casing and the well wall, the well is sealed. The sealant material is an expanding clay called Wyoming bentonite that is actually montmorillonite. A good seal is important in agricultural wells because of the danger of chemical spills near the well that could wash down the well.

Fig. 10.13 Well casing

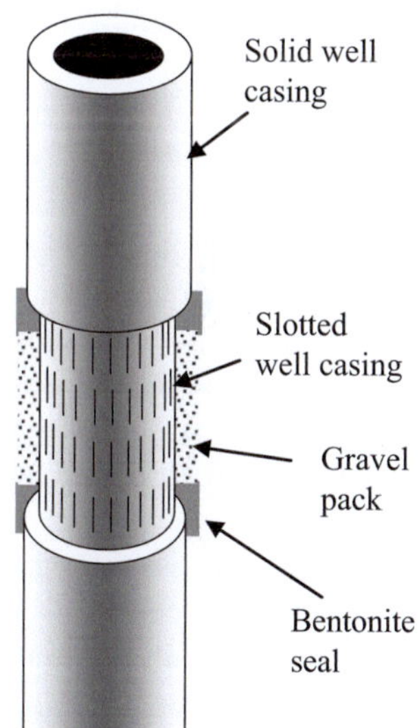

The gravel pack is poured into the well over the entire height of the water bearing aquifer. In order to allow water to freely enter the well while at the same time preventing particles from entering the well, the gravel pack has a size gradation that maintains high hydraulic conductivity but at the same time has small enough particles to filter out particles that are carried with the pumped water. Specifications for gravel pack particle size distribution are found in many handbooks.

If a well is drilled into an unstable formation such as unconsolidated sand and gravel, then the well casing must be driven into the well right behind the drill bit so that the well does not collapse. If the well is drilled into rock, then the well casing is driven into the well after the drilling is completed.

One of the criteria for selection of well casing wall thickness is the differential pressure head on the inside and outside of the well, which is generally the difference between the water elevation within the well and the piezometric surface outside the well. This may change due to the fact that the water level is drawn down within the well during pumping. Well casing is most susceptible to collapse during installation before the gravel pack around the well is installed. Shipping can sometimes damage well casing so that it is not perfectly cylindrical. A small eccentricity, even 1 %, reduces the differential head limitation.

Energy is lost as water moves through the slotted casing. This results in an additional depth of drawdown in the well. This energy loss (additional drawdown) can be calculated as a function of well flow rate.

Groundwater Pollution

Regulation of groundwater will continue to grow in importance for agriculture. The trend is to increase regulations that restrict the leaching or runoff of agricultural chemicals, pesticides and nutrients, from agriculture.

Prior to the 1970s, scientists thought that groundwater was naturally protected from contamination by the layers of soil above the groundwater. Although layers of soil may help slow or reduce pollution of the groundwater, some chemicals are not removed from water as it moves through the soil. In addition, the soil has a maximum capacity to absorb pollutants, and continued addition of pollutants to soil will overwhelm this capacity. As a result of some catastrophic environmental disasters in the 1970s such as dumping of extremely toxic chemicals in landfills within sand aquifers, Congress passed the Safe Drinking Water Act and authorized the Environmental Protection Agency (EPA) to set standards for maximum levels of contaminants in drinking water, and to regulate activities that may adversely affect water quality. The protection effort includes disposal regulations, water treatment, and water monitoring.

There are two major types of groundwater pollution: point source and nonpoint source. A point source comes from a single location such as a factory, concentrated animal feedlot operation (CAFO), or a landfill. Nonpoint source pollution comes from a broad area such as nitrate pollution from agricultural fertilizers and animal waste used for fertilization on farms. As irrigation and precipitation infiltrate into the soil, the water carries pollutants with it downward toward the groundwater. Point sources cause groundwater contamination plumes that spread as the groundwater moves below the ground surface and carries the pollutants with it. Nonpoint source pollution directly pollutes large groundwater regions.

Questions

1. What are the five major types of aquifers?
2. Describe four different types of aquifers commonly used for irrigation and give an example of each.
3. Draw a confined and unconfined aquifer.
4. Discuss groundwater recharge and discharge components for a basin
5. Discuss the impact of overpumping on stream flow in arid climates.
6. Discuss the natural condition and impact of man on the Ogallala aquifer.
7. Discuss the impact of irrigation on water table elevation and salinization.
8. Calculate the porosity, storativity, specific yield, and specific retention for an aquifer that has 40 % water below the water table, and yield of 1.2 m of water for every 5 m drop in water table elevation.
9. A coarse sand aquifer has a water table slope of 1 m/100 m. Evaluate at the upper and lower limits of coarse sand hydraulic conductivity. What is the Darcy velocity of the water in the aquifer? The cross sectional area of the aquifer is 100 m x 1,000 m. What is the volume of water flow in 1 year? Convert water volume to acre-ft. How many acres of cotton could be irrigated with this volume per year? Also calculate for a silt aquifer with a hydraulic conductivity at the lower end of silt conductivities.
10. Find the hydraulic gradient and the direction of flow with the East axis (x-axis) as zero degrees for the following three wells. Show your work (work it by hand), and check your work with the *Groundwater* program.

	East	North	Elevation
Well 1	50	600	104
Well 2	400	250	105
Well 3	200	50	108

11. If the aquifer in question 10 is a coarse sand aquifer with hydraulic conductivity equal to 100 m/day and porosity of 0.40 calculate the Darcy velocity and the velocity that a contaminant plume would travel through the aquifer.
12. What is the reason that aquifer pollution is much more difficult to correct than surface water pollution?
13. What are the primary pollutants from agriculture that have contributed to aquifer pollution?

14. What often happens to shallow aquifers when irrigation is introduced to a region?
15. Find the transmissivity and conductivity of confined aquifer in which the flow rate to a well is *400 gpm*, and observation wells at distances of 100- and 200-m from the pumping well have depths to the water table of 100-m and 98-m, respectively. The upper surface of the aquifer is 140 m below ground and the aquifer is 40 m thick. Check your calculations with the *Confined aquifer* worksheet.
16. Find the conductivity and transmissivity in an unconfined aquifer in which the flow rate to a well is *400 gpm*, and observation wells at distances of 100- and 200-m from the pumping well have depths to the water table of 100-m and 98-m, respectively. The lower boundary of the aquifer (upper surface of aquitard) is 150 m below the ground.
17. Calculate the transmissivity and storativity of a confined aquifer. The following drawdown data was collected from an observation well 100 m from the pumping well. The well flow rate was 2,000 m^3/day.

Time after initiation of pumping	Drawdown s (m)
1 min	0.44
2 min	0.55
4 min	0.66
8 min	0.77
0.01 day	0.86
0.02 day	0.97
0.04 day	1.08
0.08 day	1.19
0.16 day	1.30
0.32 day	1.41
0.64 day	1.52
1 day	1.59

18. For the aquifer in question 17, calculate the drawdown in the well for a series of points between 1 hr and 1 week. The well diameter is 50 cm (use r = 0.25 m in the *Theis s* vs. *t* worksheet). The pump flow rate is 3,000 m^3/day.
19. The well described in questions 17 and 18 has a maximum acceptable drawdown of 20 m inside the well. The pressure loss in the casing is flow rate (m^3/day)/4,600. Calculate the maximum allowable pump flow rate and the drawdown at 80 %, 60 %, 40 %, and 20 % of maximum. Plot the drawdown vs. flow rate curve. What is the shape of the curve? Convert the maximum flow rate to units of GPM and report whether this well would be considered a good well. Assume that the drawdown after 7 days is the steady state drawdown.
20. The static water table in the aquifer described in questions 17–19 is 40 m below the ground surface. The pump in the well has a pump curve as shown below. The pump has an open discharge 0.5 m above the ground surface, and the sum of minor losses (K) = 3.9 (including velocity head losses at the discharge). The pump hangs on a 12 inch pipe (Schedule 40) at an elevation 80 m below the ground surface, and there is a 2 m section of pipe above the ground surface (total 82 m pipe). The pipe has a Hazen Williams C = 100. Include the minor losses. Calculate the discharge flow rate.

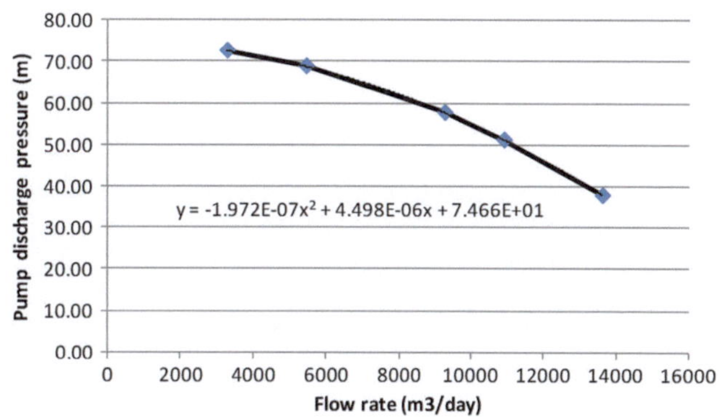

21. What are the two types of groundwater pollution? Which comes from field agriculture?

References and Resources

American Groundwater Trust. www.AGWT.org

Elhadj E (2008) Dry aquifers in Arab countries and the looming food crisis. Middle East Rev Int Aff 12(4):1–12

Huffman R, Fangmeier D, Elliott W, Workman S (2013) Soil and water conservation engineering. American Society of Agricultural and Biological Engineering, St. Joseph

JohnsonScreens. Publisher of Groundwater and Wells (the definitive resource on well design). http://www.weatherford.com/weatherford/groups/public/documents/johnsonscreens/

Lvovich MI (1979) World water resources, present and future. GeoJournal 3(5):423–433

National Groundwater Association. www.NGWA.org

USBR Groundwater manual (1995) US Department of the Interior. Bureau of Reclamation. pp 661. http://www.usbr.gov/pmts/wquality_land/GndWater.pdf

USGS Groundwater Information. http://water.usgs.gov/ogw/

11 Open Channel Flow

Open channels can range from small irrigation furrows to huge irrigation canals that are hundreds of kilometers long and supply billions of cubic meters per year for irrigation, industry, and domestic purposes. Agricultural canal categories include the irrigation district main, secondary and tertiary canals, laterals, on-farm irrigation ditches, and drainage channels. This chapter covers the structures and principles that are related to open channel delivery of water to agriculture: water diversion structures, conveyance efficiency, siphons, canal hydraulics (uniform flow, energy drop structures, and gradually varied flow), and flow measurement. The government has encouraged farmers to conserve water by lining irrigation ditches with concrete (Fig. 11.1); however, concrete channels can develop cracks and gaps that have excessive water loss. Economic analysis can determine whether lining a canal is worth the cost. Manning's equation calculates the head loss along a canal based on slope, roughness, and channel geometry. Energy dissipation structures use supercritical flow and hydraulic jumps to dissipate energy. The Froude number determines the relationship between subcritical and supercritical flow. A finite difference solution calculates water depth changes along a canal with gradually varied flow.

Water Diversion

Properly designed diversion structures are an important part of irrigation canal systems. Water is backed up behind a diversion structure such as a dam and then flows by gravity to a different field or region. Many river valleys have canal systems that take water out of the river at the upper end of the valley, and the canal, having a shallower grade than the river, is able to irrigate farms along the sides of the valley.

Many irrigation districts begin with large diversion structures that direct it into their main canal. For example, the Imperial Diversion Dam (Fig. 11.2) in southeastern California diverts water from the Colorado River into the All-American Canal, which delivers water to the Imperial Irrigation District and to San Diego. The dam increases the water surface elevation (head) and diverts water into the diversion canal. The diversion has large desilting basins, without which the canal system and farm furrows would rapidly fill with silt.

Diversions within canals generally have downstream control structures to control the upstream water level (Fig. 11.3).

The same technique is used to divert water from small streams to on-farm irrigation systems. A small dam is constructed, sometimes with a fish ladder, just downstream from the turnout (Fig. 11.4). There are many methods to divert water from irrigation ditches into fields, such as slide gates over small tubes (Fig. 11.1), slide gate turnouts or alfalfa valves over spiles (Fig. 11.5) or siphon tubes over the canal bank (Fig. 11.6).

Water also needs to be collected from the ends of fields and carried away from the field. Drainage channels and tailwater recovery channels (Fig. 11.6) are normally installed below grade in order to remove drainage water from the field by gravity flow (Fig. 11.7).

Conveyance Efficiency

Irrigation canals lose water through seepage and evaporation, which reduces conveyance efficiency. The conveyance efficiency is the amount of water delivered by a canal divided by the amount of water diverted into the canal. The water duty is the percent of water lost from the canal.

P. Waller, M. Yitayew, *Irrigation and Drainage Engineering*, DOI 10.1007/978-3-319-05699-9_11

Fig. 11.1 Square and trapezoidal irrigation ditches (Credit NRCS)

Fig. 11.2 Imperial diversion dam (Courtesy of United States Bureau of Reclamation)

Example 11.1 An irrigation district must divert 20 cm (8 in) depth of water into a lateral in order to deliver 15 cm (6 in) depth of water to a field. What are the efficiency and water duty?

$$\text{Efficiency} = 15/20(100\%) = 75\%$$
$$\text{Water duty} = 5/20(100\%) = 25\%$$

Evaporation from canals and reservoirs is approximately the same as reference ET_0. The volume of water lost per day is the product of canal water surface area and ET_0.

Fig. 11.3 Headgate on irrigation district lateral canal and downstream control structure

Fig. 11.4 Irrigation turnout upstream from dam in stream

Example 11.2 Reference evapotranspiration is 12 mm/day in summer and 4 m/year. Calculate the evaporation per day in summer and per year from a 200 m long, 1 m wide canal.

$$V_{Evap/day} = (200\,m)\ (1\,m)\ (0.012\,m/day) = 1.2\ m^3/day/1,233 = 0.001\,ac - ft/day.$$
$$V_{Evap/yr} = (200\,m)\ (1\,m)\ (4\,m/year) = 800\ m^3/year/1,233 = 0.65\ ac - ft/day.$$

Calculate the evaporation per day in summer and per year for a 540 km long, 25 m wide canal. What is the water duty if the canal delivers 1.85 billion m^3/year, and seepage losses are 3 %?

Fig. 11.5 Slide gate in farm irrigation ditch bank with energy dissipation blocks to prevent erosion (*left*) and spile (*right*)

Fig. 11.6 Siphon tubes over irrigation ditch bank (Credit NRCS Jeff Vanuga)

Fig. 11.7 Tailwater recovery channel (Credit NRCS Jeff Vanuga)

$$V_{Evap/day} = (540,000\,m)\ (25\,m)\ (0.012\,m/day) = 162,000\,m^3/day/1,233 = 131\,ac - ft/day$$
$$V_{Evap/yr} = (540,000\,m)\ (25\,m)\ (4\,m/yr) = 54,000,000\,m^3/year/1,233 = 44,000\ ac - ft/year$$

The percent evaporation is (54,000,000/1.85 × 10^9) (100 %) = 3 %.

Total water duty = % seepage losses + % evaporation losses = 3 % + 3 % = 6 %.

Canals may be lined in order to reduce seepage. Lining canals is especially important in permeable (sandy) soils. The *NRCS NEH 7* describes canal lining materials as follows:

> *Ditch lining materials include compacted soil, high expanding colloidal clay (bentonite), hand formed nonreinforced or reinforced concrete, slip formed nonreinforced concrete, pneumatic applied concrete mortar (gunnite), cold spray-applied membrane, and flexible membranes of plastic, elastomeric, or butyl rubber. Flexible membranes should be protected from physical damage and ultraviolet light by covering with aggregate or soil. Flexible membranes with concrete or aggregate protection can be installed underwater if the water velocity is less than 5 feet per second.*

Lining canals may have political ramifications if seepage to groundwater is reduced. One person's lost seepage water is another person's irrigation water. For example, the Imperial Irrigation District in California estimates that lining a 23-mile section of the All-American Canal that passes through sand dunes will save 70,000 acre-ft/yr (86 million m^3/yr). The Metropolitan Water District in Los Angeles will pay the Imperial Irrigation District to line the canal and divert the saved water for urban use. The problem with lining the canal, however, is that Mexican farmers just across the border depend on the seepage water from the All-American canal to recharge aquifers from which they pump irrigation water. They claim that this water is rightfully theirs because the canal has been seeping since it was completed in 1942, and they have developed farms that depend on the canal seepage. However, the seepage volume is not mentioned in the treaty signed in 1944 between the

Table 11.1 Canal seepage rates reported in published studies (After Leigh and Fipps (2009))

Lining/soil type	Seepage rate (gal/ft^2/day)	Seepage rate (L/m^2/day)
Unlined[a]	2.21–26.4	90–1,076
Portland cement[b]	0.52	21
Compacted earth[b]	0.52	21
Brick masonry lined[c]	2.23	91
Earthen unlined[c]	11.34	462
Concrete[d]	0.74–4.0	30–163
Plastic[d]	0.08–3.74	3.2–152
Concrete[d]	0.06–3.22	2.4–131
Gunite[d]	0.06–0.94	2.4–38
Compacted earth[d]	0.07–0.6	2.9–24
Clay[d]	0.37–2.99	15–122
Loam[d]	4.49–7.48	183–305
Sand[d]	4.0–19.45	163–792
Clay[e]	1.5	61
Silty clay loam[e]	2.24	91
Clay loam[e]	2.99	122
Silt loam earth[e]	4.49	183
Loam[e]	7.48	305
Fine sandy loam[e]	9.35	381
Sandy loam[e]	11.22	457

[a]DeMaggio (1990)
[b]U.S. Bureau of Reclamation (1963)
[c]Nayak et al. (1996)
[d]Nofziger (1979)
[e]Texas Board of Water Engineers (1946)

United States and Mexico, which guaranteed 1.5 MAF (million acre-ft) of Colorado River water per year to Mexico.

Leigh and Fipps (2009) did a literature study and compiled a table with typical canal seepage rates as a function of lining and soils (Table 11.1).

Canal seepage can be measured by closing the ends of a canal and measuring the rate of decline of the canal water surface (the best method), by measuring the inflow and outflow to a canal, or, in an earthen canal, by measuring seepage rates with an infiltrometer.

Canal seepage rates in lined canals can be higher than seepage rates in unlined canals because nonreinforced concrete lined canals crack (Fig. 11.8).

In studies in Texas conducted by Fipps, seepage losses in old concrete lined small to medium width (1–3 m wide) canals were in the range of 15–24 gal/ft^2/day (600–960 L/m^2/day); larger canals, with better construction techniques (reinforced concrete) generally had much lower seepage losses – in the range of 1–4 gal/ft^2/day (40–160 L/m^2/day). In the same study, earthen canals had seepage rates less than 2 gal/ft^2/day (80 L/m^2/day); to be fair, it should be noted that these unlined canals were constructed in less permeable soils. Another advantage to the unlined canals was that

Fig. 11.8 Cracked concrete lined canal (Credit NRCS, Ron Nichols)

silting of the canal bottoms had decreased the rate of infiltration into the soil (Fipps G, 2005, Personal communication). In the same study by Fipps (2005, Personal communication), irrigation district conveyance efficiencies in an old irrigation district ranged from 60 % to 75 %.

Lining canals in order to prevent infiltration is expensive. Lining a 75 cm deep canal with a 30 cm wide bottom can cost $60,000/km, and lining a major canal can cost $600,000/km.

Total water requirement for an irrigation district is the farm water requirement divided by the total district efficiency. The total efficiency of an irrigation district includes the storage efficiency of the reservoir, the conveyance efficiency of the canal system, flexibility factors (percent extra water) required for multiple farm delivery, irrigation delivery waste at turnouts due to lack of timing of water use and water delivery, and on-farm irrigation efficiency.

Example 11.3 Calculate the conveyance efficiency to field 1 in Fig. 11.5 from the point of water diversion to the irrigation district. The conveyance efficiency of the irrigation district up to the farm turnout is 70 %. The main concrete canal on the farm has a conveyance efficiency of 90 % and the earth-lined canal has a conveyance efficiency of 80 %

The total conveyance efficiency from the point of diversion to the irrigation district to field 1 is (Fig. 11.9)

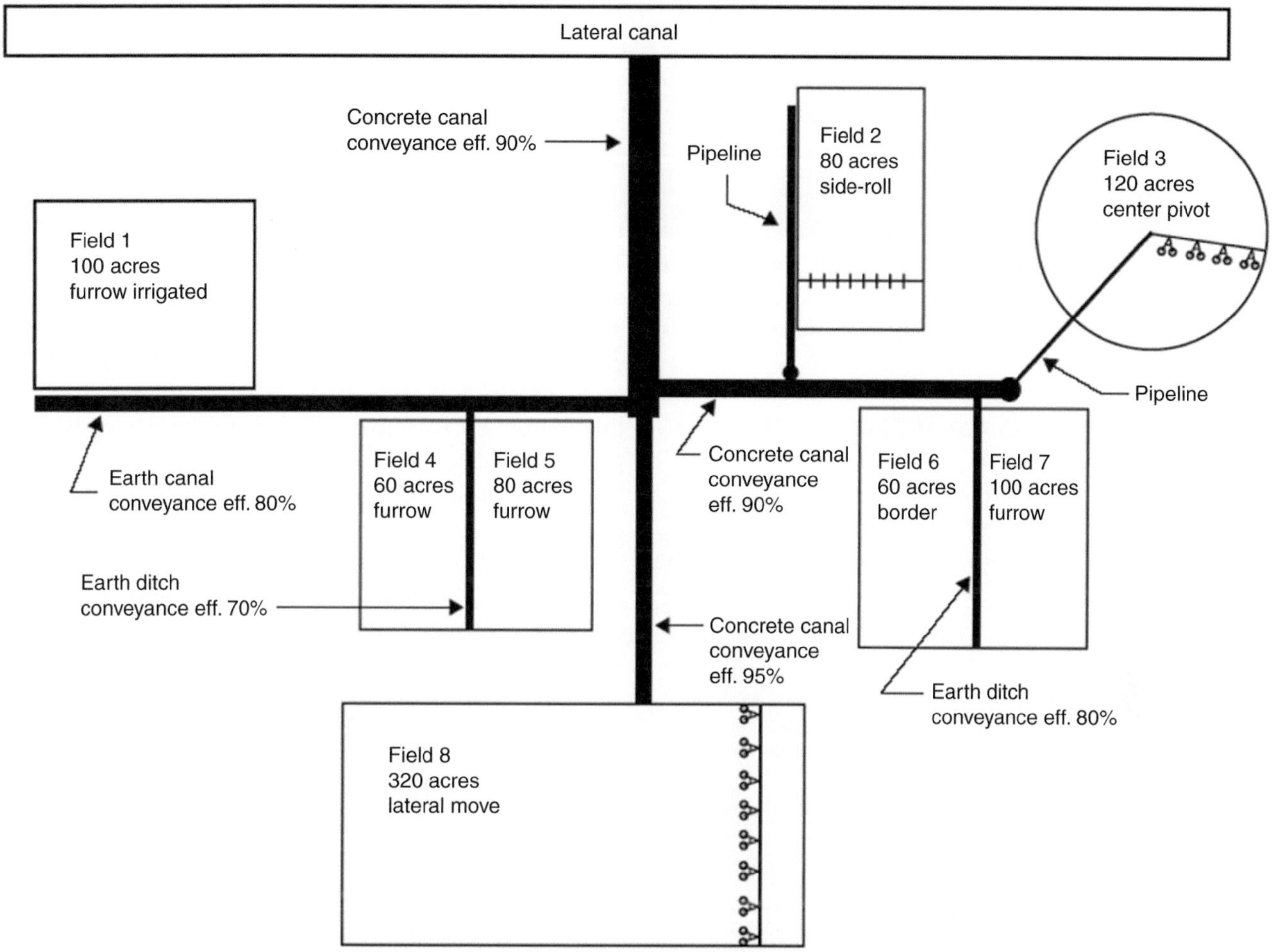

Fig. 11.9 Farm water conveyance system (Credit NRCS. National Irrigation Guide, Part 652, Chap. 7. http://www.irrigationtoolbox.com/NEH/Part652_NationalIrrigationGuide/ch7.pdf)

$$\text{Eff} = (0.7)\ (0.9)\ (0.8)\ (100\%) = 50\%$$

The wetted perimeter of a canal determines the area of infiltration (seepage). Most canals have a trapezoidal shape (Fig. 11.10). The wetted perimeter of a trapezoidal canal is

$$P = b + 2s = b + 2y\left(1 + z^2\right)^{0.5} \qquad (11.1)$$

where

P = wetted perimeter, m
z = side slope (run over rise)
b = channel bottom width, m
y = canal water depth, m.

Example 11.4 A farm irrigation ditch has a trapezoidal shape and has a 30 cm wide bottom, 2:1 side slopes (2 in the horizontal direction and one in the vertical direction, z = 2) and is 75 cm deep. It is filled with water for

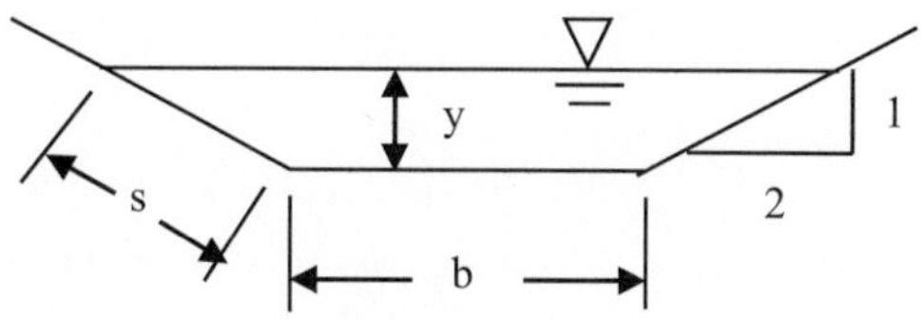

Fig. 11.10 Trapezoidal canal

60 days per year, and it is located in a silt loam soil. The irrigation ditch is 1 km long. The cost of irrigation water is \$3.27/ha-cm. The project life is 20 years, and the required rate of return is 8 %. The cost of lining the ditch is \$60,000/km. Determine whether to line the irrigation ditch.

The wetted perimeter is

$$P = b + 2{*}y{*}\left(1 + z^2\right)^{0.5} = 0.3 + 2{*}0.75{*}\left(1 + 2^2\right)^{0.5} = 3.65\ \text{m}$$

In a silt loam soil, the expected infiltration (seepage) rate is 183 L/m^2/day (Table 11.1)

$$V_{Seepage/day} = 3.65\ m*1,000\ m*183\ L/m^2/day = 660,000\ L/day/km$$
$$V_{Seepage/yr} = 660,000\ L/day*60\ days*0.001\ m^3/L*0.01ha - cm/m^3 = 400\ ha - cm/yr$$
$$\$_{Seepage/yr} = 400\ \ ha - cm/yr*\$3.27/ha - cm = \$1,300/year.$$

The present value of $1,300/yr for a 20 yr project at 8 % interest is approximately $13,000. Therefore, it is not profitable to line the irrigation ditch. However, if the ditch was always filled, then the present value of the cost of seepage would be 6 * $13,000 = $78,000, greater than the lining cost.

Pipes and Siphons in Canal Systems

Pipes can be used to get past obstructions or canyons and yet maintain the same head (elevation). For example, the Roza Irrigation District in the lower Yakima Valley in Washington State delivers water from the Roza dam on the Yakima River in the upper Yakima Valley. The canal is kept at a high elevation along the Rattlesnake Hills. In order to maintain this elevation, five siphons, having a total length of 7 km and averaging 5 m in diameter, bring the Roza irrigation water under five rivers (valleys) and preserve the elevation energy in the canal. Because the canal is at a high elevation above many of the farms it serves, the district is able to deliver water to farms through pipes and maintain a high enough pressure (>210 kPa) for farmers to run sprinklers with low pressure nozzles without a pump.

A partially completed concrete siphon on the Salt River Project Canal in Phoenix, Arizona is shown in Fig. 11.11.

Fig. 11.11 Partially completed concrete siphon on Salt River Project (Credit USBR)

Fig. 11.12 Culvert across drainage ditch

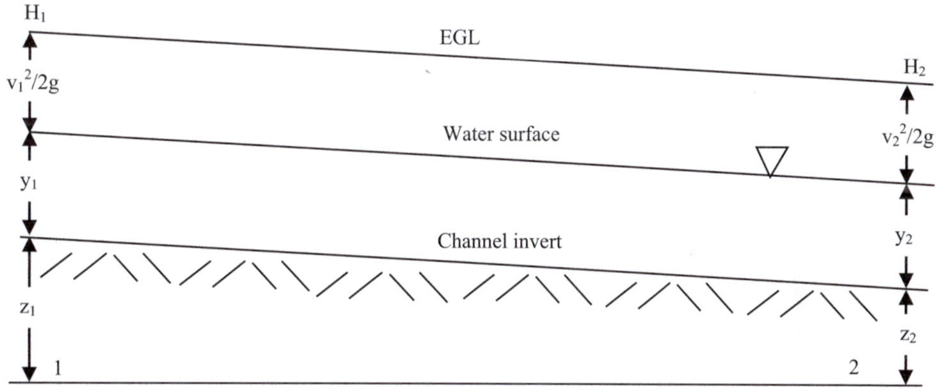

Fig. 11.13 Uniform open channel flow

Smaller siphons are also needed to transfer water past obstructions in smaller irrigation canals: under or over rivers, drainage channels, canals, and roads. Smaller canals can use a corrugated metal pipe to traverse drainage channels or roads (Fig. 11.12).

Canal Hydraulics: Steady-State, Uniform Flow

Canal flow is characterized as uniform, non-uniform, steady, and/or unsteady flow. Uniform flow is characterized by uniform depth along a channel ($y_1 = y_2$ in Fig. 11.13). Steady-state refers to the fact that flow does not change with time. Unsteady state flow is not covered.

The energy in open channel flow is the sum of the elevation of the channel, z, the depth of water flow, y (can be converted to pressure in Bernoulli's equation for a given depth in the channel), and the velocity head, $v^2/2\ g$ (Fig. 11.13). The energy grade line (EGL) is the sum of kinetic and elevation energy (Fig. 11.13).

$$H = y + \frac{v^2}{2g} + z \tag{11.2}$$

where

y = depth of flow in channel, m
v = average velocity of water flow in channel, m/sec
z = elevation of channel invert (bottom of channel), m.

Open channel flow is driven by inertial forces (momentum) and gravitational forces (differences in elevation of the water surface). For steady state (constant in time) and uniform (constant in space) flow in a channel, the water surface slope is the same as the channel slope (Fig. 11.13) and momentum is constant in time and space. Thus, the slope of the energy grade line (EGL) is equal to the slope of the channel. For uniform and steady-state flow in a channel, the flow velocity is determined by Manning's equation, which is a function of the channel roughness, energy grade line (equal to slope, S), and channel dimension (hydraulic radius).

$$\mathrm{v} = \frac{R^{2/3}S_0^{0.5}}{n} \quad Q = \frac{AR^{2/3}S_0^{0.5}}{n} \tag{11.3}$$

where

S_0 = channel slope, positive downward in direction of flow, m/m
n = Manning's roughness coefficient
v = velocity of flow in the channel, m/sec.
Q = flow in channel, m^3/sec
A = cross-sectional area of channel.

The hydraulic radius, R, is representative of the length dimension of the channel perpendicular to the direction of flow. It is calculated as the cross-sectional area of the channel divided by the wetted perimeter.

$$\mathrm{R = A/P} \tag{11.4}$$

The slope is the difference in elevation between two points (Fig. 11.13) divided by the distance between the points. For uniform flow, the difference in elevation is also equal to the friction loss, H_f.

$$S_0 = \frac{z_2 - z_1}{L\cos\theta} = \frac{H_f}{L} \tag{11.5}$$

where

z_1 = Elevation energy at upper end of channel, m
z_2 = Elevation energy at lower end of channel, m
L = Length of channel, m
θ = channel slope, degrees.

Equation 11.3 can be rearranged to solve for slope.

$$S_0^{0.5} = \frac{\mathrm{v}n}{R^{2/3}} \quad S_0 = \frac{\mathrm{v}^2n^2}{R^{4/3}} \tag{11.6}$$

Substitute Eqs. 11.5 into 11.6

$$H_f = L\frac{\mathrm{v}^2n^2}{R^{4/3}} = L\frac{\left(\frac{Q}{A}\right)^2n^2}{R^{4/3}} = L\frac{Q^2n^2}{A^2R^{4/3}} \tag{11.7}$$

Channel area is a function of the square of average channel dimension, D^2, and R is a function of D, thus

$$H_f = L\frac{Q^2n^2}{A^2R^{4/3}} = kL\frac{Q^2n^2}{D^{5.3}} \tag{11.8}$$

where

k = constant

The Manning's equation as written in Eq. 11.7 is similar to the to the Hazen-Williams equation.

Minimum, normal (design), and maximum values of the roughness coefficient, n, are tabulated in Table 11.2. The roughness coefficient increases surface roughness.

Geometric elements of typical channel cross sections are presented in Table 11.3.

When Manning's equation is solved in order to find required channel dimensions for a given flow rate, it is normally written as in Eq. 11.9. The right side of Eq. 11.9 is calculated first, and then an iterative solution is used to find geometric parameters on the left side (section factor) of the equation. After the design depth of flow is found, 25 % freeboard elevation is normally added to the channel depth.

$$AR^{2/3} = \frac{Qn}{S_0^{0.5}} \tag{11.9}$$

Example 11.5 A concrete lined irrigation ditch (Fig. 11.14) has a slope of 0.2 % = 0.002 m/m. Flow rate in the channel is 100 L/sec, the Manning's roughness coefficient, n, of the channel is 0.013. Calculate the depth of flow in the channel. The channel is a trapezoidal channel with bottom width of 0.3 m and side slopes of 1:1 (z = 1).

Solve for left side of Eq. 11.9. The following procedure is in the *Canal Depth* Worksheet.

$$\frac{Qn}{S_0^{0.5}} = \frac{0.1*0.013}{0.002^{0.5}} = 0.029$$

Use iteration to find y. First try y = 18 cm = 0.18 m.

$$AR^{2/3} = (b+zy)y\left(\frac{(b+zy)y}{b+2y\sqrt{1+z^2}}\right)^{2/3}$$
$$= (0.3+(1)(0.18))(0.18)\left(\frac{(0.3+(1)(0.18))(0.18)}{0.3+(2)(0.18)\sqrt{1+1^2}}\right)^{2/3}$$
$$= 0.01945$$

Adjust y as follows (this is a fast iteration procedure).

$$y_2 = y_1\left(\frac{(AR^{2/3})_{actual}}{(AR^{2/3})}\right)^{0.5} = 0.18\left(\frac{0.029}{0.01945}\right)^{0.5} = 0.2201\ m$$

Table 11.2 Values of manning's n for various channel and agricultural surfaces (Chow 1959)

Type of surface	Minimum	Maximum	Normal
a. Corrugated metal culvert flowing partly full	0.021	0.030	0.024
b. Lined channels			
1. Metal smooth surface – unpainted	0.011	0.014	0.012
2. Metal smooth surface – painted	0.012	0.017	0.013
3. Concrete – trowel finish	0.011	0.015	0.013
4. Concrete – float finish	0.013	0.016	0.015
5. Concrete – unfinished	0.014	0.020	0.017
6. Concrete – finished with gravel on bottom	0.015	0.020	0.017
7. Corrugated metal			
c. Earth, straight and uniform			
1. Clean, recently completed	0.016	0.020	0.018
2. Clean, after weathering	0.018	0.025	0.022
3. Gravel, uniform section, clean	0.022	0.030	0.025
4. With short grass, few weeds	0.022	0.033	0.027
d. Earth, winding and sluggish			
1. No vegetation	0.023	0.030	0.025
2. Grass, some weeds	0.025	0.033	0.030
3. Dense weeds or aquatic plants in deep channels	0.030	0.040	0.035
4. Earth bottom and rubble sides	0.025	0.035	0.030
5. Stony bottom and weedy sides	0.025	0.045	0.035
6. Cobble bottom and clean sides	0.030	0.050	0.040
e. Dragline-excavated or dredged			
1. No vegetation	0.025	0.033	0.028
2. Light brush on banks	0.035	0.060	0.050
f. Rock cuts			
1. Smooth and uniform	0.025	0.040	0.035
2. Jagged and irregular	0.035	0.050	0.040
g. Channels not maintained, weeds and brush uncut			
1. Dense weeds, high as flow depth	0.050	0.120	0.080
2. Clean bottom, brush on sides	0.040	0.080	0.050
3. Same, highest stage of flow	0.045	0.110	0.070
4. Dense brush, high stage	0.080	0.140	0.100
h. Agricultural surfaces			
1. Smooth, bare soil surfaces and furrows, SCS (1974, 1984)			0.04
2. Small-grain crops with drill rows in the direction of water flow (SCS)			0.10
3. Alfalfa, broadcast small grains, SCS (1974)			0.15
4. Dense alfalfa or alfalfa on long fields with no ditches, Clemmens (1991)			0.20
5. Dense sod crops and small grains drilled perpendicular to the flow (SCS)			0.25

Another iteration results in 0.2224. Thus, final answer is 0.22 m = 22 cm water depth.

Add 25 % freeboard elevation.

$$0.22*1.25 = 0.28 \text{ m}$$

Canal Hydraulics: Energy Dissipation

Irrigation canals and channels generally have long stretches of uniform subcritical flow in almost level canals. However, if the natural land slope is greater than the maximum subcritical flow slope, then periodic energy dissipation structures are needed. Energy dissipation structures may also be needed downstream from a diversion dam. Hydraulic jumps often take place at the bottom of energy dissipation structures as the flow transitions from supercritical to subcritical flow. This is explained by the fact that the energy in open channel flow is the sum of the elevation potential energy (y = water surface elevation) and kinetic energy (v). Flow can either have high elevation energy (y_2 in Fig. 11.15) or high kinetic energy (y_1 in Fig. 11.15). The Froude diagram shows the relationship between depth and energy (Fig. 11.15).

Table 11.3 Geometric elements of channel sections (After Cuenca (1989) from Chow (1959))

Section	Area A	Wetted perimeter, P	Hydraulic radius R	Top widthT
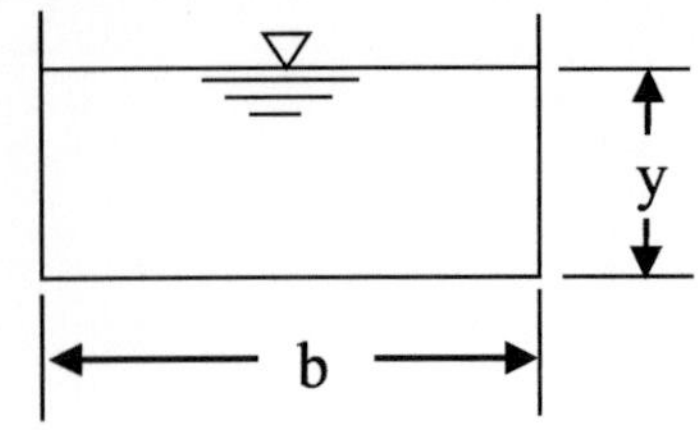 Rectangle	by	b + 2y	$\frac{by}{b+2y}$	b
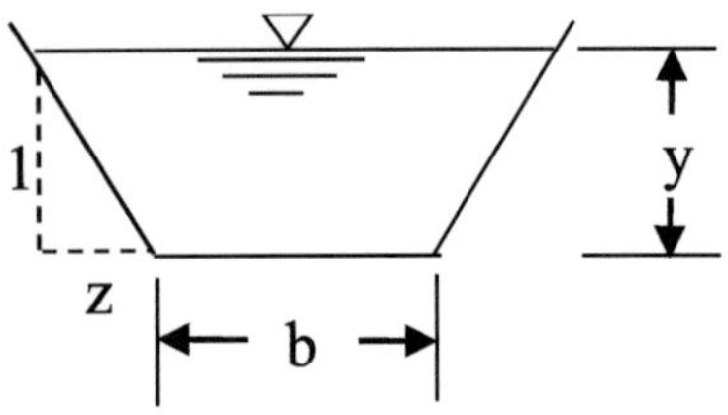 Trapezoid	(b + zy)y	$b+2y\sqrt{1+z^2}$	$\frac{(b+zy)y}{b+2y\sqrt{1+z^2}}$	b + 2zy
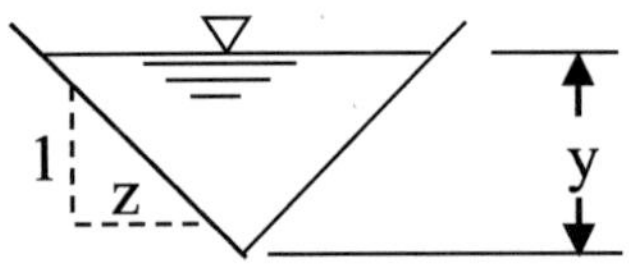 Triangle	zy^2	$2y\sqrt{1+z^2}$	$\frac{zy}{2y\sqrt{1+z^2}}$	2zy
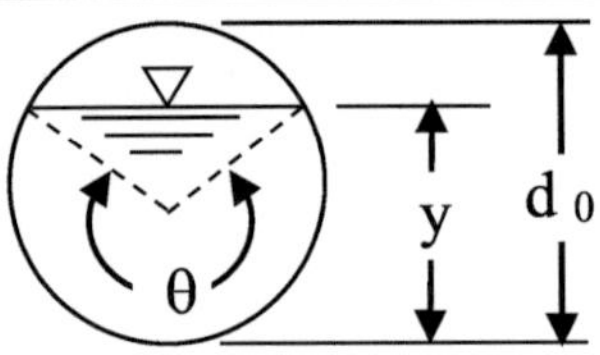 Culvert	$\frac{1}{8}(\theta - \sin\theta)d_0^2$	$\frac{1}{2}(\theta)d_0$	$\frac{1}{4}\left(1-\frac{\sin\theta}{\theta}\right)d_0$	$\left(\sin\frac{\theta}{2}\right)d_0$
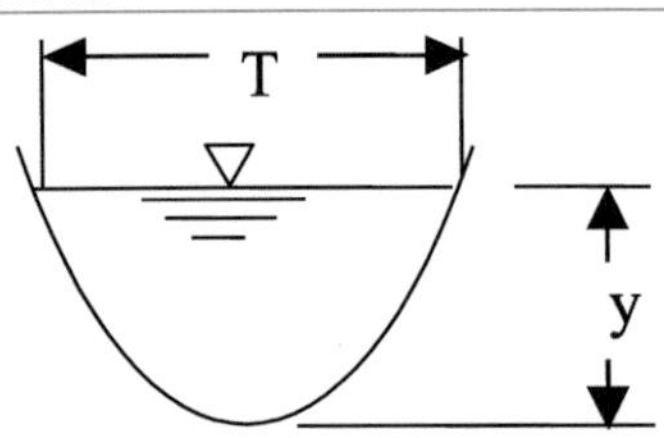 Parabola	$\frac{2}{3}Ty$	$T+\frac{8}{3}\frac{y^2}{T}$ Satisfactory for $0 < x < 1$ where $x = 4y/T$	$\frac{2T^2y}{3T^2+8y^2}$	$\frac{3A}{2y}$
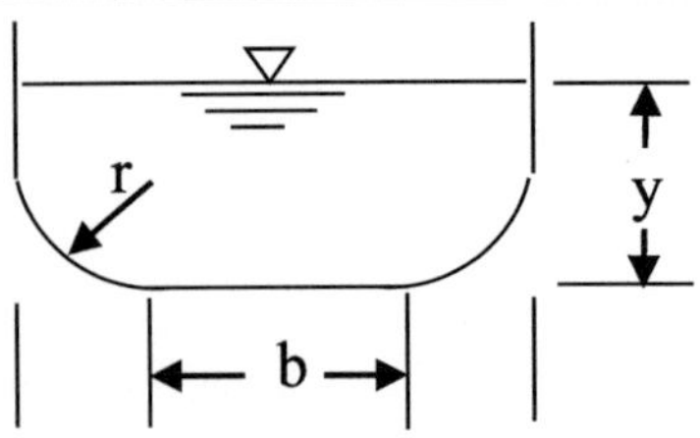 Round-cornered rectangle	$\left(\frac{\pi}{2}-2\right)r^2+$ $(b+2r)y$	$(\pi-2)r+$ $b+2y$	$\frac{\left(\left(\frac{\pi}{2}-2\right)r^2+(b+2r)y\right)}{\left((\pi-2)r+b+2y\right)}$	b + 2r

(continued)

Table 11.3 (continued)

Section	Area A	Wetted perimeter, P	Hydraulic radius R	Top widthT
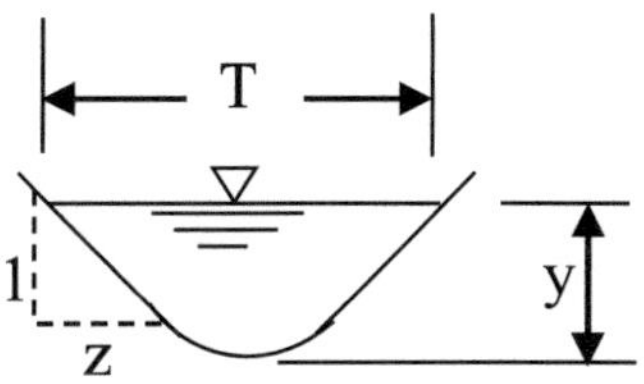 Round-bottomed triangle	$\frac{T^2}{4z} - \frac{r^2}{z}*\left(1 - z\ \cot^{-1}z\right)$	$\frac{T}{z}\sqrt{1+z^2} - \frac{2r}{z}\left(1 - z\ \cot^{-1}\ z\right)$	A/P	$2\left[z(y-r) + r*\sqrt{1+z^2}\right]$

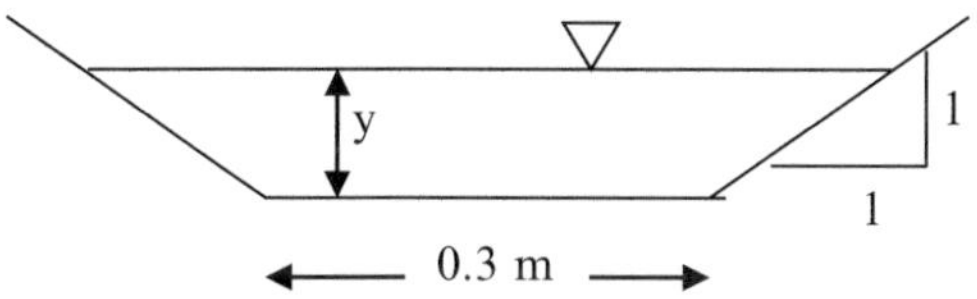

Fig. 11.14 Trapezoidal channel cross-section

$$H = y + \frac{v^2}{2g} \quad (11.10)$$

A hydraulic jump takes place when flow jumps from high kinetic energy to high elevation energy. For low slopes with low velocity and high depth (the norm in open channels), the elevation energy is greater; however, for steep slopes with low depth and high velocity, the kinetic energy is greater. If velocity is high and depth is low, then flow is called supercritical and is in the lower range of Fig. 11.15. If velocity is low and depth is high, then flow is called subcritical, and is in the upper range of Fig. 11.15.

The Froude number, Fr, is the dimensionless ratio of a momentum (velocity) force to the gravitational force. Flow is said to be critical if the Froude number is equal to one, subcritical if the Froude number is less than one, and supercritical if the Froude number is greater than one. The general form of the Froude number, which is applicable for channels with vertical sides is as follows:

$$F = \frac{\mathrm{v}}{(\alpha y g)^{0.5}} \quad (11.11)$$

where

α = velocity distribution coefficient, dimensionless.

The α term is necessary because total energy in a channel is not a linear function of velocity because of the variation in flow with distance from the channel wall. The velocity distribution coefficient in Eq. 11.11, α, is the ratio between the actual kinetic energy in a channel and kinetic energy calculated with average velocity. It is assumed to be one in irrigation channels with turbulent flow.

The Froude number can be written as follows for all open channels (non-vertical sides):

$$Fr^2 = \frac{Q^2 T}{A^3 g} \quad Fr = \frac{\mathrm{v}}{\left(^{A}/_{Tg}\right)^{0.5}} \quad (11.12)$$

where

Q = total flow in channel, m^3/s
T = channel top width, m
A = channel cross-sectional area, m^2.

If the slope of the channel is steep in the direction of flow, then velocity is high with respect to the depth, the Froude number is greater than 1, and flow is supercritical; thus, canals must be designed with a small slope so that the Froude number is less than 1, and flow is subcritical. Irrigation canals and ditches are designed to operate at subcritical flow for two reasons. In earth-lined canals and ditches, flow must be subcritical in order to prevent erosion. In concrete lined irrigation canals and ditches, it would be impossible to have energy control structures and diversions if the channel operated at supercritical flow because flow would change from supercritical to subcritical flow if the water was slowed at diversion structures.

Energy drop structures are classified as vertical (hard) drops or sloping drops (Fig. 11.16). Five energy drop structures used in irrigation canals are shown in Fig. 11.17. Problems with vertical drop structures include sedimentation and maintenance of the impact basin, and erosion

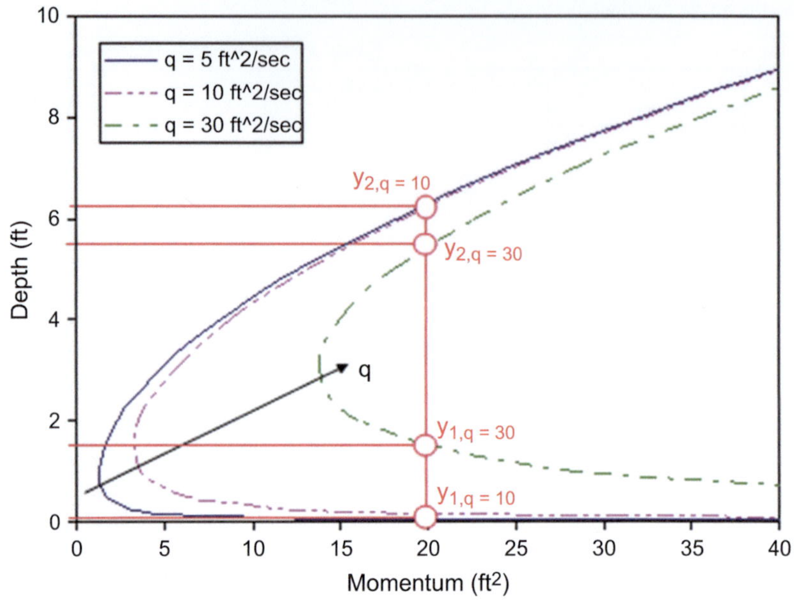

Fig. 11.15 Froude diagram (Wikipedia)

Fig. 11.16 Baffle chute energy drop structure (Wikipedia)

downstream from the impact basin. Sloped drop structures either dissipate energy with rip rap or concrete blocks along the slope or at the base of the sloped section with a hydraulic jump.

If energy is dissipated with a hydraulic jump (Fig. 11.18) at the base of a sloped drop structure, then the structure must be carefully designed. Energy is lost in the turbulence of a hydraulic jump; however, momentum is conserved. Thus, the depth of flow and velocity of flow after a hydraulic jump can be found with the conservation of momentum equation. If there is no change in flow direction, then the summation of forces before and after the hydraulic jump is the same.

$$\rho Q_1 v_1 + P_1 = \rho Q_2 v_2 + P_2 \tag{11.13}$$

$$\begin{aligned} \rho Q_1 \frac{Q_1}{A_1} + \gamma \overline{y}_1 A_1 &= \rho Q_2 \frac{Q_2}{A_2} + \gamma \overline{y}_2 A_2 \\ \rho \frac{Q^2}{gA_1} + \overline{y}_1 A_1 &= \rho \frac{Q^2}{gA_2} + \overline{y}_2 A_2 \end{aligned} \tag{11.14}$$

For a rectangular channel, Eq. 11.14 can be written in terms of flow per unit width, q.

$$\rho \frac{q^2}{gA_1} + \overline{y}_1 A_1 = \rho \frac{q^2}{gA_2} + \overline{y}_2 A_2 \tag{11.15}$$

Equation 11.15 can be rearranged and written in terms of the Froude number for a rectangular channel.

$$\frac{y_2}{y_1} = \frac{1}{2}\left(-1 + \sqrt{1 + 8Fr_1^2}\right) \tag{11.16}$$

$$Fr_1 = \frac{q}{y_1 \sqrt{g y_1}} \tag{11.17}$$

where

Fr_1 = upstream Froude number for a rectangular channel.

The amount of energy lost in the hydraulic jump depends on the upstream Froude number. A higher Froude number leads to a greater energy loss. An upstream Froude number between 4.5 and 9.0 maintains a stable hydraulic jump and high energy loss so this is the desired range of Froude number for upstream flow. Lower Froude numbers create

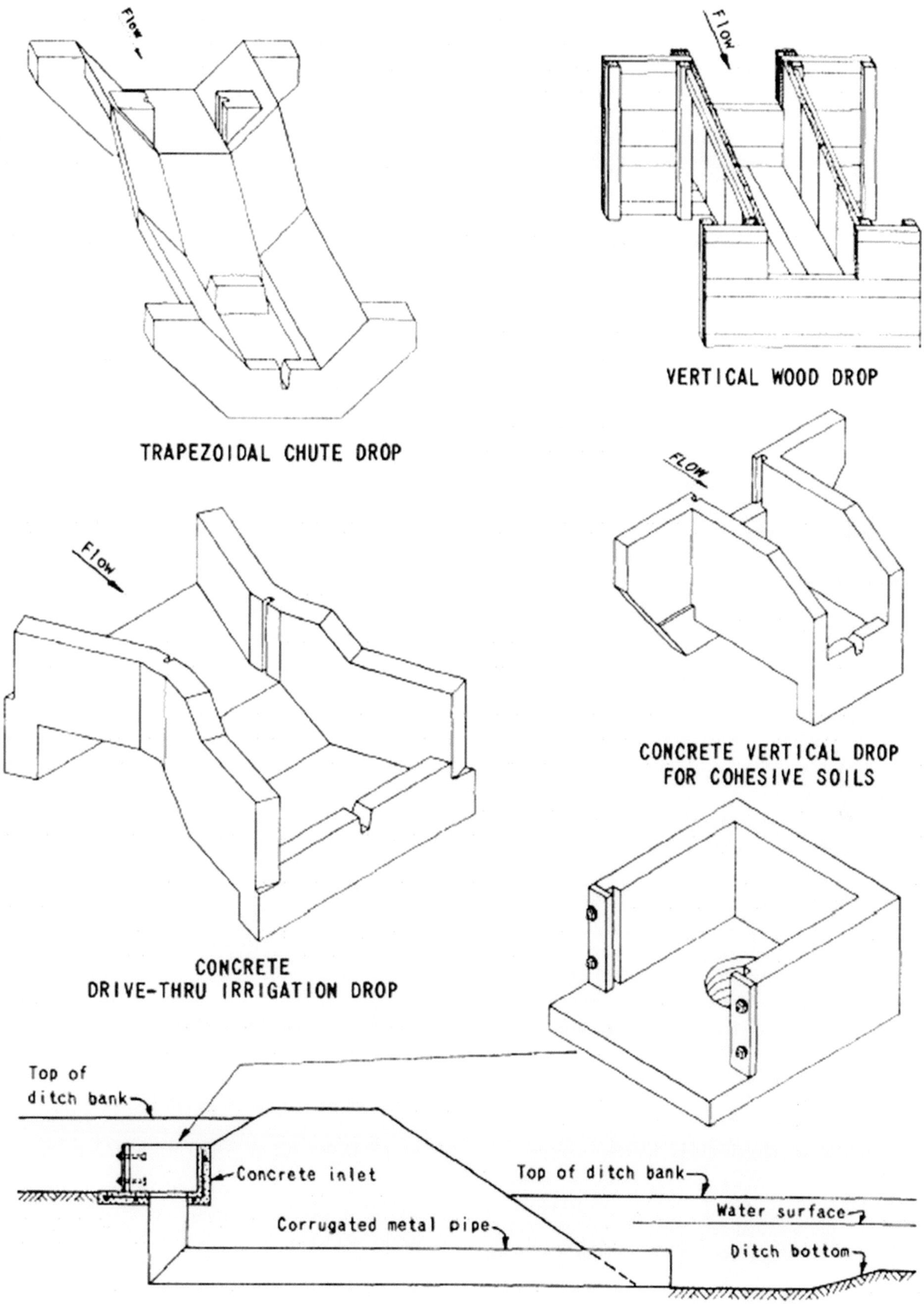

Fig. 11.17 Five energy dissipation structures (Credit NRCS)

Fig. 11.18 Hydraulic jump

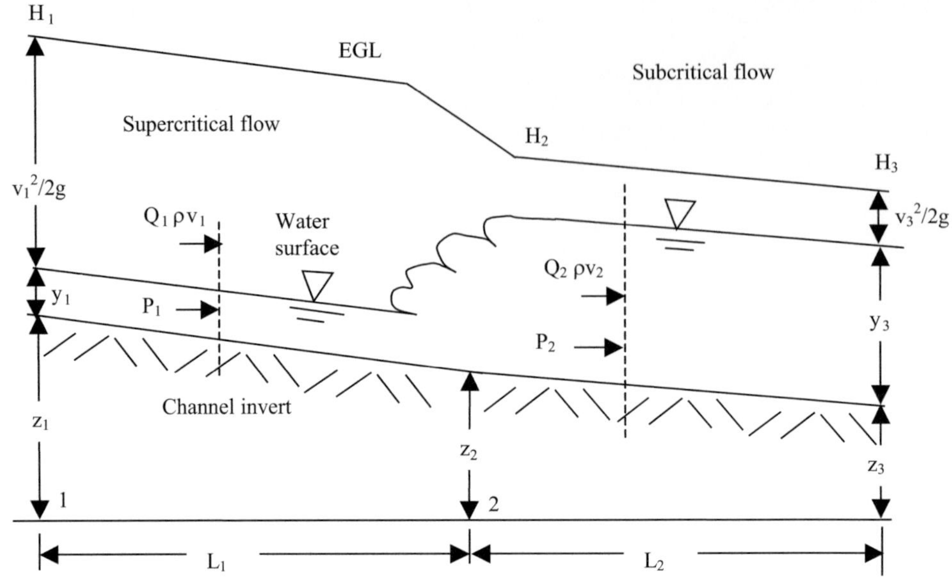

Table 11.4 Hydraulic jump characteristic versus upstream froude number (After Chin (2000))

Fr_1	Jump characteristic
1–1.7	Standing wave (undular jump); kinetic-energy loss (as a percentage of the upstream kinetic energy, $V_1^2/2$ g) < 5 %
1.7–2.5	Smooth surface rise (weak jump); kinetic energy loss 5–15 %
2.5–4.5	Unstable oscillating jump, where each irregular pulsation creates a large wave that can travel far downstream, damaging earth banks and other structures; kinetic energy loss 45–70 %
4.5–9.0	Stable jump, best performance and action, and insensitive to downstream conditions (steady jump); kinetic energy loss 45–70 %
>9.0	Rough, somewhat intermittent (strong jump); kinetic energy loss 70–85 %

downstream instability. Table 11.4 lists energy loss and hydraulic jump wave behavior vs. Froude number.

Example 11.6 A sloped drop structure is to be placed along a channel in order to dissipate energy. The channel cross section is b = 0.3 and z = 2. Manning's n is 0.013. The flow is 100 L/s. Find the slope of the drop structure required to develop a Froude number equal to 6.0 before the hydraulic jump.

The equation for top width of a trapezoidal channel is b + 2 z y (Table 11.3). The equation for channel area is (b + z y) y. Substitute into Eq. 11.12 with Fr = 6.

Iterate to solve for y: 0.059 m.

Now find the slope.

$$Q = \frac{AR^{2/3}S_0^{0.5}}{n} \Rightarrow S_0 = \left(\frac{Qn}{AR^{2/3}}\right)^2 = \left(\frac{(0.10)\ (0.013)}{(0.025)\ (0.044)^{2/3}}\right)^2 = 0.178$$

Thus, the bottom slope should be 17.8 %

One of the main reasons to carefully characterize the downstream depth of a hydraulic jump is to ensure that the channel walls are high enough to contain the jump. Equation 11.13 can be used to find downstream depth for rectangular channels as is shown in the following example.

Example 11.7 Repeat Example 11.6 for a rectangular channel with 0.3 m width. Find the downstream depth after the hydraulic jump.

$$Fr^2 = \frac{Q^2T}{A^3g}$$

$$6^2 = 36 = \frac{Q^2(b)}{(by)^3g} \Rightarrow y = \sqrt[3]{\frac{Q^2b}{36b^3g}} = \sqrt[3]{\frac{0.1^2(0.3)}{36*0.3^3\ *9.8}}$$
$$= 0.068\ m$$

The same answer is found with Eq. 11.14.

$$q = Q/b = 0.1\ m^3/s/0.3\ m = 0.333\ m^2/\sec$$

$Fr_1 = \frac{q}{y_1\sqrt{gy_1}} = 6 = \frac{0.333}{y_1\sqrt{9.8y_1}}$ y_1 is found by iteration = 0.068 m

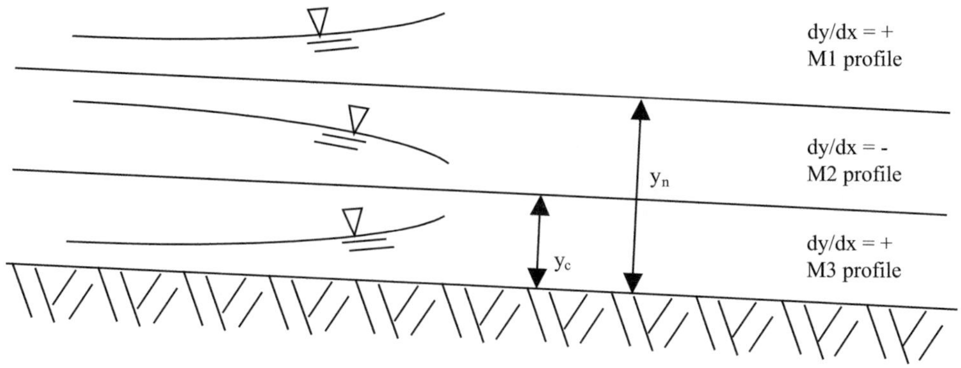

Fig. 11.19 Mild (M) channel

Find the slope

v = q/y = 0.333 m^2/ sec / 0.068 m = 4.90 m/ sec

$$v = \frac{R^{2/3}S_0^{0.5}}{n} \Rightarrow S_0 = \left(\frac{vn}{\left(\frac{by}{b+2y}\right)^{2/3}}\right)^2$$

$$= \left(\frac{4.90*0.013}{\left(\frac{0.3*0.068}{0.3+2*0.068}\right)^{2/3}}\right)^2 = 0.245$$

$$Q = \frac{AR^{2/3}S_0^{0.5}}{n} \Rightarrow S_0 = \left(\frac{Qn}{AR^{2/3}}\right)^2$$

$$= \left(\frac{0.1*0.013}{(0.3*0.068)*0.0468^{2/3}}\right)^2 = 0.241$$

In this case, the bottom slope is 24 %, which is 6 % higher than for the trapezoidal channel in Example 11.6. Thus, channel geometry has an influence on the required slope needed to generate a sufficiently high Froude number.

The downstream water surface depth is found with Eq. 11.13. for a rectangular channel.

$$y_2 = y_1\frac{1}{2}\left(-1+\sqrt{1+8Fr_1^2}\right)$$

$$= 0.068\frac{1}{2}\left(-1+\sqrt{1+8*6^2}\right) = 0.544\ m$$

Canal Hydraulics: Gradually Varied Flow

Canal velocity and depth can change gradually, rather than a hydraulic jump, upstream or downstream from an obstruction. This section gives an introduction to the calculation of gradually varying depth in a canal.

$$S_0 - S_f = \frac{\Delta\left(y+\alpha\frac{v^2}{2g}\right)}{\Delta x} \quad (11.18)$$

where

S_f = friction loss calculated with Manning's equation, also slope of EGL, m/m

Δx = distance down the channel, m.

For a rectangular channel, Eq. 11.11 can be substituted into Eq. 11.18 with Q /A substituted for v and written as follows:

$$\frac{dy}{dx} = \frac{y_2 - y_1}{x_2 - x_1} = \frac{S_0 - S_f}{1 - \alpha Fr^2} \quad (11.19)$$

Channel slopes are classified as mild (M), steep (S), critical (C), horizontal (H), and adverse (A). A mild slope is shown in Fig. 11.19. The normal depth is the depth of uniform and steady state flow. A mild slope is classified as a slope in which the normal depth is greater than the critical depth. Thus, the flow is subcritical in an M channel, which is the slope used for irrigation canals and ditches. There are three conditions for flow on a mild slope. If the depth of flow is greater than the normal depth, then the flow is categorized as M1. This would occur if there was a downstream obstruction and water was backed up in the channel. In this case, the depth of flow is increasing. If flow depth is between normal and critical depth (M2), then there is a discharge downstream and flow depth will decrease. If the flow depth is less than critical (supercritical, M3), then the upstream flow is approaching the channel from a steep slope and will increase in depth or possibly undergo a hydraulic jump.

Example 11.8 For the channel section described in Example 11.5 and an initial depth of 0.4 m, calculate the depth of flow 20-, 40-, 60-, 80-, and 100-m downstream.

Calculate the critical depth at a flow rate of 100 L/s = 0.1 m^3/sec.

$Fr^2 = 1 = \frac{Q^2T}{A^3g} = \frac{Q^2(b+2zy)}{((b+zy)y)^3g} = \frac{0.1^2(0.3+2(1)y)}{((0.3+1y)y)^3 9.8}$ By iteration $y_c = 0.182$.

Because the normal depth is greater than the critical depth, the slope is subcritical and the channel is classified as a mild slope (M) as shown in Fig. 11.16.

As calculated in Example 11.5, the normal depth, y_n, is 0.25 m. The actual depth, y = 0.4 m, is greater than the normal depth so the flow condition is M1 (Fig. 11.16). Thus, the flow depth is increasing with distance down the channel (there is a downstream obstruction such as a weir).

The section factor is calculated as follows:

$$AR^{2/3} = (b + zy)y\left(\frac{(b + zy)y}{b + 2y\sqrt{1 + z^2}}\right)^{2/3}$$

$$= (0.3 + (1)(0.4))(0.4)\left(\frac{(0.3 + (1)(0.4))0.4}{0.3 + (2)(0.4)\sqrt{1 + 1^2}}\right)^{2/3}$$

$$= 0.094$$

The friction slope is calculated with Eq. 11.19.

$$S_f = \left(\frac{nQ}{AR^{2/3}}\right)^2 = \left(\frac{0.013*0.1}{0.094}\right)^2 = 0.00019 \text{m/m}$$

Calculate the Froude number

$$\text{T} = \text{b} + 2\text{ zy} = 0.3 + 2*1*0.4 = 1.1 \text{ m}$$

$$\text{A} = (\text{b} + \text{zy})\text{y} = (0.3 + 0.4)0.4 = 0.28 \text{ m}^2$$
$$Fr^2 = \frac{Q^2T}{A^3g} = \frac{0.1^2*1.1}{0.28^3*9.8} = 0.0511$$

The downstream depth at 20 m, y_{20}, is

$$y_2 = y_1 + (x_2 - x_1)\frac{S_0 - S_f}{1 - \alpha Fr^2}$$
$$= 0.4 + (20)*\frac{0.002 - 0.00019}{1 - 1*(0.0511)} = 0.438 \ m$$

Calculate the downstream depth at 40 m. First calculate the upstream section factor.

$$AR^{2/3} = (b+zy)y\left(\frac{(b+zy)y}{b+2y\sqrt{1+z^2}}\right)^{2/3}$$

$$= (0.3 + 1*0.438)*0.438\left(\frac{(0.3+1*0.438)0.438}{0.3+2*0.438\sqrt{1+1^2}}\right)^{2/3}$$

$$= 0.11$$

The friction slope is calculated with Eq. 11.19.

$$S_f = \left(\frac{nQ}{AR^{2/3}}\right)^2 = \left(\frac{0.013*0.1}{0.11}\right)^2 = 0.00013 \text{ m/m}$$

Calculate the Froude number

$$\text{T} = \text{b} + 2\text{ zy} = 0.3 + 2*1*0.438 = 1.17 \text{ m}$$

$$\text{A} = (\text{b} + \text{zy})\text{y} = (0.3 + 0.438)0.438 = 0.321 \text{ m}^2$$
$$Fr^2 = \frac{Q^2T}{A^3g} = \frac{0.1^2*1.17}{0.321^3*9.8} = 0.035$$

The downstream depth at 40 m, y_{40}, is

$$y_2 = y_1 + (x_2 - x_1)\frac{S_0 - S_f}{1 - \alpha Fr^2} = 0.43 + (20)*\frac{0.002 - 0.00013}{1 - 1*(0.035)}$$
$$= 0.477 \text{m}$$

Flow depth at 60 m downstream is calculated similarly. Flow depths and distances are as follows:

Distance (m)	Depth, y (m)	Difference (m)
0	0.4	
20	0.438	0.0381
40	0.477	0.0388
60	0.516	0.0392
80	0.555	0.0394
100	0.595	0.0396

The flow depth increases with distance down the channel, as predicted by the M1 flow profile in Fig. 11.19. The flow profile is also concave upward as with the M1 profile. This is because the difference between flow depths increases with distance down the channel.

Iteration within each step would have improved the accuracy of the solution by using an estimate of average channel depth between x_2 and x_1 to calculate S_f. However, there is not a large difference between y_2 and y_1 in this example so the answer would not change significantly.

Flow Measurement

Flow measurement is based on the concept that volumetric flow rate is equal to the product of flow velocity and cross-sectional area. The discharge from a flow measurement device is a function of the upstream depth of water.

$$\text{Q} = \text{vA} \tag{11.20}$$

Flow from a submerged orifice is described by the orifice equation. The contraction coefficient is 1.0 for orifices with no contraction (gradual entrance into orifice) and 0.61 for sharp edged orifices (m or ft).

$$Q = C_c A\sqrt{2gh} \tag{11.21}$$

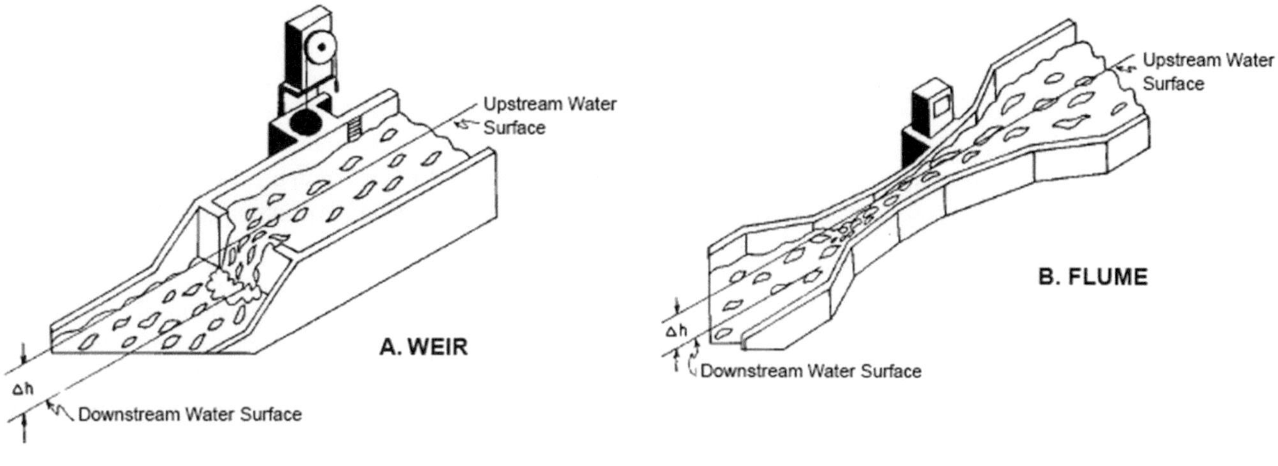

Fig. 11.20 Broad crested weir and long throated flume (Credit NRCS)

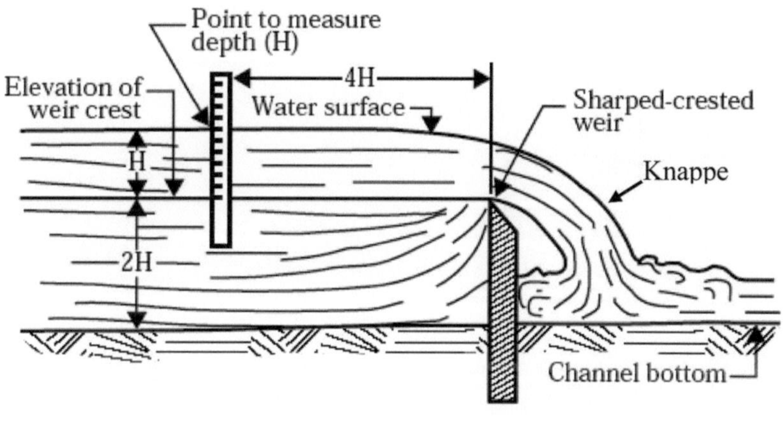

Fig. 11.21 Profile of sharp-crested weir (Credit NRCS)

where

C_c = contraction coefficient
h = depth of center of orifice below upstream water surface or pressure in pipe, m.

Sharp-crested weirs or flumes (Fig. 11.20) are used to measure subcritical flow in channels.

Sharp-crested weirs (Fig. 11.21) have a sharp edged weir blade (7–20 mm width). Discharge calculation accuracy over a sharp crested weir is within ± 2 % under best flow conditions.

For sharp-crested weirs, water approach velocity should be less than 0.15 m/sec, and tranquil flow conditions should extend 15–20 times H upstream from the weir blade. The face of the weir should be vertical and perpendicular to the direction of water flow and the upstream edge should be sharp. The depth of water over the crest should be at least 5 cm for accurate measurements. The minimum flow should be more than 2 % of the maximum flow for accurate measurement of low flows. Suppressed weirs extend completely across a channel and contracted weirs do not. The crest and edges of contracted weirs should be at least 2H away from the channel sides.

The following equation is a general equation for flow over a sharp crested weir (Fig. 11.21), where flow rate is a function of the upsteam depth of flow, H:

$$Q = C_d \frac{2}{3} \sqrt{2g}\, L_b H^{3/2} \tag{11.22}$$

where

C_d = discharge coefficient, determined by analysis and calibration tests
H = upstream elevation of water surface above weir (Fig. 11.20).

There are five types of sharp-crested contracted weirs: rectangular (Fig. 11.22), V-notch (Fig. 11.23), Cipoletti (Fig. 11.24), partially contracted rectangular weirs, and partially contracted 90-degree V-notch weirs.

The Francis equation is used to calculate flow rate over a standard contracted rectangular weir. The last term, 0.2 H, accounts for the fact that the edges of the flow contract from the sides of the weir.

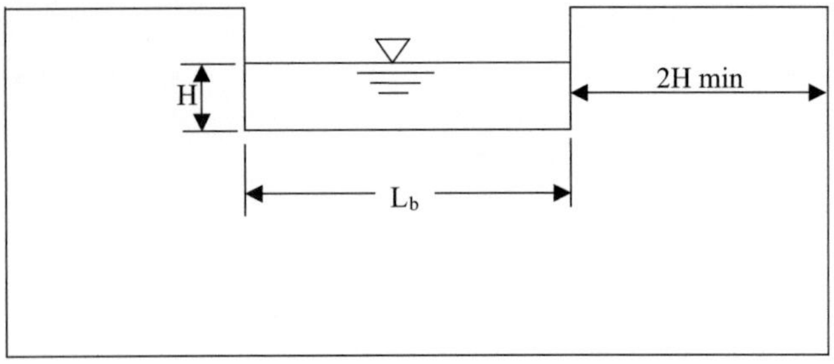

Fig. 11.22 Contracted rectangular sharp-crested weir

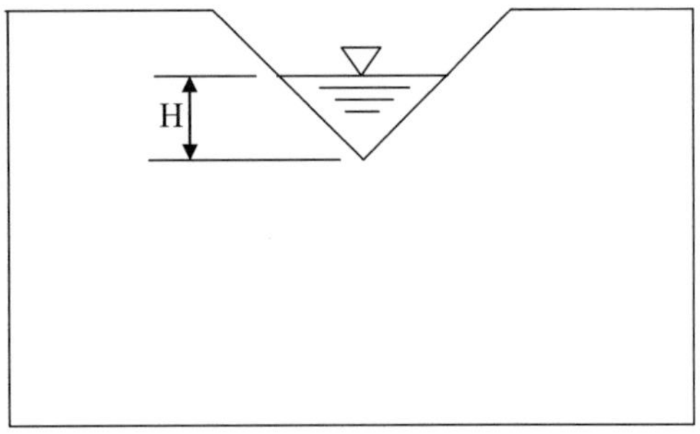

Fig. 11.23 90-degree V-notched weir

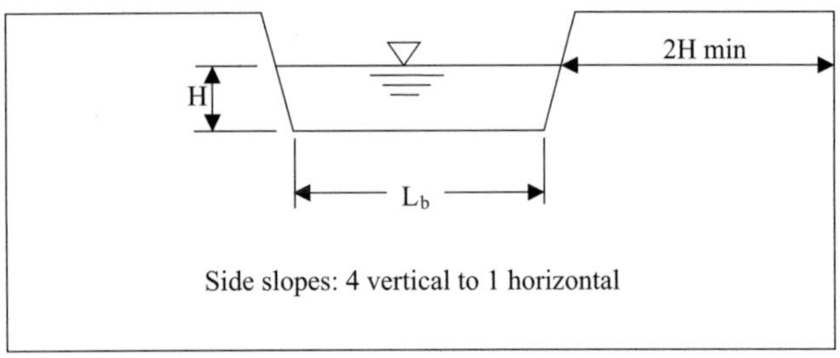

Fig. 11.24 Standard cipolletti weir

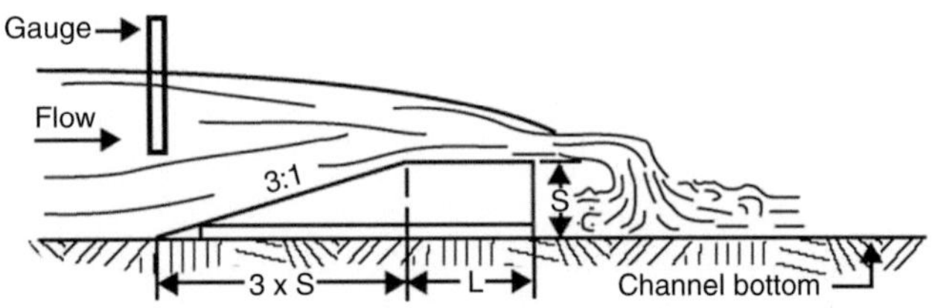

Fig. 11.25 Profile of long throated flume (Credit NRCS)

$$Q = 3.33H^{3/2}(L_b - 0.2H)(\text{ft and ft}^3/\sec)$$
$$Q = 1.74H^{3/2}(L_b - 0.2H)(\text{m and m}^3/\sec) \quad (11.23)$$

where

Q = discharge flow rate neglecting velocity of approach, ft^3/sec or m^3/sec
L_b = length of weir, ft or m
H = head on weir, ft or m.

The constant discharge coefficient, 3.33, is valid as long as the head over the weir, H, is not more than 1/3 the length of the weir, L_B. If the head over a rectangular weir is greater than 1/3 the length, then tables must be used to find the constant discharge coefficient. The USBR Water Measurement Manual (WMM) has an appendix with tables of flows over contracted weirs and contains information on proper weir design.

Equation 11.25 can be used to calculate flow over a 90-degree V-notch weir (WMM). Equations for calculation of flow over V-notch weirs with other angles as well as tables of calculations are given in the WMM.

$$Q = 2.49H^{2.48}(\text{ft and ft}^3/\sec)$$
$$Q = 1.35H^{2.48}(\text{m and m}^3/\sec) \quad (11.24)$$

In-class Exercise 11.1 Of the three weirs, rectangular, V-notch, and Cipolletti, which is best for small flows?

Which weir provides the highest sensitivity to changes in flow rate?

Which weir is best for large flow rates?

The following equation can be used to find the flow rate over a standard Cipolletti weir (Fig. 11.24). The equation is valid as long as H is no greater than 0.33 L_b.

$$Q = 3.367L_bH^{3/2}(\text{ft and ft}^3/\sec).$$
$$Q = 1.86L_bH^{3/2}(\text{m and m}^3/\sec). \quad (11.25)$$

Example 11.9 If the head, H, over a standard Cipolletti weir is 0.15 m and the weir blade length, L_b, is 0.5 m, then what is the flow rate?

$$Q = 1.86L_bH^{3/2} = 1.86*0.3*0.1^{3/2} = 0.018\ m^3/\sec.$$

A standard suppressed rectangular weir extends across the full width of a channel. This type of weir has no flow contraction from the sides. An air vent is needed below the knappe to the atmosphere in order to ensure that air pressure below the knappe (Fig. 11.21) remains the same as atmospheric pressure. The crest height above the channel bottom should be at least 3H. The equation for a suppressed rectangular weir is the same as for the contracted rectangular weir except that the 0.2H value for side contraction is left out.

$$Q = 3.33H^{3/2}(L_b)(\text{ft and ft}^3/\sec)$$
$$Q = 1.74H^{3/2}(L_b)(\text{m and m}^3/\sec) \quad (11.26)$$

Broad-crested weirs and long-throated flumes (Figs. 11.20 and 11.25) have long flat approaches. Broad-crested weirs are simply long throated flumes with no side contractions.

Fig. 11.26 Large long-throated flume under construction (Credit U.S. Water Conservation Laboratory, Phoenix, Arizona)

Flumes increase water velocity and change water flow from subcritical to supercritical as the flow is accelerated. Acceleration occurs by increasing the bottom depth (Fig. 11.26) and/or converging the sidewalls (Fig. 11.25). Flumes include a measurement of upstream flow depth, which is directly correlated with flow rate because the upstream profile is unique for each flow rate if the water passes through critical depth. A major advantage of flumes over weirs is that flumes only dissipate 25 % of the energy that is dissipated in the weir. Thus, a flume may be advantageous on shallow slopes where water energy must be conserved. Flumes are also more difficult to tamper with and alter flow readings than weirs; this prevents users from obtaining an unfair allocation of water. A disadvantage of flumes is that they cost more than weirs.

Long throated flumes are easier to install (Figs. 11.26 and 11.27) than other flume geometries. A concrete ramp is simply placed in the channel bottom. In the past, Parshall flumes were installed in many channels; however, because of difficulty of installation, they have decreased in popularity. However, in some regions, local regulations still mandate the use of Parshall flumes.

Calibration of weirs and flumes may be necessary if the geometry or upstream flow conditions are not as specified in design manuals. Water measurement manuals should be consulted for nonstandard conditions.

If upstream flow is choppy, then readings will be inaccurate. In this case, an energy dissipation structure can be installed upstream from a weir or flume as shown in Fig. 11.28. As the structure forces water under a flat surface, waves are dissipated.

When no water measurement structure is installed, water flow rate can be measured in canals or surface irrigation ditches by measuring the length of time that it takes for a floating ball or other buoyant device to travel a 30 m length of the channel. Repeat the measurement five or ten times and take the average. The water at the surface of the channel travels faster than the average velocity; thus, the calculated velocity should be multiplied by a coefficient (Table 11.5). The corrected velocity is then multiplied by the cross-sectional area of the channel in order to find the flow rate.

The velocity at the channel water surface also can be found by placing a thin and wide meter stick in the center of the channel and observing the difference in elevation of

Fig. 11.27 Large long-throated flume under operation for left to right flow (Credit U.S. Water Conservation Laboratory, Phoenix, Arizona)

Fig. 11.28 Underpass wave suppressor section (USBR Water Measurement Manual)

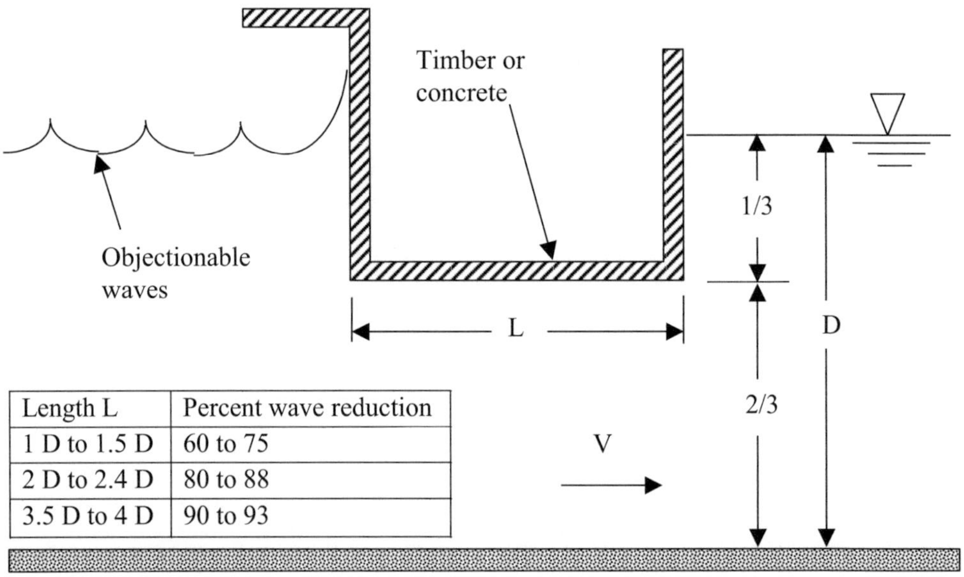

Length L	Percent wave reduction
1 D to 1.5 D	60 to 75
2 D to 2.4 D	80 to 88
3.5 D to 4 D	90 to 93

water with the sharp edge of the meter stick parallel to the flow and turned 90°. The difference in elevation is equal to the velocity head in meters at the surface of the ditch.

Example 11.10 A channel has the cross sectional area calculated in Example 11.5. The difference in head on the meter stick is 6 cm between the parallel and perpendicular to the flow positions. Calculate the flow velocity at the surface and the flow rate in the channel.

$$\mathrm{v} = (2\mathrm{gh})^{0.5} = (2*9.8*0.06)^{0.5} = 1.08\ \mathrm{m/sec}.$$

$$\mathrm{v}_{\mathrm{adj}} = 1.08*0.66 = 71\ \mathrm{m/sec}\ \text{(Table 11.5)}$$

$$A = (b + zy)y = (0.3 + 1*0.25)*0.25 = 0.14\ m^2$$

$$\mathrm{Q} = \mathrm{vA} = 0.71*0.14 = 0.1\ \mathrm{m/sec}$$

Table 11.5 Velocity coefficient versus depth of water in the ditch

Average depth (m)	Coefficient
0.3	0.66
0.6	0.68
0.9	0.70
1.2	0.72
1.5	0.74
1.8	0.76
2.7	0.77
3.6	0.78
4.5	0.79
6	0.80

Questions

1. Describe the different types of canals.
2. How is water diverted from a main canal to a lateral canal?
3. How is water diverted from a lateral canal to a field?
4. What is the reason for energy dissipation structures in canals and in canal outlets?
5. Calculate the conveyance efficiency and water duty for a canal that is 20 km long, has a wetted top width = 20 m, wetted perimeter = 26 m, and cross-sectional area = 100 m^2? Average canal flow velocity is 1 m/sec. Reference ET is 10 mm/day. Average seepage rate is 5 mm/day. In addition to reporting the water duty and efficiency, convert the seepage rate to L/m^2/day.
6. Calculate the conveyance efficiency to field 1 in Fig. 11.9 from the point of water diversion to the irrigation district. The conveyance efficiency of the irrigation district up to the farm turnout is 80 %. The main concrete canal on the farm has a conveyance efficiency of 80 %, and the earth-lined canal has a conveyance efficiency of 80 %.
7. Redo Example 11.4, but assume the ditch is in loam and sandy loam soil (Fig. 11.8) with a seepage rate of 1.5 m^3/m^2/day.
8. Describe the method used to run canal water past a road, drainage ditch, or river valley.
9. A concrete lined trapezoidal channel (Fig. 11.14) has a slope of 0.3 % = 0.003 m/m. Flow rate in the channel is 300 L/sec, and the Manning's roughness coefficient, n, of the channel is 0.015. Calculate the depth of flow in the channel. The bottom width is 1 m and side slope z = 1.5.
10. Calculate the Froude number for the previous problem, and determine whether the channel has supercritical or subcritical flow.
11. If a canal was laid on a relatively steep slope, what strategy could be used to prevent supercritical flow in the canal?
12. For the canal dimensions described in question 9, calculate the chute slope required to have a stable hydraulic jump with Froude number = 6.
13. Describe why a Froude number 6 is desirable before a hydraulic jump.
14. Calculate the flow over a standard contracted rectangular weir if the head over the weir is 0.13 m and the wier blade is 0.6 m across.

References

Chin D (2000) Water-resources engineering. Prentice-Hall, Englewood

Clemmens AJ (1991) Irrigation uniformity relationships for irrigation system management. J Irrig Drain Eng 117(5):682–699

Cuenca RH (1989) Irrigation system design: an engineering approach. Prentice Hall, Englewood Cliffs, p 552

DeMaggio J (1990) Technical Memorandum: San Luis unit drainage program project files. U.S. Bureau of Reclamation, Sacramento

Fipps G (2000) Potential water savings in irrigated agriculture for the rio grande planning region (Region M). Texas Agricultural Extension Service/Texas Agricultural Experiment Station/Texas A and M University System, College Station

Leigh E, Fipps G (2009) Measuring seepage losses from canals using the ponding test method. Publication B-6218. Texas AgriLife Extension Service, College Station

Nayak et al (1996) The influence of canal seepage on groundwater in Lugert Lake irrigation area. Oklahoma Water Resources Research Institute, Stillwater

Nofziger (1979) Profit potential of lining watercourses in coastal commands of Orissa. Environ Ecol 14(2):343–345

U.S. Bureau of Reclamation (1963) Lining for irrigation canals. U.S. Department of the Interior, Bureau of Reclamation, Denver

Te Chow V (1959) Open channel hydraulics. McGraw-Hill, New York

Texas A and M AgriLife Extension, Irrigation Technology Program website. Accessed at www.idea.tamu.edu

Texas Board of Water Engineers (1946) Seepage losses from canals in Texas. Austin. July 1

USBR Water Measurement Manual (WMM) http://www.usbr.gov/pmts/hydraulics_lab/pubs/wmm/

12 Center Pivot Irrigation Systems

A center pivot irrigation system is a movable pipe structure that rotates around a central pivot point connected to a water supply. Center pivot irrigation systems are the most popular sprinkler irrigation systems in the world because of their high efficiency, high uniformity, ability to irrigate uneven terrain (Fig. 12.1), and low capital, maintenance, and management costs. The history of center pivot irrigation systems began in Nebraska in the 1950s, and there are now hundreds of thousands of center pivot irrigation systems in the world. Center pivots are "perhaps the most significant mechanical innovation in agriculture since the replacement of draft animals by the tractor" (Splitter, Scientific American). The systems move through the field by electrically powered tractor wheels. Sprinkler flow rates increase toward the outer end of the pivot because the end of the pivot travels faster. The primary design constraint is the prevention of runoff at the end of the pivot, where application rates are highest. This chapter covers center pivot pipeline and mainline design, selection of sprinklers, and optimization of the design with respect to yield, energy requirement, components, and economics.

Center pivots irrigate large areas (50 ha or 160 ac). The huge green circles dot many agricultural regions (Fig. 12.2). As seen from the photograph, the major disadvantage of center pivots is that they don't naturally irrigate the corners of fields.

Because center pivots are circular, 21.5 % of production is lost in the unirrigated corners. If land is inexpensive and water is limited (Fig. 12.3), then loss of production in the corners may not be of concern. However, if land is limited, then there are several methods to reduce, eliminate, or irrigate corners. Corner areas can be reduced by staggering the circles (Fig. 12.3), which requires large land areas that are not previously broken into square sections. Corner areas can be reduced or eliminated with extendable arms at the end of center pivots that swing out and irrigate corners. Corner areas are eliminated with linear move irrigation systems, which are like center pivots but travel linearly; however, water is supplied to linear move system from lined concrete ditches or movable pipe systems, which are expensive and often difficult to maintain or operate. Big gun sprinklers (Fig. 12.4) at the end of the pivots can extend the entire irrigated circle of the pivot or can be turned on specifically in the corners.

Sprinkler Types

The original center pivots had widely spaced high-pressure impact sprinklers mounted on top of the center pivot pipeline (Fig. 12.5). However, in order to reduce evaporation and pressure energy, many pivots now have drop tube sprayers (Fig. 12.3), which require less pressure (less energy) than top-mounted impact sprinklers. Low energy precision application sprinklers (LEPA) require even less energy (Fig. 12.6)

Sprinklers that hang below the pipe but are elevated above the crop canopy are called Mid-Elevation Sprinkler Application (MESA) sprinklers. The original MESA sprinklers have downward-facing nozzles that spray water onto flat or serrated plates (Fig. 12.7) and have a 13 m application diameter. Newer rotating sprinklers (Fig. 12.8) have a rotating stream with a much larger diameter of application (23 m) and a more uniform application pattern. In addition, droplets are larger so there is less evaporation; however, rotating sprinklers operate at 20 PSI (140 kPa) while serrated plate sprinklers only require 10 PSI (70 kPa).

Sprinklers with a larger application diameter are less prone to ponding and runoff because the application rate is lower. Because of their larger diameter, rotating sprinklers have an average application rate 60 % less than the serrated plate sprinkler. Impact sprinklers may have a 100 ft diameter (30 m) and thus only 20 % of the application rate of serrated plates (less runoff).

Low Energy Precision Application (LEPA) bubblers hang from drop tubes and are typically spaced two rows

P. Waller, M. Yitayew, *Irrigation and Drainage Engineering*, DOI 10.1007/978-3-319-05699-9_12

Fig. 12.1 Potato fields under irrigation with center-pivot sprinklers in Idaho, USA (Credit: US Army Corps of Engineers)

apart (1.5–2 m) in order to provide uniform water application to all plants, if not to the soil volume. LEPA systems have the lowest energy costs because they save water (reduce evaporation) and can operate between 6 and 10 PSI (40–70 kPa) pressure. LEPA bubblers flood the furrow so furrows must be diked every few meters with plastic furrow inserts or with a dammer diker tractor implement.

LESA (Low Elevation Spray Application) and LPIC (Low Pressure, In-Canopy) sprayers (Fig. 12.6) are between LEPA bubblers and MESA sprinklers. These sprinklers are popular in arid climates for crops such as alfalfa that are not planted in rows. In some applications, LESA sprinklers are used to germinate the crop and then modified after germination by adding a "sock," which is a flexible plastic tube that fits over the sprayer and directs water down into the furrow. In low-profile crops such as alfalfa, LESA sprayers are placed just above the canopy in order to minimize evaporation.

Evaporation

Sprinkler droplet evaporation is a function of wind speed, vapor pressure deficit, sprinkler nozzle pressure and sprinkler nozzle diameter. Frost and Schwalen recorded the depth of evaporation from sprinklers (not center pivot sprinklers) over a range of conditions, and they developed a nomograph to calculate percent evaporation. The nomograph was later converted to the following equation by Trimmer which is accurate in the midrange of the Frost and Schwalen nomograph:

$$L_e = \begin{bmatrix} 1.98(D)^{-0.72} + 0.22(e_s - e_a)^{0.63} \\ +3.6*10^{-4}(h)^{1.16} + 0.14(U)^{0.7} \end{bmatrix}^{4.2} \tag{12.1}$$

Where

T = airtemperature, °C
RH = relativehumidity, fraction
$e_s - e_a$ = vapor pressure deficit, kPa
L_e = evaporation and wind drift, percent
D= nozzle diameter, mm
h = nozzle operating pressure, kPa
U = wind velocity, m/s.

$$e_s - e_a = 0.61\ \exp\left(\frac{17.27T}{T + 237.3}\right)(1 - RH) \tag{12.2}$$

The work of Frost and Schwalen was not calibrated for center pivots; however, the Trimmer equation is used in the

Fig. 12.2 Satellite image of circular fields in center pivot area (Credit NASA)

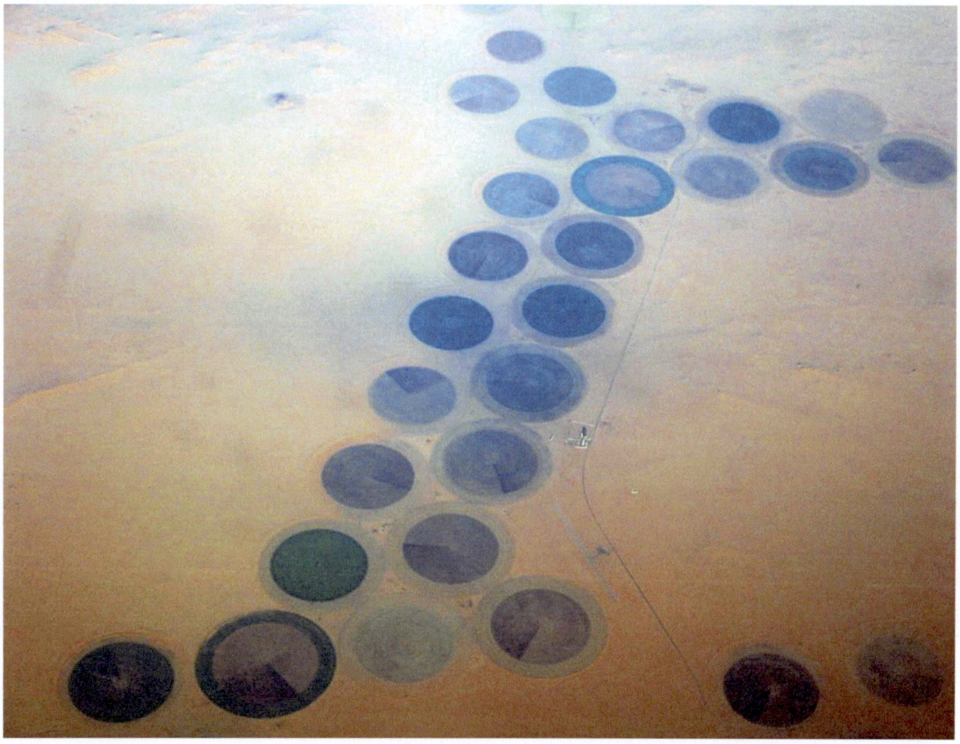

Fig. 12.3 Center pivot irrigation systems in the Sahara desert (Credit NASA)

Fig. 12.4 Big gun at end of center pivot irrigation system (Credit NRCS)

Fig. 12.5 Impact sprinklers on top of pivot

following example as a rough indication of evaporation rates from impact sprinklers mounted on top of center pivots.

Example 12.1 What is the expected evaporation fraction for the following conditions (humid and arid climates) with impact sprinklers?

Condition 1. Arid climate

$$\mathrm{RH} = 10\%, \mathrm{T} = 40^0\mathrm{C}, \mathrm{D} = 9/64\text{''}(3.57\ \mathrm{mm}), \mathrm{U} = 16\ \mathrm{km/h}(4.44\ \mathrm{m/s}), \text{and } \mathrm{h} = 70\ \mathrm{psi}(490\ \mathrm{kPa})$$

Condition 2. Humid climate

$$\mathrm{RH} = 90\%, \mathrm{T} = 25^0\mathrm{C}, \mathrm{D} = 9/64\text{''}(3.57\ \mathrm{mm}), \mathrm{U} = 8\ \mathrm{km/h}(2.22\ \mathrm{m/s}), \text{and } \mathrm{h} = 70\ \mathrm{psi}(490\ \mathrm{kPa})$$

Condition 1. Arid climate

$$e_s - e_a = 0.61\ \exp\left(\frac{17.27(40)}{40 + 237.3}\right)(1 - 0.1) = 6.63\ kPa$$

Fig. 12.6 Drop tube sprayers

Fig. 12.7 Serrated plate sprayer on drop tube (Credit NRCS)

$$L_e = \begin{bmatrix} 1.98(3.57)^{-0.72} + 0.22(6.63)^{0.63} + 3.6{*}10^{-4}(490)^{1.16} \\ + 0.14(4.44)^{0.7} \end{bmatrix}^{4.2} = 39\%$$

If the droplet remains in the air longer than in the experiments used to derive Eq. 12.3, such as for a pivot, then spray evaporation losses would be more than 39 %.

Condition 2. Humid climate

$$e_s - e_a = 0.61\ \exp\left(\frac{17.27(25)}{25 + 237.3}\right)(1 - 0.9) = 0.32\ kPa$$

$$L_e = \begin{bmatrix} 1.98(3.57)^{-0.72} + 0.22(0.32)^{0.63} + 3.6{*}10^{-4}(490)^{1.16} \\ + 0.14(2.22)^{0.7} \end{bmatrix}^{4.2} = 7.6\%$$

Evaporation losses are clearly unacceptable in the arid climate. These results show why center pivot systems were not adopted in arid environments until the advent of LEPA and LESA systems.

In addition to droplet evaporation in the air, evaporation losses also include water that stays on the canopy and

Fig. 12.8 Rotator on drop tube (Credit NRCS)

evaporates and water that evaporates from soil surface. It depends on depth of application, but a typical estimate is that 10 % of the applied water remains on the foliage. If another 12–15 % is lost to evaporation, then to total loss would be 20–25 % water lost to evaporation. In contrast to higher elevation spray systems, LEPA systems have only 3–5 % evaporation loss (in arid climates), and LESA systems have a 12–15 % total evaporation loss in a high profile crop such as corn. Although direct water evaporation from droplets or the canopy may reduce the transpiration requirement of the crop, the evaporation loss in arid climates is unacceptable for top mounted impact sprinklers.

The following sections use the *Chapter 12 Center Pivot* program. An excellent center pivot design program called CPED (Center Pivot Evaluation and Design) was developed by NRCS and is available online. One advantage of CPED is that it has many component specifications.

Center Pivot Pipeline Flow Rate

The entire center pivot flow rate is the product of gross application rate and irrigated area divided by the expected fraction of the day that the center pivot is running. The fraction is generally less than 1.0 due to repairs or crop management.

$$Q_p = 0.116 \frac{i_g A}{1 - L_{r-m}} \tag{12.3}$$

where

Q_p = center pivot flow rate, L/se
$L_{r\text{-}m}$ = fraction of down time for center pivot during peak ET, dimensionless
A = Area of field, ha
i_g = required gross application rate, mm/day

Example 12.2 A 400 m radius center pivot irrigation system operates for 22 out of 24 hours per day. Calculate the required pivot flow rate if i_g = 16.2 mm/day.

$$A = \pi\, r_{max}^2 = \pi(400^2)(1\ \text{ha}/10{,}000\ \text{m}^2) = 50.3\ \text{ha}$$

$$Q_p = 0.116 \frac{i_g A}{1 - L_{r-m}} = 0.116 \frac{(16.2\ mm)\ (50.3\ ha)}{1 - (2/24)}$$

$$= 103\ L/s$$

The *CP_infilt_RO* worksheet in the *Chapter 12 Center pivot* program calculates pivot flow in cell B5.

	A	B	
1	Pivot length (radius)	400	m
2	Pivot area	50.3	ha
3	Hours per day operation	22	hr
4	Gross application req	16.2	mm/day
5	Pivot flow rate	103.0	L/sec
6	Normal gross app 24 hr	17.7	mm/day

Infiltration Rate and Application Rate

The most common constraint in center pivot design is prevention of runoff (Fig. 12.9) at the end of the pivot. If application rate exceeds the infiltration rate + soil surface storage, then there is ponding and runoff. The reason that the end is the critical location is that application rate is high because application time is low with the high rate of travel.

Fig. 12.9 Ponding and runoff at end of pivot (Courtesy of NRCS)

The time of application, T_a, at the end of the pivot is proportional to the ratio of the sprinkler wetted diameter to center pivot circumference, $2\pi r$.

$$T_a = \left(\frac{D_w}{2\pi r}\right) T_r \tag{12.4}$$

where

D_w = sprinkler wetted diameter, m
T_r = time for one complete rotation of the pivot, hr
r = radius of the center pivot, m
T_a = application time for outer sprinkler, hr

Traditional pivots complete one rotation each day. However, new high velocity pivots complete three or more revolutions each day. The advantage of more frequent rotations and a smaller application depth per rotation is that even if infiltration rate is exceeded by application rate, the small depth of application is contained by soil surface storage capacity; thus, there is no runoff, and the water can slowly infiltrate into the soil after the pivot passes by. The disadvantage is that more evaporation takes place. For example, in Fig. 12.10 the high velocity LESA sprinkler application to alfalfa is held by surface depressions and surface crop residue. If there are no depressions, a steep slope, or no crop residue on the soil surface, then surface storage is low.

In this case, with a young crop, evaporation would be higher with multiple passes since the ground surface would be flooded for a greater percentage of time. The difference in evaporation would be more exaggerated at night since open pond evaporation continues in the night while plant evapotranspiration ceases at night. With a mature crop, there may be little difference in evaporation with multiple passes since evaporation from the ground surface is low in comparison to canopy transpiration.

Example 12.3 A drop tube LESA sprayer has a 4 m wetted diameter (D_w), and the time of rotation for the 400 m long pivot is 22 hours, calculate the time of water application for the last (outer) sprinkler on the center pivot pipeline.)

$$T_a = T_r\left(\frac{D_w}{2\pi\ r}\right) = 24\ hr\left(\frac{4\ m}{2\pi(400\ m)}\right) = 0.038\ hr$$
$$= 2.29\ \text{min}$$

This calculation is performed in cells E1:E5 in the *CP_Infilt_RO* worksheet.

Sprinkler distance	400	m
Wetted diameter	4	m
Rotation rate	24	hrs rotation
Sprinkler app time	2.29	minutes
Sprinkler spacing	2	m

The application rate is not uniform over time. It is zero when sprinkler droplets first reach a point in the field, is a maximum when the sprinkler is directly overhead, and then decreases back to zero after the sprinkler passes by. The application rate vs. time curve is often modeled as an ellipse (Fig. 12.11) even though the actual shape of the application rate curve is often different from an ellipse.

$$\frac{\left(\frac{di}{dt}\right)^2}{\left(\frac{di}{dt}\right)^2_{max}} + \frac{t^2}{\left(T_a/_2\right)^2} = 1 \tag{12.5}$$

where

di/dt = instantaneous application rate, mm/min
t = length of time after a point is first wetted by the pivot, min
di/dt_{max} = maximum application rate, mm/min.

Fig. 12.10 Surface storage in level alfalfa field (Courtesy of NRCS)

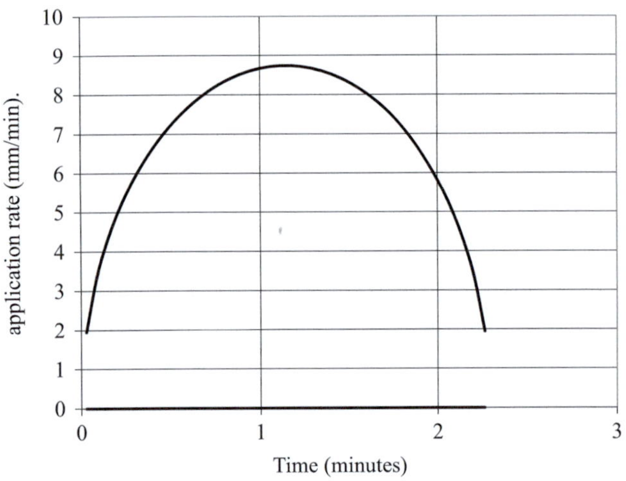

Fig. 12.11 Application rate versus time for point in the field with 2.1 minute application time

The application rate vs. time is calculated by rearranging Eq. 12.5 and solving for instantaneous application rate, di/dt (Fig. 12.11). In Eq. 12.6, t = 0 when the application rate is maximum.

$$\left(\frac{di}{dt}\right) = \sqrt{\left(\frac{di}{dt}\right)^2_{\max}\left(1 - \frac{t^2}{\left({}^{T_a}/_2\right)^2}\right)} \quad (12.6)$$

Let t = 0 when application begins by substituting $t - T_a/2$ for t.

$$\left(\frac{di}{dt}\right) = \sqrt{\left(\frac{di}{dt}\right)^2_{\max}\left(1 - \frac{\left(t - {}^{T_a}/_2\right)^2}{\left({}^{T_a}/_2\right)^2}\right)} \quad (12.7)$$

The maximum application rate, di/dt_{max} (mm/min) is found with the following equation [derived based on area of ellipse = πab: $2i_a = \pi\,(di/dt_{max})\,(T_a/2)$. The net application depth i_a is the depth of water that reaches the soil (i_g minus evaporation).

$$\left(\frac{di}{dt}\right)_{\max} = \left(\frac{4}{\pi}\right)\left(\frac{i_a}{T_a}\right) \quad i_a = i_g(1 - L_e) \quad (12.8)$$

Example 12.4 Plot the instantaneous application rate and calculate the maximum application rate for the end of the LESA pivot described in Examples 12.2 and 12.3: i_g = 17.7 mm, T_a = 2.29 min. The percent of sprinkler water that evaporates before it hits the soil is 11 %.

$$i_a = i_g(L_e) = 17.7\ \text{mm}(1 - 0.11) = 15.7\ \text{mm}$$

$$\frac{di}{dt_{\max}} = \left(\frac{4}{\pi}\right)\left(\frac{i_a}{T_a}\right) = \left(\frac{4}{\pi}\right)\left(\frac{15.73}{2.29}\right) = 8.74\ mm/\min$$
$$= 52\ cm/hr$$

The net application rate and maximum application rate are calculated in cells H1:H3 in the *CP_Infilt_RO* worksheet.

	A	B	C
1	Pivot length (radius)	400	m Hide graph
2	Pivot area	50.3	ha
3	Hours per day operation	22	hr
4	Gross application req	16.2	mm/day Show
5	Pivot flow rate	103.0	L/sec
6	Normal gross app 24 hr	17.7	mm/day
7			
8	Surface storage	4	mm
9	Intake family for infiltration functions (write intake family number in cell B7)		
10	Intake family	4	Put intake family to I
11	NRCS_a	0.4445	a and b are autc
12	NRCS_b	0.8230	
13	Include the initial depth	TRUE	This is the 0.6985 mı
14	Runoff	2.82	mm
15	Infiltration + storage	12.92	mm
16	Sprinkler application	15.75	mm
17			
18			
19	Intake families	a (cm/hr)	b
20	0.05	0.0533	0.618
21	0.1	0.062	0.661
22	0.15	0.0701	0.683
23	0.2	0.0771	0.699
24	0.25	0.0853	0.711
25	0.3	0.0925	0.72
26	0.35	0.0996	0.729

	D	E	F	G
1	Sprinkler distance	400	m	Fraction evap
2	Wetted diameter	4	m	Net app / day
3	Rotation rate	24	hrs rotation	net depth/pass
4	Sprinkler app time	2.29	minutes	Max appl. rate
5	Sprinkler spacing	2	m	Sprinkler flow

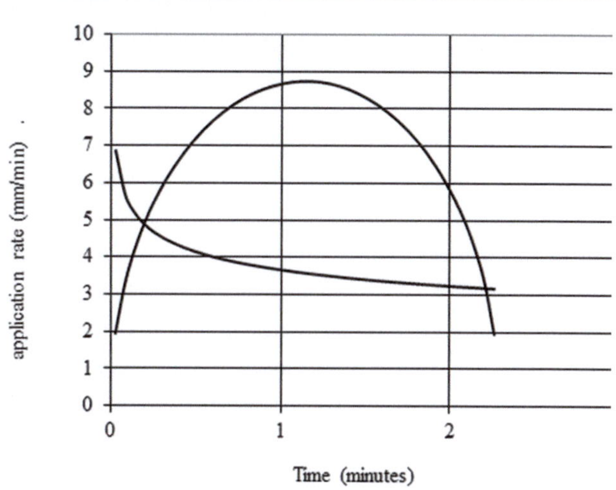

Fig. 12.12 Calculation of runoff and infiltration for intake family 4 soil in *CP_Infilt_RO* worksheet

	D	E	F	G	H	I
1	Sprinkler distance	400	m	Fraction evap	0.11	
2	Wetted diameter	4	m	Net app / day	15.73	mm/day
3	Rotation rate	24	hrs rotation	net depth/pass	15.73	mm/pass
4	Sprinkler app time	2.29	minutes	Max appl. rate	8.74	mm/min
5	Sprinkler spacing	2	m	Sprinkler flow	1.030	L/sec

Application rate vs. time (Fig. 12.11) is calculated with Eq. 12.7.

$$\left(\frac{di}{dt}\right) = \sqrt{\left(\frac{di}{dt}\right)_{\max}^{2}\left(1 - \frac{(t - {}^{T_a}/_2)^2}{({}^{T_a}/_2)^2}\right)}$$

$$= \sqrt{8.74^2\left(1 - \frac{(t - 1.145)^2}{(1.145)^2}\right)}$$

The application rate vs. time is shown in Fig. 12.11 from the *CP_Infilt_RO* worksheet.

The application rate curve can be compared with the soil's infiltration rate curve to determine whether the application rate exceeds the infiltration rate at any time during the application period. For example, an infiltration rate and application rate curve are shown in Fig. 12.12. The infiltration rate decreases as the soil fills with water. If any section of the application rate curve is greater than the infiltration rate curve, then ponding takes place.

The *CP_Infilt_RO* worksheet (Fig. 12.12) allows the user to select from the NRCS intake families listed in cells A20: C38. Enter the intake family in cell B10, and the coefficient and exponent are automatically changed in cells B11:B12. You can include the NRCS constant 0.6985 mm (see Chap. 3) by writing TRUE in cell B13. The surface storage depth is entered into cell B8, the depth of water that is stored in depressions on the soil surface during each pass. Functions in VBA then calculate the total runoff, infiltration + storage, and total depth of sprinkler application for each pass in cells B14:B16.

One way to extend the time of application, T_a, is the use of booms that extend out from the pivot pipeline with each boom holding several sprinklers. In order to prevent nonuniform application, the booms are angled diagonally from the pivot pipeline in order to offset boom sprinklers. Surface storage can be increased in order to prevent runoff. Tillage equipment can be used in increase soil surface roughness or cloddiness or crop residue can be used on the soil surface. field slope can significantly increase runoff as shown in Figure. For rough or cloddy ground, the NRCS estimates

Fig. 12.13 Infiltration and storage calculation in *CP_Infilt_RO* worksheet

	A	B	C
8	Surface storage	4	mm
9	Intake family for infiltration functions (write intake family number in cell B7)		
10	Intake family	4	Put intake family to left
11	NRCS_a	0.4445	a and b are automatically
12	NRCS_b	0.8230	updated
13	Include the initial depth	TRUE	This is the 0.6985 mm infilt
14	Runoff	2.82	mm
15	Infiltration + storage	12.92	mm
16	Sprinkler application	15.75	mm

Fig. 12.14 Application rate curve for two pivot rotations per day with LESA sprinkler and soil infiltration curve for intake family 4 soil

	C	D	E	F	G
m		Sprinkler distance	400	m	Fraction
ha	Hide graph	Wetted diameter	4	m	Net app
hr		Rotation rate	12	hrs rotation	net deptl
mm/day	Show graph	Sprinkler app time	1.15	minutes	Max app
L/sec		Sprinkler spacing	2	m	Sprinkler

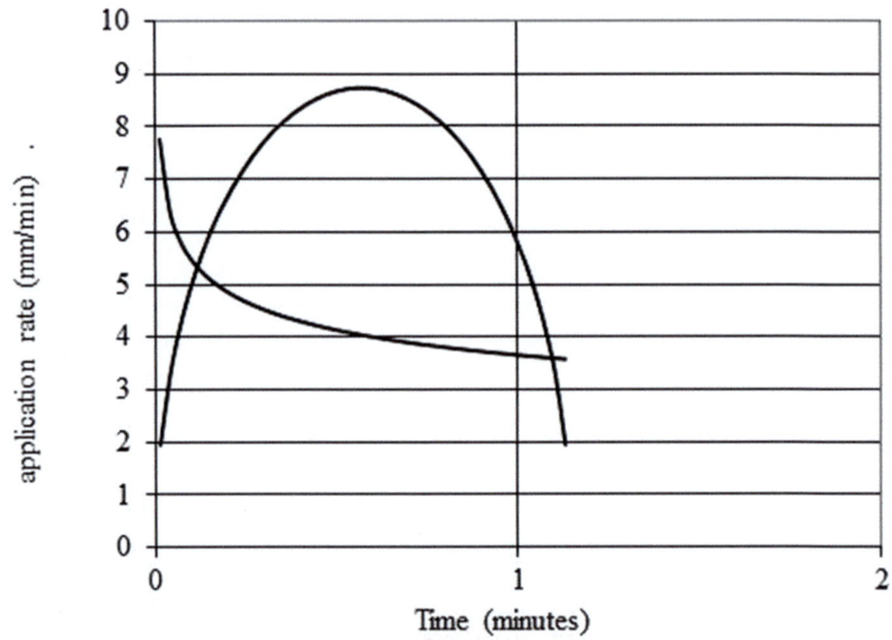

that surface storage as 1.2, 0.7, 0.25, and 0 cm for slopes of 0.5, 2, 4, and 4.5 %, respectively.

Example 12.5 Determine whether the LESA system in Examples 12.2, 12.3 and 12.4 will have ponding and or runoff in an intake family 4 soil with a single rotation per day and with two rotations per day. Surface depressions hold 4 mm/pass. Show the infiltration and sprinkler application curves.

Single rotation per day

The sprinkler application and infiltration curve data is generated in in cells D11:D50. Infiltration curves and sprinkler application curves are shown in Figs. 12.13 and 12.14. The infiltration curve does not include surface storage or the 0.6985 mm in the NRCS equation. However, the calculation of runoff (2.82 mm) in cell B14 (Fig. 12.13) does include the terms.

This example illustrates why LESA systems must rely on surface storage to hold water until infiltration is completed. Even the highest infiltration rate classification (intake family 4) has runoff with the small application diameter.

Two rotations per day (12 hours per rotation).

In this case, the application time is half, 1.14 minutes, and the application depth per pass (7.2 mm) is also half of the

Fig. 12.15 Two passes per day with no instantaneous infiltration and no surface storage results in 3.29 mm runoff

	A	B	C
8	Surface storage	0	mm
9	Intake family for infiltration functions (write intake family number in cell B7)		
10	Intake family	4	Put intake family to left
11	NRCS_a	0.4445	a and b are automatically
12	NRCS_b	0.8230	updated
13	Include the initial depth	FALSE	This is the 0.6985 mm infilt
14	Runoff	3.29	mm
15	Infiltration + storage	4.58	mm
16	Sprinkler application	7.87	mm

Fig. 12.16 Single pass per day with instantaneous infiltration TRUE and 4 mm surface storage results in no runoff

	A	B	C
8	Surface storage	4	mm
9	Intake family for infiltration functions (write intake family number in cell B7)		
10	Intake family	4	Put intake family to left
11	NRCS_a	0.4445	a and b are automatically
12	NRCS_b	0.8230	updated
13	Include the initial depth	TRUE	This is the 0.6985 mm infilt
14	Runoff	0.00	mm
15	Infiltration + storage	7.87	mm
16	Sprinkler application	7.87	mm

single pass application depth. Thus, the maximum application rate is the same as the single pass.

$$\frac{di}{dt_{\max}} = \left(\frac{4}{\pi}\right)\left(\frac{i_a}{T_a}\right) = \left(\frac{4}{\pi}\right)\left(\frac{7.85\ mm}{1.145\ \min}\right) = 8.74\ mm/\min$$
$$= 52\ cm/hr$$

$$\left(\frac{di}{dt}\right) = \sqrt{\left(\frac{di}{dt}\right)^2_{\max}\left(1 - \frac{(t - {}^{T_a}/_2)^2}{({}^{T_a}/_2)^2}\right)}$$
$$= \sqrt{52^2\left(1 - \frac{(t - 0.572)^2}{(0.572)^2}\right)}$$

The *CP_Infilt_RO* worksheet is adjusted for this scenario by changing cell E3, as shown in Fig. 12.14.

The depth of runoff without surface storage and instantaneous infiltration (0.6985 mm) is determined by setting cells B10 to zero and cell B13 to FALSE (3.29 mm in Fig. 12.15). However, with a surface storage depth of 4 mm, there is no runoff (Fig. 12.16).

Example 12.6 Determine the depth of ponding for an impact sprinkler mounted on top of a center pivot irrigating an intake family 4 soil and an intake family 0.2 soil. The wetted diameter is 35 m. The net application rate, i_a, is 14.4 mm/pass, and the pivot has a 22 hr rotation cycle. Ignore instantaneous infiltration and the 0.6985 constant in the NRCS calculation of total infiltration. The pivot diameter is 400 m diameter pivot.

$$T_a = T_r\left(\frac{D_w}{2\pi r}\right) = 22\ hr\left(\frac{35}{2*\pi*400}\right) = 0.306\ hr$$
$$= 18.4\ \min$$

$$d_{\max} = \left(\frac{4}{\pi}\right)\left(\frac{i_a}{T_a}\right) = \left(\frac{4}{\pi}\right)\left(\frac{14.4}{18.4}\right)$$
$$= 1.0\ mm/\min\left(\frac{cm}{10\ mm}\right)\left(\frac{60\ \min}{1\ hr}\right) = 6\ cm/hr$$

$$\left(\frac{di}{dt}\right) = \sqrt{\left(\frac{di}{dt}\right)^2_{\max}\left(1 - \frac{(t - {}^{T_a}/_2)^2}{({}^{T_a}/_2)^2}\right)}$$
$$= \sqrt{6^2\left(1 - \frac{(t - 9.2)^2}{(9.2)^2}\right)}$$

There is no ponding with the intake family 4 soil; however, approximately half of the water, 7.1 mm, is ponded on the intake family 0.2 soil (Fig. 12.17).

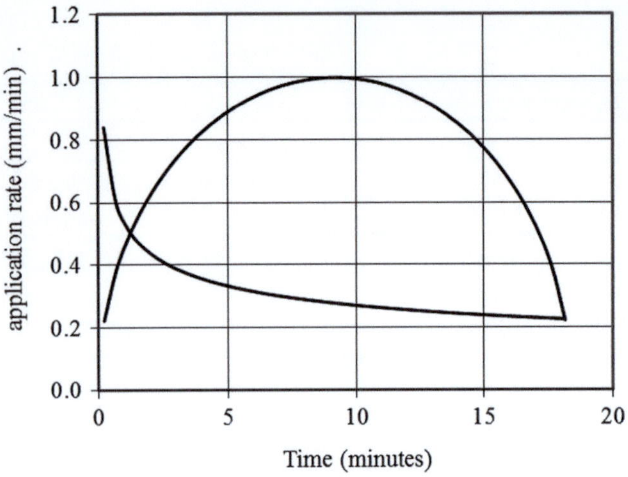

Fig. 12.17 Application rate curve and intake family curves for single rotation per day with top mounted impact sprinkler

Fig. 12.18 Sprinkler spacing

Center Pivot Sprinkler Spacing and Flow Rate

Sprinkler flow rates and spacing (Fig. 12.18) vary along the center pivot pipeline. Individual sprinkler flow rates are proportional to the ratio of land area irrigated by each sprinkler to the entire pivot flow area.

$$q_r = r(s_r)(2Q_p)/(r_{max})^2 \qquad (12.9)$$

where

q_r = sprinkler flow rate at a radial distance from the pivot point, L/sec.
s_r = distance between sprinklers at radial distance r, m
r = radial distance from pivot point, m.

Example 12.7 Sprinkler spacing is 1.5 m. Calculate the flow rate of the middle and last sprinklers for a 400 m center pivot with a 95.3 L/sec flow rate.

$$q_{400} = 400\ \text{m}(1.5\ \text{m})(2)(95.3\ \text{L/sec})/(400\ \text{m})^2$$
$$= 0.71\ \text{L/sec} = 43\ \text{L/min} = 11.3\ \text{gal/min}$$

The flow rate at the midpoint of the pivot pipeline is calculated as follows.

$$q_{200} = 200\ \text{m}(1.5\ \text{m})(2)(95.3\ \text{L/sec})/(400\ \text{m})^2$$
$$= 0.36\ \text{L/sec} = 21\ \text{L/min} = 5.7\ \text{gal/min}$$

The q_{200} flow rate is half of the q_{400} flow rate: sprinkler flow rates vary linearly along the pipeline because the area covered by each sprinkler is product of spacing and circumference, which are directly proportional to the distance to the center.

Sprinkler spacing and flow rate can be varied along a pivot in order to achieve a relatively uniform application distribution along the pipeline without having extremely large nozzles at the end. For example, the center pivot represented in Fig. 12.19 has sprinkler spacing, s_r, 5 m, 3 m, and 2 m, in the inner, middle, and outer sections of the pivot, respectively. Intervals are varied near wheels where half-circle sprinklers are used in order to avoid wetting the wheel tracks. Wetting the tracks cause the wheels to sink into the soil. It is difficult to have a uniform application rate near the pivot point (left side of Fig. 12.19) because the relative radius changes quickly. The application ratse along the rest of the pivot are relatively uniform (10 % measured CV based on nozzle flow rates).

Pivot manufacturers determine optimal sprinkler flow rates and nozzle sizes with computer programs. Farmers receive a ring of nozzles from manufacturers, and remove the nozzles from the ring one at a time in order to install them in the right order.

Sprinkler flow rate is proportional to the square root of pressure because a sprinkler is an orifice – $Q = kH^{0.5}$. If one sprinkler operates at 25 m pressure and another at 20 m pressure, then the pressure ratio is 0.8, but the flow ratio is 0.91.

$$Q_2 = Q_1\sqrt{\frac{H_2}{H_1}} = Q_1\sqrt{\frac{20}{25}} = 0.91\ Q_1$$

Pressure regulators (Fig. 12.20) solve the pressure variation problem; however, they dissipate 5 PSI (35 kPa) pressure so the system pressure requirement is higher. Pressure regulators expand a spring as pressure increases, causing the cross-sectional flow area to decrease. This decreased area increases the pressure drop across the device and maintains the design discharge operation pressure. The 30 PSI (210 kPa) pressure regulator shown in Fig. 12.20 requires 35 PSI (245 kPa) at the pressure regulator inlet.

Center pivots operating on flat ground do not require pressure regulators because the nozzles are designed to operate at the pressure along the pipeline.

Sprinkler flow rate is calculated with a modified orifice equation.

Fig. 12.19 Sprinkler flow rates, application rate and intervals between sprinklers on center pivot irrigation system

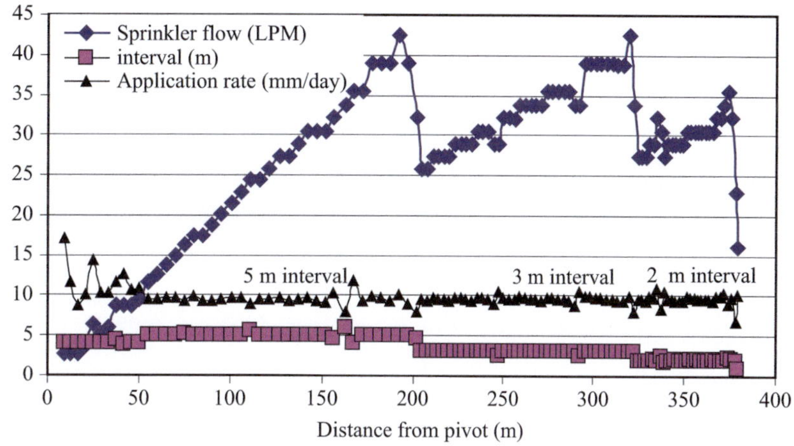

Fig. 12.20 Pressure regulator (Courtesy of Nelson Irrigation Corporation)

$$Q = 29.82\ D^2H^{0.5}C_d \qquad (12.10)$$

where

H = pressure, PSI
D = nozzle diameter, in
Q = flow rate (GPM)
C_d = sprinkler nozzle discharge coefficient

For metric units:

$$Q = 0.0666\ D^2H^{0.5}C_d \qquad (12.11)$$

where

H = pressure, kPa
D = nozzle diameter, mm
Q = flow rate (LPM).

Example 12.9 What is the flow rate of a 3.57 mm nozzle (9/64") at a pressure of 340 kPa (50 PSI)?

$$Q = 0.0666\ D^2H^{0.5}C_d = (0.0666)(3.57^2)(340^{0.5})(0.97) = 15.2\ \text{LPM} = 4\ \text{GPM}$$

If center pivots operate on a landscape with significant elevation change, then pressure regulators are required. The rule of thumb is that if energy fluctuation due to elevation change is greater than 20 % of design pressure, then pressure regulators should be used. Rather than a rule of thumb, engineering economic analysis can determine whether or not to add pressure regulators. Would the increase in uniformity and resultant decrease in water cost be greater than the capital cost of the pressure regulators and the extra cost of energy. This question is evaluated in Example 12.11.

Example 12.11 Calculate the cost of using pressure regulators to regulate pressure (3.5 m extra pressure) in a center pivot. The evapotranspiration requirement, ET_c, is 1,200 mm, and the 400 m pivot covers a 50 ha area. The water application efficiency is 85 %, and the pump energy efficiency is 85 %. The cost of energy is $0.10/kW-hr. Assume a 20 year project life and 6 % required rate of return (ROR).

Depth of water required

$$1,200\ \text{mm}/0.85 = 1,412\ \text{mm}$$

The increased pressure requirement is 5 PSI = 3.5 m.

Use Eq. 2.16 to calculate the energy per ha

$$E = \frac{0.0272(1,411\ mm)(3.5\ m)}{0.85} = 158\ kW - hr/ha$$

Calculate the energy cost per year for the 50 ha area.

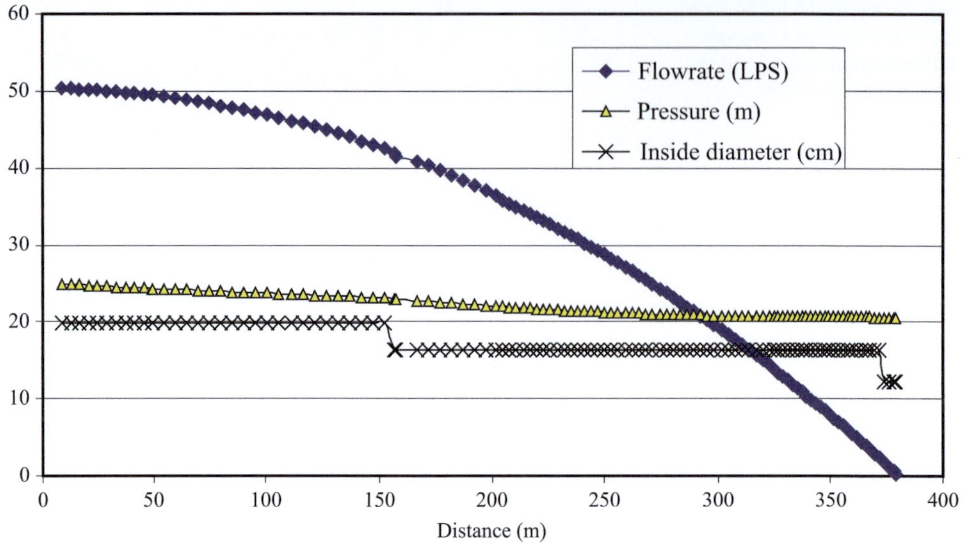

Fig. 12.21 Pipe flow rate, pressure, and pipe diameter for center pivot

$$(\$/ha)^*50\ ha = \left(\frac{158\ kW - hr}{ha}\right)\left(\frac{\$0.10}{kW - hr}\right)^*50\ ha$$
$$= \$790/year$$

Convert to the present value.

$$20\ \text{year project}, 6\%\ \text{ROR},\ \$790\ \text{payment} \rightarrow \$9,061.$$

If the increased value of yield with uniform application is greater than $9,061 over the life of the project, the pressure regulators should be used.

Center Pivot Pipeline Design

Minimizing center pivot pipe diameter reduces pipe and structure cost because the structure must carry the weight of the pipe. However, reducing pipe size increases pressure loss in the pipeline and thus increases the pressure requirement of the pump (increases energy cost).

Center pivot pressure loss can be calculated in a spreadsheet, as demonstrated in Chap. 7. Alternatively, if the entire pipeline has a single diameter, then the Christensen factor (F) for multioutlet pipelines can be used. The Christensen factor for center pivots is 0.54.

$$H_{act.} = H_{Lp}F \qquad (12.12)$$

where

H_{act} = actual pressure loss in the multioutlet pipeline, m
H_{Lp} = pressure loss in a pipeline with no outlets with inlet flow rate, m
F = Christiansen's F factor.

Example 12.10 Calculate the pressure loss along a center pivot irrigation pipeline. Inlet flow rate is 50.3 L/sec, Length is 380 m, and inside diameter of the pipe is 19.79 cm (8 inch pipe) along the entire length of the pipe. The inside surface of the pipe is galvanized steel.

For galvanized steel, assume a C value of 130.

$$H_L = kL\left(\frac{\left(\frac{Q}{C}\right)^{1.852}}{D^{4.87}}\right) = 1.22^*10^{10}(380\ m)\left(\frac{\left(\frac{50.3}{130}\right)^{1.852}}{198^{4.87}}\right)$$
$$= 5.24\ m$$

$$H_{act} = H_{Lp}F = 5.24\ m(0.54) = 2.8\ m$$

As flow decreases along the center pivot, it is possible to reduce pipe diameter. For multiple pipe diameters, the pressure loss must be calculated for each pipe section with a spreadsheet.

The center pivot pipeline represented by Fig. 12.21 has an initial pipe diameter of 19.79 cm ID. Pipe diameter is reduced to 16.3 cm ID (6 inch pipe) at 152 m and to 12.2 cm ID (4 inch pipe) at 374 m where the pipe extended beyond the last tower. The pressure change over the length of this pipeline on a level field is shown in Fig. 12.24. There is minimal pressure drop at the end even though pipe diameter is small. With the multiple pipe diameters, the pressure drops from 25 m at the beginning of the pipeline to 20.6 m at the end of the pipeline, a loss of 4.4 m. The pressure loss would be 2.8 m (Example 12.10) with a single pipe diameter.

Center Pivot Simulation in Uneven Landscapes

The *Chapter 12 Center pivot* program calculates sprinkler flow rates and pressures for center pivot movement on an uneven landscape. Sprinkler, pipe, energy, and financial

Create pivot field

Center pivot seasonal simulation

Use the data below to locate pivots in field	
Number of spans	4
Number of angular sections (arcs)	4
Number of pivots	3
Number of pixels on Worksheet covered by field	400
Keep elevations	TRUE
Make calculations below for pivot number:	1
Pressure Regulator?	TRUE
Pressure regulator loss (m)	3.5
Cost of pressure regulators ($/reg)	5
Nozzle Cd	0.97
Big Gun effective radius (m) (not active)	0
Minimum allowable pressure at sprinkler (m)	12
Center pivot pipe height above ground (m)	3
Sprinkler height above ground (m)	1
Sum elevation gain (neg.) + friction to pivot inlet	13.5

m or UTM dec. deg.	Center Latitude y	Center Longitude x	Pivot length (m)
Pivot 1	3697220.0000	342434.0000	400
Pivot 2	3697500.0000	342935.0000	400
Pivot 3	3696900.0000	342900.0000	300
Pivot 4			
Pivot 5			
Pivot 6			
Pivot 7			
Pivot 8			
Pivot 9			
Pivot 10			
Pivot 11			
Pivot 12			
Pivot 13			
Pivot 14			
Pivot 15			
Pivot 16			
Pivot 17			
Pivot 18			
Pivot 19			
Pivot 20			

Sprinkler spacing intervals	4
Distance from pivot point (m)	Space (m)
8	4
50	5
200	3
325	2

	Number of pipe diameter intervals	3
Distance	Inside diameter of steel pipe (mm)	Cost ($/m)
0	248	148
320	198	115
380	163	100
400	163	100

Pump efficiency (fraction)	0.85
Interest rate (as annual fraction)	0.08
Life(yr)	15
Energy cost ($/kW-hr)	0.1
Maintenance cost/yr as fraction of capital	0.05
Capital cost of pivot and system ($/project)	0.85
Labor cost per irrigation day ($/ha/day)	0.08

Seasonal evaluation		
Precipitation	7.5	cm
CV	0.2	
Minimum AW	64	cm
Maximum AW	74	cm
Interval AW	2	cm
Minimum AW-CWPF	20	cm

Crop growth, yield, and cost parameters		
CWPF where x = AW	a	b
$a + bx^{0.5} + cx + dx^2 + ex^3$	-3954	1067
Leach Eqn where x = AW	a	b
ae^{bx}	0.2736	0.0469
AWlimit	100	cm
ECiw	0	dS/m
ECt	7.7	dS/m
Ky	0.85	
Slope b	5.2	% / (dS/m)
Leach cost	1	$/ha-cm
Water cost	3.27	$/ha-cm
Yield price	0.92	$/kg

Fig. 12.22 Center pivot pipe and sprinkler specifications in *CP_Parameterization* worksheet

parameters are input into the *CP_Parameterization* worksheet in cells C9:C41 (Fig. 12.22). Crop parameters are input into cells F25:F44. The pipe diameters along the center pivot pipeline are specified in the *CP_Parameterization* worksheet in cells A29:C33 (Fig. 12.22). The distances in column A represent the distance from the pivot point at which each pipe diameter begins. The number of pipe diameters is specified in cell C28. The purposes of other cells in the worksheet are labeled and should be self-explanatory. They are based on concepts in this or previous chapters.

In order to allow the user to store data for more than one pivot, the positions of the pivot and the spatial cells are shown in the *Center_pivot_spatial_data* Worksheet (Fig. 12.23). Pivot cells are specified by the program according to cell number, pivot number, angular position,

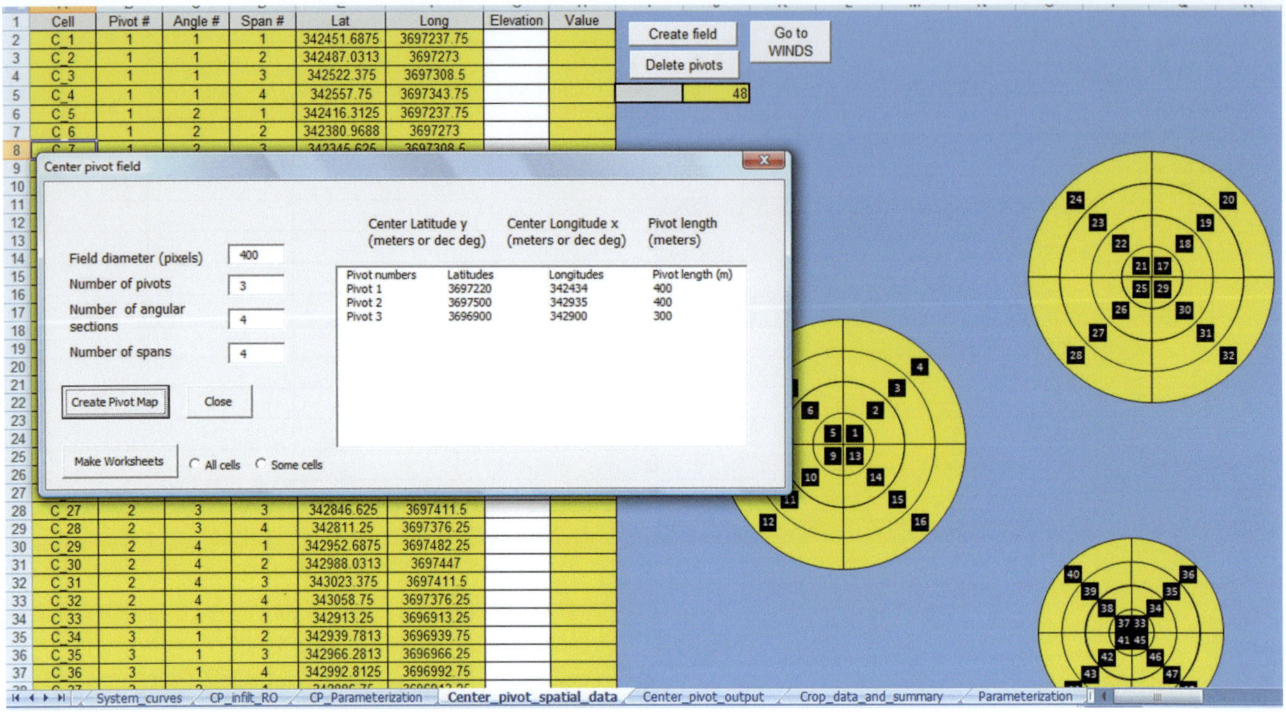

Fig. 12.23 Center_pivot_spatial_data worksheet with three pivots

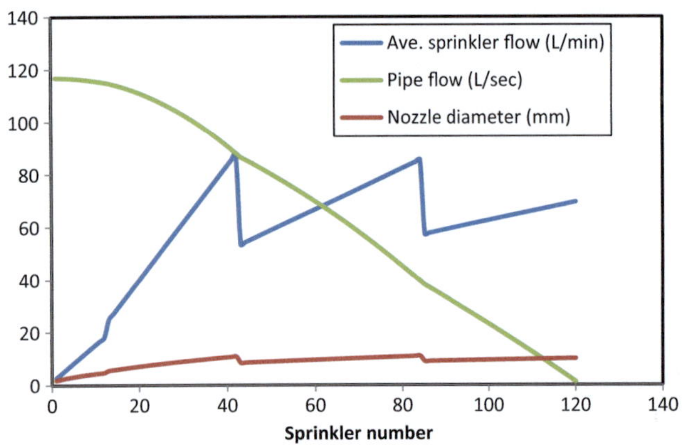

Fig. 12.24 Sprinkler flow rate, pipe flow rate, and nozzle diameter versus distance along pivot

radial position, and latitude and longitude of the centroid. Elevation of the sections are input by the user in column G (white column) of the *Center_pivot_spatial_data* Worksheet. Make sure to write TRUE in cell C6 in the *CP_Parameterization* worksheet in order to keep the elevations that you entered when the program runs. Cells E2:G20 in the *CP_Parameterization* worksheet allow the user to specify the locations and lengths of the pivots. In order to allow the user to visualize the locations of the pivots and section locations, the pivots are drawn on the *Center_pivot_spatial_data* worksheet (Fig. 12.23). The locations can also be changed in the dialog box, which is triggered by the *Create pivot field* button. The upper box in the Form allows the user to specify the dimensions of the worksheet (number of pixels) in the which the pivots are drawn.

The *Center Pivot Seasonal Simulation* button runs a simulation of the selected (cell C9) center pivot. Spatial, flow, and elevation data are written to the *CP_flow_output* worksheet. System curves (flow vs. inlet pressure) for the center pivot are written to the *CP_System_curves* worksheet. Yield data are written to the *CP_yield_output* worksheet. Summary economic data is written to the *CP_summary_data* worksheet.

Sprinkler flow rate, nozzle diameter and pipe flow rate (Fig. 12.24), elevation profile of each pipe section (Fig. 12.25), are written to the *CP_flow_output* worksheet. Pipe pressure losses are calculated based on pipe flow rates with the Hazen-Williams equation. As in Chap. 7, the calculation begins with the last sprinkler and moves backward toward the first. Sprinkler nozzle diameters are then calculated based on the resultant pipe pressure at each position.

Sprinkler pressures (Fig. 12.26) and flows (Fig. 12.27) in each quadrant are calculated based on elevation and the Hf calculation (friction loss). Flow rates are a function of nozzle diameter and pressure. Figure 12.28 shows relative application rates based on the variable sprinkler flow rates. All values are relative to the lowest application rate, which is at the end of the pivot in quadrant 2, the high point on the field.

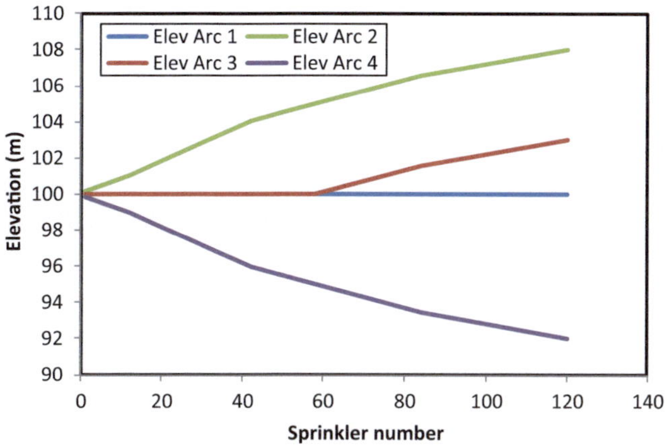

Fig. 12.25 Elevation of each quadrant versus distance along pivot

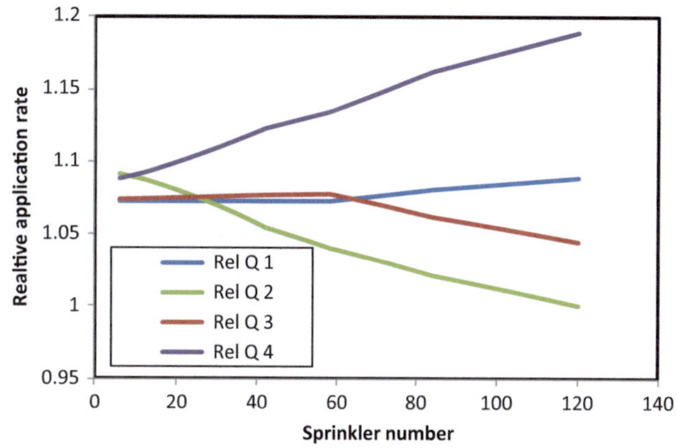

Fig. 12.28 Relative application rates in quadrants along pivot without pressure regulators

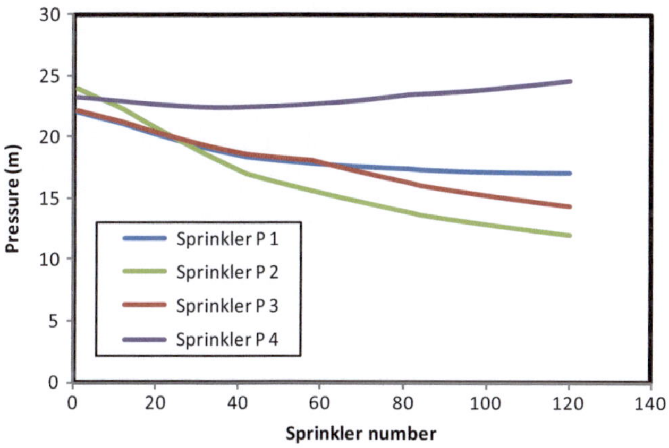

Fig. 12.26 Sprinkler pressures in each quadrant versus distance along pivot

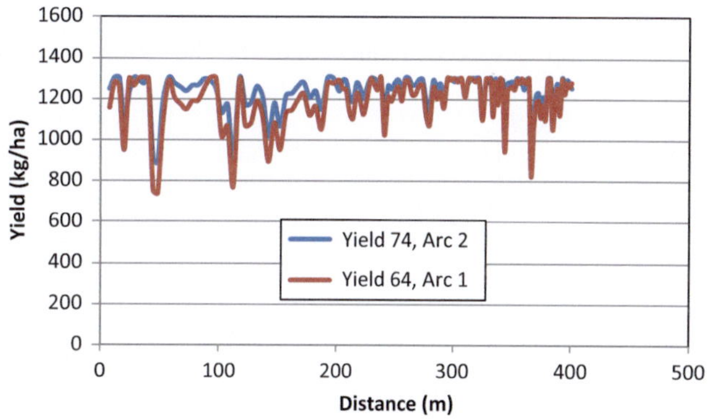

Fig. 12.29 Yields for two different quadrants and average depths

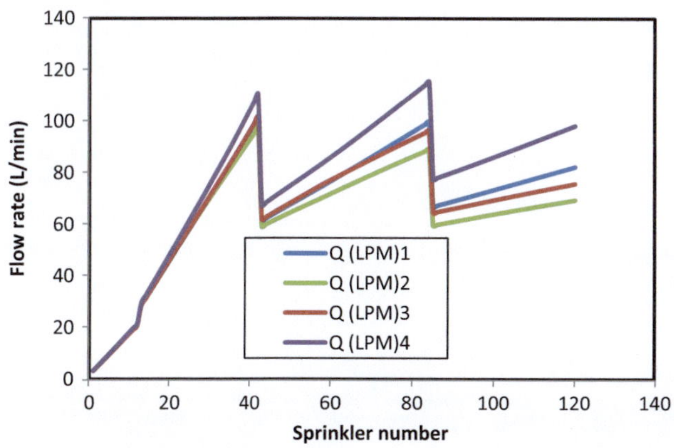

Fig. 12.27 Sprinkler flow rates in each quadrant along the pivot

Quadrant 2 has a lower overall pressure and flow rate than the other quadrants, particularly the downhill quadrant (4) as shown in Fig. 12.26. This is caused by the fact that the pivot has a system curve, which is the relationship between flow rate and pressure. Greater pressure is required for the uphill quadrant (2) so the overall flow rate goes down. The system curve calculation is conducted in VBA by running the pivot in each quadrant over a range of pressures and then developing pressure-flow rate curves for each quadrant. The minimum pressure along the pipeline is not necessarily at the end of the pipeline. There may be a high point within the field that has a low pressure. The program finds the minimum pressure as a function of inlet pressure.

Spatial variability (CV in cell F26 in *CP_Parameterization* worksheet) is evaluated for the application depths specified in cells F27 and F28 in *CP_Parameterization* worksheet. Applied water depths and yield output (Fig. 12.29) data are written to the *CP_yield_output* worksheet for each average depth applied and each quadrant. Finally, an economic summary of all evaluated depths is written to the *CP_summary data* worksheet

Center Pivot Pump and Pipe Network Design

Center pivot farms are often supplied by a central reservoir and pump station (Fig. 12.30).

The pump station should be designed to supply water flow rates that range from the water requirement for one pivot to all of the pivots on a farm. Although wells can be drilled at each pivot point, flexibility of operation and maintenance is gained when wells are networked into a central reservoir, which then supplies water to all pivots. Use of a reservoir also provides some protection against temporary water supply failures from the canal district or individual wells. A set of centrifugal pumps is connected to a single manifold. The manifold then feeds a set of stainless steel screen filters. The filters are connected to a second manifold that feeds the pipe network that supplies all of the pivots on the farm.. In this way, any pump or any screen can be shut down and repaired or cleaned while the others are running.

The pump pressure requirement is the sum of the following:

- Required sprinkler pressure
- Required pressure drop across the pressure regulator,
- Pipe friction loss from one end of the pivot to the other,
- Elevation difference between the ground surface and the sprinklers,
- Elevation difference between the pumping water surface and the maximum elevation in the field,
- Friction loss in the supply pipe network,
- Pressure loss in filters
- Friction losses in fittings and valves.

Example 12.12 Design a pump station for the pivot farm shown in Fig. 12.31. The maximum elevation in the center pivot irrigated fields 1, 2, 3, and 4 are 10 m, 4 m, 6 m, and 8 m higher than the reservoir, respectively. Pipe friction in the N-S pivot submains is 2 m, and pipe friction in the EW mainline is 4 m to the midpoint and 3 m from the midpoint to the end. Each pivot requires 95 LPS

Use 4 pumps so that 1–4 pivots can be run at any time. Field 1 is the worst case; thus, design all pumps to supply the correct pressure to pivot 1. You can't connect pumps with different discharge pressures to the same manifold.

+15 m – sprinkler pressure (20 PSI)
+3.5 m – pressure regulator required pressure difference between pipe and operating pressure
+2.8 m – pipe friction loss in center pivot
+2 m – elevation of sprinklers above ground surface.
+10 m – elevation difference between highest point in field and reservoir surface
+2 + 3 + 4 m – Pipe network
+3 m – friction loss in stainless steel screen filter
+5 m – friction loss in valves and fittings (check valve, manifold, pump suction).
49m – total pressure required.

Each of the 4 pumps should provide 49 m pressure at 95 LPS.

Fig. 12.30 Reservoir and pump station for center pivot farm. Cylindrical stainless steel ring filters and manifold in foreground. Motors on top of turbine pumps are seen behind manifold (Credit Paul Colaizzi, University of Arizona and USDA-ARS Bushland, Texas)

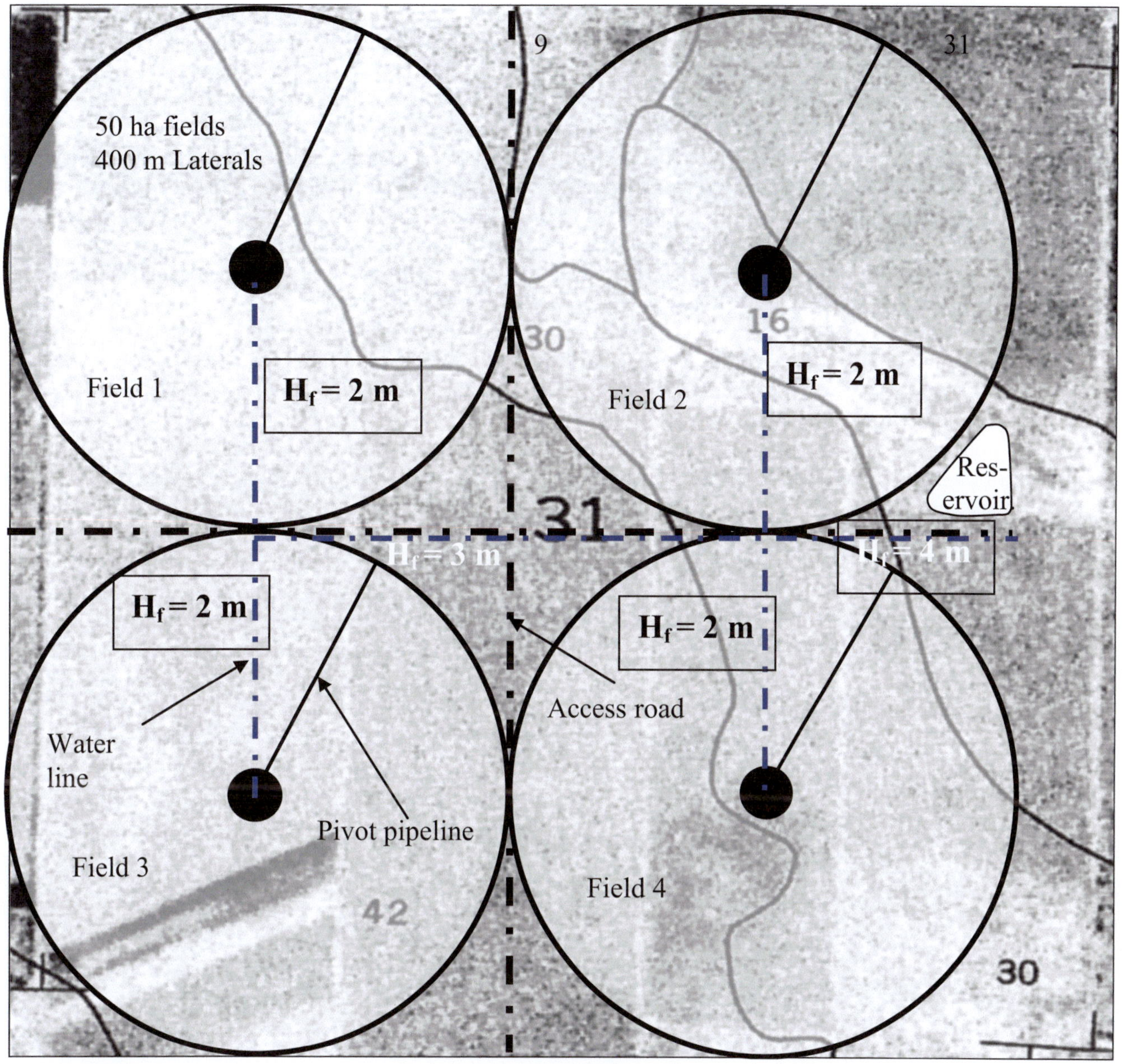

Fig. 12.31 Center pivot layout on a typical section of land (1 mile by 1 mile) with pipe friction losses for Example 12.10

Questions

1. Discuss the different types of sprinkler nozzles and systems used on center pivots. Discuss the strengths and weakness of the systems.

2. What is the difference between a linear move and a center pivot irrigation system?
3. Calculate the percent evaporation from sprinkler droplets for the parameters in Example 12.1 except that relative humidity is 50 %. If the application depth is 25 mm to a mature corn crop from overhead impact sprinklers, then what is the total depth of evaporation + canopy interception loss?
4. Calculate the percent evaporation from overhead sprinklers for the parameters in Example 12.1 except that relative humidity is 60 %. What is the depth of evaporation for an application depth of 25 mm to a mature corn crop from overhead impact sprinklers? Consider canopy interception and droplet evaporation?
5. Calculate the flow rate of a center pivot that has a length of 350 m, and gross application depth 15 mm/day. The pivot operates for 21 hours/day.
6. Calculate the maximum application rate for the parameters in question 12.5. The sprinkler wetted

diameter is 4 m and the percent evaporation is 14 %. Then, show that the maximum application rate is the same if the pivot rotates three times per day (7 hour rotation). Show calculation and explain why the maximum application rate is the same in both cases.

7. Using the parameters in question 12.6, calculate the depth of runoff and maximum application rate during each pass for an intake family 3 soil with 2 mm surface storage. Include the 0.6985 initial infiltration depth. Calculate for three revolutions per day. The percent evaporation is 14 %. Use *Chapter 12 Center pivot* program and hand calculations.
8. Repeat Question 12.7, but change to one rotation per day.
9. Calculate the flow rate and runoff at the middle sprinkler (175 m from pivot point) for the pivot described in questions 12.6–12.9 for a single 21 hour rotation per day. Sprinkler spacing is 2 m at the middle of the center pivot. Fraction evaporation is 14 %. Sprinkler wetted diameter is 3.5 m. The soil intake family is 1. Surface storage is 4 mm.
10. Derive the Christensen's F factor in Eq. 12.12 by assuming that a center pivot has four sprinklers (1/4, 1/2, 3/4, and full length down the pivot).
11. Low-pressure sprinkler nozzles at Paradise Cattle Company have a wetted diameter (Dw) of 2 m. The average rotation period of the center pivot is 8 hours. The pivot is 400 m radius and the flow rate is 95 L/sec. 5 % of water is lost to evaporation. Calculate the daily gross application rate, and plot the instantaneous application rate as a function of time at the 400 m. The Paradise Cattle Company pivots used in this and the next problem are described in the *Center pivot pump and pipe network design* section.
12. Plot the instantaneous application rate vs. time for at 200 m for the same parameters as in question 12.11.
13. Calculate the pressure loss in a center pivot that has a length of 350 m, and gross application depth applied during each pass is 5 mm. Time of rotation is 8 hours. Use 198 mm pipe. There is no down time.
14. A sprinkler has a flow rate of 5 GPM at 20 PSI. What is the flow rate at 25 PSI?
15. Use the *Chap. 12 Center pivot* model to find the optimal water application depth for CV values of 0.1 and 0.3.
16. Is it worth adding pressure regulators for the elevations shown in Fig. 12.23? Regulators cost is $5.00 per. This is the same as question 12.15, but add the regulators.
17. A center pivot irrigation system requires 200 kPa sprinkler pressure. There are five pivots each with a flow rate of 110 L/sec. Pressure regulators are used. Sprinklers are 1 m above the land surface. There is a 4 m pressure loss in the pivot pipeline, and 10 m head loss in the pipe network between the pumps and the worst case pivot. The maximum elevation of the land surface is 20 m higher than the reservoir. Make all other necessary assumptions. How many pumps are required? At what pressure and flow rate should the pumps operate?

13 Turf Sprinkler Irrigation Systems

The green industry, which includes turf and landscape irrigation, has an economic impact in the United States of $150 billion/year. Effective turf irrigation systems keep turf areas healthy and attractive and use water efficiently, which are the primary goals of turf irrigation. This chapter begins with a review of sprinkler system layout and spacing. The *Sprinkler Uniformity* program allows the user to visualize and quantify the effect of spacing on uniformity. This is followed by an economic rationale for sprinkler spacing. The last section covers sprinkler pipe network design and scheduling calculations. There is a range of turf sprinkler types. Fixed spray sprinklers cover small areas (<5 m radius). They have a relatively high flow rate with respect to their small radius of application, which results in a high application rate (30 mm/hr). Rotors and impact sprinklers cover a larger area (>5 m radius) and have a lower application rate (<25 mm/hr) than fixed spray sprinklers. Large golf course sprinklers operate at high pressure and can have 50 m wetted diameters.

Sprinkler Patterns

Turf sprinklers generally have a decreasing application rate vs. distance from the sprinkler. Ideally, the overlapped patterns from two adjacent sprinklers results in a uniform application. In general, the best superposition of the two patterns is thought to take place with head-to-head coverage, which means that sprinkler spray from one sprinkler barely reaches the next sprinkler (Fig. 13.1). Because of the decreasing application rate vs. distance, sprinklers should always be placed in corners and along the sides of turf areas.

Application rate (mm/hr) is calculated as a function of flow rate, angle of coverage, and sprinkler spacing. Equations for full, half, and quarter circle sprinklers spaced at head-to-head spacing are shown in Fig. 13.1.

The sprinkler pattern in Fig. 13.1 is called a square pattern. Figure 13.2 shows a triangular pattern, which generally has higher uniformity than a square pattern for the same lateral spacing. Arranging sprinklers in odd shaped areas is sometimes difficult, and perfection may be impossible. It is unadvisable to mix sprinkler types in order to water all parts of odd shaped areas; however, a property may require more than one type of sprinkler. Small turf areas are generally watered by fixed spray heads, and large turf areas are generally watered by rotors or impact sprinklers. Triangular patterns fit in large fields, but it is difficult to arrange sprinklers in triangular patterns in small square turf areas because of the difficulty of offsetting sprinklers at the ends of the area.

The triangular pattern consists of equilateral triangles with 60° angles. The tangent of 60° is 1.73, thus sprinklers are spaced along laterals 2/1.73 = 1.156 times farther apart than lateral row spacing. For example, if sprinkler spacing is 12.2 m, then laterals are spaced 10.56 m apart.

Sprinkler System Evaluation, Efficiency, and Uniformity

Application uniformity tends to decrease as sprinkler spacing increases beyond head to head coverage. A decrease in application uniformity can result in higher water costs over time if turf managers overirrigate in order to ensure that all parts of the turf area receive adequate water.

The economics of turf irrigation are different from that of agriculture because there is no yield. In general, turf is irrigated based on the principle that there is a minimum acceptable level of appearance. For example, an elite golf course would not allow any brown spots over the entire golf course. Thus, the golf course is over-irrigated on average in order to ensure that no part of the golf course is under-irrigated. One the other hand, a cash strapped school may irrigate just enough to keep all sections of the field barely alive. Regardless of the goal, the priority is to keep the mimimum water application depth above a threshold value.

P. Waller, M. Yitayew, *Irrigation and Drainage Engineering*, DOI 10.1007/978-3-319-05699-9_13

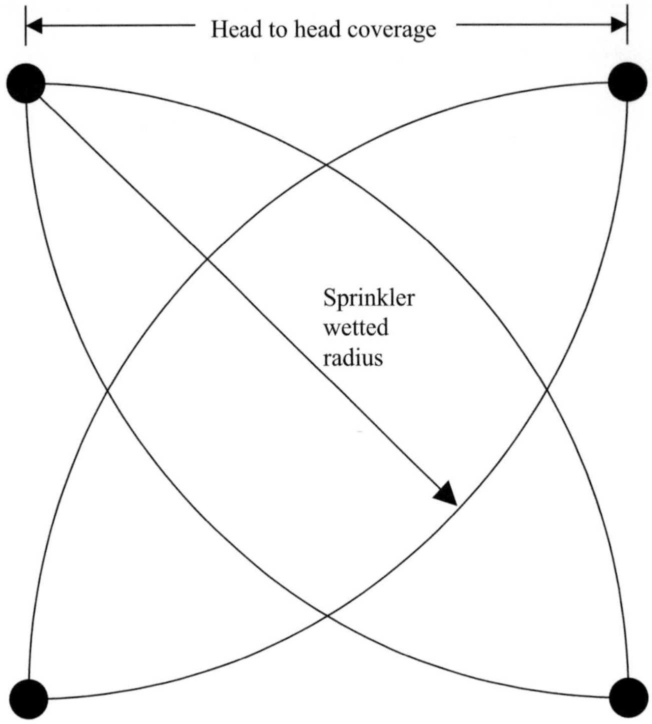

Full-circle sprinkler

$$in/hr = \frac{GPM * 96.3}{ft * ft}$$

$$mm/hr = \frac{LPM * 60}{m * m}$$

Half-circle sprinkler

$$in/hr = \frac{GPM * 193}{ft * ft}$$

$$mm/hr = \frac{LPM * 120}{m * m}$$

Quarter-circle sprinkler

$$in/hr = \frac{Q * 385}{ft * ft}$$

$$mm/hr = \frac{LPM * 240}{m * m}$$

Fig. 13.1 Sprinkler application rate equations

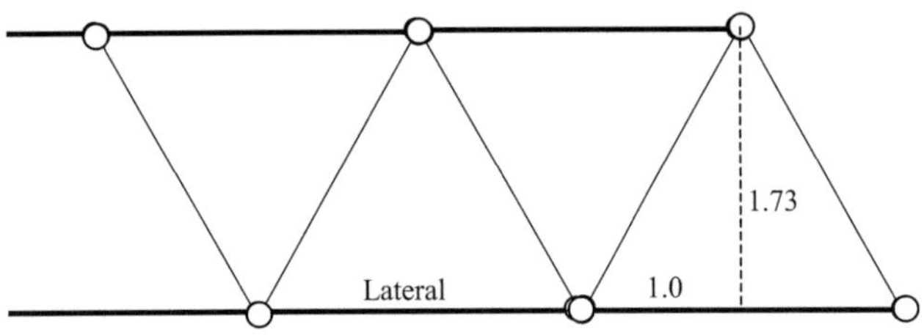

Fig. 13.2 Triangular sprinkler pattern

Based on the criterion of a minimum acceptable water depth for turf appearance or survival, the Scheduling Coefficient (SC) is a useful method of economic analysis of turf irrigation.

$$SC = \text{average depth applied/minimum depth applied.} \quad (13.1)$$

The scheduling coefficient thus calculates the fraction of wasted water. The *Chapter 13 Sprinkler Uniformity* program calculates the scheduling coefficient and the coefficient of uniformity for specified triangular or square spacings. The The first step is to input the application rate vs. distance curve for sprinklers. Distances and depths are entered into the Principal worksheet in columns G:K (Fig. 13.3).

The next step is to select the sprinkler of interest in the Main worksheet (Fig. 13.4). The final step is to input the spacing in the Triangle or Square worksheet (Fig. 13.5).

Example 13.1 Use the Sprinkler Uniformity program to calculate the CU and SC for the I-45 sprinkler for 10.5 m row spacing and 12.2 m head spacing for square and triangular pattern.

The square pattern is shown in the *Square* worksheet (Fig. 13.5). The left figure shows the application pattern for one sprinkler and the right figure shows the application pattern for square spacing. Although the CU is high, 91 %, the scheduling coefficient is also relatively high 1.32. This means that 32 % extra water is applied in order to maintain a minimum depth applied in the center of the four sprinklers. A high SC is common for square spacing because of the low application rate in the center.

The triangular CU (Fig. 13.6) is only slightly higher (92 %) than that for the square spacing (91 %); however, the SC (1.16) is much lower than the SC for square spacing. Thus, only 16 % of water is wasted for the triangular spacing, which is half of the wasted water for the square spacing.

The Economic summary worksheet finds the optimal spacing for the selected and spacing. Financial costs are entered in the worksheet. Scheduling coefficients and costs are calculated for a range of spacings in order to find the optimal design.

Example 13.2 Determine the optimal economic spacing and pattern for the I-41 sprinkler if trenching + pipe costs are $2.00 per m and each sprinkler + fittings costs $25.00

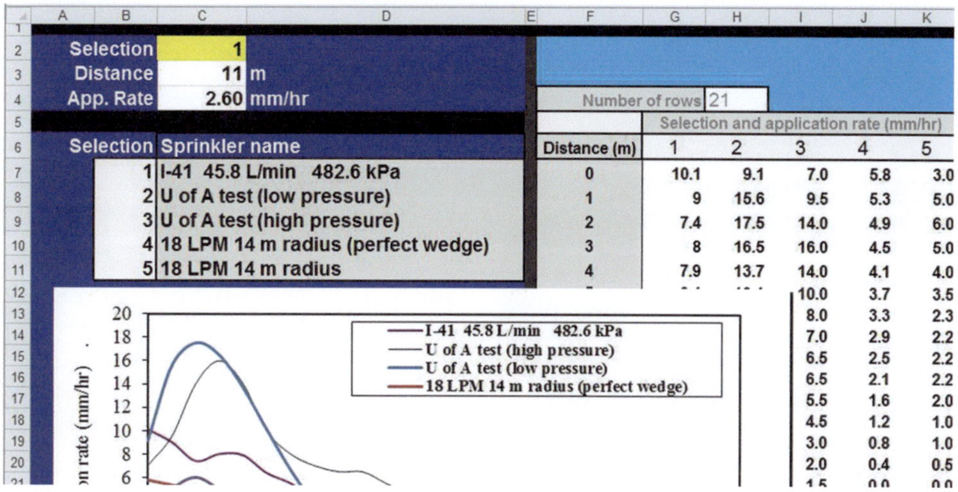

Fig. 13.3 Sprinkler depth versus distance data in *Principal* worksheet

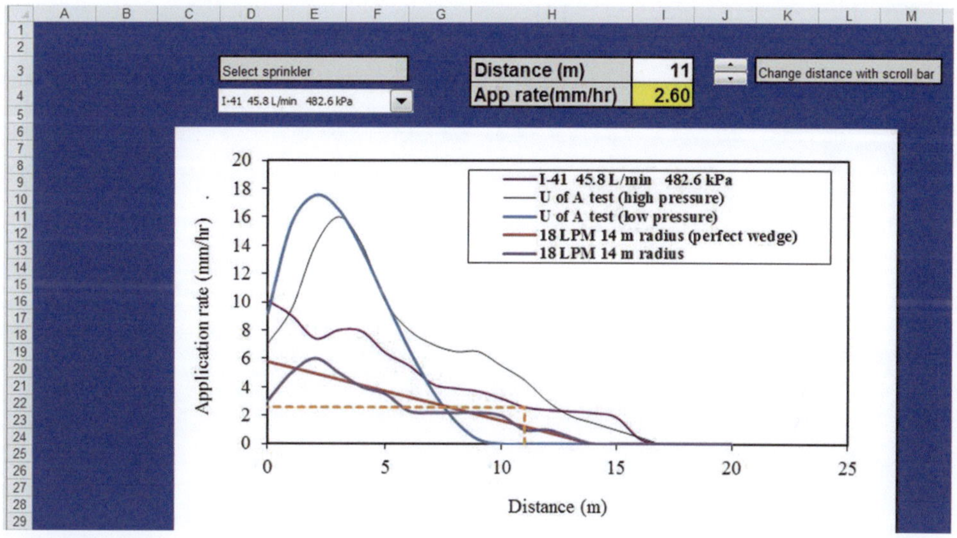

Fig. 13.4 Selection of sprinkler and distance in *Main* worksheet

per unit. The cost of irrigation water is \$0.10 per m^3. The required depth of water/year is 1.8 m (including evapotranspiration). Assume that square spacing uses equal head and row spacings and that triangular spacing uses equilateral triangles.

An equation can be established for the cost of pipe and sprinklers per ha. The length of pipe is area/row spacing.

$$\text{Pipe length/ha} = 10{,}000/s_d$$
$$\text{Number of sprinklers/ha} = 10{,}000/(s_l\ s_d)$$

where

s_l = sprinkler spacing along lateral, m
s_d = row spacing, distance between laterals, m.

The cost of materials and installation is calculated as follows:

$$\$/\text{ha} = 10{,}000(\$2/s_d + \$25/(s_l\ s_d))$$

The water cost is calculated as follows:

$$\$/\text{ha} = (1.8\ \text{m})(\text{SC})\left(10{,}000\ \text{m}^2/\text{ha}\right)\left(\$0.10/\text{m}^3\right) = 1{,}800\ \text{SC}$$

The desired spacings are entered in Column D and the economic parameters in Column B in the *Economic summary* worksheet (Fig. 13.7).

Click the *Run Economics* Button to run the VBA calculation program. Scheduling coefficients for a range of spacings

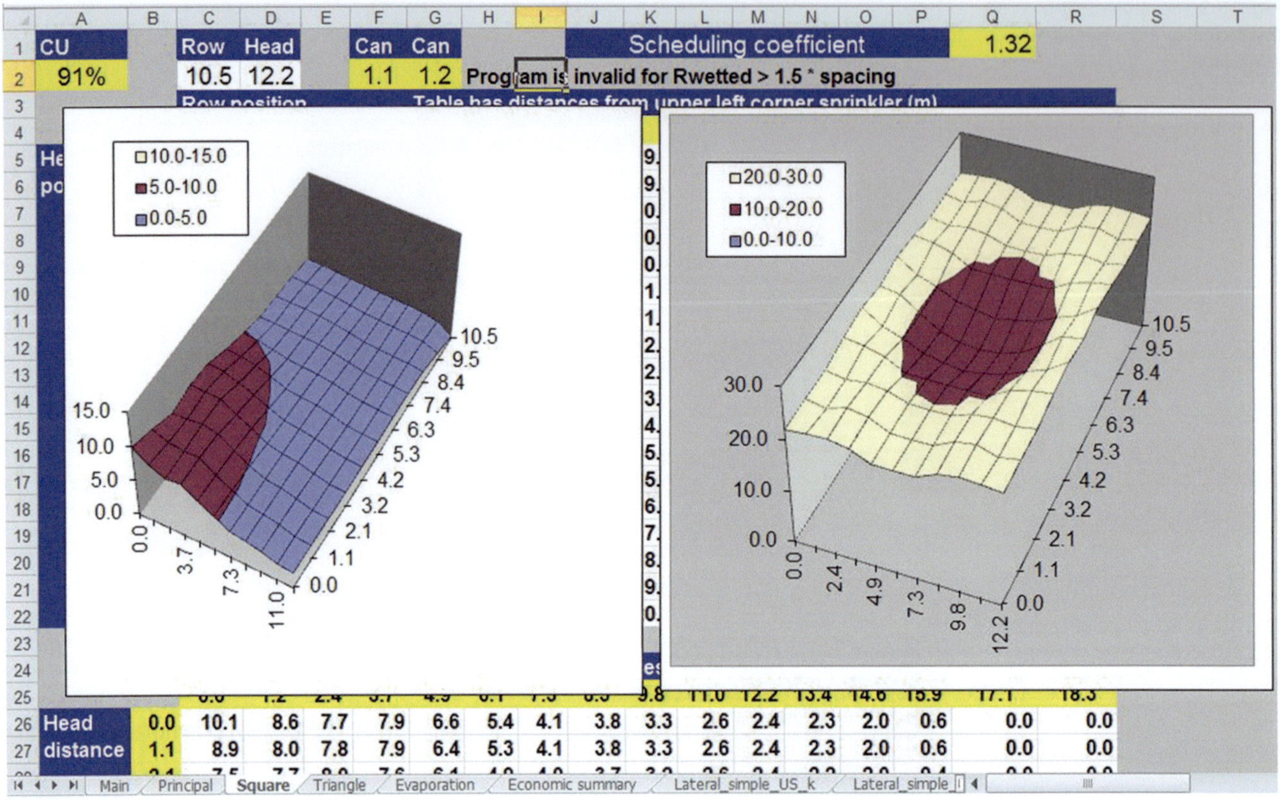

Fig. 13.5 Square spacing worksheet for *Sprinkler Uniformity* program: water distribution pattern for square spacing with Hunter I-41 at 10.5 m row spacing and 12.2 m head spacing

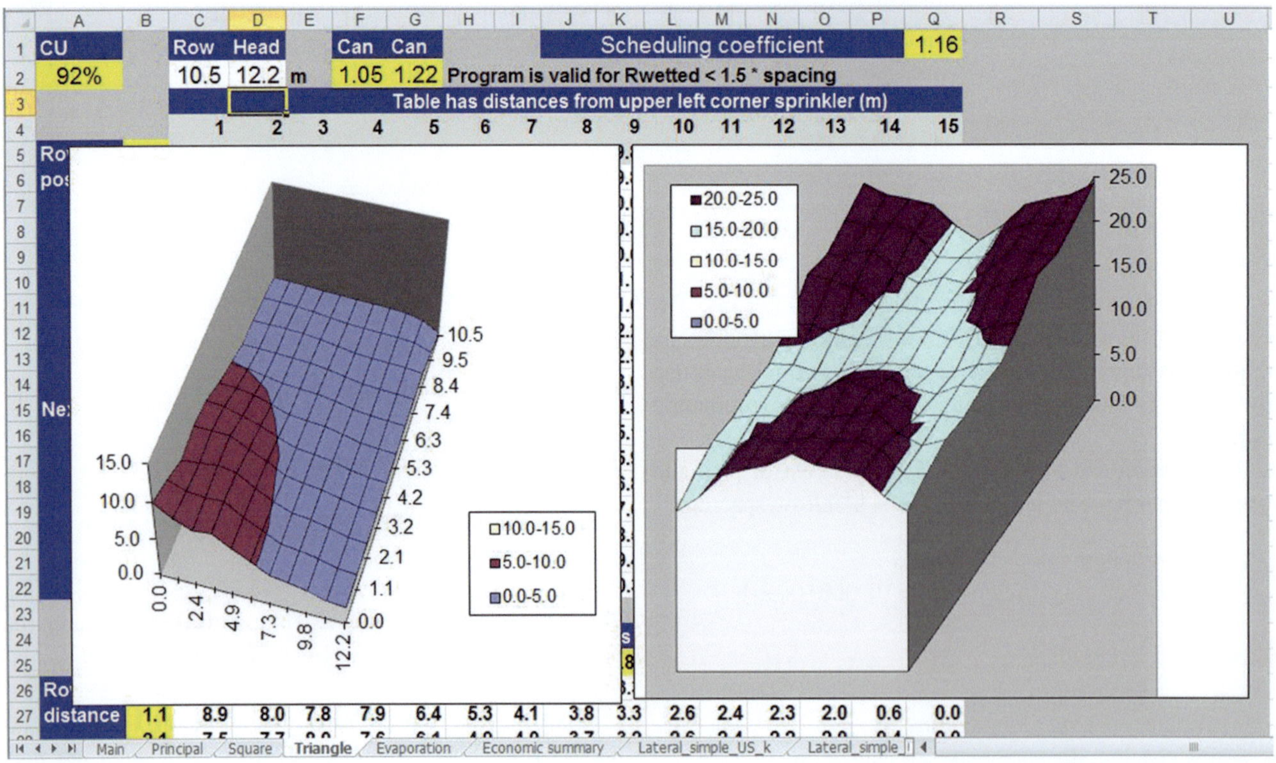

Fig. 13.6 Triangular spacing in *Triangle* worksheet for I-41 at 10.5 m row spacing and 12.2 m head spacing

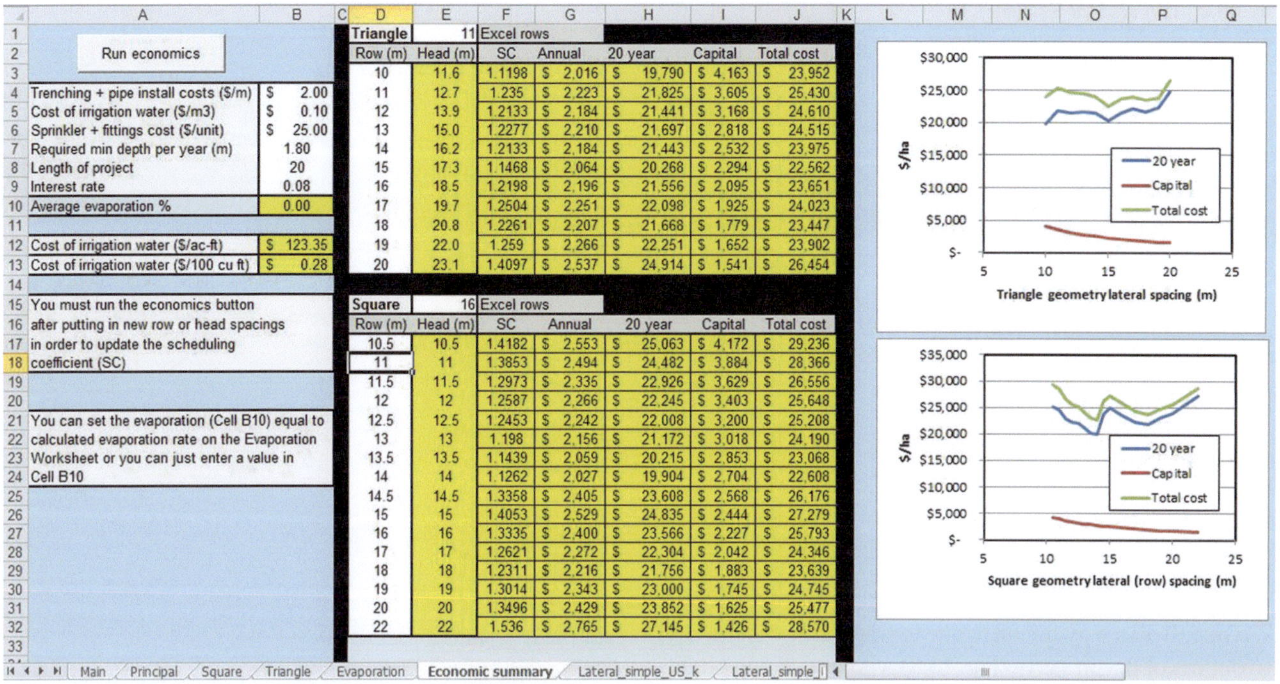

Fig. 13.7 *Economic summary* worksheet evaluation of I-41 sprinkler

are output in column F. The scheduling coefficient for triangular spacing increases between 19 m and 20 m. The square spacing sprinkler coefficient rises and falls and is less stable than the triangular spacing coefficient. The square spacing scheduling coefficient is more sensitive to minor variations in the one-dimensional application distribution. Water cost follows the same trend as the scheduling coefficient (Fig. 13.7).

The minimum present value for the triangle geometry, $22,562/ha, is at 15 × 17.3 m spacing. The minimum present value for the square geometry, $22,608/ha, is at 14 m spacing. Thus, both have comparable costs. The sprinkler used in this analysis has a wetted diameter slightly greater than 15 m. Thus, the recommended spacing is approximately equal to head to head spacing.

Once systems are installed, periodic uniformity tests should be conducted in order to evaluate system degradation. Catch cans should be placed between sprinklers and next to sprinklers. If a grid is used, then approximately 20 cans should be placed between four sprinklers in a square spacing or three sprinklers in a triangular spacing. The sprinkler system should run until at least 25 ml are collected by each can.

The *Sprinkler Uniformity* program does not consider wind. If an area is windy, then spacing is generally decreased (Table 13.1). It is best to run lateral pipes in a direction perpendicular to the general wind direction and to place sprinklers closer together along laterals. This helps avoid strips of dry ground, which are common in windy areas if sprinklers are widely spaced in the direction perpendicular to the wind direction. Triangular spacing works well in windy areas with pipes perpendicular to the wind because of better downwind overlap.

Table 13.1 Recommended sprinkler spacings for different wind speeds

Rectangular spacing			
Wind speed		Sprinkler spacing – percent of diameter	Lateral spacing – percent of diameter
(kph)	(mph)		
0–5	0–3	50	60
6–11	4–7	45	60
12–20	8–12	40	60
Square spacing			
0–5	0–3	55	55
6–11	4–7	50	50
12–20	8–12	45	45
Equilateral triangle spacing			
0–5	0–3	60	
6–11	4–7	55	
12–20	8–12	50	

System Design

There are four steps in the design of a turf sprinkler irrigation system. The first step is to perform a complete assessment of the site. Time of watering may be constrained by public use. Pressure should be measured or calculated at the expected

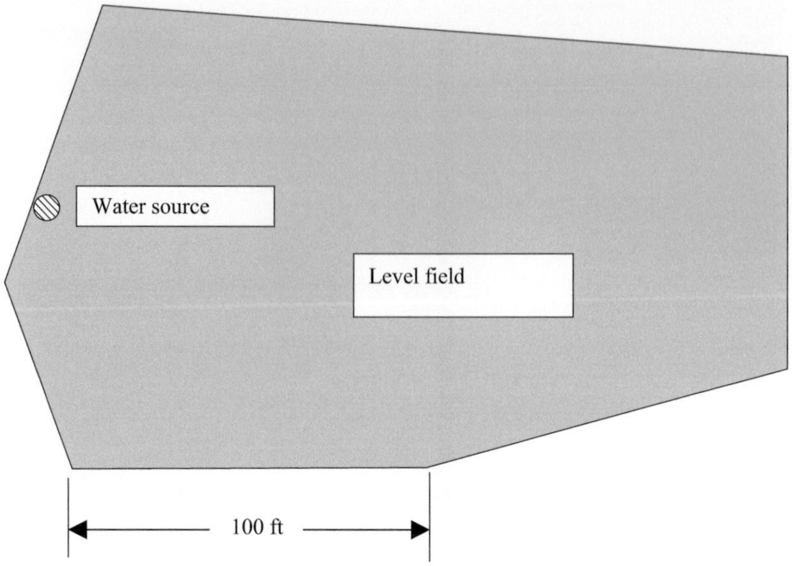

Fig. 13.8 Turf area for Example 13.3

flow rate. It also may fluctuate due to water usage in the city. An outline of the property (Fig. 13.8) with design constraints and the location of the water source is drawn. Once the map is completed, the next step is to draw the positions of all the sprinklers, which might be more art than science on an irregularly shaped property. A sprinkler should be selected that meets the spatial dimensions of the project and also is appropriate for the use; for example, a commercial sprinkler (vandal resistant) should be used on a commercial property. Then, the pipe system is laid out, grouping common types of sprinklers if possible, and keeping the total flow rate in any one zone below the maximum allowable flow rate. Pipe sizes are selected based on friction loss. The last step is to compile a report that lists the parts and pipe lengths and can be used by contractors or homeowners for cost estimation.

Site information
- Water source, ET, map, pressure, obstacles, soils, time of watering

Sprinkler selection and placement
- Spacing – flow rate

Pipe system design
- Zones, laterals, mains, valves, irrigation schedule

Preparation for bidding and installation
- Picking parts and preparing for contractor bids.

Example 13.3 Design a sprinkler system for the area shown in Fig. 13.8.

Site information for public site.

Maximum soil infiltration rate = 0.5 in/hr.
TAW = 2 inches.
Pressure at water source = 80 PSI.
Minimum sprinkler pressure = 50 PSI.

Table 13.2 Sprinkler parameters

Nozzle	Radius (ft)	gal/min	Half-circle application rate (in/hr)
8	51	7.4	0.55
6	49	5.5	0.44
4	41	3.7	0.21

Max ET_c (middle of summer) = 0.26 in/day.
Eight hours available for water application (between 10 p.m. and 6 a.m.).
Maximum system flow rate = 35 gal/min.
Average night atmospheric conditions: RH = 75 %, T = 18 °C, U = 8 km/h (2.22 m/sec).
Based on irregular geometry, assume that the scheduling coefficient is 1.5.

Sprinkler selection
Because wind speed is in the 6–11 km/hr range, choose the 50 % of diameter coverage (Table 13.1) for square spacing. The side lengths are in the range of 100 ft so select a sprinkler with approximately 50 foot wetted radius. Table 13.2 lists the flow rate and radius of an appropriate rotor sprinkler with number 4, 6, and 8 nozzles. The wetted radius is approximately 50 ft, which matches the requirements of the project. The application rates are in the range of the maximum application rate (application rate <0.5 in/hr) or less. This sprinkler is rated for commercial applications (vandal resistant with a steel cap), which is an important characteristic in public facilities: brilliant people like to do thing like shoot BB guns at the tops of sprinklers in order to turn the sprinklers into water fountains.

Sprinkler placement
The next step is to locate the sprinklers on the property (Fig. 13.9). Because the geometry of the site is irregular, not

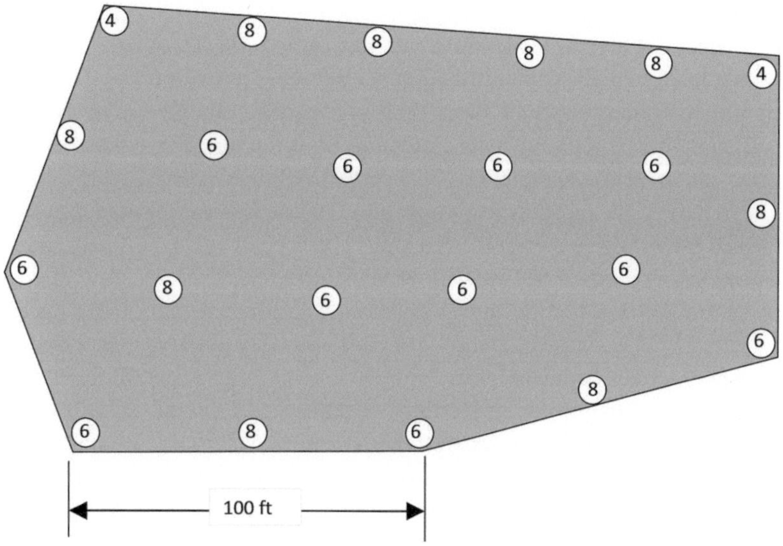

Fig. 13.9 Sprinkler locations for Example 13.3

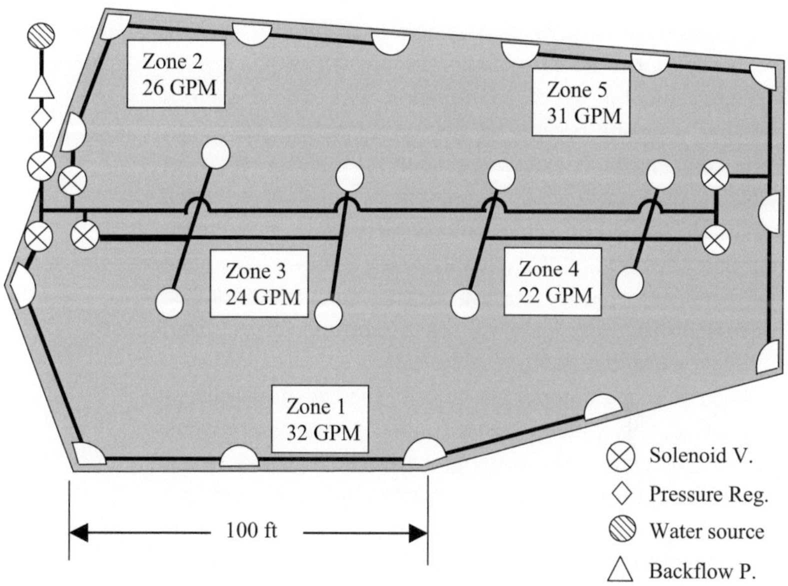

Fig. 13.10 Valves, pipes, and zones for Example 13.3

all sprinklers are spaced 50 ft apart, but they are placed as close to 50 ft intervals as possible with sprinklers in all corners.

Zones

With 35 gal/min, a maximum of five number 6 nozzles (5.5 gal/min) can be placed in each zone. A maximum of four sprinklers with number 8 nozzles (7.4 gal/min) can be placed in each zone.

The center full-circle sprinklers can be placed in two zones with four sprinklers in each zone (Fig. 13.10). Number 6 nozzles are selected for these zones; however, a number 8 nozzle is selected for the lower left sprinkler in zone 3 because it covers a larger area than the other sprinklers in zone 3.

Corner sprinklers (90+ degree arc) can be grouped in one zone or they can be incorporated into the side sprinkler groups. If corner sprinklers are grouped together, then they can use the same nozzle diameter (flow rate) as other sprinklers on the project because they are controlled by one valve and have a shorter watering time. However, extra pipe is needed to link all the corner sprinklers in one zone. In this case, the decision is made to group the corner sprinklers with the side sprinklers (180 degree arc), and to use number 4 nozzles in the corners (3.7 gal/min) and number 8 nozzles (7.4 gal/min) on the sides. The corner sprinklers with larger than 90 degree arcs are given number 6 nozzles (Fig. 13.9).

Once the sprinklers are selected and the zones are defined, then the pipe locations can be laid out. The solid lines represent sprinkler lateral lines. Even though valves are shown within the turf area, they are installed as close to the edge of the turf area as possible, or preferably outside the turf area.

Pipe system design
It is generally recommended that landscape irrigation pipes should be placed 30 cm below the ground surface. Sprinklers should be connected to the pipes with swing joints. People and equipment apply downward pressure on sprinklers; the force can break the lateral pipe if the sprinklers are installed directly over the pipe. Swing joints allow sprinklers to be pushed down without breaking the pipe. The second advantage of swing joints is that they allow adjustment of sprinkler elevation and location during installation. Complete swing joints can be purchased, or they can be made with Marlex (crystalline polypropylene) 90s.

The rule of thumb for agricultural systems is that lateral pressure loss for sprinkler systems should not exceed 20 %, resulting in 10 % flow variation. However, a more conservative approach may be appropriate for landscape irrigation systems where water costs are higher.

A backflow prevention device is placed after the water source in order to prevent the possible siphoning of chemicals from the irrigation system back into the potable water supply. In the case of this project, which has a level turf area and for which chemigation will not be used, a pressure vacuum breaker (PVB) is sufficient. However, more expensive backflow prevention devices such as reduced pressure backflow preventers or double check valves are needed on some projects, depending on topography and chemigation. City regulations should be consulted for specification of the correct backflow prevention device for each project. The 1½″ PVB selected for this project has a 5 PSI pressure loss at 30–35 gal/min.

A pressure regulator is placed after the backflow prevention device. City water pressure is not constant. If city water pressure fluctuates, then, without a pressure regulator, water application rate to the project is unknown. Although not required by law, a pressure regulator can save money by reducing water waste. On this project, a 2″ pressure regulator is selected with 6 PSI pressure loss at 30 gal/min.

Because pressure regulators have a significant pressure loss, it might not be possible to use pressure regulators on projects with marginal pressure; if sprinkler pressure drops below a threshold value (typically 40 PSI for conventional sprinklers), then a "doughnut" shaped water application pattern results with a dry spot near the sprinkler.

Solenoid valves are generally sized one size smaller than mainline pipe. Thus, 1½″ solenoid valves are selected for the 2″ mainline pipe. A 1½″ solenoid valve has a pressure loss of 3.6 PSI at 30 gal/min flow rate. An extra solenoid valve, called a master valve, is placed before the other solenoid valves on the mainline. The master valve is turned on when any of the other valves are activated. A master valve prevents water waste if one of the zone valves does not shut down properly or if the mainline is damaged and has a leak. In this example, estimate that the two solenoid valves (master and zone valve) have a pressure loss of 7.2 PSI prior to each zone.

The major design constraint for turf sprinkler systems is that the worst-case sprinkler must have adequate pressure; thus, if the worst-case zone has adequate pressure, then the rest of the sprinklers will have more than adequate pressure. On this project, Zone 1 or Zone 5 are the zones that are most likely to have the lowest pressure since they have the largest flow rate and the longest pipe networks (Fig. 13.10). Zone 5 has better hydraulic characteristics within the zone because the valve connection is in the middle of the zone; however, Zone 1 has a better position on the mainline because it is closer to the water source. In order to select the worst-case zone, evaluate the hydraulics of both zones. Sprinkler flows are added from the end to the beginning in order to find the flow rate in each pipe section. Pipe flows for zone 1 are shown in Fig. 13.11. For example, the second to last pipe supplies two sprinklers so its flow rate is 7.4 + 5.5 = 12.9 gal/min.

A sprinkler flow vs. pressure equation can be developed based on catalog flow rates and the assumption that flow vs. pressure exponent is 0.5. The coefficient k for the number 4, 6, and 8 nozzles is calculated as follows where the flow rates Q are the listed flows at 50 PSI pressure:

$$k8 = Q/Hx = 7.4/50^{\circ}.5 = 1.05$$
$$k6 = Q/Hx = 5.5/50^{\circ}.5 = 0.778$$
$$k4 = Q/Hx = 3.7/50^{\circ}.5 = 0.523$$

Pipe friction loss is calculated with the Hazen-Williams equation. On this project, use Class 200 pipe for 1″ and 1.25″, and Class 125 pipe for 1.5″ diameter and greater. Start with 1.25″ pipe in the last section in zone 1. The inside diameter of the 1.25″ Cl 200 pipe is 1.50″.

$$H_L = kL\left(\frac{\left(\frac{Q}{C}\right)^{1.852}}{D^{4.87}}\right) = 10.46(60)\left(\frac{\left(\frac{7.42}{140}\right)^{1.852}}{1.50^{4.87}}\right) = 0.375\ ft$$

where

H_L = head loss in pipe, ft
k = constant, 10.46
Q = flow rate, gal/min

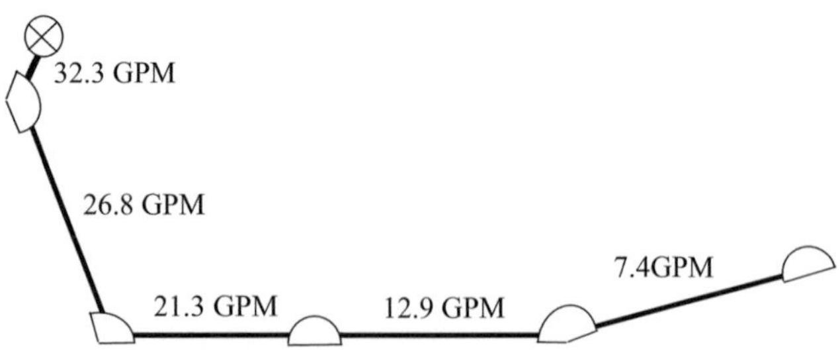

Fig. 13.11 Zone 1 pipe flow rates

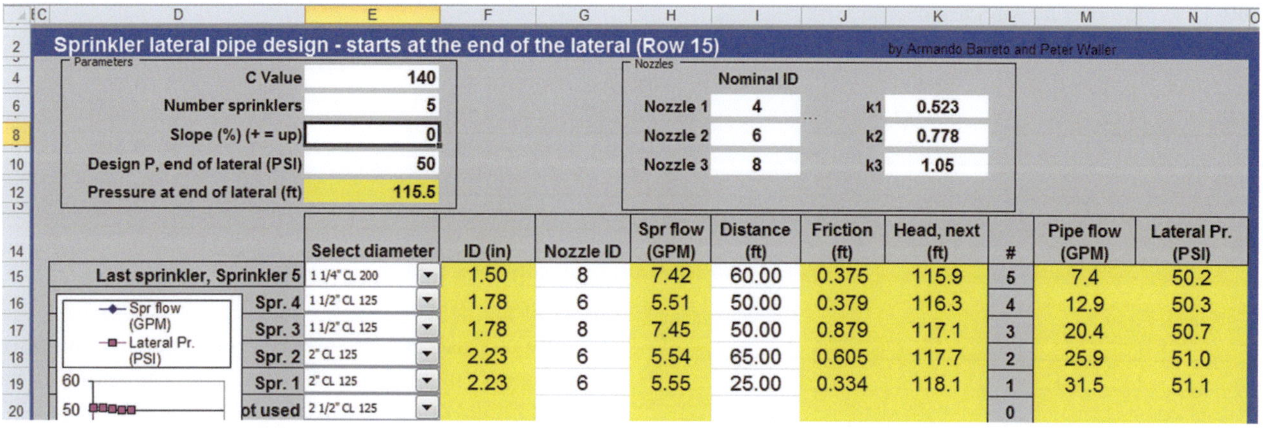

	Select diameter	ID (in)	Nozzle ID	Spr flow (GPM)	Distance (ft)	Friction (ft)	Head, next (ft)	#	Pipe flow (GPM)	Lateral Pr. (PSI)
Last sprinkler, Sprinkler 5	1 1/4" CL 200	1.50	8	7.42	60.00	0.375	115.9	5	7.4	50.2
Spr. 4	1 1/2" CL 125	1.78	6	5.51	50.00	0.379	116.3	4	12.9	50.3
Spr. 3	1 1/2" CL 125	1.78	8	7.45	50.00	0.879	117.1	3	20.4	50.7
Spr. 2	2" CL 125	2.23	6	5.54	65.00	0.605	117.7	2	25.9	51.0
Spr. 1	2" CL 125	2.23	6	5.55	25.00	0.334	118.1	1	31.5	51.1
pt used	2 1/2" CL 125							0		

Fig. 13.12 Pipe diameters and pressure losses in zone 1 in *Lateral_simple_US_k* Worksheet

C = pipe roughness coefficient, 140 for PVC pipe
D = inside diameter of pipe, inches
L = pipe length, ft.

The Sprinkler Uniformity program has pipe calculation worksheets. In this case, use the *Lateral_Simple_US_k* worksheet (Fig. 13.12). The pipe diameters in column F are linked to the selections in column E. The sprinkler flow rate coefficients (cells I6:J8) are linked to the Nozzle ID numbers in column G. With the selected pipe diameters, the total pressure loss is 1.1 PSI in the zone. Thus, the pipe pressure just downstream from the solenoid valve is 51.1 PSI.

The next step is to evaluate zone 5 (Fig. 13.13 and Table 13.3). Only the upper half of zone 5 is analyzed. The pressure drop is 0.7 PSI. The pressure loss in 200 ft of 2″ mainline pipe at a flow rate of 31.4 gal/min is 2.6 ft or 1.1 PSI. Thus, the total pipe pressure loss when zone 5 is running is 1.1 + 0.7 PSI = 1.8 PSI. This is greater than the 1.1 PSI drop in zone 1 so zone 5 is the worst-case zone.

The total pressure required for the project is calculated by summing the required sprinkler pressure plus all of the friction losses to the worst-case sprinkler, which is the upper sprinkler in zone 5. Four PSI is added as a safety factor to account for losses due to system degradation over time.

Pressure vacuum breaker	5 PSI
Pressure regulator	6 PSI
2 Solenoid valves	7.2 PSI
Mainline loss friction loss	1.1 PSI
Zone 5 friction loss	0.7 PSI
Safety factor	4 PSI
Sprinkler pressure	50 PSI
Total	74 PSI

The pressure available to the project is 80 PSI. Thus, the design is acceptable.

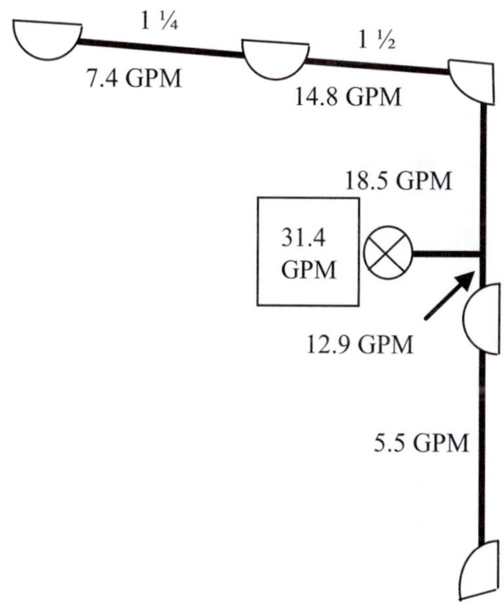

Fig. 13.13 Sprinkler flow rates for zone 5

Watering schedule

The calculation of the watering schedule begins with the evaporation calculation. The Evaporation worksheet (Fig. 13.14) uses the Frost and Schwalen algorithm (Eqs. 12.1 and 12.2) to estimate the percent evaporation during irrigation. Calculate the evaporation based on the average nozzle diameter (nozzle 6, 6/64″). The nighttime relative humidity, wind speed, and temperature are used for the calculation because the system operates at night. Thus, 11 % of gross application evaporates before it hits the soil.

With SC = 1.5, the daily depth of water required is

$$\text{Depth required} = ET_c/(1 - L_e)SC = 0.26/(1 - 0.11)1.50 = 0.44\ \text{in/day}$$

Table 13.3 Flow rates for upper half of zone 5

Pressure (PSI)	Nozzle number	Sprinkler flow (gal/min)	Pipe flow (gal/min)	Pipe ID Nominal (in)	Length (ft)	Pressure loss (ft)
50	8	7.400	7.400	1.5	60	0.376
50.16	8	7.412	14.812	1.78	50	0.492
50.38	4	3.712	18.524	1.78	50	0.745
50.70						

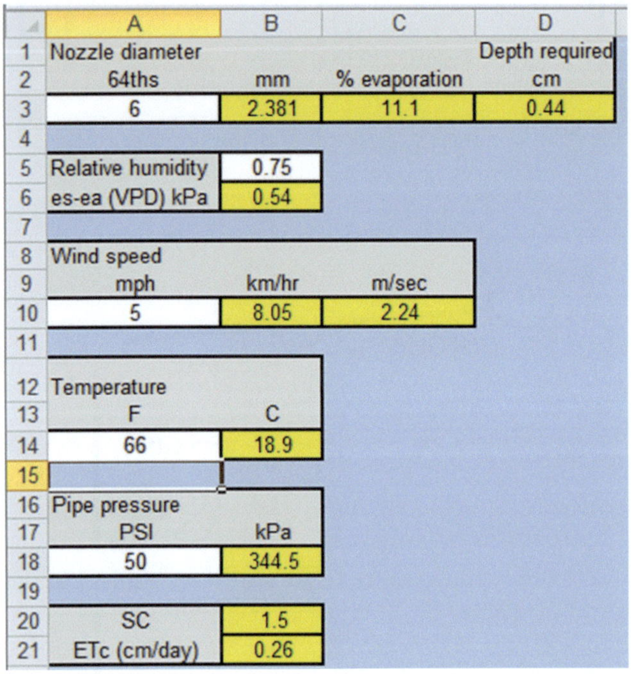

Fig. 13.14 *Evaporation* worksheet

Zones 3 and 4 have full circle sprinklers with number 6 nozzles. The full-circle application rate with number 6 nozzles is 0.22 in/h. Thus, zones 3 and 4 must operate for 2 hours/day. In general, zones 1, 2, and 5 have half-circle sprinklers with number 8 nozzles, with an application rate of 0.55 in/h. Thus, the operation time for these zones is

$$\text{Hrs/day} = (0.44\ \text{in/day})/(0.55\ \text{in/hr}) = 0.8\ \text{hrs/day}.$$

The total required watering time is (2 zones) (2 hrs/d) + (3 zones) (0.8 hrs/d) = 6.4 hours per day.

Thus, the area can be watered in the specified length of time (8 hours/day).

Example 13.4 An athletic field has a maximum summer ET rate of 0.24 in/day. The field dimensions are 240 ft by 312 ft. The water pressure is 90 PSI and maximum flow rate is 80 gal/min. The maximum infiltration rate is low, *0.22 in/hours*. The turf should be watered at night between 9 p.m. and 5 a.m. The reclaimed water source is at the southwest corner of the field

Large sprinklers are selected in order to minimize trenching and minimize application rate. The sprinkler has an 80 ft radius at 40 gpm. A 5 × 4 sprinkler grid (312 ft/4 = 78 ft and 240/3 = 80 ft) divides the field into approximately square spacing with 80 ft between sprinklers (Fig. 13.15). The sprinklers have a solenoid valve in the sprinkler so only one pipe is needed in each trench (Fig. 13.15). It is not possible to arrange the field with triangular spacing because of the small number of sprinklers and the square field dimensions.

Even with sprinklers operating alone (no adjacent sprinklers running), runoff will probably occur because of the low infiltration rate of the soil. In this case, a technique is used that is similar to the high velocity center pivots. Sprinklers are cycled on and off, which allows ponded surface water on the surface to soak into the soil before more water is applied. This process is called "cycle and soak."

If it is assumed that the scheduling coefficient is 1.3 and evaporation is 10 %, then the required depth of water per day is

$$(0.24\ \text{in/day})(1.3)/0.9 = 0.35\ \text{in/day}.$$

The application rate for the half circle sprinklers is 0.59 in/hr. Thus, the application time is

$$\textit{Calculated watering time} = \frac{\textit{ET requirement (in/day)}}{\textit{Application rate (in/hr)}} = \frac{0.35\ (in/day)}{0.59\ (in/hr)} = 0.6\ hr.$$

The full circle sprinklers require twice as much time, 1.2 hours, and the quarter circle sprinklers require half as much time, 0.3 hr.

There are three full circle zones, five half circle zones, and two quarter circle zones. The total watering time is

$$3(1.2) + 5(0.6) + 2(0.3) = 7.2\ \text{hours}$$

Thus, the area can be watered within the time constraint (8 hours).

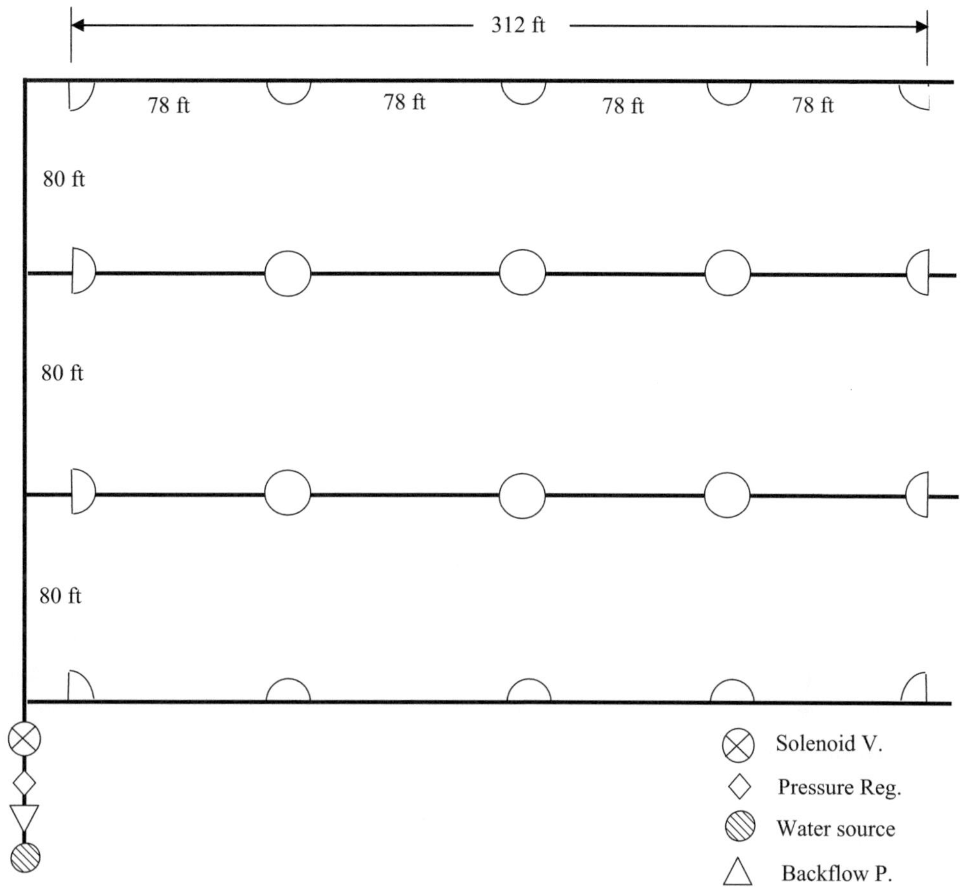

Fig. 13.15 Sprinkler locations for I-90, G900 series sprinklers

Pipes, up to the last pipe section, should be sized to carry the full 80 gal/min flow rate. Water source pressure is 90 PSI so 10 PSI can be allocated to pipe friction loss and valve losses.

The pipe friction loss is calculated for the sprinkler that is the maximum distance (worst-case) from the water source (upper right in Fig. 13.15). The distance from the water source to the second to the last sprinkler on the line is 480 ft. The distance to the last sprinkler is then 78 ft.

Flow	Distance	Diameter	Friction loss
80 gal/min	480 ft	3 inch (3.23 ID)	2.55 PSI
40 gal/min	78 ft	2 inch	0.64 PSI
Total			3.19 PSI

A backflow prevention device is not needed in this case for the reclaimed water system. If a master solenoid valve and zone solenoid valve are installed, then there is enough pressure to operate the system with a 4 PSI loss in a pressure regulating valve.

1 Solenoid valve	3 PSI
Pipe friction loss	3.2 PSI
Safety factor	4 PSI
Sprinkler pressure	80 PSI
Total	90 PSI

Small residential turf areas are irrigated with spray heads. In general, the flow meter is 5/8″ (16 mm), and the maximum flow rate is 12 gal/min (45 L/min). For these systems, use one pipe size (1″ (25 mm)), because the price of pipe and fittings is less than the cost of going to the store to purchase specific diameter parts.

Example 13.5 Irrigate a 30 ft × 30 ft turf area with spray heads.

The turf area shown in Fig. 13.16 is set up with 1 gal/min spray heads in the corners, 2 gal/min spray heads on the sides, and a 4 gal/min spray head in the middle. All of the pipes are 1″ (25 mm). The area must be broken into two zones because the upper zone has the maximum flow.

Two quarter-circle	1 + 1 gal/min
Three half-circle	2 + 2 + 2 gal/min
One full-circle	4 gal/min
Total flow rate	12 gal/min

The application rate is calculated based on the full-circle sprinkler flow rate (Fig. 13.16)

$$in/hr = \frac{(96.3)(gal/\min)}{(ft)(ft)} = \frac{96.3(4)}{(15)(15)} = 1.7\ in/hr$$

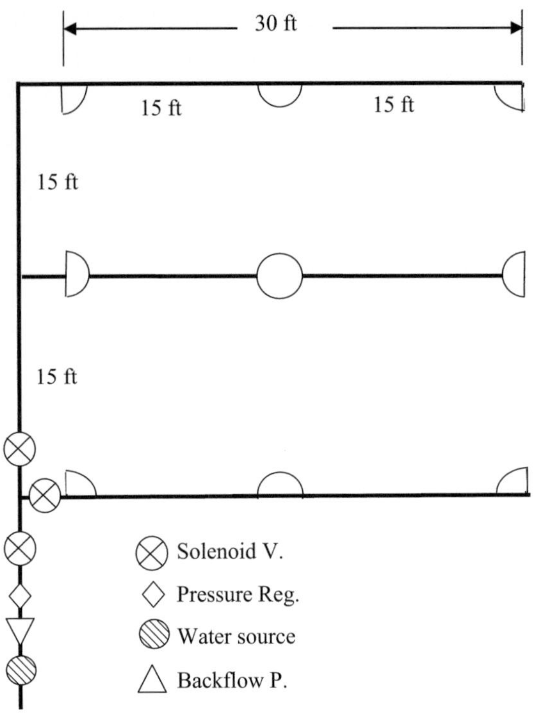

Fig. 13.16 Spray heads in turf area

Subsurface Drip Irrigation

Subsurface drip irrigation is an alternative to sprinkler irrigation that may be appropriate for odd-shaped turf areas, climates with high evaporation, and when sprinkler spray is not allowable. Drip irrigation tubing with internal emitters is laid approximately 0.15 m below the ground surface. Spacing is dependent on soil texture. The application rate is calculated as a function of emitter flow rate and emitter and line spacing. Equations are derived in Chap. 16.

Example 13.6 Design a subsurface drip system to irrigate a turf area with summer reference ET = 12 mm/day. Emitter spacing and tubing spacing are 0.3 m. Irrigation efficiency is 76 %. The AWC = 15 %. The crop coefficient, K_c, is 0.7. Soil is sandy loam. MAD is 0.6. Turf rooting depth is 0.25 m. Each emitter has an 8 L/h flow rate. The total irrigated area is 140 m^2.

The daily water volume applied per emitter is

$$\begin{aligned} V_{emitter} &= (A)(ET_0)(K_c)/Eff \\ &= (0.3\ m)(0.3\ m)(12\ mm/day)(0.7)/0.76 \\ &= 1L/day \end{aligned}$$

The soil water holding capacity for each emitter is

SWHC = (A)(Z)(AWC)(MAD) = (0.3 m) (0.3 m) (0.25 m) (0.15) (0.6) = 0.002 m^3 = 2 L

Thus, there should be a maximum of 2 days between irrigations. The emitter flow rate is 8 L/h so the daily watering time should be 0.25 hours or 15 minutes every other day in order to apply an average of 1 L/d/emitter.

Questions

1. Describe the major categories of turf sprinklers
2. Which turf sprinklers have the highest application rate?
3. What is the maximum precipitation rate in mm/hr for a coarse sandy loam with a 6 % slope?
4. Discuss the reason for placing sprinklers in corners of turf areas.
5. Describe head to head coverage.
6. Why must sprinklers with different application rates be placed on different valves?
7. What are the advantages of swing joints?
8. What is the application rate for a 1.8 m spray SRS spray head with half circle coverage? Assume 1.8 × 1.8 m spacing. Comment on whether this is a low, average, or high application rate.
9. What is the application rate for a full circle 5 GPM (19 LPM) sprinkler on 50 ft × 50 ft (15.2 m × 15.2 m) spacing? Calculate for U.S. (Imperial) and metric units.
10. What is the application rate for the same sprinkler with half circle coverage?
11. Lay out sprinklers in zones in a 45 ft × 45 ft (13.7 m × 13.7 m) turf area with the SRS spray heads with 4.6 m radius (Table 13.2). City water pressure is 40 PSI (276 kPa) and maximum flow rate for the system is 12 GPM (45.4 LPM). Divide the turf area into zones so that the maximum flow rate is not exceeded. You do not need to show pipes in the drawing. Reference ET rate is 12 mm/day and the crop coefficient for turf is 0.7. Scheduling coefficient is 1.3 and there is 7 % evaporation from sprinkler droplets. Determine the watering time per day.
12. A pipe system has 10 sprinklers that are 15 m apart and each sprinkler has a flow rate of 20 L/min. The field is level. Select appropriate pipe sizes for each section so that total pressure loss in the system is no more than 5 PSI (3.5 m). Show a hand calculation for two sections but you can use the *Lateral simple metric* Worksheet for the entire lateral design. Repeat for a 2 % uphill slope (if possible) and a 2 % downhill slope with the Simple Lateral Worksheet.
13. What is the scheduling coefficient for the U of A high pressure sprinkler in the *Chapter 13 Sprinkler Uniformity* program on the following grids? Discuss the reasons for the highest SC with the last spacing.
   14 m × 14 m square.
   13 m × 15 m triangle

16 × 16 m square
15 × 17.3 triangle
20 × 20 square
19 × 22 triangle

14. Use the *Economic summary* worksheet for the U of A high pressure sprinkler. Cost of each sprinkler head installation is \$30. Cost of trenching and pipe is \$5.00/m. Life of the project is 10 years. Interest rate is 9 %. Required water depth/year is 2 m. Evapotranspiration rate is 25 %. Cost of water is \$0.50/m^3. Determine the lowest present value for the square or triangular spacing. Which is lowest?
15. Redo Example 13.3. In this case, the city can provide 70 gal/min (fewer zones are required). Set up a spreadsheet to calculate lateral losses. Leave sprinklers in the same positions.
16. Lay out a sprinkler network for an athletic field, 120 m east × 60 m north. The potable water source is in the Northwest corner, and is 700 kPa. Maximum flow rate is 400 LPM. Maximum summer ET rate is 10 mm/day. Infiltration rate is 10 mm/hr. The turf should be watered at night between 9 p.m. and 5 a.m.

14 Agricultural Sprinkler Irrigation

Agricultural sprinkler systems include wheel-lines, hand-lines, undertree sprinklers, and microsprinklers. The design process includes selecting an appropriate sprinkler and sprinkler spacing, calculating the number and dimensions of zones, designing the pipe network, and selecting the pump. Constraints such as orchard tree spacing and premanufactured aluminum pipe lengths often limit the spacing options. Sources of nonuniformity in agricultural sprinkler systems include variability of application rates within a sprinkler wetted area and hydraulic variation of lateral pressure. Although not normally part of the design process, this chapter shows how to describe the variability of wheel-line application depth based on pressure variation and sprinkler application pattern. The instructor may decide not to include these details. An economic/environmental model is presented that optimizes seasonal application depth with respect to yield, energy and water cost, and environmental contamination.

If possible, laterals should be oriented in the downhill direction, even if that means bringing a mainline up to the top of the hill Pipe sizes are selected with the goal of maintaining uniform discharge pressure and flow rate throughout the sprinkler system. In this case, the slope of the field is an advantage, and pipes are selected such that friction energy loss and elevation pressure gain offset each other. In this way, uniform pressure is maintained along the pipe. For this reason, it is generally worthwhile, even if the water source is at the bottom of the field, to pipe the water to the top of the field and then run laterals and/or submains in the downhill direction.

Water hammer is not a hazard for sprinkler system laterals and submains because every lateral/sprinkler acts as an air vent/pressure relief valve. Thus, the 1.5 m/sec rule may be relaxed in order to meet the more important goal of maintaining uniform pressure along the laterals. This does not apply to the mainline because it is blocked from the sprinklers by a valve.

Because flow decreases with distance, sprinkler laterals generally have smaller pipes at the end and larger pipes near the inlet. The process of designing the laterals begins at the last sprinkler. The minimum acceptable pressure is assigned to the last sprinkler on the lateral. Then, the pressure loss or gain between the last and previous sprinkler is calculated. Next, the flow at the previous sprinkler is calculated and the process continues. The pressure at the previous sprinkler is calculated with the following equation.

$$H_{n-1} = H_n + H_L + (s_L)(S/100) \tag{14.1}$$

where

H_n = pressure at the nth sprinkler, m
H_{n-1} = pressure at previous sprinkler, m
H_L = friction loss in pipe, m
s_L = spacing between sprinklers along lateral, m
S = slope, negative for inlet higher than distal end, m/m.

Wheel-Lines

Wheel-lines (also known as wheelmove, sideroll, or lateral-roll) are generally used to irrigate pastures or hay in regions where farms are not large enough for center pivot irrigation systems. Many of these farms are in the northwest United States. For example, 336,000 out of 1.3 million irrigated acres in Utah are irrigated with wheel-lines (Hill 2000). Wheel lines are often used on 40-acre fields (¼ mile by ¼ mile, 1,300 m × 1,300 m), but they are also used on smaller fields.

Wheel-lines are classified as move-and-set or periodic-move irrigation systems. The farmer moves the lateral once or twice a day and allows the system to irrigate for 12 or 24 hours before it is moved to the next position. Wheel-lines have a labor advantage over hand-lines because the

P. Waller, M. Yitayew, *Irrigation and Drainage Engineering*, DOI 10.1007/978-3-319-05699-9_14

Fig. 14.1 Wheel-line (Credit NRCS)

wheel-line movement is powered by a small (~2 HP) air-cooled engine (Fig. 14.1). The engine turns the entire pipeline and wheels. The farmer disconnects the wheel-line from the mainline hydrant, turns on the motor to move the wheel-line, and then reconnects the wheel-line to the mainline at the next hydrant. This process takes 20 to 30 minutes. Wheel-lines movement requires much less labor than hand-lines where each pipe is moved by hand to the next position. However, moving hand-lines is a good means of character formation for teenagers who must get up each morning before school and carry hand-lines through a wet field for a few hours.

Hydrants are attached to a mainline that runs along one side of the field (Fig. 14.2). The mainline for wheel-line systems is generally buried PVC plastic pipe with hydrants on vertical pipes sticking out of the ground. However, the mainline can also be constructed from temporary aluminum pipe sections that lay on the ground surface. The mainline is often placed along a fence line for protection. In Fig. 14.3, a mainline runs down the center of two fields (along the fence line) with wheel-lines or hand-lines running in fields along either side.

Handlines are often used to germinate crops. In this case, hand-lines are connected to all hydrants, and water is applied to the entire field several times per day (Fig. 14.4).

Hand-lines can be rotated around a mainline (Fig. 14.2) such that the driest area is always irrigated next. However, wheel-lines cannot be moved from one side of the mainline to the other. Thus, when wheel-lines reach the end of the field, they are moved back to the other end of the field and started over in the same direction. Sometimes farmers choose to irrigate in both directions but this may lead to

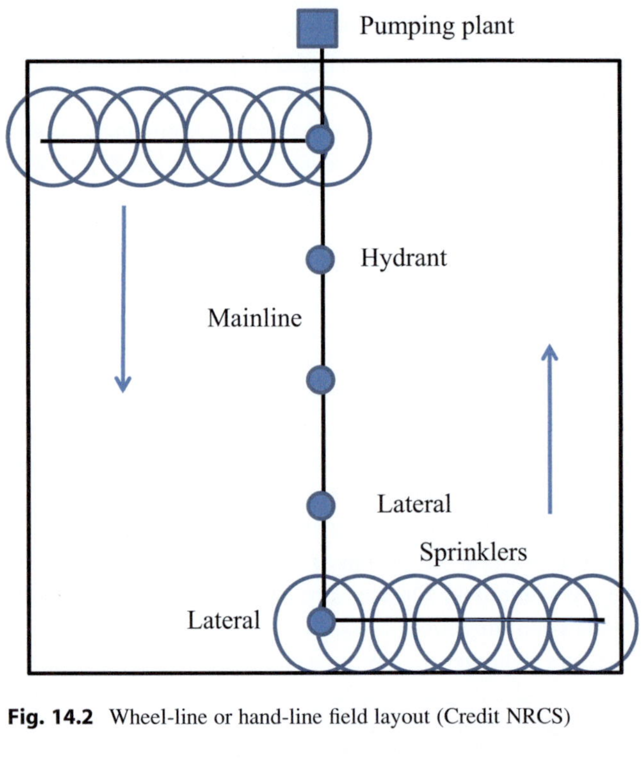

Fig. 14.2 Wheel-line or hand-line field layout (Credit NRCS)

Fig. 14.3 Wheel-line or hand-line hydrant (Credit NRCS)

Fig. 14.4 Handlines used to germinate lettuce (Credit NRCS)

excessive leaching if the soil profile is already filled by the recent. An assessment of the water-holding capacity of the soil can be made to determine if this is an acceptable practice. One technique used to lessen the effect of moving back to the beginning of the field is to skip every other hydrant in each direction (Hill 2000).

Where wheel-lines and hand-lines are popular the northwest United States, the typical evaporation rate is 0.3 inches (8 mm) per day for pasture or alfalfa irrigation (Fig. 14.2). Wheel-line sprinklers are generally brass impact sprinklers on self-leveling risers (Fig. 14.2) with a 5/32″ (3.97 mm) nozzles operating at 50 PSI (345 kPa). They typically have 40 ft (12.2 m) spacing between sprinklers on the lateral and 60 ft (18.3 m) spacing between hydrants on the mainline. The wetted radius of a 5/32″ nozzle at 50 PSI (345 kPa) is 45 ft (13.7 m). The flow rate can be calculated with Eq. 12.22

$$Q = 0.0666\, D^2H^{0.5}C_d = 0.0666\; 3.97^2\; 345^{0.5}(0.97) = 18.9\ \text{L/min} = 5.0\ \text{GPM}.$$

In order to prevent water runoff, the steady-state infiltration rate must not be exceeded by the application rate. Wheel-lines can be modeled as a line source irrigation system. If the sprinklers have a wedge-shaped application pattern and there is head-to-head spacing, then the wheel-line as a whole has a wedge shaped application pattern with a peak over the wheel-line. One can calculate the maximum application rate with an analytical equation. For a wedge-shaped application pattern, the application rate at any distance from the sprinkler is calculated by interpolation where di/dt_{max} is the peak application rate at the sprinkler.

$$\left(\frac{di}{dt}\right) = \left(\frac{di}{dt}\right)_{max}\left(\frac{r_{max}-r}{r_{max}}\right) \tag{14.2}$$

where

r_{max} = maximum wetted radius, m
r = distance from sprinkler, m.

The sprinkler flow rate, Q, is equal to the summation of the product of all application rates and incremental areas.

$$\begin{aligned} Q = \sum\left(\frac{di}{dt}\right)_n A_n = \int_0^{2\pi}\int_0^{rmax}\left(\frac{di}{dt}\right)_{max}\left(\frac{r_{max}-r}{r_{max}}\right) \\ r\ dr\ d\theta = \frac{\pi}{3}\,r_{max}^2\left(\frac{di}{dt}\right)_{max} \\ \left(\frac{di}{dt}\right)_{max} = \frac{Q}{\frac{\pi}{3}\,r_{max}^2} \end{aligned} \tag{14.3}$$

Thus, for a 5/32″ nozzle with perfect wedge-shaped application pattern, the maximum application rate at the sprinkler is

$$\begin{aligned}\left(\frac{di}{dt}\right)_{max} &= \frac{1.135\ m^3/hr}{\frac{\pi}{3}\,(13.7)^2} = 0.00576\ m/hr \\ &= 5.76\ mm/hr\end{aligned}$$

The application rate vs. distance from the sprinkler for a wedge-shaped application pattern is the straight line shown in Fig. 14.5. Actual sprinklers do not have a perfect wedge pattern, but something closer to the other pattern shown in Fig. 14.5.

The line source application rate is also a function of the sprinkler overlap. For a wedge shaped application pattern, the application rate at the sprinkler is the sum of the application rate of the sprinkler and the sprinklers on both sides.

$$\begin{aligned}\left(\frac{di}{dt}\right) &= \left(\frac{di}{dt}\right)_{max} + 2\left(\frac{di}{dt}\right)_{max}\left(\frac{r_{max}-s_L}{r_{max}}\right) \\ &= \left(\frac{di}{dt}\right)_{max}\left(1+2\left(\frac{r_{max}-s_L}{r_{max}}\right)\right)\end{aligned} \tag{14.4}$$

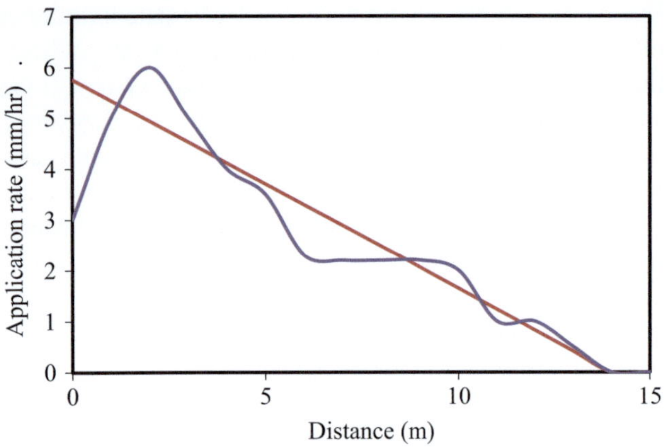

Fig. 14.5 Perfect wedge-shaped pattern and a typical sprinkler distribution pattern

For example, the sprinklers in this example have a 13.7 m application radius, but the sprinklers are only spaced 12.2 m apart. The application rate at the sprinkler is calculated as follows.

$$\left(\frac{di}{dt}\right) = \left(\frac{di}{dt}\right)_{max}\left(1 + 2\left(\frac{r_{max} - s_L}{r_{max}}\right)\right)$$
$$= 5.76\left(1 + 2\left(\frac{13.7 - 12}{13.7}\right)\right) = 7.2\ mm/hr$$

Application patterns are shown in Fig. 14.6 (from *Sprinkler Uniformity* program), with a peak application rate at the wheel-line and lower application rate between wheel-line positions. For a typical sprinkler pattern such as than shown in Fig. 14.5, with a dip near the sprinkler, some overlap is preferable because it fills in the dip near the sprinkler. This is seen in Fig. 14.6 where the application pattern directly over the wheel-line transect for the typical sprinkler (right side) is actually quite uniform.

With the normal wheel-line spacing and sprinkler nozzle, as long as the soil steady-state infiltration rate is greater than 7.2 mm/hr (0.3 in/hr), then the 5.0 GPM nozzle with wedge-shaped application pattern at 40 ft spacing is acceptable. This would correspond with the 0.3 intake family soil. It is not appropriate to rely on surface storage with wheel-lines because wheel-lines irrigate in the same position for 24 hours. In addition, many wheel-lines are placed in hilly fields so runoff is more likely than in a flat field. In Fig. 14.6, both patterns have a low application rate in the middle between laterals. The CUs for the two sprinkler patterns at 40 ft by 60 ft spacing (Fig. 14.6) are in the range of 80 %. Windy conditions would decrease the CU because droplets would not reach the midpoint between wheel-line positions. Studies at Utah State showed an average 62 % uniformity on 40 ft by 60 ft spacing. For this reason, wheel-line positions are staggered such that the wheel-line is offset to one side of the hydrant during one cycle and to the other side of the hydrant on a next cycle through the field (Fig. 14.7). This results in a 40 ft by 30 ft spacing and much higher CU (Fig. 14.8). Studies at Utah State showed 87 % uniformity for the staggered spacing.

Wheel-lines are constructed from 4″ (3.856 in or 97.9 mm ID) or 5″ (4.844 in or 123.0 mm ID) thick-wall (H-26) aluminum irrigation pipe. Five-inch pipe may be required for the first sections of long wheel-lines or on wheel-lines with high flow rate sprinklers in order to reduce friction loss and maintain pressure uniformity along the wheel-line. The larger pipe can provide higher torque for moving wheel-lines up hills.

Example 14.1 Design (select pipe size) a ¼-mile longwheel-line that has a 1 % downward slope from the inlet to the distal end. Use 5/32″ nozzles with sprinklers on 40 ft spacing. Use sample calculations for the first few sprinklers. Subsequently, evaluate the sprinkler application variability.

Derive an equation for sprinkler flow rate vs. pressure.

$$Q = 0.0666\ D^2H^{0.5}C_d = 0.0666\left(3.97^2\right)\left(H^{0.5}\right)(0.97).$$
$$Q(L/min) = 1.018\ H^{0.5}(kPa)$$

Try the smaller diameter, 4″ (97.9 mm ID) pipe, in order to burn pressure to compensate for energy gained by elevation loss.

Set the last sprinkler at the minimum design pressure of 345 kPa (50 PSI) and calculate flow rate.

$$Q = 1.018(345)^{0.5} = 18.9\ L/min$$

Calculate pressure loss in the pipe between the last sprinkler and the next to last sprinkler. The Hazen-Williams C-value for aluminum irrigation pipe is 130.

$$H_L = 1.22{*}10^{10}{*}12.2\ m\left(\left(\frac{18.9/60}{130}\right)^{1.852}/97.9^{4.87}\right)$$
$$= 0.00043\ m$$

Calculate pressure at the next to last sprinkler with Eq. 14.1.

$$345\ kPa/9.8 = 35.2\ m$$
$$H_{n-1} = H_n + H_L + (s_L)(S/100)$$
$$= 35.2 + 0.00043 + 12.2(-0.01)$$
$$= 35.08\,m = 343.8\,kPa$$

Calculate flow at the next to last sprinkler

$$Q = 1.018(343.8)^{0.5} = 18.88\ L/min$$

Flow in the pipe leading to the next to last sprinkler is

$$Q_{pipe} = Q_{n-1} + Q_n = 18.88 + 18.9 = 37.8\ L/min.$$

This process continues until the pressures in all sprinklers are calculated (Fig. 14.9). The calculations are made in the *Wheel-line* worksheet (Fig. 14.10).

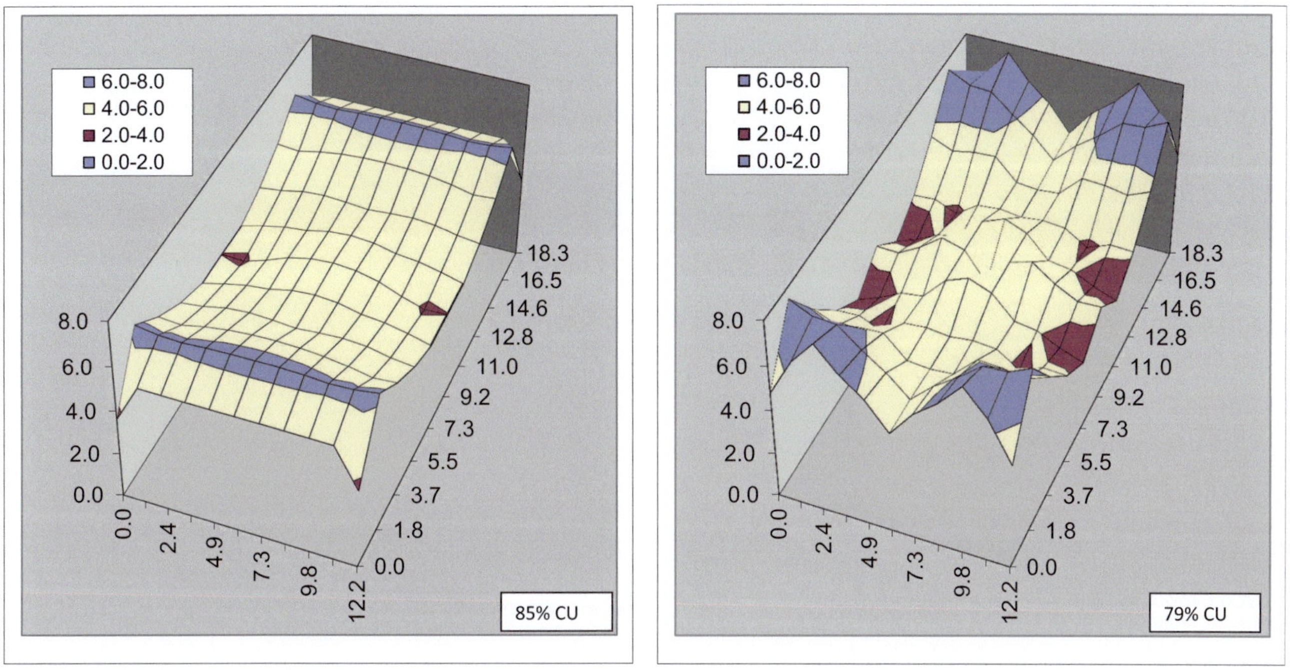

Fig. 14.6 Uniformities for 5/32″ nozzle on 40 ft × 60 f. spacing (12.2 × 18.3) with perfect wedge shape application pattern (*left*) and typical application pattern (*right*)

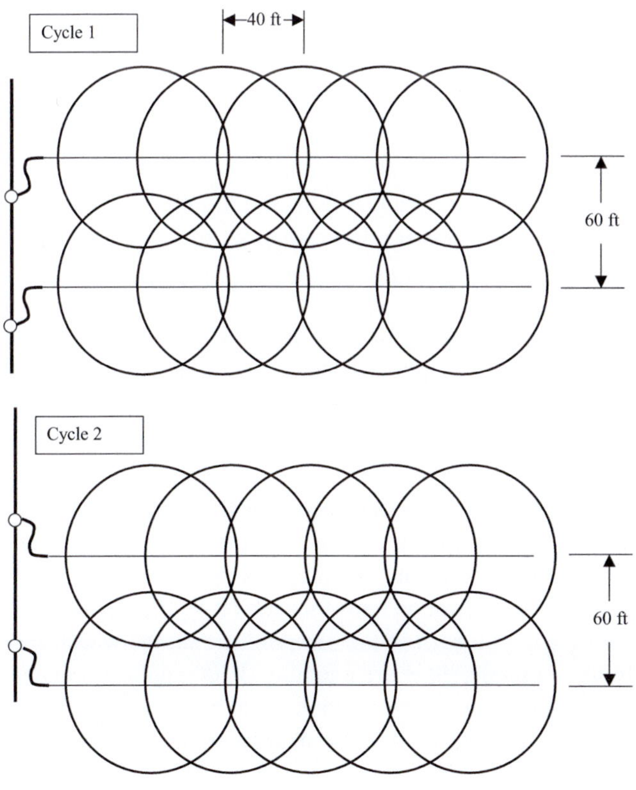

Fig. 14.7 Staggered wheel-line sets

Because of the downhill slope on the wheel-line, the maximum percent difference in flow is less than 3 % (Fig. 14.10). The energy loss due to friction becomes greater than the energy gain due to elevation over the first 120 m of the lateral. However, it is not worth switching to larger pipe in order to reduce friction loss because the pressure at the end of the pipe is greater than the inlet pressure with the 4″ (97.9 mm) pipe.

Unlike the simulation in Chap. 7, sprinkler application rates are generally not normally distributed. It is preferable to calculate the application distribution of a sprinkler zone by using the actual sprinkler application distribution (Fig. 14.8) and the hydraulic variation in sprinkler pressure (Fig. 14.9) or flow rate. In order to do this, relative application rates at all sections of the zone are calculated. In this case, the minimum sprinkler flow rate (18.47 LPM at sprinkler 11 in Fig. 14.10) is designated as the reference flow rate, 1.0 (cell J29 in Fig. 14.10). The field position (Row 0, column 6 in the middle of the right side of Fig. 14.6) minimum application rate is designated as the reference application rate, 1.0. Thus, the minimum application rate in the field is found at sprinkler 11, row 0, column 6. The application rate for any other position is the product of the ratio of sprinkler flow rate to minimum sprinkler flow rate and the ratio of application rate to minimum application rate.

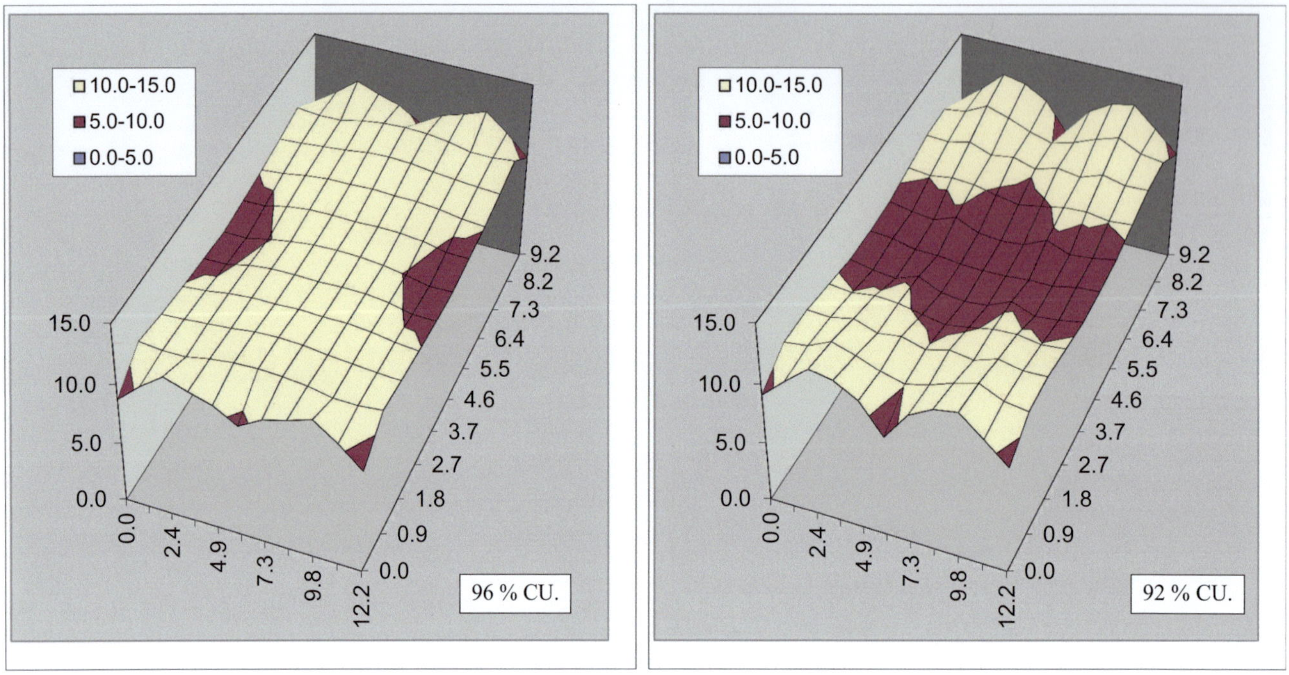

Fig. 14.8 Uniformities for 5/32″ nozzle on 40 ft × 30 f. spacing with perfect wedge shape application pattern (*left*) and typical application pattern (*right*)

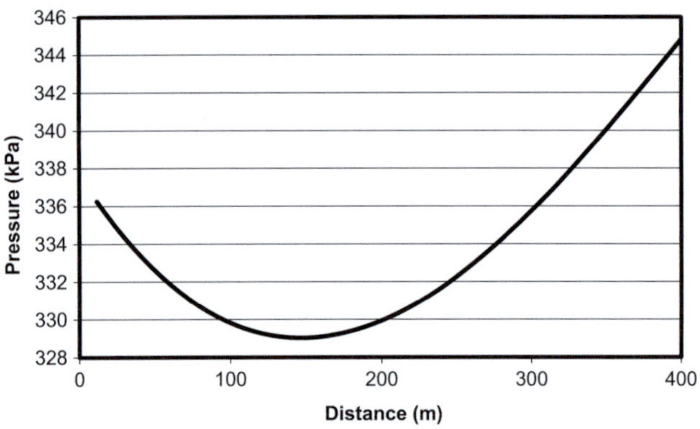

Fig. 14.9 Sprinkler pressures for quarter mile wheel-line with 5/32″ nozzles on 1 % downward slope with 4″ inch pipe

$$Rel_{j,k} = \left(\frac{Q_j}{Q_{\min}}\right)\left(\frac{\frac{di}{dtk}}{\frac{di}{dt\min}}\right) \tag{14.5}$$

The next step is to find the seasonal application depths at all positions in the zone ($AW_{j,k}$). The average gross application depth ($AW_{ave\text{-}g}$) is the total flow to the zone (cell I5 in Fig. 14.10) divided by the field area and multiplied by the seasonal irrigation time. The average net application depth is the average gross application depth minus evaporated water. The average relative application rate (Rel_{ave}) is found by summing all relative application rates ($Rel_{j,k}$) and dividing by the number of field positions. The application rate at any field position is then found with the following equation.

$$AW_{j,k} = \frac{Rel_{j,k}}{Rel_{ave}}\left(AW_{ave-g}\right)\ (1 - L_e) \tag{14.6}$$

where

$AW_{j,k}$ = net depth of water application at position j, k, cm
$\overline{Q}_{\min}$ = minimum sprinkler flow rate, L/min
$\overline{Q}_j$ = sprinkler flow rate at position j, L/h
AW_{ave-g} = average gross application depth, cm
j = sprinkler position
k = position in sprinkler area grid (Fig. 14.7)
$Rel_{j,k}$ = relative depth of application, $(di/dt)_{j,k} / (di/dt)_{min}$.
Rel_{ave} = average of relative depths of application
L_e = fraction evaporation

The program calculates the yield and environmental costs are calculated at each field position based on the applied water at each position. The program begins by calculating the relative application rates for each 10 × 10 sprinkler grid associated with each of the 33 sprinklers on the wheel-line. In this example, there are a total of 3,300 application rates. For example, the depth of application at the 11th sprinkler (cell J30 in Fig. 14.10), 2.4 m Row, 0 m Head location (Fig. 14.3) is 1.47.

$$Rel_{11,(2..4,0)} = \left(\frac{Q_{11}}{Q_{\min}}\right)\left(\frac{\frac{di}{dt(2.4,0)}}{\frac{di}{dt\min}}\right) = (1.0153)\ (1.45)$$
$$= 1.47$$

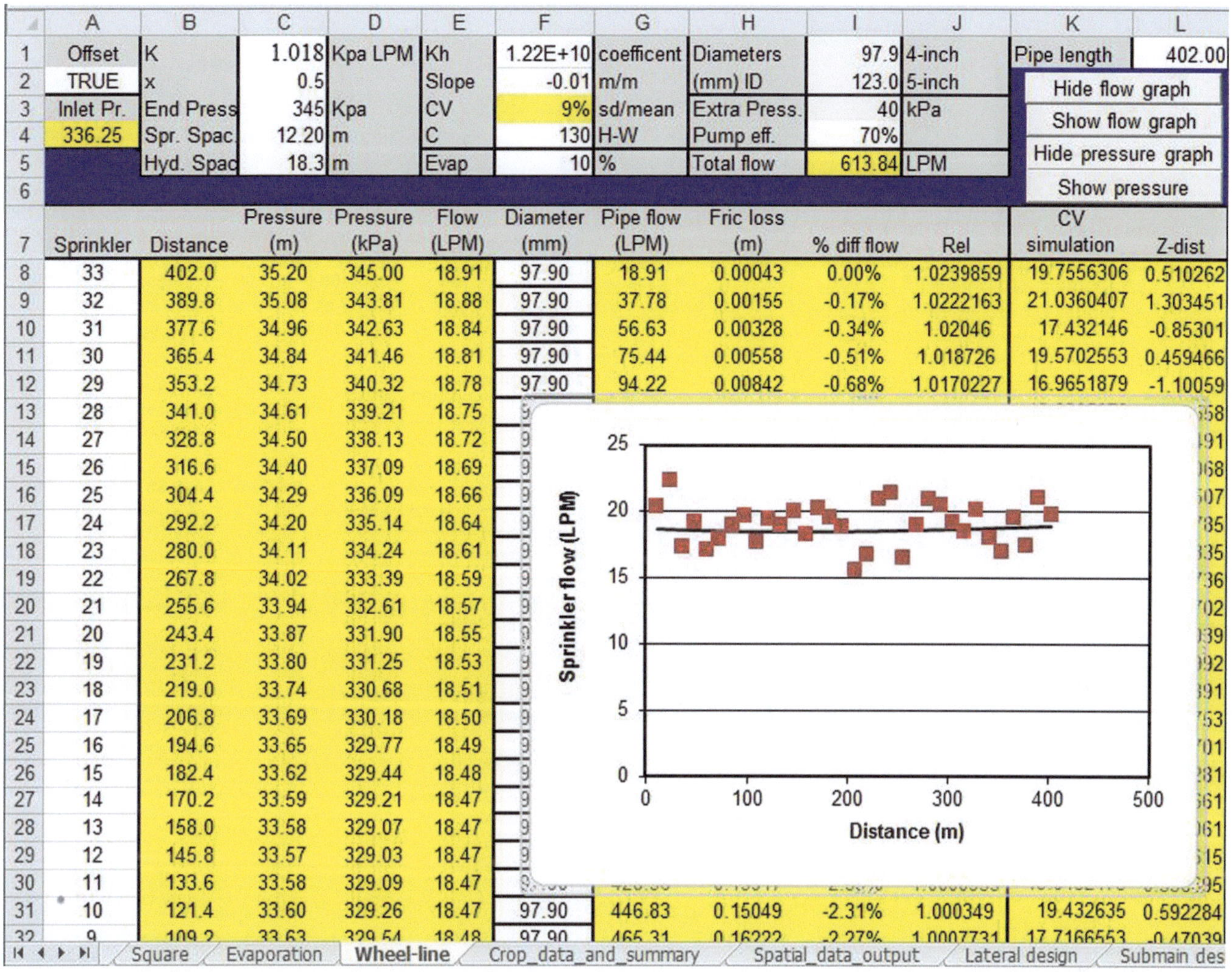

Fig. 14.10 *Wheel-line* worksheet calculations of sprinkler pressures and flows for quarter mile wheel-line with 5/32″ nozzles on 1 % downward slope with 4″ pipe

Example 14.2 Perform an economic analysis with the distribution of application rates predicted with Eqs. 14.5 and 14.6 in order to find the optimal seasonal depth of application for a ¼-mile long wheel-line with 60 ft × 40 ft spacing in a level field. Use the wheel-line parameters in Fig. 14.10. Crop and cost parameters are in white cells in Fig. 14.11. Calculate the cost of energy. The cost of energy is $0.10/kW-hr.

The crop is alfalfa hay. Up to 1 m applied water depth, the yield is 1 metric ton per 12 cm applied depth of water (coefficient = 0.083), which is a linear crop water production function (Sammis 1981). Water applied in excess of 1 m (100 cm in cell G4) is leached. Irrigation water is low in salinity so salinity is not a factor. The value of the alfalfa is $100.00/metric ton.

When the *Wheel line* Button in the *Crop_data_and_summary* worksheet is clicked, the program outputs results to rows 10–17. The average gross depth of application at which maximum profit occurs, 100 cm, is in written by the program to cell K6. Net depth is written to cell K5.

The program is activated by clicking the *Wheel line* button in the *Crop_data_and summary* worksheet (Fig. 14.11). The profits for the range of specified depths in cells C3:C4 in *Crop_data_and summary* worksheet (Fig. 14.11) are output to rows 10–17. The number of columns and rows in the "Square" worksheet is entered into cells C6 and C7. For energy calculations, the seasonal irrigation time is the average application rate divided by the zone flow rate (cell I5 in Fig. 14.10). The program calculates the overall profit (Fig. 14.11, rows 16–17) for a range of average seasonal application rates. The calculated application depths are written to the *Spatial_data_output* worksheet (Fig. 14.12).

The cost of energy per ha is calculated based on the required pressure and volume applied (see Chap. 2). The required inlet pressure is 336 kPa (cell A4 in Fig. 14.10).

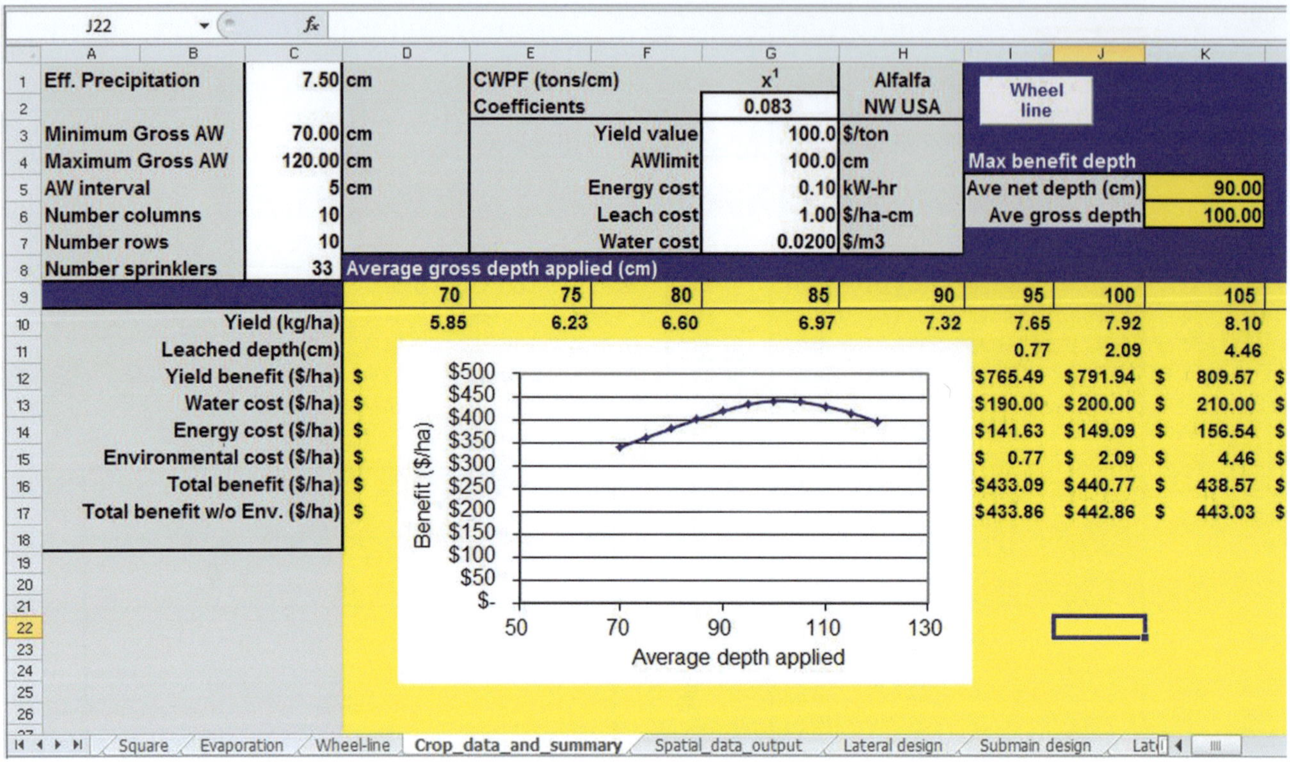

Fig. 14.11 *Crop_data_and_summary* worksheet calculations of costs and profits

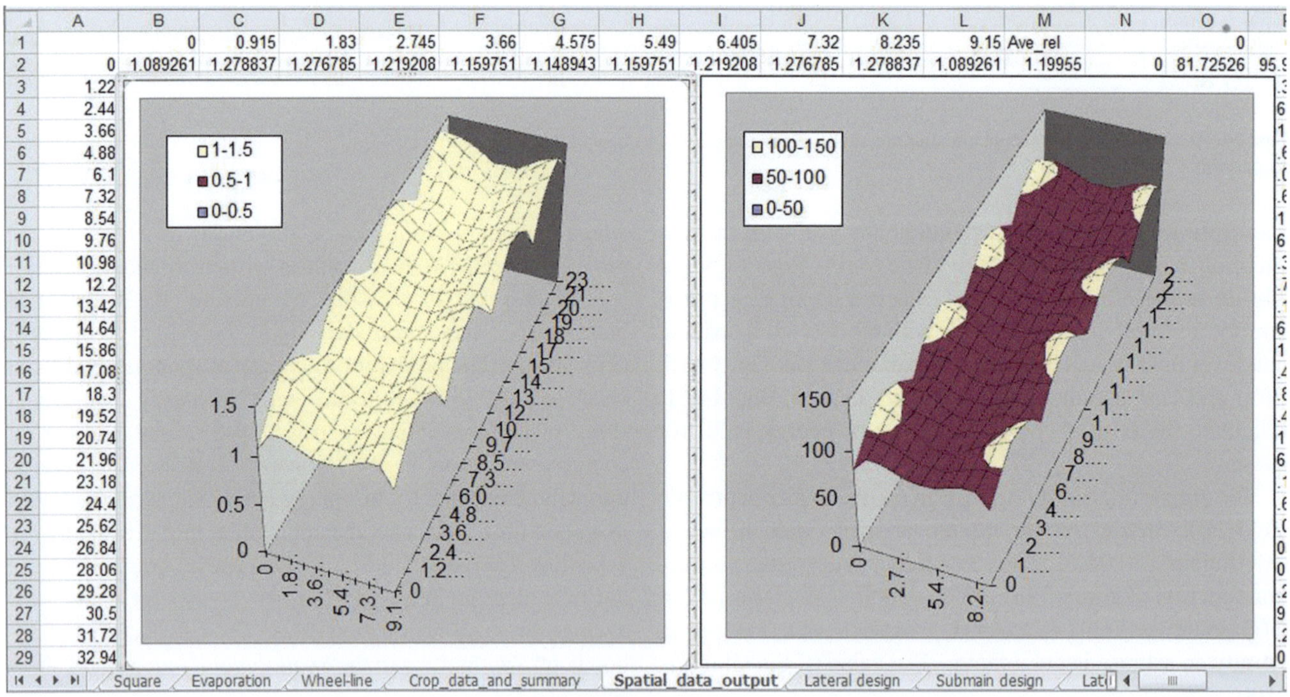

Fig. 14.12 Relative application rates and net application rates along wheel-line

Add the 40 kPa required for pump and fittings and the required pump pressure is 376 kPa (38.3 m). Pump efficiency is 70 %

$$E(kW - hr/ha) = \frac{0.0272\ (i_{mm})\ (h)}{Eff}$$
$$= \frac{0.0272\ (1050\,mm)\ (38.3\,m)}{0.7}$$
$$= 1{,}563\ kW - hr/ha$$

Thus, energy cost is $156.30/ha.

The design procedure for hand-lines is the same as for wheel-lines. Hand-lines are generally 30 ft (9 m) sections of 3″ (74.0 mm ID) thin-wall aluminum pipe, with 40 ft (12 m) spacing between hydrants on the main line. Aluminum hand lines are also sold as 20 ft sections of 2″ (48.6 mm ID) pipe. The C value is 90. They generally use a 1/8″ straight-bore nozzle and operate between 40 PSI (276 kPa) and 50 PSI (345 kPa)

Orchard Irrigation with Undertree Sprinklers

Most orchards are irrigated with undertree impacts, rotors, or microsprinklers. Overtree sprinklers on tall risers have fallen out of favor because wetted canopies tend to have disease problems. The fact that trees have large rooted areas reduces the importance of uniform application, as long as each tree receives an equal volume of water. Tree roots are generally deep so deep waterings allow a long period between irrigation events. The difficulties associated with orchard irrigation are that tree trunks block the spray, and that tree spacing constrains sprinkler spacing. Another constraint is that the direction of tractor travel through the orchard may make it necessary to slot sprinklers between trees (Fig. 14.13).

The first question that should be asked during the design process is which direction the farmer farms (drives the tractor). The sprinklers cannot be placed in the path of tractor travel or the tractor will run them over. However, due to slope or other factors, the laterals may need to run in the direction of tractor travel. If so, then the sprinklers must be slotted between the trees as shown in Fig. 14.13. This involves extra work during installation since small trenches must be dug between the location of each sprinkler and the lateral.

Orchards are often constructed on the sides of hills because hills are warmer than low areas: cold air settles in low areas. A cold front in early spring, particularly during bud break, can kill some or all of the buds and eliminate the crop, and the temperature difference between hills and valleys can prevent crop loss. As a result, many orchards are constructed on mild to steep slopes. Except for the steepest slopes, laterals should be run downhill, if possible, in order to maintain uniform pressure. As with other systems, the pipes are sized so that the energy lost due to friction is equal to the energy gained due to elevation. If the laterals run perpendicular to the direction of tractor travel, then sprinklers are placed between trees as shown in Fig. 14.14. An extra sprinkler is generally added to the windward side of the lower line because trees at the windy edge of the orchard tend to dry out.

Diagonal laterals may be convenient for some sprinkler patterns (Fig. 14.15).

Microsprinklers are generally attached to drip tubing that runs along the tree rows in the direction of tractor travel (Fig. 14.16). A Microsprinkler is generally placed near each tree.

One of the original purposes of overtree irrigation, in addition to providing water to plants, was frost protection in early spring. During cold nights, especially during bud break when the new buds are most susceptible to damage by cold, the sprinklers apply water to the plants and the process of water freezing on the plants releases heat and prevents frost damage to the buds. In the northwest United States,

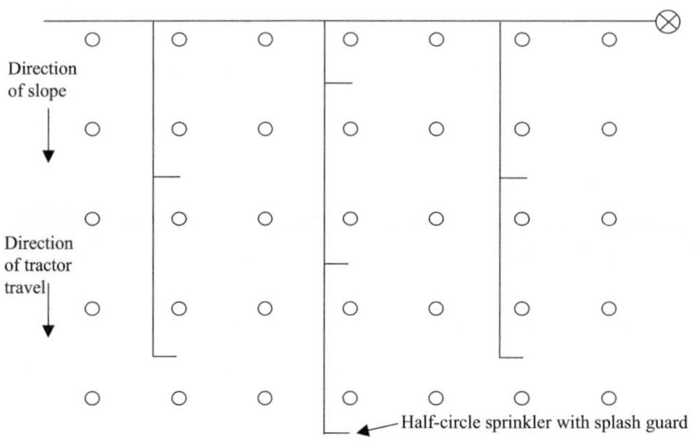

Fig. 14.13 Undertree sprinkler laterals with sprinklers slotted between trees

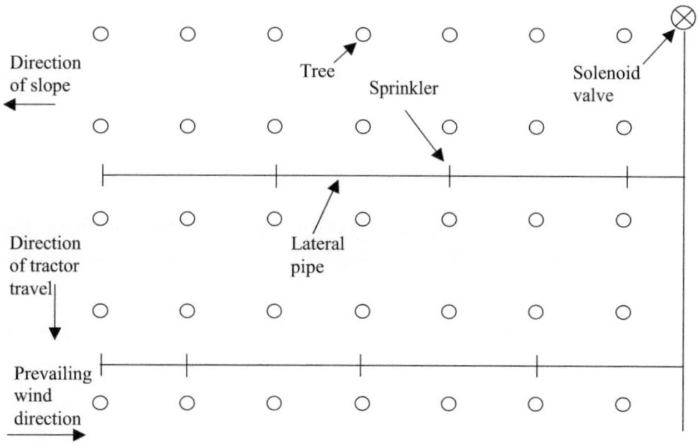

Fig. 14.14 Undertree sprinkler laterals perpendicular to direction of tractor travel

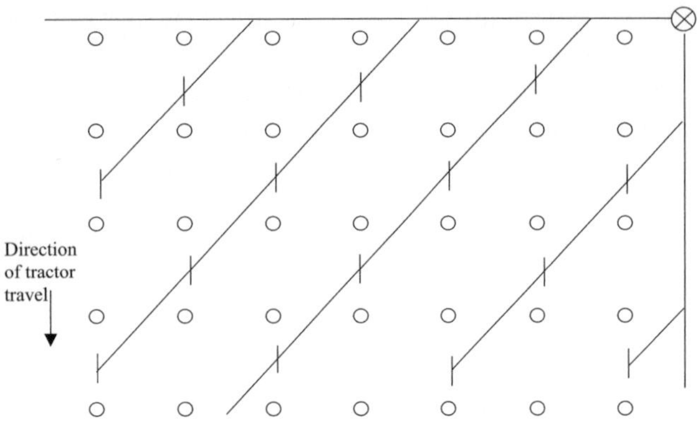

Fig. 14.15 Undertree sprinkler laterals in diagonal direction

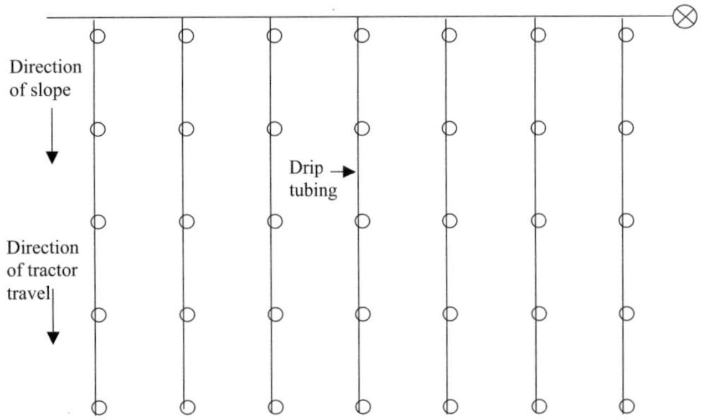

Fig. 14.16 Microsprinkler laterals (drip tubing): microsprinkler placed near each tree

where overtree irrigation was a common practice in apple orchards, a virus in the water damaged the crop when the water was applied directly to the plant. As a result, undertree irrigation is now more popular. Frost protection can still be performed with undertree irrigation because the water freezing on the ground releases heat. In extremely dry climates, the water sublimates directly to the gas phase rather than freezes. Sublimation requires heat so this actually makes the air colder and causes increased frost damage to the buds.

Undertree sprinklers are generally placed on ¾″ PVC pipe risers approximately 18″ (50 cm) above the ground surface. They keep the water below the canopy by using a low angle nozzle. It is also sometimes necessary to keep water spray off the trunk, especially for sensitive trees such as cherries. In this case, a stream splitter device may be attached to the sprinkler in order to deflect the water away from the tree.

If inadequate pressure is available to run conventional straight-bore nozzles, then low pressure nozzles with a hexagonal orifice are an option. These nozzles are designed to work at low pressures in the range of 25–30 PSI (210 kPa).

One challenge with orchard sprinkler design is finding the best sprinkler spacing. For example, trees may be spaced on a 20 × 20 ft (6 × 6 m) spacing and sprinklers have a 30 ft (9 m) radius. If sprinklers are spaced one per tree, then the spacing is 20 × 20 ft and the sprinklers are far closer than head to head coverage requires (capital cost is excessive). However, 40 ft × 40 ft would be much further apart than head to head covering. This may actually be acceptable since the goal is generally to supply each tree with an equal volume of water. Head to head coverage is not as important as making sure that every tree receives the same amount of water. Orchards with a cover crop (grass in the orchard) must be irrigated with sprinklers that wet the entire ground surface; thus, microsprinklers are not acceptable.

In many of the cooler orchard regions irrigated by impact sprinklers or rotors, a typical practice is to have ten zones in the orchard, each of which is watered for 24-hours every 10 days. The systems are designed for peak ET, and then the grower can cut back on the schedule as needed. Orchards watered with microsprinklers must be irrigated more frequently because these sprinklers only wet a fraction of the ground surface. In general, growers prefer a 12-hr or 24-hr irrigation schedule in order to make scheduling more straightforward.

Example 14.3 A cherry orchard with a cover crop in eastern Washington has trees on 18 ft × 18 ft spacing. The canopy height is 5 ft above the ground surface. Maximum soil infiltration rate is 0.3 in/hr (7.5 mm/hr). Peak summer reference ET is 0.3 in/day (7.5 mm/day). The total available water is 8 in (20 cm). The seasonal ET_c is 42 inches (1.06 m). The direction of tractor travel in the field is in the same direction as the 2 % slope. Sprinkler pressure is provided by piped flow from a higher elevation canal. Expected sprinkler operating pressure is 30 PSI (207 kPa). Expected evaporation is 5 %. Leaching is not necessary. Select a sprinkler from the following three options:

L20, 1/8″ hexagonal (low pressure) nozzle, 30 ft wetted radius, 2.58 GPM @30 PSI.

L20, 7/64″ hexagonal (low pressure) nozzle, 29 ft radius, 1.98 GPM @ 30 PSI

LF, yellow nozzle w/ olive green deflector, 35 ft radius, 1.63 GPM @ 30 PSI

The tree spacing dictates that sprinklers can either be spaced at 18 ft × 18 ft, 18 ft × 36 ft, or 36 ft × 36 ft. None of the choices are optimal. The 18 ft × 18 ft spacing is ruled out because of cost.

Because the slope is in the direction of tractor travel, the three possible options are the slotted designs shown in Fig. 14.13 (36 ft × 36 ft) and 14.19 (36 ft × 18 ft) or the diagonal design in Fig. 14.15 (36 ft × 18 ft). The diagonal spacing with one sprinkler every other tree (Fig. 14.15) is preferable because it provides every tree with the same

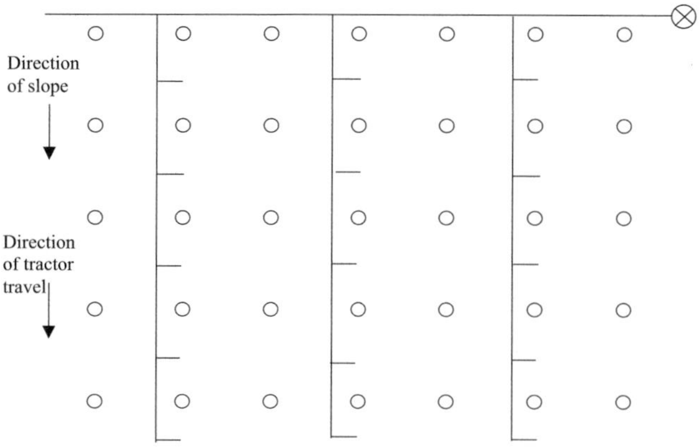

Fig. 14.17 Slotted spacing option for cherry trees

volume of water, unlike the slotted options in Figs. 14.13 or 14.17. We will evaluate the diagonal spacing (water distribution is better) and the 36 ft × 36 f. slotted design in Fig. 14.13 (lower cost).

The diagonal design is evaluated first.

Application rates (diagonal spacing, 18 ft × 36 ft)

The first step is to select the sprinkler and nozzle. For the L20 sprinkler with the 1/8″ nozzle at the diagonal 18 ft × 36 ft spacing, the application rate is

$$di/dt = \frac{(2.58)\ (96.3)}{(18)\ (36)} = 0.38\ in/hr$$

Thus the application rate with the 1/8″ nozzle is greater than the maximum application rate of 0.30 in/hr, and this design is not acceptable. Next, try the 7/64″ nozzle.

$$di/dt = \frac{(1.98)\ (96.3)}{(18)\ (36)} = 0.29\ in/hr$$

The application rate with the 7/64″ nozzle is just less than the maximum application rate. More investigation of the maximum soil infiltration rate may be advisable since the application rate is close to the specified maximum infiltration rate.

Calculate the application rate for the LF sprinkler with the diagonal 18 ft × 36 ft spacing.

$$di/dt = \frac{(1.63)\ (96.3)}{(18)\ (36)} = 0.24\ in/hr$$

Irrigation schedule (diagonal spacing, 18 ft × 36 ft)

From Table 3.3, the MAD for orchard fruit trees is 0.5. Thus, the readily available water is

$$\mathrm{RAW} = \mathrm{TAW\ MAD} = (8\,\mathrm{in})(0.5) = 4\,\mathrm{in}$$

In a 24 hour watering cycle, the expected net application depths for the 1/8″ nozzle and the LF sprinkler are.

$$\mathrm{i} = \mathrm{di/dt(time)}(1 - \mathrm{L_e})$$
$$(7/64''\ \mathrm{nozzle})\ \mathrm{i} = (0.29\ \mathrm{in/hr})\ (24\ \mathrm{hours})\ (1 - 0.05) = 6.6\ \mathrm{inches\ in\ 24\ hours.}$$
$$(\mathrm{LF\ sprinkler})\ \mathrm{i} = (0.24\ \mathrm{in/hr})(24\ \mathrm{hours})(1 - 0.05) = 5.5\ \mathrm{inches\ in\ 24\ hours.}$$

Neither design is acceptable for a 24-hour irrigation cycle because the applied water depth is greater than the RAW. Thus, the farmer would use 12-hour cycles in order to avoid leaching and apply either 3.3 inches (7/64″ nozzle) or 2.75 inches (LF sprinkler).

The reference evapotranspiration rate is 0.3 in/day. Peak K_c for cherries with a cover crop is 1.25 in arid, moderate wind conditions (the climate in eastern Washington in summer). Thus, the peak evapotranspiration would be

$$\mathrm{ET_c} = \mathrm{ET_r\ K_c} = (0.30)(1.25) = 0.375\ \mathrm{in/day.}$$

$$(7/64''\ \mathrm{nozzle})\quad (3.3\ \mathrm{in})/(0.375\ \mathrm{in/day}) = 8.8\ \mathrm{days.}$$
$$(\mathrm{LF\ sprinkler})\quad (2.75\ \mathrm{in})/(0.375\ \mathrm{in/day}) = 7.3\ \mathrm{days.}$$

Note: only 12 or 24 hour cycles were evaluated because these are generally desired by the farmer. This criterion is not as important if the farmer has hired irrigators that work 24 hours per day or if an automatic controller can accomodate other irrigation set times.

Application rate (slotted spacing, 36 ft × 36 ft)

The L20 sprinkler is not acceptable in this case. It has a 29 or 30 ft radius whereas the sprinklers are spaced 36 ft × 36 ft apart. Thus, trees in the rows between sprinkler laterals would get significantly less water than the rows of trees with sprinklers in the row. The only acceptable design is the LF sprinkler because it has a 35 ft wetted radius.

$$di/dt = \frac{(1.63)\ (96.3)}{(36)\ (36)} = 0.12\ in/hr$$

Irrigation schedule (slotted spacing, 36 ft × 36 ft)

In a 24-hr cycle, the depth of application would be 2.75 inches so the number of days between irrigations would be 7 days.

System cost

The slotted design would have much lower pipe, sprinkler, and trenching cost since laterals run directly downhill with half as many sprinklers: 70 % of the diagonal pipe length ($\cos(45^0) = 0.7$).

However, the uniformity is far superior with the diagonal design where all trees receive the same volume of water.

Sprinkler selection

If the farmer chose the diagonal design, then he/she would have to select between the LF and L20 sprinklers. The fact that the LF nozzle requires a 7 day cycle may be attractive because this might make scheduling more straightforward. The LF series sprinklers are better at deflecting water from the tree, which is important for cherry tree survival. Cherry trees are sensitive to sprinkler spray hitting the trunk. As one person familiar with cherry trees said, "Cherry trees spend 20 years trying to die and then they finally do."

Orchard Irrigation with Microsprinklers

Microsprinklers have a relatively small flow rate and wetted diameter. They are placed on stakes with a tubing connection to drip laterals. Their flow rates generally range up to 0.5 GPM (2 L/min), and the diameter of coverage ranges from 10- to 30 ft. Because they have small orifices, canal and pond water must be filtered with sand filters. Well water must be filtered with screen or disk filters. They are not suitable for arid climates because they emit a fine spray or stream, and the small drops evaporate in dry air.

One of the primary users of microsprinklers is the Florida citrus industry. The University of Florida recommends 50–75 % coverage of the soil surface area and has found that greater coverage results in higher yields (Parsons and Morgan 2004). One research project showed that two microsprinklers per tree had higher yields than one microsprinkler per tree (Parsons and Morgan 2004). Depletion levels (MAD) of 25–50 % for microsprinkler irrigation were recommended with lower depletion levels in spring (Parsons and Morgan 2004).

Example 14.4 Citrus grown in a humid climate (Florida) is spaced on 6 m by 4 m spacing. Trees on this spacing have a peak summer water use of 35 L/day/tree. There is no cover crop. Microsprinklers are placed along the row between every tree and have 1.0 L/min flow rate and a 4 m diameter wetted area. Effective depth of rooting is 0.6 m and allowable MAD in summer is 0.25. The AWC of the sandy soil is 8 %. The expected loss to evaporation is 6 %. Assume that the irrigation efficiency is 90 %. Specify an irrigation schedule for this system.

The soil water storage, S, which is the product of RAW and wetted area, can be calculated with Eq. 16.5 (Chap. 16) where D_b is the wetted diameter and Z is root zone depth.

$$S = 780 \quad Z \quad AWC \quad MAD$$
$$D_b{}^2 = 780(0.6\ m)(0.08)(0.25)\left(4^2\right) = 150\ L$$

Calculate the allowable days between irrigations

$$150\,\mathrm{L}/35\,\mathrm{L/day} = 4.2\ \mathrm{days}.$$

Thus, the irrigation system can be run every four days in order to supply the required 35 L/day.

Calculate the length of each irrigation application time

The application flow rate should be adjusted for evaporation and efficiency.

$$\begin{aligned} Q_{net} &= Q_{gross}(1 - L_e) \quad (\mathrm{Eff}) = 1\,\mathrm{L/min}(1 - 0.06)(0.9) \\ &= 0.85\ \mathrm{L/min} \end{aligned}$$

If the irrigation interval is decreased to every two days, then only 70 L must be replaced during each irrigation. The replacement volume divided by the adjusted flow rate is the application time.

Calculate the irrigation run time

$$\begin{aligned} \text{Volume ET/Application rate} &= 70\,\mathrm{L}/0.85\,\mathrm{L/min} \\ &= 82\ \text{minutes.} \end{aligned}$$

Thus, the schedule is 82 minutes every two days or 164 minutes every four days.

Sprinkler Network Design

Chapter 7 showed how to design a single sprinkler lateral. This chapter shows how to design a zone with laterals and a submain. The rule of thumb is that the entire zone should have a pressure variation no greater than 20 %, but strategic use of slopes can result in almost uniform pressure across the zone. As with previous examples in this book, the laterals are designed beginning with the end sprinkler and working backwards. The next step is the submain design. Just as laterals are designed with a k-x curve for sprinklers, the submain is designed with a k-x curve for laterals. This is accomplished by developing a system curve (pressure vs. flow rate) for laterals. As inlet pressure to the lateral is varied, the flow to the lateral varies. Lateral flow rate vs. inlet pressure is plotted, and an exponential equation is fitted to the flow vs. pressure curve.

Example 14.5 An irrigation system has 100 sprinklers arranged in ten laterals with ten sprinklers each (Fig. 14.18). All laterals are 20 m apart and sprinklers are 15 m apart along the lateral. The laterals have a downhill slope of 4 %. The submain has no slope. Sprinklers have a nominal flow rate of 16 L/min at 360 kPa. Minimum acceptable sprinkler pressure is 430 kPa. Gross average depth applied per season is 1 m. Energy cost is $0.10/kW-hr. Friction loss in pump fittings is 4 m. Select pipe diameters and find the required operating pressure and flow rate. The pump curve is:

$$H_{pump} = \left(-5.18*10^{-5}\right) Q^2 + \left(8.28*10^{-3}\right)Q + 600$$

where

Q = pump flow rate (L/min)
H = pump discharge pressure (kPa)

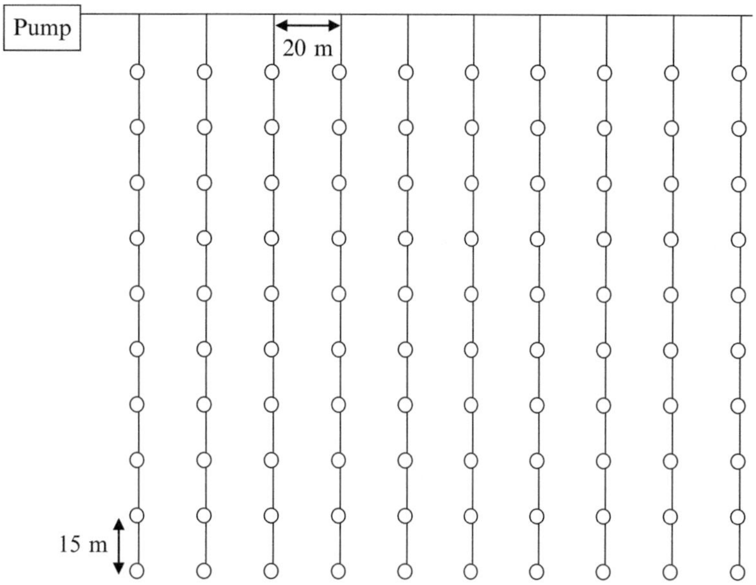

Fig. 14.18 Sprinkler system with 15 m × 15 m spacing between sprinklers

The design is performed in the *Lateral design* and *Submain design* worksheets; however, the following example calculations are given.

The sprinkler k value is calculated based on the nominal flow rate and pressure of the sprinkler.

$$k = \frac{Q}{H^{0.5}} = \frac{16}{360^{0.5}} = 0.843 \quad Q = 0.843\left(H^{0.5}\right)$$

The sprinkler flow rate at the last sprinkler is calculated as follows:

$$Q = kH^{0.5} = (0.843)\ \left(440^{0.5}\right) = 17.69\ \text{L/min} = 0.295\,\text{L/sec}.$$

Pipe pressures are calculated in units of m, the pressure at the last sprinkler is

$$440\ \text{kPa} \rightarrow 440/9.8 = 44.9\ \text{m}.$$

The head loss for 1 in (26.6 mm ID) SCH 40 pipe with pipe length 15 m is calculated:

$$H_L = 1.22{*}10^{10}(15\ m)\ \left(\left(\frac{0.295}{150}\right)^{1.852} / 26.6^{4.87}\right)$$

$$= 0.205\ m$$

Next, pressure at the next to last sprinkler can be calculated with elevation and friction loss.

$$H_{n-1} = H_n + H_L + s_L(S/100)$$
$$H_9 = 44.9 + 0.205 + 15\text{m}(-4)/100 = 44.5\text{m}$$

The *Lateral design* worksheet (Fig. 14.19) allows the user to try different pipe diameters (Fig. 14.19). The worksheet automatically adjusts the number of sprinklers based on the number of sprinklers in cell G6. Diameters are then selected by the user. Because the slope is high, the pipe diameters are relatively small in order to balance elevation energy gain with pipe friction loss.

A submain is similar to a lateral in that it has outlets with flow-pressure relationships. In order to develop an exponential equation for use in the submain design, the "Calculate lateral coefficients" triggers a VBA program that calculates flow vs. pressure data in columns L and M of the Lateral design worksheet. This is accomplished by varying the downstream pressure and recording the inlet pressure and flow rate in columns L and M. This data is then plotted and exponential equation for the lateral flow rate vs. pressure is calculated with Trendline. The lateral flow-pressure equation is shown above the graph ($Q = 6.94H^{0.531}$).

The next step is to design the submain with the lateral equation in the same way that the lateral was designed with the sprinkler equation (Fig. 14.20). The k and x values for the lateral are entered in cells E6 and E8.

The design pressure at the last lateral on the submain is assumed to be the design inlet pressure for the lateral (I24 in Fig. 14.19): 45.0 m. The flow rate into the last lateral is calculated with this pressure (kPa = H_m (9.8)).

$$Q_{lateral-n} = 6.944((45.0)\ (9.8))^{0.531} = 176.1\ \text{L/min (see cell G15)}$$

Submain friction losses contribute to the nonuniformity in a zone so it is also important to minimize pressure loss in the submain. In this example, it is more difficult to avoid pressure variation in the submain than the lateral because the submain is on level ground. Pipe diameters are selected in

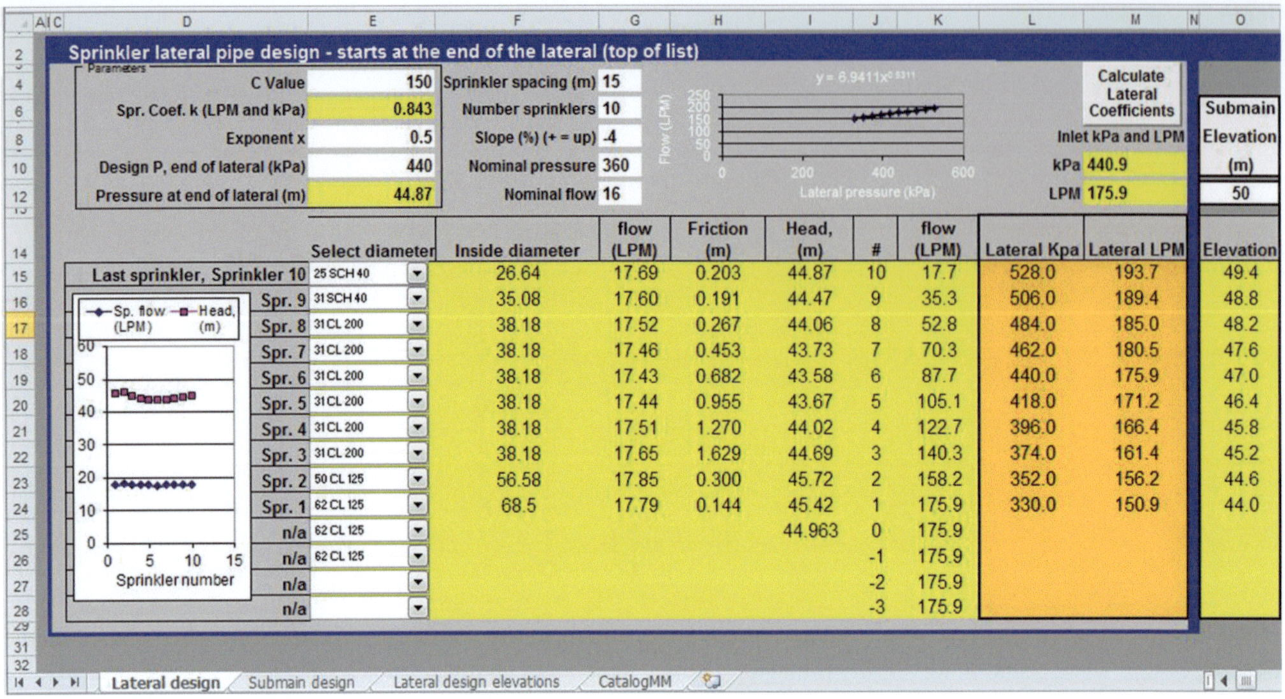

Sprinkler lateral pipe design - starts at the end of the lateral (top of list)

Parameters			
C Value	150	Sprinkler spacing (m)	15
Spr. Coef. k (LPM and kPa)	0.843	Number sprinklers	10
Exponent x	0.5	Slope (%) (+ = up)	-4
Design P, end of lateral (kPa)	440	Nominal pressure	360
Pressure at end of lateral (m)	44.87	Nominal flow	16

Calculate Lateral Coefficients — Inlet kPa and LPM: kPa 440.9, LPM 175.9. Submain Elevation (m): 50

	Select diameter	Inside diameter	flow (LPM)	Friction (m)	Head, (m)	#	flow (LPM)	Lateral Kpa	Lateral LPM	Elevation
Last sprinkler, Sprinkler 10	25 SCH 40	26.64	17.69	0.203	44.87	10	17.7	528.0	193.7	49.4
Spr. 9	31 SCH 40	35.08	17.60	0.191	44.47	9	35.3	506.0	189.4	48.8
Spr. 8	31 CL 200	38.18	17.52	0.267	44.06	8	52.8	484.0	185.0	48.2
Spr. 7	31 CL 200	38.18	17.46	0.453	43.73	7	70.3	462.0	180.5	47.6
Spr. 6	31 CL 200	38.18	17.43	0.682	43.58	6	87.7	440.0	175.9	47.0
Spr. 5	31 CL 200	38.18	17.44	0.955	43.67	5	105.1	418.0	171.2	46.4
Spr. 4	31 CL 200	38.18	17.51	1.270	44.02	4	122.7	396.0	166.4	45.8
Spr. 3	31 CL 200	38.18	17.65	1.629	44.69	3	140.3	374.0	161.4	45.2
Spr. 2	50 CL 125	56.58	17.85	0.300	45.72	2	158.2	352.0	156.2	44.6
Spr. 1	62 CL 125	68.5	17.79	0.144	45.42	1	175.9	330.0	150.9	44.0
n/a	62 CL 125				44.963	0	175.9			
n/a	62 CL 125					-1	175.9			
n/a						-2	175.9			
n/a						-3	175.9			

Fig. 14.19 *Lateral design* worksheet

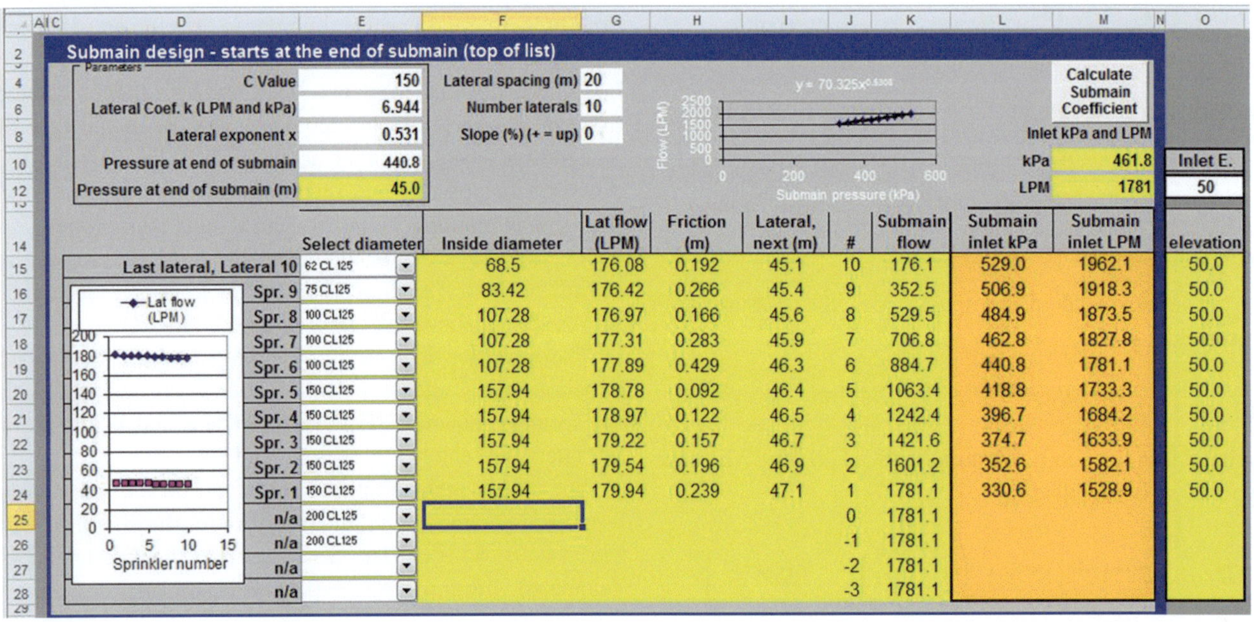

Submain design - starts at the end of submain (top of list)

Parameters			
C Value	150	Lateral spacing (m)	20
Lateral Coef. k (LPM and kPa)	6.944	Number laterals	10
Lateral exponent x	0.531	Slope (%) (+ = up)	0
Pressure at end of submain	440.8		
Pressure at end of submain (m)	45.0		

Calculate Submain Coefficient — Inlet kPa and LPM: kPa 461.8, LPM 1781. Inlet E.: 50

	Select diameter	Inside diameter	Lat flow (LPM)	Friction (m)	Lateral, next (m)	#	Submain flow	Submain inlet kPa	Submain inlet LPM	elevation
Last lateral, Lateral 10	62 CL 125	68.5	176.08	0.192	45.1	10	176.1	529.0	1962.1	50.0
Spr. 9	75 CL125	83.42	176.42	0.266	45.4	9	352.5	506.9	1918.3	50.0
Spr. 8	100 CL125	107.28	176.97	0.166	45.6	8	529.5	484.9	1873.5	50.0
Spr. 7	100 CL125	107.28	177.31	0.283	45.9	7	706.8	462.8	1827.8	50.0
Spr. 6	100 CL125	107.28	177.89	0.429	46.3	6	884.7	440.8	1781.1	50.0
Spr. 5	150 CL125	157.94	178.78	0.092	46.4	5	1063.4	418.8	1733.3	50.0
Spr. 4	150 CL125	157.94	178.97	0.122	46.5	4	1242.4	396.7	1684.2	50.0
Spr. 3	150 CL125	157.94	179.22	0.157	46.7	3	1421.6	374.7	1633.9	50.0
Spr. 2	150 CL125	157.94	179.54	0.196	46.9	2	1601.2	352.6	1582.1	50.0
Spr. 1	150 CL125	157.94	179.94	0.239	47.1	1	1781.1	330.6	1528.9	50.0
n/a	200 CL125					0	1781.1			
n/a	200 CL125					-1	1781.1			
n/a						-2	1781.1			
n/a						-3	1781.1			

Fig. 14.20 *Submain design* worksheet

order to keep the overall pressure loss low without spending an excessive amount on pipe. There is no quantitative economic analysis in this example (compare present value of cost of energy to capital cost of pipe). For the first section, a diameter is selected, 62 cm (2½″), which results in a reasonably low increase in pressure over the last section, $H_L = 0.192$ m.

The final design results in a pressure at the network inlet equal to 47.1 m. Thus, the pressure difference between the lowest and highest pressure sprinkler is 47.1 m–43.7 m = 3.4 m. Percent pressure variation is 3.4/45 * 100 % = 7.6 %. This results in a sprinkler discharge flow variation across the zone of approximately 4 %. As shown in previous examples, the variation due to sprinkler application pattern variability would be much higher than the variability due to 4 % sprinkler flow variation.

From an energy standpoint, there is a cost to the pressure variation because the system could operate at a lower pressure without the pressure variation. The cost of the pressure variation is the product of the pressure loss and the volume applied. The pressure loss is 3.8 m, the seasonal gross depth applied to the field is 1 m, and the pump efficiency is 80 %. Thus, the total cost of this energy loss is calculated as follows:

$$E(kW-hr/ha) = \frac{0.0272\ (i_{mm})\ (h)}{Eff}$$
$$= \frac{0.0272\ (1{,}000)(3.4)}{0.8}$$
$$= 115\ kW-hr/ha = \$11.5/ha/season$$

As with the lateral, a system curve is generated in the upper right corner of Fig. 14.20 for the submain by clicking the *Calculate Submain Coefficient* button:

$$Q_{system}(L/min) = 70.3\ H^{0.531}$$

The intersection of the pump curve and the system curve is the operating point for the pump. If 4 m head (40 kPa) is lost in pump valves and fittings at the pump station, then the system curve must be adjusted to account for this pressure loss. This is not the same as the 3.8 m head loss in the pipe network.

$$Q_{system}(L/min) = 70.3\left(H_{pump} - 40\right)^{0.531}$$
$$H_{pump}(kPa) = \left(Q_{system}/70.3\right)^{1/0.531} + 40$$

where

H_{pump} = pump discharge pressure, kPa.

Substitute the submain equation into the pump equation since $Q_{system} = Q_{pump}$.

$$Q = 70.3\left(\left(-5.18*10^{-5}\right)\ Q^2 + \left(8.28*10^{-3}\right)Q + 600 - 40\right)^{0.531}$$
$$0 = (70.3)\ \left(-5.18*10^{-5}\right)\ Q^2 + \left(8.28*10^{-3}\right)Q + 560)^{0.531} - Q$$

The root of this equation is Q = 1,734 L/min. Thus, the pump operating pressure is

$$H_{pump} = \left(-5.18*10^{-5}\right)\ \left(1{,}734^2\right) + \left(8.28*10^{-3}\right)(1{,}734) + 600$$
$$= 459\ kPa$$

If H_{pump} is 459, then H_{system} is 459–40 = 419 kPa

The calculated intersection point is shown in Fig. 14.21. The same solution is found with the iterative technique in the *System and pump curves* worksheet.

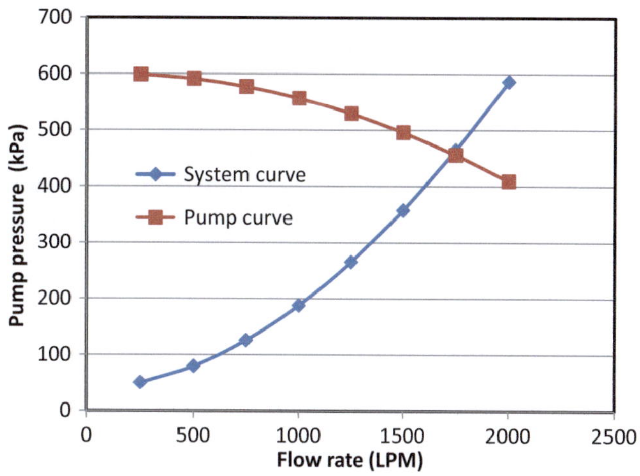

Fig. 14.21 Pump curve and system curve for Example 14.5

$$Q_{system}(L/min) = 70.3(419)^{0.531} = 1{,}735\ L/min$$

Questions

1. An orchard is on a hill. How should the mainline, submain, and laterals be positioned so that all the sprinklers on the property have nearly the same operating pressure?
2. True or false. The two main causes of nonuniformity of application in wheel-line sprinkler systems are hydraulic variation along the pipeline and spatial variability of soils.
3. True or false. Water hammer is a major problem in sprinkler laterals.
4. The last two sprinklers on a PVC pipeline have a flow rate of 20 L/min. The end pressure is 350 kPa. The distance between sprinklers is 15 m. The slope is 3 % downhill. Select pipe sizes for the last two pipe sections so that the three sprinklers have no more than a 0.2 m variation in head between them. Use a Hazen-Williams C value of 150. You can use the Worksheet to find the pipe sizes, but also make the two calculations of pipe pressure losses and change in pressure from one sprinkler to the next by hand.
5. A ¼ mile wheel-line has a downhill slope of 2 %. Use 3/16″ nozzles (ID = 4.8 mm). Calculate an equation for sprinkler flow rate vs. pressure. Then, determine whether pressure would have less variation with a 4 in (97.9 mm) or 5 in (123 mm) pipe.
6. For the parameters in Example 14.2, calculate the seasonal depth of water application at the last sprinkler, 6.1 m row position, 0 m head position. Then calculate the depth of water application at the 2.4 m row position,

0 m head position for the first sprinkler. Are these the extremes of application depth? What is the percent difference between the maximum and minimum application rates? The average seasonal depth of water application to a field is 75 cm. Calculate the application depths at the maximum and minimum positions. Evaporation rate is 10 %.

7. Redo Example 14.2, but don't offset the wheel-line positions with respect to the hydrants. Compare the total profit, and the optimal depths with those found in Example 14.2. Recalculate the pump power and the average depth applied as in Example 14.2.
8. Redo Examples 14.1 and 14.2, but use handlines. The normal design for handlines is 40 ft along the mainline and 30 ft between nozzles. Select a nozzle and flow rate from catalogs at the following web sites. The length of the run is 1/8 mile long and the handlines use 3″ aluminum pipe. The slope is 0.005 m/m in the downhill direction. Use the same evapotranspiration, precipitation, power, and production functions as in Example 14.2. Don't offset the handlines. Select sprinklers based on catalog data below. Show the variability due to hydraulics and variation in wetting due to sprinkler patterns. Maximum application rate in 0.3 in/hr. Operate the handlines at 45 PSI pressure.
   http://www.rainbird.com/ag/products/impacts/30H.htm
   http://www.rainbird.com/documents/ag/chart_20JH.pdf
9. Redo Example 14.3 with 14 ft by 14 ft tree spacing. Leave all other parameters the same.
10. Calculate a microsprinkler irrigation schedule for an orchard with 4 m × 3 m tree spacing and microsprinkler spacing. Each tree has a peak summer water use rate of 25 L/day. Microsprinklers have a 0.7 L/min flow rate and a 3 m diameter wetted area. Rooting depth is 1 m and allowable MAD is 0.35. The AWC is 12 % for a sandy loam soil, the expected loss to evaporation is 12 %, and the irrigation efficiency is 90 %. Specify an irrigation schedule for his system.
11. Repeat Example 14.5, except use 12 laterals by 12 sprinkler geometry, and the pump curve is $H = -5.18*10^{-5} Q^2 + 0.00828 * 10^{-3} Q + 900$. Find the operating point of the system

References

Hill R (2000) Wheel-line sprinkler irrigation operation and management. Utah State University Extension. Electronic Publishing. http://extension.usu.edu/files/publications/publication/ENGR_BIE_WM_08.pdf

Parsons L, Morgan K (2004) Management of microsprinkler systems for Florida citrus. HS-958, one of a series of the Horticultural Sciences Department, Florida Cooperative Extension Service, Institute of Food and Agricultural Sciences, University of Florida, Gainesville, Florida

Sammis TW (1981) Yield of alfalfa and cotton as influenced by irrigation. Agron J 73:323–329

Landscape Irrigation Systems 15

As with turf irrigation, the goal of landscape irrigation is aesthetic. Plants (trees, shrubs, groundcover, and flowers) can be irrigated to just survive or to thrive, to maintain plant biomass or to have vegetative growth. Research on landscape plant water requirements has been limited, and many systems are not adjusted to match seasonal changes or changes in plant canopy area. The largest water user in many cities is irrigation so improved irrigation management is critically important in water stressed regions. Although traditional landscape drip irrigation systems have proven to be unreliable, new systems are more reliable: multiport emitters mounted on PVC pipe, inline drip irrigation tubing, and bubbler irrigation systems. Another key to successful system performance is the proper design and installation of the control zone. The typical control zone has a ball valve, solenoid valve, filter, and pressure regulator installed in a valve box. This chapter focuses on the installation methods and components in landscape irrigation systems. This chapter includes equations on in-line emitter coil designs around trees and bubbler adjustment. The instructor may decide not to include this information since these are not normally part of the design process.

Status of Landscape Irrigation Systems

The author did a survey of landscape drip irrigation systems in Phoenix, Arizona in the early 90s. The city encourages the use of low water use plants and drip irrigation as water saving techniques – Xerigation. Although drip irrigation can be efficient in agriculture when systems are carefully maintained and scheduled by skilled farmers, drip systems in urban landscapes, with minimal flow rates under the plant canopy are generally unseen and neglected. Emitters tend to plug or develop high flow rates, and maintenance is minimal and rare. Sadly, water use in poorly managed Xerigation systems is often higher than in turf irrigation systems. Nevertheless, the great majority of landscape industry professionals and homeowners thought that drip irrigation is an excellent and efficient irrigation system. The study found that the great majority of professionals, let alone homeowners, could not calculate a drip irrigation schedule.

In-class Exercise 15.1 Could people gain a "feel" for the water requirements of plants and the length of time to water per week without actually calculating depths of water to apply?

If people cannot calculate the amount of water that plants need, then would you trust their judgment that drip irrigation is an excellent watering system?

Does it really matter if urban drip irrigation systems are efficient or uniform since plants just grow more if they get more water?

The first author and colleagues from the City of Phoenix analyzed 40 landscape drip irrigation systems, nearly all of which were professionally-maintained commercial or municipal systems. At the time, most drip systems had individual or multiport emitters plugged into polyethylene laterals with barbed fittings. If sites had components that operated correctly, then the site was given a score of 1. If the components at a site were missing or not operating correctly, then the site was given a score of zero. Number of acceptable sites was divided by total number of sites and multiplied by 10 to calculate a score between 0 and 10 (Table 15.1). Scores for each type of component are listed in Table 15.1. For example, a score of 1 meant that 4 out of 40 sites were acceptable.

One of the common problems with the systems was that landscaping tools, insects, or animals broke or chewed through the polyethylene tubing. When this happened, dirt entered the systems before or during the repair, and this led to plugging. Once emitters begin to plug, further attempts to "repair" the system often made things worse. For example, a drip system in a Phoenix park was trenched through by a fiber optic company and dirt was allowed to enter the system. In order to "clean out" the system, workers at the site

P. Waller, M. Yitayew, *Irrigation and Drainage Engineering*, DOI 10.1007/978-3-319-05699-9_15

Table 15.1 Drip irrigation component ratings

Parameter	Rating
Pressure regulators	7
Lateral hydraulics	6.6
Solenoid valves	9.6
Wire ties (not water proof rec. 0)	8
Flush valves (missing received 0)	1
Backflow preventor installation	4.2
Automatic controllers	8.7
Filters	7.2
Maintenance practices	0.25
Distribution uniformity	0
Efficiency	1.2
One-year old polyethylene laterals	9
Older polyethylene laterals	2
Old PVC laterals	9.5

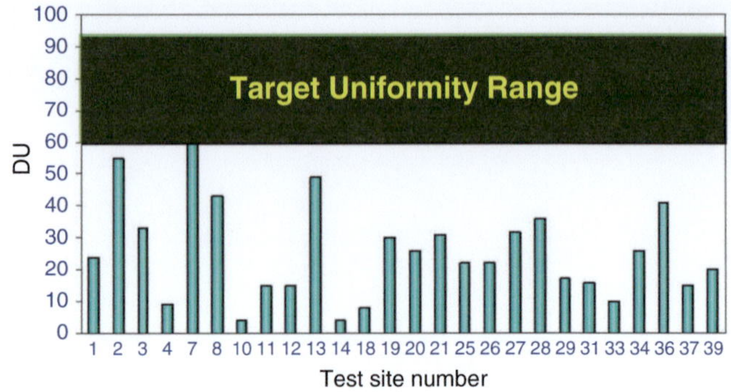

Fig. 15.1 Distribution uniformity at drip irrigation sites in Phoenix (Credit Aung Hla, University of Arizona)

removed the pressure regulators and permanently damaged the emitters by running them at 80 PSI (560 kPa) city pressure and blowing diaphragms out of emitters.

One of the problems with using compression or barbed fittings is that the connections are not as secure as PVC with glued or screwed fittings. When emitters pop out of distribution tubing or tubing is damaged, a stream of water sprays onto the landscape like a fountain.

Another common problem was that emitters were left running after plants died. It was quite common to see landscape irrigation systems watering large areas with no plants.

Approximately 2/3 of landscape drip irrigation systems had acceptable hydraulics (Table 15.1). Drip tubes can extend for hundreds of feet in landscapes without having unacceptable pressure losses. Generally, the components in the valve boxes (solenoids, filters, and pressure regulators) and controllers were in acceptable working order (Table 15.1). This is because these components must be fixed when the system breaks down or the system does not work at all.

Uniformity was measured. In landscape irrigation, the measurement of uniformity is complicated by the fact that the plants are different sizes; however, landscape irrigation uniformity was evaluated by measuring the volume applied to each plant and dividing by the canopy area. Distribution uniformity (DU) is the lowest quarter of depths divided by the average depth. Agricultural drip system DU is expected to be in the range of 90 %: all drippers flow at approximately the same rate. Distribution uniformity (DU) of landscapes in the study averaged less than 20 % (Fig. 15.1). The Irrigation Association recommends uniformities of 80 % for drip/micro irrigation. Only one site had a DU greater than 70 %. Not coincidentally, this was the only site that was managed by someone who could actually calculate irrigation requirements and conscientiously adjusted the number of drip emitters and irrigation timing to match irrigation requirements. The actual uniformities were even worse than reported because fountain emitters and completely plugged emitters were not included in the analysis. In addition, neither were areas with a multitude of emitters and no plants.

Statistical uniformity (SU) is another method of calculating irrigation uniformity.

$$SU = 1 - \frac{standard\ deviation}{mean} \tag{15.1}$$

Average statistical uniformity at the sites was negative 40 %. This means that the standard deviation of application depths was 40 % greater than the mean of the application depths. The conclusion of the study was that barbed emitters are unreliable and the PVC mounted multiport emitters, bubbler irrigation, and inline tubing are preferable alternatives.

Irrigation efficiency is the recommended rate of water use divided by the applied water. The efficiencies at the Phoenix drip study sites were measured by evaluating the controller irrigation schedules. The efficiencies (Fig. 15.2) were extremely variable, which is an indication that managers do not know how to adjust controllers. There is really no such thing as greater than 100 % efficiency as shown in Fig. 15.2; however, in this case, greater than 100 % efficiency means that less water was applied than was required.

Landscape Irrigation Control Zones

A control zone is placed at the entrance to every drip irrigation zone. It should include a manual ball valve, a solenoid valve, a filter, and a pressure regulator (Fig. 15.3). The solenoid valve, filter, and pressure regulator assembly are typically placed below the soil surface in a valve box. One to several valves (control zones) can be placed in a valve box. Fittings should be placed far enough apart so that unions can be unscrewed and the individual control zones removed for repair or replacement. Valves boxes should be installed in

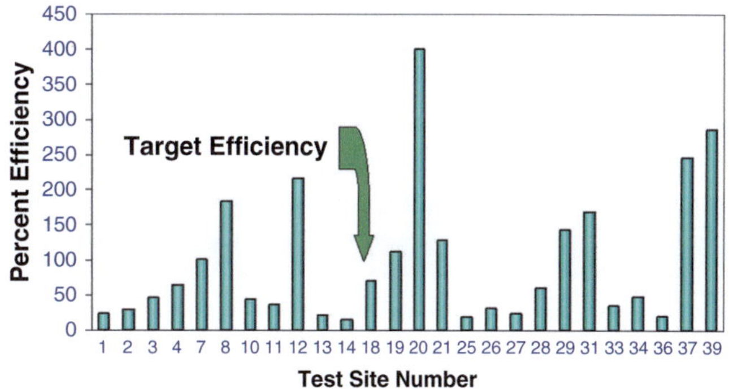

Fig. 15.2 Efficiencies of drip irrigation sites in Phoenix (Credit Aung Hla, University of Arizona)

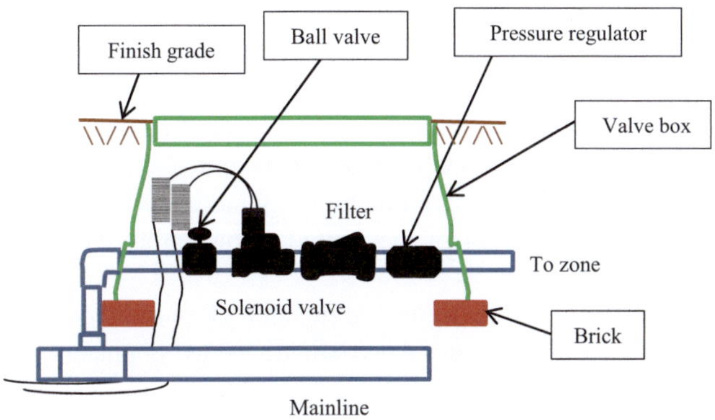

Fig. 15.3 Control zone components in valve box

such a way that they do not fill with mud. The top of the valve box should be at or above grade (Fig. 15.3). It should not be placed in a depression such that mud and water regularly flow into the valve box. The lower part of the box should be wrapped with geotextile cloth in order seal the valve box from insects and soil. The chemical WD-40 can be sprayed into the valve box in order to discourage insects. The mainline runs under the valve box and a riser connects to the control zone (Fig. 15.3). Solenoid valve wire is special coiled wire that allows the solenoid to be removed but also does not become a mess in the box. Electrical connectors that connect the controller wire to the solenoid valve are waterproof silicon-filled connectors. The bottom of the box should be filled with 75 mm depth of gravel in order to keep the valve box clean. The valve box should be leveled on bricks before gravel and backfill are added (Fig. 15.3). On many commercial projects, a mainline runs around the property, and valves are grouped along the mainline in subsurface valve boxes.

Irrigation controllers are powered by standard 110 V AC current. A hole can be drilled in the wall of the building in order to bring wires by conduit from an electrical socket or other location to the controller. The controller outputs 24 V power to the solenoid valves. One common wire (usually white) from the controller is connected to all solenoid valves and the controller. One 24 V "hot" wire is connected to each valve. Protection of electrical wiring is important. Metal conduit should be used to protect 110 V wire from the building to the controller. The 24 V valve wires should also be encased above ground. However, once the wires (i.e., #14 AWG) are in the ground they can buried without a conduit. The wires should be placed below the water pipes in the trench in order to protect them; thus, they are laid in the trench before the water pipes.

Special solenoid valves are required for low flow drip irrigation zones. The higher flow rate valves will not shut off properly in a low flow zone. If a low flow valve is used in a high flow rate sprinkler zone, then pressure loss is excessive.

Prior to the control zone, a backflow prevention device should isolate the irrigation system from the municipal water supply. If this is not done, then harmful chemicals could enter the potable water supply and cause sickness. The type of backflow prevention device depends on many factors such as whether or not chemigation is used and the topography of the property. Local municipal regulations should be consulted in order to select the correct backflow prevention device.

Drip System Design

In theory, there should be enough emitters to cover the root zone of the tree (beyond the canopy), however, the high cost of emitters can make this financially impractical. In the study of drip irrigation systems in Phoenix, Arizona, the average distance of emitters from trees was 1.3 feet and the average tree canopy width was 17 feet.

Multioutlet emitters threaded onto PVC risers and placed in valve cans (Fig. 15.4) are much easier to identify (Fig. 15.5) and maintain. Many of the problems associated with landscape drip irrigation are eliminated. The emitters are easily found and accessed in the cans and there are no unreliable polyethylene connections or barbed fittings. Distribution tubing goes out under the can to the watering points.

Another reliable landscape drip alternative is inline drip irrigation with the emitters manufactured into the tubing. There are several reasons for the surge in popularity of inline drip irrigation: (1) inline emitters have a lower cost per wetted area than single or multiport emitters, (2) it is easy to replace the entire inline lateral when the emitters or tubing wear out, (3) inline drip irrigation tubing lasts longer than point source drip emitters, (4) even if one or two emitters go bad, there are many other emitters, and (5) many inline drip emitters are self-flushing and pressure compensating.

Inline drip tubing is manufactured from polyethylene, and it can be exposed to direct sunlight. However, on commercial sites, it may be preferable to bury the tubing under mulch in order to prevent vandalism. The need to bury the tubing is most important in areas with high traffic where people might abuse the tubing. If the tubing is buried directly in the soil, then there is a possibility that some plant roots will grow into the emitters. Some emitters are impregnated with an herbicide or seal when not in use in order to prevent root intrusion.

Fig. 15.4 Multioutlet emitter in valve can

An exposed inline emitter and tubing is shown in Fig. 15.6. One inlet to the emitter provides backpressure behind a pressure compensating diaphragm, and the other is the entrance to the turbulent flow path. As pressure increases, the diaphragm closes down and restricts flow.

One of the advantages of inline tubing is that it creates a line source of water, and plants will grow roots along the tube. The increased root development is better for the plant than the restricted root development around a single point source emitter. In order to create a line source, the distances between emitters along inline tubing should be less than the wetted diameter (3/4 of the wetted diameter) as shown in Fig. 15.7.

Emitter wetted diameter is a function of emitter flow rate and soil infiltration rate. Approximate wetted diameters are shown in Table 15.2.

The salts in irrigation water remain in the soil as the water evaporates from the soil surface. In drip irrigated soils, the salts tend to be pushed to the edges of the wetted area, and rings of salt can be seen around emitters. If emitters spacing

Fig. 15.5 Multioutlet emitter valve cans in landscape

Fig. 15.6 Inline drip tubing

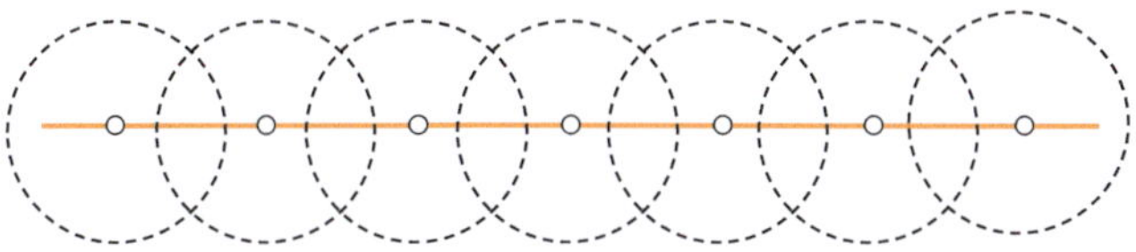

Fig. 15.7 Wetted diameter and spacing along inline drip irrigation tubing

Table 15.2 Wetted diameters (m) for drip irrigation emitters (After Benami and Ofen 1984)

Soil type	Emitter flow rate (Lph)		
	2	4	8
	Wetted diameter (m)		
Light (sand)	0.3	0.6	1.0
Medium (loam)	0.7	1.0	1.3
Heavy (clay)	1.0	1.3	1.7

is slightly wider than the wetted diameter, then salts will concentrate between emitters as they are pushed to the edges of the wetted diameter. Thus, inline tubing should be laid next to the plant as shown in in order to push the salts away from the root zone, and emitter spacing should be close enough to prevent salt accumulation between emitters. Alternating tubing from one side of the plant row to the other helps keep tubing next to plants.

In-class Exercise 15.2 What is the maximum distance between 4 LPH emitters in order to create a line source wetting pattern in a sandy soil? Would you select 30, 45, or 60 cm spacing?

One option for laying tubing around a tree that minimizes the need for fittings is to coil the tubing around the tree (Fig. 15.8). The distance between coils should be slightly less than the wetted width of the line source. The faded area

Fig. 15.8 Coil of inline drip tubing around tree

in Fig. 15.8 is the wetted area and the darker line is the drip tubing. The outer circle is the canopy diameter.

An equation can be developed that solves for the wetted area as a function of the length of tubing and the number of coil turns. If tubing begins half of the wetted diameter away from the tree, then the distance of the inner end of the tubing from the trunk center is the trunk radius + half of the wetted diameter. The coil should increase in radius by the wetted diameter for every 360° coil.

An following equation calculates the distance of the end of the coil from the center of the canopy:

$$r_{end} = r_t + \frac{B}{2} + (B)C \tag{15.2}$$

where

r_t = trunk radius, m
B = distance between successive coils of tubing, m
C = number of complete coil turns (360°), dimensionless
r_{end} = distance of the end of the tube from trunk center, m.

The change in tubing length, dL, for any change in number of coil turns, C, is

$$dL = 2\pi\ rdC \tag{15.3}$$

Rearrange and solve for r as a function of L and C.

$$r = \left(\frac{dL}{2\pi dC}\right) \tag{15.4}$$

Substitute Eq. 15.4 into Eq. 15.2.

$$\begin{aligned} \frac{dL}{2\pi dC} &= r_t + \frac{B}{2} + (B)C \\ dL &= 2\pi B\left(\frac{r_t}{B} + \frac{1}{2} + C\right)dC \end{aligned} \tag{15.5}$$

Integrate Eq. 15.5.

$$\int_0^L dL = 2\pi B \int_0^C \left(\frac{r_t}{B} + \frac{1}{2} + C\right) dC$$
$$L = 2\pi B\left(\left(\frac{r_t}{B} + \frac{1}{2}\right)C + \left(\frac{C^2}{2}\right)\right) \quad (15.6)$$
$$L = \pi B\left(\left(\frac{2r_t}{B} + 1\right)C + C^2\right)$$

The average wetted area under the canopy is calculated based on the end of the tube radius from the trunk, r_{end}, because the wetted area is r + WD/2 at the end of the coil, but it is r − WD/2 at the outer edge of the previous coil. Thus, the differences in area cancel. Rearrange Eq. 15.2 in order to solve for C as a function of r_{end}.

$$C = \frac{r_{end} - r_t - \frac{B}{2}}{B} \quad (15.7)$$

The wetted area as a function of the distance of the end of the drip tube from the trunk center is

$$A_{wetted} = \pi\, r_{end}^2 \quad (15.8)$$

The ratio of the wetted area over the entire canopy area is

$$\frac{A_{wetted}}{A_{canopy}} = \frac{r_{end}{}^2}{r_{canopy}{}^2} \quad (15.9)$$

Based on Eq. 15.9, if the ratio of r_{end} to the canopy radius is 1/2, then the ratio of the wetted area to the canopy area is ¼.

If the tubing length, L, is known, then a quadratic equation can be used to solve Eq. 15.6 for C, the number of turns of the coil.

$$-L + \pi B\left(\left(\frac{2r_t}{B} + 1\right)C + C^2\right) = 0$$

where

$$a = \pi B, \quad b = \pi B\left(\frac{2r_t}{B} + 1\right) = \pi(2r_t + B), \quad c = -L$$

$$C = \frac{-b \pm \sqrt{b^2 - 4ac}}{2a}$$
$$= \frac{-\pi(2r_t + B) \pm \sqrt{(\pi(2r_t + B))^2 + 4L\pi B}}{2\pi B} \quad (15.10)$$

Example 15.1 Find the length of coiled tubing required to wet ¼ of the canopy area of an 8 m diameter orange tree. Soil is sandy, and emitter flow rates are 4 LPH. The tree trunk is 0.3 m diameter. Determine the cost of tubing per tree if the inline tubing costs $0.76/m.

In sandy soil with approximately 4 LPH flow rate, the wetted width from the irrigation tubing is 0.6 m (Table 15.2). Thus, the coil can be placed around the tree with coils approximately 0.6 m apart.

Based on Eq. 15.9, if the ratio of wetted area to canopy area is ¼, then the ratio of the coil radius to canopy radius is ½; thus, if the canopy diameter is 8 m, then the distance from the tree center to the end of the coil, r_{end}, is 2 m. Number of coil turns is calculated with Eq. 15.7.

$$C = \frac{r_{end} - r_t - \frac{B}{2}}{B} = \frac{2 - 0.15 - \frac{0.6}{2}}{0.6} = 2.58 \ turns$$

Find the required length of tubing

$$L = \pi B\left(\left(\frac{2r_t}{B} + 1\right)C + C^2\right)$$
$$= \pi{*}0.6\left(\left(\frac{2{*}0.15}{0.6} + 1\right){*}2.58 + 2.58^2\right) = 20 \ m$$

The cost of tubing is 20 m * $0.76/m = $15.20.

Example 15.2 For the parameters in Example 15.1, find the fraction of canopy area wetted with a 10 m length of tubing.

Use Eq. 15.10 to find the number of turns in the coil.

$$C = \frac{-\pi(2r_t + B) \pm \sqrt{(\pi(2r_t + B))^2 + 4L\pi B}}{2\pi B}$$
$$= \frac{-\pi(2{*}0.15 + 0.6) \pm \sqrt{(\pi(2{*}0.15 + 0.6))^2 + 4{*}10{*}\pi{*}0.6}}{2{*}\pi{*}0.6}$$
$$= \frac{-\pi(0.9) \pm \sqrt{(\pi{*}0.9)^2 + 24{*}\pi}}{1.2{*}\pi} = 1.67 \ turns$$

The end radius with the 10 m tube can be calculated with Eq. 15.2.

$$r_{end} = r_t + \frac{B}{2} + BC = 0.15 + \frac{0.6}{2} + (0.6)(1.67) = 1.45 \ m$$

For the 8 m diameter tree, this would result in a percent wetted area of

$$1.45^2/4^2(100\%) = 1.32/16(100\%) = 13\%.$$

One technique used to encourage root growth is to move drip emitters away from the trunk toward the edge of the canopy as the tree grows. This encourages the roots to continue to move outward toward the edge of the canopy. Encouraging the roots to grow outward provides a better anchor for the tree and prevents toppling in high winds. One of the advantages of a coil is that it can easily be extended outward as the plant grows.

Inline tubing is used in beds with ground cover or flower. Parallel rows of tubing are connected to polyethylene or PVC headers at both ends of the planter. For short tubing lengths, inline emitter tubing is manufactured in 1/4″ (7 mm) diameter.

Bubbler/Water Harvesting Irrigation Systems

An alternative to drip irrigation that preceded drip irrigation is bubbler irrigation (Fig. 15.9). Bubblers operate at a much higher flow rate (0.5-, 1-, or 2-GPM; 2-, 4-, or 8-LPM) than drip irrigation systems; thus, they have larger orifices and don't require filtration. The disadvantage is that more zones are needed. They are installed on black flexible PVC pipe with glued and screwed connections. Thus, the bubblers and pipe systems are more durable than typical drip systems and are not as readily broken by landscape tools, degraded in heat, or consumed by animals or insects. However, bubblers are more difficult to install because trenched PVC pipe must be routed to each bubbler.

Fig. 15.9 Bubbler at base of shrub

Bubblers are mini flood irrigation systems. As such, they can work in concert with a water harvesting system if they are installed in a contoured basin landscape profile (Fig. 15.10). The contours direct rainwater to the plants. Contoured basins are stable and do not break down over time. The high points between plants should be approximately 50–100 mm (2–4 inches) higher than basin low points.

Bubblers are not suitable for densely planted landscapes because the water output per unit area is high. They are more appropriate for widely spaced shrubs and trees that can be surrounded by a basin; in general, it is difficult to supply more than one plant with an individual bubbler. If canopy area and basin diameter are large, then two bubblers may be required to cover the entire basin. Basins can be a several centimeters below grade, and mounds several centimeters above grade in order to avoid adding or removing soil from the site. Plants that are susceptible to trunk rot, such as pine trees, should be placed in donut shaped basins (Fig. 15.11) so that the trunk will not remain wet for an extended period.

Flexible PVC can be used to connect the bubbler to the PVC lateral pipe (Fig. 15.10). A gray ½″ electrical male adapter connects the bubbler to the flexible PVC. Orienting the lateral Tee sideways (Fig. 15.12) can minimize the amount of tubing out of the ground and allows the bubbler to be run in a shallow trench to the watering location. In this way, one lateral can be trenched in the center of a landscaped area and flexible black PVC pipe can be used to connect the bubblers to the main lateral without excessive trenching.

There are two primary types of bubblers: pressure compensating and adjustable flow. Pressure compensating bubblers have a relatively constant flow rate over a range of pressures and are generally available in 2-, 4-, 6-, and 8-LPM flow rates. One problem with some pressure compensating bubblers is that the little O-shaped rubber ring that maintains constant flow sometimes becomes dislodged and cuts off nearly all flow. Adjustable flow bubblers can be adjusted for flows between 0- and 8-LPM.

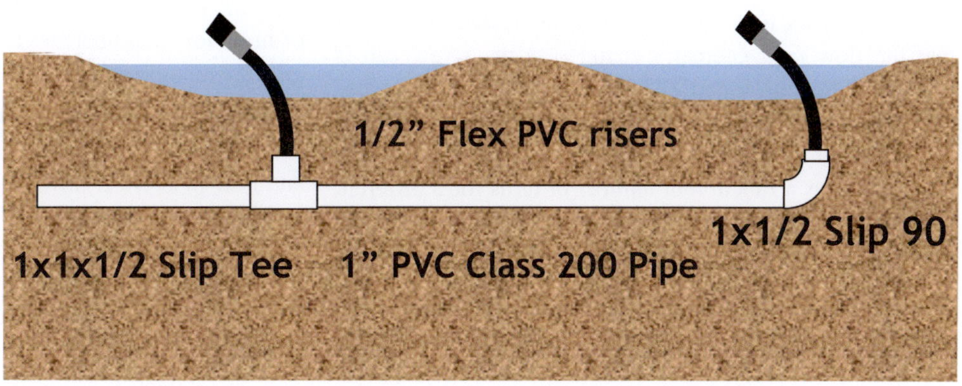

Fig. 15.10 Contoured basin design for bubbler/water harvesting systems

Fig. 15.11 Bubbler irrigation of pine trees

Fig. 15.12 Sideways Tee orientation can reduce unsightly pipe above the ground

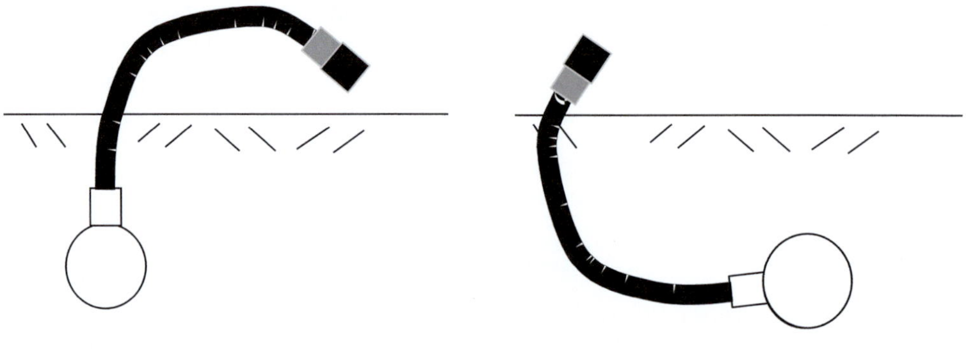

Bubbler flow rate vs. pressure can be modeled with the emitter exponent equation.

$$q = kH^x \tag{15.11}$$

where

q = bubbler flow rate, lpm
k = coefficient, dimensionless
H = pressure, kPa
x = exponent, dimensionless.

Unlike sprinklers, which have an exponent of 0.5 because flow passes through a round orifice, adjustable bubblers have a more complex flow path. This is complicated by the fact that the flow path geometry changes as the screw cap is adjusted. The author and Zhixu Yuan measured flow rates at different screw and cap positions and pressures (140 and 210 kPa) and developed equations for x (Fig. 15.13) and k (Fig. 15.14) as a function of number of turns for 5 different bubblers. The k and x values were calculated for each screw turn position as shown in Eqs. 15.12 and 15.13.

$$x = \frac{\log(Q_{140}) - \log(Q_{210})}{\log(140) - \log(210)} \tag{15.12}$$

where

Q_{140} = flow rate at 140 kPa, LPM
Q_{210} = flow rate at 210 kPa, LPM

$$k = \frac{Q_{140}}{140^x} \tag{15.13}$$

With some bubblers, flow rate is adjusted by turning a screw. Others are adjusted by turning a cap. Each brand of bubbler has a unique screw mechanism so the relationship between number of screw turns, flow rate and pressure is unique for each bubbler. Figure 15.13 shows the x value as a function of the number of 360° turns. This graph also indicates the range of turns for each bubbler. The equations next to each curve give the relationship between x and number of 360° turns, T.

The k value in Fig. 15.14 is basically a unit conversion constant. The values in Fig. 15.14 refer to pressure in kPa and flow in LPM. If flow was in GPM with pressure in kPa, then the k values in Fig. 15.14 would be divided by 3.785.

In-class Exercise 15.3 What is the exponent x for pressure compensating bubblers if flow rate does not change with pressure?

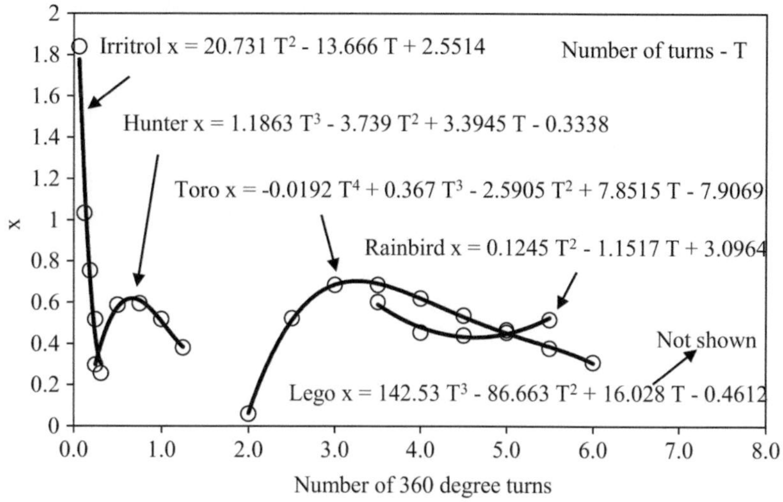

Fig. 15.13 Bubbler flow equation exponent x versus number of turns for five bubblers

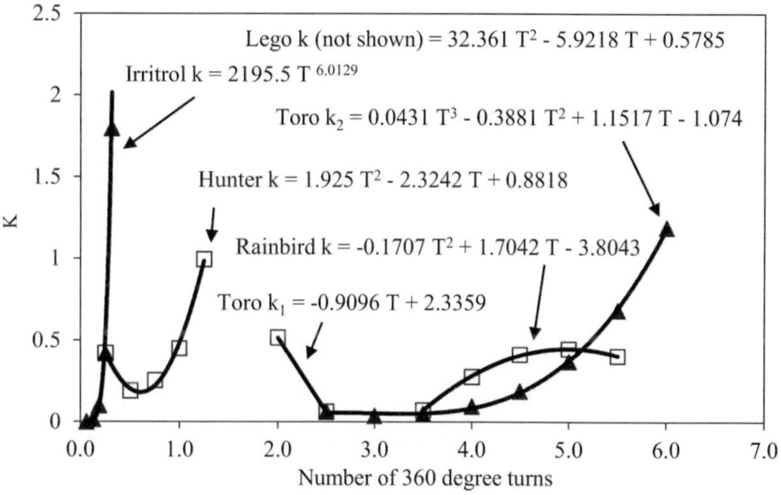

Fig. 15.14 Coefficient k versus number of turns for flow in LPM and pressure in kPa

Maintenance and Economics

Maintenance of landscape drip systems is rare because landscape companies are generally contracted to keep the site clean and pruned, most homeowners have other priorities, and drip components can be hidden by soil or plants. One study in Tucson, Arizona found that water usage decreases dramatically when drip systems are completely abandoned and people start watering with a hose. This is not at all surprising because people are actually observing the amount of water applied and the plant response to water application when they water with a hose. With automatic systems, the water application is generally unobserved.

Possibly, maintenance criteria used in large-scale irrigation projects can be used to evaluate maintenance of landscape irrigation systems. Gold-plated maintenance is maintenance to a level that virtually stops all system deterioration. Pragmatic maintenance attempts to keep a system in "good but not perfect working order." Minimal maintenance repairs those things that are absolutely necessary for major parts of the system to convey water, but is not concerned with levels of reliability, with some areas going out of service, or with diminished conveyance capacity. The great majority of landscape irrigation systems are maintained at the minimal maintenance level; for example, valves and automatic controllers are repaired because the system does not work at all without these components, but drip emitters are rarely adjusted or repaired.

If a drip irrigation system that waters 10 trees (4 emitters per tree) and 30 shrubs (1 emitter per shrub) is maintained once per year, then the following time investment might be necessary.

Measure distribution uniformity in the zones (4 minutes/emitter) (2 zones) (20 emitters/zone).

Calculate required flow rate per plant (3 minutes per plant) (40 plants).

Change the number of emitters per plant if necessary (5 minutes per plant) (40 plants).

Replace defective emitters (5 minutes per emitter) (1/5 emitters per year) (70 emitters).

Adjust emitter locations by moving distribution tubing (2 minutes per emitter) (½ emitters) (70).

The total time investment is 11 hours. If $100 replacement parts were used per visit and the cost of the worker is $40/hr, then the total cost would be $540.

Economics

The economic benefit of landscape irrigation systems can be evaluated based on the value of the landscape. One way to assign a value to residential landscaping is to determine the relationship between the selling price of the home and landscaping. Home values increase between 10 % and 17 % if landscaping around the home is rated as good or excellent vs. poor or average. Many homes in irrigated regions have values in the range between $200,000 to $700,000. Thus, the value of landscaping ranges from $20,000 to $100,000.

Example 15.3 Determine whether installing an irrigation system with landscaping is an economically wise choice. Assume a required rate of return of 8 % and a project life of eight years (owner will sell house in 8 years when the landscape is mature).

Assumptions

- Water costs for irrigation are $500/yr.
- Capital cost of installation of the irrigation system and landscaping is $5,000
- Home selling price in 8 years is expected to be $400,000.
- Irrigation system maintenance costs are $500/year
- Plant replacement costs are $200/year
- Ignore psychological benefit of looking at and interacting with landscaping, and cooling.

Costs

Annual costs are sum of water, maintenance, and plants: $500 + $500 + $200 = $1,200.

The present value of $1,200 payments for eight years at an 8 % rate of return is $6,900.

Total present value of costs is $5,000 + $6,900 = $11,900.

Benefits

The value of landscaping in eight years will be $400,000 (0.17) = $68,000.

Present value of benefits is $147,000 $(1 + 0.08)^{-8}$ = $37,000.

The present value of benefits is 3 times greater than the present value of costs.

Questions

1. Discuss the questions presented in Exercise 15.1. How would the scientific method change the results and people's perceptions? (Write at least 1 page, double-spaced). Answers will vary.
2. Describe the components in a landscape irrigation control zone and the function of each part.
3. Describe how a controller is wired and how it controls the valves.
4. Compare the advantages and disadvantages of barbed fitting emitters, line source drip irrigation and bubbler irrigation.
5. Describe how a pressure compensating in-line emitter works.
6. Answer in-class Exercise 15.2. Using Table 15.2, determine the maximum distance between 4 LPH emitters in order to create a line source wetting pattern in a sandy soil? Would you select 30, 45, or 60 cm spacing?
7. Draw out a yard that you know about or can imagine and lay out the location of sprinklers and drip emitters. You can do this on a piece of graph paper where 1 inch = 10 ft or some other appropriate scale. You could also draw it out in a computer program. Locate positions of valves and pipes as well as define zones. For bubblers, let flow rate be 2 GPM, and for emitters let flow rate be 2 GPH. Use PVC or polyethylene tubing where appropriate. Make sure to group similar emitters in zones. Select pipe and calculate pipe friction losses.
8. Evaluate the economic costs and benefits of the following system. Assume 8 % ROR and 8 year project life.
   - Water costs for irrigation are $500/yr
   - Capital cost for installation is $1,500
   - Home selling price is $200,000 and landscaping adds 17 % to the value of home. Home will be sold in 8 years and home price is expected to decrease in value by 10 % over the 8 year period.
   - Irrigation system maintenance are $250/yr
   - Plant replacement costs are $100/yr
9. Which types of irrigation devices would be appropriate for four 5 m diameter, widely spaced, trees?
10. Which types of irrigation devices would be appropriate for 10 x 50 ft planter with ground cover?
11. If you had an oleander hedge that extended for 200 ft, which type of irrigation system would be most appropriate and why?

12. If you had six 1.2 m diameter shrubs that are spaced at 3 m interval, which type of irrigation system would be most appropriate and why?
13. Calculate the flow rate of the Toro Bubbler with three turns at 140 and 210 kPa.
14. Find the length of coiled tubing required to wet ½ of the canopy area of a 10 m diameter orange tree. Soil is clay, and emitter flow rates are 3 LPH. The tree trunk is 0.3 m diameter. Determine the cost of tubing per tree if the inline tubing costs $0.76/m.
15. For the parameters in question 14, find the fraction of canopy area wetted with a 12 m length of tubing.

Reference

Benami A, Ofen A (1984) Irrigation engineering. Irrigation Engineering Scientific Publications, Israel Institute of Technology, Technion City, Israel

16 Landscape Irrigation Design and Management

Landscape irrigation design generally includes an owner interview, site survey, determination of the municipal water pressure and flow rate, calculation of plant water requirements, and zoning and hydraulics. This chapter includes two programs, one for pipe design and the other for emitter specification and irrigation scheduling. The pipe design program includes valves, pipes, fittings, and other components and allows the user to select components from a drop down list.

The scheduling program is the procedure begins with the measurement of the canopy diameter and crop coefficient for each plant in the landscape. This is followed by a calculation of the water requirement of each plant. Based on the water requirement, the program calculates the number of emitters required by each plant. Based on the reference ET, the program recommends an irrigation schedule. The procedure is similar to the calculation of an agricultural schedule except that density of planting, variable sizes of plants, and exposure to heat sources such as roads and buildings are also considered. The program also estimates the percent wasted water due to nonuniformity of application. This chapter also includes a design program for the pipe connection between the city water main and the valve box.

Owner Interview

The irrigation designer should discuss the various irrigation system options with the owner. This discussion should include life expectancy, safety features (backflow and master valve), operation and design parameters (irrigation scheduling), distribution uniformity, capital cost, cost of water, chemical injection, filtration, pressure regulation, and, if a pump is required, energy consumption and flow metering.

The property owner's level of expertise and willingness to invest time or money into management of the system should be evaluated. The designer should explain the required maintenance schedule and required level of expertise for each system option to the owner. If an owner is not able to maintain or pay someone to maintain the irrigation system, then the most reliable system should be installed.

Site Survey

A simple hand-drawn map of the landscape elements and buildings on the property should be made. For a private residence, this can be done by hand on graph paper with an appropriate scale; i.e., 1 inch = 10 ft (1 cm = 1 m). The water source, static pressure, flow meter diameter, service line, walls, concrete, turf areas, tree and shrub areas, wind speed, and wind direction should be indicated on the drawing. In the case of a commercial property or a professionally designed residential landscape, the landscape design is typically drawn with Adobe Illustrator or AutoCAD. The landscape can be faded into the background with either of these programs and the irrigation plan can be drawn over the faded landscape background. The design in Fig. 16.1 is not necessarily recommended, but it illustrates how the irrigation system can be drawn on the original landscape design.

Determining Maximum Flow Rate and Working Pressure

The water pressure and flow rate available to the irrigation system is dependent on the static water pressure, flow meter diameter, and service line diameter and length. The static water pressure can be determined with a pressure gauge attached to a hose bib when all water in the residence is turned off or from the water purveyor. For small systems, a quick check of dynamic flow rate and pressure can be obtained by turning on a hose bib that is connected to the service line and measuring the length of time required to fill a 5 gallon bucket.

P. Waller, M. Yitayew, *Irrigation and Drainage Engineering*, DOI 10.1007/978-3-319-05699-9_16

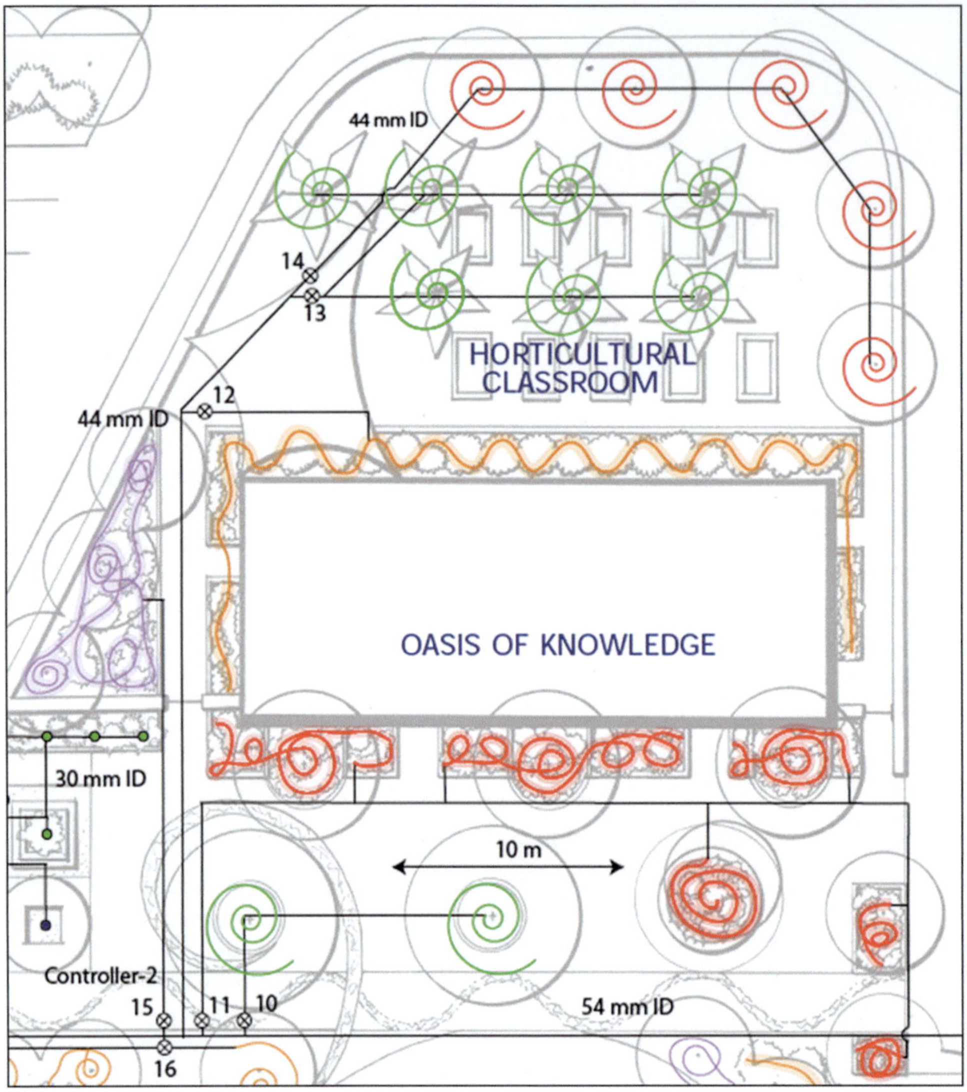

Fig. 16.1 Inline emitter tubing snaked through various types of planters

Rain Bird recommends the evaluation of the following three criteria for determining the maximum flow rate to a landscape irrigation system:

1. The pressure loss through the water meter should not exceed 10 % of the minimum static water pressure available in the city main.
2. The maximum flow through the meter for irrigation should not exceed 75 % of the maximum safe flow of the meter (maximum flow calculated in Figure 16.3).
3. Velocity of flow through the service line should not exceed 5 or 7-1/2 ft/s (1.5 to 2.3 m/s) for PVC or copper pipe, respectively. Higher velocity is allowed in copper (rated at 400 PSI).

Over the years, as pipe materials have improved, service lines have been installed that are constructed from galvanized steel, copper tubing, or PVC pipe. The Hazen-Williams equation with a design C value of 140 can be used to calculate the pressure loss in copper tubing. The inside diameters of ½, ⅝, ¾, 1, 1¼, and 1½ inch Type K (400 PSI) copper tubing are 0.527, 0.662, 0.745, 0.995, 1.245, and 1.481 in (13.4, 16.8, 18.9, 25.3, 31.6, and 37.6 mm ID), respectively. For unlined schedule 40 steel pipe (dimensions in Tables 8.1 and 8.2), assume a design C value of 100 to account for degradation of the pipe over time.

Example 16.1 A typical connection to a residence is shown in Fig. 16.2. The static pressure in the city pipeline (A) is 111 PSI (7.7 bar/0.098 = 78 m). The valve (D) and water meter (E) nominal diameter are ¾ in. The irrigation system will be connected at point F. Calculate the pressure available to the irrigation system if the 2.3 m/sec maximum velocity criterion is used to determine the maximum flow to the irrigation system. Check the other two rules.

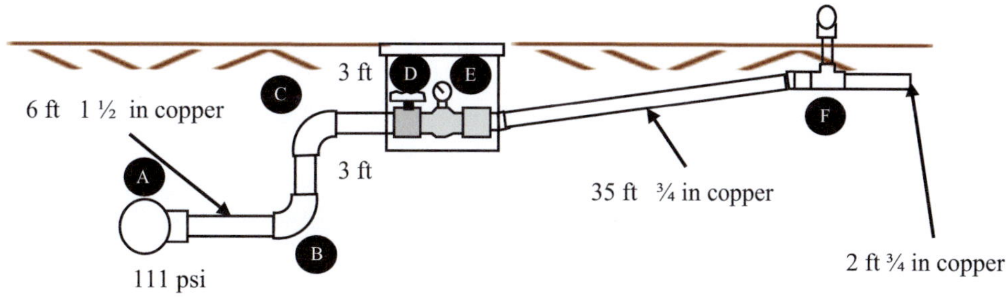

Fig. 16.2 Typical connection to residence from city mainline (After Rain Bird Corp figure)

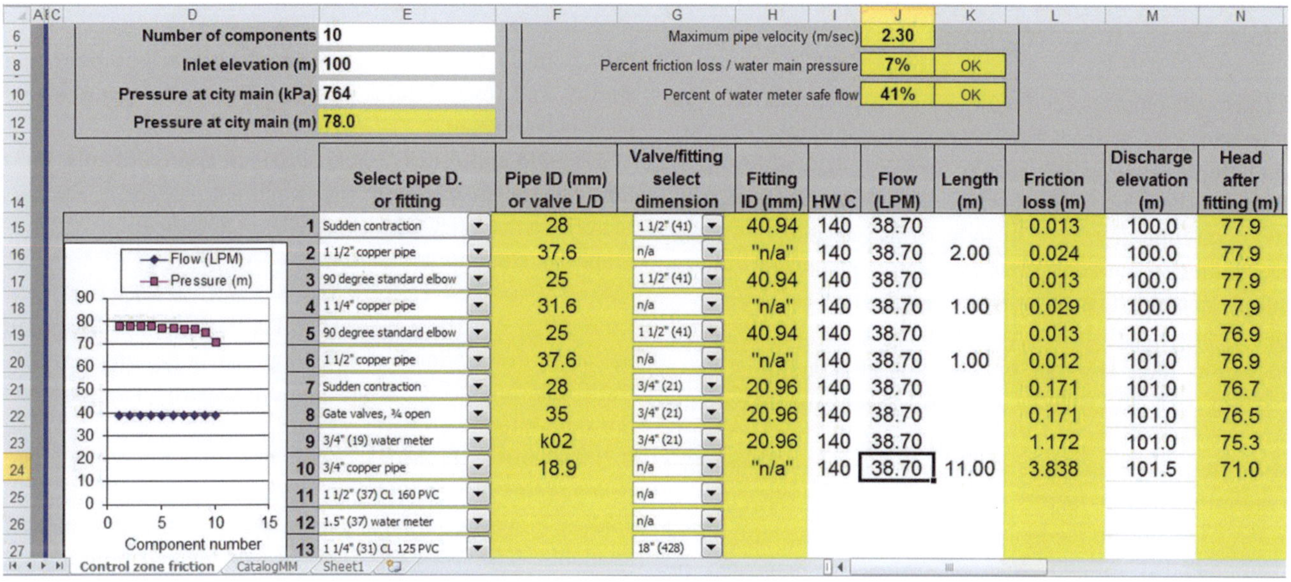

Number of components	10
Inlet elevation (m)	100
Pressure at city main (kPa)	764
Pressure at city main (m)	78.0

Maximum pipe velocity (m/sec)	2.30	
Percent friction loss / water main pressure	7%	OK
Percent of water meter safe flow	41%	OK

	Select pipe D. or fitting	Pipe ID (mm) or valve L/D	Valve/fitting select dimension	Fitting ID (mm)	HW C	Flow (LPM)	Length (m)	Friction loss (m)	Discharge elevation (m)	Head after fitting (m)
1	Sudden contraction	28	1 1/2" (41)	40.94	140	38.70		0.013	100.0	77.9
2	1 1/2" copper pipe	37.6	n/a	"n/a"	140	38.70	2.00	0.024	100.0	77.9
3	90 degree standard elbow	25	1 1/2" (41)	40.94	140	38.70		0.013	100.0	77.9
4	1 1/4" copper pipe	31.6	n/a	"n/a"	140	38.70	1.00	0.029	100.0	77.9
5	90 degree standard elbow	25	1 1/2" (41)	40.94	140	38.70		0.013	101.0	76.9
6	1 1/2" copper pipe	37.6	n/a	"n/a"	140	38.70	1.00	0.012	101.0	76.9
7	Sudden contraction	28	3/4" (21)	20.96	140	38.70		0.171	101.0	76.7
8	Gate valves, ¾ open	35	3/4" (21)	20.96	140	38.70		0.171	101.0	76.5
9	3/4" (19) water meter	k02	3/4" (21)	20.96	140	38.70		1.172	101.0	75.3
10	3/4" copper pipe	18.9	n/a	"n/a"	140	38.70	11.00	3.838	101.5	71.0
11	1 1/2" (37) CL 160 PVC		n/a							
12	1.5" (37) water meter		n/a							
13	1 1/4" (31) CL 125 PVC		18" (428)							

Fig. 16.3 *Chapter 16 Service Line* workbook

There are 4 m of 40 mm copper pipe with 2 elbows (B and C). The equivalent length of a 40 mm elbow is 1.52 m. Thus, the total length of 40 mm copper pipe is 7 m. There are 11 m of 20 mm copper pipe. The ¾ in gate valve (D) adds 0.15 m equivalent length – 11.15 m.

The inside diameter of 20 mm pipe is 18.9 mm. The maximum flow rate in the pipe is

$$Q = vA = v{*}\pi D^2/4$$

$$= 2.3\ \text{m/sec}{*}\pi(0.0189)^2/4{*}1{,}000\ \text{LPS}/(\text{m}^3/\text{sec}) = 0.645\ \text{LPS}$$

Friction loss in the 40 mm (37.6 mm ID) pipe is

$$H_L = 1.22{*}10^{10}{*}L\left(\frac{\left(\frac{Q}{C}\right)^{1.852}}{D^{4.87}}\right)$$

$$= 1.22{*}10^{10}{*}7\ m\left(\frac{\left(\frac{0.645}{140}\right)^{1.852}}{37.6^{4.87}}\right) = 0.09\ m$$

Friction loss in the ¾ in, 20 mm nominal diameter (18.9 mm ID) pipe is

$$H_L = 1.22{*}10^{10}{*}11.15\ m\left(\frac{\left(\frac{0.645}{140}\right)^{1.852}}{18.9^{4.87}}\right) = 3.9\ m$$

From Table 16.1, Friction loss in the ¾ in water meter is 0.11 bar (1 m/0.098 bar) = 1.2 m.

$$\begin{aligned}\text{Pressure(F)} &= \text{static pres.} - \text{friction} - \text{elev.}\\ &= 78 - 0.09 - 3.9 - 1.2 - 1.5\ \text{m(elevation)}\\ &= 71\ \text{m.}\end{aligned}$$

Maximum flow rate in the 20 mm valve (1.89 LPS in Table 16.1) is less than 75 % of maximum flow. Pressure loss (1.2 m) in valve is far less than 10 % of the static working pressure. OK.

The same calculations can be made in the *Chapter 16 Service Line* workbook (Fig. 16.3). Note that the final pressure in both calculations is 71 m. The worksheet has a list of all service line components and pipe diameters in column E. The diameter of valves and fittings must be selected in Column G, but no diameter is needed for pipes in Column G because pipe diameters are seqlected in Column E. The number of components, inlet elevation, and pipe diameter are selected in the upper section of Column E. Surveyors often set the datum at 100 m. Flows, Hazen-Williams coefficients and pipe lengths must be specified in Columns I, J, and K.

Table 16.1 Species factor (K_s) for different plant types (After Costello and Jones, 2014)

Vegetation	High	Average	Low
Trees	0.9	0.5	0.2
Shrubs	0.7	0.5	0.2
Ground cover	0.9	0.5	0.2
Mixture of trees, shrubs, and ground cover	0.9	0.5	0.2
Cool season turfgrass (i.e., rye)		0.8	
Warm Season Turfgrass (i.e., Bermuda)		0.6	

Plant Water Requirements

Landscape plant water requirements depend on the plant physiology as well as the watering strategy. For example, drought tolerant plants can be irrigated to survive or to thrive: to keep plants marginally alive, to keep them attractive with limited growth, or to have lush vegetative growth. In addition, some landscapes with native plants only require supplemental irrigation water during the first year or two as plants are established. External factors can influence the irrigation rate. Drought or low rainfall conditions on a local or regional level can lead to the necessity for extreme water conservation measures in some years, but very little water restriction in other years. In some cities, municipal regulations require landscapers and architects to specify low water use plants in landscape designs. Unfortunately, use of recommended drought tolerant plants, as mandated by municipal landscaping codes, may not actually save water because of the lack of knowledge of water requirements by homeowners and landscape managers.

Costello and Jones (2014), http://ucanr.edu/sites/WUCOLS/, developed the landscape coefficient method, where the ET_L (evapotranspiration rate of the landscape) is the product of K_L, the landscape coefficient, and ET_0 just as ET_c is the product of K_c and ET_0 in agriculture. The landscape coefficient is the product of three factors.

$$K_L = K_s{*}K_{mc}{*}K_d \tag{16.1}$$

where

Ks = Adjustment factor representing characteristics for a particular plant species (dimensionless)

Kmc = Adjustment factor for microclimate influence upon the plants (dimensionless)

Kd = Adjustment factor for plant density (dimensionless)

The species factor (Table 16.1) is similar to the crop coefficient in agriculture. Although research into landscape plant species factors lags behind agricultural crop coefficients, the quantification of landscape plant species factors is improving. Many studies have been performed that quantify the water requirements of various trees and shrubs, and several guidelines on plant water requirements have been written.

Costello and Jones (2014) classified approximately 3500 landscape plant species in four water use categories (species factors) in the WUCOLS (Water Use Classification of Landscape Species) guide: High (H) = 70–90 % ET_0, Moderate (M) = 40–60 % ET_0, Low (L) = 10–30 % ET_0, and Very Low (VL) = <10 % ET_0. The classification was based on the recommendations of five regional committees within California composed of horticulture experts. They divided the state into six climatic regions and classified each of the 2,200 plants as H, M, L, or VL based on the region. The designations were based on experience rather than research studies; however, for most plants, this is the best information available. The WUCOLS guide also indicates whether a plant is appropriate for a given climatic region. California has diverse climates (mountains, seaside, inland valley, and desert) so the climatic regions in California include many of the world's climatic regions. The WUCOLS guide recommends assigning species factors in Table 16.2 to the high, moderate, and low plant water use classifications. In general, low water use plant species factors range from 0.1 to 0.3 for low water use plants, 0.4 to 0.6 for medium water use plants, and 0.7 to 0.9 for high water use plants.

Drought tolerant (low water use) trees and shrubs can actually use more water than trees and shrubs from humid areas (high water use) because they are opportunistic. For example, the species factor for mesquites (drought tolerant plant) in nonlimiting water conditions can be 1.0 whereas the species factor for oaks (mesic plant) in the same nonlimiting water condition is 0.5 (Levitt et al. 1995). However, mesquites can survive on much less water than oaks. When given excessive water, drought tolerant plants have luxuriant canopy growth with high water consumption; large canopies can require extensive pruning and lead to plant toppling in high winds. Low-water use ground covers have a range of species factors; they generally range from 20 % of ET_0 to 50 % of ET_0 for acceptable appearance.

The density factor refers to the percent canopy coverage in a landscape. The density factor, K_d, is 1.0 if tree cover is greater than 70 % or if shrub canopy cover is greater than 90 % (Costello and Jones, 2014). This criterion is based on agricultural orchard studies that show that water loss from orchards does not increase as canopy cover increases from 70 % to 100 %. Below 70 % cover, however, orchard water loss declines because there is significant bare ground receiving sunlight (Costello and Jones, 2014). Agricultural studies show that orchard grass planted under orchards increases water loss by 25–80 % (Costello and Jones, 2014). Similarly, landscapes planted in tiers (ground cover under trees) will have greater water loss so K_d increases beyond 1.0 to the range between 1.1 and 1.3. The density factor may range between 0.5 and 0.9 for low-density plantings.

The microclimate factor refers to the influence of roads, buildings, or exposure to plant evapotranspiration. Microclimates can exist in landscapes. Plants near hot areas with reflected sun from buildings and roads or exposed directly to wind will have an increase in ET, and the microclimate factor, K_{mc}, can range from 1.1 to 1.4. If the landscape is close to reference evapotranspiration conditions (open area), then the microclimate factor is set to 1.0. If the plants are in shade or protected from wind, then they have a low microclimate factor: between 0.5 and 0.9. Unfortunately, there are no quantitative procedure for the calculation of microclimate factors.

Irrigation schedules can be calculated based on the water requirement of an individual plant or the water requirement of a landscaped area. Calculating an irrigation schedule requires the following steps:

- Calculate plant water use (LPD).
  LPD = Liters per day used by the plant or landscaped area.
- Calculate soil water storage (S).
  S = volume of usable water (liters) in soil.
- Calculate days between irrigation (S/LPD).
- Calculate irrigation run time (S/Q).
  Q = bubbler or emitter flow rate (LPM or LPH).

The evapotranspiration rate for a landscaped area is

$$LPD = \frac{\left(\frac{ET_L\ mm}{day}\right)\left(\frac{m}{1,000\ mm}\right)\left(A\ m^2\right)\left(\frac{1,000\ L}{m^3}\right)}{Eff} = \frac{ET_L{*}A}{Eff} \tag{16.2}$$

where

Eff = efficiency, fraction
LPD = evapotranspiration volume per emitter, liters per day
A = landscaped area, m^2
ET_L = landscape evapotranspiration rate, mm/day.

The evapotranspiration for an individual tree or shrub is:

$$LPD = \left(ET\frac{mm}{day}\right)\left(\frac{m}{1,000\ mm}\right)\left(\frac{1}{Eff}\right)\left(\frac{\pi D^2\ m^2}{4}\right)\left(\frac{1,000\ L}{m^3}\right) = ET_L{*}\frac{\pi D^2}{4(Eff)} \tag{16.3}$$

where

D = diameter of canopy, m
ET_L = evapotranspiration, mm/day.

The soil water storage is dependent on soil texture (AWC), root zone depth, and allowable soil water depletion (MAD). For a landscaped area, the volume of water readily available to the landscape is given by the following equation.

$$S = \left(A\ m^2\right)(Z\ m)(AWC)(MAD)\left(\frac{1,000\ L}{m^3}\right) = 1,000{*}Z{*}AWC{*}MAD{*}A \tag{16.4}$$

where

S = Volume of water storage in soil available to the plant, L
Z = root zone depth, m
AWC = available water holding capacity, fraction
MAD = management allowable depletion, fraction
A = *wetted* area, m^2.

For calculation of water available to individual plants that are watered by one or more bubblers or drip emitters, only the wetted area should be included in the calculation of available plant water. If it is assumed that the wetted soil volume is a cylinder below the soil surface wetted area, then S is calculated as follows:

$$S = \left(\frac{\pi D_b^2\ m^2}{4}\right)(Z\ m)(AWC)(MAD)\left(\frac{1,000\ L}{m^3}\right) = 780{*}Z{*}AWC{*}MAD{*}D_b^2 \tag{16.5}$$

where

D_b = diameter of individual bubbler basin or drip emitter wetted diameter, m.

If more than one emitter is used and wetted areas are not connected, then the wetted area calculated with Eq. 16.5 should be multiplied by the number of emitters.

Many plants concentrate roots near the soil surface. As a result, there is not as much water extraction from the lower parts of the root zone. One common plant water extraction model assumes that 40 %, 30 %, 20 %, and 10 % of the water extracted by the plant comes from the upper, 2 middle, and lower layers, respectively. If this is the case, then for calculations of available water, assume that 100 %, 75 %, 50 %, and 25 % of the readily available water is actually used by the plant in the upper, 2 middle, and lower layers, respectively. The fraction of the readily available water used by the plant is

$$(1 + .75 + .5 + .25)/4 = 2.5/4 = 0.62$$

Ideally, the irrigation schedule (days between irrigation events) should be adjusted based on the observed rate of evapotranspiration. With a computerized irrigation system and central control at a computer in the office, it is possible to adjust irrigation schedules with changes in reference evapotranspiration. It is even possible to do this automatically based on input from a weather station. However, landscape managers with on-site controllers generally do not seem to have time to adjust the irrigation timer on a weekly or even monthly basis, let alone based on daily changes in evapotranspiration. It is reasonable to ask landscape managers and homeowners to adjust their timers four times per year. In arid climates, an rule of thumb is that spring and fall reference evapotranspiration is ½ of summer reference evapotranspiration, and winter reference evapotranspiration is ¼ of summer reference evapotranspiration. The species factor may change during the year (some plants lose their leaves in winter) so the ½ and ¼ adjustments may need some fine tuning based on the type of plants in the landscape. Other climatic regions will have different relationships between warm season and cool season evapotranspiration.

Example 16.2 Calculate the watering schedule for a 3 m diameter (3 m) sweet acacia tree that is surrounded by bare soil with ET_0 in summer equal to 12 mm/day. Calculate the schedule (days between irrigations and watering time for the four seasons) for a bubbler in a 2 m diameter basin, and for four emitters. Bubbler flow rate is 8 LPM. Emitter flow rates are 4 LPH and the soil is loamy sand (AWC = 8 %); thus, emitter wetted diameter is 0.6 m. Acacia are drought tolerant trees, and the owner does not want excessive growth, but also wants the plant to look healthy. Assume that Acacia root water extraction does not follow the 40-30-20-10 pattern. Effective root zone depth is 1 m. Because acacias are drought tolerant trees and the owner does not want excessive growth, MAD is 0.8. Irrigation efficiency is 80 %

Calculate K_L.

The WUCOLS guide classifies sweet acacia (acacia farnesiana, page 63) as a low water use tree. The low range is 0.1–0.3. Since the owner wants the tree to look healthy but not have excessive growth, estimate a species factor that is on the upper end of the low range, 0.25. There is bare soil below the tree so the density coefficient is 1.0. Because the acacia is exposed on all sides to sun and wind, the evapotranspiration should be higher than if the tree was in a dense planting so assume that the microclimate coefficient is 1.2.

Calculate K_L

$$\mathrm{K_L} = \mathrm{K_S}^*\mathrm{K_{mc}}^*\mathrm{K_d} = 0.25^*1.2^*1.0 = 0.3.$$

Calculate plant evapotranspiration rate

$$\mathrm{ET_L} = \mathrm{K_L}^*\mathrm{ET_0} = 0.3^*12\ \mathrm{mm/day} = 3.6\ \mathrm{mm/day}$$

Calculate plant daily water requirement.

$$\mathrm{LPD} = \mathrm{ET_L}^*\left(\pi\mathrm{D}^2\right)/4/\mathrm{Eff} = 3.6\ \mathrm{mm/day}^*\pi/4\left(3^2\right)/0.8 = 31.8\ \mathrm{LPD}$$

Calculate soil water storage (S) for bubbler

AWC is 8 % (0.08), and MAD = 0.8.

$$S = 780^*Z^*AWC^*MAD^*D_b^2 = 780^*1.0^*0.08^*0.8^*2^2 = 200\ L$$

Calculate soil water storage (S) for 4 drip emitters

$$S = 4^*780^*Z^*AWC^*MAD^*D_b^2 = 4^*780^*1.0^*0.08^*0.8^*0.6^2 = 72\ L$$

Calculate days between irrigation (S/LPD) for bubbler

$$\mathrm{Days} = \mathrm{S/LPD} = 200\ \mathrm{L}/32\ \mathrm{LPD} = 6\ \mathrm{days}$$

Calculate irrigation run time (S/Q) for bubbler

$$\mathrm{Minutes} = \mathrm{S/LPM} = 192\ \mathrm{L}/8\ \mathrm{LPM} = 24\ \mathrm{minutes}$$

Calculate days between irrigation (S/LPD) for drip emitters

$$\mathrm{Days} = \mathrm{S/LPD} = 72\ \mathrm{L}/32\ \mathrm{LPD} = 2\ \mathrm{days}$$

Calculate irrigation run time (S/Q) for drip emitters

$$\mathrm{Hours} = \mathrm{S/LPH} = 64\ \mathrm{L}/(4^*4)\mathrm{LPH} = 4\ \mathrm{hours}$$

The schedule might be adjusted to 30 minutes every 7 days (week) in summer in order to make programming the automatic controller more straightforward. Acacia trees have foliage year-round so the species factor does not change during the year. Thus, the schedule would be adjusted to 30 minutes every 2 weeks in spring and fall, and every month in winter. The schedule could also be adjusted to 6 hours every 3.5 days (twice per week) in summer, every week in spring and fall, and every 2 weeks in winter.

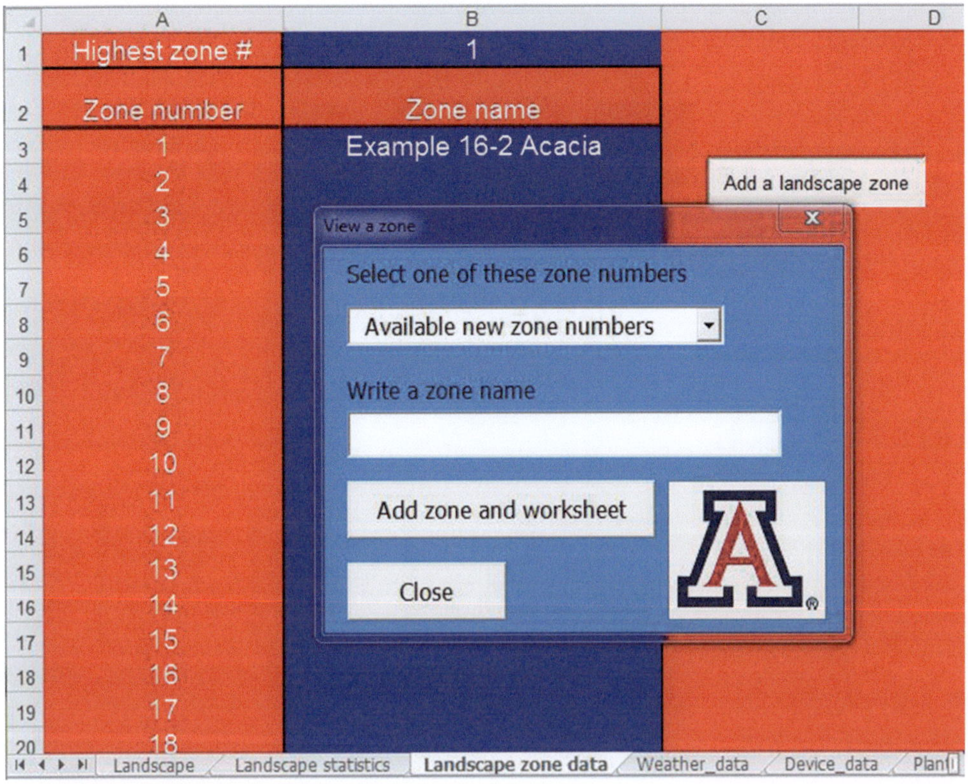

Fig. 16.4 *Landscape zone data* worksheet

The same calculation is demonstrated in the *Chapter 16 Landscape Irrigation* workbook (Figs. 16.4 and Fig. 16.12) in the following pages. The first step is to create the zone. Open the Landscape zone data worksheet and click the *Add a landscape zone* button. Create the *Example 16-2 Acacia zone* (zone 1) worksheet, as shown in Fig. 16.4.

The next step is to open the *Landscape* worksheet and click the "Step 1. Modify zone" button in order to open the dialog box and input soil and vegetation characteristics (Fig. 16.5). Although loamy sand is selected, the problem specifies that the AWC should be 8 % so this parameter is modified in the Textbox. The microclimate factor is selected as 1.2 and the density is 1.0. Then the *Write data to landscape worksheet* button is clicked.

The next step is to add the plant and bubbler. Common plants and irrigation devices in ar region can be added to the *Plant data* (Fig. 16.6) and *Device data* (Fig. 16.7) worksheets. In this case, the acacia tree is added to the plant data worksheet with the appropriate diameter and root depth. "P" in column H refers to the fact that it is a plant rather than a planted area (A) or hedge (H).

The device data worksheet includes various emitters and bubblers. The devices in this problem are pressure compensating emitters and bubblers, which means that there is no change in flow rate with pressure. These are designated by "PC". The first two devices in the *Device data* worksheet (Fig. 16.7) are pressure compensating and have the appropriate diameters and flow rates. There is no overlap so the overlap fraction is zero. The irrigation efficiency is set to 0.78 in order to agree with this problem.

The Landscape program allows for a reduction in water application in drought years (Fig. 16.8). The Thrive/survive percentage can be decreased below 100 % during drought years, when less than the required rate of water application is applied in order to conserve water. The soil depth is also specified on this page. As long as the soil depth is deeper than the plant root depth, then there is no constraint due to soil depth. In this case, the leaching fraction box is unchecked so the irrigation salinity level in Cell H3 is not considered and no extra water is applied to leach salts from the root zone.

The next step is to add the plant to the *Landscape* worksheet. Click "Step 2. Add plants" and the dialog box shown in Fig. 16.9 appears. The plant was added to the dialog box with the drop down menu and the bubbler is ready to be added. Once the bubbler is added, click "Add plant to Landscape page" and the plant is added (Fig. 16.10).

The bubbler and drip calculations for this problem cannot be run at the same time because drip and bubbler systems should not be mixed in the same zone. The flow rate of bubblers is almost two orders of magnitude greater than the flow rate of drip emitters. The next step is to click *Step 3*.

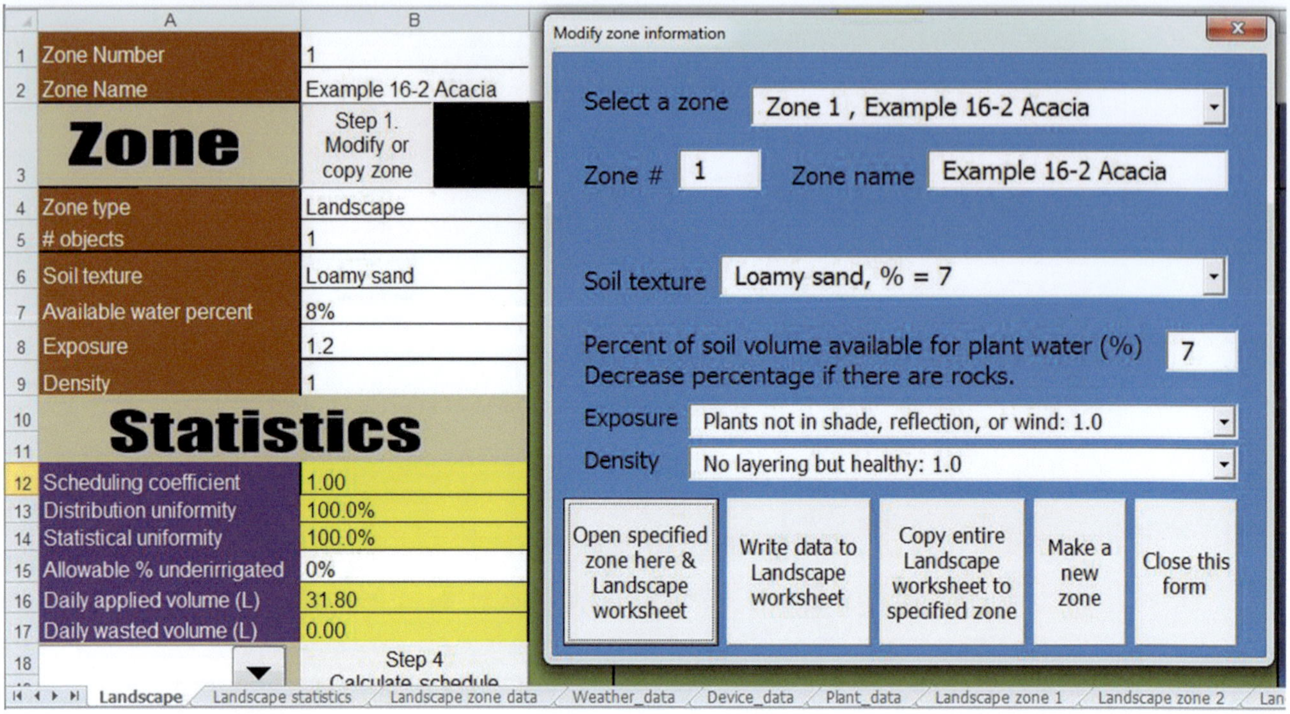

Fig. 16.5 *Landscape* worksheet with *Modify zone* dialog box open

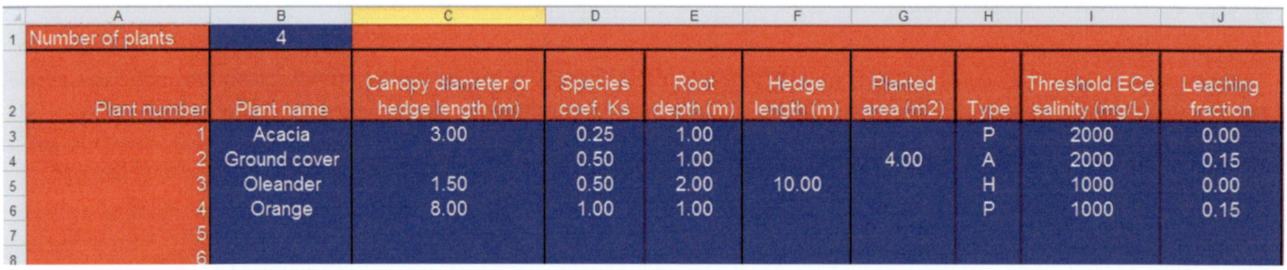

Number of plants	4								
Plant number	Plant name	Canopy diameter or hedge length (m)	Species coef. Ks	Root depth (m)	Hedge length (m)	Planted area (m2)	Type	Threshold ECe salinity (mg/L)	Leaching fraction
1	Acacia	3.00	0.25	1.00			P	2000	0.00
2	Ground cover		0.50	1.00		4.00	A	2000	0.15
3	Oleander	1.50	0.50	2.00	10.00		H	1000	0.00
4	Orange	8.00	1.00	1.00			P	1000	0.15
5									
6									

Fig. 16.6 *Plant data* worksheet

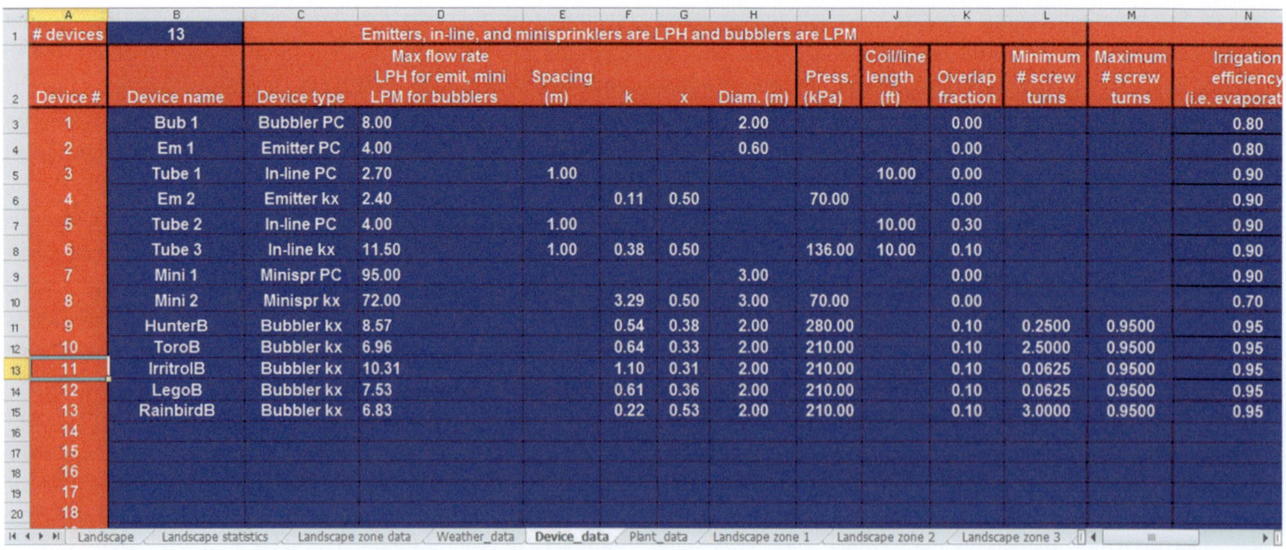

# devices	13	Emitters, in-line, and minisprinklers are LPH and bubblers are LPM											
Device #	Device name	Device type	Max flow rate LPH for emit, mini LPM for bubblers	Spacing (m)	k	x	Diam. (m)	Press. (kPa)	Coil/line length (ft)	Overlap fraction	Minimum # screw turns	Maximum # screw turns	Irrigation efficiency (i.e. evaporat
1	Bub 1	Bubbler PC	8.00				2.00			0.00			0.80
2	Em 1	Emitter PC	4.00				0.60			0.00			0.80
3	Tube 1	In-line PC	2.70	1.00					10.00	0.00			0.90
4	Em 2	Emitter kx	2.40		0.11	0.50		70.00		0.00			0.90
5	Tube 2	In-line PC	4.00	1.00					10.00	0.30			0.90
6	Tube 3	In-line kx	11.50	1.00	0.38	0.50		136.00	10.00	0.10			0.90
7	Mini 1	Minispr PC	95.00				3.00			0.00			0.90
8	Mini 2	Minispr kx	72.00		3.29	0.50	3.00	70.00		0.00			0.70
9	HunterB	Bubbler kx	8.57		0.54	0.38	2.00	280.00		0.10	0.2500	0.9500	0.95
10	ToroB	Bubbler kx	6.96		0.64	0.33	2.00	210.00		0.10	2.5000	0.9500	0.95
11	IrritrolB	Bubbler kx	10.31		1.10	0.31	2.00	210.00		0.10	0.0625	0.9500	0.95
12	LegoB	Bubbler kx	7.53		0.61	0.36	2.00	210.00		0.10	0.0625	0.9500	0.95
13	RainbirdB	Bubbler kx	6.83		0.22	0.53	2.00	210.00		0.10	3.0000	0.9500	0.95
14													
15													
16													
17													
18													

Fig. 16.7 *Device data* worksheet

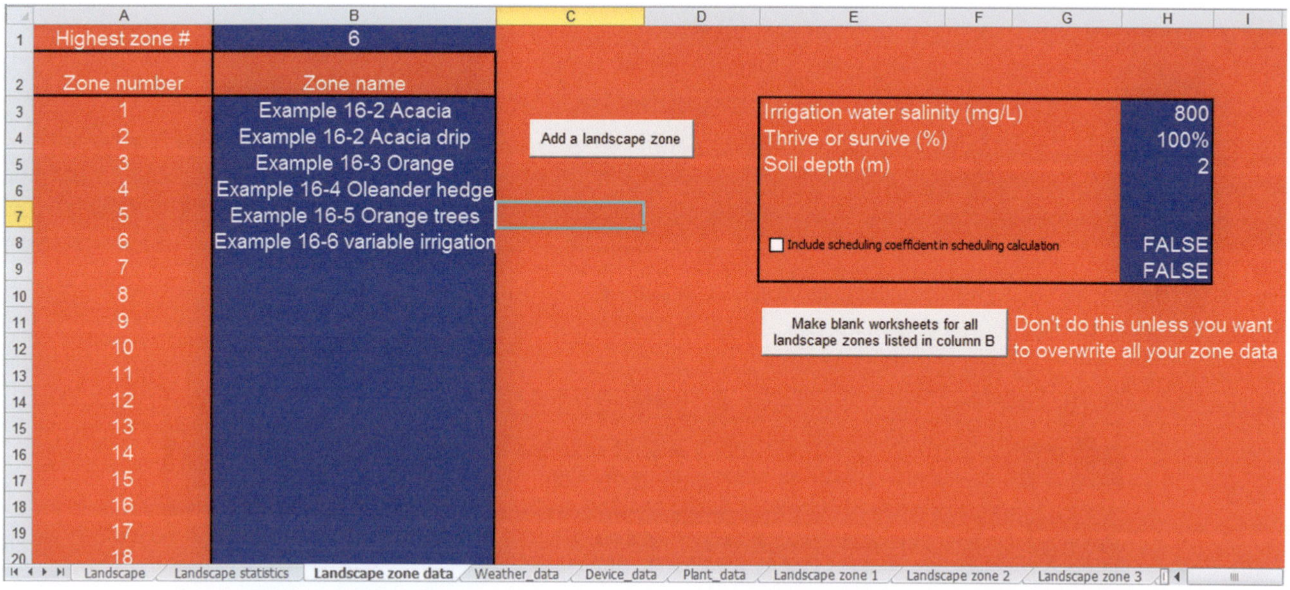

Fig. 16.8 *Landscape zone data* worksheet

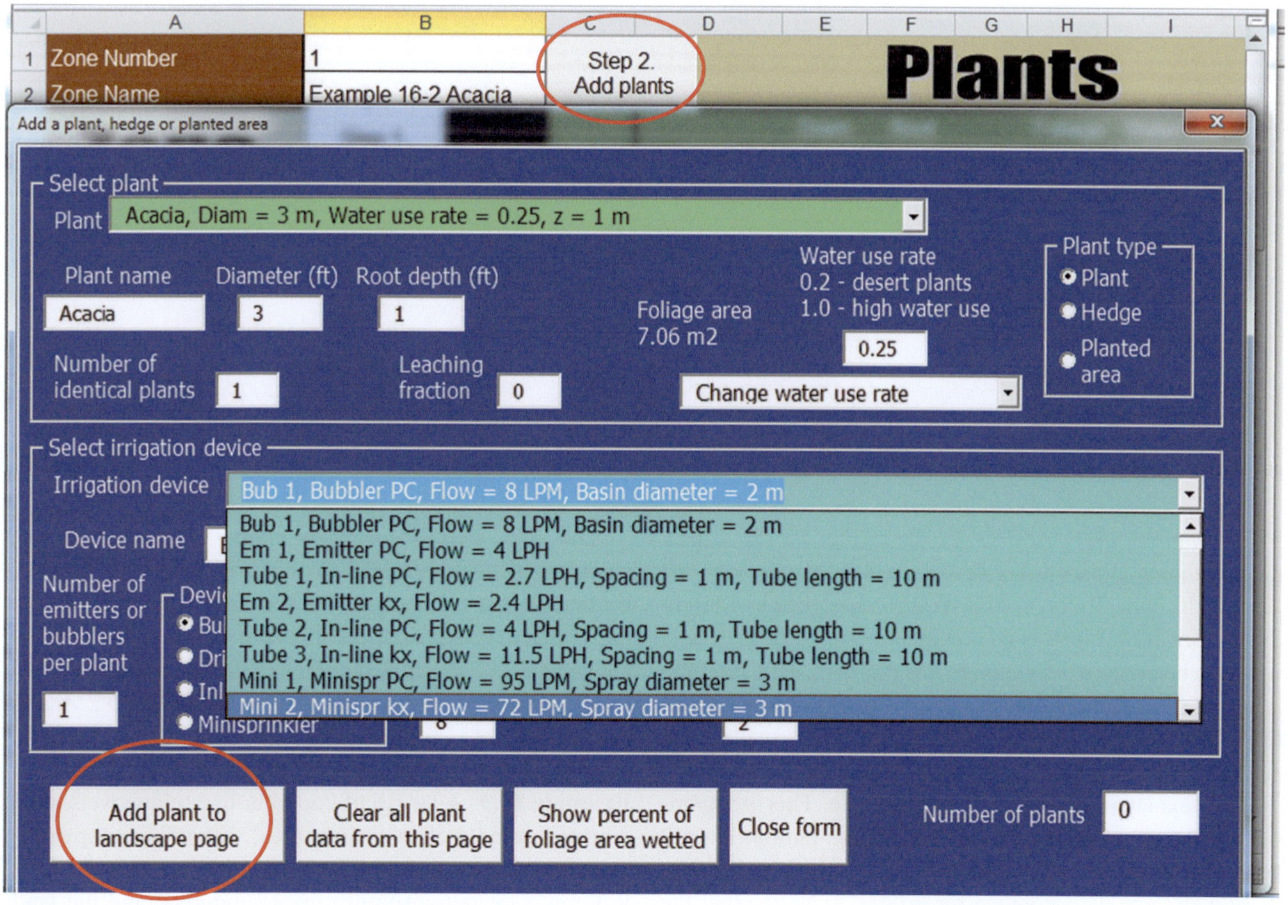

Fig. 16.9 *Add plant* dialog box

Plant number	Plant name	Diam. or width (m)	Root depth (m)	Length (m)	Foliage area (m)	Water use (0.2-desert, 1-high)	Plant type H, P, A	Number emitters/ plant	Name	Type	Flow (LPM) Bub (LPH) Em
1	Acacia	3.00	1		7.0686	0.25	Plant	1	Bub 1	Bubbler PC	8

Fig. 16.10 The acacia plant and bubbler added to the *Landscape* worksheet

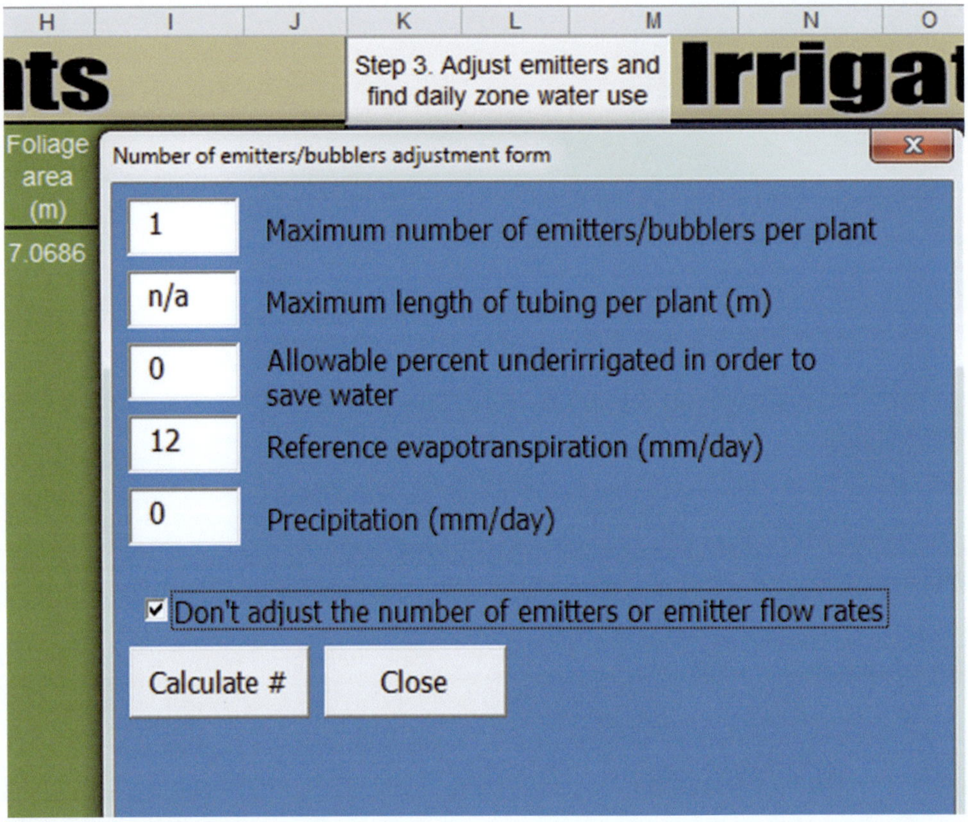

Fig. 16.11 Dialog box to calculate schedule

Adjust emitters and calculate schedule (Fig. 16.11). Check the box to not adjust emitter flow rate since there is only one plant in the zone. Reference ET has been specified as 12 mm/day in Cell B21. There is no precipitation (Cell B22). There will also be no underirrigation in this problem. This function is used if there is more than one plant and allows some plants to be underirrigated in order to avoid overirrigating the entire zone.

The next step is to calculate the schedule. This is accomplished by clicking the *Calculate schedule* button. Do not adjust the precipitation or reference ET at this point because these were already accounted for in Step 3. Notice that the irrigation period and days between irrigations are the same as manual calculations

The next step is to make a new zone for the drip calculations: *Example 16-2 Acacia drip*. Make sure to specify that there are 4 emitters. Also specify that you will input the wetted diameter and specify that it is 0.6. The schedule is the same as the calculated values (2 days and 4 hours). This zone is included in the workbook as *Example 16-2 Acadia drip* worksheet.

Example 16.3 A length of coiled in-line tubing wets a 4 m diameter area (¼ of the canopy area) within the canopy of an 8 m diameter orange tree. The total coil flow rate is 180 LPH. Reference evapotranspiration is 12 mm/day, and the landscape coefficient for orange trees is 1.0. Root zone depth is 2 m, and orange trees have a 40 – 30 – 20 – 10 water

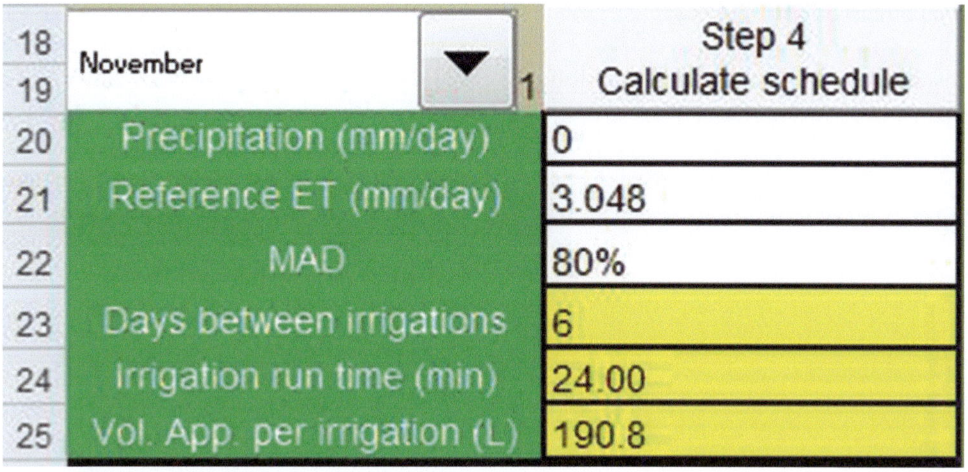

Fig. 16.12 *Calculate schedule* button and output

extraction pattern. MAD is 0.5. Soil AWC = 0.08. Calculate a watering schedule.

Determine the water requirement of the tree

$$ET_L = ET_0{*}K_L = 12\ \text{mm/day}{*}1.0 = 12\ \text{mm/day}$$

$$ET_L{*}(\pi D^2)/4/Eff = 12\ \text{mm/day}{*}\pi/4(8^2)/0.8 = 867\ \text{LPD}$$

Determine the readily available water in the wetted region (S)

Treat the wetted coil as a circular wetted area rather than an inline drip tube. Multiply root zone depth by 0.62 (40 – 30 – 20 – 10 water extraction pattern) and multiply by the area wetted. The wetted area is 4 m diameter.

$$S = 780{*}Z{*}AWC\,{*}MAD{*}D_b^2$$

$$= 780(2{*}0.62\ m)(0.08)(0.5)(4\ m)^2 = 619\ L$$

Find hours of watering required per day

$$753\ \text{LPD}/180\ \text{LPH} = 4.3\ \text{hours/day}.$$

Find the time between irrigations.

If the drip system is operated during the day, when ET is high, then the water applied would not exceed the water capacity of the soil, and water would not be wasted. A second option would be to turn on the system twice per day for 2.15 hours. In order to make the same calculations with the Chapter 15 Landscape irrigation program, the in-line drip tube is treated as a bubbler, since wetted area is not specified as a function of tube length but is a wetted circle. The flow rate is specified as 180/60 = 3 LPM. Specify that the leaching fraction is zero (Column J) in *Plant data* worksheet and Column X in *Landscape* worksheet.

For a hedge (straight, uniform row of plants) watered with in-line tubing (Fig. 16.13), the watering schedule is a function of the hedge width (evaporation rate) and the wetted width (soil water storage). Because the geometry is the same along the hedge, the schedule can be calculated for hedge interval equal to the emitter spacing.

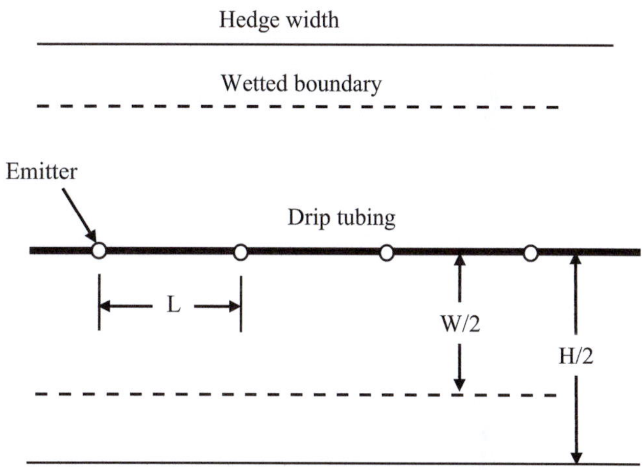

Fig. 16.13 Wetted width (W) and hedge width (H) for irrigation scheduling calculation

The evapotranspiration rate (LPD) for a hedge interval equal to the emitter spacing is calculated as follows:

$$LPD = \frac{\left(\frac{ET_L\ mm}{day}\right)\left(\frac{m}{1,000\ mm}\right)(H{*}L\ m^2)\left(\frac{1,000\ L}{m^3}\right)}{Eff} = \frac{ET_L{*}H{*}L}{Eff} \tag{16.6}$$

where

LPD = evapotranspiration for the distance between emitters, liters per day
L = length between emitters, m
H = hedge (plant) width, m

ET = depth of evapotranspiration, mm/day.

The soil water storage per emitter is calculated as follows:

$$S = \left(L^*W\ m^2\right)(Z\ m)(AWC)(MAD)\left(\frac{1,000\ L}{m^3}\right)$$
$$= 1,000^*Z^*AWC^*MAD^*L^*W \quad (16.7)$$

where

W = wetted width

Example 16.4 Calculate the watering schedule for a 10 m oleander hedge. The owner desires to have rapid hedge growth. Reference evapotranspiration is 8 mm/day. The oleander hedge is in a single layered landscape. The hedge is 1.5 m wide, the distance between emitters is 0.3 m (33 emitters), and the emitter flow rate is 4 LPH. The oleander hedge is on the south side of a building. Irrigation efficiency is 80 %. Soil is sandy and AWC is 0.08. Root depth is 2 m

Calculate K_L. Soil is sandy, and wetted width from emitters is 0.6 m. Oleander is a medium water use plant when irrigated for a lush appearance so assume that the species coefficient is 0.5. The density coefficient is 1.0 for the single layered landscape. The oleander is on the south side of a building so assume that the microclimate coefficient is 1.3. Calculate K_L and ET_L

$$K_L = K_S{}^*K_{mc}{}^*K_d = 0.5^*1.3^*1.0 = 0.65$$
$$ET_L = K_L{}^*ET_0 = 0.65^*8\ \text{mm/day} = 5.2\ \text{mm/day}$$

Calculate plant water use (LPD/emitter)

$$LPD = \frac{ET_L{}^*H^*L}{Eff} = \frac{5.2\ mm/day^*1.5\ m^*0.3\ m}{0.8}$$
$$= 2.9\ LPD/emitter$$

Calculate soil water storage (S/emitter)

Specify that MAD is 0.5 in order to maintain oleander with a thriving appearance

$$S = 1,000(Z)(AWC)(MAD)(L)(W)$$
$$= 1,000(2)(0.08)(0.5)(0.3)(0.6) = 14.4\ L$$

Calculate days between irrigation (S/LPD) and

$$\text{Days} = \text{S/LPD} = 14.4/2.9 = 5\ \text{days}$$

Calculate irrigation run time (S/Q)

$$\text{Hours} = \text{S/LPH} = 14.4\ \text{L}/4\ \text{LPH} = 3.6\ \text{hours.}$$

The watering schedule is included in the Chapter 16 *Landscape irrigation* program. To do it yourself, specify the plant and irrigation characteristics as shown in Fig. 16.14. Make sure to select *Input the wetted diameter* and specify the wetted diameter as 0.6 m. Also, change the spacing between emitters to 0.3 m. Specifying the correct number of emitters (lower left corner) is not necessary in this case since the program automatically calculates this number based on spacing and tubing length; however, it can be input as 33 in the *Landscape* worksheet (Column K) if you want to record it. Also, specify that there is no overlap (Column S) since this is a single tube; in this case, overlap specifies the overlap from the line sources (tubes) and not the overlap of the individual emitters. The plant evaporation is calculated as a function of the total hedge length and area rather than as the area over an individual emitter. Specify that the rain harvest (Column T) is zero.

In some cases, two inline drip laterals are placed on either side of the hedge. This increases the wetted area under the hedge. Dual lines can be constructed as a looped system, with water feeding from both ends. This configuration is less likely to plug in the case of a break in the pipe because water comes to the break in the pipe from both sides and prevents dirt from washing into the pipe.

Planters may have uneven shapes with various types of trees and shrubs in a multiple layers. In this case it is best to design for the least drought tolerant plant. If this is not done, then the drought tolerant plants will remove water from the soil and the less drought tolerant plant will suffer. For ET calculations, the planter can be treated as a single unit and the total irrigation flow rate can be compared to the total planter water requirement in order to develop an irrigation schedule.

Zoning

Tree water requirements are so much greater than shrub requirements that it is nearly impossible to water them in the same zone. It is best to divide trees and shrubs into different zones. However, if shrubs are directly under trees and they have similar water requirements during the year, then they can be placed in the same zone. In addition to dividing trees and shrubs into different zones, it is best to divide high water use, medium water use, and low water use plants into different zones. Some plants lose foliage in winter while others continue to use water and require winter irrigation. If these plants are on the same zone, then it is hard to meet the needs of the plant that requires water in winter without wasting water on the plant that doesn't require water in winter.

Even within zones with identical plants, the growth rates may vary. Thus, it is necessary to increase the water application rate to larger plants and vice versa for smaller plants.

The design process for a zone with plants that have different water requirements is performed as follows. This

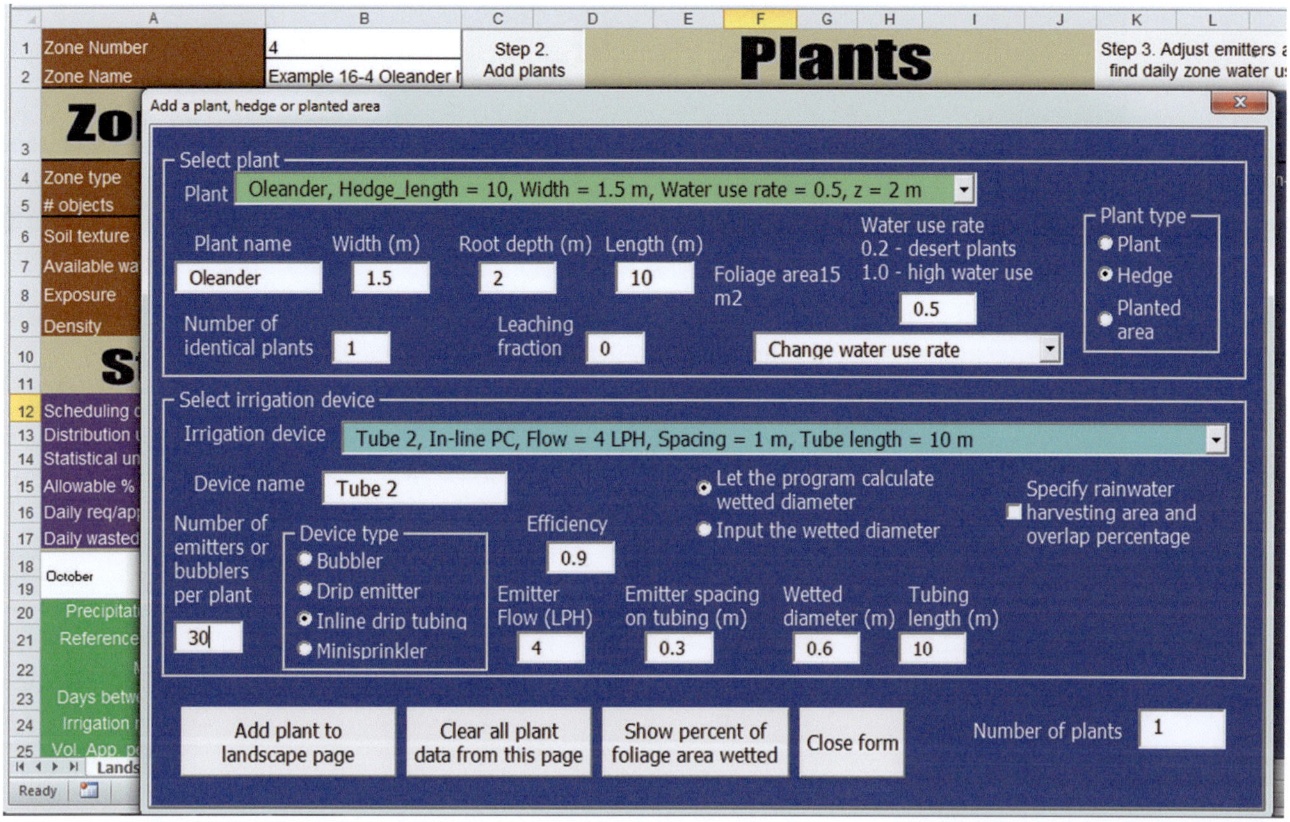

Fig. 16.14 Plant and irrigation dialog box for Oleander with in-line tubing

example applies to bubbler irrigation, but it could also apply to drip irrigation.

1. Calculate each plant's daily irrigation requirement (LPD).
2. Calculate each plant's soil water holding capacity (S).
3. Find the plant with the highest daily irrigation requirement in the irrigation zone.
4. Calculate the water requirements of all other plants in the zone as a percentage of the water requirement of the largest tree in the zone.
5. Calculate the maximum flow possible from the bubbler or set of emitters, and assign this flow rate to the plant with the highest water requirement. Calculate the flow to all other plants as the product of the percent of the maximum water requirement and the maximum bubbler or set of emitters flow rate.
6. Design the pipe system to connect the bubblers to the irrigation valve.
7. If bubblers, then calculate the number of screw turns or cap turns for each bubbler based on bubbler flow rate and irrigation pipeline pressure. If inline tubing, calculate length of tubing. If emitters, calculate number of emitters.
8. If basin volume is less than SWHC, then S = basin volume.
9. Calculate the days between irrigations as S/LPD for the largest plant.
10. Calculate the watering time as S / LPM for the largest plant (maximum flow rate).

Example 16.5 Design a bubbler irrigation system for the 4 orange trees in Fig. 16.15. Use Hunter bubblers. The pressure on the discharge side of the valve is 288 kPa. Reference evapotranspiration is 10 mm/day. Assume that orange tree rooting depth is 1.5 m. MAD = 0.5 and K_L = 1.0. Soil is sandy loam. Depth of 2 m basins is 7 cm. Irrigation efficiency is 78 %

Calculate each plant's daily irrigation requirement (LPD).

Crop evapotranspiration, $ET_c = ET_0 * K_c = 10$ mm/day * 1.0 = 10 mm/day.

$$\text{Bubbler 1}: \text{LPD} = \text{ET}_\text{L}\left(\pi \text{D}^2\right)/4/\text{Eff} = 10\left(\pi 4^2\right)/4/0.78 = 160 \text{ LPD}$$

$$\text{Bubblers } 2-4: \text{LPD} = \text{ET}_\text{L}\left(\pi \text{D}^2\right)/4/\text{Eff} = 10\left(\pi 3^2\right)/4/0.78 = 90 \text{ LPD}$$

Total LPD for the entire zone = 160 + 90 (3) = 430 LPD

Calculate each plant's soil water holding capacity (S).

All bubblers have 2 m basins.

$$S = 780{*}Z{*}AWC{*}MAD{*}D_b{}^2 = 780{*}1.5{*}0.12{*}0.5{*}2^2 = 281\ L$$

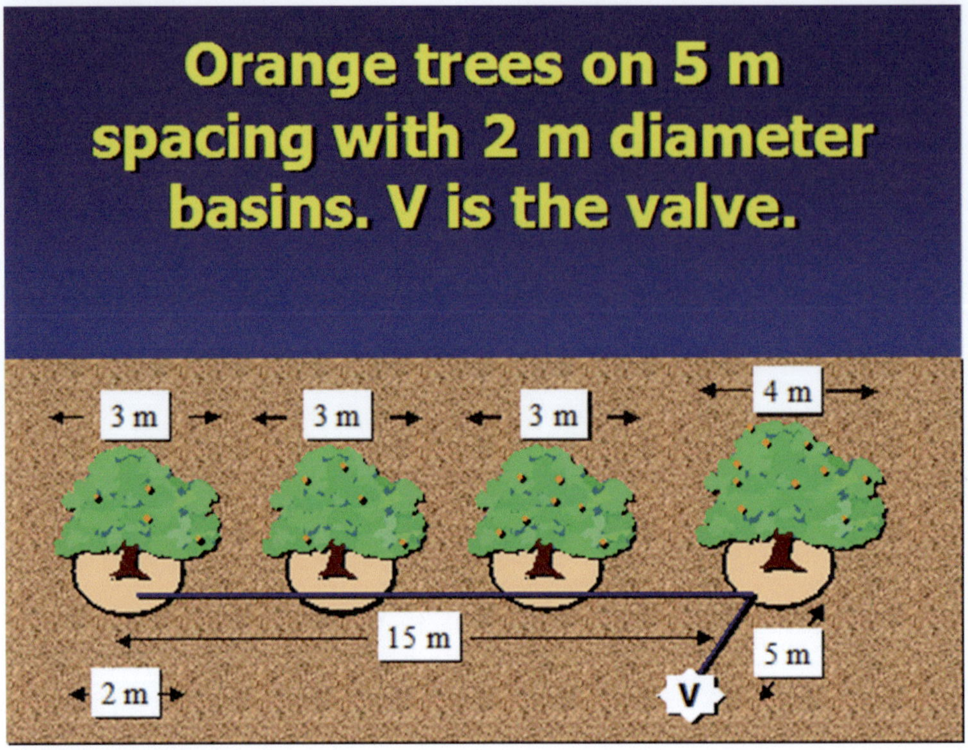

Fig. 16.15 Zone with four orange trees

Find the plant with the highest daily irrigation requirement in the irrigation zone

$$\text{Bubbler}\#1 \text{ requires } 160 \text{ LPD}$$

Calculate the water requirements of all other plants in the zone as a percentage of the water requirement of the largest tree in the zone.

$$\text{Bubblers } 2-4 \text{ require } 90/160{*}100\% = 56\%$$

*Calculate the maximum flow possible from the bubbler brand or emitters that are used, and assign this flow rate to the plant with the highest water requirement. Calculate the flow to all other plants as the product of the following: (percent of the maximum water requirement) * (maximum bubbler flow rate).*

Calculate maximum flow at system operating pressure of 280 kPa for Hunter bubblers.

Maximum number of turns for Hunter bubblers is 1.25 rotations ($360^0 * 1.25$).

From Fig. 15.13

$$x = 1.1863(T)^3 - 3.739(T)^2 + 3.3945(T) - 0.3338$$
$$x_{max} = 1.1863(1.25)^3 - 3.739(1.25)^2 + 3.3945(1.25) - 0.3338 = 0.384$$

From Fig. 15.14

$$k = 1.925(T)^2 - 2.3242(T) + 0.8818$$
$$k_{max} = 1.925(1.25)^2 - 2.3242(1.25) + 0.8818 = 0.984$$

Calculate maximum flow

$$Q = kH^x = (0.984)\left(280^{0.384}\right) = 8.67 \text{ LPM}$$

Calculate flow rate for bubblers 2–4.

$$Q = (0.56)(8.67 \text{ LPM}) = 4.85 \text{ LPM}$$

Design the pipe system to connect the bubblers to the irrigation valve.

Because the flow rate is low, <40 LPM, it is possible to use no larger than 1 inch pipe on the entire property. Don't worry about telescoping the size to ¾ and ½ inch in order to save a little money on pipe. It is more trouble to buy the different fittings than it is worth to save a little money on pipe. Usually, because residential landscape irrigation systems do not use more than 50 LPM, it is easiest to just use 1 inch (25 mm) pipe on the entire property. Using one size of pipe makes the purchase of fittings easier, and there are usually less return trips to the irrigation store. The pipe system

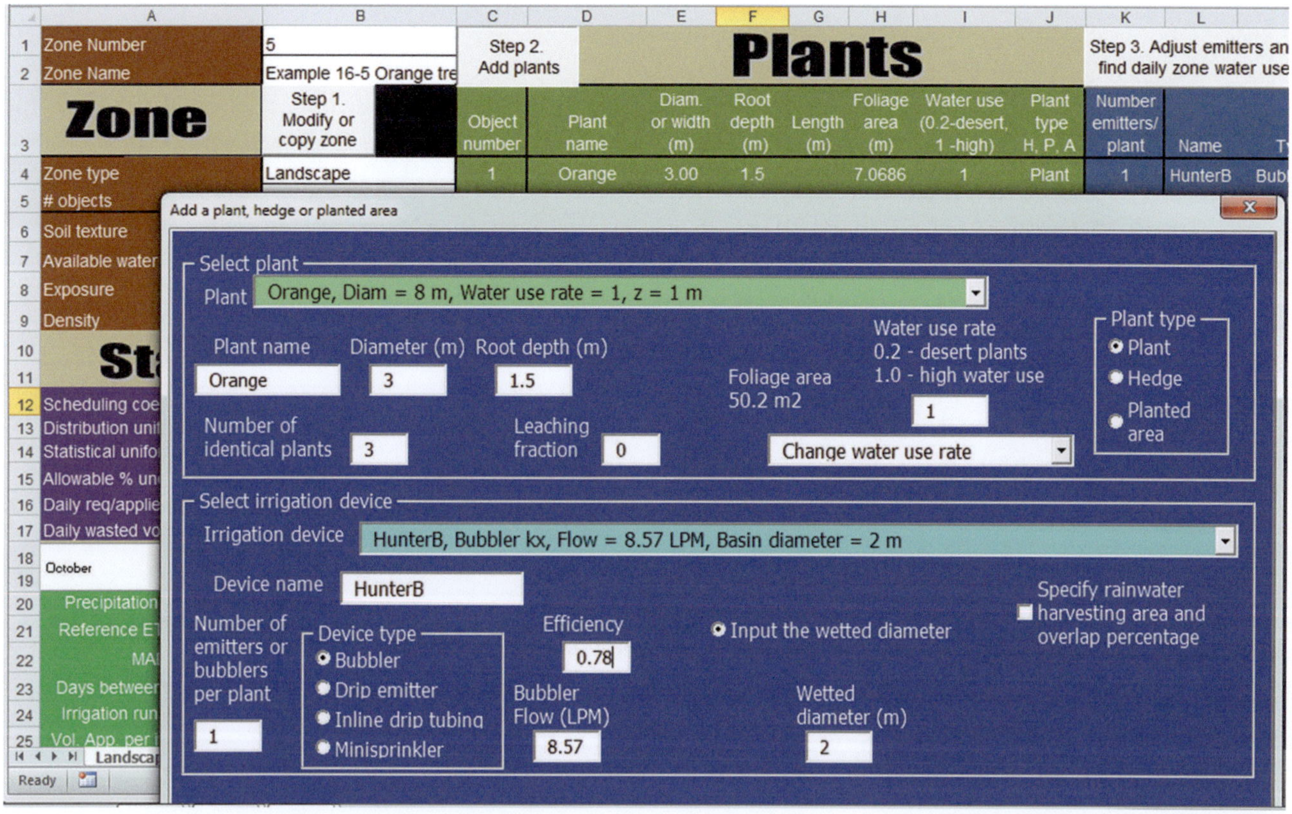

Fig. 16.16 *Add plant* dialog box for 3 orange trees in Example 16.5

includes 4 lengths of 5 m pipe, and total pressure loss in this system is trivial (<1 kPa).

Calculate the number of screw turns or cap turns for each bubbler based on bubbler flow rate and irrigation pipeline pressure.

Calculate all bubbler flow rates at 286 kPa.

$$\text{Bubbler 1 flow rate} = Q = kH^x = 0.984{*}286^{0.384} = 8.64\ \text{LPM}$$

$$\text{Bubblers } 2-4 \text{ flow rate} = 8.64{*}0.56 = 4.84\ \text{LPM}$$

Find the number of turns for bubblers 2–4 by solving the following equation ($Q = kH^x$) for T by iteration

$$Q = \left(1.925\ T^2 - 2.3242\ T + 0.8818\right){*}H\hat{}\left(1.1863\ T^3 - 3.739\ T^2 + 3.3945\ T - 0.3338\right)$$

Solution: T = 0.44 turns, and Q = 4.84 LPM.

Adjust bubblers 2–4 to 0.44 turns.

Check that basin surface volume is less than S

$$\text{Basin surf. volume} = \pi\left(D_b^2/4\right)(\text{depth of basin})\left(1{,}000\ \text{L/m}^3\right) = 3.14\left(2^2\right)(0.07\ \text{m})(1{,}000) = 879\ \text{L}$$

SWHC = 281 L so SWHC is the limiting factor, let S = 281 L

Calculate the days between irrigations as S / LPD for the largest plant.

$$\text{Days} = \text{S/LPD} = 281/160 = 1.8\ \text{days} \rightarrow 1\ \text{day}$$

At this point, the decision must be made whether to irrigate daily and stay within the water holding capacity of the soil or irrigate every two days. Daily irrigation is probably a better option since orange trees respond well to daily irrigation and because bubbler irrigation is quick and would not be stretched over a significant fraction of the ET period. Thus, the watering time is calculated as a fraction of the daily volume of ET (LPD) rather than the water holding capacity of the soil (S).

Calculate the watering time as LPD / LPM for the largest plant (maximum flow rate).

$$\text{Minutes} = \text{LPD/LPM} = 160/8.64\ \text{LPM} = 18.5\ \text{minutes}$$

The problem can also be solved with the *Chapter 16 Landscape irrigation* program. In this case, there are three identical plants (3 m diameter) so specify that there are three plants in the *Add plant* dialog box (Fig. 16.16). Select the

Object number	Plant name	Diam. or width (m)	Root depth (m)	Length (m)	Foliage area (m)	Water use (0.2-desert, 1-high)	Plant type H, P, A	Number emitters/ plant	Name	Type	Flow (LPM) Bub (LPH) Em
1	Orange	8.00	1		50.265	1	Plant	7	Bub 1	Bubbler PC	8
2	Orange	4.00	1		12.566	1	Plant	1	Bub 1	Bubbler PC	8
3	Orange	5.00	1		19.635	1	Plant	1	Bub 1	Bubbler PC	8
4	Orange	3.00	1		7.0686	1	Plant	1	Bub 1	Bubbler PC	8

Fig. 16.17 *Landscape* worksheet before the number of bubblers per plant is adjusted

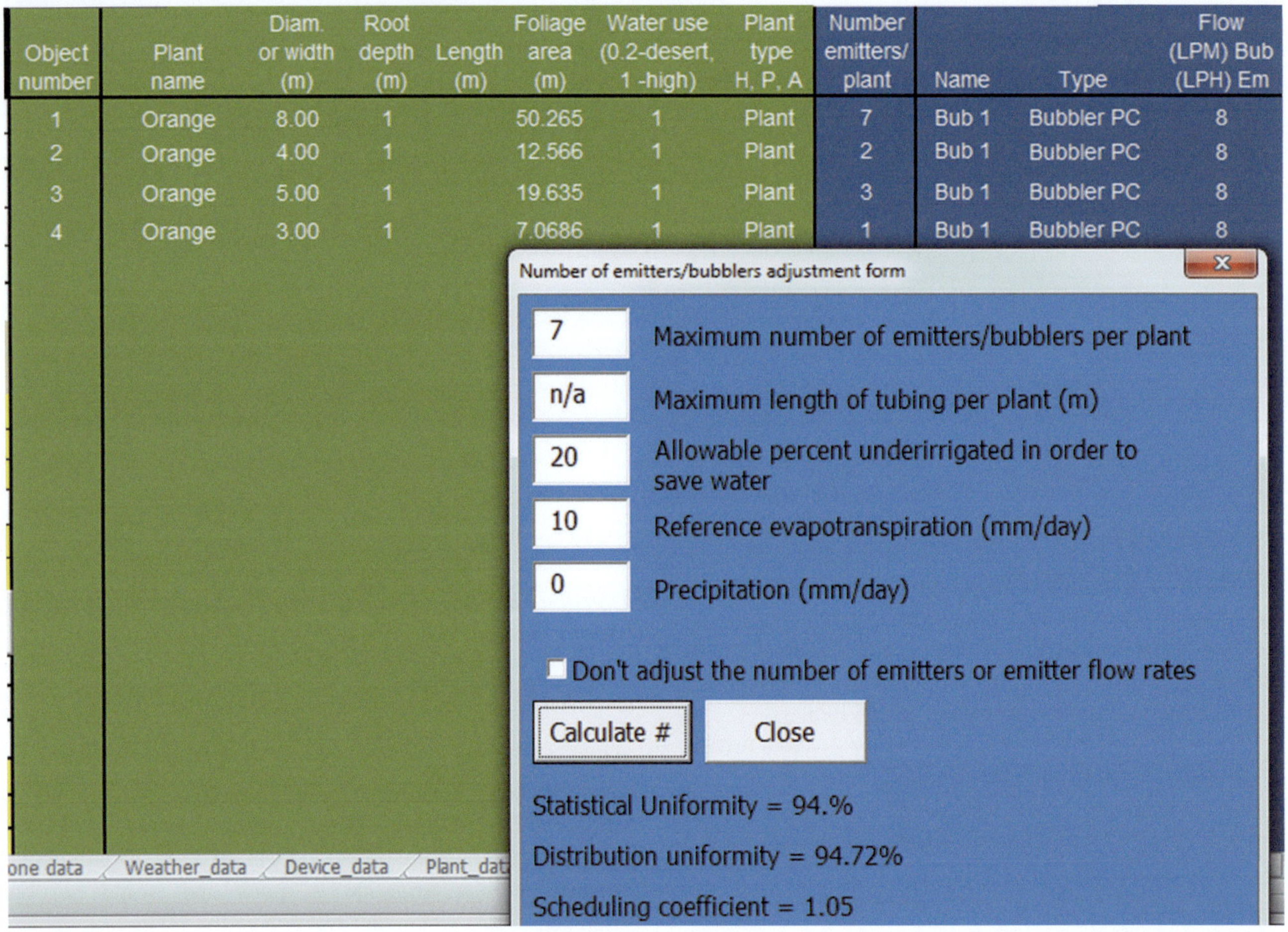

Fig. 16.18 *Landscape* worksheet after the number of bubblers per plant is adjusted

Hunter kx bubbler. The term kx refers to the fact that it is an adjustable bubbler, and that its flow rate varies with pressure. This is in contrast with the PC devices (pressure compensating), which have essentially no flow change with pressure change. The second object is then added (4 m diameter orange tree.). In this case, there is more than one object so the bubbler flow rates are adjusted to the relative sizes of the plants.

With multiple plants, the program has a complex algorithm to determine the relative amount of water applied to each plant. With adjustable emitters, the number of screw turns can be adjusted to give the correct amount to each plant. This is done for the *Example 16-5 Orange trees* worksheet zone just as it is done in Example 16.5. The calculated screw turns, application rates, and irrigation schedule are the same in the program as in the example.

With nonadjustable emitters, it is not possible to give each plant the correct application rate. Many plants will be under or over irrigated. However, the program adjusts the number of emitters in order to improve uniformity and then

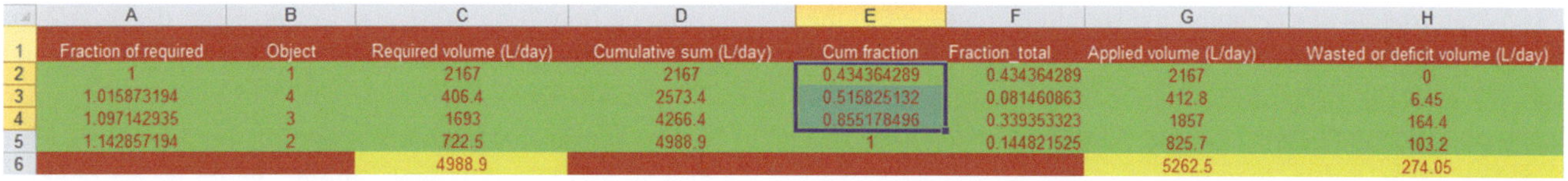

	A	B	C	D	E	F	G	H
1	Fraction of required	Object	Required volume (L/day)	Cumulative sum (L/day)	Cum fraction	Fraction_total	Applied volume (L/day)	Wasted or deficit volume (L/day)
2	1	1	2167	2167	0.434364289	0.434364289	2167	0
3	1.015873194	4	406.4	2573.4	0.515825132	0.081460863	412.8	6.45
4	1.097142935	3	1693	4266.4	0.855178496	0.339353323	1857	164.4
5	1.142857194	2	722.5	4988.9	1	0.144821525	825.7	103.2
6			4988.9				5262.5	274.05

Fig. 16.19 *Landscape statistics* worksheet

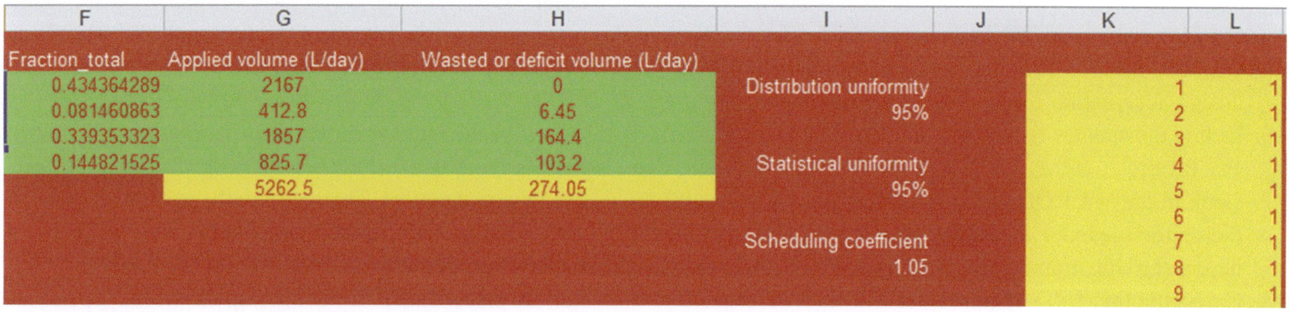

F	G	H	I	J	K	L
Fraction_total	Applied volume (L/day)	Wasted or deficit volume (L/day)				
0.434364289	2167	0	Distribution uniformity		1	1
0.081460863	412.8	6.45	95%		2	1
0.339353323	1857	164.4			3	1
0.144821525	825.7	103.2	Statistical uniformity		4	1
	5262.5	274.05	95%		5	1
					6	1
			Scheduling coefficient		7	1
			1.05		8	1
					9	1

Fig. 16.20 Distribution uniformity and statistical uniformity in *Landscape statistics* worksheet

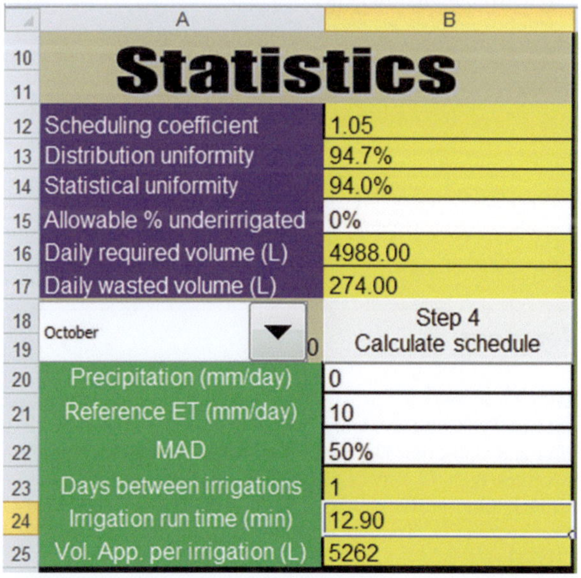

	A	B
10–11	**Statistics**	
12	Scheduling coefficient	1.05
13	Distribution uniformity	94.7%
14	Statistical uniformity	94.0%
15	Allowable % underirrigated	0%
16	Daily required volume (L)	4988.00
17	Daily wasted volume (L)	274.00
18–19	October 0	Step 4 Calculate schedule
20	Precipitation (mm/day)	0
21	Reference ET (mm/day)	10
22	MAD	50%
23	Days between irrigations	1
24	Irrigation run time (min)	12.90
25	Vol. App. per irrigation (L)	5262

Fig. 16.21 Calculated irrigation statistics from step 3 (cells B12:B17) and calculated irrigation schedule from step 4 (cells B20:B25)

allows specification of a rate underirrigation in order to reduce waste. This is demonstrated in Example 16.6.

Example 16.6 An orchard has 3 orange trees that are 8 m diameter, 4 orange trees that are 4 m diameter, 6 orange trees that are 5 m diameter, and 4 orange trees that are 3 m diameter. All are irrigated with 8 LPM nonadjustable bubblers. Find the number of bubblers per plant and the amount applied per day to each set of orange trees. The maximum number of bubblers per plant is 7. The leaching fraction is 0.15 and the irrigation efficiency is 80 %.

Set up the groups of orange trees in the zone by adding each set with the Add plant dialog box. After running Step 3. Adjust emitters and find daily zone water use, the program adjusts the number of bubblers for each group as shown in the *Landscape* worksheet in Fig. 16.17. Specify 10 percent of plants can be underirrigated in order to avoid overirrigating other zones.

The number of bubblers are adjusted in step 3. It is also specified that 20 % underirrigation is allowed, which means that 20 % of plants can be underirrigated in order to prevent overirrigation of other plants (Fig. 16.18).

The results of the adjustment are shown in the *Landscape statistics* worksheet (Fig. 16.19). The fraction of overirrigation for each set of trees is shown in column A. None of the trees is underirrigated because the group of trees with the no irrigation consumes 40 % of the water in the zone, which is more than the 20 % underirrigation criterion.

The program also calculates the distribution uniformity, statistical uniformity, and the scheduling coefficient (Fig. 16.20). Column L divides the application volume into 100 increments and shows the amount of under or overirrigation in each increment.

The next step is to run Step 4 and calculate the irrigation schedule (Fig. 16.21). The zone should be irrigated for 13 minutes every day.

Questions

1. List the 5 steps required to design a turf or landscape irrigation system
2. What issues should be discussed with the property owner before designing the irrigation system?

3. What aspects of the landscape should be included in the site plan?
4. What is the pressure loss in a "5/8" municipal water meter at 12 GPM? Use the *Service line friction loss* worksheet. Is it acceptable?
5. Redo Example 16.1 but increase flow rate to 45.5 LPM. Use 1¼ inch copper pipe and fittings prior to the valves and water meter. Then use the 5/8 inch water meter and 5/8 inch valve and 5/8 inch copper in the last section. Use the *Service line friction loss* worksheet. Is the design acceptable?
6. Define the species factor, microclimate factor, and density factor.
7. Look in the WUCOLS guide and determine the species factor for oleander in Blythe, California. Use the low desert classification.
8. Calculate the LPD requirement for a 2 m diameter oleander in Tucson, AZ Reference ET is 12 mm/day. The microclimate and density factors are 1.0. Irrigation efficiency is 85 %
9. Calculate the LPD requirement for a 5 m diameter orange tree in Blythe, CA. The landscape, microclimate, and density coefficients are 1.0. Reference ET is 14 mm/day. Irrigation efficiency is 85 %.
10. Calculate S under an emitter that has a 1.0 meter diameter wetted area with a 1.0 m deep root zone. The MAD is 0.5. Field capacity is 20 % and permanent wilting point is 10 %.
11. The orange tree in question 9 has 6 emitters with the water storage per emitter as calculated in question 10. The reference ET rate is 10 mm/day. Calculate a watering schedule for the orange tree. Emitter flow rates are 4 LPH. Note: this question demonstrates the problem with having few emitters on a large tree.
12. Calculate the watering schedule for an oleander hedge with 2 m width. Reference ET rate is 10 mm/day. Wetted width is 0.6 m and rooting depth is 2 m. Use a landscape coefficient of 0.5 and MAD = 0.5. Let distance between emitters = 0.3 m. Root depth is 2 m. Soil is sandy loam with AWC = 0.12.
13. Use Landscape Irrigation program to calculate number of screw turns for bubblers irrigating 4 orange trees with 5, 6, 7, and 8 m diameter canopy. There is one bubbler per tree. The pressure is 280 kPa (adjusted in Device data page). Determine the irrigation schedule. Soil is loamy sand, density factor is 1.0, microclimate factor (exposure) is 1.2. Leaching fraction is 0.15. There is no overlap in wetting patterns. Application efficiency is 95 %.
14. Calculate the number of bubblers per plant for 20 orange trees laid out in a 5 (NS) × 4 (EW) grid. Tree diameters are shown in meters. Then redo with spiral inline emitter configuration. Compare pipe sizes and number of valves. Application efficiency is 90 %. Leaching fraction is 0.15. There is no overlap in wetting patterns. Maximum number of bubblers per plant is five. Microclimate and density factors are 1.0. For the same soil and weather parameters as in example 13, calculate the irrigation schedule.

5	4	3.4	6
2	4	6	3
3	4	6	4
1	3	4	4
4	4	4	6

References

Costello L, Jones KS (2014) Water use classification of landscape species. WUCOLS IV. University of California Cooperative Extension and California Department of Water Resources. http://ucanr.edu/sites/WUCOLS/WUCOLS_IV_User_Manual/

Levitt DG, Simpson JR, Tipton JL (1995) Water use of two landscape tree species in Tucson, Arizona. J Am Soc Hortic Sci 120 (3):409–416

17 Agricultural Drip Irrigation

Subsurface drip irrigation saves water, improves crop yields and quality, and facilitates fertilizer application; however, system performance is dependent upon skilled management. Potential disadvantages include salt accumulation near plants, restricted root development, high system costs, and restricted crop rotation. The three primary hydraulic classifications of drip emitters are laminar, turbulent, and pressure compensating. Analysis of Reynolds equation shows the advantage of turbulent flow emitters over laminar flow emitters. Pressure compensating emitters have the best hydraulic performance and generally rely on diaphragms that reduce flow at high pressure. As with sprinkler irrigation laterals, drip irrigation laterals are multi-outlet systems. The hydraulics can be calculated with analytic equations if slope is uniform, or with spreadsheets for any field. Most drip laterals in agriculture are classified as in-line, with emitters manufactured into the tubing. Proper filtration is a key to successful drip irrigation system performance. Sand filters are a necessary pretreatment step when the water source is a pond or stream. The design of lateral length and diameter is evaluated by the emission uniformity, which is a function of number of emitters per plant, minimum and average pressure in the lateral, the emitter exponent, and the manufacturer's coefficient of variation. The most popular drip system is subsurface drip irrigation with dual feed laterals (Fig. 17.1).

Drip Emitters

Drip emitters have small openings that dissipate pressure and discharge a small flow of water in discrete drops or a continuous tiny stream. Emitters are classified as laminar flow, turbulent, orifice, vortex, partially pressure compensating, or pressure compensating. The hydraulic relationship between pressure and flow is a function of the type of emitter. This relationship is given as $q = kh^x$ where q is emitter discharge, h is operating pressure head, k is the emitter discharge coefficient, and x is the emitter discharge exponent. Laminar flow emitters (long path) are long, narrow tubes, and energy is dissipated in the tube. The emitter exponent, x, is 1.0, which indicates that flow is directly proportional to pressure. Another drawback is that tube diameter is narrow and prone to plugging. The tube geometry may be straight (spaghetti tubing) or follow a spiral flow path around a cylindrical core.

Turbulent emitters ($x = 0.57$) dissipate energy in turbulent eddies that form in tortuous paths within the emitter. Orifice emitters dissipate energy in a single orifice; thus, the diameter is extremely small (less than laminar) and these emitters are prone to plugging. The flow varies with the square root of pressure so $x = 0.5$ as with sprinklers. Vortex emitters are similar to orifice emitters except that the water passes through one turbulent eddy before exiting the orifice; the turbulent eddy decreases the emitter exponent x to 0.4. As with orifice emitters, vortex emitters have a narrow orifice and are prone to plugging.

Pressure compensating emitters have virtually no change in flow rate over a range of pressures. These are emitters with diaphragm closing as pressure increases and vice versa. Some emitters with diaphragms "pseudo pressure compensating" may be labeled as pressure compensating, but they are not truly pressure compensating. This mislabeling can cause unexpected nonuniformity if the engineer assumes that the emitters are truly pressure compensating. The engineer will assume that pressure loss along the tubing will not cause a change in flow rate and fail to design a large enough diameter tube to prevent pressure loss. This could cause lawsuits so the design team should check emitters.

Turbulent flow emitters are designed so that vortices are set up in the flow path. One advantage of turbulent flow emitters is that they have a larger diameter flow path because energy is dissipated in turbulent eddies rather than in small diameter tubes or orifices. Thus, they are less likely to plug than laminar flow or orifice emitters. Pressure compensating

P. Waller, M. Yitayew, *Irrigation and Drainage Engineering*, DOI 10.1007/978-3-319-05699-9_17

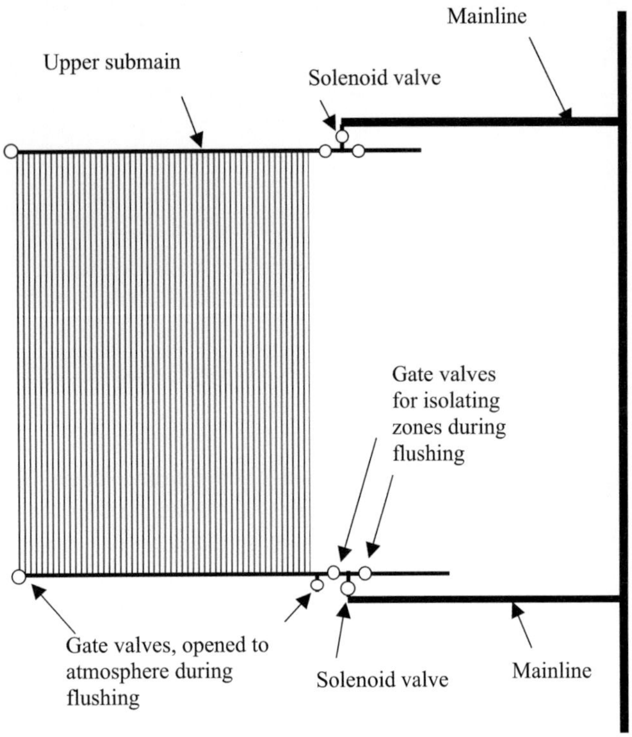

Fig. 17.1 Dual feed subsurface drip irrigation system

emitters with a diaphragm that retracts until pressurization are even less likely to have plugging than normal turbulent emitters. The flushing procedure for the turbulent emitter is the opposite of that for the diaphragm emitter.

Analysis of the Darcy-Weisbach equation shows that x is 1.0 for laminar flow emitters and 0.57 for turbulent flow emitters.

$$h_f = 6.377 fL\left(\frac{q^2}{D^5}\right) \tag{17.1}$$

The Darcy-Weisbach friction factor is 64/Re for laminar flow. Rearrange Eq. 17.1 substitute 64/Re for f.

$$\begin{aligned} q^2 &= \left(\frac{h_f D^5}{6.377 fL}\right) = \left(\frac{h_f D^5}{\left(\frac{64}{Re}\right)6.377L}\right) = \left(\frac{h_f Re D^5}{(64)\ 6.377L}\right) \\ &= \left(\frac{h_f \rho V D^6}{(64)\ 6.377\mu L}\right) = \left(\frac{h_f \rho\left(\frac{q}{A}\right)D^6}{(64)\ 6.377\mu L}\right) \\ q &= \left(\frac{h_f \rho D^6}{(64)\ 6.377\mu LA}\right) \end{aligned} \tag{17.2}$$

Thus, q is a linear function of pressure drop, h_f (equal to the pressure in the lateral, H), and the emitter exponent, x, is 1.0. The constant k in the emitter flow equation is equal to the rest of the terms on the right side of Eq. 17.2.

$$q = kh_f = kH^1 \tag{17.3}$$

q varies linearly with viscosity in Eq. 17.2; thus, laminar flow emitters are highly sensitive to viscosity changes because of temperature change. Because water can heat up in drip laterals in the field, especially on the ground surface, this fact can cause major changes in emitter flow rate from one end of the drip lateral to the other.

The Darcy-Weisbach is next used to derive the relationship between flow rate and pressure in a turbulent emitter. The Blasius friction factor for turbulent flow in smooth pipe is

$$f = 0.316\ Re^{-0.25} \tag{17.4}$$

Substitute the Blasius friction factor into the Darcy-Weisbach equation.

$$\begin{aligned} q^2 &= \left(\frac{h_f D^5}{6.377 fL}\right) = \left(\frac{h_f D^5}{\left(\frac{0.316}{Re^{0.25}}\right)6.377L}\right) \\ &= \left(\frac{h_f Re^{0.25} D^5}{(0.316)\ 6.377L}\right) = \left(\frac{h_f \rho^{0.25}\left(\frac{q}{A}\right)^{0.25} D^{5.25}}{(0.316)\ 6.377\mu^{0.25}L}\right) \\ &= \left(\frac{h_f \rho^{0.25} q^{0.25} D^{5.25}}{(0.316)\ 6.377\mu^{0.25}LA^{0.25}}\right) \\ q^{1.75} &= h_f\left(\frac{\rho^{0.25}D^{5.25}}{(0.316)\ 6.377\mu^{0.25}LA^{0.25}}\right) \\ q &= h_f^{1/1.75}\left(\frac{\rho^{0.25}D^{5.25}}{(0.316)\ 6.377\mu^{0.25}LA^{0.25}}\right)^{1/1.75} \\ &= h_f^{0.57}\left(\frac{\rho^{0.25}D^{5.25}}{(0.316)\ 6.377\mu^{0.25}LA^{0.25}}\right)^{0.57} \end{aligned} \tag{17.5}$$

Equation 17.5 shows that the value of x for turbulent flow emitters is 0.57: $q = k\ h_f^{0.57}$. Although the theoretical derivation shows that turbulent flow emitters have an emitter exponent equal to 0.57, some turbulent flow emitters have an emitter exponent in the range of 0.5 or less. Flow rate in turbulent emitters is insensitive to viscosity changes, q is proportional to $(\mu^{-0.25})^{0.57} = \mu^{-0.145}$. Because of lower possibility of plugging, less sensitivity to temperature change, and less sensitivity to pressure change, turbulent flow emitters are superior to laminar flow emitters. Figure 17.2

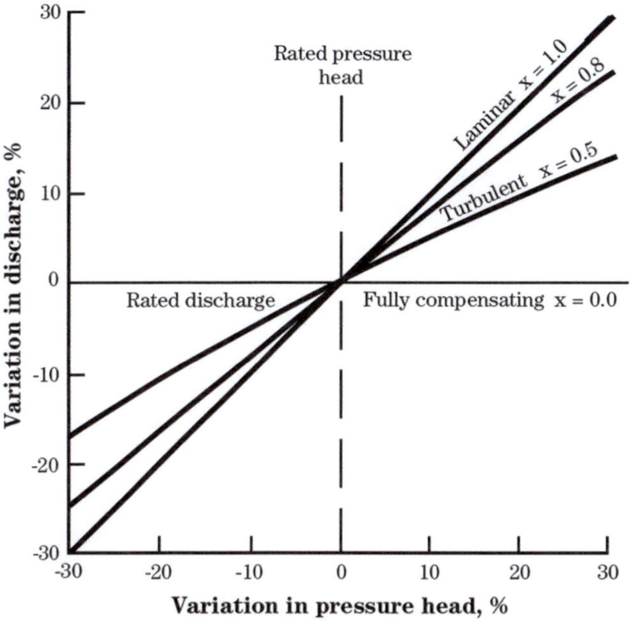

Fig. 17.2 Flow rate versus pressure variations for different emitter exponents (Credit NRCS)

shows emitter flow rate vs. pressure variations for several values of x.

Example 17.1 Calculate the flow rate for the following emitters (laminar, turbulent, and pressure compensating) at 50 and 100 kPa pressure (coefficient k for units of LPH and kPa).

$k = 0.02$ and $x = 1$ laminar
$k = 0.2$ and $x = 0.5$ turbulent or orifice
$k = 2$ and $x = 0$ pressure compensating

$x = 1$	$x = 0.5$	$x = 0$
$q = 0.02\ (50^1) =$ 1.0 LPH.	$q = 0.2\ (50^{0.5}) =$ 1.4 LPH.	$q = 2\ (50^0) =$ 2 LPH.
$q = 0.02\ (100^1) =$ 2.0 LPH.	$q = 0.2\ (100^{0.5}) =$ 2.0 LPH.	$q = 2 * (100^0) =$ 2 LPH.

Drip Tubing

There are primary three types of agricultural drip tubing. Thick wall (45 mil) polyethylene tubing with inline (integral) emitters has emitters manufactured into the tubing wall or injection molded emitters attached to the wall and is used in row crops (Fig. 17.3). Thick wall (45 mil) polyethylene tubing without inline emitters and into which single port emitters are inserted is used in vineyards and landscaping.

Thin wall polyethylene drip tape (lays flat when not pressurized) is sold with a wall thickness ranging from 4 to 15 mil and thicker for larger diameters: a mil is one thousandth of an inch. The thinner wall drip tape, 4–6 mil, is used for one season, cannot be used in soil with rocks or clay, and must be buried. Eight mil tubing may be used for 1–2 years. Fifteen mil drip tape has been shown to last in the field for 15 years when buried and very carefully managed. However, rodents, contact with farm implements, and poor chemical control can dramatically reduce the life span of tubing. Thin wall drip tape has emitter flow rates ranging from 0.5 to 2 LPH and emitter spacing ranging from 10 to 75 cm.

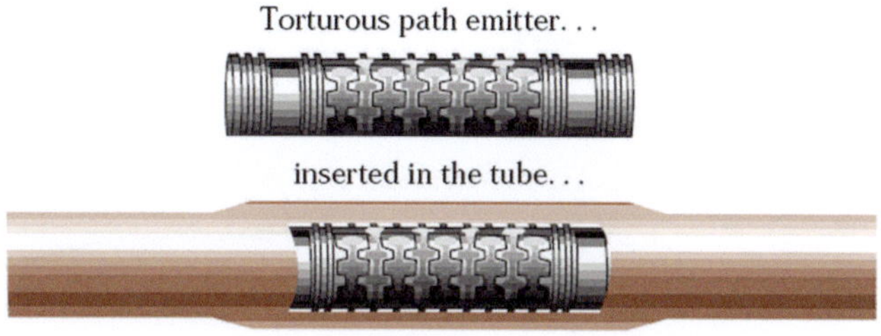

Fig. 17.3 Injection molded inline emitter attached to thick wall polyethylene tubing (Credit NRCS NEH)

Thick wall tubing (45 mil wall thickness) maintains a cylindrical shape when empty. It can be purchased in many diameters, but typical diameters used in agricultural and landscaping drip laterals are 12, 16, and 22 mm ID. Thick wall inline drip tubing has emitter flow rates ranging from 0.5 to 2 LPH and spacing up to 1.3 m. Common inside diameters of drip tape and thick wall tubing with integral emitters are 13, 16, and 20 mm ID. Some manufacturers also make 22 and 35 mm ID for longer rows.

Subsurface Drip Tape Installation

The most common type of agricultural drip irrigation is subsurface drip irrigation (SDI): drip tape with inline emitters buried approximately 15 cm below the soil surface. Subsurface drip irrigation (SDI) was developed by the USDA Agricultural Research Service in Fresno, California, and adapted to the commercial farm scale by Dr. Howard Wuertz at Sundance Farms in Coolidge, Arizona.

A tape injection unit attached to a tractor automatically installs subsurface drip tape. Spools, holding 1000 to 2000 m rolls of drip tape, feed the roll of tape into atube that trails behind a shank. Tubing is typically installed 0.15 m below the ground surface. With the advent of differential GPS, the location of the tubing is recorded as it is installed; before GPS, the tubing was installed in the center of beds, and the beds were very carefully maintained year after year so that the location of tubing was known. It is important to use high quality and properly maintained tape injection equipment because a small burr or weld in the feeder tube can rip the drip tape, which results in the need for reinstallation.

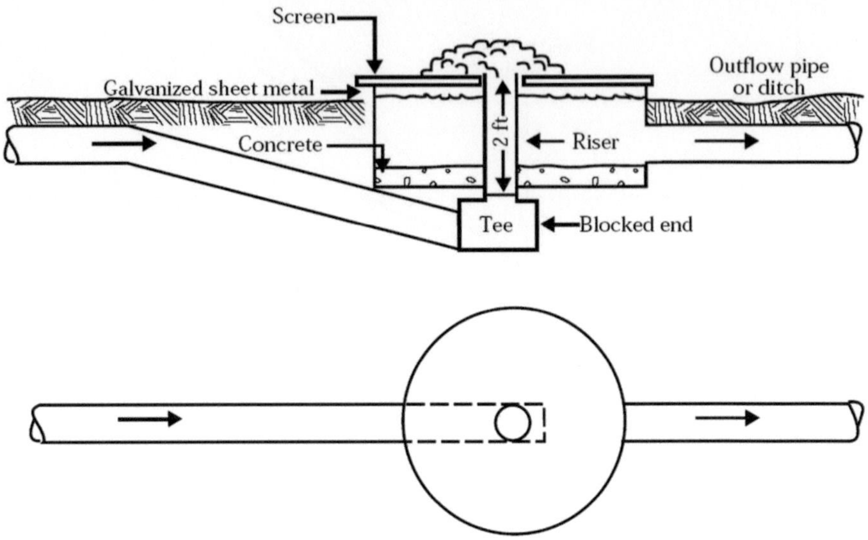

Fig. 17.4 Flat stainless steel screen filter design (Credit NRCS)

Table 17.1 Recommended stainless steel screen and riser pipe diameters for round flat filters (Credit NRCS)

Flow rate		Flow rate	Screen diameter		Riser pipe diameter	
(GPM)	(cfs)	(cms)	(m)	(in)	(m)	(in)
1	450	12.7	42	1.07	8	0.20
2	900	25.5	48	1.22	10	0.25
3	1,350	38.2	60	1.52	12	0.30
4	1,800	51.0	72	1.83	15	0.38
5	2,250	63.7	84	2.13	18	0.46

Filtration

Drip emitters have small openings and are easily plugged; thus, filtration is required to remove suspended sediment and biological materials from irrigation water. There are three primary types of filters used for drip irrigation: sand filters, mesh filters, and disc filters.

Prefiltration: Stainless Steel Screen Filters

If water has a high concentration of algae or sediment, then it will overwhelm and plug sand filters; thus, a prefiltration step, such as a settling pond for sediment or a stainless steel screen filter for algae, must be installed upstream from the filters. Algae can be removed with cylindrical stainless steel screen filters as shown in Fig. 12.30 or with round flat screen filters as shown in Fig. 17.4. Flat screen filters do not require regular cleaning because the water naturally pushes the debris off the edge of the filter. These filters can also be constructed with a linear flow path with water dropping to the filter from one side and pushing debris off the other end. Recommended screen and riser geometry for round flat filters is listed in Table 17.1.

Prefiltration: Settling Basins

Settling basins can be used as a prefiltration step for waters with high suspended sediment loads. They effectively remove silt or sand but do not remove clay particles because clay particles are small and do not settle. Water flow velocity is low in order to ensure laminar flow with no turbulent eddies. For laminar flow, the settling velocity is determined by Stokes law and can be calculated with the following equation.

$$V_p = 3.43 * 10^{-5} D^2 (SG1) \quad (17.6)$$

where

V_p = settling velocity of particle, m/min,
D = particle diameter, microns,
SG = specific gravity of particle, 2.65 for mineral (soil) particles.

Basins should be long enough so that particles of a given size settle to the bottom of the basin. Settling basin lengths should be approximately 5 times their width in order to prevent short- circuiting (a stream flowing through one side of the settling basin while other zones are stagnant). In order to distribute flow across the basin, a baffle structure may be necessary at the basin inlet. The inlet should be below the surface to prevent contaminants from floating on the water surface. Settling basin depth should be 0.6 m. The surface area of the settling basin is calculated based on the requirement that the velocity remains laminar at an average depth of

0.6 m. A safety factor of 2.0 is incorporated into the design to account for sedimentation and a decrease in depth of the settling basin. If settling basin depth is 0.6 m and side slopes are 2:1, then the water surface area of the settling basin is calculated with the following equation.

$$Area = 0.001\ F\ \left(\frac{Q}{V_p}\right) = 0.001\ (2.0)\ \left(\frac{Q}{V_p}\right) \tag{17.7}$$

Where

Area = water surface area, m^2,
F = safety factor,
Q = irrigation system flow rate, LPM
V_p = settling velocity of design particle size, m/min.

The width of the settling basin is a function of the area.

$$W = \sqrt{\frac{Area}{5}}$$

where

W = width of settling basin, m.

The length of the settling basin is a function of the width.

$$L = 5\ W$$

where

L = length of settling basin, m.

Example 17.2 Irrigation system flow rate is 1200 LPM, and the water carries a large volume of silt. What are the required dimensions of the settling basin. Assume that the design particle diameter is 30 microns (0.03 mm)

$$V_p = 3.43{*}10^{-5} D^2 (SG1) = 3.43{*}10^{-5} (30^2)(1.65) = 0.051\ m/min$$

$$Area = 0.001\ F\left(\frac{Q}{V_p}\right) = 0.001\ (2.0)\ \left(\frac{1,200}{0.051}\right) = 47\ m^2$$

$$W = \sqrt{\frac{Area}{5}} = \sqrt{\frac{47}{5}} = 3.1\ m \quad L = 3.1\ m(5) = 15.5\ m$$

Prefiltration-Centrifugal Sand Separation

Well water often has a significant fraction of sand. If water has a sand particle concentration in the range between 2 and 20 ppm, a centrifugal sand separator can be installed in the irrigation line to remove sand (Netafim Hydroyclone Sand Separators Guide, 2015). The vortex action forces sand to the outside of the tank and sand falls into the sedimentation tank as water is removed from the outlet on top of the cone. Pressure loss in a sand separator is in the range of 2–5 m. Sand separators are typically followed by an appropriate filter. A screen or disc filter may be used if the sand is removed by the sand separator.

Removal Efficiency

The removal efficiency for any filter or prefiltration step is $(1 - C_{passed})/C_{initial}$ where C_{passed} is the concentration of particles in filter discharge and $C_{initial}$ is the concentration of particles in the filter inlet. Filtration efficiency should be verified before water is injected into a drip irrigation system because filtration failure and clogging can destroy a drip system within a short time and replacement of emitters is the consequence. In addition to an initial performance evaluation, filtration efficiency should be checked regularly.

Filter Mesh Size

Sand, screen, and disc filters are classified based on a mesh size. The filter pores or openings should be approximately ¼ the size of emitter openings. Drip irrigation manufacturers specify a mesh size required for each emitter. For a screen filter, filter mesh size refers to the number of slots per inch (25.4 mm). For example, a 200 mesh filter has 200 slots per linear inch in each direction. Most drip emitter manufacturers recommend mesh sizes between 100 and 200, with 150 and 75 micron openings, respectively. Although sand filters do not have a woven mesh like a screen filter, sand filters can be classified according to mesh size. Silica sand is normally used in sand filters: crushed silica sizes 8, 11, 16, 20, and 30 have screen mesh equivalents of 70, 140, 170, 230, and 400 mesh, respectively (Kansas State University Extension Bulletin).

Filter Selection

Sand filters are required for low quality water (algae or suspended sediment) from a surface water body such as a pond, lake, canal, or stream. Even sand filters, however, may be overloaded by some water sources that are extremely high in sediment or algae. In this case, a prefiltration process is needed. Screen and disc filters have a very small surface area compared to sand filters; thus, they plug rapidly if water has a significant concentration of sediment or organic matter. Screen and disc filters are suitable for municipal water and well water. Screen or disc filters are often placed after sand filters to protect the system if the sand filter fails. If a filter has not been used with a certain water source and system application, then the filter should be tested on-site at the

Fig. 17.5 Sand media filters (Credit Yardney Corp)

expected flow rate. However, even with testing, caution is required because water quality and biological growth rates in canals and streams can fluctuate with temperature, nutrients, or other factors.

Sand Filters

Pressurized tank sand filters (Fig. 17.5) use a pressure differential to force water through the sand media. Particulate matter is caught in the pores between sand particles. The pressure differential across the filter increases as sand becomes clogged with sediment or algae. Once the pressure differential reaches a threshold value, a backflush cycle is triggered. During the backflush cycle, water flow is closed off from the inlet manifold by the backflush valve (Fig. 17.5) and flows upward through the filter from the outlet manifold. The backflush valve directs the discharge to a discharge pipe. Sand filters can be purchased in manual backflush or automatic backflush configurations. An automatic backflush system requires at least two filters installed in parallel because water must pass through one filter in order to backflush the other filter. The backflush flow rate should be set so that a small amount of sand is flushed out of the filter during each backflush cycle. This flow rate fluidizes the sand bed and cleans the bed. Even with proper cleaning, the sand filter media needs to be replaced every two to three years.

During the backflush cycle, pressure in the outlet manifold must be maintained at the pressure recommended by the manufacturer in order to maintain adequate upward flow. The required outlet manifold pressure may be as high as 30 PSI (210 kPa) (Netafim Sand Media Filter Guide, 2015). Thus, a booster pump may be needed during flushing if the required outlet manifold pressure is significantly higher than normal drip system pressure. If drip tubing has a pressure rating lower than the required backflush pressure, a pressure regulating valve is required downstream from the filter in order to prevent damage to the tubing.

Automatic backflush filters have electronic pressure sensors installed at the inlet and outlet of the sand filter. These sensors monitor the pressure loss through the sand filter, and they trigger a backflush cycle when the pressure differential reaches a threshold set by the operator. The threshold is normally 5–7 PSI (35–42 kPa) greater than the pressure differential with clean sand. Filters can also be set to backflush at set time intervals. Typical backflush intervals are every 2–4 hours and typical backflush cycle times are two minutes; however, this is dependent on water quality. Approximately 30 seconds are required between sequential filter backflushes in order to allow the sand in the filter to settle.

Pressure loss through the sand filter is the sum of losses in the backflush valve, sand media, outlets at the bottom of the sand media, and, if they are installed, secondary screen or disc filters on the discharge side of the filter. Thus, the pressure requirement of the system must include all of the filtration losses, which may be higher than the drip emitter operating pressure.

In selecting sand filters, a rule of thumb is that the maximum flow rate through sand filters should be 18 LPS per square meter of cross-sectional tank area (25 GPM/ft^2). Sand filters should be mounted on a concrete pad that is 0.1 m (4 in.) thick with 0.15 × 0.15 m (6 × 6 in.) footings for the legs. The pad dimensions should exceed the filter by 30 cm (1 ft) on all sides.

Example 17.3 Irrigation system flow rate is 300 GPM (1162 LPM). The filter system requires a 30 PSI (210 kPa) discharge manifold pressure. Select a sand filter diameter for a 2 sand filter system. Available alternatives are 24, 30, and 36 in (0.6, 0.75, and 0.9 m) diameter. Assume that the filter has a 1.5 PSI (10 kPa) pressure loss when the sand is clean. The required outlet manifold pressure is 30 PSI (280 kPa). What is the required inlet manifold pressure?

In a 2 filter system, each filter carries half of the flow, 150 GPM (580 LPM). Make an initial estimate based on the 18 LPS/m^2 rule.

$$(580\ \text{LPM})\,/(60\,\text{sec}\,/\text{min})/\left(18\ \text{LPS}/\text{m}^2\right) = 0.54\ \text{m}^2$$

$$\text{A} = \pi \text{D}^2/4 \quad D = \sqrt{\frac{4A}{\pi}} = \sqrt{\frac{4\ (0.54)}{\pi}} = 0.83\ m = 33\ \text{in.}$$

The next largest filter size should be selected, the 36 in filter.

Seven PSI should be added to the clean sand pressure loss (8.5 PSI, 59 kPa). If the required outlet manifold pressure is 30 PSI (280 kPa), then the required inlet pressure is 38.5 PSI (340 kPa).

Disc Filters

Disc filters utilize stacks of grooved donut shaped discs that capture sediment between the discs. During filtration, the stack of discs is compressed together, but during the backflush cycle, the stack is loosened as backflush water in the opposite direction through the discs. The backflush water and sediment are discharged from the filter to a discharge pipe.

Screen Filters

Screen filters are commonly used for drip systems that are supplied with relatively clean well water or municipal water. They have low surface area so they plug rapidly if there is significant algae or sediment in the water. The fine mesh can be made of several materials; however, nylon is not suitable for areas with high temperatures because nylon stretches in the heat.

Design of Drip Emitter Spacing, Flow Rate, and Schedule

The first step in the drip irrigation system design process is to make an educated estimate of the emitter spacing along the lateral, the spacing between laterals, the emitter flow rate, and lateral length. Factors that determine the selection of spacing and flow rate include plant spacing, plant rooting characteristics, soil texture, and lateral hydraulics. After a preliminary hydraulic evaluation, it may be necessary to modify the spacings and flow rate. This process continues until the economic cost is minimized and the appropriate design constraints (uniformity and application rate) are met.

Calculation of the soil water holding capacity is generally not required for drip irrigation systems because drip irrigation is a high frequency irrigation system with daily or even more frequent water application.

The gross water application rate (mm/hr) is.

$$(di/dt)_g = \left(\frac{Emitter\ flow\ rate}{area\ per\ emitter}\right) = \left(\frac{q_e}{s_l\ s_d}\right) = \left(\frac{L/hr}{m^2}\right) \tag{17.8}$$

Where

$(di/dt)_g$ = gross water application rate, mm/hr,
s_l = spacing along lateral, m,
s_d = spacing between laterals, m,
q_e = emitter flow rate, L/hr.

The net application rate is the product of the gross application rate and the water application efficiency

$$(di/dt)_a = (di/dt)_g(Eff/100) \tag{17.9}$$

The required number of watering hours per day is the evapotranspiration rate (average for the field, mm/day) divided by the net application rate (average rate over the field area, mm/hr). Divide max $ET_{c\text{-max}}$ (worst case) by the net application rate to find the required hours of application per day in each zone.

$$T_z(hr/day) = \frac{ET_{c-\max}\ (mm/day)}{(di/dt)_a\ (mm/hr)} \tag{17.10}$$

where

T_z = application time per day in each zone, hr/day.

The number of watering periods is the hours of irrigation system operation per day, T_r, divided by T_z. The number of periods must be rounded down to the next smaller whole number.

$$N_p = \frac{T_r}{T_z} \tag{17.11}$$

where

N_p = number of watering periods per day.

In-class Exercise 17.1 Emitters are spaced at 0.25 m along the lateral and 1 m between laterals, and emitter flow rate is 1 LPH. For a daily plant water requirement of 12 mm/day, calculate the application rate and the application time if plants are watered on a daily basis. Assume 90 % efficiency. System is operated for 22 hr/day. Calculate the number of watering periods per day.

Although there may be tens or even hundreds of valves on a large farm, for design purposes in this phase, assume that N_p groups of zones will be watered during the N_p periods. The area of the farm that is watered during each watering period is

$$A_p = \frac{A_f}{N_p} \tag{17.12}$$

where

A_p = area watered during each watering period, ha.
A_f = area of the farm, ha.

The number of emitters per ha is

$$E_d = \left(\frac{10,000\ m^2}{ha}\right)\left(\frac{1\ emitter}{s_L\ s_d\ m^2}\right) \tag{17.13}$$

where

E_d = emitter density, number of emitters per ha.

The required irrigation system flow rate is

$$Q_p = \frac{E_d Q_e A_p}{3,600} \tag{17.14}$$

where

Q_p = irrigation system flow rate, LPS.

Lateral Design

This chapter includes an Excel workbook that includes the examples in chapters 17 and 18, which focus on dual feed laterals. The workbook in these chapters also does not include local losses due to emitter barbed fittings or inline emitter flow constrictions. The *Chapter 21 Hydroponic irrigation* workbook includes local losses and has far fewer spreadsheets. Thus, it is easier to use for analysis of single feed laterals.

In general, the Darcy-Weisbach equation, with the Blasius or laminar friction factor (64/Re), is used to calculate drip lateral hydraulics. A spreadsheet can be used to calculate pressure and flow distribution along the lateral. The calculation processs begins at the end of the lateral since upstream flow rate and pressure loss are a function of downstream flow rate and pressure. Assume the minimum acceptable pressure at the distal end of the lateral and work backwards to the inlet: flow in each pipe section is the sum of all the downstream emitter flow rates. In this section, the additional pressure loss from barbed drippers inserted into the pipeline is ignored.

$$Q_{n-pipe} = \sum\nolimits_{n}^{N_e} Q_{n-e} \tag{17.15}$$

where

$Q_{n\text{-pipe}}$ = flow rate in pipe section n,
$Q_{n\text{-e}}$ = flow rate in emitter n

The difference in pressure from one emitter to the next is

$$H_n = H_{n+1} + h_f + S_e s_l \tag{17.16}$$

where

H_n = pressure at emitter n, m,
S_e = slope (m/m), negative for inlet higher than distal end,
s_l = distance between emitters,
n = emitter number with n = 1 at the inlet.

One additional constraint on drip lateral design is that sediments must be periodically flushed from laterals. There is disagreement as to the required flushing velocity, but a typical rule of thumb is that flow velocity must be at least 0.5 m/sec (1.5 ft/sec) at the end of laterals in order to keep particles in suspension. The ASAE (EP405) Standard for Microirrigation recommends a velocity of 1 ft/sec (0.33 m/sec) while others recommend 2 ft/sec (0.67 m/sec). Additional flow velocity may be needed to scour the sides of laterals if a slime begins to develop in tubing.

Example 17.4 Calculate the pressure and emitter flow rate along a 12 mm ID single feed lateral with emitter spacing, s_l = 0.5 m, k = 0.2 (LPH and kPa), x = 0.5, S_e = −0.002 (inlet higher than distal end) and S_e = 0.002 (inlet lower than distal end), and L (length of lateral) = 100 m. Minimum operating pressure is 100 kPa (10.2 m). Because the inlet is higher than distal end, S_e = −0.002. Make two manual calculations and then use the *Single feed lateral* worksheet.

The pressure at the last emitter is 100 kPa.

$$Q_{200} = 0.2\left(100^{0.5}\right) = 2.0\ \text{LPH}.$$

Calculate pressure loss in the lateral section before the last emitter.

$$v = \frac{Q}{A} = \left(\frac{2\ \text{L ph}}{\pi(6/1000)^2\ \text{m}^2}\right)\left(\frac{\text{m}^3}{1,000\ \text{L}}\right)\left(\frac{\text{hr}}{3,600\ \text{sec}}\right)$$
$$= 0.0049\ \text{m/sec}$$

$$\text{Re} = \frac{\text{vD}}{\nu} = \frac{0.0049\ (12\ \text{mm}/1,000\ \text{mm/m})}{1*10^{-6}} = 59$$

Flow is laminar (Re < 2000) so use the laminar flow friction factor equation.

$$f = \frac{64}{\text{Re}} = \frac{64}{59} = 1.1$$

$$h_f = 6.377 fL\left[\frac{Q^2}{D^5}\right] = 6.377\ (1.1)\ (0.5\ m)\ \left[\frac{2^2}{12^5}\right]$$
$$= 0.000056\ m$$

Friction loss is 0.000056 m between the last emitter and the next to last emitter. Pressure at the next to last emitter is calculated based on friction loss and elevation difference between the two emitters. Thus, pressure at the next to last emitter (199) is:

$$\begin{aligned} H_n &= H_{n+1} + \ \ h_f + S_e s_l \\ &= 10.2 \ \ \text{m} + 0.000056 + (-0.002 \ \ \text{m/m})(0.5 \ \ \text{m}) \\ &= 10.19906 \ \ \text{m}/0.102 = 99.99 \ \ \text{kPa}. \end{aligned}$$

Pressure is lower in the next to last emitter because friction loss is less than the elevation difference between the last two emitters. Flow from emitter 199 is

$$Q_{199} = 0.2\left(99.99^{0.5}\right) = 1.9999 \ \text{Lph}$$

Flow in the next to last pipe section of the lateral is the sum of the flows of the last 2 emitters.

$$Q = 2 + 1.9999 = 3.9999.$$

The Blasius equation for friction factor is used once the Reynolds number rises above 2000. Calculations are made in the *Single feed lateral* worksheet (Fig. 17.6). Ignore the information in the upper right corner of the worksheet.

Average flow rate in the downslope lateral is 2.105 LPH, and average pressure is 11.33 m. Figure 17.7 shows the pressure vs. flow distribution for the upslope ($S_e = 0.002$) lateral. Average flow rate in the upslope lateral is 2.124 LPH, and average pressure is 11.55 m.

The *Flushing* worksheet evaluates the flushing process. In order to maintain minimum scouring velocity, flow rate and pressure were specified at the end of the downslope lateral as 264 LPH (0.5 m/sec in 12 mm tubing) and 1.0 m, respectively. This results in an inlet flow rate of 544 LPH and inlet pressure of 12.4 m (Fig. 17.8).

Drip irrigation lateral hydraulics can also be calculated with analytic equations that use the Christensen's factor.

$$h_{ac} = h_f F \tag{17.17}$$

where

h_{ac} = actual pressure loss in the pipeline
h_f = friction loss in fully flowing pipe
F = Christensen's friction factor.

Christensen's friction factor for drip irrigation laterals is

$$F = \frac{1}{m+1} + \frac{1}{2N} + \frac{(m-1)^{0.5}}{6N^2} \tag{17.18}$$

where

N = number of outlets
m = exponent in Darcy-Weisbach equation, (2).

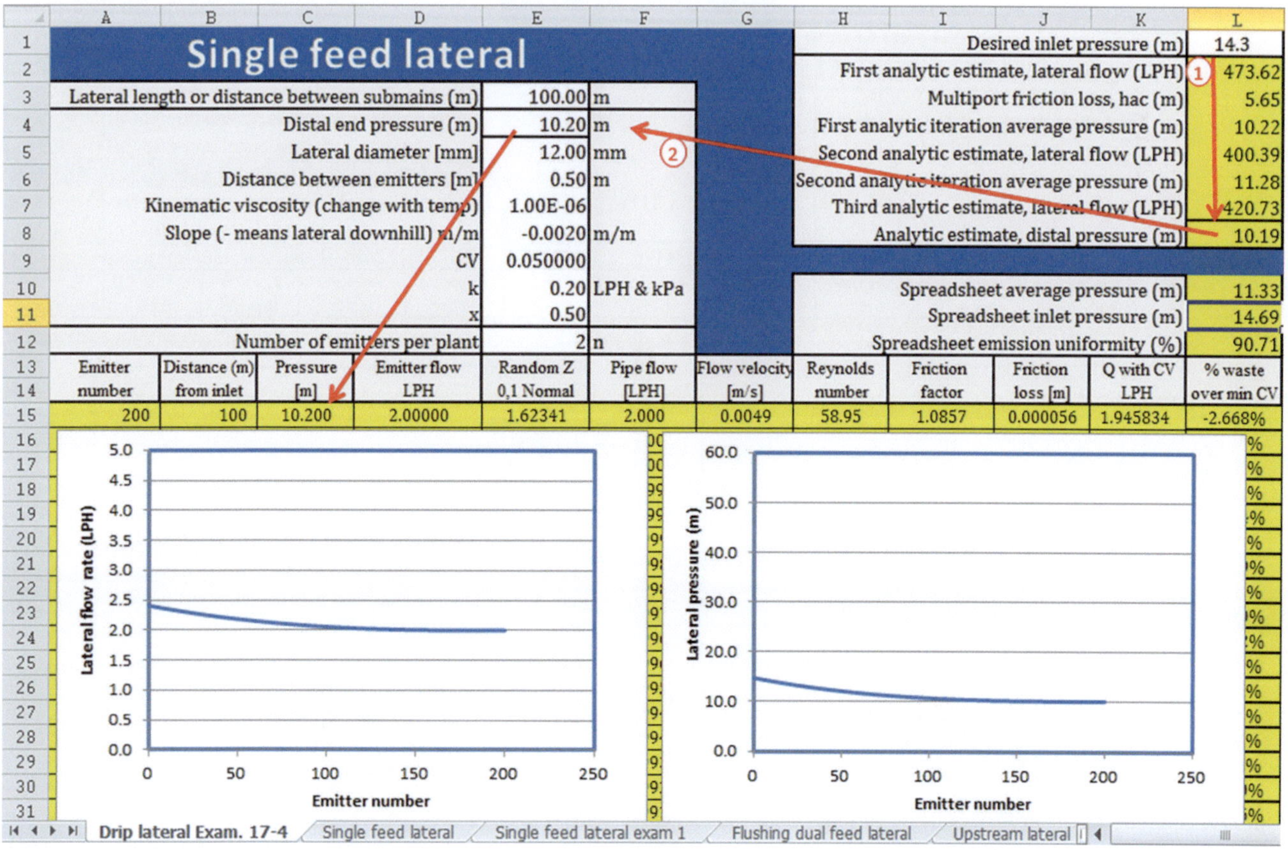

Fig. 17.6 Emitter flow rate and pressure for downhill lateral

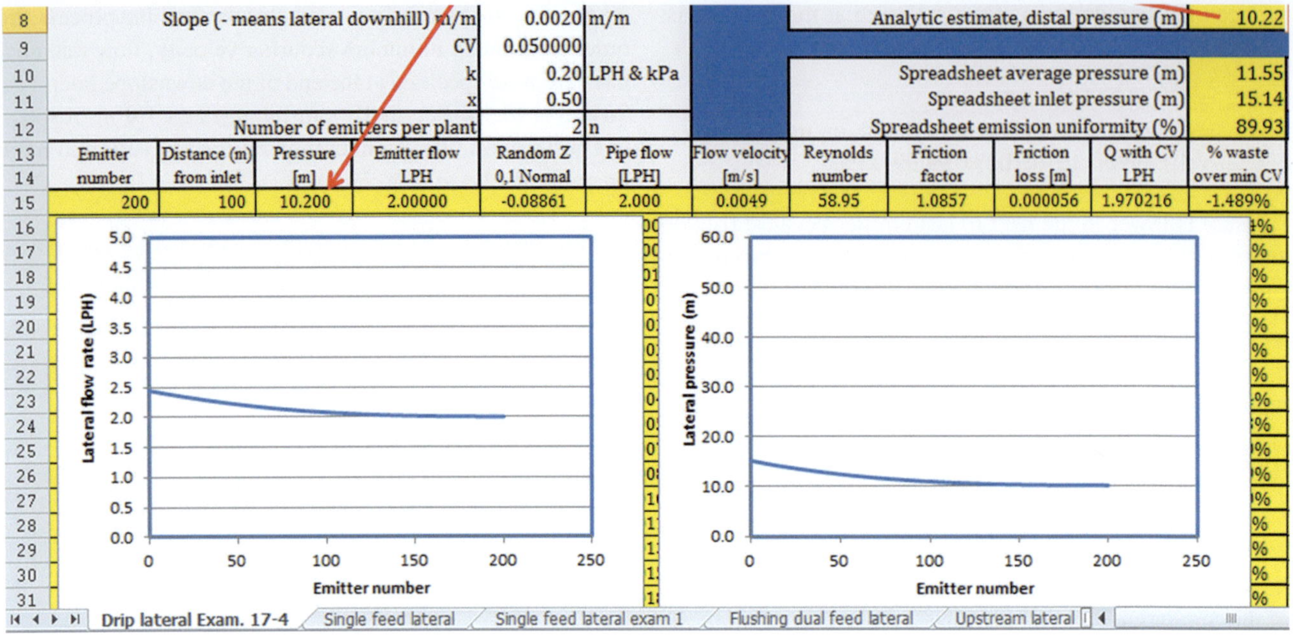

Fig. 17.7 Emitter flow rate and pressure for uphill lateral

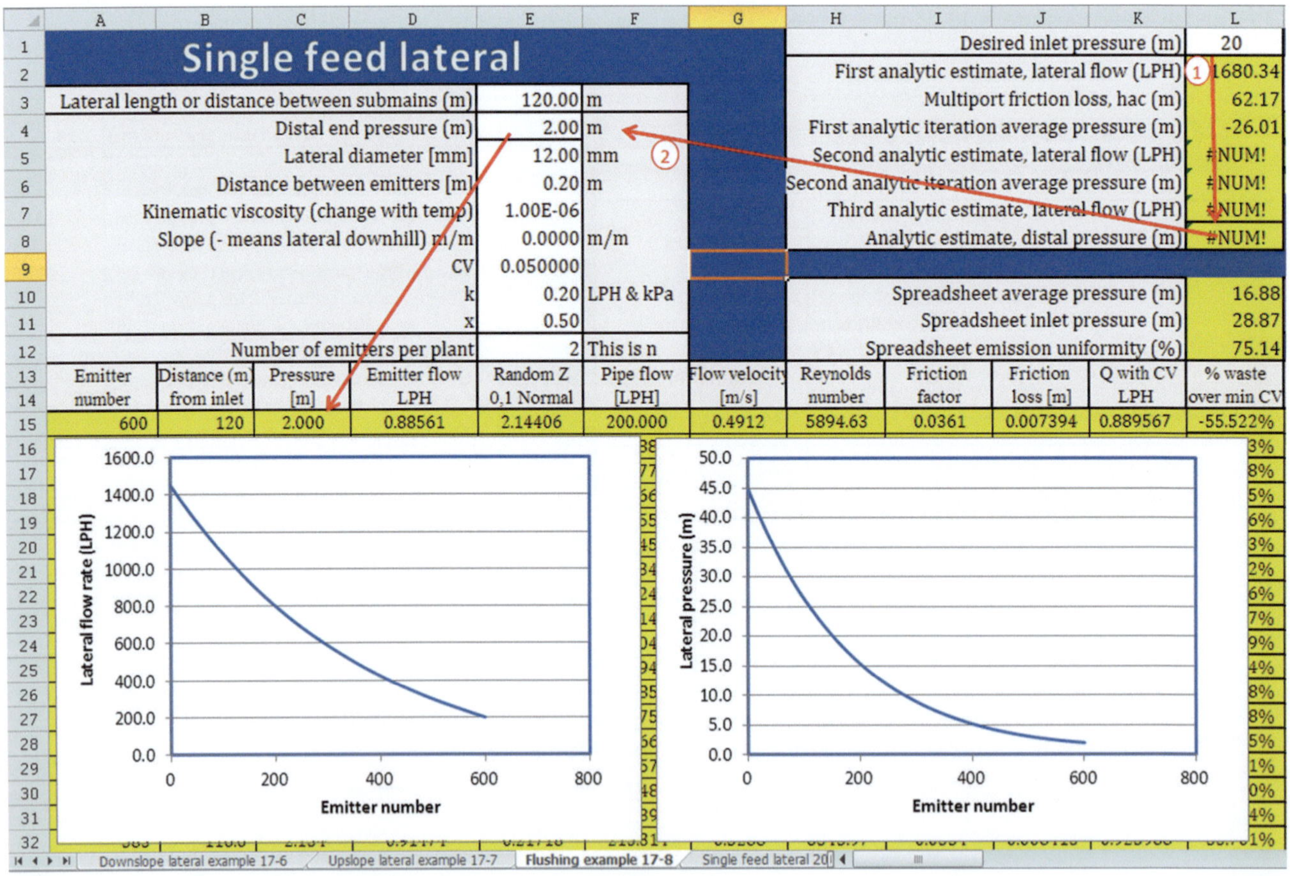

Fig. 17.8 Flushing lateral hydraulics for downslope lateral with 0.5 m/sec velocity

With pressure compensating emitters, all emitters have the same flow rate, and the lateral flow rate is the sum of all of the emitter flow rates; thus, the only design parameter is that the last emitter (lowest pressure emitter on a slope) on the lateral must operate at or above the minimum acceptable pressure. However, if emitters are not pressure compensating, then the flow rate for the lateral should be calculated based on the average pressure in the lateral. Pressures in laterals can be computed with Eqs. 17.19, 17.20, 17.21, and 17.22.

$$H_a = H_d + \frac{h_{ac}}{3.852} + \frac{S_e\ L}{2} \tag{17.19}$$

$$H_o = H_a + 0.74\ h_{ac} + \frac{S_e L}{2} \tag{17.20}$$

$$H_d = H_a - 0.26\ h_{ac} - \frac{S_e L}{2} \tag{17.21}$$

$$H_0 = H_d +\ h_{ac} + S_e L \tag{17.22}$$

where

H_a = average pressure in lateral, m,
H_o = pressure at inlet, m,
H_d = pressure at distal end, m,
S_e = slope (m/m), negative for inlet higher than distal end,
L = lateral length, m.

If the design is based on the pressure at the distal end as the minimum acceptable pressure, then Eq. 17.17 can be used directly to find the average pressure. A few iterations are necessary since h_{ac} is a function of $H_{a.}$ If the design is based on a known inlet pressure, H_0, then Eq. 17.18 is rearranged and solved for H_a after some iteration. If the design is based on H_a, then H_o and H_d are calculated directly (no iteration) with Eqs. 17.18 and 17.19.

Example 17.5 Repeat Example 17.4 with Eqs. 17.19, 17.20, 17.21, 17.22, and Christensen's F.

Calculate the Christensen's F for the Darcy-Weisbach equation with 200 emitters.

$$F = \frac{1}{m+1} + \frac{1}{2N} + \frac{(m-1)^{0.5}}{6N^2}$$
$$= \frac{1}{2+1} + \frac{1}{2\ (200)} + \frac{(2-1)^{0.5}}{6\ (200^2)} = 0.338$$

Make an initial guess for lateral flow rate is based on the flow from the last emitter.

$$0.2(100\ \text{kPa})^{0.5} = 2\ \text{LPH}.$$

There are 2 emitters per m and the length of tubing is 100 m. Thus, there are 200 emitters and the total flow rate per lateral is (2 LPH/emitter) (200 emitters) = 400 LPH.

Calculate friction loss in fully flowing pipe

$$\text{v} = \frac{\text{Q}}{\text{A}} = \left(\frac{400\ \text{Lph}}{\pi(6/1000)^2\text{m}^2}\right)\left(\frac{\text{m}^3}{1,000\ \text{L}}\right)\left(\frac{\text{hr}}{3,600\ \text{sec}}\right)$$
$$= 0.98\ \text{m/sec}$$
$$\text{Re} = \frac{vD}{\nu} = \frac{0.98\ (12\ mm)/(1,000\ mm/m)}{1*10^{-6}} = 11,800$$
$$\text{f} = \frac{0.316}{\text{Re}^{1/4}} = \frac{0.316}{11,800^{1/4}} = 0.0303$$
$$h_f = 6.377 fL\left[\frac{Q^2}{D^5}\right]$$
$$= 6.377\ (0.0303)\ (100\ m)\left[\frac{(400\ LPH)^2}{(12\ mm)^5}\right] = 12.4\ m$$
$$\text{h}_{\text{ac}} = \text{h}_\text{f}\text{F} = (12.4\ \text{m})(0.338) = 4.2\ \text{m}.$$

Pressure (m) at the distal end is known

$$\text{H}_\text{d},\ \text{is}\ 100\ \text{kPa}(0.102) = 10.2\ \text{m}$$

Calculate average pressure in the lateral (S_e = −0.002)

$$H_a = H_d + \frac{h_{ac}}{3.852} + \frac{S_e L}{2}$$
$$= 10.2 + \frac{4.2}{3.852} + \frac{-0.002\ (100\ m)}{2}$$
$$= 10.2 + 1.09 - 0.1 = 11.19\ m$$

Second iteration

Find average emitter flow rate and lateral flow rate

$$0.2(11.19\ \text{m}/0.102\ \text{m/kPa})^{0.5} = 2.095\ \text{LPH}$$
$$\rightarrow 2.095\ \text{LPH/emitter}(200\ \text{emitters}) = 419\ \text{LPH}$$

Calculate friction loss in fully flowing pipe.

$$v = \frac{Q}{A} = \left(\frac{419\ Lph}{\pi(6/1000)^2\ m^2}\right)\left(\frac{m^3}{1,000\ L}\right)\left(\frac{hr}{3,600\ \text{sec}}\right)$$
$$= 1.03\ m/\text{sec}$$
$$\text{Re} = \frac{vD}{\nu} = \frac{1.03*12/1,000}{1*10^{-6}} = 12,360$$
$$f = \frac{0.316}{\text{Re}^{1/4}} = \frac{0.316}{12,360^{1/4}} = 0.03$$
$$h_f = 6.377 fL\left[\frac{Q^2}{D^5}\right] = (6.377)(0.03)(100\ m)\left[\frac{419^2}{12^5}\right] = 13.5\ m$$

The Christensen's F for the Darcy-Weisbach equation with 200 emitters is 0.338.

$$\text{h}_{\text{ac}} = \text{h}_\text{f}\text{F} = (13.5\ \text{m})(0.338) = 4.6\ \text{m}.$$

Third iteration

Calculate average pressure in the lateral

$$H_a = H_d + \frac{h_{ac}}{3.852} + \frac{S_e\ L}{2}$$
$$= 10.2 + \frac{4.6}{3.852} + \frac{(-0.002)(100\ m)}{2}$$
$$= 10.2 + 1.19 - 0.1 = 11.29\ m$$

Recalculate lateral flow rate based on average pressure of 11.29 m

$$\text{Average emitter flow rate} = 0.2*(11.29\ \text{m}/0.102\ \text{m/kPa})^{0.5} = 2.10\ \text{LPH}$$

Calculate inlet pressure

$$H_0 = H_d + h_{ac} + S_e L$$
$$= 10.2\ \text{m} + 4.6\ \text{m} + (-0.002)(100\ \text{m}) = 14.6\ \text{m}.$$

The inlet pressure calculated with the analytic solution (14.6 m) is the same as the inlet pressure calculated with the spreadsheet (14.69 m) in Example 17.5. Average flow rate along the lateral at the average pressure is $Q_a = 0.2\ H_a^{0.5} = 0.2\ (11.29/0.102)^{0.5} = 2.104$ LPH, which is the same as the average flow rate calculated with the spreadsheet, 2.105 LPH.

Calculate for upslope (inlet is lower than distal end)

Use same initial guess for lateral flow rate = (2 LPH/emitter) (200 emitters) and $h_{ac} = 4.2$ m.

Calculate average pressure in the lateral ($S_e = 0.002$)

$$H_a = H_d + \frac{h_{ac}}{3.852} + \frac{S_e\ L}{2}$$
$$= 10.2\ m + \frac{4.2\ m}{3.852} + \frac{(+0.002)(100\ m)}{2}$$
$$= 10.2 + 1.09 + 0.1 = 11.39\ m$$

Calculate again with average pressure.

Find average emitter flow rate and lateral flow rate

$$0.2(11.39/0.102)^{0.5} = 2.11 \rightarrow 2.11\ \text{LPH/emitter*200 emitters} = 422\ \text{LPH}422\ \text{LPH} \rightarrow h_{ac} = h_f F = 13.7*0.338 = 4.6\ \text{m}.$$

Calculate average pressure in the lateral

$$H_a = H_d + \frac{h_{ac}}{3.852} + \frac{S_e\ L}{2}$$
$$= 10.2\ m + \frac{4.6\ m}{3.852} + \frac{(0.002)(100\ m)}{2}$$
$$= 10.2 + 1.19 + 0.1 = 11.49\ m$$

Average emitter flow rate = $0.2\ (11.49\ /\ 0.102)^{0.5} = 2.12$ LPH

$$H_o = H_d + h_{ac} + S_e\ L = 10.2 + 4.6 - (0.002)(100\ m)$$
$$= 10.2 + 4.6 + 0.2 = 15.0\ m$$

Again, the inlet pressure calculated with the analytic solution (15.0 m) is the same as the inlet pressure calculated with the spreadsheet (15.14 m) in Example 17.5. Average flow rate along the lateral at the average pressure is $Q_a = 0.2\ H_a^{0.5} = 0.2 * (11.49/0.102)^{0.5} = 2.122$ LPH, which is the same as the average flow rate calculated with the spreadsheet, 2.124 LPH.

A similar iteration method is used in the spreadsheet that enables the user to input a distal end pressure based on an upstream pressure (cells L1:L8), but the results are slightly different from these calculations.

The uniformity along a drip irrigation lateral is typically quantified with the emission uniformity equation:

$$U_e = 100\left[1 - \left(\frac{1.27}{\sqrt{n}}\right)C_v\right]\frac{q_{\min}}{q_{average}} \tag{17.23}$$

where

U_e = emission uniformity
n = number of emitters per plant
C_v = manufacturers coefficient of variation
q_{min} = minimum emitter discharge based on lateral hydraulic calculation
q_{ave} = average emitter discharge based on lateral hydraulic calculation.

For new systems, the coefficient of variation that is used in the emission uniformity equation is the number reported by the factory whereas older systems would develop a higher coefficient of variation.

Example 17.6 Calculate the emission uniformity for Example 17.4 (downslope and upslope). Assume that the coefficient of variation is 0.05 or 5 % and that there are 2 emitters per plant.

Downslope uniformity

$$q_{ave} = 0.2\ H_a^{0.5} = 0.2(11.29/0.102)^{0.5} = 2.104\ \text{LPH}$$
$$q_{min} = 0.2(100)^{0.5} = 2\ \text{LPH}$$

$$U_e = 100\left[1 - \left(\frac{1.27}{\sqrt{2}}\right)0.05\right]\frac{2}{2.104} = 91\ \%$$

This is the same uniformity that is calculated in the worksheet (Fig. 17.6).

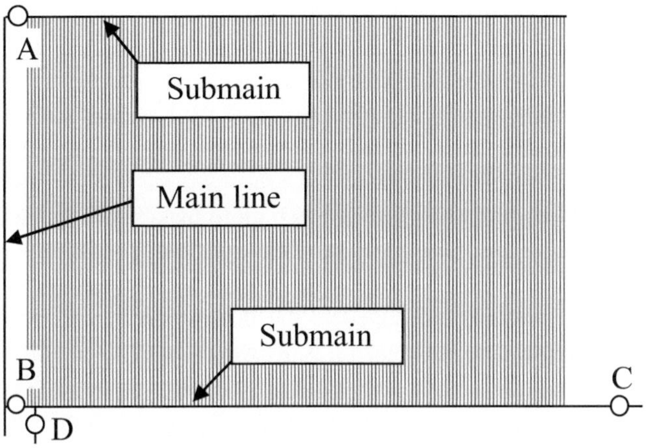

Fig. 17.9 Dual feed laterals with flushing

Upslope uniformity

$$q_{ave} = 0.2\ H_a^{0.5} = 0.2(11.49/0.102)^{0.5} = 2.122\ \text{LPH}$$
$$q_{min} = 0.2(100)^{0.5} = 2\ \text{LPH}$$

$$U_e = 100\left[1 - \left(\frac{1.27}{\sqrt{2}}\right)0.05\right]\frac{2}{2.122} = 90\ \%$$

When drip laterals are fed from both ends (Fig. 17.9) plugging is reduced, and flushing can be automated. Another advantage is that if a break in the drip lateral occurs, water enters the break from both ends, and dirt is not washed into the section of tube that is downstream from the break (the problem with single feed laterals). During irrigation, valves A and B are open and valves C and D are closed. During flushing, valves A, C, and D are open and valve B is closed. Submain diameters should be large enough to maintain a sufficient flushing flow velocity in the laterals. Use gate valves (lower pressure loss) for valves C and D in order to minimize backpressure during flushing.

For modeling a dual feed lateral laid on a slope, the point of zero flow is not midway between the submains. The zero flow point in the lateral will be closer to the lower submain. For a more accurate simulation, the water flowing into the lateral from the lower submain (lateral upslope direction) is modeled separately from the water flowing into the lateral from the upper submain (lateral downslope direction). The midpoint is shifted until the point of zero flow has the same distal end pressure in each lateral.

Example 17.7 Design the emitter spacing, lateral lengths and lateral diameter for a subsurface drip irrigation system on a 32 ha row crop farm. Use dual feed laterals. The rows are in the N-S direction so the laterals must also be N-S. Rows are 1 m apart, and there should be one lateral per row. Farm size is 32 ha with 820 m (N-S) × 400 m (E-W) boundaries. The farmer already has dirt access roads every 205 m in the E-W direction for field access. Roads are 5 m wide. There is a 0.2 % slope in N-S direction with North as the high elevation. The system efficiency is 90 %. The maximum daily plant water requirement ($ET_{c\ -\ max}$) in the middle of summer is 10 mm/day. Soil is a loam (medium texture). Assume that the system will operate for 22 hours per day during peak ET, leaving 2 hours for repairs, etc. Emitters are turbulent flow with x = 0.5 and K = 0.2 for units of kPa and LPH (2 LPH emitters). Minimum acceptable design pressure is 100 kPa. Manufacturer's CV = 0.05. Calculate emission uniformity of the lateral.

Emitter wetted diameter for a 2 LPH emitter is 0.7 m in a medium textured soil. In order to create a line source along the drip lateral, select a 0.5 m emitter spacing along the lateral.

Calculate the gross application rate

$$(di/dt)_g = \left(\frac{2\ LPH}{(0.5\ m)(1\ m)}\right) = 4\ mm/hr$$

Calculate the net application rate

$$(di/dt)_a\ = (4\ mm/hr)(90\ \%/100\ \%) = 3.6\ mm/hr$$

Calculate the required hours of application per day in each zone.

$$T_z = \frac{ET_{c-\max}}{(di/dt)_a}\ = \frac{10\ mm/day}{3.6\ mm/hr} = 2.78\ hr/day$$

Calculate the number of watering periods.

$$N_p = \frac{T_r}{T_z} = \frac{22}{2.78} = 7.91\ periods$$

Normally, the number of periods would be rounded down. However, 7.91 is close to 8 so round up to 8 periods.

Calculate the area that is watered during each period (subtract the road area).

$$A_f = \left(820 - (4\ \text{road})(5\ \text{m/road})(400\ \text{m})/\left(10{,}000\ \text{m}^2/\text{ha}\right)\right.$$
$$= 32\ \text{ha.}$$

$$A_p = \frac{A_f}{N_p} = \frac{32\ ha}{8} = 4\ ha$$

Calculate number of emitters per ha

$$E_d = \left(\frac{10{,}000\ m^2}{ha}\right)\left(\frac{1\ emitter}{s_L{*}s_d\ m^2}\right)$$
$$= \left(\frac{10{,}000\ m^2}{ha}\right)\left(\frac{1\ emitter}{(0.5\ m)(1\ m)}\right)$$
$$= 20{,}000\ emitters/ha$$

The required irrigation system flow rate is

$$Q_p = \frac{E_d Q_e A_p}{3,600}$$

$$= \frac{(20,000\ emitters/ha)(2\ LPH/emitter)(4\ ha)}{3,600\ \sec/hr}$$

$$= 44.4\ LPS$$

The next step is to select a lateral length. The laterals must run in the N-S direction, and the length of the field in the N-S direction is 800 m. Because the farmer already has E-W roads every 200 m, use 200 m long dual feed laterals. The hydraulic calculation for a 200 m long dual feed lateral during irrigation is the nearly the same as designing two 100 m long single feed laterals in the upslope and downslope direction. However, during the flushing cycle, water must travel the entire distance between submains, 200 m. Use 16 mm ID tubing to maintain uniform flow distribution during irrigation and to provide a large enough diameter to flush the 200 m laterals. During flushing, the required flow rate at the distal end of a 16 mm ID lateral is $Q = vA = v\pi r^2 = (0.5\ m/sec)(\pi)(0.008)^2 (1000\ LPS/(m^3/sec)) = 0.074$ LPS = 361 LPH. The 361 LPH flushing flow rate was added spreadsheet at the distal end of the lateral (Fig. 17.10). The required lateral inlet flushing pressure is 16.2 m.

The next step is to model the dual feed system during irrigation (Fig. 17.11). The irrigation parameters are entered into both the *Upper lateral* and *Lower lateral* worksheets, and a VBA program (*Make calcs* button) makes the hydraulic calculations. The pressure difference between the two ends of the lateral is specified in cell N2. The program adds 13 emitters are added to the downslope lateral. The average pressure in the lateral is the weighted average of the average pressures in the upslope and downslope lateral sections: $H_{a\text{-upslope}} = 10.55$ m; $H_{a\text{-downslope}} = 10.46$ m. The same numbers are also seen in the spreadsheet analysis (Fig. 17.11).

$$H_{a-average} = \frac{187\ (H_{a-upslope}) + 213\ (H_{a-downslope})}{400}$$

$$= \frac{187\ (10.55\ m) + 213\ (10.46\ m)}{2} = 10.50\ m$$

$$q_{ave} = 0.2\ H_a^{0.5} = 0.2(10.50\ m/0.102)^{0.5} = 2.029\ LPH$$

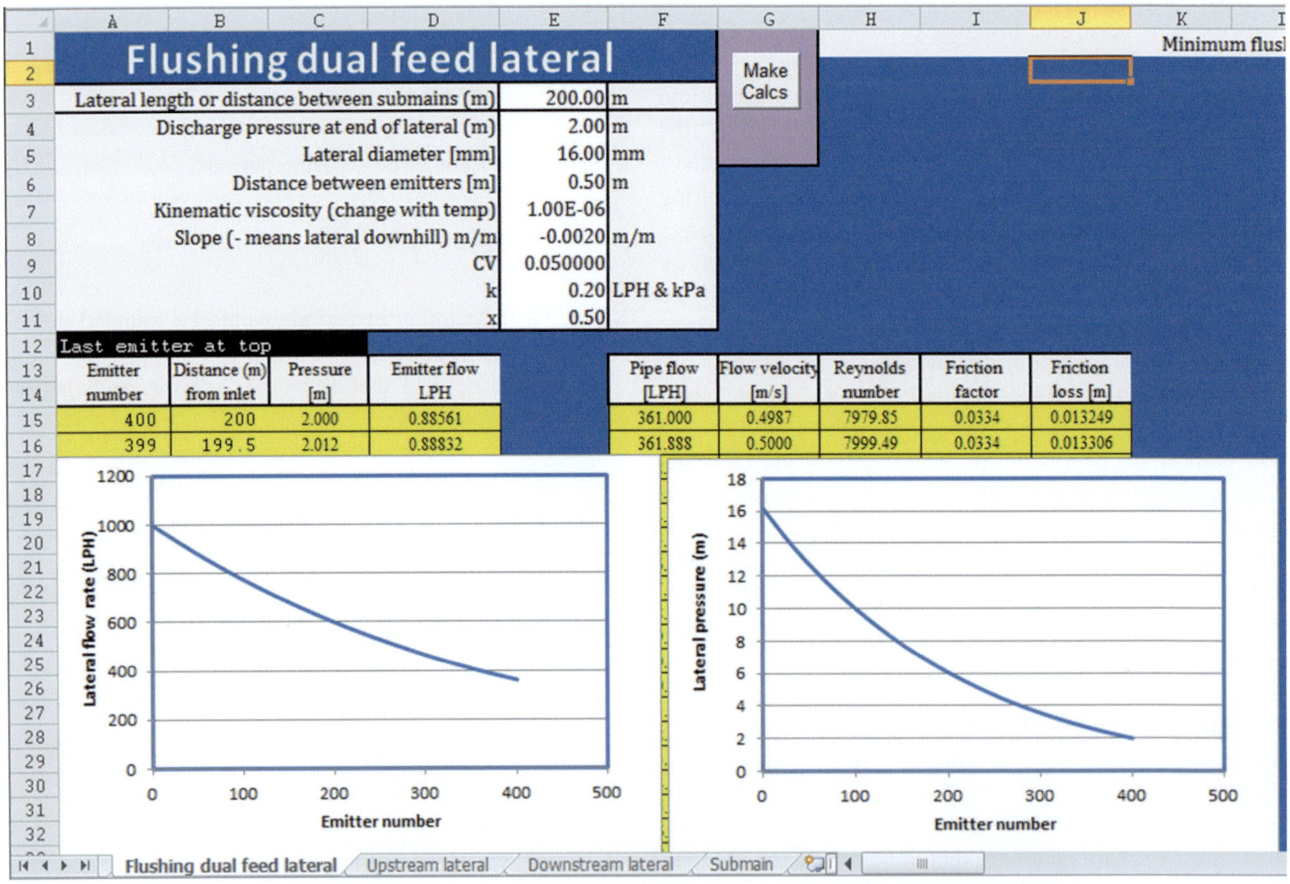

Fig. 17.10 Lateral flow rate and pressure during flushing cycle for 16 mm tubing with 2 LPH emitters at 0.5 m spacing

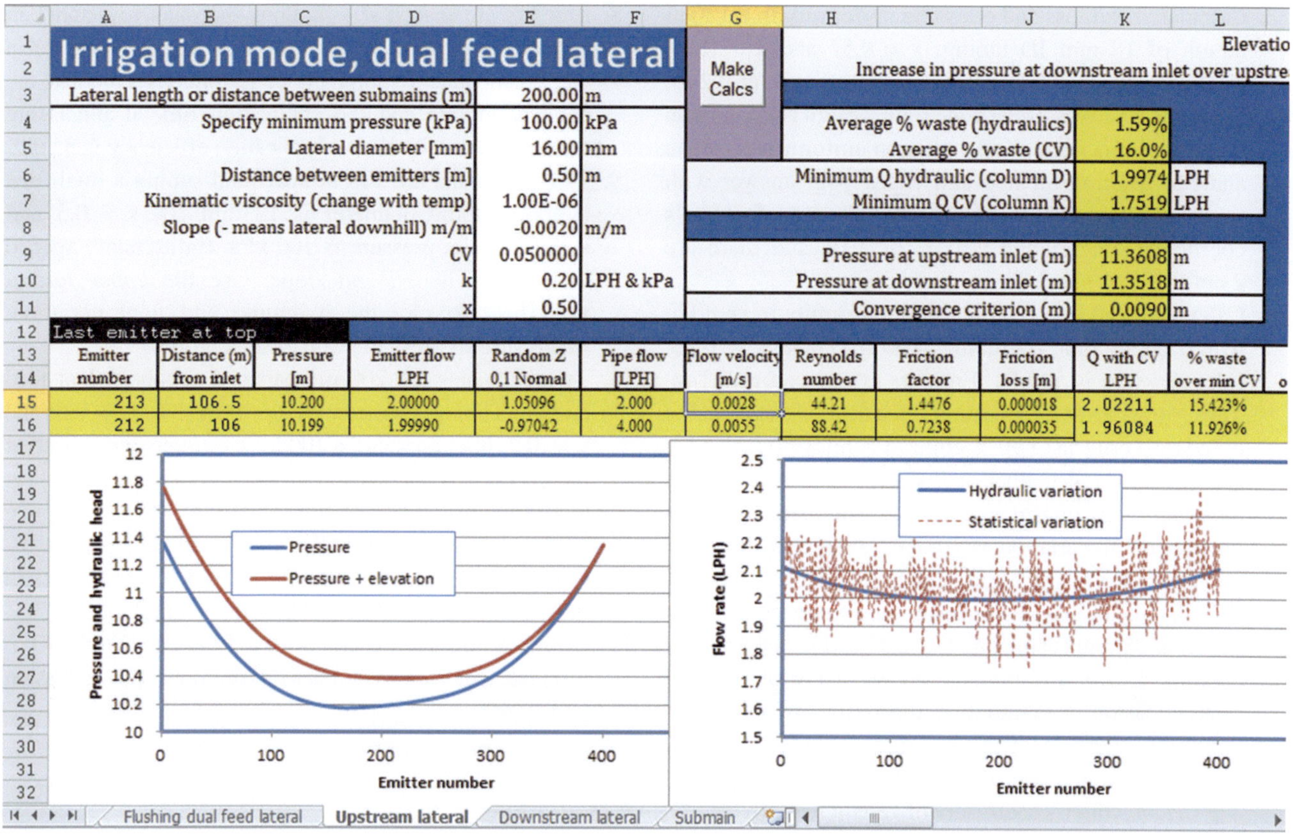

Fig. 17.11 Flow variation in dual feed irrigation lateral

Minimum pressure is 10.174 m (Fig. 17.11)

$$q_{min} = 0.2(10.17/0.102)^{0.5} = 1.997 \text{ LPH}$$

$$U_e = 100\left[1 - \left(\frac{1.27}{\sqrt{2}}\right)0.05\right]\frac{1.997}{2.029} = 94 \%$$

Connecting Drip Irrigation Laterals to PVC Submains

Subsurface drip irrigation laterals are typically placed 0.15 m below the ground surface, and submains are deeper. Polyethylene tubing tends to kink when it is bent so it is difficult to connect directly to the submain and bend the polyethylene tubing without kinking. Black flexible PVC pipe (1/2 in., 13 mm) does not kink when it is bent. It can be glued into a saddle that is glued to the top of the PVC submain (Fig. 17.12). Use of saddles or grommets and drilling holes into the submain rather than using Tees saves money – large Tees are expensive. The black flexible PVC can be connected to the drip lateral with a PVC to drip lateral connector.

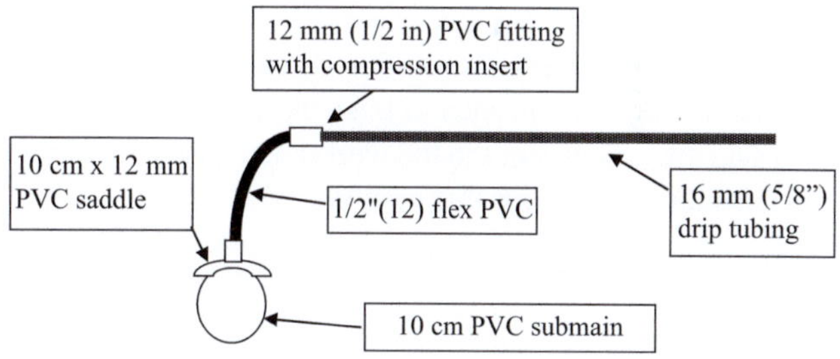

Fig. 17.12 Connection of subsurface drip irrigation lateral to PVC submain

Questions

1. How many 4 ft (1.22 m) diameter sand filters are needed for a 260 ha (640 ac) drip irrigated farm? Crop ET is 11 mm/day. Irrigation efficiency is 90 %.
2. Design particle size for a settling basin is 25 microns. Irrigation system flow rate is 1000 GPM. What are the dimensions of the settling basin?

3. Calculate head loss and emission uniformity in a 120 m length of 18 mm ID tubing. x = 0.57 and k = 0.15. Inlet pressure is 200 kPa. Emitters are spaced at 0.2 m. The crop is carrots. Determine if the design is acceptable based on a criteria of 90 % emission uniformity. Use the analytic solution method and check your answer with the *Single feed lateral* spreadsheet. The manufacturer's coefficient of variation is 0.07 or 7 %, and there are 2 emitters per plant. There is no slope.
4. Calculate head loss and emission uniformity in an 80 m length of 18 mm ID tubing. x = 0.57 and k = 0.15. Inlet pressure is 50 kPa. Emitters are spaced at 0.2 m. The crop is carrots. Determine if the design is acceptable based on a criteria of 90 % emission uniformity. Use the analytic solution method and check your answer with the *Single feed lateral* spreadsheet. The manufacturer's coefficient of variation is 0.07 or 7 %, and there are 2 emitters per plant. There is no slope.
5. Calculate head loss and emission uniformity in a 120 m length of 18 mm ID tubing. x = 0.57 and k = 0.15. Inlet pressure is 200 kPa. Emitters are spaced at 0.2 m. The crop is carrots. Determine if the design is acceptable based on a criteria of 90 % emission uniformity. Just use the *Single feed lateral* spreadsheet for the calculation. The manufacturer's coefficient of variation is 0.07 or 7 %, and there are 2 emitters per plant. Slope is 5 % downhill.
6. For the parameters in question 3, calculate the inflow rate and pressure needed to flush the 120 m length tube. Make the calculations in the worksheet. Discuss whether flushing is viable for this system.
7. For the parameters in question 5, calculate the inflow rate and pressure needed to flush the 120 m length tube. Make the calculations in the worksheet. Discuss whether flushing is viable for this system.
8. Emitters are spaced at 1 m along the plant row and 1 m between plant rows, and emitter flow rate is 2 lph. For a daily plant water requirement of 12 mm/day, calculate the application rate and the application time if plants are watered on a daily basis. Application efficiency is 85 %.
9. Two submains are 200 m apart and supply a dual feed lateral. Tubing diameter is 12 mm ID. x = 0.5 and k = 0.2. Inlet pressure is 100 kPa. Emitters are spaced at 0.5 m. There is no slope. Use the *upper lateral* worksheet to plot the hydraulic and flow variation. Make sure to press the Make calcs button.
10. Two submains are 200 m apart and supply a dual feed lateral. Tubing diameter is 12 mm ID. x = 0.5 and k = 0.2. Inlet pressure is 100 kPa. Emitters are spaced at 0.5 m. There is 1 % slope. Use the *upper lateral* worksheet to plot the hydraulic and flow variation. There is no pressure difference between the uphill and downhill inlets. Explain the difference between the red and blue lines in the pressure graph. Which line is the hydraulic head?
11. Design a dual feed lateral with a 1 % slope, and 240 m distance between submains. Determine the emission uniformity.

References

Netafim (2015) AGF sand media filters: operation, installation, and maintenance guide. Netafim USA, Fresno. http://www.netafimusa.com/files/literature/agriculture/filters/A048-AGF-Manual.pdf

Netafim (2015) Hydrocyclone sand separators: operation, installation, and maintenance guide. Netafim USA, Fresno. http://www.netafimusa.com/files/literature/agriculture/filters/hydrman-hydrocyclone-manual.pdf

ASAE Engineering Practice: ASAE EP405 – Design, Installation, and Performance of Trickle Irrigation Systems (1993) American Society of Agricultural Engineers, St. Joseph

18 Drip Irrigation System Design

This first half of this chapter focuses on the design of a dual feed subsurface drip irrigation system, which is the most common agricultural drip system. The first example demonstrates the procedure for design and economic comparison of alternative mainline and submain designs for subsurface drip irrigation. Uniformity is calculated for an entire irrigation zone rather than an individual lateral. Submains are designed with lateral flow rate vs. pressure curves just as laterals are designed with emitter flow rate vs. pressure curves. The economic analysis includes water, energy, and pipe costs. One of the major advantages of a dual feed system is the automated flushing process; the example evaluates the additional pumping installation and pipe costs associated with flushing.

The last half of this chapter provides a rationale for the selection between types of inline drip irrigation laterals. There are many alternatives with different wall thicknesses, tubing diameters, emitter spacings, emitter types, and manufacturer's coefficients of variation. How does one decide which is the best alternative? A computer program is provided that allows the evaluation of the various alternatives. It uses Monte Carlo simulation to simulate the effect of emitter degradation and tubing replacement interval on yield and cost.

Example 18.1 Two possible pipe designs are presented in Figs. 18.1 and 18.2. Because submains are split in design 1, smaller submains can be used and emission uniformity is improved. However, extra pipes are eliminated in design 2. Determine, based on economics, whether design 1 (Fig. 18.1) is better than design 2 (Fig. 18.2). Assume 20 year project life and 8 % ROR. There is a 0.2 % slope in the N-S direction with the upper end at North end. Assume normal head losses in valves and fittings. The canal water surface is 1 m below ground surface. Use the parameters specified in Example 17.7.

The cost of energy is $0.10/kW-hr, and the cost of water is $3.27/ha-cm. Annual ET_c is 1 m/y. Irrigation efficiency is 90 % and pumping efficiency is 80 %. Ignore the cost of valves and fittings because the cost of valves and fittings in the two designs is approximately the same.

Use Class 125 PVC pipe for submains and mainlines. Mainlines are defined as those pipes that are upstream from the valves. The design shown in Fig. 18.1 includes E-W submains (1–32) that are fed from the center by EW mains 33–48, which are fed by N-S mainline 49–52. The design shown in Fig. 18.2 includes submains that are fed directly from the N-S mainline 49–52.

Drip irrigation laterals are 200 m length with 16 mm tubing fed from both ends (Fig. 17.21). The cost of 16 mm ID drip tape with turbulent emitters is $0.10/m. Emitters are turbulent ($x = 0.5$), 2 L/h at 100 kPa ($k = 0.2$), spacing = 0.5 m.

For submain design, a relationship can be developed between lateral inlet pressure at the submain and lateral flow rate. The distal end pressure can be varied in the worksheet in order to generate the inlet pressure vs. flow rate curves (Fig. 18.3). As calculated in Chap. 17, the upper lateral is 106.5 m long (213 emitters) and the lower lateral is 93.5 m long (187 emitters) in the simulations represented in Fig. 18.3. The lateral flow exponents 0.502 and 0.507 are nearly the same as the emitter flow exponents; however, the *Make calcs* program varies the number of emitters in upper and lower sections as inlet pressures change (Fig. 18.4) and has a greater exponent for the upper lateral section and a lower exponent for the lower lateral section.

Design 1

The Design 1 valves and pipes for each zone are shown in Fig. 18.5. The solenoid valve supplies the entire zone (both halves with water) during an irrigation event. Isolation gate valves are used to isolate the two halves of the zone

P. Waller, M. Yitayew, *Irrigation and Drainage Engineering*, DOI 10.1007/978-3-319-05699-9_18

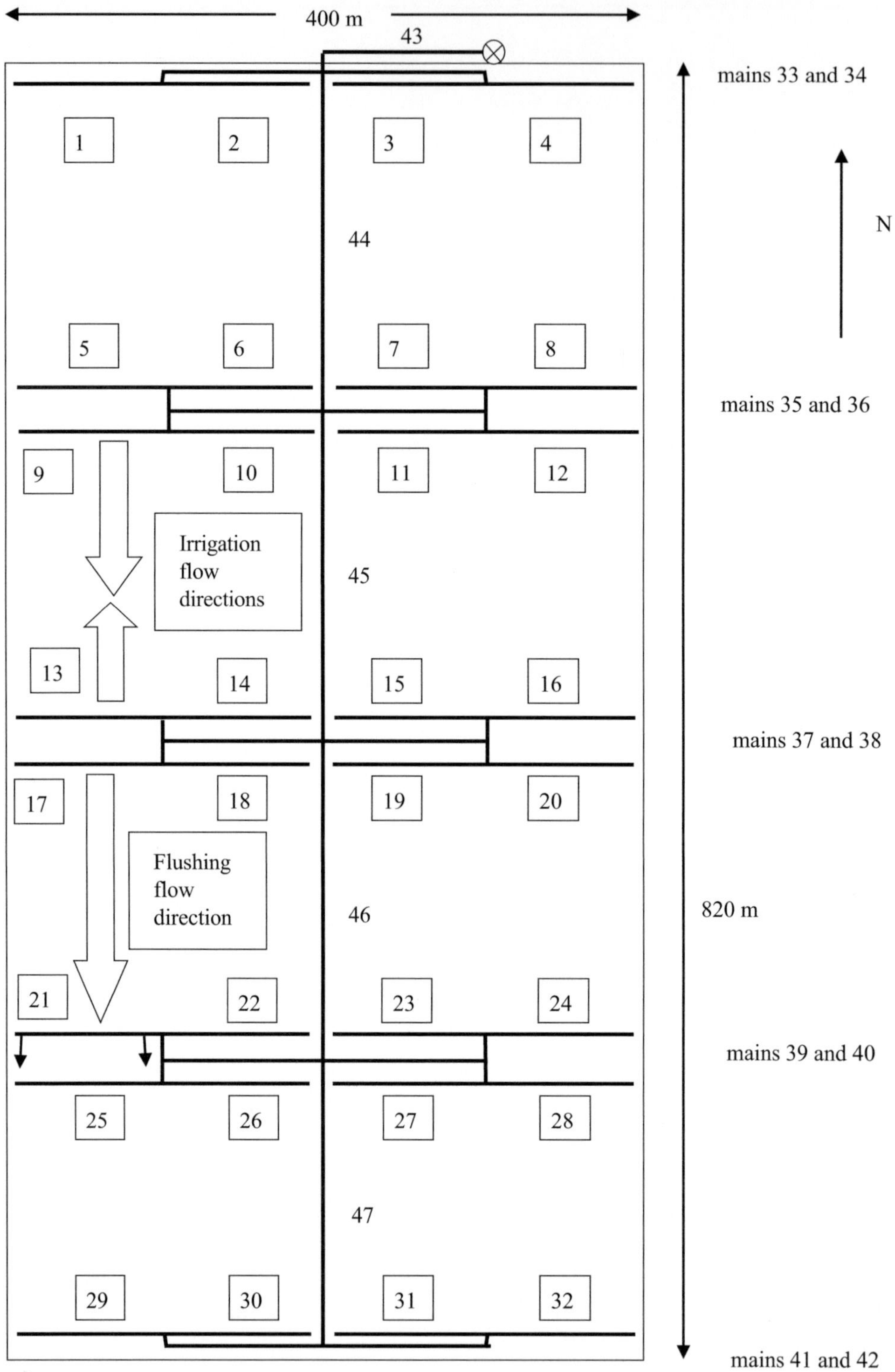

Fig. 18.1 Drip irrigation design number 1 for Example 18.1. Dark lines are PVC pipes

during flushing. Cutting the zones in half during flushing reduces the flow rate and required size of the mainline during flushing. It also enables the farmer to irrigate half of the zone, if necessary. The EW mainline supplying the submains is designed to supply the entire zone with water during irrigation or half the zone during flushing. Laterals on the right half of the zone are not shown in Fig. 18.5.

Design of Submains for Design 1

The submain design is subject to the following criteria. First, pressure loss along the submain should be minimized in order to maintain uniformity. Second, the submains should carry the flow required for flushing, which is significantly higher than the irrigation flow rate.

Fig. 18.2 Drip irrigation design number 2 for Example 18.1

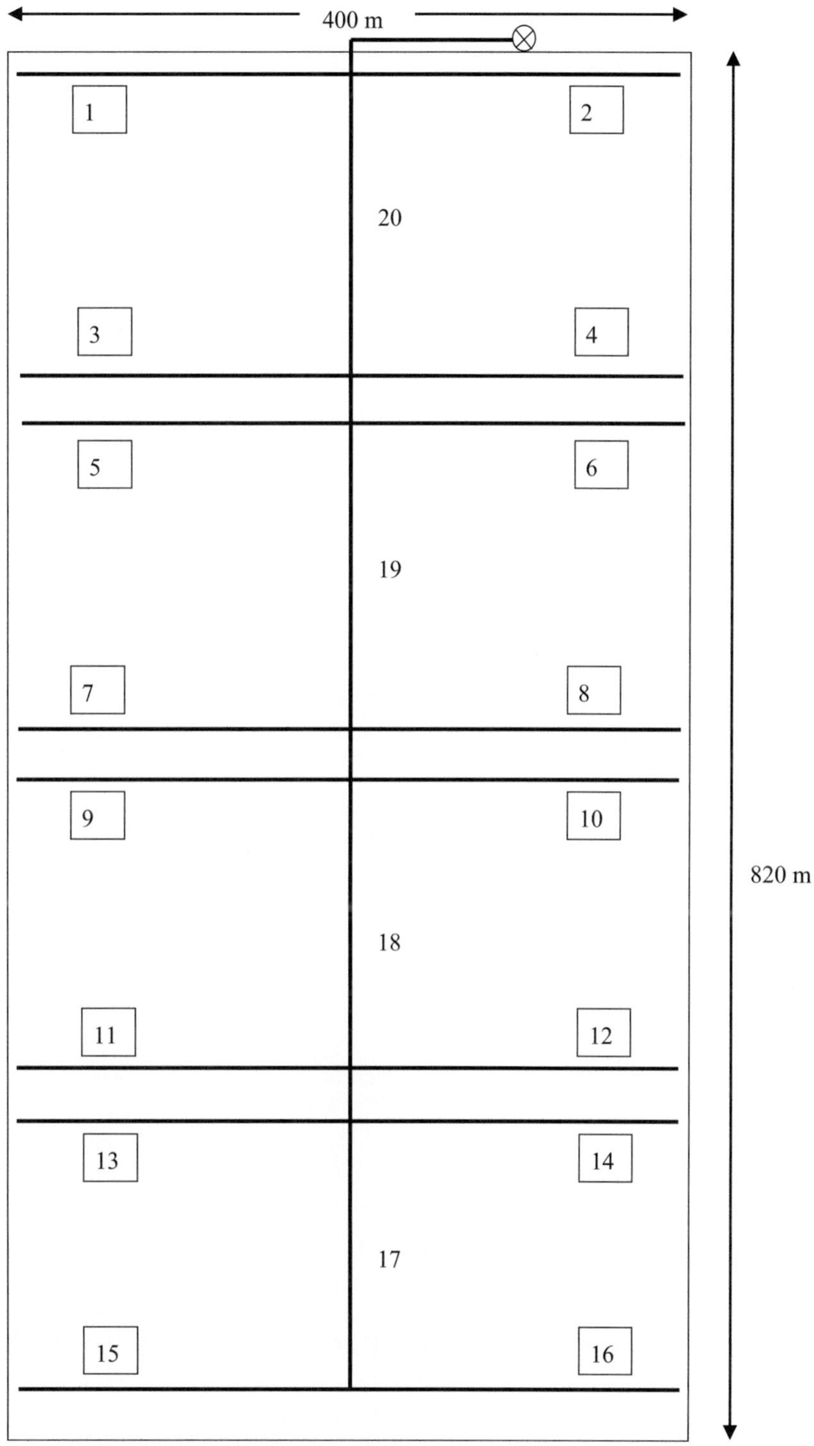

The area supplied by each submain in Fig. 18.5 is approximately 1 ha (100 m × 100 m).

$$Q_p = \frac{E_d Q_e A_p}{3,600} = \frac{20,000\ emitters/ha * 2\ LPH*1\ ha}{3,600} = 11.1\ LPS$$

The flow velocity in 83 mm ID (3 in.) Class 125 pipe at 11.1 LPS is 2.2 m/sec. The 83 mm pipe flow velocity decreases below 1.5 over the last third of the submain. The velocity in 107 mm ID (4 in.) Class 125 pipe at 11.1 LPS is 1.3 m/sec. Thus, 107 mm pipe is acceptable based on the 1.5 m/sec rule for the first two-thirds. However, the 1.5 m/sec rule might be ignored in this case because a pressure surge is dissipated by the multiple outlets to laterals.

During flushing, the lateral inlet flow rates in the upper submain are 1,000 LPH (calculated in previous chapter), and

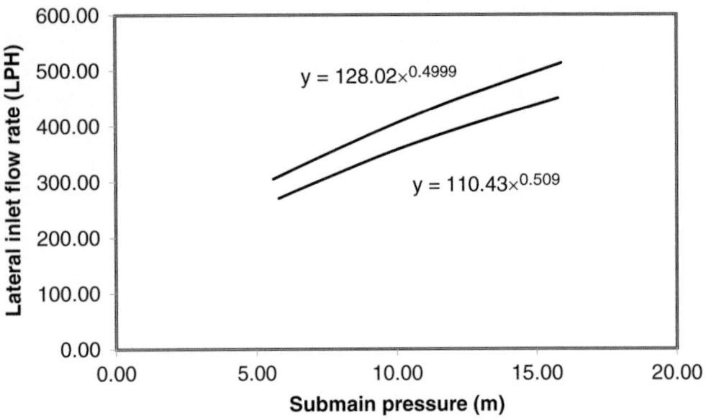

Fig. 18.3 Lateral flow rate versus pressure for upper and lower sections, keeping number of emitters constant on upper and lower sections at 213 and 187, respectively

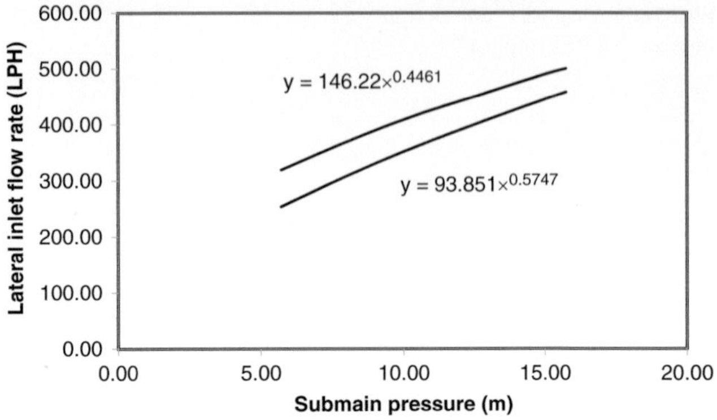

Fig. 18.4 Lateral flow rate versus pressure for upper and lower sections, keeping number of emitters constant on upper and lower sections at 213 and 187, respectively

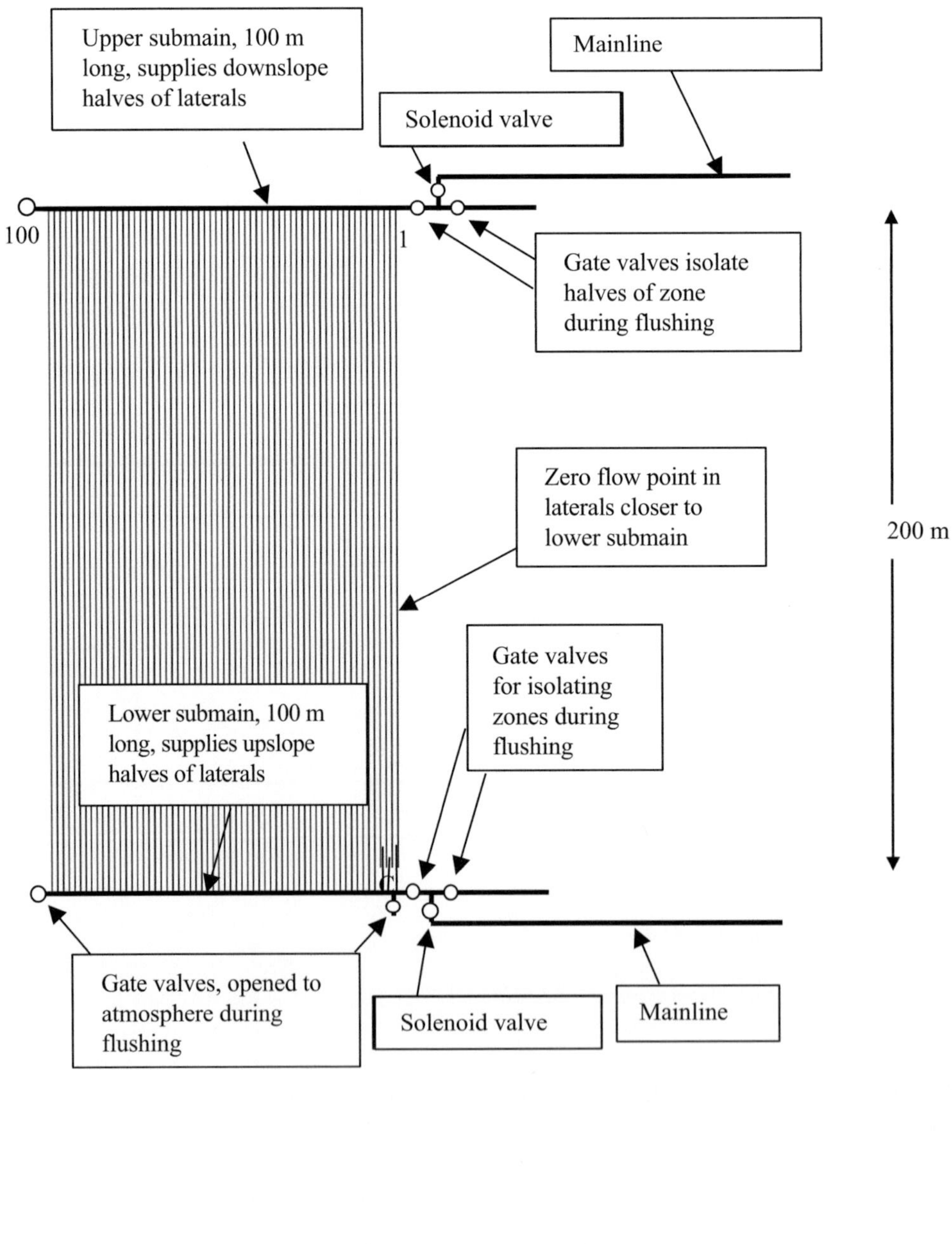

Fig. 18.5 Half of zone for design 1

lateral discharge flow rates to the lower submain are 360 LPH (Fig. 13.22). Thus, the upper submain must carry 1,000 LPH/lateral * 100 laterals = 100,000 LPH = 27.8 LPS. The lower submain divides the flushing discharge flow rate in two directions, so the largest flow rate in the lower submain is 360 LPH * 50 laterals = 18,000 LPH = 5 LPS. The flow velocity in 107 mm ID (4 in.) Class 125 pipe at 27.8 LPS is 3.1 m/sec.

If the criterion for submain diameter is based on the 1.5 m/sec rule, then a 107 mm pipe is not acceptable during flushing (3.3 m/sec velocity at 28 LPS flow rate). However, if slowly-closing 107 mm (4 in. gate valves are used as flushing outlets on the lower submain (Fig. 18.5), then water hammer will not occur because the gate valves require 10 to 20 seconds to close; the velocity will slowly decrease as the gate valves close. Sundance Farms in Arizona has a system that is similar to the design in this example: they use 107 mm submains and 107 mm gate valves, and they have never had water hammer problems during flushing.

In order to derive the lateral inlet flow rate vs. pressure curve for the flushing lateral (Fig. 18.6), the *Flushing dual feed lateral* worksheet varies downstream pressure and records the upstream flow rate and pressure in cells O6:Q15. Based on the information in these cells, the equation for lateral inlet flow rate is $Q_{lateral} = 278H^{0.46}$, where H is the pressure at the lateral inlet. The pressure at the end of the upper submain should be set to 16.2 m in order to have 2 m head at the end of the last lateral (Fig. 18.7).

The lateral inlet flow rates along the submain during irrigation are based on the equation in Fig. 18.4, $Q_{lateral\text{-}upper} = 146\ H_0^{0.45}$, and the requirement that inlet pressure

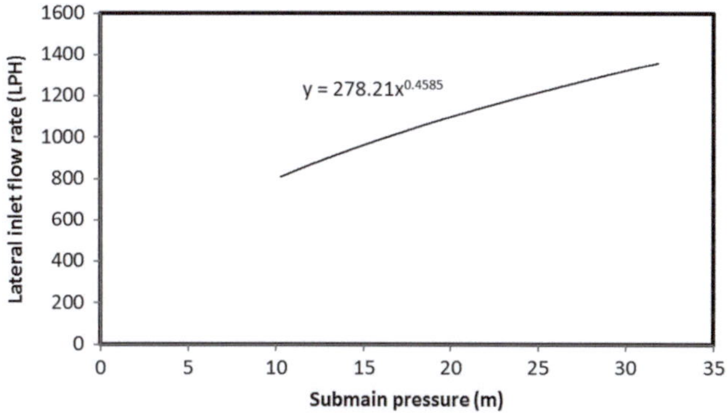

Fig. 18.6 Lateral inlet flow rate and discharge flow rate versus inlet pressure during flushing from upper submain

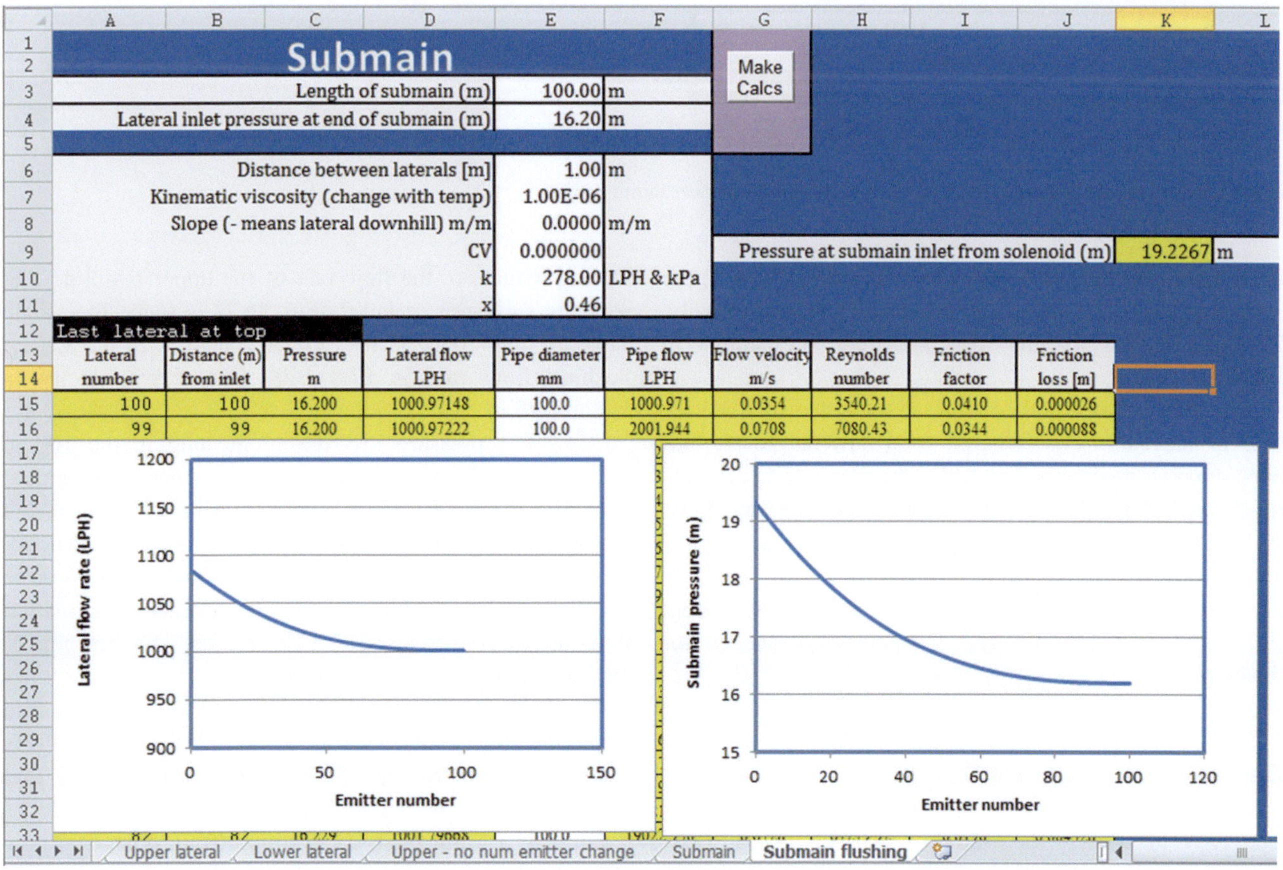

Fig. 18.7 Upper submain pressure and lateral inlet flow rate distribution during flushing for a 107 mm diameter upper submain

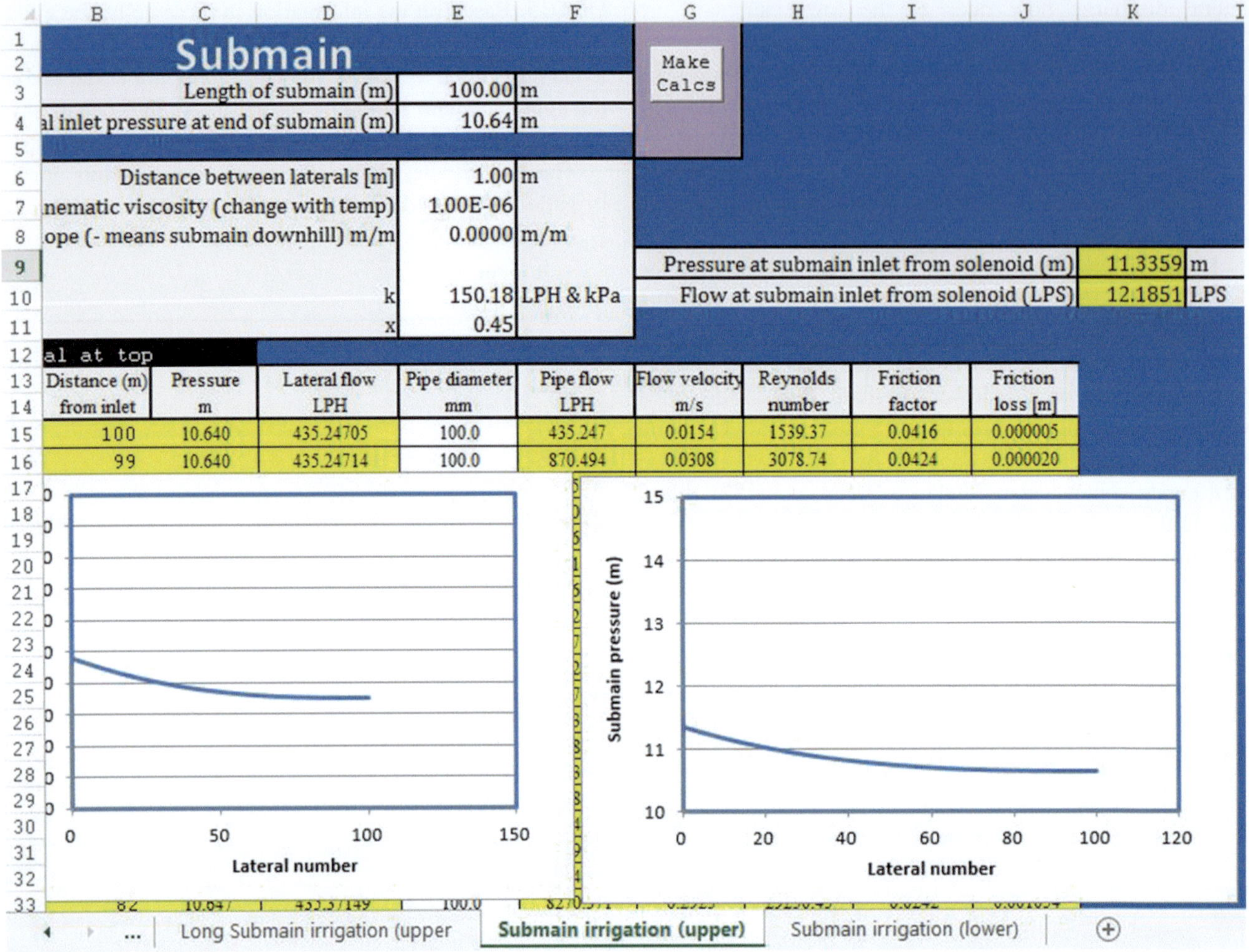

Fig. 18.8 Upper submain pressure distribution and lateral flow rates during irrigation

is 11.4 m. The inlet flow rate and pressure into the upper submain (Fig. 18.8) are 12.1 LPS and 11.3 m, respectively. The calculations for the lower submain are shown in Fig. 18.9.

The inlet pressure and flow rate in the lower submain are 11.9 m and 10.5 LPS. Thus, the total flow rate to the half zone during irrigation is 22.6 LPS. The flow rate to the zone if all emitters had the same pressure and flow rate would be 22.2 LPS (11.1 * 2); thus, the percent wasted water in the zone is (22.6 − 22.2)/(22.2) * 100 % = 1.8 %. The low percentage of wasted water might indicate that the system is overdesigned, but the large diameters are needed for flushing.

Design of Mainlines for Design 1

The E-W mainlines (for example, pipe 33 in Fig. 18.1) cannot exceed the 1.5 m/sec rule because there is a solenoid valve that could close quickly and cause water hammer. During irrigation, the flow rate of the upper mainline that supplies submains 1 and 2 (Fig. 18.1) is twice the single submain flow rate, (12 LPS) (2 zones) = 24 LPS. The flushing flow rate is 27.8 LPS. The flow velocity in 158 mm (6 in.) Cl 125 pipe at 27.8 LPS is 1.4 m/sec < 1.5 m/sec. The pressure loss during flushing in the 100 m mainline at 27.8 LPS is 0.94 m. During irrigation (24 LPS), the pressure loss is 0.73 m. All mainlines (except mains 41 and 42 in Fig. 18.1) must be designed for the upper submain flushing flow rate. Mains 41 and 42 are not used to supply water for flushing laterals; however, their irrigation flow rate of 10.6 * 2 = 21.2 LPS requires 158 mm pipe to remain below the 1.5 m/sec rule.

The N-S mainline should be designed to carry the entire irrigation flow rate (45.2 LPS) to the last zone. This design flow rate for the system is based on the fact that 8 zones are required (Chap. 17) and each of the eight blocks requires 45.2 LPS. First, try 206 mm (8 in. pipe. The flow velocity at 45.2 LPS is 1.4 m/sec, and the friction loss is 0.71 m/100 m. The last section of N-S mainline can be

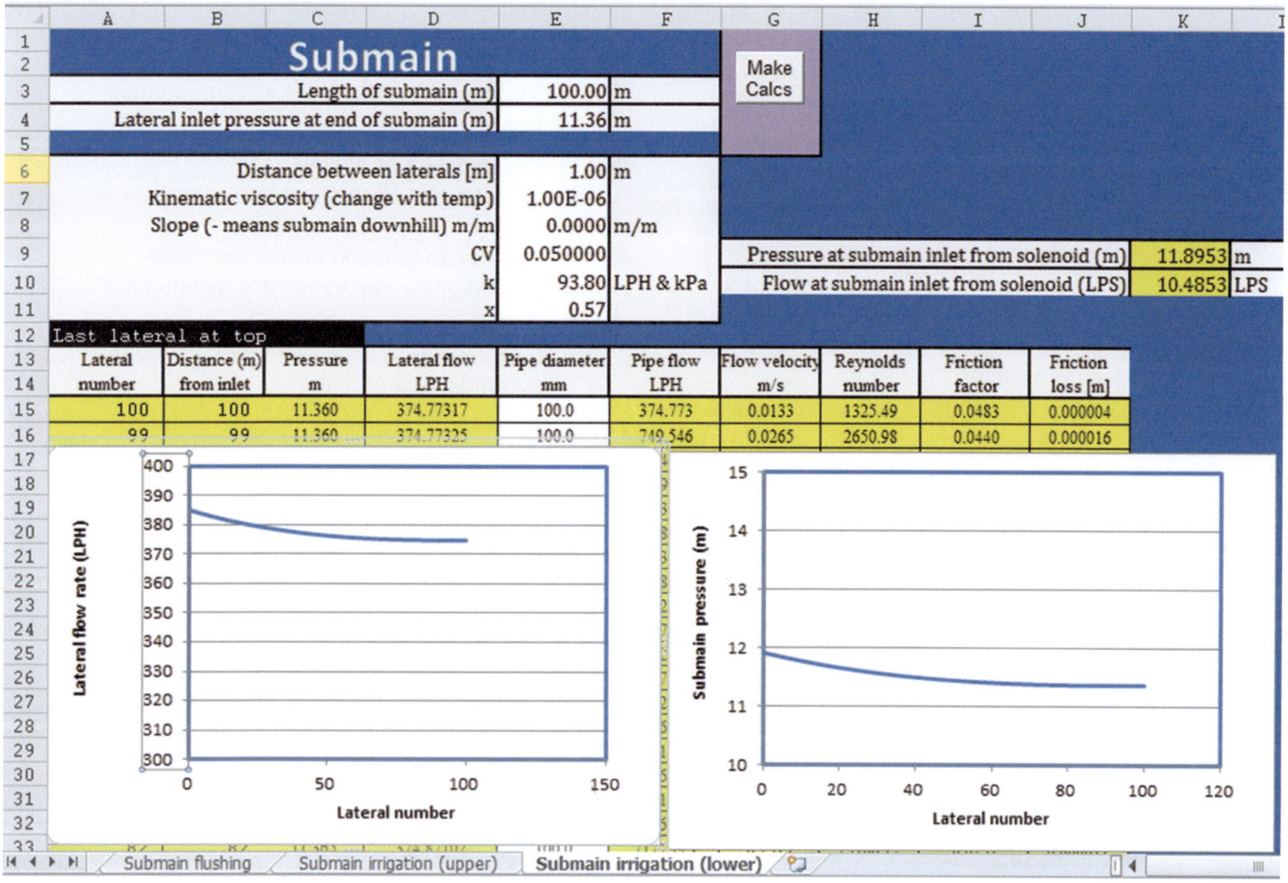

Fig. 18.9 Lower submain pressure distribution and lateral flow rates during irrigation

Table 18.1 Pipe diameters, friction loss, and pressure loss for mainline sections (Fig. 18.1) from pump to worst case mainline (number 42)

	Pipe	43–46 (700 m) 45 LPS		47 (200 m) 21 LPS		42 (100 m) 21 LPS	
Pipe size	cost	h_f (m)	ΔP (m)	h_f (m)	ΔP (m)	h_f (m)	ΔP (m)
6 (158 mm)	$5.87/m	na >1.5	na >1.5	−1.28	−0.88	−0.64	−0.64
8 (206 mm)	$9.71/m	−4.7	−3.5	−0.36	+0.04	−0.18	−0.18
10 (256 mm)	$18.6/m	−1.6	−0.4	−0.12	+0.28	−0.06	−0.06

107 mm (4 in. because mains 41 and 42 carry a maximum 21.2 LPS flow rate. The friction loss in 158 mm pipe at 21.2 m/sec is 0.64 m/100 m. Friction losses for mainline pipe diameters are shown in Table 18.1. The elevation loss between the pump and mains 39 and 40 is 600 m * 0.2 m/100 m = 1.2 m. The pressure gain due to elevation was subtracted from the friction loss in order to calculate ΔP in Table 18.1. The elevation gain is 0.4 m in section 47 and 0 m in section 42.

Slightly higher pressure will be supplied to the upper zones that are closer to the pump because of friction loss in the mainline. If one zone has a pressure that is higher than others, then gate valves can be installed and partially closed to dissipate the excess energy. Using gate valves to dissipate energy decreases their expected life; however, gate valves are relatively inexpensive compared to automatic pressure reducing valves. If flanged gate valves are used, then they are easily replaced.

The next step is to compare pipe costs and energy costs. In sections 43–46, the difference in the pump pressure requirement is 3.1 m (3.5 m and 0.4 m) between the 8 in and 10 in pipe. The depth of water required per year (described in problem statement) at 90 % irrigation efficiency is 1 m/0.9 = 1.11 m. Use Eq. 2.15 to determine the cost of energy.

$$E = \frac{0.0272(1,110\ mm)(3.1\ m)}{0.8} = 117\ kW - hr/ha$$

$$(\$/ha) = \left(\frac{117\ kW - hr}{ha}\right)\left(\frac{\$0.10}{kW - hr}\right) = \$11.70/ha$$

Difference between annual cost of 206 mm (8 in.) and 256 mm (10 in.) pipe is $11.70/ha * 32 ha = $374/yr.

Present value of annual energy savings ($374) with 256 mm (10 in.) pipe (20 year, 8 %) is $3,675

The cost of 700 m of 256 mm (10 in.) pipe (Table 18.1) is 700 * $18.60/m = $13,000

The cost of 700 m of 206 mm (8 in. pipe) is 700 * $9.71 = $6,800

The extra cost of 256 mm (10 in.) pipe is $13,000 − $6,800 = $6,200. This is more than the present value of the energy savings with 256 mm (10 in.) so use the 206 mm (8 in.) pipe in sections 43–46.

In section 47, the pressure requirement for 158 mm (6 in.) pipe is 0.88 m greater than the pressure requirement for 8 in. pipe. The cost of energy must be calculated for the entire farm unless a variable frequency pump is purchased.

$$E = \frac{0.0272(1,110\ mm)(0.88\ m)}{0.8} = 33.2\ kW - hr/ha$$

$$(\$/ha) = \left(\frac{33.2\ kW - hr}{ha}\right)\left(\frac{\$0.10}{kW - hr}\right) = \$3.32/ha$$

Total yearly cost for 32 ha = $3.32/ha * 32 = $106/yr.

Present value of energy saving with 206 mm (8 in.) pipe (20 year, 8 %) is $1,043

The cost of 200 m of 206 mm (8 in.) pipe (Table 18.1) is 200 m * $9.71/m = $1,942

The cost of 200 m of 158 mm (6 in. pipe is 200 m * $5.87 = $1,174

The extra cost of the larger pipe is $1,942 − $1,174 = $768. Thus, the energy savings ($1,043) with larger pipe in section 47 is greater than the cost difference ($768): select the larger pipe, 206 mm.

A cost analysis can also be made for E-W mainlines. The difference in pressure loss between 6 in pipe and 8 in pipe (Table 18.1) is 0.64 − 0.18 = 0.46 m. The yearly energy savings is $56, and the present value of energy savings is $545. The difference in cost of pipe for all the 100 m mains is $7,680. Thus, it is more economical to use the smaller pipe for the E-W submains.

Pump Requirement for Design 1

The system pressure requirement is based on supplying the minimum pressure to the worst-case emitter, which is located at the end of the last lateral on submain 25 (Fig. 18.1). Assume typical pressure drops in valves, filters, and fittings: 2 m head loss in the solenoid valve, 8 m head loss in the sand filter, and 5 m head loss in other valves and fittings. The pump is extracting water from a canal with water surface 1 m below the ground surface.

10.2 m	Minimum operating pressure for emitters
+1.2 m	Pressure loss in lateral
+0.4 m	Pressure loss in submain
+2 m	Pressure loss in solenoid valve
+0.7 m	Pressure loss in E-W main
+4.7 m	Pressure loss in 700 m N-S mainline
- 1.2 m	Elevation gain in 700 m N-S mainline
+8 m	Pressure loss in sand filter
+5 m	Friction loss in valves and fittings (check valve, manifold, pump suction)
+1 m	Elevation of ground surface above canal water surface
32 m	Total pump pressure requirement

Thus, the pump flow requirement is 45.2 LPS and pressure requirement is 32 m during irrigation. If pumping efficiency is 80 %, then the power requirement during irrigation is (Eq. 2.17)

$$Pow_{irrigate} = Q\rho gh/Eff = 0.0452*1,000\ kg/m^3* 9.8\ m/\sec^2*32\ m*0.001\ kW/W/0.8 = 17.7\ kW$$

During flushing, the pressure required at the submain inlet is 18.5 m (Fig. 18.7). If isolation valves are used and only half of the zone is flushed at a time, then required flow rate is 27.8 LPS. The total pressure requirement during flushing is

18.5 m	Minimum operating pressure at submain inlet (Fig. 18.7)
+2 m	Pressure loss in solenoid valve
+0.9 m	Pressure loss in E-W main
+2.0 m	Pressure loss in 700 m N-S mainline
- 1.2 m	Elevation gain in 700 m N-S mainline
+8 m	Pressure loss in sand filter
+5 m	Friction loss in valves and fittings (check valve, manifold, pump suction)
+1 m	Elevation of ground surface above canal water surface
35.9 m	Total pump pressure requirement

$$P_{flushing} = Q\rho gh/Eff = 0.0278*1,000\ kg/m^3* 9.8\ m/\sec^2*35.9\ m*0.001\ kW/W/0.8 = 12.2\ kW$$

The power requirement for flushing is less than the power required for irrigation. It would be preferable to find a pump that has both points (45 LPS @ 32 m and 27 LPS @ 37 m) on the pump curve. It is more important to optimize (highest efficiency point) the pump for the irrigation use because that is the primary energy use.

Cost of Tubing and Pipe for Design 1

Total length of drip tubing in design 1 is the field area divided by the lateral spacing.

$$32\ \text{ha}*10,000\ \text{m}^2/\text{ha}/(1\ \text{m/lateral}) = 320,000\ \text{m}$$

Drip tape	320,000 m * $0.10/m	= $32,000
Submain 4 inch	100 * 32 = 3,200 m * $2.72/m	= $8,704
Mainline 6 inch	100 * 10 = 1,000 m * $5.87/m	= $5,870
Mainline 8 inch	900 m * $9.71	= $8,739
Total		= $55,313

Design 2

The second design is not included in the *Chapter 17 and 18 lateral and submain calculations* workbook. Design 2 (Fig. 18.2) does not have E-W mains, but only has E-W submains. During irrigation, a minimum of 4 ha is irrigated at one time (44.4 LPS). During flushing, the gate valve in the center of the lower submain (Fig. 18.13) is shut, and flush valves are only turned on one side of the lower submain at a time.

The lateral design characteristics during irrigation and flushing are the same as design 1 (From Fig. 18.3, K = 127 and x = 0.502). Lateral flow rate vs. pressure was input into the upper submain spreadsheet for a 200 lateral outlet submain (Fig. 18.10). For this design, 106 mm (4 in.) is used for the last 50 m, and 158 mm (6 in.) is used over the first 150 m. The 1.5 m/s rule would allow a switch to 106 mm (4 in.) before 150 m, but energy losses and flow change were excessive with a longer length of 106 mm (4 in.) pipe. The pressure required at the upper submain connection to the N-S mainline is 12.2 m (Fig. 18.10). Total flow is 24.4 LPS.

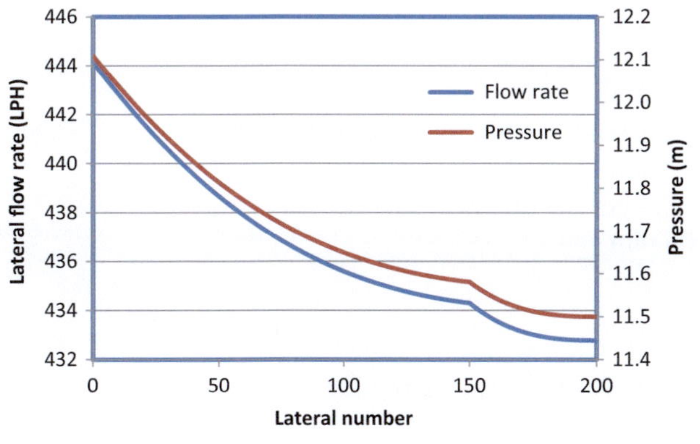

Fig. 18.10 Pressure and flow distribution in upper submain during irrigation for design 2: 106 mm (4 in.) pipe is used for last 50 m and 158 mm (6 in.) pipe is used for first 150 m

For the lower submain laterals, K = 111 and x = 0.507. The lower submain has 125 laterals at 158 mm (6 in.) diameter and the rest at 106 mm (4 in.), and the pressure requirement at the beginning of the submain is 12.5 m because water is going uphill (Fig. 18.11).

The flow rate required during irrigation in the upper submain is 24.4 LPS, and the flow rate required in the lower submain is 22.5 LPS. Thus, the total flow rate required during irrigation is 46.9 LPS (22.5 + 24.4). If the entire zone were irrigated at the design flow rate, then the flow rate would be 44.4 LPS. Thus, the percent wasted water with this design is (46.9–44.4)/44.4 = 5.5 %.

During flushing, the flow rate increases to 58.3 LPS (Fig. 18.12), and the pressure requirement at the upper submain connection to the N-S mainline is 20 m.

Thus, the power requirement during flushing is significantly higher than the irrigation power requirement, and a booster pump is required during flushing. In addition, the N-S mainline flow velocity would slightly exceed the 1.5 m/sec rule with 207 mm (8 in.) pipe at 59 LPS. It might be possible to save energy by flushing half of the zone with the configuration shown in Fig. 18.13. Half of the zone is flushed

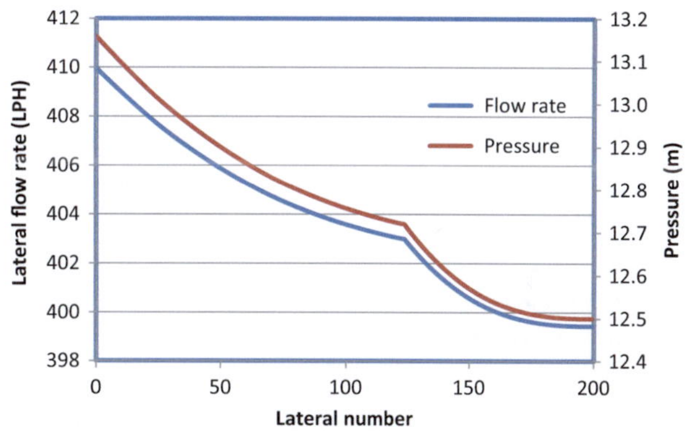

Fig. 18.11 Pressure and flow distribution in lower submain during irrigation for design 2: 106 mm (4 in.) pipe is used for last 75 m and 158 mm (6 in.) pipe is used for first 125 m

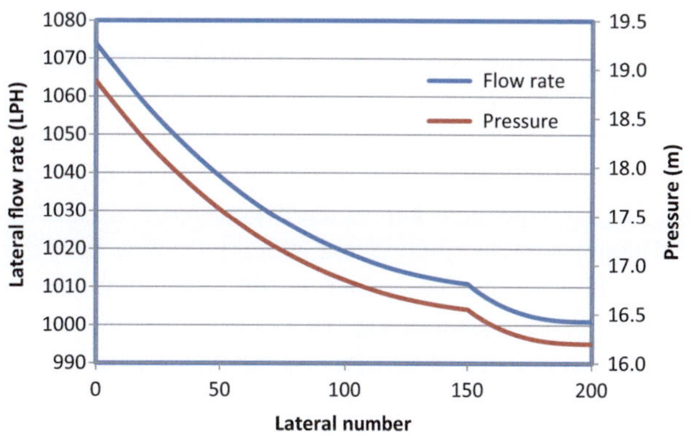

Fig. 18.12 Pressure and flow distribution in upper submain during flushing for design 2

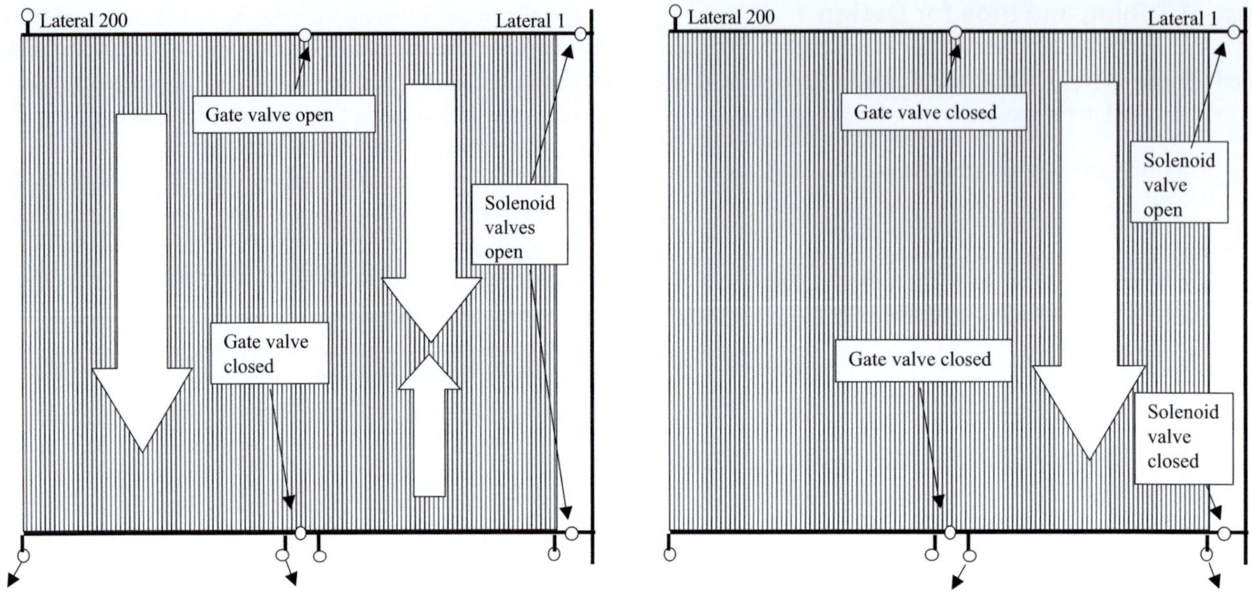

Fig. 18.13 Flow directions during flushing for the two halves of the zones in design 2

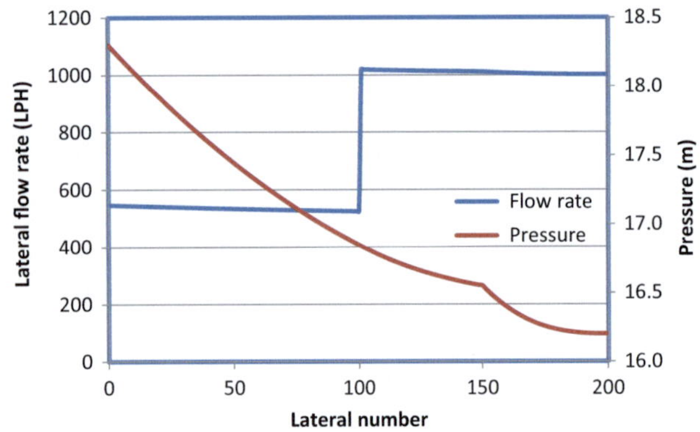

Fig. 18.14 Pressure distribution and flow rates during flushing last half of zone

at a time by dividing submains in half with gate valves. The valve positions and flow directions as the two halves are flushed are shown in Fig. 18.13. When the half of the zone farthest from the N-S mainline is flushed (left side of Fig. 18.13), the lateral flow rates in the first half of the submain (laterals 1–100) are calculated based on irrigation flow rates (Fig. 18.3) and based on flushing flow rates (Fig. 18.6) in the latter half of the lateral.

The pressure distribution and lateral inlet flow rates during flushing of the outer half of the zone (left side of Fig. 18.13) are shown in Fig. 18.14. The total flow rate in the upper submain is 43.3 LPS and the flow rate in the first half of the lower submain is 10.6 LPS (Fig. 18.8) so the total flow rate during flushing is 53.9 LPS. The 1.5 m/sec rule is slightly exceeded in the N-S mainline. The pressure required at the submain inlet is 21.3 m. This flow rate (53.9 LPS) is 7 LPS greater than the irrigation flow rate (46.9 LPS), and increased pressure through use of a booster pump during flushing would be required; thus, a booster pump would still be needed, and the design would be much more complex. Thus, the original design without gates valves is selected (59 LPS).

Pipe Costs for Design 2

Assume that the same N-S mainline pipe sizes are required.

Polyethylene tubing	320,000 m * $0.10/m	= $32,000
L. Submain 106 mm	75 * 8 = 600 m * $2.72/m	= $1,632
U. Submain 106 mm	50 * 8 = 400 m * $2.72/m	= $1,088
L. Submain 158 mm	125 * 8 = 1,000 m * $5.87/m	= $5,870
U. Submain 158 mm	150 * 8 = 1,200 m * $5.87/m	= $7,044
Mainline 206 mm	900 m * $9.71	= $8,739
Total		= $56,373

Pump Requirement for Design 2

The flow rate in both the upper and lower submains during irrigation is 45.2 LPS. The energy required at the inlet to the submains during irrigation is 11.8 m. The worst-case emitter is the emitter at the end of the last lateral on submain 11 (Fig. 18.2). Use the same pressure losses in valves, fittings, and lifting from canal as in design 1.

10.2 m	Minimum operating pressure for emitters
+1.2 m	Pressure loss in lateral
+0.8 m	Pressure loss in submain
+2 m	Pressure loss in solenoid valve
+4.7 m	Pressure loss in 700 m N-S mainline
−1.2 m	Elevation gain in 700 m N-S mainline
+8 m	Pressure loss in sand filter
+5 m	Friction loss in valves and fittings (check valve, manifold, pump suction)
+1 m	Elevation of ground surface above canal water surface
32 m	Total pump pressure requirement

Thus, the pump flow requirement is 46.9 LPS and pressure requirement is 32 m during irrigation.

If pumping efficiency is 80 %, then the power requirement during irrigation is (Eq. 2.17)

$$P_{\text{irrigation}} = Q\rho gh/Eff = 0.0469{*}1,000\ \text{kg/m}^3{*}\ 9.8\ \text{m/sec}^2{*}32\ \text{m}{*}0.001\ \text{kW/W}/0.8 = 18.4\ \text{kW}$$

If only half of the zone is flushed at a time, then the pressure required at the inlet to laterals is 18.5 m. If isolation valves are used and only half of the zone is flushed at a time, then required flow rate is 59 LPS. The total pressure requirement during flushing is

20 m	Minimum operating pressure at submain inlet (Fig. 18.7)
+2 m	Pressure loss in solenoid valve
+5.9 m	Pressure loss in 700 m N-S mainline
-1.2 m	Elevation gain in 700 m N-S mainline
+8 m	Ressure loss in sand filter
+5 m	Friction loss in valves and fittings (check valve, manifold, pump suction)
+1 m	Elevation of ground surface above canal water surface
45 m	Total pump pressure requirement

$$P_{\text{flushing}} = Q\rho gh/Eff = 0.059{*}1,000\ \text{kg/m}^3{*}\ 9.8\ \text{m/sec}^2{*}45\ \text{m}{*}0.001\ \text{kW/W}/0.8 = 32.5\ \text{kW}$$

The power requirement for flushing is much greater than the power required for irrigation. Thus, a booster pump is required during flushing. In addition, extra flow rate is required which may be difficult to add, depending on the water source. Possibly, a reservoir would need to be constructed to provide extra water during flushing. The requirement of extra flushing capacity adds costs that are not in design 1.

The Design 1 water application is has higher uniformity so yield should be greater and environmental contamination should be less. However, only capital cost, water cost, and energy cost are included in this economic comparison.

Economic Comparison of Designs 1 and 2

Pipe Design 2 has a higher pipe cost: \$56,373 – \$55,313 = +\$1,060.

Pump Estimate the extra booster pump required for flushing in design 2 costs \$4,000.

Water Cost The percent wasted water in design 1 is 1.8 % while the percent wasted water in design 2 is 5.5 %. The difference in efficiency is used to calculate the extra annual volume of water required for design 2 is

$$\Delta V == (1\ m)(32\ ha)(0.055 - 0.018) = 1.18\ ha - m$$

The annual extra cost of water for design 2 is (1.18 ha-m) (\$3.27/ha-cm) (100 cm/m) = \$387/yr. The present value (20 years, 0.08) of the extra cost is \$3,800 for the entire farm.

Energy Cost Ignore the cost of flushing because it is an insignificant fraction of total energy use. Time of operation is the volume required divided by the minimum (worst-case) application rate.

$$V/Q = (1\ \text{m}{*}32\ \text{ha}{*}10,000\ \text{m}^2/\text{ha})/(0.0444\ \text{m}^3/\text{sec})/(3,600\,\text{sec/hr}) = 2,020\ \text{hours}$$

The pump power is multiplied by the time of operation to find annual energy use.

Design 1: P*t = 17.7 kW * 2,020 hours = 35,791 kW-hrs → \$3,579/yr @ \$0.10/kW-hr
Design 2: P*t = 18.4 kW * 2,020 hours = 37,168 kW-hrs → \$3,717/yr @ \$0.10/kW-hr

The difference in power cost between design 1 and design 2 is \$158/yr. The present value (20 years, 0.08) of the extra cost of power in design 2 is \$1,600.

Difference in present value between designs 1 and 2 (+ means design 2 is more expensive):

$$+\$5,000(\text{capital}) + \$3,800(\text{water}) + \$1,600(\text{energy}) = +\ \mathbf{\$10,400}$$

Decision The present value of design 1 is less than design 2. It is also possible that an auxiliary reservoir will be needed to provide extra flow capacity for flushing for design 2. In addition, design 1 adds the flexibility of irrigating half zones. **Select design 1.**

An Economic Rationale for Selecting Between Drip Lateral Types

There are numerous types of inline drip systems. This section shows how to use economics and expected spatial variability and reliability of different drip lateral types to select between inline drip products. Factors that influence variation and cost include tubing diameter and length, emitter type, flow rate, spacing, and expected life. The Chapter 18 *Economic analysis* workbook incorporates these factors in an economic analysis. The variation is combined with crop water production functions (CWPF) to assess the effect of spatial variation on yield and profit. Present value of yields over the life of the project is compared to the present value of the cost of tubing and emitter replacement for various alternatives. In the *Drip lateral CV analysis* worksheet, columns A:I make hydraulic calculations for the lateral. Columns K:V simulate randomized application depths for a range of minimum application depths specified in Row 17 (Fig. 18.15).

If there are multiple emitters per plant, then the manufacturers coefficient of variation for single emitters is divided by the square root of the number of emitters per plant in order to find the variability of water application to plants. This standard statistical procedure is used to calculate the standard deviation of mean values based on the standard deviation of individual members of a sample group.

$$\sigma_{means} = \frac{\sigma}{N^{0.5}} \rightarrow CV_{means} = \frac{CV}{N^{0.5}} \rightarrow CV_{plant} = \frac{CV_{emitter}}{N^{0.5}} \tag{18.1}$$

where

σ_{means} = standard deviation of application rate per plant, mm

N = number of emitters per plant

CV_{means},CV_{plant} = coefficient of variation of application rate per plant, mm.

The *Drip lateral CV analysis* worksheet develops equations for yield and leaching vs. CV for each of the lateral options in the worksheet (Table 18.2). Columns Y:AJ calculate yield as a function of the application depths. Columns AL:AW calculate leaching as a function of

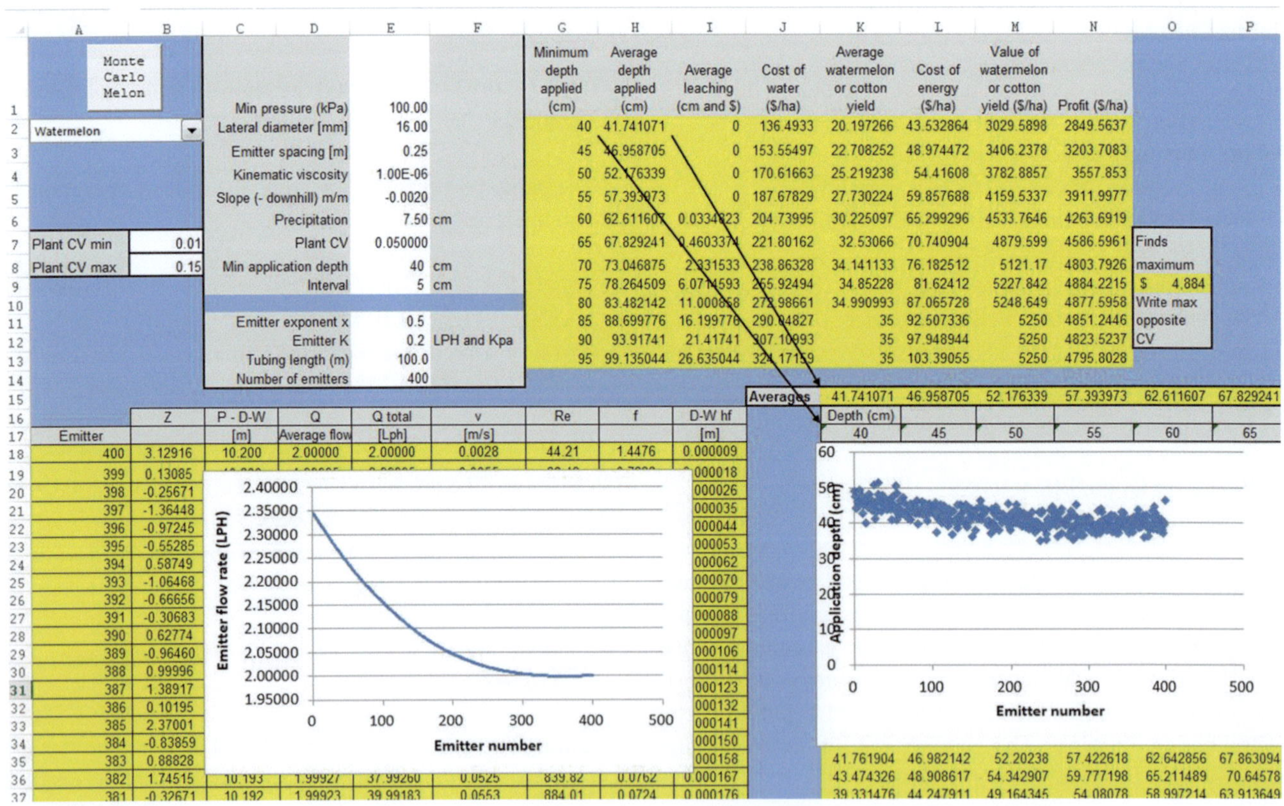

Fig. 18.15 Emitter flow rates over and variable application rates

Table 18.2 Five drip tape alternatives for subsurface drip irrigation of watermelons (*Drip lateral CV analysis* worksheet)

	AK	AL	AM	AN	AO	AP	AQ	AR	AS	AT
2										
3		Type	Cost ($/m)	Repl. (yr)	diam (mm)	k	x	spacing	Min Press (kPa)	type
4	1	8 mil tape	0.09	5	16	0.1	0.5	0.25	50	Turbulent
5	2	8 mil tape,	0.07	5	16	0.2	0.5	0.5	50	Turbulent
6	3	15 mil tape	0.12	10	12	0.2	0.5	0.5	50	Turbulent
7	4	15 mil tape	0.2	10	16	2	0	0.5	100	Pressure comp
8	5	15 mil tape	0.24	10	16	1	0	0.25	100	Pressure comp

application and plant water uptake. A VBA program is triggered with the Monte Carlo simulation button in order to simulate yield and leaching over the range of CV values specified in B7 and B8. The Monte Carlo simulation selects the best irrigation depth for each manufacturer's CV and thus finds the maximum benefit for each alternative and CV value (Fig. 18.19).

Because CV is specified for each year, the benefit vs. CV curves (Fig. 18.19) can be used in the financial analysis (*Watermelon financial calcs* and *Cotton financial calcs* worksheets) in order to calculate the value of each system for the life of the project.

The replacement interval for in-line drip irrigation tubing primarily depends on tubing wall thickness (the common units are mils). In general, 6 and 8 mil tubing are considered thin wall and are only designed to last one or two seasons while 15 mil tubing is considered thick wall and has been observed to last for 15–20 years on subsurface drip irrigated farms. New, high-quality, in-line emitters typically have a coefficient of variation (CV) in the range of 2–3 %. The CV increases over time. This type of information is compiled in Table 18.2 and is used in the Monte Carlo simulation.

Example 18.2 A subsurface drip irrigation system will be installed for irrigation of watermelon. The five drip tape alternatives are shown in Table 18.2. Select the best alternative.

There is a 10 yr project life and 8 % ROR. Initial cost of installation, other than tubing price and tape injection and connection, is the same for all types of tubing. Drip tape replacement cost is $150/ha (tape injection and connection). The cost of energy is $0.10/kW-hr. Pumping efficiency is 80 %. The cost of water is $3.27/ha-cm. The environmental penalty for leaching is $1/ha-cm. Effective seasonal precipitation is 7.5 cm. Fixed costs other than water, energy, tubing, and leaching are $3,700/ha/yr, which includes annualized capital costs (irrigation mainline, submain, and pump station installation) and annual costs such as seed, fertilizer, labor, and fuel, cultivation, and harvesting. Maximum watermelon yield is 35 metric tons/ha @ $150/metric ton.

Laterals are 100 m long and laid on a downward slope (−0.002 m/m). All emitters on 0.5 m spacing have a nominal 2 L/h flow rate. The minimum operating pressure for turbulent emitters is 50 kPa (k = 0.2828 and x = 0.5). The minimum operating pressure for pressure compensating emitters is 100 kPa (k = 2.0 and x = 0). All emitters on 0.25 m spacing have a nominal 1.0 L/h flow rate. All emitters have initial 2.5 % manufacturer's CV and 2.5 % increase in CV per year.

Watermelons are commonly planted in the center of 2 m beds with 0.5 m between plants. In order to increase wetted root volume (larger water and nutrient reservoir) for the watermelons, 3 drip laterals are installed on each bed (Fig. 18.16). All tubing is buried 20 cm below the soil surface.

There has been little research directed toward developing a watermelon CWPF. Thus, the FAO K_Y value will be used to develop a linearized CWPF for yield vs. dry stress. With respect to wet stress, there is often a decrease in watermelon yield due to over watering; however, the decrease is caused by melon splitting or other physiological responses that would be unlikely to occur with subsurface drip irrigation laterals buried at 20 cm depth. Therefore, there is no need to penalize the yield due to overwatering. The seasonal ET_c is 0.8 m, and if greater than or equal to 0.8 m is provided, then yield is the maximum potential yield, 35 metric tons/ha. Watermelon are sensitive to water stress so $K_y = 1.1$ (FAO). An equation for the watermelon yield vs. applied water depth with 7.5 cm precipitation is

$$Y_L = \left(1 - K_y\left(1 - \frac{AW + P}{AW_{req}}\right)\right)Y_{max}$$
$$= \left(1 - 1.1\left(1 - \frac{AW + 0.075}{0.8}\right)\right)35 \ metric\ tons$$

where

AW = applied depth, m.

Thus, a step function is used to calculate watermelon yield (Fig. 18.17).

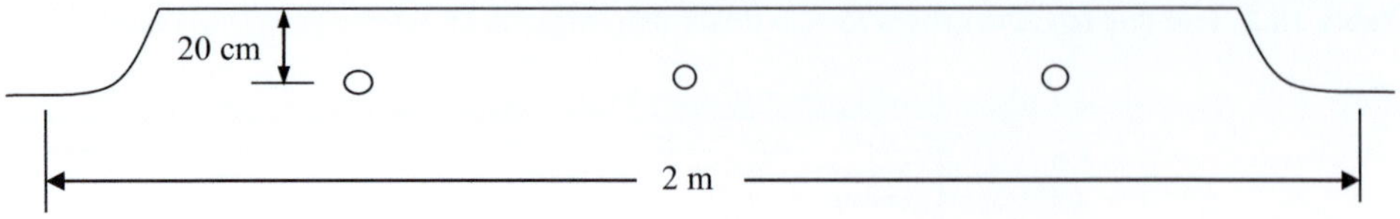

Fig. 18.16 Watermelon bed and drip lateral locations

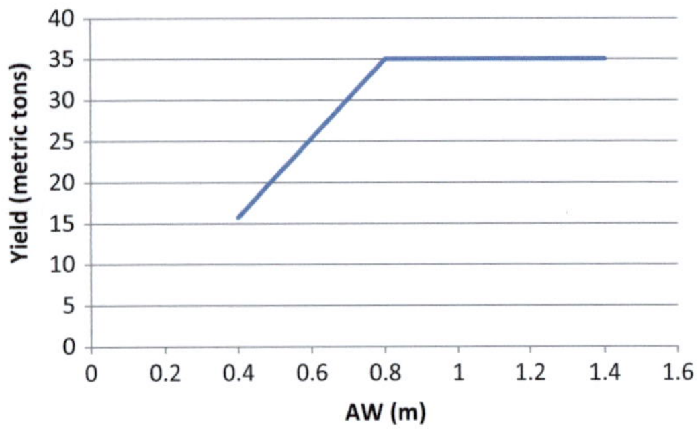

Fig. 18.17 Watermelon yield versus AW with seasonal precipitation = 0.075 m

If AW < 0.725 $Y_L = (0.00312 + 1.375 * AW) * 35$
Else $Y_L = 35$

Excess water (AW + P > 0.8 m) is leached.

Pipe installation costs include the price of the tubing (Table 18.2) and labor. The length of tubing/ha is the area divided by the spacing.

$$10{,}000\ \text{m}^2/(2\ \text{m}/3\ \text{tubes}) = 15{,}000\ \text{m tubing/ha.}$$

The cost of option 1 installation (just tubing installation, not mainline etc.) is

$$15{,}000\ \text{m}*\$0.09/\text{m} + \$150(\text{labor per ha}) = \$1{,}500/\text{ha}$$

Solution

1. Develop an equation for annual yield benefit vs. CV for each of the five tubing alternatives.

Options 2–4 have 3 emitters per plant (emitter spacing = 0.5 m, watermelon spacing = 0.5 m, 3 laterals per plant). As calculated below, the CV_{plant} during the first year is 0.0145 (1.5 %) and increases by the same amount every year, based on the problem definition. With 0.25 m spacing and 6 emitters per plant, options 1 and 5 have a CV_{plant} during the first year equal to 0.01 (1 %) and increase by the same amount each year.

$$CV_{plant} = \frac{CV_{emitter}}{N^{0.5}} = \frac{0.025}{3^{0.5}} = 0.0145$$

$$CV_{plant} = \frac{CV}{N^{0.5}} = \frac{0.025}{6^{0.5}} = 0.010$$

A VBA macro is triggered by the Monte Carlo button that automatically increases CV from the minimum to the maximum (cell B7:B8) for each tubing alternative and finds an optimal depth and profit based on the information in Fig. 18.18. The profit (column N) is calculated as yield – water cost – energy cost – leaching cost.

An equation for maximum profit vs. CV is developed for each tubing option (Fig. 18.19). Only two equations are needed in this example because there are only two different emitter configurations (3 emitters per plant and 6 emitters per plant).

The next step is to incorporate the equations in Fig. 18.19 in an economic analysis. Because. The fixed costs of $3,700 per year are subtracted from the equations derived in Fig. 18.19. The CV_{plant} for each of the tubing alternatives are shown in Table 18.3.

The next steps in the analysis are run in the *Watermelon financial calcs* worksheet. Annual profit vs. year is calculated based on the equations (Fig. 18.19) and expected annual CV values for each alternative (Fig. 18.20). The user must enter the equations into the appropriate columns in rows 2–11.

The annual fixed costs ($3,700/ha) are subtracted from the profits and are then converted to present values in the second and third tables and then summed in line 40 in the *Watermelon financial calcs* worksheet (Fig. 18.21).

Decision Select option 2 because it has the highest present value, $6,098 (the least expensive alternative). The cost of replacing option 2 at the end of the 5th year is more than paid for by the increased uniformity.

Example 18.3 Repeat Example 18.2, but for cotton. There is one drip lateral per row of plants with 1 m spacing between rows. Use the Grimes and El-Zik CWPF (Chap. 2), and the leaching fraction equation based on the Grimes and El-Zik CWPF developed in Chap. 6. Cotton price is $0.92/kg. Fixed annual costs are $600/year. Soil is sandy loam. Emitter spacing is 0.5 m.

Yield is calculated with the Grimes and El-Zik (Fig. 2.1) cotton yield function $Y_L = (-3954 + 1067\ (AW + P)^{0.5} - 54.14\ (AW + P))$. Leached depth = $0.2736e^{0.0469(AW + P)}$

With sandy loam and 2 L/h emitter flow rate, the wetted diameter (Table 15.2) is 0.7 m. Thus, the overlap is 0.35 m/ 0.5 m * 100 % = 70 %. The 1.0 L/h flow rate emitters have a 0.4 m wetted diameter in the sandy loam soil (estimated

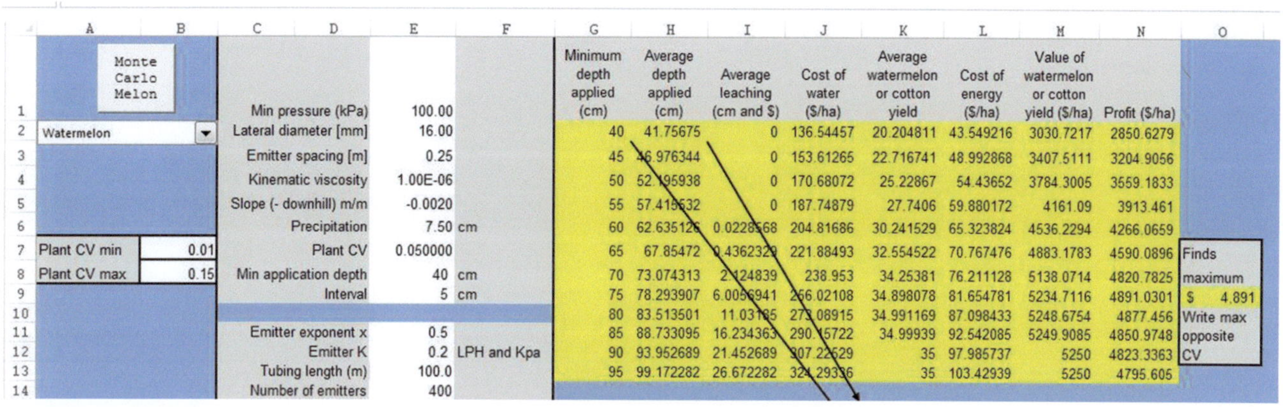

Fig. 18.18 Economic optimization for each tubing alternative for a given CV and irrigation flow characteristics

Fig. 18.19 Profit versus CV for each of the tubing alternatives

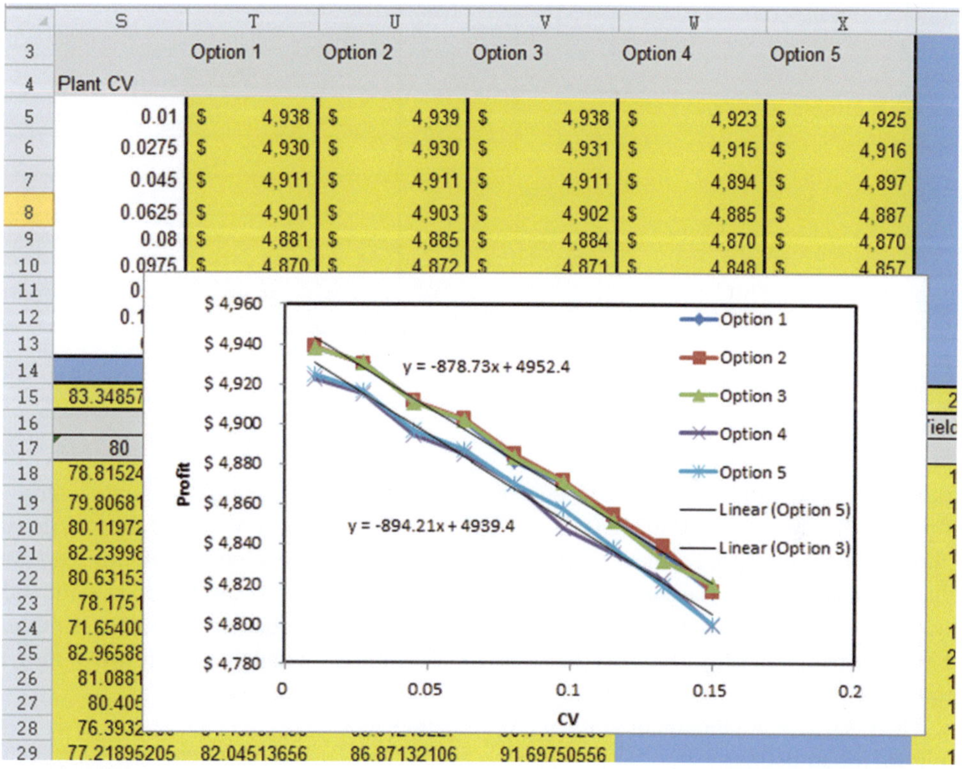

Table 18.3 Annual CV_{plant} for tubing options

yr	Option 1, .25 m	Option 2, .5 m	Option 3, .5 m	Option 4, .5 m	Option 5, .25 m
1	0.01	0.0145	0.0145	0.0145	0.01
2	0.02	0.029	0.029	0.029	0.02
3	0.03	0.0435	0.0435	0.0435	0.03
4	0.04	0.058	0.058	0.058	0.04
5	0.05	0.0725	0.0725	0.0725	0.05
6	0.01	0.0145	0.087	0.087	0.06
7	0.02	0.029	0.1015	0.1015	0.07
8	0.03	0.0435	0.116	0.116	0.08
9	0.04	0.058	0.1305	0.1305	0.09
10	0.05	0.0725	0.145	0.145	0.1

yr	Option 1		Option 2		Option 3		Option 4		Option 5	
1	0.01	4934.84	0.0145	4931.17	0.0145	4895.28	0.0145	4924.73	0.01	4928.54
2	0.02	4926.68	0.029	4919.34	0.029	4883.57	0.029	4912.47	0.02	4920.08
3	0.03	4918.52	0.0435	4907.5	0.0435	4871.85	0.0435	4900.2	0.03	4911.62
4	0.04	4910.36	0.058	4895.67	0.058	4860.14	0.058	4887.93	0.04	4903.16
5	0.05	4902.2	0.0725	4883.84	0.0725	4848.42	0.0725	4875.67	0.05	4894.7
6	0.01	4934.84	0.0145	4931.17	0.087	4836.7	0.087	4863.4	0.06	4886.24
7	0.02	4926.68	0.029	4919.34	0.1015	4824.99	0.1015	4851.13	0.07	4877.78
8	0.03	4918.52	0.0435	4907.5	0.116	4813.27	0.116	4838.86	0.08	4869.32
9	0.04	4910.36	0.058	4895.67	0.1305	4801.56	0.1305	4826.6	0.09	4860.86
10	0.05	4902.2	0.0725	4883.84	0.145	4789.84	0.145	4814.33	0.1	4852.4

	Annual benefit ($/ha)					Tubing installation costs ($/ha)				
	Opt 1	Opt 2	Opt 3	Opt 4	Opt 5	Opt 1	Opt 2	Opt 3	Opt 4	Opt 5
0	0	0	0	0	0	1500	1200	1950	3150	3750
1										0
2										0
3										0
4										0
5										0
6										0
7										0
8										0
9										0
10										0

Fig. 18.20 Annual profits (yield – water – energy – leaching cost) for each alternative in *Watermelon financial costs* worksheet

based on linear difference between doubled flow rates in Table 11.4). The overlap is 0.2 m/0.25 m * 100 % = 80 %. Thus, both emitter spacings have approximately ¾ overlap. Plants that are close to one emitter would probably have root systems with access to water from the far emitter, so assume 2 emitters per plant (Fig. 18.22) for the emission uniformity equation.

The CV_{plant} as a function of emitter CV is calculated as follows for all options.

$$CV_{plant} = \frac{CV_{emitter}}{N^{0.5}} = \frac{0.025}{2^{0.5}} = 0.017$$

Likewise, each plant receives water from 4 emitters with the 0.25 spacing. Thus, the CV_{plant} would be calculated as follows:

$$CV_{plant} = \frac{CV_{emitter}}{N^{0.5}} = \frac{0.025}{4^{0.5}} = 0.0125$$

Annual CV values for the different tubing options are shown in Table 18.4.

Using the *Drip lateral CV analysis* worksheet, profit vs. CV curves are not linear for cotton (Fig. 18.23). Thus, polynomial trendlines are used to fit the data.

The next step is to input the annual CVs and profit vs. CV equations into the *Cotton financial calcs* worksheet. Fixed annual costs are $600; thus, for option 5, the annual benefits are calculated as

$$\text{Annual benefit(option 5)} = -1074\ CV^2 + 11\ CV + 833600$$

The length of tubing/ha is the area divided by the spacing.

$$10{,}000\ \text{m}^2/(1\ \text{m/tubes}) = 10{,}000\ \text{m tubing/ha}.$$

For example, the cost of option 1 installation (just tubing installation, not mainline etc.) is

$$10{,}000\ \text{m*\$0.09/m} + \$150(\text{labor per ha}) = \$1{,}050/\text{ha}$$

The calculated benefit per year in the *Cotton financial calcs* worksheet is shown in Fig. 18.24.

	A	B	C	D	E	F	G	H	I	J	K
2	1	0.01	4934.84	0.0145	4931.17	0.0145	4895.28	0.0145	4924.73	0.01	4928.54
3	2	0.02	4926.68	0.029	4919.34	0.029	4883.57	0.029	4912.47	0.02	4920.08
4	3	0.03	4918.52	0.0435	4907.5	0.0435	4871.85	0.0435	4900.2	0.03	4911.62
5	4	0.04	4910.36	0.058	4895.67	0.058	4860.14	0.058	4887.93	0.04	4903.16
6	5	0.05	4902.2	0.0725	4883.84	0.0725	4848.42	0.0725	4875.67	0.05	4894.7
7	6	0.01	4934.84	0.0145	4931.17	0.087	4836.7	0.087	4863.4	0.06	4886.24
8	7	0.02	4926.68	0.029	4919.34	0.1015	4824.99	0.1015	4851.13	0.07	4877.78
9	8	0.03	4918.52	0.0435	4907.5	0.116	4813.27	0.116	4838.86	0.08	4869.32
10	9	0.04	4910.36	0.058	4895.67	0.1305	4801.56	0.1305	4826.6	0.09	4860.86
11	10	0.05	4902.2	0.0725	4883.84	0.145	4789.84	0.145	4814.33	0.1	4852.4
12											
13		Annual benefit ($/ha)					Tubing installation costs ($/ha)				
14		Opt 1	Opt 2	Opt 3	Opt 4	Opt 5	Opt 1	Opt 2	Opt 3	Opt 4	Opt 5
15	0	0	0	0	0	0	1500	1200	1950	3150	3750
16	1	4935	4931	4895	4925	4929	0	0	0	0	0
17	2	4927	4919	4884	4912	4920	0	0	0	0	0
18	3	4919	4908	4872	4900	4912	0	0	0	0	0
19	4	4910	4896	4860	4888	4903	0	0	0	0	0
20	5	4902	4884	4848	4876	4895	1500	1200	0	0	0
21	6	4935	4931	4837	4863	4886	0	0	0	0	0
22	7	4927	4919	4825	4851	4878	0	0	0	0	0
23	8	4919	4908	4813	4839	4869	0	0	0	0	0
24	9	4910	4896	4802	4827	4861	0	0	0	0	0
25	10	4902	4884	4790	4814	4852	0	0	0	0	0
26											
27		Yearly profit ($/ha)					Present value of yearly profit ($/ha)				
28		Opt 1	Opt 2	Opt 3	Opt 4	Opt 5	Opt 1	Opt 2	Opt 3	Opt 4	Opt 5
29	0	-1500	-1200	-1950	-3150	-3750	-1500	-1200	-1950	-3150	-3750
30	1	1235	1231	1195	1225	1229	1143	1140	1107	1134	1138
31	2	1227	1219	1184	1212	1220	1052	1045	1015	1039	1046
32	3	1219	1208	1172	1200	1212	967	959	930	953	962
33	4	1210	1196	1160	1188	1203	890	879	853	873	884
34	5	-298	-16	1148	1176	1195	-203	-11	782	800	813
35	6	1235	1231	1137	1163	1186	778	776	716	733	748
36	7	1227	1219	1125	1151	1178	716	711	656	672	687
37	8	1219	1208	1113	1139	1169	658	652	601	615	632
38	9	1210	1196	1102	1127	1161	605	598	551	564	581
39	10	1202	1184	1090	1114	1152	557	548	505	516	534
40						Total	5664	6098	5766	4749	4274

Fig. 18.21 Annual benefits and costs and present values in *Watermelon financial calcs* worksheet

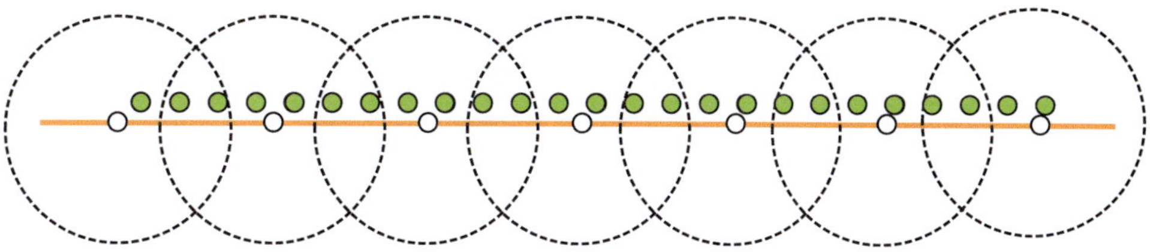

Fig. 18.22 Emitter and plant spacing for 0.5 m spaced emitters in cotton row with 2 L/h flow rate in loam soil

Table 18.4 Annual CV_{plant} for tubing options in cotton

yr	Option 1, .25 m	Option 2, .5 m	Option 3, .5 m	Option 4, .5 m	Option 5, .25 m
1	0.0125	0.0177	0.0177	0.0177	0.0125
2	0.025	0.0354	0.0354	0.0354	0.025
3	0.0375	0.0531	0.0531	0.0531	0.0375
4	0.05	0.0708	0.0708	0.0708	0.05
5	0.0625	0.0885	0.0885	0.0885	0.0625
6	0.0125	0.0177	0.1062	0.1062	0.075
7	0.025	0.0354	0.1239	0.1239	0.0875
8	0.0375	0.0531	0.1416	0.1416	0.1
9	0.05	0.0708	0.1593	0.1593	0.1125
10	0.0625	0.0885	0.177	0.177	0.125

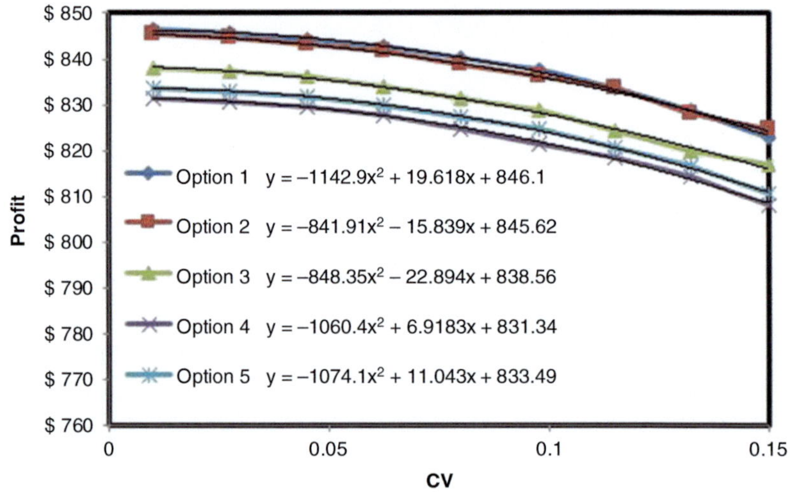

Fig. 18.23 Profit versus CV for cotton analysis in *Drip lateral CV analysis* worksheet

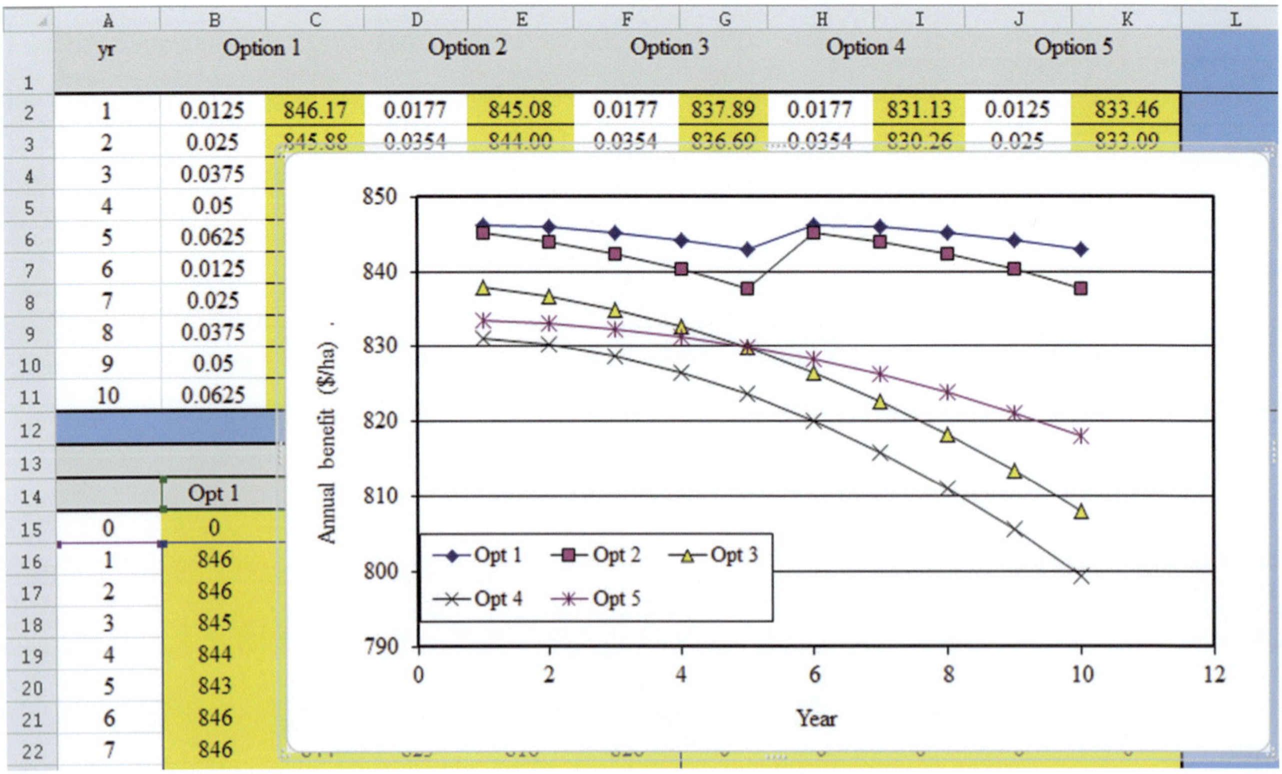

Fig. 18.24 Annual profits (yield – water – energy – leaching cost) in *Cotton financial calcs* worksheet

	A	B	C	D	E	F	G	H	I	J	K
1	yr	Option 1		Option 2		Option 3		Option 4		Option 5	
2	1	0.0125	846.17	0.0177	845.08	0.0177	837.89	0.0177	831.13	0.0125	833.46
3	2	0.025	845.88	0.0354	844.00	0.0354	836.69	0.0354	830.26	0.025	833.09
4	3	0.0375	845.23	0.0531	842.41	0.0531	834.95	0.0531	828.72	0.0375	832.39
5	4	0.05	844.22	0.0708	840.28	0.0708	832.69	0.0708	826.51	0.05	831.36
6	5	0.0625	842.86	0.0885	837.62	0.0885	829.89	0.0885	823.65	0.0625	829.98
7	6	0.0125	846.17	0.0177	845.08	0.1062	826.56	0.1062	820.12	0.075	828.28
8	7	0.025	845.88	0.0354	844.00	0.1239	822.70	0.1239	815.92	0.0875	826.23
9	8	0.0375	845.23	0.0531	842.41	0.1416	818.31	0.1416	811.06	0.1	823.85
10	9	0.05	844.22	0.0708	840.28	0.1593	813.38	0.1593	805.53	0.1125	821.14
11	10	0.0625	842.86	0.0885	837.62	0.177	807.93	0.177	799.34	0.125	818.09
12											
13		Annual benefit ($/ha)					Tubing installation costs ($/ha)				
14		Opt 1	Opt 2	Opt 3	Opt 4	Opt 5	Opt 1	Opt 2	Opt 3	Opt 4	Opt 5
15	0	0	0	0	0	0	1050	850	1350	2150	2550
16	1	846	845	838	831	833	0	0	0	0	0
17	2	846	844	837	830	833	0	0	0	0	0
18	3	845	842	835	829	832	0	0	0	0	0
19	4	844	840	833	827	831	0	0	0	0	0
20	5	843	838	830	824	830	1050	850	0	0	0
21	6	846	845	827	820	828	0	0	0	0	0
22	7	846	844	823	816	826	0	0	0	0	0
23	8	845	842	818	811	824	0	0	0	0	0
24	9	844	840	813	806	821	0	0	0	0	0
25	10	843	838	808	799	818	0	0	0	0	0
26											
27		Yearly profit ($/ha)					Present value of yearly profit ($/ha)				
28		Opt 1	Opt 2	Opt 3	Opt 4	Opt 5	Opt 1	Opt 2	Opt 3	Opt 4	Opt 5
29	0	-1050	-850	-1350	-2150	-2550	-1050	-850	-1350	-2150	-2550
30	1	246	245	238	231	233	228	227	220	214	216
31	2	246	244	237	230	233	211	209	203	197	200
32	3	245	242	235	229	232	195	192	187	182	184
33	4	244	240	233	227	231	180	177	171	166	170
34	5	-807	-612	230	224	230	-549	-417	156	152	157
35	6	246	245	227	220	228	155	154	143	139	144
36	7	246	244	223	216	226	143	142	130	126	132
37	8	245	242	218	211	224	132	131	118	114	121
38	9	244	240	213	206	221	122	120	107	103	111
39	10	243	238	208	199	218	112	110	96	92	101
40						Total	-121	196	181	-664	-1014

Fig. 18.25 Annual benefits and costs and present values in *Cotton financial calcs* worksheet

The annual fixed costs ($600/ha) are subtracted from the profits and are then converted to present values in the second and third tables and then summed in line 40 in the *Cotton financial calcs* worksheet (Fig. 18.25).

Option 1 does not yield a positive present value at the required rate of return, and options 2 and 3 are barely positive. Options 4 and 5 are highly unprofitable. If crop failure were to occur in one or more years, and this analysis was based only on normal yields, then options 2 and 3 would not be profitable in this analysis. This example shows why drip irrigation systems are usually installed for high value crops. The yield may not pay for the added cost of the drip system (typically $2,500/ha capital cost).

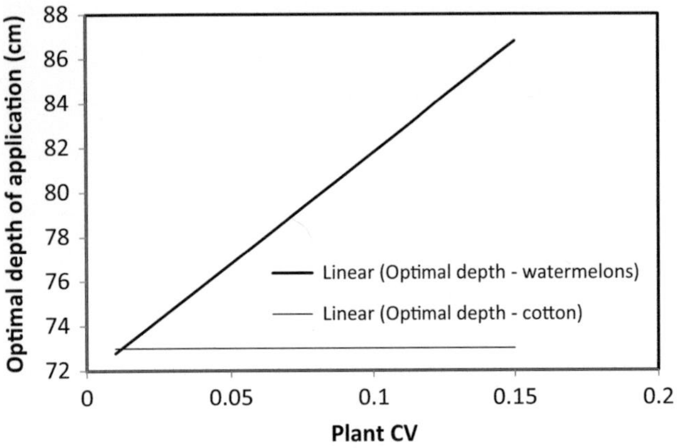

Fig. 18.26 Optimal AW_{min} versus CV_{plant} for cotton and watermelon

Examples 18.2 and 18.3 assumed only one crop per year, and the same crop every year with constant fixed costs, constant yield and crop price. However, many drip systems are installed in areas that support 2 crops per year.

A comparison of the cotton and watermelon examples demonstrates the differences in watering practices between high value and low value crops. Watering can be excessive with high value crops (Fig. 18.26) because the goal is to maintain maximum yield, but under watering of some parts of the field is economically justified with low value crops. When CV increases with cotton, there is no change in the optimal watering depth. The reason that cotton optimal watering depth does not increase with CV is that there is a relatively important financial cost associated with overwatering some parts of the field. The value of the additional water is not justified by an increased income from the crop. However, water is a relatively small cost in comparison to the profit from watermelon. Thus, it is never beneficial to underwater some parts of the field in order to save money on water.

Questions

1. A 90 m long submain supplies 12 mm ID laterals that are 90 m long. This is a single feed system. The laterals are spaced 1 m apart. Emitters are spaced every 0.3 m, $k = 0.25$, and $x = 0.41$. Slope of laterals is 1 % downhill. Manufacturer's coefficient of variation is 5 % and number of emitters per plant is 2. Verify that the lateral has at least 90 % emission uniformity. If not, then increase the pipe diameter. The submain is on level ground. Find the lateral flow rate vs. pressure equation (Fig. 18.3) and design the submain (select diameters). Because of flushing, minimum allowable size of the submain is 100 mm. The minimum acceptable pressure is 80 kPa. Find the required inlet pressure for the submain. Also, determine whether the emission uniformity is greater than 90 % for the entire zone.
2. Repeat question 1; however, there is no slope on the lateral. Evaluate emission uniformity on the individual lateral and in the zone. Is the emission uniformity above or below 90 %. Compare the exponent and the coefficient in the lateral flow pressure equation to that of question 1. Explain the differences and similarities.
3. Repeat question 2, but change to dual feed laterals with submains that are 180 m apart. For the purpose of hydraulic calculations, laterals are 90 m long to the midpoint. Emission uniformity should be greater than 90 % for any zone; thus, increase the pipe diameter to 14 mm, and determine whether this diameter results in an emission uniformity that is greater than 90 % for the entire zone. If the emission uniformity is more than 1 % greater than 90 %, then there is no need to make an additional calculation for the zone, because the zone will drop the uniformity by less than 1 %. There is no need to show all the graphs and equations. Just the results are sufficient.
4. Based on the parameters in questions 2 and 3, calculate the inlet pressure needed for the submains. Find the equation for submain inlet flow vs. pressure.
5. Based on the parameters in questions 2–4, evaluate flushing in the dual feed lateral with submains spaced on 180 m intervals. Use the *Flush dual feed lateral* worksheet. Find the inlet pressure required and the equation for lateral flow vs inlet pressure.
6. Based on the parameters in questions 2–5, evaluate flushing in the submain. There are 100 laterals spaced 1 m apart. Use the *Submain flushing* worksheet. Check that flow velocity is not excessive. Pipe flow velocity restrictions (normally < 1.5 m/sec) can be relaxed for flushing mode, as long as the owner slowly closes valves during the flushing process, keep the velocity below 2 m/sec in this question. Find the required inlet flow velocity and pressure. Find the equation for inlet pressure vs. flow rate.
7. Using the information compiled in questions 1–6, design a pump, filter, and mainline system for a level field that has dimensions 400 m $\times$ 400 m. Allow 20 m for a central road so the length of laterals (distance between submains) is 180 m. Use the 100 m $\times$ 180 m zones that you have already designed. The road travels in the EW direction, and the pump is in the NW corner. Using the structure in Fig. 18.1, specify the required pipe sizes in mains 33–38 and 43–45. The irrigation schedule allows for 8 zones so each zone is run by itself. For example, pipes 1 and 5 are activated at the same time, etc... Mains 33–36 are used for flushing because flushing originates in submains 1, 2, 3, 4, 9, 10, 11, and 12. Use

the *Chapter 18 Mainline* workbook to find flow velocities and head losses. Specify the required pump flow rate and pressure. You do not need to pick a particular pump (unless you want to). 0 year project life and 8 % ROR. The cost of energy is $0.10/kW-hr, and the cost of water is $3.27/ha-cm. Annual ETc is 1 m/y. Irrigation efficiency is 90 % and pumping efficiency is 80 %. Sand filter losses are 7 m and pump station losses are 3 m. Solenoid valve losses are 2 m. Do not worry about the flushing flow rate or pressure or the booster pump that would be required for flushing.

8. Open the *Chapter 18 Economic analysis* workbook and *Cotton Drip lateral CV analysis* worksheet. Reduce the plant CV in cell E7 to 0.05. Select cotton as the crop in cell A2. In the range E1:E14, change the tubing diameter to 12, the plant CV to 0.05, the emitter coefficient to 0.2, and the emitter exponent to 0.5. Note, the Monte Carlo simulation program changes these values during the simulation. Plot the emitter flow rates in column D vs. emitter number in column A. Plot the 40 cm application depth in column K vs. emitter number in column A. Explain why some of the application depths are less than 40. You can highlight one of the cells in Column K and look at the equation in order to find the answer. Plot the yield vs. emitter number curve for the 50 cm depth in column AA and the 75 cm depth in column AE. Explain the shapes of the yield curves.
9. Open the *Chapter 18 Economic analysis* workbook and *Cotton Drip lateral CV analysis* worksheet. Move the graph away from table T5:X13. Clear cells T5:X13. Click the Monte Carlo Cotton button in cell P1. Watch what happens in cells H2:W13. Then click the Monte Carlo Cotton button in cell A1 and watch what happens in column E. and explain how the algorithm works. How many simulations are run at each tubing option and CV value (count the number of blinks in the formula bar for each condition)?
10. Open the *Chapter 18 Economic analysis* workbook and *Cotton Drip lateral CV analysis* worksheet. Select cotton as the crop in cell A2. As shown below, change the replacement period for the 8 mil tape to 2 years (column AN), and run the Monte Carlo simulation by clicking the Monte Carlo button in cell B1. Note: the Monte Carlo simulation requires several minutes running time. Notice that the VBA program changes the parameters in the range E1:E14. Make Trendlines for each of the curves in the profit vs. CV graph in the range T1:X13. Compare with the equations in Fig. 18.23. If they are different, then explain why. Explain why options 1–2 have higher profit vs. CV than options 3–5. Explain why option 3 has higher profit than option 4.
11. In the *Cotton financial calcs* worksheet, change the CV values for the every other year replacement scheme in columns B and D for options 1 and 2, as described in question 10. Add installations costs every other in cells G15:H25. How does every other year replacement affect the annual benefit in rows 2:11 (also shown in the graph)? How does every other year replacement affect the overall profit of the system as shown in row 40? What is the only remaining option with positive

19 Chemigation

Chemigation is the application of fertilizers and pesticides to agricultural crops with irrigation water. Chemicals are also injected into microirrigation systems in order to prevent microorganisms and calcium carbonate from clogging drip emitters. Microorganisms are killed by injecting chlorine or other disinfectants. Calcium carbonate precipitation is prevented by injecting acid and lowering pH. The potential hazards associated with chemical storage, transport, and injection into irrigation water include precipitation of solids within the irrigation water and clogging of the drip system, production of poison gas, explosions, contamination of the environment, and burning or loss of eyesight from chemical splashing. This chapter describes acid injection for prevention of calcium carbonate deposition, chlorine injection for control of fungal and bacterial growth, and the application of fertilizer and non-fertilizer materials into microirrigation systems.

Use of chemicals should always be supervised by people with knowledge of the chemistry and potential reactions. The chemical supplier will have information about the chemical compatibilities of their products and should be consulted before new chemicals or formulations are injected. Water quality should always be evaluated before injecting a chemical into irrigation water because chemicals may act differently in different waters. Concentration limits for hazardous chemicals in irrigation water are dependent on label requirements, toxicity to the crop, or on corrosivity in the irrigation system. In some cases, increased salinity due to fertilizer addition can be harmful to the crop.

Injection of two or more chemicals at the same time may cause chemical reactions and the resulting precipitation of the reactants may clog microirrigation laterals or emitters. In order to pre-test for chemical precipitation, mix the chemicals and irrigation water in a transparent inert jar. These "jar tests" should be conducted at the same pH, temperature, and other conditions under which the chemicals will be applied. Safety glasses and protective clothing should always be worn when conducting jar tests.

Preventing Calcium Carbonate Deposition with Acid Injection

Calcium (Ca^{++}) and carbonate (CO_3^{-2}) tend to precipitate (form a solid) out of solution and onto the surface of drip emitters. The deposition of calcium carbonate eventually clogs the emitters and reduces flow rate. Because calcium and carbonate do not normally precipitate out of solution in the range of 6.5 pH, microirrigation managers normally lower pH to 6.5 by injecting acid into irrigation water.

Carbonate is the major species that controls pH (acidity) in water. It is a buffer against pH change because hydrogen ions in acid combine with carbonate ions (CO_3^{2-}) and form bicarbonate (HCO_3^-) and eventually carbonic acid (H_2CO_3). Thus, the amount of acid that is required to lower pH to 6.5 depends on the concentration of carbonate in the water.

The critical pH at which carbonate precipitation takes place decreases as water temperature increases. This effect is a problem in surface microirrigation systems where the water becomes warmer during the daytime as it moves through the black plastic tubing. Acid injection rates may need to be increased to compensate for heating.

Because the reaction is a function of the number of positive (H^+) and negative (CO_3^{2-} and HCO_3^-) charges, it is convenient to use units of normality (meq/L) rather than molarity. Each carbonate ion (CO_3^{-2}) has a charge of negative two, thus, the normality (number of charges) is twice the molarity (number of ions).

The pH is a measure of the acidity or alkalinity of water; it is the log of one over the hydrogen ion molarity. The brackets [ ] represent concentration in moles/L.

$$pH = \log\frac{1}{[H^+]} = -\log[H^+] \tag{19.1}$$

Hydronium ions (hydrogen), hydroxide (OH^-), and water are in equilibrium within "water". The reaction is:

P. Waller, M. Yitayew, *Irrigation and Drainage Engineering*, DOI 10.1007/978-3-319-05699-9_19

$$H_2O \leftrightarrow H^+ + OH^- \quad (19.2)$$

The product of the concentration (molarity) of hydrogen $[H^+]$ and hydroxide $[OH^-]$ ions equals a constant, the dissociation constant (also called ionization constant) for water

$$K_w = [H^+]\ [OH^-] \quad (19.3)$$

The ionization constant for water, $[H^+]\ [OH^-] = 10^{-14}$. Thus, the product of hydrogen and hydroxide concentrations is always equal to 10^{-14}.

In-class Exercise 19.1 If the molarity of hydronium ions is 10^{-6}, then what is the pH?

If the molarity of hydroxide ions is 10^{-6}, then what is the pH?

Water that is neither acidic nor alkaline has the same number of hydronium and hydroxide ions. When pH is lower than seven, water is acidic, and when pH is higher than 7, it is alkaline

In-class Exercise 19.2 The dissociation constant for water $K_w = 10^{-14}$. What is the molarity of hydronium ions that is neither acidic nor alkaline (neutral).

Acids and bases dissolved in water ionize and/or hydrolyze. A strong acid, such as sulfuric acid, will completely dissociate in water and all hydronium ions are available to combine with carbonate in order to lower pH. However, the reaction is similar to the water reaction in Eq. 19.2 for a weak acid, with the reaction proceeding in both directions.

$$HA \leftrightarrow H^+ + A^- \quad (19.4)$$

The ionization constant, K_I, for a weak acid is equal to the product of hydrogen and anion concentration divided by the unionized acid. Ionization constants for various acids are found in the CRC Handbook of Chemistry and Physics.

$$\frac{[H^+][A^-]}{[HA]} = K_I \quad (19.5)$$

where

$[H]$ = hydrogen ion concentration (moles/L)
$[A^-]$ = anion concentration (moles/L)
K_I = dissociation constant for the weak acid

When both materials are in the same solution, i.e. acetic acid dissolved in water, both equilibria (water and acid) need to be satisfied.

Example 19.1 Calculate the pH for a volume of acetic acid ($H\text{-}CH_2COOH$) dissolved in pure water such that its concentration is equal to 0.01 moles/L. K_I for acetic acid is $1.85 * 10^{-5}$.

If X is the concentration of H^+ then A^- (CH_2COOH) will also be X. The reason $A^- = H^+$ is that the concentration of H^+ added by the acid is several orders of magnitude greater than the concentration of H^+ in pure water. From Eq. 19.5 we can calculate:

$$X^2/0.01 = 1.85 \times 10^{-5} \rightarrow [H^+] = X = 4.3 \times 10^{-4}$$

We can now calculate the pH and OH^- concentration in this solution.

$$[OH^-] = 10^{-14}/4.3 \times 10^{-4} = 2.3 \times 10^{-11} \quad pH = 3.36$$

If the pH of water is known, then the concentration of HA and A^- can be calculated based on the concentration of acid added to water. Let C_T be the total concentration of a weak acid added to the solution: C_T is the sum of the nonionized acid, HA, and the ionized acid, A^-.

$$C_T = [HA] + [A^-] \quad (19.6)$$

Combining Eqs. 19.5 and 19.6, concentrations of HA and A^- as a function of final pH are:

$$[HA] = \frac{C_T[H^+]}{K_I + [H^+]} \quad (19.7)$$

$$[A^-] = \frac{C_T K_I}{K_I + [H^+]} \quad (19.8)$$

where

C_T = sum of molarities of concentration of HA and A^- in water.

Using Eqs. 19.7 and 19.8, the log of concentration of HA and A^- vs. pH for 0.01 molar acetic acid (Example 19.1) is plotted in Fig. 19.1.

In open waters, the carbonate equilibrium is more complex than that shown in Fig. 19.1 because there are five species: gaseous carbon dioxide $[(CO_2)_g]$, aqueous carbon dioxide $[(CO_2)_{aq}]$, carbonic acid (H_2CO_3), bicarbonate (HCO_3^-) and carbonate (CO_3^{2-}). In addition, Henry's Law is used to calculate the diffusion of carbon dioxide from air to water; however, water in drip irrigation tubes is not in contact with the atmosphere. Carbonic acid and aqueous CO_2 are indistinguishable in water; thus, only three species are required in the model: H_2CO_3, HCO_3^-, and CO_3^{2-}.

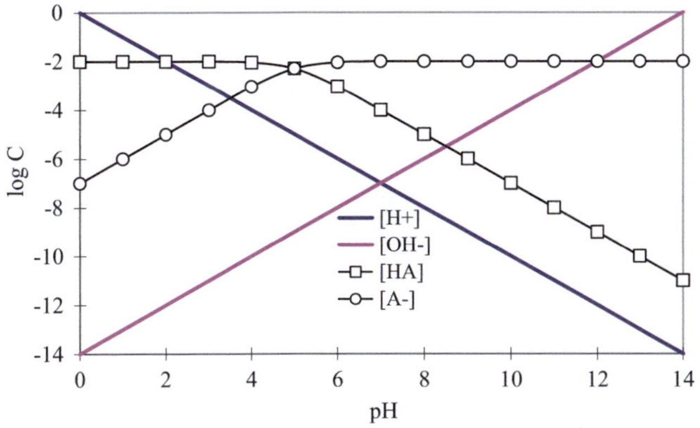

Fig. 19.1 Acetic acid concentrations vs. pH with 0.01 molar added to solution

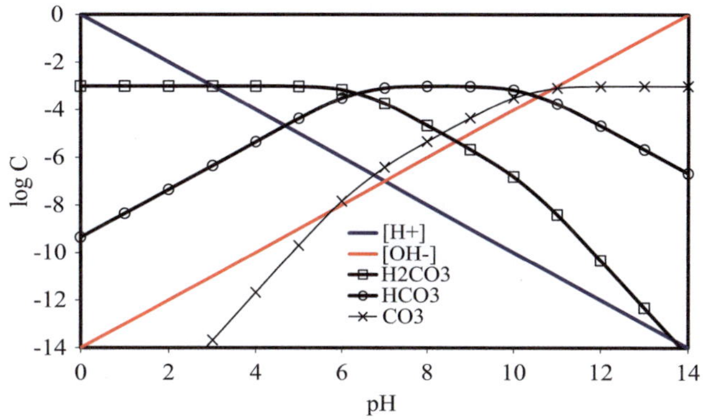

Fig. 19.2 Carbonate concentration versus pH for total concentration = 0.001 moles/L

Carbonic acid, H_2CO_3, ionizes in 2 stages so there are 2 ionization constants:

$$H_2CO_3 \leftrightarrow H^+ + HCO_3^- \tag{19.9}$$

$$\frac{[H^+][HCO_3^-]}{[H_2CO_3]} = K_1 \tag{19.10}$$

$$HCO_3^- \leftrightarrow H^+ + CO_3^{-2} \tag{19.11}$$

$$\frac{[H^+][CO_3^{-2}]}{[HCO_3^-]} = K_2 \tag{19.12}$$

Where

K_1 = dissociation constant for carbonic acid, $4.47 * 10^{-7}$
K_2 = dissociation constant for bicarbonate, $4.68 * 10^{-11}$

The 2 constants are multiplied to obtain the ionization constant for the complete dissociation of carbonic acid; $K_1 \times K_2 = 2.5 \times 10^{-17}$

$$\frac{[H^+][HCO_3^-]}{[H_2CO_3]}\frac{[H^+][CO_3^{-2}]}{[HCO_3^-]} = \frac{[H^+]^2[CO_3^{-2}]}{[H_2CO_3]} = K_1K_2 \tag{19.13}$$

Equations 19.9, 19.10, 19.11, 19.12, and 19.13 are used to derive the following equations that express the concentrations of the different carbonate species as a function of pH.

$$[H_2CO_3] = \frac{[C][H^+]^2}{[H^+]([H^+]+K_1)+K_1K_2} \tag{19.14}$$

$$[HCO_3^-] = \frac{[C][K_1][H^+]}{[H^+]([H^+]+K_1)+K_1K_2} \tag{19.15}$$

$$[CO_3^{-2}] = \frac{[C][K_1][K_2]}{[H^+]([H^+]+K_1)+K_1K_2} \tag{19.16}$$

where

C = molarity of all carbonate species.

The carbonate species concentrations vs. pH for $C_T = 0.001$ moles/L (10^{-3}) are shown in Fig. 19.2.

The term, alkalinity, refers to the capacity of negative ions (hydroxide, carbonate and bicarbonate anions) in water to combine with protons (H^+) from the acid. Typically, water analysis reports specify alkalinity, the amount of carbonate in water, as mg/L of calcium carbonate ($CaCO_3$). There are 60 mg CO_3^{2-} for every 100 mg of $CaCO_3$. The concentration units for alkalinity in Eq. 19.17 are meq/L (normality), which is the product of molarity and valence.

$$\sum\left(HCO_3^- + CO_3^{2-} + OH^-\right) \tag{19.17}$$

Normally, strong acids are injected into irrigation water because they are more effective at lowering pH and because they may be less expensive. The strong acids are HCl, HBr, HI, HNO_3, H_2SO_4, and $HClO_4$. Nearly all other acids are weak acids. Strong acids are acids for which the reaction in Eq. 19.4 only proceeds to the right. For example, the equation for sulfuric acid ionization in water is

$$H_2SO_4 \rightarrow 2H^+ + SO_4^{2-} \tag{19.18}$$

Sulfuric acid is made by compressing sulfur trioxide gas and water vapor. It is typically sold as 98 % sulfuric acid, which is 18.78 mol/L sulfuric acid.

With irrigation system injection, the acid charge flow rate (meq/s) should equal the neutralized alkalinity charge flow rate (meq/s). The charge flow rate is the product of water flow rate and normality, QC = (L/s)(meq/L) = meq/s: Q_aC_a

$= Q_wC_w$, where a is acid and w is water alkalinity. The equation can be rearranged and solved for acid flow rate, Q_a.

Example 19.2 Calculate the injection rate of 98 % sulfuric acid required to drop the pH to 6.5 for Colorado River water:

pH = 8.3
Carbonate alkalinity as $CaCO_3$ is 100 mg/L,
Irrigation system flow rate, Q_w, is 1,000 L/s

Carbonate concentration as $CaCO_3$ equal to 100 mg/L is equal to 60 mg/L CO_3^{-2} because the molecular mass of carbonate (CO_3^{-2}) is 60 g/mole, and the molecular mass of calcium carbonate ($CaCO_3^{-2}$) is 100 g/mole; thus, carbonate molarity is calculated as follows:

$$\left(\frac{100\ mg\ CaCO_3}{L\ water}\right)\left(\frac{0.6\ mg\ CO_3^{-2}}{1\ mg\ CaCO_3}\right)\left(\frac{g\ CO_3^{-2}}{1,000\ mg\ CO_3^{-2}}\right)\left(\frac{1\ mole\ CO_3^{-2}}{60\ g\ CO_3^{-2}}\right) = 0.001\ mole/L$$

The next step is to calculate the total alkalinity at pH = 8.3. Second, calculate the total alkalinity at pH = 6.5. Finally, calculate the change in alkalinity between 8.3 and 6.5.

Calculate hydroxide ion molarity at pH(8.3) : $[OH^-]$ $= 10^{-(14-8.3)} = 2*10^{-6}$mole/L.

Calculate initial HCO_3^- molarity at pH(8.3).

$$[HCO_3^-] = \frac{[C][K_1][H^+]}{[H^+]([H^+]+K_1)+K_1K_2} = \frac{[0.001]\left[4.47*10^{-7}\right]\left[5.01*10^{-9}\right]}{\left[5.01*10^{-9}\right]\left(\left[5.01*10^{-9}\right]+4.47*10^{-7}\right)+4.47*10^{-7}*4.68*10^{-11}} = 9.8*10^{-4}\ mol/L$$

Calculate the initial CO_3^{-2} molarity at pH(8.3)

$$\left[CO_3^{-2}\right] = \frac{[C][K_1][K_2]}{[H^+]([H^+]+K_1)+K_1K_2} = \frac{[0.001]\left[4.47*10^{-7}\right]\left[4.68*10^{-11}\right]}{\left[5.01*10^{-9}\right]\left(\left[5.01*10^{-9}\right]+4.47*10^{-7}\right)+4.47*10^{-7}*4.68*10^{-11}} = 9.8*10^{-4}\ mol/L$$

Add up the total alkalinity at pH = 8.3 (multiply CO_3^{2-} molarity by 2).

$$\sum\left(HCO_3^- + CO_3^{2-} + OH^-\right) = 9.80*10^{-4} + 2*9.15*10^{-6} + 2*10^{-6} = 0.00100\ \text{eq/L}$$

Hydroxide ion concentration at pH(6.5) = $10^{-(14-6.5)}$ $= 3.2 * 10^{-8}$ mole/L.

Calculate HCO_3^- molarity at pH(6.5).
Calculate the alkalinity at the final pH(6.5).

$$\left[HCO_3^{-2}\right] = \frac{[C][K_1][H^+]}{[H^+]([H^+]+K_1)+K_1K_2} = \frac{[0.001]\left[4.47*10^{-7}\right]\left[3.16*10^{-7}\right]}{\left[3.16*10^{-7}\right]\left(\left[3.16*10^{-7}\right]+4.47*10^{-7}\right)+4.47*10^{-7}*4.68*10^{-11}} = 5.86*10^{-4}\ mol/L$$

Calculate CO_3^{-2} molarity at pH(6.5)

$$[CO_3^{-2}] = \frac{[C][K_1][K_2]}{[H^+]([H^+]+K_1)+K_1K_2} =$$

$$= \frac{[0.001][4.47*10^{-7}][4.68*10^{-11}]}{[3.16*10^{-7}]([3.16*10^{-7}]+4.47*10^{-7})+4.47*10^{-7}*4.68*10^{-11}} = 8.67*10^{-8}\ mol/L$$

Calculate total alkalinity at pH(6.5).

$$\sum (HCO_3^- + CO_3^{2-} + OH^-) = 5.86*10^{-4} + 2*8.67*10^{-8} + 3.2*10^{-8} = 0.000586\ eq/L$$

Change in alkalinity = initial alkalinity – final alkalinity

$$0.001\ eq/L - 0.000586\ eq/L = 4.14*10^{-4} eq/L = 0.414\ meq/L$$

Calculate acid injection rate

$$Q_aC_a = Q_wC_w \rightarrow Q_a*37{,}560\ meq/L = 1{,}000\ L/S*0.414\ meq/L$$
$$Q_a = 0.414*1{,}000\ L/S/37{,}560 = 0.011\ L/s = 40\ L/hr$$

The same calculations are made in the *Acid injection rate* worksheet (Fig. 19.3).

Strong acids are effective, but they can corrode injection pumps, are unsafe, may have high concentrations of heavy metals, or may be unavailable at low cost. One alternative is organic acids. Most organic acids used in agriculture are forms of dicarboxylic acid. In theory, organic acids can remove salts from drip emitters even when pH remains high: organic polyacids combine with cations to form soluble salts and thus dissolve precipitated calcium and magnesium; however, weak acids are generally not as effective at preventing calcium carbonate deposition as strong acids and rarely used.

Although farmers typically lower water pH to 6.5 in order to prevent calcium carbonate precipitation, the threshold pH for precipitation may not be 6.5. The tendency of calcium carbonate to precipitate is a function of many factors: calcium and magnesium concentration, pH, bicarbonate concentration, temperature and TDS (total dissolved solids). The Langelier Saturation Index (LSI) calculates the tendency of calcium carbonate to precipitate. If the LSI is less than 0, then it is likely that no precipitation will occur.

$$LSI = pH_m - pH_c \qquad (19.19)$$

An LSI less than zero means that the calculated equilibrium pH_c of the water is greater than the measured pH_m. This means that the carbonate will come out of solution. The following method for calculating LSI is taken from the "Corrosion Doctors" website.

$$pH_c = (9.3 + A + B) - (C + D) \text{ where :} \qquad (19.20)$$

where

$A = (Log_{10}[TDS\ as\ ppm] - 1)/10$
$B = -13.12 \times Log10(°C + 273) + 34.55$
$C = Log_{10}[Ca^{2+}\ as\ CaCO_3\ (mg/L)] - 0.4$
$D = Log_{10}[alkalinity\ as\ CaCO_3\ (mg/L)]$
pH_c = calculated pH.
pH_m = measured pH

Example 19.3 Calculate LSI for the following Colorado River water sample at 25 °C and 35 °C.

Measured pH_m	7.96.
Alkalinityas $CaCO_3$	131 mg/L
Calciumas $CaCO_3$	72 mg/L
TDS	610 mg/L

Calculate the LSI at 25 °C

$$A = (Log_{10}[610] - 1)/10 = 0.1785$$
$$B = -13.12 \times Log10(25^0C + 273) + 34.55 = 2.088$$
$$C = Log_{10}[72] - 0.4 = 1.457$$
$$D = Log_{10}[131] = 2.117$$
$$pH_c = (9.3 + 0.1785 + 2.088) - (1.457 + 2.117) = 7.99$$
$$LSI = pH_m - pH_c = 7.96 - 7.99 = -0.03$$

Because the LSI is close to zero and parameters fluctuate, there might be precipitation at 25 °C. In addition, there are several other factors that may change the tendency of the water to precipitate, such as the high sulfate concentration in Colorado River water (200 mg/L). To be safe, add acid. The calculations are also made in the *LSI* worksheet (Fig. 19.4).

Recalculate for 35 °C

$$B = -13.12 \times Log10(35^oC + 273) + 34.55 = 1.90$$
$$pH_c = (9.3 + 0.1785 + 1.90) - (1.457 + 2.117) = 7.80$$
$$LSI = pH_m - pH_c = 7.96 - 7.80 = 0.16$$

	A	B	C	D	E
1	Molarity	0.001	Flow rate of water (L/sec)	1000	[H+]
2	K1	4.47E-07	Initial water pH	8.3	5.01E-09
3	K2	4.68E-11	Final water pH	6.5	3.16E-07
4	Kw	1.00E-14			
5			Species	Concentration (mol/L)	eq/L
6			Initial concentration of CO3	9.15E-06	1.83E-05
7			Initial concentration of HCO3	9.80E-04	9.80E-04
8			Initial concentration of OH	2.00E-06	2.00E-06
9			Initial total alkalinity (eq/L)		1.00016E-03
10					
11				Concentration (mol/L)	eq/L
12			Final concentration of CO3	8.67E-08	1.73E-07
13			Final concentration of HCO3	5.86E-04	5.86E-04
14			Final concentration of OH	3.16E-08	3.16E-08
15			Final total alkalinity (eq/L)		5.86E-04
16					
17			Change in total alkalinity (eq/L)		4.14E-04
18					
19			Initial hydrogen ion concentration (mol/L)		5.01E-09
20			Final hydrogen ion concentration (mol/L)		3.16E-07
21			Change in hydrogen ion concentration		3.11E-07
22					
23			Hydrogen ions (mol/L) and (eq/L) required		4.15E-04
24					
25			Acid concentration (mol/L)		18.78
26			moles H / moles acid		2
27			Acid concentration (eq/L)		37.56
28			Flow rate of acid required (L/sec)		0.0110
29			Flow rate of acid required (L/hr)		39.74
30					

Carbonate equil. Figure 19-2 | Acid inj. rate Ex. 19-2 | LSI Example 19-

Fig. 19.3 *Acid injection rate* worksheet

	A	B	C	D	E	F
1	pH_meas	7.96				
2	Temperature	25	C			
3	Ca (as CaCO3)	72	mg/L			
4	CO3 (as CaCO3)	131	mg/L			
5	TDS	610	mg/L			
6						
7	A	0.178532984				
8	B	2.088282615				
9	C	1.457332496				
10	D	2.117271296				
11	pHcalc	7.992211807				
12	LSI	-0.032211807	If positive, then precipitation			

Fig. 19.4 *LSI* worksheet

The positive LSI at 0.16 indicates that precipitation is more likely at a higher temperature.

If the pH is lowered to 6.5 by injecting acid, then the LSI becomes

$$LSI = pH_m - pH_c = 6.5 - 7.80 = -1.3$$

With a negative LSI, it is unlikely that precipitation will occur at pH 6.5.

Biological Control with Chlorine

Fungal or bacterial growth combined with possible bacterial slime within drip irrigation systems can rapidly plug drip emitters and result in disastrous crop losses. Drip irrigation systems are especially susceptible to growth of bacteria and fungi when fertigation provides a nutrient source for bacteria. Chlorine and other biocides are injected in order to kill microorganisms.

The first step is to evaluate whether biocides are necessary. Greenhouses or organic growers may not be able to use chlorine. For example, if water is from a deep well and no carbon (sugar) source (energy) for bacteria is available, then bacteria will not normally grow, and a biocide may not be necessary; however, if well water has been in contact with the atmosphere, then energy sources for bacterial growth and bacteria may diffuse into the water from the atmosphere. Well water that is high in iron may also provide an energy source for certain bacteria that oxidize iron from Fe^{2+} to Fe^{3+} in order to acquire energy. On the other hand, bacterial growth is likely if irrigation water is from an open water source such as a pond or canal.

Chlorine can be added to irrigation water in 3 forms: gaseous (Cl_2), liquid bleach (NaHOCl), or solid ($Ca(OCl)_2$). When chlorine gas is dissolved in water, the following reactions occur:

$$H_2O + Cl_2 = HOCl + H^+ + Cl^- \quad HOCl \leftrightarrow H^+ + OCl^- \tag{19.21}$$

HOCl (hypochlorous acid) and OCl^- (hypochlorite) are referred to as free available chlorine. There is more hypochlorous acid when pH is low and more hypochlorite when pH is high. Hypochlorous acid is 40–80 times more effective at killing biological organisms than hypochlorite because the negative charge on hypochlorite prevents it from entering the cell wall. At a pH of 6.5, 90 % of the chlorine is in the hypochlorous acid form. At a pH of 7, only 73 % of the chlorine is in the hypochlorous acid form, and as pH increases the reaction continues to proceed toward hypochlorite. For this reason, water is normally kept in the range of 6.0–6.5 during chlorination. The form of chlorine also makes a difference. Injection of chlorine gas lowers pH, and injection of bleach (sodium hypochlorite) raises pH.

Recommended chlorine injection rates are 1–2 ppm (elemental chlorine, Cl_2 concentration) for continuous injection, and 4–20 ppm for monthly, ½ to 1-hour, shock treatments. The injection rate depends on the amount of organic matter and other substances that react with chlorine.

Chlorine in the liquid form is the same compound as household bleach. Gaseous chlorine (much more dangerous) is delivered in gas cylinders and injected directly into the irrigation water. The change in weight of the cylinder over time can be used to calculate the mass flow rate of chlorine into the irrigation water. The percent by weight of elemental chlorine in each of the chlorine forms is shown in Table 19.1.

Example 19.4 Calculate the injection rate of chlorine into irrigation water if the desired concentration is 2 ppm (mg/L) (as elemental chlorine) and the irrigation flow rate is 44 L/s.

Table 19.1 Forms of chlorine and percent elemental chlorine (as Cl_2)

Form	Percentage (Cl_2)	Specific gravity
Gas	100	----
5 % bleach	5.1	1.07
7 % bleach	7.1	1.10
12.5 % bleach	12.5	1.19
15 % bleach	15	1.26
Calcium hypochlorite (solid)	51	

Gaseous (mass flow rate)

$$\left(\frac{44\ L\ water}{sec}\right)\left(\frac{2\ mg\ Cl_2}{L\ water}\right)\left(\frac{3{,}600\ sec}{hr}\right)\left(\frac{kg}{10^6\ mg}\right) = 0.317\ kg/hr$$

Liquid (liquid flow rate, use 15 % bleach)

$$\left(\frac{44\ L\ water}{sec}\right)\left(\frac{2\ mg\ Cl_2}{L\ water}\right)\left(\frac{100\ mg\ bleach}{15\ mg\ Cl_2}\right)\left(\frac{kg}{1*10^6\ mg}\right)\left(\frac{L}{1.26\ kg}\right)\left(\frac{3{,}600\ sec}{hr}\right) = 1.7\ L/hr$$

Solid

$$\left(\frac{44\ L\ water}{sec}\right)\left(\frac{2\ mg\ Cl_2}{L\ water}\right)\left(\frac{mg\ Ca(OCl)_2}{0.51\ mg\ Cl_2}\right)\left(\frac{3{,}600\ sec}{hr}\right)\left(\frac{kg}{10^6\ mg}\right) = 0.62\ kg/hr$$

The calculations are made in the *Chlorine* worksheet (Fig. 19.5)

Fertigation

Fertigation is the process of applying fertilizers to crops through irrigation systems. It is a subset of chemigation. One of the advantages of high frequency irrigation systems (center pivot and microirrigation) is the ability to apply fertilizer in frequent, uniform, and small quantities (spoon fed). Because nutrients are applied at the right time and the right place, plant growth is optimized and leaching is minimized. Many studies have shown that plant response to fertigation is generally higher than response to broadcast fertilization (Bar-Yosef 1999), especially on sandy soils with low cation exchange capacities. Low frequency systems such as surface irrigation systems may also by suitable for fertigation but uniformity of application may be a concern.

Effective fertigation requires calculation of nutrient requirements based on plant requirements and soil nutrients, selection of the most effective formulations, preparation of solutions for injection, and scheduling injections to ensure

Fig. 19.5 *Chlorine* worksheet

	A	B	C	D	E
1	Water flowrate	2640	L/min		
2	Chemical concentration	2	ppm		
3					
4	Chlorine gas	5280	mg/min	0.3168	kg/hr
5					
6	Liquid (bleach)	Percent OCl	L/hr		
7	Sodium hypochlorite (15 %)	15	1.6762		
8	Sodium hypochlorite (10 %)	10	2.7548		
9	Sodium hypochlorite (5.25 %)	5.25	5.6395		
10					
11	Solid (CaOCl2)	0.6212	kg/hr		

that essential nutrients are available as needed (Granberry et al. 2001). Fertilizer application rates depend on crop, soil, and time. Individual growers and agronomists use trial and error over many years to develop fertilization schemes that are specifically tailored to elicit the desired crop response, yield, and quality. As one might expect, these fertilization schemes are often kept secret.

Calculation of Plant Nutrient Requirements

Soil nutrient tests are conducted just prior to planting in order to determine the availability of soil nutrients. The seasonal nutrient requirement is the difference between the plant requirement and soil nutrient availability. Plant tissue analysis is used during the growing season to assess the current plant nutrient status and adjust fertilization rates accordingly. Distribution of roots must be considered when determining the availability of nutrients. For many drip irrigated crops, the rhizosphere (root zone) is concentrated in the vicinity of the emitter; thus, plants may have limited access to nutrients in the soil that are far from the emitter.

The nitrogen soil test reports the amount of nitrate – nitrogen (NO_3-N) in soil (as elemental N). Soil tests for nitrogen are conducted from 0- to 60-cm soil depth, and nitrogen content is reported in mg/L. This number is converted to total nitrogen amount in kg/ha by multiplying by 2.2. For example, if nitrate content in soil is reported as 13 mg/L, then there are 28 kg/ha of nitrogen in the soil. If the crop requires 147 kg/ha, then the amount of N that must be applied in fertilizer is 119 kg/ha. For nutrients other than nitrogen, such as potassium or phosphorous, extraction methods are used to determine the amount of available nutrient rather than the total amount of the nutrient in the soil.

General Considerations

The distribution uniformity of chemical application follows the distribution uniformity of water application. Because a well-designed microirrigation system should have a distribution uniformity in the range of 90 % or greater, it is reasonable to expect that the distribution uniformity of chemical application should also be 90 % or greater. Whenever possible, fertilizer injections should be made in small, frequent doses (spoon feeding) that fit within regularly scheduled microirrigation events (designed to match plant water use) to help avoid unnecessary leaching.

Because concentration of injected chemicals in irrigation water is not constant due to time delays in the movement of the water-chemical mixture through the piping system, the one-fourth rule is used: chemigation should start after one-fourth of the total irrigation set time, injection should occur during the middle two-fourths, and the lines flushed with clean water during the last one-fourth of an irrigation event. Fertilizers and other agrochemicals (except chlorine) should never be left in the pipeline when the system is not operating.

Advantages and Disadvantages of Fertigation and Pestigation

Microirrigation potentially offers numerous benefits for chemical application because of its flexible delivery arrangement. Some of the advantages over tractor application of chemicals include fuel and labor savings, less soil compaction, constant nutrient concentration in the root zone, improved productivity, placement of the chemical in the root zone, and decreased worker exposure to chemicals. The downside to chemigation includes higher required management expertise, high capital costs, restricted pesticide use for chemigation, risk of environmental contamination through misapplication, need to coordinate chemical and irrigation applications, and chemical corrosion of metal components.

Fertilization Requirements

Crop nutrient requirements change during the season depending on the growth stage. Total seasonal nutrient requirements for various crops are shown in Table 19.2; the values are based on research by Bar-Yosef (1999) for crops grown in a Mediterranean climate.

The grower makes decisions about applications during each stage based on desired crop response, soil nutrient reserves, available fertilizer formulations, preplant fertilization, irrigation schedules, criteria for maximum nutrient concentration in irrigation water, and other factors. Fertilizer formulations are generally not exactly in the proportions that match relative plant uptake of different nutrients. Thus, the grower application schedule may not be perfectly in sync with the plant uptake of nutrients during the growing season.

Example 19.5 Develop a nutrient application schedule for cantaloupe (cantaloupe). The growing season will be split into five stages: (0–25 days), (25–50 days), (50–75 days), (75–100 days), and (100–120 days) (Table 19.3). Soils are not limited in potassium, so no potassium fertilizer is required. Antecedent nitrogen in the soil is 28 kg/ha. The grower adds phosphorous in stage 3 (Table 19.4).

From Table 19.2, total N, P, and K required are 148 kg/ha, 25 kg/ha and 389 kg/ha. The stage requirements were calculated (left columns in Table 19.4) and the nutrient application schedule was assigned (right columns in Table 19.4). Nitrogen application was allowed to be behind schedule because of antecedent nitrogen in the soil. All phosphorous was applied during the 3rd stage. Potassium was not needed.

Table 19.2 Seasonal nutrient uptake amounts for various crops (based on Bar Yosef)

Crop	N (kg/ha)	P (kg/ha)	K (kg/ha)
Muskmelon (cantaloupe)	148	25	389
Carrot	277	72	592
Sweet corn	237	40	323
Broccoli	183	25	255
Chinese cabbage	137	38	281
Celery	151	36	588
Lettuce	113	22	244
Cotton	233	45	186
Potato	166	n/a	n/a
Eggplant	296	33	399
Bell pepper	206	30	372
Fresh tomato	248	25	367
Greenhouse tomato	431	49	709
Processing tomato	395	56	517

Fertilizer Selection and Calculation of Injection Rates

Fertilizers for fertigation systems are selected based on plant response, solubility, cost, effect on soil pH, formulation, and potential for reaction with other chemicals within the distribution system. Liquid fertilizers can be purchased in liquid form or stock solutions can be mixed on-farm. Fertilizer is often available in premixed liquid form for fertigation systems. Transportation costs are higher for liquid fertilizers than for solid forms. However, premixed fertilizer often save time and labor, and avoids problems associated with poorly prepared on-farm mixes (Granberry et al. 2001). Solid fertilizers used in making liquid fertilizer formulations should be in the granular form. However, some granular fertilizers are not completely soluble in water and should not be injected into microirrigation systems. The maximum amount of a solid that can be dissolved in water is the solubility (kg/L). Solubility changes with temperature and with concentration of other ions in solution. Fertilizers should not be mixed at the maximum solubility because temperature change can cause chemical precipitation and to mixing time is extensive.

The type of nitrogen that is used by the plant has an effect on both the plant and the soil environment. Plants can take

Table 19.3 Percent of seasonal N, P, and K requirements for a muskmelon crop in Yuma, Arizona

Stage	Nitrogen N (kg/ha) % required	Phosphorus P (kg/ha) % required	Potassium K (kg/ha) % required
1 (0–25 days)	3	4	1
2 (25–50 days)	16	16	14
3 (50–75 days)	52	36	43
4 (75–100 days)	22	36	38
5 (100–120 days)	7	8	4

Table 19.4 N, P, and K application rate for a muskmelon crop in Yuma, Arizona.

Stage	Nitrogen N (kg/ha)		Phosphorus P (kg/ha)		Potassium K (kg/ha)	
	Required	Applied	Required	Applied	Required	Applied
1 (0–25 days)	5	5	1	0	4	0
2 (25–50 days)	24	35	4	0	55	0
3 (50–75 days)	77	35	9	25	166	0
4 (75–100 days)	32	35	9	0	150	0
5 (100–120 days)	9	9	2	0	14	0

Table 19.5 Conversions of relevant nutrient units of measurements (After Burt et al. 1995)

Unit	Multiplier		Result	
NO_3-N (mg/L) (elemental N)	×	4.42	∋	NO_3 (mg/L)
NH_4-N (mg/L) (elemental N)	×	1.29	∋	NH_4 (mg/L)
NH_4 (mg/L)	×	0.78	∋	NH_4-N (mg/L)
NO_3 (mg/L)	×	0.226	∋	NO_3-N (mg/L)
P_2O_5 (mg/L)	×	0.67	∋	PO_4 (mg/L)
P_2O_5 (mg/L)	×	0.44	∋	PO_4-P (mg/L)
PO_4-P (mg/L)	×	3.07	∋	PO_4 (mg/L)
K_2O (mg/L)	×	0.83	∋	K (mg/L)

up nitrogen as ammonium, NH_4^+, or as nitrate, NO_3^-. Nitrogen applied as ammonium, a cation, is less likely to be leached from soil than nitrogen applied as nitrate (anion) because cations are adsorbed by the clay particles in the soil. Because the plant must maintain electroneutrality (same number of anions and cations), plants tend to secrete cations (hydrogen ions) when most of the nitrogen is taken up as ammonium. This process tends to acidify the soil. Acidification of some soils may be a problem with acid-based fertilizers, and alternatives such as calcium nitrate should be considered. A balanced fertilization program may require supplemental foliar applications of micronutrients.

Fertilizers are often applied as ammonium-N with the expectation that nitrifying bacteria will slowly convert the ammonium to nitrate during the growing season; typically, plants prefer to take up nitrogen in the nitrate form. At higher pH, the ammonium (NH_4^+)/ammonia (NH_3) equilibrium goes toward ammonia, and thus volatilization of ammonium is more likely at high pH. For surface drip irrigation, ammonia volatilization may result in a significant loss of nitrogen to nitrous oxide or nitrogen gas.

Typically, soil tests report phosphorus (P), and potassium (K) availability in soils as the oxidized forms, P_2O_5 and K_2O, respectively. Fertilizer formulation labels for nitrogen (N), phosphorous (P), and potassium (K) are based on percent N, P_2O_5 and K_2O, respectively (Rosen and Eliason, 1996). Applicable conversion factors are given in Table 19.5. Characteristics of various fertilizers are listed in Table 19.6.

Urea sulfuric acid can be mixed with other solutions such as phosphoric acid, muriate of potash, and sulfate of potash. Urea sulfuric acid mixtures should always be mixed in the correct order and only with approved fertilizers as specified by the manufacturer (Burt et al. 1995).

Proper mixing procedures should always be followed with all fertilizer formulations. Mixing procedures (after Burt et al. 1995) are listed as follows:

1. Always fill the mixing container with 50–75 % of the required water to be used in the mix.
2. Always add the liquid materials to the water before adding the dry, soluble fertilizers.
3. Always add the dry ingredients slowly with circulation or agitation to prevent the formation of large, insoluble, or slowly soluble lumps.
4. Always add acid into water, not water into acid.
5. Never mix an acid or acidified fertilizer with chlorine (gas, liquid, or solid).
6. Do not mix concentrated fertilizer solutions with other concentrated fertilizer solutions.
7. Because fertilizer solutions are applied in very small dosages, and if injected at different locations in the irrigation line, many incompatibility problems are avoided.

Example 19.6 A grower wants to apply 4 kg/ha of potassium (K) as K_2O and 2 kg/ha of nitrate-N to a 20 ha field during a 1-week period. Calculate the mass of KNO_3 fertilizer that must be dissolved and the amount of solution required for the 1-week period. Irrigation schedule is 4 hr/day, every day.

Total potassium required as K_2O is 4 kg/ha × 20 ha = 80 kg.

Total potassium required as K is found by taking the ratio of K to K_2O (0.83 from Table. 19.4)

$$\text{Mass K} = 80 \times 0.83 = 66.4 \text{ kg}$$

Now solve for the number of moles of K required using information from Table 3.1.

$$\begin{aligned}\text{Moles K} &= 66.4 \text{ kg} \times 1000 \text{ g/kg}/(39.1 \text{ g/mole}) \\ &= 1,698 \text{ moles K}\end{aligned}$$

Solve for the number of moles of nitrogen required.

$$\begin{aligned}\text{Moles N} &= 2 \text{ kg/ha} \times 20 \text{ ha} \\ &\quad \times 1,000 \text{ g/kg}/(14 \text{ g N/mole}) \\ &= 2,857 \text{ moles N}\end{aligned}$$

The number of moles of N and number of moles of K that are applied must be the same because there is a one-to-one ratio of NO_3 and K in the fertilizer. Thus, excess potassium will be applied.

Solve for the mass of fertilizer required.

$$\begin{aligned}&\text{Molecular weight of fertilizer} \\ &\quad = \text{Mol.wt. } NO_3 + \text{Mol wt. K} = 62 + 39.1 \\ &\quad = 101.1 \text{ g/mol.} \\ &\text{Mass of fertilizer} \\ &\quad = 2,857\text{moles} \times 101.1\text{g/mol.} \times 0.001 \text{ kg/g} = 289 \text{ kg.}\end{aligned}$$

The solubility of potassium nitrate is 0.133 kg/L (Table 19.5). Volume of water required based on solubility of potassium nitrate.

$$289 \text{ kg}/(0.133 \text{ kg/L}) = 2,172 \text{ L}$$

Table 19.6 Types of fertilizer and suitability for fertigation through microirrigation systems (after Burt et al. 1995 and originally presented in Evans and Waller 2006)

Fertilizer	Compound	Precaution	Characteristics
Anhydrous ammonia	82-0-0	Never inject into microirrigation Sys	n/a
Aqua ammonia	20-0-0	Never inject into microirrigation systems	n/a
Ammonium nitrate solution (AN-20) NH_4NO_3 *H_2O	20-0-0	Never mix solution with concentrated acids	Density – 1.29 kg/L Commonly used for fertigation
Urea ammonium nitrate (UAN32) $(NH_2)_2CO$*NH_4NO_3	32-0-0	Don't mix with CAN17 or calcium nitrate solutions	Density – 1.33 kg/L Commonly used liquid fertilizer with highest N.
Calcium ammonium nitrate (CAN17) $Ca(NO_3)_2$*NH_4NO_3	17-0-0-8.8 Ca	Do not combine with solutions containing sulfates/thiosulfate	Density – 1.55 kg/L Effective for obtaining high fruit quality in some crops.
Ammonium phosphate $(NH_4)H_2PO_4$	8-24-0	Can precipitate if injected at high rates or into hard water	Density – 1.26 kg/L Not commonly used for fertigation.
Ammonium polyphosphate $(NH_4)_2H_2P_2O_7$	9-30-0, 10-34-0, 11-37-0	Can precipitate with carbonates in high pH water	9-30-0, Density – 1.36 kg/L 10-34-0, Density – 1.37 kg/L 11-37-0, Density – 1.41 kg/L
Ammonium polysulfide $(NH_4)_2S_x$	20-0-0-45	Dangerous hydrogen sulfide gas results from contact with acid	Density – 1.05 kg/L Solution has very high pH, but can act as soil acidifying agent.
Ammonium thiosulfate $(NH_4)_2S_2O_3$	12-0-0-26	Do not mix with acids or apply to low pH soils	Density – 1.33 kg/L Acidifying agent, ideal for treatment of calcareous soils.
Metal chelates	Most metal micronutrients must be chelated with EDTA, DTPA, or EDDHA to prevent precipitation		Chelated metals are very effective in alkaline soils.
Phosphoric acid H_3PO_4	0-54-0	Never mix phosphoric acid with calcium fertilizer	Density – 1.69 kg/L Green acid is most common, but is less pure than white acid.
Potassium chloride KCl	0-0-60 Dry form	Not recommended for Cl sensitive crops	Solubility – 0.347 kg/L Highest used potassium fertilizer.
Potassium nitrate KNO_3	13-0-44	None	Solubility – 0.133 kg/L Second most popular source of potassium fertilizer.
Potassium phosphate KH_2PO_4	0-52-34	None	Density – 1.26 kg/L Used in greenhouse and nursery
Potassium sulfate K_2SO_4	0-0-50	None	Solubility – 0.12 kg/L Popular for fertigation as good source of sulfur
Potassium thiosulfate (KTS) $K_2S_2O_3$	0-0-25-17 0-0-22-23	KTS blends should have pH less than 6	Contains 0.36 kg/L K_2O and 0.25 kg/L S
Sulfuric acid H_2SO_4	0-0-0	Very Hazardous. Injection w/ CAN17 results in clogging	Density – 1.83 kg/L
Urea solid $(NH_2)_2CO$	46-0-0	Never mix urea with sulfuric acid	Solubility – 1 kg/L
Urea phosphate $(NH_2)_2CO$*H_3PO_4	17-44-0		The acidity prevents ammonia volatilization
Urea sulfuric acid $(NH_2)_2CO$*H_2SO_4	10-55-18 19-49-16 28-19-9	55 Sulfuric acid 18 S 49 Sulfuric acid 16 S 15 Sulfuric acid 9 S	10-55-18 Density – 1.54 kg/L 19-49-16 Density – 1.53 kg/L 28-19-9 Density – 1.42 kg/L

It is easier to dissolve potassium nitrate at a concentration less than the maximum concentration (0.133 kg/L). The grower doubles the volume of liquid used to dissolve the fertilizer to approximately 5,000 L.

If irrigation occurred on daily basis and fertilizer injection was conducted for 2 h each irrigation, then the injection rate for the 20 ha block would be calculated as follows.

$$5{,}000\ \text{L}/(2\ \text{h/day*7 days}) = 357\ \text{L/h}$$

The grower could adjust the relative amounts of N and K by mixing a fertilizer that is higher in nitrogen and lower in potassium. One possible combination would be ammonium nitrate and potassium nitrate.

Table 19.7 Nutrient application amounts and irrigation schedule per stage for cantaloupe crop

Stage	Nutrients -kg/block/stage					
	N	P	K	Ca	Irr. events per stage	Fertigation hours/stage
1 (0–25 days)	75	0	0	0	6	12
2 (25–50 days)	525	0	0	0	12	24
3 (50–75 days)	525	375	0	0	16	32
4 (75–100 days)	525	0	0	300	16	32
5 (100–120 days)	135	0	0	300	16	32

Example 19.7 A grower plans to meet the nutrient application requirements specified in Table 19.7. The 75 ha field is divided into 5 blocks of 15 ha each. In addition to N and P fertilization, the grower also adds 20 kg/ha calcium during the fourth and fifth stages. Nutrient application rates (Table 19.7) per block are the product of kg/ha for each stage (Table 19.4) and the area of each block. Water is Colorado River water with pH = 8. Irrigation water is applied at a rate of 4 mm/hr.

The grower injects the fertilizer during the middle 2 h of each 4 h irrigation event. Thus, there are 2 fertigation hours per irrigation event as shown in Table 19.7. In order to save time and avoid possible problems, the grower decides to purchase premixed liquid formulations. Urea-ammonium nitrate (UAN32) is selected during the first 2 stages because only nitrogen is required. For the third stage, liquid formulations of UAN32 and ammonium polyphosphate (10-34-0) are selected and will be injected at two ports into the irrigation pipeline. Ammonium polyphosphate would not normally be selected for injection into Colorado River water (pH ~ 8) because of the danger of precipitation of calcium phosphate ($CaPO_4$). However, the grower injects sulfuric acid in order to lower irrigation water pH to 6.5 at an injection port upstream from the two fertilizer injection ports. Thus, the grower can avoid precipitation of fertilizers with Ca in the Colorado River water source. The soil is alkaline so soil acidification is not a hazard, and lowered pH may improve calcium availability to the plant. For the last two stages, the grower decides to use CAN17 as the calcium source. CAN17 cannot be injected with urea-ammonium nitrate so the grower decides to use ammonium nitrate as the nitrogen source if CAN17 does not supply the nitrogen required during the fourth and fifth stages.

UAN32 will supply 75 kg N per block over 12 h of fertigation (Table 19.6) during the first stage. The density of UAN32 is 1.33 kg/L and the formulation is 32-0-0 (Table 19.5).

Volumetric N content of UAN32
$= 1.33\ \text{kg/L} \times 0.32 = 0.425\ \text{kg/L of N}$
UAN32 volume per block during stage 1
$= 75\ \text{kg}/0.425\ \text{kg/L} = 176\ \text{L UAN32}$
Injection rate $= 176\ \text{L}/12\ \text{h of fertigation} = 14.7\ \text{L / h}.$
Mass of UAN32 applied $= 176\text{L} * 1.33\text{kg / L}$
$= 234\ \text{kg UAN32/block}$

UAN32 will supply 525 kg N per block over 24 h of fertigation during stage 2.

UAN32 volume per block during stage 2
$= 525\ \text{kg}/0.425\ \text{kg / L} = 1,235\ \text{L UAN32}$
Injection rate $= 1,235\ L/24\ \text{h of fertigation} = 51.5\ \text{L / h}.$
Mass of UAN32 applied $= 1,235\ \text{L}*1.33\ \text{kg/L}$
$= 1,642\ \text{kg UAN32/block}$

Ammonium polyphosphate will supply 375 kg P per block over 32 h of fertigation during the third stage. The density of ammonium polyphosphate (APP) fertilizer is 1.37 and the formulation is 10-34-0 (Table 19.5). Thus, 34 % of the mass of the fertilizer is phosphorous as P_2O_5. P_2O_5 is multiplied by 0.44 (Table 19.4) to convert to elemental P (PO_4-P).

Volumetric P content of APP
$= 1.37\ \text{kg / L} \times 0.34 \times 0.44 = 0.205\ \text{kg / L of P}$
APP volume per block during stage 3
$= 375\ \text{kg}/0.205\ \text{kg/L} = 1,829\ \text{L of APP}$
APP Injection rate $= 1,829\ \text{L}/(32\ \text{h of fertigation})$
$= 57\ \text{L / h of APP}$
Mass of APP applied $= 1,829\ \text{L}*1.37\ \text{kg / L}$
$= 2,500\ \text{kg APP / block}$

The nitrogen supplied by the ammonium polyphosphate during stage 3 is calculated as follows:

Volumetric N content of APP $= 1.37\ \text{kg/L} \times 0.10$
$= 0.137\ \text{kg / L of N}$
Mass N applied as APP $= 1,829\ \text{L} \times 0.137\ \text{kg / L}$
$= 250\ \text{kg of N}$

If 250 kg N is applied as ammonium phosphate, then the remainder of the nitrogen requirement (525–250 = 275 kg/block) during stage 3 must be supplied by UAN32.

UAN32 volume per block during stage 3
$= 275\ \text{kg}/(0.425\ \text{kg / L}) = 647\ \text{L of UAN32}$
Injection rate $= 647\ \text{L}/(32\ \text{h of fertigation})$
$= 20\ \text{L / h of UAN32}$
Mass of UAN32 applied $= 647\ \text{L}*1.33\ \text{kg / L}$
$= 860\ \text{kg UAN32 / block}$

Calcium ammonium nitrate (CAN17) will supply 300 kg Ca per block over 32 h of fertigation during the fourth stage and fifth stages. The density of CAN17 fertilizer is 1.55 and the formulation is 17-0-0-8.8 Ca (Table 19.5). Thus, 8.8 % of the mass of the fertilizer is calcium.

$$\begin{aligned}&\text{Volumetric Ca content for CAN17}\\&\quad = 1.55\ \text{kg} / \text{L} \times 0.088 = 0.136\ \text{kg} / \text{L of Ca}\\&\text{CAN 17 volume per block during stage 3}\\&\quad = 300\ \text{kg}/(0.136\ \text{kg/L}) = 2,205\ \text{L of CAN17}\\&\text{Injection rate of CAN17} = 2,205\ \text{L}/(32\ \text{h of fertigation})\\&\quad = 69\ \text{L/h of CAN17}\\&\text{Mass of CAN17 applied} = 2,205\ \text{L}*1.55\ \text{kg} / \text{L}\\&\quad = 3,418\ \text{kg CAN17} / \text{block}\end{aligned}$$

The amount of nitrogen supplied by CAN17 during stages 4 and 5 is calculated as follows.

$$\begin{aligned}&\text{Volumetric N content of CAN17} = 1.55\ \text{kg} / \text{L} \times 0.17\\&\quad = 0.264\ \text{kg} / \text{L of N}\\&\text{Mass N applied as CAN17}\\&\quad = 2,205\ \text{L CAN17} \times 0.264\ \text{kg} / \text{L} = 582\ \text{kg of N}\end{aligned}$$

Thus, CAN17 will supply slightly more N per block, 582 kg, than is required during the fourth stage, 525 kg, and much more than is required during the fifth stage, 135 kg. As a result, the grower may decide to apply less N during the third stage and the application of UAN32 during the third stage may not be necessary.

The final step in assessing a fertigation strategy is to check the concentration of nutrients in irrigation water and the impact on the root zone salinity concentration.

The total depth of irrigation water applied to the field is

$$4\ \text{mm} / \text{h}*66\ \text{irrigations}*4\ \text{h} / \text{irrigation} = 1,056\ \text{mm}.$$

The total volume of irrigation water applied to the 75 ha field is

$$1.056\ \text{m}*75\ \text{ha}*10,000\ \text{m}^2/\text{ha} = 792,000\ \text{m}^3$$

The total mass of fertilizer (salts) applied is the sum of the fertilizer mass applied during all stages * 5

$$\begin{aligned}&5*(234 + 1,642 + 2,500 + 860 + 3,418 + 3,418)\\&\quad = 60,360\ \text{kg salts}\end{aligned}$$

The average concentration in irrigation water is

$$60,360\ \text{kg}/792,000\ \text{m}^3 = 0.076\ \text{kg} / \text{m}^3 = 76\ \text{mg} / \text{L}$$

Extra leaching will be required in order to remove the additional 76 mg/L salts in the irrigation water (leaching fraction in Chap 4). The calculations are made in the *Fertigation* worksheet (Fig. 19.6). Note that white cells are user input and yellow cells are calculated.

Although not necessarily a nutrient, farmers add acid to irrigation water in order to lower the pH of soils because soil pH can have an effect on nutrient availability. One reason that plants may appear nutrient deficient is not a lack of nutrients in the soil but rather a lack of availability of nutrient to the plant due to pH.

Chemigation of Non-Fertilizer Materials

Only chemicals that are labeled for chemigation should be applied with irrigation systems. The pesticide application may also be subjected to additional state regulations above label specifications. The United States Environmental Protection Agency has established criteria for evaluating the leaching potential of pesticides. Pesticides are considered less likely to leach if they have a short life in the soil before degrading, are not very soluble in water, and are likely to be adsorbed onto soils and organic matter.

Pesticide injection rate can be calculated based on irrigation system area and time of application. A typical pesticide application might include one h of water application before pesticide is injected, 3 h of pesticide injection, and then one h of water application after pesticide application in order to flush all pesticides from the irrigation system. Flushing is especially important with the application of soil fumigants in order to prevent crop damage when irrigation is resumed after planting.

The concentration of the pesticide in the irrigation water must also be checked in order to ensure that the concentration of pesticide in the irrigation system is not corrosive to irrigation system components.

Example 19.8 A grower wants to apply Telone EC at a rate of 45 L/ha over a period of 3 h to an irrigation zone that covers 10 ha. The irrigation water flowrate to the irrigation zone is 5,000 L/min. The density of Telone EC is 1.1 kg/L.

The required injection rate is

$$\begin{aligned}\text{Injection rate, L} / \text{h} &= 45\ \text{L} / \text{ha} \times 10\ \text{ha}/3\ \text{h} = 150\ \text{L} / \text{h}\\&= 2.5\ \text{L} / \text{min}\end{aligned}$$

It has been found that Telone EC at concentrations greater than 1,500 mg/L is corrosive to irrigation system components. In order to avoid exceeding one-half of the corrosive concentration (the rule of thumb), the concentration of Telone EC in irrigation water should not exceed 750 mg/L.

$$\begin{aligned}&\text{Telone EC Concentration}\\&\quad = 2.5\ \text{L/min Telone}/5,000\ \text{L/min water}\\&\quad \times 1.1\ \text{kg} / \text{L Telone} \times 1 \times 10^6\text{mg} / \text{kg}\\&\quad = 550\ \text{mg} / \text{L Telone}\end{aligned}$$

Thus, at an injection rate of 2.5 L/min, the concentration of Telone in irrigation water is less than 750 mg/L and is acceptable.

	A	B	C	D	E	F
1	Nitrogen table	Total N required kg/ha	148			
2	Stage	0-25 days	25-50 days	50-75 days	75-100 days	100-125 days
3	Fertilizer type	UAN32	UAN32	UAN32	CAN17	CAN17
4	kg N per L	0.425	0.425	0.425	0.136	0.26
5	fraction of total	0.03	0.16	0.52	0.22	0.07
6	kg per ha required	4	24	77	33	10
7	kg per ha applied	5	35	35	35	9
8	area/block (ha)	15	15	15	15	15
9	kg/block/stage required	75	525	274	-57	-447
10	Irrigation days per stage	6	12	16	16	16
11	Fertilization run time/irr (hr)	2	2	2		
12	Fertigation hours per stage	12	24	32		
13	Volume fert / stage	176	1235	646		
14	Injection rate	14.7	51.5	20.2		
15	Fertilizer density	1.33	1.33	1.33		
16	Mass of fertilizer applied	235	1643	859		

	A	B	C	D	E	F
19	Other elements requirement	0	0	25	20	20
20	Stage	0-25 days	25-50 days	50-75 days	75-100 days	100-125 days
21	Element and Fertilizer type	none	none	Phosphorous from APP	Calcium from CAN17	Calcium from CAN17
22	kg element per L			0.205	0.136	0.136
23	fraction of total			1.00	1.00	1.00
24	kg per ha required			25	20	20
25	kg per ha applied			25	20	20
26	area/block (ha)			15	15	15
27	kg/block/stage			375	300	300
28	Irrigation days per stage			16	16	16
29	Fertilization run time/irr (hr)			2	2	2
30	Fertigation hours per stage			32	32	32
31	Volume fert / stage			1829	2206	2206
32	Injection rate			57.2	68.9	68.9
33	Fertilizer density			1.37	1.55	1.55
34	Mass of fertilizer applied			2506	3419	3419
35	kg N per L			0.137	0.264	0.264
36	Mass of N applied (kg)	0	0	251	582	582
37	Mass of fertilizer applied (kg)	0	0	2506	3419	3419
38						
39	Total mass of salt applied	235	1643	3365	3419	3419
40						
41	Total mass of salt/season (kg)	60403				
42	Water application rate (mm/hr)	4				
43	Depth applied (mm) (4 hr/event)	1056				
44	Vol applied (m3) (5 blocks)	792000				
45	Increase in salinity (mg/L)	76				

Fig. 19.6 *Fertigation worksheet*

The Chapter 19 Chemigation workbook has several worksheets that are not reviewed in this chapter: *Units, Mixing,* $CaCO_3$ *solubility, Sulfuric acid, Hardness, and Gypsum.*

Questions

1. Determine a fertilizer injection schedule for N, P, and K for bell peppers for the following table. All other assumptions, areas, and fertilizer types are as in the chapter 19 examples.

Table HW11-1 Nutrient application rates and times for Bell Peppers

Stage	Nitrogen N (kg/ha)		Phosphorus P (kg/ha)		Potassium K (kg/ha)	
	Required	Applied	Required	Applied	Required	Applied
1 (0–25 days)	17	17	2	0	4	0
2 (25–50 days)	36	35	5	0	55	0
3 (50–75 days)	50	40	9	23	166	0
4 (75–100 days)	65	50	9	0	150	0
5 (100–120 days)	39	36	5	0	14	0

2. Plot carbonate, bicarbonate, and carbonic acid concentration as a function of pH for alkalinity as calcium carbonate equal to 120 mg/L.
3. Calculate the amount of 98 % sulfuric acid required to drop the pH from 7.9 to 6.8 for water with alkalinity as calcium carbonate equal to 120 mg/L.
4. Calculate the LSI for the following water analysis at 25 °C and 35 °C. Measured pH is 7.96.
   - alkalinity as calcium carbonate equal to 120 mg/L.
   - calcium as calcium carbonate equal to 65 mg/L
   - TDS equal to 1,000 ppm
5. Calculate the injection rate of 10 % chlorine bleach required in order to develop a concentration of 3 ppm elemental chlorine in an irrigation system with a 200 LPM flow rate.

References

The Evans and Waller document is the original source of much of this chapter. Dr. Evans was a Federal employee so there is no copyright.

Bar-Yosef B (1999) Advances in fertigation. Adv Agron 65:2–77. Academic Press

Burt CM, O'Connor K, Ruehr T (1995) Fertigation. Irrigation Training and Research Center, California Polytechnic State University, San Luis Obispo, p 320

Evans and Waller (2006) Chemigation. (Ed) Freddy Lamb. Americaan Society of Agricuitural and Biological. Engineers, Published ASABE Irrigation Monograph, St. Joseph

Granberry DM, Harrison KA, Kelley WT (2001) Drip chemigation: injecting fertilizer, acid and chlorine, vol 1130, Bull. University of Georgia, College of Agricultural and Environmental Science, Cooperative Extention Service, Athens, p 13

Rosen CJ, Eliason R (1996) Nutrient management for commercial fruit & vegetable crops in Minnesota. University of Minnesota Ext. Pub. BU-05886. 61 pp. http://www.extension.umn.edu/distribution/cropsystems/DC5886.html

Surface Irrigation

20

Surface irrigation methods include furrow (Fig. 1.15), border, and basin irrigation. This chapter shows how to use volume balance methods in spreadsheets and the WinSRFR program, http://www.ars.usda.gov/services/software/download.htm?softwareid=250, to design furrow surface irrigation systems. The same procedures are used in the design of other surface irrigation methods. Volume balance methods compare the applied volume to surface and subsurface storage volumes in order to calculate parameters such as infiltration rate or the rate of water advance down the field. The primary design objectives in surface irrigation are to maximize application efficiency and uniformity. The primary design variables are inlet flow rate, and application time. Field length and slope are not easily changed without earthmoving equipment. Soil permeability, soil surface roughness, furrow geometry, water temperature, wind direction and speed, and surface crusting can vary during the growing season and thus change the infiltration rate and flow characteristics. Variability in these parameters and spatial variability of soil properties can make it difficult to optimize surface irrigation systems.

Surface irrigation events have three phases: advance, storage, and recession. During the advance phase, the wetting front (Fig. 20.1) moves down the field. The advance time is the length of time required for the wetting front to reach the end of the field. If the advance time is long, then the upstream end of the field receives more water than the downstream end; thus, water application is not uniform. Higher flow rates result in faster advance of the wetting front and uniform infiltration over the length of the field. However, flow velocity must be lower than the erosive flow velocity, and high flow velocity may result in excessive runoff from the end of the field if the flow rate is not reduced once the wetting front reaches the end of the field.

The second phase of surface irrigation is the storage phase (Fig. 20.2). After the advance reaches the end of the field, the water must remain ponded for a sufficient length of time for the end of the field to receive the required depth of water. The length of the storage phase depends on the required depth of infiltration, and the soil infiltration rate. It may last from several hours to 24 hours. If the storage phase is long, then a significant quantity of water may run off the end of the field. Also, significant leaching may occur at the upstream end. With respect to irrigation efficiency (under or over irrigation), the length of the storage phase is the most important design criterion for surface irrigation. In furrow irrigation systems, the length of the storage phase at the end of the field is the time of cutoff minus the advance time. The time of cutoff is the time when the irrigation flow to the field is shut off.

After irrigation water is turned off at the time of cutoff, the recession phase begins: ponded water infiltrates or moves down the furrow and the upper end dries (Fig. 20.2). In furrow irrigation systems, the upper end of the furrow dries immediately after the time of cutoff, and then the dried section increases as the water infiltrates and moves off the end of the field. However, recession does not begin immediately for border and level basin systems because there is a much greater ratio of water on the field surface to wetted soil area than in furrows (Fig. 20.3).

Surface Irrigation Infiltration

Infiltration rate determines the rate of advance, and the required length of the storage phase. Surface irrigation systems cannot be placed in certain fields because the soil infiltration rate prevents successful operation of surface irrigation systems. For example, in sandy soils, the wetting front may never reach the end of the field due to excessive infiltration. In sloping clay soils, the required storage time may be extremely long and cause excessive runoff. Thus, determination of the infiltration rate should be made before deciding to install a surface irrigation system or selecting between different surface irrigation systems.

The soil infiltration rate is a key parameter in surface irrigation models. If the soil infiltration parameters are in

P. Waller, M. Yitayew, *Irrigation and Drainage Engineering*, DOI 10.1007/978-3-319-05699-9_20

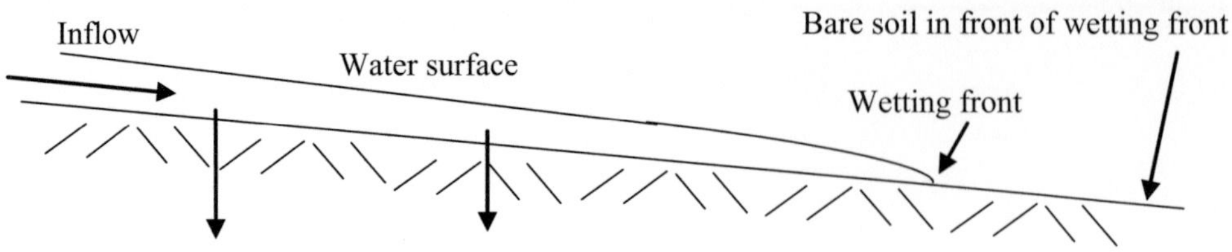

Fig. 20.1 Advance phase during surface irrigation

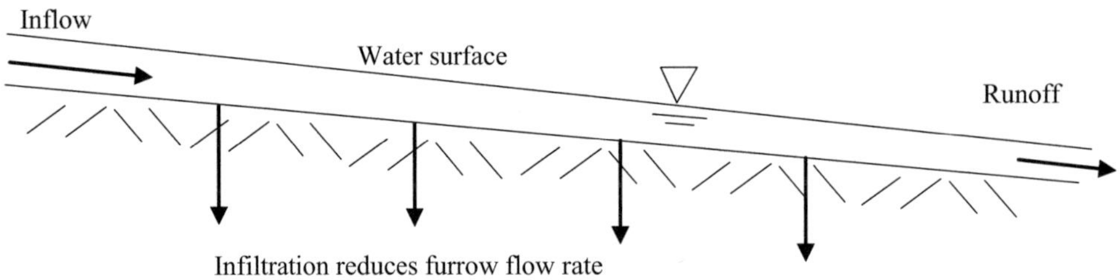

Fig. 20.2 Storage phase during surface irrigation of open-ended field

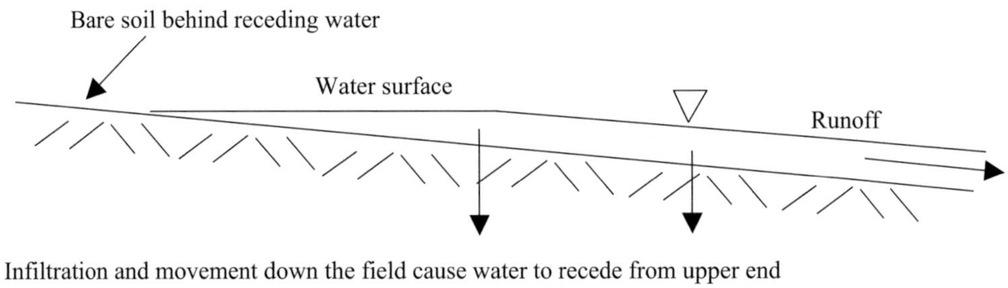

Fig. 20.3 Recession phase during surface irrigation of open-ended field

error, then no matter which model is used to design the irrigation system, the design will be in error. Many methods are used to determine average field infiltration rate. (1) Observe the rate of advance and use the two-point volume balance method to determine the infiltration parameters. (2) Observe a surface irrigation event and perform a mass balance by measuring the inflow, outflow, rate of advance, and time of recession. (3) Same as 2 with the addition of an infiltrometer test. (4) Same as 2 with the addition of measuring flow depth at intervals in the field. (5) Same as 2, except with the assumption of Phillips' infiltration equation. (6) Use several infiltrometers to measure the rate of infiltration. (7) Determine the intake family from an NRCS soils map.

Although it is preferable to use direct measurements of infiltration (observation of advance and recession times during an irrigation event or use of infiltrometers), infiltration rate can be estimated based on soils maps and NRCS intake families (method 7 listed above). The infiltration equation as typically written for surface irrigation modeling is similar to but uses different terms than the NRCS intake family equations.

$$d = at^b + c \quad \text{(NRCS form)} \tag{20.1a}$$

$$d = kt^a + bt + c \text{ (as typically written for surface irrigation models)} \tag{20.1b}$$

where

d = infiltrated depth, mm,
k, b, and c = constants,
a = exponent, dimensionless.

The bt term in Eq. 20.1 is not in the NRCS intake family equation, but is often added to the surface irrigation infiltration equation in order to better represent steady infiltration due to the downward force of gravity. The first term, kt^a or at^b, represents the decreasing rate of infiltration due to decreased capillary suction and larger distance between the soil surface and the infiltration wetting front (Fig. 3.12). The third term, c, represents initial rapid infiltration into cracks.

In cracking clay soils, the c term can constitute the majority of infiltration. Some furrow irrigation models use the $kt^a + c$ term during the initial infiltration and $bt + c$ during the final stages of infiltration Clemmens et al. (2006).

Many furrow irrigation models only use the first term, kt^a, which is called the Kostiakov equation. Merriam and Clemmens

Table 20.1 Comparison of NRCS (Eq. 20.1a $d = at^b$) and Merriam and Clemmens (1985) infiltration equation ($d = kt^a$) with time in units of hours in infiltration depth in units of mm

NRCS ($at^b + c$)				Infiltration (100 mm) time (hr)	Merriam and Clemmens (kt^a)	
Intake family	a	b	c		k	a
0.05	6.69	0.618	7.0	70.70	30.09	0.28
0.1	9.28	0.661	7.0	32.66	29.18	0.35
0.15	11.5	0.683	7.0	21.29	30.08	0.39
0.2	13.5	0.699	7.0	15.87	31.33	0.42
0.25	15.7	0.711	7.0	12.27	32.88	0.44
0.3	17.7	0.720	7.0	10.03	34.45	0.46
0.35	19.7	0.729	7.0	8.44	36.06	0.48
0.4	21.6	0.736	7.0	7.27	37.69	0.49
0.45	23.6	0.742	7.0	6.36	39.33	0.5
0.5	25.5	0.748	7.0	5.64	41	0.52
0.6	29.3	0.757	7.0	4.59	44.29	0.53
0.7	33.2	0.766	7.0	3.85	47.62	0.55
0.8	36.9	0.773	7.0	3.31	50.91	0.56
0.9	40.7	0.779	7.0	2.89	54.19	0.58
1	44.4	0.785	7.0	2.56	57.49	0.59
1.5	60.2	0.799	7.0	1.73	71.12	0.62
2	75.3	0.808	7.0	1.30	84.31	0.65
3	103	0.816	7.0	0.88	109.05	0.69
4	129	0.823	7.0	0.67	132.88	0.71

(1985) assessed numerous infiltration curves measured throughout the Western United States and developed a relationship between a and k, based on the time to infiltrate 100 mm – a typical depth of application per irrigation event. Equation 20.2 calculates a, and subsequently, k can be calculated.

$$a = 0.675 - 0.2125\ \mathrm{LOG}_{10}(T_{100}) \tag{20.2}$$

where

T_{100} = time to infiltrate 100 mm, hr.

Example 20.1 Find a and k if the time to infiltrate 100 mm is 15.9 hours.

First, determine a with Eq. 20.2.

$$\begin{aligned} a &= 0.675 - 0.2125\ \mathrm{LOG}_{10}(T_{100}) \\ &= 0.675 - 0.2125\ \mathrm{LOG}_{10}(15.9) = 0.42 \end{aligned}$$

Find k by rearranging Eq. 20.1.

$$k(\mathrm{mm/hr}) = d/t^a = 100/15.9^{0.42} = 31.3 \quad d = 31.3t^{0.42}$$

A comparison between NRCS intake family infiltration constants and Merriam-Clemmens calculated a and k values for equivalent times to infiltrate 100 mm are shown in Table 20.1. The two equations have dramatically different exponents with NRCS intake family exponents approximately 0.25 lower than Merriam-Clemmens intake family exponents.

Example 20.2 Compare the infiltration curve calculated with the Clemmens a and k values to the NRCS infiltration curve for a soil requiring 10 hour to infiltrate 100 mm.

The soil with a 10 hour infiltration time in Table 20.1 is the 0.3 intake family.

$$d = at^b + c = 17.7\ t^{0.72} + 7 = 17.7(600)^{0.72} + 7$$

Calculate the equivalent Clemmens infiltration exponents, a and k, for the case where 10 hours are required to infiltrate 100 mm.

$$\begin{aligned} a &= 0.675 - 0.2125\ \mathrm{LOG}_{10}(T_{100}) \\ &= 0.675 - 0.2125\ \mathrm{LOG}_{10}(10) = 0.462 \\ k(\mathrm{mm/hr}) &= d/t^a = 100/10^{0.462} = 34.4 \\ d &= 34.4\ t^{0.462} \quad (\mathrm{thours}) \end{aligned}$$

A comparison of infiltration curves is shown in Fig. 20.4.

With lower exponents, the Merriam-Clemmens intake families predict higher infiltration at the beginning of the irrigation event, which results in a slower calculated advance time. The NRSC intake family curve predicts greater infiltration at the end of the irrigation event, which results in a greater calculated depth of infiltration if the irrigation event

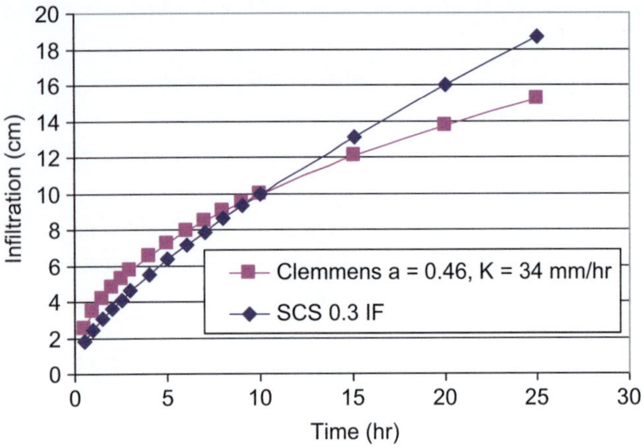

Fig. 20.4 Comparison of infiltration curves for Clemmens a and k and NRCS intake family 0.3 – soil requires 10 hours to infiltrate 100 mm

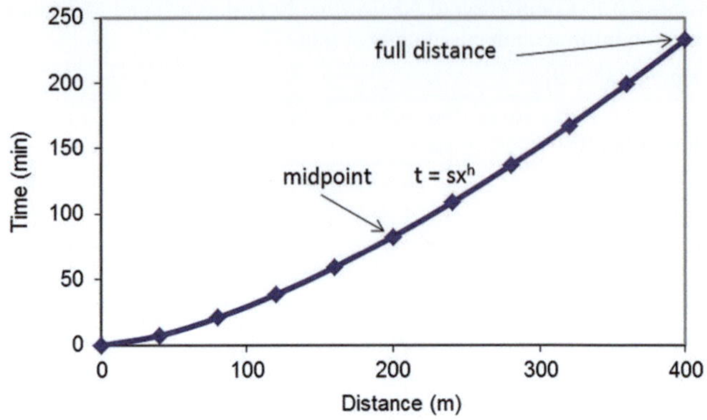

Fig. 20.5 Advance curve parameters for two-point volume balance method

is longer than 10 hours. Thus, the Merriam-Clemmens intake families provide a more conservative estimate of long-term infiltration (predict less) and advance time (takes longer).

The six methods for calculating infiltration are described in the following pages. Each of them has strengths and weaknesses and are appropriate for different applications.

1. Observe the position of the wetting front over time and plot the advance curve (Fig. 20.5). Regression can then be used to fit an exponential equation to the advance curve. The infiltration equation parameters, *k* and *a*, are adjusted by iteration until the observed advance curve matches the calculated advance curve. The inputs to the two point volume balance method are the inflow rate and the advance times to half of the length of the field and the entire length of the field. If bt is included in the infiltration Eq. (20.1), then a three point volume balance method could be used to find *k, a,* and *b*. As stated previously, the drawback with this method is that it only measures infiltration during advance. Thus, it is a poor estimator of long-term infiltration during the storage phase. This method should only be used if the rate of soil infiltration is high and the storage phase is very short in comparison to the advance time. Use of the exponential advance curve to calculate infiltration is only valid if the slope of the field is not flat, such that water does not back up (becomes deeper at the beginning of the field), and if the inflow remains on during the entire advance phase.
2. Observe a surface irrigation event and perform a mass balance by measuring the inflow, outflow, rate of advance, and time of recession. This method can be used to determine an average rate of infiltration vs. the average time of ponding for the area irrigated. The volume infiltrated is found by subtracting the total outflow volume from the total inflow volume. The average depth, $d_{average}$, infiltrated is the volume infiltrated divided by the area irrigated. The average time of infiltration is found be subtracting the advance curve from the recession curve (Fig. 20.6) at several positions and finding the intake opportunity time (IOT_i in Fig. 20.6). Then, the average intake opportunity time, $t_{average}$, is found be taking the weighted average of the intake opportunity time at the different field positions. Then $t_{average}$ and $d_{average}$ are used to find the intake family.

$$d_{average} = kt_{average}^{a} + bt_{average} + c \tag{20.3}$$

This method is likely to result in the correct calculation of the average infiltration in a field; however, because of the uncertainty of the intake family exponents, it is likely that the advance rate will be under or overestimated. For example, with the infiltration equation as written in Eq. 20.3, there are four unknowns (*k, a, b,* and *c*). Thus, an infinite assortment of a, k, b, and c values would solve the equation. However, if it is assumed that b and c are zero, and that there is a direct relationship between intake opportunity time and exponent *a* (as in Eq. 20.2), then there is only one solution. Unfortunately, as is shown in Table 20.1, there is considerable uncertainty with respect to the value of exponent *a*, for a given infiltration depth vs. time (the intake families have dramatically different exponents).
3. If a single infiltrometer test at a point in the field is combined with the irrigation event evaluation described in method 2, then the exponent *a* can be derived from the infiltrometer test, and the *k* value can be calculated based on the average infiltration rate in the field. The assumption is that the shape of the infiltration curve (represented by *a*) is relatively constant in the field, but the magnitude of infiltration varies. This method solves the problem of the uncertainty over the *a* value, and the rest of the steps in method 2 are the same. This is the method used by the NRCS for irrigation evaluations.

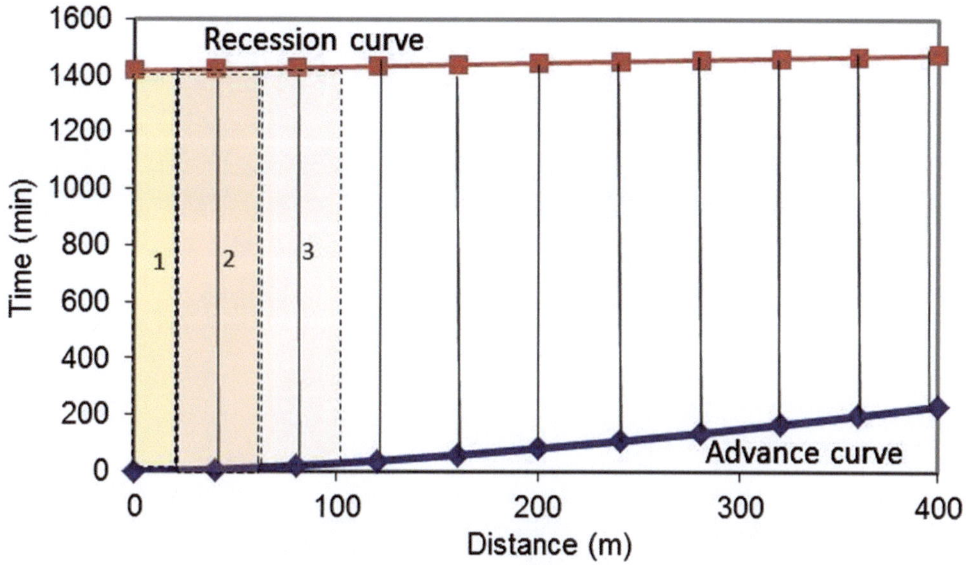

Fig. 20.6 Calculation of average intake opportunity time (IOT) by averaging depth infiltrated at each field section

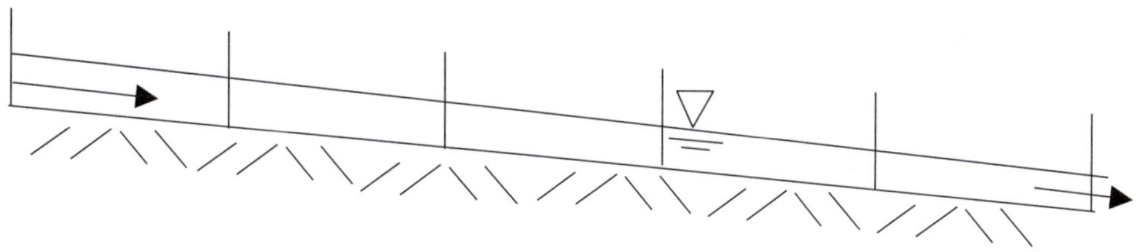

Fig. 20.7 Stakes for measuring depth of flow in furrows

4. Same as two with the addition of measuring flow depth at intervals in the field. If measuring stakes are placed in the field, and flow depth is measured in the field over time (Fig. 20.7), then the volume of infiltration can be calculated during the entire irrigation event (inflow – outflow – depth of surface storage). The stakes must be surveyed with respect to absolute elevation (not just the elevation of the stake with respect to the soil). This method is more time intensive than the other methods, and is more apropriate in a research project.

In-class Exercise 20.1 Draw a graph of the shape of the infiltrated depth vs. time curve in method 4 as it would appear on a log-log graph in Excel. How would you find a?

5. Same as two, except with the assumption of Phillips' infiltration equation. Philips derived the infiltration equation from theory. He found that the exponent, *a,* should theoretically be 0.5 and that the saturated hydraulic conductivity, K_s, should be used to model constant infiltration.

$$d = kt^{0.5} + K_s \quad (20.4)$$

The Philips infiltration equation yields reasonably accurate infiltration curves, which result in irrigation model data having a reasonably accurate agreement with field data. It is probably the best choice when limited data is available. The only unknown is k.

6. Use multiple infiltrometers (Fig. 20.8) to measure spatially varying infiltration. This method is extremely time consuming and has only been conducted in a few research studies. Recent studies show that the coefficient of variation of infiltration due to varying soil properties is in the range of 25 %.

Graded Furrow Infiltration

Graded furrows have a slope and an open ended discharge at the lower end. They are one of the most common irrigated methods and are used for row crops such as cotton, tomatoes, and cantaloupes. Water is introduced to the furrows in small streams from gated pipe, siphon tubes (Fig. 20.9), plastic risers, and other methods.

Because only a fraction of the soil surface is ponded with furrow irrigation, the average infiltration rate over the field

Fig. 20.8 Double ring infiltrometer

Fig. 20.9 Graded furrow irrigation supplied by siphon tubes from head ditch (Credit NRCS. Rick Schlegel)

area is less than infiltration in a level field. However, the infiltration rate per wetted soil area is greater than level field infiltration because water infiltrates in two directions in the furrow (Fig. 20.10).

Small flumes can be used at the upstream and downstream ends of furrows to measure flow rate (Fig. 20.11).

A furrow infiltrometer is constructed by blocking two points in a short section of the furrow. A flowing furrow

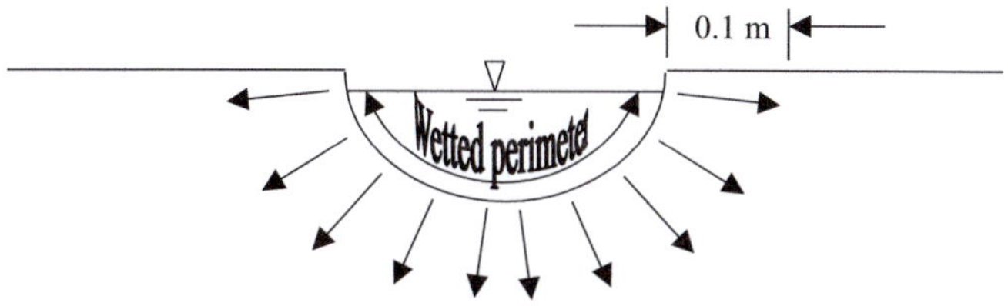

Fig. 20.10 Furrow infiltration geometry

Fig. 20.11 Furrow flume (Credit USDA-ARS David Bjorneberg)

infiltrometer (Fig. 20.12) measures inflow and outflow with flumes, orifices, or other flow measurement devices. A nonflowing infiltrometer uses a float valve that triggers a pump which replaces infiltrated water from a tank, and the water level in the tank is monitored over time. The advantage of a furrow infiltrometer is that infiltration rate is measured directly. The disadvantage is that infiltration over a short section of the furrow may not be representative of the field average, due to spatial variation of soil properties. However, even calculating infiltration based on an irrigation event over an entire furrow may not be representative of the field. Typically, the advance rate from one furrow to the next in a field varies dramatically.

If it is not possible to conduct an infiltrometer test or to monitor advance and recession during an irrigation event, then the NRCS intake families for level surfaces can be adjusted to furrows infiltration rate as follows.

- Add 0.213 m (0.1 m on each side of the furrow) to the wetted perimeter (Fig. 20.5) of the furrow to account for lateral infiltration to the center of the beds.
- Multiply the infiltration rate for a level field by the following ratio (wetted perimeter + 0.213 m)/(width between furrows).

Fig. 20.12 Blocked furrow infiltrometer (Credit NRCS)

The Two-Point Volume Balance Method

The two-point volume balance model calculates the position of the wetting front during advance based on a comparison of the applied volume to the surface and subsurface storage. The surface storage is the volume stored in the furrow and is

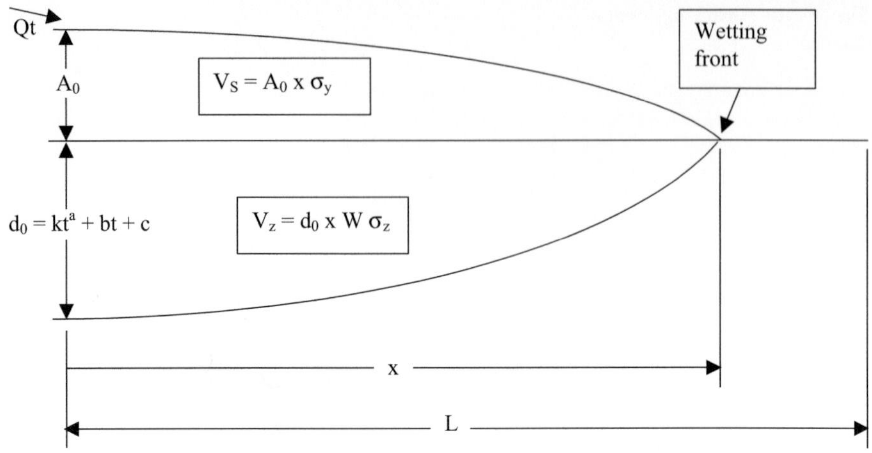

Fig. 20.13 Surface and subsurface storage for volume balance method

the product of cross-sectional area of flow at the upper end of the furrow, furrow spacing, distance to the wetting front, and a shape factor. The subsurface storage is the volume stored in the furrow and is the product of the depth infiltrated at the upper end of the furrow, distance to the wetting front, and a shape factor (Fig. 20.13).

The total storage is the sum of surface and subsurface storage.

$$V_T = V_z + V_s. \tag{20.5}$$

where

V_z = volume of subsurface storage (infiltrated volume), m^3,
V_s = volume of surface storage (water in furrow), m^3.
V_T = volume of total storage

The inflow volume is the product of inflow rate and time.

$$V_{in} = 3.6Qt \tag{20.6}$$

where

Q = inflow rate, L/sec,
t = time since irrigation water was turned on, hr,
V_{in} = volume applied to furrow since water was turned on, m^3,

The cross-sectional flow area at the upper end of the furrow can be calculated with Manning's equation. Furrow shape is typically described by a power function or a trapezoid. In this text, the furrow shape is modeled as a trapezoid. As with canals, the side slope, z, in furrows is specified as the run over the rise. The cross sectional area, A, and the wetted perimeter, P, of a trapezoidal furrow are given in Table 10.3. Typical Manning's n values for agricultural surfaces are given in Table 11.2. In smooth furrows, Manning's n is 0.04.

Example 20.3 For a 1.0 L/s flow rate, calculate the depth of flow in a trapezoidal furrow with z = 2 and a bottom width, b, of 0.1 m. The furrow has a slope of 0.002 m/m, and the Manning's n is 0.05. Finally, calculate the Clemmens *k* and *a* values for the upper end of the furrow if the time to infiltrate 100 mm is 10 hours (as in Example 20.2). Calculations made in *Furrow* worksheet.

Calculate the section factor (Chap. 10)

$$\frac{Qn}{S_f^{1/2}} = \frac{(0.001\ m^3/\sec)\ (0.05)}{0.002^{1/2}} = Section\ factor = 0.001118$$

$AR^{2/3}$ is calculated with the following equation.

$$AR^{2/3} = A(A/P)^{2/3}$$
$$= (b + zy)y\left(((b + zy)y)/\left(b + 2y\left(1 + z^2\right)^{\wedge 0.5}\right)\right)^{2/3}$$

Make an initial guess, y = 0.04 m.

$$0.001118 = AR^{2/3} = (0.1 + (2)(0.04)(0.04)\left(\left(0.1 + (2)(0.04)\right)(0.04)/\left(0.1 + (2)(0.04)\left(1 + 2^2\right)^{\wedge 0.5}\right)\right)^{2/3}$$
$$0.001118 = 0.00063$$

Iterate by inputting the required section factor (0.01118) and the first guess for a section factor into the following equation. This equation is designed for rapid convergence for this iteration.

$$y = \text{initial } y\left(SF/AR^{2/3}\right)^{0.5} = 0.04(0.001118/0.00063)^{0.5}$$
$$= 0.0533 \text{ m.}$$

$$0.001118 = (0.1 + (2)(0.0533)(0.04)\left(\left(0.1 + (2)(0.0533)\right)(0.0533)/\left(0.1 + (2)(0.0533)(1+2^2\right)^{\wedge 0.5}\right)\right)^{2/3}$$
$$0.001118 = 0.001124 \rightarrow y = 0.0533.$$

Calculate wetted perimeter

$$P = b + 2y\left(1+z^2\right)^{0.5} = 0.1 + (2)(0.0533)\left(1+2^2\right)^{0.5}$$
$$= 0.338 \text{ m}$$

Adjust the infiltration rate for the equivalent furrow infiltration width.

$$\frac{0.338 \text{ m} + 0.213 \text{ m}}{1.0\ m} = 0.551$$

Clemmens *k* and *a*, from Example 20.2, are 34.45 and 0.462, respectively. Adjust *k* by multiplying by 0.551. No change is required for a.

$$\text{Adjusted}k = (0.551)(34.4 \text{ mm/hr}) = 19.0(\text{mm/hr})$$
$$d = 19.0\ t^{0.462}$$

The surface shape factor is based on the shape of the water profile down the furrow. If the entire furrow was filled to the same depth, then the shape factor would be 1.0. However, the depth of water in the furrow decreases with distance down the furrow. The surface shape factor is the fraction of water in a furrow in comparison to a furrow with constant depth. It is usually assumed that the surface shape factor, σ_y, is 0.75 during the advance phase. Thus, the volume of water stored in the furrow is calculated as follows:

$$V_s = \sigma_y A_0 x \tag{20.7}$$

where

σ_y = surface shape factor, 0.75,
A_0 = cross sectional area at the upper end of the furrow at time t_x, m^2,
x = advance distance down the furrow at time t, m,
V_s = volume of surface storage at advance distance x, m^3.

The subsurface storage for furrow irrigation is a function of infiltration and field parameters. If b and c in Eq. 20.1 are zero, then the total volume of subsurface storage (infiltration) is

$$V_z = d_0\sigma_z Wx = k\,t^a\ \sigma_z Wx \tag{20.8}$$

where

W = distance between furrows, m,
d_0 = infiltration at upper end of furrow (as in Example 20.2), m,
σ_z = subsurface shape factor (Eq. 20.9),
t = time since the beginning of the irrigation event, hr,
a = exponent in infiltration equation.

If b and c are not zero, then the total volume of infiltration down the furrow is

$$V_z = d_0\sigma xW = \left(c + \sigma_z k\,t^a + \frac{h}{1+h}bt\right)xW\sigma_z \tag{20.9}$$

where

h = exponent in advance equation.

The position of the wetting front during advance is modeled with the advance equation.

$$t = sx^h \tag{20.10}$$

where

s = advance equation coefficient.

The subsurface shape factor is a function of a and h.

$$\sigma_z = \frac{h + a(h-1) + 1}{(1+a)(1+h)} \tag{20.11}$$

The relationship between the advance exponent h, and the advance times to two points in the field can be found by taking the logarithm of Eq. 20.11.

$$\log t = \log s + h \log x \tag{20.12}$$

Rearrange the equation and solve for log s

$$\log s = -h \log x + \log t \tag{20.13}$$

Equation 20.13 follows the form, y = mx + b. For two log x and log t points, the slope of the line between them is h. If the two distances (points x_1 and x_2) are the full length of the furrow, x_L, and half the length of the furrow, $x_{L/2}$, then x_2 is twice the length of x_1 and the following equation is used to calculate h.

$$h = \frac{\log t_2 - \log t_1}{\log x_2 - \log x_1} = \frac{\log\left(\frac{t_2}{t_1}\right)}{\log\left(\frac{x_2}{x_1}\right)} = \frac{\log\left(\frac{t_L}{t_{L/2}}\right)}{\log\left(\frac{2}{1}\right)} \quad (20.14)$$

The first step in the two-point volume balance method is to find the normal depth of flow at the given inflow rate. Next, an initial guess is made for t_L and $t_{L/2}$, and these values are used in Eq. 20.14 to make an initial calculation for h. Then, the subsurface shape factor is calculated (Eq. 20.11), and the surface and subsurface storage are calculated. Finally, $t_{L/2}$ is adjusted based on the inflow volume and storage at time $t_{L/2}$, and t_L is adjusted based on the inflow volume and storage at time t_L with Eq. 20.5. Next, the iteration procedure is repeated with the new estimates of t_L and $t_{L/2}$. During the iteration procedure, the time required for water to advance to a certain point in the furrow is adjusted by the ratio of inflow volume, V_{in}, to total storage volume, V_T.

$$t_{m+1} = t_m\left(\frac{V_T}{Q\ t_m}\right)^{1.4} \quad (20.15)$$

where

m = iteration number

Example 20.4 Calculate s and h and plot the advance curve for a 400 m long furrow with 1 m spacing between furrows. Inflow rate is 1.0 L/s. Ten hours are required to infiltrate 100 mm in level soil. Manning's roughness n is 0.05, slope is 0.002 m/m, bottom width is 0.1 m, z is 2, and the upstream flow depth is 0.0533 m (as calculated in Example 20.3). Use a convergence criterion of less than 1 min difference between iterations.

Calculate cross-sectional area at the furrow inlet, A_0.

$$A_0 = (b + z\ y)y = (0.1\,m + (2)(0.0533\ m))(0.0533\ m) = 0.011 m^2$$

From Table 20.1, if 10 hours are required to infiltrate 100 mm in level soil, then a = 0.46 and k = 34.4. Adjusted k for furrows (as calculated in Example 20.3) is 19.0.

The design procedure starts with a guess for the advance time to ½ the field length and to the end of the field. For this example, we arbitrarily guess 100 and 250 minutes, respectively. For a flow rate of 1.0 L/s, calculate inflow volumes during these two periods.

$$V_{L/2} = Qt = 100\ min(60\ sec/min)(0.001 m^3/sec) = 6\ m^3$$
$$V_L = Qt = 250\ min(60\ sec/min)(0.001\ m^3/sec) = 15\ m^3$$
$$h = \log(t_{L/2}/t_L)/\log(1/2) = \log(100/250)/\log(1/2) = 1.32$$

The next step is to calculate the subsurface shape factor.

$$\sigma_z = \frac{h + a(h-1) + 1}{(1+a)(1+h)} = \frac{1.32 + 0.46(1.32 - 1) + 1}{(1 + 0.46)(1 + 1.32)} = 0.73$$

The next step is to calculate subsurface storage. The infiltrated depths at the upper end of the field at 100 and 250 minutes are calculated.

$$d_{L/2} = kt^a = 19.0(100/60)^{0.46} = 24.0\text{ mm infiltrated.}$$
$$d_L = kt^a = 19.0(250/60)^{0.46} = 36.6\text{ mm infiltrated.}$$

Calculate subsurface storage at $t_{L/2}$ (time to reach L/2) and t_L

$$Vz_{L/2} = d_0 s_z W\ x = (24.0/1{,}000\ mm)(0.73)(1.0\ m)(200\ m) = 3.50\ m^3$$
$$Vz_L = d_0 s_z W\ x = (36.6/1{,}000\ mm)(0.73)(1.0\ m)(400\ m) = 10.68\ m^3$$

Calculate surface storage at $t_{L/2}$ and t_L.

$$V_{s\ L/2} = s_y A_0 x = (0.75)(0.011\ m^2)(200\ m) = 1.65\ m^3.$$
$$V_{s\ L} = s_y A_0 x = (0.75)(0.011\ m^2)(400\ m) = 3.29\ m^3.$$

Calculate total storage at $t_{L/2}$ and t_L.

$$V_{T\ L/2} = 1.65 + 3.50 = 5.15\ m^3$$
$$V_{T\ L} = 3.29 + 10.68 = 13.97\ m^3$$

The advance times are adjusted with Eq. 20.16

$$t_{m+1} = t_m\left(\frac{V_T}{Q\ t_m}\right)^{1.4} = 100\left(\frac{5.15}{6}\right)^{1.4} = 81\ \min$$
$$t_{m+1} = t_m\left(\frac{V_T}{Q\ t_m}\right)^{1.4} = 250\left(\frac{13.97}{15}\right)^{1.4} = 226\ \min$$

The procedure is then repeated for the next iteration with $t_{L/2}$ = 81 min, and t_L = 226 min. Subsequent iterations are shown in Table 20.2.

The solution has met the convergence criteria since there is less than one-minute change in advance times from the previous iteration. Solve for s (hours and meters) and write the advance equation; the advance time is 231/60 = 3.85 h.

$$t = sx^h \quad s = t/x^h \quad s = 231/400^{1.49} = 0.0305$$
$$t = 0.0305\ x^{1.49}$$

The advance curve is shown in Fig. 20.14.

Table 20.2 Iteration steps for two point volume balance method

	t_1	t_2	h	σ_z	$V_{in-t\,L/2}$	$V_{in-t\,L}$	$V_{TL/2}$	V_{TL}
Initial guess	100	250	1.32	0.729	6	15	5.15	14.0
Iteration 1	80.7	226	1.49	0.747	4.84	13.6	4.90	13.7
Iteration 2	82.0	230	1.49	0.747	4.92	13.8	4.92	13.8
Iteration 3	82.1	231	1.49	0.747	4.92	13.8	4.93	13.8

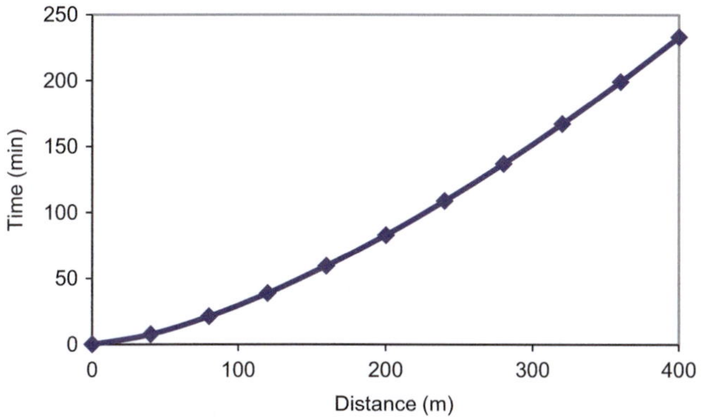

Fig. 20.14 Advance curve for Example 20.5

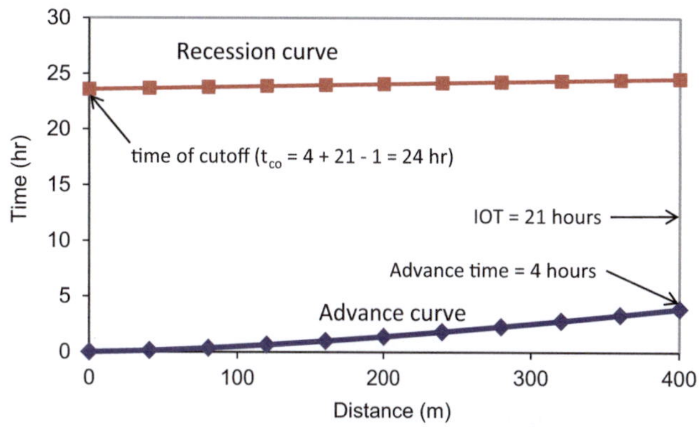

Fig. 20.15 Advance and recession curves for Example 20.6

After cutoff (the time when the water is turned off), recession time in a furrow is found by dividing volume stored in the furrow at the time of cutoff by the inflow rate (Clemmens et al. 2006). Volume stored in the furrow at the time of cutoff is calculated with a new σ_y:

$$V = A_0 L \sigma_y \quad (20.16)$$

$$t_{rec} = V/Q \quad (20.17)$$

where

t_{rec} = time of recession, min,
Q = flow rate, m^3/min,
σ_y = surface shape factor at time of cutoff, 0.8.

Example 20.5 Calculate the time of recession for Example 20.4 where Q = 1.0 L/s, A_0 = 0.011 m^2, and L = 400 m.

$$Q = \left(\frac{1.0\ L}{\text{sec}}\right)\left(\frac{m^3}{1,000\ L}\right)\left(\frac{60\ \text{sec}}{\text{min}}\right) = 0.06\ m^3/\text{min}$$

$t_{rec} = V/Q = (0.011\ m^2)(400\ m)(0.8)/0.06 = 58$ minutes (see Fig. 20.15)

Recession is assumed to be linear so linear interpolation can be used to plot the recession curve between the time of cutoff and the time of recession.

Furrow Irrigation Scheduling and Evaluation

Furrow irrigation systems generally require manual labor to cut off water to furrows and to initiate inflow to other furrows. In general, farmers prefer to have irrigation sets of 12 or 24 hours because the irrigation can be changed at the same time each day. However, if farmers have employees running irrigation systems for 24 hours per day, then dividing the irrigation schedule into days or half days is not as important since irrigation sets can be changed at any time.

If the minimum ponding time is at the end of the furrow, then the time that irrigation is cut off can be found by subtracting the recession time from the advance time + required ponding time at the end of the furrow.

$$t_{co} = t_{adv} + IOT_{req} - t_{rec} \quad (20.18)$$

where

t_{co} = time of cutoff, min,
t_{adv} = advance time, min,
IOT_{req} = required intake opportunity time, min

The length of time between irrigation events is the depth of ponding divided by the evapotranspiration rate.

Example 20.6 Adjust flow parameters and depth of infiltration so that the farmer has a 12 or 24-hour irrigation cycle. Make an initial guess that the required depth of infiltration is 100 mm. Use soil and furrow flow parameters as calculated in

Examples 20.3, 20.4 and 20.5. Determine the time of cutoff and the length of time between irrigation events if the maximum ET_c + efficiency losses in summer is 11 mm/day.

The time required to infiltrate 100 mm is calculated with the kt^a equation.

$$t = \left(\frac{d}{k}\right)^{1/a} = \left(\frac{100}{19}\right)^{1/0.46} = 37\ h$$

The required infiltration time exceeds 24 hour, which may not be acceptable to the grower. If recession time is one hour, and advance time is 4 hours, then the time available for the storage phase on a 24 hour irrigation cycle is 21 hours. Calculate the depth that infiltrates after 21 hours.

$$d = kt^a = (19)(21\ \text{hr})^{0.46} = 77\ \text{mm}.$$

The required ponding time at the end of the furrow is 21 hours, the advance time is 231 min (4 h), and the recession time is 58 min (1 h). Confirm that the time of cutoff is 24 hours.

$$t_{co} = t_{adv} + IOT_{req} - t_{rec} = 4 + 211 = 24\ \text{hr}$$

The advance and recession curves are shown in Fig. 20.15.

One concern with furrow systems is the wetting pattern. Furrows might need to be closer together in sandy soils if complete wetting is needed.

WinSRFR

WinSRFR (Fig. 20.16) *Event Analysis World* (Fig. 20.17) calculates infiltration k & a with the two-point method.

Select *Furrow* and *Elliot-Walker Two-point* method (Fig. 20.17) in *Event Analysis World*, Enter information on the System Geometry (Fig. 20.18) page: furrow is 400 m long, furrows are 1 m apart, one furrow per set, and 0.002 m/m slope. Select a trapezoidal furrow with bottom width 100 mm.

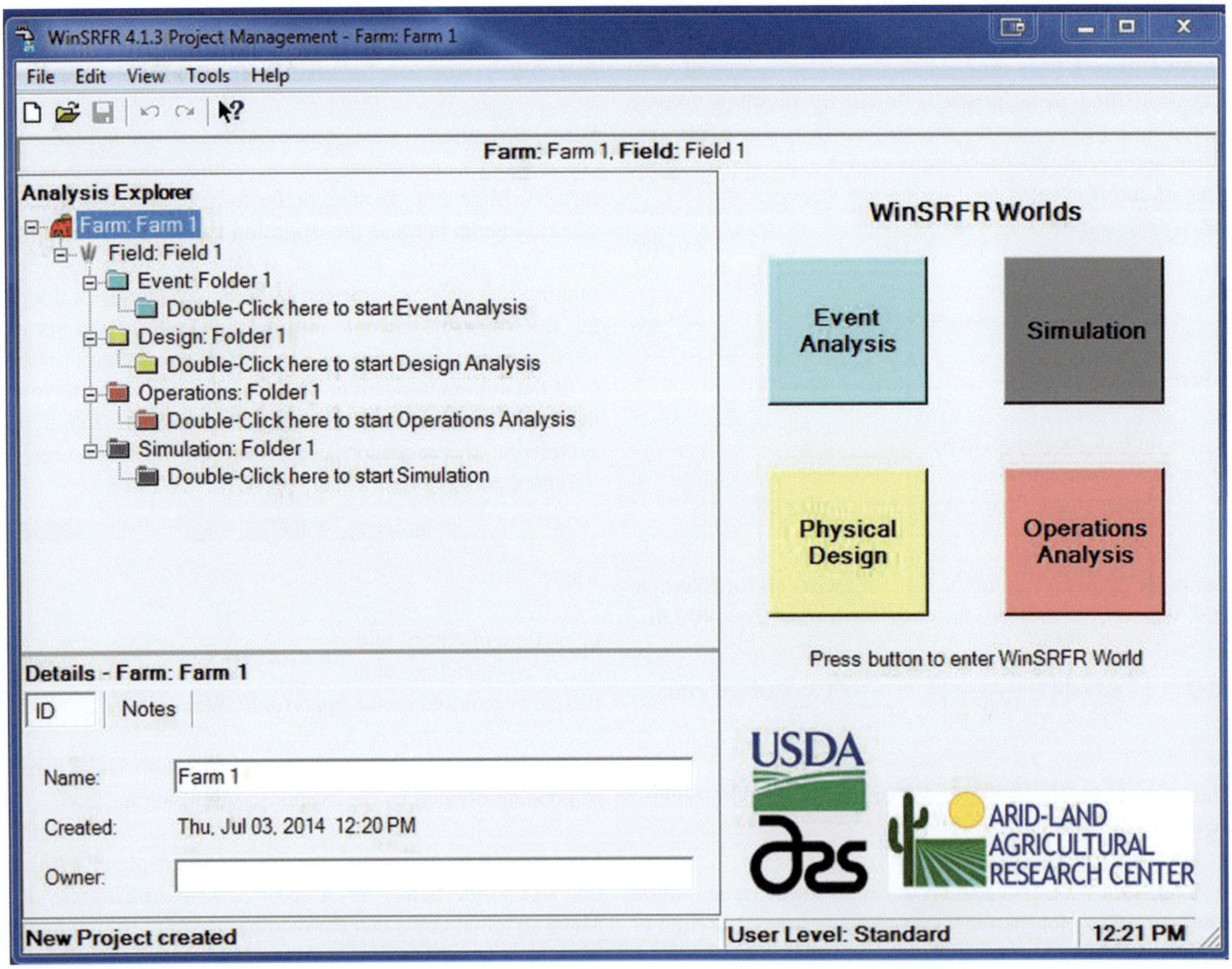

Fig. 20.16 WinSRFR front page

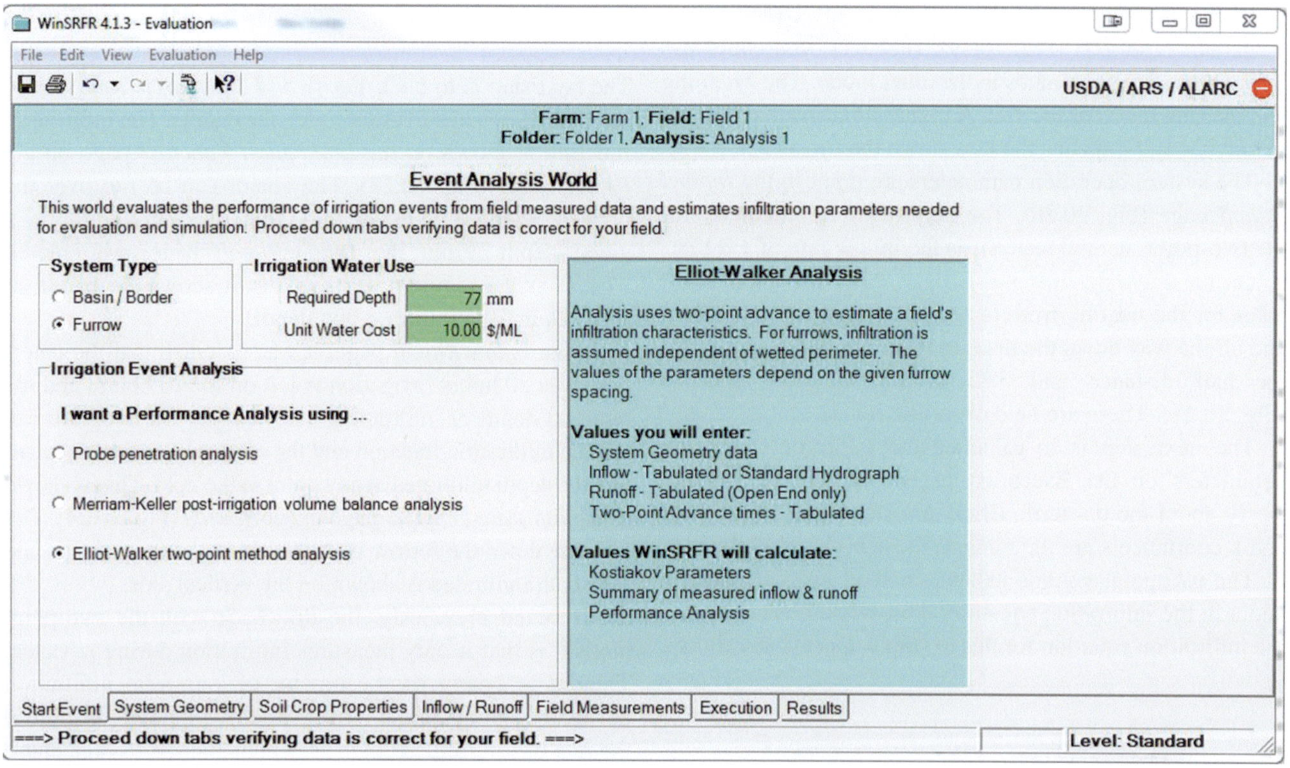

Fig. 20.17 WinSRFR event analysis world

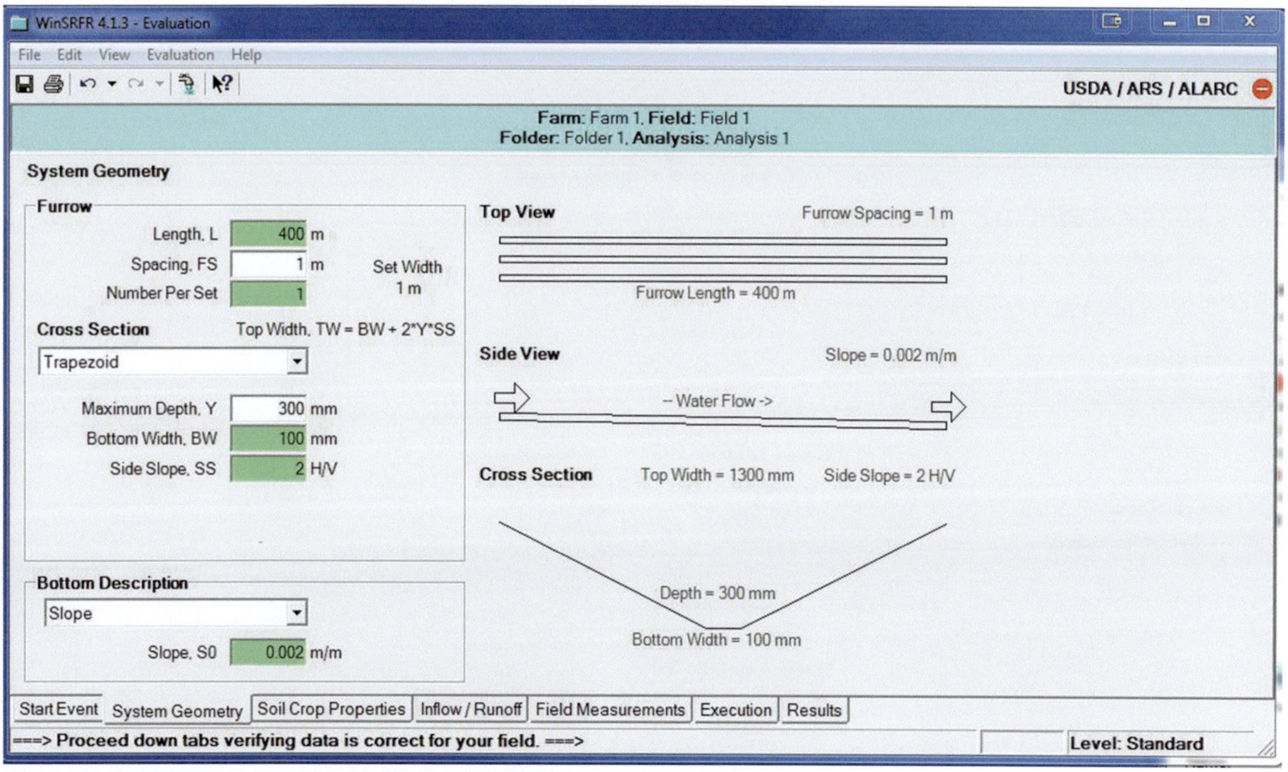

Fig. 20.18 *System geometry* page in WinSRFR event analysis world

In the *Soil/Crop Properties* page, select *User entered value* and type 0.05 (Fig. 20.19). However, you could use one of the defined n values in the other mode. The Manning n is the roughness of the soil. A higher n means that the soil is rougher and impedes the flow down the furrow.

The system operation parameters are input in the *Inflow/Runoff* page (Fig. 20.20). The only required parameter for the two-point method analysis is the inflow rate of 1.0 L/s.

On the *Field Measurements* page, enter the time that it takes for the wetting front to reach halfway down the field and all the way down the field. In this case, enter 1.35 hr for the half distance and 3.85 hr for the full distance (Fig. 20.21). These are field observed values.

The next step is to calculate the infiltration equation parameters on the Execution page (Fig. 20.22). Do not worry about the b – term. Click *estimate a and k* and the a and k coefficients are the same as those in the worksheet.

The two point volume balance method just calculates k and a in the infiltration equation; thus, b and c are zero and the infiltration equation for this example appears as follows.

$$d = kt^a + bt + c = 18.79\, t^{0.47}$$

The next step is to click *Verify and Summarize Analysis* on the Execution page in order to see the results. The infiltration function is shown on the *Infiltration Function* page on the *Results* page (Fig. 20.23). The cursor can be run over the page in order to find the ponding time required to infiltrate a given depth. From the *Inflow/Runoff* page, we entered 77 mm as the required depth. Thus, it shows the length of time required to infiltrate that depth.

In the *Inflow/Runoff* page, it was specified that the cutoff time was 20 hours (irrigation is left on for 20 hours) and the required depth of infiltration was 77 mm. The program can use the infiltration function and the other parameters to calculate the depth infiltrated at any point in the field. This is shown in the *Infiltrated Depths* page under *Results* (Fig. 20.24). The distance down the furrow is shown on the horizontal axis and the depth infiltrated is shown on the vertical axis.

As stated previously, the drawback with the two-point method is that it only measures infiltration during advance. Thus, it is a very poor estimator of long-term infiltration during the storage phase. This method should only be used if the rate of soil infiltration is very high and the storage phase is very short in comparison to the advance time. In addition, the exponential advance curve is only valid if the slope of the field is not flat, such that water does not back up (becomes deeper at the beginning of the field), and if the inflow

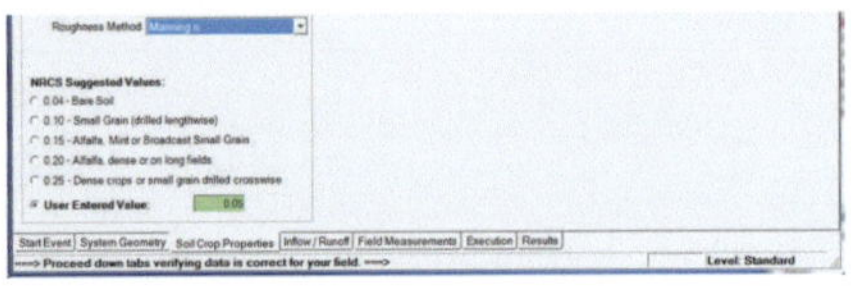

Fig. 20.19 WinSRFR *soil/crop properties* page

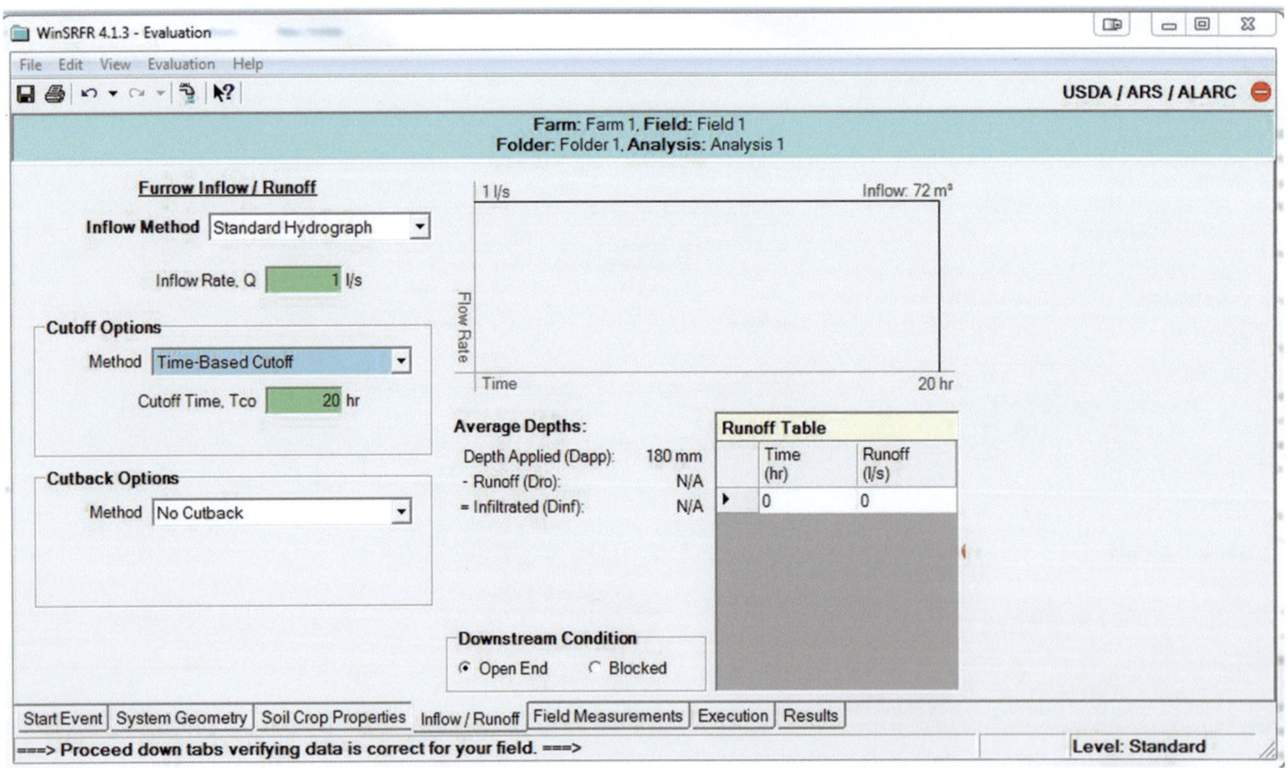

Fig. 20.20 WinSRFR inflow/runoff page

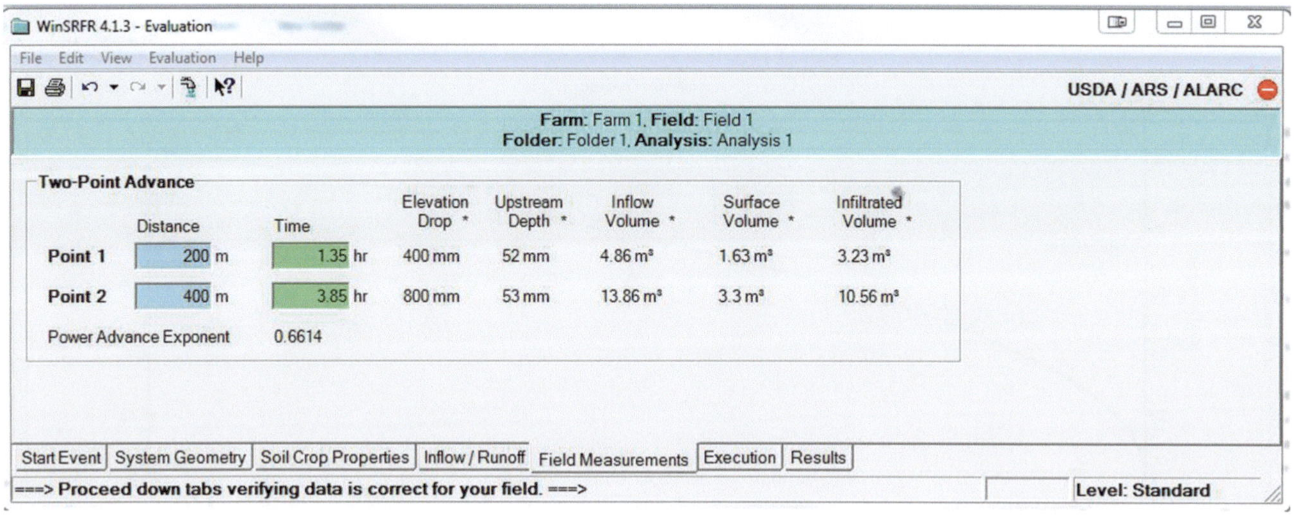

Fig. 20.21 WinSRFR two-point advance page

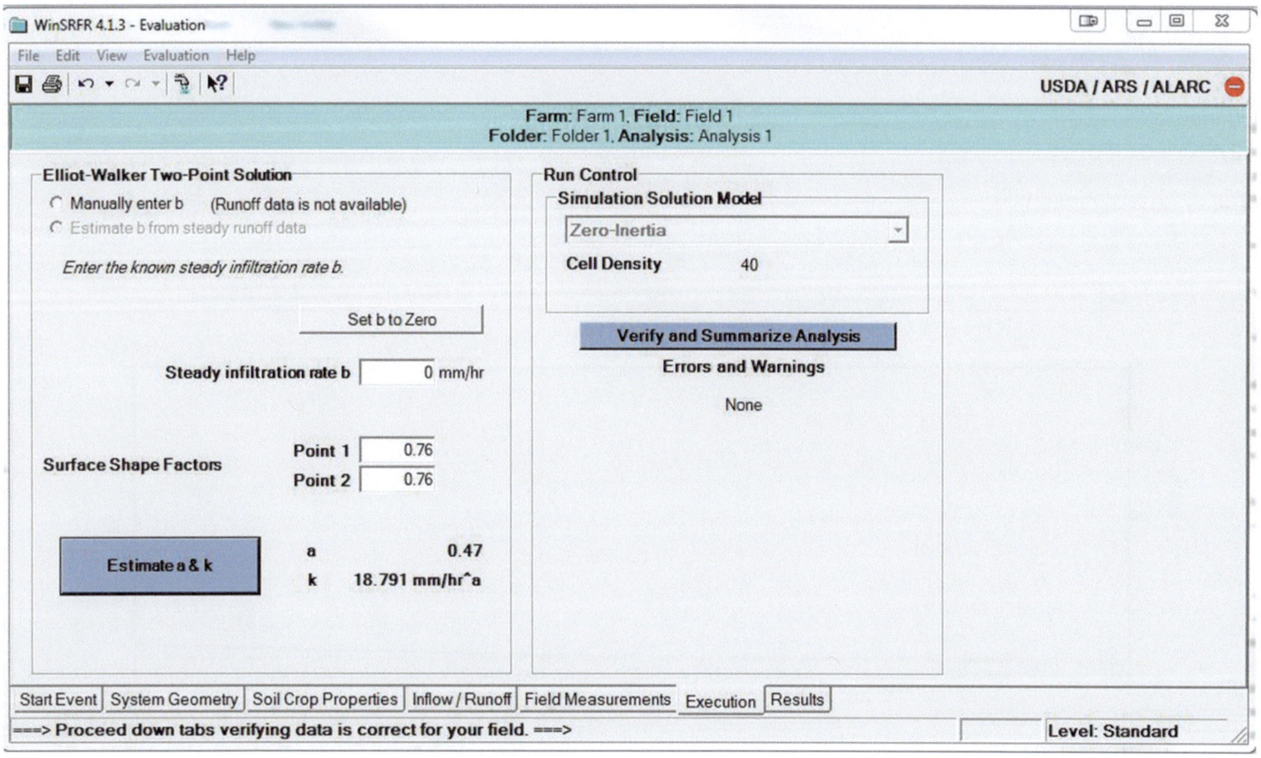

Fig. 20.22 Execution page in WinSRFR event analysis world

remains on during the entire advance phase. For other situations, the *Merriam-Keller post-irrigation volume balance analysis* is more accurate.

Distribution uniformity, DU, and efficiency are the most common parameters used for evaluation of surface irrigation systems. DU is the average of the low quarter infiltrated depths divided by the average infiltration depth over the field. Efficiency can be calculated as follows.

$$Eff = \frac{V_{in} - V_{RO} - V_{DP}}{V_{in}} \tag{20.19}$$

where

V_{RO} = volume of water that runs off the end of the field, m^3,
V_{DP} = volume of water infiltrated that is not used for leaching or ET, m^3.

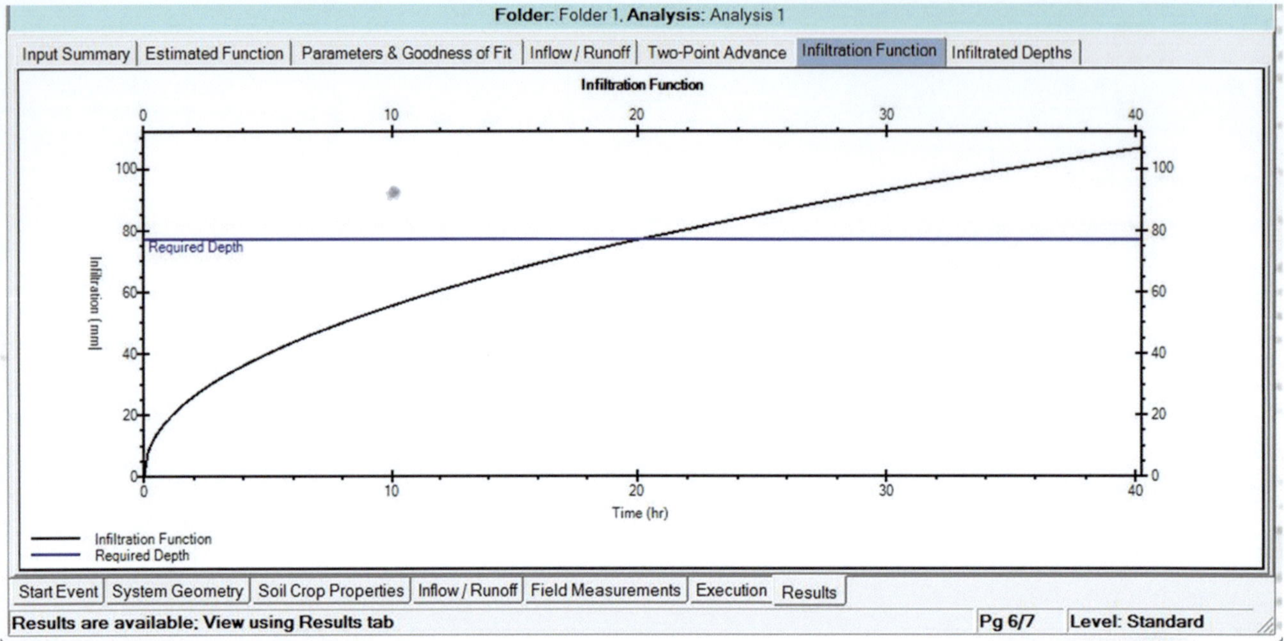

Fig. 20.23 WinSRFR *infiltration function* in *results* page

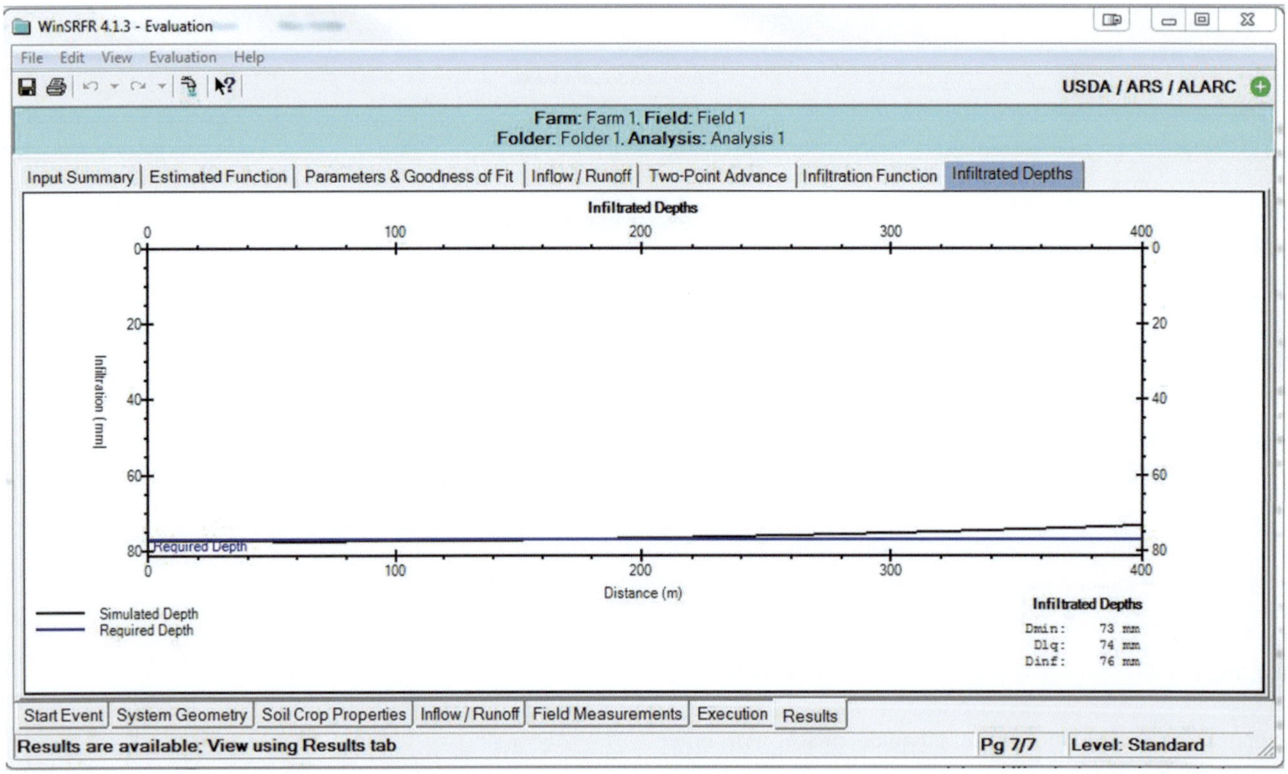

Fig. 20.24 WinSRFR *infiltrated depths* in the *results* page

Typically, the design depth of infiltration is applied to the end of the furrow, and upper sections are overirrigated. Calculation of DU and efficiency require calculation of the volume of

$$V_j = W \Delta x_j d_j \tag{20.20}$$

where

Δx_j = length of furrow represented by point j, m,

d_j = depth of infiltration at point j, m,
V_j = volume of infiltration in furrow section j, m^3.

The intake opportunity time (IOT_j as shown in Fig. 20.6) in any section of the furrow is calculated by subtracting the advance time from the recession time

$$IOT_j = t_{co} + x_j/L\ t_{rec} s x_j^h \tag{20.21}$$

where

x_j = distance to position j from inlet, m.

The depth infiltrated at position j is

$$\begin{aligned} d_j &= kt^a + bt + c \\ &= k\left(t_{co} + x_j/L\ t_{rec} s x_j^h\right)^a \\ &\quad + b\left(t_{co} + x_j/L\ t_{rec} s x_j^h\right) + c \end{aligned} \tag{20.22}$$

The volume of infiltration over the entire furrow is the summation of infiltrated volumes in furrow segments

$$V_{furrow} = \sum_{j=1}^{n} V_j \tag{20.23}$$

where

V_{furrow} = volume infiltrated into the entire furrow, m^3.

The required volume of water infiltration per furrow, V_{req}, is the product of required depth, length, and width of the furrow

$$V_{req} = W\ L\ d_{req} \tag{20.24}$$

where

d_{req} = design irrigation depth, usually depth applied at end of furrow, m,
V_{req} = volume of water required for infiltration at design irrigation depth, m^3.

If the required depth is infiltrated at the end of the furrow (no underirrigation), then the volume of deep percolation for the entire furrow, V_{DP}, is

$$V_{DP} = \sum_{j=1}^{n} V_j - V_{req} \tag{20.25}$$

The runoff volume is the volume applied – volume of infiltration

$$V_{RO} = V_{in} - V_{furrow} \tag{20.26}$$

where

V_{RO} = volume of runoff, m^3.

In this case, short furrow sections will be evaluated for yield and leaching. The volume of water required in each furrow section is

$$V_{j-req} = W\ \Delta x_j\ d_{req} \tag{20.27}$$

The volume of deep percolation in each furrow section, V_{j-dp}, is

$$\begin{matrix} V_j > V_{j-req} & V_{j-DP} = V_j - V_{j-req} \\ V_j < V_{j-req} & V_{j-DP} = 0 \end{matrix} \tag{20.28}$$

Example 20.7 Calculate the distribution uniformity and irrigation efficiency for the furrow described in Example 20.6. A more accurate estimate of the advance time is 3.85 h and the recession time is 0.97 h. The required depth of infiltration is 77 mm.

Equation 20.22 can be used to find the infiltrated depths at each of the 11 positions along the furrow (Fig. 20.25). For example, the infiltrated depth and volume (Eq. 20.20) at position 2 (80 m from the inlet representing the segment from 60 m to 100 m) are calculated as follows.

$$\begin{aligned} d_{80} &= k\left(t_{co} + x_j/L\ t_{rec} s x_j^h\right)^a \\ &= 19\left(24\ h + (80/400)(0.97\ h) - \left(0.0305{*}80^{1.49}\right)/60\right)^{0.46} \\ &= 81.4\ mm \end{aligned}$$
$$V_{80} = W \Delta x_{80}\ d_{80} = 1m{*}40m{*}81.4/(1{,}000\ mm/m) = 3.26\ m^3$$

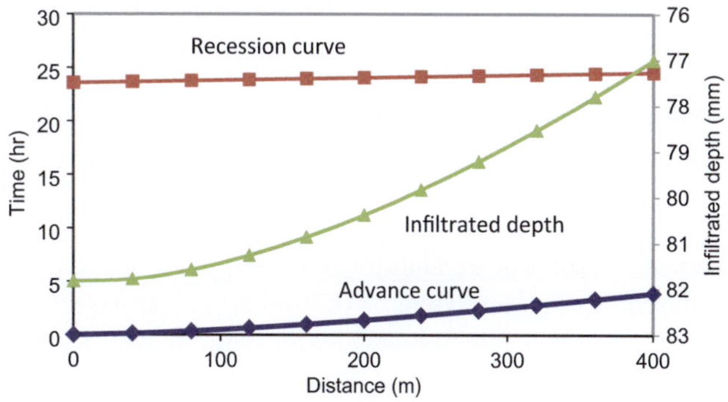

Fig. 20.25 Infiltrated depth versus distance

Table 20.3 Intake opportunity time and infiltrated depth along the furrow

Location (j)	Distance (m)	Advance time (hr)	Recession time (hr)	Opportunity time (hr)	Infiltrated depth (mm)	Inf. volume (m^3)
0	0	0	23.8	23.8	81.7	1.63
1	40	0.12	23.9	23.8	81.6	3.27
2	80	0.35	24.0	23.7	81.4	3.26
3	120	0.64	24.1	23.5	81.1	3.25
4	160	0.98	24.2	23.2	80.7	3.23
5	200	1.37	24.3	22.9	80.3	3.21
6	240	1.80	24.4	22.6	79.7	3.19
7	280	2.26	24.5	22.2	79.2	3.17
8	320	2.76	24.6	21.8	78.5	3.14
9	360	3.29	24.7	21.4	77.8	3.11
10	400	3.84	24.8	20.9	77.0	1.54

Fig. 20.26 Irrigated furrows with multiple siphon tubes (Credit Bert Clemmens, USDA-ARS)

Intake opportunity times, infiltrated depths, and infiltrated volumes in furrow segments are shown in Table 20.3. Infiltrated depths along the furrow are shown in Fig. 20.26.

The low quarter is the average of the infiltrated depths over the last quarter of the field. The following equation can be used where the field is divided into 10 sections:

$$\begin{aligned} d_{LQ} &= (d_8 + d_9 + 0.5\, d_{10})/2.5 \\ &= (78.5\text{ mm} + 77.8\text{ mm} + (77.0\text{ mm})(0.5))/2.5 \\ &= 77.9\text{ mm} \end{aligned}$$

The last term d_{10} is multiplied by 0.5 because it only represents half of a section.

The average depth infiltrated for the entire furrow, d_{ave}, is

$$d_{ave} = (0.5\, d_0 + d_1 + d_2 + d_3 + d_4 + d_5 + d_6 + d_7 + d_8 + d_9 + 0.5 d_{10})/10 = 80.0\text{ mm}$$

Distribution uniformity is the low quarter divided by the furrow average

$$DU = d_{LQ}/d_{ave}(100\%) = 77.9\text{ mm}/80.0\text{ mm} = 97\%.$$

Total volume of infiltration is the total depth infiltrated over the length of the furrow. The sum of volumes, V_j, in the last column in Table 20.3 is 32.0 m^3.

Calculate deep percolation

$$\begin{aligned} V_{req} &= W\,L\,d_{req} = (1\text{ m})(400\text{ m})(77\text{ mm}/1{,}000\text{ mm/m}) \\ &= 30.8\text{ m}^3 \\ V_{DP} &= \sum_{j=1}^{n} V_j - V_{req} = 32.0\text{ m}^3 30.8\text{ m}^3 = 1.2\text{ m}^3 \end{aligned}$$

Calculate applied volume

$$V_{in} = Q\,t_{co} = (0.001\text{ m}^3/\text{ sec})(24\text{ hr})(3{,}600\text{ sec}/\text{hr}) = 86.4\text{ m}^3$$

Calculate runoff volume

$$V_{RO} = V_{in} - V_{furrow} = 86.4 - 32.0 = 54.4\ m^3$$

Calculate the deep percolation percentage

$$DP\% = V_{DP}/V_{in}*100\% = 1.2/86.4*100\% = 1.4\%$$

Calculate the runoff percentage

$$RO\% = V_{RO}/V_{in}*100\% = 54.4/86.4 = 63\%$$

The high runoff percentage will lead to wasted water unless one of two solutions is implemented: cutback irrigation or tailwater reuse.

Flow rate can be reduced (cut back) once the wetting front reaches the end of the furrow in order to reduce runoff at the end of the furrow. For example, furrows irrigated with two siphon tubes can be cut back to one siphon tube after flow has reached the end of the furrow This technique is called cutback irrigation.

One constraint on cutback irrigation is that the cutback flow rate must be great enough to satisfy all of the infiltration needs of the furrow. If infiltration over the furrow is greater than cutback flow rate, then water recedes from the end of the furrow. The volume infiltration rate at the time that the water reaches the end of the furrow can be calculated as in the following equation (Clemmens et al. 2006); however, in the example in this text, a spreadsheet is used to calculate the infiltration at each position in the field.

$$Q_L = WL\left[b + h^{(a-1)}kt_L{}^{(a-1)}\right]/3,600 \qquad (20.29)$$

where

Q_L = volume infiltration rate, L/s.
t_L = time of advance to end of furrow, hr,
W = furrow spacing, m,
L = furrow length, m.

Equation 20.29 can also be used to find the volume infiltration rate at any time during the advance phase by substituting x for L and t_x for t_L where x is the distance from the furrow inlet to the wetting front.

As with calculation of uniformity and efficiency, numerical integration can be used to find the average infiltration rate over the field during the storage phase: volume infiltration rates are calculated for field segments and summed. The infiltration rate at any point, j, at time t is

$$i_j = ak\left(tsx_j^h\right)^{a-1} + b \qquad (20.30)$$

Once the flow is cut off, the depth of water in the furrow is reduced so the wetted perimeter and thus the infiltration rate decreases. Thus, the required intake opportunity time at the end of the furrow or depth of infiltration must be adjusted.

Example 20.8 Determine the cutback time and flow rate for Example 20.7 with 1.0 LPS initial flow rate if the cutback flow rate is half of the initial flow rate. Determine depth of infiltration.

Half of the initial flow rate (1.0 L/s) is 0.5 L/s. However, the volume infiltration rate is 0.82 L/s so cutback cannot take place at the time that water reaches the end of the furrow. Volume infiltration rate can be calculated during the storage phase using numerical integration in a spreadsheet (Table 20.4). For example, the average infiltration rate at the 80 m position 33 minutes after advance has reached the end of the field is

$$i_{80} = ak(t - sx_{80}{}^h)^{a-1} + b = 0.46 * 19((231+33-0.0305 * 80^{1.49})/60)^{0.46-1} + 0 = 4.11 \text{ mm/h}$$

Table 20.4 Volume infiltration rate vs. time after advance reaches end of furrow.

Distance (m)	Advance time (min)	IOT at time t_L (min)	t_L	Infiltration rate (mm/h): 10 min after t_L	33 min after t_L	62 min after t_L	80 min after t_L	86 min after t_L
0	0	231	Calculate Q_L at time t_L (this column) with equation 20-29. Columns to the right (times after t_L) calculated with numerical integration in spreadsheet	4.13	3.93	3.71	3.60	3.56
40	7	223		4.20	3.99	3.76	3.64	3.61
80	21	210		4.34	4.11	3.86	3.73	3.69
120	38	192		4.53	4.28	4.00	3.86	3.82
160	59	172		4.80	4.51	4.19	4.03	3.98
200	82	149		5.17	4.81	4.43	4.24	4.19
240	108	123		5.69	5.22	4.75	4.52	4.46
280	136	95		6.46	5.81	5.19	4.90	4.81
320	165	65		7.73	6.70	5.82	5.42	5.30
360	197	34		10.39	8.28	6.79	6.19	6.02
400	231	0		23.00	12.11	8.56	7.47	7.20
Ave infiltration in furrow (mm/hr)				6.69	5.57	4.89	4.61	4.52
Total infiltration in furrow (L/s)			0.82	0.74	0.62	0.54	0.51	0.50

Thus, the flow can be cut back 86 minutes after water reaches the end of the furrow (231 + 86 minutes = 5.28 hr. The new cross-sectional area of flow and the wetted perimeter are calculated with Manning's equation. The new flow depth is 3.77 cm, and the new wetted perimeter is

$$P = b + 2\,y(1+z^2)^{0.5}$$
$$= 0.1\text{ m} + (2)(0.0377\text{ m})(1+2^2)^{0.5} = 0.269\text{ m}.$$

Adjust the infiltration rate. $\dfrac{0.269\text{ m} + 0.213\text{ m}}{1.0\ m} = 0.482$

Clemmens *k* and *a*, from Example 20.2, are 34.45 and 0.462, respectively. Adjust *k* by multiplying by 0.482. No change is required for a.

$$\text{Adjusted } k = 0.482(34.4) = 16.6(\text{mm/hr}).$$

Thus, if water is not backed up at the end of the furrow in order to maintain wetted perimeter, then there is a potential miscalculation associated with cutback irrigation. The ratio of the reduced infiltration rate to the infiltration rate at the original flow (1.0 L/s) is 16.6 / 19 (100 %) = 87 %. Thus, the depth of infiltration at the end of the furrow is only 87 % of the original calculated depth.

$$77\text{ mm }(0.87) = 67\text{ mm}.$$

Thus, water should be slightly ponded at the end of the furrow in order to avoid this problem with reduced infiltration. Assuming that infiltration is sufficient at the end of the furrow, then the volume of flow/furrow for cutback operation, would be (5.3 hr + 18.7/2) /24 hr * 100 % = 61 % of the no cutback volume. Power costs would also be 61 % of the no cutback power. If infiltration is reduced, then the *cutback* worksheet has an option to calculate the cutback time based on the reduced wetted perimeter in cells M1:S5. These cells are based on equations that are not included in this document.

Reuse Systems

An effective method for improving irrigation efficiency is to use a runoff reuse system. Runoff water is pumped from a tail water recovery point (Fig. 20.27) back up to the head ditch (Fig. 20.28) and reused. In this way, water that was lost to runoff is reused.

The number of furrows that can be irrigated with a reuse irrigation system is

$$F = N\left(\frac{Q_s\left(^{E}ff_{ds}/_{100}\right)}{Q_f} + Q_r\left(\frac{^{E}ff_{dr}/_{100}}{Q_f}\right)\right) = N(n_s + n_r) \tag{20.31}$$

where

N = number of sets,
F = number of furrows,
Q_s = supply flow rate, L/s,
Q_f = individual furrow flow rate, L/s,
Q_r = pumped flow rate in reuse pipeline, L/s,
n_s = number of furrow irrigated with supply water,
n_r = number of furrows irrigated with reuse water.
Eff_{ds} = supply delivery system efficiency, %,
Eff_{dr} = reuse delivery system efficiency, %,

The calculation of reuse flow rate based on supply flow rate is

$$Q_r = Q_s(Eff_{ds}/100\%)(Eff_{rcs}/100\%)(RO\%/100\%) \tag{20.32}$$

where

Eff_{rcs} = efficiency of reuse collection system and reservoir, %.

Substitute Eq. 20.32 into Eq. 20.31.

$$F = N\frac{Q_s\left(^{E}ff_{ds}/_{100}\right)}{Q_f}\left(1 + \left(\frac{RO\%}{100}\right)\left(^{E}ff_{dr}/_{100}\right)\left(^{E}ff_{rcs}/_{100}\right)\right) \tag{20.33}$$

It is more convenient to operate a runoff reuse system with a large reservoir size that holds a one to two set (zone or daily irrigation volume) supply of water. Then, the reuse pump can operate continually. For design purposes, the reservoir should be considered full at the beginning and end of the irrigation event. This is an appropriate assumption during the peak irrigation season, when the irrigation cycle must begin again as soon as the last cycle is completed. For periods of the year when less frequent irrigation is required, then the reservoir may be empty at the beginning and end of the irrigation cycle. In this case, an extra day is required for the irrigation cycle because the first day of the cycle is only irrigated with the supply source water, Q_s, as the reservoir is filled. The last day is only irrigated with the reuse water, Q_r, as the reservoir is drained.

Side slopes of on-farm reservoirs should be 2:1–2.5 to 1 with a very shallow slope (5:1) at one end for cleaning and

Fig. 20.27 Tailwater recovery reservoir with pumping unit (Credit USDA-ARS)

as an emergency exit. The reservoir should be designed with 30 cm extra depth because sediment builds up in the bottom of the reservoir between cleanings.

Example 20.9 Part 1: Design a runoff recovery system, and determine the required supply flow rate, Q_s, to the head ditch for the furrows described in Example 20.6 and 20.7 (s = 0.0305, h = 1.49). Assume that there is no cutback during the irrigation event. The field width is 800 m wide. The ET_c is 11 mm/day. The delivery efficiency of the head ditch is 90 %. The collection system efficiency of the runoff recovery system, prior to water reentering the head ditch, is 87 % (3 % lost to evaporation and 10 % lost to seepage in the tailwater ditch and reservoir). Assume that the reservoir is full at the beginning and end of the irrigation cycle. The field requires 1 m depth of irrigation per year.

Part 2: Determine whether a reuse system should be installed if the required rate of return is 8 %, and the project length is 20 years. Compare to the cost of cutback (Example 20.8) and no reuse system.

Supply source water is pumped from a well with a dynamic water table depth of 50 m, and the well pump efficiency is 80 %. Reuse water is pumped a distance of 1,200 m, and the difference in elevation between the head ditch inlet and the reuse reservoir water surface is 3 m. Reuse pump efficiency is 70 %

There is an environmental cost of $1/ha-cm for runoff water discharged to surface streams. Assume that cutback runoff percentage (RO%) is 35 % and the runoff with no cutback is 63 %.

Cost of energy is $0.10/kW-hr.

Assume that the cost of the reuse pump and valves is $5,000. However, the cost of the well pump decreases by $2,000 because a lower flow rate is required for the reuse system. The cost of reservoir excavation is $1.00/m^3. The cost of trenching for the reuse pipe is $1.00/m.

Part 1 Solution : Design of reuse system.

The furrow is designed to apply 77 mm, d_{req} at end of furrow, per irrigation. Thus, the interval between irrigation events is 7 days, and the field should be divided as shown in Fig. 20.18, except that all of the zones would be the same size. The number of zones is

$$N = (77\ \text{mm})/(11\ \text{mm/day}) = 7\ \text{sets}$$

The field is 800 m wide, and furrows are 1 m wide so there are 800 furrows, F.

The reuse delivery efficiency, Eff_{dr}, is the same as the pumped water delivery system efficiency since reuse water is added to the head ditch at the same point, 0.9. The reuse collection system efficiency, Eff_{rcs}, is 0.87. The required system flow rate, Q_s, with the reuse system is found by rearranging Eq. 20.33.

$$Q_s = \frac{FQ_f}{N\left({}^{E}ff_{ds}/_{100}\right)\left(1 + \left(\frac{RO\%}{100}\right)\left({}^{E}ff_{dr}/_{100}\right)\left({}^{E}ff_{rcs}/_{100}\right)\right)}$$

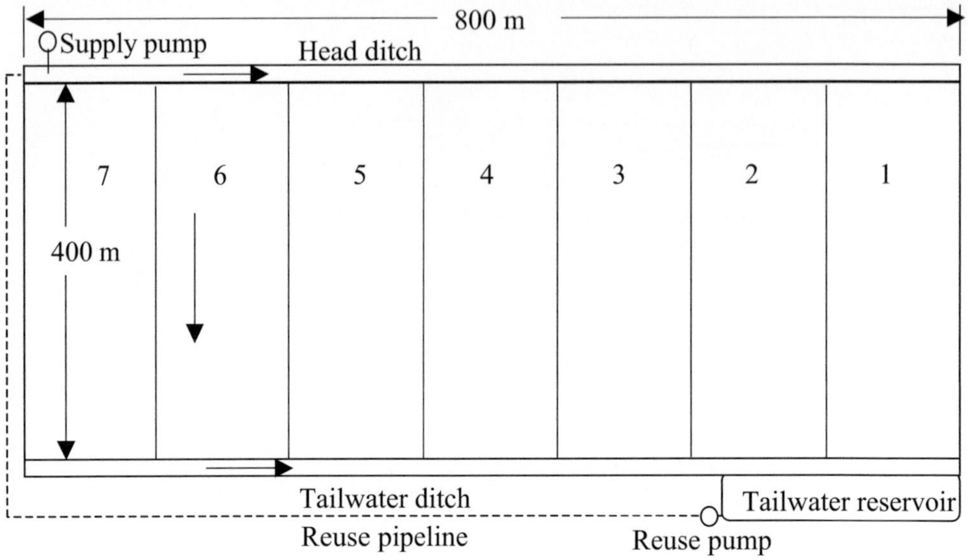

Fig. 20.28 Tailwater recovery system schematic

$$Q_s = \frac{800*1.0\ L/s}{7(0.9)(1+(0.63)(0.9)(0.87))} = 85\ L/s$$

$$Q_r = Q_s(Eff_{ds}/\ 100\%)(Eff_{rcs}/100\%)\ (RO\%/100\%) = 85(0.9)(0.87)(0.63) = 42\ L/s$$

Check results with Eq. 20.31

$$F = 7\left(\frac{85(0.9)}{1} + \frac{42(0.9)}{1}\right) = 800\ furrows$$

Number of furrows irrigated per day is 800 furrows/7 sets = 114 furrows per irrigation. Supply system flow rate, Q_s, without reuse is 114 furrows (1 L/s-furrow)/0.9 = 127 L/s.

Let the reservoir hold a 1-day water supply (48 = 42/0.87).

$$48\ L/s\ (24\ hr)(3,600\ sec/hr)/1,000\ L/m^3 = 4,000\ m^3$$

If the reservoir is 2 m deep, then the average reservoir storage area is 2,000 m^2. Average dimensions of 100 m x 20 m would be adequate. If the side slopes are 2:1, and the bottom width is 16 m, then the top (water surface) width would be 24 m, for an average width of 20 m. Likewise, the bottom length should be 93 m and the top length (water surface) should be 107 m for an average length of 100 m. The top of the reservoir should be 0.3 m higher to account for filling of the bottom and an additional 0.3 m for freeboard. Thus, the width of the reservoir with a 2:1 side slope would increase by 1.2 m to 25.2 m. The length would increase by 0.3 m (2) + 0.3 m (5) = 2.1 m. Thus, the length of the reservoir is 109.1 m.

Part 2: Cost

Present value analysis (based on present value of energy costs and capital cost of pipe) was used to select a 206 mm (8 in. pipe. A low pressure rating (Cl 100) can be used because the maximum pressure in the pipe is below 15 m, because the velocity is below 1.5 m/s, and because water hammer will not occur if there is not a valve at the discharge end. The cost of 206 mm pipe is $8.00 per m so the cost of the 1,200 m reuse pipeline is $9,600. The pressure loss at 42 LPS flow rate is 8.57 m.

The irrigation time per year is found by dividing the annual required depth by the depth delivered at the end of the furrow by each irrigation

$$1,000\ mm/(77\ mm/irrigation) = 13\ irrigation\ cycles$$
$$(13\ irrigation\ cycles)(7\ irrigations/cycle)(24\ hours/irrigation) = 2,184\ hr$$

Pumping costs:

There is a lift of 3 m between the reservoir surface and the head ditch. The present value of reuse system pumping (Q_r = 42 L/s) is

$$P_r = Q_r\rho gh/Eff$$
$$P_r = 0.042(1,000kg/m^3)(9.8m/\sec^2)(3+8.57m)(0.001kW/W/0.7) = 6.8kW$$
$$E_r = 6.8kW(2,184hr) = 14,480kW-hr \rightarrow \$1,448/yr \rightarrow PV = \$14,200$$

The present value of the supply system pumping during reuse (85 L/s) is

$$P_s = Q_s \rho gh/Eff$$
$$P_s = 0.085(1,000kg/m^3)(9.8m/\sec^2)(50m)$$
$$(0.001kW/W/0.8) = 52kW$$
$$E_s = 52kW(2,184hr)$$
$$= 113,700kW - hr \rightarrow \$11,370/yr \rightarrow PV = \$111,600$$

The present value of pumping without reuse or cutback (Q_s = 127 L/s) is

$$P_s = Q_s \rho gh/Eff$$
$$P_s = 0.127m^3/s(1,000kg/m^3)(9.8m/\sec^2)(50m)$$
$$(0.001kW/W/0.8) = 79kW$$
$$E_s = 79kW(2,184hr)$$
$$= 170,000kW - hr \rightarrow \$17,000/yr \rightarrow \$166,900$$

The present value of pumping with cutback flow is 61 % of the total without cutback.

$$Cutbackcost = \$166,900*0.61 = \$102,000$$

Environmental costs:
Cost of runoff water from the no reuse, no cutback system is

$$V_s = 0.127m^3/s(3,600s/hr)(24hr/irrigation)(13irrigations)$$
$$(0.01m^3/ha - cm) = 1,426ha - cm$$
$$V_{RO} = Q_s(Eff_{ds})(RO\%) = 1,426ha - cm(0.9)(0.63)$$
$$= 809ha - cm/yr \rightarrow \$809/yr \rightarrow PV = \$7,900$$

The cost of runoff with the cutback system (1/2 of no cutback) is PV = \$7,900/2 = \$3,950.

Capital cost:
The capital cost of the reuse system cost includes the cost of the reuse pump (minus the \$2,000 saved on the supply pump), pipeline, trenching, and excavation of the reservoir. The volume of the reservoir is 2.3 m/2 m * 4,000 m^3 = 4,600 m^3. The cost of excavation is

$$\$1.00/m^3*4,600m^3 = \$4,600.$$

The length of the pipeline is 1,200 m. At \$1/m, the cost of trenching the pipeline is \$1,200. Total capital cost of the reuse pipe system, pump, and reservoir is

$$\$4,600 + \$1,200 + \$9,600 + \$3,000 = \$18,400.$$

Labor costs:
Labor costs will increase with the cutback system because water will be cutback after 5 hours. When water is cutback from two furrows, then another furrow is started. Thus, irrigation sets will be staggered. It will also be more critical to monitor the end of the furrows in order to determine whether the runoff rate is adequate. Assume that an irrigator will be required for 2 hours per day with the reuse and no cutback/no reuse systems (2 hr * 13 irrigations * 7 zones * \$10/hr = \$1,820/yr → PV = \$17,900). Assume that an irrigator is required for 4 hr/day for the cutback system (4 hr * 13 irrigations * 7 zones * \$10/hr = \$3,640/yr →PV = \$35,700

Present values of costs are compared in Table 20.5

The cutback system has the lowest cost; however, the reuse system may be preferable since there is uncertainty regarding the infiltration rate at the end of the furrow with cutback flow.

Erosive Flow Velocity

In the case of Example 20.7, the distribution uniformity was high because the advance ratio (ratio of advance to storage phase) was low. However, in fields with high infiltration rates and long furrows, the advance ratio, the distribution uniformity will be low. One alternative is to increase the furrow flow rate so that the advance time is reduced. However, increasing furrow velocity can cause erosion. Higher velocity flow has more shear stress at the soil surface and more intense turbulent eddies. Shear stress pick up particles, and eddies keep particles suspended in the stream. The relative magnitude of eddy drag forces and gravitational forces on particles determines whether particles will remain entrained in the stream or settle to the furrow bottom. The maximum nonerosive flow velocity for erosive and non-erosive soils is 8 and 13 m/min, respectively. Sandier soils that do not form soil aggregates are more easily eroded by the furrow irrigation stream. Thus, soils that have a significant percentage of sand (sandy soils) tend to be classified as erosive. Other soils, such as unconsolidated silty soils, are known for severe erosion problems. With these soils, addition of polyacrilymide (PAM) to the irrigation water has been shown to onsolidate soil particles at the soil surface and prevent erosion.

Surge Irrigation

In high infiltration rate soils, one technique that has been used successfully to decrease the depth of water applied during the advance phase is surge irrigation. A surge valve is used to divide the water between two halves of the field. The surge valve cycles water between one side of the field and the other. With surge irrigation, water advances down a furrow and then infiltrates after the water is turned off. In medium to coarse textured soils, the surface tends to consolidate and seal up once the water is turned off. When water is

Table 20.5 Present value of costs for different irrigation systems

System	No cutback, no reuse	Cutback	Reuse
Energy	\$166,900	\$102,000	\$111,600 + \$14,200
Labor	\$17,900	\$35,700	\$17,900
Environment	\$7,900	\$3,950	0
Capital	0	0	\$18,400
Total	\$192,700	\$141,650	\$162,100

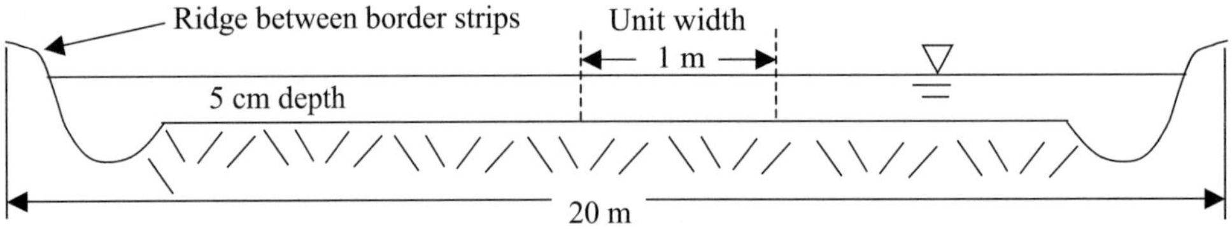

Fig. 20.29 Typical dimensions for border strip in direction perpendicular to flow

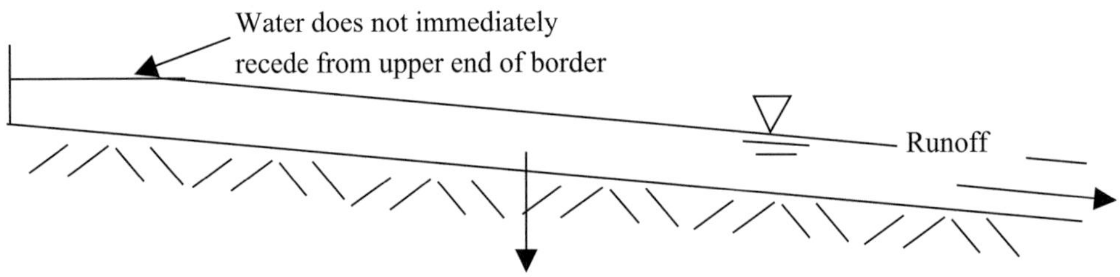

Fig. 20.30 Ponding on border strip after cutoff

reintroduced in the furrow, infiltration is low, and the advance phase is fast. With surge irrigation, the advance time can be nearly as fast as continuous flow, but use half as much water.

Border Strip Irrigation

Graded border strips are strips of flat planted crops, which are level in the direction perpendicular to flow and sloped in the direction of flow. They are confined by small ridges on either side (Fig. 20.29). Water runs down the border as a sheet. WINSRFR has a border strip option.

The advance curve for border strip irrigation is calculated in the same way as for furrow irrigation with $\sigma_y = 0.7$. However, calculation of the recession curve is more complex because, unlike furrows, recession does not occur immediately in a border strip (Fig. 20.30).

The upper end remains ponded because there is a large volume of water on top of the soil that does not immediately infiltration. As with furrows, recession is caused by the combined effects of infiltration and runoff. After cutoff, the water volume of water continues to move down the border strip such that the wetting front continues to move toward the end of the border. Thus, the time of cutoff can take place before the wetting front reaches the end of the border. Because water continues to move down the border after cutoff, the infiltration distribution is often better than that observed with furrow irrigation. In fact, the border strip can be designed so that the recession curve is parallel to the advance curve such that infiltration over the border strip is nearly uniform. Nearly level borders are known for high efficiency.

Nonuniformity in soil surface slope and sideslopes in the border strip can dramatically reduce uniformity. Just a few centimeters variation in slope can lead to dramatic variation in infiltrated depth because low areas receive ponded water for a longer period. A sideslope can cause water to primarily run down one side of the border such that one side of the border receives very little water.

Level Basin and Level Border Irrigation

Level basins are large flat areas. As such, the goal is to move the water across the border as fast as possible and then to let the water pond for the required length of time. A typical design flowrate for a square 15 acre (7 ha) level basin is 5 cfs

($8.5\ m^3$/min). The water is then allowed to infiltrate over time. Well-designed level basins can have irrigation efficiencies of 90 %.

Level borders are 1.0 mile (1.6 km) long at the J.G. Boswell Ranch in California. Huge axial pumps (12,000 LPM) mounted on caterpillar tractors pump canal water over levees. The water sits in the border and excess can be returned to the canal. The reason that that these long border strips have uniform irrigation is that the soil is swelling clay. Once the cracks close after approximately 5 minutes, infiltration essentially stops.

One type of level border system that is becoming more popular due to high efficiency and low labor requirement is the drain back system. In these systems, level borders are fed by a common canal/ditch at one end, when one border is finished water is allowed to drain back into the ditch, and this water is used to irrigate the next borders down the ditch.

Questions

1. Describe the three phases of surface irrigation.
2. Answer the following questions true or false.
   (a) Uniformity is generally high if the storage phase is relatively small in comparison to the advance phase.
   (b) The advantage of the two-point method is that the infiltration rate during the storage phase can be extrapolated from the infiltration rate calculated during the advance phase.
   (c) The Kostiakov equation includes steady state infiltration.
   (d) The two-point volume balance method can be expanded to find the coefficient b in the steady state term by using another point.
   (e) The vertical infiltration rate as calculated from a double ring infiltrometer can be adjusted for furrow irrigation by taking the width of the furrow divided by the distance between furrows.
3. Using the Merriam and Clemmens approach, calculate k and a if the time to infiltrate 100 mm is 8 hours. Make the calculation by hand and in the *Infiltration* worksheet. Compare to the closest NRCS (SCS) curve number: the curve that is closest to 100 mm over 8 hours.
4. Flow rate = 2.0 L/s, z = 1.7, b = 0.2 m, Manning's n = 0.05, furrow slope = 0.001 m/m. Calculate the depth of flow by hand and with the *Furrow* worksheet. Using the Clemmens k and a values from question 6, adjust the k value for the wetted perimeter and a furrow spacing of 1 m. Make calculations by hand and in furrow worksheet.
5. Using the information from question 4, calculate s and h and plot the advance curve for a 500 m long furrow with 1 m spacing between furrows. Inflow rate is 2.0 L/s. Use a convergence criterion of less than 1 min difference for advance time between iterations. Make calculations by hand and with the furrow worksheet.
6. Find the time of cutoff for the parameters in question 5. The depth required is 76 mm. You should calculate a total irrigation time of approximately 12 hr.
7. Use the Furrow times worksheet to find the depth of infiltration every 50 m down the furrow for the parameters in questions 5–6. Calculate the depth of infiltration at 150 m by hand and compare to the worksheet. Calculate the DU LQ by hand and compare to the value in cell G33. By hand, calculate the DP%, RO%, and efficiency based on applied volume and infiltrated volume reported in the worksheet. In this question, do not use cutback irrigation. Adequately irrigate the entire field (minimum required depth is applied to end of furrow).
8. For the parameters in questions 5–7, observe how the parameters for cutback irrigation are calculated in the cutback worksheet. Can cutback take place as soon as water reaches the end of the furrow? What will be the cutback flow rate (half of initial flow rate). Find the following parameters and report the cell number in which they are found: the average infiltration at the end of the furrow when water reaches the end of the furrow, the volume infiltration rate when water reaches the end of the furrow (LPS), the average infiltration rate (mm/hr) 10 minutes after water reaches the end of the furrow, and the time when cutback can take place. Compare the total depth applied to the depth applied without cutback, and the RO%, DP%, and efficiency
9. Calculate an irrigation schedule for the parameters in previous questions for peak ETc 10 mm/day. Leaching fraction is 10 %.
10. Design a runoff-recovery system for the parameters in questions 3–7 and 9. Do not worry about economic analysis or comparison to cutback irrigation unless asked by instructor. Determine the required supply flow rate, Q_s, to the head ditch. The field width is 800 m wide. The delivery efficiency of the head ditch is 90 %. The collection system efficiency of the runoff recovery system, prior to water reentering the head ditch, is 87 % (3 % lost to evaporation and 10 % lost to seepage in the tailwater ditch and reservoir). Assume that the reservoir is full at the beginning and end of the irrigation cycle.
11. The required intake opportunity time is 17 hours, advance is 2 hours, and recession time is one hour. What is the t_{co}?
12. Volume of deep percolation is 20 m^3, volume of runoff is 25 m^3, and volume used in the soil profile is 60 m^3. What is the irrigation efficiency, inflow volume, deep percolation percentage, and runoff percentage?

13. The average depth of infiltration in a field is 100 mm, and the average depth of infiltration over the last 25 % of the field is 90 mm. What is the DU LQ?
14. Why can't one just increase the flow rate to any velocity in order to get water across the field as quickly as possible?
15. Why does surge irrigation improve irrigation efficiency in some soils?
16. This question and the following questions use WinSRFR. Download WinSRFR onto your computer and copy a screen showing that it is open on your computer.
17. Put the data in Figs. 20.16, 20.17, 20.18, 20.19, 20.20, 20.21 and 20.22 into WinSRFR and get the same result shown in the chapter. Copy the data from the Estimated Function page and paste to this document. Include Two point advance per furrow data and Kostiakov a and b data.
18. Modify the previous question by putting in an advance time to 2 hours to the midpoint in the field and 5 hours to the end of the field. Copy the data from the Estimated Function page.
19. Modify the previous questions by changing the furrow bottom width to 200 mm and the slope to 0.003. Copy the data from the Estimated Function page.
20. Modify the furrow side slope to 1 and do a screen capture of the System geometry page, showing the modified furrow shape.
21. Change the furrow shape to the Power Law shape and use exponent M = 0.4. Do a screen capture of the System Geometry page, showing the modified furrow shape.
22. Change the Manning n on the Soil/Crop Properties page to that for alfalfa, mint, or broadcast small grain. Write down this value as the answer.

References

Clemmens AJ, Walker WR, Fangmeier DD, Hardy LA (2006) Design of surface systems, chapter 14. In: Glenn H (ed) ASABE irrigation monograph. American Society of Agricultural and Biological Engineering, St. Joseph

Merriam JL, Clemmens AJ (1985) Time rated infiltrated depth families. In: Development and management aspects of irrigation and drainage systems. ASCE, pp 67–74

21 Hydroponic Irrigation Systems

Hydroponic irrigation systems generally utilize drip irrigation with soilless growing media such as sand, rock wool or coco-coir. The drip systems have online barbed drip emitters and deliver water through distribution tubing to the media. With their closely spaced barbed fittings, the friction loss calculation for hydroponic drip laterals must include the local losses due to barbed fittings. Irrigations are generally triggered by a light sensor placed above the greenhouse. The duration of the watering cycle is a few minutes. The irrigation cycle not only adds water, but also adds oxygen to the growing media. All nutrients, including micronutrients and oxygen, must be supplied by irrigation. Fertigation calculations for hydroponic systems are generally made on a concentration basis. Chemical is either mixed into the irrigation water in large tanks, or proportional injectors add fertilizer solution to the irrigation water. This chapter presents a calculation method and worksheet for the five macronutrients required in hydoponic fertigation systems. The chapter also includes a VBA/Excel program that solves for fertilizer amounts for both macronutrients and micronutrients. VBA/Excel programs also calculate friction loss in laterals and in the injection system.

The Growing Media

The advantage of hydroponic (soil-less) media is that it allows growers to rapidly change nutrient concentrations in order to "steer" the plant toward a desired endpoint, such as high sugar in fruit or more reproductive (more fruit rather than leaves) growth. The media can range from sand to rock wool. A sand culture system (Figs. 21.1 and 21.2) has several inches of sand over the entire floor area.

The most common growing media for hydroponic irrigation is rock wool or ground coconut husks (coco-coir) in plastic bags (Fig. 21.3).

Hydroponic Drip Systems

Hydroponic drip irrigation systems are adapted to the unique requirements of hydroponic irrigation. Controllers are generally triggered by a light sensor on top of the greenhouse. Unlike conventional controllers, they have the capability of turning the system on several times per hour. They may also be linked to an expert system that controls all greenhouse components.

Solenoid valves used in greenhouse irrigation systems must be rated for high salinity water. Many solenoid valves will degrade in saline water. The diaphragms degrade and prevent full shut off or not open correctly.

Submains supply water to polyethylene laterals that are laid on the greenhouse floor. Buried submains are generally white PVC while submains on the ground surface should be black polyethylene pipe.

Emitters are typically 2 LPH online emitters with barbed fittings inserted into the lateral (Fig. 21.4). Spaghetti tubing brings water from drip emitters to plant containers, and plastic stakes fix the tubing to plant containers (Fig. 21.4). Pressure compensating emitters are normally used in greenhouse hydroponic systems (Fig. 21.5). They should have an extremely small change in flow rate under the range of pressures in the greenhouse. As with solenoid valves, one of the dangers is that plastic or rubber materials in emitters might degrade in the high salinity water. Thus, emitters must be rated for hydroponic irrigation.

Each slab (bag) supports six cubes, each cube having one plant. Thus, there are six emitters supplying water to each slab. Because of the high hydraulic conductivity of the slabs, the flow from the six emitters mixes in the bag. Thus, the emission uniformity equation should use six emitters per plant since the flow from the six emitters is averaged in each bag.

Greenhouses have high nutrient loads in the water. Discharging the water to groundwater can cause pollution

P. Waller, M. Yitayew, *Irrigation and Drainage Engineering*, DOI 10.1007/978-3-319-05699-9_21

Fig. 21.1 Sand culture system with lettuce (Courtesy of Murat Kacira (Controlled Environment Agriculture Center, The University of Arizona))

Fig. 21.2 Sand culture system with tomatoes (Courtesy of Murat Kacira (Controlled Environment Agriculture Center, The University of Arizona))

Fig. 21.3 Hydroponic irrigation system with rock wool in plastic bags (Courtesy of Gene Giacomelli (Controlled Environment Agriculture Center, The University of Arizona))

Fig. 21.4 Water and nutrient delivery system for hydroponic irrigation (Courtesy of Gene Giacomelli (Controlled Environment Agriculture Center, The University of Arizona))

Fig. 21.5 Rows of plants by dripper and trough (Courtesy of Gene Giacomelli (Controlled Environment Agriculture Center, The University of Arizona))

of the groundwater. Nutrient and water recycle reduces the environmental impact of hydroponic irrigation. An efficient return system begins with gutter systems (Fig. 21.5). The major difficulty is bringing the water back into the greenhouse. The fertilizer concentration in the return water must be monitored and adjusted to the required nutrient recipe. Other required procedures for recycle include sand filtration and disinfection. In order to avoid adding chlorine or other biocides to water, some recycle systems include UV light for disinfection.

Friction Loss in Drip Irrigation Laterals

Most hydroponic systems have online emitters with barbed insert fittings, which cause additional friction loss in drip irrigation laterals. The friction loss associated with emitters is referred to as a local loss. Barbed fittings restrict the flow area due to the barbed insert fitting. Thus, the emitters contract and then expand the flow, which causes turbulence and additional pressure loss. Inline emitters also have local losses, but the losses are generally less than barbed fittings and they are spaced further apart in conventional agriculture. Thus, local losses were not discussed in Chap. 17; however, inline emitters that are welded to the inside of the tube or inserted into the tube also cause contraction and then expansion of flow area. Local losses are of particular concern in hydroponic systems because of the normal practice of closely spaced emitters with barbed fittings inserted into the drip lateral. The following paragraphs discuss local losses associated with both inline and online emitters.

Demir et al. (2007) summarized the research on local losses due to emitters. The following information on local losses is from that paper, and much more information on local losses in various types of emitters can be found in the same paper. Bagarello conducted an extensive study of friction loss caused by various barbed emitters. He developed an obstruction index that accounted for the reduced flow area due to emitters. The term α is the ratio of friction loss in pipe with emitters to friction loss in pipe without emitters. Bagarello calculated α, which is the same as the minor loss coefficient for fittings, as follows:

$$\alpha = 1.68\left[\left(\frac{A_p}{A_g}\right) - 1\right]^{1.29}$$

where

H_s = friction loss between two emitters, m,
α = local losses as a fraction of kinetic height ($V^2/2$ g),
A_p = cross-sectional area of pipe without emitter,
A_g = cross-sectional area of pipe with emitter,

An equation for local losses with inline emitters was developed by Provenzano and Pumo (2004).

$$\alpha = 0.116\left[\left(\frac{D}{d}\right)^{13.87} - 1\right]$$

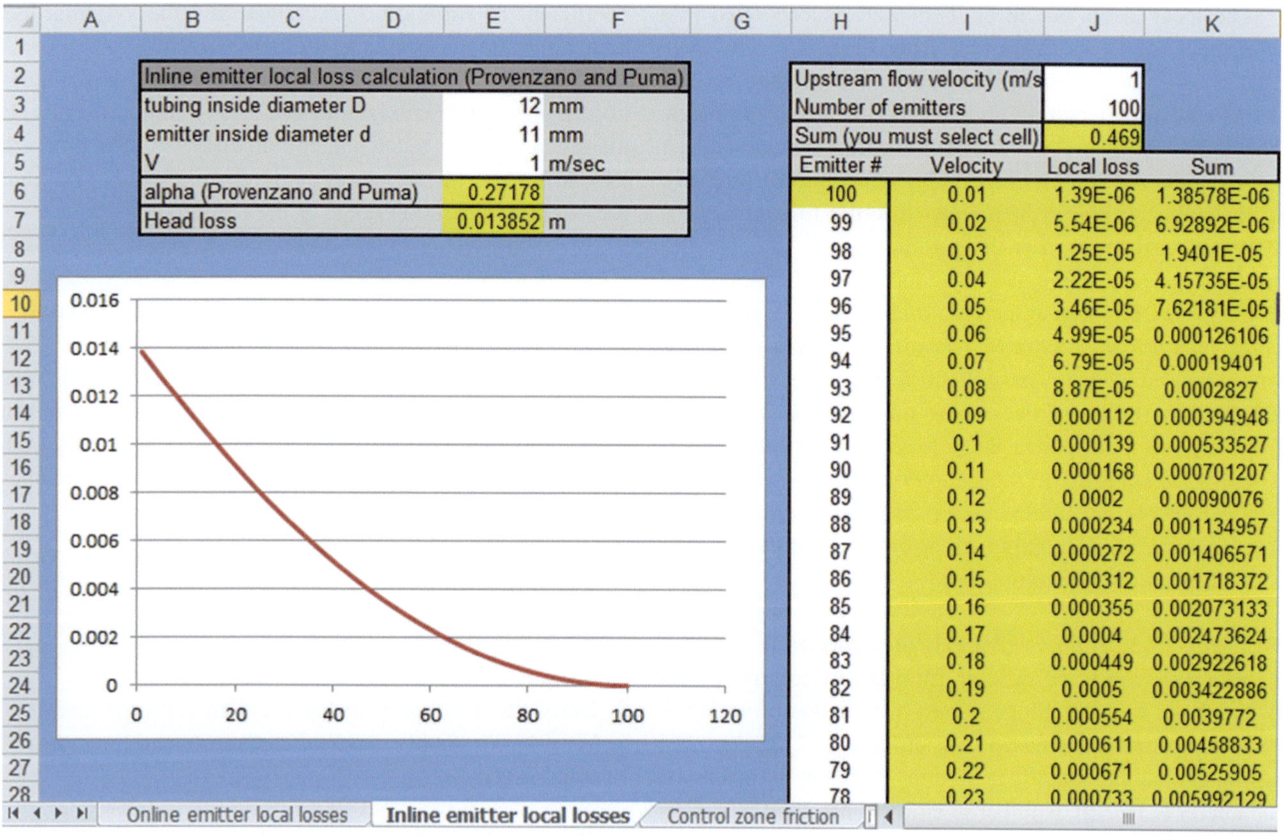

Fig. 21.6 Local losses (not pipe friction loss) calculation for inline emitters in Example 21.1

where

D = pipe inside diameter
d = inline emitter inside diameter

Example 21.1 A 12 mm diameter irrigation lateral has pressure compensating inline emitters that reduce the internal diameter of the pipe to 11 mm. Flow velocity is 1 m/sec. Calculate the friction loss due to 1 emitter. Use the *Inline emitter local losses* worksheet to calculate the friction loss due to 100 emitters, where flow is reduced proportionally vs. distance along the lateral.

$$\alpha = 0.116\left[\left(\tfrac{D}{d}\right)^{13.87} - 1\right] = 0.116\left[\left(\tfrac{12}{11}\right)^{13.87} - 1\right] = 0.27$$

$$\Delta H_s = \alpha\frac{V^2}{2g} = 0.27\frac{1^2}{2g} = 0.0139\,m$$

Friction loss due to 1 emitter at 1 m/sec flow rate is 0.0139 m.

The calculation for 100 emitters is made in the *Inline emitter local losses* worksheet. Note that this is just the sum of the local losses and does not include pipe friction loss (see Fig. 21.6).

Example 21.2 A 12 mm diameter irrigation lateral has an online emitter with barbed fitting, which blocks the flow path such that the equivalent diameter of the pipe is 11 mm Maximum flow velocity is 1 m/sec. Calculate the friction loss due to 100 emitters in the *Online emitter local losses* worksheet.

$$\alpha = 1.68\left[\left(\tfrac{A_p}{A_g}\right) - 1\right]^{1.29} = 1.68\left[\left(\tfrac{113}{95}\right) - 1\right]^{1.29} = 0.196$$

$$\Delta H_s = \alpha\frac{V^2}{2g} = 0.196\frac{1^2}{2g} = 0.010\,m$$

Friction loss due to 1 emitter at 1 m/sec flow rate is 0.010 m.

The calculation for 100 emitters is made in the *Online emitter local losses* worksheet. Total loss due to local losses is 0.339 m.

Equations have been developed by researchers and companies for specific types of emitters. In order to develop an accurate prediction of local losses due to the type of emitters that is used in an application, it is best to consult research papers and company tests that focus on specific emitter types and specific emitters, respectively.

The local losses due to emitters were incorporated in a worksheet for lateral head loss from Chap. 17. The *Drip lateral w/o local losses* worksheet is the same as the worksheets from Chap. 17 and does not include local losses. The *Drip lateral w local losses* worksheet includes minor losses in the friction loss calculations in column J. The two worksheets can be run in parallel in order to evaluate the effect of local losses. For the parameters in Chap. 17 with emitters spaced at 0.5 m, the effect local losses, as calculated in Example 21.1, was to increase total friction loss over the entire lateral from 4.49 m to 5.47 m. Thus, the friction loss was increased by approximately 20 %. Including local losses also had the effect of decreasing the emission uniformity from slightly greater than 90 % to slightly less than 90 %. These worksheets are in the *Chapter 21 Hydroponic irrigation* workbook. For inline emitter systems in which there is not a contraction in tubing area at the point of the emitter, local losses would be minimized.

Example 21.3 Emitters (2 LPH) are spaced 0.17 m in a 12 mm pipe that is 50 m long. Emitters are turbulent with k = 0.2 and x = 0.5. Slope is zero. Coefficient of variation is 5 %. Full tubing area without emitters is 113 mm^2 and tubing area at the emitters is 85 mm^2. Pressure at the end of the lateral is 10.2 m. Number of emitters per plant is 2. Calculate emission uniformity and head loss. Include one calculation with local losses and one calculation without local losses.

The alpha coefficient is calculated.

$$\alpha = 1.68\left[\left(\frac{A_p}{A_g}\right) - 1\right]^{1.29} = 1.68\left[\left(\frac{113}{85}\right) - 1\right]^{1.29} = 0.40$$

Calculations are made in the *Example 21-3 wo local losses* worksheet and the *Example 21-3 w local losses* worksheet. The pressure loss in the pipe without considering local losses is 14.81 m and with local losses is 19.64 m. Thus, the head loss increases from 4.6 to 9.4 m with local losses. The head loss due to local losses is 4.8 m, which is greater than the head loss due to pipe friction.

Considering the local losses decreases the emission uniformity from 90.4 % and is 86.5 %. This example demonstrates the importance of including local losses in calculations of lateral performance, particularly in hydroponic systems with closely spaced barbed emitters.

Because of the high value of greenhouse products and the need to carefully regulate water and nutrient application to plants, the standard for emission uniformity in hydroponic systems is higher than in conventional agriculture. For many growers, an emission uniformity of 90 % is not acceptable. For this reason, hydroponic systems normally have high quality pressure compensating emitters with manufacturer's coefficient of variation 3 % or less, and the systems thus have emission uniformity in the range of 97 %.

Even with pressure compensating emitters, miscalculation of pressure loss in laterals can lead to nonuniformity. If the pressure loss is much higher than expected, then pressure loss at the end of the lateral could drop below the optimal pressure compensation range of the emitters.

Hydroponic Irrigation System Maintenance

System maintenance is especially critical in hydroponics because the water-holding capacity of the growing media is low. In general, plants die after 2 days without irrigation. As with other drip irrigation systems, the following maintenance procedures are generally necessary.

1. Injection of biocides
2. Injection of acids to prevent precipitation of salts
3. Flush drip laterals regularly
4. Filtration
5. Daily check of pump station pressure and operation
6. Weekly or monthly check of uniformity and flow rates of drip emitters

Once bacterial growth begins in hydroponic irrigation systems, high nutrient concentration in hydroponic irrigation water causes rapid growth. The problem might be compounded by the fact that some of the most effective biocides are not acceptable in greenhouses. For example, greenhouse growers do not normally inject chlorine because chlorine can harm plants and chlorine gas (toxic) can escape into the greenhouse air. Alternative disinfection methods include hydrogen peroxide (a weak biocide), UV light, and ozonation.

The problem with disinfection is why greenhouse growers search for water that is unlikely to support microbial growth. Water from wells that are deeper than 60 m (200 ft) will probably not have appreciable amounts of bacteria or organic carbon, and disinfection is probably not necessary; however, the well water must not contact the atmosphere. Holding well water in an open tank between the well and the irrigation system might introduce light (an energy source for phototrophic organisms) and bacteria into the water.

Although calcium and carbonate can precipitate and form a scale on drip emitters, greenhouse growers normally add acid and lower pH to the range of 5.5–6.5 in order to make nutrients more available to the plant; thus, calcium carbonate deposition is unlikely. The most common acid in greenhouses is phosphoric acid, but nitric acid and sulfuric acid are also used. Nitric acid is most expensive, but it is appropriate for nitrogen tolerant tomato varieties. Phosphoric acid has the added benefit of increasing the phosphorous – potassium ratio.

In general, greenhouse growers use high quality well water so screen or disk filters are adequate. However, when water is recycled, then bacteria are in the recycled water, and sand filters are needed. In addition, organic fertilizers can have particulates and also can cause slimes to grow in tanks, tubing, and emitters.

Irrigation Scheduling

Hydroponic systems are normally scheduled based on solar radiation intensity. Pyranometers are mounted outside the greenhouse and measure solar radiation. A typical (inexpensive) pyranometer measures total electromagnetic radiation energy between 400 and 1,100 nanometers, approximately 90 % of total solar radiation. The normal schedule is one irrigation application per 80 J/cm^2. The plants can be steered toward the desired growth characteristics by decreasing or increasing the frequency. Decreasing the frequency can increase salinity in the growing media, which increases plant stress, thus increasing fruit sugar and reproductive growth (more fruit rather than leaves).

The scheduling strategy is adjusted based on time of day. For tomatoes, there is normally no irrigation for the first 1 ½ to 2 hours after sunrise and the last 1½ to 2 hours before sunset in order to prevent fruit cracking. If heat pipes are used during the night to warm the greenhouse, then two irrigations during the night are recommended. Midday frequency can be increased in order to decrease plant stress. The benefit of frequent irrigations during midday is that oxygen concentration in the growing media increases.

Researchers have found that the application of 100 ml per emitter per irrigation event maximizes the intake of oxygen from the air and water into the growing media. In addition, dissolved oxygen is contained in the irrigation water. The minimum acceptable level of dissolved oxygen in hydroponic irrigation water is 6 mg/L. Recycled water and warm water often has less oxygen than water from wells because it is warmer, and oxygenation or cooling of the water may be necessary.

In order to supply 100 ml per irrigation, a 2 LPH emitter is left on for (100 ml)/(2,000 ml/hr) = 3 minutes. With high solar radiation, the irrigation system may need to cycle as often as once every 10 minutes. Greenhouse designers have found that a maximum of two zones comfortably fit within the required cycle time (with a margin of safety): (2 zones) (3 minutes/zone) + valve cycle time = 6 minutes + valve cycle time.

Extra water must be applied to leach salts out of the rock wool slabs. Drip irrigated vegetable crops normally have between 30 % and 40 % overdrain percentage. Overdrain percentages vary based on plant type, plant maturity, time of day, and season of the year. Outflow/inflow is the overdrain percentage. The daily inflow can be measured by placing a drip emitter in its own container. The outflow can be measured by placing a container below the gutter. The overdrain percentage is the daily drainage volume (outflow) divided by the daily emitter volume. For example, if the emitter inflow is 1 LPD, and drainage overflow is 1/3 LPD, then the overdrain fraction is 1/3: (1/3)(100) = 33 % overdrain percentage.

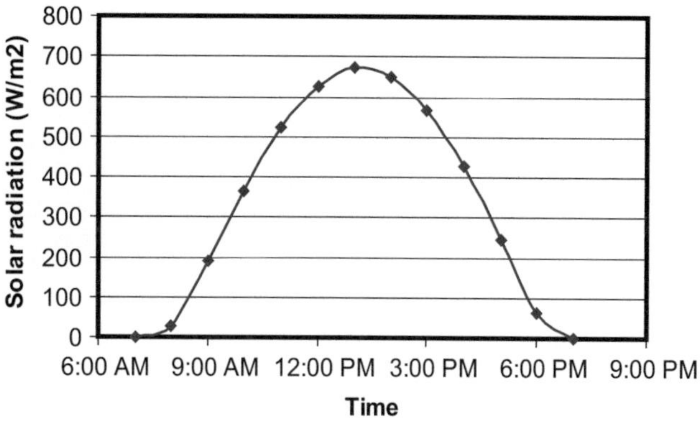

Fig. 21.7 Solar radiation intensity during winter day in Tucson, Arizona

Example 21.4 Develop an irrigation schedule based on the solar radiation data shown in Fig. 21.7.

The unit conversion from power to energy accumulation per hour is calculated as follows:

$$\left(\frac{29\ W}{m^2}\right)\left(\frac{1\ J/\sec}{W}\right)\left(\frac{3,600\ \sec}{hr}\right)\left(\frac{1\ m^2}{10,000\ cm^2}\right)$$
$$= 10.4\ J/cm^2$$

Calculate the number of irrigations per hour by dividing the hourly energy that reaches the greenhouse by 80 J/cm^2. Results are shown in Table 21.1.

While irrigation amount in field agriculture is normally calculated as depth of water applied per unit area, greenhouse growers normally calculate application rate as volume applied per plant per day. Volume application rate can be converted to depth application rate. Typical tomato plant spacing is 0.2 m along the row and 1.5 m between rows. Thus, area per plant is approximately (0.2 m) (1.5 m) = 0.3 m^2/plant.

Example 21.5 Calculate depth applied per day in the greenhouse. The volume applied per plant is 2 Liters/day, which is the same as 0.002 m^3/day. The plant spacing is 0.3 m^2

$$\text{Volume/Area} = \left(0.002\text{m}^3/\text{day}\right)/\left(0.3\text{m}^2\right) = 0.006\text{m/day}$$
$$= 6\text{mm/day}$$

Table 21.1 Solar energy and irrigation schedule during winter in Tucson

Time	Average power (W/m^2) reaching greenhouse	Energy (J/cm^2) reaching greenhouse	Number of irrigations
6:00–7:00 AM	0	0	
8:00 AM	29	10.4	
9:00 AM	190	68.4	1
10:00 AM	365	132	2
11:00 AM	521	188	3
12:00 PM	626	225	3
1:00 PM	671	242	3
2:00 PM	650	234	3
3:00 PM	568	204	3
4:00 PM	427	154	2
5:00 PM	247	88.9	
6:00 PM	62	22.2	
10:00 PM			1
2:00 PM			1

Fertigation

The ability to precisely manage nutrient concentrations is one of the major advantages of hydroponic irrigation. Nutrient recipes vary with crop, variety, maturity, climate, and desired results, such as increased sugar. Growers often guard their nutrient strategies carefully because it might give them an advantage over other growers. On the other hand, many standard recipes are available from research or trade organizations.

In 1804, Justus Liebig determined that plants required certain mineral elements. In 1860, two German plant physiologists (J. Sachs and W. Knopf) attempted to determine the amount of nutrients required by plants by growing the plants in water. Their experiments were hampered by the fact that they did not realize that the roots needed air as well as water, and they didn't supply oxygen to their plant root systems. The first plant nutritionist in the United States, Dennis R. Hoagland, developed a nutrient solution (Table 21.2) required for growth of plants in hydroponic culture (in water). Formulations used in modern hydroponic greenhouses are similar (Table 21.3).

As shown in Table 21.3, nutrients can be divided into the categories of macronutrients and micronutrients:

- Macronutrients: N, P, K, Ca, Mg, S
- Micronutrients: Fe, Mn, B, Zn, Cu, Mo, Cl.

Nutrients in water can be monitored with lab tests, inexpensive color chart tests or plant evaluation. Growers normally just observe plant leaf color or other plant characteristics to detect plant nutrient deficiency. Nitrogen deficiency causes chlorosis or yellowing of the leaves. Phosphorous deficiency results in stunted plants and delayed maturity. Potassium deficiency causes weak stems and stalks (purple color). Calcium deficiency causes loss of flowers. Most growers do not conduct water analyses on a regular basis.

Table 21.2 Hoagland solution

Fertilizer	Molarity, mg/l (ppm)
KNO_3.	0.006
$Ca(NO_3)$	0.004
$NH_4H_2PO_4$	0.001
$MgSO_4 * 7H_2O$	0.002
Mixture of 0.5 % FeSO4 and 0.4 % tartaric acid: 0.6 ml/l added 3 times per week.	
$MnCl_2 * 4H_2O$	0.5 Mn; 6.5 Cl
H_3BO_3	0.5 B
$ZnSO_4 * 7H_2O$	0.05 Zn
$Cu SO_4 * 5H_2O$	0.02 Cu
$H_2MoO_4 * H_2O$	0.05 Mo

Table 21.3 Recommended nutrient and tank concentrations from BC Agriculture

Elements to be supplied (Argus Greenhouse Nutrient Management brochure)			*Irrigation water C (mg/L) (no injection)*	
		ppm		
Nitrogen	(N)	200	*** *A tank:*	
Phosphorus	(P)	40	Calcium Nitrate	947
Potassium	(K)	200	Iron Chelate	15
Calcium	(Ca)	180	Potassium Nitrate	206
Magnesium	(Mg)	35		
Sulfur	(S)	46	*** *B tank:*	
Iron	(Fe)	2.0	Magnesium Sulfate	350
Manganese	(Mn)	0.7	Monopotassium Phosphate	174
Boron	(B)	0.5	Potassium Sulfate	206
Zinc	(Zn)	0.5	Manganese Chelate	5.8
Copper	(Cu)	0.5	Solubor	2.4
Molybdenum	(Mo)	0.05	Zinc Chelate	3.5
			Copper Sulfate	2.0
			Sodium Molybdate	0.13

Plant nutrient requirements change during the season. For example, the ratio of potassium to nitrogen supplied to tomatoes is nearly double early in the first half of the growing season and nearly the same during the second half of the growing season. Ratios of nutrients are also variety specific.

Because of electroneutrality, plants are only able to absorb a fixed number of cations. An excess of one cation in the nutrient solution can result in a deficiency of another cation in the plant. Maintaining the correct balance of cations and anions in the nutrient solution is important. For example, increased magnesium will decrease the amount of calcium taken up by the plant.

It is important to test a water supply before mixing the nutrient solution. The nutrient concentrations used at the Controlled Environment Agricultural Center (CEAC) at the University of Arizona for production of tomatoes is shown in Table 21.4. The rationale for this recipe is as follows:

- Lower nitrogen at the beginning of season (early fall) prevents rapid vegetative growth in hot weather.
- High potassium weeks 6–12 to enhance flower production
- Recipe after week 12 maintains balance between vegetative and reproductive growth.

The tomatoes at the CEAC are planted at the end of the summer. During the early phase of the growing season (early fall), nitrogen concentration in the hydroponic solution is low, 95 ppm, in order to prevent excessive vegetative growth. The small plants require approximately 0.75 L/day. During the latter part of the growing season, the nitrogen concentration is 189 ppm, High nitrogen prevents blight and helps the plants maintain high yields. The plant water requirement is also high late in the season (>4 L/day). Calcium is decreased late in the season, but it cannot become so low that blossom end rot takes place (Fig. 21.8).

Table 21.4 Nutrient concentrations used at the CEAC (Credit Pat Rorabaugh)

Nutrient	Week 0–6	Week 6–12	Week 12+
	PPM (mg/L)	PPM (mg/L)	PPM (mg/L)
N	95	145	189
P	47	47	39
K	281	351	341
Ca	212	190	170
Mg	65	60	48

Because fertilizer is applied at a constant rate in the irrigation water, greenhouse growers calculate fertilizer requirements in terms of concentration in the irrigation water (ppm or mg/L) rather than in kg/ha as for field crops. Fertilizer in the greenhouse is typically injected into the irrigation water from 2 tanks called the A and the B solution (Table 21.3). The A solution is calcium nitrate, iron, and sometimes potassium nitrate and the B solution is all of the other nutrients. Calcium nitrate is separated from the other nutrients because of the tendency for calcium to precipitate out with the sulfate in the B solution and form gypsum.

If injector pumps are used to pump fertilizer into the irrigation line, then the tank concentrations should be multiplied by the ratio of the irrigation flow rate to the fertilizer injection flow rate. For example, if the fertilizer injection rate is 1/100 of the irrigation flow rate, then multiply the concentration of fertilizer in Table 21.3 by 100 in order to calculate the tank concentration.

Example 21.6 Calculate the concentration of potassium nitrate in the B tank (Table 21.3) for an injection flow rate that is 1/100 of the irrigation flow rate. Check to make sure that the higher concentration does not exceed the maximum solubility.

Fig. 21.8 Blossom end rot (Credit Wikipedia)

Multiply the concentration of potassium nitrate by 100

$$206\text{mg/L}*100 = 20,600\text{mg/L} = 0.0206\text{kg/L}.$$

The maximum solubility of potassium nitrate (Chap. 19) is 0.133 kg/L, which is much higher than the fertilizer tank concentration.

Pat Rorabaugh at the U of A CEAC developed a sequential fertilizer calculation method (Example 21.7) that begins with calcium and continues with nitrate, potassium, phosphorous, and magnesium.

Example 21.7 Calculate the concentrations of the five macronutrients that are shown in the "Week 12+" column in Table 21.3. Assume that a 1,000 L tank is mixed, and that the water flow rate is 200 X greater than the fertilizer injection rate.

First, calculate the required amount of calcium nitrate in the A tank to supply the required calcium.

- Assume that source water contains 29 mg/L Ca
- Need 170 mg/L – 29 mg/L = 141 mg/L
- Calcium fertilizer is calcium nitrate
- $Ca(NO_3)_2$ is 19 % by weight Ca
- 141 ppm / 0.19 = 742 mg/L $Ca(NO_3)_2$
- Increase concentration by 200 = 742 × 200 = 148,400 mg/L $Ca(NO_3)_2$
- Convert to kg – 0.148 kg/L $Ca(NO_3)_2$.

Next, the amount of nitrate-nitrogen (NO_3-N) added by the calcium nitrate is calculated along with the NO_3-N that must be added with other fertilizers.

- Calcium nitrate is 15.5 % N
- 742 mg/L $Ca(NO_3)_2$ * 0.155 = 115 mg/L N
- Need 189 mg/L N
- The difference is 189–115 = 74 mg/L

Add remaining NO_3-N with potassium nitrate

- KNO_3 is 13.85 % N
- 74 mg/L/0.1385 = 534 mg/L
- Increase concentration by 200 = 534 × 200 = 106,800 mg/L KNO_3
- Convert to kg – 0.1068 kg/L KNO_3

Calculate concentration of potassium, K, still needed.

- Potassium nitrate is 38.67 % K
- 534 mg/L KNO_3 * 0.3867 = 206.6 mg/L K
- Need 341 mg/L K
- The difference is 341–206.6 = 134.4 mg/L K

Add potassium and phosphorous w/ KH_2PO_4

- KH_2PO_4 is the only source for phosphorous, so calculate phosphorous requirement first.
- Monopotassium phosphate is 22.78 % P
- 39 mg/L/0.2278 = 171 mg/L KH_2PO_4
- Increase concentration by 200 = 171 × 200 = 34,200 mg/L KH_2PO_4
- Convert to kg – 0.0342 kg/L KH_2PO_4

Calculate concentration of potassium, K, still needed (again).

- Monopotassium phosphate is 28.75 % K
- 171 mg/L KH_2PO_4 * 0.2875 = 49.2 mg/L K
- Need 341 mg/L K
- The difference is 341-(206.6 + 49.2) = 85.2 mg/L K still needed

Add remaining potassium with K_2SO_4

- Potassium sulfate is 44.87 % K
- 85.2 mg/L/0.4487 = 189.9 mg/L K_2SO_4
- Increase concentration by 200 = 189.9 × 200 = 38,000 mg/L K_2SO_4
- Convert to kg – 0.038 kg/L K2SO4

Add magnesium w/ $MgSO_4$ * $7H_20$

- Magnesium sulfate is 9.8 % Mg
- 48 mg/L/0.098 = 489.8 mg/L $MgSO_4$ * $7H_20$
- Increase concentration by 200 = 490 × 200 = 98,000 mg/L $MgSO_4$ * $7H_20$
- Convert to kg – 0.098 kg/L $MgSO_4$ * $7H_20$

Calculate the mass of calcium nitrate added to the 1,000 L fertilizer tank. Other masses are shown in Table 21.5.

$$0.148\text{kg/L}Ca(NO_3)_2*1,000\text{L} = 148\text{kg}.$$

Table 21.5 Mass of fertilizer added to 1,000 L tank

Fertilizer	Concentration (kg/L)	Mass (kg)
$Ca(NO_3)_2$	0.148	148
KNO_3	0.107	107
KH_2PO_4	0.0342	34
K_2SO_4	0.038	38
$MgSO_4$* $7H_20$	0.098	98

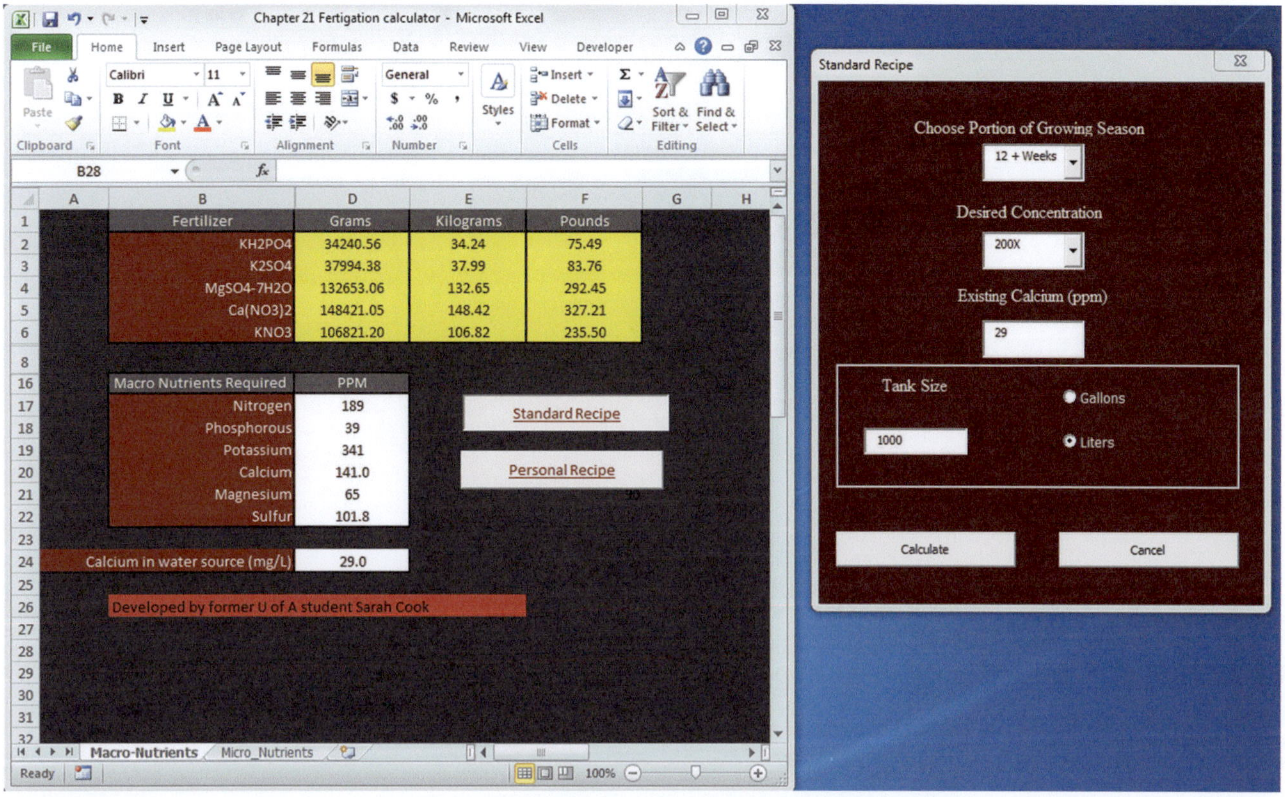

Fig. 21.9 *Chapter 21 Fertigation calculator*, *Macronutrients* worksheet (Courtesy of Sarah Cook. The University of Arizona)

A calculator for fertilizer concentrations was developed by former student Sarah Cook at the University of Arizona. The *Chapter 21 Fertigation calculator* calculates the same values (Fig. 21.9). The calculator uses linear programming in Solver to solve for the optimal mix of fertilizers.

The only difference between the calculations in the example and the worksheet is that the magnesium sulfate concentration is higher because of the added sulfur requirement in the calculator. The *Chapter 21 Fertigation calculator* also calculates micronutrient requirements (Fig. 21.10).

Growers can use salinity stress to increase the concentration of sugar in the fruit. It can also be used to steer the plant away from vegetative growth and toward reproductive growth. Thus, salinity is a fertigation management parameter. Growers use electrical conductivity (EC) meters to monitor and adjust fertilizer and salt concentration in irrigation water. Normal EC levels for tomatoes are 2.8 dS/m in the irrigation water and 3.5 dS/m in the bag. If salinity is too low, growers add salts to the fertigation solution.

The Pump Station and Nutrient Delivery System

In order to inject nutrients from the A, B, and acid tanks, hydroponic injection systems (Fig. 21.11) have three positive displacement injectors that inject water from the tanks into the irrigation system. Positive displacement pumps provide precise injection flow rate control. Proportional injection pumps, such as those shown in Fig. 21.10, are adjusted based on the ratio of fertilizer solution injection flow rate to irrigation flow rate. Other injection pumps, such as diaphragm pumps, are set to inject at a given rate. Venturi injectors are not suitable for greenhouse irrigation because they cannot be adjusted precisely. Dr. Stephen Poe set up the plywood with unistrut injection board shown in Fig. 21.11. Unitstrut clamps hold the pipes firmly in place. Parts are listed in Table 21.6.

The pressure loss in each of the components in the irrigation board is calculated in the *Injection board friction* worksheet (Fig. 21.12). The components upstream or downstream from the board are included in the upper part of the worksheet.In this case, a sand filter, pump station losses, and solenoid losses total 12 m pressure loss. The required pressure at the board assembly discharge is 11.3 m. The "Head at board inlet" is adjusted until the "Act. end pressure" is equal to the "Req. end pressure."

As shown in Fig. 21.11, the Dosatron injectors are placed in parallel in the circuit. This is the recommended geometry from the manufacturer. Some users mistakenly place them in series on the same pipe. The pressure loss in this arrangement is excessive and is shown in Fig. 21.12. The injectors are installed with PVC unions in order to facilitate easy removal. When the systems are not used, they should be removed, filled with water, and sealed in order to prevent drying of

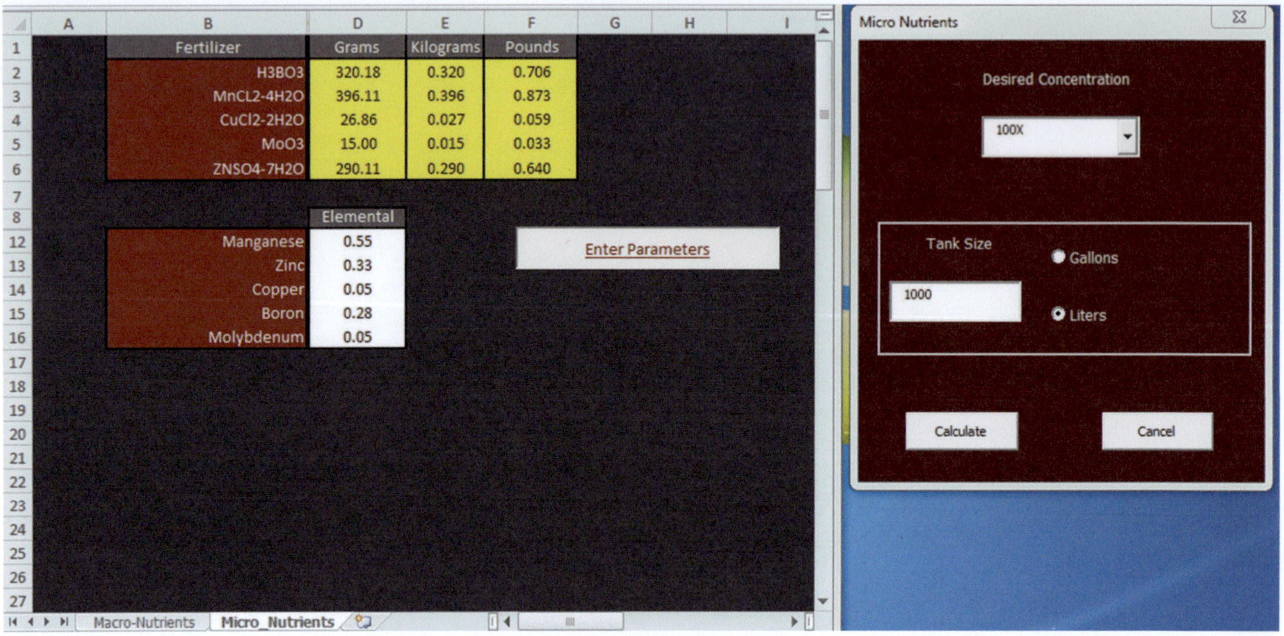

Fig. 21.10 Micronutrients worksheet in *Chapter 21 Fertigation calculator* (Courtesy of Sarah Cook, the University of Arizona)

Fig. 21.11 Greenhouse fertigation board (Courtesy of Dr. Stephen Poe. University of Arizona)

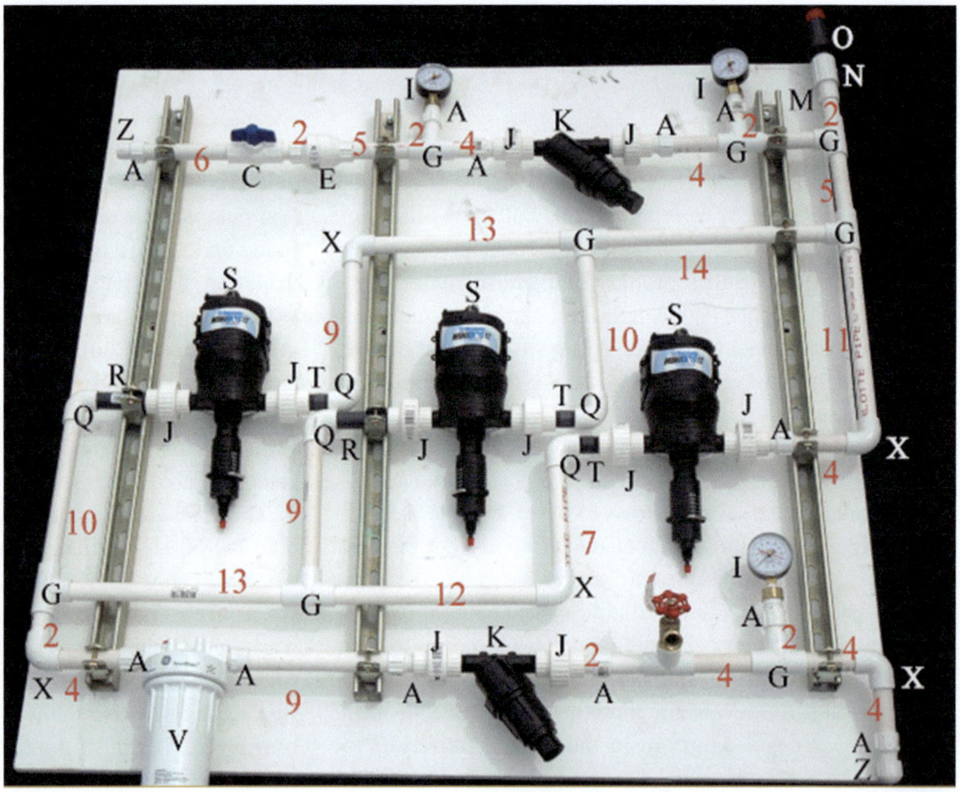

internal components. They also must be removed and stored if temperature drops below freezing. A filter before the injector prevents particulates from entering the system. The filter after the injectors protects the drip emitters in case chemicals react and form precipitates. A check valve is included in order to prevent chemical backflow into the water source. Consult local regulations for the type of check valve that is required for specific applications and water sources.

The injection board required end pressure (Fig. 21.12) is dependent on the drip tubing hydraulics, which is calculated with the *Drip lateral* worksheet (Fig. 21.14) for parameters in Example 21.3. The inlet pressure to the drip system from the *Drip lateral* worksheet is equal to the "Req. end pressure" of the board (cell K10 in Fig. 21.12). The "Act. end pressure" is set equal to the "Req. end pressure" by adjusting the "Head at board inlet."

The lengths of pipe in the injection board were left out of the calculation in Fig. 21.12 because they represent a small fraction of the total head loss in the board. However, they could be included in the calculation by adding more rows to the calculation area in the worksheet. This could be done by increasing the number of fittings or pipes in cell E6 in Fig. 21.12 and adding rows to the calculation area. Adding more combo boxes in column E can be done by copying an existing combo box and moving it to the new row. Then right click on the new combo box and make the cell link refer to the cell that it is placed in. The combo boxes refer to the *Catalog MM* worksheet. If a part is not represented, then you can add it to the table in the *Catalog MM* worksheet.

The submain pressure loss is also included in the calculation of total head loss and added to the pressure requirement in Fig. 21.12. A *Submain* worksheet is included in the *Chapter 21 Hydroponic irrigation* workbook. In general, hydroponic irrigation systems have two zones so flow is alternated between two submains.

Table 21.6 Greenhouse fertigation board parts (Courtesy of Dr. Stephen Poe)

A. ¾″ MALE ADAPTER – 12	M. ¾″ MIXING BOWL (FILTER REMOVED) – 1
B. ¾″ BALL VALVE SxS – 1	N. 1″ COUPLER (OR ¾) – 1
C. ¾″ CHECKVALVE SxS – 1	P. ¾″ 0–100 PSI PRESSURE GAUGE – 3
D. ¾″ CAP – 2	T. ¾″ TEE SxSxS – 6
E. 1″ THREADED UNION – 4	U. ¾″ THREADED UNION – 4
F. ¾″ IN-LINE Y FILTER – 2	V. 1″ AIR VALVE – 1
H. ¾″ HOSEBIB – 1	X. 1″ MALE TO – ¾″ S ADAPTER – 4
L. ¾″ PVC ELBOW SxS – 12	Y. ¾″ TEE SxTHDxS – 1
	Z. 1″ INJECTOR

Example 21.8 For the parameters in Example 21.3, change to 2 LPH pressure compensating emitters Let the minimum required pressure for the pressure compensating emitters equal 15 m. Submains are 50 m long with 1.5 m spacing between laterals. Diameter of submains is 75 mm. There are two zones, each supplied by one of the submains. Chemical injection is from diaphragm pumps, which cause no pressure

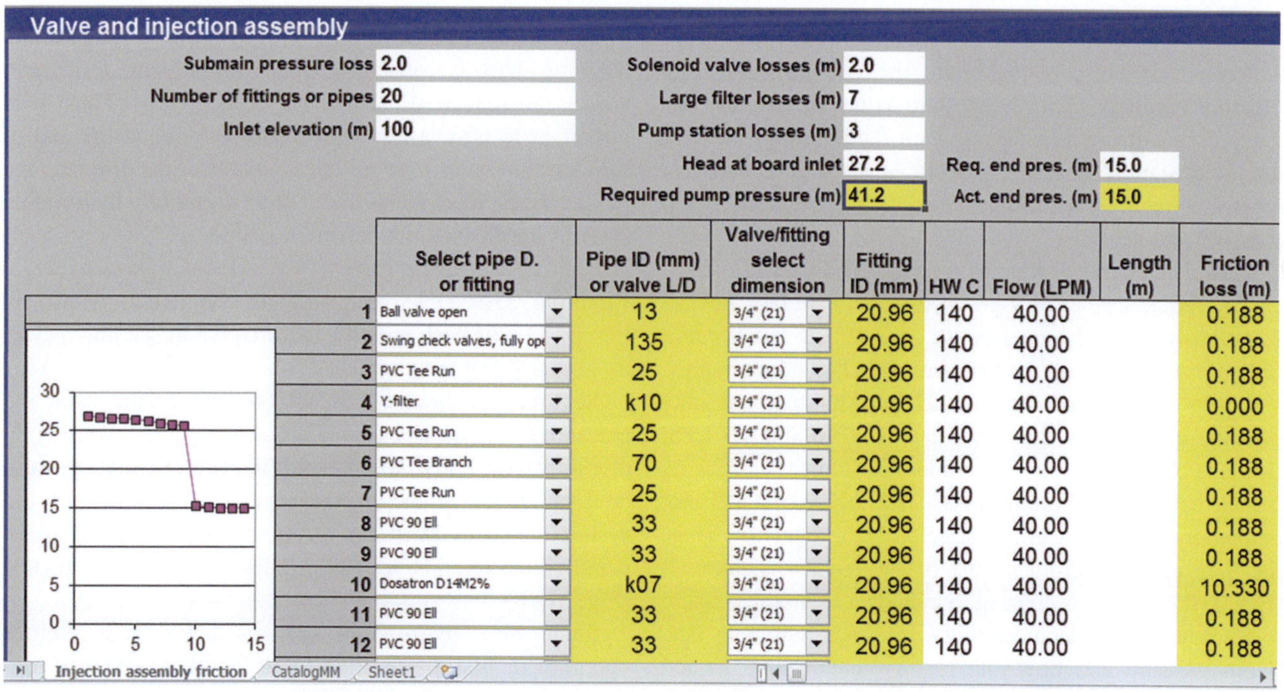

	Select pipe D. or fitting	Pipe ID (mm) or valve L/D	Valve/fitting select dimension	Fitting ID (mm)	HW C	Flow (LPM)	Length (m)	Friction loss (m)
1	Ball valve open	13	3/4" (21)	20.96	140	40.00		0.188
2	Swing check valves, fully op	135	3/4" (21)	20.96	140	40.00		0.188
3	PVC Tee Run	25	3/4" (21)	20.96	140	40.00		0.188
4	Y-filter	k10	3/4" (21)	20.96	140	40.00		0.000
5	PVC Tee Run	25	3/4" (21)	20.96	140	40.00		0.188
6	PVC Tee Branch	70	3/4" (21)	20.96	140	40.00		0.188
7	PVC Tee Run	25	3/4" (21)	20.96	140	40.00		0.188
8	PVC 90 Ell	33	3/4" (21)	20.96	140	40.00		0.188
9	PVC 90 Ell	33	3/4" (21)	20.96	140	40.00		0.188
10	Dosatron D14M2%	k07	3/4" (21)	20.96	140	40.00		10.330
11	PVC 90 Ell	33	3/4" (21)	20.96	140	40.00		0.188
12	PVC 90 Ell	33	3/4" (21)	20.96	140	40.00		0.188

Fig. 21.12 *Injection board friction* worksheet with parallel Dosatron injectors

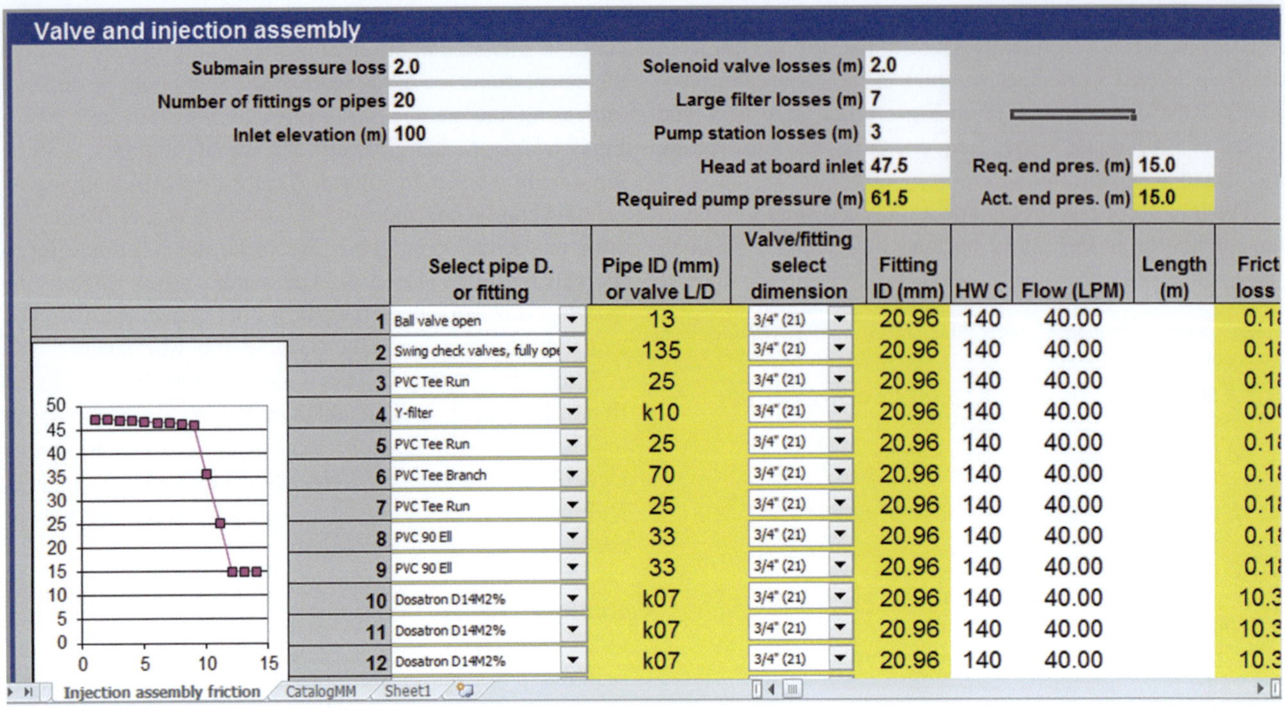

Fig. 21.13 *Injection board friction* worksheet with Dosatron injectors installed in series

loss in the irrigation pipeline. A sand filter is used to filter the water, which has 6 m maximum pressure drop. A large pressure regulator controls pressure to the system and has 3 m pressure drop across the regulator. Total fittings losses at the pump station are 3 m. Solenoid valves to each submain have a 2 m pressure drop. Calculate the required pump pressure and flow rate.

Calculate lateral hydraulics (in Figs. 21.14 and 21.15).

The required inlet pressure to the lateral is 18.4 m. The equation for lateral inlet flow vs. pressure is flat because pressure compensating emitters are used. The submain pressure loss is calculated in Fig. 21.16. The required inlet pressure and flow to the submain is 18.6 m and 4.3 LPS. Because of the way that the problem is defined, there is no need to calculate the fittings losses. The only valve that is not defined in the worksheet is a large pressure regulator. Thus, the 3 m pressure loss can just be written into column L (Fig. 21.17). There is no need to include a submain pressure loss because that is included in the 18.6 m requirement. The pump pressure and flow rate requirement is 33.6 m and 4.3 LPS.

One of the major concerns with hydroponic irrigation is flushing. This is especially true with organic fertilizers, which can induce slime and bacterial growth. Thus, it is important to have a flow velocity of 0.5 m/s at the end of the lateral in order to scour the sidewalls of the drip laterals. Unlike dual feed subsurface drip irrigation, hydroponic systems are flushed one lateral at a time.

Example 21.9 Determine whether the lateral shown in Fig. 21.14 will flush with the required 0.5 m/sec flow velocity at the end of the lateral.

Set the distal end flow rate to 200 LPH and the distal end pressure to 0.1 m. Because it is open discharge, there is no pressure at the end of the lateral. The required inlet pressure is 9 m. Thus, at the normal operating pressure of the system, which is 18 m, the flushing process can take place if individual laterals or small groups of laterals are flushed (Fig. 21.18).

Some new hydroponic systems have large emitter openings or just distribution tubing alone in order to prevent plugging, particularly in the case of organic fertilizers.

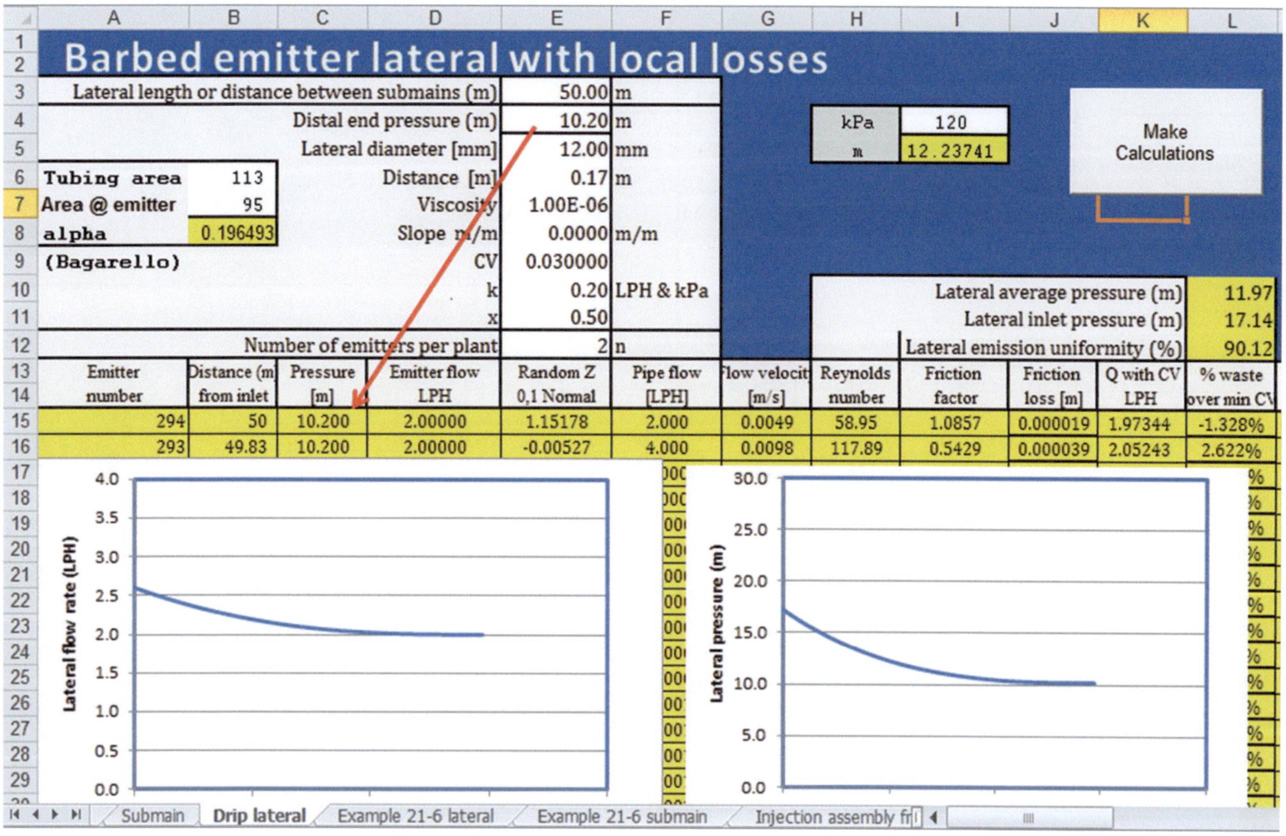

Fig. 21.14 *Drip lateral* worksheet with local losses calculated with Bagarello equation

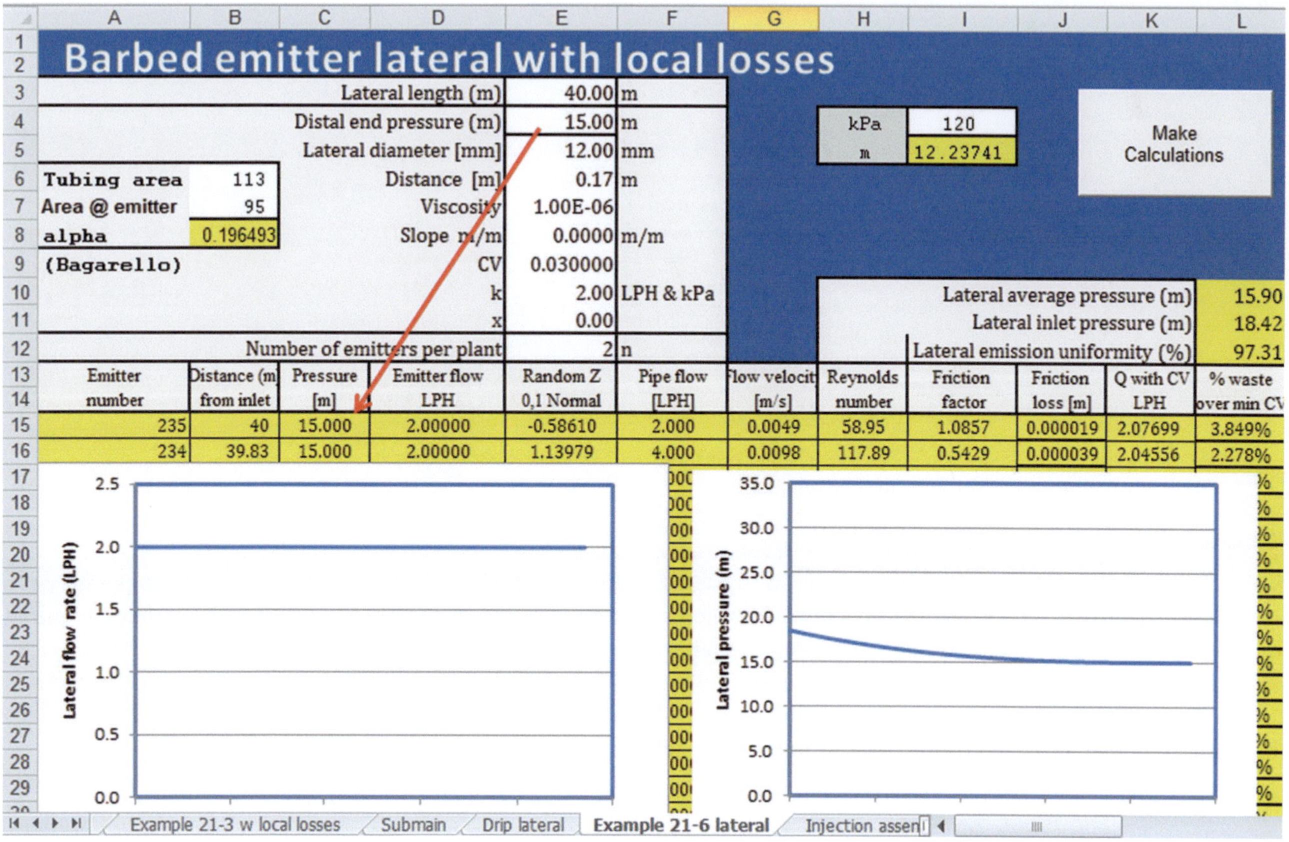

Fig. 21.15 Drip lateral for Example 21.8 with pressure compensating emitters

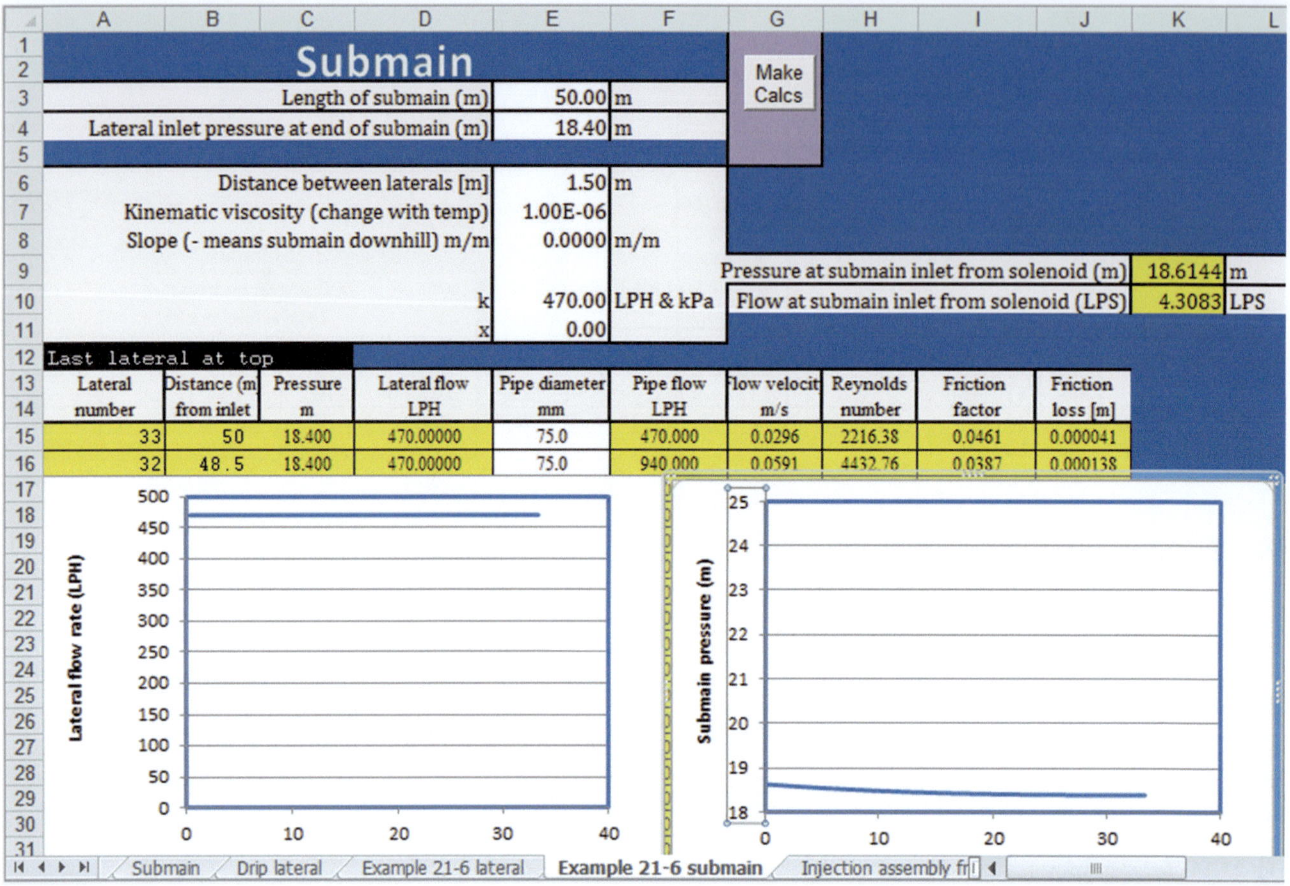

Fig. 21.16 Submain hydraulics for Example 21.8

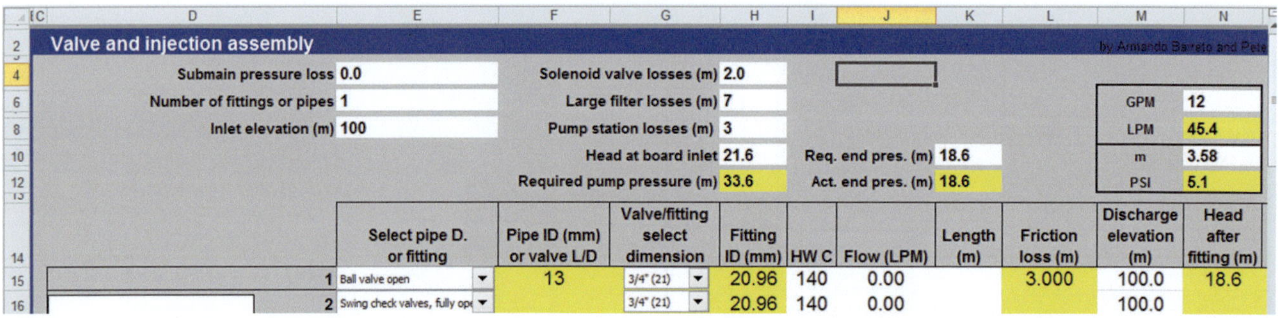

Fig. 21.17 Pump station pressure calculations for Example 21.8

Some are similar to the bubbler irrigation systems discussed in the Chap. 22. Two advantages of these systems are that no emitters are required and the pressure requirement is lower. However, the major advantage is that they are not as prone to plugging. One disadvantage is that flow is higher in laterals; thus, lateral pipes may be required; however, the expense of extra pipe may be more than offset by the labor that is sometimes required to fix plugged drip emitters in greenhouses. Even worse, plugged emitters can lead to total crop loss since there is often only a few days of irrigation water in the slabs or bags.

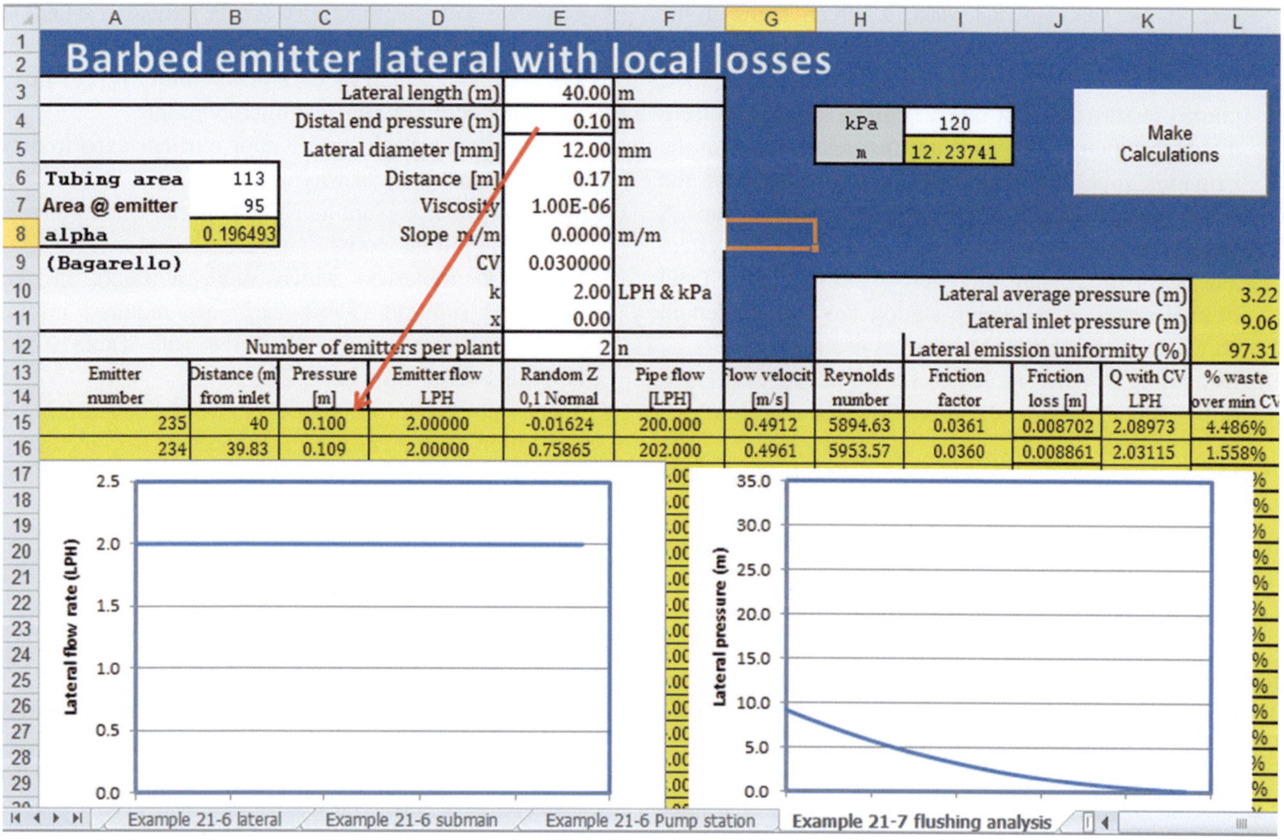

Fig. 21.18 Flushing analysis for Example 21.9

Questions

1. Turbulent flow emitters are cheaper than pressure compensating emitters. The emitter flow rate flow rate is 2 L/hr, and emitter spacing is 0.17 m. The emitter k is 0.2 and x is 0.5. Tubing diameter is 12 mm. Manufacturer's CV is 3 %. Barbed fittings reduce the cross-sectional area of the tube to 95 mm^2. There is no slope. Include minor losses due to emitter barbs. Inlet pressure to the lateral is 120 kPa. Calculate whether the emission uniformity exceeds a 95 % threshold for hydroponic drip irrigation. Should the grower switch to pressure compensating emitters? Use the Drip lateral worksheet to make the calculation.
2. For the parameters in question 1, switch to pressure compensating emitters with 2 LPH flow rate. In order to do this, change the values in cells E10:E11. Calculate the emission uniformity. Make a screen copy of the worksheet.
3. A 12 mm diameter irrigation lateral has pressure compensating inline emitters that reduce the internal diameter of the pipe to 10 mm. Flow velocity is 1.2 m/sec. Calculate the friction loss due to 1 emitter by hand. Use the *Inline emitter local losses* worksheet to calculate the friction loss due to 100 emitters, where flow is reduced proportionally vs. distance along the lateral. Compare to the loss calculated in Example 21.4. Make a screen copy of the worksheet.
4. A 12 mm diameter irrigation lateral has pressure compensating online emitters that reduce the internal diameter of the pipe to 90 mm^2. Flow velocity is 0.5 m/sec. Calculate the friction loss due to 1 emitter by hand. Use the *Online emitter local losses* worksheet to calculate the friction loss due to 100 emitters, where flow is reduced proportionally vs. distance along the lateral. Compare to the loss calculated in Example 21.4. Make a screen copy of the worksheet.
5. Emitters (4 LPH) are spaced 0.34 m in a 12 mm pipe that is 50 m long. Emitters are turbulent with k = 0.4 and x = 0.5. Slope is zero. Coefficient of variation is 5 %. Full tubing area without emitters is 113 mm^2 and tubing area at the emitters is 85 mm^2. Calculate emission uniformity and head loss. Include one calculation with local losses and one calculation without local losses. Compare to the friction loss due to local losses in Example 21.5. Make sure to press the *Make calculations* button. Make a screen copy of the worksheet with local losses.
6. A submain supplies the lateral described in question 5. The submain is 100 m long and laterals are spaced every 1.5 m. The inside diameter of the submain is 100 mm. Using the *Submain* worksheet, find the pressure

loss in the submain and make a screen copy of the worksheet. Make sure to press the *Make calcs* button.

7. For the parameters in question 6, calculate the required pump pressure and flow rate. Chemical injection is from diaphragm pumps, which cause no pressure loss in the irrigation pipeline. Ring filters are used to filter the water, which have 4 m maximum pressure drop. A large pressure regulator controls pressure to the system and has 4 m pressure drop across the regulator. Total fittings losses at the pump station are 5 m. Solenoid valves to each submain have a 3 m pressure drop.
8. Determine whether the lateral shown in Fig. 21.14 will flush with the required 0.5 m/sec flow velocity at the end of the lateral.
9. List the maintenance procedures for hydroponic drip irrigation.
10. Why is it desirable to have water from a deep well for irrigation of greenhouse crops?
11. Describe typical irrigation practices at dawn and dusk in a tomato greenhouse.
12. Develop an irrigation schedule for the following solar radiation intensity. The units for energy are those given by the Arizona agricultural weather network. You will need to convert to J/cm^2.

Time (hours)	Solar radiation (MJ/m^2)
7	0.15
8	0.41
9	0.94
10	1.61
11	2.99
12	3.33
13	3.4
14	3.3
15	2.7
16	2.44
17	0.65
18	0.5
19	0.06

13. Calculate daily depth applied for the irrigation schedule in question 8. Each water application volume is 100 ml. The area represented by each plant/emitter is 0.27 m^2.
14. List the macronutrients and micronutrients
15. Why were the Sachs and Knopf nutrient experiments unsuccessful? Why or why not?
16. Who was the first plant nutritionist in the United States, and where did he work?
17. Based on molarity, which three nutrients in the Hoagland solution (Table 21.2) are required in the greatest amount? Does this agree with Table 21.3? Does this agree with Table 21.4?
18. T/F. Testing water for nutrients prior to determining a fertigation regime in the greenhouse is not necessary because greenhouse plants require so much fertilizer anyway?
19. Calculate the amount of manganese chelate in the B tank (Table 21.3) if the injection rate is 1/50th of the greenhouse irrigation flow rate.
20. T/F. Adding too much of one cation can decrease the uptake of other cations.
21. Calculate the concentrations of the fertilizers required to formulate the nutrient mix shown in the "Week 6–12" column in Table 21.4. Mix enough fertilizer to dissolve in a 500 L tank when the water flow rate is 200 X greater than the fertilizer injection rate. The source water contains 20 mg/L Ca. Perform calculations by hand and check answer with fertigation calculator.

Reference

Demir V, Yurdem H, Degirmencioglu A (2007) Development of prediction models for friction losses in drip irrigation laterals equipped with integrated in-line and on-line emitters using dimensional analysis. Biosyst Eng 96(4):617–631. doi:10.1016/j.biosystemseng.2007.01.002

22 Low-Head Gravity Bubbler Irrigation System

Introduction

Low-head gravity bubbler irrigation systems apply water to the soil as a small stream from a small diameter tubes known as delivery hoses. To avoid runoff, small basins around the plant are sometimes needed when the application rate from the delivery hoses exceed the infiltration rate of the soil. The system is particularly well-suited for irrigation of perennial tree crops such as orchards, olives, vines, and palm trees. Traditional irrigation systems, such as furrows, can be easily converted into bubbler systems. As the name implies the systems are based on gravity where the flowrate through the delivery tube is dependent on the elevation difference between the source and the outlet of the tube. For that reason, the systems can operate at a pressure head as low as 1 meter (3.3 ft) and do not require mechanical pumps. Because the emmision point of the delivery tubes are large compared to other microirrigation emitters, the system does not require an elaborate filtration system. Keller (1990) ranked bubbler systems in the same low-risk category as surface irrigation systems, since both systems are based on gravity flow and they do not require mechanical pumps or filtration systems. The initial cost of installing bubbler systems is comparable to or lower than the cost of most other microirrigation systems that include pumps and filters. Operation and maintenance costs for bubbler systems are substantially less than for other systems due to lower energy requirements and no mechanical breakdowns.

One of the disadvantages of bubbler systems is possible air locks occuring within the pipelines systems located on level ground. Air locks within the system can prtially or entirely block the flow of water, and thereby significantly decrease the water application uniformity of the system.

Despite their simplicity and advantages, bubbler systems are not extensively used. Speculations are that bubbler systems are not used because no manufactured components are needed and micro-irrigation companies would rather sell and install systems requiring component parts such as emitters, pressure regulators, and filters. In addition, the low use of bubbler systems may be related to the suitability of these systems mainly for limited crops mainly orchard and tree crops and to the possibility of air locks occurring in the delivery hoses and laterals that block or restrict water flow through the system.

For developing countries, the energy advantages of bubbler systems are more important because many of these countries have energy shortage and plastic emitters are not readily available.

System Components

Low-head gravity bubbler irrigation systems consist of a mainline connected to a water source, a constant head device, manifolds, laterals, and small-diameter delivery hoses, Figs. 22.1 and 22.2. The laterals are laid midway between two rows of trees, and small-diameter hoses (called delivery hoses or tubes) are inserted in the laterals to deliver water to the trees. Hoses are anchored to a tree or stake, and hose heights are adjusted so that water flows out from all hoses at equal rates.

The distinguishing feature of bubbler systems is the flexible delivery hoses, Fig. 22.3, attached to the lateral in contrast to small manufactured emitters commonly used with other microirrigation systems. These hoses allow greater rates of water to discharge into the small basins, and they do not require a filtration system because of their large orifice openings. The basins are usually circular or rectangular in shape, and are bordered by low embankments, or levees, so that water is uniformly distributed over the root zone.

Installation of the systems require dynamic calibration which is a procedure by which errors in friction loss calculations can be evenly distributed along the lateral by adjusting the elevation of each delivery hose.

P. Waller, M. Yitayew, *Irrigation and Drainage Engineering*, DOI 10.1007/978-3-319-05699-9_22

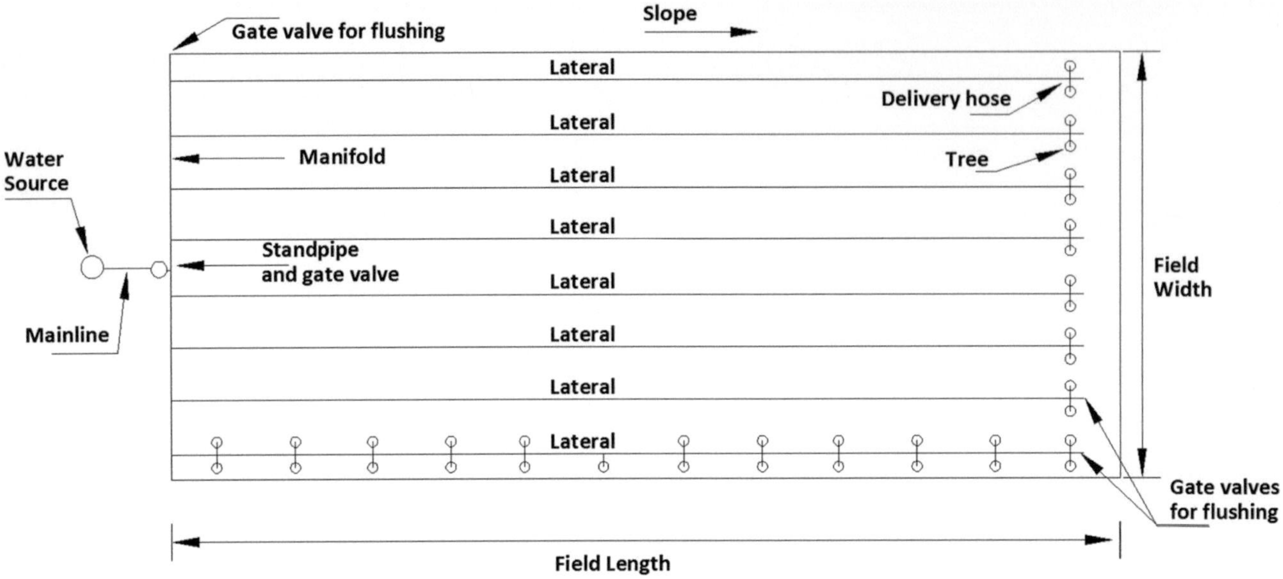

Fig. 22.1 A typical bubbler irrigation layout

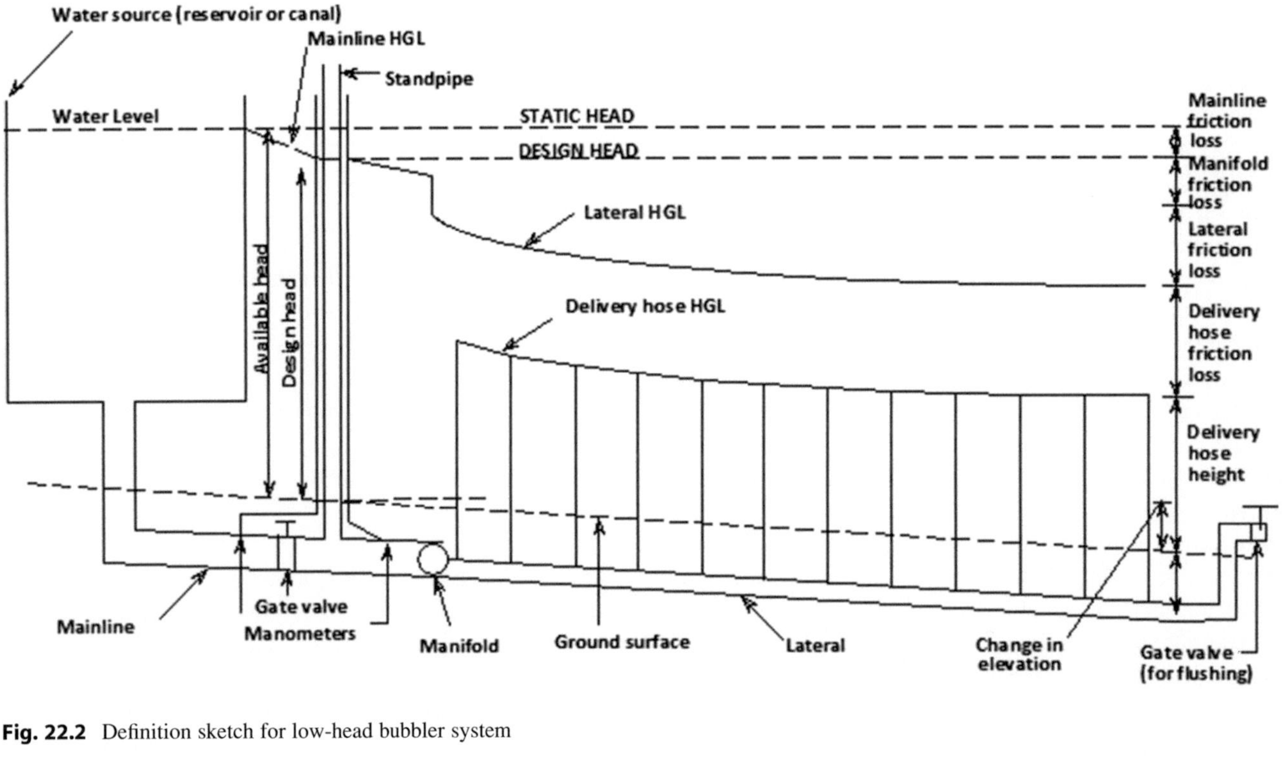

Fig. 22.2 Definition sketch for low-head bubbler system

Development of Low-Head Gravity Bubbler System

Rawlins (1977) designed and installed the first system what became known as a bubbler system when he converted a low-head furrow irrigation system into a closed-conduit gravity-flow system at Tacna, Arizona. The goal was to reduce the amount of excess irrigation runoff by increasing irrigation efficiency and, thereby, reduce salt levels in irrigation runoff.

The systems consisted of 76-mm (3-in) corrugated polyethylene (PE) laterals buried 0.6 m (2 ft) deep and connected to an existing low-head concrete pipeline. The laterals were laid midway between the rows of trees and replaced the furrows that traditionally irrigated the citrus orchard.

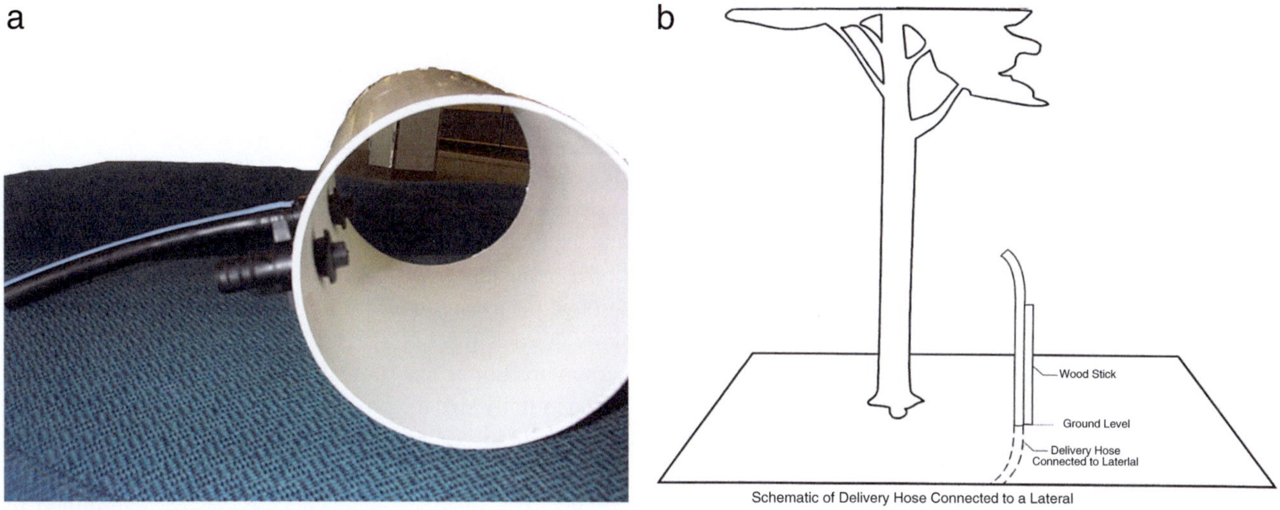

Fig. 22.3 Delivery hose connection on lateral Pipe (**a**) and support schematics (**b**)

Others, mainly Thornton and Behoteguy (1980), Carr and Kay (1980), and Roth (1992) later designed and installed bubbler systems after learning about these systems by visiting one of the pilot projects started by Rawlins. Thornton and Behoteguy (1980) followed Rawlin's example at Riverside, California and installed a bubbler system at Thermal, California, to irrigate 300 trees along a windbreak. Their lateral design was more complex than Rawlins', since their lateral was 732 m (2400 ft) long. They used the same design procedure as Rawlins, but their lateral was decreased in diameter several times in the direction of flow because of the long lateral length. Thornton and Behoteguy stated that some of their major problems were due to the lack of manufactured water-tight fittings; animals chewing and insects entering the delivery hoses; and high labor requirements for installation. Carr and Kay (1980) first learned about bubbler irrigation systems during a visit at Riverside, California, and then designed and installed a bubbler system to irrigate 320 trees at the National Agricultural Engineering College at Silsoe, England. This same system is also described by Hull (1981).

All of the bubbler systems described by Rawlins (1977), Thornton and Behoteguy (1980), Carr and Kay (1980), and Hull (1981) were similar in design since they were all situated on fields with considerable ground slope; design head was 1 meter (3.3 ft); laterals were made of corrugated PE tubing; delivery hoses were made of smooth 9.5-mm (3/8-in) diameter PE tubing; design flows ranged from 0.032 to 0.067 lps (0.5–1.0 gpm); and the Manning and Darcy-Weisbach equations were used to size the lateral and delivery hose diameters, respectively. Carr and Kay mentioned that bubbler irrigation systems can operate at a minimum design head of 1 meter (3.3 ft) and a maximum head of 5 meters (16.4 ft), and Hull stated bubbler systems are restricted to slopes of 1–3 %. However, systems of lesser slopes were installed later by Roth (1992) in Arizona.

During the 1980s, Roth (1992) installed bubbler systems at three different University of Arizona Research Stations located at Yuma, Maricopa, and Waddell, Arizona, USA. All three systems are installed on level ground and operated successfully. Roth had first learned about bubblers during the 1970s at Tacna, Arizona, and he designed the Yuma system in the early 1980s to irrigate slightly over 2 acres of various deciduous trees. The second system Roth designed and installed was in 1987 for the Citrus Agriculture Research Center at Waddell to irrigate over 12 acres of citrus trees. Roth's third system was installed at Maricopa in 1989 and is similar in size to the Yuma system. The bubbler system installed at Maricopa was also documented by Waheed (1990).

The systems at Yuma and Maricopa are different from other documented bubbler systems in that the laterals are not corrugated PE pipes and the systems are located on level ground. Roth(1992) used polyvinylchloride (PVC) pipes for laterals because commercial pipe fittings were available. PVC pipes also helped to reduce friction head losses, a critical factor for systems located on level ground, and facilitated connecting the delivery hoses to the laterals. However, PVC pipes increased capital costs slightly. Roth also modified the supports of the delivery tubes by inserting the delivery hoses through 19-mm (3/4-in) diameter PVC pipes, Fig. 22.3; this eased adjusting delivery hose height and helped to protect the hoses from animals, birds and insects.

Waheed (1990) listed the total cost of the Maricopa bubbler system at $1620 (1993 Cost Index) per acre, with approximately 50 % of the cost for labor and 50 % for materials. The major portion of material costs was in the laterals and manifold.. Yitayew and Reynolds (1997) also compared the cost of bubbler system and microspray systems and showed the clear cost advantages of bubbler over the microspray systems.

Available literature on bubbler systems until the 1990s describes various design aspects, but does not include a comprehensive design procedure. In addition, the dissemination of bubbler design criteria has largely occurred by personal contact and by site visits to existing bubbler systems. A site visit is helpful to enable a designer to visualize the simple hydraulic principles of bubbler systems and to understand the various design intricacies. However, the fact that site visits seem to be a prerequisite to design bubbler systems probably confirms that the available literature does not sufficiently explain bubbler irrigation design or convey the simplicity of such systems. To address this problem Reynolds (1993), Reynolds et al. (1995), Yitayew et al. (1995), developed design procedures and presented a computer program for deign of bubbler irrigation systems,Didan et al. (1996) Yitayew et al. (1999).

Hydraulics Considerations

Low-head gravity bubbler systems are different from other microirrigation systems as they are based on gravity-flow and do not require external energy to push the water. The energy equation is the primary equation used for bubbler system design and analysis and is given by:

$$\frac{p_1}{\gamma} + z_1 + \frac{V_1^2}{2g} = \frac{p_2}{\gamma} + z_2 + \frac{V_2^2}{2g} + \sum h_f + \sum h_{ml} \quad (22.1)$$

Where,

h_f = friction head loss in pipes, m (ft)
h_{ml} = minor losses at pipe fittings, m (ft)
v = flow velocity of water in pipes, mps (fps)
p = pressure within the pipe, N/m^2 (lb/ft^2)
z = elevation of pipe centerline with respect to a reference datum, m (ft)
γ = specific weight of water, N/m^3, (lb/ft^3)
g = gravitational constant, 9.81 m^2/s, (32.2 ft^2/s)

The terms $V^2/2$ g, p/γ, and z are commonly known as the velocity, pressure, and elevation heads, respectively. The velocity head is the head that accelerates the water from rest to the velocity at that point. The pressure head is equal to the pressure at that point divided by the unit weight of water. The elevation head is the vertical distance of a given point in the system from an arbitrary datum taken at the centerline of the pipe. The sum of the pressure head and the elevation head is termed the piezometric head ($H = p/\gamma + z$). A plot of the piezometric head is the hydraulic grade line (HGL). A plot of the piezometric plus the velocity head is the energy grade line (EGL).

The energy equation is useful to size the pipe diameters of a bubbler system by determining the piezometric heads for the upstream and downstream ends of the system. The piezometric heads at the upstream and downstream ends of the bubbler system are determined from the elevation of the water source and the field layout. The difference between the upstream and downstream piezometric heads is the total allowable head loss through the system used for design purpose. The friction loss through each pipe of the bubbler system will comprise a certain amount of the total allowable head loss.

To initially size the pipe diameters, the velocity heads and minor head losses are assumed zero, Fig. 22.4, but they will be accounted for later when calculating the delivery hose heights. The diameter of each pipe can be determined by using the flow rate, known pipe length, and calculated allowable head loss for each pipe component into the friction loss equation and solving for the diameter. For sprinkler and microirrigation design, the Hazen-Williams friction loss equation is the most commonly used formula because of its simplicity. It is expressed as:

$$h_f = K\frac{\left(\frac{Q}{C}\right)^{1.852}}{D^{4.87}}L \quad (22.2)$$

Where,

K = 1.22 × 10^{10} for metric units, (10.46, for English units)
C = Hazen-Williams friction coefficient
Q = flow of water, lps, (gpm)
D = inside pipe diameter, mm, (in.)
h_f = friction head loss in the pipe, m (ft)
L = length of lateral, m (ft)

It should be recognized that the Hazen-Williams equation was originally developed for flow in water distribution networks where Reynolds numbers are greater than 100,000. For low-pressure systems, the Reynolds number is rarely greater than 100,000 for pipe diameters less than 128-mm (5-in). For Reynolds numbers less than 25,000, a C-value of 150 commonly used for smooth plastic pipes seriously underestimates pipe friction losses, (Watters and Keller 1978). Because bubbler systems tend to have low Reynolds numbers, the Hazen-Williams equation traditionally has not been used for bubbler design.

The Darcy-Weisbach equation is the most universal formula used for computing head loss in all types of pipes and is given by:

$$h_f = f\frac{L}{D}\frac{V^2}{2g} \quad (22.3)$$

in which the friction factor, f, is obtained from the Moody diagram or computed by the equations listed below.

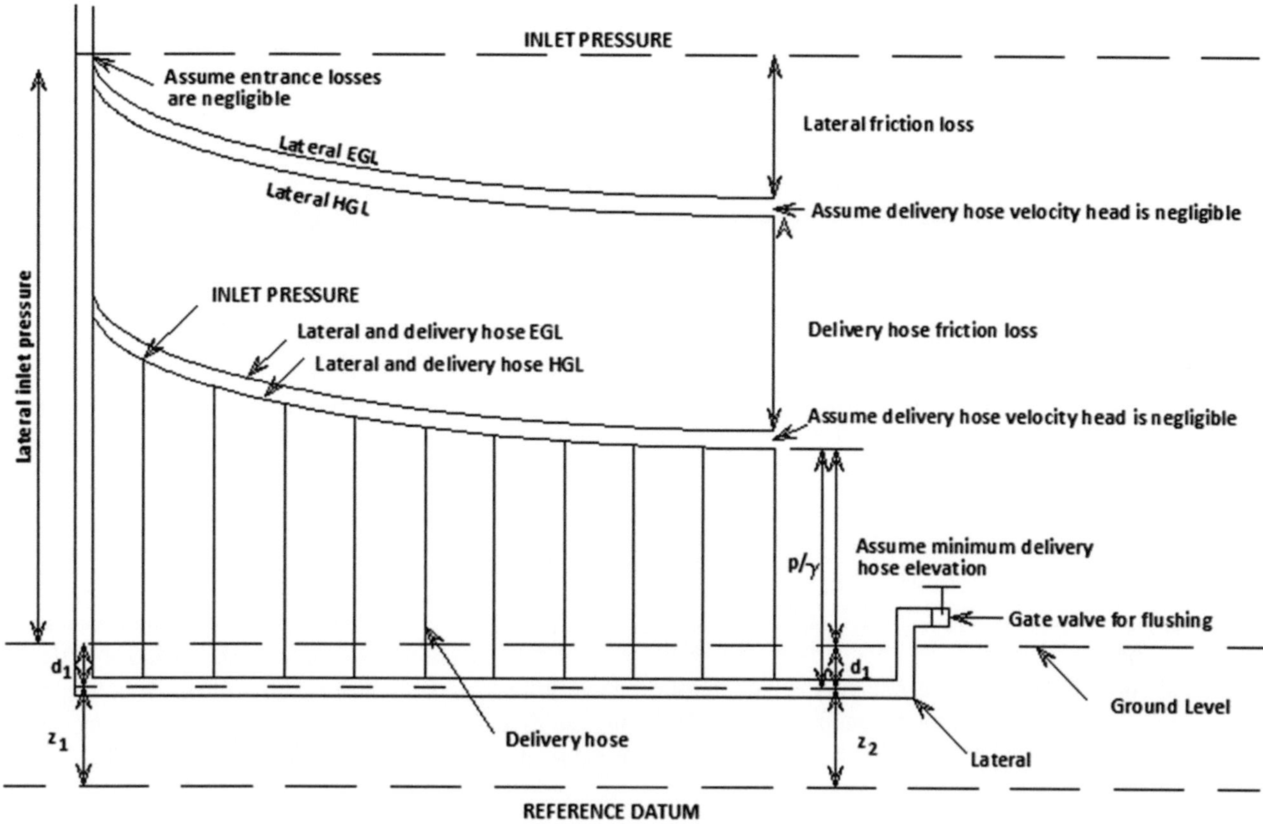

Fig. 22.4 Piezometer diagram for initial sizing of pipe diameters

For smooth plastic pipes, the friction coefficient, f, for laminar (Eq. 22.4), transitional (Eq. 22.5), and turbulent (Eqs. 22.6 and 22.7) flow conditions are:

$$\text{For} \quad Re < 2000, \quad f = \frac{64}{R_e} \tag{22.4}$$

$$\text{For } 2000 < R_e < 4000, \quad f = 3.42 \times 10^{-5} R_e^{0.85} \tag{22.5}$$

$$\text{For } 4000 < R_e < 10^5, \quad f = \frac{0.316}{R_e^{0.25}} \tag{22.6}$$

$$\text{For } 10^5 < R_e < 10^7, \quad f = \frac{0.13}{R_e^{0.172}} \tag{22.7}$$

Keller and Bliesner (1990) and Boswell (1984) recommend Eqs. 22.4, 22.6 and 22.7 in microirrigation design, with Eq. 22.6, the Blasius equation, having a Reynolds number lower limit. The Reynolds number lower limit for the Blasius equation is typically 3000–4000, however, for desk-top calculations, Eq. 22.5 as defined by Wu and Fangmeier (1974) can be ignored by setting the lower limit for the Blasius equation at 2000.

By combining the Darcy-Weisbach Eq. 22.3 and the Blasius Eq. 22.6, an equation for small diameter pipes, similar in form to the Hazen-Williams equation, can be obtained:

$$\text{For } 2000 < R_e < 10^5, \quad h_f = K \frac{Q^{1.75}}{D^{4.75}} L \tag{22.8}$$

where,

D = inside diameter, mm (in), for pipes less than 128-mm (5-in) diameter.

K = 7.89 × 10^5 for metric units (0.00133, English units), and for water temperature at 20 °C (68 °F).

For smooth pipes larger than 128-mm (5-in), Keller and Bliesner (1990) note that the Reynolds number will typically be larger than 100,000. By incorporating Eq. 22.7 into the Darcy-Weisbach equation, a friction loss equation for large smooth pipes is obtained:

$$\text{For } 10^5 < R_e < 10^7, \quad h_f = K \frac{Q^{1.828}}{D^{4.828}} L \tag{22.9}$$

where,

D = inside diameter, mm (in), for pipes greater than 128-mm (5-in) diameter.

K = 9.58 × 10^5 for metric units, (0.001 for English units), and for water temperatures at 20 °C (68 °F).

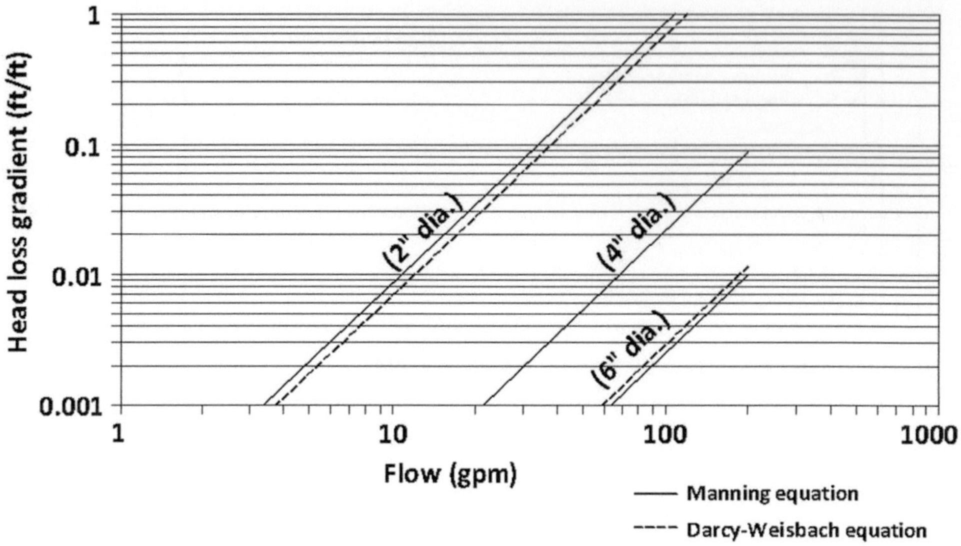

Fig. 22.5 Comparison between Manning and Darcy_Weisbach equations (After Reynolds 1993)

For sizing the corrugated polyethylene laterals, Rawlins (1977), Thornton and Behoteguy (1980), Carr and Kay (1980), and Hull (1981)) used the Manning equation with an n-value of 0.016. By setting the hydraulic radius to D/4 for pipes, the Manning equation is:

$$h_f = K\, n^2 \frac{Q^2}{D^{\frac{16}{3}}} L \qquad (22.10)$$

where,

K = 1.02×10^{11} for metric units, (13.12 for English units)
n = Manning's roughness coefficient

The Darcy-Weisbach equation can also be used to size the diameter of corrugated polyethylene pipelines by assuming an f-value for corrugated polyethylene tubing. To find which f-value to use, the Darcy-Weisbach equation can be set equal to the Manning equation, and the following relationship obtained:

$$F = \frac{K\, n^2}{D^{\frac{1}{3}}} \qquad (22.11)$$

where,

K = 124.5 for metric units, (185 for English units)
D = inside pipe diameter, m (ft)

From Eq. 22.11, an n-value of 0.016 will give an f-value of 0.068, or 0.07, for an inside pipe diameter of 102-mm (4-in). An f-value of 0.07 for a 102-mm (4-in) diameter polyethylene pipe also agrees with experiments conducted by Hermsmeier and Willardson (1970). Figure 22.5 compares the Darcy-Weisbach equation with an f-value of 0.07 and the Manning equation with an n-value of 0.016 for three different pipe diameters. From these data, it appears that an f-value of 0.07 is approximately the same as an n-value of 0.016 for corrugated polyethylene pipe ranging from 51 to 154 mm (2–6 in) in diameter.

By substituting an f-value of 0.07 into Eq. 22.3, the Darcy-Weisbach equation for corrugated plastic tubing is:

$$h_f = K \frac{Q^2}{D^5} L \qquad (22.12)$$

where,

K = 5.78×10^6 for metric units, (0.00218 for English units).

Hence, Eq. 22.8 for small smooth pipes, (22.9) for large smooth pipes, and (22.12) for corrugated PE pipes, are the final turbulent flow formulas recommended for calculating friction losses in bubbler systems when using the Darcy-Weisbach equation.

Three friction head loss equations are commonly used for the design of closed-conduit, micro-irrigation systems: the Hazen-Williams, Manning, and Darcy-Weisbach equations. These turbulent flow equations are also commonly used to calculate friction losses in the downstream reaches of irrigation laterals and manifolds, because the flow in these reaches is laminar and friction losses are essentially negligible. Previous researchers in bubbler design sized the diameters for the corrugated PE laterals by using the Manning equation with an n-value of 0.016, while the diameters for the delivery hoses were sized by using the Darcy-Weisbach and Blasius equations. Tables 22.1 and 22.2 presents summary of the

Table 22.1 Friction_loss_equations

	Mainline, manifolds, and laterals		Delivery hoses
	Smooth PVC	Corrugated PE	Smooth PE
(a) Pipe diameter equal to or less than 128 mm (5 in)			
h_f, (metric)	$7.89 \times 10^5 F \frac{Q^{1.75}}{D^{4.75}} L$	$5.78 \times 10^6 F \frac{Q^2}{D^5} L$	$7.89 \times 10^5 \frac{Q^{1.75}}{D^{4.75}} L$
h_f, (English)	$0.00133\ F \frac{Q^{1.75}}{D^{4.75}} L$	$0.00218\ F \frac{Q^2}{D^5} L$	$0.00133 \frac{Q^{1.75}}{D^{4.75}} L$
(b) Pipe diameter greater than 128 mm (5 in)			
h_f, (metric)	$9.58 \times 10^5 F \frac{Q^{1.828}}{D^{4.828}} L$	$5.78 \times 10^6 F \frac{Q^2}{D^5} L$	N/A
h_f, (English)	$0.001\ F \frac{Q^{1.828}}{D^{4.828}} L$	$0.00218\ F \frac{Q^2}{D^5} L$	N/A

Water temperature at 20 °C (68 °F)

Table 22.2 Unit constant, K, values for friction head loss equations

Temperature (°F)	Viscosity (ft^2/sec)	Constant, K[a] (English)
40	1.664×10^{-5}	0.00148
50	1.410×10^{-5}	0.00142
60	1.217×10^{-5}	0.00137
68	1.084×10^{-5}	0.00133
70	1.059×10^{-5}	0.00132
80	9.300×10^{-6}	0.00128
90	8.260×10^{-6}	0.00124
Temperature (°C)	Viscosity (m^2/sec)	Constant, k (metric)
5	1.519×10^{-6}	8.738×10^5
10	1.308×10^{-6}	8.418×10^5
15	1.141×10^{-6}	8.135×10^5
20	1.007×10^{-6}	7.885×10^5
25	8.970×10^{-7}	7.660×10^5
30	8.040×10^{-7}	7.453×10^5

[a]Use Eq. 22.8 to compute the unit constant, K,

friction loss equations applied in bubbler irrigation design and the unit constants, K, in the friction loss equations respectivel

$$h_f = K \frac{Q^{1.75}}{D^{4.75}} L \quad (22.13)$$

Where,

$K = 7.89 \times 10^5$ at 25 °C or 0.00133 at 68 °F.

Equation 22.8 converts to:

$$K = \frac{8\ (0.32)\ (v)^{0.25}}{4^{0.25} \pi^{1.75} (32.2)} \frac{(12)^{4.74}}{(448.8)^{1.75}}, \text{(English)}$$

$$K = \frac{8\ (0.32)\ (v)^{0.25}}{4^{0.25} \pi^{1.75} (9.81)} \frac{(1000)^{4.74}}{(1000)^{1.75}}, \text{(metric)}$$

Table 22.3 Christiansen reduction coefficient, F, for equally-spaced outlets along manifolds and laterals

Number of outlets	F (end)[a]	F (mid)[b]	Number of outlets	F (end)	F (mid)
1	1.00	1.00	8	0.42	0.38
2	0.64	0.52	9	0.41	0.37
3	0.54	0.44	10–11	0.40	0.37
4	0.49	0.41	12–15	0.39	0.37
5	0.46	0.40	16–20	0.38	0.36
6	0.44	0.39	21–30	0.37	0.36
7	0.43	0.38	>31	0.36	0.36

After Keller and Bliesner (1990)
[a]First outlet is a full space from pipe inlet
[b]First outlet is one-half space from pipe inlet

Christiansen Reduction Coefficient

The Christiansen (1942) reduction coefficient, F, is commonly used to calculate head losses in multiple outlet pipes and to initially size the diameters of mainlines, manifolds, and laterals. Applying the Christiansen reduction coefficient to head loss equations simplifies calculations for multiple-outlet pipes by estimating the friction loss along the entire length of the multiple-outlet pipe. The Christiansen factor is based on the assumptions that all the water is carried to the end of the line and that the multiple outlets are evenly spaced with equal discharge.

When using the Christiansen coefficient, the total friction loss for a multiple outlet pipe is expressed as:

$$h_f = F^* h_{f'} \quad (22.14)$$

where,

h_f = friction head loss between the upstream and downstream ends of a multiple outlet pipe, m (ft).
F = Christiansen reduction coefficient that depends on the number of outlets along the multiple-outlet pipe, Table 22.3.
h_f' = friction loss in a length of pipe assuming no outlets along the pipe, m (ft).

As Table 22.3 indicates, Christiansen factors for half spacing from the submian on the laterals for the first outlet are smaller than for full length spacing, thereby reducing the total friction loss in the lateral. For the design of bubbler systems, any savings of head is critical, and designing half-space for the first outlet is recommended to minimize head loss and to utilize the field area more efficiently for orchard crops.

System Design Considerations

The goal of bubbler irrigation design is to size the diameters of the pipelines to deliver an equal volume of water to each tree while keeping the total friction and minor head losses less than the available head at the water source. To permit flexibility in determining the best design, a thorough understanding of the hydraulic design principles of bubbler systems is required, and ingenuity must always be employed to keep costs low and designs flexible. Generally, the simpler the design, the less the material will cost. For example, minimizing the number of manifolds and laterals, which comprise a major portion of total material costs, will reduce costs.

Corrugated polyethylene (PE), smooth PE, and polyvinylchloride (PVC) pipes are the most common pipe materials used for the laterals, and care should first be taken to design the bubbler system with internal pipe diameters that are commercially available in-country. In addition, using corrugated PE pipe may be impractical for designs with very low heads on level ground, because the friction losses for corrugated PE pipes are much larger and small pipe undulations may cause air locks in the laterals.

Another important material consideration is the type of connection used to connect the delivery hoses to the laterals. The type of connection chosen is often subject to the availability of materials at a particular locality because commercial connections are usually not available.

Because bubbler systems are based on gravity flow, the slope of the field is another crucial factor in design. The design for systems located on level ground and those on gradual slopes differ slightly because their maximum and minimum delivery hose heights occur at different points along the lateral, as illustrated by comparing Figs. 22.6 and 22.7. Also, systems on gradual slopes will gain energy down-slope, and this allows laterals to be longer than on level ground and permits greater diversity in design for a given available head. Correspondingly, designs on level ground do not gain energy; lateral lengths cannot be as long as on sloping ground with a given head; and design flexibility is limited.

Another major difference between design of bubblers and other microirrigation systems is that design flows are assumed at the beginning of the design, and these flow rates, if run continuously, are typically much greater than the peak water requirement of any crop. In contrast, sprinkler and drip systems are typically designed to operate continuously for 90 % of the time during the peak consumptive period and their design flow rates are calculated at the beginning of design. Therefore, bubbler systems tend to operate less frequently than sprinkler or drip irrigation systems because their design flow rates are much greater.

This section below discusses the steps required to design a bubbler system, and the associated design parameters. Design examples and criteria for systems located on gradual slopes is presented later. Irrigation frequency and duration of

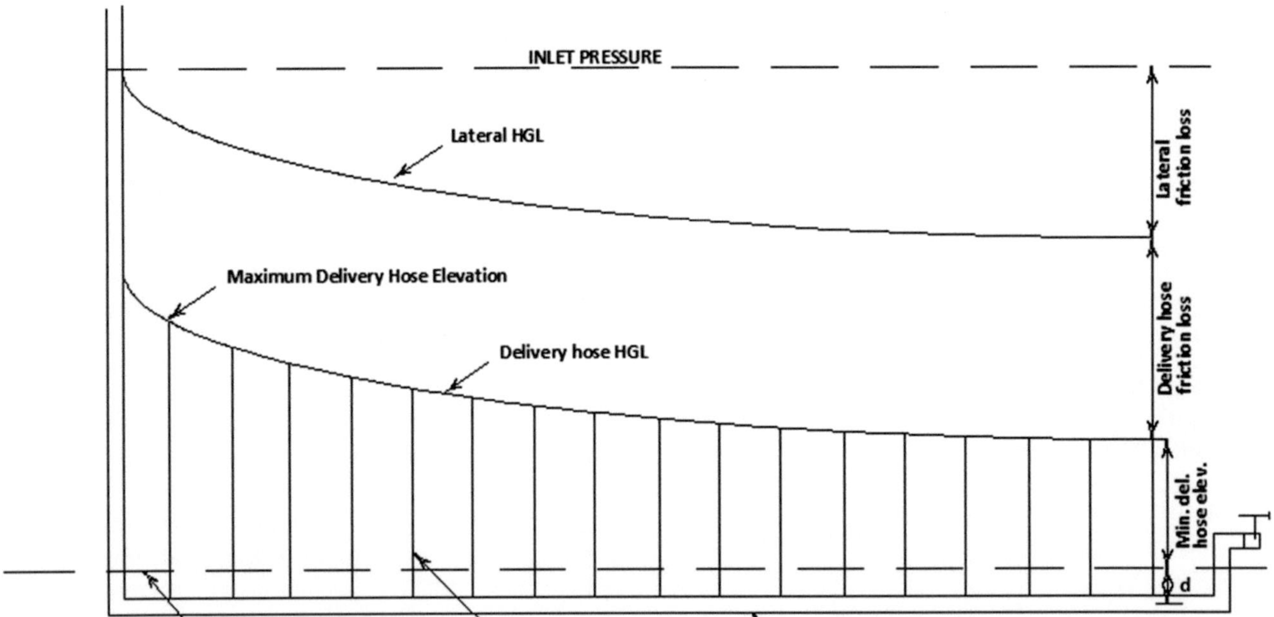

Fig. 22.6 Definition sketch for level ground designs

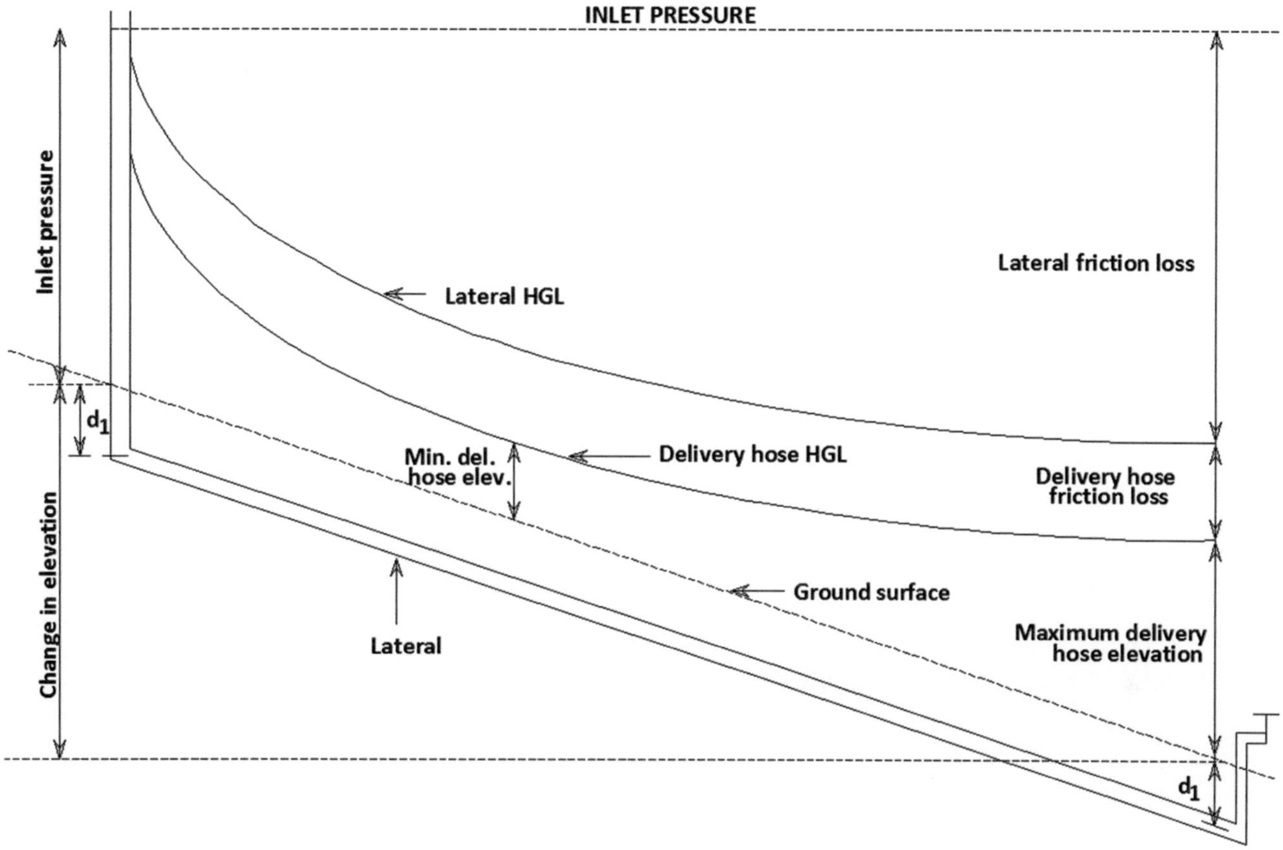

Fig. 22.7 Definition sketch for gradual slope designs

water application to meet the specific crop water requirements for a particular soil and climate are determined following the principles presented in earlier chapters.

The major steps and associated design parameters for bubbler design are summarized as follows:

Step 1: Determine field layout.
Step 2: Design constant head device.
Step 3: Select design criteria.
Step 4: Calculate length and number of pipes.
Step 5: Calculate design flow rates for pipes.
Step 6: Select pipe diameters.
Step 7: Calculate delivery hose elevations and check spreadsheet results.
Step 8: Plot piezometric diagram.
Step 9: Determine time of application
Step 10: Install system and calibrate delivery hose elevations.

A more detailed description of each design step and the associated design parameters follows.

Steps in the Design Process

Determination of Field Layout

The following parameters are determined from the field layout (Fig. 22.8):

Plan and profiles of the piping system.
Field length, L.
Field width, W.
Slope of field, S_L.
Cross-slope of field, S_W.

The field plan and profiles should be surveyed accurately. Elevation surveys with an error of a few inches will affect delivery hose elevations by a few inches and also will affect the uniformity of flow through the system. Typical field layouts of bubbler systems are shown in Fig. 22.9. For level ground designs, standpipes located in the middle of

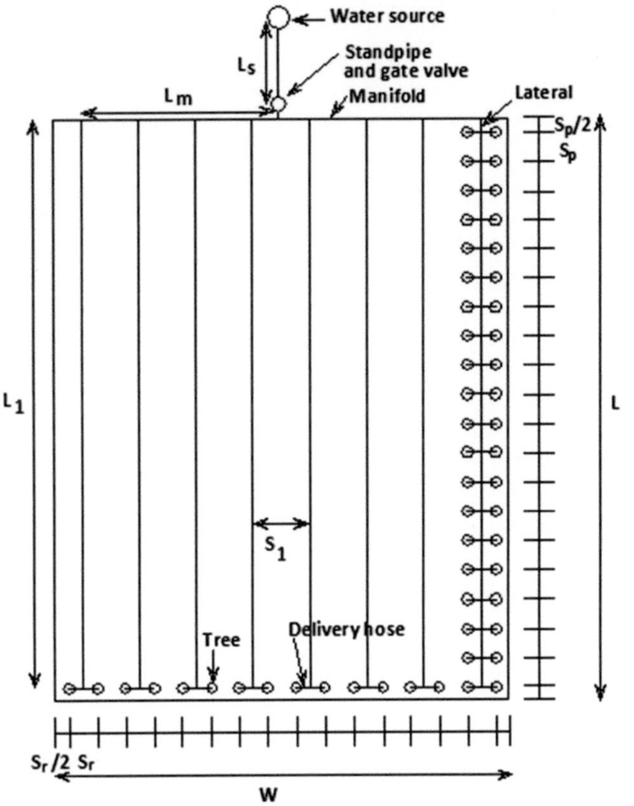

Fig. 22.8 Typical field layout for bubbler system

the field, as shown in Fig.22.9d, will reduce accumulation of friction loss errors since the laterals are shorter, and the design head can more easily be distributed equally to each section of the field.

Design a Constant Head Device

If the water source (reservoir or canal) does not operate at a constant head, a constant head device should be installed near the water source or somewhere along the mainline to a provide constant design head during operation. The constant head device can simply be a standpipe and gate valve. Other types of constant head devices could be considered by a designer, depending on site conditions. A standpipe can also serve as both an air release and sediment trap, as shown in Fig. 22.10.

For systems with a standpipe, two manometers should be installed (one upstream of the gate valve and one downstream) and attached permanently to the standpipe. The purpose of the upstream manometer is to monitor the reservoir elevation, and the purpose of the downstream manometer is to monitor the operating level of the bubbler system. A permanent reference mark should be made on the downstream manometer to indicate the system design head.

One of the critical factors affecting the uniformity of flow in bubbler systems is operating the system at heads different from the design head (the permanent reference mark). Such operation will result in different flow rates for each delivery hose and will severely affect the uniformity of the entire system. Fluctuations above or below the design head should be kept minimal to maintain a higher level of flow uniformity.

The position of the constant head device determines the available head and the design head. The design head is defined as:

$$H_d = H_s - h_s \tag{22.15}$$

where,

H_d = design head, m (ft).
H_s = available head, m (ft).
h_s = friction head loss between the water source and the constant head device, m (ft).

The available head is measured at the water source elevation, and the design head is measured at the constant head device. The available head less the mainline head loss is the design head. The distinguishing differences between the two is that the design head will always remain constant and the available head will fluctuate.

Selection of Design Criteria

The following design criteria must be selected before the design can proceed and are subject to the designer's preference, within certain limits:

1. Type of crop and spacing.
   Spacing per plant, S_p, m, (ft).
   Spacing per tree row, S_r, m, (ft).
   Spacing between laterals, S_l, m, (ft).
   Spacing between delivery hoses, S_e, m, (ft).
   Spacing for first outlet, $S_p/2$ or S_p, m, (ft).
2. Depth of buried lateral, d_l, m, (ft).
3. Maximum and minimum delivery hose elevations, H_{max} and H_{min}, respectively, m, (ft).
4. Delivery hose design flow, q_{dh}, lps (gpm).

Spacings

Plant spacings are dependent on the type of crop, and lateral and delivery hose spacings must conform to the plant spacings. For example, a common citrus tree spacing is

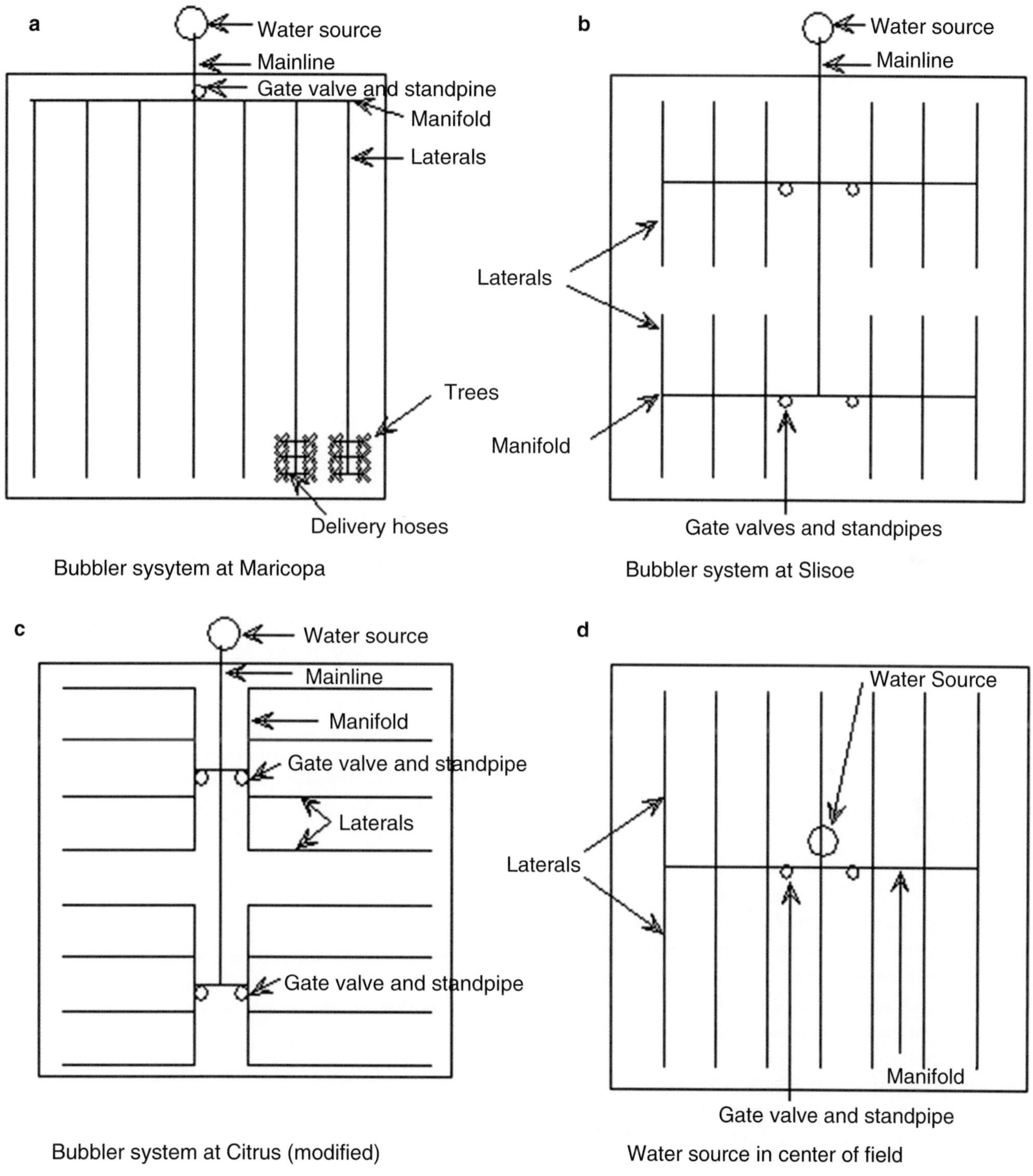

Fig. 22.9 Field layout examples

6.7 × 6.7 meters (22 × 22 ft). So a lateral laid mid-way between tree rows will have a lateral spacing, S_l, that is equal to twice the row spacing, or 13.4 meters (44 ft). The outlet spacing, S_e, is equal to the plant spacing, or 6.7 meters (22 ft). Note that this set-up assumes two delivery hoses are placed next to one another to act as one outlet and to water trees on both sides of the lateral. The first delivery hose outlet is recommended to be located at one-half the tree spacing to minimize friction losses in the lateral.

The number of outlets and the location of the first delivery hose are the two factors required before the Christiansen reduction coefficient can be determined. If two delivery hoses are assumed to act as one outlet, the delivery hose spacing, S_e, is then equal to the plant spacing, S_p, and all

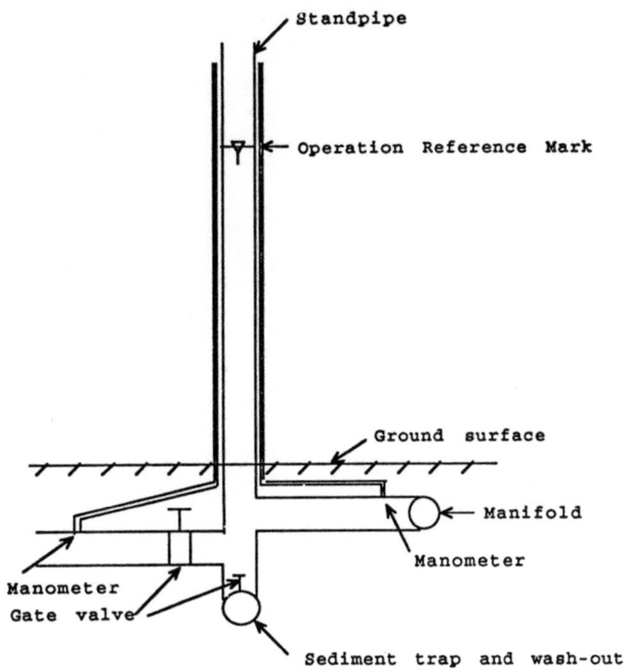

Fig. 22.10 Sketch of a constant head device

delivery hose trenches will be perpendicular to the lateral. However, if one delivery hose is the outlet, than the delivery hose spacing, S_e, is one-half the plant spacing, $S_p/2$, and alternate delivery hose trenches must be dug at less than right angles from the lateral to the tree.

Lateral Burial Depth and Delivery Hose Elevations

The recommended depth of the buried lateral is 0.46 m (1.5 ft) since deeper depths will increase trenching costs and lesser depths may not anchor the pipe adequately or protect it from heavy loads. If heavy machinery will be operated in the orchard, burial depths greater than 0.46 m (1.5 ft) may be required to protect the pipelines.

For practical purpose all delivery hoses in a system are assumed to have the same length. The maximum delivery hose heights are assumed to be greater than or equal 1 m and the minimum hose height should not be less than 0.3 meters. Delivery hoses set at elevations lower than 0.3 meter (1 ft), risk damage from ponded water or trampling by workers or animals. Delivery hose heights could be set at heights higher than 1 meter (3.3 ft), but the falling water would increase soil erosion at the point of impact. One way to increase delivery hoses heights without increasing soil erosion is to place a tee at the point of discharge of the hose, and run a delivery tube from the side of the tee down to the basin. The upper end of the tee is exposed to atmospheric pressure and is set at the calculated delivery hose height to obtain the desired flow.

Delivery Hose Design Flow

Design flow rates for the delivery hoses, q_{dh}, are a function of the minimum hose diameter required to prevent excessive friction losses and clogging; maximum orifice diameter to ensure uniformity of discharge; and flow velocity to flush air out of pipelines

The delivery hose diameter is chosen based on customary orifice diameters of 6–10 mm (1/4–3/8 in). Common delivery hose flow rates of 0.032 or 0.047 lps (0.5 or 0.75 gpm) for 10-mm (3/8-in) diameter tubes have been used in the past bubbler designs. These design flows have velocities greater than the minimum flushing velocities required to prevent air locks from occurring in 10-mm (3/8-in) diameter delivery hoses. For systems where available head is greater than 2 meters (6.6 ft), delivery hoses of 6-mm (1/4-in) diameter might be considered, and deliver flow rates of 0.013–0.063 lps (0.20–1.0 gpm) are required to prevent air locks.

Determination of Length and Number of Pipes

Component parts of a bubbler system include the water source, constant head device, mainline, manifolds, laterals, and delivery hoses. The layout of these components must be determined before pipe diameters can be chosen, and the most critical factors in determining the layout are location of the water source and slope of the field.

The following parameters can be determined by simple geometry from the field layout as illustrated in Fig. 22.8:

1. Length of pipes
   L_s, is the distance between the water source and desired location of the standpipe, m, (ft).

$$L_m = W/2 - S_r \tag{22.16}$$

$$L_l = L - S_p/2 \tag{22.17}$$

$$L_{dh} = S_r/2 + d_l + H_{max} \tag{22.18}$$

2. Number of pipes and trees.

$$N_e = 2L/S_e \tag{22.19}$$

$$N_l = W/2S_r \tag{22.20}$$

$$N_t = N_e * N_l \tag{22.21}$$

where,

L_s = Mainline length, m (ft).
L_m = Manifold length, m (ft).

L_l = Lateral length, m (ft).
L_{dh} = Delivery hose length, m, (ft).
W = Field width, m (ft).
L = Field length, m, (ft).
S_r = Row spacing, m, (ft)
S_p = Plant spacing, m, (ft).
d_l = Depth of lateral burial, m, (ft).
H_{max} = Max. delivery hose height, m (ft).
N_e = Number of delivery hoses per lateral.
S_e = Delivery hose spacing, m (ft).
N_l = Number of laterals.
N_t = Total number of trees.

Calculation of Pipe Design Flows

The pipe design flow rates are calculated from the assumed delivery hose flow rate:

$$q_l = q_{dh} {*} N_e \tag{22.22}$$

$$q_m = q_l {*} N_l \tag{22.23}$$

$$Q_s = N_t {*} q_{dh} \tag{22.24}$$

where,

q_{dh} = Delivery hose flow rate, lps, (gpm).
q_l = Lateral flow rate, lps (gpm).
q_m = Manifold flow rate, lps, (gpm).
Q_s = Mainline flow rate, lps, (gpm).
N_e = Number of delivery hoses along the lateral.

Selection of Pipe Diameters

Keller and Bliesner (1990) note that various designers use different methods to size components of sprinkler and micro-irrigation pipeline networks, including;

1. Economic method – selecting the most economical pipe size when based on long-term pumping costs.
2. Headloss gradient or Unit head loss method – setting a limit on the head loss per unit length of pipe.
3. Velocity method – setting a limit on the flow velocity.
4. Percent head loss method – setting a limit on the friction head loss for each pipe.

The velocity method and the percent head loss method are often used to size pipelines for bubbler systems. The velocity method is useful for sizing the larger diameter pipes such as the mainline and manifolds, because long-term sediment deposition in the large diameter pipelines is a major concern. When using the velocity method, the continuity equation is used:

$$Q = V\ A = \frac{V\ D^2}{K} \tag{22.25}$$

where,

K = 1273 for metric units, (0.4058, for English units)
Q = flow rate, lps (gpm)
V = velocity of flow in pipe, mps (fps)
D = inside diameter of pipe, mm (in)
A = inside cross-sectional area of pipe, mm^2 $(in)^2$

The value of 0.6 mps (2 fps) in the laterals was recommended by Rawlins (1977) and Worstell (1975) to prevent deposition of silt-laden sediment from irrigation waters in the laterals. Whether this velocity is sufficient to transport the sediment as suspended load in a given case depends on the particle size distribution of the material.

Pipes sometimes must be designed with flow velocities less than 0.3 mps (2 fps) to prevent excessive friction loss. Under such conditions, it may be necessary to construct a sediment trap along the mainline or to flush out the system regularly to maintain design capacity. In addition, gate valves or caps should be placed at the end of each manifold and lateral for flushing purposes.

The headloss gradient method is commonly used in microirrigation design. Figures 22.11 and 22.12 are examples of typical design charts for systems of PE, PVC, and corrugated PE pipe which are also applicable for low-head gravity systems.

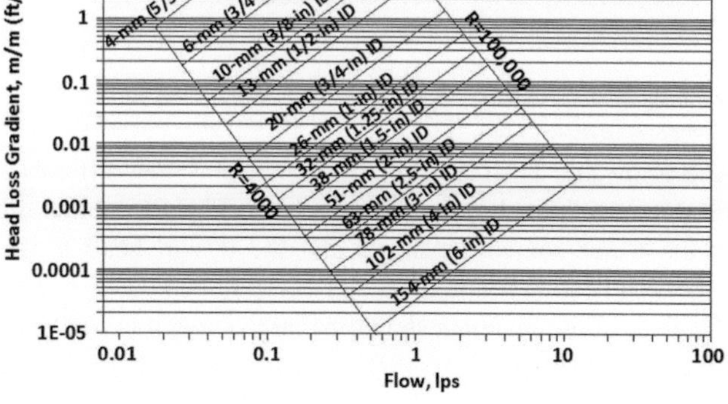

Fig. 22.11 Head loss gradient chart for smooth (PE and PVC) pipes (water temprature = 20 °C (68 °F)) (After Reynolds 1993)

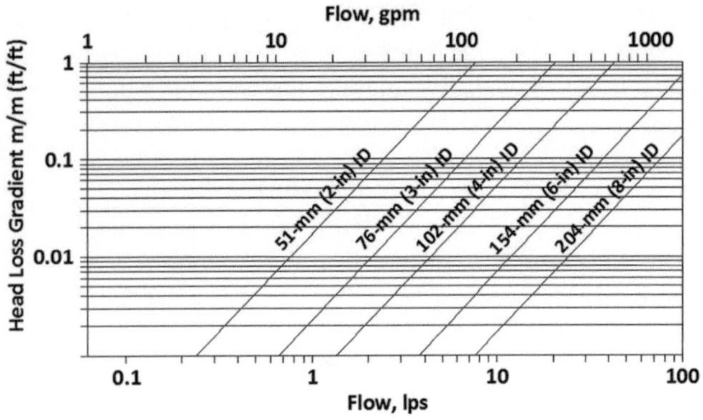

Fig. 22.12 Head loss gradient for corrugated PE pipe (f = 0.07, (water temprature = 20 °C (68 °F)) (After Reynolds 1993)

Allowable Head Loss

The allowable head loss gradient is determined by the following equation for both level ground and gradual-slope designs:

$$\frac{h_f}{L} = \frac{(H_u - H_{do}) - \Delta z}{FL} = \frac{(h_f)_a}{FL} \tag{22.26}$$

Where,

h_f/L = head loss gradient, m/m (ft/ft).
H_u = pressure head upstream, m (ft).
H_{do} = pressure head downstream, m (ft).
Δz = change in elevation between upstream and downstream and is negative for downslopes, m (ft).
$(h_f)_a$ = Allowable head loss in the pipe, m (ft)

The design procedure for bubbler systems usually proceeds starting from the water source to the distal end of the field. First, the mainline is sized with respect to the available head and the design head. Next, the manifold is sized as between 5 % and 20 % of the design head. Finally, the laterals and delivery hoses are sized simultaneously by the percent head method.

Sizing Mainlines

The allowable head loss in the mainline is the difference between available head and design head. The available head is the water source elevation, and the design head is the operating head at the standpipe or constant head device. The diameter of the mainline should be large enough to maintain flushing velocities of 0.6 mps (2 fps) to prevent long-term sediment deposition problems. If the available head is not great enough to maintain velocities of 0.6 mps (2 fps), the diameter should be sized according to the difference between available head and the desired design head. In the later case, it may be desirable to install a small sediment trap near the intake or near the constant head device.

Sizing Manifolds

The design of manifolds is easier than the design of laterals since the delivery hoses do not have to be sized at the same time. Keller and Karmelli (1974) indicate that when a percent head method is used to size manifolds for pressurized systems, the most economical design is with approximately 55 % of the head loss in the laterals and 45 % in the manifold. However, it is important to note that this design criteria for pressurized systems is not applicable to bubbler systems because the pressure gradient is small for low-head bubblers and the delivery hoses will consume a large percent of the design head. Therefore, the allowable head loss in the bubbler manifold must be less than 45 % of the design head.

Past bubbler designs has shown that the allowable manifold head loss, h_m, should be 20 % or less of the design head at the standpipe. For manifolds sized with a pressure variation of 20 % of design head, friction losses in the manifold must be calculated to determine the inlet pressure for each lateral. In this case, the delivery hose elevations will have to be calculated separately for each lateral because inlet pressure heads for each lateral will be different.

A simple way to minimize calculations for delivery hose elevations is to assume the manifold friction losses to be 5 % or less of the design head. This assumption will increase the manifold diameter, but manifold friction losses can be neglected, allowing all laterals along the manifold to have the same inlet pressure. Thus, calculations of delivery hose elevations for one lateral can be applied to all laterals connected to the same manifold, and the number of hose elevation calculations will be reduced by the number of laterals connected to the manifold. This assumption should be avoided for long manifolds with five or more laterals, because the increase in manifold diameter will be prohibitively expensive.

Sizing Laterals and Delivery Hoses

Table 22.4 presents comparative (1993) prices of PVC and corrugated PE pipe materials and is a useful guide for sizing pipe diameters. Costs are approximately the same for rigid PVC pipe and corrugated PE pipe for lateral diameters of 63 mm (2.5 in) and less. Therefore, PVC pipes are recommended for diameters of 63 mm (2.5 in) and smaller to minimize friction losses.

For diameters greater than 63 mm (2.5 in), corrugated PE laterals are cheaper, but this pipe material is unsuitable under certain conditions due to excessive head losses on low-head systems. In addition, using corrugated PE pipes

Table 22.4 Common pipe diameters and cost for corrugated PE tubing, rigid PVC pipe, and flexible PE tubing (1993)[a]

Internal diameter (in)	Equivalent units (mm)	Current cost($/ft)/comments
(a) Corrugated PE tubing (LIDCO, manufacturers)		
2	52	0.24
3	78	0.33
4	102	0.39
6	154	0.98
8	204	1.43
10	254	Available from manufacturer upon special request
12	306	
(b) Rigid PVC pipes (common sizes for laterals)		
1.25	32	0.13
1.5	38	0.16
2	52	0.24
2.5	63	0.36
3	78	0.52
4	102	0.95
6	154	1.68
(c) PE tubing (Hardie irrigation, manufacturers)		
5/32	4	0.03
1/4	6	0.04
3/8	10	0.05
1/2	13	Not recommended for delivery hoses due to low flow uniformity
9/16	15	
5/8	16	
11/16	18	
3/4	20	
1	26	

[a]Prices quotes from Irrigation and Sprinkler Supply, Inc.; Tucson, Arizona; March, 1993

may provide only minimal savings because Figs. 22.13 and 22.14 indicate that the flow capacity for a given diameter corrugated PE pipe is the same as for the next smaller diameter PVC pipe available. Table 22.4 also shows that the next smaller diameter PVC pipe that is available typically costs approximately the same as the larger corrugated PE pipe. Since the amount of flow and head loss gradient are the governing factors in bubbler design, a smaller diameter PVC pipe will, in most cases, cost the same and transmit the same flow per unit head loss as the next larger diameter corrugated PE pipe that is commercially available.

Calculation of Delivery Hose Elevations

Total Head Losses

After all pipe diameters have been determined, the elevations of the delivery hoses can be calculated by determining the total head loss from the constant head device to the point of discharge. The total head loss is the sum of the friction and minor losses.

$$h_l = \sum h_f + \sum h_{ml} \tag{22.27}$$

where,

h_l = Total head loss in pipes, m (ft)
h_f = Friction loss in pipes, m (ft)
h_{ml} = Minor losses in pipes, m (ft)

Friction loss equations from Table 22.1 must be used to estimate friction losses in the pipes. The head loss gradient charts in Figs.22.11 and 22.12 cannot be used to calculate delivery hose elevations because interpolation from these charts will not be sufficiently accurate to determine friction losses in the laterals and delivery hoses.

Minor Losses

All minor losses between the water source and the point of discharge must be accounted for, including pipe entrances, elbows, tees, reducers, valves, etc.,

$$h_{ml} = C_L \frac{V^2}{2g} \tag{22.28}$$

where,

C_L = Loss coefficient in various pipe fittings

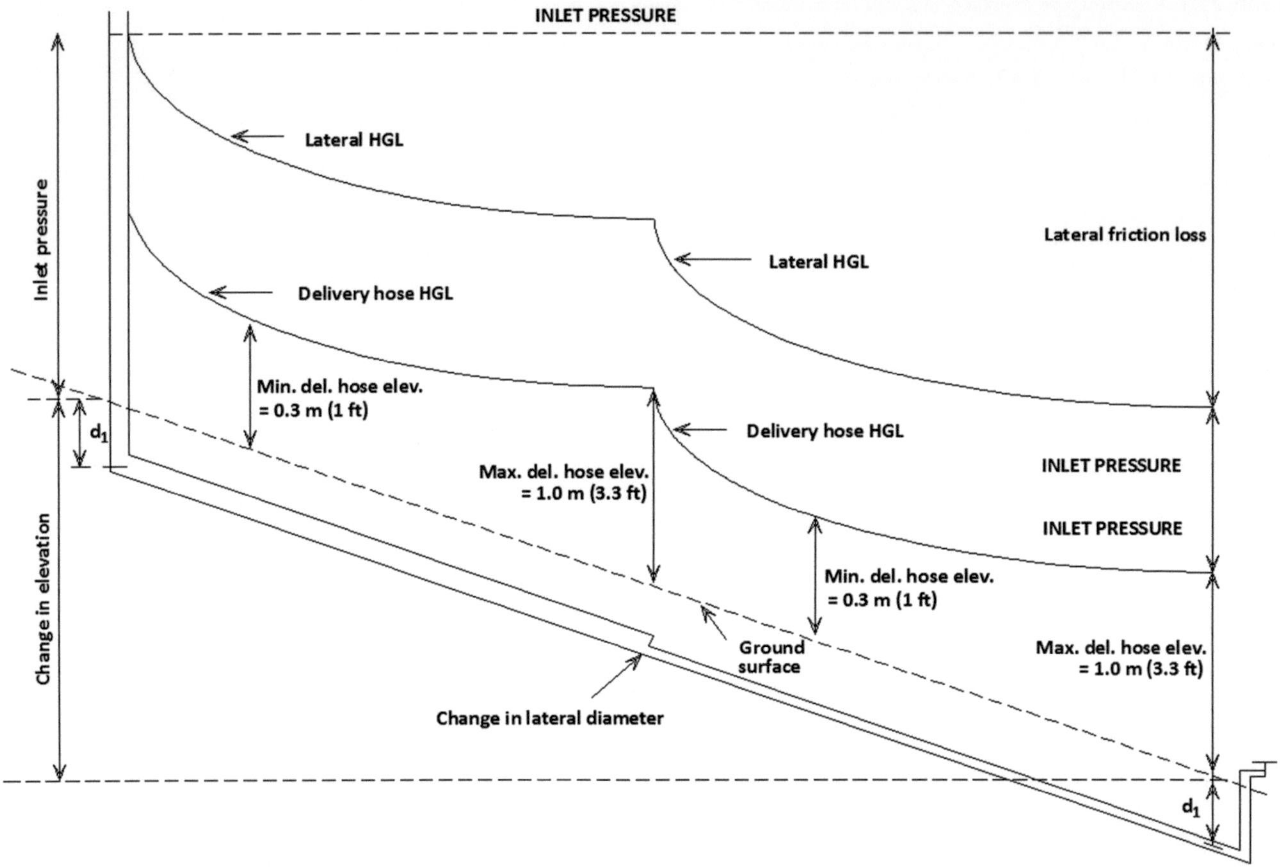

Fig. 22.13 Change in lateral diameter

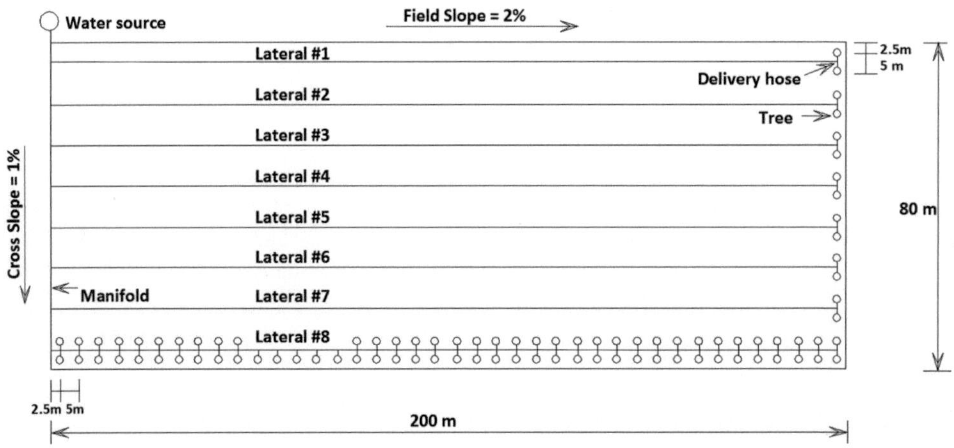

Fig. 22.14 Field layout for gradual slope design

Keller and Bliesner (1990) give a complete list of loss coefficients, and some common C_L values for bubbler systems are listed in Table 22.5.

A simple method to use to account for head losses in bubbler systems is to compute or assume friction and minor losses from the constant head device to the lateral inlet junction. Then account separately in the spreadsheet for friction and minor losses from the lateral inlet to the point of discharge.

In general, minor losses from the lateral inlet to the point of discharge will include losses due to:

1. The delivery hose projection into the lateral.
2. The change in lateral diameter.

Table 22.5 Common loss coefficients, C_L, for bubbler systems[a]

Fitting	C_L
Tees	0.9
Elbows	0.9
Square entrance	0.5
Projected entrance	0.8 – 1.0
Pipe contraction	$0.7[1 - (D_{sm}/D_{la}]^b$
Gate valve, fully open	0.2
Gate valve, 3/4 open	1.0
Gate valve, 1/2 open	5.6
Gate valve, 1/4 open	24

[a]From Albertson et al. (1960) and Keller and Bliesner (1990)
[b]When determining the velocity head loss at a reducer, the diameter and flow that gives the highest head should be used. D_{sm} is the smaller diameter pipe and D_{la} is the larger diameter pipe

The velocity head term from the energy equation must also be accounted for and can be combined easily with the minor losses when calculating the delivery hose elevations in the spreadsheet tabulations. The velocity head may be computed by using:

$$\frac{V^2}{2g} = K\frac{Q^2}{D^4} \qquad (22.29)$$

where,

K = 8.26 × 10^4 for metric units (2.59 × 10^{-3} for English units)
Q = flow rate, lps (gpm)

Plot Piezometric Diagram

A piezometric diagram for the laterals and delivery hoses can be constructed from the above delivery hose elevation table by plotting the hose elevations on the lateral elevation profile. Spreadsheet software such as Excel can be used for plotting an energy diagram of each lateral to help ensure the design is correct.

Install System and Calibrate Delivery Hose Elevations

Once the piezometric diagram is drawn, the bubbler system can be installed. The design head elevation marked on the constant head device (standpipe) is recommended as the best reference datum for installing the system to prevent ground undulations from affecting the delivery hose elevation. After all delivery hoses have been set at the calculated elevations, the system should be turned on and dynamically calibrated, as described earlier, to ensure nearly equal flow rates at each delivery point.

Design Examples

Gradual Slope Design

The major difference between designs for laterals on gradual slopes and on level ground is that energy is gained on sloping ground and the lateral diameter must be sized to keep the lateral hydraulic grade line (HGL) approximately parallel to the ground slope. Also, the minimum delivery hose elevation of 0.3 m (1 ft) is no longer located at the end of the lateral, but will occur somewhere upstream, Figs. 22.7 and 22.13.

For gradual slope designs, care must be taken to analyze the delivery hose HGL so that it is not less than the minimum delivery hose elevation or greater than the maximum delivery hose elevation. However, designs within ±0.15 m (0.5 ft) of the delivery hose design elevation limits may be acceptable since these limits were chosen arbitrarily and are not fixed.

If the delivery hose HGL is not located between approximately the maximum and minimum delivery hose elevations, the lateral diameter must be reduced somewhere along the lateral to allow the HGL to run more parallel to the ground slope. For bubbler systems with long laterals and steep slopes, the lateral diameter may have to be reduced several times to keep the lateral HGL approximately parallel to the ground surface. Pressure regulators can be used for lateral designs on steep slopes, but this will increase material costs.

Calculations to determine where to decrease lateral diameters down-slope for multiple outlet pipes are based on the following equation, (Keller and Bliesner 1990):

$$L_{sm} = L_t\left[\frac{(h_f)_a - h_{la}}{h_{sm} - h_{la}}\right]^{\frac{1}{2.75}} \qquad (22.30)$$

where,

L_{sm} = length of the smaller pipe, m (ft).
L_t = total length of the lateral, m (ft).
$(h_f)_a$ = allowable head loss in the lateral, m (ft).
h_{la} = friction loss for larger diameter pipe along the entire lateral length, m (ft)
h_{sm} = friction loss for smaller diameter pipe along the entire lateral length, m (ft).

The above equation is very useful for determining where to make a single reduction of the lateral diameter (two pipe sizes). For longer laterals, the pipe diameter may have to be reduced several times, and Keller and Bliesner (1990) describe in detail several procedures for multiple diameter reductions.

For very long laterals, note that calculation errors for friction head losses will accumulate and that the f-value used for design is more prone to change over time due to scum build-up on the pipe walls or from sediment deposition in the laterals. Rawlins (1993) recommends that delivery hose elevations be adjusted by dynamic calibration from time to time to compensate for these changes, and he also recommends placing gate valves at the nodes of each lateral reduction to help control the flow through the pipeline and to help compensate for accumulation of friction loss errors for long laterals over time. While gate valves can help to control the flow, they will not solve completely the problem of accumulated friction loss errors when determining the delivery hose elevations. Also, placement of valves on each node will increase material costs. Therefore, the maximum length of laterals on gradual slopes is limited by the accumulation of friction loss errors, and Thornton and Behoteguy (1980) recommended that lateral lengths longer than 457 m (1500 ft) be avoided

Gradual Slope Design Example

Example 22.1 Design a bubbler irrigation system to irrigate a deciduous orchard with tree spacings of 5 m × 5 m (16.4 ft × 16.4 ft). The bubbler system will replace a furrow system to improve irrigation uniformity and efficiency.

The field has a slope of 2 %, with a cross-slope of 1 %, and field dimensions are 80 m × 200 m (26.2 ft × 656 ft). Reduce the diameter of the laterals and manifold each once (two pipe sizes).

Assume a design head of 1.1 meter (3.6 ft) can be maintained at the standpipe and the maximum and minimum delivery hose elevations are 1 m (3.3 ft) and 0.3 m (1 ft), respectively.

Calculations Follow the design procedure oulined earlier.

Step 1: Determine field layout.

Refer to Fig. 22.14 for the field layout.

Given:

Field length = L = 200 m (656 ft)

Field width = W = 80 m (26 ft)

Slope of field = $S_L = 0.02$

Cross slope of field = $S_W = 0.01$

Step 2: Design constant head device.

Constant head device is standpipe with gate valve.

The given design head = $H_d = H_s = 1.1$ m (3.6 ft)

Step 3: Select design criteria.

Given:

Tree spacing = $S_p = S_r = 5$ m (16 ft)

Maximum delivery hose elev. = $H_{max} = 1.0$ m (3.3 ft)

Minimum delivery hose elev. = $H_{min} = 0.3$ m (1.0 ft)

Assume:

Lateral is laid mid-way between two rows of trees and two delivery hoses comprise an outlet.

Del. hose spacing $= S_e = S_p = 5$ m (16 ft)

Lateral spacing $= S_l = 2S_r = 10$ m(32 ft)

Depth of lateral $= d_l = 0.46$ m(1.5 ft)

Delivery hose flow $= q_{dh} = 0.032$ lps(0.5 gpm)

Step 4: Calculate length and number of pipes.

Length of pipes:

Length of mainline = L_s = not applicable

Length of manifold = L_m = W-S_r

= 80–5 = 75 m (246 ft)

Length of lateral = $L_l = L - S_p/2$

= 200–5/2 = 197.5 m (648 ft)

Length of del. hose = $L_{dh} = S_r/2 + d_l + H_{max}$

= (5 m)/2 + 0.46 m + 1 m

= 4.0 m (13 ft)

Number of pipes and trees:

Number of del. hoses per lateral = $N_e = 2\ L/S_e$

=(2)200 m/5 m = 80

Number of laterals = $N_l = W/2S_r$ = 80 m/10 m = 8

Total number of trees = $N_t = N_e * N_l$ = 640 trees

Step 5: Calculate design flow rate for pipes.

Flow in lateral = $q_l = q_{dh} * N_e$

= (0.03155)(80) = 2.52 lps (40 gpm)

Flow in manifold = $q_m = q_l * N_l$

= (2.52)(8) = 20.16 lps (320 gpm)

Flow in mainline = $Q_s = q_{dh} * N_t$

= (0.03155)(640)

= 20.16 lps (320 gpm)

Step 6: Select pipe diameters.

Manifold Diameter

The manifold has a 1 % slope; therefore the pressure head at each lateral inlet will be slightly different. The manifold diameter should be reduced once (two pipe sizes) to maintain an inlet pressure head of nearly 1.1 meter (3.6 ft) for each lateral.

The allowable head loss can be obtained from Eq. 22.25,

$$(h_f)_a = (H_u - H_{do}) - \Delta z = (1.1 - 1.1) - (-0.01(75)) = 0.75\text{m}(2.46\text{ft})$$

From Table 22.3, the Christiansen reduction coefficient, F, is 0.42 for 8 outlets, with the first outlet one full spacing from the inlet. By using the above values into Eq. 22.25, the manifold head loss gradient is:

$$h_f/L = (h_f)_a/L_m = 0.75/(0.42)(75) = 0.023$$

From Fig. 22.15, a design flow of 20.19 lps (320 gpm) and a head loss gradient of 0.023 will give a corrugated PE pipe with diameter of 204-mm (8-in).

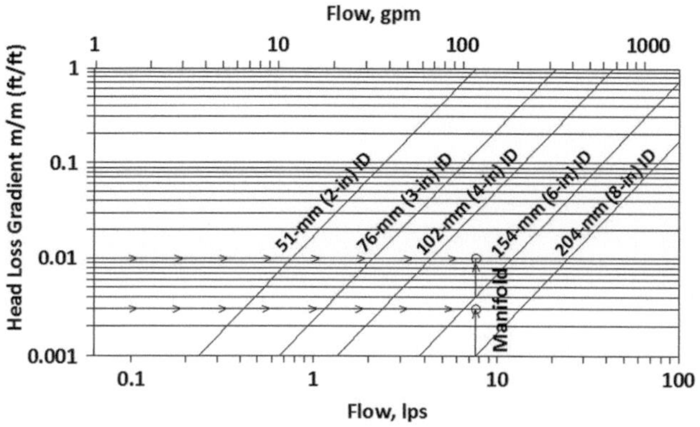

Fig. 22.15 Sizing the manifold and lateral with corrugated PE pipe (from Fig. 22.12)

To reduce the manifold to a smaller size, the head loss for water to flow the entire length of the smaller and larger pipes must be determined. From Fig. 22.15, the head loss gradient for the 204-mm (8-in) corrugated PE pipe is 0.007. From Eq. 22.25, the total head loss running along the entire length for the 204-mm (8-in) pipe is equal to:

$$h_{la} = FL(h_f/L) = (0.42)(75m)(0.007m) = 0.22m(0.72ft)$$

To reduce the pipe to a smaller size, a pipe size of 102-mm (4-in) is chosen because reducing to a 154-mm (6-in) pipe diameter does not reduce the lateral inlet pressures along the manifold very much. By using Fig. 22.15, the head loss gradient for a 102-mm (4-in) pipe is 0.23, and the total head loss running along the entire length is:

$$h_{sm} = FL(h_f/L) = (0.42)(75)(0.23) = 6.9m(22.7ft)$$

From Eq. 22.29, the length of the smaller diameter pipe is:

$$L_{sm} = 75\left[\frac{0.75 - 0.22}{6.9 - 0.22}\right]^{\frac{1}{2.75}} = 29.8 \text{ m } (98 \text{ ft})$$

A pipe length of 30 m (98.4 ft) is chosen for the 102-mm (4-in) section since this is the closest length to a lateral node on the manifold.

The piezometric diagram tabulation for the manifold is shown in Table 22.6 and illustrated in Fig. 22.16. Friction loss (Eq. 22.12), obtained from Table 22.1, was used to calculate column (6) of Table 22.6.

Lateral Diameter

The allowable head loss for the first lateral is obtained from Eq. 22.25 and is;

$$(h_f)_a = (H_u - H_{do}) - \Delta z = (1.084 - 1) - (-0.02)(197.5) = 4.03 \text{ m}(13.23 \text{ ft})$$

Where,

H_u = lateral inlet pressure for lateral # 1 (from Column 9, Table 22.6).

From Table 22.3, the reduction coefficient, F, is 0.36 for 40 multiple outlets, with the first outlet being one-half space from the inlet. By substituting the above values into Eq. 22.25, the head loss gradient for the lateral is:

$$h_f/L = (h_f)_a/FL_l = 4.03/(0.36(197.5)) = 0.057$$

To determine if a corrugated PE or smooth PVC pipe should be used for the laterals, a flow of 2.52 lps (40 gpm) and head loss gradient of 0.057 from Fig. 22.15 will give a 76-mm (3-in) corrugated PE pipe; and from Fig. 22.15, a 51-mm (2-in) PVC pipe . From Table 22.4, the 51-mm (2-in) PVC and 76-mm (3-in) corrugated PE pipes costs the same, however the 51-mm (2-in) PVC pipe is chosen because it is rigid.

To reduce the diameter of the PVC lateral one pipe size, the next available diameter is found from Table 22.4 to be 38-mm (1.5-in) PVC pipe. From Fig. 22.17, the head loss gradients for the larger and smaller pipes are 0.03 and 0.15, respectively, and the head losses along the entire length for 38-mm (1.5-in) and 51-mm (2-in) diameter pipes are:

$$h_{la} = FL_l(h_f)_a = (0.36)(197.5m)(0.03) = 2.13m(7.0ft)$$
$$h_{sm} = FL_l(h_f)_a = (0.36)(197.5m)(0.15) = 10.7m(35.1ft)$$

From Eq. 22.29, the approximate length of the smaller diameter pipe is:

$$L_{sm} = 197.5\left[\frac{4.03 - 2.13}{10.7 - 2.13}\right]^{\frac{1}{2.75}} = 114.9 \text{ m}(376.9 \text{ ft})$$

A pipe length of 115 m (377 ft) is chosen for the 38-mm (1.5-in) diameter section along the lateral since the lateral nodes along the manifold occur at 5 m (16.4 ft) spacings.

Delivery Hose Diameter

The allowable delivery hose friction head loss for gradual slopes is the upstream pressure head less the minimum delivery hose height:

$$h_{dh} = 1.084 - 0.3m = 0.78m(2.6ft).$$

Table 22.6 Manifold piezometer tabulation (gradual slope design) From BUBBLER program

Manifold (two pipe sizes)									
Larger diameter				=	204 mm (8 in)				
Length				=	45 m (147.6 ft)				
Flow range				=	20.16–10.1 lps (320–160 gpm)				
Smaller diameter				=	102 mm (4 in)				
Length				=	30 m (98.4 ft)				
Flow range				=	7.56–2.52 lps (120–40 gpm)				
Out. Sect.	Flow	Section Length	Total Length	Ground Elevation	Manifold Hf	Minor Losses	Total Hf	Manifold HGL	Lat. Inlet Presuure
(i, j)	(lps)	(m)	(m)	(m)	(m)	(m)	(m)	(m)	(m)
(1)	(2)	(3)	(4)	(5)	(6)	(7)	(8)	(9)	(9)
0	0	0	0	0.00	0.000	0.00	0.000	1.100	0.000
1	20.160	5	5	−0.05	0.033	0.03	0.066	1.034	1.084
2	17.640	10	15	−0.15	0.051	0.03	0.084	0.983	1.133
3	15.120	10	25	−0.25	0.037	0.03	0.070	0.945	1.195
4	12.600	10	35	−0.35	0.026	0.03	0.059	0.919	1.269
5	10.080	10	45	−0.45	0.017	0.03	0.050	0.903	1.353
6	7.560	10	55	−0.55	0.299	0.04	0.339	0.597	1.147
7	5.040	10	65	−0.65	0.133	0.04	0.173	0.464	1.114
8	2.520	10	75	−0.75	0.033	0.04	0.073	0.430	1.180
Manifold friction head loss				=	0.630 m				

and the head loss gradient is:

$$h_f/L_{dh} = 0.78m/4m = 0.195m/m(ft/ft)$$

From Fig. 22.17, a head loss gradient of 0.195 and a design flow of 0.0315 lps (0.5 gpm) will give a delivery hose diameter, D_{dh}, of 10 mm (3/8 in).

Step 7: Calculate delivery hose elevations

After the lateral and delivery hoses are determined, the delivery hose elevations can be calculated and tabulated as presented in Table 22.7. The friction loss equations used for columns (6) and (8) are from Table 22.1. The same analysis (steps 6 and 7) must be done for each lateral because the inlet pressure at each lateral is different.

Step 8: Plot piezometric diagram

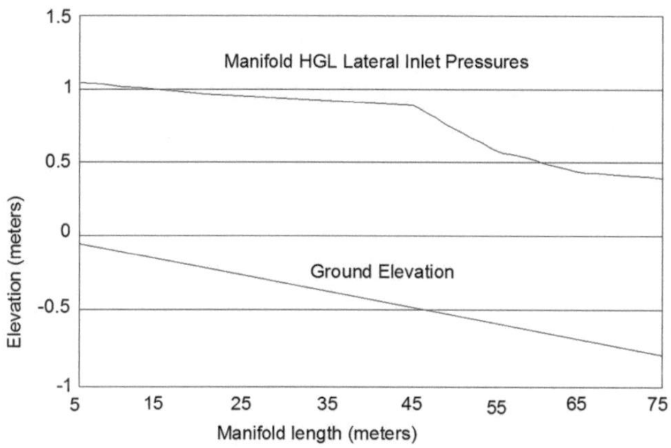

Fig. 22.16 Manifold piezometric diagram for gradual slope design

Note that the delivery hose elevations in column (12) of Table 22.7 are slightly lower than the minimum delivery hose design height of 0.3 m (1 ft). Since the values are within ±0.15 m (0.5 ft) of the delivery hose design limits, the HGL curve can be considered acceptable.

The piezometric diagram for the lateral and delivery hoses is shown in Fig. 22.18.

Step 9: Determine time of application (Previous chapters)

Step 10: Install system and calibrate hose elevations.

Follow the same procedure as previously described.

Finally from Fig. 22.17 the following diameters for laterals and delivery hoses are obtained.

$$D_l = 63mm(2.5in)$$
$$D_{dh} = 10mm(3/8in)$$

Step 7: Calculate Delivery Hose Elevations

After the lateral and delivery hose diameters are determined, the elevation of the delivery hoses can be calculated and tabulated.

Step 8: Plot Piezometric Diagram

From column (12) of Table 22.7, the delivery hose elevations are between the maximum and minimum delivery hose elevations. A piezometric diagram for the laterals and the delivery hoses are constructed from Table 22.7 by plotting column (5), (7), and (11) on the lateral elevation profile, as shown in Fig. 22.18.

A piezometric diagram for the laterals and delivery hoses are constructed from Table 22.7 by plotting columns (5), (7), and (11) on the lateral elevation profile, as shown in Fig. 22.18.

Fig. 22.17 Sizingof the lateral and delivery hose with smooth PE or PVC pipe (From Fig. 22.11)

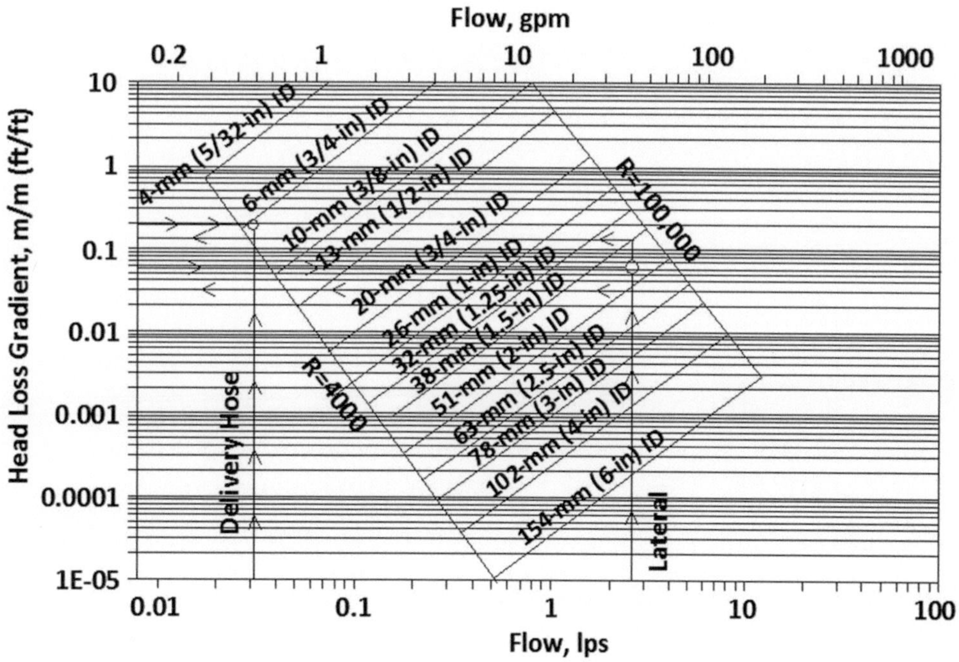

Table 22.7 Delivery hose elevation tabulation (gradual slope design) from BUBBLER program

Lateral # 1 (with two pipe sizes)		
Larger lateral diameter	=	51 mm (2 in)
Larger pipe length	=	82.5 m (270.7 ft)
Larger pipe flow range	=	2.52–1.5 lps (40–24 gpm)
Smaller lateral diameter	=	38 mm (1.5 in)
Smaller pipe length	=	115 m (377.2 ft)
Smaller pipe flow range	=	0.045–0.063 lps (23–1 gpm)
Delivery hose diamater	=	10 mm (3/8 in)
Delivery hose length	=	4 m (13.12 ft)
Delivery hose flow rate	=	0.0315 lps (0.5 gpm)

Out/ Sect. (i,j)	Flow Qj	Section Length Lj	Total Length (Ll)j	Ground Elev. (z)j	Lat.F/loss (hlat)j	Lat. HGL (HGL)i	Hose F/loss (hdh)i	V/Head + Minor losses	Total losses (hl) i	Hose HGL (HGL)i	Hose Elev. (z)	Refer Datum (y)
	(lps)	(m)	(m)	(m)	(m)	(m)	(m)	(m)	(m)	(m)	(m)	(m)
(1)	(2)	(3)	(4)	(5)	(6)	(7)	(8)	(9)	(10)	(11)	(12)	(13)
0			0.0	0.00		1.084				0.935		0.00
1	2.520	2.5	2.5	−0.05	0.077	1.007	0.133	0.016	0.226	0.858	0.908	−0.24
2	2.457	5.0	75	−0.15	0.147	0.860	0.133	0.016	0.296	0.711	0.861	−0.39
3	2.394	5.0	12.5	−0.25	0.141	0.719	0.133	0.016	0.290	0.570	0.820	−0.53
4	2.331	5.0	17.5	−0.35	0.134	0.585	0.133	0.016	0.283	0.436	0.786	−0.66
5	2.268	5.0	22.5	−0.45	0.128	0.456	0.133	0.016	0.277	0.307	0.758	−0.79
6	2.205	5.0	27.5	−0.55	0.122	0.335	0.133	0.016	0.271	0.186	0.736	−0.91
7	2.142	5.0	32.5	−0.65	0.116	0.219	0.133	0.016	0.265	0.070	0.720	−1.03
8	2.079	5.0	37.5	−0.75	0.110	0.109	0.133	0.016	0.259	−0.040	0.710	−1.14
9	2.016	5.0	42.5	−0.85	0.104	0.004	0.133	0.016	0.253	−0.145	0.706	−1.24
10	1.953	5.0	47.5	−0.95	0.099	−0.094	0.133	0.016	0.247	−0.243	0.707	−1.34
11	1.890	5.0	52.5	−1.05	0.093	−0.187	0.133	0.016	0.242	−0.336	0.714	−1.44
12	1.827	5.0	57.5	−1.15	0.088	−0.275	0.133	0.016	0.237	−0.424	0.726	−1.52
13	1.764	5.0	62.5	−1.25	0.083	−0.357	0.133	0.016	0.231	−0.506	0.744	−1361
14	1.701	5.0	67.5	−1.35	0.077	−0.435	0.133	0.016	0.226	−0.584	0.766	−1.68
15	1.1638	5.0	72.5	−1.45	0.072	−0.507	0.133	0.016	0.221	−0.656	0.794	−1.76
16	1.575	5.0	77.5	−1.55	0.068	−0.575	0.133	0.016	0.217	−0.724	0.826	−1.82
17	1.512	5.0	82.5	−1.65	0.063	−0.638	0.133	0.016	0.212	−0.787	0.863	−1.89
18	1.449	5.0	87.5	−1.75	0.237	−0.875	0.133	0.017	0.386	−1.040	0.727	−2.14
19	1.386	5.0	92.5	−1.85	0.219	−1.093	0.133	0.017	0.368	−1.259	0.608	−2.36
20	1.323	5.0	97.5	−1.95	0.202	−1.295	0.133	0.017	0.351	−1.461	0.506	−2.56
21	1.260	5.0	102.5	−2.05	0.185	−1.480	0.133	0.017	0.335	−1.646	0.421	−2.75
22	1.197	5.0	107.5	−2.15	0.169	−1.650	0.133	0.017	0.319	−1.816	0.351	−2.92
23	1.134	5.0	112.5	−2.25	0.154	−1.804	0.133	0.017	0.303	−1.970	0.297	−3.07
24	1.071	5.0	117.5	−2.35	0.139	−1.943	0.133	0.017	0.289	−2.109	0.258	−3.21
25	1.008	5.0	122.5	−2.45	0.125	−2.069	0.133	0.017	0.275	−2.234	0.233	−3.33

26	0.945	5.0	127.5	−2.55	0.112	−2.180	0.133	0.017	0.261	−2.346	0.221	−3.45
27	0.882	5.0	132.5	−2.65	0.099	−2.280	0.133	0.017	0.249	−2.446	0.221	−3.55
28	0.819	5.0	137.5	−2.75	−0.087	−2.367	0.133	0.017	0.236	−2.533	0.234	−3.63
29	0.756	5.0	142.5	−2.85	0.076	−2.443	0.133	0.017	0.225	−2.608	0.258	−3.71
30	0.693	5.0	147.5	−2.95	0.065	−2.508	0.133	0.017	0.214	−2.674	0.293	−3.77
31	0.630	5.0	152.5	−3.05	0.055	−2.563	0.133	0.017	0.204	−2.729	0.338	−3.83
32	0.567	5.0	157.5	−3.15	0.046	−2.809	0.133	0.017	0.195	−2.774	0.393	−3.87
33	0.504	5.0	162.5	−3.25	0.037	−2.645	0.133	0.017	0.187	−2.812	0.455	−3.91
34	0.441	5.0	167.5	−3.35	0.030	−2.675	0.133	0.017	0.179	−2.841	0.526	−3.94
35	0.378	5.0	172.5	−3.45	0.023	−2.698	0.133	0.017	0.172	−2.864	0.603	−3.94
36	0.315	5.0	177.5	−3.55	0.016	−2.714	0.133	0.017	0.166	−2.880	0.687	−3.98
37	0.252	5.0	182.5	−3.65	0.011	−2.725	0.133	0.017	0.160	−2.891	0.776	−3.99
38	0.189	5.0	187.5	−3.75	0.007	−2.732	0.133	0.017	0.156	−2.898	0.869	−4.00
39	0.126	5.0	192.5	−3.85	0.003	−2.735	0.133	0.017	0.153	−2.901	0.966	−4.00
40	0.063	5.0	197.5	−3.95	0.001	−2.736	0.133	0.017	0.150	−2.902	1.065	−4.00
Total lateral friction loss =					3.820 m							

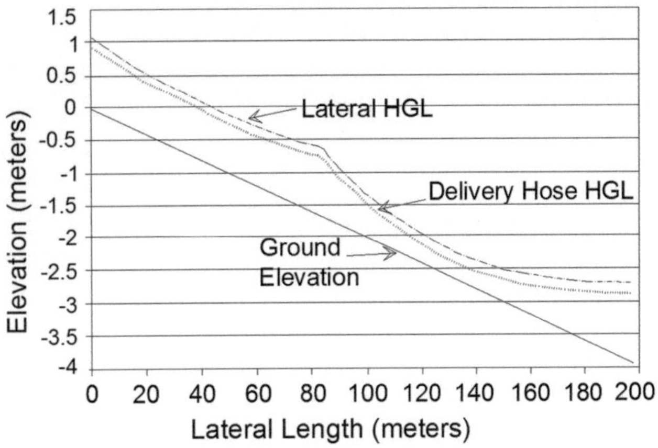

Fig. 22.18 Piezometric diagram for gradual slope design (two pipe sizes)

Questions

1. What are the advantages and disadvantages of low-head gravity bubbler system?
2. What is the cause of airlocks in low-head gravity bubbler systems
3. What are the effects of airlocks?
4. How do you avoid airlocks?
5. What is the main cause of non-uniformity of flow along the lateral in bubbler systems?
6. The Hazen-Williams empirical equation for flow in pipes is given by $Q = 0.849CAR^{0.63}\left(\frac{h_f}{L}\right)^{0.54}$ Metric system Where Q is flow rate, A is cross-sectional area, R is the hydraulic radius, L is length of pipe, and h_f is the friction head loss. The friction head loss h_f can be written from the above relationship as $h_f = K\left[\frac{Q}{C}\right]^{1.852} D^{-4.87}L$ For h_f in meter, D in meter, L in meter, and Q in cubic meters per second determine the value of K.
7. For microirrigation design we use the Hazen-Williams equation more commonly than the Darcy-Wiesbach equation even though the later is more accurate for different fluid and flow conditions and the Hazen-Williams equation is easier to use for water. a) Develop a relationship between Darcy-Wiesbach resistance coefficient (f) in Equation – and the Hazen-Williams (C) in Equation – and b) develop a relationship between C and Reynolds number. Assume water at 20°C.
8. The friction drop ratio a for a microirrigation lateral can be given by $R_x = \frac{h_{fx}}{h_{fL}}$ where h_{fx} is the head loss due to friction from the head end to any point x along the lateral and h_{fL} is the head loss for the entire length of the lateral. It can also be shown that R_x can be calculated by $R_x = 1 - \left(1 - \frac{x}{L}\right)^{m+1}$ where x is the distance from the head end to any point along the lateral, L is the length of the lateral, and m is the exponent for the velocity or flow varying in the friction loss equation Ex. m = 2 for Darcy-Wiesbach equation and 1.852 for Hazen-Williams equation. The pressure head, H_x at any given point x on the lateral neglecting the velocity head term can be computed as $H_x = H_o - h_{fx} - h_{Zx}$ where H_x is the pressure head at a point located at a distance x from the pipeline inlet H_0 is pressure head at the pipeline inlet and H_{Zx} is difference in elevation between point x, and pipeline inlet Using these relationships show that the average pressure head, H_a, can be calculated using $H_{av} = H_0 - \frac{m+1}{m+2}h_{fL} - \frac{H_{ZL}}{2}$
9. Determine the size of a PVC manifold for low-head gravity bubbler system with each lateral to carry a flow of with 2 l/s flow to meet the crop demand and is to be laid in a field 100 m wide and 200 m long. The field is going to be used to grow orchard with tree spacing is 5 m by 5 m. The standpipe is to be at the middle of the manifold. The allowable head loss in the manifold is 0.2 meters.
10. Design a bubbler system to irrigate a citrus orchard with tree spacing of 6 m by 6 m. The field is level and has a dimension of 120 m by 96 m. The water source is a low-head pipeline located at the edge of the field. Assume the design head at the constant head device is 1.2 m and the maximum and minimum delivery hose elevations are 1 m and 0.3 m respectively. Also assume the laterals are laid midway between two rows of trees. To prevent air locks, assume the delivery hose flow rate of 0.047 L/s.

References

Albertson ML, Barton JR, Simons DB (1960) Fluid mechanics for engineers. Prentice-Hall, New Jersey

Boswell MJ (1984) James Hardie micro-irrigation design manual. James Hardie Irigation, El Cajon

Carr M, Kay MG (1980) Bubbler irrigation. Hortic Ind, March:11–12

Christiansen JE (1942) Irrigation by Sprinkling. California agriculture experiment station bulletin 670. University of California, Davis

Didan K, Reynolds C, Yitayew M (1996) Bubbler users manual: a computer program for designing low-head gravity flow bubbler irrigation system for trees, vines, and orchard crops. Technical bulletin 268, Agricultural Experiment Station, College of Agriculture, The University of Arizona, Tucson 85721

Hermsmeier LF, Willardson LS (1970) Friction fac-tors for corrugated plastic tubing. ASCE J Irrig Drain Div 96(No IR3):265–271

Hull PJ (1981) A low pressure irrigation system for orchard tree and plantation crops. The agriculture engineer. Summer 36(2):55–58

Keller J (1990) Modern irrigation in developing countries. In: Proceedings of the 14th international congress on irrigation and drainage. Rio de Janeiro, Brazil, International Commission on Irrigation and Drainage No 1-E, pp 113–138

Keller J, Bliesner RD (1990) Sprinkle and trickle irrigation. Van Nostrand Reinhold, New York

Keller J, Karmelli D (1974) Trickle irrigation design parameters. Trans ASAE 17(4):678–684

Rawlins SL (1977) Uniform irrigation with a low-head bubbler system. Agric Water Manag 1(2):167–178

Reynolds CA (1993) Design and evaluation of bubbler irrigation systems. MS thesis The University of Arizona, Tucson

Reynolds CA, Yitayew M, Petersen M (1995) Low- head bubbler irrigation systems Part I: design. Agric Water Manag 29:1–24. Elsevier Science B. V

Roth RL (1992) Personal communication at Maricopa, Arizona

Thornton JR, Behoteguy D (1980) Operation and installation of a bubbler system. ASAE technical paper no 80-2059, St. Joseph

Waheed SI (1990) Design criteria for low head bubbler systems. Master's thesis, University of Arizona

WattersGZ, Keller J (1978) Trickle irrigation tubing hydraulics. ASAE technical paper no 78-2015, St. Joseph

Worstell RV (1975) An experimental buried multiset irrigation system. ASAE technical paper no 75-2540, St. Joseph

Wu IP, Fangmeier DD (1974) Hydraulic design of twin chamber irrigation laterals, technical bulletin no. 216, The agricultural experiment station. University of Arizona, Tucson

Yitayew M, Reynolds C (1997) Cost savings: the low-head gravity-flow bubbler irrigation system advantages. Paper No 97184 presented at the 1997 annual international meeting of ASAE, Minneapolis Convention Center, Minneapolis, 10–14 Aug 1997

Yitayew M, Reynolds CA, Sheta AE (1995) Bubbler irrigation system design and management. Presented and published in the proceeding of the fifth international microirrigation congress, Orlando, 2–6 Apr 1995

Yitayew M, Didan K, Reynolds C (1999) Microcomputer based low-head gravity -flow bubbler irrigation system design. Int J Comput Electron Agric 22:29–39

23 Wastewater Contaminants and Treatment

With the utilization of irrigation systems for wastewater treatment and disposal, engineers should be aware of common contaminants, methods of on-farm wastewater treatment, and methods of risk evaluation. The primary concerns with wastewater treatment and reuse are prevention of disease and prevention of nutrient contamination and eutrophication of surface and subsurface water resources. Regulations and treatment processes have been set up as a barrier between contaminants and people. In addition, best management practices have been established for optimal and safe utilization of waste. Epidemiological studies and quantitative risk assessment methods can help to establish the risk to human health of wastewater treatment, containment, and reuse practices. Based on risk assessment, regulators establish rules for wastewater treatment and disposal. The last wall of defense against disease and death is the immune system.

Pathogens

Disease causing organisms are called pathogens. Pathogens include nonliving viruses and organisms from four of the kingdoms of life.

Prokaryotes – do not have nucleus or organelles in cells.

- Monera (bacteria). All pathogens labeled as bacteria are part of the monera kingdom. The monera kingdom also includes organisms that are important components of wastewater treatment: They degrade waste, convert ammonia to nitrate and ultimately nitrogen gas, and kill harmful pathogens in wastewater

Eukaryotes – include a nucleus and organelles

- Protista. All of the eukaryotes that don't fit in the animal, plant, or fungi kingdoms. This kingdom includes the protozoa, some of which consume bacteria in latter phases of wastewater treatment, and others cause waterborne diseases.
- Fungi. Responsible for primary decomposition of organic waste. Fungi grow roots into organic matter, including human internal organs, and suck out the nutrients. Some waterborne diseases are caused by fungi.
- Animalia – helminth worms live in wastewater and cause disease.

Some pathogens are spread through wastewater by the fecal-oral route. The process begins when pathogens multiply in the intestinal systems of humans and animals and are then excreted. If left untreated, sewage can enter water systems and contaminate drinking water or food.

Of the pathogens transmitted through water: the most notorious are cholera and typhoid fever. There are an estimated 16 million cases of typhoid fever and 600,000 deaths per year in the world. In addition to typhoid fever, 10 million deaths per year are blamed on waterborne diseases, and between 30 % and 40 % of worldwide illnesses are transmitted by ingestion of contaminated drinking water. Lack of treatment infrastructure and regulations in many developing nations leads to a direct connection when sewage and drinking water.

Pathogens vary in size. Viruses are between 10 and 100 nm in diameter. Mycoplasma and bacteria are between 100 nm and 10 μm. Protozoa are between 1 and 100 μm. Helminth worms are the largest. The size of pathogens determines the level of filtration needed to physically remove them from water.

Viruses

Viruses are not living organisms and only function as obligate parasites: they replicate inside other cells by removing the membrane around the DNA and causing the DNA to produce mRNA that is coded to produce new viruses. Some

P. Waller, M. Yitayew, *Irrigation and Drainage Engineering*, DOI 10.1007/978-3-319-05699-9_23

cause the host human cell to blow up and scatter the new viruses to infect new cells. Others cause the host cell to continually produce new viruses. Viruses are small (10–100 nm) and difficult to remove by filtration. Toxic viruses transmitted from human to human in wastewater include poliovirus, coxsackie virus, echovirus, hepatitis A, hepatitis E, caliciviruses, rotaviruses, and reoviruses.

The viruses in the gastrointestinal tract are transmitted by the fecal-oral route and are thus of concern in wastewater. One of the deadliest fecal-oral viruses is the coxsackie virus. It replicates in the intestinal lining (epithelial cells), and then it enters the bloodstream. Once it enters the bloodstream, it may cause myocarditis in the heart, meningitis or paralysis in the central nervous system, or encephalitis in the brain. The Norwalk virus (a calicivirus) causes an estimated 23 million cases of diarrhea per year in the United States. Of those cases, food and recreational water activities each cause approximately 7 million cases per year and 3.5 million cases per year are caused by consumption of drinking water.

Bacteria

Bacterial pathogens feed on organic and chemical nutrients. Most fecal-oral toxic bacteria are anaerobic heterotrophs (consume other organisms or organic matter) that grow in the intestine. They generally cause diarrhea, and some emit toxins that may puncture the intestinal lining. Some of the most harmful gastrointestinal bacteria are salmonella, campylobacter, shigella, E. coli O157:H7, and vibrio cholera

Salmonella typhi are bacteria that are spread in water and cause typhoid fever. *They* colonize on organ surfaces. Their favorite location is the gall bladder. Typhoid fever can causes fever, rashes, headache, joint pain, and even death, and has one of the highest fatality rates of waterborne diseases. Because *Salmonella typhi* are in the intestinal tract, the disease is spread through the feces and urine of infected people. Thus, if sewage water is not treated before entering the potable water system in a city, then the disease spreads rapidly. Open sewage systems are also a hazard because flies can land on the sewage and then land on food.

Infected food preparation workers can spread typhoid fever through the fecal-oral route. As they wipe their bottoms, the bacteria stick to their hands, and then they spread the bacteria during food preparation. The most famous case was Typhoid Mary in New York: Mary Mallon was a cook for rich families. One of the families, which contracted typhoid fever, hired a private investigator to determine the cause of a typhoid outbreak in their family. The private investigator found that 22 people contracted typhoid fever at 7 different jobs where Mary worked from 1900 to 1906. Mary refused to believe that she was the source of typhoid fever. In fact, when officials asked to take urine and stool samples, Mary attacked them with a carving knife. Eventually, public health authorities confined her to Brother Island. She was released after 2 years with the understanding that she would give up cooking. However, an outbreak of typhoid fever (25 people) occurred 5 years later at a location where a Mrs. Brown was the cook. Mrs. Brown turned out to be Mary Mallon. New York authorities then confined her to Brother Island for the last 23 years of her life.

Protozoa

Protozoan pathogens feed off the human gut. Some of the most harmful protozoan pathogens are giardia lambia, cryptosporidium, cyclospora, microsporidia, and toxoplasma gondii. Cryptosporidium, which can originate in runoff into streams from feedlots and dairies, is especially resistant to chlorination during the oocyst phase (Fig. 23.1).

Animal and Human Waste

Diseases associated with animal waste include salmonella, anthrax, tuberculosis, tetanus, hog cholera, and foot and mouth disease. Diseases are introduced into water bodies through runoff from fields or cattle confinement facilities. In order to prevent animal borne pathogens from entering surface water bodies, regulations often specify vegetated riparian zones next to water bodies in order to intercept pathogens before storm water runoff enters streams or ponds.

Rather than test for specific pathogens in wastewater, it is much cheaper to test for the presence of indicator organisms or surrogates that are produced in large quantities by humans or animals. It is assumed that if an indicator organism such as fecal coliform is absent, then other pathogens are absent. The ratio of different coliforms can indicate the source of water pollution (animal or human) because the ratio of fecal coliform to fecal streptococcus varies between animals and humans. Animals have a low fecal coliform to fecal streptococcus ratio (<0.6) while humans have a high ratio (4.4). Because fecal coliform is an indicator organism, maximum fecal coliform limits are established for water bodies. Public water supplies before water treatment should be less than 2,000 fecal coliforms/100 ml (MPN/100 ml). Swimming locations should have less than 200 fecal coliforms/100 ml.

Removal of Pathogens by Biological Treatment and Oxidation

Wastewater treatment removes most pathogens from water. The goal is to lower the pathogen count to an acceptable level of risk. The concentration of pathogens in the incoming

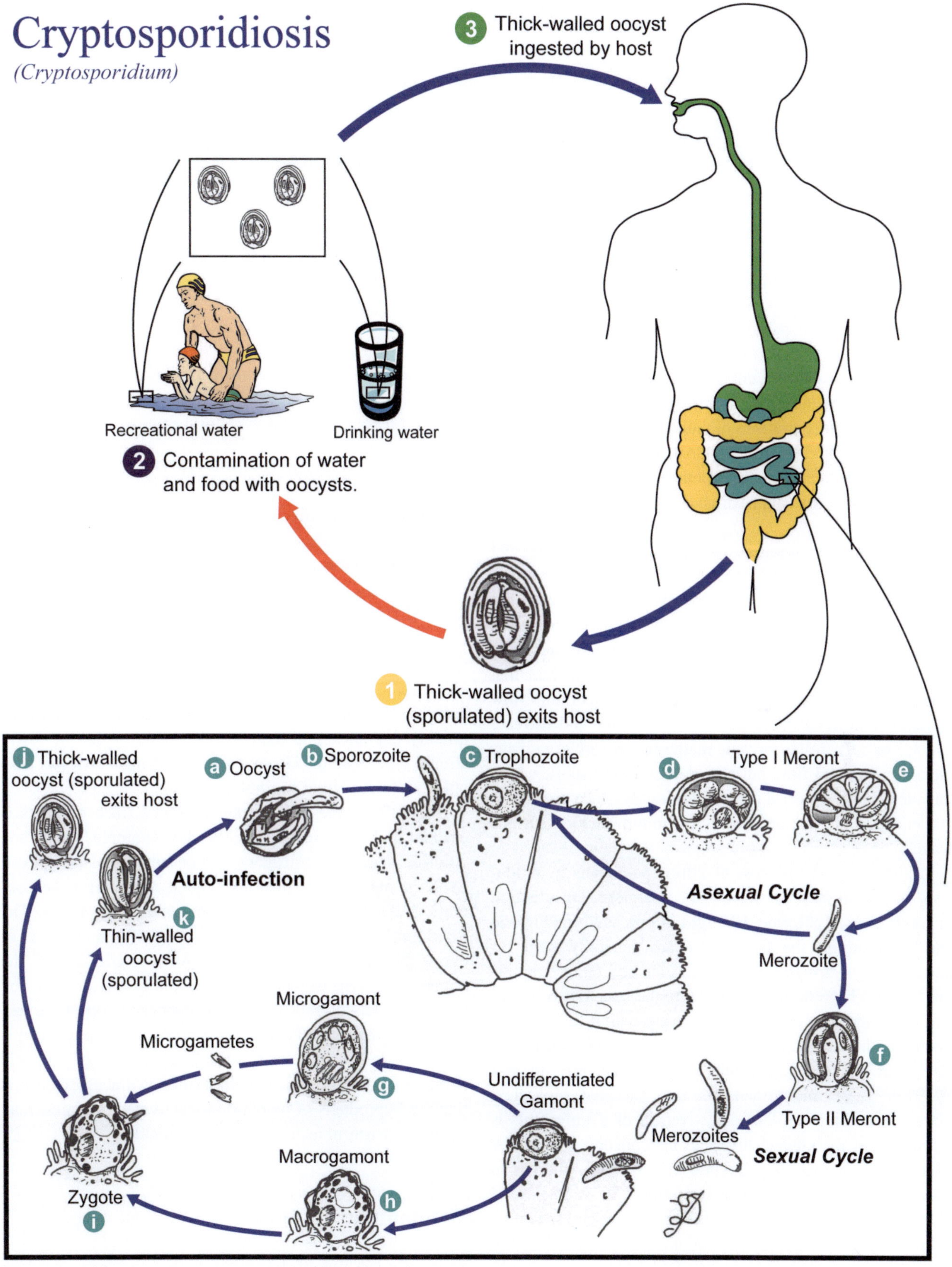

Fig. 23.1 Cryptosporidium life cycle (Credit CDC/Alexander J. da Silva)

Table 23.1 Typical concentrations of organisms in raw and treated (prior to chlorination) wastewater (organisms/100 ml) (Source: Charles Gerba, University of Arizona)

Organisms	Raw sewage	Treated
Enteroviruses	10–100	1–10
Salmonella	100–10,000	10–10,000
Coliforms	10^7–10^9	10^6–10^8
F-RNA coliphages	10^6–10^7	10^5–10^6
Somatic coliphages	10^6–10^7	10^5–10^6
Bacteroides phages	10^4–10^5	10^3–10^4

wastewater stream is one factor that determines the effluent pathogen concentration. In general, developing countries have much higher pathogen counts in raw wastewater because there is a higher incidence of disease. Other factors that influence pathogen concentrations in wastewater are socioeconomic status, per capita water use, and time of year. Typical concentrations of pathogens in raw and treated wastewater are shown in Table 23.1. In general wastewater treatment decreases the concentrations of many microorganisms by approximately 10 times or "1 log removal."

The ability of pathogens to survive in the environment determines the required treatment time. Survival time is influenced by environment, type of organism, and life phase. For example, the cyst and oocysts phases in giardia and cryptosporidium, respectively, are much more resistant than the latter phases of life. Environmental factors that decrease pathogen survival time are high temperature, low water content in soils, antagonistic microflora, and extreme pH (<3 or >9). Soil or organic matter that adsorb organisms increases survival time but also decreases transport, which may decrease the hazard to the environment since organisms are not leached to transported to groundwater or surface water.

Disinfection, the process of killing pathogens by oxidation, is a final step in wastewater treatment. Disinfection methods include chlorine, UV light, ozone, chlorine dioxide and chloramines; the latter three are generally used for disinfection of drinking water. Chlorine and other oxidants kill harmful bacteria by entering into the bacterial cell wall and oxidizing the cell wall, or entering within the cell and disrupting biochemical processes such as protein formation. The goal of the disinfection step is generally to kill 99–99.9 % of microorganisms that survived biological wastewater treatment. The effectiveness of disinfectants is a function of the contact time and the concentration of the disinfectant in water (C* t) where C is in mg/L and t is in minutes. For example, if contact time is 30 minutes, and chlorine concentration is 1 mg/L, then Ct = 30. Typical Ct values for chlorine disinfection are 7,200 for helminth worms, 0.1–1.0 for viruses, and 0.001–0.01 for E. coli. Viruses, protozoa, and helminth worms are the most resistant to disinfection.

Historically, waterborne diseases were common until the introduction of chlorine as a disinfectant for public water systems; the incidence of typhoid fever in the United States dropped from an average of 25 incidences per 100,000 people per year to approximately 400–500 cases per year in the entire nation (less than 0.2 incidences per 100,000 people). Thus, the per capita disease frequency dropped by 100 times due to the introduction of chlorine into public water supplies (Fig. 23.8).

As chlorine gas is dissolved in water, it reacts with water to form hypochlorous acid (HOCl), hydrogen (H^+) and chloride ions (Cl^-):

$$Cl_2 + H_2O \Leftrightarrow HOCl + H^+ + Cl^-.$$

Chlorine exists in two forms within water, hypochlorous acid (HOCl) and hypochlorite (OCl^-). The hypochlorous acid and hypochlorite, OCl^-, equilibrium in water is

$$HOCl \Leftrightarrow H^+ + OCl^-.$$

Hypochlorous acid (free chlorine) is 80 times more effective at killing bacteria than hypochlorite because the hypochlorite has a charge and is repelled from the cell. Hypochlorous acid is the primary form of chlorine in acidic waters, pH < 6.5. Thus, if the initial pH is high, acid must often be added in concert with chlorine in order to drop the pH to 6.5.

Ozone is a strong oxidizer and can be used in place of chlorine; it is not used with chlorine because it oxidizes the chlorine. The problem with ozone disinfection is that ozone quickly degrades, and there is no residual ozone in downstream pipe systems. However, it is a stronger oxidizer than chlorine (more lethal to some organisms).

The effectiveness of oxidants against trace organics is oxidant and contaminant specific. For example, ozone and chlorine deactivate estrogens and antimicrobials; however, ozone is much better than chlorine at removing psychosomatic and similar drugs. Concentration also is a factor. The Southern Nevada Water Authority tried different levels of ozone between 3 and 8 mg/L. At 8 mg/L, all of the compounds were deactivated. They found that degradation is most closely correlated with the concentration of the free residual, rather than the total mass dose. The free residual is what is left over after initial oxidation of organic compounds. One problem with oxidizing trace organics is that the byproducts of oxidation (deactivated organics) may be just as harmful as the original chemicals.

The Immune System

There is a last line of defense if wastewater treatment fails and pathogens enter the body, the human immune system. The immune system produces antibodies that are tailored to fight specific pathogens that invade the body. The key factor in the fight against pathogens is the relative time required to produce antibodies or other defensive functions in comparison to the reproductive rate of the pathogen. The concentration of pathogens that are ingested and their rate of multiplication, as well as the health of the immune system, determine the resistance to disease.

The immune system is constantly looking for foreign cells in the body by looking for foreign markers on the surfaces of cells, called epitopes. When an invading cell is recognized, stem cells are triggered to become two types of immune cells, lymphoid cells and myeloid cells. B cells are a lymphocyte (lymphoid cell) that is produced after the body recognizes an antigen. The B cells produce plasma cells that produce antibodies that will fight one specific type of pathogen. The antibodies attach themselves to a pathogen and trigger a series of biochemical reactions that destroy the pathogen. Macrophages are myeloid cells that consume pathogens and break them down into peptide chains that are then displayed on the surface of the macrophage. Another lymphocyte, helper T cells, records the shape of the peptide chain and then signals B cells to start producing plasma cells.

Cytotoxic T cells (also known as killer T cells) attach to pathogens and kill them with granules filled with chemicals. Phagocytes are white cells that consume pathogens. Unlike antibodies, these cells do not recognize specific pathogens; they just consume any foreign cell. Other complex systems are designed to kill specific types of pathogens and operate in specific parts of the body such as the mucus membranes.

The body builds up immunity to disease because some of the T cells and B cells used in the initial defense remain after the battle as memory cells. The next time that the pathogen is encountered, the immune system is ready to respond. Vaccines are inactive pathogens that trigger an immune response without the danger of disease. Thus, T cells and B cells remain and will respond quickly to the real pathogens when they are encountered.

Normally, the body is able to manufacture immune cells and antibodies quickly enough to stop the multiplication of pathogens within the body. However, some pathogens can overcome the immune system. For example, when *Salmonella typhi* bacteria are ingested into the digestive tract, they are consumed by monocytes in the bloodstream. However, rather than the monocytes killing the *Salmonella typhi*, the bacteria multiply within the monocyte for a period of 1–2 weeks. Then the bacteria spill out into the bloodstream and can overwhelm the immune system. The large numbers of bacteria in the bloodstream cause the body to develop a high fever. Eventually, the *Salmonella typhi* enter the digestive tract and other organs. Once in the intestines and other organs, the bacteria give off a toxin that causes inflammation and can even puncture the lining of the intestines or cause bursting or leakage in other organs. Victims of typhoid fever are often sick for several months and may die. However, if people have been inoculated against typhoid fever, then the process is aborted before it has a chance to begin.

The rate of infection depends on the organism and the host human. For example, only half of normal healthy adults exposed to enteric (intestinal) viruses (polio and Coxsackie) become infected while 100 % of people exposed to measles become infected. Healthy adults are less likely to become infected than small children, the elderly, and those with compromised immune systems. In addition, the level of infection can vary between persons. Some infected people do not show any sign of disease while others may have a mild illness, severe disease, or death.

Risk Assessment

No human activity is absolutely safe; however, the goal is to make the activity as safe as possible. The 10 most dangerous activities in America, in order, according to a survey of experts, are smoking, drinking alcoholic beverages, driving a car, handguns, surgery, motorcycles, x-rays, electric power, and swimming. Drinking water is not even listed among the top 30 most dangerous activities. Water systems are designed with the criterion that no more than one out of 10,000 people each year become infected with a waterborne pathogen (USEPA Surface Treatment Rule). In comparison, one out of every 2 people in America become sick each year from a food borne pathogen (food poisoning).

Epidemiological studies look at past outbreaks of disease and associate the outbreaks with given practices. In quantitative risk assessment, the possibility of each of the steps in disease transmission is multiplied in order to arrive at the total probability of disease transmission. The process of risk analysis includes three steps: risk assessment – determining the probability that an adverse event will occur and its magnitude, risk management – considers various regulatory options to minimize the risk, and risk communication – transfer of risk information to experts and non-experts.

Quantitative risk assessment has four steps:

1. Hazard Identification – identifying the contaminant (i.e. Salmonella)
2. Dose–response Assessment – relationship between the number of organisms ingested and the probability of becoming infected (i.e. how many does it take to make you sick)
3. Exposure Assessment – Determining the concentration of a pathogen in the water and estimating amount of contact and possibly ingestion.
4. Risk Characterization – Estimating the potential impact (infection, disease) of a pathogen based on the severity of its effects.

Hazard identification is the process of identifying the disease carrying organism, finding the transmission pathway, determining the population exposed to the disease, and quantifying the health effects of the disease.

The dose–response relationship is a function of the type of pathogen and the human immune system. The risk of infection increases as the ingested quantity of pathogen increases. In theory, a single microorganism can multiply and cause disease, but for some pathogens, a minimum quantity of pathogens is required to overwhelm the immune system. The response in a dose–response study is also called an endpoint. Endpoints in a study could be visible symptoms, disease, or death.

The infection rate is the percent of people exposed to the disease who become infected.

Exposure assessment is the product of the assumed concentration of the pathogen, the amount ingested per unit of time, the amount of time of exposure.

In the following example, the risk of disease transmission is calculated for a landscape irrigated with reclaimed wastewater by making the following calculations:

1. Concentration of pathogens in irrigation water
2. Concentration in soil after irrigation
3. Duration of human exposure to soil
4. Rate of ingestion of soil
5. Dose–response

Example 23.1 Partially treated wastewater from an overloaded wastewater lagoon is used for landscape irrigation in a water-stressed city. Calculate the risk of exposure to *Salmonella typhi* bacteria for 50 children who play for 8 hours in the landscaped area. Water content in the soil is 15 % by volume. Soil bulk density is 1.15 mg/ml. The concentration of Salmonella in the lagoon is 10^3 MPN/100 ml. MPN refers to the most probable number. It is found by counting bacterial colonies on plates or gels in the lab. Salmonella can survive in soil for an extended period, so assume that the concentration of salmonella in the soil water is the same as concentration in the irrigation water. Assume that the salmonella are not preferentially absorbed in the upper range of soil by organic matter or clay particles. Children playing in a soil normally ingest approximately 500 mg (size of a peanut) of soil per 8 hours of playtime.

The concentration of salmonella in soil is calculated as follows:

$$1{,}000\ \text{MPN}/100\ \text{ml water} * 15\ \text{ml water}/100\ \text{ml soil} = 150\ \text{MPN}/100\ \text{ml of soil.}$$

The volume of soil ingested is the mass of the soil divided by the bulk density. Assume that the.

$$500\ \text{mg soil ingested}/1.15\ \text{mg/ml soil} = 435\ \text{ml soil}$$

$$435\ \text{ml soil}*150\ \text{MPN}/100\ \text{ml soil} = 650\ \text{Salmonella bacteria}$$

If this were a real study, statistical methods would be used to determine the distribution of ingestion rates; however, in this case, assume that 10 children ingest a greater amount than the average: between 1,000 and 10,000 salmonella bacteria. Thus, 40 children ingest between 0 and 1,000 bacteria. Studies have shown that 1,000 of bacteria must be swallowed at one time in order to cause infection by salmonella. Thus, assume that the children that ingest less than 1,000 salmonella bacteria will probably not become infected. Based on a salmonella dose–response curve, approximately 20 % of children (2 out of the 10) exposed to between 1,000 and 10,000 Salmonella bacteria will become infected. Thus, approximately 2 children out of the total of 50 (4 %) will become infected.

Biochemical Oxygen Demand (BOD)

In addition to causing disease, wastewater can have other detrimental effects. Waste is a food source for microorganisms. In order to consume the waste (respiration), microbes require oxygen. Thus, releasing wastewater into streams can deplete oxygen in the stream and kill wildlife. If wastewater with a heavy organic load is added to surface waters, the dissolved oxygen concentration can decrease from the normal 5–7 mg/L to 2 mg/L or less, a level at which fish die. In order to prevent oxygen depletion, regulations restrict the oxygen demand of wastewater released into streams.

Biochemical Oxygen Demand (BOD) is defined as the amount of oxygen required for the bacterial decomposition (oxidation) of organic matter under aerobic conditions at a standard temperature and incubation time. The amount of oxygen required varies with concentration of organic matter, concentration of bacteria, nature of organic matter, and type of bacteria. Organic matter undergoes two primary transformations in the process of BOD exertion: oxidation and synthesis and endogenous respiration

Oxidation and Synthesis

$$\underset{\text{(Organic Matter)}}{\text{COHNS}} + O_2 + \text{Nutrients} \xrightarrow{\text{Bacteria}} CO_2 + NH_3 + \underset{\text{(New Bacterial Cells)}}{C_5H_7NO_2} + \text{Other End Products} \quad (23.1)$$

Fig. 23.2 Oxygen demand and uptake of organic waste and microorganims

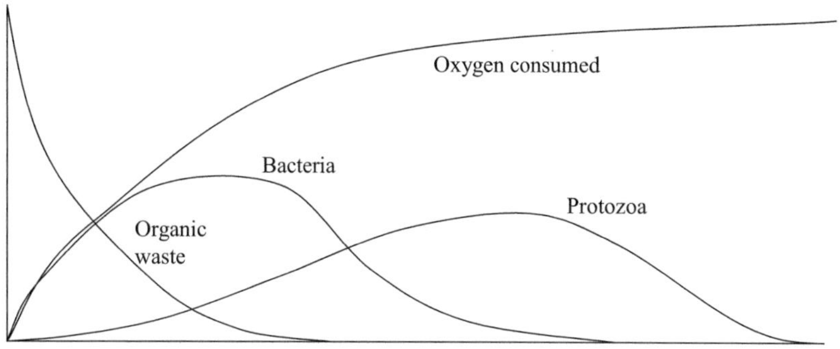

Endogenous Respiration

$$C_5H_7NO_2 + 5O_2 \xrightarrow{\text{Protozoa and bacteria}} 5CO_2 + 2H_2O + NH_3 + \text{Energy} \quad (23.2)$$

When organic matter (CHONS) is oxidized (Eq. 23.1), the resulting bacterial cells have the formula: $C_5H_7NO_2$. Dead bacterial cells are then consumed by protozoa in a process called endogenous respiration. The result of endogenous respiration is stable, nontoxic byproducts.

The test for BOD is a well-established procedure. The BOD of a sample is measured by adding a sample aliquot to a bottle and filling the bottle with dilution water, which is distilled water supplemented with buffering agents, nutrients, and trace metals and aerated to raise the concentration of dissolved oxygen to saturation. The initial dissolved oxygen concentration (DO_i) of the sample is measured, and it is then incubated in the dark at a constant temperature, usually 20°C. At the end of the test, the dissolved oxygen concentration (DO_f) is measured.

$$BOD = \left(DO_i - DO_f\right)\frac{V_b}{V_s} = \Delta DO\ (DF) \quad (23.3)$$

where

DO_i = initial dissolved oxygen concentration, mg/L,
DO_f = final dissolved oxygen concentration, mg/L,
DF = dilution factor: volume of the bottle divided by volume of the sample,
V_b = volume of the bottle, ml,
V_s = volume of sample added to the bottle, ml,

The traditional BOD test, developed in England, is 5 days long: it took 5 days for water in the Thames River to travel from London (source of sewage) to the ocean. The 5-day test is called BOD_5, and typical wastewater plant effluent standards are 30 mg/L BOD_5. In the early stages of oxidation of organic matter, oxygen uptake is approximately equal to decrease in BOD (Fig. 23.2). This process generates new bacteria cells. In the initial phase, the protozoans contribute to organic removal but are not as efficient as the bacteria in competing for substrate (organic matter). As food supplies for microorganisms become scarce, bacteria begin to die off or predate on other living and dead bacteria. Eventually, the number of protozoans increases as they consume bacteria. During these processes, the organic matter is recycled through a number of organisms, with oxygen being consumed as the organic matter is metabolized for each biomass.

The BOD_5 value for raw domestic sewage is 200–300 mg/L while that for animal waste effluent ranges from 6,000 (dairy and beef) to 13,000 (swine) mg/L. Anaerobic lagoon or anaerobic digestion systems can reduce animal waste BOD to a range between 200 mg/L to 4,000 mg/L, depending on the source and the detention time. Wetlands can further decompose the waste; however, animal waste effluent from anaerobic lagoons must be diluted in order to reduce the toxic ammonia concentration that would kill the wetlands plants.

The primary products of the initial breakdown of organic matter are CO_2, H_2O, and NH_3 (ammonia). Oxidation of ammonia to nitrate (nitrification) constitutes a second stage BOD, known as nitrogenous BOD (NBOD) whereas the first stage is often called carbonaceous BOD (CBOD). NBOD occurs simultaneously with the latter stages of CBOD. Nitrification (nitrogenous oxidation) stoichiometry is as follows:

$$\begin{array}{lll} NH_3 + \frac{3}{2}O_2 & \xrightarrow{\text{Nitrosomonas}} & NO_2^- + H^+ + H_2O \\ NO_2^- + \frac{1}{2}O_2 & \xrightarrow{\text{Nitrobacter}} & NO_3^- \\ \hline NH_3 + 2O_2 & \longrightarrow & NO_3^- + H^+ + H_2O \end{array} \quad (23.4)$$

The oxidation of ammonia begins after 7 or 8 days of wastewater treatment have passed. Thus, the BOD_5 primarily measures the CBOD and does not include the NBOD.

Nutrients and Waste in Soils and Wetlands

Although nutrients are needed for crop growth, they are pollutants when discharged into the environment from agriculture or wetlands. The major nutrients of concern in wastewater are nitrogen and phosphorous. Excess concentrations of these nutrients cause algal blooms in streams (eutrophication), and nitrate in water can cause blue baby disease – methemoglobimia:

the nitrate replaces oxygen in the blood stream and the baby turns blue. Thus, nitrate discharge concentration in surface water and ground water is limited to 10 mg/L NO_3-N. This is the same as 42 mg/L NO_3^-, as it is specified in Europe. Wastewater effluent standards may include nitrate, ammonia, and/or total nitrogen and are generally in the range of 10 mg/L. Often, design requirements for nitrogen are lower in summer than in winter for natural wastewater systems because bacteria are more active at higher temperature.

Nitrogen exists in many forms in water: organic nitrogen, amines, ammonia (NH_3), ammonium (NH_4^+), nitrite (NO_{23-}), nitrate (NO_3^-). Total nitrogen, TN, includes all nitrogen forms. Total ammonia includes ammonia and ammonium ion. Nitrogen also exists as nitrogen gas (N_2), and nitrous oxide gas (NO_2) in the atmosphere. The goal in wastewater treatment of nitrogen is to convert the total nitrogen in water to nitrogen gas or to remove the nitrogen by plant uptake.

Bacteria sequentially transform nitrogen from one form to another in soils and wetlands. The nitrogen cycle includes anaerobic (no oxygen) bacteria, aerobic (oxygen) bacteria and facultative (anaerobic or aerobic) bacteria.

The steps from conversion of organic N to NO_3^- (nitrate) are called mineralization, which includes ammonification and nitrification.

organic matter → amines → ammonium → nitrite → nitrate → nitrous oxide or nitrogen gas
anaerobic — aerobic — anaerobic or aerobic
ammonification — nitrification — denitrification

In soils, nitrate is a much greater groundwater pollution hazard than ammonium. The reason for this is that clay particles in soil have a negative surface charge; thus, they tend to absorb positively charged ammonium. In contrast, nitrate is not absorbed, and is washed directly through the soil to the groundwater when water leaches through the soil.

In some cases, farming practices release nitrate at inopportune times. For example, in the Midwestern United States, nitrogen in plant matter that is disked into the soil during fall mineralizes in winter (becomes nitrate). Then, spring rains wash the nitrate through the soil and into the groundwater, drains, and streams. In irrigated areas, excess irrigation or storms after a heavy fertilization can leach large quantities of fertilizer through the soil. Thus, this is a problem that has little relation to fertilization but primarily to drainage and farming practices.

Application of manure, biosolids, and other waste products is a common method of fertilization and waste disposal. The timing of waste application is important. For example, nitrogen in fresh manure is approximately 60–80 % in the organic form (NRCS AWMFH). Anaerobic lagoon effluent has an organic fraction of approximately 20–30 %. Organic N is not available to crops and is immobile in soil. When organic N is applied to soils, it takes 4–5 months for 40–90 % of the organic N to be converted to ammonia and then to nitrate in cold climates. However, in warmer climates, total mineralization may take place in 2 months (NRCS AWMFH).

Ammonia (NH_3) is the unionized form and ammonium ion (NH_4^+) is the ionized form. Ammonia is not available to plants, but ammonium ion is available. Ammonia and ammonium ion are in equilibrium in water, with the percent of unionized ammonia dependent on temperature and pH. Although generally not a concern in soils, ammonia is a concern in wetlands and water bodies because it kills plants and fish. Ammonia is so toxic to fish that the EPA's maximum recommended limit for unionized ammonia in surface waters is 0.02 mg/L.

The nitrogen cycle is complex in wetlands, with many sources, sinks, and processes (Fig. 23.3). There are three sources of ammonia in wetlands: hydrolysis of urea, bacterial decomposition of fecal matter, and oxidation of cells (Fig. 23.3). In addition, denitrification and degradation of wetlands plants can result in formation of ammonia. Ammonia is lost by volatilization to the atmosphere if the total ammonia (unionized) concentration in a wetland is greater than 20 mg/L. If ammonia concentration is low, then the primary ammonia removal mechanism in wetlands is absorption to soil and organic particles or nitrification. Nitrification of ammonia in wetlands is limited by the rate that wetlands plants add oxygen to the water.

In wetlands, negatively charged nitrate does not absorb to soil particles and must be removed by plant uptake or bacterial denitrification. Denitrification takes place in anaerobic conditions so anaerobic zones are often added to wetlands in order to complete the nitrogen cycle. As with nitrification, denitrification is a two-step process that involves the creation and destruction of nitrite (NO_2^-). Nitrate is also taken up by plants so nitrate removal by the wetland can be increased if plants are periodically harvested.

Herbaceous aquatic plants pump oxygen into the water (Fig. 23.4) and facilitate nitrification and denitrification by

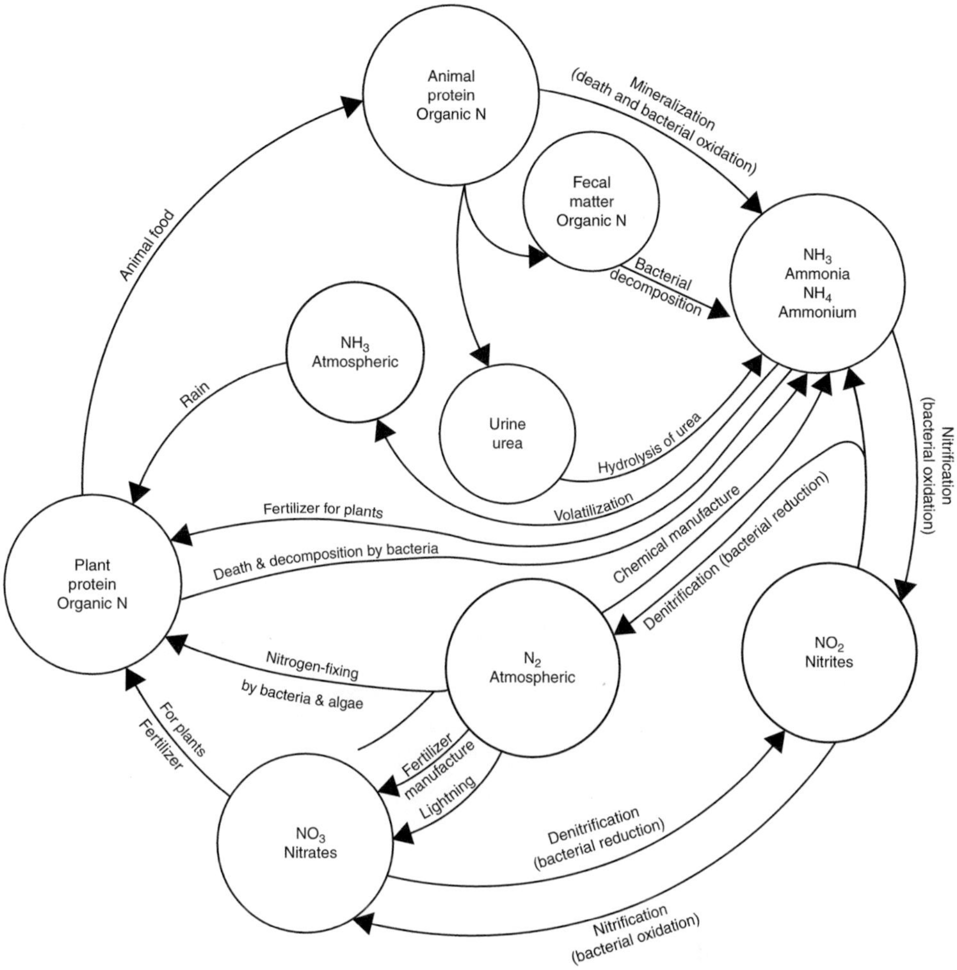

Fig. 23.3 The nitrogen cycle in wetlands (Credit NRCS, NEH 637–03)

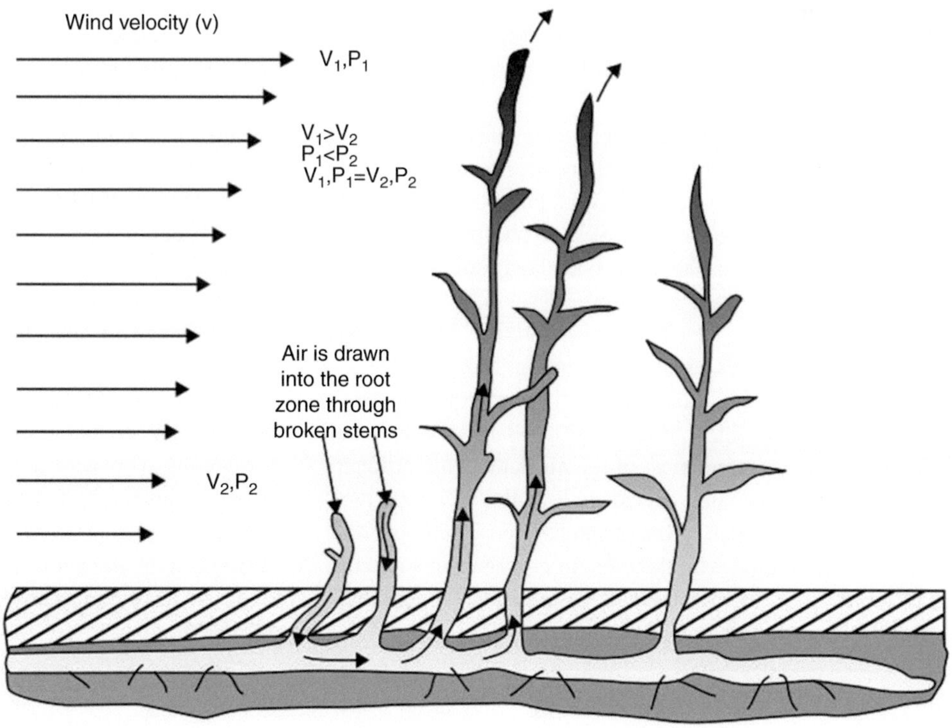

Fig. 23.4 Oxygen flow into wetlands induced by herbaceous aquatic plants

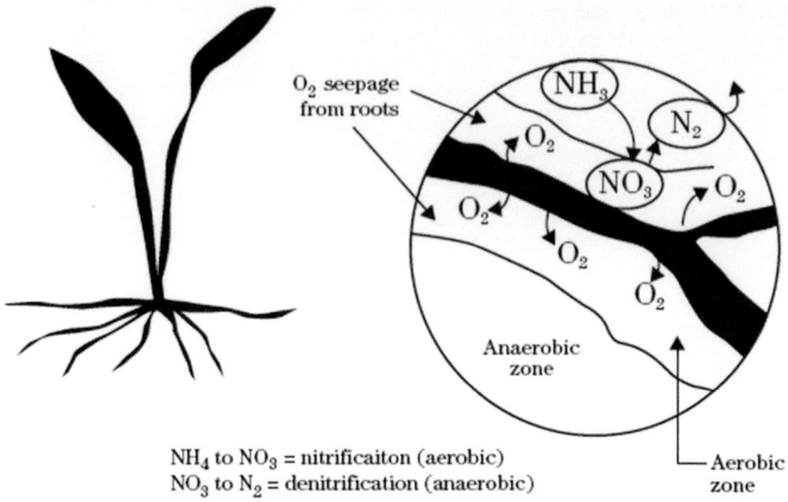

Fig. 23.5 Microenvironments with high oxygen near plant roots (Credit NRCS, NEH 637–03)

creating microenvironments with high oxygen near the roots (Fig. 23.5).

The soil in wetlands is an important part of the treatment process. Plants suck wastewater into soil as they draw in water for evapotranspiration. As the wastewater moves through the soil, it is filtered and treated by microorganisms. Organic nitrogen associated with suspended solids can be removed via sedimentation and incorporated into the soil matrix.

Dead plant litter within the water is an important substrate for microbial growth. Plant litter, along with nondegradable solids, and dead microorganisms settle on the soil surface and form a peat layer. This layer absorbs metals and phosphorous. In fact, 90 % of wetlands treatment takes place on microbes that are attached to plant litter (Kadlec and Knight 1996).

In addition to nitrogen, the other primary nutrient of concern is phosphorous. Phosphorous exists in three forms – organic, soluble, and attached – with 70 % of phosphorous in fresh animal waste in the organic form. Soluble phosphorous is used by plants. Attached phosphorous is bound to positively charged cations such as calcium. For example, gypsum, calcium phosphate ($CaPO_4$) is attached phosphorous. Phosphorous transformations in soils are as shown in Fig. 23.6.

If not overloaded, agricultural soil is an effective nutrient and contaminant filter (Fig. 23.7). Plant uptake and microorganisms remove most of the nutrients from effluent. However, if soils are overloaded with effluent, then contaminants can leach to the ground water. The upper soil region is the location with the most microorganisms and the most rapid waste treatment. Once waste reaches the groundwater, little waste treatment takes place.

The NRCS lists the following animal waste practices (Fig. 23.8) that can lead to environmental contamination.

1. Contaminated well: Well water contaminated by bacteria and nitrates because of leaching through soil.
2. Waste storage structure: Poisonous and explosive gases in structure.
3. Animals in poorly ventilated building: Ammonia and other gases create respiratory and eye problems in animals and corrosion of metals in building.
4. Waste applied at high rates: Nitrate toxicity and other N-related diseases in cattle grazing cool-season grasses; leaching of NO3 and microorganisms through soil, fractured rock, and sinkholes.
5. Discharging lagoon, runoff from open feedlot, and cattle in creek: (a) Organic matter creates low dissolved oxygen levels in stream; (b) Ammonia concentration reaches toxic limits for fish; and (c) Stream is enriched with nutrients, creating eutrophic conditions in downstream lake.
6. Runoff from fields where livestock waste is spread and no conservation practices on land: P and NH4 attached to eroded soil particles and soluble nutrients reach stream, creating eutrophic conditions in downstream lake.
7. Eutrophic conditions: Excess algae and aquatic weeds created by contributions from items 5 and 6; nitrite poisoning (brown-blood disease) in fish because of high N levels in bottom muds when spring overturn occurs.
8. Leaching of nutrients and bacteria from poorly sealed lagoon: May contaminate ground water or enter stream as interflow.

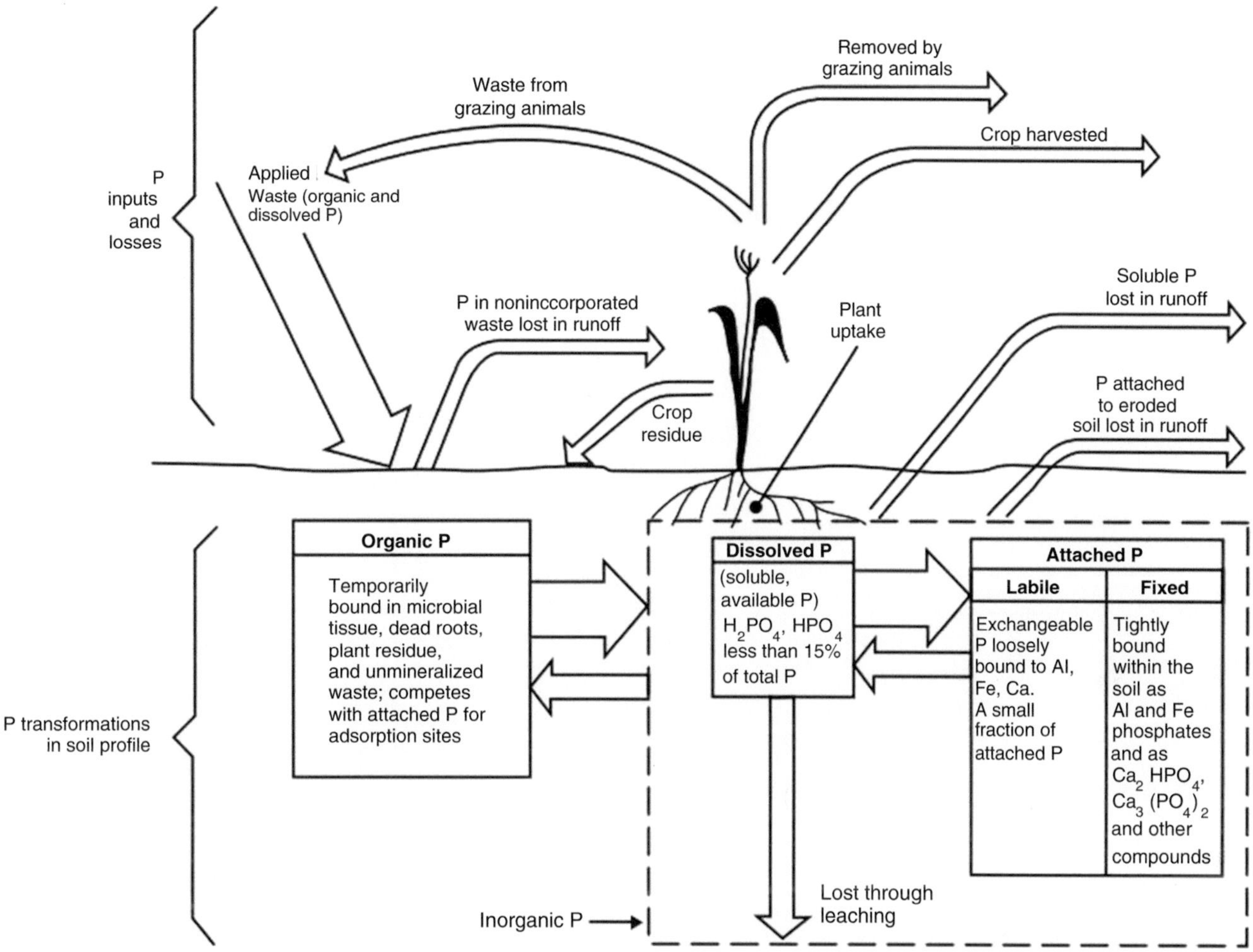

Fig. 23.6 Phosphorus inputs and losses at a waste application site and phosphorus transformation within the soil profile (abbreviated phosphorus cycle) (Credit NRCS AWMFH)

In addition to ground water pollution, surface water runoff from agriculture impacts streams and other water bodies. Approximately 20,000 miles of streams and rivers in the US are negatively impacted by animal waste effluent BOD and nutrients.

Trace Chemicals

Pesticides and other agricultural chemicals are facing increasing scrutiny and regulation. Pesticides such as atrazine, diuron, and hexachlorocyclohexane have been found in drinking water. Animal feeding operation chemicals, such as growth stimulants, are a major concern. Other trace chemicals that may be in soil include selenium and other metals. Soils and wetlands have been found to remove many of the trace chemicals that originate in cities.

Monitoring trace contaminants is a major challenge. Many of the tests are extremely expensive. Even if it was possible to monitor all of the chemicals in wastewater, a multitude of chemical test results confuses industry, regulators, and the public. As a result, the use of surrogates or indicators to detect the general presence of hazardous chemicals is more practical.

- Surrogates: bulk measurements of parameters such as total organic carbon (TOC) that indicate the presence of organic chemicals.
- Indicators: specific measurements of individual chemicals that are representative of other chemicals within a group of similar chemicals.

There are many surrogates: TOC, BDOC (biologically degradable organic carbon), DOC (dissolved organic carbon

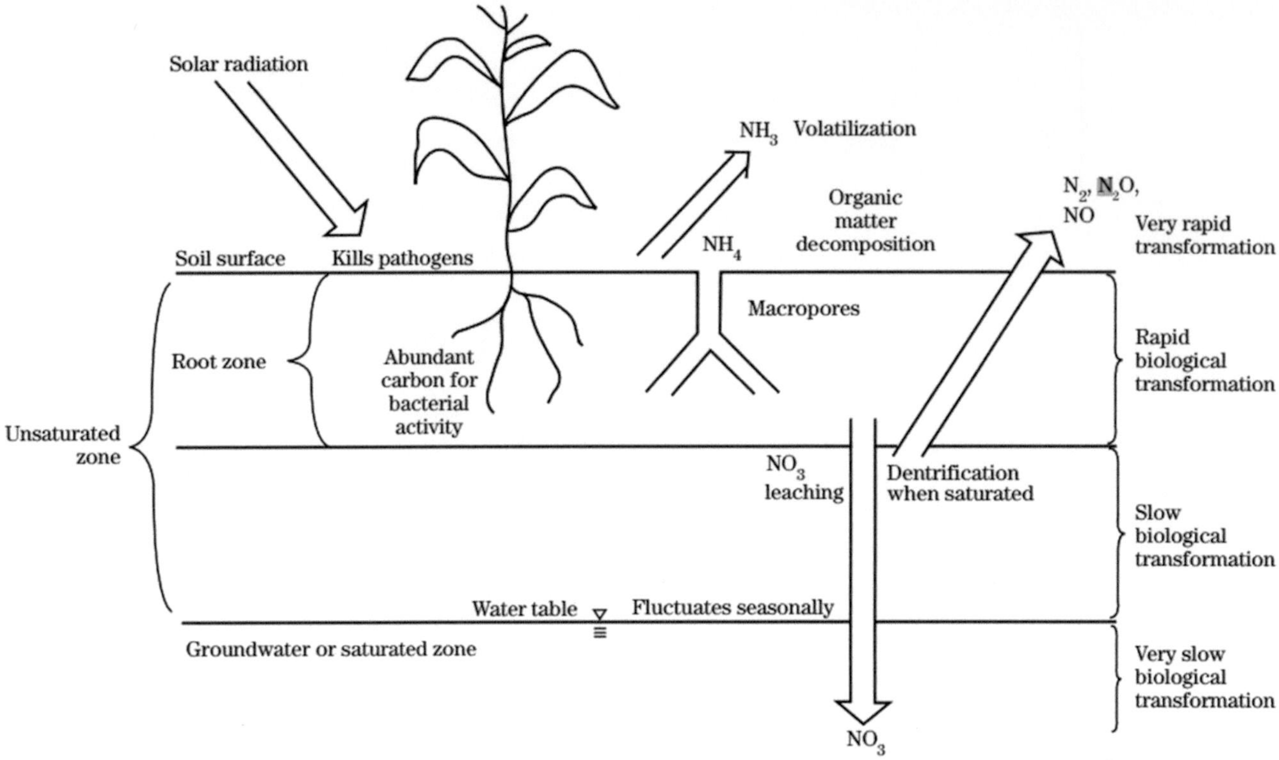

Fig. 23.7 Waste treatment in soil (Credit: NRCS AWMFH)

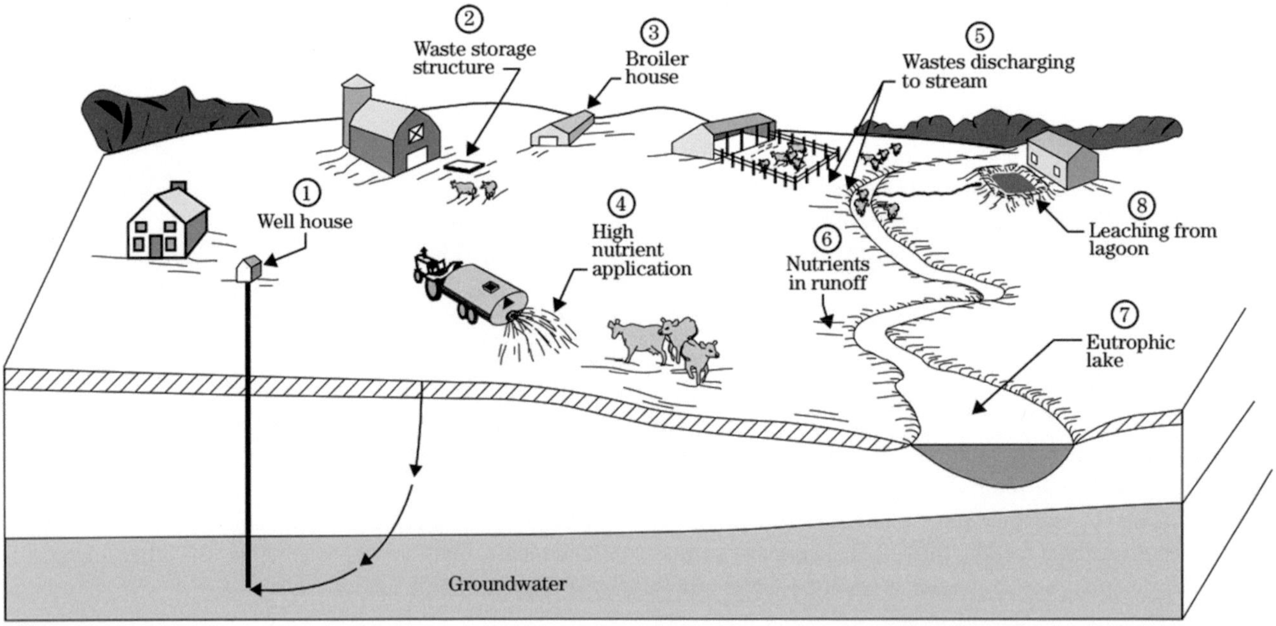

Fig. 23.8 Animal waste management practices that lead to environmental contamination (Credit: NRCS AWMFH)

in clean water after sediments have settled), color, COD (chemical oxygen demand, no microorganisms), BOD (biological oxygen demand) UV absorbance/fluorescence, particle size distribution, absorption analysis, and physical characteristics. The selection of indicators and/or surrogates depends on the types of wastewater treatment processes and the organics of concern. It is important select representative indicators of the different groups of chemicals. The suite of indicators should avoid duplication or gaps. Due to the concern with trace organics, the surrogates used for water testing have changed. The old view was to test for the following surrogates: BOD (organisms consuming oxygen), COD (chemical oxygen demand, no living organisms), TKN (total Kjeldahl nitrogen), VSS (ammonia and other solids that volatilize), pesticides, and solvents. New tests often include DOC, DON (dissolved organic nitrogen), TOC, DBP, endocrine activity (EDC), and PCPP (pharmaceutical chemicals).

Degradation rate is a major factor in the government's assessment of the level of hazard of chemicals. Even within some classes of compounds, slight differences in chemistry can change the degradation rate. For example, among steroids, those with an attached OH group degrade much more slowly than those without the OH group. Even though a chemical degrades, there is more to the story. The metabolites (degradation products) may be more harmful than the original chemical.

In countries with unregulated industrial discharge into sewage systems, there may be high concentration of heavy metals in irrigation water (Cd, Cu, Zn, Ni, Pb). These are generally not removed by conventional wastewater treatment. Chemicals may become concentrated in agricultural crops irrigated by this water. This is of concern because heavy metals cause many diseases when they are in high concentrations. Alternative treatment systems such as wetlands and soil aquifer treatment have in some cases been found to be more effective at removing heavy metals than conventional wastewater treatment systems.

Physical Characteristics

In addition to chemical and biological characteristics, four physical properties are used to describe waste streams: weight, volume, total solids, and moisture content. Total solids are divided into volatile solids and fixed solids. Wastes that are more than 95 % liquid are called liquid waste; wastes with less than 75 % liquid are a solid; wastes with a moisture content between 75 % and 95 % are called a slurry. Physical, as well as chemical and biological waste characteristics, of human and animal waste, as excreted, are given in Table 23.2.

Although physical and chemical characteristics of excrement from humans and animals are similar, raw human sewage (from developed countries) and animal waste effluent may have dramatically different characteristics because of high dilution in human sewage systems (Table 23.3). In a water short country or in low water flow facilities such as mobile home parks, per capita waste flow rate is low, and, as a result, concentration of human waste in sewage is higher.

Table 23.2 Waste characteristics – as excreted (Data from NRCS AWMFH)

		Human – adult	Beef – mature steer	Dairy – lactating cow	Swine – grower	Chicken – layer
Mass/day	kg/d/1,000 kg	30	59	80	63	60
Volume/day	L/d/1,000 kg	34	59	81	62	58
Moisture	%	89	88	88	90	75
TS (total solids)	% w.b.	10.9	11.6	12.5	10	25
TS (total solids)	kg/d/1,000 kg	3.3	6.8	10	6.3	15
VS (volatile solids)	kg/d/1,000 kg	1.9	6.0	8.5	5.4	11
FS (fixed solids)	kg/d/1,000 kg	1.4	0.74	1.5	0.94	4.3
COD (carbonaceous oxygen)	kg/d/1,000 kg	3	6. 1	8.9	6.1	14
BOD_5	kg/d/1,000 kg	1.3	1.4	1.6	2.1	3.7
N (elemental nitrogen)	kg/d/1,000 kg	0.2	0.31	0.45	0.42	0.83
P (elemental phosphorous)	kg/d/1,000 kg	0.02	0.11	0.07	0.16	0.31
K (elemental potassium)	kg/d/1,000 kg	0.07	0.24	0.26	0.22	0.34
C:N ratio			11	10	7	7

Table 23.3 Characteristics of human and dairy cow wastewater (Data from NRCS AWMFH)

Component	Units	Raw human WW	Secondary human WW	Raw dairy (Milk house + parlor)	Anaerobic dairy lagoon effluent (MH + MP)	Aerobic dairy lagoon (MH + MP)
Volume (L/d/1000 kg)		5600	5300	37	37	37
L/d/person (50 kg/person)		280	260			
Moisture	%	99.95	99.95	99.4	99.75	99.95
TS (total solids)	mg/L	500	500?	6,000	2500	500
VS (volatile solids)	mg/L	350	0	4,200	1100	200
FS (fixed solids)	mg/L	150	0	1,800	1400	300
COD (carbonaceous)	mg/L	450	0	5,000	1500	150
BOD_5	mg/L	200	25	1,000	350	35
N (elemental)	mg/L	30	20	200	200	20
$NH_4 - N$	mg/L	0	10		120	12
P (elemental)	mg/L	10	10	100	58	10
K (elemental)	mg/L	10	12	300	500	0

Questions

1. What is the fecal-oral pathway of disease transmission?
2. Pathogens come from four kingdoms of life: list and describe them.
3. List viruses that cause waterborne disease.
4. List bacteria that cause waterborne disease.
5. List protozoan parasites that cause waterborne disease.
6. Look up three of the diseases listed in Table 23.1 on the Internet and describe the symptoms.
7. Why are indicator organisms used and how do they distinguish between different sources of disease?
8. List the factors that determine wastewater treatment plant discharge pathogen concentration.
9. Calculate the required contact time in order to remove 99.9 % of remaining E-coli at a chlorine concentration of 3 mg/L
10. What was the key factor that reduced typhoid fever in the United States?
11. Why aren't ozone and chlorine used together?
12. What environmental factors influence pathogen fate in the environment?
13. Give a brief summary (one paragraph) of how the body fights pathogens.
14. How do vaccines help the body fight disease?
15. Why must acid be injected at the same time as chlorine during disinfection?
16. Redo the risk assessment problem. The expected number of salmonella bacteria in wastewater is 10^5 MPN/100 ml, and 200 children are expected to play for 16 hours in the landscaped area.
17. What are the 3 steps of Risk Analysis?
18. What are the 4 steps of Risk Assessment?
19. What is BOD and why is high BOD detrimental when wastewater is discharged to streams?
20. Describe the difference between oxidation and synthesis and endogenous respiration.
21. Calculate the BOD_5 if the dilution factor is 20 to 1, the initial dissolved oxygen concentration is 6 mg/L, and the final dissolved oxygen concentration is 2 mg/L.
22. List the steps in the nitrogen cycle in soils.
23. How is the nitrogen cycle carried to completion in wetlands with herbaceous aquatic plants?

References and Resources

Centers for Disease Control and Prevention (CDC): Water related diseases. http://www.cdc.gov/ncidod/diseases/water/drinking.htm

Kadlec RH, Knight R (1996) Treatment wetlands. CRC Press, Boca Raton

National Resource Conservation Service (NRCS) Waste management tools and agricultural waste management field handbook

Waste Treatment in Wetlands and Agriculture 24

Municipal and animal waste can be treated in ponds, wetlands, and agricultural fields. The sizing of ponds, wetlands, and field application areas is based on waste characteristics and volume. Wetland and pond areas are calculated based on standard treatment times, such as 5 day hydraulic retention time, or based on decay rate equations. The decay rate constant is primarily a function of temperature. Application of animal waste to soils by solids spreading (vehicle), liquid spreading (vehicle) or through a sprinkler system requires a series of calculations: the nutrient content in the wastewater, nutrient needs of the crop, the degradation and volatilization of nutrients in storage, mineralization, denitrification, and plant uptake in soils. The rate of application by sprinklers is dependent on percent solids in the waste and soil texture. This chapter includes an NRCS example calculation of dairy waste application to cropped fields. The soil salinization hazard from animal waste is calculated from the manure salinity load. Finally, an example shows how to calculate the blended ratio of wastewater to fresh water when wastewater concentration of nitrogen exceeds the crop nitrogen requirements.

Wetlands

This section includes rate equations, biological and hydraulic loading rates, and strategies for nitrogen removal. Microbial degradation of contaminants in wetlands over time can be modeled with a first-order reaction equation.

$$\frac{dC}{dt} = -kC \tag{24.1}$$

Equation 24.1 can be integrated and solved for final concentration, C_2.

$$C_2 = C_1 \ e^{-kt} \tag{24.2}$$

where

C_2 = contaminant concentration in effluent, mg/L,
C_1 = influent concentration, mg/L,
k = temperature dependent rate constant, 1/days,
t = hydraulic retention time, days.

Subsurface flow and free water surface wetlands are generally designed based on plug flow (no mixing) through the wetland. Thus, the time of reaction, T, is equal to the hydraulic detention time (HDT) of water in the wetland (the length of time that water remains in the wetland).

$$C_{out} = C_{in} \ e^{-kt} \tag{24.3}$$

where

C_{out} = the contaminant concentration in the effluent, mg/L,
C_{in} = the influent concentration, mg/L,

The rate equation can be modified to account for natural processes that may add contaminants to the flow. For example, nitrogen is continuously added to the wetland by plant decomposition, animal waste and other contributions. Thus, first-order decay equation includes the irreducible background concentration.

$$C_{out} - C^* = e^{(-kt)} \left(C_{in} - C^*\right) \tag{24.4}$$

where

C^* = irreducible background concentration (mg/L)

The reaction rate constant is commonly based on the wetland area divided by the flow rate rather than the hydraulic detention time. This is based on the assumption of a typical average depth of water in the wetland. If a wetland

P. Waller, M. Yitayew, *Irrigation and Drainage Engineering*, DOI 10.1007/978-3-319-05699-9_24

Table 24.1 Constants at 20 °C for FWS and SSF wetlands (Credit (Knight et al. 1995), Arizona Department of Environmental Quality Wetlands Design Manual)

Surface Flow (FWS)	BOD	TSS	NH_4^+	NO_3^-	TN	TP
k_{20} (m yr^{-1})	35	1000	18	35	22	12
θ_k	1.0	1.0	1.04	1.09	1.05	1.0
Irred. C_{20} (mg/L)	6	4.7 + .09 C_1	0.0	0.0	1.5	0.02
θ_c	1.0	1.065	–	–	1.0	1.0
Subsurface Flow	BOD	TSS	NH_4^+	NO_{24-}	TN	TP
k_{20} (m yr^{-1})	180	1000	34	50	27	12
θ_k	1.0	1.0	1.04	1.09	1.05	1.0
Irred. C_{20} (mg/L)	6	4.7 + .09 C_1	0.0	0.0	1.5	.02
θ_c	1.0	1.065	–	–	1.0	1.0

has a dramatically different depth than conventional wetlands, then area-based rate constants are not applicable.

$$C_{out} - C^* = e^{\left(\frac{-kA}{Q}\right)}\left(C_{in} - C^*\right) \qquad (24.5)$$

where

A = wetland area, m^2,
Q = flow rate, m^3/yr.

The rate constant, k, is a function of temperature because microbial metabolism generally increases with temperature. The rate constant, k, is typically given for 20 °C water and then modified based on the difference between water temperature and 20 °C.

$$k = k_{20}\theta_k^{(T-20)} \qquad (24.6)$$

where

k_{20} = rate constant at 20 °C, d^{-1}
θ_k = temperature coefficient
T = Temperature, °C.

The irreducible background concentration is also a function of temperature and is modeled with a similar equation.

$$C^* = C_{20}\,\theta_k^{(T-20)} \qquad (24.7)$$

where

C_{20} = background concentration at 20 °C,

Reaction rate and background concentration constants for six contaminants in FWS and SSF wetlands are given in Table 24.1.

Hydraulic detention time is also the time in the reaction rate equations. It is calculated as wetland volume divided by the product of area and average depth. Subsurface (SSF) wetland volume is also a function of porosity:

$$V = Ad\phi \quad t = V/Q \qquad (24.8)$$

where

A = wetland area, m^2,
d = depth of water, m,
ϕ = porosity of gravel media,
V = wetland volume, m^3.

Example 24.1 Effluent is discharged from septic tanks into a subsurface flow wetland. Wetlands area is 1,070 m^2. Total nitrogen in septic tank effluent is 36 mg/L. Hydraulic loading rate is 25,000 m^3/yr. Water temperature in the wetland is 20 °C. Calculate total nitrogen, TN, in wetlands effluent with Eq. 24.4. Recalculate for 4 °C. Determine whether the wetland is large enough to comply with a maximum allowable TN equal to 10 mg/L in wetland effluent. Next, convert the area-based constant in Table 24.1 to a time-based constant with the assumption that the area-based constant is calibrated for a void fraction in the gravel equal to 0.33 and a depth of flow equal to 0.4 m. Then calculate effluent quality with a 5 day detention time.

Calculation of effluent concentration at 20 °C

The irreducible background concentration for total nitrogen is not temperature dependent and is a constant, 1.5 mg/L. The area-based rate constant, k_{20}, at 20 °C is 27 m/yr (Table 24.1).

$$C_{out} = C^* + e^{\left(\frac{-kA}{Q}\right)}\left(C_{in} - C^*\right)$$
$$= 1.5 + e^{\left(\frac{-27\,*\,1,070}{25,000}\right)}(36 - 1.5) = 12\ mg/L\ TN$$

The wetland is inadequate because the treatment goal of 10 mg/L of total nitrogen is exceeded. Thus, the size should be increased.

Calculation of effluent concentration at 4 °C

$$k = k_{20}\theta_k^{(T-20)} = 27{*}1.05^{(4-20)} = 12.37\ m/yr$$
$$C_{out} = C^* + e^{\left(\frac{-kA}{Q}\right)}\left(C_{in} - C^*\right) = 1.5 + e^{\left(\frac{-12*1,070}{25,000}\right)}(36 - 1.5)$$
$$= 22\ mg/L\ TN$$

The wetland is even more inadequate at a lower temperature. It is not surprising that the wetland is inadequate given the fact that the hydraulic detention time is 2 days:

$$V_{cv} = d\ A\ \theta = 0.4\ m(1,071\ m^2)(0.33) = 141\ m^3$$
$$T = V/Q = 141\ m^3/25,000\ m^3/yr(365\ day\ /\ yr)$$
$$= 2\ days$$

Calculate an equivalent time-based rate constant (e^{-kT}) *from the area-based rate constant.*

$$\frac{-k_{area}A}{Q} = -k_{time}T = \frac{-k_{time}V_{cv}}{Q}$$
$$k_{time} = \frac{k_{area}A}{V_{cv}} = \frac{k_{area}}{d\theta} = \frac{(27\ m/yr)}{0.4\ (0.33)}\left(\frac{1}{365\ day/yr}\right)$$
$$= 0.56\ d^{-1}$$

Calculate effluent TN with 5-day detention time and 20 °C (A = 1,070 (5/2) = 2,680 m^2)

$$C_{out} = C^* + e^{(-kT)}\left(C_{in} - C^*\right) = 1.5 + e^{(-5*0.56)}(36 - 1.5)$$
$$= 3.6\ mg/L\ TN$$

Calculate effluent TN with 5-day detention time and 4 °C (A = 2,678 m^2)

$$k = k_{20}\theta_k^{(T-20)} = 0.56\left(1.05^{(4-20)}\right) = 0.26\ m/yr$$
$$C_{out} = C^* + e^{(-kT)}\left(C_{in} - C^*\right) = 1.5 + e^{(-5\ *\ 0.26)}(36 - 1.5)$$
$$= 11\ mg/L\ TN$$

The 5-day detention time at 4 °C almost complies with the effluent standard of 10 mg/L.

In Table 24.1, the area-based rate constant is approximately 5 times higher in the SSF than the FWS for BOD. The increased decay rate means that SSF wetlands can have a much smaller area than FWS wetlands if the criterion is BOD treatment. FWS wetlands require approximately 20 m^2/person for secondary treatment, while SSF wetlands require only 2.5 m^2/person for secondary treatment. The improvement in BOD treatment is due to the filtration capacity of the gravel medium. However, the improved filtration also leads to the main problem with SSF wetlands, clogging and loss of hydraulic conductivity.

Biological and Hydraulic Loading Rates

Wetlands can be designed based on a maximum biological loading rate where loading refers to kg of contaminant (BOD_5 loading rate) per ha of wetland surface area per day. The *maximum* loading rate is calculated by estimating the oxygen transfer capacity of the wetland vegetation. The BOD should not overwhelm the capability of the wetland to produce oxygen (EPA 1988). The BOD loading rate should also not be extremely low (reason below). The required oxygen (kg/d) to meet the total demand of the waste load can be calculated as:

$$\text{Required oxygen(kg/d)} = 1.5\ BOD_5(kg/d) \qquad (24.9)$$

The available oxygen is the product of oxygen transfer rate and wetlands area.

$$\text{Available oxygen(kg/d)} = TrO_2\ A \qquad (24.10)$$

where

T_rO_2 = oxygen transfer rate for the vegetation, kg/(ha-d)
A = wetlands surface area, ha.

Herbaceous aquatic plants transfer oxygen at rates between 50 and 450 kg/ha/day in FWS wetlands. A typical transfer rate is 200 kg/ha-d. The BOD_5 loading rate for a wetland can be found by setting the available oxygen equal to the required oxygen for BOD removal.

$$1.5\ BOD_5 = (T_rO_2)(A)$$
$$\frac{BOD_5}{A} = \frac{200\ kg/ha/d}{1.5} = 133\ kg/ha/d$$

As a safety factor, the BOD_5 oxygen load should be less than the available oxygen. Thus, the EPA (1988) recommends a maximum loading rate of 112 kg BOD_5/ha-d for FWS wetlands. In practice, hydraulic loading rates for FWS wetlands typically range from 150 to 500 m^3/ha/day with a loading rate of 200 m^3/ha/day resulting in the greatest treatment efficiency (EPA 1988).

The minimum acceptable BOD_5 loading rate is based on the amount of carbon required for denitrification. If inadequate carbon is available, then the microbial reaction that causes denitrification does not occur. This is also why denitrification does not take place in soils with low organic carbon.

Alternation of Aerobic and Anaerobic Treatment Zones for Nitrification and Denitrification

With high waste levels and limited land area, supplemental oxidation may be needed. One alternative is floating aquatic plant (FAP) wetlands, which have mechanical air injection and alternating aerated and nonaerated zones.

Peter Livingstone developed an innovative design with 4 anaerobic/aerobic cycles (Fig. 24.1) at the Desert Museum

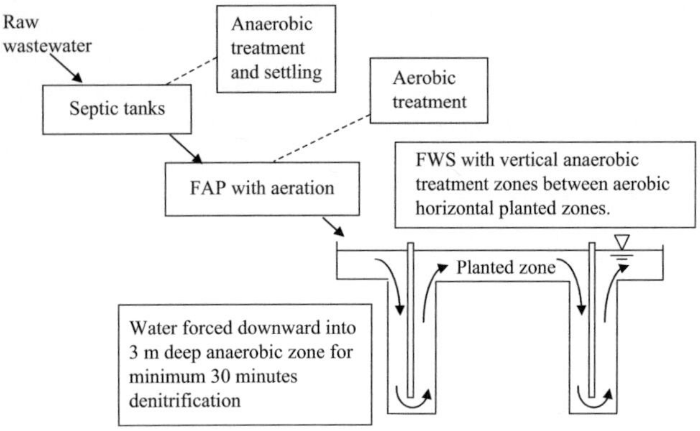

Fig. 24.1 Desert Museum wetlands in Tucson Arizona (Designed by Peter Livingstone)

in Tucson, Arizona. This wetland, with 4 vertical denitrification zones (only two are shown) has undetectable nitrogen levels at the end of the wetland; however, before the vertical zones were installed, nitrogen in effluent exceeded the acceptable discharge concentration (>10 mg/L). The vertical zones were relatively easy to install: a trench and barrier is placed across the wetland that forces the water downward into the anaerobic zone.

Water Balance in Wetlands

Evapotranspiration can vary dramatically with the seasons, and a wetland must operate in summer and winter in order to keep the plants alive. A volume balance for water flow in a wetland must be conducted in order to ensure that sufficient water is allocated to the wetland during dry periods. Seepage, evapotranspiration, and precipitation depths are multiplied by wetlands surface area to calculate rate of volume change from each of these processes.

$$\Delta d_{cv}(A) = Q_{in}t - Q_{out}t - d_{seepage}(A) - ET(A) + P(A) \tag{24.11}$$

where

Q_{in} = inflow rate, m^3/day,
t = time, day,
Q_{out} = effluent flow rate, m^3/day,
ΔV = change in water volume within the wetland, m^3,
ET = evapotranspiration, m,
A = wetlands surface area, m^2,
P = depth of precipitation, m,
$d_{seepage}$ = depth of infiltration below the wetland, m,
Δd_{cv} = depth of water in the wetland, m.

Example 24.2 The wastewater inflow rate of the Pinetop-Lakeside wetland near Pinetop, Arizona is 1,500 m^3/d during summer (Table 5.2, Arizona Department of Environmental Quality manual). The area (A) of the wetland is 38 ha. The evapotranspiration rate (ET) is 1.14 cm/day in summer, and precipitation is effectively zero. Assume that the average percentage of the wetland that is filled with plant material is 3 %. Neglect seepage. What is the change in water surface elevation during a 1-day period if no supplemental water is added to the wetland?

$$V_{ET} = A\ ET = (38\ \text{ha})(1.14\ \text{cm})(10{,}000\ \text{m}^2/\text{ha})(1\text{m}/100\ \text{cm}) = 4{,}300\ \text{m}^3$$
$$V_{inflow} = 1{,}500\ \text{m}^3$$
$$V_{outflow} = V_{seepage} = 0.$$
$$\Delta V_{cv} = V_{inflow} - V_{ET} = 1{,}500 - 4{,}300 = -2{,}800\ \text{m}^3$$

Depth change (Δd) during a 1-day period without supplemental water is ΔV_{cv} over area.

$$\Delta d = \frac{\Delta V_{cv}}{A(1 - \%plants)} = \frac{-2{,}800\ m^3}{380{,}000\ m^2(1 - 0.03)} = -0.0076\ m = -0.76\ cm$$

Thus, supplemental water may be required in summer.

Waste Application to Crops

Animal Waste

Animal waste management is surprisingly important for the environment and for farm economics. Animal waste production in the United States is 130 times greater than the human waste production; a single 50,000 unit hog farm produces as much waste as the entire city of Los Angeles (Lusk 1998). Nitrogen containing compounds in animal waste are the primary water quality concern associated with animal waste (Lusk 1998). Nutrients can lead to eutrophication in streams and ultimately to large dead zones in the ocean. Many counties in the US have more nitrogen in animal waste than crop uptake capability. In this case, treatment and removal is necessary to prevent water pollution. In theory, the loop is closed if nutrients in animal waste are applied to crops, and crops take up the waste; however, misapplication of waste, volatilization of greenhouse gases, or inefficient irrigation can lead to environmental contamination.

There are numerous methods of animal waste disposal and utilization. Waste can be applied to fields as dry manure, compost, slurry, or relatively clear treated liquid. There are advantages and disadvantages associated with each method.

Untreated animal waste is hazardous and toxic to plants. Because of the extremely high contaminant load, animal waste effluent is treated for long periods in anaerobic lagoons prior to introduction into wetlands or soils. Animal

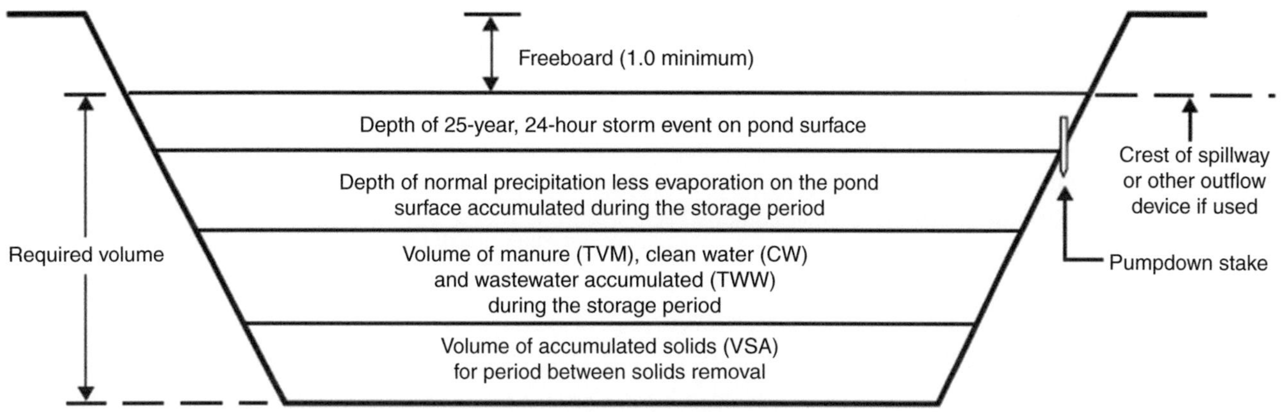

Fig. 24.2 Waste storage pond volume (without a watershed) (Credit NRCS AWMFH)

waste lagoons and wetlands require different treatment times than urban wastewater treatment facilities and wetlands because of higher contaminant concentrations in animal waste.

Animal waste has high salinity and can lead to salinization of soils and ponds. If animal waste is treated in lagoons and recirculated for flushing waste treatment facilities, the high level of evaporation in arid regions can lead to even higher salinity in lagoon water. The salinity can kill the microbes that are supposed to treat the waste.

Wet manure is typically high in ammonia and/or ammonium, which can be toxic to plants both in wetlands and in agriculture. Germination of seeds is particularly susceptible to high ammonium. Ammonium, a cation, can also block the uptake of calcium by mature plants. Thus, manure must be strategically applied, for example as a side dressing away from seeds, and in limited quantities. Another alternative is manure pretreatment to remove ammonium. Pretreatment of manure by drying in open feedlots in arid regions before application to fields, or by treatment in an anaerobic lagoon in humid regions, can reduce ammonia to acceptable levels; however, one problem with ammonia removal by drying of waste or in anaerobic lagoons is that much of the ammonia is volatilized to the atmosphere, along with methane. Ammonia and methane are potent global warming gases. Methane digestion, also known as anaerobic digestion, reduces greenhouse gas emissions from animal waste treatment. It also produces fuel for electricity generation or heating. However, many of these systems have failed, and they are generally only suitable in large animal feeding operations.

The combination of waste storage ponds or lagoons and free water surface wetlands can be effective at removing much of the nutrient load from animal waste, although volatilizing greenhouse gases into the atmosphere. The percent removal of ammonia, organic N, phosphate and BOD_5 can approach 90 % but is often in the range of 50 % (AWMFH). The storage capacity of animal waste storage ponds (Fig. 24.2) must include the volume of sludge, manure and wastewater, as well as enough storage for the 24-hour 25-year storm plus a 1 ft (0.3 m) freeboard. Additional volume should be added if the storage pond is connected to a watershed (water can run into the pond).

Animal waste effluent wetlands areal loading rates can be calculated based on the desired effluent nitrogen concentration (*AWMFH*).

$$LR = 0.68(C_{out}) - 7.88 \tag{24.12}$$

where

LR = areal TN loading rate, (kg TN)/ha/day,
C_{out} = desired wetland effluent TN concentration, mg/L.

The water surface area of the wetland is the TN mass flow rate divided by the loading rate.

$$A = TN / LR \tag{24.13}$$

where

TN = total nitrogen loading rate, kg/day
A = water surface area of free water surface wetland, ha.

Although Eqs. 24.12 and 24.13 provide a general guideline for agricultural wetlands design, the specific quality of wastewaters varies, and the equations should be refined for local conditions, during the design phase if possible, but if not, during wetlands operation.

Soil type has a direct impact on the hazard level of applying manure or animal waste effluent to soils. Sandy soils with little absorption capability and high hydraulic conductivity are more likely to leach contaminants to groundwater. In addition to land application, leaking tanks, unlined lagoons, and surface streams on farms can lead to aquifer contamination by animal waste. Once contaminants pass below the root zone, it is unlikely that they will be treated before reaching the groundwater so it is important to keep animal waste nutrients within the root zone in order to allow them to be taken up by the crop or degraded by microbes.

Composting animal waste adds cost, but compost is chemically stable and is a good soil amendment that improves soil structure. Raw manure is much more reactive, has a much higher pathogen concentration, and is more odiferous; however, if the goal is to economically add fertilizer, then manure is a better option. Compost production requires high labor input: regular turning over, adding water, and possibly adding carbon. In addition nutrients are lost during in the process. One disadvantage of applying wet manure or municipal sludge to fields is that pathogen concentrations are high. Growing edible crops is not recommended within 18 months of sludge or manure application in arid regions and within 36 months in humid regions in order to allow for pathogen reduction in soils.

The number of animals per unit land application area and the nutrient production per animal can be balanced with the crop nutrient requirement. If the nutrient load is excessive, then nutrient load can be lowered by reducing the number of animals or by extending the waste treatment time. Wetland size can be tailored to treat the wastewater to the point that it matches the nutrient load required by the farm, as shown in Eq. 24.12. Alternatively, disposal capacity can be increased by increasing land area or changing to crops that have higher nutrient requirements.

Animal waste might contain high levels of copper, zinc, and aluminum, which can be toxic to plants, humans, or animals consuming the plants. Another hazard is that animal waste often contains hormones and other chemicals that are fed to animals in order to increase production.

Animal waste that is treated in lagoons or wetlands can be dried and then applied to soil as a solid by a solids spreader (vehicle). It can also be applied as a liquid through a liquid spreader (vehicle) or through a sprinkler system. Wetlands supernatant (clear discharge with no solids), can be handled by any type of surface or sprinkler irrigation system, However, solids in wastewater limits the options, and special high-pressure pumps with chopping blades are required for pumping. Up to a solids content of 12 % (by weight) for swine and poultry manure and up to 7 % solids for cattle manure, liquid waste and slurries from lagoons or waste storage ponds can be applied to land by big gun sprinklers. Conventional sprinklers can apply cattle manure with up to 4 % solids and swine or poultry manure with up to 7 % solids.

Percent solids (wet basis) and percent moisture (wet basis) are calculated as follows:

$$P = Percent\ solids = \frac{Dry\ weight}{Wet\ weight}$$
$$Percent\ moisture = \frac{Wet\ weight - Dry\ weight}{Wet\ weight} \quad (24.14)$$

Waste may need to be diluted in order to pump waste through a sprinkler system. The dilution ratio is calculated as follows:

$$DR = \frac{P_0 - P_d}{P_d} \quad (24.15)$$

where

DR = dilution ratio, (volume water)/(volume manure)
P_0 = original solids content, g/g,
P_d = diluted solids content, g/g.

Example 24.3 If cattle manure has 20 % solids, then what dilution ratio is required to apply it through a sprinkler system at a solids content of 5 %. What is the ratio of water to manure?

$$DR = \frac{P_0 - P_d}{P_d} = \frac{20 - 5}{5} = 3$$
$$\text{DR} = (\text{volume water})/(\text{volume manure})$$
$$\text{Ratio} = \text{DR} \rightarrow 3 \text{ parts water to 1 part manure.}$$

When applying waste, uniform application prevents pollution and uneven distribution of nutrients in the field. Proper design of big gun spacing or traveling big guns increases uniformity. The design procedure for big gun application of lagoon effluent was described by Fulhage and Pfost from Missouri Extension (http://extension.missouri.edu/p/EQ327). In order to prevent runoff, the maximum application rates in Table 24.2 should not be exceeded.

Solids in irrigation water reduce infiltration rate because they clog soil pores. The reduction coefficient is the ratio of reduced infiltration with high solids content to infiltration with clean water (Table 24.3). Application rates in Table 24.2 must be multiplied by coefficients in Table 24.3 to obtain the solids adjusted application rate. Notice that the infiltration rate of fine-textured soils is not affected at all by solids in wastewater since the soil particles are smaller than the wastewater particles. Runoff to streams is more likely in sloping fields so it is especially important to ensure that infiltration rate is not exceeded in sloping fields.

Pipelines used for delivering wastewater have unique hydraulic requirements and characteristics. Average pipe velocity in all pipe sections should be kept greater than

Table 24.2 Maximum application rate, in/hr and cm/hr, as a function of total application depth (Credit NRCS AWMFH)

Total application depth (in)	0.25	0.5	0.75	1	1.25	1.5	2
Soil texture	Maximum application rate (in/hr)						
Sand	6	6	6	6	6	6	6
Loamy sand	6	6	4.83	4.22	3.86	3.62	3.32
Sandy loam	4.91	2.97	2.32	1.99	1.8	1.67	1.51
Loam	3.11	1.69	1.21	0.98	0.84	0.74	0.62
Silt loam	2.7	1.45	1.03	0.82	0.7	0.61	0.51
Sandy clay loam	1.74	0.96	0.69	0.56	0.48	0.43	0.37
Clay loam	1.27	0.68	0.48	0.39	0.33	0.29	0.24
Silty clay loam	1.09	0.57	0.4	0.32	0.26	0.23	0.19
Sandy clay	0.61	0.33	0.23	0.19	0.16	0.14	0.12
Silty clay	0.84	0.44	0.3	0.24	0.2	0.17	0.14
Clay	0.39	0.21	0.14	0.11	0.09	0.08	0.07
Total application	Maximum application rate (mm/hr)						
depth (mm)	6.3	12.7	19.0	25.4	31.7	38.1	50.8
Sand	15.2	15.2	15.2	15.2	15.2	15.2	15.2
Loamy sand	15.2	15.2	12.3	10.7	9.8	9.2	8.4
Sandy loam	12.5	7.5	5.9	5.1	4.6	4.2	3.8
Loam	7.9	4.3	3.1	2.5	2.1	1.9	1.6
Silt loam	6.9	3.7	2.6	2.1	1.8	1.5	1.3
Sandy clay loam	4.4	2.4	1.8	1.4	1.2	1.1	0.94
Clay loam	3.2	1.7	1.2	0.99	0.84	0.74	0.61
Silty clay loam	2.8	1.4	1.0	0.81	0.66	0.58	0.48
Sandy clay	1.5	0.84	0.58	0.48	0.41	0.36	0.30
Silty clay	2.1	1.1	0.76	0.61	0.51	0.43	0.36
Clay	0.99	0.53	0.36	0.28	0.23	0.20	0.18

Table 24.3 Reduction coefficient for infiltration determined by % solids (Credit NRCS AWMFH)

Soil texture	Percent solids (by wt)						
	0.5	1.0	2.0	3.0	5.0	7.0	10.0
Sand	0.88	0.55	0.31	0.22	0.13	0.10	0.07
Loamy sand	0.7	0.54	0.37	0.28	0.19	0.14	0.10
Sandy loam	0.87	0.77	0.63	0.53	0.40	0.32	0.25
Loam	0.97	0.93	0.88	0.83	0.74	0.67	0.59
Silt loam	0.98	0.95	0.91	0.87	0.81	0.75	0.68
Sandy clay loam	0.99	0.97	0.95	0.92	0.87	0.83	0.78
Clay loam	0.99	0.99	0.98	0.97	0.94	0.92	0.89
Silty clay loam	1.00	1.00	0.99	0.99	0.98	0.97	0.86
Sandy clay	1.00	1.00	1.00	1.00	0.99	0.99	0.99
Silty clay	1.00	1.00	1.00	1.00	1.00	1.00	1.00
Clay	1.00	1.00	1.00	1.00	1.00	1.00	1.00

0.6 m/sec in order to prevent sedimentation. Friction loss is higher for slurries than for clean water (Table 24.4).

The nutrient balance calculation for application of waste to crops includes the fraction that is lost to the atmosphere in lagoons and wetlands, fraction mineralized in the soil, fraction denitrified in the soil, fraction lost to deep seepage or runoff, and fraction taken up by plants. Many of these terms have a range of values.

Nutrient retention percentages for various waste storage facilities are shown in Table 24.5. Note in the last and third to last row that a higher dilution ratio results in lower nutrient retention (higher nutrient loss) in the storage pond.

Mineralization is the process of converting organic nitrogen to nitrate. Not all organic nitrogen is mineralized immediately. Rates of mineralization for different waste sources

Table 24.4 Friction loss ratio, slurries vs. clean water (15–25 cm ID) (Credit NRCS AWMFH)

	Percent solids					
Velocity (m/s)	4	5	6	7	8	10
0.3	1.1	1.5	2.1	2.9	4	5.3
0.45	1	1.2	1.5	2.1	2.5	4
0.6	1	1	1	1.6	1.9	3.3
0.75	1	1	1	1.3	1.6	2.9
0.9	1	1	1	1.2	1.5	2.7
1.05	1	1	1	1.1	1.3	2.5
1.2	1	1	1	1	1	2.4
1.35	1	1	1	1	1	2.3
1.5	1	1	1	1	1	2.2
1.65	1	1	1	1	1	2.1
1.8	1	1	1	1	1	2
1.95	1	1	1	1	1	2
2.1	1	1	1	1	1	2

Table 24.5 Range of nutrient retention in waste treatment storage facilities (NRCS AWMFH (Table 11.5))

	Beef			Dairy			Poultry			Swine		
	N	P	K	N	P	K	N	P	K	N	P	K
Manure stored in open lot in cool, humid region	5570	7080	5570	7085	8595	8595				5570	6580	5570
Manure stored in open lot in hot, arid region	4060	7080	5570	5570	8595	8595						
Manure, liquids, and solids stored in a covered, essentially watertight structure	7085	8595	8595	7085	8595	8595				7585	8595	8595
Manure, liquids, and solids stored in an uncovered, essentially watertight structure	6075	8090	8090	6575	8090	8090				7075	8090	8090
Manure, liquids, and solids (diluted less than 50 %) held in a waste storage pond				6580	8095	8095						
Manure stored in pits beneath slatted floor	7085	8595	8595	7085	9095	9095	8090	9095	9095	7085	9095	9095
Manure treated in anaerobic lagoon or stored in waste storage pond after being diluted more than 50 %	2035	3550	5065	2035	3550	5065	2030	3550	5060	2030	3550	5060

are given in Table 24.6. Rate of degradation affects the amount of N available to the crop.

Nitrogen is lost by ammonia volatilization to the atmosphere if manure or animal waste effluent is applied to soil surface. The percent loss is a function of the type of application system, length of time that waste remains on the soil surface, temperature of the soil, and moisture content of the soil. For sprinkler application, approximately 25 % of nitrogen is volatilized (Table 24.7).

Denitrification (conversion of nitrate to atmospheric nitrogen by microorganisms) requires a carbon source. Thus, soils low in carbon have low denitrification rates. Denitrifying organisms have higher growth rates (denitrification rates) in anaerobic soils. Denitrification rates are greatest in organic (high carbon) poorly drained (anaerobic) soils. Dentrification rates per year as a function of drainage class and organic matter content are shown in Table 24.8. No-till fields have higher denitrification rates so the AWMFH recommends using one class wetter drainage for those fields. For manure N, double all values. For tile drained (subsurface drainage) soils, use one class better drainage. For irrigation or humid climates, use values at upper end of the range. For arid or semiarid non-irrigated sites, use values at lower end of the range. For soils with a compacted, very low permeability layer below plow depth, but above 1.2 m depth, use one class wetter drainage.

The leaching index, as defined by the NRCS, is the depth of water (inches) leached below the root zone. This index has been correlated with the percent of inorganic N (NO_3^-) that is leached below the soil root zone. The percent leaching losses for leaching indexes of <2, 2–10, and >10 are 5 %, 10 %, and 15 %, respectively (*AWMFH*). Higher leaching

Table 24.6 Mineralization rates in soils for wastes from different animal waste treatment facilities (Credit NRCS AWMFH)

	Nitrogen			Phosphorus			Potassium		
	Years after initial application								
	1	2	3	1	2	3	1	2	3
Fresh poultry manure	90	92	93	80	88	93	85	93	98
Fresh swine or cattle manure	75	79	81	80	88	93	85	93	98
Layer manure from pit storage	80	82	83	80	88	93	85	93	98
Swine or cattle manure stored in covered storage	65	70	73	75	85	90	80	88	93
Swine or cattle manure stored in open structure or pond (<50 % dil)	60	66	68	75	85	90	80	88	93
Cattle manure with bedding stored in roofed area	60	66	68	75	85	90	80	88	93
Effluent from lagoon or diluted waste storage pond (>50 % dilution)	40	46	49	75	85	90	80	88	93
Manure stored in open lot, cool-humid	50	55	57	80	88	93	85	93	98
Manure stored in open lot, hot-arid	45	50	53	75	85	90	80	88	93

Table 24.7 Percent of nitrogen in applied manure still potentially available to the soil after surface volatilization (Willrich, 1974) (Credit NRCS AWMFH)

	Percent of nutrients retained		
Injection	95		
Sprinkler	75		
Days between application by truck and incorporation	Warm dry	Warm wet	Cool wet
1	70	80	100
4	60	70	95
7 or more	50	60	90

Table 24.8 Approximate N dentrification rates for various soils. Percent of inorganic N denitrified (Credit NRCS AWMFH from Meisinger and Randall (1991))

	Soil drainage classification				
Soil organic matter content (%)	Excessively well drained	Well drained	Moderately well drained	Somewhat poorly drained	Poorly drained
<2	2–4	3–9	24–14	6–20	10–30
2–5	3–9	24–16	6–20	10–25	15–45
>5	24–12	6–20	10–25	15–35	25–55

rates result in higher N leaching. Application timing influences the depth leached; for example, application of waste or fertilizer just before spring rains or heavy irrigation would result in maximum nutrient loss.

The ratio of nutrients (N, P, and K) in manure is not the same as the ratio of N, P, and K requirements of plants: Phosphorous is generally overapplied if the manure application is based on supplying the N requirement. Overapplication of phosphorous leads to environmental contamination in some regions; in this case, waste should not be applied based on the phosphorous requirement. The potential for phosphorous pollution is greatest an acidic soils and when there are adjacent water bodies. Phosphorous contamination is generally not a potential hazard in arid regions because soils are alkaline, and there are no adjacent river bodies; thus, the waste application rate can be solely based on the N requirement. Worksheets are available from the NRCS that specify whether waste application should be based on nitrogen, phosphorous, or potassium.

Example 24.4 This example is from *NRCS AWMFH*, but it has been converted to metric units.

Given: A dairy farm has 200 lactating dairy cows in Central Wisconsin. Average mass/cow is 500 kg. The N, P, and K nutrient loads per cow are 0.45-, 0.07-, and 0.26-kg/d/1000 kg, respectively. The wastewater from the milking house effluent has N, P, and K concentrations equal to 200-, 100-, and 300-mg/L, respectively. All milking house – milking parlor waste and manure are added to an anaerobic waste storage pond. The bottom of the pond dimensions are 18 × 60 m and the maximum operating depth is 3.6 m. Side slopes are 2:1 (2 horizontal by 1 vertical). The 25-year,

224-hour storm is 15 cm. Milking parlor + milking house wastewater volume = 19 L/cow/day. No runoff water from holding areas or adjoining fields is allowed to run into the pond. Mean annual precipitation is 81 cm, annual evaporation from the pond surface is 30 cm. Dilution of manure by rainfall and milk house and milk parlor cleaning water is approximately 50 %. The climate has moderate rainfall and humidity. Phosphorous pollution is a hazard in the region.

Manure has been applied for several years every spring and is incorporated into the soil by plowing within 1 day of application.

Soils are moderately well-drained silt loams and have a leaching index of 6, which means that 15 cm (6 in.) percolates below the root zone. The organic matter content is 3 %. The rock fraction is 10 %. The soils are subject to frequent flooding. Soil slopes range up to 10 %.

The crop is corn, with a rooting depth of 1.5 m. Required yearly nutrient additions for crop growth (grain corn) are 132 kg/ha nitrogen (N), 52 kg/ha phosphorous (P_2O_5), and 39 kg/ha potassium (K_2O). Manure is applied by truck in spring just before planting when soil is wet. Soil temperature is between warm and cool (Table 24.7) but closer to cool (needed for denitrification rate calculation)

Required: Calculate manure and wastewater application rates for truck application and wastewater application rate for sprinkler application.

Answer the following questions:

1. What is the required application rate?
2. Estimate the land area required for maximum nutrient utilization.
3. What is the application rate in metric tons per hectare for the area that would provide maximum nutrient utilization?
4. For an irrigation system design, determine the total depth of wastewater application based on nitrogen requirements and for maximum nutrient utilization.

Solution:

Step 1. Estimate the total nutrients (NPK) in the excreted manure.

Annual nutrient load in excreted manure (rates from Table 2.6)

$$N = \frac{(200\ cows)(500\ kg/cow)(0.45\ kg/d)\ (365\ d/yr)}{1{,}000\ kg} = 16{,}400\ kg$$

$$P = \frac{(200\ cows)(500\ kg/cow)\ (0.07\ kg/d)\ (365\ d/yr)}{1{,}000\ kg} = 2{,}550\ kg$$

$$K = \frac{(200\ cows)(500\ kg/cow)\ (0.26\ kg/d)\ (365\ d/yr)}{1{,}000\ kg} = 9{,}500\ kg$$

Step 2. Calculate annual nutrient load in the wastewater

This waste calculation includes some of the manure (as excreted) from the previous calculation because it dissolves in the water; thus, the amount of manure applied will be slightly overestimated. Annual nutrient load added to wastewater:

$$N = (200\ cows)(19\ L/d/cow)(200\ mg/L)\left(10^{-6}\ kg/mg\right)(365\ d) = 277\ kg$$

$$P = (200\ cows)(19\ L/d/cow)(100\ mg/L)\left(10^{-6}\ kg/mg\right)(365\ d) = 138\ kg$$

$$K = (300\ cows)(19\ L/d/cow)(300\ mg/L)\left(10^{-6}\ kg/mg\right)(365\ d) = 416\ kg$$

Total nutrients produced:

$$\text{Total N} = 16{,}400\ \text{kg} + 277\ \text{kg} = 16{,}700\ \text{kg}$$
$$\text{Total P} = 2{,}550\ \text{kg} + 138\ \text{kg} = 2{,}700\ \text{kg}$$
$$\text{Total K} = 9{,}500\ \text{kg} + 416\ \text{kg} = 9{,}900\ \text{kg}$$

Fertilizer requirements for crops are reported as the oxidized form of potassium (K_2O) and phosphorous (P_2O_5) and the elemental form for N. Convert P and K to the form used in crop requirement calculations. Elemental P is multiplied by 2.29, and element K is multiplied by 1.21 to calculate the amount of the oxidized forms of P and K.

$$\text{Total N} = 16{,}700\ \text{kg}$$
$$\text{Total P as } P_2O_5 = 2.29(2{,}700\ \text{kg}) = 6{,}200\ \text{kg}$$
$$\text{Total K as } K_2O = 1.21(9{,}900\ \text{kg}) = 12{,}000\ \text{kg}$$

Step 3. Subtract nutrients lost during storage.

The nutrients are held in a waste storage pond with 50 % dilution. From Table 24.5, use the maximum nutrient depletion for the waste storage pond (65 %, 80 %, and 80 % retained for N, P, and K, respectively) since the dilution (50 %) is on the high side for the low dilution category (<50 %) in Table 24.5.

$$\text{Nutrients after storage losses} = (\text{fraction retained})(\text{production})$$

$$N = 0.65(16{,}700) = 10{,}900\ \text{kg}$$
$$P_2O_5 = 0.8(6{,}200) = 4{,}960\ \text{kg}$$
$$K_2O = 0.8(12{,}000) = 9{,}600\ \text{kg}$$

Step 4. Determine the plant available nutrients gained by mineralization in the soil.

Mineralization rates in soil are 55 %, 90 % and 93 % (Table 24.6) for nitrogen, phosphorous, and potassium, respectively for the third year in a waste storage pond. Phosphorous and potassium values are for both >50 % and <50 % waste storage ponds, and nitrogen is an intermediate value between 49 % (>50 % dilution) and 68 % (<50 % dilution).

$$\text{Nutrients available to plant} = (\text{fraction})(\text{application rate})$$

$$N = 0.55(10{,}900) = 6{,}000\ \text{kg}$$
$$P_2O_5 = 0.9(4{,}960) = 4{,}500\ \text{kg}$$
$$K_2O = 0.93(9{,}600) = 8{,}900\ \text{kg}$$

Step 5. What are the plant nutrient requirements? This data is available from farmers, extension personnel, crop advisors, seed companies, or fertilizer companies.

$$N_{-\text{plant}} = 132 \text{ kg/ha}$$
$$P_2O_{5-\text{plant}} = 52 \text{ kg/ha}$$
$$K_2O_{-\text{plant}} = 39 \text{ kg/ha}$$

Adjust the plant N requirement based on denitrification, leaching, and volatilization. We have already adjusted the waste N loading rate mineralization rate under step 4. Thus, mineralization is ignored under step 5.

Add denitrification losses to the plant nitrogen requirement.

From Table 24.7, a moderately well drained soil with an organic matter content of 3 % has an annual denitrification rate between 6 % and 20 %. Because the climate is moderate, estimate the midpoint of this range, 13 %. The values in Table 24.7 must be doubled for manure application so estimate the denitrification rate as 26 %. The plant nitrogen requirement is 132 kg/ha so the field nitrogen required is

$$N_{-\text{required-denitrification}} = 132 \text{ kg/ha}/0.74 = 178 \text{ kg/ha}$$

Add leaching losses to the plant nitrogen requirement

The plant nitrogen requirement must be increased to replace anticipated leaching losses. A leaching index of 6 (6 inches of annual percolation below the root zone), results in an annual nitrate loss of 10 %.

$$N_{-\text{required-denitrification and leaching}} = 178 \text{ kg/ha}/0.9 = 200 \text{ kg/ha}$$

Add application (volatilization) losses to plant nitrogen requirement.

Because the manure will be incorporated within 1 day, the expected nitrogen retention after volatilization is between 80 % (warm wet) and 100 % (cool wet), but since soil is closer to cool than warm, estimate percent retained as 95 %

$$N_{-\text{required-denitrification, leaching, and volatilization}} = 200 \text{ kg/ha}/0.95 = 210 \text{ kg/ha}$$

The answer to question 1 is that the required application rate is

$$N_{-\text{plant}} = 210 \text{ kg / ha}$$
$$P_2O_{5-\text{plant}} = 52 \text{ kg / ha}$$
$$K_2O_{-\text{plant}} = 39 \text{ kg / ha}$$

Step 6. Compute the field area to which manure can be applied.

First, try calculating the required land area using the nitrogen requirement.

$$\text{Area} = (6{,}000 \text{ kg/year N})/(210 \text{ kg/ha N}) = 29 \text{ ha} = 72 \text{ acres}$$

Calculate the application rate of P and K with 29 ha application area

$$\text{Application rate } P_2O_5 = 4{,}500 \text{ kg}/29\text{ha} = 155 \text{ kg/ha}$$
$$\text{Application rate } K_2O = 8{,}900 \text{ kg}/29\text{ha} = 308 \text{ kg/ha}$$

Thus, application rate of phosphorous is 3 times greater than plant requirement and application rate of potassium is 8 times greater than plant nutrient requirements, if application rate is based on nitrogen.

For maximum nutrient utilization and also to prevent phosphorous pollution, calculate the land application area based on the potassium requirement, and add other nutrients with fertilizer.

$$\text{Area} = (8{,}900 \text{ kg/year } K_2O)/(39 \text{ kg/ha } K_2O) = 228 \text{ ha} = 560 \text{ acres}$$

In some cases, manure may be overapplied with respect to potassium and phosphorous if the risk of pollution by these nutrients due to overapplication is low. That is not the case here. In this case, only 13 % of the N requirement is applied (29/228) in manure.

Step 7a. Calculate the application rate and volume for truck application. Truck applies both wastewater and manure. First calculate the pond volume.

The waste storage pond contains the manure produced by the 200 cows plus the milk parlor wastewater. Precipitation and evaporation must be considered to obtain the total volume of stored material.

Manure excreted per day = 81 $L/day/1{,}000\ kg$ (Table 2.6).

$$Volume\ manure/year = \frac{(81\ L/d)(200\ cows)(500\ kg/cow)(365)}{1{,}000\ kg} = 3{,}000{,}000\ L/year$$

$$Volume\ wastewater/year = 19\ L/day/cow{*}200\ cows{*}365 = 1{,}400{,}000\ L/year$$

Volume of water added by precipitation

The working depth of the pond is 3.6 m. This depth includes the expected precipitation – evaporation. An additional safety factor should be added, which includes the expected 25-year 224-hour storm + 0.3 m freeboard. The 25 year 24 hour storm is 0.15 m. Thus, the distance from the bottom of the pond to the top is 3.6 m + 0.3 + 0.15 = 4.05 m.

With 2:1 side slopes, the pond surface dimensions are calculated.

$$2(4.05 \text{ m}) + 60\text{m} = 68 \text{ m}$$
$$2(4.05 \text{ m}) + 18\text{m} = 26 \text{ m}$$

Thus, volume of water added by precipitation is

$$\begin{aligned}&(\text{Precipitation})(\text{Pond area at top of pond})\\&= 68\ \text{m}(26\ \text{m})(0.81\ \text{m})(1,000) = 1,430,000\ \text{L}\end{aligned}$$

The volume of water lost to evaporation is calculated based on the average working pond area. Maximum working depth is 3.6 m so average depth is 1.8 m.

$$\begin{aligned}2(1.8\ \text{m}) + 60\ \text{m} &= 64\ \text{m}\\2(1.8\ \text{m}) + 18\ \text{m} &= 22\ \text{m}\end{aligned}$$

Thus, volume of water lost to evaporation is

$$\begin{aligned}&(\text{Evaporation rate})\ (\text{Pond area at mid depth})\\&= 64\ \text{m} \times 22\ \text{m}(0.3\ \text{m})(1,000) = 420,000\ \text{L}\end{aligned}$$

Total water added to the pond per year (precipitation + wastewater volume – evap.) is

$$1,400,000 + 1,430,000 - 420,000 = 2,400,000\ \text{L}$$

The dilution ratio is (volume water)/(volume water + manure)

$$2,400,000/(2,400,000 + 3,000,000) = 44\%$$

The manure, as excreted, contains 12.5 % solids. The percent solids of the diluted wastewater is

$$\begin{aligned}&(\text{Original solids content})(\text{dilution ratio})\\&(12.5\%)(2,400,000\ L/(3,000,000\ L + 2,400,000\ L))\\&= 5.5\%\ \text{solids.}\end{aligned}$$

The density of dairy manure with 7 % solids is 0.96 kg/L (manure bulk density can be found by weighing a known volume of manure).

The mass of the stored material is $(0.96\ kg/L)(5,400,000\ L) = 5,200,000\ kg = 5,200$ metric tons. The application rate is $5,200$ metric tons$/228\ ha = 23$ metric tons$/ha$.

The volume of manure per ha is found by dividing by the bulk density.

$$23\ \text{metric tons/ha}/0.96 = 24\ m^3/\text{ha}.$$

Step 7b. Application of waste by sprinklers (calculated based on N and K)

For application of waste by sprinklers, the nitrogen requirement should be recalculated based on surface volatilization expected from sprinkler application (Table 24.7).

$$\begin{aligned}&N_{-\text{required}-\text{denitrification, leaching, and volatilization}}\\&= 200\ \text{kg/ha}/0.75 = 267\ \text{kg/ha}\end{aligned}$$

Leaching and denitrification may also increase under sprinkler irrigation because of possible addition of more water and wetter soil. However, those possible losses are ignored here. Use the N mass/year after mineralization is accounted for (6,000 kg/yr N).

$$\begin{aligned}&\text{Area for full nitrogen application}\\&= (6,000\ \text{kg/year}\ N)/(267\ \text{kg/ha}\ N) = 22\ \text{ha}\\&= 55\ \text{acres}\end{aligned}$$

Dilute the manure such that total solids is 4 % in order to apply dairy manure through the irrigation equipment,

$$DR = \frac{P_0 - P_d}{P_d} = \frac{7-4}{4} = 0.75$$

The volume of combined manure and wastewater in the waste storage pond after rainfall and evaporation is 5,400,000 L. The required addition of water to reduce the solids ratio to 4 % is the product of 5,400,000 and the DR.

$$\begin{aligned}&\text{Volume dilution water/volume original wastewater} = 0.75\\&\text{Volume of added water} = 0.75(5,400,000\ \text{L}) = 4,000,000\ \text{L}\end{aligned}$$

The total volume of water to be applied to the field is 5,400,000 + 4,000,000 L = 9,400,000 L

The depth of water to be applied (ponded depth) to the field is the volume/area. If the application rate is based on the potassium requirement, then the area of application should be 228 ha (2,280,000 m^2). The depth applied is volume/area.

$$\begin{aligned}\text{Depth applied} &= \text{Volume/Area} = 9,400/2,280,000\\&= 0.0042\ \text{m} = 0.4\ \text{cm}.\end{aligned}$$

Check the solution. The concentration of K_2O in the wastewater is 8,900 kg/9,400 m^3 ~ 1 kg/m^3. The volume applied to a hectare with a 0.4 cm application depth (0.004 m) is

$$\begin{aligned}&1\ \text{kg/m}^3 {*} 0.004\ \text{m} {*} 10,000\ \text{m}^2/\text{ha}\\&= 40\ \text{kg}\ \text{K}_2\text{O/ha(correct)}\end{aligned}$$

If the rate was calculated based on the nitrogen requirement, then the area of application would be

$$\begin{aligned}\text{Area} &= (6,000\ \text{kg/year N})/(267\ \text{kg/ha N}) = 22.5\ \text{ha}\\&= 225,000\ \text{m}^2\end{aligned}$$

The calculated depth of application for nitrogen would be

$$\begin{aligned}\text{Depth applied} &= \text{Volume/Area} = 9,400/225,000\\&= 0.042\ \text{m} = 4\ \text{cm}.\end{aligned}$$

Thus, 10 times greater depth is applied with nitrogen if the rate is based on the nitrogen requirement. In this case, based on Table 24.2, the maximum application rate by sprinklers would 1.8 mm/hr with a total application depth of 4 cm on silt loam soil. With 4 % solids in irrigation water, the infiltration rate is 0.85 times as great as it would be for clear water on silt loam soil (Table 24.3). This is between the 3 % and 5 % values. Thus, the maximum application rate is 1.8 * 0.85 = 1.5 mm/hr. The sprinkler system should be designed to apply water at no more than 1.5 mm/hr if the entire depth is applied at one time.

Calculation of Salinity Load from Animal Waste

Animal waste carries a high salt load. Wastewater salinity in storage ponds can range up to 12 dS/m or even higher with recirculating systems. Farmers in arid regions have found that they must limit the application of manure to fields in order to avoid soil salinization.

The salinity level in manure and wastewater depends on the original concentration of salinity in the manure, the amount of wash water, and the depths of precipitation and evaporation. There is a wide range of salinity levels in manure. Average and extreme salinity levels of manures tested in Colorado are shown in Table 24.9. Manure salinity can vary depending on the level of salinity in feed and other factors. Because of the wide range of salinity levels in manure and other wastes, the EC or concentration of ions should be tested when calculating the maximum allowable waste application rate with respect to salinity. Even on a single farm, researchers have found that 20 to 40 samples must be collected in order to adequately characterize the average salinity in manure. Field management practices such as leaching and crop drought tolerance (salinity tolerance) also determine the maximum waste application rate.

Total salinity percentage can be estimated based on the percentages of the major cations (NRCS AWMFH) as specified in Table 24.11. The justification for Eq. 24.16 is that the cations comprise half of the total salinity: the other half is anions.

$$\%\text{salts} = 2(\%\text{K} + \%\text{Ca} + \%\text{Na} + \%\text{Mg})\,(\text{by mass in dry manure}) \tag{24.16}$$

Recommended manure application rates depend on the crop and the climatic region (Table 24.10). Dryland application rates recommendations can be almost 10 times less than warm irrigated regions. This assumes that irrigation will be used to leach salts from the soil. Typical cation concentrations in different manures are listed in Table 24.11.

Example 24.5 Calculate the salinity in the waste storage pond of Example 24.4, and calculate the salinity added to the field (kg/ha) if 0.4 cm is applied by sprinkler irrigation. Use cation concentrations from Table 24.11.

The total percentage of salinity in dry dairy manure is calculated with Eq. 24.16 using the cation concentrations specified for dairy manure in Table 24.11.

$$\%\text{salts} = 2(\%\text{K} + \%\text{Ca} + \%\text{Na} + \%\text{Mg}) \quad (\textit{by mass in dry manure})$$
$$\%\text{salts} = 2(2.86 + 1.2 + 0.47 + 0.55) = 2(5.08) = 10.16\%$$
$$= 0.10\ \text{kg/kg}$$

Fresh dairy manure has a solids content of 12.5 %. Thus, multiply by 0.125 to find mass of salts in wet manure.

$$(0.1016\ \text{kg salts/kg solids})(0.125\ \text{kg solids/kg manure}) = 0.0127\ \text{kg salts / kg manure}$$

Lactating dairy cows produce 80 kg/d/1,000 kg manure

$$\textit{Mass manure/year} = \frac{(200\ \textit{cows})(500\ \textit{kg/cow})(80\ \textit{kg/d})(365\ \textit{d})}{1{,}000\ \textit{kg}}$$
$$= 2{,}920{,}000\ \textit{kg manure}$$

Mass of salts produced per year

$$(2{,}920{,}000\ \text{kg manure})(0.0127\ \text{kg salts/kg manure}) = 37{,}000\ \text{kg salts}$$

Volume of manure produced per year

$$\textit{Volume manure/year} = \frac{(200\ \textit{cows})(500\ \textit{kg/cow})(81\ \textit{L/d})(365\ \textit{d})}{1{,}000\ \textit{kg}}$$
$$= 2{,}960{,}000\ \textit{L manure}$$

Calculate the concentration of salts in fresh manure

$$(37{,}000\ \text{kg salts})\left(1{*}10^6\text{mg/kg}\right)/2{,}960{,}000\ \text{L manure} = 12{,}500\ \text{mg/L}$$

The volume of manure is diluted by 2,400,000 L as precipitation and washwater. Calculate the diluted concentration.

$$(12{,}500\ \text{mg/L})(2{,}400{,}000\ \text{L}/(3{,}000{,}000\ \text{L} + 2{,}400{,}000\ \text{L})) = 5{,}600\ \text{mgL} = 0.0056\ \text{kg/L}$$

The depth of water application is 0.4 cm so (0.004 m) (10,000 m^2/ha) = 40 m^3 is applied to the field. Thus, the rate of salt application is (40 m^3) (1,000 L/m^3) (0.0056 kg/L) = 224 kg/ha.

Table 24.9 Salinity levels for fresh manure (Credit Colorado State Extension Service)

Manure source	Average EC dS/m	Minimum	Maximum
Beef	28.2	8.4	42.5
Horse	6.2	3.3	10.2
Sheep	23.4	9.4	42.8
Chicken	23.7	16	40.7
Dairy	18.8	9	29.5
Llama	5.5	4.1	6.9
Hog	34.8	34.8	34.8
Dairy compost	24.5	12.8	43.6
Turkey compost	35.8	2.4	42.2

Table 24.10 Recommended animal manure application rates (dry rate basis) on cultivated land. Source: Fuller and Warrick (1985)

Region	Optimum	Maximum loading rate (metric tons)
Warm irrigated (arid) regions Livestock manure Poultry manure	324–56 t/ha 11–22 t/ha	80 t/ha (to salt tolerant crops)
Cold irrigated (arid) regions Livestock manure Poultry manure	324–45 t/ha 9–20 t/ha	50 t/ha (to salt tolerant crops)
Dryland regions Livestock manure Poultry manure	6.7–11 t/ha Half of irrigation region	10 t/ha (to salt tolerant crops)
Humid regions Livestock manure Poultry manure Slurry (5 % cattle solids)	324–45 t/ha 11–22 t/ha 4 cm depth in 30 days	324–45 t/ha 11–22 t/ha 4 cm depth in 30 days

Table 24.11 Typical percentages of cations (elemental) in dry animal manures (Credit Chen et al. (2003))

	Cattle			
Cation	Dairy	Beef	Feedlot	
Calcium (Ca)	1.2	1.06	0.69	
Magnesium (Mg)	0.55	0.3	0.34	
Sodium (Na)	0.47	0.25	0.12	
Potassium (K)	2.86	1.44	0.92	
	Poultry			
	Chick starter	Pullet grower	17–40 weeks	Post-molt diet
Calcium (Ca)	4	4	9.6	6.9
Magnesium (Mg)	0.66	0.68	0.91	0.96
Sodium (Na)	0.77	0.64	0.72	0.6
Potassium (K)	2.7	2.5	3.9	3.8
	Swine			
	Nursery	Grower	Finisher	
Calcium (Ca)	1.6	4.2	2.9	
Magnesium (Mg)	0.61	0.86	0.62	
Sodium (Na)	0.28	0.28	0.15	
Potassium (K)	1.6	1.6	1.2	

Example 24.6 Calculate the salinity added to a field (kg/ha) if 80 metric tons/ha by dry weight of dairy manure is applied to the field.

EC is 18.8 dS/m (Table 24.11).

$$(18.8\ \mathrm{dS/m})\ (640\ \mathrm{mg/L/dS/m}) = 11{,}500\ \mathrm{mg/L}$$
$$= 1.15\%\ \text{by weight salts.}$$

Fresh dairy manure is 12.5 % solids. Thus, the fraction of salts in dry dairy manure is

$$(0.0115 \text{ kg salts/kg manure})/(0.125 \text{ kg solids/kg manure}) = 0.092 \text{ kg salts/kg solids}$$

An application rate of 80 metric tons/ha is the same as 80,000 kg/ha.

$$80,000 \text{ kg/ha dry weight}*0.092 \text{ kg salts/kg solids} = 7,400 \text{ kg/ha}.$$

Irrigation of Crops with Municipal Wastewater

Wastewater cannot have excessive nutrients for some crops. In this case, wastewater should be blended with fresh water in order to prevent excessive nutrient application.

Example 24.7 Calculate the required blend of groundwater and municipal treated wastewater in order to provide the nitrogen requirement of sorghum, which is 50 kg/ha. The treated wastewater nitrogen concentration is 24 mg/L elemental N with 75 % as ammonium and 25 % as nitrate. Organic N is less than 10 % of the nitrogen load and is ignored in this calculation. The groundwater has a nitrate concentration of 3 mg/L. Previous studies have shown that ammonium in this wastewater is effectively converted to nitrate within 1–2 days after application to soils. Because the water pH is too low for ammonia formation and volatilization, the ammonia is in the ammonium ion form. Thus, it will not volatilize to the atmosphere. The depth of water application is 0.9 m/year. The crop nutrient requirement is 100 kg/ha, and 30 kg/ha of the nitrogen requirement has been applied as biosolids. Calculate the required dilution rate of the wastewater with groundwater.

The volume of water application per ha is calculated as follows:

$$(0.9 \text{ m/year})(10,000 \text{ m}^2/\text{ha}) = 9,000 \text{ m}^3/\text{ha}$$

30 kg/ha is applied as biosolids; thus, 70 kg/ha must be applied in wastewater. The required nitrogen concentration in wastewater is calculated as follows:

$$70 \text{ kg/ha}/9,000 \text{ m}^3/\text{ha} = 0.0078 \ kg/m^3 = 7.8 \text{ g/m}^3 = 8 \text{ mg/L N}$$

The NO_3 –N concentration in the wastewater is 24 mg/L, three times greater than what is required. Calculate the fraction of wastewater as follows:

$$C_T = C_{WW}F_{WW} + C_{GW}F_{GW}$$
$$8 = C_{WW}F_{WW} + C_{GW}(1 - F_{WW}) = 24 \ F_{WW} + 3(1 - F_{WW})$$
$$8 = 23 \ F_{WW} + 3 - 3 \ F_{WW} \rightarrow F_{WW} = 5/20 = 1/4$$

where

$C_{applied}$ = application concentration, mg/L
C_{WW} = wastewater concentration, mg/L
F_{WW} = wastewater fraction,
C_{GW} = groundwater concentration, mg/L
F_{GW} = groundwater fraction.

The fraction of wastewater is ¼; thus, three parts ground water is required for every part wastewater.

Questions

1. Explain the meaning of Eq. 24.1 in a sentence.
2. Integrate Eq. 24.1 and derive Eq. 24.2.
3. Explain the similarity between Eqs. 24.2 and 24.3, and explain the justification for using Eq. 24.2 for calculating effluent concentration from a wetland (Eq. 24.3).
4. Show how Eq. 24.5 is derived from Eq. 24.4.
5. Explain the meaning of Eq. 24.6.
6. Repeat the area-based calculations of Example 24.1, but use 25 °C and 10 °C instead of 20 °C and 4 °C.
7. Treated wastewater has a BOD_5 of 100 mg/L, and flow rate is 200 L/min. What is the required area of a FWS wetland? Calculate based BOD load and hydraulic load. Design for an effluent BOD of 12 mg/L.
8. Why is there a minimum acceptable BOD_5 loading rate for wetlands?
9. Calculate the size of a FWS wetland required to treat an animal waste effluent stream. The desired TN concentration on the discharge side of the wetland is 30 mg/L, the waste flow rate is 50 m^3/day, and the TN concentration in the dairy waste effluent is 200 mg/L.
10. A storage pond is expected to receive 10,000 m^3 of manure, clean water, and wastewater during a 6 month storage period (winter) (50 m^3/day * 200 days). The pond is only pumped once every 2 years and is expected to have settling of 600 m^3 of solids during the 2 year interval. The normal depth of evaporation during the winter is 0.5 m, and the expected precipitation during the winter is 0.25 m. The depth of the 25-year 224-hour storm is 10 cm. The dimensions of the base of the pond are 20 × 20 m. The side slope of the pond is 2.5:1. (2.5 run × 1 rise). Calculate the depth of the pond.
11. A dairy farm has 500 lactating dairy cows in Central Arizona. Average mass/cow is 500 kg. Cows are kept in open lots, and 95 % of manure is dried (not put in ponds). The 5 % of manure from the milking parlor is washed into an anaerobic waste storage pond with 6 L water/day/cow wash water. The maximum pond operating depth is 4 m. Side slopes are 2.5:1 (2 horizontal by 1 vertical). The 25 year – 24 hour storm is 7 cm. Annual precipitation is 15 cm, and mean annual evaporation is 200 cm. Soils on the sites for waste application are well drained sandy loams and have a leaching index of 6 (6 inches (15 cm)

percolates below the root zone). The organic matter content is $< 2\ \%$. The soils are flood irrigated. Soil slopes are close to dead level. Crop is cotton. Required annual nitrogen addition is 150 kg/ha nitrogen (N). Assume that no extra phosphorous or potassium are required, but that there is no environmental hazard associated with overapplication of phosphorous or potassium in these soils and in this hydrologic setting with no adjacent surface water bodies. Manure is applied by truck in spring just before planting when soil is warm and dry. Assume that the storage period is 365 days and use average evaporation rate during the year. Manure has been applied for several years every spring and is incorporated into the soil by plowing within 1 day of application. Calculate application rates for truck application of dried manure and for sprinkler application of liquid manure from the waste storage pond. Assume that 150 m^3/yr remain on the bottom of the pond each year.

12. Calculate the dry manure salinity application rate (kg/ha) for question 11.
13. Calculate the salinity application rate if 20 t/ha poultry manure (dry weight basis) is applied to a field.
14. Calculate the required blend of groundwater/wastewater to provide 200 kg/ha. The municipal wastewater concentration after secondary treatment is 15 mg/L N with 75 % ammonia in an arid region. The crop requires 1.2 m depth of water. Groundwater has 6 mg/L nitrate.

References

Chen et al. (2003) Value added chemicals from animal manure. Department of Energy Pacific Northwest National Laboratory and Washington State University. www.pnl.gov/main/publications/external/technical_reports/PNNL-14495.pdf

Fulhage, C, Pfost D (2000) Calibration of lagoon irrigating equipment. Missouri Extension. Department of Agricultural Engineering. http://extension.missouri.edu/p/EQ327

Fuller WH, Warrick AW (1985) Soils in waste treatment and utilization, vol 1. CRC Press, Boca Raton

Knight RL et al (1995) Arizona guidance manual for constructed wetlands for water quality improvement. Arizona Department of Environmental Quality, Phoenix

Lusk P (1998) Methane recovery from animal manures. The current opportunities casebook. Prepared by Resource Development Associates for the National Renewable Energy Laboratory. http://www.nrel.gov/docs/fy99osti/25145.pdf

Meisinger JJ, Randall GW (1991) Estimating nitrogen budgets for soil-crop systems. In: Managing nitrogen for groundwater quality and farm profitability, vol 5, pp 85–124

Natural Resources Conservation Service, Agricultural Waste Management Field Handbook (AWMFH) (1996). Part 651. http://www.nrcs.usda.gov/wps/portal/nrcs/detail/national/water/quality/?cid=stelprdb1044732

U.S. Environmental Protection Agency (1988) Design manual. Constructed wetlands and aquatic plant systems for municipal wastewater treatment. Office of Research and Development. Center for Environmental Research Information. http://water.epa.gov/type/wetlands/upload/design.pdf

WINDS Salinity and Nitrogen Algorithms 25

This chapter introduces the *WINDS* (Water-use, Irrigation, Nitrogen, Drainage, and Salinity) model, which simulates water, salts, and nitrate in agricultural fields. The *WINDS* model uses daily time steps and allows for up to 13 soil layers and multiple field positions in simulations. The focus of this chapter is the nitrogen and salinity models in WINDS. The input salinity parameters include initial concentration in each soil layer, date of manure application, manure salinity, crop salinity equation threshold and slope, and irrigation water salinity. Unlike the salinity model, the nitrate model includes reaction terms: mineralization, denitrification, and plant uptake. Because nitrate is the primary form of nitrogen used by plants, the mass balance for nitrogen focuses on the sources and sinks for nitrate. The rate of microbial activity (mineralization and denitrification) is strongly dependent on temperature; thus, this chapter presents algorithms for calculating annual and diurnal temperature variation in soils. The input nitrate parameters include initial concentration in layers (mg/kg), organic matter distribution with depth, fertilization application dates and depths, and irrigation water nitrate concentration. Reaction terms include mineralization rate constants, Michaelis-Menton uptake coefficients, seasonal nitrate requirement, optimal soil nitrate concentration, fraction of plant nitrogen uptake as nitrate, nitrogen dissolution rate, and nitrogen stress factors.

The fact that nearly all of the salinity terms in the total salinity mass balance (Eq. 4.12) are ignored was justified in Chap. 4. The effect of salinity on crop growth was also described in Chap. 4. The *WINDS* salinity model simply calculates the movement of salts between layers and the daily effect of salinity on crop growth.

The *WINDS Salinity* dialog box is shown in Fig. 25.1. In this case, five soil layers plus an evaporation layer (Chap. 6) are included in the simulation.

The dissolution rate is the length of time that it takes for manure or other salinity source to break down and release the salinity into the soil solution.

All data in *WINDS* input dialog boxes are written to the Active Data Worksheet (Fig. 25.2). In this case, the *Salinity* dialog box data is written to rows 250–298.

DRAINMOD (Skaggs 1980) is the most popular drainage model in the world. Breve et al. developed a nitrate simulation component for DRAINMOD called DRAINMOD-N. In the following derivation of the nitrate mass balance, the DRAINMOD-N approach is followed (Breve et al. 1997). Because the plant primarily utilizes the nitrate form of nitrogen and because nitrate is a pollutant, the DRAINMOD-N program only performs a mass balance on the nitrate form of nitrogen.

There are several sources and sinks for nitrogen in soils. For nonlegumes (legumes fix their own nitrogen from the atmosphere), the two major sources of nitrogen are fertilizer application and mineralization of organic nitrogen to nitrate. Sinks include plant uptake, denitrification, and leaching. The following sections describe how to model these processes.

Fertilizer Application

Farmers typically calculate fertilizer application rates or requirements in terms of kg/ha (lb/ac) of nitrogen. The change in nitrate concentration in the soil due to fertilization is calculated with Eq. 25.1 (Breve et al. 1997).

$$\Gamma_{fer} = K_{fer}\frac{A_{fer}}{D_{fer}} \tag{25.1}$$

where

Γ_{fer} = addition of N by fertilizer, mg NO_3-N per liter of soil,
K_{fer} = dissolution rate, d^{-1},
A_{fer} = amount of fertilizer present in D_{fer}, kg/ha,
D_{fer} = depth of incorporation, m.

P. Waller, M. Yitayew, *Irrigation and Drainage Engineering*, DOI 10.1007/978-3-319-05699-9_25

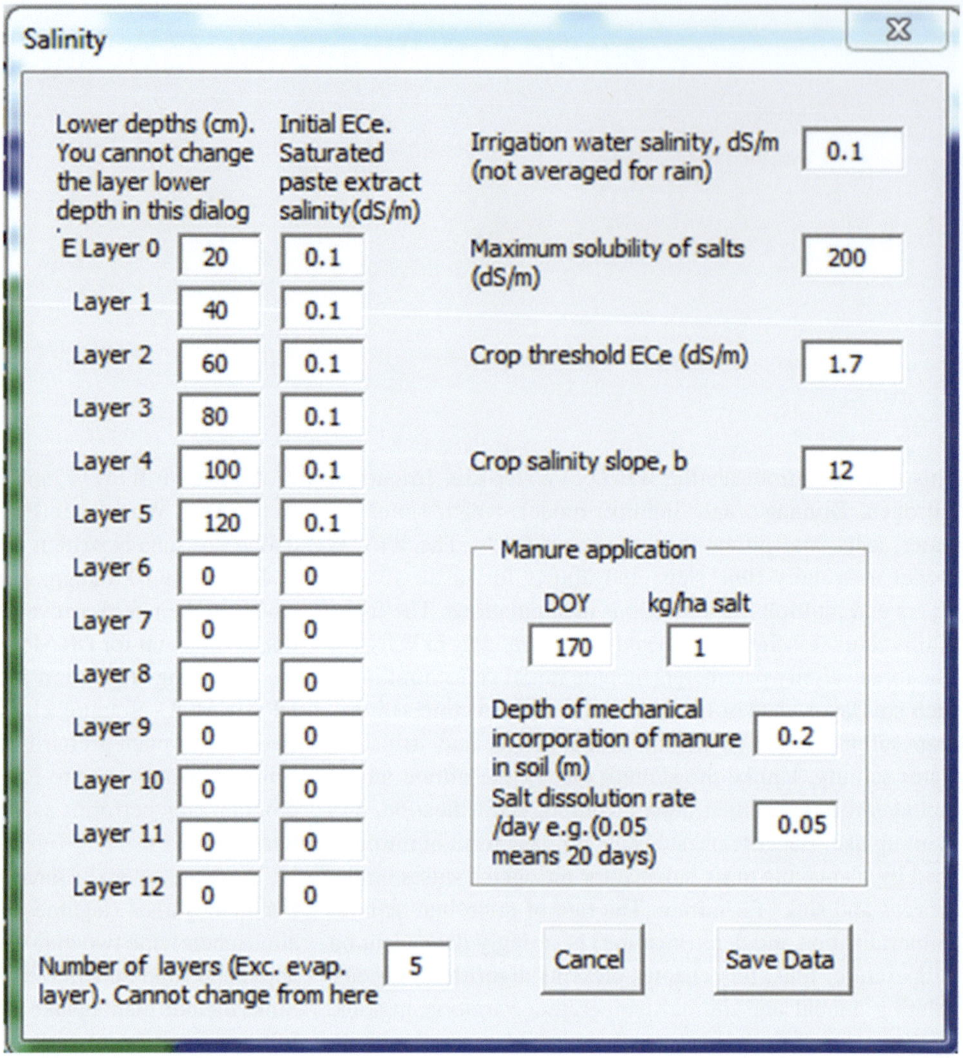

Fig. 25.1 Salinity dialog box in *WINDS* model

The fertilizer dissolution rate, K_{fer}, is the fraction of applied fertilizer that is dissolved into the soil-water solution per day. The dissolution rate is 1.0 if the fertilizer completely dissolves within a day. The dissolution rate is less than 1.0 if dissolution takes place over many days. For example, K_{fer} would be 0.05 per day if dissolution took place over 20 days

Almost all nitrate in soil is dissolved in soil water. Nitrate concentration in soil water is the concentration in the soil volume divided by the volumetric water content:

$$N_{water} = N_{soil}/\theta \tag{25.2}$$

where

N_{soil} = nitrate concentration in soil volume, mg/L_{soil},
N_{water} = nitrate concentration in soil water, mg/L.

Example 25.1 A fertilizer is applied at a rate 150 kg/ha of NO_3-N and is incorporated to a depth of 0.1 m. The dissolution rate is 1.0: all of the fertilizer is dissolved on the day that it is incorporated. Calculate the soil and water nitrate concentrations.

$$\begin{aligned}\Gamma_{fer} &= K_{fer}\frac{A_{fer}}{D_{fer}} \\ &= 1.0\left(\frac{150\ kg/ha}{0.1\ m}\right)\left(\frac{ha}{10,000\ m^2}\right)\left(\frac{1\ m^2}{1,000\ L}\right)\left(\frac{10^6\ mg}{kg}\right) \\ &= 150\ mg/L\ soil\end{aligned}$$

If water content is 20 %, then change in nitrate concentration in water in the upper 0.1 m is calculated as follows.

$$\Delta N_{water} = \Gamma_{fer}/\theta = 150/0.2 = 750\ \text{mg/L water}.$$

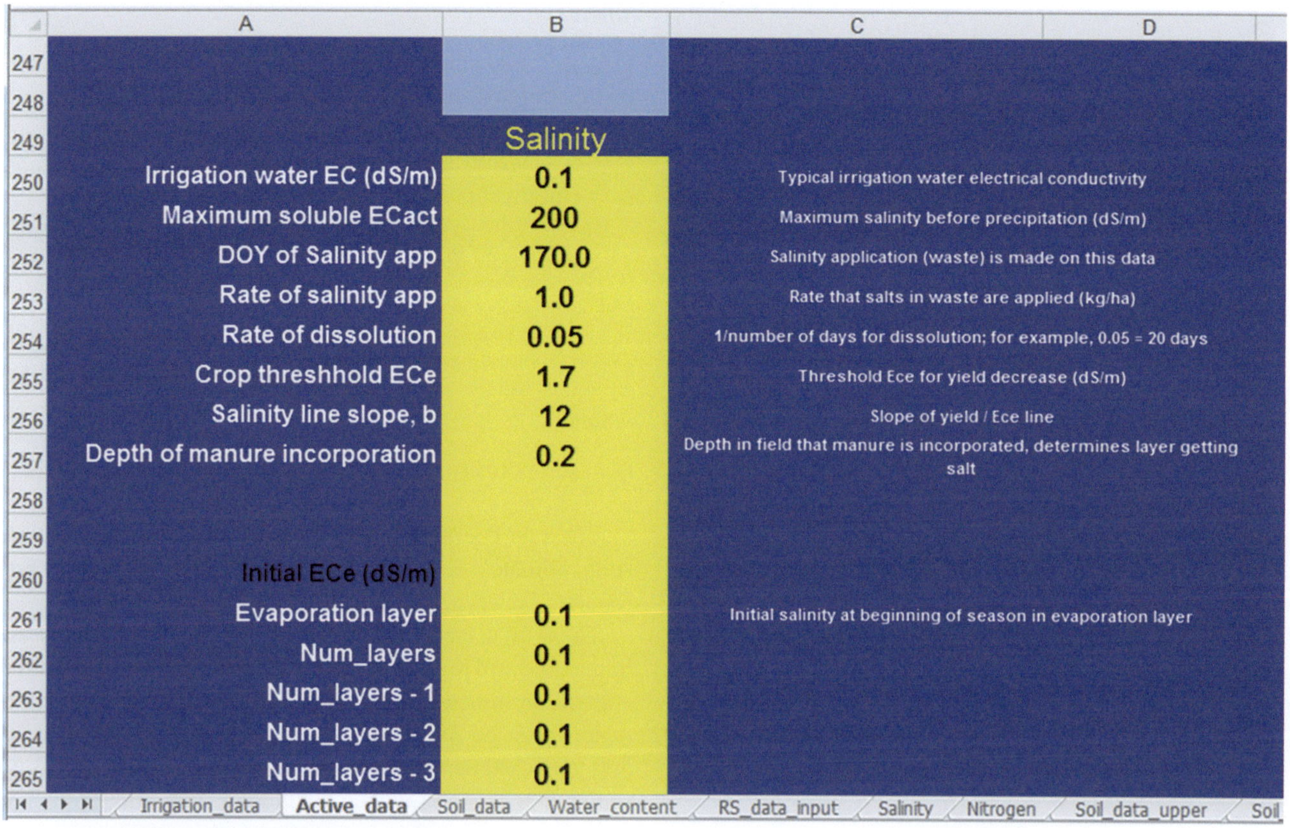

Fig. 25.2 Salinity information in *Active Data* worksheet in *WINDS* model

Based on Example 25.1, it is easy to see why leaching of nitrate can be an environmental hazard, especially after a heavy fertilizer application: the nitrate maximum contaminant level (MCL) in groundwater is 10 mg/L. If a large storm or wasteful irrigation takes place immediately after a nitrate application, then nitrate will be leached to the groundwater because nitrate, with a negative charge, is not adsorbed onto negatively charged clay particle surfaces. If nitrogen is applied as ammonium ion NH_4^+ or as organic nitrogen, the danger of leaching may not be as great because ammonium has a positive charge and is adsorbed onto negatively charged clay surfaces; thus, it tends not to leach during a storm or overirrigation. However, this is not the case if it mineralizes quickly to nitrate, as in hot and relatively dry climates.

Mineralization (Source of Nitrate)

In the *WINDS* model, the conversion of soil organic matter to nitrate (dead plant material → amino compounds → ammonification → nitrification) is lumped into one process – mineralization. Other potential sources of nitrate in agricultural soils are irrigation water, fertilizer, and manure. Each source of nitrogen may have an associated mineralization or nitrification rate. The mineralization rate is a function of temperature, water content, and soil organic matter concentration. Microbial activity in soils increases with temperature (up to a limit), and peaks at mid-range water contents: microbes involved in mineralization reactions work fastest when there is both water and oxygen.

Equation 25.3 (Breve et al. 1997) calculates the mineralization rate of organic matter in the soil as a function of organic matter concentration, temperature, and water content. Because mineralization rate is dependent on the organic matter content and not on the nitrate concentration in the soil, the equation is zero-order (no dependence on nitrate concentration in soil) model:

$$\Gamma_{mnl} = K_{mnl}\, f_{mnl\theta}\, f_{temp}\, \rho O_n \qquad (25.3)$$

where

Γ_{mnl} = net mineralization rate, μg/cm^3/day = mg/L soil/day,
K_{mnl} = rate coefficient for net mineralization, d^{-1},
ρ = soil bulk density, g/cm^3,
O_n = concentration of organic nitrogen present, μg/g soil,
$f_{mnl\theta}$ = soil water content adjustment factor, dimensionless,
f_{temp} = temperature adjustment factor, dimensionless.

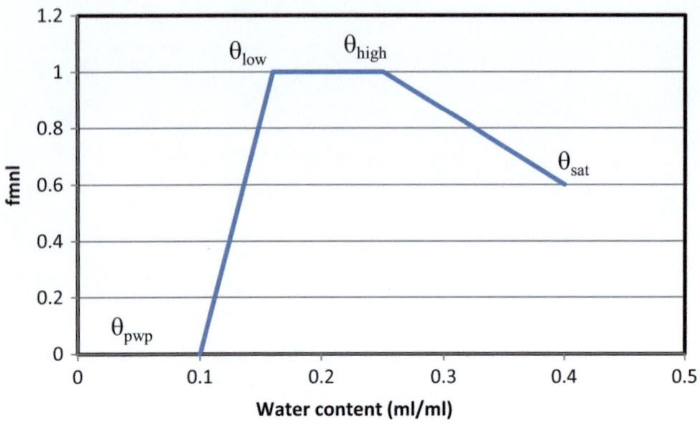

Fig. 25.3 Three part step function for soil water content adjustment factor for nitrate mineralization

The phases in Fig. 25.3 can be modeled (Breve et al. 1997) with a three-part step function:

$$f_{mnl\theta} = \left(\frac{\theta - \theta_{pwp}}{\theta_{low} - \theta_{pwp}}\right) \quad \cdots\cdots\cdots \quad \theta_{wp} \le \theta \le \theta_{low}$$
$$f_{mnl\theta} = 1 \quad \cdots\cdots\cdots\cdots\cdots\cdots\cdots \quad \theta_{low} \le \theta \le \theta_{high}$$
$$f_{mnl\theta} = 0.6 + 0.4\left(\frac{\theta_{sat} - \theta}{\theta_{sat} - \theta_{high}}\right) \quad \cdots\cdots \quad \theta_{high} \le \theta_{sat} \tag{25.4}$$

For example, if θ_{pwp}, θ_{low}, θ_{high}, and θ_{sat} are 0.1, 0.16, 0.25, and 0.4, respectively, then the shape of the $f_{mnl\theta}$ curve is as shown in Fig. 25.3.

The temperature adjustment factor used in *WINDS* has the same form as the wetlands temperature adjustment factor described in Chap. 24. The similarity is expected because the same microbes are involved in respiration in soils and wetlands. Q_{10} is typically 3 (Breve et al. 1997).

$$f_{temp} = Q_{10}^{\left(\frac{T - t_b}{10}\right)} \tag{25.5}$$

where

Q_{10} = the rate of change associated with a 10 °C change in soil temperature
T = actual soil temperature, °C,
t_b = base temperature for which microbial activity is defined, 20 °C.

Concentration of organic nitrogen as a function of soil depth can be modeled with an exponential equation (Breve et al. 1997):

$$O_n = O_{n_{max}} e^{-\alpha\ z} \tag{25.6}$$

where

$O_{n\text{-}max}$ = maximum organic nitrogen content in the top layer of soil, µg/g,
O_n = organic matter concentration at depth z, µg/g,
z = distance of layer (cell) from surface, cm,
α = empirical constant (0.02–0.05), dimensionless.

Soil particle density, ρ_p, is 2.65 g/cm^3. Soil bulk density, ρ, is

$$\rho = \rho_p(1 - \phi) \tag{25.7}$$

where

ϕ = soil porosity

The rate coefficient, K_{mnl}, varies dramatically between soils, climates, and source of organic material. Values are reported between 10^{-4} and 10^{-5} (Breve et al. 1997).

There is a large difference in organic matter between humid and arid regions. Desert soils have low organic matter content so mineralization rates are low. The total organic matter near the soil surface in desert soils is approximately 500 µg/g (500 ppm) soil. In contrast, organic matter concentration in topsoil in Lumbee sandy loam in North Carolina is approximately 3,300 µg/g soil (Breve et al. 1997).

Of the total organic matter fraction, approximately 60 % of organic matter (by mass) is carbon (300 µg C/g soil). Nitrogen concentration in organic matter is typically about 1/10 the concentration of carbon (30 µg N/g soil). Thus, the mass of nitrate added by mineralization each year at the soil surface in an arid region is almost trivial (30 µg N/g soil) (0.05) = 1.5 µg N/g soil/year.

Temperature Model

Soil temperature is one of the most important parameters in the mineralization equation (Eq. 25.3). Temperature change can be modeled with a finite difference energy balance model, or it can be simulated with sin curves, based on observed data from weather stations. The *WINDS* model uses the sin curve approach. Agricultural weather stations typically have temperature probes at 2 soil depths (5 and 10 cm). As with air temperature, diurnal (daily) soil temperature follows as a sin wave. The range or amplitude of diurnal (daily) variation of soil temperature decreases exponentially with depth and becomes insignificant at a depth of 1 m. Figures 25.4, 25.5, and 25.6 show typical diurnal temperature waves. Although

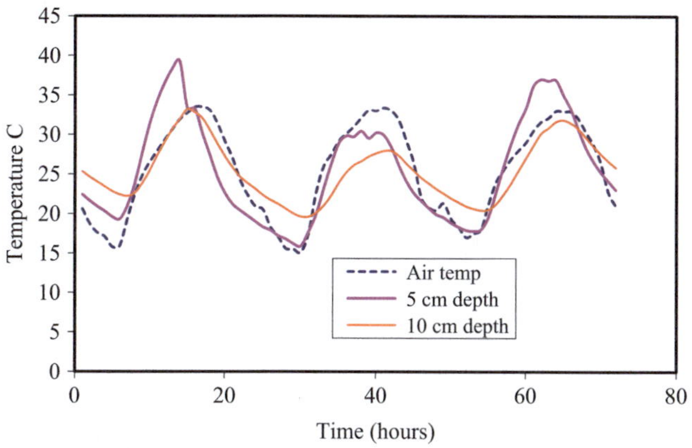

Fig. 25.4 Temperature changes in air and at 5 and 10 cm depth (Data from AZMET, Maricopa, Arizona weather station, days 160–162 (summer), 1998)

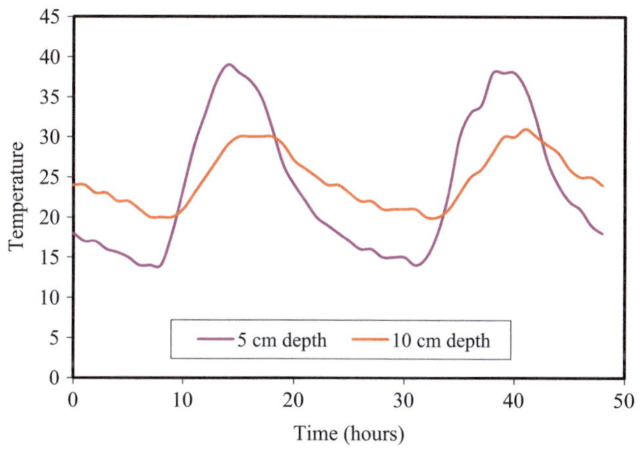

Fig. 25.5 Soil temperature changes for a dry soil

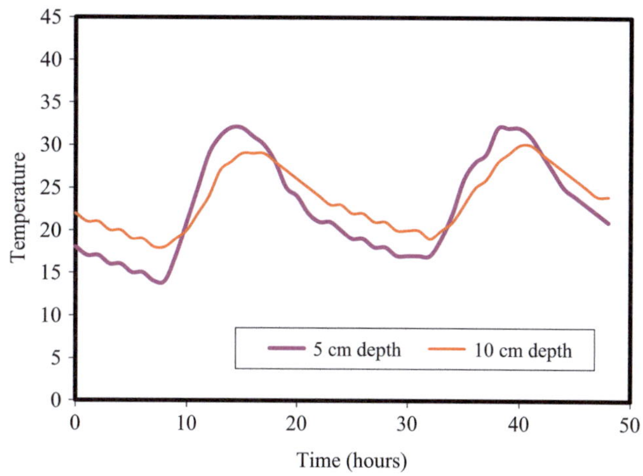

Fig. 25.6 Soil temperature changes for a wet soil

the temperature change in soil is not a perfect sin wave, it is a reasonable approximation for the purpose of modeling soil microbial processes. Although not all microbes respond identically, microbial activity in the soil generally increases up to 40 °C but then decreases or shuts down at higher temperatures. Soil temperature can become much higher in the upper soil layers in arid climates. At this point, the *WINDS* model only considers average daily temperature in the profile and does not calculate diurnal fluctuation rates or high and low diurnal temperatures. The primary reason for this is that *WINDS* is a daily time step model. The microbial rates could be calculated at high and low temperatures in each layer and averaged, but this capability is not currently in the *WINDS* model.

One of the factors that influence temperature fluctuation is water content. The difference between temperature change in a dry soil and a wet soil is observed in Figs. 25.5 and 25.6, which were collected on the same day and in the same soil type, but the soils had different moisture contents. The dry soil (Fig. 25.5) temperature range is 25 °C at 5 cm depth, but it is 17 °C in a wet soil (Fig. 25.6) at the same depth. However, the *WINDS* model currently uses average daily temperature to compute microbial activity so this difference has no effect on the simulation.

Soil temperature is a function of depth. The propagation of the thermal wave in soils is a function of thermal diffusivity, D (m^2/s). Thermal diffusivity is the thermal conductivity divided by the heat capacity of soil. The thermal diffusivity of the soil is calculated based on the measured amplitude of temperature change, A_1 and A_2, at two depths in the soil, z_1 and z_2.

$$D = \frac{\pi}{P}(z_2 - z_1)^2 \left[\ln\frac{A_1}{A_2}\right]^{-2} \quad (25.8)$$

where

P = period of one complete sin wave fluctuation (24 hours)
z = depth in soil, m,
A = temperature fluctuation (T_{max} − T_{min})/2 at depth z
P = period, yr or hr.

In order to develop a sinusoidal model of temperature, it can be assumed that the average daily soil temperature with depth and time is constant for a series of days or a series of years.

$$\overline{T}_{(z,t)} = Const \quad (25.9)$$

where

T = temperature
t = time, hr.

Surface temperature, T(0,t), is a sinusoidal function of time and mean temperature.

$$T(_{0,t}) = \overline{T} + A_0 \sin\left[\frac{2\pi}{P}(t - t_{-m})\right] \quad (25.10)$$

Temperature at any depth and time, T(z,t) is

$$T(z,t) = \overline{T_z} + A_0\ e^{\left(\frac{-z}{d}\right)} \sin\left(\frac{2\pi}{P}(t - t_m) - \frac{z}{d}\right) \quad (25.11)$$

where

$\overline{T}$ = average soil temperature at depth z during period, P, °C,
t_m = time when T = $\overline{T}$ at the soil surface on the rising sin curve, hr,
t = time, hr,
$A_o = (T_{max} - T_{min})/2$ for the period, P, at the soil surface, °C,
z = depth, m,
d = damping depth of the thermal wave (no change), m,

The damping depth, d, is

$$d = (PD/\pi)^{1/2} \quad (25.12)$$

The temperature change at a given depth usually lags behind the temperature change in the soil above it. The lag time between two layers (1 and 2) is

$$t_{lag} = \frac{z_2 - z_1}{2}\left(\frac{\pi}{PD}\right)^{0.5} \quad (25.13)$$

Amplitude at any depth is a function of surface amplitude and damping depth. Typically, surface temperature is not measured so it is calculated based on temperature at 5 cm depth.

$$A_z = A_0 e^{-z/d} \quad (25.14)$$

Example 25.2 Use information from Fig. 25.4 to calculate the thermal diffusivity and damping depth. Calculate the diurnal temperature curves at 5 cm increments down to 40 cm depth.

Amplitude at 5 and 10 cm: use approximate values since there is variability between days

$$A_5 = (T_{max} - T_{min})/2 = (30 - 16)/2 = 7^0C$$
$$A_{10} = (T_{max} - T_{min})/2 = (28 - 20)/2 = 4^0C$$

Calculate average soil thermal diffusivity, D, with data from the 5 cm and 10 cm depths.

$$D = \frac{\pi}{P}(z_{10} - z_5)^2\left[\ln\frac{A_5}{A_{10}}\right]^{-2} = \frac{3.14}{24}(0.1 - 0.05)^2\left(\ln\frac{7}{4}\right)^{-2}$$
$$= 0.001\ m^2/hr$$

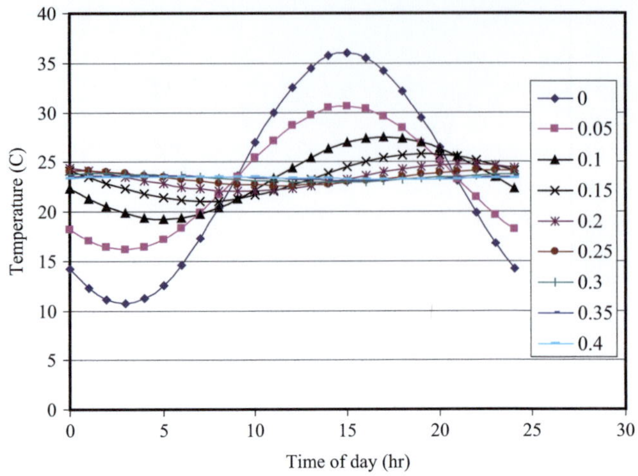

Fig. 25.7 Modeled temperatures every 5–40 cm depth

Calculate damping depth d:

$$d = (PD/\pi)^{0.5} = ((24)(0.001)/3.14)^{0.5} = 0.09\ \text{m}$$

Assume that average surface temperature is the same as the average temperature measured at the 5 cm depth. $\overline{T}_5 = \overline{T}_0 = 23°C$. From Fig. 25.4, the temperature is typically about 23 °C at 9 am; thus, t_{m-5} = 9 hrs. Next find t_m at the soil surface by determining the lag time between the surface and the 5 cm depth.

$$t_{lag} = \frac{z_2 - z_1}{2}\left(\frac{\pi}{PD}\right)^{0.5} = \frac{0.05 - 0}{2}\left(\frac{\pi}{24{*}0.001}\right)^{0.5} = 0.27\ hr$$
$$t_{m-surface} = t_{m-5} - t_{lag} = 9 - 0.2 = 8.8\ \ hr$$
$$A_0 = A_5/(\exp(-z/d)) = 7/(\exp(-0.05/0.09)) = 13^0C$$

Terms are plugged into Eq. 25.11, and modeled temperature data every 0.05 m to 0.4 m depth is shown in Fig. 25.7.

$$T(z,t) = 23 + 13\ e^{\left(\frac{-z}{0.09}\right)} \sin\left(\frac{2\pi}{24}(t - 8.8) - \frac{z}{0.09}\right)$$

For many regions, an annual temperature curve (Fig. 25.8) is reasonably approximated by a sin curve and can be modeled with Eq. 25.11 with z set equal to zero. On an annual basis, there is little difference between average daily temperatures at different depths. Example 25.3 shows how to approximate the annual sin curve. Because *WINDS* uses a time step of 1 day, the simple equation used in Example 25.7 is the current method of daily soil temperature calculation in *WINDS* for all layers; however, it would be more accurate to model microbial processes with the hourly temperature change or to calculate maximum and minimum temperatures at each depth and use an averaging technique. Annual temperature parameters are input in the Active data worksheet in cells B400:B402.

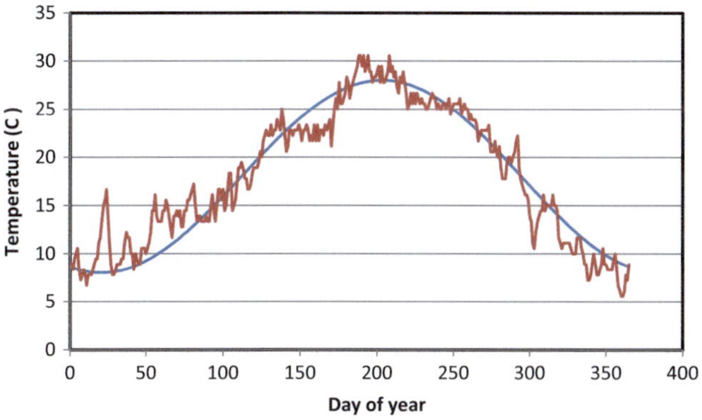

Fig. 25.8 Annual temperature wave at 10 cm depth: Tucson, AZ

Example 25.3 Fit a sin curve to the temperature data shown in Fig. 25.9.

Average temperature is the average temperature for the year and is the same at all depths, $\overline{T}_5 = \overline{T}_0 = 18$ °C, and there is no lag time between depths for daily average temperature. Thus, there is no need to consider damping depth. The rising sin curve crosses 18 °C at approximately 130 days. The temperature fluctuation is ±10 °C.

$$T(t) = 18 + 10\sin\left(\frac{2\pi}{365}(t - 130)\right)$$

Mineralization Model

This section shows examples of mineralization calculations.

Example 25.4 Find the mineralization rate for a sandy loam soil in a humid climate. θ_{pwp} θ_{low}, and θ_{sat} are equal to 0.044, 0.14, and 0.305, respectively. Temperature is 28 °C. The water content is 0.0778. Average layer depth is 20 cm. Surface organic matter content is 3,000 μg/g, K_{mnl} = 0.00005 d^{-1}, $\alpha = 0.025$.

Find the water content adjustment factor. Water content is between θ_{pwp} and θ_{low}

$$f_{mnl\theta} = \left(\frac{\theta - \theta_{pwp}}{\theta_{low} - \theta_{pwp}}\right) = \left(\frac{0.0778 - 0.044}{0.14 - 0.044}\right) = 0.35$$

Find the temperature adjustment factor. The temperature adjustment factor at 28 °C is

$$f_{temp} = Q_{10}^{\left(\frac{T-t_b}{10}\right)} = 3^{\left(\frac{28-20}{10}\right)} = 2.41$$

Find the average organic matter concentration.

$$O_n = O_{n_{max}} e^{-0.025z} = 3{,}000\ e^{-0.025(20)} = 1{,}819\ \mu g/g$$

Find the mineralization rate and the change in concentration in cell water

The saturated water content is 0.305. Thus, the porosity is approximately 0.305. Soil bulk density is.

$$\rho = \rho_p(1 - \phi) = 2.65(1 - 0.305) = 1.84.$$

$$\begin{aligned} \Gamma_{mnl} &= K_{mnl} f_{mnl\theta} f_{temp} \rho O_n \\ &= 0.00005(.35)(2.41)(1.84)(1{,}819) \\ &= 0.14\ \text{mg/L soil /d} \end{aligned}$$

$$dN_1 = \Gamma_{mnl}/\theta = 0.14/0.0778 = 1.81\ \text{mg/L/day}$$

Example 25.5 Find the mineralization rate constant for a desert soil. Assume that the average yearly temperature is 20 °C and average soil water content is 0.16 ($f_{mnl\theta} = 1$ and $f_{temp} = 1$). Soil porosity is 0.45.

$$\Gamma_{mnl} = 1.5\ \mu g/g/year/365 = 0.0041\ \mu g/g/day$$
$$\rho = \rho_p(1 - \phi) = 2.65(1 - 0.45) = 1.46\ \text{g/cm}^3$$

Rearrange Eq. 25.8 and solve for the mineralization rate constant

$$K_{mnl} = \frac{\Gamma_{mnl}}{f_{mnl\theta} f_{temp} \rho O_n} = \frac{0.0041}{1{*}1{*}1.46{*}500} = 0.000018$$

Example 25.6 Calculate the mineralization of organic nitrogen for a 1-day period in a desert soil at 7.5 cm and 20 cm depth for the following parameters. Water content at both depths is 0.12 cm^3/cm^3

θ_{pwp}	0.056 cm^3/cm^3
θ_{low}	0.11 cm^3/cm^3
θ_{high}	0.14 cm^3/cm^3
θ_s	0.30 cm^3/cm^3
K_{mnl}	0.00002 d^{-1}
temp	14.3 °C
O_{nmax}	500 μg/g

7.5 cm depth

Calculate $f_{mnl\theta}$. Water content is less than θ_{low}.

$$f_{mnl\theta} = \left(\frac{\theta - \theta_{wp}}{\theta_{low} - \theta_{wp}}\right) = \left(\frac{0.12 - 0.0516}{0.14 - 0.0516}\right) = 0.77$$

Calculate f_{temp}:

$$f_{temp} = Q_{10}^{\left(\frac{T-t_b}{10}\right)} = 3^{\left(\frac{15-20}{10}\right)} = 0.577$$

Calculate organic matter content, O_n.

$$O_n = 500e^{-0.025(7.5)} = 414\ \mu g/g$$

Calculate mineralization rate

$$\Gamma_{mnl} = K_{mnl} f_{mnl\theta} f_{temp} \rho O_n$$
$$= 0.00002(0.77)(0.577)(1.35)(414)$$
$$= 0.0049 \ mg/L \ soil/day$$

Calculate the change in concentration of nitrate in soil water.

$$(0.0048 \ \text{mg}/L \ \text{soil/day})(L \ \text{soil}/0.12 \ \text{L water}) = 0.04 \ \text{mg/L/d water.}$$

Repeat for 20 cm.

The $f_{mnl\theta}$ and the f_{temp} adjustment factors are the same.
Calculate organic matter content, O_n.

$$O_n = 500e^{-0.025(20)} = 303 \ \mu g/g$$

Calculate mineralization rate

$$\Gamma_{mnl} = K_{mnl} f_{mnl\theta} f_{temp} \rho O_n$$
$$= 0.00002(0.77)(0.577)(1.35)(303)$$
$$= 0.0036 \ mg/L \ soil/d$$

Calculate the change in concentration of nitrate in soil water.

$$(0.0036 \ mg/L \ soil/day)(L \ soil/0.12 \ L \ water) = 0.03 \ mg/L/d \ water.$$

Thus, 0.03 mg/L of NO_3-N (low) is added to soil water on a daily basis at 20 cm depth.

Denitrification (Nitrate Sink)

Denitrification is the conversion of nitrate to nitrogen gas or nitrous oxide gas. Denitrifiers require oxygen or another electron acceptor as they break down organic molecules in order to obtain energy from the carbon bonds in the molecules. Microorganisms can use oxygen or nitrate as an electron acceptor (oxidizer) during respiration. Based on chemical thermodynamics, nitrate is the first alternative electron acceptor after oxygen (before iron, sulfur, or carbon). If oxygen is in the soil (aerobic conditions), then denitrification is not likely to occur, but if the soil is slightly anaerobic, then nitrate will be used as an electron acceptor, and denitrification will take place.

Rates for denitrification range from 0.004 to 1.08 d^{-1}. A carbon source is also needed for denitrification. Because denitrification occurs in the liquid phase in the soil, the source of carbon must be soluble organic carbon. Thus, the nitrification reaction is more directly dependant on soluble organic carbon than on total organic matter in the soil. The concentration of soluble organic carbon is included in the denitrification rate constant, K_{den}. Because of low soluble organic carbon in desert soils, denitrification rates in desert soils are extremely low and often negligible. However, in sludge-amended desert soils, denitrification rate may be higher until the carbon is lost by microbial processes to the atmosphere.

Breve et al. (1997) calculated the net denitrification rate with the following equation:

$$\Gamma_{den} = K_{den} f_{den\theta} f_{temp} f_z \theta \ N \tag{25.15}$$

where

Γ_{den} = denitrification rate, mg/L_{soil}/d,
K_{den} = denitrification rate constant, d^{-1},
θ = water content for last time step, ml/ml,
N = nitrate concentration in soil water, mg/L,
$f_{den\theta}$ = soil water content adjustment factor for denitrification, dimensionless,
f_{temp} = same temperature adjustment factor used in mineralization calculation,
f_z = depth adjustment factor to account for fraction of carbon in soil.

Denitrification will not occur when water content in the soil is less than the lower threshold water content, θ_{den}. The lower threshold (Breve et al. 1997) for denitrication is 60 % of water filled pore space ($\theta_{den} = 0.6\ \theta_{sat}$). Above the lower threshold, the soil water content adjustment factor increases parabolically up to 1.0 at saturated water content.

$$f_{den\theta} = 0 \ \text{for} \ \theta < \theta_{den}$$
$$f_{den\theta} = \left(\frac{\theta - \theta_{den}}{\theta_{sat} - \theta_{den}}\right)^2 \quad \text{for} \ \theta > \theta_{den} \tag{25.16}$$

The depth adjustment factor is the same as the organic matter adjustment factor for mineralization.

$$f_z = e^{-0.025z} \tag{25.17}$$

where

z = depth in soil, cm.

Example 25.7 Calculate the denitrification rate in the sandy loam soil described in Example 25.4 at 20 cm depth. Nitrate concentration in soil water is 10 mg/L, and the denitrification rate constant is 0.002 d^{-1}. Change the water content to 0.27. Calculate the change in concentration within the soil volume and change in concentration within the soil water.

Calculate the water content adjustment factor.

$$f_{den\theta} = \left(\frac{\theta - \theta_{den}}{\theta_{sat} - \theta_{den}}\right)^2 = \left(\frac{0.27 - 0.183}{0.305 - 0.183}\right)^2 = 0.51$$

Calculate the temperature adjustment factor.

$$f_{temp} = Q_{10}^{\left(\frac{T-t_b}{10}\right)} = 3^{\left(\frac{28-20}{10}\right)} = 2.41$$

Calculate depth adjustment factor, f_z.

$$f_z = e^{-0.025*20} = 0.61$$

Calculate net loss of nitrate due to denitrification.

$$\begin{aligned}\Gamma_{den} &= K_{den} f_{den\theta} f_{temp} f_z \theta N \\ &= 0.002(0.51)(2.4)(0.61)(0.27)(10) \\ &= 0.004 \ mg/L_{soil}/day\end{aligned}$$

Calculate change in concentration within soil water.

$$\Delta N = mg/L_{soil}/day/\theta = 0.004/0.27 = 0.014 \ mg/L$$

Plant Uptake (Sink)

The seasonal mass of nitrogen that is physically removed by plants from soils can be calculated by multiplying the plant mass by the percent of nitrogen in the plant. The change in nitrogen concentration with respect to soil volume during a growing season is (Breve et al. 1997)

$$\Gamma_{upt} = \frac{Y \ fN}{10 \ Rz} \quad (25.18)$$

where

Γ_{upt} = seasonal plant uptake, mg/L_{soil},
Y = yield (dry matter), kg/ha,
fN = fraction of nitrogen present in the crop dry matter, dimensionless,
Rz = root length, m,

The number 10 in Eq. 25.18 can be found with dimensional analysis.

$$\frac{Y \ fN}{Rz} \Rightarrow \left(\frac{kg}{ha - season}\right)\left(\frac{1}{m}\right)\left(\frac{10^6 \ mg}{kg}\right)\left(\frac{ha}{10^4 \ m^2}\right)\left(\frac{m^3}{10^3 \ L}\right) = \frac{1}{10} mg/L_{soil}/season$$

Nitrogen consumption varies with plant growth rate and vegetative growth. Certain phases of the growth cycle have rapid plant mass accumulation or vegetative growth. In order to account for this variable growth rate, the season is divided into phases with fractional nutrient uptake defined for each fraction of the growing season. If no other information is available, it is assumed that the nitrogen uptake follows the crop coefficient curve for evapotranspiration.

Example 25.8 Calculate the nitrogen removed from the root zone during the growing season. Plant dry matter is 10,000 kg/ha. Fraction of nitrogen in dry matter is 1.55 %. Assume that the root zone is 0.8 m deep. Calculate daily uptake 60 days after planting if maturity is reached 125 days after planting, and significant nitrogen uptake begins 25 days after planting.

$$\begin{aligned}\Gamma_{upt} &= \frac{Y \ fN}{10 \ dz} = \frac{10,000 \ (0.0155)}{10 \ (0.8)} \\ &= 19.4 \ mg/L_{soil}/season\end{aligned}$$

Daily uptake

Period of nitrogen uptake is 125 days − 25 days = 100 days
$(19.4 \ mg/L/season)/(100 \ days \ uptake/season) = 0.194 \ mg/L_{soil}/day$

Some of the plant nitrogen is taken up as ammonium. Because the *WINDS* model only considers the nitrate form of nitrogen, only the nitrate uptake is considered in the mass balance. Thus, the plant nitrogen uptake in the model must be partitioned between ammonium and nitrate uptake. This is specified as 0.75 in one of the lower right textboxes in Fig. 25.9. In this case, the nitrate uptake is ¾ of the total plant nitrogen requirement.

The ratio of nitrate and ammonium in the soil solution influences the percentage of nitrogen taken uptake as nitrate. Plants have range of preference for nitrate over ammonium; however, the overall trend is that the amount of nitrate and ammonium absorbed is similar to the ratio of the dissolved concentrations in the soil water. Temperature and solution pH can influence the relative amounts of ammonium and nitrate absorption. Ammonium uptake increases with pH. Nitrate is absorbed more readily above 25°. Warncke and Barber (1973) conducted an experiment in which they exposed corn seedlings to different ratios of nitrate and ammonium in the soil water and then measured the ratio of uptake of ammonium and nitrate in different treatments (Table 25.1).

Insufficient nitrate/nitrogen results in crop stress. A general rule of thumb for cotton and grain crops is that 30 mg $N0_3$-N/kg soil (soil dry weight) is required in order to have adequate nitrate uptake with no yield reduction. If soil nitrate concentration is in excess of the optimal range, then some crops may undergo luxury consumption of nitrogen with corresponding rank (excessive) growth. Excess nitrogen often results in insufficient reproductive (grains, fruits, etc.) growth and excessive vegetative (shoots and leaves) growth. Rank growth might also weaken the plant and cause it to fall over in high winds. The *WINDS* model stress factors for excess or insufficient nitrogen are specified in the *Nitrogen* dialog box.

The denitrification equation calculates the removal from nitrate (mg/kg soil) and thus is a function of soil bulk density (Eq. 25.19). It can be assumed that all nitrate in soils is dissolved in soil water. The unit conversion from nitrate in soil-water to mg/kg should be calculated as follows:

$$N_{soil-mg/kg} = N\left(\frac{\theta}{\rho_b}\right) \quad (25.19)$$

Table 25.1 Influence of NH_4^+/NO_3^- ratio and nitrogen concentration of the nutrient solution on the ratio of NH_4^+/NO_3^- absorbed during corn seedling growth from 13 to 18 days (After Warncke and Barber (1973))

Mean N concentration in soil solution (μmol/L)	Solution NH_4^+/NO_3^- ratio				
	8.40	2.46	1.05	0.49	0.17
	NH_4^+/NO_3^- absorbed				
16	6.82	2.88	1.11	0.56	0.17
67	9.40	2.78	0.97	0.50	0.15
303	4.91	2.38	1.45	0.70	0.28
1,507	3.71	2.05	1.48	0.68	0.33
6,105	3.50	2.09	1.04	0.69	0.31

where

Nsoil-mg/kg = nitrate concentration in soil, mg/kg soil.

Example 25.9 A soil contains 30 mg NO_3-N /kg soil. Soil porosity is 45 %, and soil water content is 20 %. Calculate the concentration of NO_3-N in soil water.

Bulk density and soil-water nitrate are calculated as follows:

$$\rho_b = (1 - \varphi)\rho_p = (1 - 0.45)2.65 = 1.46\ kg/L$$
$$N = N_{soil-mg/kg}\left(\frac{\rho_b}{\theta}\right) = 30\left(\frac{1.45}{0.2}\right) = 217\ mg/L$$

This calculated nitrate concentration is approximately the same as that used in greenhouse hydroponic solutions.

Plants have three modes of nutrient uptake, advection, diffusion, and root interception.

- *Advection*: solutes are carried to the roots with the water that is drawn into the roots.
- *Diffusion*: Roots selectively move solutes out of the soil and into the roots. This process depletes solutes near the roots and a concentration gradient is established. The concentration gradient draws more nutrients toward the roots and the process continues. Diffusive flux of nutrients can generally occur over a length of from 0.1 to 15 mm. Thus, the mass of nutrients in the soil solution that is available for diffusion into roots is a function of the density of roots in the soil.
- *Root interception*: As roots grow in the soil, solutes are intercepted and drawn into roots.

Advection generally supplies all of a plant's micronutrients because water uptake brings sufficient micronutrients to the plant. Macronutrient uptake is primarily by diffusion because water uptake does not transport sufficient macronutrients to the plant. The plant creates a zone of low concentration around the root in order to draw these nutrients into the root. For this reason, the rate of nitrogen uptake is not calculated as the product of the volume of water uptake and the concentration of nitrogen in soil water; instead, the Michaelis-Menton Eq. (25.20) is used to calculate the rate of nitrate uptake. It models the rate of diffusion as a function of nitrogen concentration. The Michaelis-Menton equation is generally written in terms of root surface area and nitrate concentration, which is the most accurate approach; however, if root density is unknown, then the equation can be written in terms of nitrate concentration in soil and constants adjusted based on experiments, historical plant uptake data, or other methods.

$$m_{upt} = \frac{N_{\max}(N_{ave-soil} - N_{\min})}{Km + N_{ave-soil} - N_{\min}} \tag{25.20}$$

where

m_{upt} = mass of NO_3-N taken up, kg/ha/day,
N_{max} = Maximum plant NO_3-N uptake rate, kg/ha/day,
Km = Michaelis-Menten constant; adjusted so that required uptake takes place at optimal soil concentration,
$N_{ave\text{-}soil}$ = NO_3-N concentration in soil, mg/kg soil,
N_{min} = Concentration at which influx = efflux and net m_{upt} = 0, mg/kg,

As described previously, the required nitrate uptake is the product of the total plant nitrogen requirement and the fraction of nitrogen taken up as nitrate.

$$N_{req} = Nitrogen_{req} {*} Frac\_NO_3 \tag{25.21}$$

where

$Nitrogen_{req}$ = nitrogen required by the plant, kg/ha/day
N_{req} = nitrate required by the plant, kg/ha/day
$Frac\_NO_3$ = fraction of nitrogen uptake taken up as nitrate.

The Michaelis-Menten constant Km can be adjusted based on the assumption that the required nitrate uptake takes place at the optimal soil nitrate concentration. This is accomplished by rearranging Eq. 25.20 and adjusting Km and/or N_{min} so that the shape of the curve matches the expected or experimental nitrate uptake curve.

$$Km = \left(N_{opt-soil} - N_{\min}\right)\left(\frac{N_{\max}}{N_{req}} - 1\right) \tag{25.22}$$

where

$N_{opt\text{-}soil}$ = nitrate concentration at which uptake is optimal, mg/kg.

The ratio of actual uptake to required nitrate uptake is used in Chap. 29 to calculate a nitrogen stress coefficient and corresponding yield reduction.

Example 25.10 Plot the uptake rate vs. soil nitrogen concentration, and calculate the uptake of nitrate at a soil nitrate concentration of 25 mg/kg. $\text{Nitrogen}_{\text{req}} = 2.3$ kg/ha/d, $N_{\min} = 2$ mg/kg soil, and $N_{\max} = 2.5$ kg/ha. The fraction of nitrogen taken up as nitrate is 0.8. The optimal level of nitrate uptake takes place when the soil nitrate concentration is 30 mg/kg.

The optimal nitrate uptake rate is

$$N_{\text{req}} = \text{Nitrogen}_{\text{req}}(\text{Frac\_NO}_3) = 2.3\ \text{kg/ha/d}(0.8)$$
$$= 1.84\ \text{kg/ha/d}$$

Adjust Km so that optimal nitrate uptake takes place when soil nitrate concentration is 30 mg/kg.

$$Km = \left(N_{opt-soil} - N_{\min}\right)\left(\frac{N_{\max}}{N_{req}} - 1\right)$$
$$= (30 - 2)\left(\frac{2.5}{1.84} - 1\right) = 9.83$$

The uptake at $N_{\text{ave-soil}} = 25$ mg/kg soil is

$$m_{u\,pt} = \frac{N_{\max}(N_{ave-soil} - N_{\min})}{Km + N_{ave-soil} - N_{\min}}$$
$$= \frac{2.5\ kg/ha/d*(25\ mg/kg - 2\ mg/kg)}{9.83\ mg/kg + 25\ mg/kg - 2\ mg/kg} = 1.75\ kg/ha/d$$

Thus, the ratio of actual nitrate uptake to optimal nitrate uptake is 1.75/1.84 = 0.95.

The nitrate parameters used in the *WINDS* model are entered into the *Nitrate* dialog box (Fig. 25.9). The data is written to cells B300:B360 in the *Active data* worksheet.

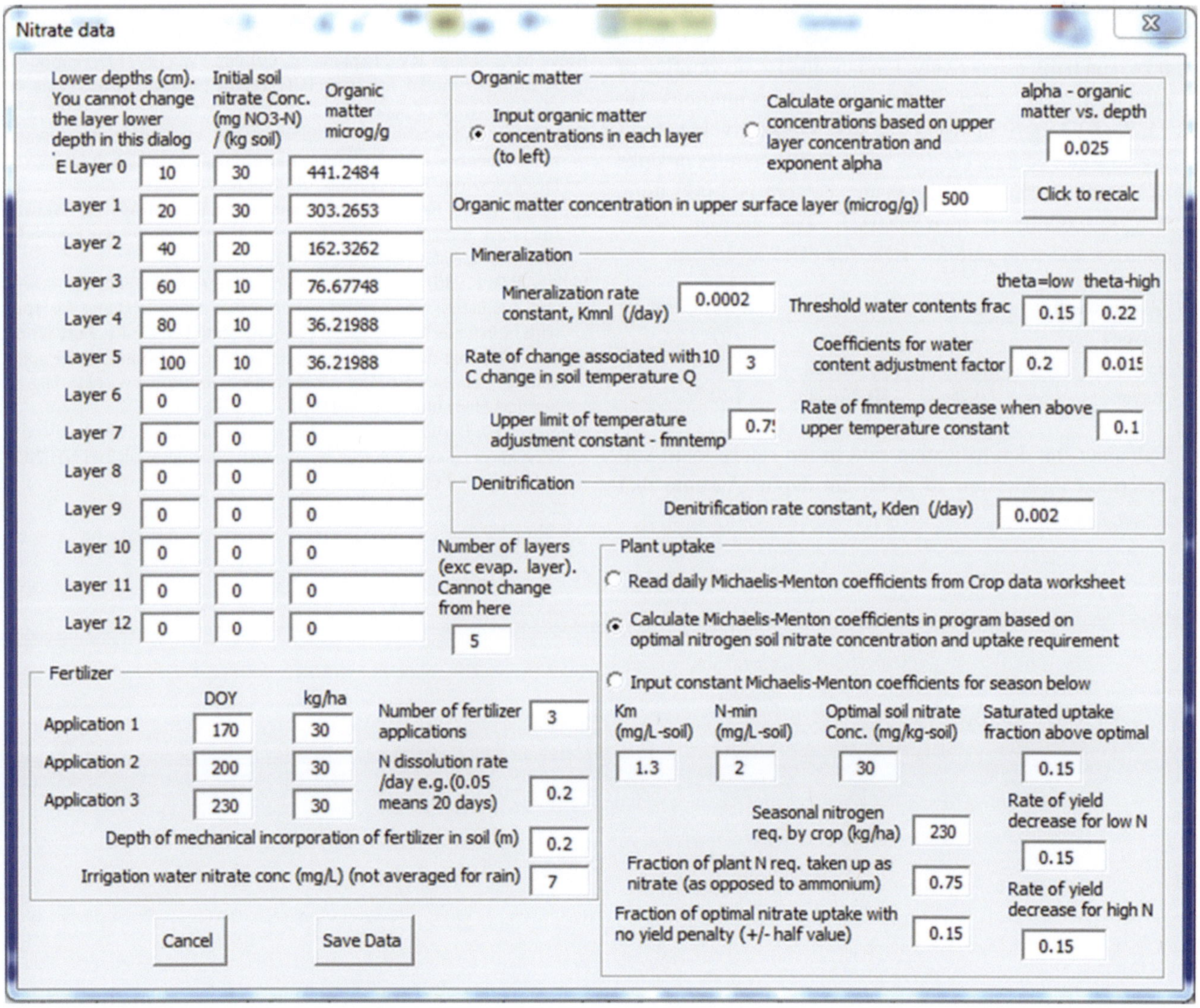

Fig. 25.9 *Nitrate* dialog box in *WINDS* model

Questions

1. An application of 100 kg/ha of NO_3-N is incorporated to a depth of 0.15 m. The 100 refers to only the N portion of nitrate. All of the fertilizer is dissolved on the day that it is incorporated. Calculate the change in the concentration of fertilizer in the upper 0.15 m per soil volume and in soil water. Water content is 0.3 L/L.
2. Annual Tucson temperature data at 10 cm and 50 cm depth is available in the *Chapter 25 WINDS salinity and nitrogen* workbook. Develop an annual sin wave Eq. (25.16) based on this data, and plot temperatures at 5, 10, 50, and 100 cm.
3. Hourly data is available in the *Chapter 25 WINDS salinity and nitrogen* workbook for 3 days in January and 2 days in June. Develop a diurnal sin wave Eq. (25.16) based on this data, and plot temperatures at the surface and at 5, 10, 50, and 100 cm for January and June. Compare the lag times calculated with the diffusivity to the lag times observed in the figures. Compare the equations that were derived based on data for the two different seasons.
4. Find the mineralization over a 2 day period in a sandy loam soil with θ_{pwp} θ_{low}, θ_{high}, and θ_{sat} are equal to 0.1, 0.17, 0.26, and 0.42, respectively. Find the final concentration of nitrate in soil water if the initial concentrations are 15, 8, and 3 mg/L in layers 1, 2, and 3 respectively assuming no water content changes and no other sources or sinks of nitrate. Surface organic matter content is 2,000 μg/g, $K_{mnl} = 0.00005\ d^{-1}$, $\alpha = 0.021$. The average water contents and temperatures in the top three layers are

Water content	Cell elevations		Temperature
Layer 1	0.16	0–40 cm,	27 °C
Layer 2	0.28	40–80 cm,	22 °C
Layer 3	0.25	80–120 cm,	20 °C

5. Calculate the denitrification rate in the sandy loam soil described in question 14 at 60 cm depth. Assume that initial nitrate concentration in soil water is 8 mg/L, and that the denitrification rate constant is 0.002 d^{-1}. Let the depth adjustment factor equal 0.021. Calculate the change in concentration within the soil volume and change in concentration within the soil water. Consider both the denitrification and mineralization to calculate the final concentration in layer 2 after 2 days.
6. Determine the seasonal removal of nitrate from the root zone (mg/L_{soil}) for a crop that has a yield of 8,000 kg/ha and has a nitrogen percentage of 1.6 %. Root zone depth is 1.5 m.
7. $N_{req} = 1.4$ kg/ha, $N_{min} = 2$ mg/kg soil, and $N_{max} = 1.8$ kg/ha. The fraction of nitrogen taken up as nitrate is 1.0. The optimal level of nitrate in the soil is 30 mg/kg. Find Km, plot the uptake rate vs. soil nitrogen concentration, and calculate the uptake of nitrate at a soil nitrate concentration of 20 mg/kg.

References

Breve MA, Skaggs RW, Parsons JE, Gilliam JW (1997) Drainmod-N, A nitrogen model for artificially drained soils. ASAE Trans 40 (4):1067–1085, 2 papers

Crop evapotranspiration – Guidelines for computing crop water requirements – FAO Irrigation and drainage paper 56. http://www.fao.org/docrep/X0490E/X0490E00.htm

Kimberly Idaho Research and Extension Center FAO 56 Website (includes spreadsheets by Richard Allen). http://www.kimberly.uidaho.edu/water/fao56/index.html

Skaggs RW (1980) Drainmod. Reference report. Methods for design and evaluation of drainage-water management systems for soils with high water tables. North Carolina State University, Fort Worth

USDA-ARS. Diagnosis and improvement of saline and alkali soils. United States Salinity Laboratory http://www.ars.usda.gov/Services/docs.htm?docid=10155

Warncke DD, Barber SA (1973) Ammonium and nitrate uptake by corn (Zea mays L) as influenced by nitrogen concentration and NH/NO3 ratio. Agron J 65:950–953

Water and Solute Mass Balance Models

26

This chapter describes the *WINDS* tipping bucket approach to water, salt, and nitrate transport in soils. The tipping bucket model is based on the law of conservation of mass (Fig. 26.1). The model runs quickly because rapid changes due to irrigation or storm events are simulated with conservation of mass, which allows for daily time steps. The algorithms march forward in time with Euler's finite difference method. Because the model is fast, it can simulate daily changes in water, salinity, and nitrogen at hundreds of locations within fields over a growing season.

The continuity equation can be written for any control volume (Fig. 26.1):

$$\Delta m_{cv} = m_{in} + m_{IR} - m_{out} \qquad (26.1)$$

where

m_{in} = mass entering system, kg,
m_{out} = mass leaving system, kg,
m_{IR} = mass generated or lost within system by internal reaction, kg,
Δm_{cv} = change of mass within control volume, kg.

Water Mass Balance

The conservation of mass equation is written in terms of volume of water rather than mass because the density of water is constant in the range of soil temperature.

$$\Delta V_{cv} = V_{in} + V_{IR} - V_{out} \qquad (26.2)$$

where

V_{in} = volume entering system, m^3,
V_{out} = volume leaving system, m^3,
V_{IR} = volume generated within system by internal reaction, m^3,
ΔV_{cv} = change of volume within control volume, m^3.

Equation 26.2 is used to calculate daily water volume change in the root zone. The change in volume is the difference between daily water inflow, outflow, and internal change.

The sources and sinks for the entire soil profile are shown in Fig. 26.2. The soil can be divided into layers with a volume balance written for each layer. In the tipping bucket model, it is assumed that water only flows downward through the root zone and freely drains below the soil profile. This assumption is appropriate for soils in which water redistribution takes place within a few days. Subsequent chapters describe upward and downward flow of water in the soil due to energy gradients. The tipping bucked model is used for irrigation or large storm events. Each soil layer is a control volume that is treated like a "bucket" from which water spills into the next layer when the "bucket" is full.

The soil is divided in layers in Fig. 26.3. Each of the layers is treated as a bucket that can hold a certain volume of water (Fig. 26.3). The volume of each bucket is the product of layer volume and field capacity. When the volume of the bucket is exceeded during an irrigation event, then excess water drains to the next layer. The internal reaction term is evapotranspiration.

Evapotranspiration can be treated like an internal reaction term ($V_{ET} = V_{IR}$) for soil water.

$$V_{in} - V_{out} - V_{ET} = \Delta V_{cv} \qquad (26.3)$$

where

V_{ET} = volume lost to evapotranspiration during time step.

Layer water volume is the product of depth of water in the layer (equivalent to ponded depth but distributed within pores) and layer horizontal area, $\Delta x\ \Delta y$.

$$d_{in}\Delta x\ \Delta y - d_{out}\Delta x\ \Delta y - d_{ET}\Delta x\ \Delta y = \Delta d_{cv}\Delta x\ \Delta y \qquad (26.4)$$

P. Waller, M. Yitayew, *Irrigation and Drainage Engineering*, DOI 10.1007/978-3-319-05699-9_26

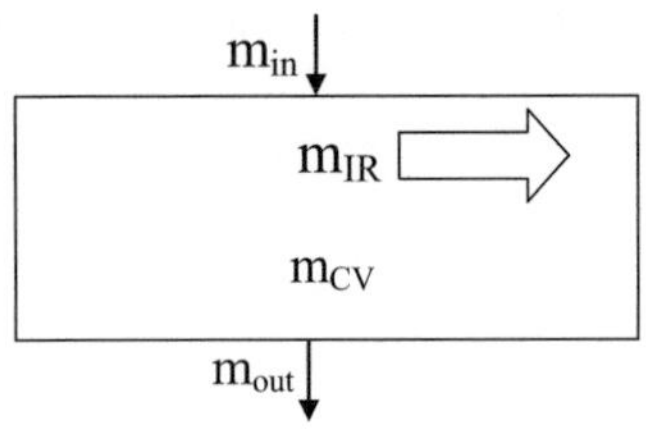

Fig. 26.1 Control volume for conservation of mass

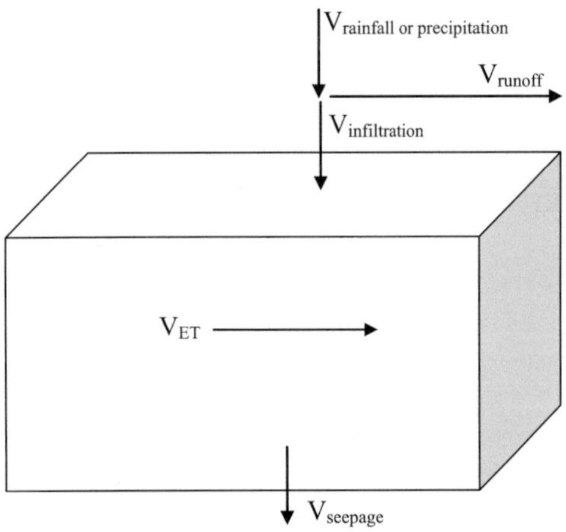

Fig. 26.2 Sources and sinks for water in a single layer

Divide Eq. 26.4 by Δx Δy

$$d_{in} - d_{out} - d_{ET} = \Delta d_{cv} \qquad (26.5)$$

where

d_{in} = depth of water that flows into the layer, m,
d_{out} = depth of water that flows out of the layer, m,
d_{ET} = depth of water lost to evapotranspiration, m,
Δd_{cv} = change in depth of water within the layer, m,

The d_{in} term at the upper boundary (soil surface) is infiltration from storms or irrigation events. Infiltration is either specified in the *WINDS* program in the *Irr* worksheets, is calculated with the Green-Ampt model (Chap. 29), or is calculated with the NRCS curve number method. If a runoff flume is used at a site, then infiltration can be calculated based on precipitation, field area, and runoff volume.

$$d_{RO} = V_{RO}/A_{field} \qquad (26.6)$$

$$i = P - d_{RO} \qquad (26.7)$$

where

i = infiltrated depth, m,
V_{RO} = volume of runoff from a field, m^3.
A_{field} = field area, m^2,
d_{RO} = depth of runoff, m.
P = precipitation depth, m.

Fig. 26.3 Layers in the soil profile for tipping bucket model

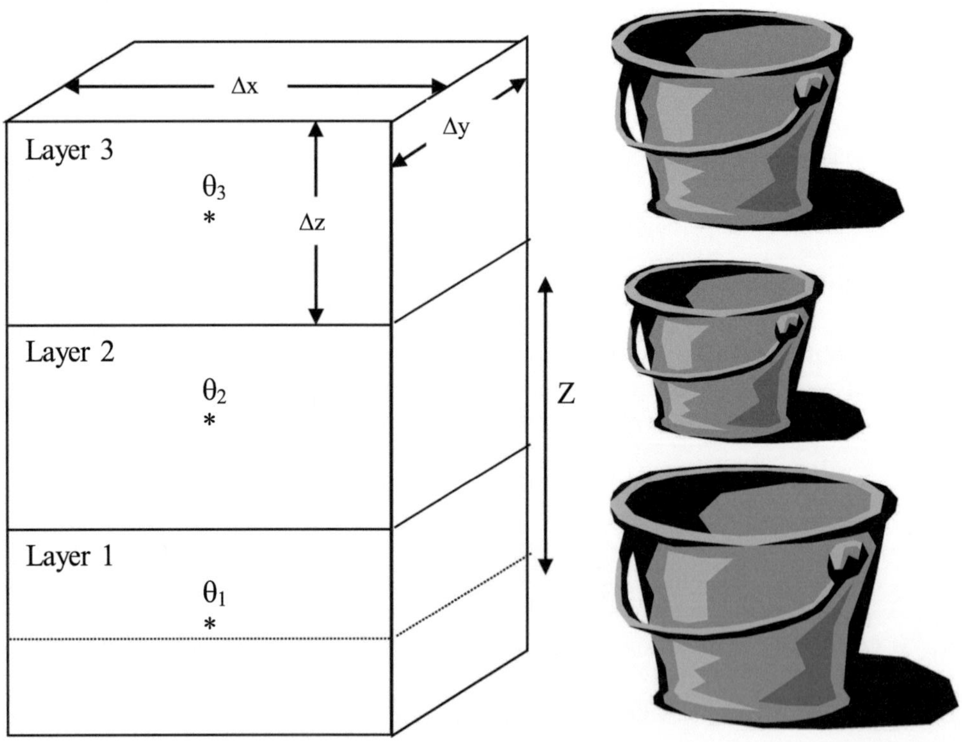

The water exiting across the lower boundary of the layer is seepage. Substitute water balance parameters into Eq. 26.5 to calculate change in the depth of water in the layer.

$$\Delta d_{cv} = i - d_{seepage} - d_{ET} \quad (26.8)$$

where

$d_{seepage}$ = depth leached below the layer, m,

The maximum depth of water that a layer can hold is the product of field capacity and layer depth.

$$d_{cv-max} = \theta_{fc}\Delta z \quad (26.9)$$

where

d_{cv-max} = Maximum depth of water in each layer, m,
θ_{fc} = Water content at field capacity, dimensionless.

The tipping bucket approach is sequential, beginning in the upper layer. If there is excess water in a layer after an infiltration event (exceeding field capacity), then the water is routed to the next layer. Then, the process repeats until there is no more seepage or the lower boundary is reached.

If there is seepage, then the final water content in the layer is field capacity. The calculation of seepage and final water content is calculated in three steps for each layer.

Step 1: Calculate initial depth of water in layer.

$$d_{initial} = \theta_{initial}\Delta z \quad (26.10)$$

where

$d_{initial}$ = initial depth of water in layer, m.
$\theta_{initial}$ = initial water content, ml/ml.
Step 2: Calculate sum of initial depth, infiltration, and ET.

$$d_{sum} = d_{initial} + i - d_{ET} \quad (26.11)$$

where

d_{sum} = antecedent water content + infiltrated depth, m.
Step 3: Calculate final water content and seepage

$$d_{sum} \geq d_{cv-max} \rightarrow \theta_{final} = \theta_{fc} \quad d_{seepage} = d_{sum} - d_{cv-max} \quad (26.12)$$

$$d_{sum} < d_{cv-max} \rightarrow \theta_{final} = d_{sum}/\Delta z \quad d_{seepage} = 0 \quad (26.13)$$

where

θ_{final} = final water content, m.

Example 26.1 During a two-day period, there are two storms. The first high intensity, storm has 3 cm precipitation depth. Runoff during and after the first storm is measured in a flume at the lower end of the 38 ha field: 3,800 m^3. A second low intensity storm adds 4 cm water to the soil profile, and there is no runoff. Model the soil as one layer that is 1.2 m deep. Average water content in the soil profile before the storm is 17 %. Field capacity is 20 %. The water table is well below the soil profile. Assume that evapotranspiration between the storms is negligible. Perform the water balance for each storm.

Calculate the maximum depth of water that the soil profile can hold

$$d_{cv-max} = \theta_{fc}\Delta z = 0.2(1.2m) = 0.24m$$

Find the depth of runoff during the first storm

$$d_{RO} = V_{RO}/A_{field} = 3{,}800m^3/\left(38ha\left(10{,}000m^2/ha\right)\right) = 0.01m$$

Find the depth of infiltration during the first storm

$$i = P - d_{RO} = 3cm - 1cm = 2cm = 0.02m.$$

Step 1: Calculate the initial depth of water in the soil profile.

$$d_{initial} = \theta_{initial}\Delta z = 0.17(1.2m) = 0.204m$$

Step 2: Calculate sum of initial depth and infiltration.

$$d_{sum} = d_{initial} + i = 0.204\,m + 0.02\,m = 0.224\,m$$

Step 3: Calculate final water content and seepage.

$$d_{sum} < d_{cv-max} \rightarrow \theta_{final} = d_{sum}/\Delta z = 0.224/1.2m = 0.187ml/ml \quad d_{seepage} = 0$$

Repeat calculation for second storm with $i = 0.04$.

Step 1: Calculate initial depth of water in soil profile.

$$d_{initial} = d_{initial} = \theta_{initial}\Delta z = 0.187(1.2m) = 0.224m$$

Step 2: Calculate sum of initial depth and infiltration.

$$d_{sum} = d_{initial} + i = 0.224m + 0.04m = 0.264m$$

Step 3: Calculate final water content and seepage.

$$d_{sum} \geq d_{cv-max} \rightarrow \theta_{final} = \theta_{fc} = 0.2ml/ml$$
$$d_{seepage} = d_{sum} - d_{cv-max} = 0.264 - 0.24 = 0.024m$$

If the soil is divided into layers (Fig. 26.3), then the three steps can be repeated for each layer. The *WINDS* model numbers layers from the bottom of the soil profile to the top. Thus, with two layers, the upper layer is layer number 2, and the mass balance is calculated as follows, where $d_{seepage\text{-}2}$ is the water that passes from layer 2 to layer 1.

$$\Delta d_{cv-2} = i - d_{seepage\ 2} - d_{ET-2} \quad (26.14)$$

The volume balance is written as follows for layers below the upper layer with seepage from the layer above entering the layer.

$$\Delta d_{cv-j} = d_{seepage\ j+1} - d_{seepage\ j} - d_{ET-j} \quad (26.15)$$

where

j = layer number.

Evapotranspiration is the product of total ET and fraction ET from each layer.

$$d_{ET-j} = ET_{frac-1} d_{ET} \quad (26.16)$$

$$\Delta d_{cv-j} = d_{seepage\ j+1} - d_{seepage\ j} - ET_{frac-j} d_{ET} \quad (26.17)$$

where

$ET_{frac\text{-}j}$ = fraction of ET that evaporates from layer j.

Example 26.2 Redo Example 26.1 with 3 layers that are 0.4 m deep. Initial water content in all layers is 0.17 and field capacity is 0.20.

The maximum water depth held by each layer is

$$d_{cv-max} = \theta_{fc} \Delta z = 0.20(0.4m) = 0.08m$$

Storm 1. Layer 3 (Upper layer)

Step 1. Calculate initial water depth.

$$d_{initial-3} = \theta_{initial-3} \Delta z = 0.17(0.4m) = 0.068m$$

Step 2. Calculate sum of initial depth and infiltration.

$$d_{sum-3} = d_{initial-3} + i = 0.068m + 0.02m = 0.088m$$

Step 3. Calculate final water content and seepage.

$$d_{sum-3} \geq d_{cv-max} \rightarrow \theta_{final-3} = \theta_{fc} = 0.2ml/ml$$
$$d_{seepage-3} = d_{sum-3} - d_{cv-max} = 0.088 - 0.08 = 0.008m$$

Storm 1. Layer 2

Step 2: Calculate sum of initial depth and infiltration.

$$d_{sum-2} = d_{initial-2} + d_{seepage-3} = 0.068m + 0.008m = 0.076m$$

Step 3: Calculate final water content and seepage.

$$d_{sum-2} < d_{cv-max} \rightarrow \theta_{final-2} = d_{sum-2}/\Delta z = 0.076m/0.4 = 0.19ml/ml \; d_{seepage-2} = 0$$

There is no seepage to the third layer and no change in water content. After the first storm, the depth of water in the first layer is 0.080 m and the depth of water in the second layer is 0.076 m. No water infiltrates to the third layer.

Storm 2. Layer 3

The second storm infiltration is 4 cm (0.04 m). The first layer is already at field capacity so the entire 0.04 m drains from layer 3 to layer 2.

Storm 2. Layer 2.

Step 1. Calculate initial water depth.

$$d_{initial-2} = \theta_{initial-2} \Delta z = 0.19(0.4m) = 0.076m$$

Step 2. Calculate sum of initial depth and infiltration.

$$d_{sum-2} = d_{initial-2} + d_{seepage-3} = 0.076m + 0.04m = 0.116m$$

Step 3. Calculate final water content and seepage.

$$d_{sum-2} \geq d_{cv-max} \rightarrow \theta_{final-2} = \theta_{fc} = 0.2ml/ml$$
$$d_{seepage-2} = d_{sum-2} - d_{cv-max} = 0.116 - 0.08 = 0.036m$$

Storm 2. Layer 1

Step 1. Calculate initial water depth.

$$d_{initial-1} = \theta_{initial-1} \Delta z = 0.17(0.4m) = 0.068m$$

Step 2. Calculate sum of initial depth and infiltration.

$$d_{sum\text{-}1} = d_{initial\text{-}1} + d_{seepage\text{-}2} = 0.068\ m + 0.036\ m = 0.104\ m$$

Step 3. Calculate final water content and seepage.

$$d_{sum-1} \geq d_{cv-max} \rightarrow \theta_{final-1} = \theta_{fc} = 0.2\text{ml/ml}$$
$$d_{seepage-1} = d_{sum-1} - d_{cv-max} = 0.104 - 0.08 = 0.024\text{m}$$

Example 26.3 A crop with a 1 m root zone is planted in a soil with two layers.

$$\text{Layer1(Upperlayer)}: \theta_{FC} = 0.20,$$
$$\theta_{PWP} = 0.10, \quad \text{layer depth } \Delta z = 0.6\text{m}, \quad ET_{frac} = 0.6$$
$$\text{Layer2(Lowerlayer)}: \theta_{FC} = 0.26,$$
$$\theta_{PWP} = 0.11, \quad \text{layer depth } \Delta z = 0.4\text{m}, \quad ET_{frac} = 0.4$$

Total ET (d_{ET}) is 0.01 m/day. Four days after a storm filled the soil to field capacity, a second storm adds 3 cm to the soil. Calculate water contents and percent depletions before and after the second storm

Calculate water status before second storm based on ET between storms and initial θ_{fc}.

Layer 2 (Upper layer) $d_{cv\text{-}max} = 0.2(0.6) = 0.12$ m (four day time step)

Step 1. Calculate initial water depth.

$$d_{initial-2} = \theta_{initial-2}\Delta z = 0.2(0.6) = 0.12\text{m}$$

Step 2. Calculate sum of initial depth, infiltration, and ET.

$$\begin{aligned} d_{sum-2} &= d_{initial-2} + i - ET_{frac-2}d_{ET} \\ &= 0.12\text{m} + 0\text{m} - (0.01\text{m/day})(0.6)(4\text{days}) \\ &= 0.096\text{m} \end{aligned}$$

Step 3: Calculate final water content and seepage.

$$\begin{aligned} d_{sum-2} < d_{cv-max} \rightarrow \theta_{final-2} &= d_{sum-2}/\Delta z \\ &= 0.096\text{m}/0.6\text{m} = 0.16\text{ml/ml} \ d_{seepage-2} = 0 \end{aligned}$$

$$\text{Depletion} = (0.01\text{m/day})(0.6\text{m})(4\text{days}) = 0.024\text{m}$$
$$\text{TAW} = \left(\theta_{fc} - \theta_{pw\,p}\right)\Delta z = (0.2 - 0.1)0.6\text{m} = 0.06\text{m}$$
$$\begin{aligned} \%\text{depletion} &= D_r/\text{TAW}(100\%) = 0.024\text{m}/0.06\text{m}(100\%) \\ &= 40\% \end{aligned}$$

Layer 1 (Lower layer) $d_{cv\text{-}max} = 0.26\ (0.4) = 0.104$ m

Step 1. Calculate initial water depth.

$$d_{initial-1} = \theta_{initial-2}\Delta z = 0.26(0.4\text{m}) = 0.104\text{m}$$

Step 2. Calculate sum of initial depth, infiltration, and ET.

$$\begin{aligned} d_{sum-1} &= d_{initial-1} + d_{seepage-2} - ET_{frac-1}d_{ET} \\ &= 0.104\text{m} + 0 - (0.01\text{m/day})(0.4)(4\text{days}) \\ &= 0.088\text{m} \end{aligned}$$

Step 3: Calculate final water content and seepage.

$$\begin{aligned} d_{sum-1} < d_{cv-max} \rightarrow \theta_{final-1} &= d_{sum-1}/\Delta z \\ &= 0.088\text{m}/0.4\text{m} = 0.22\text{ml/ml} \end{aligned}$$
$$d_{seepage-2} = 0$$

$$\text{Depletion} = (0.01\text{cm/day})(0.4\text{m})(4\text{days}) = 0.016\text{m}$$
$$\text{TAW} = \left(\theta_{fc} - \theta_{pw\,p}\right)\Delta z = (0.26 - 0.11)0.4\text{m} = 0.06\text{m}$$
$$\%\text{depletion} = D_r/\text{TAW}(100\%) = 0.016\text{m}/0.06\text{m}(100\%)$$
$$= 26\%$$

Calculate water status after second storm based on infiltration.

Layer 2 (Note that the time step is just the time of the storm)

Step 1. Calculate initial water depth.

$$d_{initial-2} = \theta_{initial-2}\Delta z = 0.16(0.6\text{m}) = 0.096\text{m}$$

Step 2. Calculate sum of initial depth, infiltration, and ET.

$$\begin{aligned} d_{sum-2} &= d_{initial-2} + i - d_{ET-2} = 0.096\text{m} + 0.03\text{m} - 0\text{m} \\ &= 0.126\text{m} \end{aligned}$$

Step 3: Calculate final water content and seepage.

$$d_{sum-2} \geq d_{cv-max} \rightarrow \theta_{final-2} = \theta_{fc} = 0.2\text{ml/ml}$$
$$d_{seepage-2} = d_{sum-2} - d_{cv-max} = 0.126 - 0.12 = 0.006\text{m}$$

$$\text{Depletion} = 0\text{m}$$
$$\%\text{depletion} = 0\%$$

Layer 1

Step 1. Calculate initial water depth.

$$d_{initial-1} = \theta_{initial-1}\Delta z = 0.22(0.4\text{m}) = 0.088\text{m}$$

Step 2. Calculate sum of initial depth, infiltration, and ET.

$$\begin{aligned} d_{sum-1} &= d_{initial-1} + d_{seepage-2} - d_{ET-1} \\ &= 0.088\text{m} + 0.006\text{m} - 0\text{m} = 0.094\text{m} \end{aligned}$$

Step 3: Calculate final water content and seepage.

$$d_{sum-1} < d_{cv-max} \rightarrow \theta_{final-1} = \theta_{sum-1}/\Delta z$$
$$= 0.094m/0.4m = 0.235ml/ml$$
$$d_{seepage-2} = 0$$

$$\text{Depletion} = 0.104m - 0.094m = 0.01m$$
$$\%\text{depletion} = D_r/TAW(100\%) = 0.01m/0.06m(100\%) = 17\%$$

Salinity Mass Balance Model

Although there are many salinity sources and sinks that could be incorporated into the salinity mass balance model, it was shown in Chap. 4, that the sources and sinks can be ignored in a typical model of salinity in agricultural soils. All of the cations and anions can be lumped into one concentration term representing all forms of salinity, EC or C. The approach in the *WINDS* model is to calculate maximum solubility of the irrigation water and if salinity exceeds the maximum solubility, then salts are removed from the water phase. However, this level of salinity is not normally encountered in agricultural fields. The maximum solubility is a function of temperature, activity coefficients of ions in solution and numerous other parameters. Programs such as HYDRUS, available through the USDA-ARS Salinity Laboratory, consider salinity transformations in soil.

The mass of salt within a layer (m_{salt}) is the product of volume of the layer (V_{layer}), fraction of soil that is occupied by water (θ), and the concentration of salt within the soil-water (C).

$$m_{salt} = V_{layer} C\, \theta \tag{26.18}$$

where

C = salt concentration, mg salt /L water,
θ = initial water content in the layer, L water /L soil,
V_{layer} = volume of the layer, L.
m_{salt} = mass of salt within the layer, mg/L_{soil} m^3.

The change of salt mass within the control volume during a time step is

$$\Delta m_{salt} = \Delta(C\, V_{cv}\, \theta) = C_{final} V_{layer} \theta_{final} - C_{initial} V_{layer} \theta_{initial} \tag{26.19}$$

where

Δm_{salt} = change in mass of salt, mg/L_{soil} m^3.
$C_{initial}$ = initial salt concentration, mg salt /L water,
C_{final} = final salt concentration, mg salt /L water,
$\theta_{initial}$ = initial water content in control volume, L water/L soil,
θ_{final} = final water content in control volume, L water / L soil.

The salt influx is the product of water influx (V_{in}) and average salt concentration. Assuming piston flow through the layer and no water passing completely through the layer, the salt drained is the product of outflow (V_{out}) and average salt concentration in the layer's water.

$$m_{in} = V_{in} C_{in} \quad (L)(mg/L) = mg/L_{water} m^3 \tag{26.20}$$

$$m_{out} = V_{out} C_{initial} \quad (L)(mg/L) = mg/L_{water} m^3 \tag{26.21}$$

where

C_{in} = salt concentration in water entering the layer, mg salt / L water
$C_{initial}$ = Initial salt concentration in layer, mg salt /L water
V_{in} = Volume of water entering the layer, m^3,
V_{out} = Volume of water exiting the layer, m^3.

Plant uptake of salinity is neglected, and the change in mass of salt is simply the difference between influx and outflow.

$$\Delta m_{layer} = m_{in} - m_{out} \tag{26.22}$$

Substitute Eq. 26.32 into Eq. 26.29

$$C_{final} V_{layer} \theta_{final} - C_{initial} V_{layer} \theta_{initial} = V_{in} C_{in} - V_{out} C_{initial} \tag{26.23}$$

With a finite difference soil solute model of a layered soil, it is convenient to express the total mass of salt based on concentration and depth, mg/L_{soil} m, or mg/L_{water} m. Salt concentration is expressed as mg/L_{water}.

$$C_{final} \Delta z\, \theta_{final} - C_{initial} \Delta z\, \theta_{initial} = d_{in} C_{in} - d_{out} C_{initial} \tag{26.24}$$

Equation 26.24 can be rearranged to solve for final salt content.

$$C_{final} = \frac{d_{in}C_{in} - d_{out}C_{intial} + C_{intial}\,\Delta z\,\theta_{intial}}{\Delta z\,\theta_{final}} \quad (26.25)$$

Equation 26.25 is used in the *WINDS* model to calculate the concentrations in soil layers. Dimensional analysis shows that the units in Eqs. 26.24 and 26.25, mg/L_{soil} m_{soil}, or mg/L_{water} m_{water} (C d_{in}) can be converted to kg/ha by multiplying by 10:

$$C\Delta z\theta = \left(\frac{mg}{L_{water}}\right)\left(\frac{m}{1}\right)\left(\frac{L_{water}}{L_{soil}}\right) = \left(\frac{mg\ m}{L_{soil}}\right)$$

$$\text{kg/ha} = \left(\frac{mg}{L_{soil}}\right)\left(\frac{1\ kg}{10^6\ mg}\right)\left(\frac{10^3\ L}{m^3}\right)\left(\frac{10^4\ m^2}{ha}\right)$$

$$(soil\ depth\ (m)) = 10\left(\frac{mg}{L_{soil}}\right)(soil\ depth)$$

Concentration, C, in Eq. 26.35 can be replaced by EC.

$$EC_{final} = \frac{d_{in}\ EC_{in} - d_{out}\ EC_{initial} + EC_{initial}\,\Delta z\,\theta_{initial}}{\Delta z\,\theta_{final}} \quad (26.26)$$

The units for the mass of salts in the layers is now (dS/m m), which is equal to mg/L m/640. Equation 26.26 can be applied to individual soil layers. With the tipping bucket model, the sequence of calculation for each time step is to first calculate the water movement beginning with the upper layer. Second, the salinity concentrations are calculated with Eq. 26.26 beginning with the upper layer and then moving downward through the soil profile from one layer to the next. The input to each layer is the seepage from the layer above. For the upper layer, Eq. 26.26 is written as

$$EC_{final} = \frac{i\ EC_{iw} - d_{seepage}\ EC_{initial} + EC_{initial}\,\Delta z\,\theta_{initial}}{\Delta z\,\theta_{final}} \quad (26.27)$$

For layers (j) below the upper layer, Eq. 26.26 is written as

$$EC_{final} = \frac{d_{seepage\text{-}j+1}\ EC_{j+1} - d_{seepage\text{-}j}\ EC_{initial} + EC_{initial}\,\Delta z\,\theta_{initial}}{\Delta z\,\theta_{final}} \quad (26.28)$$

If infiltration from above exceeds the available water storage capacity in the layer, (field capacity – water content) (layer depth), then Eq. 26.28 is invalid. In this case, the existing water and infiltrating water from above is mixed; the weighted average is (i EC_{iw} + $EC_{initial}$ $\theta_{initial}$ Δz)/(i + $\theta_{initial}$ Δz + Rain – ET). This average concentration is then the final layer concentration and the concentration of the seepage water below the layer.

Example 26.4 Using the parameters from Example 4.6, perform a transient analysis of salinity in 4 layers that are 0.5 m depth with a root zone depth that is 2.0 m deep. An irrigation is applied on the third day and every 7 days thereafter. Make hand calculations for the first 3 days for the upper two layers and verify that they agree with the *WINDS* model. Run the *WINDS* model simulation for 100 days beginning with DOY 1. Initial salinity of the saturated extract, EC_e, is 1.034 dS/m in all layers. Irrigation water salinity is 1.0 dS/m. $\theta_{FC} = 0.2$, $\theta_{PWP} = 0.1$, and $\theta_{sat} = 0.387$. MAD = 0.35. ET is 10 mm/day during the entire 100 days, and 40-, 30-, 20-, and 10-% of ET is extracted from the upper, next, next, and lower layers, respectively. Initial water content in all layers is field capacity, 0.2. First, calculate for no leaching, and then calculate for 15 % LF. The comparison is complicated by the fact that the *WINDS* model has an evaporation layer. This can be ignored by reducing the evaporation layer thickness to 1 cm, eliminating evaporation and transpiration from the evaporation layer, and setting the EC in the evaporation layer to the same concentration as the concentration in the irrigation water. Because the evaporation layer is 1 cm thick, the upper layer (below the evaporation layer) thickness in the *WINDS* model is 49 cm. In the following calculations, the calculations are more straightforward by letting the thickness or the upper layer equal 0.5 m. Note that layers are numbered from the bottom in the *WINDS* model; thus, the upper layer (below the evaporation layer) is layer 4.

In order to calculate the irrigation frequency and amount, calculate RAW.

$$\text{AWC} = \theta_{FC} - \theta_{PWP} = 0.2 - 0.1 = 0.1$$
$$\text{TAW} = \text{AWCz} = 0.1(2.0\text{m}) = 0.2\text{m}$$
$$\text{RAW} = \text{TAW MAD} = 0.2\text{m}(0.35) = 0.07\text{m} = 70\text{mm}.$$

Thus, with no leaching, irrigation takes place every 7 days and adds 70 mm to the root zone.

Calculate initial soil water salinity if the field is at field capacity and the saturated paste extract EC is 1.034 dS/m.

$$EC = EC_e\frac{\theta_{sat}}{\theta_{FC}} = (1.034)\left(\frac{0.387}{0.2}\right) = 2\ dS/m$$

The ET from each layer is calculated based on the percent removal from each layer. For example, d_{ET} from layer 4 is $ET_{frac-4}\ d_{ET}$ = 0.4 (0.01 m/d) = 0.004 mm/d.

The problem specified two scenarios: zero leaching fraction and 1.15 leaching fraction. The following calculations are made for zero leaching fraction.

Day 1: Layer 4 (upper layer)

Calculate water content at the end of day 1

$$\Delta d_{cv-4} = i - d_{seepage-4} - ET_{frac-4} d_{ET}$$
$$= 0 - 0 - 0.4(0.01m) = -0.004m$$
$$\Delta\theta = \Delta d_{cv-4}/\Delta z = -0.004m/0.5m = -0.008$$
$$\theta_{final} = \theta_{initial} + \Delta\theta = 0.2 - 0.008 = 0.192$$

Calculate EC after the first day

$$EC_{final} = \frac{i\ EC_{iw} - d_{seepage\text{-}4}\ EC_{initial} + EC_{initial}\ \Delta z\, \theta_{initial}}{\Delta z\ \theta_{final}}$$
$$EC_{final} = \frac{0 - 0 + 2.0\ (0.5)\ (0.2)}{0.5\ (0.192)} = 2.083\ dS/m$$

Day 1: Layer 3

Calculate water content at the end of day 1

$$\Delta d_{cv-3} = d_{seepage-4} - d_{seepage-3} - ET_{frac-3} d_{ET}$$
$$= 0 - 0 - 0.3(0.01m) \ = -0.003m$$
$$\Delta\theta = \Delta\theta_{cv-3}/\Delta z = -0.003m/0.5m = -0.006$$
$$\theta_{final} = \theta_{initial} + \Delta\theta = 0.2 - 0.006 = 0.194$$

Calculate EC after the first day

$$EC_{final} = \frac{d_{seepage\text{-}4}\ EC_4 - d_{seepage\text{-}3}\ EC_{initial} + EC_3\ \Delta z\, \theta_{initial}}{\Delta z\ \theta_{final}}$$
$$EC_{final} = \frac{0 - 0 + (2.0)\ (0.5)\ (0.2)}{(0.5)\ (0.194)} = 2.062\ dS/m$$

Day 2: Layer 4 (upper layer)

Calculate water content at the end of day 2

$$\Delta d_{cv-4} = i - d_{seepage-4} - ET_{frac-4} d_{ET}$$
$$= 0 - 0 - 0.4(0.01m) = -0.004m$$
$$\Delta\theta = \Delta d_{cv-4}/\Delta z = -0.004m/0.5m = -0.008$$
$$\theta_{final} = \theta_{initial} + \Delta q = 0.192 - 0.008 = 0.184$$

Calculate EC

$$EC_{final} = \frac{i\ EC_{iw} - d_{seepage\text{-}4}\ EC_{initial} + EC_{initial}\ \Delta z\ \theta_{initial}}{\Delta z\ \theta_{final}}$$
$$EC_{final} = \frac{0 - 0 + 2.08\ (0.5)\ (0.192)}{0.5\ (0.184)} = 2.17\ dS/m$$

Day 2: Layer 3

Calculate water content at the end of day 2

$$\Delta d_{cv-3} = d_{seepage-4} - d_{seepage-3} - ET_{frac-3} d_{ET}$$
$$= 0 - 0 - 0.3(0.01m) = -0.003m$$
$$\Delta\theta = \Delta d_{cv-3}/\Delta z = -0.003m/0.5m = -0.006$$
$$\theta_{final} = \theta_{initial} + \Delta\theta = 0.2 - 0.006 = 0.194$$

Calculate EC

$$EC_{final} = \frac{d_{seepage\text{-}4}\ EC_4 - d_{seepage\text{-}3}\ EC_{initial} + EC_3\ \Delta z\ \theta_{initial}}{\Delta z\ \theta_{final}}$$
$$EC_{final} = \frac{0 - 0 + (2.063)\ (0.5)\ (0.194)}{(0.5)\ (0.188)} = 2.13\ dS/m$$

Day 3: Layer 4

Irrigation depth 0.07 m is applied on this day. The infiltration depth far exceeds the capacity of layer 4 so most of the water infiltrates to layer 3. The mass balance equation for salts must be modified when water from a large irrigation event or storm passes completely through the layer. As stated previously, the rationale is that all nitrate from the irrigation water or storm is mixed with the nitrogen in the layer, and the mixed concentration is the final concentration in the layer.

The depth of seepage from layer 4 to layer 3 is calculated as follows:

$$i - ET_{layer-4} - (\Delta z\ \theta_{FC} - \ \Delta z\ \theta_{initial})$$
$$= 0.07m - 0.004m - (0.2(0.5m) - 0.184(0.5m))$$
$$= 0.058m$$

Calculate EC

$$EC_{final} = \frac{\text{mass of salts}}{\text{volume of water}} = \frac{i\ EC_{iw} + EC_{initial}\ \Delta z\ \theta_{initial}}{i - ET_{layer} + \Delta z\ \theta_{initial}}$$
$$EC_{final} = \frac{(0.07\ \text{m})\ (1\ \text{dS/m}) + 2.17{*}0.5\ (0.184)}{0.07m - (10\ mm)\ (0.4)/(1000\ mm/m) + 0.5\ (0.184)}$$
$$= 1.71\, dS/m$$

Day 3: Layer 3

As with layer 4, the modified equation must be used. In this case, the final salinity of layer 4 is the seepage concentration instead of the initial salinity of layer 4.

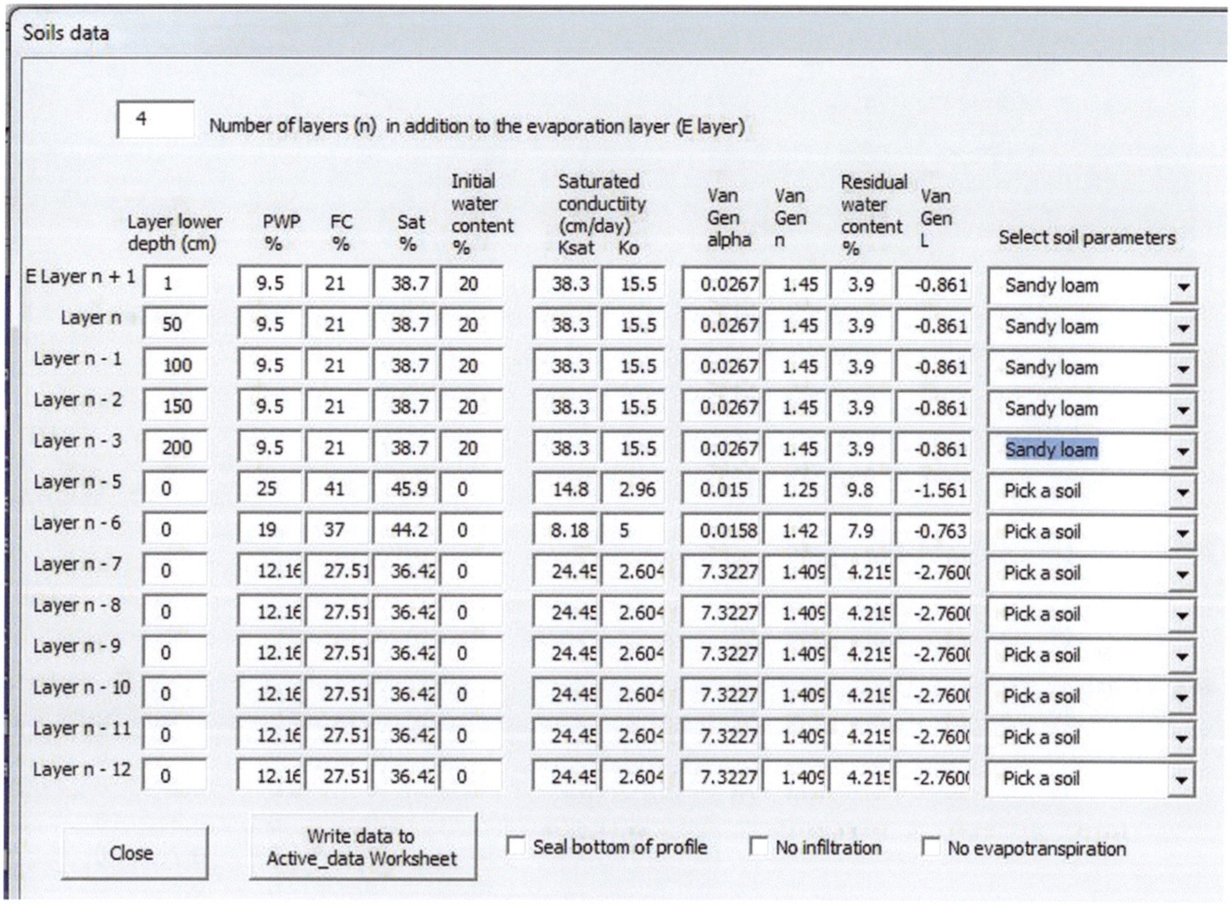

Soils data

4 Number of layers (n) in addition to the evaporation layer (E layer)

	Layer lower depth (cm)	PWP %	FC %	Sat %	Initial water content %	Saturated conductiity (cm/day) Ksat	Ko	Van Gen alpha	Van Gen n	Residual water content %	Van Gen L	Select soil parameters
E Layer n + 1	1	9.5	21	38.7	20	38.3	15.5	0.0267	1.45	3.9	-0.861	Sandy loam
Layer n	50	9.5	21	38.7	20	38.3	15.5	0.0267	1.45	3.9	-0.861	Sandy loam
Layer n - 1	100	9.5	21	38.7	20	38.3	15.5	0.0267	1.45	3.9	-0.861	Sandy loam
Layer n - 2	150	9.5	21	38.7	20	38.3	15.5	0.0267	1.45	3.9	-0.861	Sandy loam
Layer n - 3	200	9.5	21	38.7	20	38.3	15.5	0.0267	1.45	3.9	-0.861	Sandy loam
Layer n - 5	0	25	41	45.9	0	14.8	2.96	0.015	1.25	9.8	-1.561	Pick a soil
Layer n - 6	0	19	37	44.2	0	8.18	5	0.0158	1.42	7.9	-0.763	Pick a soil
Layer n - 7	0	12.16	27.51	36.42	0	24.45	2.604	7.3227	1.409	4.215	-2.7600	Pick a soil
Layer n - 8	0	12.16	27.51	36.42	0	24.45	2.604	7.3227	1.409	4.215	-2.7600	Pick a soil
Layer n - 9	0	12.16	27.51	36.42	0	24.45	2.604	7.3227	1.409	4.215	-2.7600	Pick a soil
Layer n - 10	0	12.16	27.51	36.42	0	24.45	2.604	7.3227	1.409	4.215	-2.7600	Pick a soil
Layer n - 11	0	12.16	27.51	36.42	0	24.45	2.604	7.3227	1.409	4.215	-2.7600	Pick a soil
Layer n - 12	0	12.16	27.51	36.42	0	24.45	2.604	7.3227	1.409	4.215	-2.7600	Pick a soil

Close | Write data to Active_data Worksheet | Seal bottom of profile | No infiltration | No evapotranspiration

Fig. 26.4 *Soils* dialog box with 4 layers plus evaporation layer parameters

$$EC_{final} = \frac{d_{seepage\text{-}4}\ EC_4 + EC_{initial}\ \Delta z\ \theta_{initial}}{i - ET_{layer} + \Delta z\ \theta_{initial}}$$

$$EC_{final} = \frac{(0.058\ \mathrm{m})\ (1.71\ \mathrm{dS/m}) + 2.13{*}0.5\ (0.188)}{0.058\ m - 10\ mm\ (0.3)/(1000\ mm/m) + 0.5\ (0.188)}$$
$$= 2.01\ dS/m.$$

The same calculations are made in the *WINDS* model. The process of setting up the *WINDS* model begins by defining the soil parameters. The *Soil_data* worksheet has typical soil parameters for different soil textures. A button on the *Soil_data* worksheet triggers the *Soils* dialog box (Fig. 26.4). In this case, field capacity, permanent wilting point, and saturated water content are similar to sandy loam so the sandy loam soil is selected and then slightly modified to have the parameters in this example. Only the upper four layers and the evaporation layer are defined. Only the left five parameters are used in this example. The right six parameters will be required in later chapters for water movement in response to energy gradients. In order to ignore the evaporation layer, the evaporation layer thickness is set at 1.0 cm (small).

Click the *Write data to Active_data worksheet* button. Soil parameters are stored in cells B40:B229 in the *Active_data* worksheet. The plant growth phases are defined in cells B3:B18 (Fig. 26.5). For this example, the root zone is defined as 1.0 m depth from the beginning to the end of the growing season. Kcb is also defined as 1.0 for the entire growing season because the crop coefficient is 1.0 for the entire period of the example.

Once the soils, plant, and salinity information are defined, all information in column B can be written to one or more field locations in the *Crop_data* worksheet. The target locations are entered in cells G13:G14. The information is written to locations 1 and 2 by clicking the *Copy data to Crop_data* button.

The fraction of evapotranspiration removed from each layer is defined in the *ET_fractions* worksheet (Fig. 26.6). In this case, there are four layers (in the root zone) during the

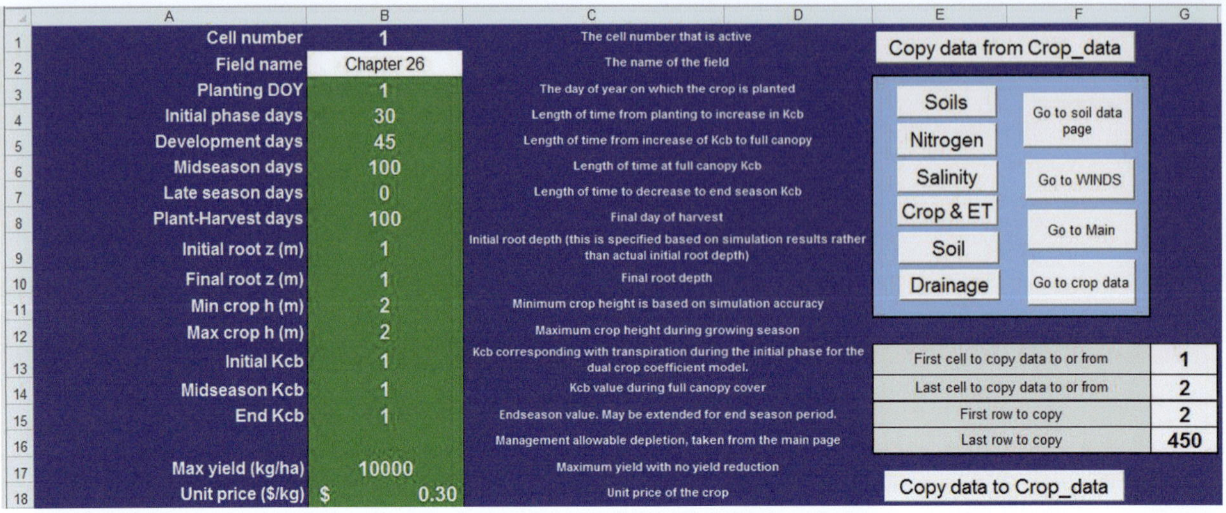

Fig. 26.5 *Active data* worksheet plant information

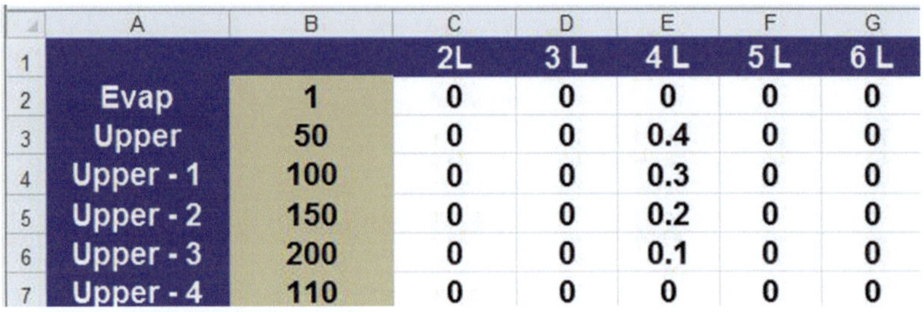

		2L	3 L	4 L	5 L	6 L
Evap	1	0	0	0	0	0
Upper	50	0	0	0.4	0	0
Upper - 1	100	0	0	0.3	0	0
Upper - 2	150	0	0	0.2	0	0
Upper - 3	200	0	0	0.1	0	0
Upper - 4	110	0	0	0	0	0

Fig. 26.6 ET_fractions worksheet with fractions defined for a 4 layer soil

entire simulation so only the 4 L column needs to be defined. No water is removed from the evaporation layer.

The reference evapotranspiration is set at 10 mm/day for all days in column D in the *Active_year_weather* worksheet (Fig. 26.7). All other columns are ignored in this simulation.

The irrigation rate is defined in the *IRR_01* worksheet (Fig. 26.8), which defines irrigation days and depths for each section of a field. In this case, the reference irrigation depth is 70 mm every 7 days. The field location Sec_1 (location 1), which has no leaching fraction, has the same application depth as the reference depth. Thus, the ratio entered in column C is 1.0. However, section 2 (Column D) has a value of 1.176 to represent the 15 % leaching fraction, as calculated in Example 4.6. It is easy to imagine that this approach can be used to specify many infiltration depths in an agricultural field.

The simulations are run from the *Main* worksheet (Fig. 26.9). Field locations in the simulation are defined in cells E2 and G2 (field positions 1 and 2). Dates are defined in row 3. Cell C5 is set to FALSE in order to ignore rainfall. Cell E5 is set to TRUE in order to include a salinity simulation. The irrigation strategy is defined as 1 (defined dates and depths in *IRR_01* worksheet) in the combobox in cell A9. The crop coefficient calculation method is defined as 5 (crop coefficient always equals 1.0 with no evaporation from the evaporation layer) in the I18 combobox. Cell I37 is defined as TRUE in order to prevent ET_fraction adjustment based on water content differences between layers, and cell I39 is defined as TRUE in order to prevent water movement between layers between irrigation events based on energy differences (Richards equation).

Once all parameters are set, the Run button is clicked. The *Get data* combobox located at cell L5 (Fig. 26.9) retrieves the simulation data from the *C01* worksheet (Section 1) and places it in the *Water content, Nitrogen* and *Salinity* worksheets, where the data can be viewed in predefined graphs. Click the *View water content* button in order to go to the *Water content* worksheet. There are many graphs available in the *Salinity, Nitrogen* and *Water content* worksheets. A dialog box (Fig. 26.10) allows the user to select between graphs. The Water content graph is selected. The dialog box also allows the user to specify the axis limits and to select which layers are displayed. The water content in the upper four layers are shown in Fig. 26.10.

			Chap 26	simulation	1		
DOY	RHmin	WS	ET0 (mm)	Rain (mm)	Tmax ©	Tmin ©	Rain time (hr)
1	28.9	1	10	0	18.8	2.5	1
2	50	1.6	10	3.3	17	8.6	1
3	47.2	2	10	10.16	19.8	10	1
4	52.1	2.2	10	1.02	15.2	6.8	1
5	39.3	1	10	0	12.5	1.5	1
6	31	1.5	10	0	14.4	-1.6	1
7	14.4	1.3	10	0	19.2	0.2	1
8	23.9	1.4	10	0	21.6	4.6	1
9	29.1	1	10	0	21.4	2.6	1
10	19.6	0.45	10	0	22.3	4.8	1
11	15.3	2.2	10	0	21.5	10.4	1
12	38.3	3.3	10	0	11.7	3.2	1
13	21.5	1.1	10	0	16	-2.3	1
14	13.2	1	10	0	20.4	3.1	1
15	11.1	1	10	0	21.7	-1.9	1
16	9.6	0.9	10	0	25.6	-1.2	1

1.00

Go to WINDS Go to Main

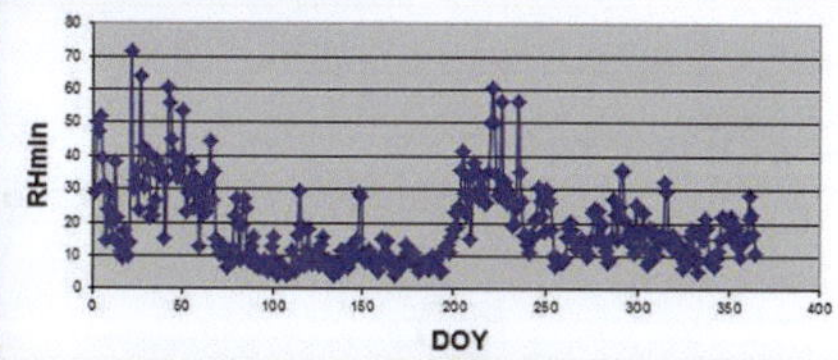

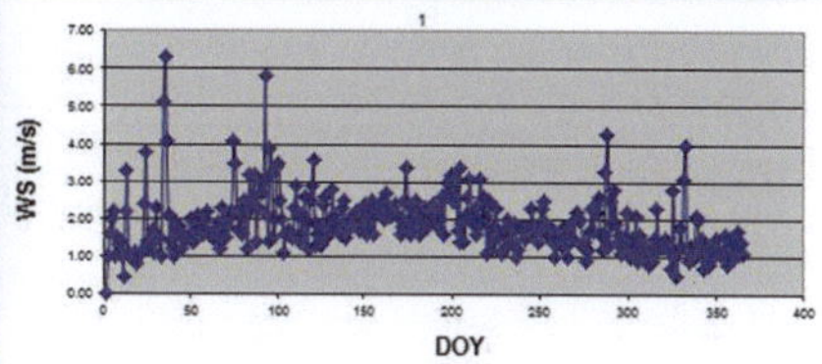

Fig. 26.7 Active year weather worksheet with 10 mm/d reference ET specified in column D

Fig. 26.8 The *IRR_01* worksheet with irrigation dates and depths defined

DOY	Ref (mm)	Sec_1	Sec_2	Sec_3	Sec_4
1					
2					
3	70	1	1.15		
4					
5					
6					
7					
8					
9					
10	70	1	1.15		
11					
12					
13					
14					
15					
16					
17	70	1	1.15		

The water contents vs. time can be found in columns AB:AF in the *Water content* worksheet. The calculated water contents are the same as the manually calculated values in this example in layers 3 and 4 for the first 3 days. The evaporation layer water content (column AF) remains at field capacity because the evaporation and evapotranspiration rates were set to zero for the evaporation layer. Thus, irrigation water just passes through the evaporation layer. Because the initial soil water salinity in the evaporation layer was set to be the same as the irrigation water salinity, the evaporation layer has no effect on the solution in this problem. The 15 % leaching fraction simulation (worksheet *C02*) has the same pattern of water content vs. time as the no leaching alternative (Fig. 26.10 and worksheet *C01*). The reason for this is that the same evapotranspiration pattern takes place and the soil is filled back to field capacity during irrigation events in both cases. However, there is more seepage water with the 15 % leaching fraction.

The salinity parameters can be specified in the *Salinity* dialog box accessed from the Salinity button in the *Active_data* worksheet (Fig. 26.11). The salinity data is written to cells B250:B288 in the *Active_data* worksheet. The initial salinity is set at 1.033 dS/m saturated paste extract salinity for this example. Because saturation is

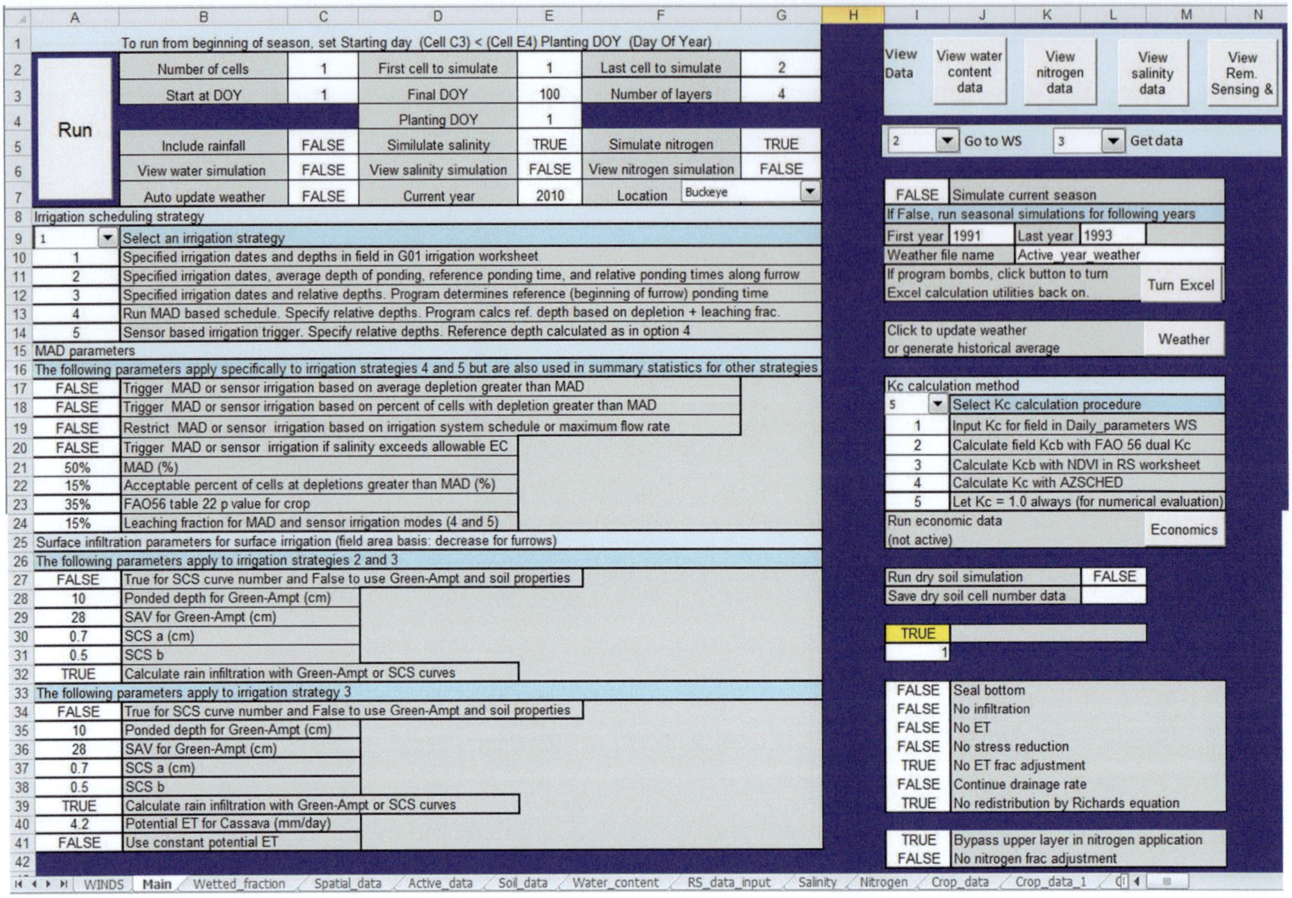

Fig. 26.9 The *Main* worksheet set up for Chapter 26 simulations

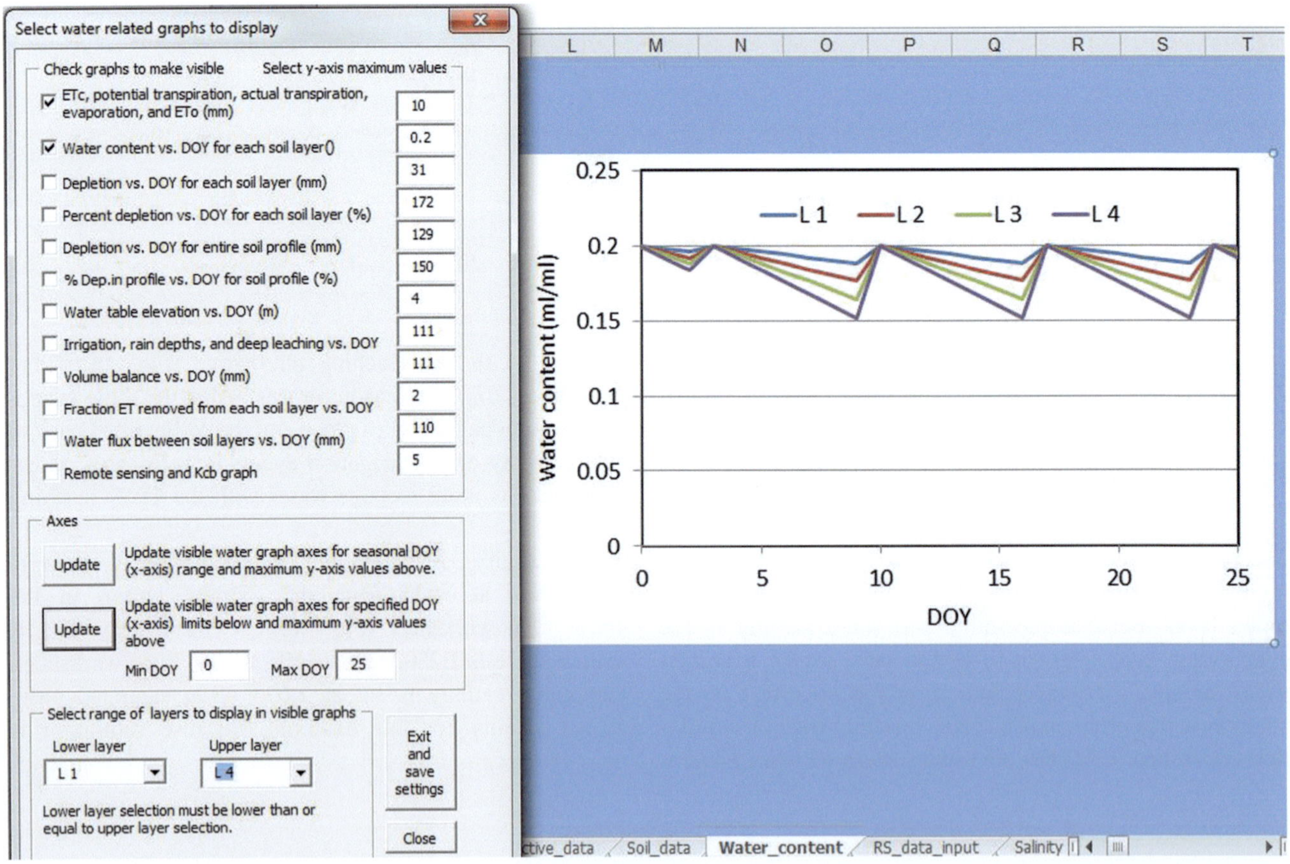

Fig. 26.10 Water Content worksheet with dialog box and water content graph

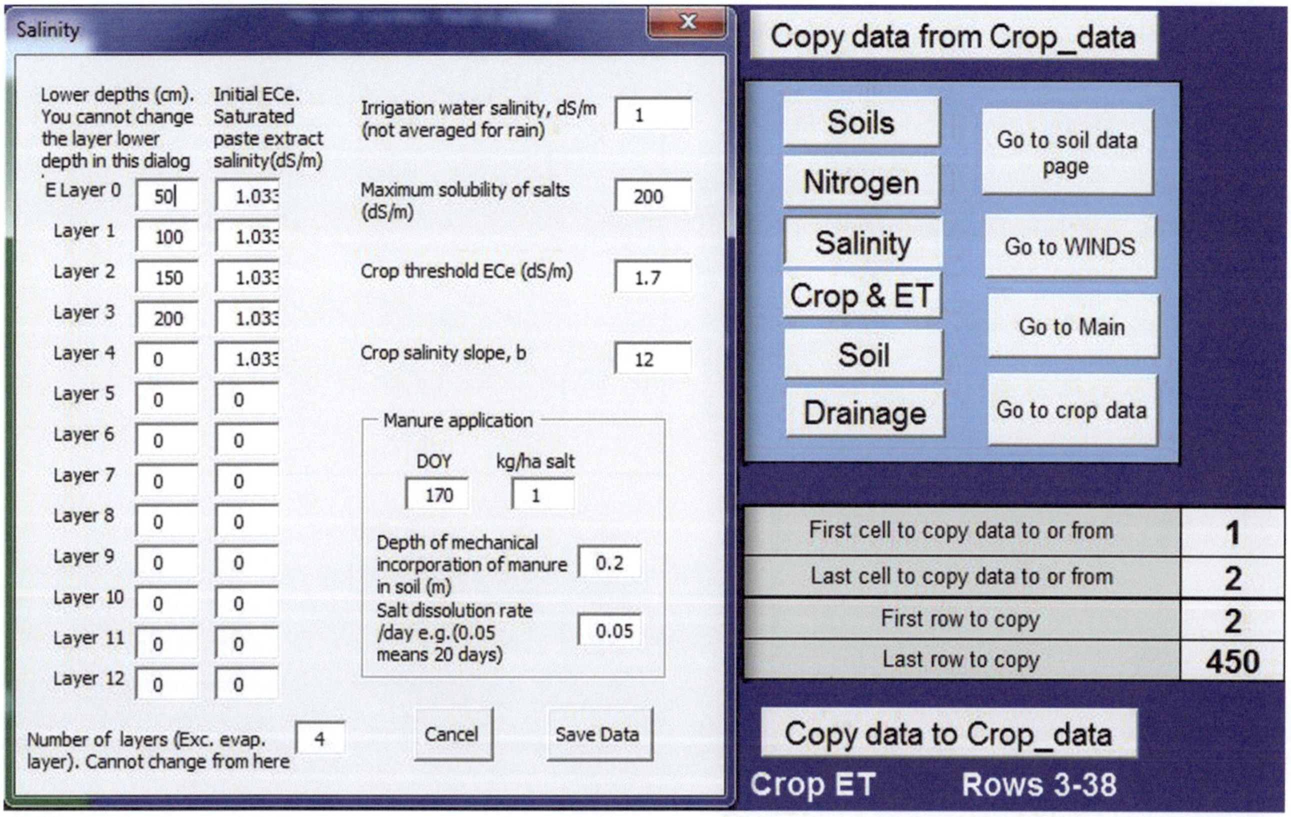

Fig. 26.11 Salinity dialog box with parameters for Example 26.5

38.7 % and field capacity is 20 %, this sets the initial soil water salinity at 2.0 dS/m.

Click the *Go to Main* button and then click the *View salinity* button in the *Main* worksheet in order to go to the *Salinity* worksheet. The Salinity worksheet also has a dialog box that allows the user to select between graphs (Fig. 26.12). The salinities after 100 days are 2-, 3.5-, 6.5-, and 7-dS/m in layers 1 to 4, respectively. The salinity increases during each cycle as the water content decreases. This is the result of the fact that the water is extracted but the salinity remains in the soil.

This example illustrates the fact that soil salinity can increase significantly within 100 days. In order to look at the 15 % leaching fraction scenario, go to the Main worksheet and click 2 in the *Get_data* combobox. The calculated salinity values are the same as the manual calculations in the first part of this example. The salinity curves for 15 % leaching fraction are shown in Fig. 26.13.

The steady state salinity (Example 4.6) values with 15 % leaching fraction (Example 26.4) were 1.5-, 2.5-, 4.3-, and 6.7-dS/m. The upper 3 layers have reached these equilibrium concentrations after 100 days while the lower layer (layer 1) is still approaching the final steady state concentration.

One of the most important checks for accuracy in simulation models is the mass balance. Even if some processes are modeled with some incorrect assumptions, a validated mass balance is the first step towards a reasonable solution; on the other hand, letting mass vanish into thin air guarantees an inaccurate solution. The mass balance verification in the *WINDS* model is based on the fact that the sum of sources and sinks should equal the change in storage. The volume balance for water is found in the *Water content* worksheet and can be accessed with the dialog box in the *Water content* worksheet. Figure 26.14 shows that the sum of sources and sinks equals the change in storage so the volume balance is verified. The reason that the sum of sources and sinks is less than 70 on irrigation days is that some water was lost to evapotranspiration.

The salinity mass balance also shows that salts are conserved in the model (Fig. 26.15). The salt balance is negative on the third day because irrigation was conducted before it was required; thus, leaching greatly exceeded 15 %.

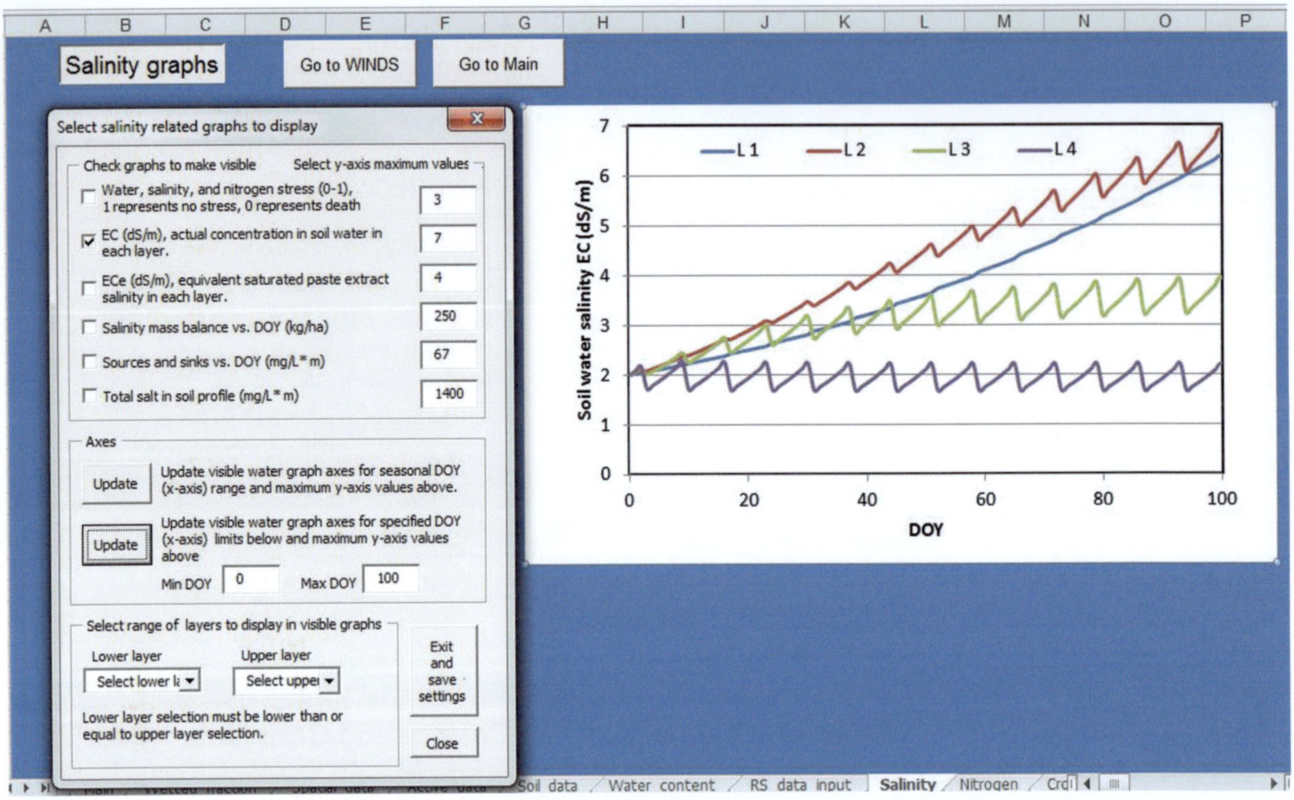

Fig. 26.12 *WINDS* model simulation of salinity for 100 days with no leaching in Example 26.5

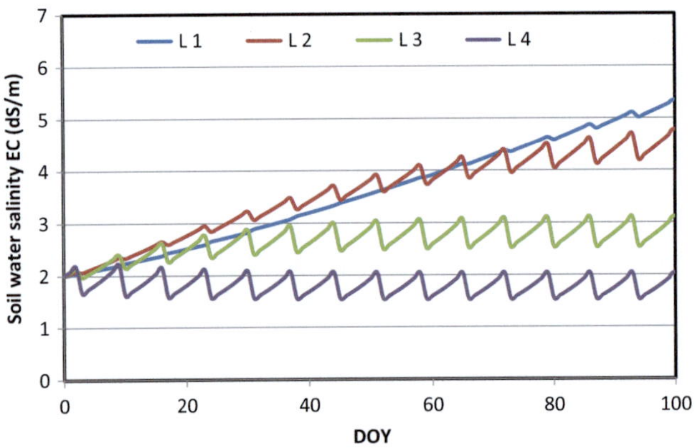

Fig. 26.13 *WINDS* model simulation of Example 26.5 salinity for 100 days with 15 % LF in Example 26.5

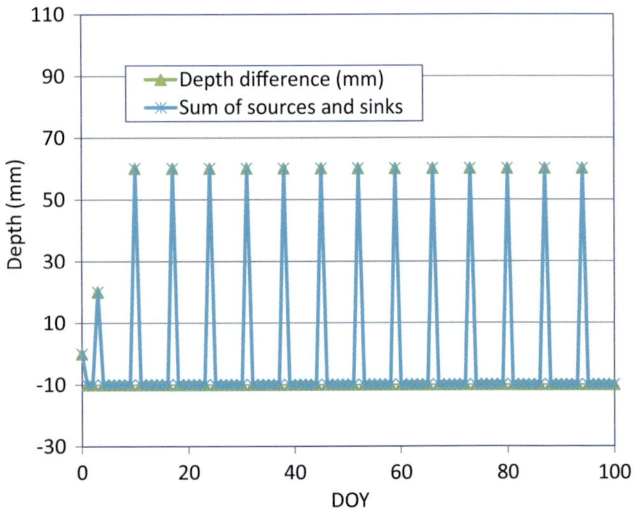

Fig. 26.14 Mass balance for water content in the *WINDS* model for Example 26.5

Nitrate Mass Balance Model

Nitrate has a negative charge; thus, we can assume that nitrate is not absorbed by soil particles and is in the soil water solution. The following derivation of the soil nitrate model is similar to the soil salinity model; however, there are reactions terms.

The sources, sinks, and flux of nitrate in a layer are shown in Fig. 26.16. The nitrate mass is typically reported as only the mass of elemental nitrogen rather than the entire nitrate molecule. This is commonly called nitrate-N (NO_3-N) and is represented by the letter N in this chapter.

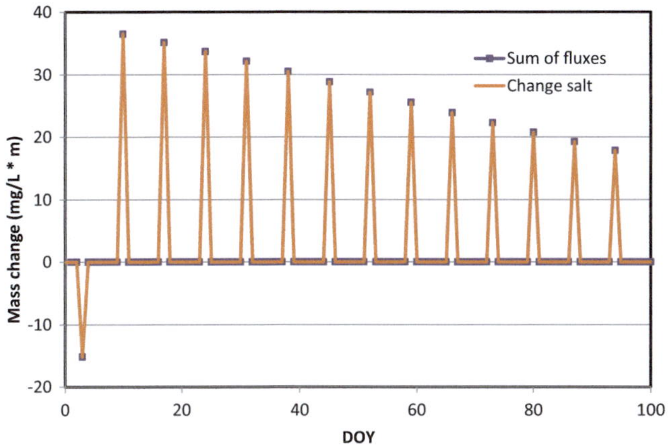

Fig. 26.15 Mass balance for salinity in the *WINDS* model for Example 26.5

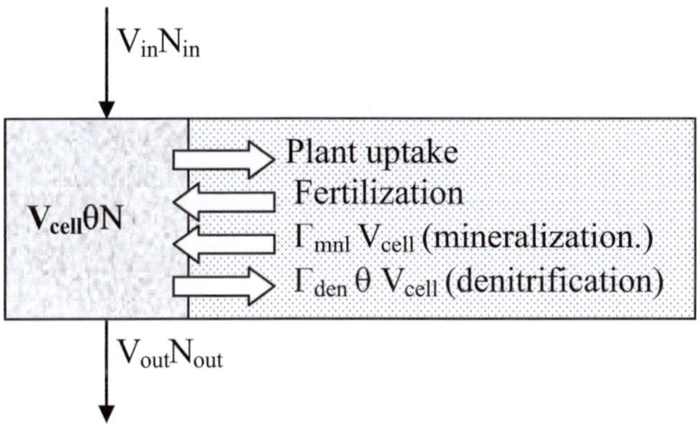

Fig. 26.16 Nitrate (NO_3-N) mass balance in root zone

As with salinity, the mass of N in the water in a layer can be calculated as mg/L m.

$$m_{N-layer} = \Delta z N\theta \tag{26.29}$$

where

$m_{N\text{-layer}}$ = mass of NO_3-N within the layer, mg/L m.

The change of N mass within the control volume is

$$\begin{aligned} dm_{N-layer} &= \Delta(N\ \Delta z\ \theta) \\ &= N_{final}\Delta z\ \theta_{final} - N_{initial}\ \Delta z\ \theta_{initial} \end{aligned} \tag{26.30}$$

where

$dm_{N\text{-layer}}$ = change in mass of NO_3-N, mg/L_{soil} m.
$N_{initial}$ = initial NO_3-N concentration, mg/L_{water},
N_{final} = final NO_3-N concentration, mg/L_{water},

The nitrate flux terms are the product of water flux and the concentration.

$$m_{in} = d_{in}N_{in} \quad mg/Lm \tag{26.31}$$

$$m_{out} = d_{out}N_{initial} \quad mg/Lm \tag{26.32}$$

where

N_{in} = NO_3-N concentration in water entering the layer, mg/L_{water}
$N_{initial}$ = Initial NO_3-N concentration in layer, mg/L_{water},

The reaction terms include mineralization, denitrification, plant uptake, and fertilization. The change in nitrate mass due to mineralization is the product of the mineralization rate (Γ_{mnl} from Eq. 25.3) per volume of soil, time, and the depth of the layer.

$$\begin{aligned} m_{min} &= \Delta z\ \Gamma_{min}\ \Delta t = (m)(mg/L_{soil}/d)(d) \\ &= mg/Lm \end{aligned} \tag{26.33}$$

where

Γ_{min} = rate of reaction, mg/L_{soil}/d.
m_{min} = mass added to the layer by mineralization, mg/L_{soil} m
Δt = time step, days.

The change in nitrate mass due to denitrification is calculated with Γ_{den} from Eq. 25.15.

$$m_{den} = \Delta z\ \Gamma_{den}\Delta t \tag{26.34}$$

The change in nitrate mass due to fertilization is calculated by dividing the kg/ha application rate by 10.

$$\begin{aligned} m_{fer} &= \frac{App\ rate(kg/ha)}{10} \\ &\Rightarrow \left(\frac{kg}{ha}\right)\left(\frac{10^6\ mg}{kg}\right)\left(\frac{ha}{10^4\ m^2}\right)\left(\frac{m^3}{10^3\ L}\right) \\ &= mg/L_{soil}\ m \end{aligned} \tag{26.35}$$

The change in nitrate mass due to plant uptake is calculated by dividing the kg/ha plant nitrate uptake by 10.

$$m_{upt} = \frac{N_{upt}(kg/ha)}{10} \Rightarrow mg/L_{soil}\ m \tag{26.36}$$

where
m_{upt} = mass of nutrients taken up by crop, mg/L m,
N_{upt} = Nitrate removed by the crop, kg/ha.
The continuity equation for mass of NO_3-N within a layer is

$$\begin{aligned} dm_{N-layer} &= d_{in}N_{in} - d_{out}N_{initial} + m_{min} - m_{den} \\ &+ m_{fer} - m_{upt} \end{aligned} \tag{26.37}$$

Substitute Eq. 26.37 into Eq. 26.30

$$N_{final}\ \Delta z\ \theta_{final} - N_{initial}\Delta z\ \theta_{initial} = d_{in}N_{in} - d_{out}N_{initial} + m_{min} - m_{den} + m_{fer} - m_{u\,pt} \tag{26.38}$$

Equation 26.38 is rearranged to solve for final nitrogen content within the soil profile.

$$N_{final} = \frac{d_{in}\ N_{in} - d_{out}\ N_{initial} + m_{min} - m_{den} + m_{fer} - m_{u\,pt} + N_{initial}\ \Delta z\ \theta_{initial}}{\Delta z\ \theta_{final}} \tag{26.39}$$

As with the salinity model, if infiltration into the layer from above exceeds the available water storage capacity in the layer (the product of field capacity and layer depth), then Eq. 26.39 is not valid. In this case, the weighted average concentration of the layer water and the infiltrating water from the above layer leaches to the next layer. Thus, the final NO_3-N concentration as well as the leachate NO_3-N concentration would be

$$N_{final} = (i\ N_{iw} + N_{initial}\ \theta_{initial}\Delta z + m_{min} - m_{den} + m_{fer} - m_{u\,pt})/(i + \theta_{initial}\Delta z - ET - E). \tag{26.40}$$

Example 26.5 For the parameters in Example 26.4 (no leaching fraction), make sample calculations of nitrate concentration during the first 3 days for the upper layer (below the 1 cm evaporation layer). Use the following nitrogen model parameters. Then use *WINDS* to calculate nitrate concentration for the first 100 days of the year. Apply 20 kg/ha fertilizer on DOY 1, to be incorporated into the upper layer. Irrigate on DOY 3, and then irrigate every 7 days thereafter. Ignore the 1 cm evaporation layer. Plant uptake of nitrogen is 50 kg/ha, with no uptake during the first 30 days and uniform uptake during the last 70 days of the growing seasons. The *WINDS* model parameters are set as shown in Fig. 26.17.

Parameter	Value
K_{mnl}	0.0002
θlow	0.15
θhigh	0.22
Q_{temp}	3
On-max	3000
α	0.025
K_{den}	0.002
Frac NO3 uptake	0.75
Fertilizer dissolution rate	1.0/day
Nitrate conc. in irrigation	10 mg/L
tm	100 days
Tbar	21.0 C
A0	17.1 C
Nitrogen requirement	50 kg/ha (no uptake until development phase is over)

(continued)

Initial nitrate concentrations (mg/kg soil) are

Evap layer	1.2
Layer 4	35
Layer	30
Layer 2	25
Layer 1	20

Day 1: Upper layer (layer 4)

Calculate initial nitrate concentration in layer 1 by converting from mg/kg soil to mg/L water. The initial water content is 20 %

$$N = N\_soil/\theta(1 - \theta_{SAT})(2.65) = 35mg/kg/0.2(1 - 0.387)(2.65) = 284mg/L$$

Calculate average soil profile temperature on day 1.

$$T(1) = 21.1 + 17.1 \sin\left(\frac{2\pi}{365}(t - 100)\right) = 4.15\ C$$

Calculate NO_3-N mineralization in layer 1
$\theta = 0.20$, $\theta_{low} = 0.15$, thus $f_{mn\theta} = 1.0$

$$f_{temp} = Q_{10}^{\left(\frac{T - t_b}{10}\right)} = 3^{\left(\frac{4.15 - 20}{10}\right)} = 0.175$$

$$\rho = \rho_p(1 - \phi)\ = 2.65(1 - 0.387) = 1.62$$

$$O_n = O_{n_{max}} e^{-0.025z} = 3{,}000\ e^{-0.025\ *25} = 1{,}606\ \mu g/g$$

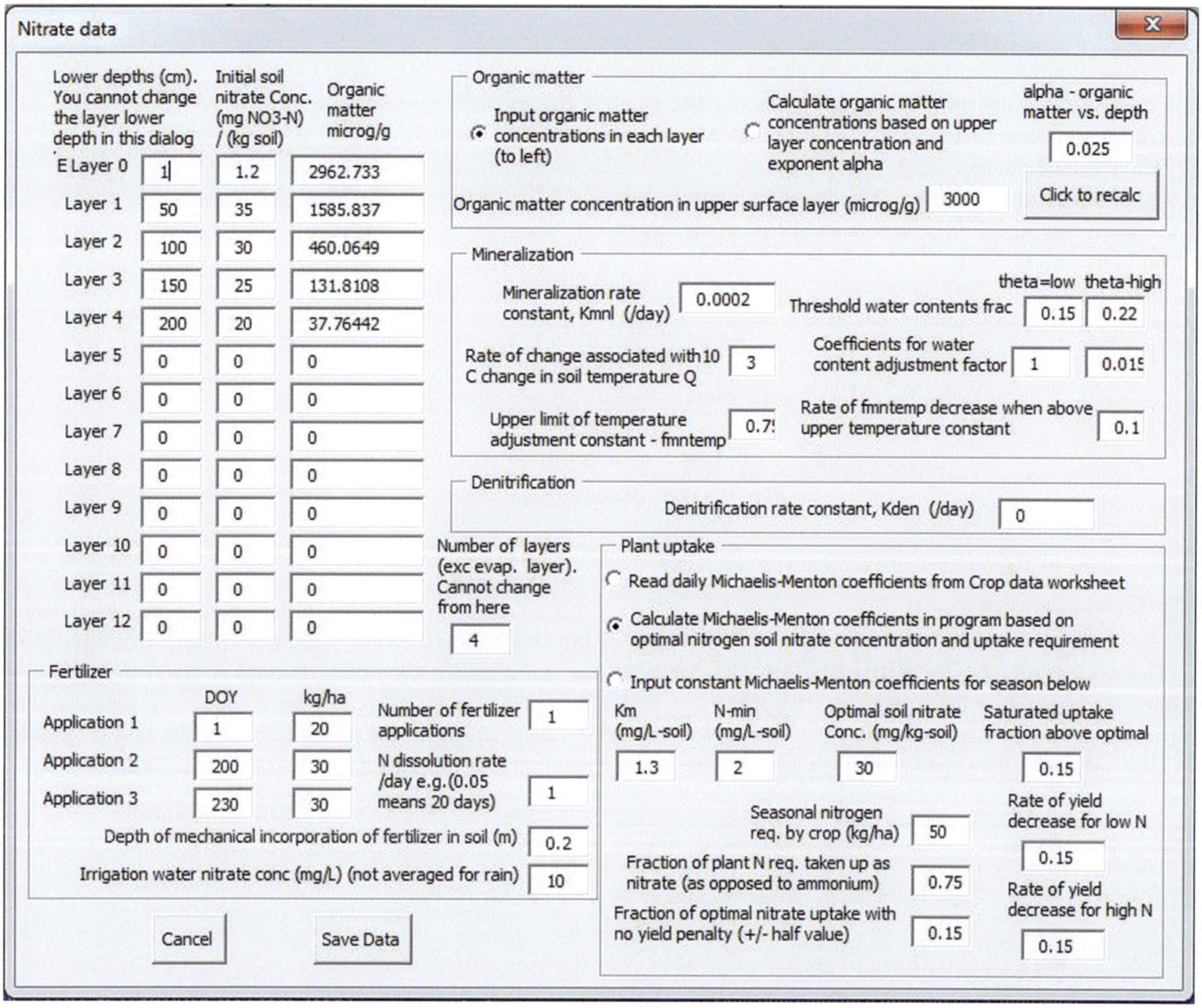

Fig. 26.17 Nitrogen specifications in *WINDS* model for Example 26.5

$$\begin{aligned}\Gamma_{mnl} &= K_{mnl}\, f_{mnl\theta}\, f_{temp}\, \rho O_n \\ &= 0.0002(1.0)(0.175)(1.62)(1,606) \\ &= 0.09106\ \text{mg}/\text{L}_{soil}/\text{day}\end{aligned}$$

$$\begin{aligned}m_{mnl} &= \Gamma_{mnl} \Delta z\ \Delta t = 0.9106(0.5\ \text{m})(1\ \text{day}) \\ &= 0.0455\ \text{mg}/\text{L}_{soil}\ \text{m}\end{aligned}$$

Calculate denitrification

$\theta = 0.20$ and $\theta_{sat} = 0.387$. Because no denitrification takes places is less than 60 % of pore space is filled with water, denitrification does not take place

Calculate fertilizer addition. The dissolution rate is 1.0. This means that all fertilizer will be added to the soil on the day after fertilization in the *WINDS* model.

There is no plant nitrogen uptake at the beginning of the season.

The water content decreases to 0.192

Calculate new concentration

$$N_{final} = \frac{d_{in}\ N_{in} - d_{out}\ N_{initial} + m_{min} - m_{den} + m_{fer} - m_{upt} + N_{initial}\ \Delta z\ \theta_{initial}}{\Delta z\ \theta_{final}}$$

$$N_{final} = \frac{(0)N_{in} - (0)N_{initial} + 0.0455 - 0 + 0 - 0 + 284\ (0.49)\ (0.2)}{0.49\ (0.192)} = 296\ mg/L$$

Day 2: Upper layer (layer 4)

Temperature remains the same; thus, NO_3-N mineralization remains the same since it is a zero order equation (not dependent on nitrate concentration), $m_{mnl} = 0.0455$

Denitrification remains zero.

Calculate fertilizer addition. The dissolution rate is 1.0. This means that all fertilizer applied on the previous day is added to the soil this day.

$$m_{fer} = \frac{App\ rate(kg/ha)}{10} = \frac{20\ kg/ha}{10} = 2\ mg/L_{soil}\ m$$

Calculate new concentration

$$N_{final} = \frac{d_{in}\ N_{in} - d_{out}\ N_{initial} + m_{min} - m_{den} + m_{fer} - m_{upt} + N_{initial}\ \Delta z\ \theta_{initial}}{\Delta z\ \theta_{final}}$$

$$N_{final} = \frac{(0)N_{in} - (0)N_{initial} + 0.0455 - 0 + 2 - 0 + 296\ (0.49)\ (0.192)}{0.49\ (0.184)} = 331\ mg/L$$

Day 3: Upper layer (layer 4)

This day has a 0.07 m depth irrigation. Most water passes through since initial water content is 0.184, and 0.2 is field capacity; thus, the soil can only hold 0.016*0.5 m = 0.008 m → 0.07 – 0.008 = 0.062 passes through layer 1. The nitrate concentration in irrigation water (and in the evaporation layer) is 10 mg/L. When water from a large irrigation event or storm passes completely through a layer, the following mass balance equation can be modified as follows. This equation calculates the mixed concentration, and is the concentration of nitrogen in the water that seeps to the next layer.

$$N_{final} = \frac{d_{in}\ N_{in} + m_{min} - m_{den} + m_{fer} - m_{upt} + N_{initial}\ \Delta z\ \theta_{initial}}{d_{in} - ET_{layer} + \Delta z\ \theta i_{initial}}$$

$$N_{final} - \frac{(0.07\ m)(10\ mg/L) + 0.0455 - 0 + 0 - 0 + 331\ (0.49)\ (0.184)}{0.07m - 10\ mm^{*}0.4/(1000\ mm/m) + 0.49\ m\ (0.184)} = 196\ mg/L$$

The 196 mg/L in layer 4 is the same concentration reported in the *WINDS* model Nitrogen page in cell AE3. The *WINDS* model nitrogen page also has many graphs and a dialog box to select from them. The mass balance graph shows that nitrate is conserved. The fertilizer addition on day 2 resulted in a gain of total nitrate, and the drainage event on day 4 resulted in a loss. The mineralization increased later in the season when the temperature increased. The irrigation events added nitrate every seven days (Fig. 26.18). A closer look shows that mineralization increased during the season, but that plant uptake of nitrogen began at DOY 30 and decreased the overall increase in nitrogen per day (Fig. 26.19).

The leaching of nitrate to the lower layers over time (Fig. 26.20) is similar to the leaching of salinity over time (Fig. 26.12).

Although the total average nitrogen in the soil profile increases, much of it is leached below the upper layers.

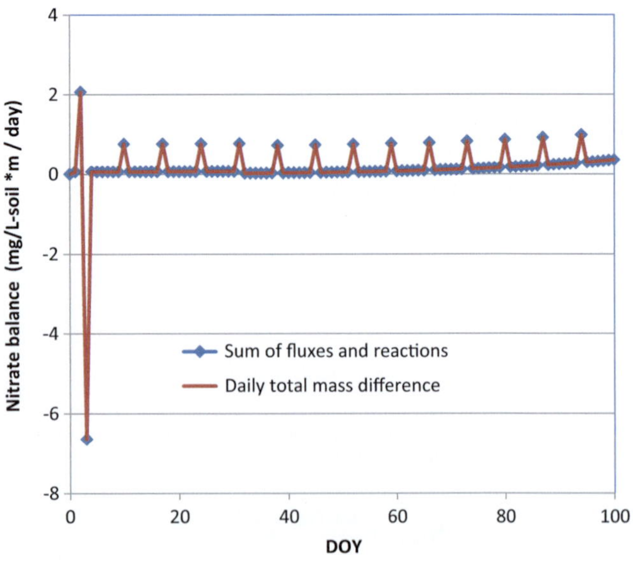

Fig. 26.18 Nitrate mass balance for Example 26.6

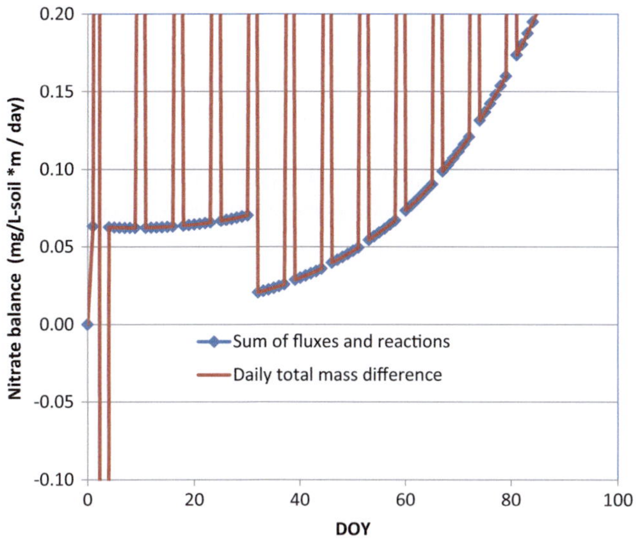

Fig. 26.19 Nitrate mass balance for Example 26.6

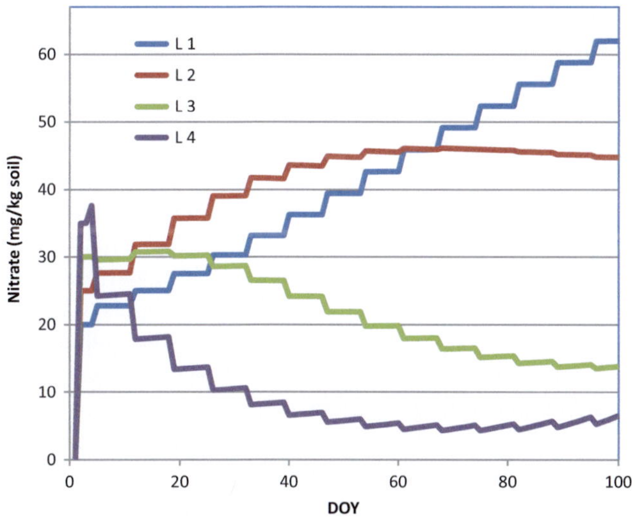

Fig. 26.20 Nitrate concentration in layers for Example 26.6

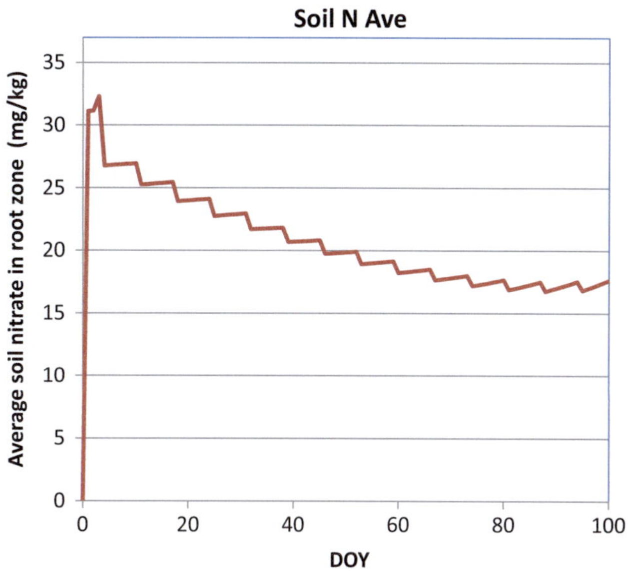

Fig. 26.21 Average weighted nitrate concentration versus time with weights based on fraction of water extraction (root density) in each layer, for Example 26.6

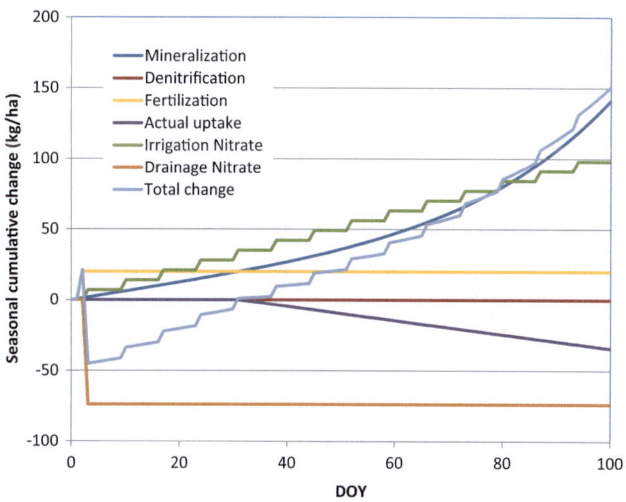

Fig. 26.22 Cumulative change versus time for all nitrate sources and sinks in Example 26.6

Thus, the weighted average nitrate concentration decreases if the weights are based on the fraction of water extraction in each layer (Fig. 26.21). Most of the nitrate in the soil profile has been leached to the lower layers (Fig. 26.20).

The cumulative contribution to the nitrate mass balance of all sources and sinks is shown in Fig. 26.22. The cumulative change due to plant uptake is only 34 kg/ha while the plant nitrogen requirement was 50 kg/ha. This was caused by the fact that only 75 % of the plant nitrogen uptake was nitrate (37.5 kg/ha) and by the uptake being penalized by low nitrate concentration in the upper layers by the Michaelis-Menton equation. The optimal concentration was specified as 30 mg/kg, but the weighted average nitrate concentration dropped below 20 mg/kg; thus, the uptake was reduced. Of course, these types of numbers would need to be generated by experiment for a particular crop. By far, the biggest loss was due to the irrigation event on the third day, when the soil water content was higher than it normally would be just before an irrigation event. This demonstrates the environmental hazard associated with overapplication or mistiming of irrigation events and nitrogen fertilization.

Questions

1. Repeat Example 26.1, but change the infiltration from the first storm to 3 cm, and the field capacity to 0.25. As before, infiltration from the second storm is 4 cm.
2. Redo question 1, but divide the soil into three layers of 0.4 m depth.
3. Redo Example 26.3, but change the upper layer FC is 0.26, and the lower layer FC to 0.24. Also, 70 % of ET is removed from the upper layer and 30 % from the lower layer.
4. Redo Example 26.4 with the WINDS model and by hand, but lower the leaching fraction to 0.05. Make calculations for the irrigation on the third day for the upper two layers by hand. Next, use the WINDS model to calculate EC for 100 days. There are only two field sections in the *WINDS Chapter 26* workbook. The sections are organized with respect to their irrigation zones in the *spatial data* worksheet. Add another G01 section in column C and write "3" in the same row in column. In cell K7, specify that the number of cells is 3 and click the *Make new sections* button. This process adds the C_3 worksheet to the end of the workbook. The next step is to populate the date in the *Crop_data* worksheet for section 3. You can do this in the A*ctive Data* worksheet or just copy the cells from section 2 (column C) to section 3 (column D) in the *Crop_data* worksheet. If you use the *Active_data* worksheet, then the copy the information from section 2, "Copy data from crop data," and then copy rows 3 to 450 to section 3 (specified in cells G13:G16) and click the "Copy data to crop data" button. After calculating the required application depth for 0.05 leaching fraction, add the calculated fraction of baseline irrigation to the section 3 column in the G01 worksheet. Go to the Main worksheet. In cell G2, specify that three sections will be evaluated. After clicking Run, select position 3 in the Get Data combo box (upper right side of the worksheet). Find the "Water content" graph and the "Irrigation, rain depth, and leaching" graph with the Selection form. If rainfall appears in the graph, remove the rainfall from the *Active year weather* page for the first 100 days. Find the soil water salinity graph in the *Salinity* worksheet. Compare to the salinity levels in Example 26.4. Copy and paste the worksheets or graphs into this document. Use the graphs to assess the processes.
5. Calculate leaching fraction for irrigation water salinity 2 dS/m and required EC_e 1.5 dS/m?
6. What are the ratios EC_e/EC_{ave}, and EC_e/ EC_{dw} in Example 26.4? EC_{dw} is the leachate salinity. Discuss the importance of understanding these ratios with respect to crop management decisions?
7. Redo Example 26.5, but change the fertilizer application to 40 kg/ha on the first day application and change nitrate concentration in the irrigation water to 20 mg/L. Make a new hand calculation of the changes due to fertilizer application and irrigation during the first 3 days in the upper cell. Run the WINDS simulation for 100 days with the higher irrigation water nitrate concentration and higher fertilization rate on day 3. The irrigation rate will be the same as question 4. You can change the nitrogen data in the Active_data worksheet and copy it to section 3 in the *Crop_data* worksheet. Make sure that cell G5 in *Main* worksheet is marked True. Run the simulation from the *Main* worksheet. Select 3 in the *Get_data* combobox in the Main worksheet. Click the *View Nitrogen* data button. Copy the following graphs into your homework document: *Nitrate* (mg/kg) in layers, *Irrigation and drainage nitrate* (you might need to update both x and y axes from the selection form or from the axes), *Reactions*, and *Cumulative leaching, nitrate and reactions*. Assess the processes by looking at the graphs.
8. A soil has three 0.4 m layers, numbered 1, 2, and 3 from the bottom, with field capacity in all layers equal to 0.25. The initial water salinity in layers 1, 2, and 3 is, 23-, 7-, and 5-dS/m, respectively. ET is 10 mm/day with 20 %, 30 %, and 50 % of ET in layers 1, 2, and 3 respectively. Irrigation water salinity is 2 dS/m. The initial water content on the previous day in layers 1, 2, and 3 is 0.18, 0.15, and 0.10, respectively. Soil porosity is 0.4. An irrigation event adds 11 cm water to the soil in the morning. Compare to the final water content, actual salinity, and saturated paste extract salinity before the morning irrigation event. Compare the changes in water salinity and saturated paste extract salinity.
9. During a one day period, the upper layer of soil, 0.4 m depth, has a mineralization rate of 0.1 mg/L * m, a denitrification rate of 0.05 mg/L*m, and plant uptake of 1 kg/ha. One cm (average for the field) depth of water is added to the layer by drip irrigation and the irrigation water has a nitrate concentration of 20 mg/L. Transpiration removes 1.4 cm from the layer. No water leaches to the next layer. The initial water content is 0.18, and the initial nitrate concentration in the soil water is 15 mg/L. Calculate the final water content and nitrate concentration in the water. Calculate the kg/ha nitrate in the layer at the end of the day.

Water and Energy Relationships in Soils

27

This chapter begins with the derivation of the conservation of energy equation. Water is driven through the soil by energy gradients while friction between water molecules in small pores dissipates energy and restricts flow. Hydraulic conductivity is the ratio of water velocity to energy dissipation, and Darcy's Law calculates velocity of water as the product of hydraulic conductivity and the energy gradient. Soil energy diagrams include gravitational potential energy and matric potential and can help to determine energy gradients. The chapter shows that vertical movement through the soil profile, horizontal flow through subsurface wetlands, or flow in any direction through any media, is calculated with the same energy gradient principle. Hydraulic conductivity decreases in unsaturated soils. Equations that describe the relationships between unsaturated hydraulic conductivity, matric potential, and water content have been developed by Brooks-Corey and by van Genuchten.

Conservation of Energy Equation

The conservation of energy equation is similar to the conservation of mass equations for water, salts, and nitrogen except that the mass (Vρ) is multiplied by an intensive property, energy per unit mass, E (Joules/kg). The product, EVρ, has units of energy, (Joules/kg)(kg/m^3)(m^3) = Joules. The energy balance equation can be applied to a control volume such as a soil or pipe as shown in Fig. 27.1.

where

E_{in} = energy of inflow water, Joules/kg,
V_{in} = inflow volume, m^3,
ρ = density of water, kg/m^3
Heat = heat dissipation due to friction, Joules
V_{cv} = volume within the control volume, m^3,
V_{out} = outflow volume, m^3,
E_{out} = energy of outflow water, Joules/kg,
$V_{flux\text{-}ave}$ = Average of V_{in} and V_{out}, m^3,
E_f = energy lost to heat due to friction, Joules/kg.
ΔE_{cv} = change in energy within the control volume, Joules/kg.

The conservation of energy equation includes the same terms as the conservation of mass equation: in, out, internal reaction, and change in storage (Fig. 27.2).

Total energy includes mechanical (pressure, elevation, and kinetic), osmotic, thermal, chemical, and other forms of energy, but the derivation in this chapter only includes mechanical energy. The internal reaction terms include friction and energy added by a pump. The flux terms include energy that moves across the control volume boundary.

Engineers often refer to the energy of water in terms of head and use units of length. The symbol H is used to refer to head. This length term refers to the elevation of the water or the pressure (energy) that would be found at an equal depth of water.

Ten meters head is approximately equal to one atmosphere and is equal to 100 kPa:

$$P = z\rho g = 10\ \text{m}(1{,}000\ \text{kg/m}^3)(9.8\ \text{m/sec}^2) = 1*10^5\text{Pa} = 100\ \text{kPa} = \text{atmospheric pressure.}$$

where

z = Elevation, m
ρ = density, kg/m^3
g = acceleration due to gravity.
P = pressure, N/m^2

Length units are converted to Joules/kg by multiplying by gravity.

$$E = H*g = m\frac{m}{\sec^2} = \frac{kg}{kg}m\frac{m}{\sec^2} = \frac{kg*m/\sec^2*m}{kg} = \frac{N*m}{kg} = \frac{Joules}{kg}$$

P. Waller, M. Yitayew, *Irrigation and Drainage Engineering*, DOI 10.1007/978-3-319-05699-9_27

where

H = energy (head) in units of m of water, m.

Substitute Hg for E in the energy equation in Fig. 27.2.

$$H_{in}V_{in}\rho g - h_f V_{flux\text{-}ave}\rho g - H_{out}V_{out}\rho g = \Delta H \ V_{cv}\rho g \quad (27.1)$$

where

h_f = energy lost due to friction, m.

Divide both sides by ρg.

$$H_{in}V_{in} - h_f V_{flux-ave} - H_{out}V_{out} = \Delta H \ V_{cv} \quad (27.2)$$

The change in internal energy over time is zero for steady-state flow.

$$H_{in}V_{in} - h_f V_{flux-ave} - H_{out}V_{out} = 0 \quad (27.3)$$

$V_{in} = V_{out} = V_{flux\text{-}ave}$ for steady-state flow:

$$H_{in} - h_f - H_{out} = 0 \quad (27.4)$$

$$h_f = H_{in} - H_{out} \quad (27.5)$$

Equation 27.5, the basis of the Bernoulli equation, is based on the assumption of steady state flow.

The Bernoulli equation kinetic energy, potential energy, and pressure, and the sum of the terms is represented by H.

$$H = \frac{v^2}{2g} + \frac{P}{\rho g} + z \quad (27.6)$$

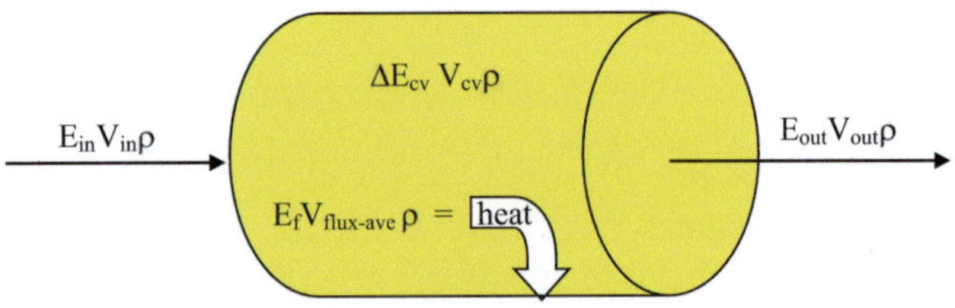

Fig. 27.1 Conservation of energy diagram

where

P = pressure, Pa,
v = velocity, m/sec,

Friction loss is the energy difference between two locations in the flow field. Substitute Eq. 27.5 into Eq. 27.4. Let $H_1 = H_{in}$ and $H_2 = H_{out}$

$$h_f = \frac{{v_1}^2}{2g} + \frac{P_1}{\rho g} + z_1 - \frac{{v_2}^2}{2g} - \frac{P_2}{\rho g} - z_2 \quad (27.7)$$

For flow in porous media (i.e., soils), the flow rate and water velocity are typically low; thus, the kinetic energy term (velocity) is negligible and it is dropped from the Bernoulli equation.

$$h_f = \frac{P_1}{\rho g} + z_1 - \frac{P_2}{\rho g} - z_2 = h_{c-1} + z_1 - h_{c-2} - z_2 \quad (27.8)$$

where

h_c = matric potential, m.

Hydraulic Conductivity

Hydraulic conductivity is the rate at which soil conducts water at a given hydraulic gradient or energy gradient. It is much higher for coarse soils (sandy) than for fine soils (Fig. 27.3).

Figure 27.4 shows the relationship between water velocity (Darcy velocity) and energy gradient for a soil with hydraulic conductivity equal to 2 m/d. Darcy's law describes the relationship between hydraulic gradient (ΔH/L) and water movement in soils (v).

$$v = -K\frac{\Delta H}{L} \quad (27.9)$$

where

L = length of the flow path, m,
v = Darcy velocity, cm/hr,
K = hydraulic conductivity of the soil material, cm/hr,

Fig. 27.2 Terms in the conservation of energy equation

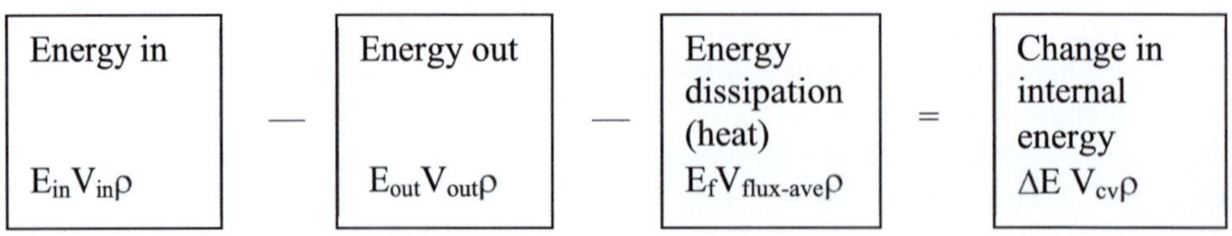

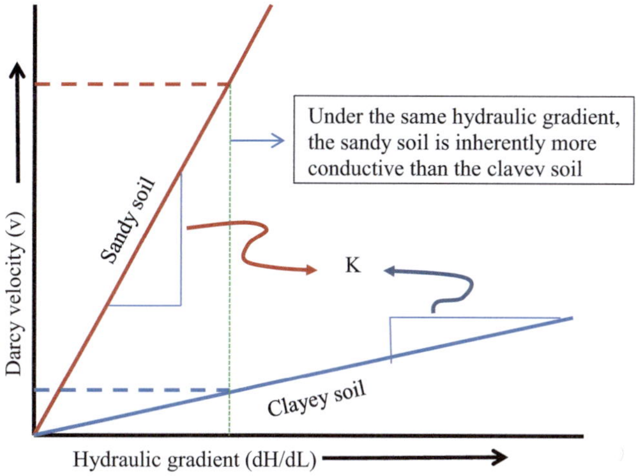

Fig. 27.3 Hydraulic conductivity: rate of flux/hydraulic gradient (After NRCS)

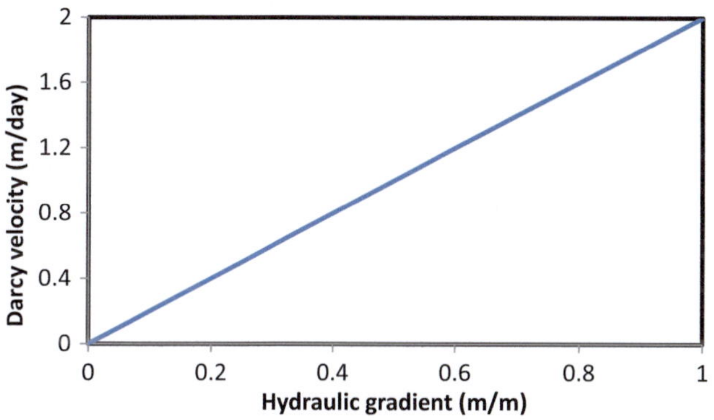

Fig. 27.4 Water velocity versus energy gradient for hydraulic conductivity equal to 2

ΔH = energy potential difference between the end points, m.

Darcy's law can be rewritten in terms of Bernoulli's equation where h_f is equal to ΔH:

$$h_f = \Delta H = \frac{vL}{K} \tag{27.10}$$

The velocity used in Darcy's equation is not the actual velocity of flow through the soil. Rather, Darcy's velocity is Q, the total volume flow rate entering the soil volume, divided by A, the cross-sectional area of flow. The actual velocity of flow in the soil is faster: Q/(A ϕ) where ϕ is soil porosity.

Example 27.1 A sand column (Fig. 27.5) has hydraulic conductivity = 5 cm/hr and a diameter of 7 cm. The water column above the sand is 0.4 m. Porosity is 0.38. Calculate the flow rate through the column, the Darcy velocity, and the average actual flow velocity in the column.

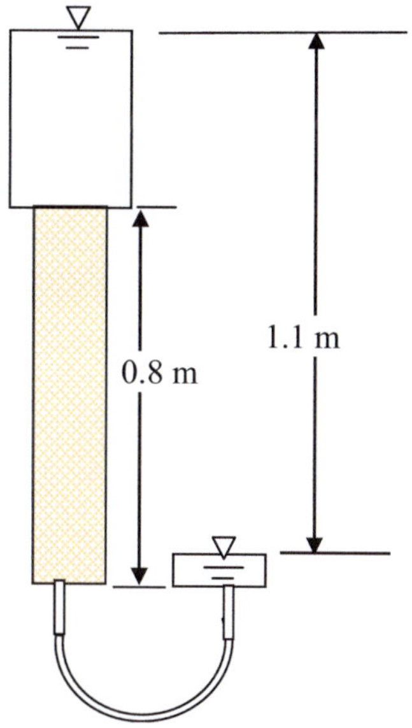

Fig. 27.5 Sand column

Define the boundaries of the control volume as the upper surfaces of the inlet and outlet tanks.

The flow distance through the sand, L, is 0.8 m.

The energy difference, ΔH, is the difference in elevations between the ponded surfaces.

$$\Delta H = (z_1 - z_2) + \frac{P_1}{\rho g} - \frac{P_2}{\rho g} = 1.1 + 0 - 0 = 1.1\ m$$

Calculate the velocity of flow with Darcy's law.

$$\mathrm{v} = k\frac{\Delta \mathrm{H}}{\mathrm{L}} = 5\ cm/hr\frac{1.1\ m}{0.8\ m} = 6.9\ cm/hr$$

Calculate the flow rate Q through the sand column

$$\mathrm{Q} = \mathrm{vA} = \mathrm{v}\left(\pi \mathrm{D}^2/4\right)$$
$$= (6.9\ \mathrm{cm/hr})\ \pi\ \left(7^2/4\right)\left(0.001\mathrm{L/cm}^3\right) = 0.26\ \mathrm{L/hr}.$$

If the porosity of sand is 0.38, then the actual velocity of water through the filter is

$$\mathrm{v}_{\mathrm{actual}} = \mathrm{v}_{\mathrm{Darcy}}/\phi = 6.9\ \mathrm{cm/hr}/0.38 = 18\ \mathrm{cm/hr}.$$

Energy Diagrams

Energy diagrams are used in this chapter to show the relationship between matric potential, elevation, and total energy.

$$H = h_c + z \tag{27.11}$$

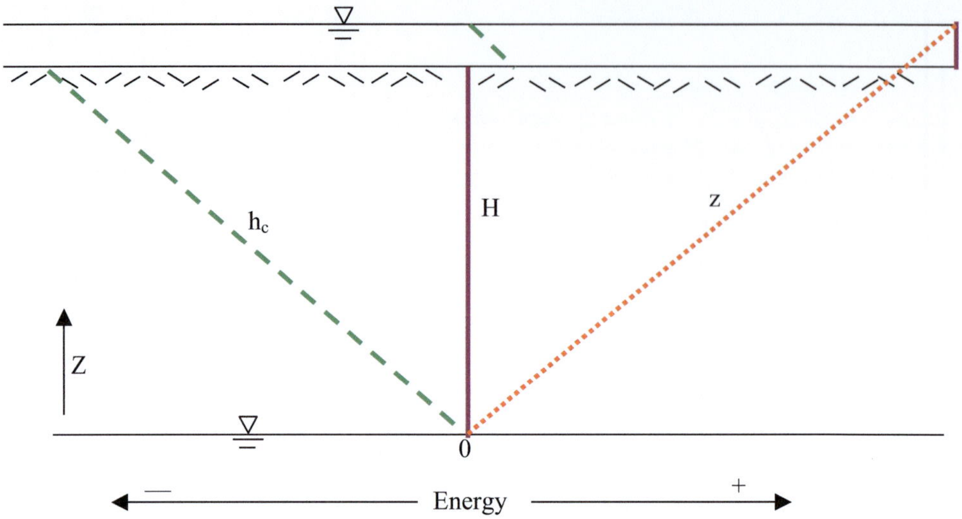

Fig. 27.6 Energy lines for soil with water table, no water movement below the soil surface, and water that has just been ponded at the water surface

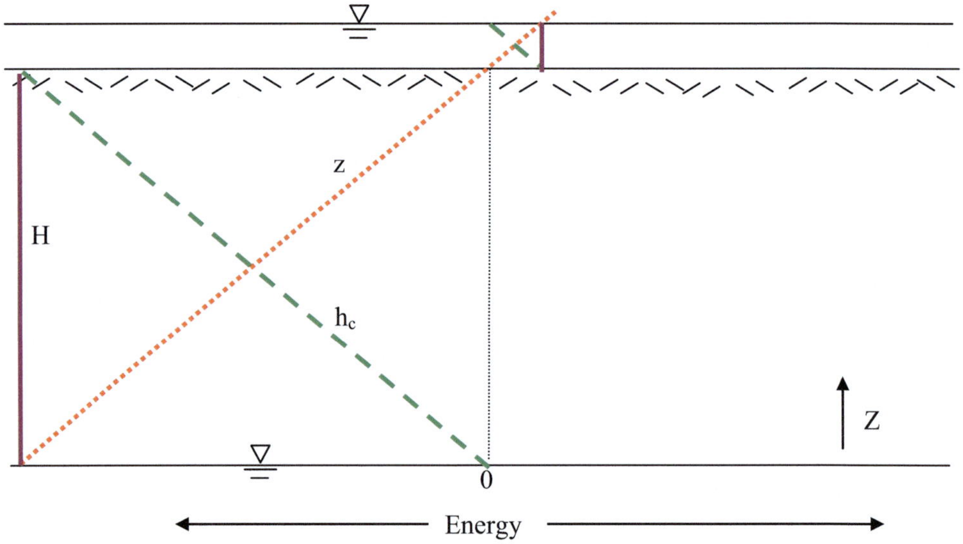

Fig. 27.7 Energy lines with datum at the soil surface and with ponded water before infiltration

H = total energy, m,
h_c= matric potential energy, m,
z = elevation energy, m.

Figure 27.6 is an energy diagram of a soil with a water table at the datum and no water movement. The datum is defined as the point at which the elevation is zero. The Z axis begins at the datum and is positive upwards. The energy axis is in the horizontal direction and is positive to the right. The water table is defined as the elevation in a soil or aquifer at which the matric potential energy (h_c) is zero. The matric potential energy, also be called capillary potential, pressure, or suction (may be designated by symbols such as h, h_m, ψ_c, and ψ_p). Water has just been ponded at the soil surface in Fig. 27.6, but it has not begun infiltrating into the soil.

The elevation energy is zero at the datum and increases at a 1:1 ratio with elevation. Thus, the elevation line is drawn at a 45^0 angle beginning at zero energy at the datum. Matric potential, h_c, at the water table is also zero by definition. Because $H = h_c + z$, H at the water table is zero. Because water in the soil is static, the total energy, H, in the soil is constant. Thus, H is a vertical solid line from the water table to the soil surface. Because $h_c = H - z$, and H is vertical, the dashed line, h_c, has the opposite slope of z.

In the ponded water, the pressure at the ponded water surface is zero and increases at a 1:1 ratio with depth (Fig. 27.6). The elevation line continues to increase from the soil surface to the ponded water surface. The total energy line, H, in the ponded water is vertical (h_c + z) because the water is not moving. Even if the water was moving in ponded water, the total energy line would be vertical because there is no energy loss.

The three energy lines can also be drawn with the datum at the soil surface (Fig. 27.7). The pressure line is still in the same position as it was when the datum was at the water table because the magnitude of pressure is not dependant on

Fig. 27.8 Energy lines for water flow in soil column

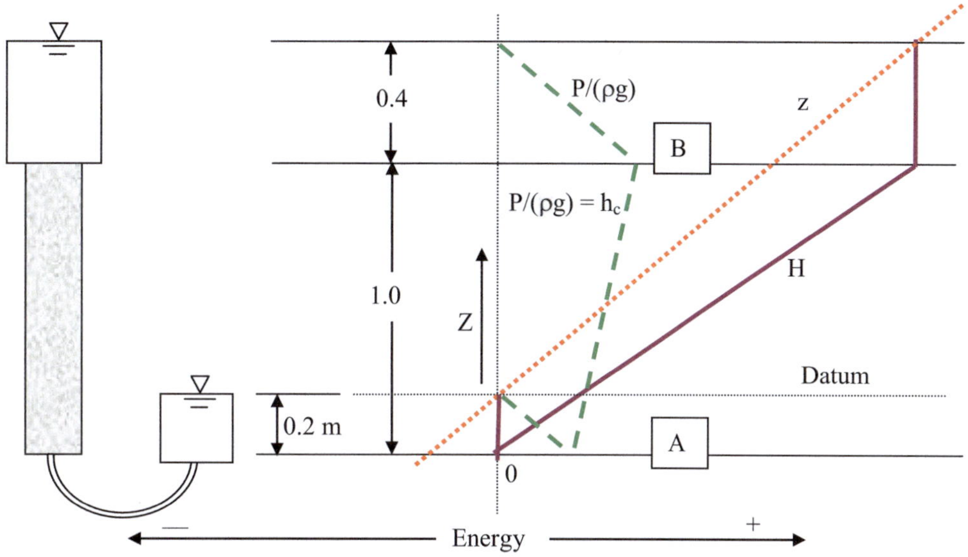

the position of the datum. However, the elevation and total energy lines shift to the left by an amount equal to the difference in the elevation of the datums in Figs. 27.6 and 27.7.

Unlike Figs. 27.6 and 27.7, water is moving through the soil column in Fig. 27.8. Thus, there is a total energy gradient and H is not a vertical line. In this case, the datum is set at the surface of the lower tank, and the elevation line is drawn at a 45^0 angle beginning at zero energy. Pressure P/ρg at the upper tank's water surface is zero and increases linearly with depth between the water surface and the top of the soil. Likewise, pressure at the top of the lower tank's water surface is zero, and increases linearly with depth between the water surface and the bottom of the soil column. If the soil in Fig. 27.12 is homogenous, then the slope of energy gradient dH/dZ does not change with elevation. If dH/dZ and dz/dZ are constant, then dh_c/dZ is a constant. Thus, a straight line representing h_c in the soil can be drawn between the known pressures at the top and bottom of the soil column (points A and B). The total energy line is the sum of the elevation energy and pressure energy lines. Because dH/dZ is linear in the soil column, connect the dots and draw the solid line for total energy, H, in the soil column.

Example 27.2 If the hydraulic conductivity of the soil column in Fig. 27.8 is 0.02 m/hr, and the porosity of the soil in the column is 0.48, then what is the actual average velocity (in the direction of water flow) of water flow through the column? If the cross sectional area of the soil column is 0.1 m^2, then what is the flow rate through the column in L/hr?

Total energy at the top of the soil column is 1.2 m and total energy at the bottom of the soil column is 0.0 m. Thus, H_f = 1.2–0.0 = 1.2 m. The length of the soil column, L, is 1.0 m.

$$v_{Darcy} = K\frac{\Delta H}{L} = 0.02\frac{1.2}{1.0} = 0.024\ m/hr$$

$$v_{actual} = v_{Darcy}/\phi = 0.024/0.48 = 0.05\ \mathrm{m/hr.}$$
$$Q = v_{Darcy}A = 0.024\ \mathrm{m/hr}*0.1\ \mathrm{m^2}*(1000\ \mathrm{L/m^3}) = 2.4\ \mathrm{L/hr.}$$

For flow in soil layers with different conductivities, the energy gradient changes from layer to layer. The effective hydraulic conductivity (average), K_e, is calculated first. Then, the Darcy velocity through the entire column is calculated. The Darcy velocity is then used to find the energy losses in each layer, which are then used to calculate the head changes through the column. Finally, elevation potential is subtracted from total energy to find matric potential in each layer. For a layered soil, effective hydraulic conductivity is calculated with a formula that is similar to that used for calculation of electrical current through a series of resistors.

$$K_e = \frac{total\ D}{{}^{D_1}/_{K_1} + {}^{D_2}/_{K_2} + {}^{D_n}/_{K_n}} \tag{27.12}$$

where

K_n = hydraulic conductivity of each soil layer.
D_n = depth of each layer

Example 27.3 The column shown in Fig. 27.9 has three soil layers. Layers 1, 2, and 3 have saturated hydraulic conductivities of 1, 2, and 3 cm/hr, respectively. The datum is at the bottom of the column. Calculate the energy, elevation, and pressure potential at points A, B, C, and D.

Calculate the flow velocity (Darcy velocity) in the column and energy loss in each layer?

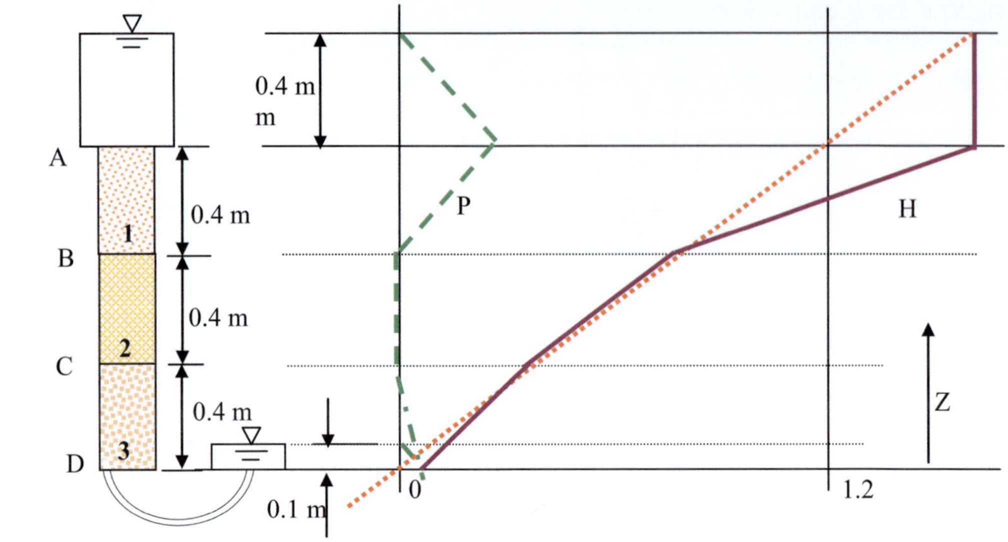

Fig. 27.9 Soil column with 3 layers used in Example 27.4

Table 27.1 Calculation table for energies in Fig. 27.10

Position	Total energy (H)	Elevation energy (z)	Matric potential (h or P)
A	1.6 m	1.2 m	1.6 − 1.2 = 0.4 m
B	1.6 − 0.82 = 0.78 m	0.8 m	0.78 − 0.8 = −0.02 m
C	0.78 − 0.41 = 0.37 m	0.4 m	0.37 − 0.4 = −0.03 m
D	0.37 − 0.27 = 0.1 m	0 m	0.1 − 0.0 = 0.1 m

$$K_e = \frac{1.2}{^{0.4}/_1 + ^{0.4}/_2 + ^{0.4}/_3} = 1.64\ cm/hr$$

$$v = K_e \frac{\Delta H}{L} = 1.64\ cm/hr \frac{1.5\ m}{1.2\ m} = 2.05\ cm/hr$$

$$\Delta H_1 = \frac{vL}{K} = \frac{2.05{*}0.4}{1} = 0.82\ m$$

$$\Delta H_2 = \frac{vL}{K} = \frac{2.05{*}0.4}{2} = 0.41\ m$$

$$\Delta H_3 = \frac{vL}{K} = \frac{2.05{*}0.4}{3} = 0.27\ m$$

The energies can be found sequentially by start with known energy at position A and then calculating energies downwards sequentially through the column as shown in Table 27.1.

Horizontal hydraulic conductivity controls the lateral movement of water in soils, such as the movement of water to subsurface drains. The mathematical formula is similar to the mathematical formula for parallel resistors. This formula will be used in Chap. 10 for calculation of lateral drainage.

$$K_e = \frac{K_1 D_1 + K_2 D_2 + K_n D_n}{total\ D} \tag{27.13}$$

Example 27.4. A tank is filled with the soils (same hydraulic conductivities) described in Example 27.3. However, inlets and discharge are located to the sides and water flows horizontally. Pressure differential across all layers is 1.5 m (Fig. 27.10). Calculate the average flow velocity through the tank and the flow velocity through each layer.

$$K_e = \frac{1{*}0.4 + 2{*}0.4 + 3{*}0.4}{1.2} = 2\ cm/hr$$

$$v = K_e \frac{\Delta H}{L} = 2\ cm/hr \frac{1.5\ m}{1.2\ m} = 2.4\ cm/hr$$

$$v_1 = K_1 \frac{\Delta H}{L} = 1\ cm/hr \frac{1.5\ m}{1.2\ m} = 1.25\ cm/hr$$

$$v_2 = 2\ cm/hr \frac{1.5\ m}{1.2\ m} = 2.5\ cm/hr$$

$$v_3 = 3\ cm/hr \frac{1.5\ m}{1.2\ m} = 3.75\ cm/hr$$

Horizontal Flow Through Gravel Wetlands

Water passes horizontally through the porous media in subsurface flow wetlands,. Pressure is atmospheric at the surface of both ends of the wetland, and velocity is negligible so the energy at either end of the wetland is the elevation of the water surfaces. Thus, the energy equation for the wetland is

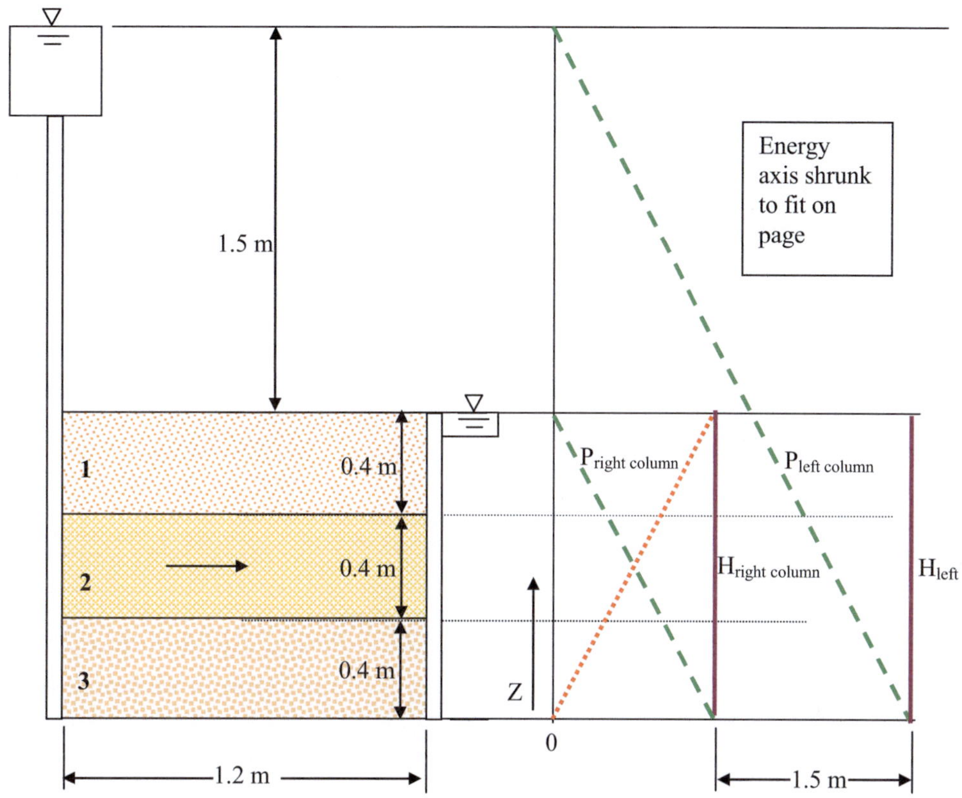

Fig. 27.10 Soil tank with 3 layers with lateral flow (Example 27.4)

$h_f = z_{in} - z_{out}$ (the water elevation difference from one end to the other).

The flow rate, Q, through the wetlands equals the cross-sectional area of the wetland * Darcy velocity.

$$Q = Av = A(K\ \Delta H/L) \quad (27.14)$$

where

K = hydraulic conductivity of the wetland fill material, m/d,
A = cross-sectional area of the wetland, m^2,
$\Delta H = z_1 - z_2$, m,
L = length of the wetland in direction of water flow, m,
Q = flow rate through the wetland, L/hr.

The volume of water in the wetland is

$$V_{cv} = A\ L\ \phi \quad (27.15)$$

Hydraulic conductivity in SSF wetlands depends on the gravel media characteristics and the amount of plugging of the media with particulate matter. Hydraulic conductivities of course gravel with little sand, sandy gravel, and fine to medium sand are estimated as 10^{-2}, 10^{-4}, and 10^{-5} m/sec, respectively. Drury and Mainzhauzen (2000) found that the hydraulic conductivities of gravel only and gravel with 20 % compost were 600 ft/day (2 * 10^{-3} m/sec) and 200 ft/day (0.67 * 10^{-3} m/sec), respectively, just after installation; however, conductivities decreased by an order of magnitude over 2 years. Filter fabric clogging and inlet clogging were the primary factors that decreased hydraulic conductivity. They also hypothesized that chemical precipitation from mine tailings water contributed to clogging.

SSF wetlands are generally installed with a level surface but with a ½ to 1 percent bottom slope.

Example 27.5 A SSF wetland has a gravel bed 0.6 m deep, 10 m wide, and 20 m long. The hydraulic conductivity of the gravel in the wetland is 10^{-3} m/sec (360 cm/hr), and the porosity is 0.38. The impermeable liner has a 0.75 % slope in the direction of flow. What are the Darcy velocity, flow rate, pore volume, and hydraulic detention time within the wetland?

The difference in impermeable liner elevation between the beginning and end of the wetland is 0.0075 * 20 m = 0.15 m. Calculate Darcy velocity,

$$v = K\frac{\Delta H}{L} = 360\frac{0.15}{20} = 2.7\ cm/hr$$

$$Q = vA = (2.7\ \text{cm/hr})(0.01\ \text{m/cm})(10\ \text{m})(0.6\ \text{m})$$
$$(1000\ \text{L/m}^3) = 162\ \text{L/hr}$$

$$V_{cv} = A\ L\ \phi = 10{*}0.6{*}20{*}0.38 = 46\ \text{m}^3$$

$$HDT = \frac{V}{Q} = \frac{46\ m^3}{162\ L/h}\left(\frac{1,000\ L}{m^3}\right)\left(\frac{day}{24\ hr}\right) = 12\ days$$

Calculation of Flow Through Any Media with Finite Differences

The flow of water through porous media is often calculated with finite difference models. When properties such as hydraulic conductivity are heterogeneous, it is appropriate to divide the region into smaller cells and make energy difference and flow calculations for each cell. These model calculations can take place in an Excel worksheet, but models written in programming languages, such as VBA, have more flexibility. The WINDS model is a one-dimensional finite difference model for vertical flow through the soil profile. This section begins with the general energy and mass flux derivation and then focuses on energy of water in soils and wetlands.

The general equation for mass flow per unit area as a function of energy is

$$\dot{m}_u = -K\frac{\Delta E}{Lg} \tag{27.16}$$

where

$\dot{m}_u$ = mass rate of flow per unit area, kg/sec/m^2,
ΔE = energy difference between cells, J/kg,
K = conductivity, m/sec,
L = distance between cells, m,
g = gravity, m/sec^2

Multiply by cell cross-sectional area to calculate the mass rate of flow.

$$\dot{m} = -AK\frac{\Delta E}{Lg} \tag{27.17}$$

where

$\dot{m}$ = mass rate of flow, kg/sec.

Density (kg/m^3) is constant and velocity is negligible for saturated water flow in porous media; thus, Eq. 27.17 can be written as Darcy's law.

$$Q = -AK\frac{\Delta H}{L} \tag{27.18}$$

The change in volume across a control volume boundary is the product of flow rate and time.

$$\Delta V = -AK\frac{\Delta H}{L}\Delta t \tag{27.19}$$

where

ΔV = change in volume within the cell, m^3.
Δt = length of time step, hr.

1	2	3	4	5

Fig. 27.11 Discretized cells for SSF wetland

A horizontal control volume, such as an SSF wetland, can be divided into cells (Fig. 27.11). In this example, it is assumed that the left side of cell 1 and the right side of cell 5 have no flow.

Equation 27.20 calculates the volume change in cell 2. Because the cell has flow from both sides, Eq. 27.20 includes flux across both boundaries. The horizontal axis is positive to the right; thus, the first half of the equation is negative because a negative energy gradient between cells 1 and 2 adds water to cell 2.

$$\Delta V_2 = \underbrace{-AK\frac{H_2 - H_1}{L}\Delta t}_{\text{left side flux}} + \underbrace{\text{AK}\frac{H_3 - H_2}{L}\Delta t}_{\text{right side flux}} \tag{27.20}$$

Equation 27.21 calculates the change in volume in cell 1. There is no flux across the left boundary of cell 1 so only the flux between cells 1 and 2 is included.

$$\Delta V_1 = +AK\frac{H_2 - H_1}{L}\Delta t \tag{27.21}$$

There is no flux across the right boundary of cell 5 so only the flux from cell 5 to cell 4 is included.

$$\Delta V_5 = -AK\frac{H_5 - H_4}{L}\Delta t \tag{27.22}$$

In each cell, the final volume at the end of the time step is the initial concentration + change in concentration

$$V_{final} = V_{init} + \Delta V \tag{27.23}$$

Combine Eqs. 27.20 and 27.23

$$V_{2=final} = V_{2=initial} - AK\frac{H_2 - H_1}{L}\Delta t + AK\frac{H_3 - H_2}{L}\Delta t \tag{27.24}$$

Similarly, a general equation for mass change (Eq. 27.16) can be written as follows.

$$m_{2=final} = m_{2=initial} - AK\frac{E_2 - E_1}{gL}\Delta t + AK\frac{E_3 - E_2}{gL}\Delta t \tag{27.25}$$

A program to compute change in mass in response the energy gradients, as calculated in Eq. 27.25, *Mass-Energy*, is written in Excel/VBA and is available with the textbook.

Example 27.6 This example demonstrates the use of finite difference model. Using Eq. 27.25, calculate the final mass after one second and after 2 seconds in a sequence of six cells with initial masses equal to 3-, 4-, 5-, 6-, 7-, and 8-kg. The energy in each cell (J/kg) is equal to 2 * mass (kg). The conductivity between cells is 6 m/sec, and the cross-sectional area of cells, A, is 0.1 m^2. The length of cells is 0.4 m. No water enters across the left boundary of the left cell or the right boundary of the right cell. Continue the calculations in the *Mass-Energy* VBA/Excel program until the system reaches equilibrium.

Time step 1

Cell 1 (left cell). Energy in cell 1 is 2 * 3 = 6, and energy in cell 2 is 2 * 4 = 8

$$m_{1=final} = m_{1=initial} + AK\frac{E_2 - E_1}{gL}\Delta t$$

$$= 3 + 0.1*6*\frac{2*4 - 2*3}{9.8*0.4}*1 = 3.31 \ kg$$

Cell 2

$$m_{2=final} = m_{2=initial} - AK\frac{E_2 - E_1}{gL}\Delta t + AK\frac{E_3 - E_2}{gL}\Delta t$$

$$m_{2=final} = 4 - 0.1*6\frac{8-6}{gL}*1 + 0.1*6\frac{10-8}{gL}*1 = 4 \ kg$$

Cells 3.5 all remain the same for the first time step as did cell 2.

Cell 6

$$m_{6=final} = m_{6=initial} - AK\frac{E_6 - E_5}{gL}\Delta t = 8 - 0.1*6\frac{16-14}{9.8*0.4}$$

$$= 7.69 \ kg$$

Time step 2.

Cell 1

$$m_{1=final} = m_{1=initial} + AK\frac{E_2 - E_1}{gL}\Delta t$$

$$= 3.31 + 0.1*6*\frac{2*4 - 2*3.31}{9.8*0.4}*1 = 3.52 \ kg$$

Cell 2

$$m_{2=final} = 4 - 0.1*6\frac{8 - 2*3.31}{gL}*1 + 0.1*6\frac{10-8}{gL}*1$$

$$= 4.09 \ kg$$

Cells 3 and 4 remain the same. Cell 5 mass is 6.91 kg and Cell 6 mass is 7.48 kg at the end of time step 2.

The *Mass-Energy* program was run for the parameters in this problem and results are shown in Fig. 27.12. The system reaches equilibrium (all cells have the same mass) after approximately 30 seconds.

For lateral flow phenomena through a subsurface flow wetlands, the cross-sectional flow area (transverse to flow direction) changes as the water elevation changes. In the following example, the changing wetlands flow area is incorporated into the calculation of water flow.

Example 27.7 Parameters for the SSF wetland are described in Example 27.5. Begin with steady state flow and then stop the discharge. Calculate the change in water surface elevation over time. The slope, b, of the wetland is 0.15 m/20 m = 0.0075 m/m (b = b–0.0075). Divide the

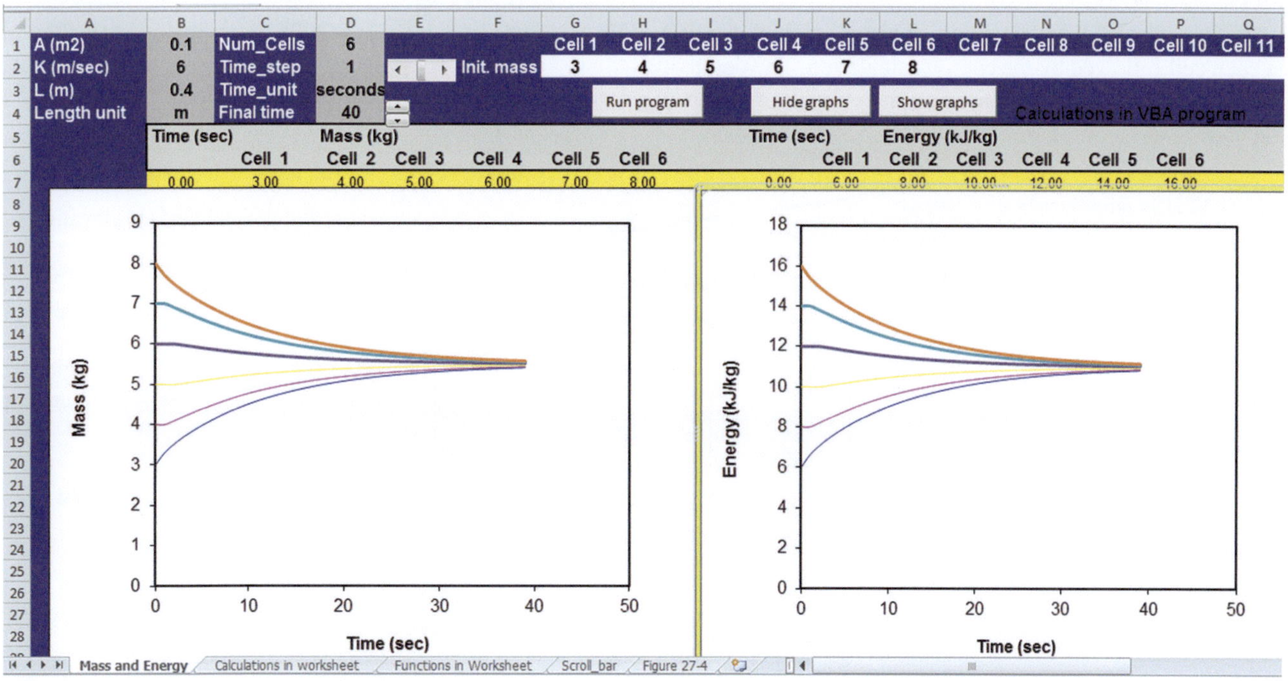

Fig. 27.12 *Chapter 27 Mass_and_Energy* program

SSF flow wetland into n = 5 cells that are 4 m in length and 10 m wide. The initial depth of flow in all cells as the water is flowing is 30 cm. Use 0.1 hour time steps. The centers of cells 1, 2, 3, 4, and 5 are 2-, 6-, 10-, 14-, and 18-m from the inlet, respectively. The distance, X, of cell midpoints from the inlet as a function of cell number j is

$$X_j = (j - 1 + 0.5)^*L_{cell}$$

where L_{cell} = L/n, L is the total wetland length, and n is the number of cells.

The following equation calculates average bottom surface elevation of any cell j when the datum is located at the bottom surface elevation at the inlet.

$$Y_j = Y_0 + b^*X_j$$

where Y_j is the average bottom surface elevation of cell j, and Y_0 is the bottom surface elevation at the inlet.

The total energy (elevation potential) of water in each cell is equal to the depth + bottom surface elevation.

$$H_j = d_j + Y_j$$

The initial depth of water in cells can be calculated based on the steady-state flow water surface elevation.

$$d_{j\text{-initial}} = H_{j\text{-initial}} - Y_j$$

The initial volume of water in cells is calculated based on the depth.

$$V_{j\text{-initial}} = d_{j\text{-initial}}{}^*L_{cell}{}^*\, W^*\phi$$

As with Example 27.5, the volume of water is calculated first in each cell followed by the energy calculation. The depth of flow in a cell is calculated based on the volume of water in the cell.

$$d_j = V_j/(L_{cell}{}^*W^*\phi)$$

where d_j is the depth at the midpoint of cell j, and V_j is the volume of water in cell j.

The mass flow rate is a function of the conductivity and cross-sectional area. The cross-sectional area of flow in cell j, A_j, is

$$A_j = d_j{}^*W.$$

Time step 1 (i = 1)

The initial average water surface elevations of cells 1 through 5 are 28.5-, 25.5-, 22.5-, 19.5-, and 16.5-cm, respectively (Fig. 27.13).

Cell 1(j = 1)

Initial parameterization

$$X_1 = (j - 1 + 0.5)^*L_{cell} = (1 - 1 + 0.5)^*4 = 2\ \text{m}$$
$$Y_1 = Y_0 + b^*X_1 = 0 + (-0.0075)^*2 = -0.015\ \text{m}$$
$$d_{1\text{-initial}} = H_{1\text{-initial}} - Y_j = 0.285 - (-0.015) = 0.3\ \text{m}.$$
$$V_1 = d_1{}^*L_{cell}{}^*W^*\phi = 0.3^*4^*10^*0.38 = 4.56\ \text{m}^3.$$
$$A_1 = d_1{}^*W = 0.3\ \text{m}^*10\ \text{m} = 3\ \text{m}^2$$

Calculation of parameters at end of time step (K = 3.6 m/hr)

$$V_{1-final} = V_{1-initial} + AK\frac{H_2 - H_1}{L_{cell}}\Delta t$$
$$V_{1-final} = 4.56\ m^3 + 3^*3.6\frac{0.255 - 0.285}{4}{}^*0.1$$
$$= 4.56 - 0.0054 = 4.5519\ m^3$$
$$d_{1-\text{final}} = V_{1-\text{final}}/(L_{cell}{}^*W^*\phi) = 4.5519/(4^*10^*0.38)$$
$$= 0.2995\ \text{m}$$
$$H_{1-\text{final}} = d_{1-\text{final}} + Y_j = 0.2995 + (-0.015) = 0.2845\ \text{m}$$

Cell 2(j = 2)

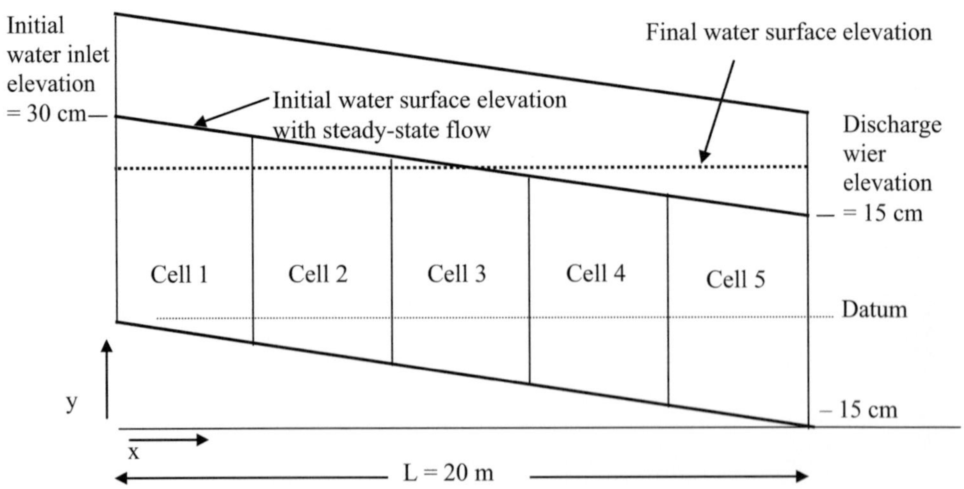

Fig. 27.13 Discretized wetland

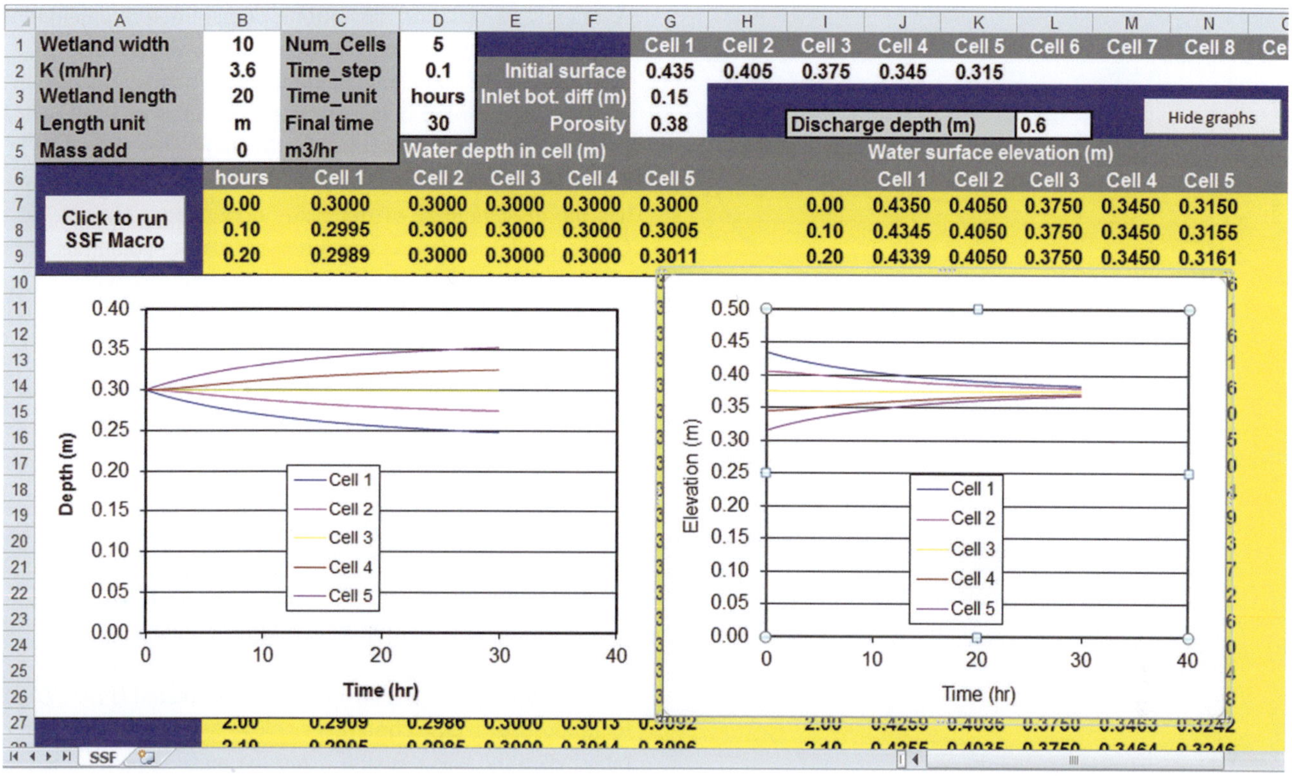

Fig. 27.14 *SSF* spreadsheet solution for Example 27.7

There is no change in water surface elevation for cell 2 during the first time step because the energy difference on both sides of the cell is the same.

$$\text{Time step } 2(\mathrm{i}=2)$$
$$\text{Cell } 1(\mathrm{j}=1)$$

The average area between cells 1 and 2 is

$$A = (0.2995 + 0.3000)/2{*}10 = 2.9975 \ \mathrm{m}^2$$
$$V_{final} = 4.5519 \ m^3 + 2.9975{*}3.6\frac{0.255 - 0.2845}{4}{*}0.1$$
$$= 4.56 - 0.0054 = 4.5439 \ m^3$$
$$\mathrm{d}_{\mathrm{final}} = \mathrm{V}/(\mathrm{L}{*}\mathrm{W}{*}\phi) = 4.5439/(4{*}10{*}0.38) = 0.2989$$

$$\text{Cell } 2(\mathrm{j}=2)$$

The average area between cells 2 and 3 is 3 m^2

$$V_{2\text{-}final} = V_{2\text{-}initial} - AK\frac{H_2 - H_1}{L}\Delta t + AK\frac{H_3 - H_2}{L}\Delta t$$
$$V_{2\text{-}final} = 4.56 - 2.9975{*}3.6\left(\frac{0.255 - 0.2845}{4}\right)0.1$$
$$+ 3{*}3.6\left(\frac{-0.03}{4}\right)0.1 = 4.55986 \ m^3$$
$$\mathrm{d}_{2\text{-}final} = \mathrm{V}_{2\text{-}final}/(\mathrm{L}{*}\mathrm{W}{*}\phi) = 4.55986/(4{*}10{*}0.38)$$
$$= 0.299991 \ \mathrm{m}$$

The other cells are calculated in the same manner for each time step. The *SSF* VBA/Excel program included with the text calculates the change in water surface over time (Fig. 27.14).

Water Content and Matric Potential Energy in Soils

Relationships between water content and matric potential (Fig. 27.15) and hydraulic conductivity and matric potential (Fig. 27.16) have been defined by the Brooks and Corey (1964) and van Genuchten (1980). Water content is a function of matric potential (water characteristic curve) because water drains from successively smaller soil pores as matric potential in the soil becomes more negative. Van Genuchten modeled the shape of the water characteristic curve (Fig. 27.15) with the following equation:

$$\theta = \theta_r + (\theta_s - \theta_r)[1 + (-\alpha \ {*}h_c)^n]^{-m} \qquad (27.26)$$

where

$\mathrm{h_c}$ = soil water matric potential (negative), cm,
n = dimensionless parameter,
m = dimensionless parameter equal to 1-1/n
α = dimensionless parameter,

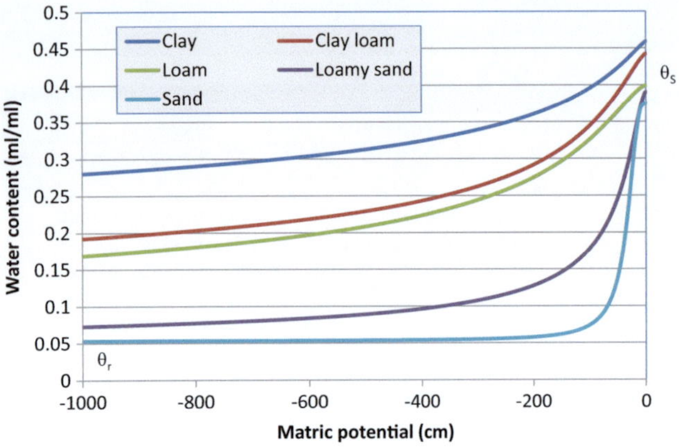

Fig. 27.15 Water characteristic curves with van Genuchten model

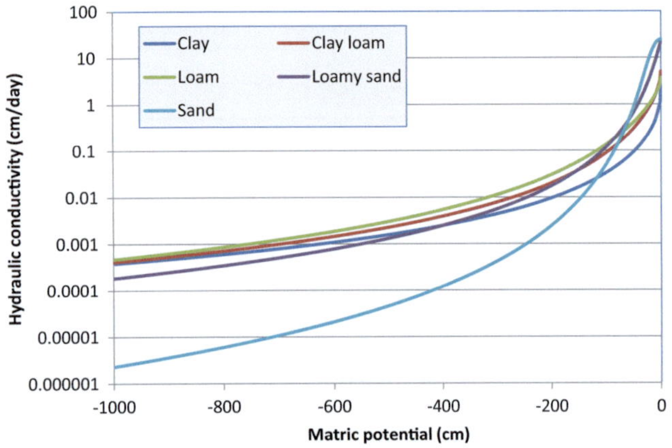

Fig. 27.16 Hydraulic conductivity versus matric potential with van Genuchten model

θ_r = residual water content, dimensionless,
θ_s = saturated water content, dimensionless
θ = actual water content, dimensionless

For example, n, α, θ_r, and θ_s for sand are 3.18, 0.0352, 0.053, and 0.375, respectively (Table 27.2). The water content for sand at – 40 cm matric potential (Fig. 27.15) is

$$\theta = \theta_r + (\theta_s - \theta_r)[1 + (-\alpha * h_c)^n]^{-m} = 0.053 + (0.375 - 0.053)\left[1 + (-0.0352*(-40))^{3.18}\right]^{-(1-1/3.18)} = 0.178$$

The Rosetta model was derived at the United States Salinity Laboratory and is available from the University of Arizona. It calculates the van Genuchten equation parameters from soil properties. The more data that is provided, the better the calculation. Possible input parameters include percentages of sand, silt, and clay, bulk density, and water content at field capacity and wilting point.

Average and standard deviations of parameters calculated for different textural classifications are shown in Table 27.2.

Although the van Genuchten model is currently more popular, the relationship between matric potential and soil water content was first described by Brooks and Corey (Fig. 27.17). Unlike the van Genuchten model, the Brooks-Corey equation has the disadvantage of not being continuous. The Brooks-Corey model assumes saturation at matric potentials less negative than what is called the bubbling pressure. The Brooks-Corey equation can be written as follows.

$$\theta = (\theta_s - \theta_r)\left(\frac{h_b}{h_c}\right)^{\lambda} + \theta_r \quad \text{for } h_c \text{ more negative than } h_b$$
$$\theta = \theta_s \quad \text{for } h_c \text{ less negative than } h_b \tag{27.27}$$

where

h_b = bubbling pressure, m (or same units as h_c),
λ = pore size distribution index, dimensionless.

The water characteristic curve can be rotated (Fig. 27.18) to model the water content vs. elevation above the water table when the water is static, and the matric potential is the negative of the elevation above the water table (Fig. 27.6). Soil is saturated between the water table and the bubbling pressure elevation, h_b, and water content decreases with elevation above the bubbling pressure elevation according to the Brooks-Corey model. A similar water content vs. elevation relationship could be generated with the van Genuchten model.

Brooks and Corey defined the "bubbling pressure", h_b, as corresponding to the maximum pore-size forming a continuous network of flow channels within a soil medium: the first pore to drain. In their model, the soil remains saturated if the matric potential is less negative than the bubbling pressure (Eq. 3.5). In actual soils, the saturated water content, θ_s, is slightly lower than the soil porosity and there is not an abrupt change in water content at the bubbling pressure.

Both the Brooks-Corey (left) and van Genuchten (right) models can be written in terms of effective water content (Eq. 27.28). Effective water content is zero when $\theta = \theta_r$, the residual water content. Effective water content is 1.0 when $\theta = \theta_s$, the saturated water content.

$$\theta_e = \frac{\theta - \theta_r}{\theta_s - \theta_r} = \left(\frac{h_b}{h_c}\right)^{\lambda} \qquad \theta_e = [1 + (-\alpha * h_c)^n]^{-m} \tag{27.28}$$

where

θ_e = effective water content, m^3/m^3.

Table 27.2 Moisture retention and hydraulic conductivity data (Schaap, 1999)

Texture class	Num samp	θ_r cm^3/cm^3	θ_s cm^3/cm^3	α (1/cm)	n	Ks (cm/day)	Ko (cm/day)	L
		Mean						
Clay	84	0.098	0.459	0.0150	1.25	14.8	2.96	−1.561
Clay loam	140	0.079	0.442	0.0158	1.42	8.18	5.00	−0.763
Loam	242	0.061	0.399	0.0111	1.47	12.1	3.70	−0.371
Loamy sand	201	0.049	0.39	0.0348	1.75	105	24.3	−0.874
Sand	308	0.053	0.375	0.0352	3.18	643	24.5	−0.93
Sandy clay	11	0.117	0.385	0.0334	1.21	11.4	4.34	−3.665
Sandy clay Lo	87	0.063	0.384	0.0211	1.33	13.2	6.93	−1.28
Sandy loam	476	0.039	0.387	0.0267	1.45	38.3	15.5	−0.861
Silt	6	0.05	0.489	0.00658	1.68	43.8	3.34	0.624
Silty clay	28	0.111	0.481	0.0162	1.32	9.62	3.17	−1.287
Silty clay Loa	172	0.09	0.482	0.00839	1.52	11.1	2.23	−0.156
Silt loam	330	0.065	0.439	0.00506	1.66	18.2	1.75	0.365
		Standard deviations						
Clay		0.107	0.079	4.8	1.17	8.3	1.82	1.39
Clay loam		0.076	0.079	4.9	1.32	12.3	1.70	0.9
Loam		0.073	0.098	5.4	1.35	8.3	1.62	0.84
Loamy sand		0.042	0.07	3.0	1.45	4	1.7	0.59
Sand		0.029	0.055	1.8	1.51	4	1.7	0.49
Sandy clay		0.114	0.046	3.7	1.15	7.8	2.19	1.8
Sandy clay Lo		0.078	0.061	5.1	1.32	7.1	1.74	0.99
Sandy loam		0.054	0.085	3.6	1.29	4.6	1.6	0.73
Silt		0.041	0.078	2.0	1.35	1.9	2.09	1.57
Silty clay		0.119	0.08	4.4	1.26	3.7	1.86	1.23
Silty clay loa		0.082	0.086	3.9	1.35	5.8	1.82	1.23
Silt loam		0.073	0.093	3.7	1.38	5.5	1.82	1.42

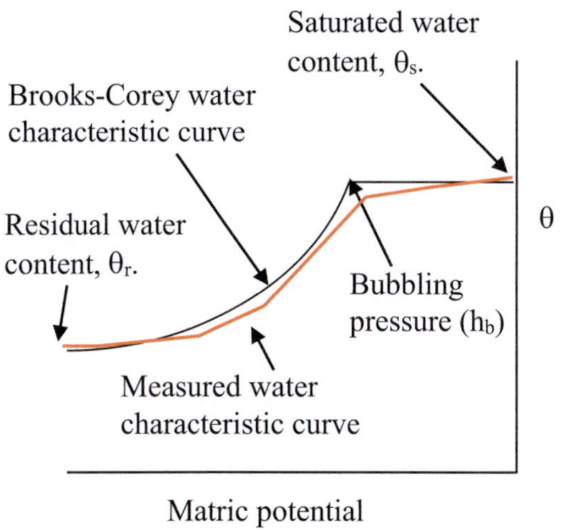

Fig. 27.17 Brooks-Corey model

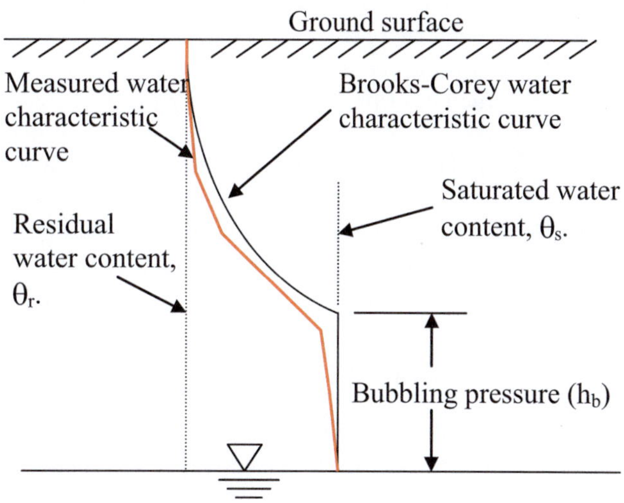

Fig. 27.18 Vertical Brooks-Corey model of water content versus height above the water table

The Brooks-Corey and van Genuchten can also be written in terms of saturation, S. Saturation is the fraction of pore space that is filled with water: volume of water/volume of pores. Effective saturation, S_e, is the fraction of voids filled between residual and saturated water content ($S = (\theta - \theta_r)/(\theta_s - \theta_r)$). As with the effective water content, θ_e, the effective saturation, S_e, ranges from 0 to 1. A graph of saturation and effective saturation vs. matric potential is shown in Fig. 27.19.

$$S_e = \frac{S - S_r}{S_s - S_r} = \left(\frac{h_b}{h_c}\right)^{\lambda} \quad S_e = [1 + (-\alpha * h_c)^n]^{-m} \tag{27.29}$$

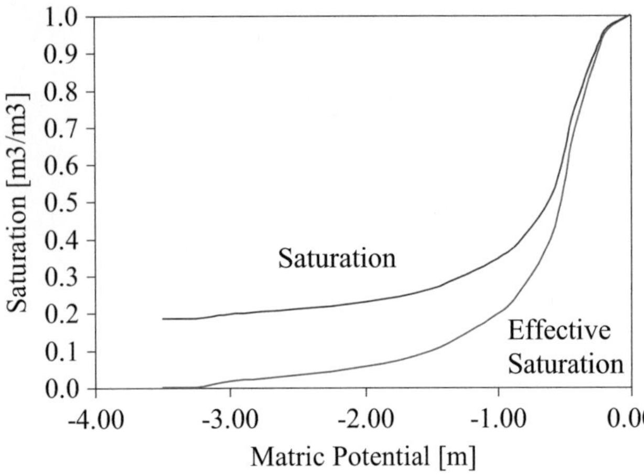

Fig. 27.19 Effective saturation for Wagram loamy sand

where

S_e = effective saturation, dimensionless.

The pore size distribution index, λ, determines the shape of the Brooks-Corey curve. It is a function of the distribution of pore sizes within the soil matrix because pore drainage is a function of pore size. A method to calculate the pore size distribution index, λ, and the bubbling pressure, h_b, was derived by Brooks and Corey. They found that log-log plots of effective saturation, S_e vs. capillary pressure, h_c formed a straight line for most soils. The intercept of the line with $S_e = 1$ is the bubbling pressure, h_b. The slope of the line on the log-log plot is the pore size distribution index, λ.

The construction of the log-log plot requires two steps. First, a log-log plot of effective saturation with an initial estimate of residual water content is constructed (estimate residual water content as the lowest measured water content). Continue to adjust residual water content until the S_e vs. h_c curve is as straight as possible. Then, calculate the slope, λ, of the straight line.

Example 27.8 Find λ and h_b for Wagram loamy sand (Fig. 27.19).

As a first guess, estimate residual water content as 0.056 (the minimum measured water content). The corresponding residual saturation is 18 %. The saturated water content ($h_c = 0$) is 30.5 % (see Table 27.3). Saturation and effective saturation vs. matric potential are calculated in Table 27.3 and plotted in Fig. 27.20. A sample calculation for effective saturation at matric potential = −1 m is shown below.

$$S_e = \frac{S - S_r}{1 - S_r} = \frac{0.344 - 0.184}{1 - 0.184} = \frac{\theta - \theta_r}{1 - \theta_r} = \frac{0.105 - 0.056}{0.31 - 0.056}$$
$$= 0.197\ m^3/m^3$$

Not all of the points on the curve line up. Adjust θ_r until all of the points along the *lower part* of the curve form a straight line. The points line up if the residual water content is adjusted to 0.044 (Fig. 27.21).

The slope of the line should be calculated with two representative points on the log-log plot. In this case, the logarithms of points at matric potentials of −0.33 and −1.5 are used to calculate the slope

$$\lambda = -\frac{\log(S_1) - \log(S_2)}{\log(h_1) - \log(h_2)} = -\frac{\log(0.81) - \log(0.138)}{\log(0.33) - \log(1.5)}$$
$$= 1.27$$

The intercept at the $S_e = 1.0$ line is 0.3 m. Thus, the bubbling pressure, h_b, is 0.3 m. The curve calculated with $\lambda = 1.27$, $h_b = 0.3$, and $\theta_r = 0.044$ is compared to the actual water characteristic curve in Fig. 27.22.

Example 27.9 A soil has the following Brooks-Corey parameters: $h_b = -0.3$ m, $\lambda = 1.27$, $\theta_s = 0.31$, $\theta_r = 0.044$. Calculate the water content at a matric potential of – 3.3 m (field capacity).

$$\theta = (\theta_s - \theta_r)\left(\frac{h_b}{h_c}\right)^{\lambda} + \theta_r$$
$$= (0.31 - 0.044)\left(\frac{-0.30}{-3.3}\right)^{1.27} + 0.044 = 0.057$$

Example 27.10 After a rainstorm, the soil water is in equilibrium, with matric potential equal to elevation above the water table. The water table is 1.7 m below the soil surface and the datum is 2 m below the soil surface. Calculate the water content at elevations of 0.3-, 0.27-, 1.1-, and 1.5-m above the water table. Use the same parameters as in Example 27.10

Calculate the water content at 0.6 m (0.3 m above the water table). At the elevation of the bubbling pressure, 0.3 m, the soil is saturated based on the Brooks-Corey model.

$$\theta = \theta_S = 0.31$$

Elevation 1.0 m
$$\theta = (\theta_S - \theta_r)\left(\frac{h_b}{h_c}\right)^{\lambda} + \theta_r$$
$$= (0.31 - 0.044)\left(\frac{-0.30}{-0.7}\right)^{1.27} + 0.044 = 0.135$$

Elevation 1.4 m
$$\theta = (\theta_S - \theta_r)\left(\frac{h_b}{h_c}\right)^{\lambda} + \theta_r$$
$$= (0.31 - 0.044)\left(\frac{-0.30}{-1.1}\right)^{1.27} + 0.044 = 0.095$$

Elevation 1.8 m
$$\theta = (\theta_S - \theta_r)\left(\frac{h_b}{h_c}\right)^{\lambda} + \theta_r$$
$$= (0.31 - 0.044)\left(\frac{-0.30}{-1.5}\right)^{1.27} + 0.044 = 0.078$$

The water content vs. elevation above the datum is shown in Fig. 27.23.

Table 27.3 Experimental data (Left two columns) and calculated values for first guess for residual saturation (0.056) of Wagram loamy sand

h_c	θ	$θ_e$	S	S_e	h_c	θ	$θ_e$	S	S_e
0.00	0.305	1.000	1.000	1.000	−0.80	0.124	0.273	0.407	0.273
−0.17	0.295	0.960	0.967	0.960	−1.00	0.105	0.197	0.344	0.197
−0.24	0.280	0.900	0.918	0.900	−1.50	0.080	0.096	0.262	0.096
−0.33	0.255	0.799	0.836	0.799	−2.00	0.070	0.056	0.230	0.056
−0.44	0.220	0.659	0.721	0.659	−2.50	0.064	0.032	0.210	0.032
−0.50	0.190	0.538	0.623	0.538	−3.00	0.060	0.016	0.197	0.016
−0.60	0.155	0.398	0.508	0.398	−3.25	0.056	0.010	0.184	0.000

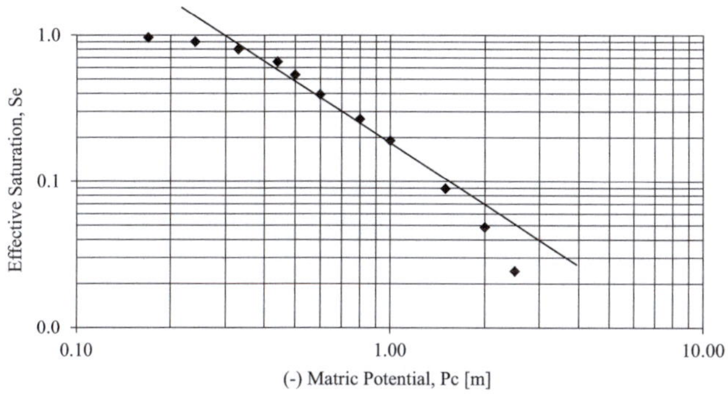

Fig. 27.20 Effective saturation as a function of capillary head (first guess)

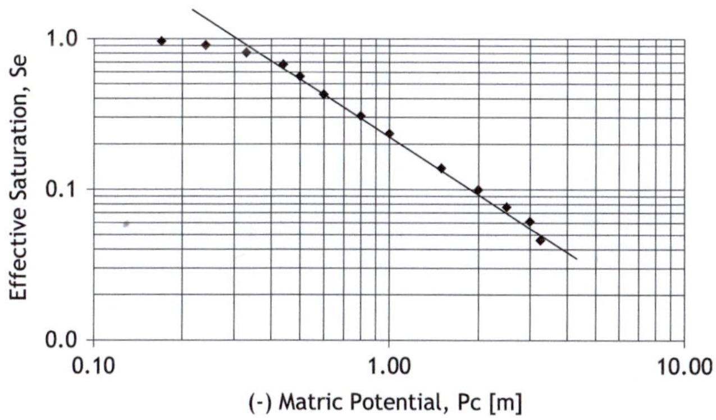

Fig. 27.21 Modified log-log plot with $θ_r = 0.044$

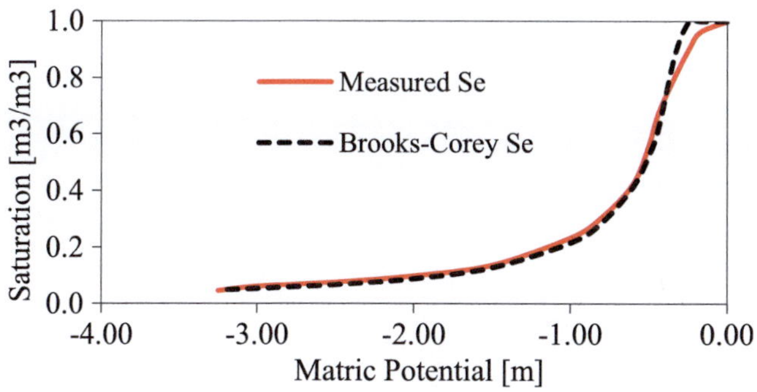

Fig. 27.22 Measured and Brooks-Corey Se comparison

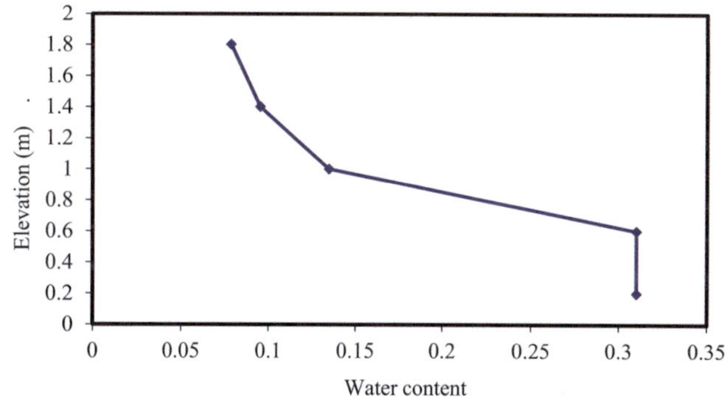

Fig. 27.23 Water content versus elevation above the water table

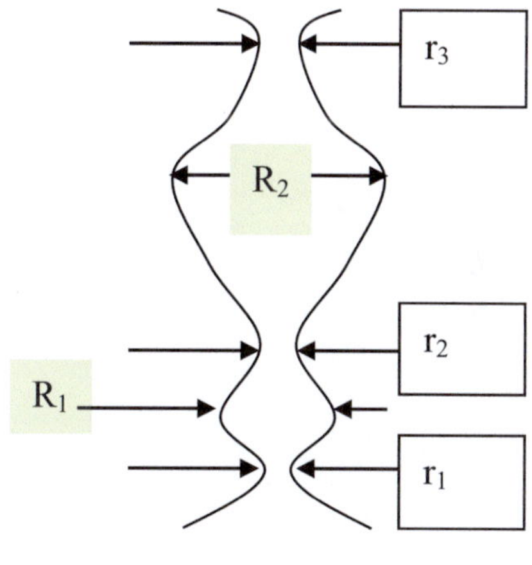

Fig. 27.24 Hysteresis diagram

The water characteristic curves (water content vs. matric potential) during soil drainage and soil filling are different. This phenomenon is called hysteresis. Hysteresis is caused by the fact that **smallest diameter** of the smallest pores prevents water from **draining** out of soil pore networks. However, the **largest** diameter of the largest pores prevents water from **filling** soil pore networks. For example, in Fig. 27.24, consider the case that the pore network is initially filled with water. Capillary forces at diameter r_1 (the smallest pore diameter in the network), will block the water in the pore network from draining. Now consider the case that the entire system is initially dry. The soil pore network will remain dry until the soil water potential is less negative (smaller) than the capillary pressure at the maximum diameter of the largest pore, R_2. During filling, when the water table is raised, water enters the soil pores from below due to capillary suction. The largest pore prevents filling until h_c is less negative than the capillary potential.

Questions

1. How is energy lost as water flows through pipelines, soils, and channels?
2. Convert 10 m head to units of kPa, atmospheres, bars, J/kg, ft head, and PSI.
3. Water flows through a 1 m long column at a rate of 2 m/d and the pressure differential from one end of the column to the other is 1 m. Calculate the hydraulic conductivity of the media in the column. What would the flow rate be if the pressure differential was 100 kPa?
4. Calculate the kinetic energy of the water in question 4 for the case of 1 m pressure differential. Is calculation of kinetic energy necessary in soil water calculations?
5. Draw the energy diagram for Example 27.1 (part 1)
6. For Example 27.1 (part 1), place the lower water surface at the same elevation as the upper boundary of the sand (0.4 m below the elevation of the upper tank water surface). Draw the energy diagram.
7. For Example 27.4, change the conductivities in cells 1, 2, and 3 to 5-, 4-, and 3-cm/hr, and change the heights of the cells to 0.3, 0.4, and 0.5-m, respectively. Draw the energy diagram.
8. For Example 27.5, change the conductivities in cells 1, 2, and 3 to 5-, 4-, and 3-cm/hr, and change the heights of the cells to 0.3, 0.4, and 0.5-m, respectively. Draw the energy diagrams at the midelevation for each of the three cells in the direction parallel to flow (rotate the axes).
9. A column has two soil layers. Layers 1 and 2 have saturated hydraulic conductivities of 1 and 2 cm/hr, respectively. The datum is at the bottom of the column. Calculate the energy, elevation, and pressure potential at points A, B, and C. Draw the energy potential lines. You do not need to draw any lines below the datum. Calculate the flow velocity (Darcy velocity) in the column and energy loss in each layer?

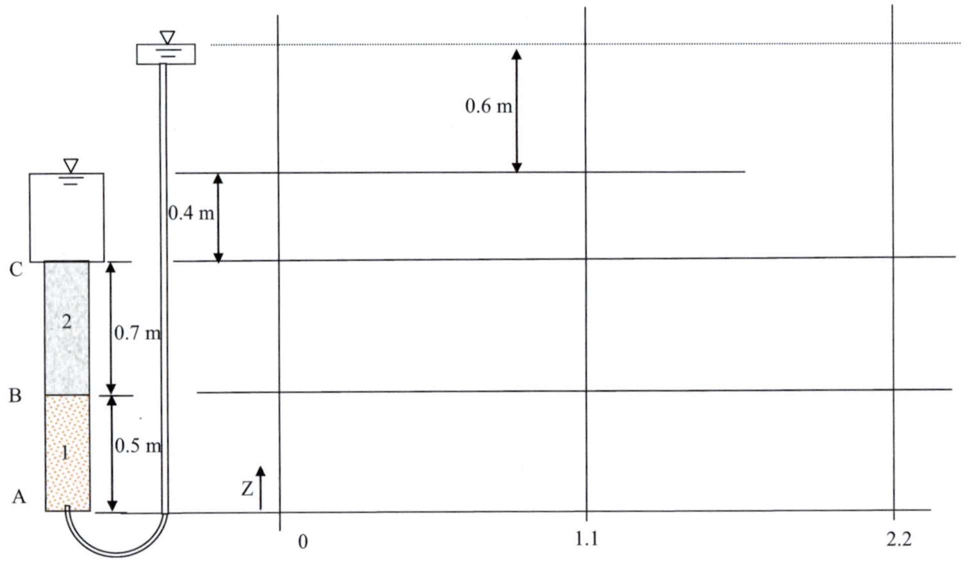

10. Draw the energy potential lines for this soil column, and calculate the total and matric potential energy at points A, B, and C. The conductivity of layer 1 is 2 cm/day, and the conductivity of layer 2 is 0.5 cm/day.

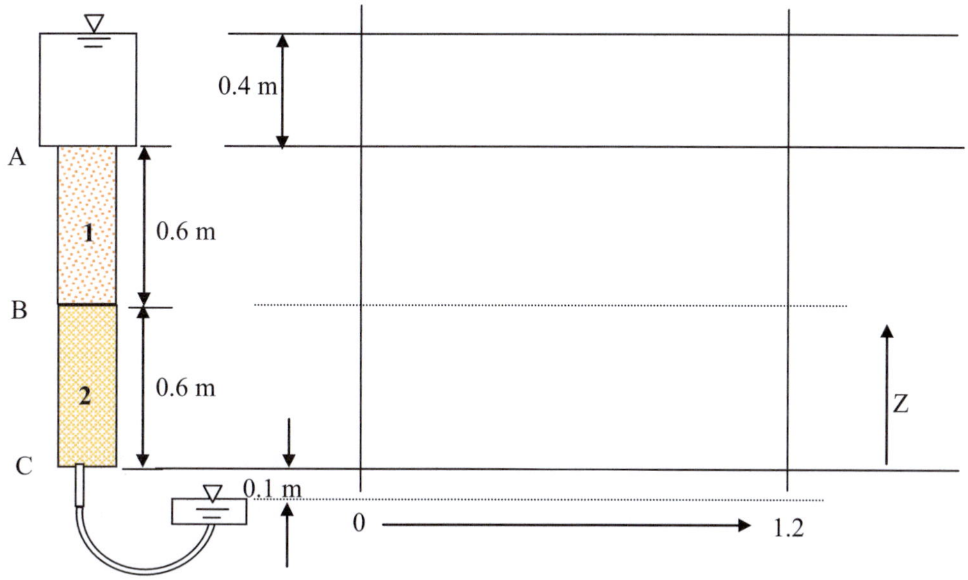

11. Using Eq. 27.25, calculate the change in mass in cell 2, which has an energy of 4 J, For this example, E (meters) = 2 * mass (kg). Cell 1 E = 2 J and Cell 3 E = 3 J. The conductivity between cells is 6 m/sec, and the cross-sectional area of cells, A, is 0.1 m^2. The length of cells is 0.4 m. The length of time steps is 1 second. Calculate final mass and energy.
12. For Example 27.6, let the initial masses be equal to 1-, 4-, 5-, 6-, 8-, and 10-kg. Print out the energy graph and discuss results.
13. Repeat question 12, but change the time step to 10 seconds. Print out the energy graph and discuss results.
14. Change the *Calculations in worksheet* so that 1 kg mass is continually added each second to the left of the control volume (cell 1). Make the time step 0.2 seconds. Then make a second modification so that 0.4 kg is removed from the right side whenever the mass on the left side is greater than 8. Simulate for 60 seconds and show in a graph. Show the equations that you used. Make sure that the A,K, and L values are the same as those in the Mass and Energy worksheet: (A = 0.1, k = 6, and L = 0.4).
15. Repeat question 14 in *Functions in Worksheet*. Show the equations/functions as well as the graph. You will need to add the Right_mass() function as well as make the modification on removed mass. Also, let the criteria be that the left mass must be greater than 10.
16. Repeat question 14 but make the changes in the VBA Mass_Energy subroutine so that the results are shown in the Mass and Energy worksheet. Show the changed code. The graph should have the same pattern as graphs in questions 14 and 15.
17. A SSF wetland has a gravel bed 0.5 m deep, 10 m wide, and 20 m long. The hydraulic conductivity of the gravel in the wetland is 10^{-3} m/sec (360 cm/hr), and the porosity is 0.38. The impermeable liner has a 0.5 % slope in the direction of flow. What are the Darcy velocity, flow rate, pore volume, and hydraulic detention time within the wetland? Calculate the percent reduction in flow from that of the wetland described in Example 27.5.
18. Evaluate your answer in question 17 with the SSF workbook. Determine whether 54 L/hr (0.054 m^3/hr) results in steady 0.3 m depth flow. Inlet flow rate is specified in cell B5. Set all cells at an initial depth of 0.3 m. Set discharge depth (cell L4) equal to 0.3 m. Inlet bot. difference (cell G3) is 0.1 m.
19. With the initial conditions described in question 17, investigate the effect of increasing the discharge depth to 0.5 m and increasing the flow rate to 90 L/hr with the *SSF* model. Run for 200 hours. Verify with the equations in question 16 that this flow depth results in steady state depth = 0.5 m.
20. With the initial conditions described in question 17, investigate the effect of keeping the discharge depth at 0.3 m and increasing the flow rate to 90 L/hr with the *SSF* model. Run for 200 hours. Show the graph and describe the final conditions after 200 hours. Look at the hydraulic gradient and discuss its effect on flow rate. Discuss the ability of SSF wetlands to handle variations in flow rate.
21. Plot the water characteristic curves for sandy loam and clay soils with the van Genuchten equation for the terms in Table 27.2. Make three graphs for each soil. 0–50 cm matric potential with linear scale, 0–15 bar matric

potential with linear scale, and 0–15 bar matric potential with logarithmic scale on the x-axis. For logarithmic scale, the matric potentials must be expressed as positive values. Describe the differences in the curves. Make sure that the water content scale on all graphs begins at zero in order to make the comparison.

22. Find λ and hb for the Goldsboro Sandy Loam water content vs. matric potential in the table below. Show initial log-log plot, adjusted log-log plot, and show final graph with a comparison of the experimental water characteristic curve and the Brooks-Corey water characteristic curve. (Note: the graph does not align quite as well as the example in the book).

hc (m)	θ
0.00	0.365
– 0.10	0.355
– 0.30	0.320
– 0.50	0.275
– 0.60	0.260
– 0.80	0.235
– 2.00	0.188
– 2.50	0.182
– 3.00	0.178
– 3.50	0.170

23. Replace the simplistic relationship between mass and energy in the Mass and Energy worksheet VBA code with the Van Genuchten equation for water content (mass) and matric potential (energy) relationship. Also, include the Van Genuchten relationship between matric potential (energy) and conductivity instead of the constant conductivity that is included in the problem. In order to simplify the problem, you can assume that the cells are horizontal in order to avoid the calculation of elevation energy. The solution is basically the VBA code used in the next chapters, but without the vertical change in elevation energy.

References and Resources

Brooks RH, Corey AT (1964) Hydraulic properties of porous media, Hydrology Papers. Colorado State University, Fort Collins

Drury WJ, Mainhauzen K (2000) Hydraulic characteristics of subsurface flow wetlands. In: Proceedings of the billings land reclamation symposium. Billings, 20–24 Mar 2000

Schaap M (1999) Rosetta model. United States Salinity Laboratory and University of Arizona. http://www.ars.usda.gov/Services/docs.htm?docid=8953

van Genuchten MT (1980) A closed form equation for predicting the hydraulic conducitivity of unsaturated soils. Soil Sci Soc Am 44(5):892–898

Modeling Soil Moisture

28

Richards equation is a model of water content and soil water flux in response to energy gradients. It is incorporated into the *WINDS* model. Water, energy, and conductivity relationships are calculated with the van Genuchten equations. Determining the rate of water table movement in response to water gain or loss requires calculation of specific yield. The WINDS model utilizes integrated forms of the Brooks-Corey and van Genuchten equations to calculate water volume in the soil profile above a water table and to calculate the specific yield of the water table. This approach requires that some soil layers in the model are in hydraulic equilibrium with the water table while flow in upper layers is calculated based on energy differences and hydraulic conductivity. Lastly, this chapter discusses upward movement of water from the water table and compares Anat's equation to a discretized solution.

Combined Energy and Volume Balance Equation

As with Chap. 26, the vertical axis, z, is defined in the positive upward direction, the datum is placed at the bottom of the control volume, and layer numbers increase from the bottom to the top of the control volume (Fig. 28.1).

The Darcy velocity, v, is calculated with the Darcy equation with the energy gradient including both elevation and matric potential.

$$\mathrm{v} = -K_e \frac{H_j - H_{j+1}}{\Delta z} \tag{28.1}$$

where

H = h_c + z
Δz = height of layer, m,
H = total energy at center of layer, m,
h_c = matric potential, m,
z = elevation above datum, m.

The soil physics equation that is used to model water flow in soils is the Richard's equation.

$$\frac{\delta\theta}{\delta t} = \frac{\delta}{\delta z}\left[K(h)\frac{\delta h}{\delta z} + 1\right] \tag{28.2}$$

Thermal and osmotic energy gradients influence water movement in soils, but they are ignored in this chapter and in many soil-water models.

In the following derivation, j refers to the layer, j − 1 refers to the layer below, j + 1 refers to the layer above, and n refers to the number of layers. Z_{L-1}. The depth of water in the layer is θ Δz. The change in depth of water is the product of the change in water content and the thickness of the layer, Δθ Δz.

Equation 28.3 is derived from Eq. 27.20. Although the cross-sectional area between layers remains constant in soils, the hydraulic conductivity changes dramatically with water content so the effective conductivity, K_e, calculated as the geometric or arithmetic mean of the conductivities in the adjacent layers is used in Eq. 28.3.

$$\Delta\theta_j \Delta z = \left(-K_{e(j\&\,j-1)}\frac{H_j - H_{j-1}}{\Delta z} + K_{e(j\&\,j+1)}\frac{H_{j+1} - H_j}{\Delta z}\right)\Delta t \tag{28.3}$$

Divide both sides by Δz and solve for change in water content.

$$\Delta\theta_j = \frac{1}{\Delta z}\left(-K_{e(j\&\,j-1)}\frac{H_j - H_{j-1}}{\Delta z} + K_{e(j\&\,j+1)}\frac{H_{j+1} - H_j}{\Delta z}\right)\Delta t$$
$$\Delta\theta_j = \frac{\Delta t}{\Delta z^2}\left(-K_{e(j\&\,j-1)}\left(H_j - H_{j-1}\right) + K_{e(j\&\,j+1)}\left(H_{j+1} - H_j\right)\right) \tag{28.4}$$

P. Waller, M. Yitayew, *Irrigation and Drainage Engineering*, DOI 10.1007/978-3-319-05699-9_28

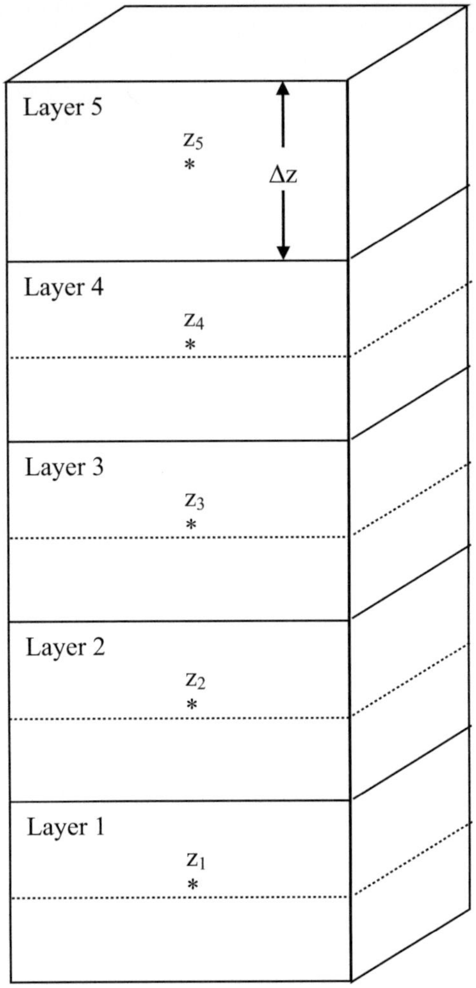

Fig. 28.1 Soil model with 5 layers beginning with layer 1 at the bottom of the control volume

Final water content is the initial water content + change in water content.

$$\theta_{j\text{-final}} = \theta_{j\text{-initial}} + \Delta\theta_j$$

$$\theta_{final} = \theta_{initial} + \frac{\Delta t}{\Delta z^2}\left(-K_{e(j\&\, j-1)}\left(H_j - H_{j-1}\right) + K_{e(j\&\, j+1)}\left(H_{j+1} - H_j\right)\right) \quad (28.5)$$

The total energy, H is the sum of matric potential energy for the layer and elevation of the midpoint of the layer above the datum. If the layers have the same thickness, Δz, then Δz can be substituted for the difference in elevation potential energy between two layers as follows.

$$H_j - H_{j-1} = \left(h_{cj} - h_{c\ j-1} + \Delta z\right)$$

$$\theta_{j\text{-}final} = \theta_{initial} + \frac{\Delta t}{\Delta z^2}\left(-K_{e(j\&\, j-1)}\left(h_j - h_{j-1} + \Delta z\right) + K_{e(j\&\, j+1)}\left(h_{j+1} - h_j + \Delta z\right)\right) \quad (28.6)$$

In the case of a confining layer at the lower end of the soil profile with no water movement below the root zone, the final water content in the lower layer (layer 1) is

$$\theta_{final} = \theta_{initial} + \frac{\Delta t}{\Delta z^2}\left(K_{e(1\,\&\,2)}\left(H_2 - H_1\right)\right) = \theta_{initial} + \frac{\Delta t}{\Delta z^2}\left(K_{e(1\,\&\,2)}\left(h_2 - h_1 + \Delta z\right)\right) \quad (28.7)$$

The final water content in the upper layer (layer n) is

$$\theta_{n\text{-}final} = \theta_{n\text{-}initial} + \frac{\Delta t}{\Delta z^2}\left(K_{e(n\,\&n-1)}\left(H_n - H_{n-1}\right)\right) = \theta_{n\text{-}initial} + \frac{\Delta t}{\Delta z^2}\left(K_{e(n\,\&\,n-1)}\left(h_n - h_{n-1} + \Delta z\right)\right) + \frac{i}{\Delta z} \quad (28.8)$$

where

i = infiltration from the soil surface (positive downward), m.

The new matric potential energy (Eq. 28.9) is calculated at the beginning of each time step, and substituted into Eqs. 28.3, 28.4, 28.5, 28.6, 28.7, and 28.8 in order to calculate flux and change in water content.

Brooks-Corey van Genuchten

$$h_c = \frac{h_b}{\left(\frac{\theta - \theta r}{\theta_s - \theta_r}\right)^{1/\lambda}} \qquad h_c = \frac{\left(\theta_e^{-1/m} - 1\right)^{1/n}}{\alpha} \quad (28.9)$$

The effective hydraulic conductivity between two layers, j and j + 1, can be calculated as the simple average or calculated as the geometric mean.

$$K_e = \frac{\Delta z}{\frac{\Delta z/2}{K_j} + \frac{\Delta z/2}{K_{j+1}}} = \frac{2}{\frac{1}{K_j} + \frac{1}{K_{j+1}}} \quad (28.10)$$

Neither the simple nor the geometric average is correct since the changes in water content and conductivity between two points are nonlinear. In the WINDS model, a conditional statement is used: the average mean is used if the water content is greater than a threshold value and if the water content in the upper layer is greater than the water content in the lower layer. The hypergeometric mean is used in if the either of the two conditions are not true.

Hydraulic conductivity is dramatically reduced with a reduction in water content because pore area available for water flow increases with water content and because the water path through the soil becomes more tortuous as water content decreases. The ratio of unsaturated hydraulic conductivity to saturated hydraulic conductivity is called the relative permeability, K_r.

$$K_r = \frac{K(\theta)}{K_S} \qquad (28.11)$$

where

$K(\theta)$ = conductivity at water content θ, cm/day
K_S = saturated hydraulic conductivity, cm/day
K_r = relative permeability

Permeability is a property of the soil matrix whereas conductivity is a function of both the soil matrix geometry and the water viscosity. The relative permeability, K_r, is the ratio between the permeabilities of the saturated and unsaturated soil conditions. Through experiments and mathematical derivation, Brooks-Corey found the following relationship between relative permeability and matric potential with the term η equal to $2 + 3\lambda$:

$$K_{rw} = \left(\frac{h_b}{h_c}\right)^{2+3\lambda} \quad K_{rw} = \left(\frac{h_b}{h_c}\right)^{\eta} \quad K(h) = K_s K_{rw} = K_s\left(\frac{h_b}{h_c}\right)^{\eta} \qquad (28.12)$$

Example 28.1 Find the relative permeability and effective saturation of Wagram loamy sand ($h_b = -30$ cm $= -0.3$ m, $\lambda = 1.27$) at $h_c = -100$ cm.

$$K_{rw} = \left(\frac{-30}{-100}\right)^{2+3(1.27)} = 0.0009$$

The hydraulic conductivity at −100 cm capillary pressure is approximately 1,000 times less then the saturated hydraulic conductivity. The effective saturation at −100 cm capillary potential is 0.21. Thus, the effective saturation is approximately 5 times lower than for a fully saturated soil, but the conductivity is approximately 1,000 times lower.

$$S_e = \left(\frac{h_b}{h_c}\right)^{\lambda} = \left(\frac{-30}{-100}\right)^{1.27} = 0.21$$

Example 28.2 Calculate the Darcy velocity between two points in an unsaturated Wagram loamy sand with matric potential values of −100 cm and −136 cm and elevations of 40 and 70 cm, respectively. The saturated hydraulic conductivity of Wagram loamy sand is 0.6 cm/hr. The distance between the two points is 50 cm. Use the Brooks-Corey model for hydraulic conductivity and the hypergeometric mean to calculate the effective conductivity.

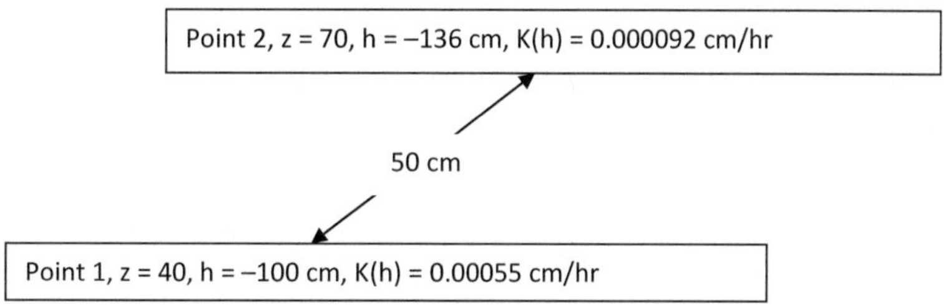

The relative conductivity, K_{rw}, at −100 cm was calculated in Example 28.1. Calculate the relative conductivity at −136 cm and the calculate actual hydraulic conductivity at each point.

$$K_{rw} = \left(\frac{-30}{-136}\right)^{2+3(1.27)} = 0.00015$$

$K(h_{100}) = K_S K_{rw\ 100} = 0.6$ cm/hr*0.00091 $= 0.00055$ cm/hr
$K(h_{136}) = K_S K_{rw\ 136} = 0.6$ cm/hr*0.00015 $= 0.000092$ cm/hr

Calculate the hypergeometric mean conductivity.

$$K_e = \frac{L_1 + L_2}{\frac{L_1}{K_1} + \frac{L_2}{K_2}} = \frac{25 + 25}{\frac{25}{0.000092} + \frac{25}{0.00055}} = 0.00016 \text{ cm/hr}$$

Calculate energy difference, h_f, between the points.

$$h_f = \Delta H = h_2 + z_2 - h_1 - z_1 = -136 + 70 = (-100) - 40 = -6 \text{ cm}$$

The fact that the energy gradient is negative means that the energy at point 1 is higher than the energy at point 2. Thus, flow is from point 1 to point 2. Calculate Darcy velocity, v.

$$v = -K_e\frac{\Delta H}{L} = -0.00016\frac{-6}{50} = 0.000019\ cm/hr$$

Van Genuchten derived an expression for relative permeability, which is used in the *WINDS* model, in terms of effective saturation.

$$K(h) = K(\theta) = K_0 \theta_e^{\,L}\left[1-\left(1-\theta_e^{\,1/m}\right)^m\right]^2 \qquad (28.13)$$

where

K(θ) = conductivity at water content θ, cm/day
K_0 = matching hydraulic conductivity at saturation, cm/day
L = curve fitting parameter, dimensionless number theoretically = 0.5 but not in actual practice (Table 7.2).

Equation 28.13 is written in terms of the matching hydraulic conductivity at saturation rather than the saturated hydraulic conductivity in order to match experimental curves more closely with the model. The matching and saturated hydraulic conductivity terms are similar but not the same (Table 27.2). The matching hydraulic conductivity is used instead of the hydraulic conductivity for the purpose of matching the mathematical curve to the experimental curve. The matching conductivity is in the *Soils* dialog box with the column heading Ko (Fig. 26.4).

Modeling Flux and Energy in Unsaturated Soils

The sections shows how *WINDS* models soil water content and flux in unsaturated soils with the Richards equation and the van Genuchten soil parameters. Example 28.3 shows that equilibrium is reached when total energy is the same in all layers.

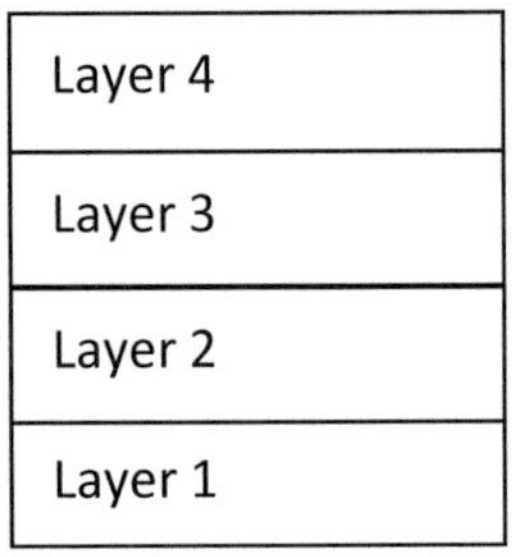

Example 28.3 Four layers have 0.5 m depth. Saturated water content = 0.45. Residual water content = 0.067, n = 1.41, α = 0.02, L = 0.5, and K_0 = 10.8 cm/day. Initial water content in all layers is 30 %. There is no infiltration and no seepage of water below the control volume. Use daily time steps. Calculate effective hydraulic conductivity between layers with different water contents with the geometric mean. Make manual calculations for two time steps and compare with *WINDS* model output. Run the *WINDS* model for 200 days. The information for this simulation is in the #1 position in the *Crop_data* worksheet.

Calculate m.

$$m = 1 - 1/n = 1 - 1/1.41 = 0.29$$

Time step 1 (i = 1). Layer 1. (j = 1)

Calculate initial effective water content and matric potential.

$$\theta_e = \frac{\theta-\theta_r}{\theta_s-\theta_r} = \frac{0.30-0.067}{0.45-0.067} = 0.608$$

Calculate initial matric potential.

$$h_c = -\frac{\left(\theta_e^{-1/m}-1\right)^{1/n}}{\alpha} = -\frac{\left(0.608^{-1/0.29}-1\right)^{1/1.41}}{0.02}$$
$$= -145.85\ cm = -1.4585\ m$$

Calculate hydraulic conductivity.

$$K(\theta) = K_0\theta_e^{\,0.5}\left[1-\left(1-\theta_e^{\,1/m}\right)^m\right]^2$$
$$= 10.8^*(0.608)^{0.5}\left[1-\left(1-0.608^{1/0.29}\right)^{0.29}\right]^2$$
$$= 0.0268\ cm/d = 0.000268\ m/d$$

Calculate final water content in layer 1.

$$\theta_{final} = \theta_{initial} + \frac{\Delta t}{\Delta z^2}\left(K_{e(1\,\&\,2)}(h_2-h_1+\Delta z)\right)$$
$$\theta_{final} = 0.30 + \frac{1}{0.5^2}(0.000268(-1.4585-(-1.4585)+0.5))$$
$$= 0.30054$$

Layer 2. (j = 2) Fluxes from top and bottom are the same so there is no change in water content.

Day 2

Calculate new effective water content in layer 1 (layers 2 and 3 remain the same)

$$\theta_{e-1} = \frac{\theta-\theta_r}{\theta_s-\theta_r} = \frac{0.30054-0.067}{0.45-0.067} = 0.6098$$

Calculate new matric potential in layer 1.

$$h_{c-1} = -\frac{\left(\theta_e^{-1/m}-1\right)^{1/n}}{\alpha} = -\frac{\left(0.6098^{-1/0.29}-1\right)^{1/1.41}}{0.02}$$
$$= -144.85\ cm = -1.4485\ m$$

Calculate new hydraulic conductivity in layer 1 (layers 2 and 3 remain the same).

$$K(\theta)_1 = K_0\theta_e^{0.5}\left[1-\left(1-\theta_e^{1/m}\right)^m\right]^2$$
$$= 10.8*(0.608)^{0.5}\left[1-\left(1-0.608^{1/0.29}\right)^{0.29}\right]^2$$
$$= 0.0273\ cm/d = 0.000273\ m/d$$

Calculate effective hydraulic conductivity between layers 1 and 2 with the geometric mean.

$$K_{e-1-2} = \frac{2}{\frac{1}{K_j}+\frac{1}{K_{j+1}}} = \frac{2}{\frac{1}{0.000268}+\frac{1}{0.000273}}$$
$$= 0.00027049\ m/day$$

Calculate final water content in layer 1.

$$\theta_{1-final} = \theta_{2-initial} + \frac{\Delta t}{\Delta z^2}\left(K_{e(2\&\ 1)}(h_2 - h_1 + \Delta z)\right)$$
$$\theta_{1-final} = 0.30054 + \frac{1}{0.5^2}(0.0002705*\ (-1.4585 - (-1.4485) + 0.5)) = 0.3011$$

Calculate the final water content in layer 2.

$$\theta_{2-final} = \theta_{2-initial} + \frac{\Delta t}{\Delta z^2}\left(-K_{e(2\&\ 1)}(h_2 - h_1 + \Delta z) + K_{e(2\,\&\,3)}(h_3 - h_2 + \Delta z)\right)$$
$$\theta_{2-final} = 0.3 + \frac{1}{0.5^2}(-0.0002705,*(-1.4585,-(-1.4485)\ +0.5) + 0.000268(0.5)) = 0.300006$$

The calculated values agree with the WINDS model calculations on the first two days in layers 1 and 2. The following text explains the calculations in written form.

Water content in layers 2 and 3 = 0.3 m/m
Energy difference between layer 3 and layer 2 = 0.5
Average conductivity and flow in vertical direction = 0.000268 m/day
Water content change in layer 2 = 0.000268 m/0.5 m = +0.000536 m/m
Updated water content in layer 2 = 0.300536 – the same as the layer 1 water content

Energy difference = 0.5 + (−0.014585 − (−0.014485)) = 0.49
Average conductivity and flow in vertical direction = (0.000268 + 0.000273)/2 = 0.0002705
Water content change in layer 2 = 0.49*0.0002705/0.5^2 = −0.0005302 m/m
Updated water content in layer 2 = 0.300536 − 0.0005302 = 0.300006

	I	J	K	L	M
34	TRUE	No infiltration			
35	TRUE	No ET			
36	FALSE	No stress reduction			
37	TRUE	No ET frac adjustment			
38	FALSE	Continue drainage rate			
39	FALSE	No redistribution by Richards equation			
40	TRUE	Turn off field capacity restriction			
41	TRUE	Bypass upper layer in nitrogen application			
42	FALSE	No nitrogen frac adjustment			

Fig. 28.2 Main page true-false values for Example 28.3

In order to run the model in WINDS, the True-False values in column I must be set as shown in Fig. 28.2. Turn off salinity and nitrogen calculations in row 5. The van Genuchten parameters are set in the *Soils* dialog box (Fig. 28.3). With the sensitivity of energy-driven water flux calculations to discontinuities, the evaporation layer cannot be ignored as it was in Chap. 26. It becomes the upper layer in this example, but with no evaporation.

Water nearly reaches equilibrium during the 200 day simulation period (Fig. 28.4). Equilibrium is reached when the total energy (Fig. 28.5), which is the elevation energy + matric potential energy (Fig. 28.6), is the same in all layers.

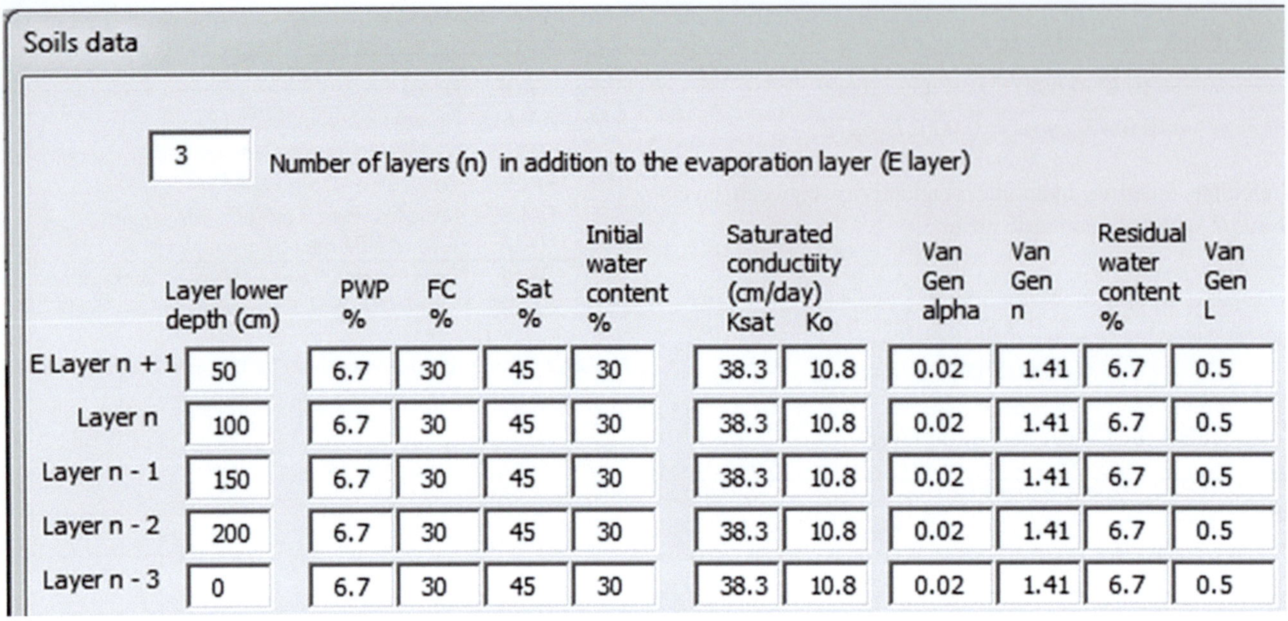

Soils data

3 Number of layers (n) in addition to the evaporation layer (E layer)

	Layer lower depth (cm)	PWP %	FC %	Sat %	Initial water content %	Saturated conductiity (cm/day) Ksat	Ko	Van Gen alpha	Van Gen n	Residual water content %	Van Gen L
E Layer n + 1	50	6.7	30	45	30	38.3	10.8	0.02	1.41	6.7	0.5
Layer n	100	6.7	30	45	30	38.3	10.8	0.02	1.41	6.7	0.5
Layer n - 1	150	6.7	30	45	30	38.3	10.8	0.02	1.41	6.7	0.5
Layer n - 2	200	6.7	30	45	30	38.3	10.8	0.02	1.41	6.7	0.5
Layer n - 3	0	6.7	30	45	30	38.3	10.8	0.02	1.41	6.7	0.5

Fig. 28.3 Soil parameters for Example 28.3

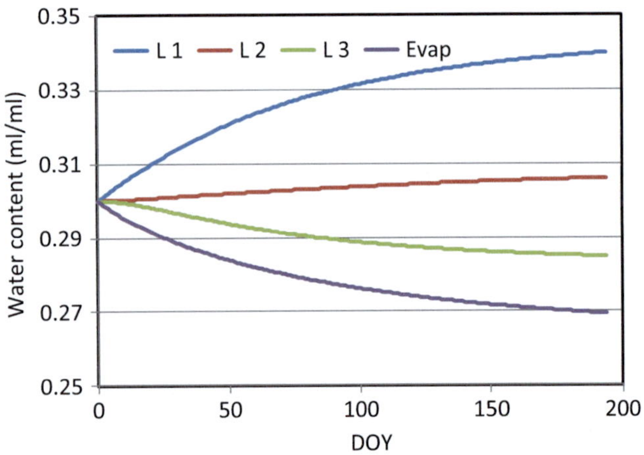

Fig. 28.4 Water content versus time from *WINDS* model for Example 28.3

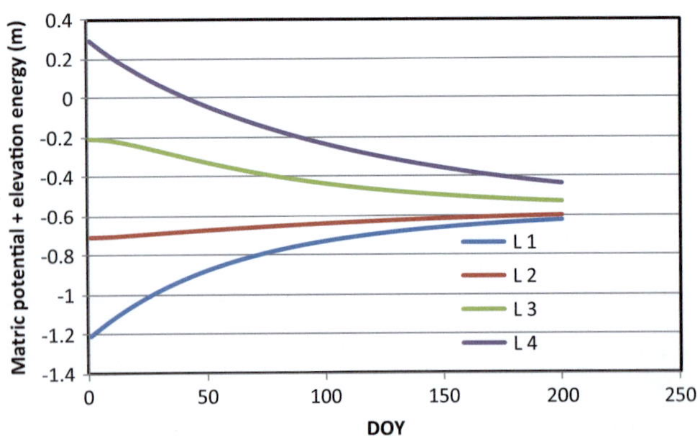

Fig. 28.5 Total energy versus time from *WINDS* model for Example 28.3

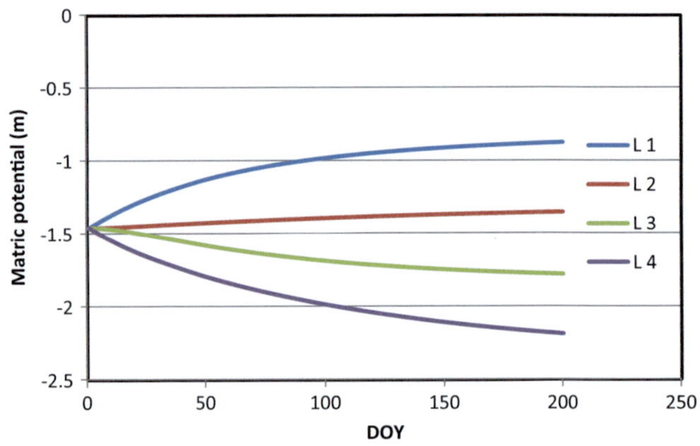

Fig. 28.6 Matric potential versus time from *WINDS* model for Example 28.3

Example 28.4 Leaving other parameters the same as in 28.3, change the van Genuchten parameters to $n = 1.75$ and $\alpha = 0.078$. Run the WINDS model for 300 days. Results are shown in Fig. 28.7. The information for this simulation is in the #2 position in the *Crop_data* worksheet.

Example 28.4 demonstrates the dramatically different results that can be obtained with two soils with the same textural classification but with different properties. The soils in Examples 28.3 and 28.4 are silt loams. Layer 1 becomes saturated after 50 days (there is no leaching below layer 1) in Example 28.4.

Normally, there is not an impermeable layer directly below the root zone (as in Example 28.4), and water either pools in a water table or seeps to a lower soil layer. For soils without a water table, the same matric potential can be assigned to the region below the lower layer in order to allow the lower layer to drain by gravity (difference in

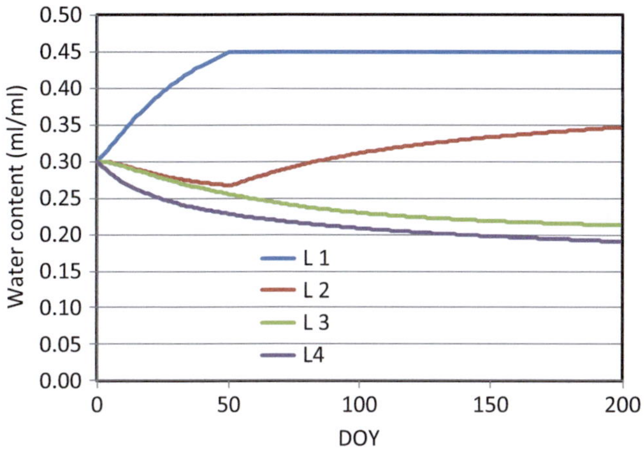

Fig. 28.7 Water content versus time from WINDS model for Example 28.4

head equal to Δz). This is accomplished by rewriting the water balance equation for layer 1 (Eq. 28.6) as

$$\theta_{final} = \theta_{initial} + \frac{\Delta t}{\Delta z^2}\left(K_{e(1\,\&\,2)}(h_2 - h_1 + \Delta z) - K_1 \Delta z\right) \quad (28.14)$$

where K_1 Δt is the flux below layer 1. Equation 28.14 is used at the lower boundary in Example 28.5. For saturated soils with a water table, seepage rate is generally a function of water table elevation, which is described in the following sections.

Water and Energy Distribution with a Water Table

When a lower confining layer such as in Example 28.4 prevents the rapid downward movement of water, the lower soil layers reach saturation. In soils with a water table, the soil can be divided into the saturated zone below the water table and the vadose zone above the water table. The saturated zone may lose water slowly to a drainage system or to a low permeability aquitard below the soil. The water just above the water table tends to remain in hydraulic equilibrium with the water table. However, at some distance above the water table, matric potentials become more negative due to water extraction by evapotranspiration, and the layers disconnect hydraulically from the water table. Thus, *WINDS* models soils with a water table in two parts: the upper section in which water moves from layer to layer by matric potential gradients caused by ET, and the lower section, which remains in equilibrium with the water table.

Specific yield is the depth of water drained from the soil per change in depth of the water table. The concept is used in both aquifer and soil drainage calculations to estimate the amount of water that is removed from an aquifer for a given change in water table elevation. The United States Bureau of Reclamation calculates specific yield of a soil as the total water drained when the water table drops from the soil surface to a given depth.

$$SY = \frac{d_d}{DTWT} \quad (28.15)$$

where

SY = Specific yield, dimensionless,
d_d = Depth drained, m,
DTWT = Depth to the water table from the soil surface, m.

Example 28.5 20 cm depth of water is drained from the soil as the water table is lowered from the soil surface to 2 m. Calculate the specific yield.

$$SY = \frac{d_d}{DTWT} = \frac{0.2\ m}{2\ m} = 0.1 = 10\%$$

Specific yield can also be defined as the incremental change in volume drained per incremental change in water table depth.

$$SY = \frac{\Delta d_d}{\Delta DTWT} = -\frac{\Delta d_d}{\Delta z_{WT}} \quad (28.16)$$

where

Δz_{WT} = change in water table elevation above a datum (z is positive upward), m.

Example 28.6 The water table drops from 1.5 to 1.6 m below the ground surface, and the incremental specific yield at 1.5 m depth is 20 %. Calculate the depth of water drained. Calculate in terms of the elevation above the datum instead of the depth to the water table to maintain consistency with later equation development.

Assume that the datum is at 2 m depth. Initial water table elevation is 0.5 m above the datum and final water table elevation is 0.4 m.

$$\Delta z_{WT} = z_{WT-2} - z_{WT-1} = 0.4\ \text{m} - 0.5\ \text{m} = -0.1\ \text{m}$$
$$d_d = -SY^{*}\Delta z_{WT} = -0.2^{*}(-0.1) = 0.02\ m$$

Specific yield can be calculated with the van Genuchten or Brooks-Corey equations. For the Brooks-Corey equation, if the water table is lowered from the soil surface elevation, z_t, to an elevation, z_{WT}, then the total depth drained, d_d, is the area of the hatched region in Fig. 28.8. The total depth of water in the soil above the water table is the integral of the equations from the water table to the soil surface. Depth drained from the soil when the water table drops from the soil surface to elevation, z_{WT}, is calculating by subtracting the drained depth

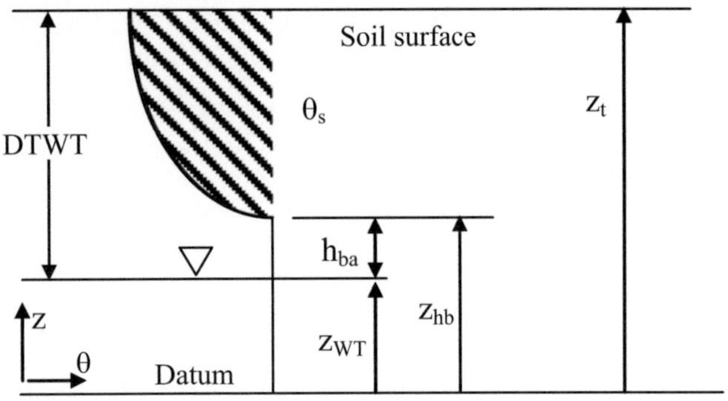

Fig. 28.8 Schematic for calculation of volume drained

from the total depth in the control volume. The drained depth is the integral of the hatched area in Fig. 28.8 with limits of integration as the absolute value of the matric potential at the lower (h_{ba}) and upper ($z_t - z_{hb} + h_{ba}$) boundaries.

z_{WT} = elevation of water table
z_{hb} = elevation of bubbling pressure
z_t = elevation of surface above datum.
h_{ba} = Abs. value of bubbling pressure

All units in m.

The area of the hatched region in Fig. 28.7 is calculated as follows:

$$d_d = \theta_s(z_t - z_{hb}) - \int_{hba}^{Zt-ZWT} \theta(z)dz \tag{28.17}$$

where

z_t = elevation of soil surface above the datum, m,
z_{hb} = elevation of bubbling pressure point above datum ($z_{WT} + h_b$), m,
d_d = total depth drained from profile, m,
$\theta(z)$ = water content at elevation z above the datum, m/m,

The Brooks-Corey equation is substituted for $\theta(z)$

$$\theta(z) = \theta_r + (\theta_s - \theta_r)\left(\frac{h_b}{h_c}\right)^{\lambda} = \theta_r + (\theta_s - \theta_r)\left(\frac{h_{ba}}{z}\right)^{\lambda}$$

$$d_d = \theta_s(z_t - z_{hb}) - \left(\int_{hba}^{Zt-ZWT}\left(\theta_r + (\theta_s - \theta_r)\left(\frac{h_{ba}}{z}\right)^{\lambda}\right)dz\right)$$

$$d_d = \theta_s(z_t - z_{hb}) - (z_t - z_{hb})\theta_r - \frac{(\theta_s - \theta_r)h_{ba}^{\lambda}}{-\lambda + 1}\left((z_t - z_{WT})^{-\lambda+1} - (h_{ba})^{-\lambda+1}\right)$$

$$d_d = (\theta_s - \theta_r)(z_t - z_{hb}) - \frac{(\theta_s - \theta_r)h_{ba}^{\lambda}}{-\lambda + 1}\left((z_t - z_{WT})^{-\lambda+1} - (h_{ba})^{-\lambda+1}\right)$$

$$d_d = (\theta_s - \theta_r)\left(z_t - z_{hb} - \frac{h_{ba}^{\lambda}}{-\lambda + 1}\left((z_t - z_{WT})^{-\lambda+1} - (h_{ba})^{-\lambda+1}\right)\right) \tag{28.18}$$

Example 28.7 Calculate the depth of water drained from the soil profile and specific yield if the water table drops from the soil surface to a depth of 1 m in Wagram loamy sand. Use Eq. 28.18. Assume that the datum is 2 m below the soil surface.

Water table depth = 1.0 m, bubbling pressure, h_{ba} = 0.3 m, Saturated water content, θ_s = 0.305,

Residual water content, θ_r = 0.044, Pore size distribution index, λ = 1.27

$$z_{hb} = z_{WT} + h_{ba} = 1 + 0.3 = 1.3 \text{ m}$$

$$d_{d(1.0\ meter)} = 0.261\left(2 - 1.3 - \frac{0.3^{1.27}}{-1.27 + 1}\left((2-1)^{-1.27+1} - 0.3^{-1.27+1}\right)\right) = 0.1022\ m$$

Thus, 10 cm of water is drained from the soil profile when the water table drops from the soil surface to 1 m depth. The specific yield is calculated as follows:

$$SY = \frac{d_d}{DTWT} = \frac{0.10}{1} = 0.1 = 10\%$$

The incremental specific yield is the depth drained for a given change in water table depth from one position to the next. This is found by taking the derivative of d_d with respect to z_{hb}. Fortunately, terms cancel in the derivative so Eq. 28.19 is relatively short.

$$SY = \frac{d(d_d)}{d(z_{WT})} = -(\theta_s - \theta_r)D_{zhb}\left(z_t - z_{hb}, -\frac{h_{ba}^{\lambda}}{-\lambda + 1}\left((z_t - z_{WT})^{-\lambda+1} - (h_{ba})^{-\lambda+1}\right)\right)$$

$$SY = \frac{d(d_d)}{d(z_{WT})} = -(\theta_s - \theta_r)\left[-1 + \left(\frac{h_{ba}}{z_t - z_{WT}}\right)^{\lambda}\right] = -(\theta_s - \theta_r)\left[-1 + \left(\frac{h_{ba}}{DTWT}\right)^{\lambda}\right] \tag{28.19}$$

Example 28.8 For Wagram loamy sand, find the incremental specific yield if the water table is at 1 m depth.

$$\frac{d(d_d)}{d(z_{WT})} = -(0.305 - 0.044)\left[-1 + \left(\frac{0.3}{1}\right)^{1.27}\right]100\% = S = 20.5\%$$

Thus, for every 1 cm change in water table depth, 0.205 cm is drained from the soil. For example, if the water table depth increased from 1 to 1.02 m (z_{wt} decreased from 1 to 0.98 m), then 0.02 m * 20.5 = 0.0041 m would be drained from the soil. Depth drained would be 0.1022 + 0.0041 = 0.1063. Check the results by calculating the total volume drained at 1.02 m with Eq. 28.18.

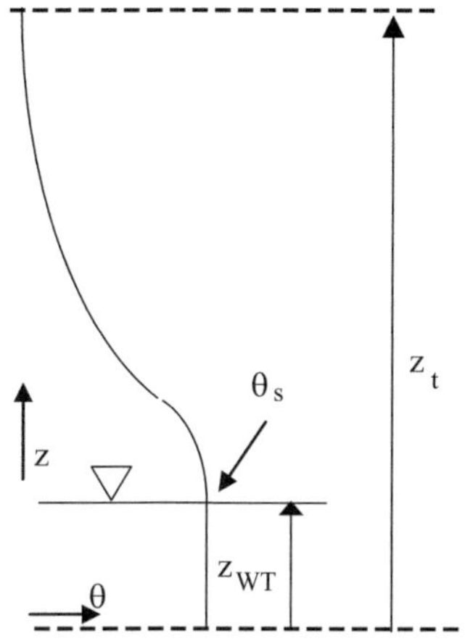

Fig. 28.9 Water profile above datum for soil with a water table for Van Genuchten model

$$d_{d(1.02\ meter)} = 0.261\left(2-, 1.28, -\frac{0.3^{1.27}}{-1.27+1}\left(\left(1.02\right)^{-1.27+1} - 0.3^{-1.27+1}\right)\right) = 0.1063\ m$$

Equation 28.17 can be modified to find the total depth of water in the soil profile: this technique is used in the *WINDS* model for the lower layers that are in equilibrium with the water table.

$$d_{total} = \theta_s(z_{hb}) + \int_{hba}^{Zt-ZWT} \theta(z)dz \qquad (28.20)$$

where

d_{total} = total depth of water in the soil profile, m,

Integrate Eq. 28.20 to find the total depth of water in the soil profile.

$$d_{total} = \theta_s(z_{hb}) + \theta_r(z_t - z_{hb}) + \frac{(\theta_s - \theta_r)h_{ba}^{\lambda}}{-\lambda+1}\left(\left(z_t - z_{WT}\right)^{-\lambda+1} - \left(h_{ba}\right)^{-\lambda+1}\right) \qquad (28.21)$$

The depth of water in a soil profile with a water table can also be found with the van Genuchten equation (Fig. 28.9). Write Eq. 28.20 in terms of van Genuchten parameters as a function of elevation z (matric potential). Substitute z, elevation above the water table, for – h_c. This is the water table algorithm in the *WINDS* model.

$$d_{total} = \theta_s(z_{WT}) + \int_0^{Zt-Zwt} \theta(z)dz \qquad (28.22)$$

$$\theta = \theta_r + (\theta_s - \theta_r)[1 + (-\alpha\ {*}h_c)^n]^{-m}$$

$$d_{total} = \theta_s(z_{WT}) + \int_0^{Zt-Zwt} \theta_r + (\theta_s - \theta_r)[1 + (\alpha\ {*}z)^n]^{-m}dz \qquad (28.23)$$

The right side of the integrand in Eq. 28.23 is the Gauss hypergeometric function, ${}_2F_1(a, b, c, w)$ where a = 1/n, b = 1 – 1/n, c = 1 + 1/n, and $w = -((\alpha\ (z_t - z_{WT}))^n$. Note that w is substituted for z because z represents elevation.

$$\int_0^{Zt-Zwt} (1 + (\alpha\ z)^n)^{(1/n-1)} = z\ * {}_2F_1\left(\frac{1}{n},\ 1-\frac{1}{n};\ 1+\frac{1}{n};\ -((\alpha\ (z_t - z_{WT}))^n)\right) \qquad (28.24)$$

The hypergeometric function (Equation 28.24) does not converge if the fourth term is greater than 1. In these water table calculations, the fourth term is greater than 1 if (z_t – z_{WT}) is large and/or α is large. A transformation can be made such that w in the hypergeometric function is always less than 1.

$${}_2F_1(a, b, c, w) = (1-w)^{-b}{}_2F_1(c-a,\ b;\ c, w/(w-1))$$
where $w = -((\alpha(z_t - z_{WT}))^n)$

With the transformation, the integral of Eq. 28.23 is

$$d_{total} = \theta_s(z_{WT}) + \theta_r(z_t - z_{WT}) + (\theta_s - \theta_r)(z_t - z_{WT})(1-w)^{1/n-1}\ {}_2F_1\left(1,\ 1-\frac{1}{n};\ 1+\frac{1}{n},\ \frac{w}{w-1}\right) \qquad (28.25)$$

It is not possible to find the derivative of Eq. 28.25 in order to find the incremental specific yield. Thus, the incremental specific yield must be found by solving for d_{total} at two different water table elevations, and then calculating the incremental specific yield with the two points as shown in Example 28.9.

Example 28.9 Calculate the incremental specific yield at a water table depth below the soil surface of 0.45. Van Genuchten parameters are n = 1.75, α = 0.078, z_t = 0.55 m, z_{WT} = 0.1 m, θ_s = 0.45, and θ_r = 0.08 (Figure 28.10).

Because the van Genuchten parameters are listed in terms of cm, w must be calculated in units of cm so multiply by 100.

$$w = -((\alpha(z_t - z_{WT}))^n)$$
$$w = -\left((0.078{*}100{*}(0.55 - 0.1))^{1.75}\right) = -9.35$$

The hypergeometric function is passed the following parameters

$$hg(1, 1 - 1/1.75, 1 + 1/1.75, -9.35/(-9.35 - 1))$$

The hypergeometric function returns a value of 1.67

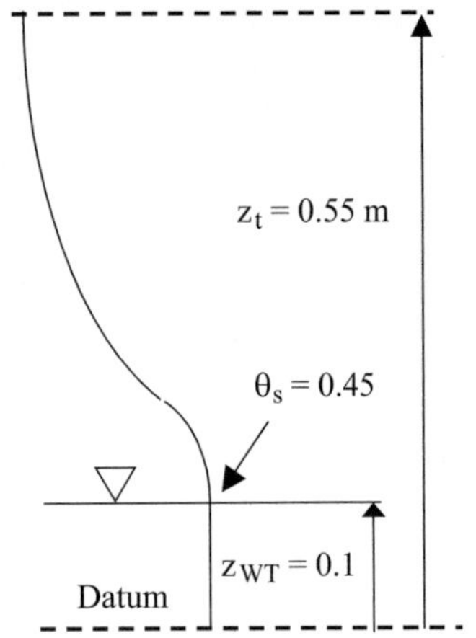

Fig. 28.10 Parameters for solution of Example 28.9

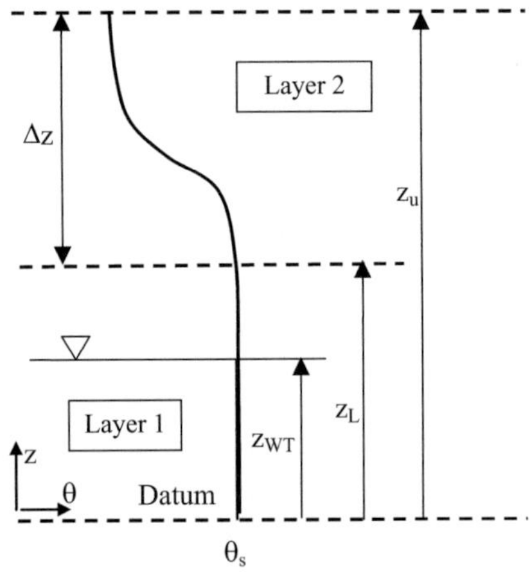

Fig. 28.11 Parameters for calculation of depth of water in a soil layer

The total depth of water in the soil is calculated as follows.

$$d_{total} = \theta_s(z_{WT}) + \theta_r(z_t - z_{WT}) + (\theta_s - \theta_r)(z_t - z_{WT})\,(1 - w)^{1/n-1}\,{}_2F_1\left(1,\ 1 - \frac{1}{n};\ 1 + \frac{1}{n},\ \frac{w}{w-1}\right)$$

$$d_{total} = 0.45(0.1) + 0.08(0.55 - 0.1) + (0.45 - 0.08)\,(0.55 - 0.1)(1 - (-9.35))^{(1/1.75-1)} * 1.67 = 0.18\ m$$

Calculate the incremental specific yield by also calculating the total depth of water in the soil profile at $z_{WT} = 0.09$ m → 0.1821 m. SY = (0.1844–0.1821)/(0.1–0.09) = 0.23 → 23 %.

The equations can be modified to find the depth of water within an individual layer that contains the water table (Layer 1 in Fig. 28.11).

The water depth below the water table is added to the water depth above the water table, where z_L and z_u are not the same as the z_L and z_u shown in Fig. 28.11 but refer to the lower and upper limits of Layer 1.

$$d_{layer} = \theta_s(z_{WT} - z_L) + \int_0^{Zu-Zwt} \theta(z)dz \qquad (28.26)$$

where

z_u = elevation of upper boundary of layer, m,
z_L = elevation of lower boundary of layer, m.
d_{layer} = depth of water within the layer, m.

Integrate Eq. 28.26.

$$d_{layer} = \theta_s(z_{WT} - z_L) + \theta_r(z_u - z_{WT}) + (\theta_s - \theta_r)\,(z_u - z_{WT})(1 - w)^{1/n=1}\,{}_2F_1\left(1,\ 1 - \frac{1}{n};\ 1 + \frac{1}{n},\ \frac{w}{w-1}\right) \qquad (28.27)$$

where

$$w = -((\alpha(z_u - z_{WT}))^n)$$

If a layer is completely above the water table such as Layer 2 in Fig. 28.11, and the layer is in hydraulic equilibrium with the water table, then the depth of water in the layer is found as follows.

$$d_{layer} = \int_{Z_L - Z_{WT}}^{Zu - Z_{WT}} \theta(z)dz$$

$$d_{layer} = \theta_r(z_u - z_L) + (\theta_s - \theta_r)(z_u - z_{WT})(1 - w_u)^{1/n-1}\,{}_2F_1\left(1,\ 1 - \frac{1}{n};\ 1 + \frac{1}{n},\ \frac{w_u}{w_u - 1}\right) - (\theta_s - \theta_r)(z_L - z_{WT})(1 - w_L)^{1/n-1}\,{}_2F_1\left(1,\ 1 - \frac{1}{n};\ 1 + \frac{1}{n},\ \frac{w_L}{w_L - 1}\right) \qquad (28.28)$$

where

$$w_u = -((\alpha(z_u - z_{WT}))^n)$$
$$w_L = -((\alpha(z_L - z_{WT}))^n)$$

The downward rate of seepage from the soil profile (from Layer 1 to below Layer 1) is controlled by the permeability of the confining layer below or by the resistance to flow to a subsurface drain system and by the depth of the water table. The drainage rate is a function of this permeability, or resistance, and the elevation of the water table.

$$d_{drainage} = f(K_d, z_{WT}, \Delta t) \qquad (28.29)$$

where

K_d = conductivity of the drainage system, m/day,
Δt = time, days,
$d_{drainage}$ = depth of drainage below the control volume, m.

The equation for downward flux may be as simple as the following equation or it may be very complex and depend on the geometry of the subsurface flow to drains.

$$d_{drainage} = K_d * z_{WT} * \Delta t \quad (28.30)$$

One of the challenges with this drainage model is that the model crashes if the theoretical framework is inaccurate. This mass transfer between layers must be calculated with algorithms that have some level of theoretical validity. If only a single layer was connected to the water table, then the specific yield values would be unrealistic. This is complicated by the fact that the water table rises and falls; thus, the boundary between the hydraulically connected lower layers and hydraulically disconnected upper layers also rises and falls. The following example demonstrates how the *WINDS* model transfers water from the upper soil region to a lower region that is in equilibrium with the water table.

Water content in those layers that are the in equilibrium with the water table are all adjusted according to Eqs. 28.26 and 28.27 after each time step. Those layers that are not connected are only adjusted based on the energy difference and conductivity between layers. This technique prevents the upward movement of water from the water table connected layers into layers that are dried by evapotranspiration. The user can specify the approximate number layers in equilibrium with the water table by adjusting the fraction of saturation that causes connection of layers to the water table.

Example 28.10 A soil has four layers, 0.5 m thickness (Fig. 28.12). The soil parameters are the same as in Example 28.4 except that saturated water content is 40 % instead of 45 %. The rate of subsurface drainage below layer 1 is a function of the water table elevation, z_{WT}: $d_{drainage} = z_{WT}$ * 0.01 m/day * 1 day. The initial elevation of the water table is 1.5 m, and all layers are in equilibrium with the water table at the beginning of the first day (Fig. 28.13). Afterward, the saturation fraction required for equilibrium is 0.68. There is no irrigation or evapotranspiration, Run the *WINDS* model for 100 days. The information for this simulation is in the #3 position in the *Crop_data* worksheet. The drainage parameters are in Fig. 28.13. They are also in cells B231: B248 in the *Active_data* sheet.

The elevation of the water table vs. time is shown with the red line in Fig. 28.14. The equilibrium layer (blue line) descends with the water table. The change in water content vs. time is shown in Fig. 28.15. A comparison of Figs. 28.14

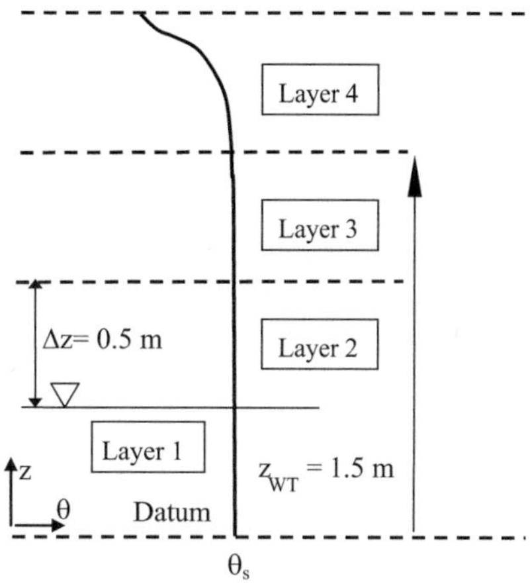

Fig. 28.12 Geometry for Example 28.10

and 28.15 reveals that once layers are not in equilibrium with the water table, the rate of water content change decreases dramatically. The number of layers in equilibrium with the water table would be changed by changing the fraction of saturation number (increase from 0.68) in the *Drainage* dialog box (Fig. 28.13).

The mass balance (Fig. 28.16) shows that mass is conserved. The change in water depth in the soil profile corresponds with the amount of water lost to drainage. In addition, the drainage rate is as expected (15 mm/day at 1.5 m water table elevation, etc...). The smoothness of curve is the result of including several layers in the water table change calculation. Otherwise, specific yield fluctuates with proximity to layer boundaries. Seepage (infiltration) below each layer vs. time is shown in Fig. 28.17. All layers with saturated lower boundaries have the same seepage rate. Once the upper end of a layer begins to become unsaturated, the amount of water entering the upper end of the layer does not match the amount of water exiting the lower end of the layer.

Example 28.11 A soil has four layers, 0.5 m thickness (Fig. 28.18). Calculate the initial and final water contents, and the change in water table depth over one day for the soil parameters in Example 28.4; however, change the saturated water content is 40 % instead of 45 %. The rate of subsurface drainage below layer 1 is a function of the water table elevation, z_{WT}: $d_{drainage} = z_{WT}$ * 0.01 m/day * days). The initial elevation of the water table is 0.5 m. Let all layers be in equilibrium with the water table for the initial distribution of water, but only the lower two layers are in equilibrium with the water table afterwards. Evapotranspiration removes 0.5 cm/day from each of layers 3 and 4. Compare calculations with the WINDS model. The information for this simulation is in the #4 position in the *Crop_data* worksheet.

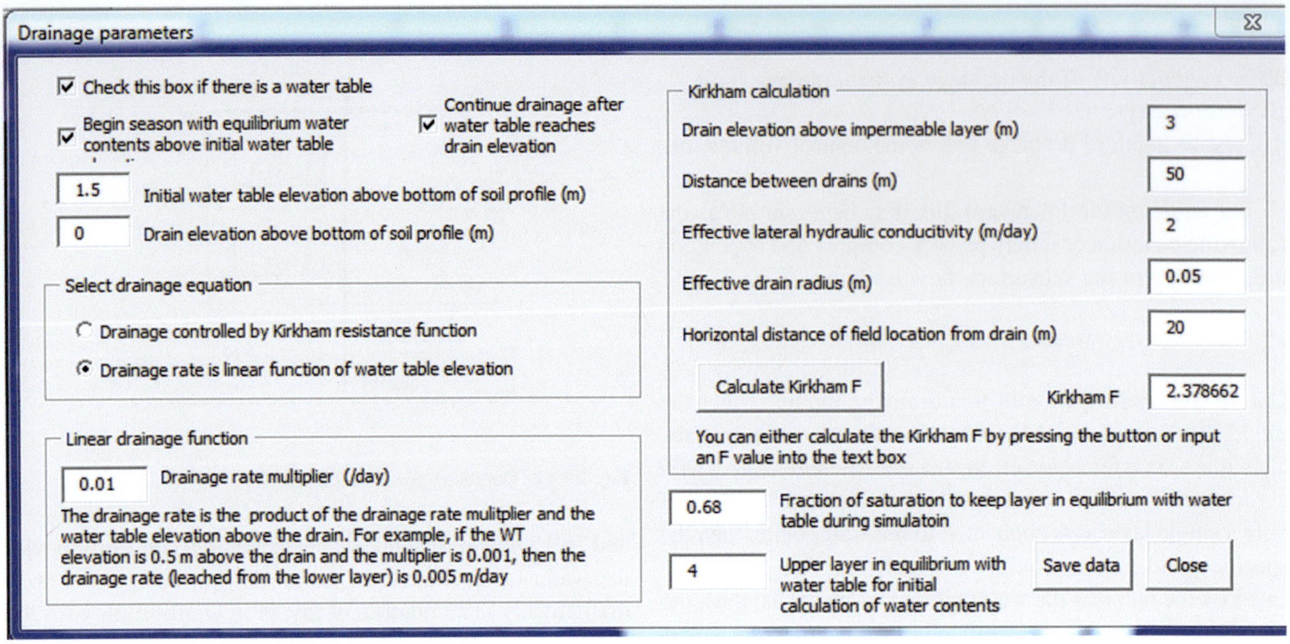

Fig. 28.13 *Drainage* dialog box for Example 28.10

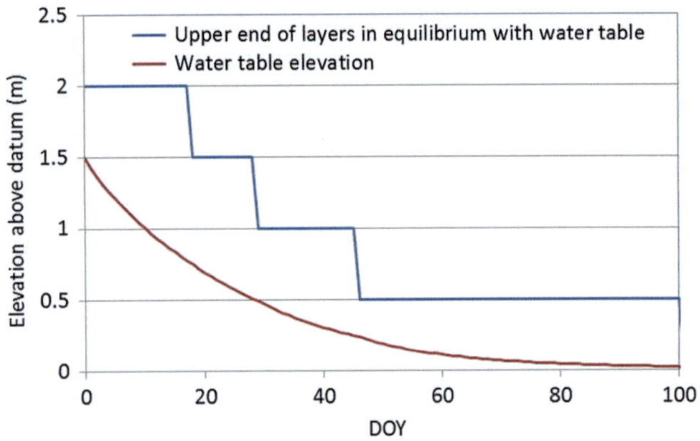

Fig. 28.14 Position of water table and cells in equilibrium with the water table versus time for Example 28.10

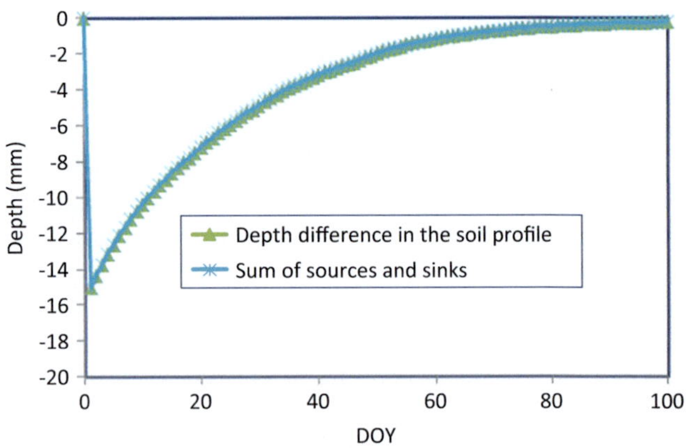

Fig. 28.16 Volume balance for Example 28.10

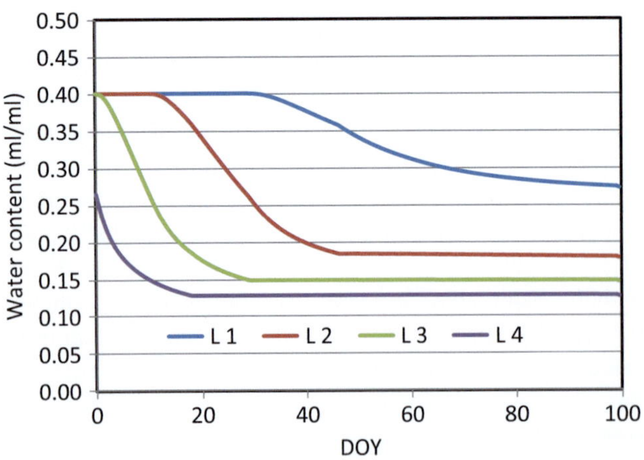

Fig. 28.15 Change in water content versus time for Example 28.10

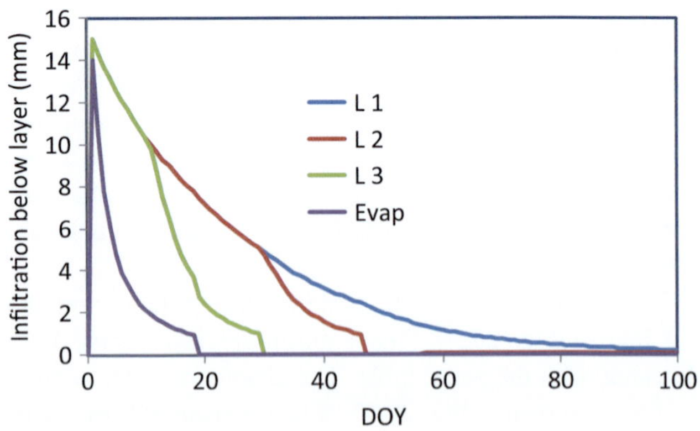

Fig. 28.17 Seepage below each layer for Example 28.10

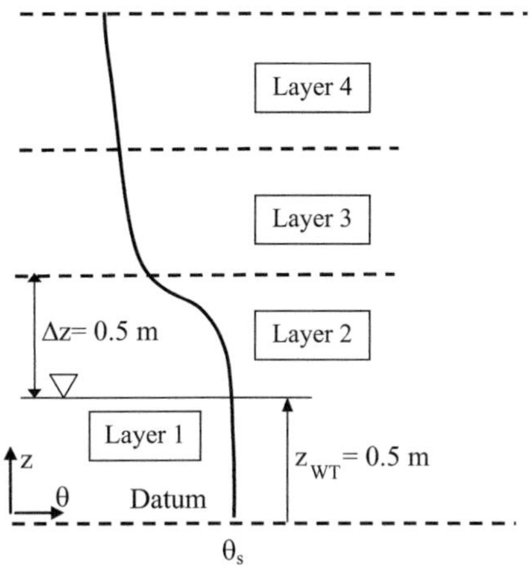

Fig. 28.18 Geometry for Example 28.11

In order to run this example, the Kcb values are set to 1.0 in cells B13:B15 in the *Active_data* worksheet, and the No_ET value in cell I35 in the *Main* worksheet must be set to False, and cells C2:C3 in the *ET_fractions* worksheet must be set to 0.5 to extract water from the upper two layers in which the root zone occupies 2 layers. Set the root zone depth = 0.99 m so that WINDS does not extract water from layer 2, as it would if the roots touched layer 2 (1.0 m).

The water table is at the upper boundary of layer 1 so layer 1 is saturated (water content = 40 %).

$$d_{layer} = 0.4\ m/m(0.5\ m) = 0.2\ m$$

The average water content in layer 2 can be found with Eq. 28.28 with the water table elevation $z_{WT} = z_L$ for layer 2 (layer 2 is in equilibrium with the water table).

$$\begin{aligned} d_{layer} &= \theta_s(z_{WT} - z_L) + \theta_r(z_u - z_{WT}) \\ &+ (\theta_s - \theta_r)(z_u - z_{WT})(1 - w)^{1/n=1} \\ &{}_2F_1\left(1,\ 1 - \frac{1}{n};\ 1 + \frac{1}{n},\ \frac{w}{w-1}\right) \\ w &= -((\alpha(z_u - z_{WT}))^n) = -\left((0.078*100(0.5))^{1.75}\right) \\ &= -10.823 \\ d_{layer} &= 0.4(0) + 0.067(0.5) + (0.33)(0.5) \\ &(1 - (-10.823))^{1/1.75=1} \\ &{}_2F_1\left(1,\ 1 - \frac{1}{1.75};\ 1 + \frac{1}{1.75},\ \frac{-10.823}{-10.823 - 1}\right) \\ d_{layer} &= 0.4(0) + 0.067(0.5) + (0.333)(0.5)(0.347)(1.715) \\ &= 0.1325\ m \\ \theta_2 &= d_{layer}/\Delta z = 0.1325\ m/0.5\ m = 0.265 = 26.5\%. \end{aligned}$$

The average water content in layer 3 is found with Eq. 28.29.

$$w_u = -((\alpha(z_u - z_{WT}))^n) = -\left((0.078*100(1.5 - 0.5))^{1.75}\right) = -36.405$$
$$w_L = -((\alpha(z_L - z_{WT}))^n) = -\left((0.078*100(1.0 - 0.5))^{1.75}\right) = -10.823$$

$$\begin{aligned} d_{layer} &= \theta_r(z_u - z_L) + (\theta_s - \theta_r)(z_u - z_{WT})(1 - w_u)^{1/n-1} \\ &{}_2F_1\left(1,\ 1 - \frac{1}{n};\ 1 + \frac{1}{n},\ \frac{w_u}{w_u - 1}\right) \\ &- (\theta_s - \theta_r)(z_L - z_{WT})(1 - w_L)^{1/n-1} \\ &{}_2F_1\left(1,\ 1 - \frac{1}{n};\ 1 + \frac{1}{n},\ \frac{w_L}{w_L - 1}\right) \end{aligned}$$

$$d_{layer} = 0.067(0.5) + (0.333)(1.0)(1 - (-36.405))^{1/1.75-1}\ (1.99) \\ -(0.333)(0.5)(1 - (-10.823))^{1/1.75-1}\ (1.715) = 0.0747\ m$$

$$\theta_3 = d_{layer}/\Delta z = 0.07477\ m/0.5\ m = 0.1495 = 15.0\%.$$

The average water content in layer 4 is also found with Eq. 28.29 and is equal to 11.5 %.

The total depth of water in the soil profile is (0.4 + 0.265 + 0.150 + 0.115) * 0.5 = 0.465 m.

Equation 28.23 can be used to verify that the total depth of water in the entire soil profile is equal to the sum of the depths in each soil layer.

$$\begin{aligned} d_{total} &= \theta_s(z_{WT}) + \theta_r(z_t - z_{WT}) + (\theta_s - \theta_r) \\ &(z_t - z_{WT})(1 - w)^{1/n=1} \\ &{}_2F_1\left(1,\ 1 - \frac{1}{n};\ 1 + \frac{1}{n},\ \frac{w}{w-1}\right) \\ w &= -((a(z_t - z_{WT}))^n) = -\left((0.078*100(1.5))^{1.75}\right) \\ &= -74.016 \\ d_{total} &= 0.4(0.5) + 0.067(1.5) + (0.333)(1.5) \\ &(1 - (-74.016))^{1/1.75=1}\ (2.096) = 0.465\ m \end{aligned}$$

Next, find the water contents after the first day. Calculate the water lost to drainage.

$$\begin{aligned} d_{drainage} &= z_{WT}*0.02\ m/day*1\ days = 0.5*0.02 \\ &= 0.01\ m/day \end{aligned}$$

The water table movement is based on the specific yield. The specific yield at any water table elevation can be found by calculating the total depth of water in the soil profile at water table elevations 0.05 m above and 0.05 m below the original water table elevation. The depths must be calculated for only the lower 2 layers since only those layers are in equilibrium with the water table in this example.

$$\begin{aligned} d_{profile}(z_{wt} &= 0.495) = 0.33148\ m \\ w &= -\left((\alpha(z_t - z_{WT}))^n\right) = -\left((0.078*100\left(1 - 0.495\right))^{1.75}\right) \\ &= -11.0135 \\ d_{total} &= 0.4(0.495) + 0.067(0.505) + (0.333)(0.505) \\ &(1 - (-11.0135))^{1/1.75-1}\ (1.72) = 0.33148\ m \\ d_{profile}(z_{wt} &= 0.505) = 0.33367\ m \end{aligned}$$

The specific yield is then calculated with low and high water contents

$$SY = (0.33367 - 0.33148)/(0.505 - 0.495) = 0.2188$$

The new water table elevation is calculated as follows (j represents the previous time step).

$$z_{wt}(j) = z_{wt}(j-1) - drainage/SY = 0.5 - 0.01/0.2188 = 0.454 \text{ m}$$

The *WINDS* model uses the *Find_wt* function to find the height of the water table. The calculated value is 0.455 m. *Find_wt* uses an iterative procedure that moves the water table until the total depth of water in the equilibrium layers is equal to the depth of water in the cells.

The new water content in layer 1 is calculated with the d_layer_WT function in the *WINDS* model.

$$\theta_1 = \text{d\_layer\_WT}(ts, tr, zu, zl, zwt(j), alpha, n)/dz$$
$$\theta_1 = \text{d\_layer\_WT}(0.4, 0.067, 0.5, 0, 0.455, 0.078, 1.75)/0.5 = 0.3999 \text{ ml/ml}$$

The new water content in layer 2 is calculated with the d_layer_Eq function in the *WINDS* model

$$\theta_2 = \text{d\_layer\_Eq}(ts, tr, zu, zl, zwt(j), alpha, n)/dz$$
$$\theta_2 = \text{d\_layer\_Eq}(0.4, 0.067, 1, 0.5, 0.455, 0.078, 1.75)/0.5 = 0.246 \text{ ml/ml}$$

The new water contents in layers 3 and 4 are calculated based on the amount of water removed by evapotranspiration since they are no longer in equilibrium with the water table. There is no flux from layer to layer because all layers are in equilibrium at the beginning of the time step. Thus, the only water loss is due to evapotranspiration.

$$\theta_3 = 0.1498 - 0.005 \text{ m}/0.5 \text{ m} = 0.1398 \text{ ml/ml}$$
$$\theta_4 = 0.115 - 0.005 \text{ m}/0.5 \text{ m} = 0.105 \text{ ml/ml}$$

The flux between the upper layers is negligible (order 10^{-6} m/d) and does not significantly change the solution. Likewise, the flux from layer 3 into layer 2 is negligible.

The next example demonstrates the fluctuation of the water table with evapotranspiration and weekly irrigations. The *WINDS* model move the equilibrium layer to the top of the soil profile or to the maximum specified layer when water content increases in the upper layers beyond field capacity.

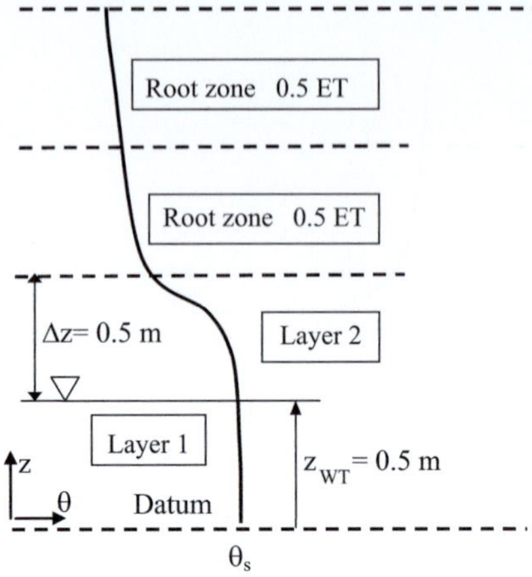

Fig. 28.19 Geometry for Example 28.12

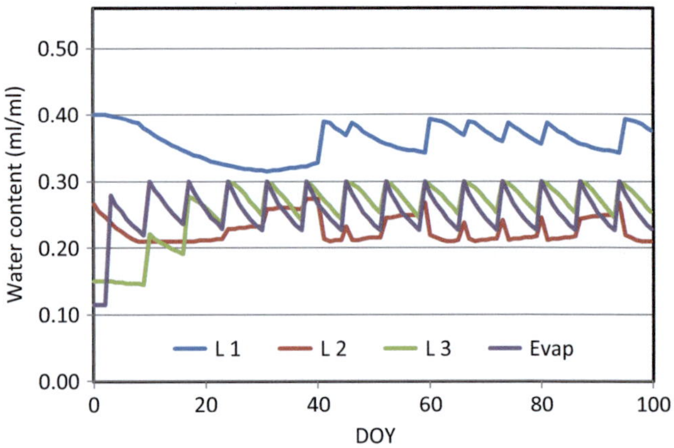

Fig. 28.20 Water content versus time for Example 28.12

Example 28.12 A soil has four layers, 0.5 m thickness (Fig. 28.19). Soil parameters are the same as the previous two examples; however, change the permanent wilting point to 14 %. The rate of subsurface drainage below layer 1 is a function of the water table elevation, z_{WT}: $d_{drainage} = z_{WT}$ * 0.01 m/day *days). The initial elevation of the water table is 0.5 m. The fraction of saturation for the equilibrium max placement is 0.75. Evapotranspiration is 1 cm/day, with the nominal fraction removed from each of layers 3 and 4 equal 50 %. However, the fraction can be adjusted based on water content, and the total evapotranspiration can be reduced based on overall soil water content in layers 3 and 4. Set the root zone depth = 0.9 m for the entire season. The information for this simulation is in the #5 position in the *Crop_data* worksheet. Apply 70 mm/week with 15 % extra as a leaching fraction. Run *WINDS* for 100 days.

Cells I43:I44 on the Main worksheet are set to True and 2, respectively, limiting the equilibrium zone to the lower two layers. Water content changes are shown in Fig. 28.20. The water table (red line) and upper limit of the equilibrium zone (blue line) are shown in Fig. 28.21.

The *WINDS* model has an algorithm, described in Chap. 29, to limit evapotranspiration if there is not sufficient water in the root zone. For example, both layers 3 and 4 begin the season at less than permanent wilting point (Fig. 28.20); thus, there is no evapotranspiration on the first two days (Fig. 28.22). The potential ET in this example

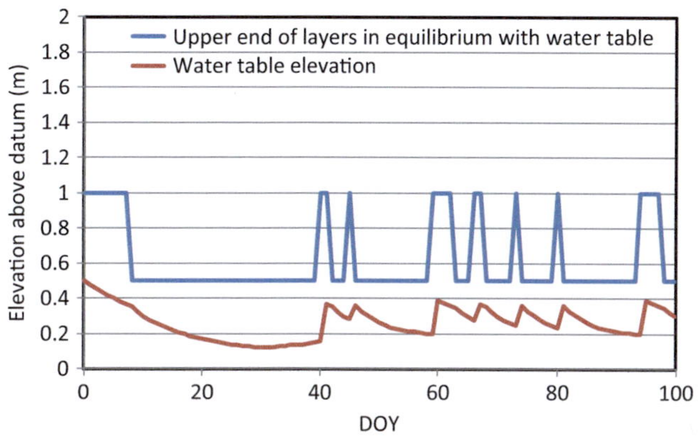

Fig. 28.21 Water table and equilibrium zone versus time for Example 28.12

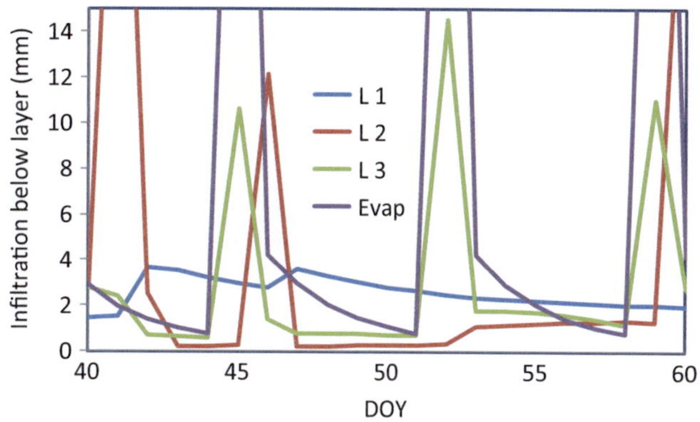

Fig. 28.24 Downward seepage between layers from DOY 40–60 for Example 28.12

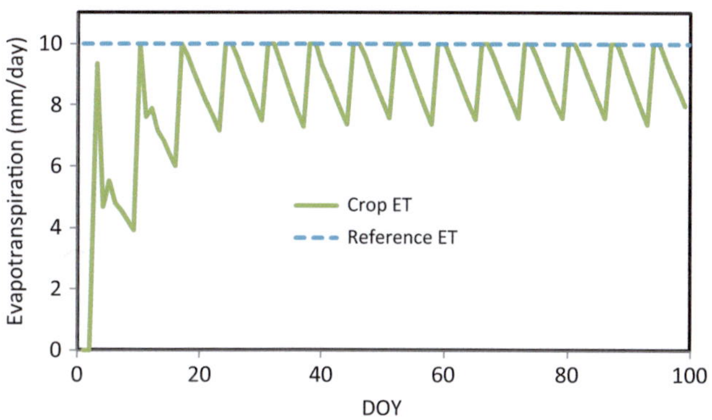

Fig. 28.22 Evapotranspiration versus time for Example 28.12

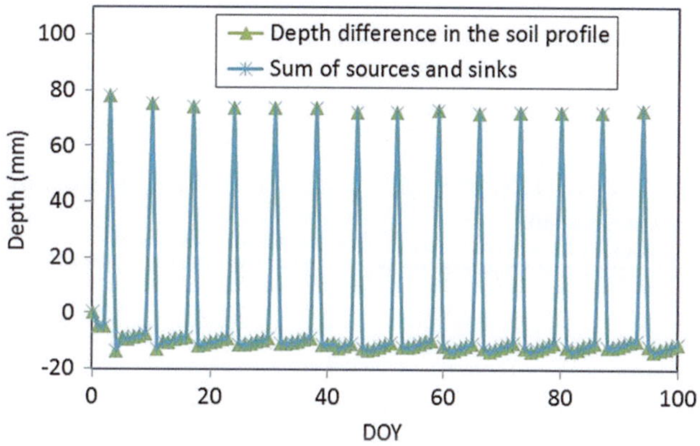

Fig. 28.25 Volume balance for Example 28.12

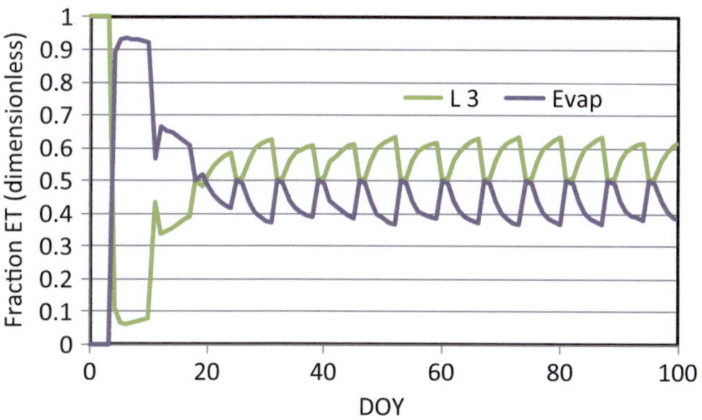

Fig. 28.23 Fractions of total ET removed from each layer for Example 28.12

is a constant 10 mm/day; however, the actual evapotranspiration (ET) is less than 10 mm/day on days (Fig. 28.22) when water content is below the MAD value.

The WINDS model shifts the majority of the ET to the root zone layers with the highest water content. Although the specified fraction of ET is 50 % each from layers 3 and 4, the actual fractions change with water content as shown in Fig. 28.23.

The reason that more water is extracted from layer 3 is that there is a slow downward seepage of water from layer 4 to layer 3 after the irrigation events. In Fig. 28.24, the top of the irrigation infiltration peaks are cut off in order to focus in the infiltration between layers during the interval between irrigation events. The volume balance shows agreement between change in storage and sum of sources and sinks (Fig. 28.25).

Upward Flux

A high water table can supply water to the plant root zone if the water table is close to the bottom of the root zone (Fig. 28.26). The phenomenon of upward flux from the water table to the bottom of the root zone can be modeled with an analytic solution derived by Anat.

$$v = \frac{K_s\left[h_b + \frac{1.89\ h_b}{\eta^2+1}\right]^\eta}{y^\eta} \tag{28.31}$$

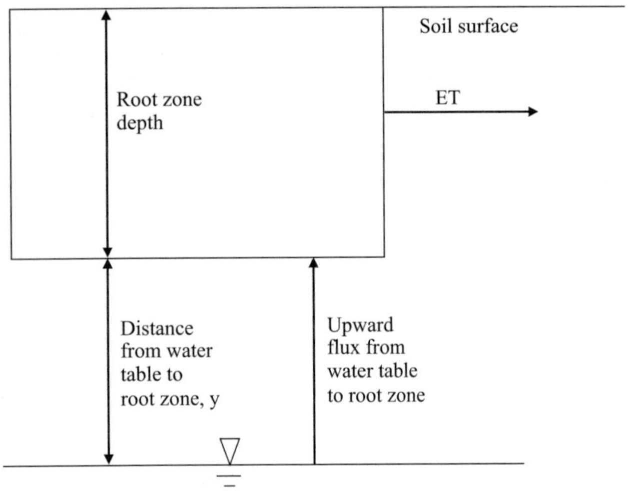

Fig. 28.26 Upward flux from water table to root zone and evapotranspiration

where

y = distance from the water table to the bottom of the root zone, m.
v = Darcy velocity, m/day.

Upward flux from the water table to the root zone or to the soil surface can also be calculated with a discretized solution of Darcy's law, which is rearranged to solve for change in matric potential (Eq. 28.32) from layer to layer (Fig. 28.27). The solution starts at the top layer with an assumed matric potential. The procedure continues until the matric potential equals zero, which would be the depth of the water table The final solution is not sensitive to the magnitude of the assumed matric potential in the upper layer because hydraulic conductivity becomes very low at very negative matric potentials.

$$\begin{aligned} v &= -K_{e(j\&j-1)}\frac{H_j - H_{j-1}}{dz} = -K_{e(j\&j-1)}\frac{h_j + z_j - h_{j-1} - z_{j-1}}{dz} \\ &= -K_{e(j\&j-1)}\frac{h_j - h_{j-1} + \Delta z}{\Delta z} = K_{e(j\&j-1)}\frac{h_{j-1} - h_j}{\Delta z} - K_{e(j\&j-1)} \\ h_{j-1} &= h_j + \frac{v\ \Delta z}{K_{e(j\&j-1)}} + \Delta z \end{aligned} \quad (28.32)$$

Example 28.13 The potential evapotranspiration is 0.01 m/day. Calculate the maximum distance between the water table and the bottom of the root zone for subirrigation in Wagram loamy sand. Compare the discretized solution (Eq. 28.32) to Anat's Eq. (28.31). Assume that saturated hydraulic conductivity of Wagram loamy sand is 0.144 m/day and that matric potential at the bottom of the soil root zone is −3.3 m (field capacity). Let the layers have a height (dz) of 0.01 m.

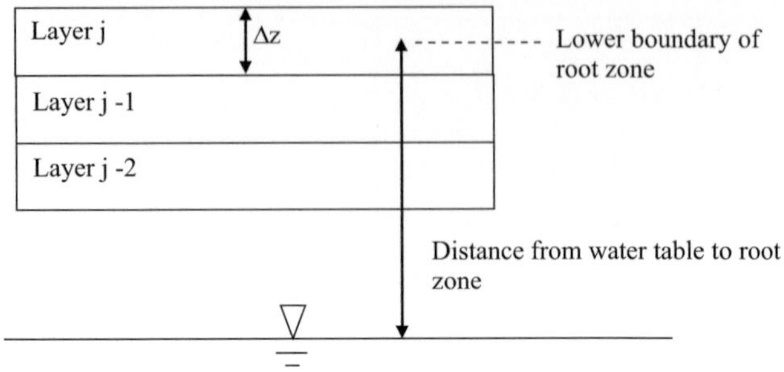

Fig. 28.27 Layer numbering for upward flux calculation

Start by finding the conductivity of the upper layer, j, assuming that matric potential is −3.3 m. Next calculate the matric potential of layer j − 1. Then, based on the new calculated j − 1 matric potential, calculate a new conductivity of layer j − 1, and so on.

$$\begin{aligned} K_j &= K_s K_{rw} = K_S\left(\frac{h_b}{h_j}\right)^{2+3\lambda} = 0.144\left(\frac{-0.3}{-3.3}\right)^{2+3(1.27)} \\ &= 1.28{*}10^{-7}\ m/day \\ h_{j-1} &= h_j + \frac{v\ dz}{K_{e(j\&j-1)}} + dz \\ &= -3.3\ m + \frac{(0.01\ m/day)(0.01\ m)}{1.28{*}10^{-7}\ m/day} + 0.01\ m \\ &= 776\ m \end{aligned}$$

The low hydraulic conductivity caused to solution to blow up. Try a less negative matric potential (h_j = −0.7 m) in the upper layer so that the solution does not blow up.

$$\begin{aligned} K_j &= K_s K_{rw} = K_S\left(\frac{h_b}{h_j}\right)^{2+3\lambda} = 0.144\left(\frac{-0.3}{-0.7}\right)^{2+3(1.27)} \\ &= 0.001\ m/day \\ h_{j-1} &= -0.7\ m + \frac{(0.01\ m/day)(0.01\ m)}{0.001\ m/day} + 0.01\ m \\ &= -0.594\ m \end{aligned}$$

Now, solve for matric potential in layer j-2.

$$\begin{aligned} K_j &= K_s K_{rw} = K_S\left(\frac{h_b}{h_j}\right)^{2+3\lambda} = 0.144\left(\frac{-0.3}{-0.594}\right)^{2+3(1.27)} \\ &= 0.0027\ m/day \\ h_{j-1} &= -0.594\ m + \frac{(0.01\ m/day)(0.01\ m)}{0.0027\ m/day} + 0.01\ m \\ &= -0.548\ m \end{aligned}$$

Continue to solve until $h_c = 0$ (Fig. 28.28). The matric potential reaches zero at a depth of 0.47 m below the bottom of the root zone. Thus, if the root zone is less than 0.47 m above the water table, then the crop will receive adequate water from the water table.

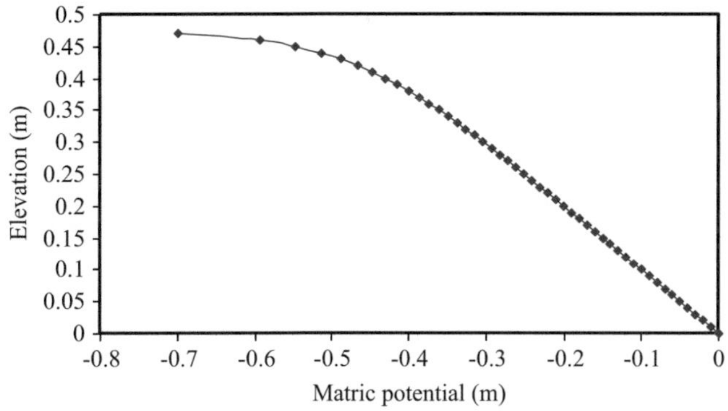

Fig. 28.28 Matric potential versus elevation above the water table

Find the upward flux with Anat's equation with a water table at a depth of 0.47 m below the root zone.

$$\text{v} = \frac{K_s\left[h_b + \frac{1.89\ h_b}{\eta^2+1}\right]^\eta}{y^\eta} = \frac{0.144\left[0.3 + \frac{1.89\ *0.3}{5.81^2+1}\right]^{5.81}}{0.47^{5.81}}$$
$$= 0.0144\ m/day$$

Anat's equation calculates an upward flux rate that is 44 % greater than the finite difference solution. However, at a slightly greater distance from the water table to the root zone, 0.52 m, the flux predicted by Anat's equation is 0.008 m/day, which is 20 % lower than the finite difference calculation.

Questions

1. Derive Eq. 28.3 from Eq. 27.20
2. Derive a finite difference equation from Eq. 28.3 that has the same structure as Eq. 28.2.
3. Include osmotic potential as a function of soil water EC in Eq. 28.5. Leave only the change in water content ($\Delta\theta$) on the left side of the equation. Use Eq. 5.2 to calculate osmotic potential based on soil water EC.
4. For the three cells listed below, write the equation derived in question 3 in terms of cell 4. Calculate the change in water content in cell 4 with and without the influence of salinity. Use a one day time step.

   $z_3 = 1.0$ m $h_3 = -2$ m $EC_{w-3} = 4$ dS/m
   $z_4 = 1.4$ m $h_4 = -3$ m $EC_{w-4} = 3$ dS/m
   $z_5 = 1.8$ m $h_5 = -5$ m $EC_{w-5} = 6$ dS/m

   The effective hydraulic conductivity between cells 3 and 4 is 0.01 m/day
   The effective hydraulic conductivity between cells 4 and 5 is 0.001 m/day
5. Calculate the water content, effective water content, effective saturation and hydraulic conductivity of Wagram loamy sand at −1 bar matric potential.
6. Calculate the Darcy velocity between two points in unsaturated Wagram loamy sand with matric potential values of −1,000 cm (point 1) and −2,000 cm (point 2) and elevations of 10 cm (point 1) and 20 cm (point 2), respectively. Assume that the saturated hydraulic conductivity of Wagram loamy sand is 0.6 cm/hr. The distance between the two points is 50 cm. Use the Brooks Corey model for hydraulic conductivity. Use the geometric mean to calculate effective hydraulic conductivity.
7. Three layers numbered 1 through 3 from bottom to top have 0.5 m depth. Use the van Genuchten equations to calculate water contents after 1 day and 2 days. Initial water content in all layers is 37 %. $n = 1.31$, $\theta_r = 0.095$, $\theta_S = 0.41$, $\alpha = 0.019$, $L = 0.5$, and $K_0 = 6.24$ cm/day. There is no infiltration and no seepage of water below the control volume. Calculate effective conductivity between two layers with the geometric mean. Calculate water contents after the first day and after the second day. It is impossible for any layer to have greater than the saturated water content.
8. Repeat question 7 but allow water to drain below the lower layer. Assume that there is no water table and use Eq. 28.27 for layer 1.
9. Repeat question 7 but restrict the downward movement of water to 0.1 cm/day from layer 1 (as with subsurface drainage and a water table).
10. Thirty cm is removed from the soil as the water table is lowered from 2.5 m to 0.5 m. Calculate the incremental specific yield between 0.5 m and 2.5 m.
11. The water table drops from 1.0 to 1.5 m below the ground surface, and the specific yield of the soil is 10 %. Calculate the depth of water removed from the soil profile.
12. What do the right and left terms on the right side of Eq. 28.17 represent? What is the hatched area in Fig. 28.8? What does $\theta(z)$ represent in the Eq. 28.17? Why is the Brooks-Corey calculation of $\theta(z)$, which is based on matric potential substituted into the equation derivation? Explain why the upper limit of integration is $z_t - z_{WT}$ and the lower limit is h_{ba}.
13. Calculate the depth of water drained and specific yield for a water table that drops from the soil surface to 0.8 m above the datum in Wagram loamy sand. Place the datum 1.5 m below the soil surface.
14. If the question 13 water table dropped from 0.7 m below the surface to 0.8 m below the surface, then how much water would you expect would be drained from the soil during this 0.1 m drop in water table elevation?

Calculate the depth drained in two ways: use the specific yield that you calculated in question 13 and also calculate the actual depth drained at 0.8 m depth with Eq. 28.18. Then use the two drained depths to calculate the incremental specific yield between 0.7 and 0.8 m depth below the soil surface.

15. Use Eq. 28.19 to find the incremental specific yield at 0.75 m depth below the soil surface for Wagram loamy sand and compare to the incremental specific yield that was calculated in question 14.
16. Find the depth drained if the water table drops from the soil surface to 50 cm depth, from the soil surface to 100 cm depth, and from the soil surface to 150 cm depth. Use the Brooks-Corey Goldsboro sandy loam parameters that you calculated in question 27.22. Set the datum at 3 m. Find the specific yield associated with the water table drop from the surface to each depth
17. Derive Eq. 28.21 from Eq. 28.20
18. Use the integrator at http://integrals.wolfram.com/index.jsp to integrate the left hand side of Eq. 28.24.
19. Find the hypergeometric function page at the Wolfram website listed in question 18 and find the transformation shown between Eqs. 28.24 and 28.25. List the line number of the transform on the Wolfram page to show that you found the transform. Find the series solution to the hypergeometric function on the same page and list the line number. Explain how you would implement the series solution in a computer code or spreadsheet.
20. A soil has the following parameters: $n = 1.5$, $\alpha = 0.06$, $z_t = 2$ m, $z_{WT} = 1$ m, $\theta_s = 0.45$, and $\theta_r = 0.08$. Calculate the a, b, c, and w terms for the transformed Gauss hypergeomtric function as shown in Eq. 28.25. A function is included in the Van Genuchten Excel/VBA program called hg that calculates the hypergeometric series solution. Calculate the series solution with the function hg by calling it from a worksheet with the following: "=hg(a,b,c,w/(w-1))". Make sure to write z in units of cm in the calculation of w since α has units of 1/cm. Finally, calculate the total depth of water in the soil profile from the datum to the soil surface.
21. Redo question 20, but place the water table at 0.1 m above the datum. Calculate the specific yield if the water table drops from 1 to 0.1 m above the datum.
22. Find the average water content in two layers. The upper layer has lower and upper boundaries that are 1 and 1.2 m above the datum. The lower layer has lower and upper boundaries that are 0.8 and 1.0 m above the datum. Both layers are in hydraulic equilibrium with the water table. The matric potential, h_c, at the center of the upper layer is – 0.5 m. Use the Brooks-Corey parameters for Wagram loamy sand.
23. Calculate the average water content in a layer that has upper and lower limits that are 1.4 m and 1.2 m above the datum. Use the parameters listed in question 20.
24. A crop requires 0.006 m/day evapotranspiration. Calculate the maximum distance between the water table and the bottom of the root zone for sub irrigation without water stress. Use Wagram loamy sand parameters. Compare the discretized solution (Eq. 28.32) to Anat's Eq. (28.31). The saturated hydraulic conductivity of Wagram loamy sand is 0.144 m/day.
25. Repeat Example 28.3; however, use the parameters for sandy loam in the *WINDS* model. Make sure to change the initial water content in the Active Data page to field capacity (0.21) before copying the data to Crop_data. Make sure that the final DOY is set to 200 days (cell E3) in order to run the simulation for the entire period. Show the water content, matric potential, and matric potential + elevation graphs. The Matric potential graph is in the *Matric potential* worksheet. Discuss results.
26. Repeat Example 28.3; however, use the parameters for sand in the *WINDS* model. Make sure to change the initial water content in the Active Data page to field capacity (0.1) before copying the data to Crop_data. Show the water content, matric potential, and matric potential + elevation graphs. The Matric potential graph is in the *Matric potential* worksheet. Discuss results. Is there reason to believe that the field capacity estimate may be too high?
27. Repeat Example 28.10; however, use the parameters for sand in the *WINDS* model. Show the water table graph and the water content graph. Evaluate drainage rate multipliers 0.01 and 0.02. The multiplier can be changed in the *Drainage* form, which is accessed from the *Active Data* worksheet.
28. Repeat Example 28.10; however, use the parameters for clay in the *WINDS* model. Show the water table graph and the water content graph. Evaluate drainage rate multiplier 0.02. The multiplier can be changed in the *Drainage* form, which is accessed from the *Active Data* worksheet. Explain the results.
29. Repeat Example 28.12; however, use the parameters for sandy loam in the *WINDS* model. Increase the drainage rate multiplier to 0.03. Set the initial elevation of the water table at 0.2 m. Show the water content graph and the water table elevation graph.

WINDS Agricultural Simulations

29

This chapter demonstrates the use of *WINDS* for modeling crops, ET, and water content in agriculture. The chapter describes the *WINDS* algorithms for infiltration, evapotranspiration, crop stress, dry and wet zones in fields, and remote sensing. Infiltration can be specified as a fixed depth or calculated with the Green-Ampt infiltration model inside the *WINDS* model. Evapotranspiration is calculated with the dual component evapotranspiration model. Yield reduction is based on water stress and the sensitivity to water stress during the growing season. Fields with drip irrigation or alternative furrow irrigation can be modeled as partially wetted fields, which can provide a better estimate of available water capacity in some cases. Two case studies are presented. The *WINDS* model is used to schedule irrigations in a cotton remote sensing experiment in Maricopa, Arizona, and it is used to evaluate farming practices for a rainfed farm in a tropical region. Finally, this chapter reviews the stress algorithms for salinity and nitrogen in the *WINDS* model.

Evapotranspiration and Crop Stress

The evapotranspiration algorithm in *WINDS* is the FAO56 dual component evapotranspiration model that was described in Chap. 6. In *WINDS*, the relationship between actual evapotranspiration and potential evapotranspiration is determined by the water stress, Ks. The relationship between water stress and yield reduction is the crop sensitivity to water stress, Ky. This value is generally much higher during the flowering and fruiting stage than at the end of the season when the crop may be maturing or even drying in preparation for harvest. The crop sensitivity to water stress during each growth phase, as defined by the crop coefficient curve, is input in Cells B19:B23 in the *Active_data* Worksheet.

In the *WINDS* model, the daily reduction in yield is weighted based on the following equation:

$$\left(\frac{Y_a}{Y_m}\right)_{water} = 1 - \frac{\sum_{j=Planting\_Date}^{Harvest}\left(K_y(j)(1 - K_{s-water}(j))\right)}{\sum_{j=Planting\_Date}^{Harvest}\left(K_y(j)\right)} \tag{29.1}$$

where

Y_a = actual seasonal yield, kg/ha
Y_m = potential seasonal yield, kg/ha,
$K_{s-water}$ (j) = water stress coefficient on DOY j,
K_y(j) = crop sensitivity to water stress on DOY j.

Example 29.1 During a period from DOY 186 to 191 (Table 29.1), crop transpiration falls below the potential transpiration. The sum of daily Ky values during a season is 150. Calculate the fraction of seasonal yield lost due to stress from DOY 185 to 192.

Substitute values into Eq. 29.1. There is a 0.55 % seasonal yield reduction due to stress during this period.

$$\left(\frac{Y_a}{Y_m}\right)_{water} = 1 - \frac{\sum_{185}^{192}(0 + 0.09 + 0.17 + 0.21 + 0.16 + 0.21 + 0.02)}{\sum_{j=Planting\_Date}^{Harvest}(150)}$$

$$= 1 - \frac{0.83}{150} = 1 - 0.0055 = .9945$$

The water stress coefficient is calculated as a function of the cumulative effect of stress caused by water depletion in all layers in the root zone. If average percent depletion in layers in the root zone is greater than MAD (p in FAO 56),

P. Waller, M. Yitayew, *Irrigation and Drainage Engineering*, DOI 10.1007/978-3-319-05699-9_29

Table 29.1 Evapotranspiration and stress parameters for Example 29.1

DOY	Actual ET_C mm	Potential ET_C mm	Ky – Crop sensitivity to water stress	$(1 - K_{s-water})(Ky_j)$
185	7.28	7.28	1.1	(1 − 7.28/7.28) 1.1 = 0
186	6.19	6.71	1.1	(1 − 6.19/6.71) 1.1 = 0.09
187	5.54	6.51	1.1	(1 − 5.54/6.51) 1.1 = 0.17
188	5.18	6.59	1.0	(1 − 5.18/6.59) 1.0 = 0.21
189	5.86	6.95	1.0	(1 − 5.86/6.95) 1.0 = 0.16
190	5.65	7.22	1.0	(1 − 5.65/7.22) 1.0 = 0.21
191	7.15	7.27	1.0	(1 − 7.15/7.27) 1.0 = 0.016
192	7.63	7.63	1.0	(1 − 7.63/7.63) 1.0 = 0

then overall reduction in evapotranspiration ($K_{s-water}$) is calculated as follows. The fraction p is analogous to MAD (management allowable depletion), but it is adjusted based on soil type and evaporation rate.

$$K_{s-water} = \frac{\sum_{j=Bottom\,layer}^{n}(TAW_j) - \sum_{j=Bottom\,layer}^{n}(Depletion_j)}{(1-p)\sum_{j=Bottom\,layer}^{n}(TAW_j)} \quad (29.2)$$

where

Bottom layer = bottom layer in the root zone,
n = total number of layers,
TAW_j = total available water in layer j, m,
$Depletion_j$ = depletion, D_r, in layer j, m.
p = threshold percent depletion (similar to MAD)

The next step is to calculate the modified fraction of transpiration removed from each layer. The specified fractions in the *ET_fractions* worksheet are first reduced by the fraction of depletion below the MAD level. If the sum of fractions does not equal 1.0, then all fractions are divided by the initial sum of fractions in order to make the final sum equal 1.0. The entire process is demonstrated in Example 29.2.

Example 29.2 Calculate the depths of water extracted and the final water content for four layers (Table 29.2) in the root zone. Management allowable depletion (p or MAD) is 0.5, ET_0 is 10 mm. The basal crop coefficient, K_{cb}, is 1.0, and there is no evaporation due to full canopy cover. Field capacity is 0.23 and permanent wilting point is 0.11 in all layers.

Table 29.2 Layer characteristics for Example 29.2

Layer number	Lower boundary (m)	Initial water content	Fraction transpiration
4	0.1	0.09	0.05
3	0.4	0.14	0.5
2	0.8	0.17	0.3
1	1.2	0.19	0.15

Solution:
The first step is to adjust the total daily ET based on the average depletion in the root zone. The water stress coefficient, $K_{s-water}$, is calculated with Eq. 9.6. The depletion and TAW summations used in Eq. 9.6 are calculated below. The threshold water content at p = 50 % (MAD = 50 %) is

$$\theta_t = \theta_{FC} - (\theta_{FC} - \theta_{PWP})\,p = 0.23 - (0.23 - 0.11)*0.5 = 0.17$$

Calculate the summations in Eq. 9.6.

$$\begin{aligned}\text{Depletion} = D_r &= (\theta_{FC-1} - \theta_1)\ \Delta z_1 + (\theta_{FC-2} - \theta_2)\,\Delta z_2 + (\theta_{FC-3} - \theta_3)\,\Delta z_3 + (\theta_{FC-4} - \theta_4)\,\Delta z_4 \\ &= (0.23 - 0.19)\ 0.4 + (0.23 - 0.17)\ 0.4 + (0.23 - 0.14)\ 0.3 + (0.23 - 0.09)\ 0.1 \\ &= 0.081\text{m}.\end{aligned}$$

$$TAW = (\theta_{FC-1} - \theta_{PWP-1})\,\Delta z_1 + (\theta_{FC-2} - \theta_{PWP-2})\,\Delta z_2 + (\theta_{FC-3} - \theta_{PWP-3})\,\Delta z_3 + (\theta_{FC-4} - \theta_{PWP-4})\,\Delta z_4$$

All layers have the same soil type

$$TAW = (0.23 - 0.11)\ 1.2 = 0.14\text{m}.$$

Average percent depletion is 0.081/0.144 * 100 % = 56 %

Percent depletion, 56 %, is greater than p so Eq. 9.6 is used to calculate the reduction in evapotranspiration due to water stress, $K_{s-water}$

$$K_{s-water} = \frac{\sum_{j=Bottom\,layer}^{n}(TAW_j) - \sum_{j=Bottom\,layer}^{n}(Depletion_j)}{(1-p)\sum_{j=Bottom\,layer}^{n}(TAW_j)} = \frac{0.144 - 0.081}{(1-0.5)\ 0.144} = 0.88$$

Calculate the total transpiration for the soil profile

Transpiration = ET_0 $K_{s-water}$ K_{cb} = 0.01 m/day * 0.875 * 1.0 = 0.00875 m = 8.75 mm

Adjust evaporation fractions and then calculate transpiration and final water contents (Table 29.3). Note that the layer thicknesses are not required in the adjustment.

Table 29.3 Calculations of water contents for Example 29.2

Layer	Nominal fraction	Adjustment	First estimate		Final fraction	
4	0.05	(0.09 − 0.11)/(0.17 − 0.11) < 0	0		0	
3	0.5	(0.14 − 0.11)/(0.17 − 0.11) = 0.5	0.25		0.25/0.7 = 0.357	
2	0.3	(0.17 − 0.11)/(0.17 − 0.11) = 1	0.3		0.3/0.7 = 0.429	
1	0.15	(0.19 − 0.11)/(0.17 − 0.11) > 1	0.15		0.15/0.7 = 0.214	
			Sum	0.7	Sum	1.0
Layer	Fraction	Transpiration (mm)	Final water content (ml/ml)			
4	0	8.75(0) = 0 mm	0.09 − (0/1000)/0.1 = 0.09			
3	0.385	8.75(0.357) = 3.12 mm	0.14 − (3.12/1000)/0.3 = 0.130			
2	0.462	8.75(0.429) = 3.75 mm	0.17 − (3.75/1000)/0.4 = 0.158			
1	0.163	8.75(0.214) = 1.87 mm	0.19 − (1.87/1000)/0.4 = 0.183			

Partially Wetted Fields

Drip systems and alternate furrow irrigation only wet a fraction of the soil area. In these cases, it is unrealistic to calculate the water holding or nutrient holding capacity of the soil based on the entire soil volume. The *WINDS* model has the capability to model partially wetted fields. This section describes the modifications of the water balance and salinity balance models for partially wetted fields. The same type of modification is used for the nitrogen model.

The wetted fraction of each soil layer is specified. In Fig. 29.1, there are 12 soil layers and the evaporation layer, and half of each layer is wetted until DOY 160, after which the field is completely wetted. This might be the case if alternate furrow irrigation were used until DOY 160 and every furrow was used until the end of the season, as in Example 9.2. If a greater fraction of the soil were wetted below the soil surface, such as with a bulb-shaped wetting pattern, then the middle layers would have a greater fraction than the top layer.

In the partially wetted case, all fluxes are expressed on a field average basis. Thus, the depth of irrigation, i, is divided by the wetted area in order to find the depth of irrigation in the wetted fraction. Plant transpiration is also expressed on a field average basis and is divided by the wetted fraction in order to calculate the water content change.

The *WINDS* model ignores the dry zone. This can be a source of error. For example, if a significant fraction of nitrate is in the dry zone and the field suddenly becomes fully wetted and roots grow into the dry zone, then this nitrogen source would not be accounted for.

In theory, there are three causes of evaporation reduction: the top of the soil profile is dry, the canopy covers the soil surface, and/or the evapotranspiration is limited by weather parameters. The *WINDS* model modifies the soil wetting constraint for partially wetted fields. The minimum of the following two K_e values is selected.

$$K_r = \frac{TEW - D_r}{TEW - REW} \quad K_e = (K_{c-max} - K_{cb})\ K_r F_w \quad (29.3)$$

$$c = \left(\frac{K_{cb} - 0.15}{K_{c\text{-}max} - 0.15}\right)^{(1+0.5\ Crop\ height)} \quad K_e = K_{c-max}(1 - c) \quad (29.4)$$

Minimum of

$K_e = K_{c-max}(1 - c)$ or $K_e = (K_{c-max} - k_{cb})\ K_r F_w$

where

F_w = fraction wetted in evaporation layers, dimensionless.

Evaporation is then calculated as $E = K_e\ ET_0$ and all of the evaporation is allocated to the wetted area. The water balance for the evaporation layer in the wetted part of the soil profile is calculated as follows.

$$\Delta d_{cv-n+1} = \left(i - d_{seepage-n+1} - ET_{frac-n+1} d_{ET} - E\right)/F_{wn+1} \quad (29.5)$$

The water balance equation for all layers is constructed in the same format.

$$\Delta d_{cv-j} = \left(d_{seepagej+1} - d_{seepagej} - ET_{frac-j} d_{ET-j}\right)/F_{w-j} \quad (29.6)$$

Then, divide by layer thickness, Δz, and wetted fraction to calculate Δθ.

$$\theta_{final} = \theta_{initial-j} + \left(d_{seepagej+1} - d_{seepagej} - ET_{frac-j} d_{ET-j}\right)/F_{w-j}/\Delta z_j \quad (29.7)$$

Rainfall and evaporation are added to the equation for the upper layer

$$\theta_{final} = \theta_{initial-j}\Delta z_j + \left(i/\Delta z_{n+1} - ET_{frac-j} d_{ETn+1}/\Delta z_{n+1} + Rain\left(R_{weight}\right)/\Delta z_{n+1} - E/\Delta z_j\right)/F_{w-n+1} \quad (29.8)$$

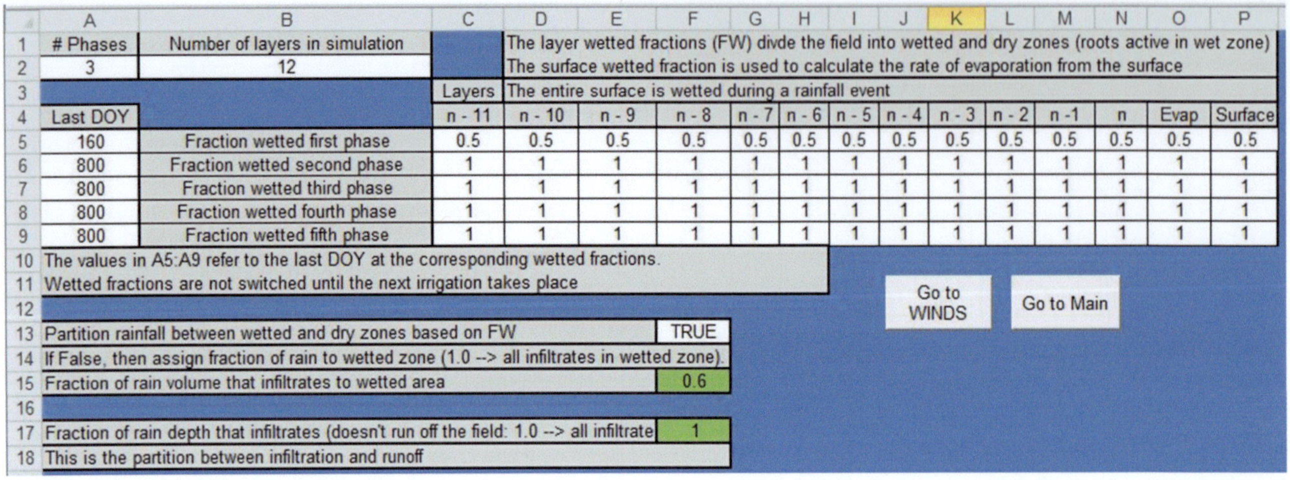

	A	B	C	D	E	F	G	H	I	J	K	L	M	N	O	P
1	# Phases	Number of layers in simulation		The layer wetted fractions (FW) divde the field into wetted and dry zones (roots active in wet zone)												
2	3	12		The surface wetted fraction is used to calculate the rate of evaporation from the surface												
3			Layers	The entire surface is wetted during a rainfall event												
4	Last DOY		n - 11	n - 10	n - 9	n - 8	n - 7	n - 6	n - 5	n - 4	n - 3	n - 2	n -1	n	Evap	Surface
5	160	Fraction wetted first phase	0.5	0.5	0.5	0.5	0.5	0.5	0.5	0.5	0.5	0.5	0.5	0.5	0.5	0.5
6	800	Fraction wetted second phase	1	1	1	1	1	1	1	1	1	1	1	1	1	1
7	800	Fraction wetted third phase	1	1	1	1	1	1	1	1	1	1	1	1	1	1
8	800	Fraction wetted fourth phase	1	1	1	1	1	1	1	1	1	1	1	1	1	1
9	800	Fraction wetted fifth phase	1	1	1	1	1	1	1	1	1	1	1	1	1	1
10	The values in A5:A9 refer to the last DOY at the corresponding wetted fractions.															
11	Wetted fractions are not switched until the next irrigation takes place										Go to WINDS		Go to Main			
12																
13	Partition rainfall between wetted and dry zones based on FW					TRUE										
14	If False, then assign fraction of rain to wetted zone (1.0 --> all infiltrates in wetted zone).															
15	Fraction of rain volume that infiltrates to wetted area					0.6										
16																
17	Fraction of rain depth that infiltrates (doesn't run off the field: 1.0 --> all infiltrate					1										
18	This is the partition between infiltration and runoff															

Fig. 29.1 *Wetted fraction* worksheet

where

R_{weight} = weighting factor to preferentially infiltrate water into wetted zone

If θ_{final} is greater than θ_{fc}, then the following equation can be used to calculate the seepage to the next layer based on the excess initially calculated for the cell:

$$d_{seepage j} = (\theta_{final} - \theta_{FC})\ \Delta z_j F_w \tag{29.9}$$

For days between irrigations, water movement is calculated based on the energy gradient between cells. Average seepage over the field area is a function of matric potential, the distance between cell midpoints, and fraction wetted. In this case, the fraction wetted in cell j is used to calculate the seepage (field average basis) to cell j − 1.

$$d_{seepage j} = \left(h_{c-j} - h_{c-j-1} + \Delta z\right)/(\Delta z) {}^{*} K_{eff} F_{w j} \tag{29.10}$$

The transpiration reduction and associated stress is calculated based on the available water and depletion (field average) in the wetted fraction.

$$K_s = \frac{TAW\ (F_w) - D_r}{(1-p)TAW} \tag{29.11}$$

The salinity mass balance equations are modified for partial wetting in the evaporation layer.

$$EC_{final\ n+1} = \frac{i/F_{w\ n+1}\ EC_{iw} - d_{seepage\ n+1}/F_{w\ n+1} EC_{initial} + EC_{initial}\ \Delta z\ \theta_{initial}}{\Delta z\ \theta_{final}} \tag{29.12}$$

As with the water balance, seepage and irrigation are the averaged values for the field and are divided by the fraction wetted to find the depth infiltrated into the wetted portion. The equation for cells (j) below the upper cell is as follows:

$$EC_{final\ j} = \frac{d_{seepage\ j+1}/F_{w j}\ EC_{j+1} - d_{seepage\ j}/F_{w j} EC_{initial} + EC_{initial}\ \Delta z\ \theta_{initial}}{\Delta z\ \theta_{final}} \tag{29.13}$$

For the nitrate mass balance, the biochemical reaction terms must be multiplied by Fw in order to obtain the field average increase in nitrate in the wetted zone

As with fully wetted fields, if seepage from above exceeds the available water storage capacity in the cell, then the following modified equation must be used in which all infiltrated water and antecedent water in the layer are mixed and then the mixed concentration passes to the next layer.

$$EC_{final} = \frac{\frac{i\ EC_{iw}}{F_w} + m_{waste} + EC_{initial}\ \Delta z\ \theta_{initial}}{\frac{i}{F_w} + \Delta z\ \theta_{initial} + Rain\left(R_{weight}\right) - E - \frac{ET_{frac}}{F_w}} \tag{29.14}$$

If there is upward movement of water in the wetted zone, then the final salinity is calculated as follows (positive seepage is downward).

$$EC_{final} = \frac{-\frac{d_{seepage\ j+1} EC_j}{F_{w\ j}} + \frac{d_{seepage j} EC_{j-1}}{F_{w\ j}} + \frac{m_{waste}}{F_{w\ j}} + EC_{initial}\ \Delta z\ \theta_{initial}}{\Delta z\ \theta_{final}} \tag{29.15}$$

The *WINDS* model cannot use the partially wetted equations in the case of subsurface drainage.

WINDS Remote Sensing Algorithm

Although the crop coefficient can be calculated based on predefined schedules from FAO56 or other sources, the weather varies between seasons and locations so the crop growth rate also varies. It is possible to use heat units to account for weather variation, but it is also possible to use real time information from remote sensing. In addition, remote sensing information is spatial, which detects variations in growth rate within a field. The *WINDS* model has the capability to update the crop coefficient in each field section with remote sensing measurements of the normalized difference vegetation index (NDVI). The NDVI is a measure of the ratio of soil to plants and is based on the ratio of red to near infrared radiation. The technique was developed by Dr. Hunsaker at the USDA-ARS Arid Lands Agricultural Research Center. The technique was used in ARS experiments for irrigation scheduling in a 14 acre cotton field. Weekly remote sensing measurements updated the NDVI. The *WINDS* crop coefficient algorithm used regression to fill in the points between the weekly measurements and to forecast into the future.

The set remote sensing measurements are placed in column B in the *RS_data_input* worksheet (Fig. 29.2). The measurements are then distributed to each of the field sections. Then, a regression was run for each field section with the recent measurements in order to predict the NDVI for upcoming days. The crop coefficient for cotton in Arizona was calculated with the following equation:

$$Kcb_{calc} = a + b\,NDVI + c\,NDVI^2 + d\,NDVI^3 \quad (29.16)$$

where

$a = 0.223$
$b = 1.74$
$c = -1.69$
$d = 0.975$

Figure 29.3 shows the NDVI measurements (column G), regression (column H), and crop coefficient (column I). In this case, remotely sensed NDVIs were taken with helicopter and a tractor mounted Crop Circle sensor and averaged, if necessary, in column G. The first three values in column G are hypothetical values to begin the season because four values are needed for the regression algorithm in *WINDS*. Then the regression algorithm calculates the corresponding Kcb values with Eq. 29.16 or a similar equation that is calibrated for the particular crop and remote sensing device.

Example 29.3 This example demonstrates the use of the *WINDS* model to simulate the 14 acre cotton field in Maricopa Arizona during the 2009 growing season. In this example, the *WINDS* model calculates the crop coefficient

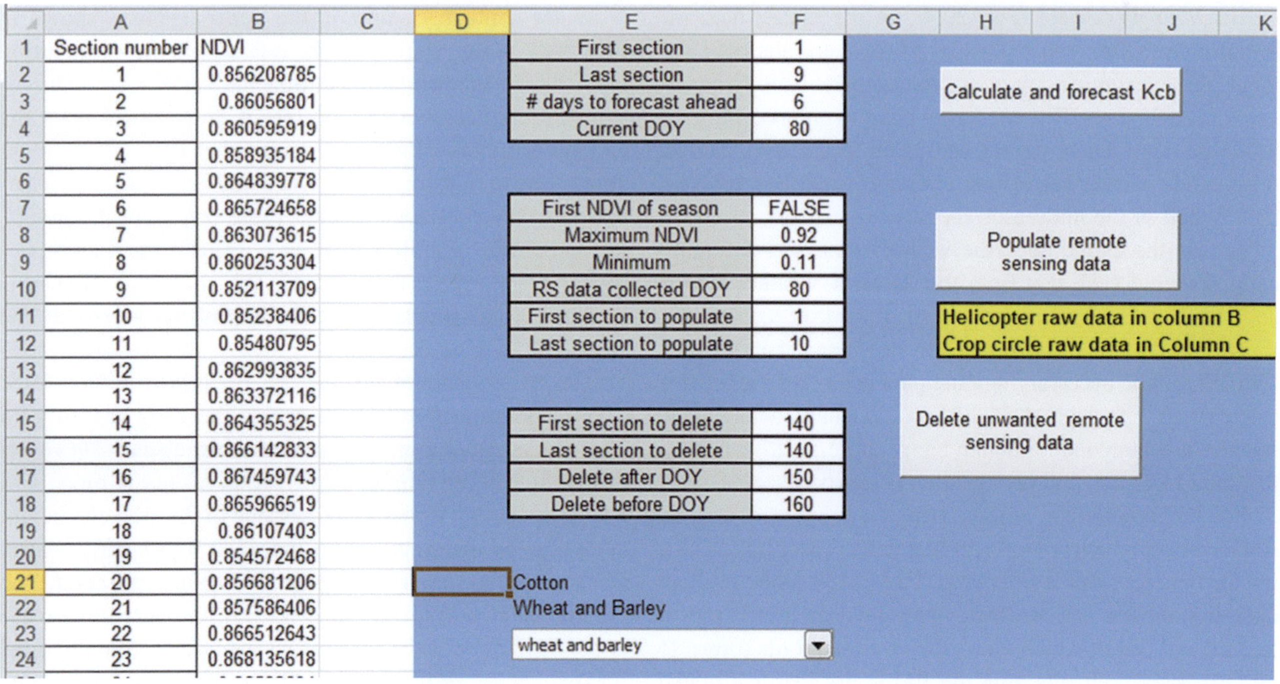

Fig. 29.2 Remote sensing page in *WINDS* model

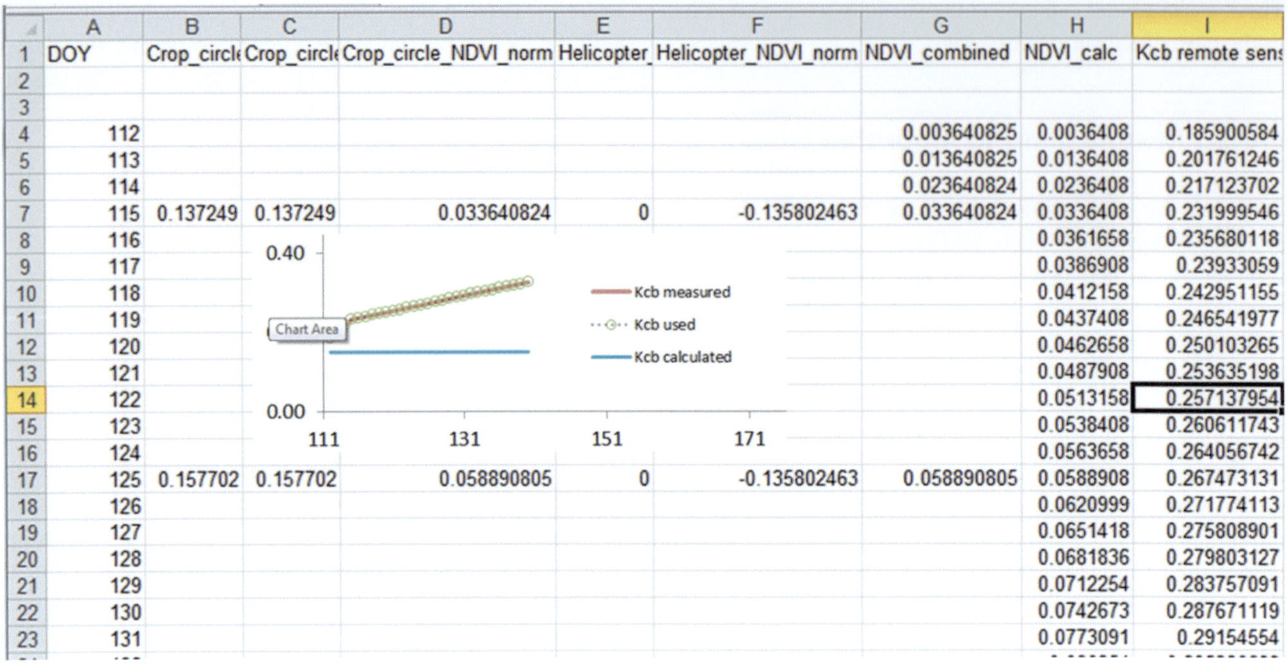

	A	B	C	D	E	F	G	H	I
1	DOY	Crop_circl	Crop_circl	Crop_circle_NDVI_norm	Helicopter_	Helicopter_NDVI_norm	NDVI_combined	NDVI_calc	Kcb remote sens
2									
3									
4	112						0.003640825	0.0036408	0.185900584
5	113						0.013640825	0.0136408	0.201761246
6	114						0.023640824	0.0236408	0.217123702
7	115	0.137249	0.137249	0.033640824	0	-0.135802463	0.033640824	0.0336408	0.231999546
8	116							0.0361658	0.235680118
9	117							0.0386908	0.23933059
10	118							0.0412158	0.242951155
11	119							0.0437408	0.246541977
12	120							0.0462658	0.250103265
13	121							0.0487908	0.253635198
14	122							0.0513158	0.257137954
15	123							0.0538408	0.260611743
16	124							0.0563658	0.264056742
17	125	0.157702	0.157702	0.058890805	0	-0.135802463	0.058890805	0.0588908	0.267473131
18	126							0.0620999	0.271774113
19	127							0.0651418	0.275808901
20	128							0.0681836	0.279803127
21	129							0.0712254	0.283757091
22	130							0.0742673	0.287671119
23	131							0.0773091	0.29154554

Fig. 29.3 Remote sensing data and calculated values in C_1 worksheet

based on the FAO curve. This example focuses on just one location in the field. In the experiment, the infiltrated depths at each field position were input into the *WINDS* model. Planting took place on DOY 112 and the irrigation switched from alternate furrow irrigation to every furrow irrigation on DOY 160.

The crop information was specified in the Crop and ET dialog box. The crop data is stored in cells B2:B23 in the *Active_data* worksheet (Fig. 29.4). In the dialog box, the initial phase (Kcb = 0.15) lasted 31 days and the development phase (crop canopy grows to complete canopy coverage) lasted 52 days. The midseason phase with maximum ET lasted 50 days. Then, the ET decreased for 23 days. Unlike most crops, cotton has a long endseason that has approximately half of the midseason ET rate.

During the experiment, the weather data was automatically acquired each day from the AZMET weather station and stored in the *Active_year_weather* worksheet; however, for this example, the entire year was downloaded at one time (Fig. 29.5). The necessary weather parameters for the *WINDS* calculations with the FAO56 dual component model are wind speed, maximum and minimum temperature, relative humidity, solar radiation, and precipitation (Fig. 26.7).

For the experiment, a soils map was generated with an EM38 probe, which is an electromagnetic sensor pulled by a tractor that develops a soils map. The EM38 probe has two operation modes, one of which views the deep profile (down to 2 m) and one of which views the shallow profile (down to 60 cm). Measurements with the EM38 are then calibrated with laboratory measurements of soil cores from 13 statistically selected locations in the field. With this measurement technique, each field location has one soil type specified for the upper 60 cm and another for all cells below 60 cm (Fig. 29.6). In this simulation, 12 soil layers + evaporation layer were included in the model.

The Main worksheet is set up to just run the water simulation (no salinity or nitrogen) with planting on DOY 112 (Fig. 29.7).

Water content vs. time in the upper layers is shown in Fig. 29.8 Water content drops quickly in the evaporation layer after each irrigation during the early season due to evaporation. The evaporation rate is much lower once the canopy is mature. Water content in other layers drops more slowly due to transpiration. Layer 9 and below have no water loss until the root zone reaches those layers. In the model, the soil profile is refilled after each irrigation to field capacity, and the excess seeps below the root zone. Water is depleted in all soil layers at the end of the season as the crop is allowed to dry out.

The environmental parameters are shown in Fig. 29.9. The evaporation rate is high during the early season with high peaks in evaporation after wetting events. However, evaporation after rainfall or irrigation events is much less when the canopy is mature. The shape of the crop coefficient curve can be distinguished in Fig. 29.9 where early season ET is much less than the reference evapotranspiration. The volume balance (Fig. 29.10) is correct for every day except at the beginning of the season and the switch from alternate furrow and every furrow on DOY 160. Both discrepancies were caused by the fact that the soil water holding capacity

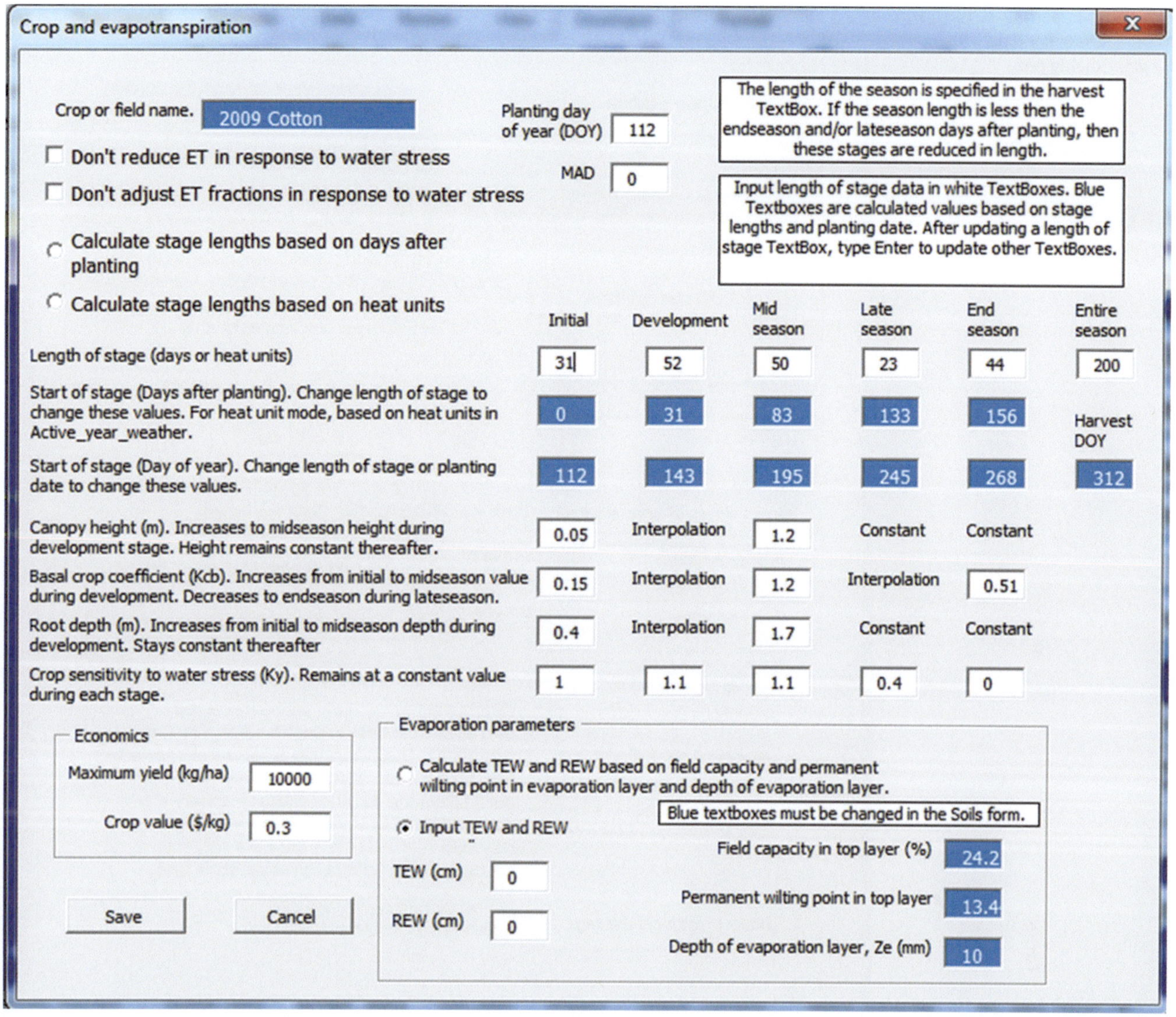

Fig. 29.4 Crop and ET dialog box in *WINDS* model

halves or doubles because of the initiation of partial wetting or the switch from partial wetting. The irrigation, rainfall, and leaching graph (Fig. 29.11) shows that the third irrigation resulted in excess leaching. However, this was the first irrigation with every furrow so much of the water was probably used to replenish the excessively depleted dry zone.

When the profile reaches the MAD (FAO p) level, then irrigation should take place. The percent depletion vs. time is shown in Fig. 29.12. The typical cotton MAD is 50 % so it appears in the model that the maximum percent depletion was exceeded several times; however, there was not observed stress. There are several possible sources of error or misrepresentations in the model. The early season percent depletion possibly appears high because of the high depletion in the evaporation layer. Its effect is magnified because it occupies a large part of the early season root zone. Another possibility is that this simulation was made with the FAO 56 crop coefficient curve, and not with the remotely sensed Kcb values. Another (Allen et al. 1998) possibility is that the crop drew water from the entire profile and not just from what was assumed to be the wetted fraction.

The fraction of evapotranspiration removed from each layer is shown in Fig. 29.13. The upper layers have a high fraction early in the irrigation cycle, and then the depletion switches to the lower layers at the end of the cycle.

The crop stress, crop sensitivity to water stress, and relative yield are output to the *Stress* worksheet (Fig. 29.14). The relative yield is calculated for the season with Eq. 29.1 and output to cell G2.

WINDS has the capability to neglect the evaporation layer in the percent depletion calculation. Write TRUE in cell I45 in the Main worksheet to compute the percent depletion in

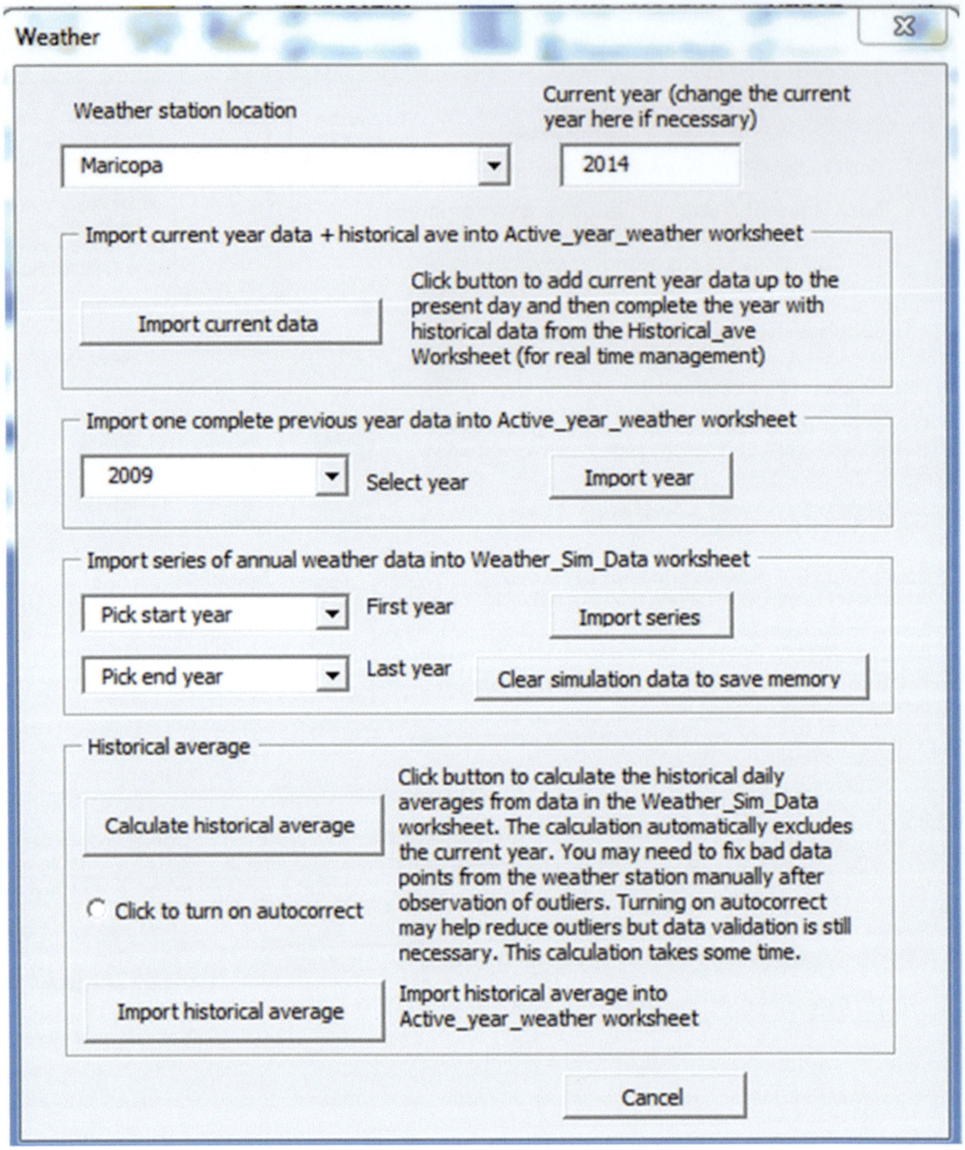

Fig. 29.5 Importing 2009 Maricopa weather data with *Weather* dialog box

all root zone layers except the evaporation layer. In this case, the early season percent depletion is much lower (Fig. 29.15) than in Fig. 29.12. With lower percent depletion, the relative yield increases from 70 % to 78 %. Nevertheless, the early season percent depletion before the first irrigation is still unrealistic. Letting the wetted fraction equal 1.0 for the entire season decreases the early season percent depletion to the unstressed range (Fig. 29.16), which is reflective of the nonstressed early season appearance of the crop.

This example demonstrates the importance of using the correct assumptions and calibrating the model to field conditions. The plant roots probably penetrated and utilized water in the dry fraction even though it was not irrigated until after DOY 160.

One of the advantages of a two component evapotranspiration model, where evaporation and transpiration are calculated independently, is that the model accounts for different soil evaporation rates for different irrigation methods. For example, ET rates were simulated for a center-pivot irrigated cotton crop in Maricopa, Arizona in 2009, with frequent and light irrigations (Fig. 29.17). There is excessive evaporation due to frequent and light irrigations in the early and late season. Once the canopy covers the soil, the difference in irrigation frequency is not as important with respect to soil evaporation; however, if the water is sprayed on the canopy, the water may evaporate from the crop. Because of the shallow nature of the root zone under center pivot irrigation, it may not be reasonable to neglect the evaporation layer in the percent depletion calculation. This comparison emphasizes the importance of correctly modeling the mode of irrigation water application (frequent light irrigations or infrequent deep irrigations), especially in hot, arid, and windy climates with high soil evaporation.

Soils data

12 Number of layers (n) in addition to the evaporation layer (E layer)

	Layer lower depth (cm)	PWP %	FC %	Sat %	Initial water content %	Saturated conductiity (cm/day) Ksat	Ko	Van Gen alpha	Van Gen n	Residual water content %	Van Gen L
E Layer n + 1	10	13.45	24.2	38.1	24.2	25.9	12.5	3	1.37	5.57	-1.35
Layer n	20	13.45	24.2	38.1	24.2	25.9	12.5	3	1.37	5.57	-1.35
Layer n - 1	40	13.45	24.2	38.1	24.2	25.9	12.5	3	1.37	5.57	-1.35
Layer n - 2	60	13.45	24.2	38.1	24.2	25.9	12.5	3	1.37	5.57	-1.35
Layer n - 3	80	12.6	22.9	37.9	22.9	26.3	12.42	3.02	1.37	5.64	-1.36
Layer n - 4	100	12.6	22.9	37.9	22.9	26.3	12.42	3.02	1.37	5.64	-1.36
Layer n - 5	120	12.6	22.9	37.9	22.9	26.3	12.42	3.02	1.37	5.64	-1.36
Layer n - 6	140	12.6	22.9	37.9	22.9	26.3	12.42	3.02	1.37	5.64	-1.36
Layer n - 7	160	12.6	22.9	37.9	22.9	26.3	12.42	3.02	1.37	5.64	-1.36
Layer n - 8	180	12.6	22.9	37.9	22.9	26.3	12.42	3.02	1.37	5.64	-1.36
Layer n - 9	200	12.6	22.9	37.9	22.9	26.3	12.42	3.02	1.37	5.64	-1.36
Layer n - 10	250	12.6	22.9	37.9	22.9	26.3	12.42	3.02	1.37	5.64	-1.36
Layer n - 11	300	12.6	22.9	37.9	22.9	26.3	12.42	3.02	1.37	5.64	-1.36

Fig. 29.6 Soil properties for Example 9.2

	A	B	C	D	E	F	G
1		To run from beginning of season, set Starting day (Cell C3) < (Cell E4) Planting DOY (Day Of Year)					
2	Run	Number of cells	1	First cell to simulate	1	Last cell to simulate	1
3		Start at DOY	112	Final DOY	360	Number of layers	12
4				Planting DOY	112		
5		Include rainfall	TRUE	Similulate salinity	FALSE	Simulate nitrogen	FALSE
6		View water simulation	FALSE	View salinity simulation	FALSE	View nitrogen simulation	FALSE
7		Auto update weather	FALSE	Current year	2014	Location	Maricopa

Fig. 29.7 *Main* worksheet for Maricopa simulation

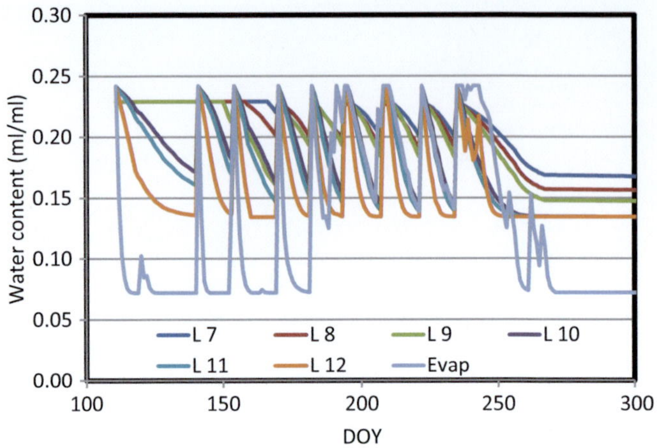

Fig. 29.8 *WINDS* simulation of water contents in upper six layers and evaporation layer for cotton crop in Maricopa, Arizona during 2009

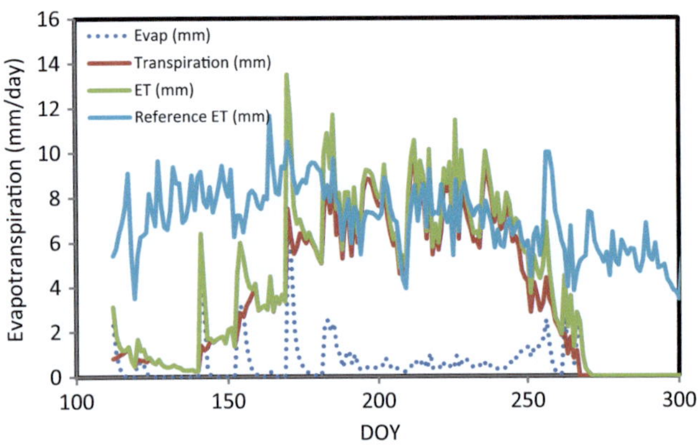

Fig. 29.9 *WINDS* simulation of evaporation and transpiration for cotton crop in Maricopa, Arizona during 2009

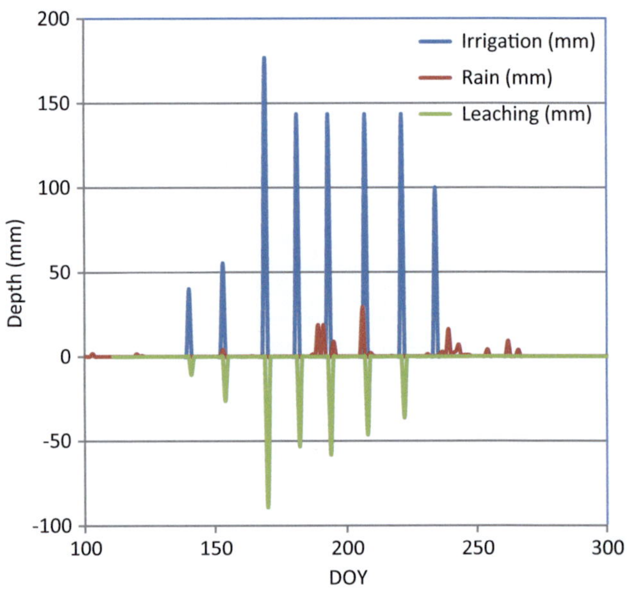

Fig. 29.11 Irrigation, rainfall, and leaching for Maricopa, Arizona cotton simulation during 2009

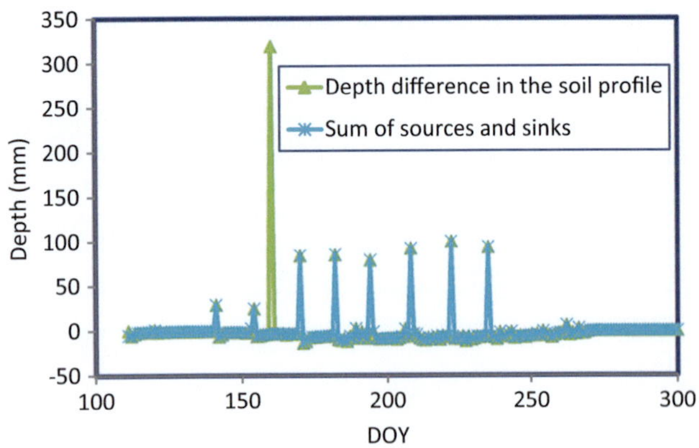

Fig. 29.10 Volume balance for cotton simulation in Maricopa, Arizona during 2009

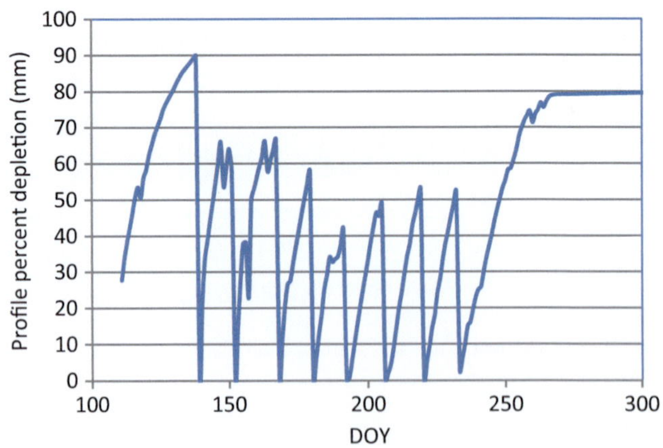

Fig. 29.12 Percent depletion for Maricopa, Arizona cotton simulation during 2009

The percent depletion (Fig. 29.18) has much less variation than with surface irrigation (Fig. 29.16), which is expected for frequent irrigations.

Green-Ampt Infiltration Model

Instead of the empirical approach of the NRCS equations, the Green-Ampt model is based on soil physics and accounts for varying available pore space due to antecedent moisture conditions. Because it is based on theory, it can account for antecedent water content, and storm or irrigation intensity and duration. The Green-Ampt model is one-dimensional and is not able to model the two-dimensional or three-dimensional infiltration geometry that takes place with furrow and drip irrigation. Thus, the Green-Ampt model is normally used to simulate infiltration in humid climates where the primary water input is precipitation and antecedent moisture before storms varies with storm frequency and depth.

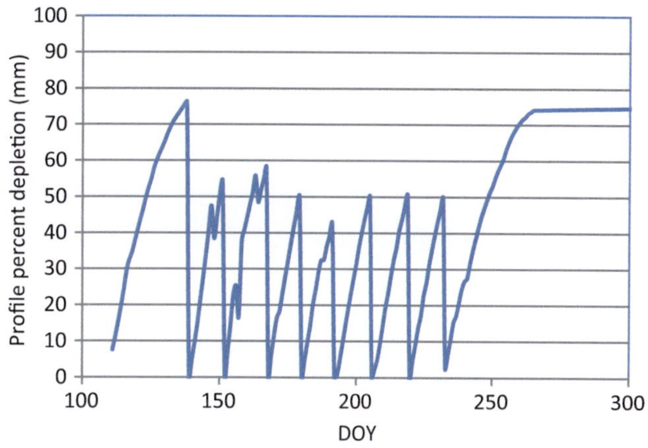

Fig. 29.15 Percent depletion, neglecting the evaporation layer, for Maricopa, Arizona cotton simulation during 2009

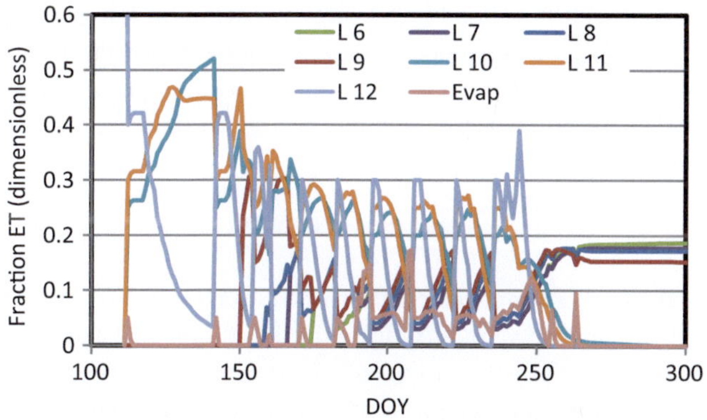

Fig. 29.13 Fraction of transpiration from each layer for Maricopa, Arizona cotton simulation during 2009

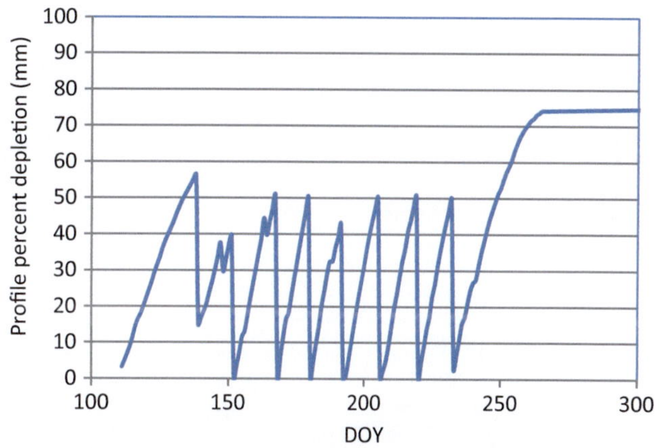

Fig. 29.16 Percent depletion, neglecting the evaporation layer and letting wetted fraction = 1.0 for entire season, for Maricopa, Arizona cotton simulation during 2009

	A	B	C	D	E	F	G	H
1	DOY	water stres	salt stress	nitrogen st	Ky	MAD (p)	relative yield	
2	112	1	1	1	1	0.3	0.780095	
3	113	1	1	1	1	0.27769		
4	114	1	1	1	1	0.375616		
5	115	1	1	1	1	0.426089		
6	116	1	1	1	1	0.444806		
7	117	1	1	1	1	0.453382		
8	118	0.977827	1	1	1	0.452749		
9	119	0.885446	1	1	1	0.44633		
10	120	0.878973	1	1	1	0.471128		
11	121	0.95466	1	1	1	0.481596		

Fig. 29.14 Crop stress and relative yield for Maricopa, Arizona cotton simulation during 2009

Green and Ampt (1911) assumed that an infiltration event could be described as a wetting front with saturated soil above the wetting front and antecedent moisture conditions below the wetting front. The Green-Ampt infiltration model can be described by an energy diagram. The progression of the wetting front is illustrated in Figs. 29.19 and 29.20.

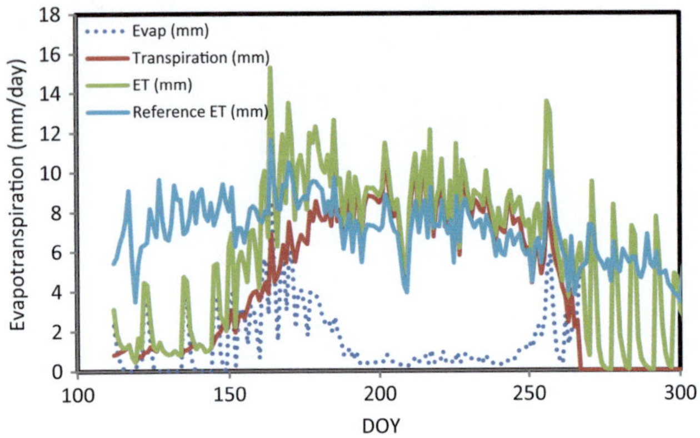

Fig. 29.17 Center pivot irrigation ET rates for 2009 cotton crop in Maricopa, Arizona (Credit Saman Rahimian. The University of Arizona)

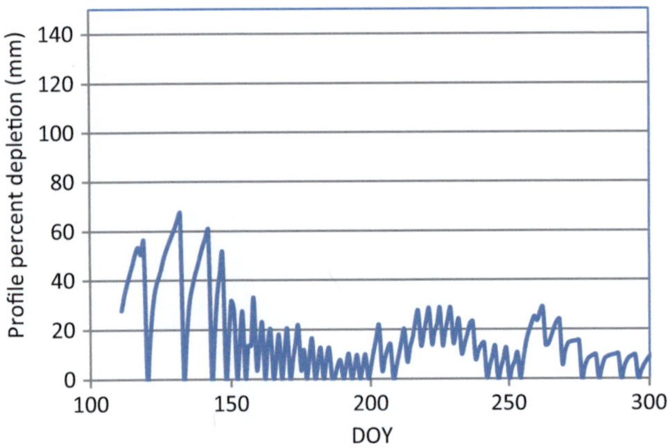

Fig. 29.18 Percent depletion for 2009 cotton crop in Maricopa, Arizona irrigated with center pivot irrigation

Elevation energy, z, at the datum is zero, and increases linearly with elevation. The pressure at the soil surface is equal to the depth of ponded water. The matric potential at the wetting front is a constant suction value based on soil properties, S_{av}. Total energy, H, at the soil surface is the ponded depth H_0 plus the elevation of the soil surface. Total energy at the wetting front is the suction (S_{av}) + elevation. The difference between the two is divided by the distance of the wetting front from the soil surface in order to calculate the total energy gradient, dH/dZ. Infiltration rate, v_{Darcy}, is then the product of the energy gradient, dH/dZ, and the saturated hydraulic conductivity, K_s. Note that dH/dZ is larger in Fig. 29.19 than in Fig. 29.20, in which the wetting front is closer to the water table. The difference in energy, H, between the soil surface and the wetting front remains constant, but the distance between the soil surface and the wetting front, L_f, increases. Because the energy gradient, dH/dZ, is highest when infiltration begins and decreases over time, the infiltration rate also decreases over time.

Although it seems logical that the suction at the wetting front would be dependent on the elevation above the water table (Figs. 29.19 and 29.20), this is not true. Brakensiek (1977) found that the suction, S_{av}, is a function of the bubbling pressure, h_b.

$$S_{av} = -0.76\, h_b \tag{29.17}$$

Brakensiek's simple calculation of S_{av} may be understood in light of the upward flux discretized solution in Chap. 28, which was insensitive to the assumed value of matric potential at the bottom of the root zone due to the decrease in hydraulic conductivity at low matric potential values: there is also a rapid change in matric potential vs. depth at the wetting front and a corresponding dramatic reduction in conductivity at low matric potential.

Fig. 29.19 Energy lines with wetting front near the soil surface during early part of infiltration event

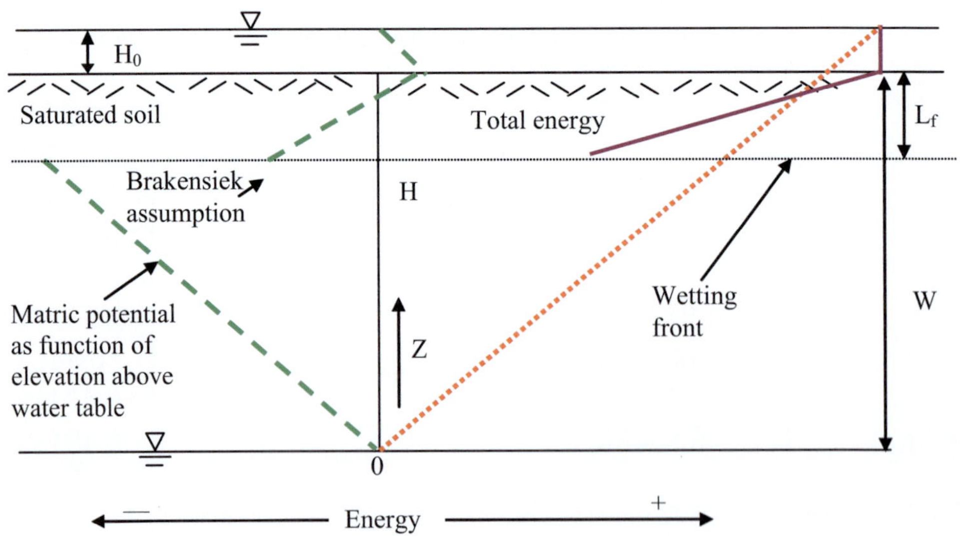

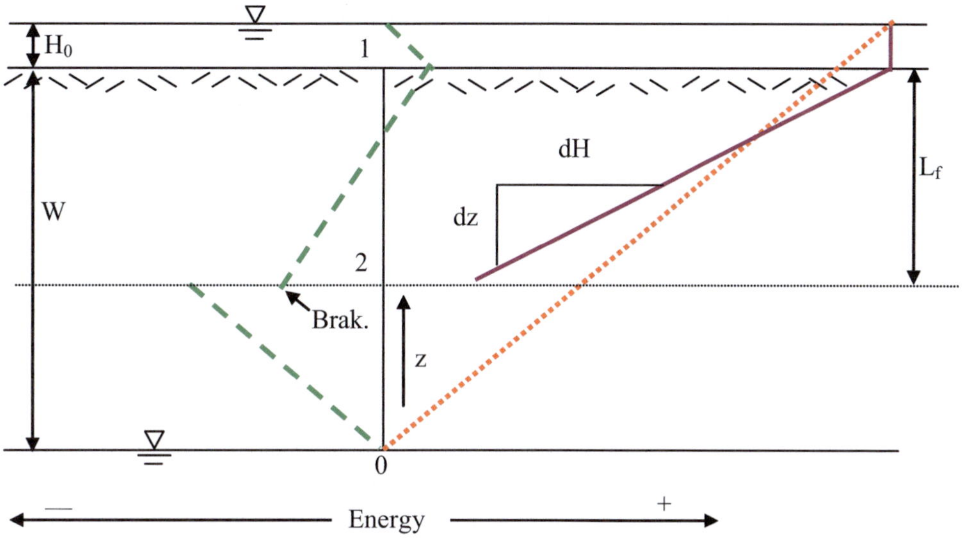

Fig. 29.20 Energy diagram during latter part of infiltration event

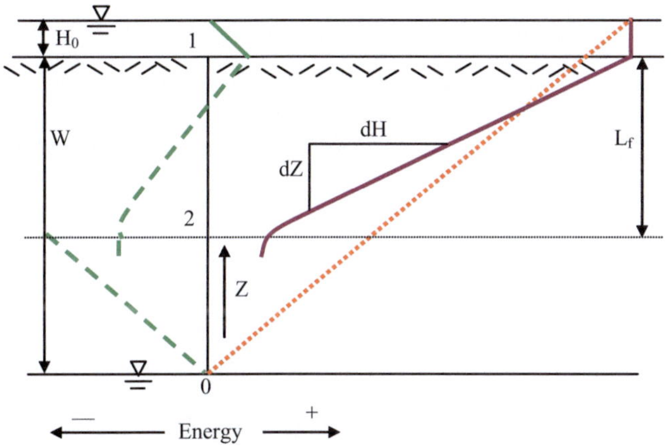

Fig. 29.21 Energy lines with diffusion at the wetting front

Diffusion occurs at the wetting front so there is not an abrupt change in water content or matric potential at the wetting front. Likewise, the energy lines at the wetting front do not make an abrupt turn as indicated by the energy graphs above. A better representation of the energy gradient at the wetting front is shown in Fig. 29.21.

In order to derive the Green-Ampt equation, let the control volume (Fig. 29.19) be bounded by the soil surface (1) and the wetting front (2). Although the boundary of the control volume moves, the energy balance can be calculated at any point in time.

$$h_f = H_1 - H_2 = h_1 + z_1 - h_2 - z_2 \qquad (29.18)$$

where

H = total potential, m,
h = matric potential, m,
z = elevation, m.

If the datum is set at the water table depth, then the elevation potential at the soil surface equals the water table depth, W.

$$\begin{aligned} h_f &= H_1 - H_2 = h_1 + z_1 - h_2 - z_2 \\ &= H_0 + W - S_{av} - (W - L_f) \\ &= H_0 - S_{av} + L_f \end{aligned} \qquad (29.19)$$

Substitute Eq. 29.19 into Darcy's equation in order to derive the Green-Ampt equation.

$$\begin{aligned} v_{Darcy} &= K_s \frac{h_f}{L_f} = K_s \frac{H_0 + S_{av} + L_f}{L_f} \\ &= \frac{(H_0 + S_{av})K_s}{L_f} + K_s \end{aligned} \qquad (29.20)$$

where

H_0 = ponded depth,

The length term in Darcy's equation is L_f, the distance from the soil surface to the wetting front. The saturated conductivity, K_s, is used in the Green-Ampt equation because the soil is wetted above the wetting front. $H_0 + S_{av} + L_f$ is difference in energy between the soil surface and the wetting front. Note that the infiltration rate, v_{Darcy}, is a function of L_f, the depth of the wetting front, and is not a function of the time since infiltration began because time is not in the equation. This fact can be used to advantage in the Green-Ampt model because the maximum rate of infiltration is not dependent on storm intensity. The antecedent water content (before infiltration) is designated as θ_I, and the water content after infiltration is saturated water content, θ_S. The difference is called the drainable porosity, m. For uniform with depth antecedent soil moisture conditions, the infiltrated depth is equal to the product of the change in water content and the depth of the wetting front.

$$i = (\theta_S - \theta_I)L_f = m\,L_f \qquad (29.21)$$

where

θ_S = saturated water content, ml/ml,
θ_I = initial antecedent water content, ml/ml.

Rearrange Eq. 29.21 and substitute into Eq. 29.20.

$$v_{Darcy} = \frac{di}{dt} = \frac{K_S(\theta_S - \theta_I)(S_{av} + H_0)}{i} + K_s \qquad (29.22)$$

Equation 29.22 is only valid if precipitation exceeds infiltration rate, which depends on the intensity of precipitation and soil hydraulic conductivity. In the case where precipitation does not exceed the infiltration rate, then infiltration is equal to precipitation. If precipitation is greater than the maximum infiltration rate, then runoff takes place; however, surface storage (see Chaps. 3 and 12) is filled before water runs off the field.

Example 29.4 A storm has an intensity of 2 cm/hr for 3 hours. Soil is Wagram loamy sand (Chap. 28). Initial water content in the soil profile is 0.044, Ks = 0.6 cm/hr, and depth of ponded water is zero. Bubbling pressure is 0.3 m. Surface storage is 0.5 cm. Calculate infiltration rate vs. time. Saturated water content = 0.305.

Calculate drainable porosity and wetting front suction

$$\theta_S - \theta_I = 0.305 - 0.044 = 0.261$$
$$S_{av} = 0.76{*}h_b = 0.76{*}0.30m = 0.23m = 23cm$$

Because the initial infiltration rate is constrained by the precipitation rate in this problem, it is not necessary to know the initial maximum infiltration rate. In the case of ponded infiltration, the early infiltration rate must be calculated. However, the calculated rate of infiltration is infinity at the first time step when the previously infiltrated depth is zero. Thus, an initial depth of infiltration must be assumed such that the calculated infiltration rate during the first time step lines up on the same curve as infiltration rates during subsequent time steps. For example, if an initial depth of 0.1 cm is assumed, then the infiltration rate during the first time step is calculated as follows.

$$\frac{di}{dt} = \frac{0.6\ cm/hr{*}0.261{*}23\ cm}{0.1\ cm} + 0.6\ cm = 36.3\ cm/hr$$

Because the precipitation rate (2 cm/hr) is lower than the maximum rate of infiltration, 36.3 cm/hr, the initial infiltration rate equals the precipitation rate. Once infiltration rate drops below the precipitation rate, then the infiltration rate is

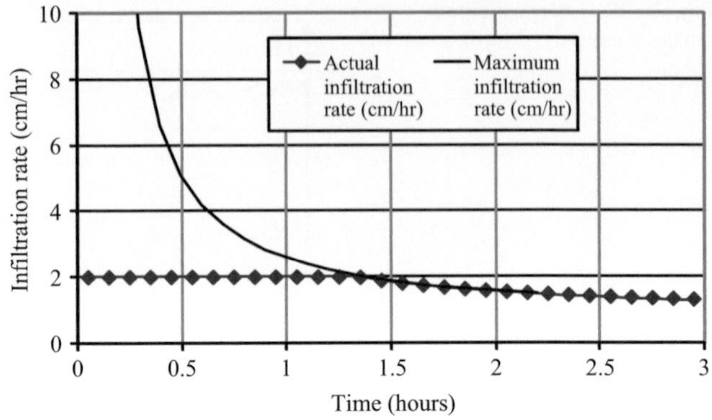

Fig. 29.22 Maximum and actual infiltration rate versus time for Example 29.4

determined by Eq. 29.22. Rearrange Eq. 29.22 to find the depth of infiltration when the maximum infiltration rate is 2 cm/hr.

$$i = \frac{K_s(\theta_S - \theta_I)\ S_{av}}{\frac{di}{dt} - K_s} = \frac{0.6{*}0.261{*}\ 23}{2 - 0.6} = 2.57\ cm$$

At a precipitation rate of 2 cm/hr, 2.57 cm would infiltrate after 1.29 hours, after which time, infiltration rate is calculated with Eq. 29.22. Uninfiltrated water is first added to surface storage, and runoff begins when surface storage is exceeded (Fig. 29.22). The maximum infiltration rate should be calculated based on the depth infiltrated rather than the time since the beginning of the storm.

The *WINDS* model uses the Green-Ampt infiltration model for multilayered soil in two VBA functions: wetting_front() and wetting_time(). The wetting_front() function calculates the depth of infiltration for a given time of ponding or precipitation. The wetting_time() function calculates the required time of ponding for a required depth of infiltration.

The wetting_front() function first calculates the wetting front suction, SAV, and drainable porosity, m, for each layer. Then, it calculates the total available air-filled pore space (above the water table), which is the sum of m * dz in each layer. The model first checks to make sure that sufficient pore space is available; otherwise, infiltration depth is set equal to available pore space. The model then checks to determine which layers have available pore space and only includes those layers in the Green-Ampt calculation. The model then solves for infiltration rate as a function of depth infiltrated. The suction at the wetting front, S_{av}, is calculated based on the soil properties of the layer that contains the wetting front. If there is precipitation, then the model checks to see whether infiltration rate is greater than precipitation rate at each time step.

Two of the limitations on rain fed agriculture in tropical climates are that wet soils in the rainy season prevent tractors from entering the field and limited rain during the dry season limits crop growth. In order to calculate the soil water balance, the rainfall must be accurately portioned between infiltration and runoff. The Green-Ampt model is ideally suited for this purpose.

Example 29.5 The *WINDS* model was used to evaluate soil moisture content in order to schedule farming operations and calculate crop stress in an economic evaluation of a large proposed farming operation in a tropical climate. Climate records were limited to daily rainfall, and maximum and minimum temperature. Daily reference ET was estimated as 4.2 mm/day for the entire year. In this climate, rainfall events are typically in the afternoon and last approximately 4 hours. Minimum relative humidity was estimated as 60 % for the entire year. Other weather parameters are in the *Active_year_weather* worksheet. The soil was divided into 5 layers, each 0.2 m depth, down to 1 m, with a 0.1 m evaporation layer. The soil is clay. There was no subsurface drainage or high water table. There was no irrigation. The crop was semi-perennial and the season was estimated as 1 year in this example. The crop could survive extended dry periods without significant loss of biomass, but growth reduction was proportionate to water stress (percent depletion). Thus, crop sensitivity to water stress was estimated as 1.0 for the entire year. There was no salinity stress because it was rainfed agriculture. Neither was nitrogen of concern. Other parameter values can be viewed in the *WINDS Chapter 29 tropical* workbook.

With clay soils, much of the precipitation does not infiltrate into the soil within the 4 hr storm event. Although storm depths are as high as 100 mm (Fig. 29.23), the four hour storm events do not allow more than 5 mm infiltration (Fig. 29.24) for a soil with a maximum infiltration rate of 1.25 mm/hr (clay). This results in reduced evapotranspiration (Fig. 29.25) and low soil water content during dry periods (Fig. 29.26).

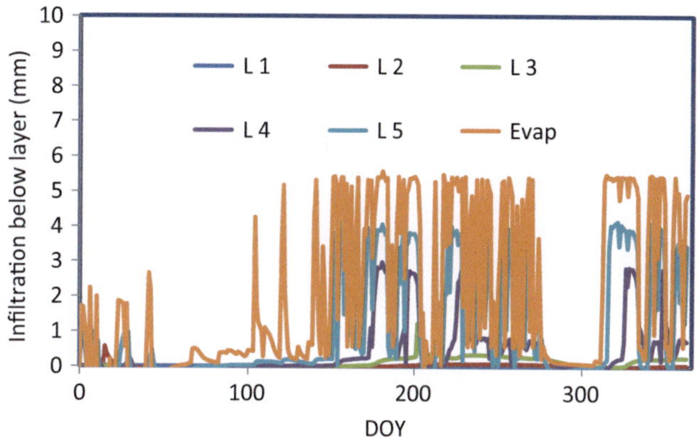

Fig. 29.24 Flux between layers in tropical climate for Example 29.5

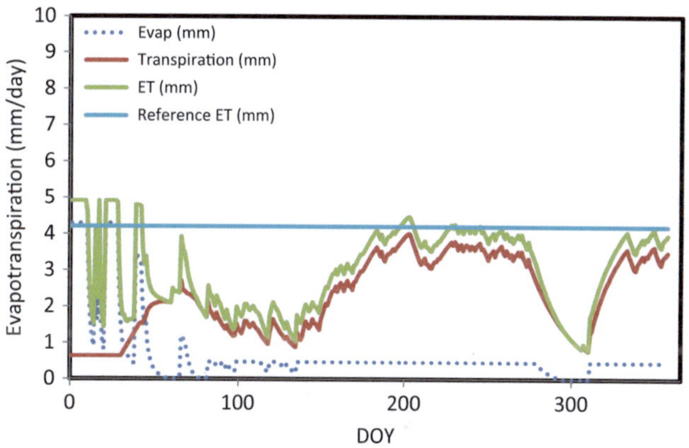

Fig. 29.25 Evapotranspiration in clay soil in tropical climate for Example 29.5

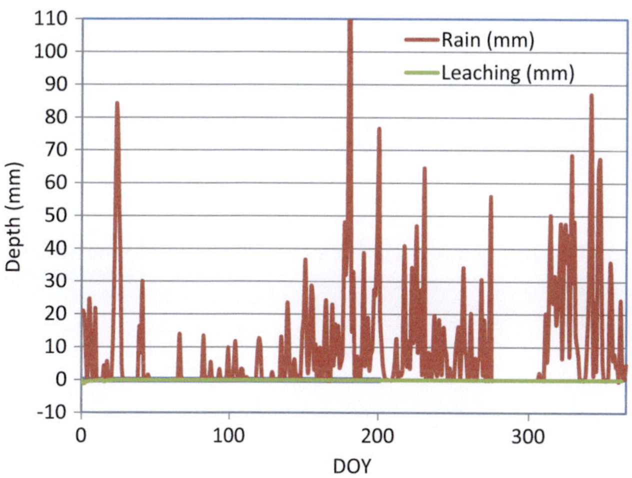

Fig. 29.23 Rainfall and leaching (none) in tropical climate for Example 29.5

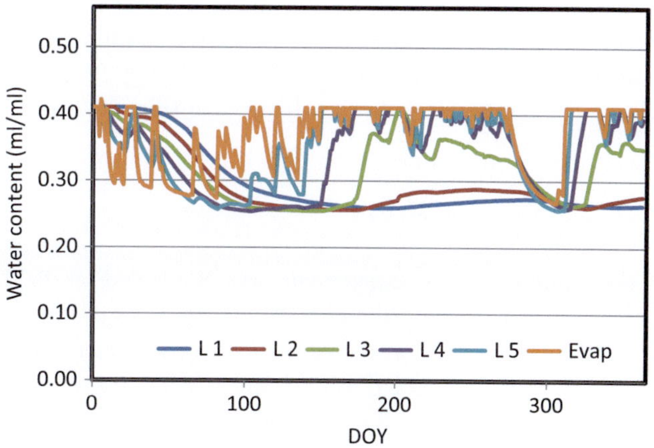

Fig. 29.26 Water content in clay soil in tropical climate for Example 29.5

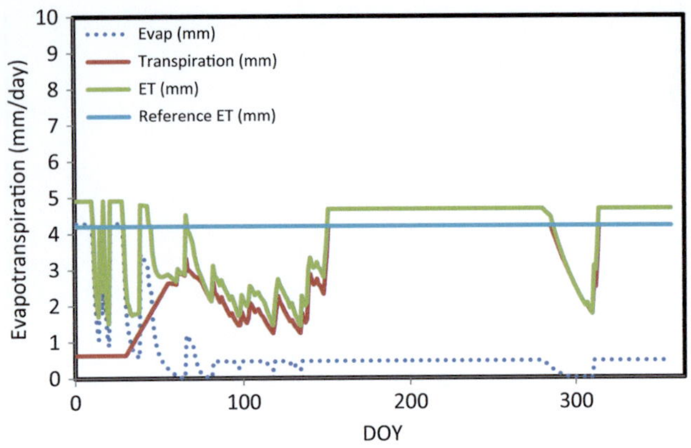

Fig. 29.27 Evapotranspiration in clay soil with water harvesting in tropical climate for Example 29.5

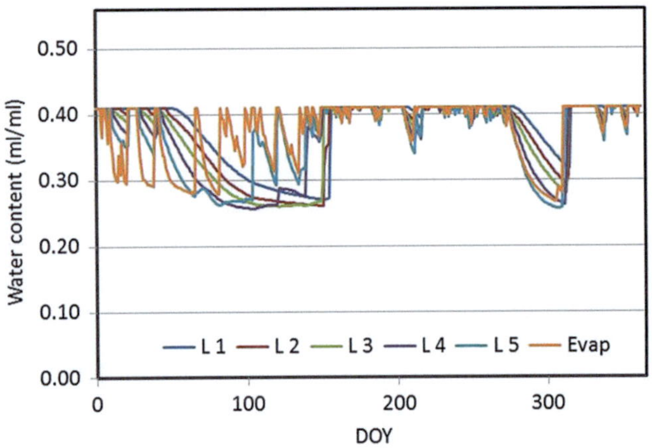

Fig. 29.28 Water content in clay soil with water harvesting in tropical climate for Example 29.5

In order to simulate a greater depth of infiltrated water, it was assumed that furrows were diked for water harvesting. In this case, it was assumed that the entire storm would infiltrate into the soil. This can be accomplished in the *WINDS* model by writing FALSE in cell A32 on the *Main* worksheet, which bypasses the calculation of infiltration by Green-Ampt or NRCS curves. In this case (Fig. 29.27), there was much less stress with water harvesting (Fig. 29.28).

WINDS Salinity Stress

Most salinity stress experiments are conducted at field capacity, and the corresponding b and $EC_{e\text{-threshold}}$ values are calculated based on the corresponding saturated paste extract salinity. In reality, salinity stress increases as water content decreases. The *WINDS* model calculates soil water salinity during each time step and then converts it to the corresponding saturated paste extract EC_e. In addition, if the soil water content is less then field capacity, then the soil salinity is greater than would be indicated by the calculated saturated paste extract salinity. Conversely, if the soil water content is greater than field capacity, then the plant salinity stress will be less than it would be at field capacity. Thus, salinity stress is calculated in the *WINDS* model based on actual water salinity and the corresponding saturated paste extract salinity at field capacity. Salinity stress, $K_{s\text{-salt}}$, is calculated with the following modified EC_e from each soil layer.

$$EC_e(k) = EC(k)*\frac{\theta_{FC-k}}{\theta_{sat-k}} \quad (29.23)$$

where

EC_e (k) = equivalent saturated paste extract salinity in layer k, dS/m,
EC(k) = actual water salinity in layer k, dS/m,
$\theta_{FC\text{-}k}$ = field capacity in layer k, ml/ml,
$\theta_{sat\text{-}k}$ = saturated water content in layer k, ml/ml.

Salinity varies in the soil profile, and the average soil profile salinity must be calculated in order to calculate salinity stress. The effective EC_e for the soil profile is calculated based on the weighted average of EC_{e-eff} in each layer k. In the *WINDS* model, this weighted average does not include the upper soil layer because of its normal dryness and lack of plant water uptake. Because the plant does not remove water from the upper layer when the soil is dry, including the upper layer in the weighted average would not result in a weighted average salinity that is representative of the extraction profile by the roots.

$$EC_{e-eff}(j) = \frac{\sum_{k=bottom\_layer}^{Num\_layers}\left(ET_{frac-k} *EC_e(k)\right)}{\sum_{k=bottom\_layer}^{Num\_layers}\left(ET_{frac-k}\right)} \quad (29.24)$$

where

bottom_layer = lower layer in the root zone
$EC_{e\text{-eff}}$ (j) = soil profile effective saturated paste extract salinity, dS/m,
j = day of year.

The salinity stress coefficient is calculated for each day.

$$K_{s-salt}(j) = 1 - \frac{b}{100*K_y(j)}\left(EC_{e-eff}(j) - EC_{e-t}\right) \quad (29.25)$$

where

$K_{s\text{-salt}}(j)$ = salinity stress coefficient on DOY j, dimensionless,
$K_y(j)$ = crop sensitivity to water stress on DOY j, %/dS/m,

Finally, the total daily crop transpiration is calculated:

$$ET = ET_0 K_{s-water} K_{s-salt} K_{cb} \qquad (29.26)$$

Although the crop reduction due to salinity stress could be calculated independently, the *WINDS* model calculates total stress during each daily time step as the product of water stress and salinity stress.

WINDS Nitrate Stress

The WINDS nitrate model has not been used in the field or approved by agronomists; nevertheless, the model is ready for testing. The nitrate mass balance parameters, as described in Chaps. 25 and 26, are implemented in the *WINDS* nitrate calculations. The daily nitrate uptake from layers in the root zone is calculated as follows. First, the seasonal nitrate requirement is divided between days. Second, the weighted average nitrate concentration in the root zone is found. Third, the Michaelis-Menton equation is used to calculate the actual daily nitrate uptake based on the weighted average nitrate concentration. Finally, the total nitrate uptake is divided between soil layers, based on the weighted nitrate contents in the layers.

There is a high correlation between daily nitrogen requirement and daily crop transpiration for many crops. Thus, the *WINDS* model determines the fractions of seasonal soil nitrogen requirement with the basal crop coefficient curve. In the model, there is no nitrogen uptake during the early and late season, which generally agrees with nitrogen uptake models. Alternatively, daily potential nitrate uptake rates or a crop nitrate coefficient curve could be specified in the model.

The Michaelis-Menton equation that was described in Chap. 25 includes the average soil nitrate concentration, mg NO_3-N/kg soil, as an input. Because soil nitrate varies in the soil profile, the weighted average soil nitrate concentration is weighted as the product of two parameters in all layers: nitrate concentration and ET fraction.

$$N_{ave\ mg/kg}(j) = \frac{\sum_{k=bottom\_layer}^{Num\_layers+1} \left(ET_{frac-k} * N_{mg/kg}(k)\right)}{\sum_{k=bottom\_layer}^{Num\_layers+1} \left(ET_{frac-k}\right)} \qquad (29.27)$$

where

$N_{ave\ mg/kg}$ (j) = average nitrate concentration in soil profile on DOY j, mg/kg soil,
$N_{mg/kg}(k)$ = average nitrate concentration in layer k, mg/kg soil,
bottom_layer = lower layer in the root zone.

The weighted average nitrate concentration is implemented in the Michaelis-Menton equation in order to calculate the nitrate uptake on DOY j.

$$m_{u\,pt-total}(j) = \frac{N_{max}(j) * \left(N_{ave-mg/kg} - N_{min}\right)}{Km + N_{ave-mg/kg} - N_{min}} \qquad (29.28)$$

where

$N_{max}(j)$ = maximum nitrate uptake rate on DOY j, kg/ha,
$m_{upt-total}(j)$ = nitrate uptake rate on DOY j, kg/ha/day.

Once the total nitrate uptake is calculated with Eq. 29.28, the uptake must be partitioned between soil layers. The calculation of nitrate uptake from each layer k (m_{upt-k}) is based on a weighting scheme that is calculated as follows, with the constraint no nitrogen is removed from a layer with a water content less than or equal to the permanent wilting point:

$$m_{u\,pt-k} = \frac{m_{u\,pt-total}}{10} \frac{ET_{frac-k} * N_{mg/kg}(k)}{\sum_{m=bottom\_layer}^{Num\_layers+1} \left(ET_{frac-m} * N_{mg/kg}(m)\right)} \qquad (29.29)$$

where

m_{upt-k} = nitrate uptake from layer k, mg/L_{soil}/day.

The ratio of actual to optimal nitrate uptake can be used to calculate a daily nitrogen stress coefficient. If nitrogen uptake is greater than the optimal nitrogen uptake, as would be the case if nitrate concentration is in excess of the recommended soil nitrate concentration, then yield may decrease in some crops because excess nitrogen uptake may lead to rank growth and decrease yield. For these crops, the daily nitrogen stress coefficient should less than 1.0 for excessive uptake of nitrate. The nitrogen stress index in the *WINDS* model is a step function:

$$K_{s-nit} = 1 - K_{Nlow}\left(N_{low} - m_{u\,pt-total}\right)/N_{low} \qquad m_{u\,pt-total} < N_{low}$$
$$K_{s-nit} = 1 \qquad N_{low} < m_{u\,pt-total} < N_{high}$$
$$K_{s-nit} = 1 - K_{Nhigh}\left(m_{u\,pt-total} - N_{high}\right)/N_{high} \qquad N_{high} < m_{u\,pt-total}$$

where

K_{s-nit} = nitrate stress coefficient, dimensionless,
K_{Nlow} = yield sensitivity to reduced nitrate uptake, %ΔY/(%Δkg/ha),
K_{Nhigh} = yield sensitivity to excessive nitrate uptake, %ΔY/(%Δkg/ha),
N_{low} = lower limit of optimal range of nitrate uptake rate, mg/kg,

N_{high} = upper limit of optimal range of nitrate uptake rate, mg/kg.

The nitrate requirement varies during the season so K_{Nlow}, K_{Nhigh}, N_{low} and N_{high} could be specified for each day of the growing season. Otherwise, it is assumed in the WINDS model that K_{Nlow} and K_{Nhigh} are 1.0 and that there is a ±10 % range of optimal nitrate uptake with no yield decrease.

Seasonal reduction in crop yield due to nitrogen stress is calculated in the same manner as seasonal reduction of yield due to salinity and water stress. The nitrogen stress for each day is weighted as a function of the daily required nitrogen.

$$\left(\frac{Y_a}{Y_m}\right)_{nit} = 1 - \frac{\sum_{j=Planting\_Date}^{Harvest}\left((1-K_{s-nit}(j))^*N_{opt}(j)\right)}{\sum_{j=Planting\_Date}^{Harvest}\left(N_{opt}(j)\right)} \quad (29.30)$$

The nitrate, water, and salinity yield fractions could be multiplied in order to find the seasonal fraction of maximum yield.

$$\frac{Y_a}{Y_m} = \left(\frac{Y_a}{Y_m}\right)_{water}\left(\frac{Y_a}{Y_m}\right)_{salinity}\left(\frac{Y_a}{Y_m}\right)_{nit} \quad (29.31)$$

The *WINDS* model outputs the daily water, salinity, and nitrogen stress factors to the *Stress* worksheet.

Example 29.6 Calculate K_{s-nit} and the nitrate uptake in layer 10 in a 12 layer simulation for a soil with the nitrate concentrations and ET fractions listed below. The sensitivity to reduced nitrate uptake is 0.7 and the sensitivity to excessive nitrate uptake is 0.5. The Michaelis-Menton terms (defined in Chap. 25) are Km = 10 mg/kg, N_{min} = 2 mg/kg, and N_{max} = 2.5. The optimal range is ±15 %. Layer 13 is the upper evapotranspiration layer. Exclude the upper layer from the weighted average calculation because water content is below permanent wilting point. The fraction of nitrogen taken up as nitrate is 90 % and the required nitrogen uptake is 2.3 kg/ha/day.

Layer	N_soil (mg/kg)	ET frac
4	9.74	0.02
5	9.87	0.03
6	9.57	0.03
7	24.0	0.04
8	1.72	0.05
9	18.7	0.05
10	14.9	0.2
11	7.57	0.25
12	4.43	0.3
13	16.0	0.03

The weighted average soil nitrate concentration (excluding evaporation layer) is

$$N_{ave\ mg/kg}(j) = \frac{\sum_{k=bottom\_layer}^{Num\_layers+1}\left(ET_{frac-k}\ {}^*N_{mg/kg}(k)\right)}{\sum_{k=bottom\_layer}^{Num\_layers+1}\left(ET_{frac-k}\right)} = \frac{8.96}{0.97} = 9.23\ mg/kg$$

The actual nitrate uptake from the soil profile is

$$m_{u\ pt-total} = \frac{N_{max}(N_{ave-soil} - N_{min})}{Km + N_{ave-soil} - N_{min}} = \frac{2.5\ kg/ha/d\ \ (9.23\ mg/kg - 2\ mg/kg)}{10\ mg/kg + 9.23\ mg/kg - 2\ mg/kg} = 1.05\ kg/ha/day$$

The optimal nitrate uptake is

$$N_{opt} = N_{req}Frac_{\_NO3} = 2.3(0.9) = 2.07 kg/ha/day$$

There is a 15 % optimal uptake range so the upper and lower limits are.

$$N_{low} = N_{opt}{}^*(1-0.075) = 2.07(1-0.15/2) = 1.91 kg/ha/day$$
$$N_{high} = N_{opt}{}^*(1+0.075) = 2.07(1+0.15/2) = 2.23 kg/ha/day$$

The uptake is below the optimal range; thus, K_{s-nit} is calculated as follows:

$$K_{s-nit} = 1 - K_{Nlow}\left(N_{low} - m_{u\ pt-total}\right)/N_{low} = 1 - 0.7(1.91 - 1.05)/1.91 = 0.68$$

The uptake from layer 10 is

$$m_{u\ pt-k} = \frac{m_{u\ pt-total}}{10}\frac{ET_{frac-k}\ {}^*N_{mg/kg}(k)}{\sum_{m=bottom\_layer}^{Num\_layers+1}\left(ET_{frac-m}\ {}^*N_{mg/kg}(m)\right)} = \frac{1.05}{10}{}^*\frac{14.9{}^*0.2}{8.96} = 0.035\ mg/L{}^*m$$

Thus, approximately 1/3 of total nitrate uptake is from layer 10.

Questions

1. Based on Fig. 29.9, estimate the midseason crop coefficient (around DOY 200). Also estimate the maximum soil evaporation coefficient early in the season and during the midseason.

2. Why would the dual crop coefficient provide a better comparison between two irrigation methods such as infrequent surface irrigation of a crop and frequent minisprinkler irrigation?
3. Based on Fig. 29.9, when would the use of a dual crop coefficient be superior to a single crop coefficient, during early season or midseason? How could you adjust a single crop coefficient to reflect the frequency of rainfall and irrigation events during the early season?
4. Calculate crop evapotranspiration. Reference evapotranspiration is 10 mm/day. The basal crop coefficient is 0.8 and the soil evaporation coefficient is 0.1. The water stress factor 0.9.
5. Use the following data to calculate the percent of maximum potential yield due to water stress. There is no salinity stress. The sum of Ky values during the season is 150.

DOY	Actual transpiration (mm/day)	Potential transpiration (mm/day)	Crop sensitivity to water stress
185	8	8	1.5
186	7	8	1.5
187	6	8	1.5
188	5	9	1.5
189	4	9	1.5
190	3	10	1.5
191	7	9	1.5
192	8	8	1.5

6. Calculate the depths of water extracted and the final water content for the four layers in the table. MAD is 0.4, ET_0 is 6 mm, K_{cb} is 0.6, and K_w is 0.05. Field capacity is 0.19 and permanent wilting point is 0.10. Assume that there is no evaporation due to full canopy cover.

Layer number	Lower boundary (m)	Initial water content	Fraction transpiration
0	0.1	0.13	0.05
1	0.4	0.17	0.6
2	0.8	0.18	0.35
3	1.2	0.22	??

7. Run the *WINDS Chapter 29 surface irrigation model.* Use the data from Example 29.3, but redo the irrigation schedule (IRR_01 worksheet) and depths such that no more than 14 mm water is lost to leaching after each irrigation (use the irrigation leaching graph in the Water Content worksheet or column CX) and no stress occurs (MAD is 0.5, percent depletions are listed in column DF and the *percent depletion* graph) before DOY 240. Let the fraction wetted (*ET_fractions worksheet*) be 1.0 for the entire growing season. Print out the *ETc*, *irrigation and leaching*, and *water content* graphs from the *Water_content* worksheet. Discuss how you might automate this adjustment process in the computer program. Discuss whether the required number of irrigations is practical. Consider whether the stress is overestimated in the early season. Also consider the effect of frequent irrigations on ETc (see ETc graph).
8. Run the *WINDS Chapter 29 surface model*. Change the soil type to clay loam. Print out the percent depletion graph from the *Water_content* worksheet. Compare to question 7 and discuss the effect of soil type on irrigation scheduling and deep leaching. How could the deep leaching have been prevented with the same large irrigation depths?
9. Initial water content in the soil profile before flood irrigation is 0.15, Ks = 0.2 cm/hr, and depth of ponded water is 10 cm. Plot the infiltration rate vs. time curve for ponded infiltration. Calculate infiltration rate every 0.1 hours with a spreadsheet. Calculate the infiltration rate and depth for 3 hours. Also plot the cumulative infiltration.
10. An irrigation event is 12 hours long. The depth of ponded water is 5 cm. Bubbling pressure = 0.3 m and saturated water content = 0.305. The antecedent water content is 15 %. Saturated hydraulic conductivity = 0.2 cm/hr. Calculate by hand for the first two time steps, Use time steps = 0.1 hr. Run the *1 layer ponded* worksheet to find the depth infiltrated after 12 hours. You may need to adjust the columns in order to make it work correctly. One of the challenges of Green-Ampt infiltration is to find an assumed depth of infiltration before the first time step, which results in a smooth overall curve. Adjust the initial infiltrated depth until the curve is smooth. Investigate the sensitivity of the estimate of total infiltration over the entire irrigation event to the initial assumed depth of infiltration. Print out the infiltration rate in cumulative infiltration curves. Based on the graph, what term in the Green-Ampt equation does the final rate of infiltration approach.
11. For the parameters in question 10, change the depth of ponding to 10 cm. Evaluate whether increasing the depth of ponding plays a significant role in infiltration in the case of typical depths used in surface irrigation events.
12. For the parameters in question 10, change the initial water content to 10 % and 20 %, and 25 %. Evaluate whether antecedent water content plays a significant role in infiltrated depth and depth of the wetting front.
13. A soil has two layers, both 20 cm depth. The upper layer (0–20 cm) has a drainable porosity equal to 0.2. The lower layer (20–40 cm) has a drainable porosity equal to 0.10. The suction at the wetting front in both layers is 20 cm, and the depth of ponding is 10 cm. The saturated

hydraulic conductivity in both layers is 0.6 cm/hr. Plot the infiltration rateand cumulative infiltration for 12 hours of ponding. Compare to a uniform drainable porosity in the entire soil profile of 0.2. Is there a significant different in infiltration rate once the wetting front reaches the 2nd layer due to the changed drainable porosity?

14. Redo question 13, but let the conductivity of the lower layer equal 0.3 cm/hr and $M = 0.1$. Calculate effective hydraulic conductivity as the geometric mean and show a sample calculation of the weighted conductivity. Plot the infiltration rate and the depth of the wetting front vs. time. Evaluate the effect of decreased conductivity in the lower layer on cumulative infiltration and how this might influence irrigation practices.
15. Redo Example 29.5 (*WINDS Chapter 29 tropical*), but use a sandy loam soil. Print the ET/transpiration graph and the water content graph. Compare the crop water stress during the season with and without water harvesting. Make sure to change the initial water content to field capacity.

References

Allen RL, Pereira DR, Smith M (1998) Crop evapotranspiration – guidelines for computing crop water requirements. FAO Irrigation and drainage paper 56. FAO – Food and Agriculture Organization of the U.N. http://www.fao.org/docrep/X0490E/x0490e00.htm

Brakensiek DL (1977) Estimating the effective capillary pressure in the Green and Ampt infiltration equation. Water Resour Res 13(3):680–682

Green WH, Ampt GA (1911) Studies on soil physics, 1. The flow of air and water through soils. J Agric Sci 4(1):1–24

30 Subsurface Drainage Design and Installation

Subsurface drainage systems remove excess groundwater below the ground surface. Perforated plastic drain tubes are placed between 1 and 2 m below the soil surface. The technique was originally called tile drainage because tile cylinders were laid end to end in a trench. Spacing and depth of drain tubes as well as lateral hydraulic conductivity of soil layers determine the rate of water removal from the field. Lateral hydraulic conductivity is generally measured with auger hole tests. Drainage can impact downstream water quality because it changes the timing of water and chemical leaching through the soil. Drainage planning and design require extensive analysis of hydrology, soil structure and texture, soil chemistry, crop rotations, field equipment, topography, waterways, and construction materials. The sizing of drainage pipes is based on the land slope and expected flow rate to pipes. Pipe layout and slope is a function of land slope and discharge location in the field. If the discharge waterway is higher than the drain outlet, then a sump and pump must be installed. Gravel envelopes or fabrics protect drain tubes from sedimentation.

Environmental Considerations

Drainage has an impact on the environment and is thus under increasing scrutiny by regulators and environmentalists.

The impact of the proposed drainage system on the community, landowner, neighbors, agricultural developments, natural resource conservation, biodiversity, wildlife habitat, water quality, economics, health, and social considerations are all be considered in the decision making and regulatory processes (NRCS Part 650 EFH 14, 2001).

Drainage water normally carries a high nitrate load because nitrate concentration is normally high in agricultural soils. However, the issue of whether subsurface drainage systems have a positive or negative effect on water pollution is a subject of debate. Fogiel and Belcher (1991) listed the following positive environmental impacts of subsurface drainage (EFH 14): reduced overland flow and erosion, reduced sediment transport and attached phosphorous and potassium off the farm, improved surface nutrient management, improved ability to monitor percolating flow, and reduction of nonpoint source pollution. The reason that erosion is reduced is that the soil generally has less water and infiltration is increased. The major disadvantage is increased discharge of nutrients and other chemicals in the soil to surface water bodies. For example, high nutrients in Mississippi River flow have caused eutrophication and dead zones in huge areas of the Gulf of Mexico. Much of this pollution comes from drainage water that is loaded with nitrate and other nutrients and agricultural chemicals from farms. Exacerbating this problem, many natural wetlands that used to flank the Mississippi River and purify the water before it entered the river have been removed by agricultural and urban development.

The NRCS recommends several techniques for the reduction of the impact of drained water from farmland on the environment. Install on-farm detention ponds or wetlands that reduce nitrate concentration in drainage water before it leaves the farm, retrofit drainage systems to reduce drainage during critical periods and manage the water table such that the nutrient load from subsurface drainage systems is reduced. Implement cropping systems that reduce nitrate losses. Improve fertilizer and animal manure management and timing.

Drainage System Planning

The first step in designing a drainage system is a general reconnaissance survey: needed items include maps, soil survey reports, a hand level, and a soil auger. Reconnaissance should begin with a walkover through the site soon after a heavy rainfall event in order to find drainage problems. Local wells and a few borings should be made in order to determine the depth to the water table. Wetlands areas

P. Waller, M. Yitayew, *Irrigation and Drainage Engineering*, DOI 10.1007/978-3-319-05699-9_30

should be delineated. Soils or vegetation types can indicate the presence of wetlands. Pacing off distances and a hand level can be used to determine the approximate slope of waterways and drainage paths. Surveying equipment can be used in locations where higher survey accuracy is needed. The NRCS (EFH 14, 2001) lists the following features that should be included in a reconnaissance survey.

- Location and extent of any wetlands.
- The areas in which crops show damage, as pointed out by the farmer, indicated by the aerial photograph, or noted in personal observations.
- Personal observations of unique landscape features, ecologically significant areas, land use patterns, operation (land management) aspects, and site visibility.
- Topography and size of the watershed area.
- Size, extent, and ownership of the area being considered for drainage.
- Location of the drainage outlet and its condition.
- Location, condition, and approximate size of existing waterways.
- Presence of cultural resources.
- Potential impacts outside the area being evaluated.
- General character of soil throughout the area needing drainage, including land capability, land use, crops and yields, and salinity or sodicity.
- High-water marks or damaging floods and dates of floods.
- Utilities, such as pipelines, roads, culverts, bridges, and irrigation facilities and their possible effect on the drainage system (see NEM part 503).
- Sources of excess water from upslope land or stream channel overflow and possible disposal areas and control methods.
- Condition of areas contributing outside water and possible treatment needed in these areas to reduce runoff or erosion.
- Condition of any existing drainage system and reasons for failure or inadequacy. Old subsurface drainage systems that have failed because of broken or collapsed sections may well be the cause of a wet area.
- Estimate of surveys needed.
- Type and availability of construction equipment.
- Feasibility.

Existing drainage pipelines are often difficult to find. Infrared remote sensing can detect the presence of existing subsurface drainage pipes by detecting the temperature of the soil. In general, the soil is warmer over subsurface drainage pipes. In addition to infrared maps, aerial photographs can show the natural drainage ways in a region.

Topographic information is important since flow is driven by gravity. The USBR recommends a scale of 1 in = 4,000 ft (2.54 cm = 1,200 m) topographic maps for reconnaissance studies and 1 in = 400 ft (2.54 cm = 120 m) for detailed design of drainage systems. A 2 m (5 ft) contour interval is recommended for preliminary studies, and a 0.6 m (2 ft) contour interval is recommended for drainage design. For large, nearly level areas, a 0.3 m (1 ft) contour interval is required. If land is relatively flat such as is common in surface irrigated agriculture, a topographic map is not adequate for planning drainage systems, and the path of the drain system must be surveyed. The NRCS recommends collecting survey information at 30–90 m (100–300 ft) intervals on flat land. The profile of existing ditch cross sections, details of utilities, and construction along ditches should also be provided to the drainage system designer.

The elevation of the drainage outlet can constrains the maximum slope of the drain pipe. The end of the drain can be no lower than the minimum elevation of the drainage outlet. If the potential drain slope in a field is unacceptable, then a sump may be required. Minimum drain depth at the upper end of the drain can be restricted by government regulations, the drainage design depth (Chap. 10), or by tillage practices. The combination of minimum depth of installation and a shallow outlet can severely restrict the slope on drainage lines. If drains are installed at a nearly level grade, then there is a possibility of air locks and sedimentation in the drainage pipeline. Dips fill with mud, and rises can have airlocks and prevent drainage. The USBR recommends that the minimum grade (slope) on a drain pipe should be 0.1 m/100 m. This is an extremely shallow grade and is susceptible to installation errors that may lead to air locks and blockages by siltation. In this case, extreme care must be taken during drainage installation to make sure that drains are installed to grade. If not installed to grade, causing the drain not flow, then the drains will need to be reinstalled at the expense of the installation company. This situation usually does not result in happy farmers, engineers, or contractors.

Laser levelers on modern trenching and drain pulling equipment provide elevation and direction control during drain installation. If laser equipment is not used, then surveyors mark elevations on stakes along the drain trench every 50 ft (15 m). A string is then tied at the marked elevations. Then a dogleg (upside down L) is held by a worker against the string and the bottom of the trench. This worker tells the trencher operator to raise or lower the depth of the trenching equipment as the trench is excavated.

Soils Analysis

Extensive soil hydraulic conductivity tests must be connected on the farm. A typical practice is to collect soil cores to 2 m depth every hectare. These can be used to develop hydraulic conductivity maps and soil profile maps for drainage design. Then, the methods presented in Chap. 31 can be used to determine the optimal drain spacing and depth.

A shallow hardpan or clay layer can influence drainage system design. If a drain is placed below an impermeable layer, then water may not be able to reach the drain.

Artesian water is a potential problem. If an aquifer is fed from below at sufficient pressure, then the capacity of the drainage system will need to be increased in order to drain the excess water from the field.

Pipe Network Layout and Elevations

Once the drain spacing is determined, a map of the drainage pipe layout can be drawn. Drains should follow the natural slope of the land if the field is relatively flat. A herringbone pattern or other geometries may be preferable depending on land slope and the outlet location. Elevation calculations and installation should begin at the outlet since that is the common point at which all branches meet.

Pipe Diameter

Drainage systems are designed to carry water from the field during the period of peak drainage. The design flow rate is often referred to as the drainage coefficient. The NRCS recommends that drain pipes should never flow under pressure because water flowing out of the pipe disturbs the envelope or soil around the drain. Drain pipes must be large enough to carry the peak flow during a storm or irrigation without becoming pressurized. The USBR bases the drainage coefficient on the rate that water flow to the drain and uses the following equation to calculate peak drainage requirement (design flow rate) for drains above an impermeable layer.

$$q_d = \frac{2\pi K m_0 D}{86,400L} \quad (30.1)$$

where

q_d = discharge from two sides per unit length of drain, m^3/sec/m,
m_0 = maximum height of water table above drain invert, m,
K = weighted average hydraulic conductivity of soil profile between maximum water table elevation, y_0, and impermeable layer.
D = average flow depth, d + $y_0/2$, m,
d = distance from drain to barrier, m,
L = drain spacing, m.

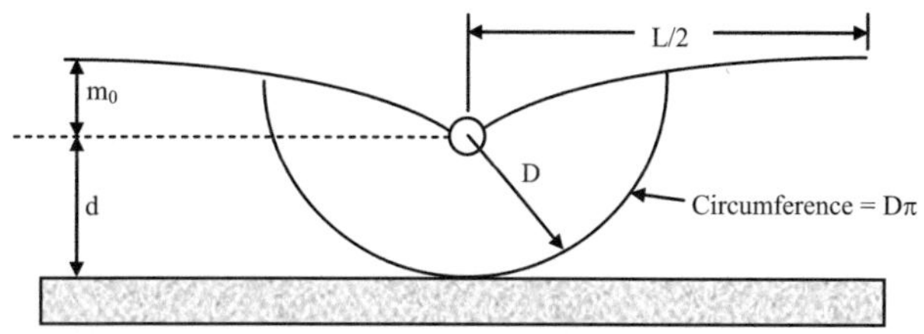

Fig. 30.1 Geometry for derivation of drainage coefficient

Equation 30.1 is derived from Darcy's law based on the geometry of Fig. 30.1 where the cross-sectional area of flow is Dπ, the energy difference is m_0, and the length is L/2.

$$q = KA\frac{\Delta H}{\Delta x} = KD\pi\frac{m_0}{L/2} \quad (30.2)$$

where

m_0 = energy difference, m,
L = distance between drains, m,
A = cross-sectional area of flow, Dπ, with D as a radius.

$$q_d = KD\pi\frac{m_o}{\frac{L}{2}} = \frac{2\pi K m_o D}{L} \quad (30.3)$$

Equation 30.3 is divided by 86,400 to convert flow rate from m^3/d/m to m^3/s/m.

$$q_d = \frac{2\pi K m_o D}{L}\left(\frac{m^3/m/\sec}{86,400\ m^3/m/day}\right) = \frac{2\pi K m_o D}{86,400\ L} \quad (30.4)$$

The drainage coefficient (Fig. 30.1 and Eq. 30.3) is not meant to represent the flow rate that occurs when the soil directly over the drain is saturated and water flows into the drain from above: In this case, the energy gradient is 1.0 rather than $m_0/(L/2)$.

The drainage coefficient can be expressed in units of mm/day by dividing by L, the distance between drains, and multiplying by 1,000 to convert from m/day to mm/day.

$$D_c = \frac{q_d}{L} = \frac{2\pi K m_o D}{L^2}\left(\frac{1,000\ mm}{m}\right) = \frac{2,000\pi K m_o D}{L^2} = \left(\frac{mm}{day}\right) \quad (30.5)$$

General drainage coefficients (Table 30.1) based on how quickly the soil must be dewatered have been developed by the NRCS (EFH 14, 2001) based on the constraint that surface depressions and the majority of the root zone should be drained within 24–48 hours after a storm.

Table 30.1 General drainage coefficients (Credit NRCS, EFH 14, 2001)

Soil	Field crops (grains, orchards, etc.)	Truck crops (vegetables)
Mineral	⅜ – ½ inch (8–12 mm)	½ – ¾ inch (12–18 mm)
Organic	½ – ¾ inch (12–18 mm)	¾ – 1½ inch (18–36 mm)

The USBR designs pipe drains that have no surface inlets to run full, and pipe drains that have surface inlets to run half full (larger pipe for fields with direct surface inlets to drains). Alternatively, the following adaption of the Manning equation calculates the minimum ID of corrugated plastic pipe drains with diameter less than 30 cm (Huffman et al. 2013) based on field area and drainage coefficient.

$$ID = 51.7(D_c{*}A{*}n)^{0.375}s^{-0.1875} \quad (30.6)$$

where

ID = inside drain diameter, mm,
D_c = drainage coefficient, mm/d,
A = drainage area, ha,
n = coefficient of roughness, 0.015 for corrugated plastic drain pipe,
s = drain slope, m/m.

Pipe flow rates as a function of drainage coefficient and drained area can be found in ASABE Standards. Pipe size as a function of flow rate, grade, and velocity is also found in ASABE Standards.

Example 30.1 Calculate the required diameter of a subsurface drain based on Eqs. 30.5 and 30.6 with the following parameters. Let Manning's n = 0.015.

Minimum depth to water table	0.48 m
Depth to impermeable layer	5 m
Drain depth:	2 m
Drain slope	s = 0.2 m/100 m
Hydraulic conductivity	K = 2.5 cm/hr
Drain spacing	L = 40 m
Length of drain pipe	L_d = 200 m
Maximum WT height above drains:	m_0 = 2 m − 0.48 m = 1.52 m
Distance from drains to impermeable layer	d = 5 m − 2 m = 3 m
Average depth of flow to drain	D = d + m_0/2 = 3 + 1.52/2 = 3.76 m
Hydraulic conductivity	K = 2.5 cm/hr = 0.6 m/day

Calculate drainage coefficient.

$$D_c = \frac{2{,}000\pi K m_o D}{L^2} = \frac{2{,}000\pi{*}0.6{*}1.52{*}3.76}{40^2}$$
$$= 13.5\ mm/day$$

Calculate pipe diameter

$$ID = 51.7(D_c{*}A{*}n)^{0.375}s^{-0.1875}$$
$$= 51.7(13.5{*}200{*}40/10{,}000{*}0.015)^{0.375}{*}0.002^{-0.1875}$$
$$= 84\ mm$$

The drain diameter, 84 mm, is between 3 and 4 inches diameter. Round up to the next largest manufactured size, 4 inches (102 mm ID and 114.5 mm OD).

Filters (Envelopes)

In many soils, a filter is required around the drain in order to prevent sedimentation of the drain by the base soil. Soils with low hydraulic conductivity do not tend to have problems with soil washing into the drain because the water does not flow fast enough to carry soil particles with it. In general, all soils with low clay content are considered problem soils with respect to sedimentation of drainage pipes (NRCS, EFH 14, 2001).

There are two primary types of envelope materials: (1) geotextile, and (2) sand and gravel. Organic materials, such as wood chips can be packed around the drains, but they may not be durable. In addition, organic materials may lead to slime formation in drains in some soils (NRCS, EFH 14, 2001). The soil textural triangle can be divided into sections that delineate base soils that need no filter, textile filters, or sand and gravel filters (Fig. 30.2).

Pulled drains can only use textile filters that are added to the drain during the manufacturing process. The filter is a sock around the drain tube that is pulled into the ground with the drain tube. The geotextile fabric may be constructed from polyester, nylon, polypropylene, polyamide, or polystyrene. These fabrics vary in mesh size, fiber size, and weight: no single fabric is suitable for all soils. The NRCS recommends that the flow capacity of the sock be 10 times that of the soil in order to avoid restricting flow to drain perforations. The recommended permittivity is 1 $m^3/(s * m^2 * m)$, which means one cubic meter per second should flow through a square m of fabric given a 1 m head differential from one side of the fabric to the other.

As large soil particles are stopped by the geotextile filter, they in turn become a filter and stop smaller particles from washing into the drain. In order to catch the larger particles, the following criteria is used by the NRCS. If less than 50 % of the soil particles pass the Number 200 sieve, then the apparent opening size (AOS) of the fabric should be the same size as the number 30 sieve (0.59 mm). On the other hand, if more than 50 % of the soil particles pass the number 200 sieve, then the apparent opening size of the fabric should be smaller (Number 50 sieve, 0.297 mm). For silty soils, sand and gravel envelopes are recommended (Fig. 30.2).

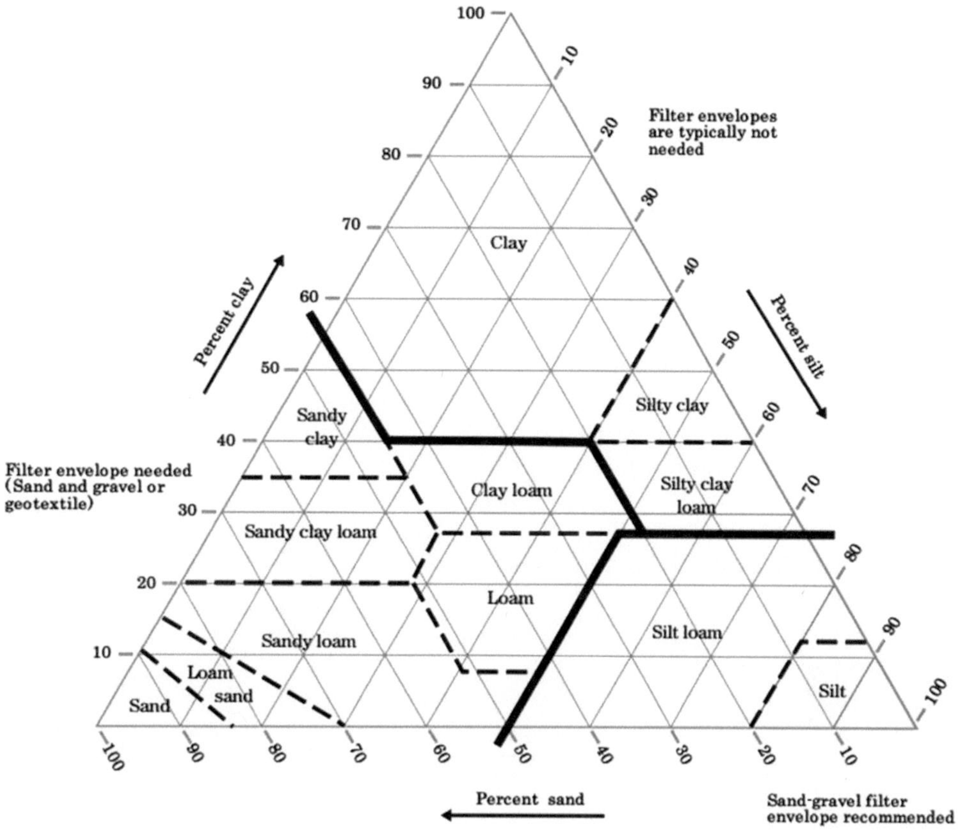

Fig. 30.2 Recommended filters for different soil textures (Credit NRCS, Part 650 Engineering Field Handbook, Chap. 14, 2001)

It is generally recommended that trenched drains have a sand and gravel filter envelope on all sides. The USBR Drainage Manual (1993) recommends that gravel envelopes should be designed with a 10 cm distance from the pipe to the surrounding soil; however, the NRCS recommends a minimum 7.5 cm (3 in) distance. Thus, a 4 in (10 cm) drain would be laid in a 25–30 cm width trench. Consult NRCS or USBR drainage manuals for further information on sand and gravel filter geometry. Sand and gravel filters increase hydraulic conductivity around the drain as well as prevent soil from washing into the drain. The envelope should be well graded, free of vegetable matter, clay and other deleterious substances that could change the hydraulic conductivity of the envelope: a well-graded envelope material has particles in all specified size ranges. Larger particles help increase conductivity and support the drain while smaller particles help reduce the movement of fines from the base material (original soil surrounding the trench) into the envelope material. Proper gradation usually requires machine sorting.

Velocity at the interface between the base material and the envelope must be low enough to prevent fine textured material from moving into the envelope. In order to meet the requirements for conductivity and filtration, there must an even distribution (gradual size gradation) of particle sizes within the envelope. If there is too much size difference between each particle size gradation in a filter, then the filter is called a gap graded filter. In this case, particle segregation may occur during placement (*NRCS NEH 633.2603*).

When sand and gravel filters are designed with the correct size gradation of particles, eroding soil is caught at the surface of the filter and these soil particles fill in developing cracks in the soil, causing subsequent flow to pass through soil pores (*NRCS NEH 633.2603*). This prevents the development of channeling or cracks in the soil and the resulting sedimentation of drains (*NRCS NEH 633.2603*). Envelope material requirements for silt-loam, sandy clay loam and loam base materials are more flexible than for fine sand and very fine sand because the finer textured soils (clays and loams) have a low velocity at the envelope – base material interface, and particles are not likely to be carried into the envelope.

The USBR recommends sampling with a soil auger every 180 m along trenches in order to characterize the base material. If the base material changes significantly along the trench system, then envelope gradation requirements should be changed.

The USBR uses two coefficients to assess envelope material gradation: coefficient of uniformity and coefficient of curvature. The coefficient of uniformity determines whether there is a sufficient range of sizes in the gravel or sand envelope material to prevent wash in of particles from the

Table 30.2 Gradation relationship between base material and diameters of envelope material

	Gradation limitations for envelope (diameter of particles, mm)										
	Lower limits, percent retained (i.e., $D_{10} \leftrightarrow 90$ %)						Upper limits, percent retained				
Base material, 40 % retained (diameter of particles, mm)	0	40	70	90	95	100	0	40	70	90	100
0.02–0.05 Silt	9.52	2.0	0.81	0.33	0.3	0.07	38.1	10.0	8.7	2.5	0.59
0.05–0.1 VFS	9.52	3.0	1.07	0.38	0.3	0.07	38.1	12.0	10.4	3.0	0.59
0.1–0.25 FS	9.52	4.0	1.30	0.40	0.3	0.07	38.1	15.0	13.1	3.8	0.59
0.25–1.0 Sand	9.52	5.0	1.45	0.42	0.3	0.07	38.1	20.0	17.3	5.0	0.59

base material. The coefficient of uniformity has two criteria: one for sand and another for gravel, where the minimum size of gravel (maximum size of sand) is 2 mm. The coefficient of curvature shows whether most of the change in size distribution occurs over a narrow range of sizes or gradually over the entire range. A coefficient of curvature between 1 and 3 reflects a relatively constant gradation of particle sizes.

Coefficient of uniformity (C_U): well graded if $C_U > 4$ for gravel and $C_U > 6$ for sand

$$C_U = D_{60}/D_{10} \tag{30.7}$$

Coefficient of curvature: between 1 and 3 for both gravel and sand

$$C_C = (D_{30})^2/(D_{10}{*}D_{60}). \tag{30.8}$$

where

D_x = Diameter (mm) where "x" is percent of particles smaller than the diameter.

For all envelopes, 100 % of the envelope material should pass the 38.1 mm sieve and 5 % or less should pass the 0.3 mm sieve. Based on experience and Eqs. 30.7 and 30.8, the USBR recommends the upper and lower limits of percent of particles that are smaller than the given sizes in Table 30.2.

For example, an envelope material in very fine sand (VFS) base material should have a sieve analysis with a gradation curve that lies between the upper and lower limits shown in Fig. 30.3. Each point on the curve is the percentage of material that is larger (is retained on the sieve) than the sieve. This percentage is the opposite of the D_x criterion (100 − x). For example, at least 30 % of the material (D_{30}) should pass the 10.4 mm sieve, but no more than 30 % should pass the 1.07 mm sieve.

Example 30.2 Evaluate the C_C and C_U of the upper and lower gradation limits for very fine sand (VFS) base material in Fig. 30.3.

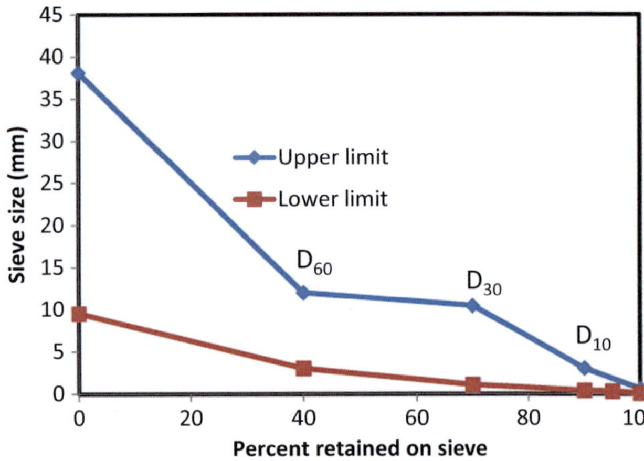

Fig. 30.3 Upper and lower gradation limits for very fine sand (VFS) base material

Upper Limit

$D_{60} = 12$ mm, $D_{30} = 10.4$ mm, $D_{10} = 3.0$ mm

$$C_u = \frac{D_{60}}{D_{10}} = \frac{12}{3} = 4 \quad C_c = \frac{(D_{30})^2}{D_{10}D_{60}} = \frac{10.4^2}{12{*}3} = 3$$

The minimum size of gravel is 2 mm; thus, almost all of the envelope material is gravel. The criterion for gravel gradation is that the C_u must be greater than 4, and the upper limit gradation C_u is exactly equal to 4. The coefficient of curvature is 3, the maximum acceptable level for both gravels and sands.

Lower Limit

$D_{60} = 3$ mm, $D_{30} = 1.07$ mm, $D_{10} = 0.38$ mm

$$C_u = \frac{D_{60}}{D_{10}} = \frac{3}{0.38} = 7.9 \quad C_c = \frac{(D_{30})^2}{D_{10}D_{60}} = \frac{1.07^2}{3{*}0.38} = 1$$

By definition, the larger size particles that are specified in the lower limit gradation are between the sizes of very coarse sand and very fine gravel. The criteria for sand gradation are that the C_u must be greater than six, and the lower limit gradation C_u is 7.9 – an acceptable value. The coefficient of

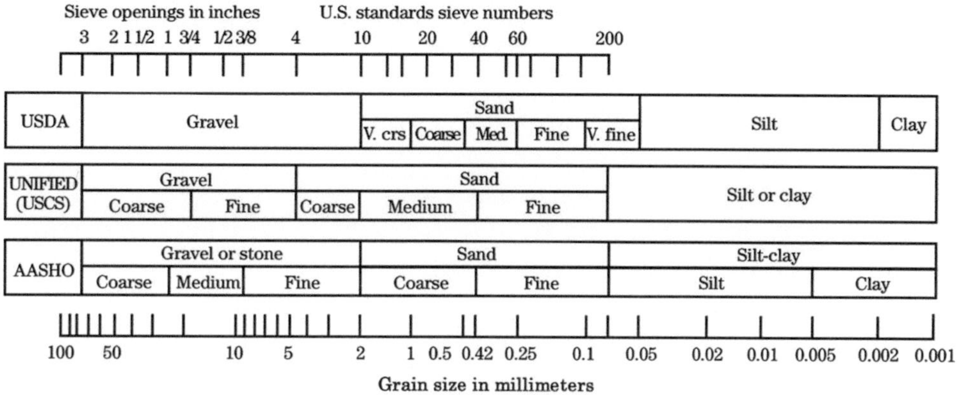

Fig. 30.4 Sieve sizes and associated particle grain diameters (Credit NRCS, Part 650 Engineering Field Handbook, Chap. 14, 2001)

curvature is 1.0, the minimum acceptable level for both gravels and sands.

Thus, both the coefficient of curvature and the coefficient of uniformity criteria are satisfied by the upper and lower limit curves on Fig. 30.3.

Envelope materials are evaluated with standard sieves. Standard sieve sizes as well as associated particle sizes are shown in Fig. 30.4.

The conductivity of the envelope material should be ten times greater than the hydraulic conductivity of the base material (USBR Drainage Manual) in order to not restrict the flow of water to drain openings; in general, if all particles in the envelope material are retained on the number 30 sieve (coarse sand), then the hydraulic conductivity of envelope should be adequate. In cases where it is suspected that the hydraulic conductivity is too low, the hydraulic conductivity of the material can be checked by filling a short piece of pipe with the envelope material and measuring the flow rate and energy difference between the two ends of the pipe. The procedure for this test is outlined in the USBR Drainage Manual.

The effectiveness of the envelope can be reduced by improper installation procedures. Care must be taken to avoid mixing the base material with the envelope during installation in saturated soil. If possible, the contractor should wait for a time of year when the water table is below the designed drain elevation. If this is impossible, then the soil may need to be dewatered before drainage installation if the base material is unstable. With trenchers that install pipe and backfill gravel in one operation, the risk of mixing base material with the envelope is less than in an open trench.

Based on conservation of energy, flow rate through the envelope material and into the drain is dependent in the energy difference between the outside of the envelope material and the interior of the drain and the envelope conductivity. The envelope dimensions (simplified as a circle) are shown in Fig. 30.5.

The average energy difference between the inside of the drain and the outside of the envelope for corrugated plastic drain pipe can be calculated with Eq. 30.9 (USBR), with the assumption that the water table intersects the upper surface of the envelope.

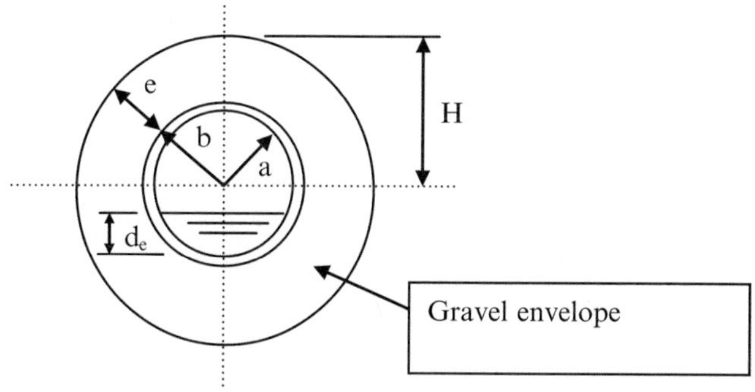

Fig. 30.5 Drainage envelope parameters

$$\overline{H} = H\left[1 - \frac{d_e}{160H}\left(22 + 29\frac{d_e}{a}\right)\right] \tag{30.9}$$

where

d_e = depth of flow in the drain, mm,
H = b + e (outside drain radius and width of envelope), mm,
a = inside drain radius, mm,
$\overline{H}$ = average energy difference between outside envelope and inside pipe, mm.

Calculation of the flow rate into the drain is then made with Eq. 30.10.

$$q_d = b\overline{H}K\phi \tag{30.10}$$

where

b = outside diameter of drain, m,
K = hydraulic conductivity of the envelope material, m/day,
q_d = flow rate into the drain, m^3/d/m,
ϕ = adjustment factor.

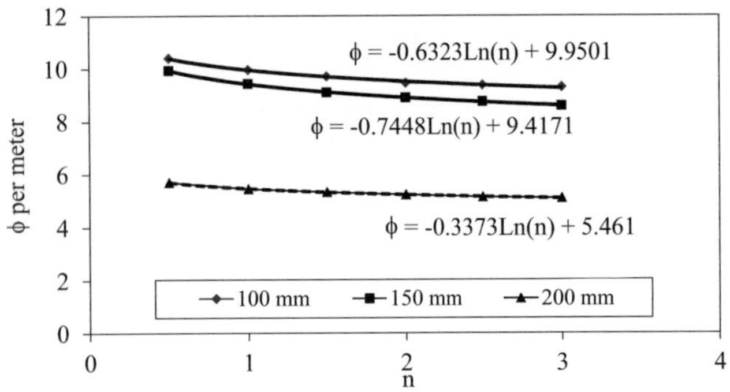

Fig. 30.6 n versus ϕ relationship

The term, ϕ, is found in Fig. 30.6 (Fig. 5.25 in the USBR drainage manual) for different sizes of drain pipe. Use of Fig. 30.6 requires the parameter n, which is equal to e/b.

$$n = e/b \tag{30.11}$$

If the drainage coefficient and the pipe diameter are already known based on drainage requirements as calculated above, then the only unknown parameters in Eqs. 30.9 and 30.10 are the width and conductivity of the envelope material.

Example 30.3 Calculate a minimum acceptable envelope width for Example 30.1 that will carry the required flow to the drain. The envelope material has a conductivity of 9.6 m/day, and the drain is running 3/4 full. The drainage coefficient is 13.5 mm/day. The outside pipe diameter is 114 .4 mm, and the internal diameter is 102 mm (4 inch pipe).

Required flow rate into the drain is the product of the drainage coefficient and the drain spacing.

$$q_d = D_c L = \left(\frac{13.5\ mm}{day}\right)\left(\frac{m}{1000\ mm}\right)(40\ m)$$
$$= 0.54\ m^3/day/m$$

Calculate a and b

$$a = 102/2 = 51\text{mm}.$$
$$b = 114.4/2 = 57.2\text{ mm}$$

Calculate d_e based on the assumption that the drain runs ¾ full.

$$d_e = 0.75{*}102\text{ mm} = 76.5\text{ mm}$$

Assume the envelope dimension, e, is 100 mm, in order to find initial n and ϕ values.

$$n = \frac{e}{b} = \frac{100\text{ mm}}{57.2\text{ mm}} = 1.75$$

$$\phi = -0.6323{*}\text{Ln}(n) + 9.9501$$
$$= -0.6323{*}\ \text{Ln}(1.75) + 9.9501 = 9.60$$

Calculate $\overline{H}$

$$\overline{H} = \frac{q_p}{bK\phi} = \frac{0.54}{0.0572{*}9.6{*}9.60} = 0.102\ m = 102\ mm$$

Calculate H (iterate to find answer)

$$H = \frac{\overline{H}}{\left[1 - \frac{d_e}{160H}\left(22 + 29\frac{d_e}{a}\right)\right]} = \frac{102}{\left[1 - \frac{76.5}{160H}\left(22 + 29\frac{76.5}{51}\right)\right]}$$
$$\Rightarrow 133\ mm$$

Calculate envelope dimension, e:

$$e = H - b = 133 - 57.2 = 76\text{ mm (3 inches)}.$$

Example 30.4 Redo Example 30.3, but assume that envelope dimension, e, is 100 mm. Find minimum hydraulic conductivity of the envelope material. Assume that the water table intersects the top of the envelope material.

$$\overline{H} = H\left[1 - \frac{d_e}{160H}\left(22 + 29\frac{d_e}{a}\right)\right]$$
$$= 100\left[1 - \frac{76.5}{160{*}100}\left(22 + 29\frac{76.5}{51}\right)\right] = 69\ mm$$

$$K = \frac{q_p}{b\overline{H}\phi} = \frac{0.54}{0.0572{*}0.69{*}9.77} = 1.4\ m/day$$

The required hydraulic conductivity is six times less for the 100 mm width envelope material than for the 76 mm width envelope material. An economic decision may need to be made on whether to increase the size of the envelope or increase the conductivity of the envelope material. For example, if the gravel pit is near the job site, then it may be more economical to increase the envelope size than perform mechanical gradation of the envelope material.

In addition to the hazard of physical drain clogging by sediments, chemical or biological drain clogging can take place in certain soils. Iron or magnesium oxides can chemically react and seal the pores. Ochre is a red sludge produced by iron oxidizing bacteria. Sandy or peat soils with organic lenses are susceptible. If sewage sludge or cannery waste are applied to soils, then they can be an iron source. Ground water can have high ferrous iron concentration (>0.2 ppm). The ochre problem can take place just in the first few months after drain installation or it can be a long-term problem. Important factors that influence ochre formation are pH and temperature.

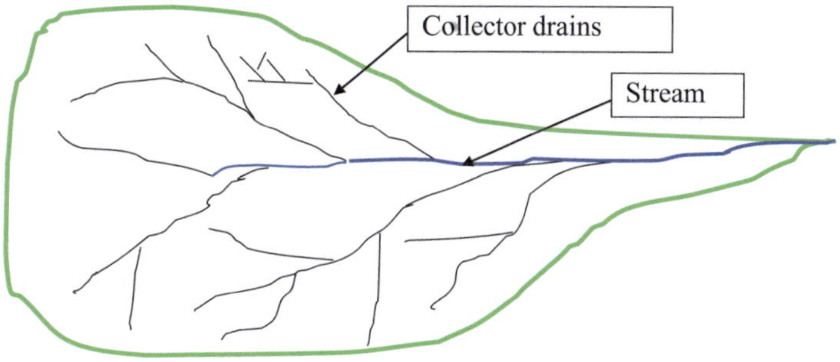

Fig. 30.7 Drains connecting low areas to river

If a gravel envelope is not used, then the bottom of the trench must support at least one quarter of the circumference of the drain pipe, in order to improve the structural strength of the drain. This can be accomplished by shaping the bottom of the trench at a 90^0 V angle. Maximum trench depths are based on no more than 20 % deflection of the drain by the overburden pressure + expected vehicle load (NRCS, EFH 14, 2001). Maximum depths range from 6 to 12 feet (2–4 m), depending on drain diameter and trench width

Field Drainage System Geometry

Three types of drainage patterns are shown in Figs. 30.7 and 30.8: random, herringbone, and parallel. In fields with undulating topography, a random pattern (actually not random but following the topography) is used to drain low areas into streams. In depressions, vertical open drain inlets can facilitate rapid drainage of ponded water after storms.

A herringbone pattern may be used if it provides advantageous slope to drains; for example, if the central collector drain is laid in the downslope direction, then a herringbone pattern can give a downslope orientation to the drains, and the drains can be laid at a constant depth in the soil rather than start shallow and end deep.

Drainage Structures

Drainage structures include surface inlets, junction boxes, silt traps, vents, and outlets. Direct surface inlets are common in regions with undulating topography and ponded depression areas. These inlets rapidly drain ponded water. However, they are not recommended by the NRCS. One of the primary hazards associated with surface inlets is drain siltation. If silt is a problem, then a blind surface inlet can be constructed, which includes highly porous material above the drain but no direct flow to the drain. Refer to NRCS design manual for drawings of these structures.

Junction boxes are required where two or more drains are joined together (Fig. 30.9). Junction boxes facilitate drain cleaning. Within the field, the junction box upper surface should be at least 18 in (45 cm) below the ground surface. Silt traps are similar to junction boxes except that the box bottom is far below the inlet and outlet. Silt traps are placed downstream from silt sources, such as an open inlet.

Plastic pipe has high tensile strength but very little resistance to collapse against an internal vacuum. However, corrugated drain pipe has higher resistance to collapse. An internal vacuum can develop if water drains rapidly from a downstream pipe while the upstream pipe has water moving more slowly down the pipe. This can occur where the drainage pipe changes from a flat slope to a steep slope. Placement of a vent (vertical pipe) at the point where slope changes from flat to steep can relieve internal vacuum and prevent collapse.

On the other hand, when pipe slope changes from steep to flat, pressure can build up in the pipe if a slug of water approaches the flat section. This is especially likely in the case of an open surface inlet. An air relief valve can be placed at these locations in order to allow air to breathe out of the pipe under high pressure. Vents and air relief valves should be placed along fence lines or other locations where they will not be in the way of farm equipment.

Drain outlets should corrugated metal pipe or other rigid pipe. The drain outlet should be protected from intrusion by rodents by a swinging gate or iron grating. Where surface outlets are submerged, the outlet should be protected by a swing gate. The outlet pipe should be fireproof if burning of weeds is expected. If surface water flows toward the outlet from above, then the drain outlet should be protected by a headwall or earth berm (NRCS, EFH 14, 2001).

Drainage Sumps and Pumps

If it is impossible to provide a gravity outlet for a drainage system, then a drainage sump and pump system may be required. A subsurface drainage sump (Fig. 30.10) can be constructed from stacked vertical, large diameter, concrete pipe culvert sections. A float valve with stop and start collars turns the pump on and off when the water level inside the sump rises and falls, respectively. It is better to spend extra money on a sump than deal with extremely shallow grades and risk siltation or air locks.

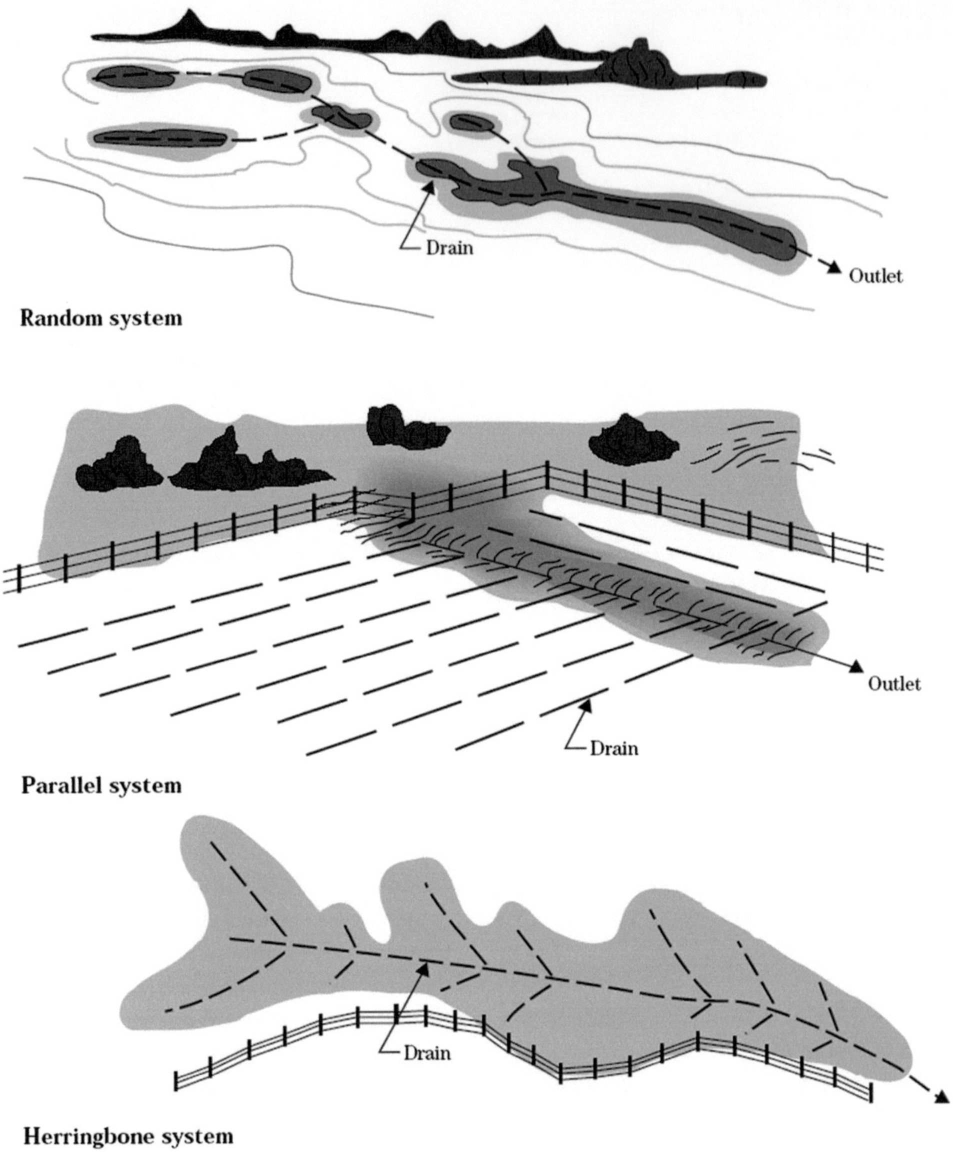

Fig. 30.8 Field subsurface drainage systems (Credit NRCS, Part 650 Engineering Field Handbook, Chap. 14, 2001)

The pump flow rate should be designed to meet the expected maximum drainage flow rate.

$$P = I \tag{30.12}$$

where

P = Pumping rate, LPM,
I = Drainage inflow rate, LPM,

During periods when the pump flow rate exceeds the drainage flow rate, the pump turns on when the water level in the sump reaches the start level and turns off when the water level reaches the stop level. The number of pump cycles per hour should not exceed 10; otherwise, the electrical efficiency of the pump decreases and the pump wears out quickly. The relationship between number of cycles, inflow and pumping rate and sump volume is given by the following equations.

$$\frac{60}{N} = \frac{S}{I} + \frac{S}{P - I} \tag{30.13}$$

where

N = number of cycles per hour
S = Storage volume, Liters.

The sump volume is based on the worst-case scenario: the maximum number of pump cycles per hour occurs when the inflow rate, I, is equal one-half of the pumping rate, P. If I = 0.5P is substituted into Eq. 30.13,

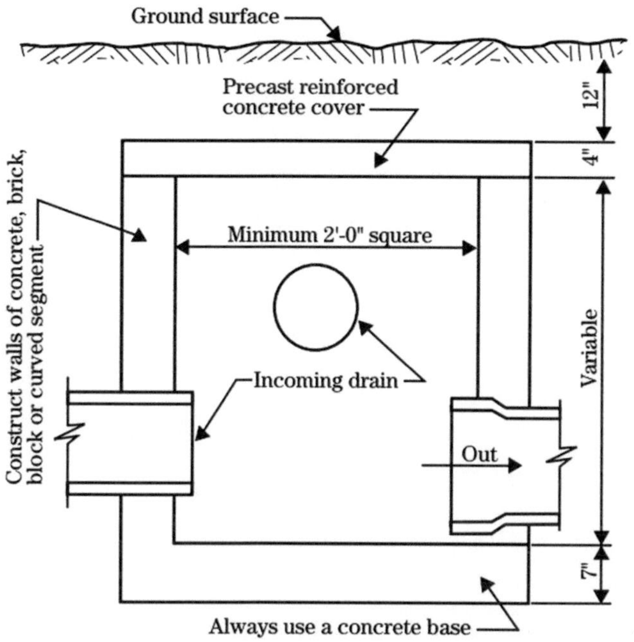

Fig. 30.9 Junction box (Credit NRCS, Part 650 Engineering Field Handbook, Chap. 14, 2001)

$$\frac{60}{N} = \frac{S}{I} + \frac{S}{P-I} = \frac{S}{0.5P} + \frac{S}{0.5P} = \frac{S}{P} \rightarrow S = \frac{60P}{N} \quad (30.14)$$

If the maximum number of cycles per hour is 10, then the sump storage volume should be (60/10) P = 6P.

Example 30.5 Design a drainage pump and sump for a maximum expected drainage rate of 200 LPM, and a design maximum number of cycles per hour of 5.

Pump flow rate = maximum drainage rate = 200 LPM.
Sump volume = 60 P / 5 = 12 (200) = 2,400 L = 2.4 m^3.

Measurement of Lateral Hydraulic Conductivity

Lateral hydraulic is measured in an auger hole. The hole is normally drilled below the water table. Water is pumped out of the hole, and then the rate that water rises in the hole is measured. The geometrical parameters in a standard auger hole test are shown in Fig. 30.11.

The auger hole should be drilled approximately 30 cm (1 ft) below the elevation of the water table at the expected elevation of the subsurface drain pipe. It may be necessary to wait until a large storm raises the water table to the necessary elevation above the drain elevation. If this is not possible, then there are auger hole tests for unsaturated soils.

- H = depth of hole below the ground water table (cm)
- r = radius of auger hole (cm)
- y = distance between ground water level and the average level of water in the hole (in) for the time interval t (s)
- Δy = rise of water (cm) in auger hole during Δt
- t = time interval (s)
- G = depth of the impermeable layer below the bottom of the hole (cm). Impermeable layer is defined as a layer that has the permeability of no more than a tenth of the permeability of the layers above.
- d = average depth of water in auger hole during test (cm)

The auger hole is evacuated with a small pump (double diaphragm or stirrup pump) or bail bucket (*NRCS EFH 14, 2001*). The level of water in the hole can be measured with a stick attached to a float or automatic depth sensor. The NRCS recommends a lightweight bamboo fishing rod with length markings attached to a cork or can (Fig. 30.12).

Ernst (1950) developed two equations for calculation of auger-hole hydraulic conductivity. The first equation is used if G is zero, and the bottom of the auger hole corresponds with the elevation of the top of the impermeable layer.

$$K_s = \frac{15,000(r^2)}{(H+10r)\left(2-\frac{y}{H}\right)y}\frac{\Delta y}{\Delta t} \quad (30.15)$$

where

K_s = hydraulic conductivity (cm/hr)

If the auger-hole is greater than 0.5H above the impermeable layer (G > 0.5H) then significant upward flow into the auger hole is expected and the following equation (Ernst 1950) is used.

$$K_s = \frac{16,667*r^2}{(H+20r)\left(2-\frac{y}{H}\right)y}\frac{\Delta y}{\Delta t} \quad (30.16)$$

The NRCS recommends the following for use of Eqs. 30.15 and 30.16. The hole diameter, 2r, should be between 6 and 14 cm. The static depth of water in the hole, H, should be greater than 25 cm and less than 200 cm. The average depth of water in the hole, y, should be greater than 0.2 H. The change in depth, Δy, should be less then ¼ y_0, the initial depth at the beginning of test.

During a conductivity test, the distance from the reference point to the water surface (Fig. 30.13) is measured over time. The difference between this distance (R) and the distance from the static water surface to the reference point (B) is plotted (Fig. 30.13) and is equal to Δy/Δt.

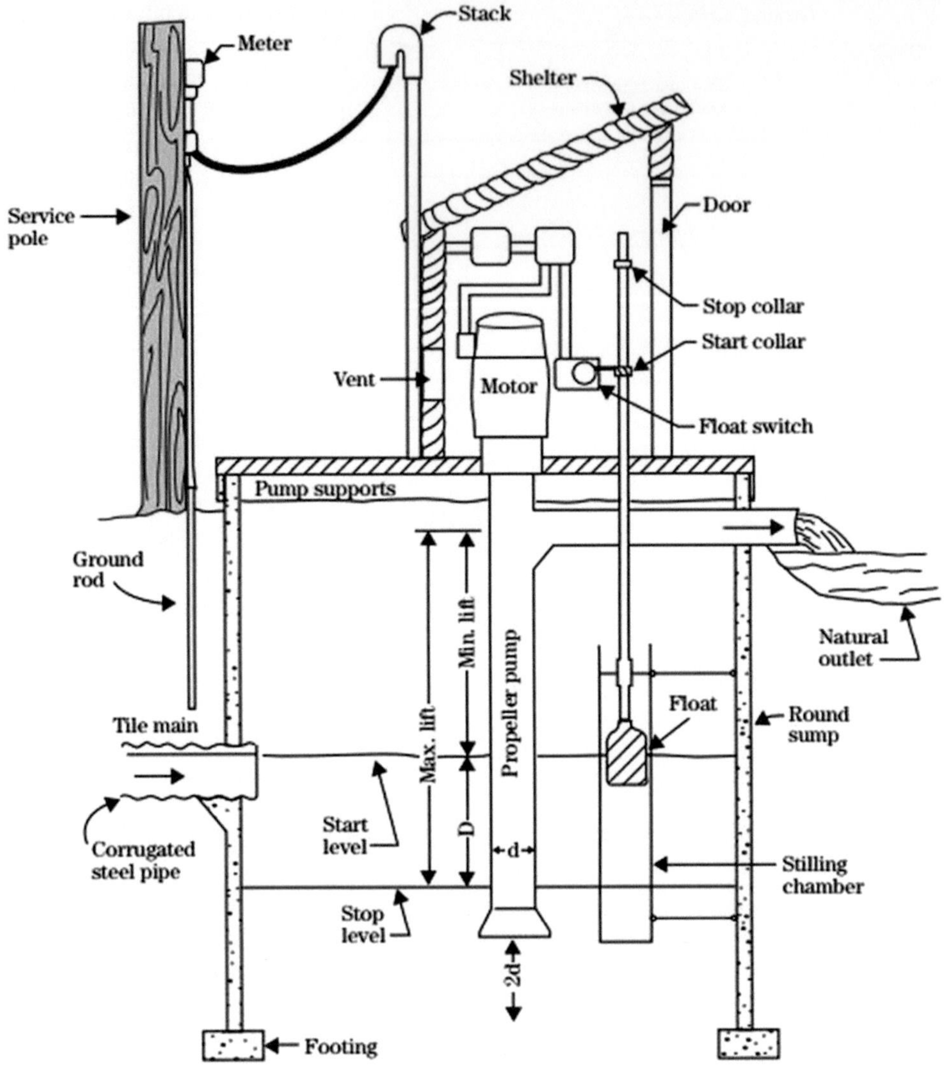

Fig. 30.10 Drainage pumping plant (Credit NRCS, Part 650 Engineering Field Handbook, Chap. 14, 2001)

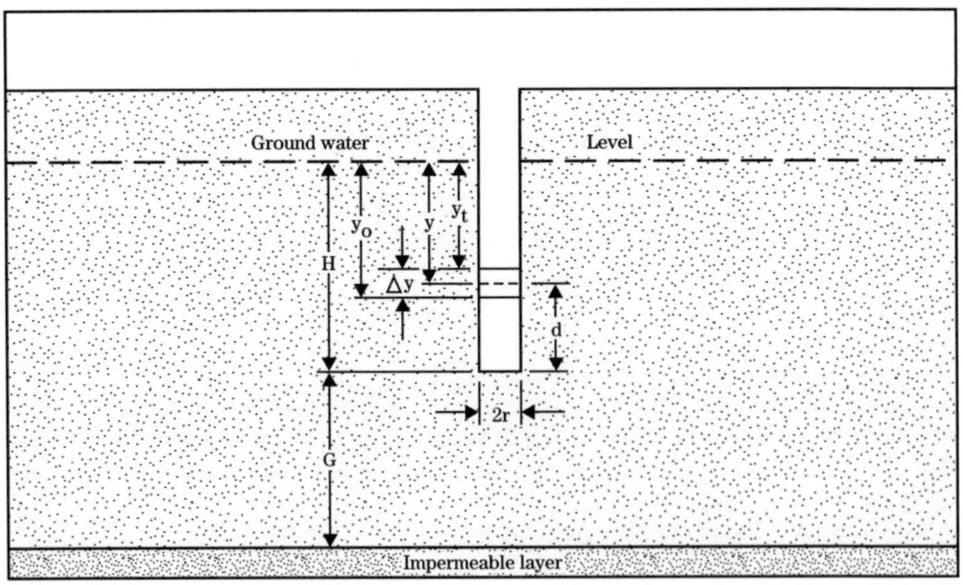

Fig. 30.11 Auger hole test parameters (Credit NRCS)

Example 30.6 Calculate hydraulic conductivity with Fig. 30.13 and the following parameters. The diameter of the auger hole is 10 cm, and the bottom of the auger-hole corresponds with the top of the impermeable layer.

Radius of auger-hole (5 cm = 2 in)	r	5 cm
Distance from bottom of auger hole to reference	D	236 cm
Distance from bottom of auger hole to static water table	H	127 cm
Distance from bottom of auger hole to average depth during test	d	41 cm
Change in elevation, R-B, during test	Δy	9.5 in = 24 cm
Length of test	Δt	150 sec

$$y = H - d = 127\ \text{cm} - 41\ \text{cm} = 86\ \text{cm}$$

$$K_s = \frac{15{,}000{*}r^2}{(H + 10r)\left(2 - \frac{y}{H}\right)y}\frac{\Delta y}{\Delta t}$$

$$= \frac{15{,}000{*}5^2}{(127 + 10{*}5)\left(2 - \frac{86}{127}\right){*}86}\frac{24\ cm}{150\ \ sec} = 3.0\ cm/hr$$

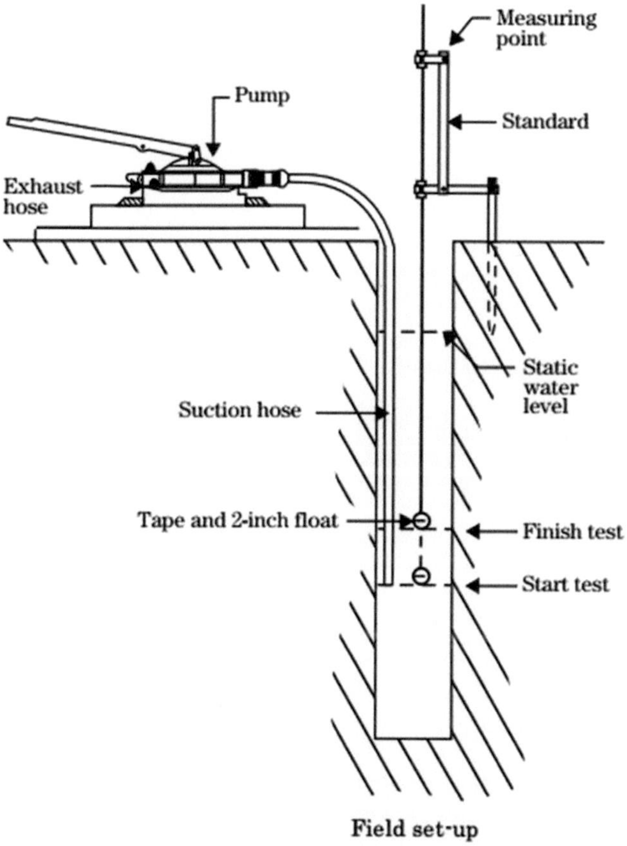

Fig. 30.12 Auger hole test equipment (Credit NRCS, Part 650 Engineering Field Handbook, Chap. 14, 2001)

Example 30.7 Calculate hydraulic conductivity based on the following measurements. The bottom of the auger hole is above the impermeable layer.

Radius of auger-hole (5 cm = 2 in)	r	5 cm
Distance from bottom of auger hole to static water table	H	102 cm
Distance from bottom of auger hole to average depth during test	y	30.5 cm
Change in elevation, R-B, during test	Δy	0.81 cm
Length of test	Δt	10 sec

The bottom of the auger-hole is greater than 0.5 H above the bottom of the impermeable layer, so use Eq. 30.16.

$$K_s = \frac{16{,}667{*}r^2}{(H + 20r)\left(2 - \frac{y}{H}\right)y}\frac{\Delta y}{\Delta t}$$

$$= \frac{16{,}667{*}5^2}{(102 + 20{*}5)\left(2 - \frac{30.5}{102}\right){*}30.5}\frac{0.81\ cm}{10\ \ sec} = 3.2\ cm/hr$$

Fig. 30.13 Data for auger hole test. Y-axis is (R-B) in cm (Credit NRCS, Part 650 Engineering Field Handbook, Chap. 14, 2001)

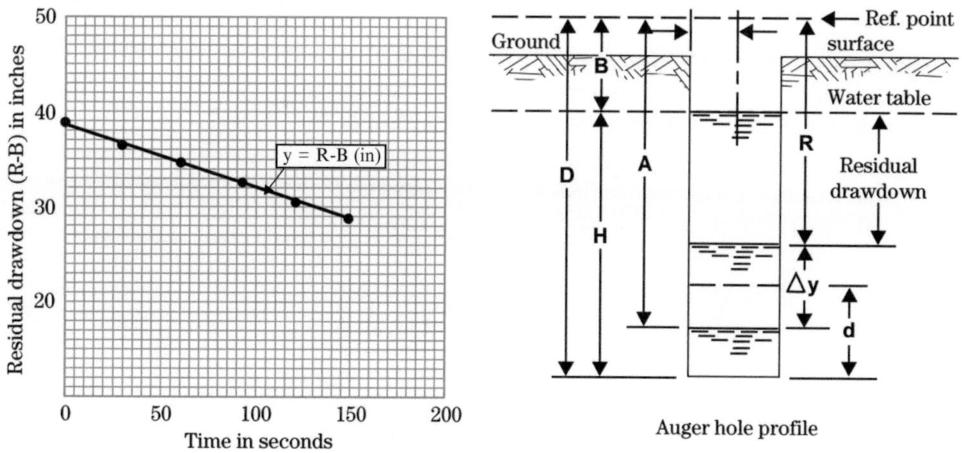

Section	Area A	Wetted perimeter, P	Hydraulic radius R	Top width T
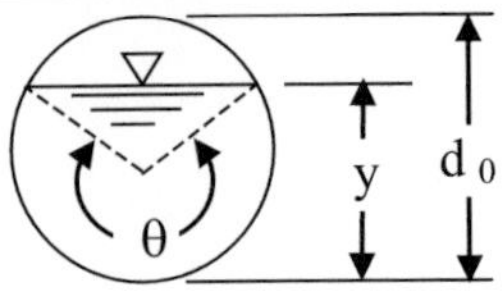 Culvert	$\frac{1}{8}(\theta - \sin\theta)d_0^2$	$\frac{1}{2}(\theta)d_0$	$\frac{1}{4}\left(1 - \frac{\sin\theta}{\theta}\right)d_0$	$\left(\sin\frac{\theta}{2}\right)d_0$

If it is not possible to conduct auger-hole tests, an alternative technique is to use soil properties to estimate hydraulic conductivity. If soil texture is known, then conductivity can be calculated with the Hydraulic Properties Calculator referenced in Chap. 3.

Questions

1. List the positive and negative environmental aspects of subsurface drainage.
2. What measures are recommended by the NRCS to reduce the impact of drained water from farmland on the environment.
3. List the features that should be noted in a drainage reconnaissance survey.
4. Based on the geometry of Fig. 30.1 and the derivation of Eq. 30.1, are drains designed for the peak flow with the water table directly over the drain after a rainstorm?
5. Derive a drainage coefficient equation for flow with the water table directly over the drain. Compare the ratio of flow rates based on your equation and Eq. 30.4. Discuss why drains are not designed based on the flow rate that takes place when the water table is directly over the drain.
6. Derive Eq. 30.6 from Manning's equation assuming that the pipe is full and half full and determine whether the equation assumes that the drain is flowing half full of full.
7. A 5 ha area has a drainage coefficient of 19.1 mm/day. The drain slope is 0.3 %. Calculate the required drain size.
8. Calculate the required diameter of a subsurface drain for the following parameters.

Manning's n	0.017
Drain elevation above impermeable layer	3 m
Drain slope	$s = 0.2$ m/100 m
Hydraulic conductivity	$K = 2$ m/day
Drain spacing	$L = 60$ m
Length of drain pipe	$L_d = 500$ m
Maximum WT height above drains:	$m_0 = 2$ m $-$ 0.48 m $= 1$ m

9. Can a geotextile filter be used in a loam soil? What mesh is recommended ?

References

ASABE Standards (2015) Design and construction of subsurface drainage systems on agricultural lands in humid areas. ASAE EP260.5. February 2015. American Society of Agricultural and Biological Engineering

Ernst LF (1955) A new formula for the calculation of the permeability factor with the auger hole method. TNO, Groningen

Fogiel A, Belcher H (1991) Research literature review of water table management impacts on water quality. Agricultural Engineering Department, Michigan State University/Land Improvement Contractors of America, East Lansing, p 111

Huffman R, Fangmeier D, Elliott W, Workman S (2013) Soil and water conservation engineering. American Society of Agricultural and Biological Engineering, St. Joseph, Mich

Natural Resources Conservation Service (2001) Part 650 Engineering field handbook chapter 14 water management (Drainage). http://directives.sc.egov.usda.gov/OpenNonWebContent.aspx?content=17551.wba

Natural Resources Conservation Service. Section 16, National engineering handbook. chapter 4. Drainage of agricultural land. http://directives.sc.egov.usda.gov/OpenNonWebContent.aspx?content=18365.wba

USBR (1993) Drainage manual: a water resources technical publication. US Department of the Interior, Bureau of Reclamation. pp 1–59, 147–174. http://www.usbr.gov/pmts/wquality_land/DrainMan.pdf

31 Subsurface Drainage Modeling

The original subsurface drainage model is the Hooghoudt equation, which is a one-dimensional steady-state simplification of the two-dimensional transient flow to parallel drains. It calculates the midpoint water table elevation between drains. Bower and van Schilfegaarde modified the Hooghoudt equation for transient analysis. The Bureau of Reclamation also developed drainage equations for transient analysis of midpoint water table elevation. Kirkham developed a Laplace analytic solution for the two-dimensional subsurface drainage geometry. He also adapted this solution for transient analysis with the concept of fixed streamtubes along the path of water flow. The advantage of this approach is that water table height can be simulated as a function of distance from the drain rather than just the midpoint water table elevation. The Kirkham streamtube approach is used in the *WINDS* drainage model, which enables *WINDS* to model water, salinity, and nitrogen in the soil profile as a function of distance from the drain. The chapter also includes an example of the economic analysis of drain spacing and depth.

Subsurface field drainage pipes are corrugated, black, high-density, polyethylene (HDPE) plastic pipe that is extruded in rolls with slits cut in the sides of the pipe to allow water to enter the pipe (Fig. 31.1). The corrugations support the walls of the pipe and prevent physical collapse. In regions with unstable soils in which particles can be carried by flowing water to the drain pipe, the pipes are wrapped with a fabric sleeve (sock) at the factory in order to prevent sediment from entering into the pipe during drainage.

Flow to subsurface drains takes two forms. Just after a heavy storm or irrigation, the entire water table may rise to the soil surface. Immediately after the event, the primary flow to the drains is from the region above the drains (Fig. 31.2, left). If the drain pipes are full due to excess flow in the system, then the water table above the drains will not decline and may even rise in portions of the field where drains have positive internal pressure (Fig. 31.2, right). Conventional drainage systems continue to remove water from the field until the water table is lowered to the elevation of the drain; this can result in insufficient water for plants during dry periods.

The water table above the drain generally declines within a day or two of the storm or irrigation event until the water table intersects the drain (Fig. 31.3). Subsequently, the water table forms an elliptical shape and flow to the drains is driven by the slope (energy gradient) of the water table.

The flow lines are not always toward the drain. Conventional (Fig. 31.4a) subsurface drainage systems can be modified to manage the water table elevation by installing structures in the drainage network such as flashboards or stoplogs that prevent water from draining through the pipe network. This is called controlled drainage (Fig. 31.4b). If a dry period is expected, then plant growth is maximized by preventing drainage and maintaining a high water table in order to carry the plant through the dry period. Subirrigation (Fig. 31.4c) adds water to the soil through the drainage system; water in the drainage canal and water flows into the drainage system.

Simulation of Yield Reduction

Wet stress and yield reduction take place during periods with an excessively high water table. Drainage simulation models typically report the average daily height of the water table at the midpoint between drains as a representative indication of the water content in the root zone. Figure 31.5 shows the daily water table depth computed by the DRAINMOD model, a popular drainage simulation program developed by Wayne Skaggs at North Carolina State University. During rainstorms, the elevation of the water table increases sharply as the soil is suddenly filled with water. Subsequently, the elevation of the water table slowly decreases as water is removed from the soil by plant roots and by subsurface drainage. The program calculates corn yield reduction based on the number of days when the water

P. Waller, M. Yitayew, *Irrigation and Drainage Engineering*, DOI 10.1007/978-3-319-05699-9_31

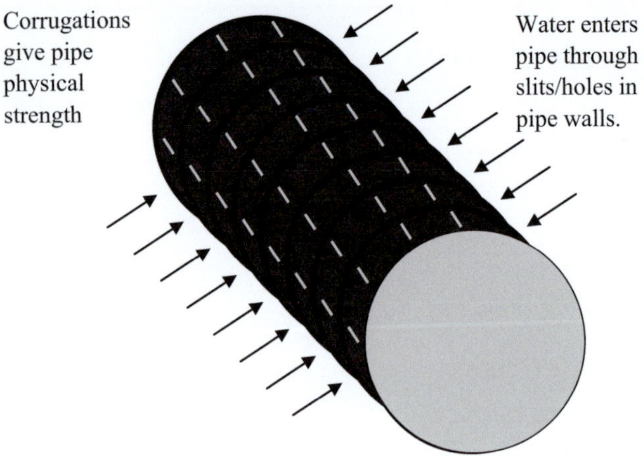

Fig. 31.1 Corrugated plastic drain pipe

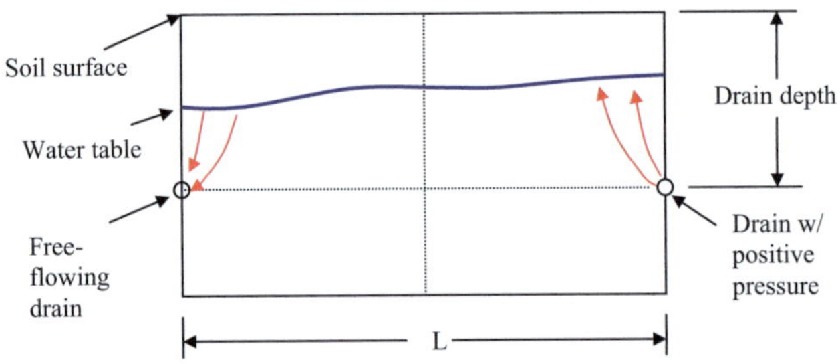

Fig. 31.2 Flow to drains just after a storm with water table above drain

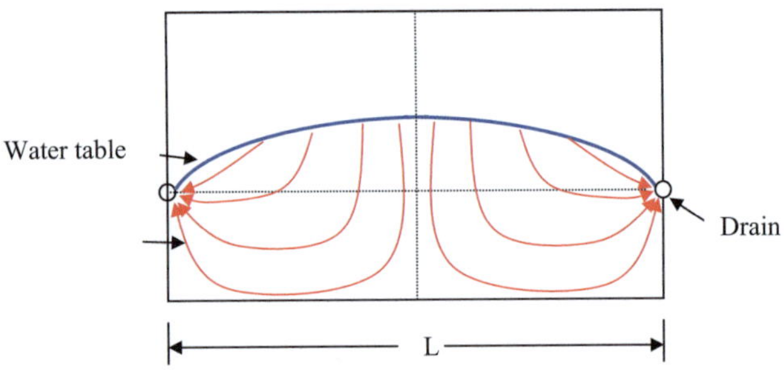

Fig. 31.3 Flow to subsurface drains between storms

table is within 30 cm of the ground surface (horizontal line in Fig. 31.5). This is done by calculating a wet stress coefficient and a weighted water table elevation.

In DRAINMOD, the wet stress during the growing season is computed as the weighted (greater weight when the crop is more susceptible to wet stress) sum of the distance from the 30 cm depth to the water table on each day that the water table is higher than 30 cm depth. The distance from the 30 cm depth to the water table is called the SEW30

$$\mathrm{SEW30} = 30 - \mathrm{DTWT} \tag{31.1}$$

where

DTWT = depth to water table, cm.

The crop susceptibility to wet stress changes during the growing season and is quantified by the crop susceptibility factor, CS, in DRAINMOD. The crop susceptibility factor is typically higher during early to middle parts of the growing season. The summation of CS * SEW-30 is called the stress day index, SDI

$$SDI = \sum_{i=1}^{n} CS_i * SEW30_i \quad \text{for DTWT} < 30 \tag{31.2}$$

Finally, yield reduction, YR, for the growing season is calculated based on the stress day index and the susceptibility of the crop to wet stress, expressed as DSLOPE.

$$\mathrm{YR} = \mathrm{YRDMAX} - \mathrm{DSLOPE} * \mathrm{SDI} \tag{31.3}$$

where

YR = percent of maximum yield
YRDMAX = yield intercept
DSLOPE = ratio between SDI and YR

Example 31.1 Calculate the SDI for a corn crop. The planting date is DOY 115 (after spring rains). The time when the water table depth is less than 30 cm is between DOY 194 and DOY 209 (Table 31.1).

DOY 194–209 is approximately 90 days (yield formation) after planting and the crop susceptibility factor at that time is 0.08. The susceptibility is low because the crop is not very susceptible to wet stress during yield formation. The SDI is

$$\mathrm{SDI} = (30 + 18 + 30 + 12 + 10 + 30 + 8 + 2) * 0.08 = 11$$

For corn, YRDMAX = 102 and DSLOPE = 0.75

$$\mathrm{YR} = 102 - 0.75 * 11 = 94\%$$

Thus, the crop has a seasonal yield reduction of 6 % due to wet stress during this part of the growing season.

If there is no outlet control for the drainage system (Fig. 31.4), then the drainage system must be designed for average conditions during the year. A well-designed depth and spacing of drains keeps the water table out of the critical root zone for the majority of growing seasons, but does not cause the water table to drop so low that dry stress becomes a problem. No drainage system can prevent the water table from ever reaching the ground surface during a heavy storm, but drainage systems will rapidly reduce the elevation of the water table after the storm as shown in Fig. 31.5.

After the water table directly over the drain drops to the drain level, the water table has an ellipse pattern and flow pattern to the drains is as shown in Fig. 31.6. The rate of flow to drains at the midpoint between drains is much slower than

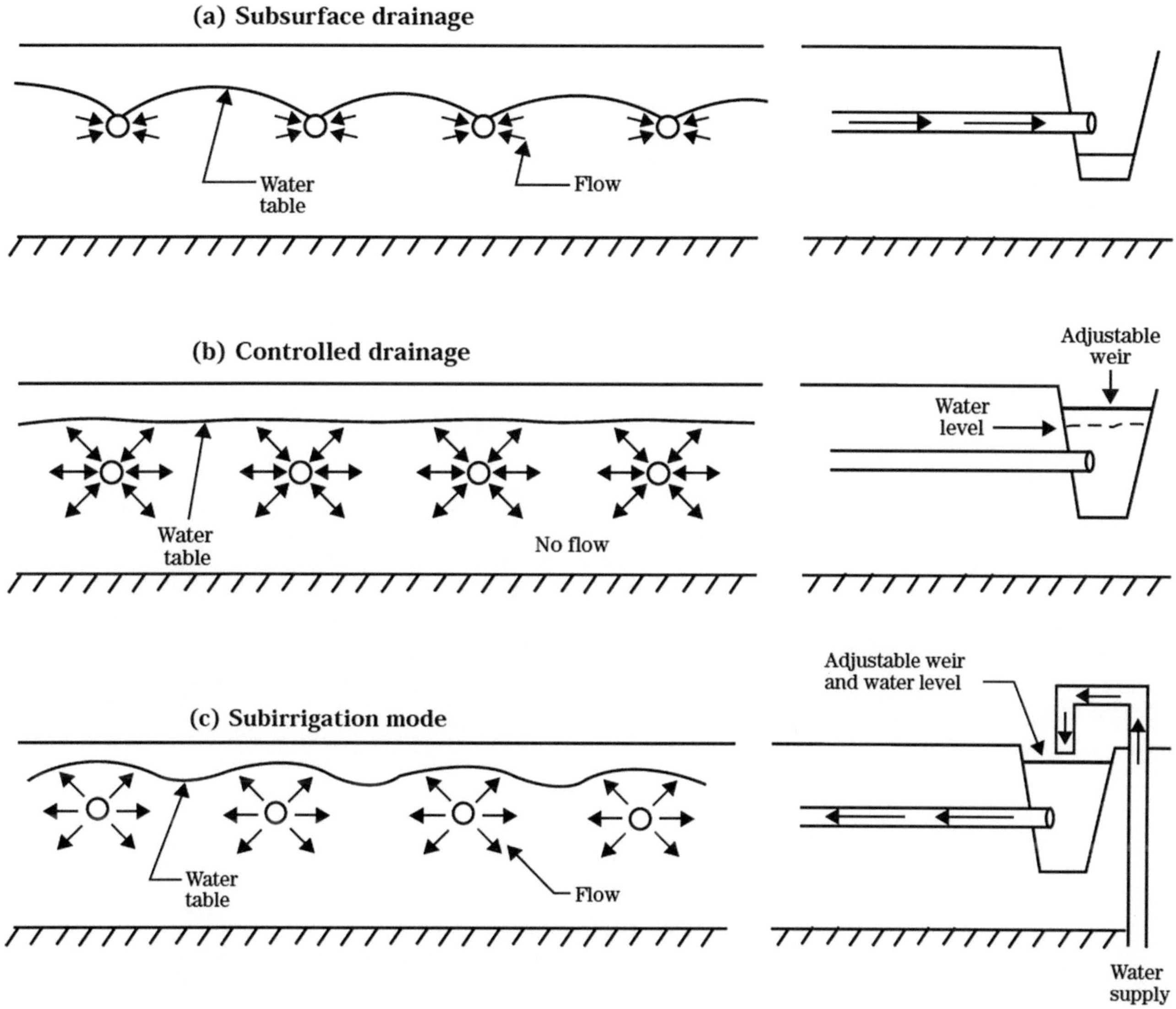

Fig. 31.4 Drainage alternatives (Credit NRCS, Part 650 Engineering Field Handbook National Engineering Handbook, Chapter 14)

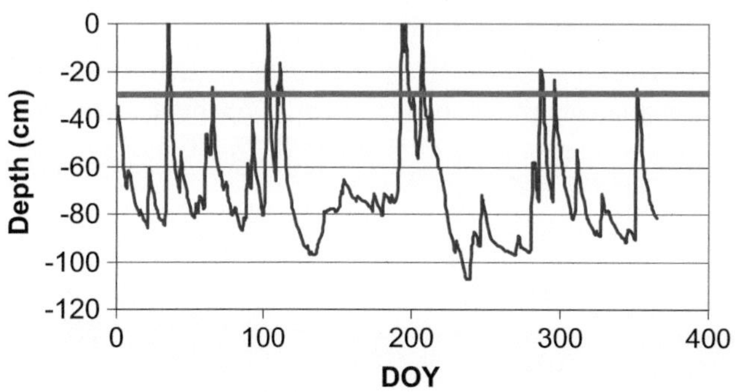

Fig. 31.5 DRAINMOD calculation of water table depth versus time

in the region close to the drain. This explains why the water table height, average soil water content, and salinity are normally higher between drains than near drains.

Drainage simulation models are based on a water volume balance for the region between drains (Fig. 31.7). The infiltration is normally assumed to be uniform across the region between drains; however, if water content and drain elevation are higher between drains, then infiltration may decrease with distance from the drain.

Hooghoudt Equation

The Hooghoudt equation was developed to calculate the midpoint water table elevation between drains with the assumption of steady state rainfall during the year: all of the precipitation for an entire year is averaged to calculate the average rate of precipitation per day. Steady state rainfall is not an unrealistic assumption in Hooghoudt's country, Holland. In order to develop a one-dimensional equation

Table 31.1 DTWT, SEW-30, and CS for data in Fig. 31.8

DOY	DTWT	SEW-30	CS	DOY	DTWT	SEW-30	CS
194	0	30	0.08	202	34	0	0.08
195	12	18	0.08	203	44	0	0.08
196	0	30	0.08	204	51	0	0.08
197	18	12	0.08	205	56	0	0.08
198	20	10	0.08	206	37	0	0.08
199	31	0	0.08	207	0	30	0.08
200	36	0	0.08	208	22	8	0.08
201	31	0	0.08	209	28	2	0.08

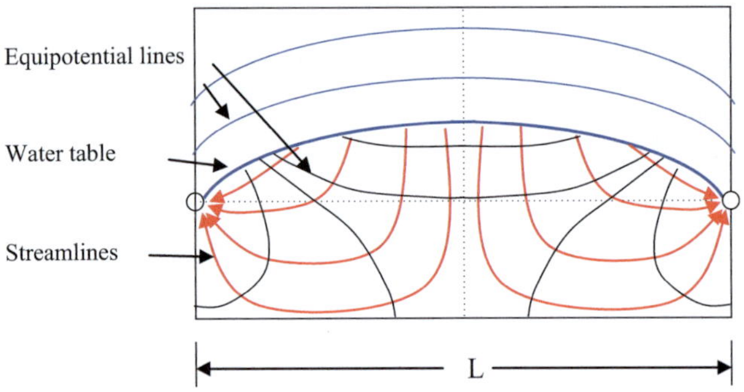

Fig. 31.6 Streamlines and equipotential lines for two-dimensional flow to parallel subsurface drains

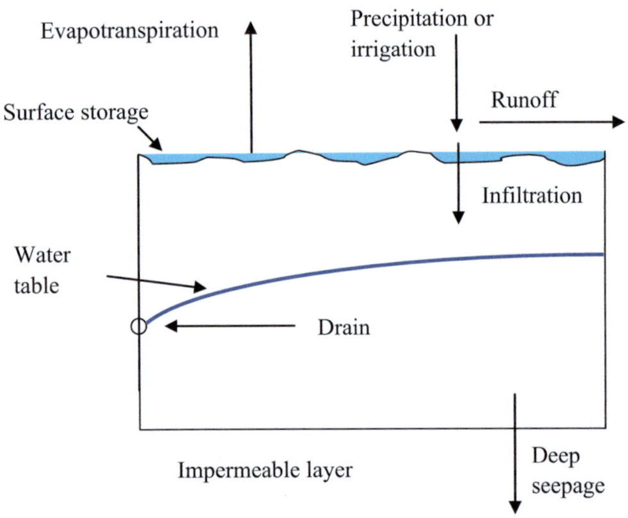

Fig. 31.7 Drainage control volume and components of drainage volume balance

for horizontal flow to drains, Hooghoudt assumed that all flow to drains below the water table was horizontal (the Dupuit-Forchheimer (D-F) assumption).

In-class Exercise 31.1 On a piece of paper, draw streamlines and equipotential lines below the water table for Dupuit-Forchheimer flow.

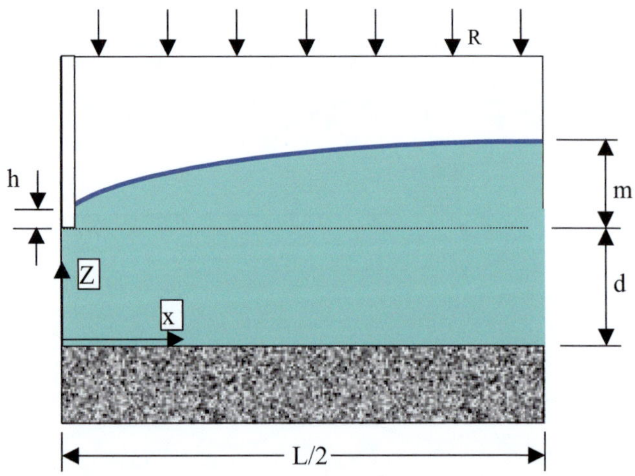

Fig. 31.8 Drainage geometry for Hooghoudt equation

Hooghoudt first derived an equation for flow to parallel ditches, which is simpler than calculating flow to subsurface drainage tubes. The parameters in the parallel ditch version of the Hooghoudt equation are shown in Fig. 31.8.

The law of conservation of mass can be used to find the total horizontal flow at any horizontal distance, x, from the drain. The flow toward the drain at distance x, q_x, is equal to the cumulative rainfall to the right of x. The direction of flow is against the positive x direction to the right.

$$\left(\frac{L}{2} - x\right)R = -q_x \tag{31.4}$$

where

q_x = flow rate per unit length of the drain at distance, x, from the drain, m^2/day,
L = drain spacing, m,
R = precipitation rate, m/day

Flow rate into the drain per unit length of drain tubing is the sum of flow into both sides of the drain

$$q_d = 2\frac{L}{2}R = LR \quad (31.5)$$

where

q_d = flow rate per unit length of drain tubing, m^2/day.

Total drain pipe flow rate is the product of flow rate per unit length and length of the drain.

$$Q = q_d{*}drain\ tube\ length\ (m) \quad (31.6)$$

where

Q = total drain flow, m^3/day

Example 31.2 Rainfall is 0.01 m/day, and drain spacing is 30 m. Calculate flow rate q_5, at a distance of 5 m from the drain. Find the flow rate into the drain per unit length of drain tubing.

$$q_x = \left(\frac{L}{2} - x\right)R = \left(\frac{30}{2} - 5\right)0.01\ m/day = 0.1\ m^2/day$$

The total flow rate per unit length, q_p, into the drain is

$$q_d = LR = 30{*}0.01 = 0.3\ m^2/day$$

The slope of the water table is the energy gradient that drives horizontal flow to the drain: the total energy at a distance x from the drain is the elevation of the water table, z, above the datum. Thus, the energy gradient along the x-axis, dH/dx, is dz/dx. The flow rate, q_x, is equal to the product of the Darcy velocity and the cross sectional area of flow, which is the elevation of the water table.

$$q_x = v_{Darcy}z = -K\frac{dz}{dx}z \quad (31.7)$$

where

z = elevation of water table above impermeable layer, m.

Substitute Eq. 31.7 into Eq. 31.4.

$$\left(\frac{L}{2} - x\right)R = Kz\frac{dz}{dx} \quad (31.8)$$

Separation of variables gives:

$$\frac{L}{2}Rdx - xRdx = Kzdz \quad (31.9)$$

Integrate Eq. 31.9

$$\int_0^{L/2}\frac{L}{2}Rdx - \int_0^{L/2}xRdx = \int_{h_d+d}^{m+d}Kzdz$$

$$\left[\frac{RL}{2}x\right]_0^{L/2} - \left[\frac{R}{2}x^2\right]_0^{L/2} = \left[\frac{K}{2}z^2\right]_{h_d+d}^{m+d} \quad (31.10)$$

$$L^2 = \frac{4K\left(m^2 - {h_d}^2 + 2dm - 2dh_d\right)}{R}$$

where

m = maximum elevation of water table above drain, m.
d = elevation of drain above impermeable layer, m,
h_d = depth of water in drain, m,
K = hydraulic conductivity, m/day.

Subsurface drain tubes can be modeled with the Hooghoudt equation for open ditches (Eq. 31.10) with the assumption that h, the depth of water in the ditch, is zero. Parameters in the Hooghoudt modification for flow into subsurface drain pipes are shown in Fig. 31.9.

If h is zero, then Eq. 31.10 is written as

$$L^2 = \frac{4Km(m + 2d)}{R} \quad (31.11)$$

If the conductivity varies between layers in the soil, then the equivalent lateral conductivity must be used in Eq. 31.11. Only half of the elevation between the drain and the mid-point water table elevation, m, should be used in the equivalent conductivity calculation.

The Dupuit-Forchheimer assumption of horizontal flow is valid far from the drain, but flow converges and forms a radial

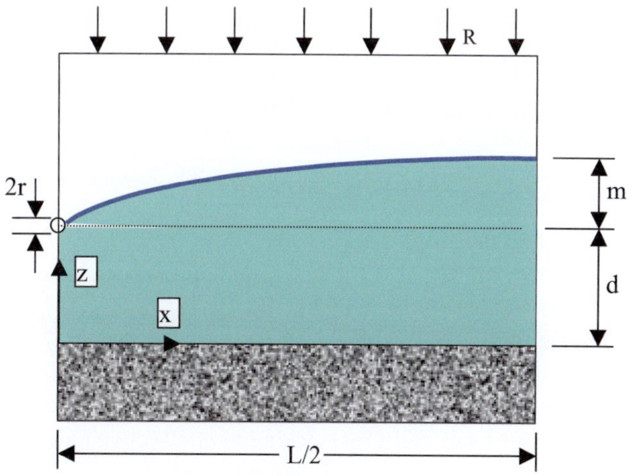

Fig. 31.9 Subsurface drain tube geometry

pattern near the drain (Fig. 31.6). Energy loss increases near the drain because the entire subsurface flow, q_p, to the drain must pass through a small area near the drain tube. Increased energy loss near the drain is accounted for in the Hooghoudt equation by reducing the depth, d, in Eq. 31.11. This reduced depth is called the equivalent depth, d_e. Equations 31.12 and 31.13 are used to calculate the equivalent depth. These equations include the effective diameter of the drain, r_e, which is less than the actual radius of the drain tube because water must move laterally along the drain into small holes in the drain. Skaggs reported that for conventional 11.4 cm OD drain tubing (4 in nominal diameter) the effective radius is 0.51 cm (DRAINMOD Manual). Selection between Eq. 31.12 or 31.13 depends on the ratio of aquifer depth (d in Fig. 31.9) to the distance between drains, L.

For $d/L < 0.3$

$$d_e = \frac{d}{1 + \frac{d}{L}\left(2.55\ln\left(\frac{d}{r_e}\right) - c\right)} \qquad c = 3.55 - 1.6\frac{d}{L} + 2\left(\frac{d}{L}\right)^2 \tag{31.12}$$

Where

r_e = the effective radius of the drain, m.

For $d/L > 0.3$

$$d_e = \frac{L}{\left(2.55\ln\left(\frac{L}{r_e}\right) - 2.93\right)} \tag{31.13}$$

Equation 31.11 is thus modified to include equivalent depth, d_e, and effective conductivity, K_e.

$$L = \sqrt{\frac{4K_e m(m + 2d_e)}{R}} \tag{31.14}$$

The procedure for calculating the water table elevation with the Hooghoudt equation is as follows:

- Specify the drain elevation above the impermeable layer, d.
- Specify the required midpoint water table elevation, m, based on the crop.
- Make an initial guess for the correct drain spacing, L.
- Solve for d_e with Eq. 31.12 or Eq. 31.13.
- Solve for a new L with Eq. 31.14.
- Calculate a new d_e by using the actual d and the new guess for L in Eq. 31.12 or Eq. 31.13: selection between Eqs. 31.12 and 31.13 for calculation of effective radius is always based on the new L and the original d.
- If the new iteration of d_e and the previous iteration have a greater than 5 % difference, then continue the iterative procedure. If not, then the solution has converged and the final L is the required drain spacing.

Example 31.3 Calculate the required spacing between drains if the farmer wants to maintain the water table at least 0.5 m below the soil surface. Yearly rainfall is 1.8 m/yr. The impermeable layer is 3.6 m below the soil surface. Drains are standard 4 in (10 cm) nominal diameter drains. Hydraulic conductivity within 1.1 m of the soil surface is 1.5 m/day, and hydraulic conductivity below the 1.1 m depth is 3.0 m/day. The drains are installed at 1.1 m depth below the soil surface. Effective drain radius for a 4 in (10 cm) nominal diameter drain is 0.51 cm = 0.0051 m. Drainage parameters are shown in Fig. 31.10.

$$\text{Average rainfall} \quad (1.8\ \text{m/year})/(365\ \text{days/year}) = 0.005\ \text{m/day}.$$

The drains are 1.1 m below the soil surface. Thus, the water table elevation above the drain, m, is

$$m = 1.1\ \text{m} - 0.5\ \text{m} = 0.6\ \text{m}.$$

The elevation of the drain, d, above the impermeable layer is

$$d = 3.6\ \text{m} - 1.1\ \text{m} = 2.5\ \text{m}.$$

Find the effective conductivity with Eq. 7.12. Let D_1 be equal to half of the elevation, m, above the drain, 0.3 m.

$$K_e = \frac{K_1D_1 + K_2D_2}{D_1 + D_2} = \frac{1.5{*}0.3 + 3.0{*}2.5}{0.3 + 2.5} = 2.84\ m/day$$

Make an initial guess of L = 40 m and solve for d_e with Eq. 31.12 because d/L is much less than 0.3.

$$c = 3.55 - 1.6\frac{d}{L} + 2\left(\frac{d}{L}\right)^2 = 3.55 - 1.6\frac{2.5}{40} + 2\left(\frac{2.5}{40}\right)^2$$
$$= 3.46$$

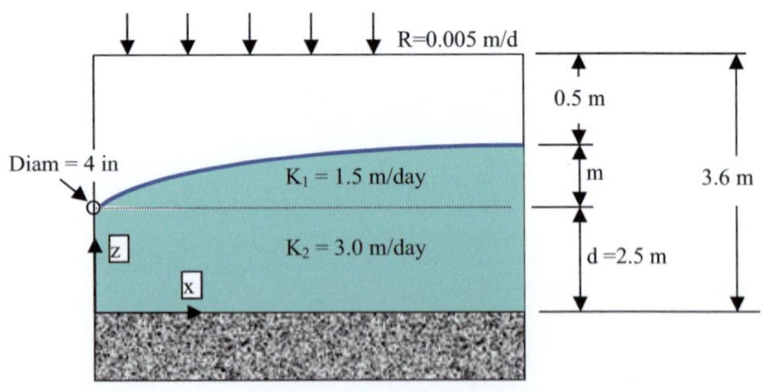

Fig. 31.10 Drainage dimensions for Example 31.3

$$d_e = \frac{d}{1 + \frac{d}{L}\left(2.55\ln\left(\frac{d}{r_e}\right) - c\right)}$$

$$= \frac{2.5}{1 + \frac{2.5}{40}\left(2.55\ln\left(\frac{2.5}{0.0051}\right) - 3.46\right)} = 1.41$$

Solve for L with Eq. 9.11

$$L = \sqrt{\frac{4K_e m(m + 2d_e)}{R}} = \sqrt{\frac{4(2.84)(0.6)(0.6 + 2(1.41))}{0.005}}$$
$$= 68.3\ m$$

The new d_e with L = 68.3 and d = 2.5 is 1.72 m.

Solve for L with the new equivalent depth

$$L = \sqrt{\frac{4K_e m(m + 2d_e)}{R}} = \sqrt{\frac{4(2.84)(0.6)(0.6 + 2(1.72))}{0.005}}$$
$$= 74.3\ m$$

The new d_e with L = 74.3 and d = 2.5 is 1.77 m.

Solve for L with the new equivalent depth

$$L = \sqrt{\frac{4K_e m(m + 2d_e)}{R}} = \sqrt{\frac{4(2.84)(0.6)(0.6 + 2(1.77))}{0.005}}$$
$$= 75.1\ m$$

The change in d_e was less than 5 % (1.72–1.77 m) so the solution has converged and the final drain spacing, L, is 75 m.

If the conductivity above the drain is different from the conductivity below the drain (as in Example 31.3), then Eq. 31.14 can be modified to account for the varying conductivity.

$$L = \sqrt{\frac{4K_1 m^2}{R} + \frac{8K_2 m d_e}{R}} \tag{31.15}$$

where

K_1 = conductivity above the drain, m/day,
K_2 = conductivity below the drain, m/day.

Example 31.4 Redo Example 31.3 with Eq. 31.15. Let initial d_e = 1.41 (initial L = 40).

Solve for L

$$L = \sqrt{\frac{4K_1 m^2}{R} + \frac{8K_2 m d_e}{R}}$$
$$= \sqrt{\frac{4(1.5)(0.6^2)}{0.005} + \frac{8(3.0)(0.6)(1.41)}{0.005}} = 67.1\ m$$

After several iterations, the solution converges to 74 m, approximately equal to the 75 m calculated in Example 31.3.

$$L = \sqrt{\frac{4K_1 m^2}{R} + \frac{8K_2 m d_e}{R}}$$
$$= \sqrt{\frac{4(1.5)(0.6^2)}{0.005} + \frac{8(3.0)(0.6)(1.76)}{0.005}} = 74.2\ m.$$

Transient Drainage Models

This section presents two different models of transient water table response to storm events and dry periods between storms: the Bower and Van Schilfegaarde (1963) modification of the Hooghoudt equation and the United States Bureau of Reclamation (USBR) transient drainage model. The Bower and Von Schilfegaarde model is used in the DRAINMOD model.

Bower and Van Schilfegaarde rearranged Eq. 31.14, and solved for rainfall based on a known drain depth and spacing. Rainfall is replaced by unit flow rate to the drain, q_d, divided by L.

$$R = \frac{4K_e m(m + 2d_e)}{L^2} \tag{31.16}$$

$$\frac{q_d}{L} = \frac{4K_e m(m + 2d_e)}{L^2} \quad q_d = \frac{4K_e m(m + 2d_e)}{L} \tag{31.17}$$

Bower and Van Schilfegaard added a constant, C, to account for the difference between the assumption of uniform infiltration in the Hooghoudt equation and the varying infiltration observed in actual drainage response to storm and irrigation events.

$$q_d = \frac{4K_e m(m + 2d_e)}{CL} \tag{31.18}$$

Even though the C term is probably necessary for an accurate solution, the DRAINMOD model assumes that C is 1.0 because no values for the C term for different soils are available in the literature. Therefore, C is dropped from the equation.

For different hydraulic conductivities above and below the drain, Eq. 31.18 becomes:

$$q_d = \frac{4K_1 m^2}{L} + \frac{8K_2 m d_e}{L} \tag{31.19}$$

The depth of drainage in the field during time Δt is

$$d_{drainage} = \frac{q_d}{L}\Delta t \tag{31.20}$$

Table 31.2 Two-week water table depths

Time (days)	Initial m (m)	q_d (m^2/day)	$d_{drainage}$ (m)	Δm	m_{final} (m)	DTWT (m)
0	0.9	0.58	0.0078	– 0.078	0.82	0.2
1	0.82	0.52	0.0070	– 0.070	0.75	0.28
2	0.75	0.47	0.0064	– 0.064	0.69	0.35
3	0.69	0.43	0.0058	– 0.058	0.63	0.41

where

Δt = time, days.
$d_{drainage}$ = average depth of drainage, m.

For the case of two-dimensional drainage geometry, the ratio of the decline of the midpoint water table elevation, d_{mid} to the drainage depth, $d_{drainage}$, can be defined as the specific yield.

$$SY_{mid} = -\frac{d_{drainage}}{\Delta m} \quad (31.21)$$

where

SY_{mid} = specific yield during decrease in midpoint water table elevation,
Δm = change in midpoint water table elevation, m.

Example 31.5 For the drainage design in Examples 31.3 and 31.4 (d = 2.5 m and L = 74.2 m), calculate the decrease in the midpoint water table elevation over time. The initial depth of the water table at the midpoint between drains is 0.2 m. The specific yield, SY_{mid}, is 10 %. Assume that C = 1.0 (equal drainage over the length, L, between drains).

Note that the effective depth, d_e, is not a function of water table elevation, m, so d_e is constant if the depth and spacing of the drains are fixed. The drain depth is 1.1 m, and the midpoint water table is 0.2 m below the soil surface.

$$d_e = 1.76 \text{ m} \quad m = 1.1 - 0.2 = 0.9 \text{ m}.$$

Find the volume drained (per unit length of drain) during the first day

$$q_d = \frac{4K_1m^2}{L} + \frac{8K_2md_e}{L}$$
$$= \frac{4(1.5)(0.9^2)}{74.2} + \frac{8(3.0)(0.9)(1.76)}{74.2} = 0.58\ m^2/day$$

The depth drained during the first day, decline in elevation and final height are calculated.

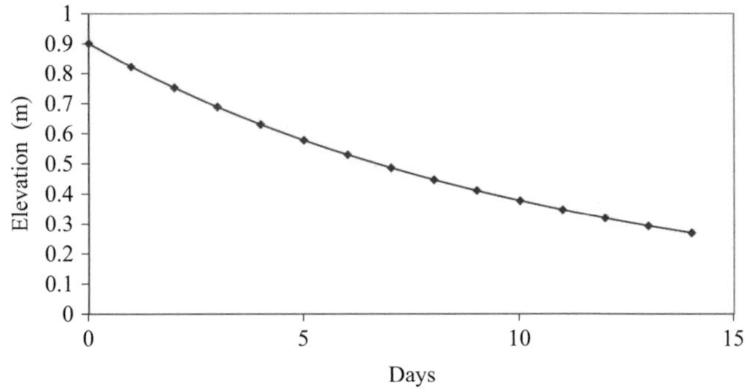

Fig. 31.11 Midpoint water table elevation versus time

$$d_{drainage} = \frac{q_d}{L}t = \frac{0.58}{74.1}(1\ day) = 0.0078\ m$$

$$\Delta m = -\frac{d_{drainage}}{SY_m} = -\frac{0.0078}{0.1} = -0.078\ m = -7.8\ cm$$

$$m_{final} = m_0 + \Delta m = 0.9 \text{ m} - 0.078 \text{ m} = 0.822 \text{ m}$$

The depth of the water table below the soil surface after the first day is 1.1 – 0.822 = 0.278 m. The process is repeated with 0.822 m as the initial height of the water table above the drain at the beginning of the second day. Based on the SEW-30 rule, the crop yield would decrease on days 0 and 1, but there would be no yield decrease on the following day with the depth to the water table at 0.349 m. Water table depths and other parameters for a two-week period are shown in Table 31.2.

The water table height, m, is shown for a two-week period in Fig. 31.11.

The USBR developed large irrigation and water resources projects in the western United States. The assumption of steady rainfall was not appropriate in the Western climate with periodic irrigations and storm events. In the 1950s, the USBR developed drainage equations and charts that allowed engineers to model the transient behavior of the water table and resultant flow to drains during the year. The USBR calculations were observed to generally fall within 10 % of observed field data.

The USBR method calculates the ratio of final and initial midpoint water table elevations, m/m_0, based on hydraulic conductivity, time, water table depth, specific yield and drain spacing. Parameters for the USBR equations are

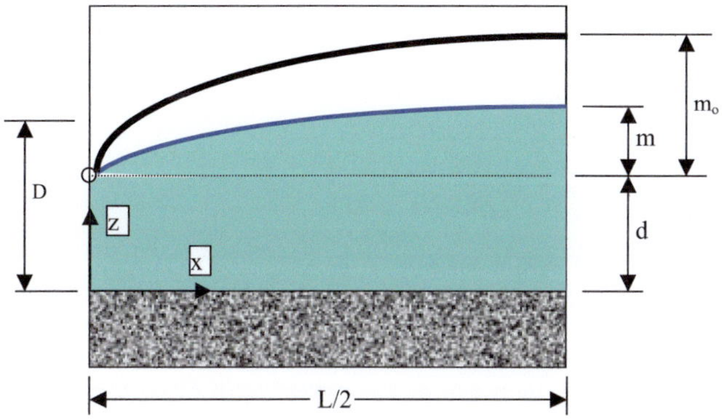

Fig. 31.12 Bureau of reclamation drainage parameters

Table 31.3 Deep percolation percentage from surface irrigation as a function of soil infiltration rate (Table 5.1 in USBR drainage manual)

Infiltration rate (mm/hr)	Deep percolation percent (f_{dp}) 100 %	Infiltration rate (mm/hr)	Deep percolation percent (f_{dp}) (100 %)
1.3	3	25	20
2.5	5	32	22
5	8	38	24
7.6	10	51	28
10	12	64	31
13	14	76	33
15	16	102	37
20	18		

shown in Fig. 31.12. Equation 31.22 is a regression fit of the "Drain above barrier" curve in Fig. 5.4 from the USBR Drainage Manual (http://www.usbr.gov/pmts/wquality_land/DrainMan.pdf). This equation calculates the ratio of the initial and final midpoint water table elevation over a specified period. In order to remain consistent with the previous derivation, m and m_0 are used instead of the USBR parameters, y and y_0.

$$\frac{m}{m_0} = 10^{0.01569-4.335\frac{K_e D t}{SY_{FC-S}L^2}} \tag{31.22}$$

where

K_e = effective hydraulic conductivity, m/day,
m_0 = initial depth of the water table above the drain, m,
m = final depth of the water table above the drain, same as m_{final}, m,
D = weighted average elevation of the water table (d_e + m_0/2), m,
$SY_{FC\text{-}S}$ = specific yield defined as difference between field capacity and saturation,
t = time, days,
L = distance between drains, m.

The depth, D, in Eq. 31.22 is the average elevation of the water table above the impermeable layer and is the sum of the equivalent depth and m/2.

$$D = d_e + m_0/2 \tag{31.23}$$

Equation 31.22 is valid when the ratio of d/m_0 is greater than 0.8. When d/m_0 is less than 0.1, drains near impermeable layer, then Eq. 31.24 is used. For in-between values, $0.1 < d/m_0 < 0.8$, an approximation is made between the two equations.

$$\frac{Z}{H} = f\left(\frac{K_e H t}{SY_{FC-S}L^2}\right) \tag{31.24}$$

where

H = Initial midpoint water table elevation above the impermeable layer.
Z = Final midpoint water table elevation above the impermeable layer.

In most irrigated areas, irrigation is the major water input to the soil. Because of poor irrigation efficiency or because of the need to leach salts below the root zone, some of the water applied during irrigation leaches below the root zone and contributes to build up of the water table. This water is called deep percolation. Only the wasted water lost below the root zone to deep percolation is added to the USBR water table model (unlike the DRAINMOD model). Average values of deep percolation as a function of soil infiltration rate during surface irrigation events were established by the USBR (Table 31.3) through field evaluations.

Deep percolation percentage increases with infiltration rate because more water is wasted to deep percolation at the upper end of fields while water advances to the ends of the field. On the other hand, excessive water can be wasted to runoff from a field with a very low infiltration rate; however, this fact is not reflected in Table 31.3 because it would run off the farm and not add to the water table. The depth of deep percolation during an irrigation event is

$$d_{dp-i} = i(1 - f_{RO})(f_{dp}) \tag{31.25}$$

where

i = gross depth of irrigation, m,
$d_{dp\text{-}i}$ = depth of deep percolation from irrigation event, m,
f_{RO} = fraction of runoff,
f_{dp} = fraction of deep percolation from Table 31.3.

The rise in the midpoint water table elevation in response to deep percolation from irrigation is

$$\Delta m = \frac{d_{d\,p-i}}{SY_{FC=S}} \tag{31.26}$$

Specific yield, SY_{FC-S}, as defined by the USBR is the difference between the saturated water content and field capacity and is estimated based on hydraulic conductivity with a graph in the USBR Drainage Manual. An equation that approximates the graph is on the *USBR relationships* page in *Chapter 31 Drainage*: SY =10^(−0.175*(LOG10 (A3))^2 + 0.7214*LOG10(A3) + 0.7188)

The USBR also developed a relationship between rainfall and infiltrated water as a function of magnitude of precipitation. This equation is also included in the *USBR relationships* worksheet:=IF(A6>2.5,IF(A6< 80,−0.007673*A6^2+1.1755*A6−2.2401,42.5),0).

The change in depth of the water table due to deep percolation from rainfall is the calculated in the same way as for irrigation (Eq. 31.26)

$$\Delta m = \frac{d_{d\,p-r}}{SY_{FC=S}} \tag{31.27}$$

where

$d_{dp\text{-}r}$ = deep percolation due to rainfall, m.

Example 31.6 Use the USBR equations to calculate the water table depth vs. time for a farm that is surface irrigated with an average irrigation depth per application of 180 mm. Average surface runoff during irrigation is 25 %. Soils have a silty clay loam texture with an infiltration rate of 12.7 mm/hr (1.27 cm/hr, 0.3 m/day with hydraulic conductivity of the soil equal to 0.3 m/day). Average precipitation per rainfall event is 31 mm. The depth from the soil surface to the impermeable layer is 5 m. The equivalent drain radius, r_e, is 0.02 m. The drain depth is 1.2 m. Drain spacing is 40 m. Minimum acceptable depth to the water table is 0.48 m.

The following irrigation, snowmelt, and rainfall events are expected based on historical records.

Snowmelt: April 22
Irrigation dates: May 28, Jun 6, Jun 20, Jul 2, Jul 14, Jul 26, Aug 10, Aug 22.
Rainfall events: Nov 1, Dec 15, Mar 1, Mar 15

Seasonal snowmelt has the same depth of deep percolation as one irrigation event.

Calculate the equivalent depth, d_e.

Calculate depth, d, the distance from the drain to the impermeable layer

$$d = 5\ \text{m} - 1.2\ \text{m} = 3.8\ \text{m}.$$

The ratio of depth to distance, d/L = 3.8/40 = 0.10: use Eq. 31.12 because d/L < 0.31. The effective radius of the drain is 0.02 m.

$$c = 3.55 - 1.6\frac{3.8}{40} + 2\left(\frac{3.8}{40}\right)^2 = 3.41$$

$$d_e = \frac{3.8}{1 + \frac{3.8}{40}\left(2.55\ln\left(\frac{3.8}{0.0051}\right) - 3.41\right)} = 1.95\ m$$

Calculate the increase in the water table elevation during irrigation events.

The fraction of deep percolation for soil with infiltration rate 12.7 mm/hr is 0.14 (Table 31.3).

$$d_{d\,p-i} = i^*(1 - f_{RO})^*(f_{d\,p}) = 0.18\ \text{m}^*(1 - 0.25)(0.14) = 0.0189\ \text{m}.$$

Specific yield, $SY_{FC\text{-}S}$, is calculated in the *USBR relationships* worksheet. For K = 1.27 cm/hr, $SY_{FC\text{-}S}$ = 7 %. Change in water table elevation during each irrigation event is

$$\Delta m = \frac{d_{d\,p-i}}{SY_{FC=S}} = \frac{0.0189}{0.07} = 0.27\ m$$

Thus, the midpoint water table elevation increases by 0.27 m during irrigation events.

Calculate the increase in the water table elevation during rainfall events.

If rainfall depth is 31 mm, then infiltration (deep percolation depth) = 25 mm (*USBR relationships* worksheet).

Deep percolation depth = 0.025 m.

$$\Delta m = \frac{d_{d\,p-r}}{SY_{FC=S}} = \frac{0.025}{0.07} = 0.357\ m$$

Calculate drainage position vs. *time of year. Start with August 22 as the first day.*

Because the water table height is typically a maximum at the end of the irrigation season, start the simulation with the water table at the maximum acceptable height at the end of the growing season. The minimum acceptable depth to the water table (from soil surface) is 0.48 m.

Calculate initial m_0.

$$1.2\text{m} - 0.48\text{m} = 0.72\text{m}.$$

Thus, on the day just after the last irrigation of the growing season, August 22, the water table elevation above the drains, m_0 = 0.72 m.

Calculate the initial D

$$D = d_e + m_0/2 = 1.95 + 0.72/2 = 2.31\text{m}$$

Calculate

$$\frac{K_e Dt}{SY_{FC-S}L^2} = \frac{0.30*2.31*1}{0.07*40^2} = 0.00629$$

Calculate

$$\frac{m}{m_0} = 10^{0.01569-4.335(0.00629)} = 0.974$$

Calculate water table depth after 1 day: m = (m/m_0) (m_0) = (0.974) (0.72 m) = 0.701 m.

The process is repeated for each day of the year. For days with water table buildup such as rainfall and irrigation days, the deep percolation is added instantaneously at the beginning of the day, and the calculation for change in water table height is conducted with the increased water table elevation. For example, on November 1, the water table elevation before the storm is 0.157 m, and the water table elevation after the storm is 0.154 + 0.357 = 0.511 m. The change in water table elevation is based on the 0.511 m water table elevation.

$$D = D + m_0/2 = 1.95 + 0.511/2 = 2.208\text{m}$$

Calculate

$$\frac{K_e Dt}{SY_{FC-S}L^2} = \frac{0.30*2.208*1}{0.07*40^2} = 0.00601$$

Calculate

$$\frac{m}{m_0} = 10^{0.01569-4.335(0.00601)} = 0.976$$

$$m = m/m_0 * m_0 = 0.976*0.511 = 0.499\text{m}.$$

Water table elevation fluctuation during the year is shown in Fig. 31.13.

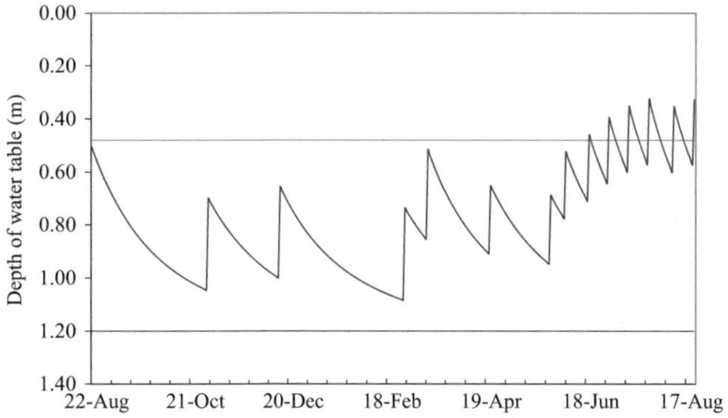

Fig. 31.13 Water table elevation versus time for L = 40 m and drain depth = 1.2 m

The water table (Fig. 31.18) elevation at the end of the growing season exceeds the criteria of 0.48 m distance from the soil surface during the growing season for 26 days, and the water table is higher at the end of the year than at the beginning of the year. Thus, the drains are spaced too far apart.

Economic Analysis

The following parameters can be incorporated into engineering economic analysis for drainage systems:

1. Capital investment
2. Maintenance
3. Crop yield or improvement in crop yield with the new drainage system
4. Environmental cost or benefit

In the following example, the two primary drainage installation methods are compared: trenching and pulling. Trenched drains are placed in a trench with a gravel pack around the drain tube. The gravel pack increases the conductivity of the region around the drain and as a result, the effective radius is increased. The increased effective radius increases the equivalent depth (Eq. 31.12) and makes it possible to have wider drain spacing. Pulled drains are installed by a large tractor that pulls a shank through the field. The drain tube is threaded down the shank and inserted in the soil at the shank depth. Soil is often compressed around the shank, and there is no gravel pack around the drain, so the effective radius is small. The maximum depth of pulled drain installation is 1.2 m. The current practice is to pull drains into the soil because trenched drains are more expensive and are only installed when the required drain depth is greater than is possible to install by pulling.

Example 31.7 Select between the following three drainage alternatives, and determine the correct spacing for each alternative. Use soil and weather parameters from Eq. 31.6. Use 4 inch nominal diameter drain tubing. Effective radius of the tubing for pulled in drains is 0.02 m and is 0.183 m for trenched drains with gravel pack. Place the USBR equations into an Excel spreadsheet in order to solve the problem.

Pulled drain at 1.2 m depth
Trenched drain at 2 m depth
Trenched drain at 2.4 m depth.

Design criteria:

The depth to the water table may not be less than 48 cm for more than 14 days during the growing season. The water

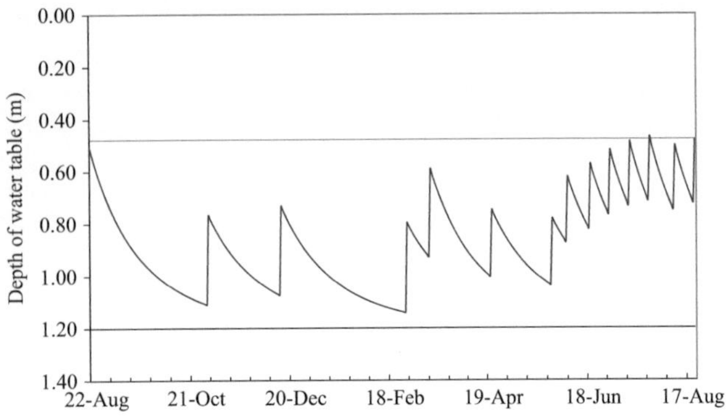

Fig. 31.14 Water table elevations versus time for L = 36.5 m and drain depth = 1.2 m

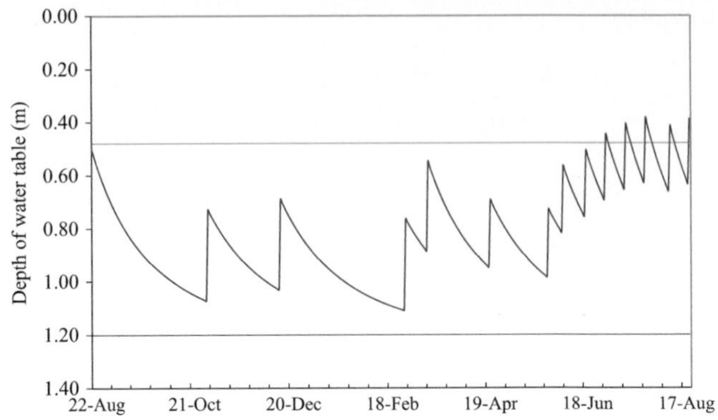

Fig. 31.15 Water table elevations versus time for L = 38.7 m and drain depth = 1.2 m

table elevation immediately following the last irrigation of the growing season cannot be higher than the water table elevation on the same day during the previous year.

Material costs and maintenance:

The cost of the collector drains and drainage structures is \$200/ha for both trenched drains and pulled drains. For this analysis, assume that the cost of drain maintenance is equal to the cost of field maintenance without drainage so it is not necessary to include it in the analysis. The cost of the 4 in drain tubing is \$3.00/m.

Economics

The price of corn is \$2.64 per bushel. Without drainage, the farmer has an average yield of 250 bushels per hectare. With a drainage system, the expected average corn yield is 320 bushels per hectare. The required rate of return is 10 % and project life is 20 years.

Trenching costs

Trenched drains: \$4.84/m of drain tube length for depths less than 2 m. There is an additional cost of \$.10/cm for trenching deeper than 2 m. Thus, there is no cost saving for trenching shallower than 2 m. The gravel pack around the drain is \$1.84/m.

Pulled drains: The cost of pulling in the drains at 1.2 m depth is \$1.69/m.

Pulled drain at 1.2 m depth.

The 40 m drain spacing was not satisfactory (Example 9.5) because the final water table depth at the end of the period (August 22) was less than the initial depth.

An iterative solution or Goal Seek in Excel can be used to find the correct solution, L = 36.5 m, where the water table elevation, m_0, is 0.72 m at the end of the growing season. At a drain spacing of 36.5 m, the water table only rises to within 48 cm of the soil surface for 1 day, July 26 (Fig. 31.14).

Goal Seek was also used to find a drain spacing that results in the water table not entering the root zone more than 14 days during the growing season, L = 38.7 (Fig. 31.15). The water table elevation at the end of the year is 0.81 m. Although the criteria for water not entering the root zone for more than 14 days is met, the first criterion, that the water table is the same elevation at the end of the year at the beginning of the year is not satisfied. Thus, a drain spacing of 36.5 m is selected in order to satisfy both criteria.

Trenched drain at 2 m depth

Desired water table depth is 0.48 m. Thus, initial m_0 is 2 m − 0.48 m = 1.52 m.

$$d = 5\text{m} - 2\text{m} = 3\text{m}.$$

Test a drain spacing of 50 m.

The ratio of depth to distance, d/L = 3/50 < 0.31. The effective radius of the drain is 0.183 m.

$$c = 3.55 - 1.6\frac{3}{50} + 2\left(\frac{3}{50}\right)^2 = 3.46$$

$$d_e = \frac{3}{1 + \frac{3}{50}\left(2.55\ln\left(\frac{3}{0.183}\right) - 3.46\right)} = 2.46\ m$$

Example calculation of water table elevation at the end of the day on August 23:

Initial m_0 = 1.52 m

$$D = d_e + m_0/2 = 2.46 + 1.52/2 = 3.22\text{ m}$$

Calculate

$$\frac{K_e D t}{S Y_{FC-S} L^2} = \frac{0.30{*}3.22{*}1}{0.07{*}50^2} = 0.00561$$

Calculate

$$\frac{m}{m_0} = 10^{0.01569 - 4.335(0.00561)} = 0.980$$

Calculate water table depth at the end of the day.

$$m = m/m_0{*}m_0 = 0.980{*}1.52 = 1.49\text{m}.$$

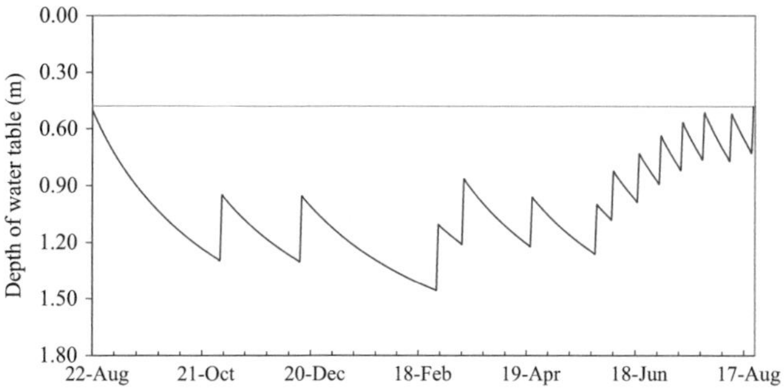

Fig. 31.16 Water table elevations versus time for L = 52.6 m and drain depth = 2 m

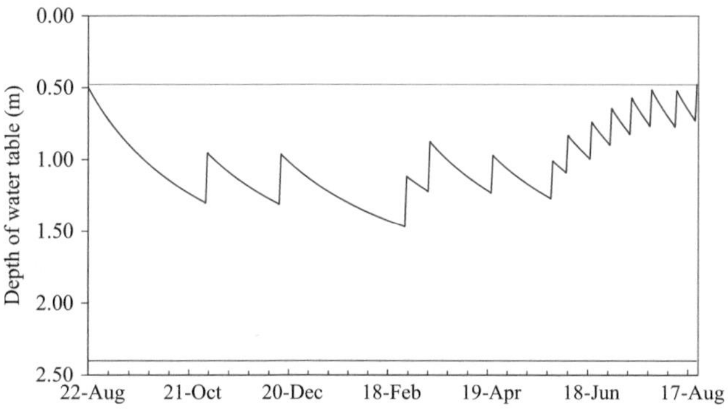

Fig. 31.17 Water table elevations versus time for L = 54.1 m and drain depth = 2.4 m

Goal seek was used to find the optimal distance, L = 52.6 m (Fig. 31.16) based on the criteria that the water table at the end of the period not exceed the water table depth at the beginning of the period.

Goal Seek was used to find the optimal distance for the 2.4 m drain depth: 54.1 m (Fig. 31.17).

Economic analysis

Economic analysis of drainage alternatives is performed in two steps. First, the drainage alternative with the lowest cost is selected. Second, a present value cost/benefit analysis of the lowest cost drainage alternative is made. If benefits are greater than costs, then the project is approved.

Comparison of the three drainage alternatives:

Costs are calculated on a per hectare basis. The length of drain tubing installed per hectare (m/ha) is the area (m^2/hectare) divided by the drain tube spacing (m). For the drains installed at 1.2 m depth, the spacing L is 36.5 m.

$$(\mathrm{m/ha}) = 10{,}000/36.5 = 274\mathrm{m}$$

The cost per m of drain tubing is the sum of materials and installation costs:

$$\begin{aligned}\$/\mathrm{m} &= \mathrm{pullingcost} + \mathrm{draintubecost} = \$1.69/\mathrm{m} + \$3.00/\mathrm{m} \\ &= \$4.69/\mathrm{m}\end{aligned}$$

The cost per ha is the product of m/ha and $/m

$$\$/\mathrm{ha} = 274{*}4.69 = \$1{,}285/\mathrm{ha}$$

The installation cost of the other two alternatives was calculated in the same manner except that the cost of gravel was added for the trenched alternatives (Table 31.4).

The lowest cost alternative is the pulled drain at 1.2 m depth.

The present value of installation of the drain tubes ($1,285) and collector drains and structures ($200/ha) is

$$\$1{,}285/\mathrm{ha} + \$200/\mathrm{ha} = \$1{,}485$$

The benefit of the drainage system is increased yield. The yield without drainage is 250 bushels/ha and the yield with drainage is 320 bushels/ha. Thus, the increase in yield is 320 − 250 = 70 bushels/ha. The amount of money received each year because of drainage is the product of the increased yield due to drainage and the crop price:

$$70\mathrm{bushels/ha/yr}{*}\$2.64/\mathrm{bushel} = \$184.80/\mathrm{ha/year}$$

The Excel present value formula, PV, can be used to find the present value of 20 annual payments of $184.80/yr at 10 % interest rate: = PV(0.1, 20, 184.8). The calculated present value with the PV formula is $1,573/ha.

The present value of the benefits of drainage ($1,573) are higher than the cost ($1,485/ha) at a 10 % rate of return. Thus, the decision is made to invest in a drainage system.

The value of a crop will have a significant influence on whether the decision is made to invest in drainage; thus, the NRCS states that drain spacings closer than 40 ft (12 m) can be justified economically for high value crops. Another key factor in the economics of drainage design is possible work at the discharge end. If a channel must be deepened in order to carry the drainage flow from a subsurface drainage network, then that would add cost to the system.

Streamtube Model

The previous models presented in this chapter modeled the two-dimensional flow geometry below the water table with one-dimensional equations. Finite element, finite difference, or analytic solutions of the La Place equation are needed to model the flow in two dimensions. Kirkham developed a two-dimensional analytic solution of the La Place equation that calculates the position of the water table over time. He simplified the geometry by assuming that the soil above the drain has infinite conductivity. Thus, he only modeled the rectangular region between the impermeable layer and the drain. He assumed that the water above the drain flowed

Table 31.4 Costs of drainage alternatives

Alternative.	Spacing, L (m)	Length (m/ha)	Pipe cost ($/m)	Installation ($/m)	Gravel ($/m)	Total cost ($/ha)
Cost – pulling 1.2 m	36.5	274	3.00	1.69	–	1,284.93
Cost – 2 m trench	52.6	190	3.00	4.84	1.84	1,840.30
Cost – 2.4 m trench	54.1	185	3.00	8.84	1.84	2,528.65

vertically downward to the drain elevation in streamtubes. The Kirkham (1958) solution to the LaPlace equation calculates the elevation of the water table over the drain vs. distance away from the drain for a given steady state recharge rate, R.

$$h = \frac{LR}{K\pi}\left[\begin{array}{l} \ln\dfrac{\sin(\pi x/L)}{\sin(\pi r/L)} \\ +\displaystyle\sum_{m=1}^{\infty}\frac{1}{m}\left(\cos\frac{m\pi r}{L/2} - \cos\frac{m\pi x}{L/2}\right)\frac{e^{-m\pi d/(L/2)}}{\sinh(m\pi d/(L/2))} \end{array}\right] \quad (31.28)$$

where

h = height of water table above drain at any distance x from the drain, m,
r = effective radius of the drain, m,
m = the number of terms in the series solution.

The formula inside the parenthesis is labeled as F(x,0) by Kirkham and is solely dependant on the geometry of the region between the drain and the midpoint between drains.

A Visual Basic program was written in the *Chapter 31 Drainage* program that calculates the Kirkham F. Although the sinh(z) function is not available in VBA, it is equal to (exp(z) − exp(−z))/2, as in the following code.

```
Function Kirkham_F(num_series, x, L, re, d) As
Single
    Dim sum, m, f
    f = Log(Sin(3.14159 * x / (L)) / Sin(3.14159 * re
    / (L)))
    sum = 0
    For m = 1 To num_series
      sum = sum + (1 / m) * (Cos(m * 3.14159 * re / (L /2))
      - Cos(m * 3.14159 * x / (L / 2))) * (Exp(-m *
      3.14159 * d / (L / 2)) / (Exp(m * 3.14159 * d /
      (L / 2)) - Exp(-m * 3.14159 * d / (L / 2))))
    Next m
    f = (f + sum) / 3.14159
    Kirkham_F = f
End Function
```

Kirkham developed a method based on the streamtube model to calculate the rate of fall of the water table at each distance between drains.

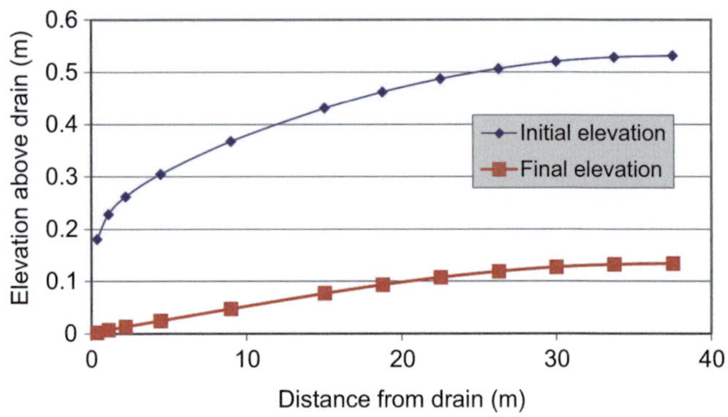

Fig. 31.18 Initial and final water table elevations versus distance from the drain calculated with Kirkham solution (0 and 14 days after equilibrium)

$$\frac{db}{dt} = \frac{-K/f}{1 + LF/b} \quad (31.29)$$

where b is the elevation of the water table at any distance x and *f* is water table elevation at that distance over specific yield (b/SY).

Example 31.8 Calculate the water table elevation vs. distance from the drain using the parameters in Example 31.3. Calculate the rate of fall of the water table specified in Example 31.8. Compare the rate of fall of the midpoint water table elevation to the Bower and Von Schilfegaarde solution in Example 31.5. For the Bower and Von Schilfegaarde solution, assume that the initial water table elevation is 0.53 m. Plot the rate of fall at each point between drains. Plot the initial and final water table positions. Show a sample calculation at 0.375 m from the drain during the first time step.

Distance between drains, L	75	m
Elevation of drain above impermeable layer, d	2.5	m
Effective drain radius, re	0.0051	m
Rainfall rate	0.005	m/d
Number in series approximation, m	20	
Hydraulic conductivity	2.84	m/day

The Kirkham water table elevation vs. distance from the drain is shown in Fig. 31.18 (blue line is initial elevation). At the same steady state recharge rate as in Example 31.3, the elevation at the midpoint is 0.53 m, which is approximately

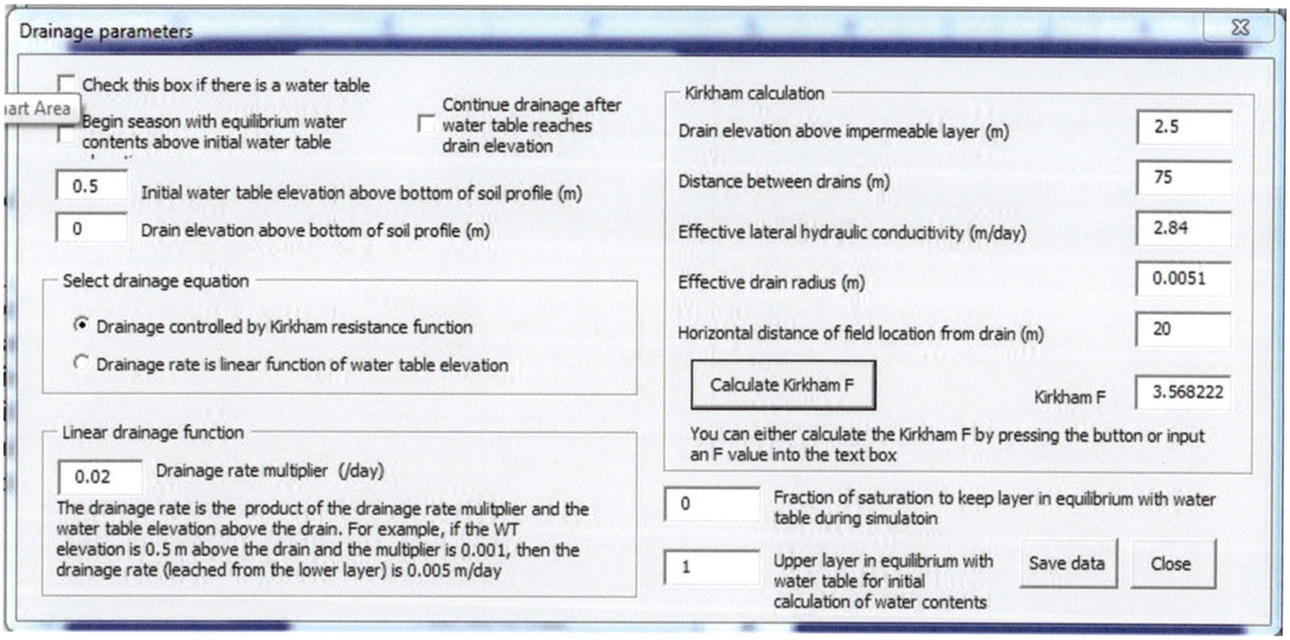

Fig. 31.19 *WINDS* model *Drainage* dialog box for drain parameters in Example 31.8

10 % lower than the elevation calculated with the Hooghoudt equation; however, there is close agreement between the Kirkham and Bower and Von Schilfegaarde (Example 31.3) solutions for the change in midpoint water table elevations over time.

The F value is calculated as 1.37 at 0.375 m from the drain. The initial water table elevation is 0.181 m at 0.375 m from the drain. The new water table elevation is calculated as follows.

$$b_{final} = b_{initial} + \frac{db}{dt} = b_{initial} + \frac{-K/f}{1 + LF/b}$$

$$= 0.181 + \frac{-2.84/0.1}{1 + 75*1.37/0.181} = 0.131\ m$$

The calculations for each time step and distance are made in the Streamtube worksheet. The initial and final elevations of the water table are shown in Fig. 31.18.

The benefit of the Kirkham model over the Bower and van Schilfgaarde model is that the water table as a function of distance from the drain can be evaluated for its effect on water, nitrogen, and salinity, and crop growth. There are significant trends vs. distance from the drain in many fields, which is why farmers often decide to install drains between drains. For example, salinity is often high in the region between drains but low near the drains. The Kirkham model allows the analysis of a number of points between drains in order to see the effect of drain spacing on crop growth and other processes. This technique could be used to improve best management practices for reducing nitrate output from drains and into surface waters.

The *WINDS* model uses the Kirkham algorithm at each field position to compute water table position vs. time. For example, the parameters from Example 31.8 can be entered into the *Drainage* dialog box (Fig. 31.19). In this case, the field position is 20 m from the drain, and the drain is at the bottom of the soil specified soil profile, which is 1.4 m deep, a typical drain depth. The program calculates the Kirkham F and sets up the model to run simulations of water table elevation vs. time at the specified location.

In the following examples, the rainfed clay soil example from Chap. 29 (Example 29.5) will be modified for the presence of a water table. Because the soil is clay, and there is little change in water content with matric potential, the "fraction of saturation to keep layer in equilibrium with water table during simulation" value (Chap. 28) must be high (0.95); otherwise, all layers will remain in equilibrium with the water table during the entire simulation. This would result in evaporation (water loss) from the upper layers rapidly dropping the water table when in reality the upper layers would disconnect hydraulically from the water table.

Because clay has an extremely high field capacity, almost saturation, it is interesting to evaluate whether an underlying water table would reduce water stress during dry periods. It would seem that the water remains in the soil whether there is an underlying water table and restriction on downward water movement or not. Evaluation with the *WINDS* model indicates that with or without (Fig. 31.20) water harvesting,

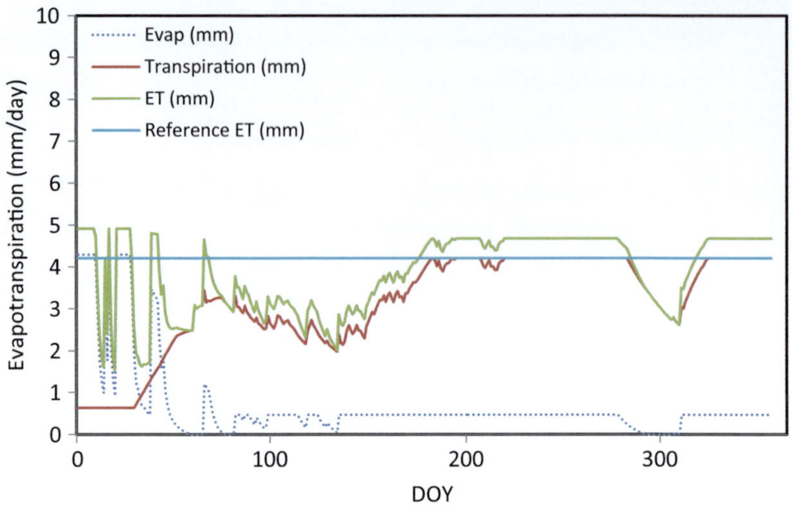

Fig. 31.20 *WINDS* model evapotranspiration for drain parameters in Example 31.8 and soil and hydrology parameters (without water harvesting) from Example 29.5

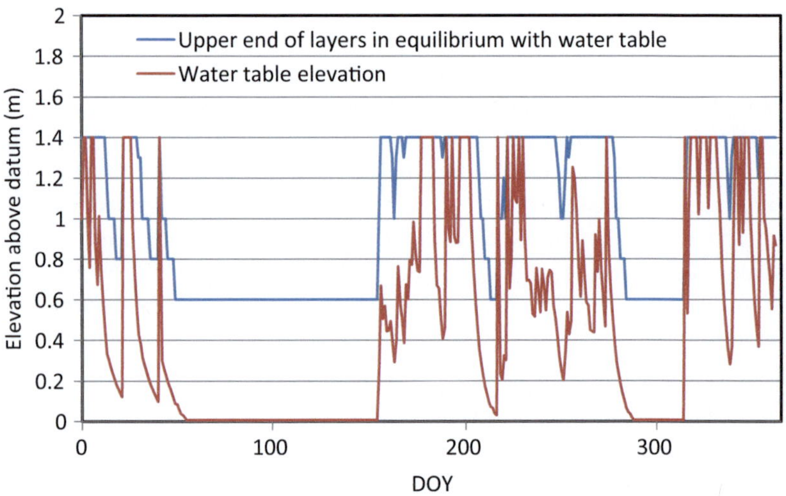

Fig. 31.21 *WINDS* model *Drainage* dialog box for drain parameters in Example 31.8

the stress is reduced with an underlying water table (compare Fig. 29.24 to Fig. 31.20).

Without water harvesting, the water table quickly falls to the bottom of the soil profile and remains there. In spite of this, the underlying water table (0.4 m below the root zone) apparently adds water to the root zone during dry periods. With water harvesting, the water table frequently rises up into the root zone (Fig. 31.21).

Questions

1. Give two reasons for installing subsurface drainage in a field and discuss its importance.
2. Describe the two different types of flow to subsurface drains
3. List the water sources and sinks for drained soils.
4. Describe the Dupuit Forchheimer assumption and draw flow lines to a subsurface drain according to the Dupuit-Forchheimer assumption.
5. Discuss the following drainage systems: open parallel ditches, conventional drainage, water table control, and subirrigation.
6. Calculate the yield reduction for corn with the following water table data. Assume that CS is 0.12 from DOY 135–143 and is 0.10 from DOY 144–160. YRDMAX = 102 and DSLOPE = 0.75. When does the storm occur?

DOY	DTWT	DOY	DTWT
135	5	143	5
136	15	144	15
137	24	145	24
138	32	146	32
139	39	147	39
140	44	148	44
141	46	149	46
142	0	150	47

7. Why does the water table elevation in Fig. 31.5 increase rapidly after a rainstorm and then decrease more and more slowly over time? Discuss the influence of energy gradients.
8. Rainfall is 0.02 m / day, and drain spacing is 40 m. Calculate flow rate q_5, at a distance of 10 m from the drain. Find the flow rate into the drain per unit length of drain tubing.
9. A drain is placed in the field in the shape of a circle with radius 20 m. Derive equations for the flow rate to a unit length of the drain (m^2/day) at any distance from the drain. Let R represent the radius of the drain circle, r = distance from the center of the circle, D = distance from the drain, and P represent the precipitation rate. Only derive an equation for the flow on the inside of the circle.
10. Calculate the required spacing between drains. The farmer wants to maintain the water table at least 0.7 m below the soil surface. Yearly rainfall is 1 m/yr. The impermeable layer is 2.5 m below the soil surface. Hydraulic conductivity of the soil is 1 m/day. The conventional practice in the region is to install drains at 1.1 m depth below the soil surface. Drains are standard 4 in (10 cm) diameter drains. Effective drain radius for the 4 in drain is 0.51 cm
11. Set up a finite difference solution in a spreadsheet in order to solve for water table elevation vs. distance from the drain for the conditions in question 10. Solve for Δz with the finite difference solution at each position by rearranging Darcy's law and solving for Δz, and sequentially calculate the increasing drain elevation beginning at the drain and working toward the midpoint between drains. For the location close to the drain (horizontal distance from drain is less than distance from drain to impermeable layer), set up the flow to the drain as a quarter circle (or slightly more as water table increases with distance from the drain), and calculate head loss as a function of distance from the drain. Where it intersects the drain, let the water table be at the midpoint elevation of the drain. Use Dupuit-Forchheimer solution far from the drain. You should have less than 10 % difference between your solution and the 0.4 m elevation m at the midpoint between drains calculated in question 10.
12. Redo problem 11, but include a sloping geometry ($\theta = 10^0$) for the impermeable layer as shown below. Let the elevation of the drain be 1.1 m above the impermeable layer at the position of the drain. Assume Dupuit-Forchheimer flow in the region far from the drain. Compare the midpoint water table elevation to that calculated in question 11.

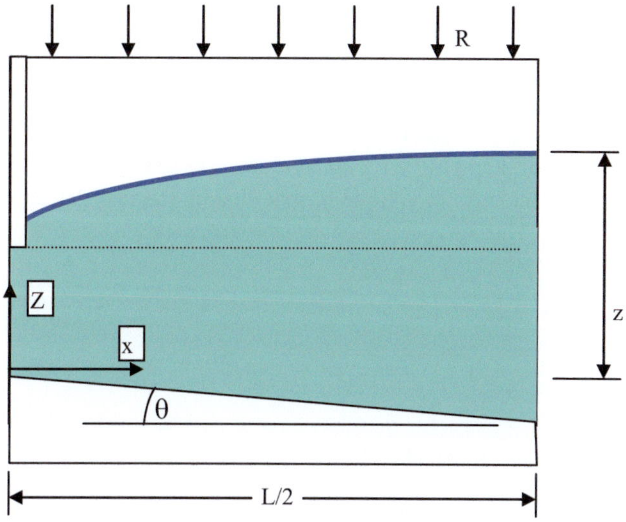

13. Redo question 12, but let $\theta = -4^0$ which means that the impermeable layer slopes upward from the drain rather than down. Compare to the midpoint water table elevation calculated in question 12.
14. Redo question 10 but let hydraulic conductivity of the soil within the top 1.1 m equal 0.3 m/day and below the drain equal 1 m/day.
15. With information from question 10, calculate the change in water table elevation over two weeks with the modified Hooghoudt equation. The initial elevation is 0.5 m, and the specific yield, SY_{mid}, is 15 %.
16. With parameters from question 10 and question 15 ($d_e = 1.0$), calculate the change in water table elevation over time with the USBR equation. The initial elevation is 0.5 m, and the specific yield, SY_{mid}, is 15 %. Remember to calculate D with the effective depth. Plot results and compare to the Hooghoudt transient graph from question 15. There is very little agreement because the KDt/SL^2 value is at the limit of the USBR curve from which it is derived.
17. Redo question 16 but let the equivalent depth, de = 3 for both the Hooghoudt and USBR solutions. The solutions are close because the USBR equation is in the central part of the USBR curve.
18. A field has an infiltration rate of 38 mm/hr. An irrigation with a gross application depth (volume applied divided

by field area) of 10 cm is applied, and the fraction of runoff is 0.2. Calculate the change in water table elevation for a drainage system with $d_e = 3.0$ m and $K = 0.5$ m/day. Plot the water table elevation (m) vs. time for four weeks of irrigation events that take place once each week. Remember to include the specific yield in the calculation of change in water table elevation and also to include drainage on the days that irrigation water is added to the water table.

19. Redo question 18, but add 20 cm gross application depth during each irrigation event. Plot the water table elevation vs. time (m).
20. Redo Example 31.9, but call drainage companies for drainage alternatives and prices in your region.
21. Redo Example 31.10, but evaluate a field soil and drainage scenario in your area, specified by the instructor.

References and Resources

Bouwer H, van Schilfgaarde J (1963) Simplified method of predicting fall of water table in drained land. Trans ASAE 6(4):288–296

Jury WA (1975) Solute travel-time estimates for tile-drained fields: I theory. Soil Sci Soc Am Proc 39:1020–1023

Kirkham D (1958) Seepage of steady rainfall through soil into drains. Am Geo U Trans 39:892–908

Kirkham D (1964) Physical artifices and formulas for approximating water table fall in tile drained land. Soil Sci Soc Am Proc 28:585–590

Kirkham D, Gaskell RE (1951) The falling water table in tile and ditch drainage. Soil Sci Soc Am Proc 15:37–42

Skaggs RW (1980) Drainmod. Reference report. Methods for design and evaluation of drainage-water management systems for soils with high water tables. North Carolina State University. Chapters 2, 3, and 5. http://www.bae.ncsu.edu/soil_water/drainmod/

van Schilfgaarde J (1963) Design of tile drainage for falling water table. J Irrig Drain Div ASCE 89(2):1–11

Solutions

Chapter 1: Solution

1. List the components of a farm water balance.

Evaporation, irrigation, canal seepage, operational waste, evapotranspiration, rainfall, runoff, deep percolation, usable return flow

2. Describe the major four irrigation methods (NRCS description).

   Surface, sprinkler, micro and sub irrigation

3. Describe the different types of irrigation efficiency

Irrigation efficiency has been defined in many ways. In general, efficiency is the amount used divided by the amount applied; however, irrigation efficiency can be thought of in different ways. Conveyance efficiency is the amount of water reaching a field divided by the amount diverted from the irrigation water source. Application efficiency is the amount of water stored in the root zone divided by the amount of water applied to the field. Storage efficiency is the amount of water stored in the root zone during a single irrigation divided by the total water-holding capacity of the root zone. Seasonal irrigation efficiency is the water volume beneficially used by the crop (including leaching) divided by the seasonal amount of water applied.

4. In general, pressurized irrigation systems are considered to be more efficient than surface irrigation systems. What are some factors that may decrease the irrigation efficiency of some sprinkler systems to below that of some surface irrigation systems?

However, loss of water by evaporation or wind, poorly maintained sprinklers or emitters, and poor uniformity due to spacing or pipe hydraulics decrease application efficiency.

5. How do irrigation systems fail?

If not regulated, water resources naturally become stretched beyond their capability to supply demand. As people begin to develop irrigated farms in a region, well drilling is generally not regulated; thus, the rate of pumping soon exceeds the rate of recharge in a successful farming region. As the groundwater table recedes to hundreds of feet below the ground surface, the cost of pumping increases, and the aquifer yield decreases until farmers eventually start going out of business.

Another source of failure is salinization of soils due to poor irrigation management or lack of drainage.

A third cause of failure is drought.

6. Summarize the virtual water concept.

Crops grown in one region and shipped to a water stressed region are essentially a shipment of water to the water stressed region.

7. What increases the value of the irrigation engineering?

An increased value of water resources.

Chapter 2: Solution

1. Why does increasing the depth of irrigation water applied per season reach a point of diminishing returns?

   There are at least three answers to this question.
   (a) Agricultural plants only transpire as much water as they need. Applying more water than the required water does not increase plant growth.
   (b) Excessive water or water in the root zone can decrease oxygen uptake from the root zone for respiration.
   (c) Some processes in some plants, such as tomato production, are enhanced when there is some water stress (dry stress) because the plants put their energy into reproduction in the event of plant death due to water depletion.

P. Waller, M. Yitayew, *Irrigation and Drainage Engineering*, DOI 10.1007/978-3-319-05699-9

2. Why is there a nearly linear relationship between evapotranspiration (ET) and yield and a nonlinear relationship between applied water and yield (Fig. 2.1)?

As water application increases, water waste due to evaporation and leaching increases. Thus, applied water is wasted, but transpiration does not change significantly.

3. How are the crop water production function and engineering economic analysis used to assess or manage irrigation systems?

The crop water production function is used to determine the optimal depth of irrigation water application. Engineering economic analysis is used to make decisions, such as which irrigation system to select or whether a project is economically profitable.

4. Find the profit with the following parameters for the Grimes and El-Zik CWPF for drip irrigated cotton: 15 cm depth of precipitation, cost of water is $0.02/m³. The selling price of cotton is $0.90/kg. Depth of applied water, $AW_{CWPF} = 62$ cm. Show your work.

$$Y_a = \left(-3954 + 1067(AW + 15\ \text{cm})^{0.5} - 54.14(AW + 15\ \text{cm})\right)$$
$$Y_a = \left(-3954 + 1067(62 + 15\ \text{cm})^{0.5} - 54.14(62 + 15\ \text{cm})\right) = 1{,}240\text{kg/ha}$$

$$\text{Crop\_price} = (Y_a)(\$/\text{kg}) = 1{,}240\ \text{kg/ha}(0.9) = \$1{,}116$$

$$WC = (\$/\text{m}^3)(100)(\text{AWcwpf(cm)}/(1 - (\text{CWPF\_Eff} - \text{Irr\_Eff})) + \text{Pre})$$
$$WC = (\$0.02)(100)(62/(1 - (0.9 - 0.9)) + 0) = \$124$$

$$\Pr = \$1{,}116 - \$124 = \$992$$

5. For the parameters in question 4, calculate the optimal depth of water application, AW_{CWPF}. Do this problem in Excel and turn in the graph that shows the yield, cost of water, and profit vs. AW_{CWPF}.

The maximum profit is found at 74 cm applied water depth, at which point the profit is $1016.

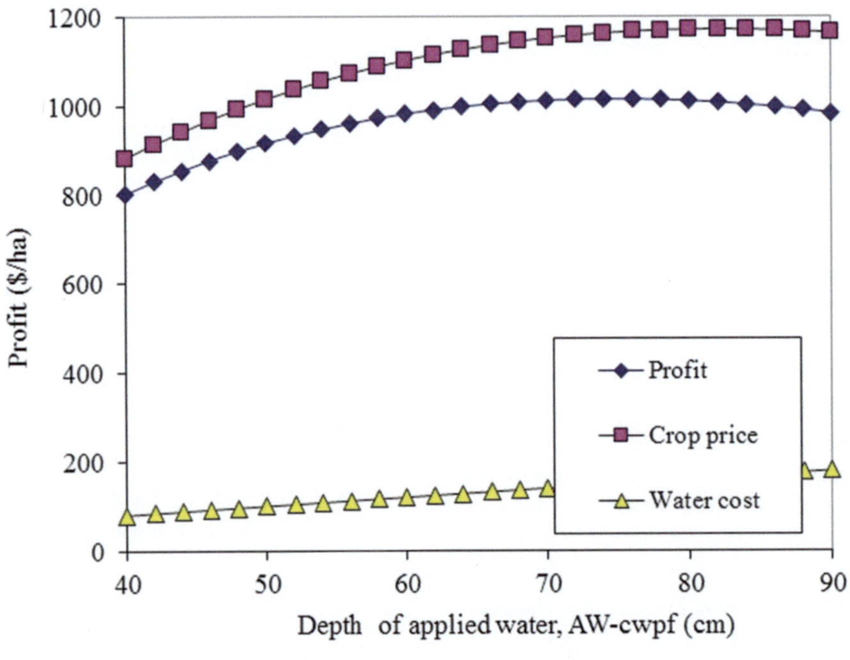

6. For the parameters in question 4, calculate the profit for a surface irrigation system with 70 % efficiency at $AW_{CWPF} = 70$ cm. Preirrigation is 45 cm. The water from preirrigation provides no benefit for crop growth in this case. Do not consider erosion. Show your work.

$$Y_a = \left(-3954 + 1067(AW + 15\ \text{cm})^{0.5} - 54.14(AW + 15\ \text{cm})\right)$$
$$Y_a = \left(-3954 + 1067(62 + 15\ \text{cm})^{0.5} - 54.14(70 + 15\ \text{cm})\right) = 1{,}240\ \text{kg/ha}$$

$$\text{Crop\_price} = (Y_a)(\$/\text{kg}) = 1{,}240\ \text{kg/ha}(0.9) = \$1{,}116$$

$WC = (\$/m^3)(100)(AWcwpf(cm)/(1 - (CWPF\_Eff - Irr\_Eff)) + Pre)$
$WC = (\$0.02)(100)(70/(1 - (0.9 - 0.7)) + 45) = \$265$

$$Pr = \$1,116 - \$124 = \$888$$

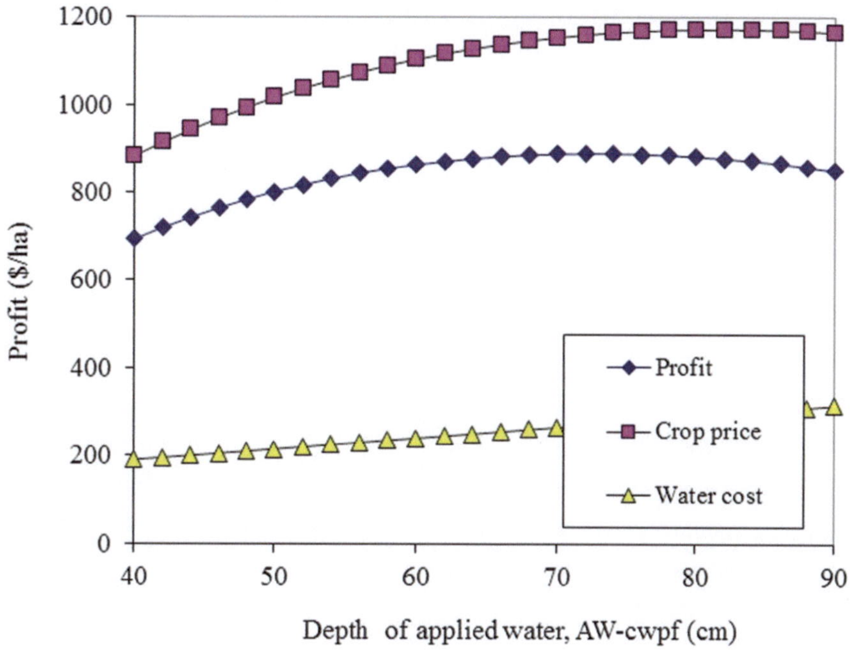

7. For the parameters in question 6, find the optimal depth of water application, AW_{CWPF}. Do this problem in Excel, and turn in the graph that shows the yield, cost of water, and profit vs. AW_{CWPF}.

The optimal depth of water application is 72 cm, at which point the profit is $889

8. For the parameters in question 6, calculate the profit per ha, but include erosion. Calculate erosion with Eq. 2.7. The cost of erosion is $0.4/kg. Show your work.

$$Y_a = \left(-3954 + 1067(AW + 15\ \text{cm})^{0.5} - 54.14(AW + 15\ \text{cm})\right)$$
$$Y_a = \left(-3954 + 1067(62 + 15\ \text{cm})^{0.5} - 54.14(70 + 15\ \text{cm})\right)$$
$$= 1,240\ \text{kg/ha}$$

$$Crop\_price = (Y_a)(\$/kg) = 1,240\ \text{kg/ha}(0.9) = \$1,116$$

$WC = (\$/m^3)(100)(AWcwpf(cm)/(1 - (CWPF\_Eff - Irr\_Eff)) + Pre)$
$WC = (\$0.02)(100)(70/(1 - (0.9 - 0.7)) + 45) = \$265$

$$i = 70/(1 - (0.9 - 0.7)) + 45$$
$$= 70/(1 - (0.9 - 0.7)) + 45 = 132.5\ \text{cm}$$

$$Sediment = \left(4.62*10^{-5}*(132.5)^3 - 0.00784*(132.5)^2 + 0.477*(132.5) - 9.4449\right)*10 = 236\ \text{kg/ha}$$

$$EnvC = 236\ \text{kg/ha}*\$0.4/\text{kg} = \$94.4/\text{ha}$$

$$Pr = \$1,116 - \$265 - \$94 = \$794$$

9. For the parameters in question 8, find the optimal depth of water application, AW_{CWPF}. Do this problem in Excel, and turn in the graph that shows the yield, cost of water, cost of erosion, and profit vs. AW_{CWPF}.

The optimal depth of water application is 62 cm, at which point the profit is $806

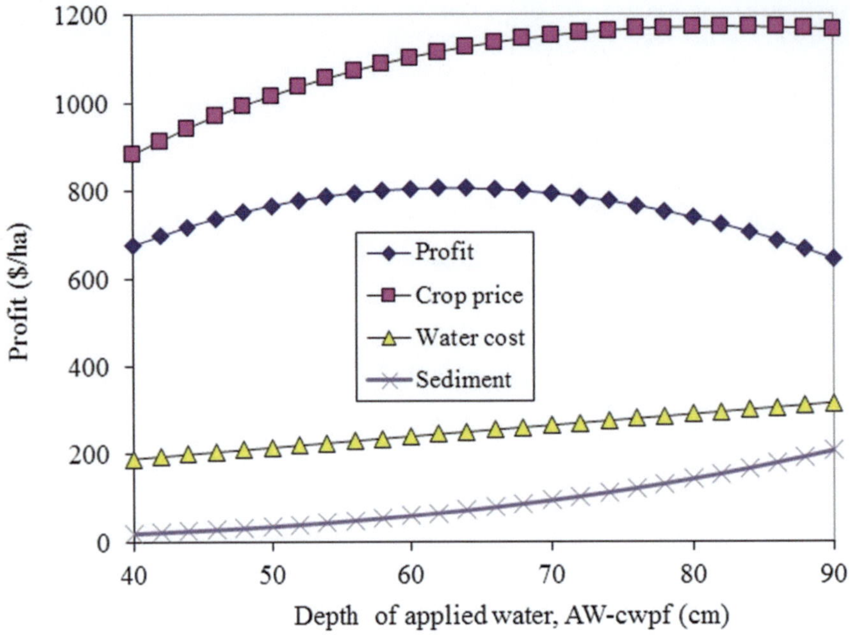

10. Use the *Variable water application* worksheet. Keep all other parameters the same, but double the cost of water. Find the optimal depth of water application.

Vary the value in cell F1 until the profit is highest in cell F9. The peak is at 57 cm.

	A–E	F	G	H–J	K
1	Target AW-CWPF	57		Average AW-CWPF (cm)	58.09
2	Fraction of Target depth at end of furrow (specified)	0.91		Average leached depth (cm)	2.32
3	Runoff percentage (specified)	20%		Average useful depth (cm)	55.78
4	Precipitation	15	cm	Total applied (AW - cm)	117.62
5	Price of crop	0.9	$/kg	Efficiency (w/o preirrigation)	77%
6	Preirr depth	45	cm	Efficiency (including preirrigation	47%
7	Price of water	0.04	$/m3	Cost of water ($/ha)	$ (470.47)
8	Cost of sediment	0.4	$/kg	Cost of sediment ($/ha)	$ (53.51)
9	Benefits - costs	563.1628	$/ha	Average yield benefit ($/ha)	$ 1,087.14

10	Distance (m)	Fraction of end depth	Fraction of target depth	AW-cwpf	Excess Leached depth (cm)	Yield ($/ha)
11	40	1.20	1.092	62.24	5.24	1117.55
12	120	1.18	1.07	61.21	4.21	1111.23
13	200	1.14	1.04	59.13	2.13	1097.42
14	280	1.08	0.98	56.02	0.00	1073.65
15	360	1.00	0.91	51.87	0.00	1035.86

Hide chart

View chart

11. Use the *Variable water application* worksheet but change the number of furrow sections to 6 with the following multipliers of end furrow application. 1.5, 1.45, 1.33, 1.25, 1.13, and 1. Keep all other parameters the same as the original version. Find the optimal depth of water application.

	A–E	F	G	H–J	K
1	Target AW-CWPF	49		Average AW-CWPF (cm)	56.93
2	Fraction of Target depth at end of furrow (specified)	0.91		Average leached depth (cm)	8.66
3	Runoff percentage (specified)	20%		Average useful depth (cm)	48.27
4	Precipitation (cm)	15		Total applied (AW - cm)	116.16
5	Price of crop	0.9	$/kg	Efficiency (w/o preirrigation)	68%
6	Preirr depth	45	cm	Efficiency (including preirrigation)	42%
7	Price of water	0.04	$/m3	Cost of water ($/ha)	$ (464.63)
8	Cost of sediment	0.4	$/kg	Cost of sediment ($/ha)	$ (50.36)
9	Benefits - costs	553.8945		Average yield benefit ($/ha)	$ 1,068.89

10	Distance (m)	Fraction of end depth	Fraction of target depth	AW-cwpf	Excess Leached depth (cm)	Yield ($/ha)
11	33.3	1.5	1.365	66.89	17.89	1141.26
12	100	1.45	1.32	64.66	15.66	1130.78
13	166.7	1.33	1.21	59.30	10.30	1098.63
14	233.3	1.25	1.14	55.74	6.74	1071.30
15	300	1.13	1.03	50.39	1.39	1020.55

Hide chart

View chart

12. Should government charge farmers with surface irrigation systems that discharge sediment to streams for the cost of dredging and for sediment removal from public drinking water supplies? Why or why not?

Difficult question. The government has chosen to pay farmers in sensitive regions to convert to pressurized irrigation systems in order to reduce sediment discharge to streams. This practice also reduces water use and allows the water to go to other uses such as new farms, cities, or industries.

Why:

Nobody should be able to pollute a public resource. Those who degrade the environment should have to pay for the consequences of their actions. The public should not allow certain practices or industries to degrade public resources.

Why not:

The government has a vested interest in supporting production agriculture. At the most basic level, national food security prevents extortion by other countries. If the government charged farmers for every environmental consequence of farming, then it would be difficult for farmers to continue to farm and to compete against farmers in other countries, who receive direct or indirect subsidies from their government.

13. Should the government pay farmers to convert to pressurized irrigation if a farming region has erosive soils?

It seems like a win-win since water is saved and erosion is decreased. The government actually does not have to pay for the entire irrigation system in order to induce farmers to convert to pressurized irrigation. Other benefits of pressurized irrigation such as reduced labor and water cost also factor into the economic decision to install a new irrigation system. Thus, a practice such as just paying for the mainline can be enough to cause farmers to make the switch. Economists can determine the level at which assistance can cause farmers to convert to pressurized irrigation systems. The alternative is to close down farms or pay for lawyers to prevent erosion through the legal system. The cost of legal action may be just as high as the cost of paying for mainlines.

14. At the government policy level, should economic analysis of irrigation systems include all environmental degradation due to irrigation?

At the policy level, environmental contamination and sustainability should be part of the analysis of irrigation systems and irrigation projects. The government is responsible for the public welfare; thus, the government should consider the impact of irrigation practices on the environment.

15. Calculate cotton yield if the required depth of applied water in a region is 120 cm, actual applied water depth is 80 cm, and the maximum yield is 1200 t/ha. Look at Fig. 2.6. Is your answer realistic?

$$Y_a = Y_{\max}\left(1 - \frac{\%\Delta Y}{\%\Delta AW}\frac{AW_{req} - AW_{CWPF}}{AW_{req}}\right)$$
$$= 1,200\left(1 - 0.75\frac{120 - 80}{120}\right) = 900\ \ kg/ha$$

16. There is 10 % yield loss due to pest stress in addition to the loss due to water stress calculated in question 15. What is the expected yield?

The reduced yield is multiplied by the pest stress

$$900\ \ \mathrm{kg/ha}(0.9) = 810\ \ \mathrm{kg/ha}$$

17. The expected annual income from a project is \$600/yr for 5 years, and the project requires a \$2,000 investment. Determine whether or not to invest in the project at a 9 % required rate of return.

The present value of the payments is 2,334 so the decision is made to invest in the project.

18. Determine the present value of the income from cotton 4 years from now for full water application and a required rate of return of 7 %. (1) Assume no inflation. (2) Assume inflation of 5 % per year in costs and no inflation in the selling price of cotton. Current yearly income is \$854.25, and current yearly expenses are \$721.16

The present value of year 4 profit with no inflation is \$451. The present value of year 4 with inflation is -\$75.

19. Calculate the cost of energy to pump water for sprinkler irrigation alfalfa for 100 ha of land with an irrigation system that requires a 30 m pressure head, and a pump that operates at 80 % efficiency. The cost of energy is \$0.10/kW-hr. Assume that the sprinkler irrigation system water application efficiency is 75 %. Assume that alfalfa requires 5 m of water per year.

The irrigation system must apply 5000 mm/0.75 = 6,667 mm

$$E = \frac{0.0272(6,667)(30\ m)}{0.8}(100\ ha)$$

$$= 680,000\ kW - hr(0.1/\text{kW} - \text{hr}) = \$68,000/\text{season}$$

20. Develop a rule of thumb for pump flow rate requirement for an area that has a max ET = 10 mm/d in terms of gpm/acre? The expected irrigation efficiency is 70 %, and the expected downtime is 10 %. What pump flow capacity will be required for a 10 acre farm? If the farm is divided into five irrigation zones, then what pump size is required?

ET rate	10	mm/day
	100	m3/day-ha
	69.44444	L/min/ha
	7.425143	gal/min/acre
Irr eff	0.7	
Rule of thumb	10.60735	gal/min/acre

For a 10 acre farm, the pump should produce 106 gal/min

The same pump size is required for any number of zones.

21. For the flow rate in question 20, what motor horsepower is required and wattage is required if the pump operates at 100 ft pressure and is 85 % efficient?

$$\text{HP} = \text{GPM*FT}/(3960\text{*Eff}) =$$

Acres	10	
Pump flow	106.0735	gpm
Pressure	100	ft
Pump eff	0.85	
HP	3.151321	

22. Using the *Cash flow diagram* worksheet, redo the analysis for alfalfa planted every year.

	A	B	C	D	E	F	G	H	I	J	K	L
10	0	-	-		-	-3000	0	0	0	0	\$ (3,000.00)	\$ (3,000.00)
11	1	Alfalfa	Alfalfa	1000	0.07	-300	1070	-208	-108	-210	\$ 244.00	\$ 225.93
12	2	Alfalfa	Alfalfa	1000	0.07	-300	1145	-216	-117	-221	\$ 291.44	\$ 249.86
13	3	Alfalfa	Alfalfa	1000	0.07	-300	1225	-225	-126	-232	\$ 342.57	\$ 271.95
14	4	Alfalfa	Alfalfa	1000	0.07	-300	1311	-234	-136	-243	\$ 397.67	\$ 292.30
15	5	Alfalfa	Alfalfa	1000	0.07	-300	1403	-243	-147	-255	\$ 457.03	\$ 311.05
16	6	Alfalfa	Alfalfa	1000	0.07	-300	1501	-253	-159	-268	\$ 520.96	\$ 328.29
17	7	Alfalfa	Alfalfa	1000	0.07	-300	1606	-263	-171	-281	\$ 589.79	\$ 344.14
18	8	Alfalfa	Alfalfa	1000	0.07	-300	1718	-274	-185	-295	\$ 663.89	\$ 358.68
19	9	Alfalfa	Alfalfa	1000	0.07	-300	1838	-285	-200	-310	\$ 743.63	\$ 372.00
20	10	Alfalfa	Alfalfa	1000	0.07	-300	1967	-296	-216	-326	\$ 829.43	\$ 384.19
21	11	Alfalfa	Alfalfa	1000	0.07	-300	2105	-308	-233	-342	\$ 921.73	\$ 395.31
22	12	Alfalfa	Alfalfa	1000	0.07	-300	2252	-320	-252	-359	\$ 1,021.00	\$ 405.45
23	13	Alfalfa	Alfalfa	1000	0.07	-300	2410	-333	-272	-377	\$ 1,127.74	\$ 414.67
24	14	Alfalfa	Alfalfa	1000	0.07	-300	2579	-346	-294	-396	\$ 1,242.49	\$ 423.02
25	15	Alfalfa	Alfalfa	1000	0.07	-300	2759	-360	-317	-416	\$ 1,365.84	\$ 430.57
26	16	Alfalfa	Alfalfa	1000	0.07	-300	2952	-375	-343	-437	\$ 1,498.40	\$ 437.37
27	17	Alfalfa	Alfalfa	1000	0.07	-300	3159	-390	-370	-458	\$ 1,640.83	\$ 443.47
28	18	Alfalfa	Alfalfa	1000	0.07	-300	3380	-405	-400	-481	\$ 1,793.84	\$ 448.91
29	19	Alfalfa	Alfalfa	1000	0.07	-300	3617	-421	-432	-505	\$ 1,958.20	\$ 453.74
30	20	Alfalfa	Alfalfa	1000	0.07	-300	3870	-438	-466	-531	\$ 2,134.70	\$ 458.00
31												\$ 4,448.88

Chapter 3: Solution

1. What is the textural class name of a soil that has 40 % clay, 20 % sand, and 40 % silt?

This is a point between the four different classifications

2. What is the textural class name of a soil that has 35 % clay, 15 % sand, and 50 % silt?

   Silty clay loam

3. Download the Soil Water Characteristics calculator from the website listed in the References, and calculate the field capacity and permanent wilting point for the soil described in question 2. Use the field capacity and permanent wilting point values to calculate AWC.

   Field capacity = 38 %
   Permanent wilting point = 22 %
   AWC = 16 %

4. Use the Soil Water Characteristics Calculator to determine whether Field Capacity or Permanent Wilting Point changes more with soil compaction and explain why. What is the percent change from Loose to Hard.

Field capacity changes more, typically by 4 %. This is because larger macropores are lost by compaction.

5. Use the Soil Water Characteristics Calculator to evaluate changes in Field Capacity or Permanent Wilting Point from zero to 8 % organic matter. Make sure your salinity is below 5 dS/m. Is the change greater for a sandy loam or a clay? Does the change increase the AWC?

There is almost no change in PWP or FC for a clay because clay has similar pore sizes as organic matter. Change in PWP is approximately 6 % and change in FC is approximately 13 % for a sandy loam with 8 % increase in organic matter. Adding organic matter does not increase the AWC for clay, but it increases by 7 % for a sandy loam.

6. A soil sample is removed from the field and weighed (130 g). The soil is then dried and the weight is 100 g. What is the gravimetric water content?

$$\theta_{grav} = \frac{m_{water}}{m_{dry\ soil}} = \frac{30\ g}{100\ g} = 0.30 = 30\ \%$$

7. If gravimetric water content θ_{grav} is 30 % and bulk density ρ_b is 1.30 g/cm^3, then what is the volumetric water content? What is the porosity?

$$\theta_V = \theta_{grav} \times \rho_b = 0.3 \times 1.3 = 0.39 = 39\ \%$$

$$\phi = 100 - \left(\frac{\rho_b}{\rho_p}\right)*100 = 100 - \left(\frac{1.3}{2.65}\right)*100 = 51\ \%$$

8. What is the FC, PWP, and AWC of a sandy loam? Use the Soil Water Characteristics Calculator (0 % and 4 % organic matter, salinity = 3 dS/m, gravel = 0 %, and compaction is normal). Discuss the impact of properties other than soil texture on hydraulic properties.

   FC = 0.06
   PWP = 0.14
   $AWC = \theta_{FC} - \theta_{PWP} = 0.14 - 0.06 = 0.08$

There is a dramatic change in hydraulic properties with organic matter, salinity, and compaction. This is why it is necessary to know more than just the soil texture.

9. What is the depth of readily available water (RAW) for sandy loam (4 % organic matter) if the effective root zone depth is 1.5 m and MAD = 0.4?

Estimate AWC for a sandy loam as 0.10 based on the Soil Water Characteristics Calculator.

$$\text{TAW} = \text{AWC*z} = 0.10*1.5 = 0.15\ \text{m}$$

$$\text{RAW} = \text{TAW*MAD} = 0.15\ \text{m}*0.4 = 0.06\ \text{m} = 6\ \text{cm}.$$

10. Define MAD and answer the following questions. What is meant by 40 % MAD? Does 40 % MAD have a water content closer to PWP or FC?

MAD is defined as the management allowed depletion and is the maximum allowable percent depletion of the AWC. A 40 % MAD refers to the fact that a maximum of 40 % of the AWC can be depleted before irrigation must take place. Thus, 40 % MAD has a water content closer to field capacity.

11. What is the percent depletion if measured water content is 19 %, field capacity is 25 % and permanent wilting point is 10 %? If the MAD is 50 %, at what water content must the next irrigation take place? If the root zone depth is 1.5 m, then what depth of available water remains for plant use before the next irrigation? If evapotranspiration rate is 1 cm/day, then what is the maximum length of time before the next irrigation?

$$\%Dep = \frac{(\theta_{FC} - \theta)}{(\theta_{FC} - \theta_{PWP})}*100 = \frac{(25 - 19)}{(25 - 10)}*100 = 40\ \%$$

The next irrigation must take place before the soil reaches a water content of 17.5 %.

$$\text{TAW} = \text{AWC*z} = 0.15*1.5 = 0.225 \text{ m}$$

$$\begin{aligned}\text{RAW} &= \text{TAW*MAD} = 0.225 \text{ m}*0.5 = 0.113 \text{ m} \\ &= 11.3 \text{ cm.}\end{aligned}$$

$$\begin{aligned}\text{Available water} &= \text{RAW} - (\theta_{FC} - \theta)*z \\ &= 0.113 - (0.25 - 0.19)*1.5 = 0.023 \text{ m} \\ &= 2.3 \text{ cm}\end{aligned}$$

There are 2.3 cm remaining and ET is 1 cm/day, so the irrigation should take place within 2 days.

12. How much irrigation water (ft) should be applied in the next irrigation if porosity is 50 %, field capacity is 27 %, and permanent wilting point is 12 % in all layers? Measured soil water content in the upper 4 ft of soil (root zone) is as follows: 0–1 ft = 21 %, 1–2 ft = 22 %, 2–3 ft = 17 %, and 3–4 ft = 22 %. Assume that the irrigation efficiency is 100 %. Redo assuming that the irrigation efficiency is 80 %.

$$\begin{aligned}D_r = (&1\text{ft}(.27 - .21) + 1\text{ft}(.27 - .22) + 1\text{ft}(.27 - .17) \\ &+ 1\text{ft}(.27 - .22)) = .65 \text{ ft}\end{aligned}$$

If irrigation efficiency is 100 %, then 0.65 ft of water should be added.

If irrigation efficiency is 80 %, then 0.65/0.8 = 0.81 ft of water should be added.

13. What is the depth of available water in the root zone if the readily available water in the root zone 1 week ago was 10 cm and the rate of evapotranspiration was 1 cm/day? During this time, a storm added 2 cm water to the soil. When should the next irrigation take place?

$$\text{Soil water available} = 10 \text{ cm} - 7*1 \text{ cm} + 2 \text{ cm} = 5 \text{ cm.}$$

The next irrigation should take place within 5 days.

14. Use the Web Soil Survey (WSS) to find a soil at the agricultural field station in your area (or a location specified by the instructor) and repeat Example 3.8 for your soil. First, go to the WSS URL listed in the References and click "Start Web Soil Survey" in the upper right corner. There are four tabs at the top of WSS. Find your location under the "Area of Interest (AOI)" tab. You can make the scaling process faster by using your mouse and outlining the location you are interested in. Then outline one field at the research station with the red area of interest rectangle tool (Second button from right along top). After outlining the area of interest, press the Soil Map tab at the top. The soils in the AOI are listed on the left. Click on the soil name for a short description of that soil. For a more extensive description, go to the Natural Resources Conservation Service Soil Series Descriptions at the URL listed in the References. Type in only the name of the series, but not the texture. Define the field capacity and permanent wilting with the Soil Water Characteristics Calculator Assume an appropriate crop in your area. Then calculate the RAW of the soil based on the MAD (Table 3.3) and root zone depth of the crop. Find the root zone depth in Table 3.2.

Answers will vary

15. Estimate the long-term ponded steady-state infiltration rate for a sandy clay loam with the Soil Water Characteristics Calculator.

Assume that the wetting front depth goes to infinity; thus, the steady state infiltration rate is equal to the saturated hydraulic conductivity for sandy loam soil, 25 mm/hr.

16. For the following infiltration data, determine the SCS intake family as shown in Example 3.3.

Time (min)	Infiltrated depth (cm)
0	0.6985
5	1.33
10	1.79
50	4.55
100	7.33
150	9.82
200	12.13

First, subtract out the initial infiltration into cracks.

Intake without cracks	
Time (min)	cm
0	0
5	0.63
10	1.09
50	3.85
100	6.64
150	9.12
200	11.43

Take the natural log values of the data

Natural log of time	Natural log of infiltrated depth
1.61	−0.459
2.30	0.085
3.91	1.348
4.61	1.892
5.01	2.211
5.30	2.437

The graph of natural log of depth vs. natural log of time is constructed and trendline is used to find the slope.

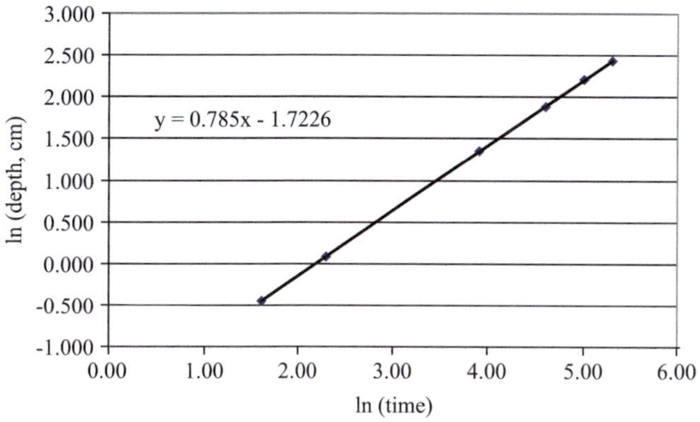

The slope is 0.785, which is equal to b.

Coefficient a is found by substituting 0.785 into the modified form of Eq. 3.19 as follows and selecting a time with a corresponding depth.

$$a = \frac{i}{t^b} = \frac{6.64}{100^{0.785}} = 0.1787$$

Thus the equation for this infiltration curve is

$$\mathrm{i} = \mathrm{at^b} + \mathrm{c} = 0.1787\ \mathrm{t}^{0.785} + 0.6985$$

This equation corresponds with the intake family 1.0 curve.

17. Calculate the depth of infiltration and infiltration rate over time and plot the two curves for an intake family 1.0 soil. Plot your infiltration rate curves in terms of cm/hr and in/hr and calculate out to 1000 minutes. At what time is the intake rate equal to 1.0 in/hr? Is this the steady state intake rate?

The cumulative infiltration is found with the following equation for the intake family 1.0 soil.

$$\mathrm{i} = \mathrm{at^b} + \mathrm{c} = 0.1786\ \mathrm{t}^{0.785} + 0.6985$$

The intake rate is calculated with the derivative.

$$\mathrm{di/dt} = \mathrm{ab(t)}^{(\mathrm{b}-1)} = 0.785{*}0.1786{*}\mathrm{t}^{(0.785-1)} = 0.14\ t^{-0.215}$$

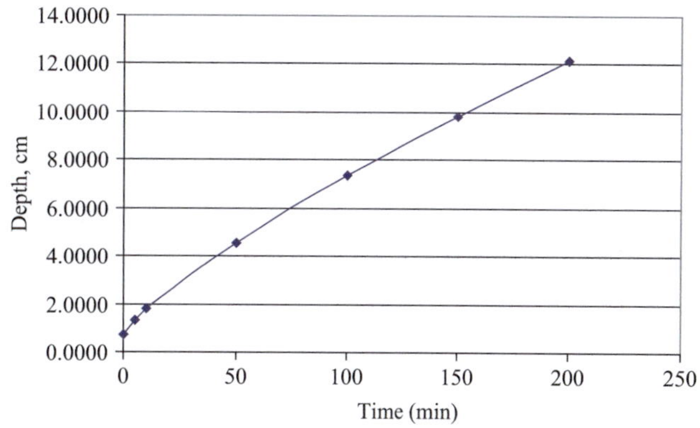

Cumulative infiltration vs. time for intake family 1.0 soil.

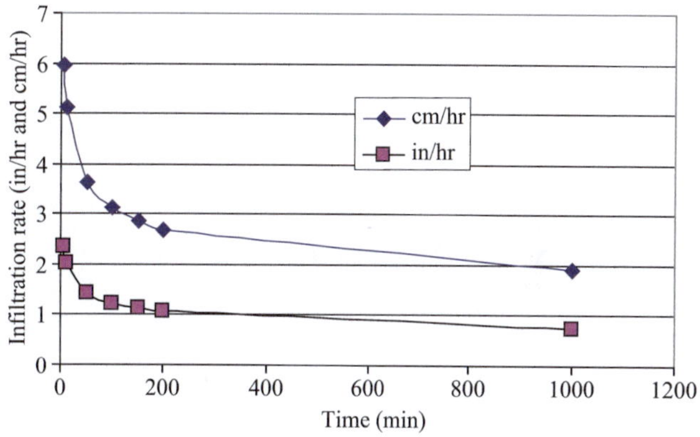

Infiltration rate vs. time for intake family 1.0 soil.

The intake rate is equal to 1.0 in/hr at approximately 200 minutes. This appears to be a steady state intake rate. The fact that the model shows a decreasing intake rate as time increases is a relic of the equation and not necessarily the behavior of the soil. The equation is probably accurate up to 200 minutes when 12 cm has infiltrated into the soil, a typical depth of irrigation.

18. Calculate the moisture contents in Table 3.11 in the upper four layers if the calibration slope is changed to 0.2. If permanent wilting point is 11 % in the upper four layers, calculate the total available water in the upper four layers assuming that the neutron measurements were taken a few days after irrigation. Calculate the readily available water in the upper four layers if MAD is 0.45.

Depth	Reading	FC	PWP	FC – PWP	Thickness	TAW
40 cm	13965	0.175582	0.11	0.065582	0.5	0.032791
60 cm	14962	0.190566	0.11	0.080566	0.2	0.040283
80 cm	16963	0.220637	0.11	0.110637	0.2	0.055319
100 cm	18509	0.243871	0.11	0.133871	0.2	0.066936

The assumption is that soil water content a few days after irrigation is equal to FC.

Sum of TAW is 0.195

$$RAW = TAW(MAD) = 0.195(0.45) = 0.09\ \text{m}.$$

Chapter 4: Solution

1. What is the molecular mass (g/mole) of calcium carbonate ($CaCO_3$)?

$$\begin{aligned}&Ca - 40.1\ g/mole\\&C - 12\ g/mole\\&O - 16\ g/mole\\&\therefore\text{Molecular mass of } CaCO_3 = 14.1 + 12 + 3{*}16\\&\qquad = 100.1\ g/mole\end{aligned}$$

2. What is the molarity and ppm of Na^+ and Cl^- if 0.02 g NaCL is dissolved in 4 L of water?

The molecular weight for each component is:

$$\begin{aligned}&Na : 22.99\ g/mole\\&Cl : 35.43\ g/mole\\&NaCL : 58.42\ g/mole\end{aligned}$$

Calculate number of moles

$0.02\ g/58.42\ g/mole = 0.0034235$ moles of NaCl are in the water.

Calculate molarity

$$0.00345\ moles/4L = 8.56E - 04\ mol/L$$

Molarity is same for Na and Cl because there is a 1:1 ratio of Na and Cl to NaCl

Calculate ppm (mg/L) Cl

$$\left(\frac{0.000856\ moles\ Cl^-}{Liter}\right)\left(\frac{35.43\ g}{mole}\right)\left(\frac{1{,}000\ mg}{g}\right) = 303\ mg/L$$

Calculate ppm (mg/L) Na

$$\left(\frac{0.000856\ moles\ Cl^-}{Liter}\right)\left(\frac{22.93\ g}{mole}\right)\left(\frac{1{,}000\ mg}{g}\right) = 196.3\ mg/L$$

3. What is the concentration of salts (mg/L) in water with $EC_{iw} = 2.4$ dS/m?

$$\begin{aligned}&C_w = EC_{iw} \times 640\\&\therefore C_w = 2.4 \times 640 = 1{,}500\ mg/L\end{aligned}$$

4. What is the soil salinity (mg/L) at saturation if the saturated paste extract EC_e is 4 dS/m?

mg/L = 4 dS/m * 640 = 2,500 mg/L.

5. How many moles of sodium chloride "NaCl" are required in 4 L of water to develop a solution that has 1000 mg/L sodium? What is the concentration of chloride in the water?

$$\text{Molecular mass of NaCl} = 23 + 35.5 = 58.5\ g/mole$$

If the concentration of sodium = 1000 mg/L

$$\begin{aligned}\therefore\text{Mass of Sodium} &= \text{concentration(mg/L)*volume of solution(L)}\\&= 1000\ mg/L{*}4L = 4000\ mg = 4\ g - Na\end{aligned}$$

$$\text{No. of moles of sodium} = 4/23 = 0.174\ \text{mole} - Na$$

∵1 mole of sodium ions is produced for every mole of sodium chloride "NaCl"

$$\therefore\text{No. of moles of sodium chloride "NaCl"} = 0.174\ \text{mole} - NaCl$$

And, No. of moles of chloride $= 0.174\ \text{mole} - Cl = 0.174 {*}35.5 = 6.177\ g$

$$\begin{aligned}\text{Concentration of chloride} &= \text{mass of solute (mg)/}\\&\quad\text{volume of solution (L)}\\&= 6177/4 = 1544.3\ mg/L\end{aligned}$$

6. How many grams of NO_3 are dissolved in 4 L of water with a nitrate concentration of 20 mg/L?

$$\begin{aligned}\text{Mass of nitrate (mg)} &= \text{concentration (mg/L)*volume of solution (L)} = 20\ (mg/L){*}4(L)\\&= 80\ mg = 0.08\ g\end{aligned}$$

7. Calculate the mass of ammonium nitrate NH_4NO_3 dissolved in 200 L water to obtain a nitrate concentration of 30 mg/L in water.

$$\begin{aligned}&\text{Molecular mass of "}NO_3\text{"} = 14 + 3{*}16 = 62\ g/mole\\&\text{Molecular mass of "}NH_4NO_3\text{"} = 14 + 4{*}1 + 14 + 3{*}16\\&\qquad = 80\ g/mole\end{aligned}$$

$$\begin{aligned}\text{Mass of nitrate "}NO_3\text{" in mg} &= \text{concentration (mg/L)*}\\&\quad\text{volume of solution (L)}\\&= 30(mg/L){*}200\ (L)\\&= 6000\ mg = 6\ g - NO_3\end{aligned}$$

$$\text{No. of moles of } NO_3 = 6/62 = 0.097\ \text{moles} - NO_3^-$$

$\because$1 mole of nitrate NO_3^- ions is produced for every mole of ammonium nitrate NH_4NO_3

$\therefore$No. of moles of ammonium nitrate NH_4NO_3 = 0.097 moles – NH_4NO_3

Mass of ammonium nitrate NH_4NO_3 in g = molecular mass (g/mole)*no. of moles = 80 g/mole*0.079 moles = 7.76 g – NH_4NO_3

8. If saturated paste extract EC_e is 2 dS/m, then what is the osmotic potential ψ_s of the water if the water content θ is 15 % and the saturated water content θ_s is 45 %?

$$\psi_s = -3.6 \times EC_e \times \frac{\theta_s}{\theta}$$
$$\therefore \psi_s = -3.6 \times 2 \times \frac{0.45}{0.15} = -21.6\text{m} = -2.16\text{atmosphere}$$

9. Explain the effect of salinity in Fig. 4.1.

Salinity increases the negative potential for a given water content. Especially for a salt sensitive plant where water stress begins at a less negative water potential, there is very little available water in the soil.

10. Explain the function of the xylem and the phloem in the plant.

The xylem and phloem constitute the plant vascular system.

- Xylem: Transfer water and nutrients upward from the roots to the leaf surfaces.
- Phloem: Phloem carries sugar downward, which plants create through photosynthesis, from the leaves to the rest of the plant and to the roots.

11. Calculate the Gibb's free energy of water "G" in a cell if the concentration of sucrose in the cell is 0.3 mole sucrose/L water at a temperature of 20 °C. Express your answer in J/mole and J/kg (a.k.a. kPa).

$$mole/Lwater = \left(\frac{1000g}{L}\right)\left(\frac{mole}{18g}\right) = 55.56 mole/Lwater.$$

Mole fraction of water = 55.56/(55.56 + 0.3) = 0.995.

$$G = RT\ln(C) = 8.314 \times 293 \times \ln(0.995) = -13.12 J/mole.$$

Multiply by 55.56 to obtain the answer in J/kg (kPa)

$$G = RT\ln(C) = 55.56 \times 8.314 \times 293 \times \ln(0.995) = -728.83 J/kg(kPa) = -7.29 atm$$

12. Calculate yield for cotton for a growing season if average values of EC_e and water content during the growing season are 12 dS/m and 13.5 %, respectively. $\theta_{fc} = 0.2$, $\theta_{pwp} = 0.1$. $K_y = 0.85$. Max yield = 1,000 kg/ha. MAD = 0.55. Threshold EC_e is 7.7 dS/m and b is 5.2. $K_y = 0.85$ for cotton.

$$AWC = \theta_{FC} - \theta_{PWP} = 0.2 - 0.1 = 0.1$$
$$TAW = AWC \times z = 0.1 \times 1.5 = 0.15$$
$$D_r = TAW \times \%depletion = 0.15 \times 0.65 = 0.0975$$
$$K_{s-water} = \frac{TAW - D_r}{(1 - MAD)TAW} = \frac{0.15 - 0.0975}{(1 - 0.55) \times 0.15} = 0.778$$
$$K_{s-salt} = 1 - \frac{b}{100 \times K_y}(EC_e - EC_{e-t})$$
$$= 1 - \frac{5.2}{100 \times 0.85}(12 - 7.7) = 0.737$$
$$K_s = K_{s-water} \times K_{s-salt} = 0.778 \times 0.737 = 0.573$$
$$Y_a = \left(1 - K_y(1 - K_s)\right)Y_m$$
$$= (1 - 0.85 \times (1 - 0.573)) \times 1000 = 637\ kg/ha$$

13. Why does high sodium ruin some soils? What types of soils are most vulnerable?

Excess sodium (Na+) can lead to breakdown of clay particle structure and can clog the soil and reduce infiltration rate to nearly zero. The most vulnerable type of soils is clay because sodium cations are loosely attracted to negative charged clay layers and they maintain a hydration shell of approximately ten water molecules around them. The water molecules in the shells of hydration will force the clay layers apart.

14. Irrigation water has 230 mg/L sodium (Na+), 60.15 mg/L calcium (Ca++), and 24.3 mg/L magnesium (Mg++). If irrigation water salinity is 1,000 ppm, then what level of hazard is presented by sodicity?

Equivalent masses are given in Table 4.1. Calculate meq/L for each cation.

460 mg/L Na^+/23 mg/meq = 20 meq / L Na^+
20 mg/L Na^+/20.05 mg/meq = 1 meq / L Ca^{++}
12 mg/L Na^+/12.15 mg/meq = 1 meq/L Mg^{++}

$$SAR = \frac{|Na^+|}{\sqrt{\frac{|Ca^{++}|+|Mg^{++}|}{2}}} = \frac{20}{\sqrt{\frac{1+1}{2}}} = \frac{20}{1} = 20$$

The EC_{iw} of the irrigation water is 960 ppm/640 (dS/m/ppm) = 1.5 dS/m. From Table 4.4, there is a moderate to severe (more severe than moderate) salinity hazard.

15. Fill in the missing total potential values and show direction of water flow. Are these total potentials more likely to occur in the day or the night?

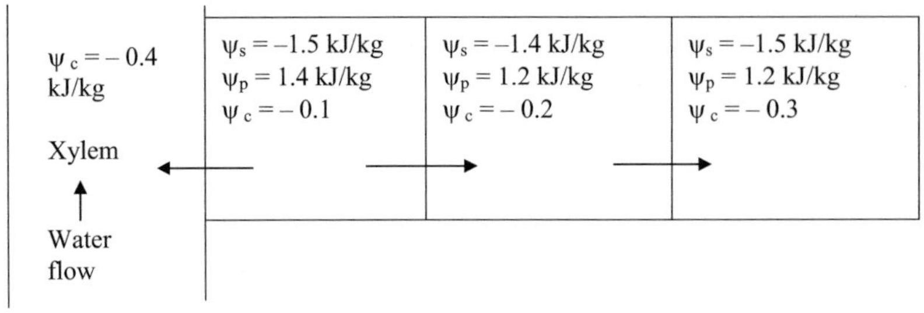

These total potential values are more likely to occur at night because turgor pressure is high and total potential is close to zero.

16. Determine the leachate salinity. Irrigation water salinity (EC_{iw}) = 2 dS/m. Applied water depth (d_{in}) = 1,250 mm/season. There is no precipitation during the growing season. Crop water demand (ET_c) = 1,000 mm/season. Average soil moisture content is the same at the beginning and end of the growing season.

$$LF = \frac{i - ET_c}{i} = \frac{1,250 - 1,000}{1,250} = 0.2$$

$$EC_{dw} = \frac{EC_{iw}}{LF} = \frac{2}{0.2} = 10\ dS/m$$

17. Irrigation water salinity (EC_{iw}) = 2 dS/m. Applied water depth (d_{in}) = 1,300 mm/season. Crop water demand (ET_c) = 900 mm/season and assume that plants extract 40 %, 30 %, 20 %, and 10 % of their water from the upper quarter, next quarter, next, and lowest quarter of the root zone, respectively. Calculate the salinity at the bottom of the root zone by treating the root zone as a whole and calculate salinity at the bottom of each of the 4 layers. Calculate the average salinity in the root zone by assuming that the field capacity is half of the saturated water content (divide average salinity in half for EC_e). Then calculate the leaching fraction that would be required to have this average salinity in the root zone with Eq. 4.20 and compare with the leaching fraction in this problem.

Calculate EC_{dw} treating the entire root as a whole:

$$LF = \frac{d_{in} - ET}{d_{in}} = \frac{1300 - 900}{1300} = 0.31$$

$$EC_{dw} = \frac{EC_{iw}}{LF} = \frac{2}{0.31} = 6.45\ dS/m$$

Calculate EC_{dw} by calculating the salinity at the bottom of each of the four quarters of the root zone:

$$LF_1 = \frac{1300 - 0.4 \times 900}{1300} = \frac{940}{1300} = 0.72$$

$$\Rightarrow \therefore EC_1 = \frac{EC_0}{LF_1} = \frac{2}{0.72} = 2.8\ dS/m$$

$$LF_2 = \frac{940 - 0.3 \times 900}{940} = \frac{670}{940} = 0.71$$

$$\Rightarrow \therefore EC_2 = \frac{EC_1}{LF_2} = \frac{2.8}{0.71} = 3.9\ dS/m$$

$$LF_3 = \frac{670 - 0.2 \times 900}{670} = \frac{490}{670} = 0.73$$

$$\Rightarrow \therefore EC_3 = \frac{EC_2}{LF_3} = \frac{3.9}{0.73} = 5.3\ dS/m$$

$$LF_4 = \frac{490 - 0.1 \times 900}{490} = \frac{400}{490} = 0.82$$

$$\Rightarrow \therefore EC_4 = \frac{EC_3}{LF_4} = \frac{5.3}{0.82} = 6.46\ dS/m$$

The calculated EC_{dw} is exactly the same for both calculations.

Calculate the average EC in the root zone for the calculation of the required leaching fraction.

$$EC_{ave} = (2 + 2.8 + 3.9 + 5.3 + 6.46)/5 = 4.1\ \text{dS/m}.$$

If the field capacity is approximately 50 % of saturated water content, then, the average EC_e in the soil is:

$$EC_{e(sat)} = EC_{ave} * (\theta_{fc}/\theta_{sat}) = 4.1 * 0.5 = 2.05\ \text{dS/m}$$

Calculate the leaching fraction required to have an average EC_e equal to 2.05 dS/m.

$$LF = \frac{EC_{iw}}{5(EC_e) - EC_{iw}} = \frac{2}{5(2.05) - 2} = 0.24$$

This calculated leaching fraction, 0.24, is less than the actual leaching fraction, 0.31, but it is reasonably close given all of the assumptions that were made.

18. Calculate the depth of irrigation water required (average for the field), IR, for cotton based on Eq. 4.23. The MAD is 50 %, the irrigation system efficiency is 60 %, the irrigation water EC_{iw} is 2 dS/m, and the TAW is 20 cm.

Max. soil salinity in the saturated paste extract (Table 4.4) for cotton with no yield reduction = 7.7 dS/m.

$$LF = \frac{EC_{iw}}{5(EC_e) - EC_{iw}} = \frac{2}{5(7.7) - 2} = 0.054$$

$$IR = \frac{100}{IE(1 - LF)} RAW = \frac{100}{60(1 - 0.054)} (0.5*20\ cm) = 17.6\ cm$$

Chapter 5: Solution

1. Atmospheric pressure is 100 kPa and the atmosphere is 2 % water by volume. What is the partial pressure of water in the atmosphere?

2 kPa

2. Why is relative humidity an important factor in evapotranspiration?

The energy gradient due to the vapor pressure difference between the leaf and the atmosphere drives the evaporated water from the plant canopy to the atmosphere

3. Calculate the relative humidity if the partial pressure, e_a, of water in the air is 1.1 kPa and T = 40 °C.

First, calculate the saturated vapor pressure $e_s = \exp\left(\frac{16.78(30)-116.9}{30+237.3}\right) = 4.25\ kPa$

Calculate the ratio of partial pressure of water to saturated vapor pressure.

$$RH = e_a/e_s = 1.1/7.4*100\% = 15\%$$

4. The density of water in the atmosphere at sea level is 0.018 kg/m^3, atmospheric pressure is 101 kPa, and temperature is 40 °C. What is the relative humidity?

The molar volume (mol/m^3) is

$$n = (18g/m^3)/(18g/mol) = 1.0 mol/m^3$$

Rearrange the ideal gas law and solve for partial pressure of water in the atmosphere

$$p = nRT/V = 1(mol/m^3)*8.314*313 = 2,600\ Pa = 2.6\ kPa.$$

First, calculate the saturated vapor pressure $e_s = \exp\left(\frac{16.78(30)-116.9}{30+237.3}\right) = 4.25\ kPa$

$$RH = 2.6/7.4*100\% = 35\%$$

5. Calculate the total resistance to water vapor transfer and the water vapor transfer conductance for a well-watered turf crop. The wind speed at 2 m elevation is 2 m/sec and the bulk surface resistance is 70 s/m. Calculate the maximum depth of water vapor transfer during 1 hour if the relative humidity is 60 %, temperature is 30 °C, and elevation is sea level.

$$r_s = 70\ s/m$$
$$r_{av} = 208/U_2 = 208/2 = 138\ s/m$$
$$r_{total} = r_s + r_{av} = 70 + 138 = 208\ s/m$$

$$h_v = \frac{1}{r_{total}} = \frac{1}{208} = 0.0048\ m/s$$

Calculate canopy vapor pressure.

$$e_c = e_s = \exp\left(\frac{16.78(30) - 116.9}{30 + 237.3}\right) = 4.25\ kPa$$

The atmospheric vapor pressure is the product of saturation vapor pressure and relative humidity

$$e_a = e_s*RH = 4.25\ kPa*0.6 = 2.54\ kPa$$

Calculate ET.

$$\begin{aligned} ET_0 &= \rho_a h_v \frac{(e_c - e_a)0.622}{P} \\ &= 1.29*0.0048 \frac{(4.25 - 2.54)0.622}{101.3} \\ &= 0.00017\ kg/(m^2 - \sec) \\ &= 0.00017\ mm/\sec = 0.61\ mm/hr \end{aligned}$$

6. Write a *sentence* describing each of the four terms in the crop evapotranspiration energy balance equation.

Latent heat of vaporization, LE is the energy used to convert liquid water to water vapor. Soil heat flux, G, is the quantity of energy that enters the soil. Net radiation, R_n is the total solar radiation minus the sum of the reflected short wave radiation and the long wave infrared radiation. Sensible heat flux, H, is the energy transfer from the crop to the atmosphere.

7. Calculate cumulative net radiation (MJ/m^2) over turf for 2 hours if average hourly radiation energy input (R_S) is 50 Langleys (cal/cm^2).

Calculate average incoming solar radiation (W/m^2). First convert Langleys to $MJ/m^2/hr$

$$50\ \mathrm{cal/cm^2/hr}{*}4.18\ \mathrm{Joules/calorie} = 2.1{*}10^2\mathrm{J/cm^2} = 2.1{*}10^6\mathrm{J/m^2} = 2.1\ \mathrm{MJ/m^2}$$

Calculate net radiation over turf (MJ/m^2) for 2 hr.

$$\mathrm{R_n} = 277.8{*}(-0.3 + 0.77{*}2.1) = 366\ \mathrm{W/m^2}{*}7200\ \mathrm{sec}/2\ \mathrm{hr}{*}10^{-6}\mathrm{J/MJ} = 2.6\ \mathrm{MJ/m^2}.$$

8. The plant canopy of a transpiring plant is generally cooler than the atmosphere above it. Why?

When water evaporates from the leaf, it removes heat from the leaf due to the latent heat of vaporization.

9. How would the height of the crop influence the 2 m wind speed measurement? Would a calculation of aerodynamic resistance based on 2 m wind speed without accounting for roughness elevation be in error? Why isn't this a problem for reference evapotranspiration calculations?

The wind speed would begin to increase at the roughness elevation. Thus, the wind speed would be zero for corn that is greater than 2 m tall. This factor is not a problem for reference evapotranspiration calculations because the reference crop is turf, which has a 10 cm (low) canopy height.

10. Explain why there is a difference between the stable wind speed and the unstable wind speed profile.

As with pipe flow, under turbulent conditions (unstable atmosphere) the velocity is distributed more uniformly in the radial direction in the pipe. Laminar flow has a large difference between the velocity of flow near the edges of the pipe and the velocity in the center of the pipe. The stable atmosphere has more laminar flow. The fact that the stable atmospheric condition is shown with a higher velocity is not significant, it is just how the picture was drawn in order to show a steeper gradient for the laminar flow.

11. Explain why all of the curves in Fig. 5.6 have the same aerodynamic resistance at 3 m/sec wind speed. Why does the −2 °C curve have a higher aerodynamic resistance than the +2 °C curve?

The lowest resistance to energy and vapor transfer is observed when surface/canopy temperatures are higher than air temperature (Ts − Ta = +2 °C): a buoyant condition with intensive mixing. There is no decrease in resistance as wind speed increases because maximum mixing is already occurring. The maximum resistance to energy and water vapor transfer occurs when wind speed is low and air temperature is higher than surface/canopy temperature; there are very few eddies, and energy and vapor must transfer by molecular diffusion.

12. Conduct dimensional analysis of Eqs. 5.15, 5.17, 5.19, and 5.20, and 5.22 and make sure that the units in the equations are consistent.

The units for Eq. 5.15 are consistent if the units for latent heat of vaporization are J/kg (J/mm-m), evaporation are m/s, and radiation are J/m^2. The units are W/m^2 on both sides

$$\lambda \mathrm{ET_0} = \mathrm{R_n} - \frac{\rho_a c_p h_a}{\Delta}[(e_{sc} - e_{sa})]$$

$$\frac{J}{mm\ m^2}{*}\frac{m}{s} = \frac{J/s}{m^2} - \frac{\left(\frac{kg}{m^3}\right)\left(\frac{J}{kg\ {}^\circ C}\right)\left(\frac{m}{s}\right)(kPa)}{\left(\frac{kPa}{{}^\circ C}\right)} = \frac{W}{m^2} \tag{5.15}$$

The units for Eq. 5.24 are consistent if the units for latent heat of vaporization are J/kg (J/mm-m) and the units for ET are mm/s.

$$\frac{\gamma \lambda ET_0}{\rho_a c_p h_v} = (e_c - e_a)$$

$$\frac{\left(\frac{kPa}{{}^\circ C}\right)\left(\frac{J}{mm\ m^2}\right)\left(\frac{mm}{s}\right)}{\left(\frac{kg}{m^3}\right)\left(\frac{J}{kg{}^\circ C}\right)\left(\frac{m}{s}\right)} = kPa \tag{5.17}$$

The units for Eq. 5.19 are consistent.

$$\lambda ET_0 = \frac{(R_n - G)\Delta + \rho_a c_p h_a[(e_{sa} - e_a) - (e_{sc} - e_c)]}{\left(\Delta + \gamma \frac{h_a}{h_v}\right)}$$

$$\left(\frac{J}{mm\ m^2}\right)\left(\frac{mm}{s}\right) = \frac{\left[\left(\frac{J/s}{m^2}\right) - \left(\frac{J/s}{m^2}\right)\right]\left(\frac{kPa}{{}^\circ C}\right) + \left(\frac{kg}{m^3}\right)\left(\frac{J}{kg{}^\circ C}\right)\left(\frac{m}{s}\right)kPa}{\left(\frac{kPa}{{}^\circ C}\right) + \left(\frac{kPa}{{}^\circ C}\right)} = \tag{5.19}$$

The units for Eq. 5.22 are consistent if the units for h_a are mm/s.

$$\lambda ET_0 = \frac{\Delta}{(\Delta+\gamma)}(R_n - G) + \frac{0.622\lambda h_a \gamma (e_{sa} - e_a)}{P(\Delta+\gamma)}$$

$$\left(\frac{J}{mm\ m^2}\right)\left(\frac{mm}{s}\right) = \frac{\left(\frac{kPa}{°C}\right)}{\left(\frac{kPa}{°C}\right)}\left[\left(\frac{J/s}{m^2}\right) - \left(\frac{J/s}{m^2}\right)\right] + \frac{\left(\frac{J}{mm\ m^2}\right)\left(\frac{mm}{s}\right)\left(\frac{kPa}{°C}\right)kPa}{kPa\left(\frac{kPa}{°C}\right)} =$$

Equation 5.23 is consistent if the units for the wind transfer function are W/(m^2 kPa)

$$\lambda ET_0 = \frac{\Delta}{(\Delta+\gamma)}(R_n) + \left(\frac{\gamma}{\Delta+\gamma}\right)f(u)(VPD)$$

$$\left(\frac{J}{mm\ m^2}\right)\left(\frac{mm}{s}\right) = \frac{\left(\frac{kPa}{°C}\right)}{\left(\frac{kPa}{°C}\right)}\left(\frac{J/s}{m^2}\right) + \frac{\left(\frac{kPa}{°C}\right)}{\left(\frac{kPa}{°C}\right)}\left(\frac{J/s}{m^2\ kPa}\right)kPa$$

The units for Eq. 5.24 are not consistent.

$$ET_{sz} = \frac{0.408\ \Delta\ (R_n - G) + \gamma\frac{C_n}{T+273}U_2(e_s - e_a)}{\Delta + \gamma(1 + C_d U_2)}$$

$$\left(\frac{mm}{d}\right) = \frac{\left(\frac{kPa}{°C}\right)}{\left(\frac{kPa}{°C}\right)}\left[\left(\frac{J/s}{m^2}\right) - \left(\frac{J/s}{m^2}\right)\right] + \frac{\left(\frac{kPa}{°C}\right)\left(\frac{1}{°C}\right)\left(\frac{m}{s}\right)kPa}{\left(\frac{kPa}{°C}\right)} =$$

13. In Eq. 5.19, the radiation term on the left and the aerodynamic term on the right are independent. Can Eq. 5.20 be broken down in the same way?

$$ET_{sz} = \frac{0.408\ \Delta\ (R_n - G) + \gamma\frac{C_n}{T+273}U_2(e_s - e_a)}{\Delta + \gamma(1 + C_d U_2)}$$

The radiation term is

$$\frac{0.408\Delta\ (R_n - G)}{\Delta + \gamma(1 + C_d U_2)}$$

The aerodynamic term is

$$\frac{\gamma\frac{C_n}{T+273}U_2(e_s - e_a)}{\lambda(1 + C_d U_2)}$$

The radiation term in equation includes wind speed in the denominator. The fact that the radiation term includes wind speed would imply that energy loss due to radiation is partially a function of wind speed. The aerodynamic term includes a dependence on temperature so this would imply that the aerodynamic losses are partially a function of temperature.

14. Compare the results of Examples 5.4 and 5.6. Discuss the reason for the difference.

The Hargreaves-Samani equation predicts a value that is 25 % lower than the standardized Penman equation.

$$(6.65 - 5.2)((6.65 + 5.2)/2)*100\% = 25\%$$

The most likely factor is wind speed. Lowering the wind speed to 1 m/sec yields the same result for the standardized Penman and the Hargreaves-Samani equations.

15. Use data from April 10, 2009, Tucson weather station data (http://ag.arizona.edu/azmet/data/0109eh.txt) to make hand calculations of daily ET with the equations. Then use the *Chapter 5 ET calculator* program and compare with your results. The latitude of Tucson is 33.95°N. Elevation is 655 m. Maximum temperature is 28 °C and minimum temperature is 8 °C. Relative humidity minimum is 7 % and relative humidity maximum is 57 %. Measured solar radiation is 27 MJ m^{-2} day^{-1}. Average wind speed at 3 m elevation is 2.8 m/sec.

Max T	28	C	Latitude	33.95	degrees	Elevation	655	m
Min T	8	C	DOY	100	days	Measured Rs	27	MJ/m2/
Tave	18.00	C	Phi	0.59		solar radiation		
es Max T	3.78	kPa	dr	1.00		Wind speed	2.8	m/s
es Min T	1.07	KPa	delta	0.13		Wind z	3	m
es-ave	2.43	kPa	w	1.66		Wind, 2 m	2.58	
RH max	0.57	fraction	Ra	35.24	MJ/m2/d			
RH min	0.07	fraction	Rs/Rso	1.004				
ea RH max	0.61							
ea RH min	0.26				Calculation of daily ET with standardized Penman equation			
ea-ave	0.44	kPa	DP ea	0.67	Input data required in white cells			
VPD (es-ea)	1.99		ave VPD	0.69				
Rso	26.89	MJ/m2/d	AZMET unit conversion			Max T	82	F
Rnl	8.82	MJ/m2/hr				Max T	27.78	C
Rns	20.79	MJ/m2/d	RS	641.7	Lang	Min T	47	F
Rn	11.97	MJ/m2/d	RS	26.82	MJ/m/d	Min T	8.33	C
P	93.79	kPa	Dewpt	34.3	F	Wind	6.3	MPH
DELTA	0.13		Dewpoint ea	1.28	C	Wind	2.82	m/s
gamma	0.06							
U2	2.58	m/s						
G	0	MJ/m2/d	USE AZMET AVERAGES for VPD				FALSE	
R(f) rad. Fn.	**0.634**							
w(f) wind Fn.	**0.989**							
denominator	**0.247**							
ETO	**6.5747**	**mm/day**						
ETO	**0.2588**	**in/day**						

16. Use hand calculations to calculate hourly ET over turf in Tucson, Arizona on DOY 100. Solar radiation is 2.84 MJ m^{-2} hr^{-1}. Average temperature is 27 °C, and wind speed is 4.2 m/s at 3 m elevation. Elevation is 655 m, relative humidity is 9 %. Longitude of Tucson is 113.18°. Tucson is in the Mountain Time Zone (Lm = 105). The longitude of the Mountain Time meridian is 105°. The period is from 2 to 3 p.m. Check your results against the spreadsheet calculations.

Tave	27	C	Latitude	33.95		Elevation	655	m
es	3.57	kPa	DOY	100		Rs measured	2.84	MJ/m2/
RH	0.09		Phi	0.593				
ea	0.32	kPa	dr	0.995		Wind speed	4.2	m/s
			delta	0.133		Wind z	3	m
Rso	2.43	MJ/m2/hr	w	1.661				
Rnl	0.53	MJ/m2/hr				Calculation of hourly ET during da		
			b	0.33		Input data required in white cells		
Rns	2.19	MJ/m2/hr	Sc	-0.027				
Rn	1.66	MJ/m2/hr	Lz	113.183	station degrees			
P	93.8	kPa	Lm	105	time zone meridian degrees			
DELTA	0.209		end time	15	end of hour (military time)			
gamma	0.062		w	0.790				
U2	3.87	m/s	w1	0.659				
G	0.166	MJ/m2/hr	w2	0.92		D		
ET	**0.679**	**mm/hr**	sin(q)	0.170		Dew		
			Ra	3.19	MJ/(m2 hr)			

17. Use average *monthly* values of ET_0, maximum temperature, and minimum temperature from a weather series in your area to calculate Fourier series coefficients for maximum and minimum temperature and ET_0.

Answers will vary.

Chapter 6: Solution

1. Reference ET_0 is 10 mm/day for 1 week, and the crop coefficient is 0.5. What is the depth of water required during 1 week?

The crop uses 5 mm/day or 35 mm/week.

2. Reference ET_0 is 7 mm/day, pan evaporation is 8 mm/day, and measured crop evapotranspiration is 5 mm, as measured by a lysimeter. What are the pan and reference ET_0 crop coefficients?

$$K_p = 5/8 = 0.62$$
$$K_m = 5/7 = 0.71$$

3. Explain the difference between the single crop coefficient and the dual crop coefficient.

The single crop coefficient is defined as the ratio between ET_c (crop evapotranspiration) and ET_o (reference evapotranspiration); thus, it combines transpiration and evaporation. On the other hand, the dual crop coefficient separates the basal crop coefficient (transpiration) and the soil evaporation coefficient (evaporation from the soil surface).

4. Explain the spikes in transpiration rate in Fig. 6.4 and the average crop coefficient in Fig. 6.5.

During the initial and development states, irrigation effects are greatest when the soil is wetted. As a result, evaporation spikes due to wet soil and irrigation during the early season, and can be as great as midseason ET. In addition, Fig. 6.5 indicates that average crop coefficient, which is the sum of crop basal coefficient and soil evaporation coefficient is greater than the basal crop coefficient during the early season because of the influence of evaporation.

5. Find the Kcb values for winter wheat (dual crop coefficient) in FAO56 Table 17, and adjust $K_{cb\text{-}mid}$ for average minimum relative humidity during mid-season and late season growth stages equal to 20 % and 30 %, respectively. Average wind speed at 2 m elevation is 2 m/sec during mid and late season growth stages. There is less than 10 % ground cover during the initial phase. There is high grain moisture at harvest. The crop is grown in California. Plot the linearized crop coefficient curve for the dual component model.

Because there is less than 10 % ground cover during the initial phase, $k_{cb\ ini} = 0.15$

$$k_{cb-mid} = k_{cb(table\ 17)} + [0.04(u_2 - 2) - 0.004(RH_{min} - 45)]\left[\frac{h}{3}\right]^{0.3}$$

From Table 17:

$$K_{cb\ mid} = 1/15$$

From Table 12:

$$h = 1$$

$$k_{cb\ mid} = 1.15 + [0.04(2-2) - 0.004(20-45)]\left[\frac{1}{3}\right]^{0.3}$$
$$\boldsymbol{k_{cb\ mid} = 1.22}$$

From Table 17, kcb-end for high moisture content at harvest is 0.30

$$k_{cb\ end} = 0.30 + [0.04(2-2) - 0.004(30-45)]\left[\frac{1}{3}\right]^{0.3}$$
$$\boldsymbol{k_{cb\ end} = 0.343}$$

Lengths of stages are given in FAO Table 11 for California.

Initial	20 days
Development	60 days
Midseason	70 days
Late	30 days

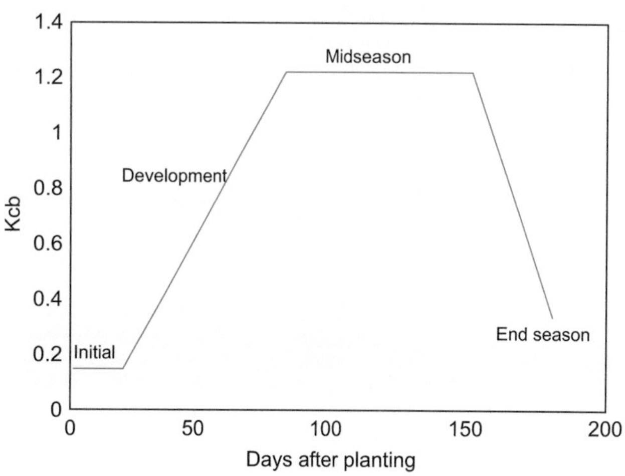

6. Repeat question 5, but plot the single crop coefficient curve (FAO Table 12) for winter wheat. Average depth of irrigation during the initial phase is 50 mm. Average number of days between irrigations during the initial period is 20 days. Minimum relative humidity is 20 % and average wind speed is 2 m/sec. The crop is machine harvested (late season K_c). During the early season, the crop is irrigated every 10 days, irrigation depths are greater than 40 mm, and the ET_0 is 7 mm/day. Also plug values into the Example 6.10 worksheet in order to solve the problem.

The stage lengths are found in FAO56 Table 6.11. The $K_{c\text{-}ini}$ values are found in FAO56 Table 6.12. The final Kc value is 0.25 because the crop is machine harvested. For early season K_{c-ini}, from Figure 30, the 10 day, 7 mm/day value is 0.21. Because the wetting is greater than 40 mm/irrigation, the value in Figure 30 is used, and the value from Figure 29 is not required. The kcb-mid is the same as in Example 5 since the equation adjustment is the same, which is reasonable since a full canopy cover has little to no evaporation and full transpiration. Thus Kc is 1.22.

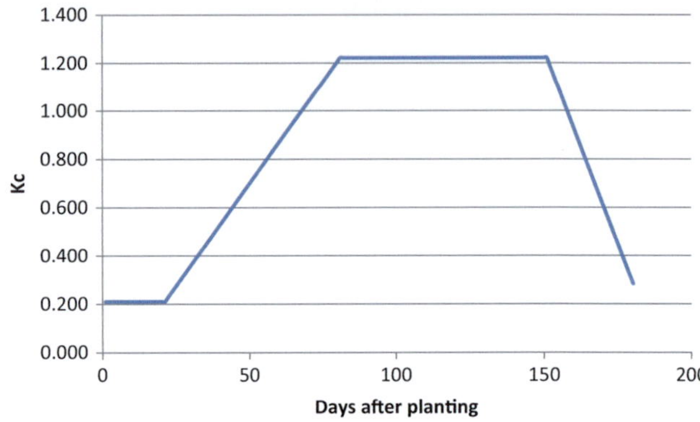

The spreadsheet solutions are in yellow

S	T	U	V	W	X
Winter wheat in California desert			Plant on Jan 1		
	Stage lengths and Kc from Tables 11 and 12 in FAO 56				
	stage length	Days after planting	Kc	Root (m)	Crop height
Initial	20	20	0.21	0.40	0.10
Dev	60	80			
Mid	70	150	1.15	0.80	1.00
End	30	180	0.25		

Adjust Kc ini based on the following values with Table 29 and 30	
Average depth applied during initial phase (mm)	50.00
Average number of days between irrigations	10.00
Maximum gross depth applied per day or irrigation (mm)	100.00
Typical midseason minimum relative humidity %	20.00
Typical midseason average daily wind speed (m/s)	2.00
Adjusted midseason crop coefficient	1.22

7. Calculate the single day crop basal transpiration for winter wheat 70 days after planting. Reference ET_0 is 7 mm/day and the crop stress coefficient is 0.8. Relative humidity is 20 %, and wind speed is 3 m/sec at 3 m elevation.

After 70 days of planting, winter wheat is in the midseason state, then:

$$k_{cb} = k_{cb(table\ 17)} + [0.04(u_2 - 2) - 0.004(RH_{min} - 45)]\left[\frac{h}{3}\right]^{0.3}$$

Calculate wind speed at 2 m, U_2.

$$U_2 = U_z \frac{4.87}{\ln(67.8z - 5.42)}$$

$$U_2 = 3\frac{4.87}{\ln(67.8*3 - 5.42)} = 2.76\ m/sec$$

From Table 17:

$$k_{cb\ mid} = 1.10$$

From Table 12:

$$h = 1$$

$$k_{cb} = 1.10 + [0.04(2 - 2.76) - 0.004(20 - 45)]\left[\frac{1}{3}\right]^{0.3}$$
$$\boldsymbol{k_{cb\ mid} = 1.16}$$
$$ET_c = ET_o * k_a$$
$$k_a = k_s\, k_{cb} + k_e$$

Then, assuming that k_e is neglected:

$$k_a = 0.8*1.16$$
$$k_a = 0.93$$
$$ET_c = 7mm/d*0.94$$
$$\boldsymbol{ET_c = 6.54\ mm/day}$$

8. Plot the winter wheat root growth curve as defined in FAO56. Initial depth is 0.4 m and the final depth is 0.8 m.

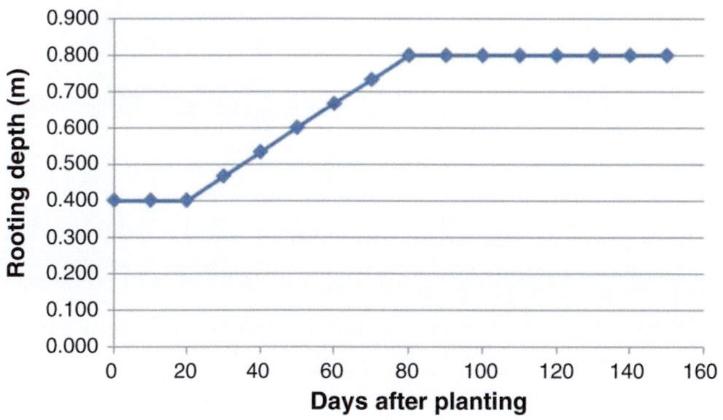

9. What is the reason that soil evaporation + basal transpiration cannot exceed K_{c-max}?

K_{c-max} is constrained by the energy available for evaporation; thus, the basal crop coefficient plus evaporation coefficient cannot exceed k_{c-max}.

10. Calculate K_{c-max} if relative humidity is 20 %, crop height is 0.5 m, and wind speed is 2 m/sec.

$$k_{c-max} = 1.20 + [0.04(u_2 - 2) - 0.004(RH_{min} - 45)]\left[\frac{h}{3}\right]^{0.3}$$
$$k_{c-max} = 1.20 + [0.04(2 - 2) - 0.004(20 - 45)]\left[\frac{0.5}{3}\right]^{0.3}$$
$$\boldsymbol{k_{c-max} = 1.25}$$

11. Explain the difference between REW and TEW.

The evaporation of REW during stage 1 is unrestricted by the soil. However, TEW takes place as the soil increasingly resists water transfer to the atmosphere. Thus, at the threshold between REW and TEW, the evaporation rate begins to decrease.

12. Field capacity is 20 % and permanent wilting point is 10 %. The depth of the surface layer is 0.11 m. REW = 7 mm. Calculate TEW. Calculate K_r for the surface layer depletion equal to 4 mm.

$$TEW = 1{,}000(\theta_{FC} - 0.5\theta_{PWP})Ze$$
$$TEW = 1{,}000(0.20 - 0.5*0.10)*0.11$$
$$TEW = 16.5\ mm$$

k_r = 1.0 because D_r does not exceed REW

13. Alfalfa has a low ET just after cutting and high ET just before cutting. For arid conditions with moderate wind, calculate alfalfa evapotranspiration just before and just after cutting if reference ET_0 is 10 mm/day. Use FAO 56 Table 12.

From Table 12 from FAO 56:
K_c = 0.4 after cutting
K_c = 1.15 before cutting

$$ET_c = ET_o * k_c$$
$$ET_c = 10mm/d*0.4$$
$$\boldsymbol{ET_c = 4\ mm/day}$$
$$ET_c = 10mm/d*1.15$$
$$\boldsymbol{ET_c = 11.5\ mm/day}$$

14. During the first week after planting watermelons, t_{max} and t_{min} are 32 °C and 8 °C, respectively. Calculate the number of growing degree days accumulated after 1 week.

THR and UPP for watermelon are 10 and 30, respectively.

Average ambient temperature does not exceed UPP so GDD is calculated as follows.

$$(t_{max} + t_{min})/2 - THR = (32 + 8)/2 - 10 = 10.$$

GDD accumulation after 7 days would be 7 * 10 = 70 days °C.

15. Calculate heat unit Fourier series K_c and ET_c curves for watermelon and corn for (Instructor selects year and location) weather data. You can download this data with the *Chapter 5 ET Calculator – Active Year Weather* worksheet from a city in Arizona or use weather from your home state. Click the *Run Weather Form* button. Calculate the Fourier W coefficients for the Tmax, Tmin and ETo curves in the *Chapter 5 ET Calculator – Fourier T and ETo* worksheet. Then, insert your W coefficients for Tmax, Tmin and ETo into the *Chapter 6 Crop ET and scheduling – Fourier T and ETo* worksheet. Then, copy your Tmax, Tmin and ETo values into the *Chapter 6 Crop ET and scheduling – ETo and temp – ch 5* worksheet. Soil is heavy-textured.

Results for Arizona:

The corn MAD will be adjusted based on the following equation with the expectation that the ETc for corn averages around 8.0 mm/day.

The p value (threshold for water depletion due to crop stress in Table 22) for corn is 0.5. Adjust with the following equation.

$$p = p_{Table\ 22} + 0.04(5 - ET_c) = 0.5 + 0.04(5 - 8) = 0.5 - 0.12 = 0.38$$

Adjust MAD downward for heavy-textured soil to 0.33

$$MAD = 0.38 - 0.05 = 0.33.$$

The following average monthly averages were acquired from AZMET in order to calculate the W coefficients. Note that the ET STD values at the bottom of each summary page are monthly sums in English units that must be divided by 30 and multiplied by 25.4 in order to calculate mm/day. Likewise, temperatures must be converted to metric.

The monthly averages (white columns) are averages of monthly data for each parameter
Spreadsheet calculates sin P and cos P, etc.., in yellow columns based on P in the blue column
Follow instructions below to perform regression on one of the three parameters

P = 2*PI()*DOY/365

Date	ETo	Tmax	Tmin	DOY	P	cos P	sin P	cos 2P	sin 2P	cos 3P	sin 3P	AZME ET (in
15-Jan	3.07	17.78	2.22	15	0.258	0.967	0.255	0.87	0.494	0.715	0.699	2.6
15-Feb	3.78	20.56	2.78	46	0.792	0.703	0.712	-0.01	1	-0.72	0.693	3.2
15-Mar	6.61	24.44	5.00	75	1.291	0.276	0.961	-0.85	0.531	-0.74	-0.67	5.6
15-Apr	8.86	28.33	7.22	105	1.807	-0.23	0.972	-0.89	-0.46	0.652	-0.76	7.5
15-May	10.04	30.56	11.67	136	2.341	-0.7	0.718	-0.03	-1	0.738	0.674	8.5
15-Jun	11.46	38.33	18.33	166	2.858	-0.96	0.28	0.843	-0.54	-0.66	0.753	9.7
15-Jul	9.57	36.11	22.78	197	3.391	-0.97	-0.25	0.878	0.479	-0.73	-0.68	8.1
15-Aug	8.62	35.56	22.22	228	3.925	-0.71	-0.71	0.004	1	0.703	-0.71	7.3
15-Sep	7.91	34.44	19.44	259	4.458	-0.25	-0.97	-0.87	0.486	0.69	0.724	6.7
15-Oct	6.38	31.11	10.56	289	4.975	0.26	-0.97	-0.87	-0.5	-0.71	0.706	5.4
15-Nov	4.13	24.44	6.11	319	5.491	0.703	-0.71	-0.01	-1	-0.72	-0.69	3.5
15-Dec	2.60	18.89	2.78	350	6.025	0.967	-0.26	0.87	-0.49	0.715	-0.7	2.2

You may need to add the Data Analysis Tool from Tools --> Add-ins
Multiple regression (Tools → Data Analysis → Regression (Excel 2003) or Data, Data Analysis, and Tools in Excel 200

Pivot water requirement | Fourier_Crop_data | number 16 | Fourier T and ETo | Number 16 corn schedule | Numb

These W coefficients were substuted into the Fourier T and ETo Worksheet as shown below. Note that the ET and Tmax curves aren't symmetrical and reflect the monsoon activity in JJ. It is more interesting that the Tmin curve does not seem to be affected by the monsoon.

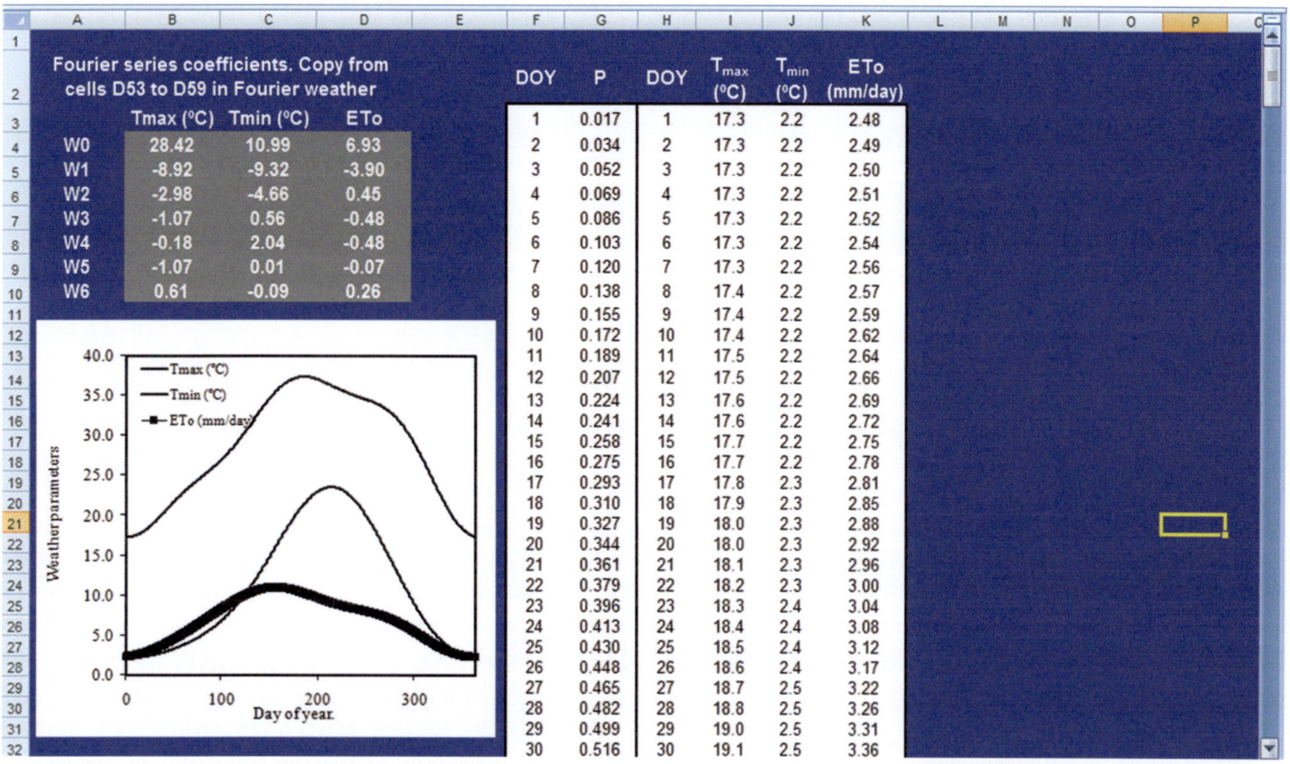

Fourier series coefficients. Copy from cells D53 to D59 in Fourier weather

	Tmax (°C)	Tmin (°C)	ETo
W0	28.42	10.99	6.93
W1	-8.92	-9.32	-3.90
W2	-2.98	-4.66	0.45
W3	-1.07	0.56	-0.48
W4	-0.18	2.04	-0.48
W5	-1.07	0.01	-0.07
W6	0.61	-0.09	0.26

DOY	P	DOY	T_{max} (°C)	T_{min} (°C)	ETo (mm/day)
1	0.017	1	17.3	2.2	2.48
2	0.034	2	17.3	2.2	2.49
3	0.052	3	17.3	2.2	2.50
4	0.069	4	17.3	2.2	2.51
5	0.086	5	17.3	2.2	2.52
6	0.103	6	17.3	2.2	2.54
7	0.120	7	17.3	2.2	2.56
8	0.138	8	17.4	2.2	2.57
9	0.155	9	17.4	2.2	2.59
10	0.172	10	17.4	2.2	2.62
11	0.189	11	17.5	2.2	2.64
12	0.207	12	17.5	2.2	2.66
13	0.224	13	17.6	2.2	2.69
14	0.241	14	17.6	2.2	2.72
15	0.258	15	17.7	2.2	2.75
16	0.275	16	17.7	2.2	2.78
17	0.293	17	17.8	2.3	2.81
18	0.310	18	17.9	2.3	2.85
19	0.327	19	18.0	2.3	2.88
20	0.344	20	18.0	2.3	2.92
21	0.361	21	18.1	2.3	2.96
22	0.379	22	18.2	2.3	3.00
23	0.396	23	18.3	2.4	3.04
24	0.413	24	18.4	2.4	3.08
25	0.430	25	18.5	2.4	3.12
26	0.448	26	18.6	2.4	3.17
27	0.465	27	18.7	2.5	3.22
28	0.482	28	18.8	2.5	3.26
29	0.499	29	19.0	2.5	3.31
30	0.516	30	19.1	2.5	3.36

The environmental data was copied from DOY 70 into the Irrigation schedule Worksheet, the MAD, LF and IE efficiency data were added, and the following graphs were generated for corn.

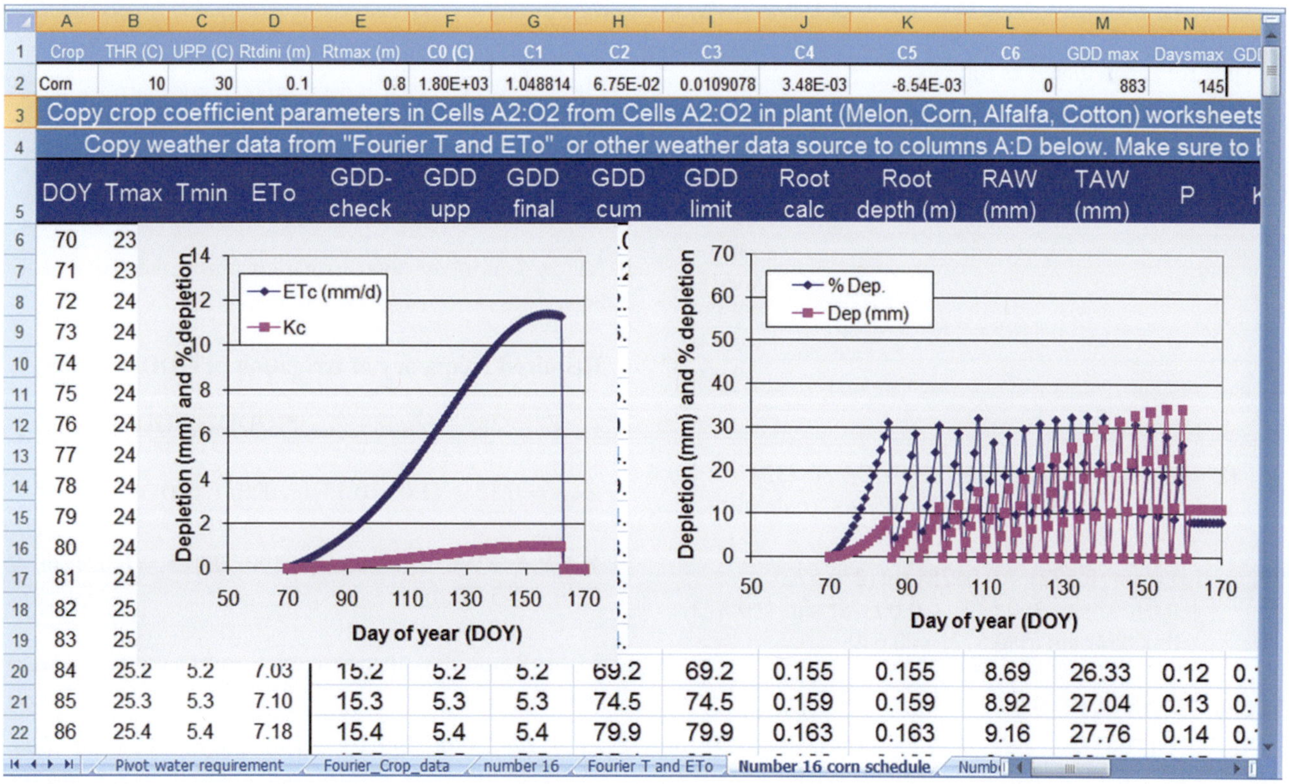

Crop	THR (C)	UPP (C)	Rtdini (m)	Rtmax (m)	C0 (C)	C1	C2	C3	C4	C5	C6	GDD max	Daysmax
Corn	10	30	0.1	0.8	1.80E+03	1.048814	6.75E-02	0.0109078	3.48E-03	-8.54E-03	0	883	145

Copy crop coefficient parameters in Cells A2:O2 from Cells A2:O2 in plant (Melon, Corn, Alfalfa, Cotton) worksheets

Copy weather data from "Fourier T and ETo" or other weather data source to columns A:D below. Make sure to

DOY	Tmax	Tmin	ETo	GDD-check	GDD upp	GDD final	GDD cum	GDD limit	Root calc	Root depth (m)	RAW (mm)	TAW (mm)	P
70	23												
71	23												
72	24												
73	24												
74	24												
75	24												
76	24												
77	24												
78	24												
79	24												
80	24												
81	24												
82	25												
83	25												
84	25.2	5.2	7.03	15.2	5.2	5.2	69.2	69.2	0.155	0.155	8.69	26.33	0.12
85	25.3	5.3	7.10	15.3	5.3	5.3	74.5	74.5	0.159	0.159	8.92	27.04	0.13
86	25.4	5.4	7.18	15.4	5.4	5.4	79.9	79.9	0.163	0.163	9.16	27.76	0.14

The depth of irrigation and ETc were 1149 and 654 mm, respectively, for corn.

Environmental data was copied from DOY 80 for the melons. The following figure was generated.

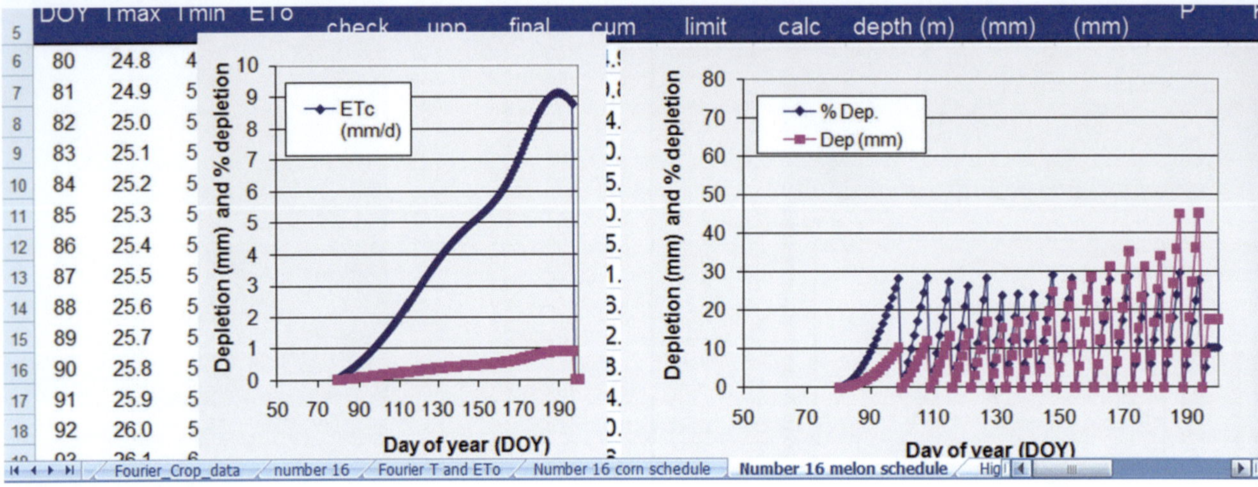

The depth of irrigation and ETc were 1121 and 588 mm, respectively, for melons.

16. As in Example 6.6, calculate corn K_c and ET_c 50 days after planting by hand (with a calculator) and show that calculated K_c and ET_c in the spreadsheet agree with the equations.

The corn GDD 50 days after planting (DOY 120) is 328, and C_0 is 1800.

Thus Kc is calculated as follows.

$$P = \pi^* GDD_{cum}/C_0 = \pi^* 328/1,800 = 0.572$$

C values for corn are in the following table.

C_1	1.05
C_2	0.0675
C_3	0.0109
C_4	0.00348
C_5	−0.00854
C_6	0

$$K_c = C_1{}^* \sin(P) + C_2{}^* \sin(2P) + C_3{}^* \sin(3P) + C_4{}^* \sin(4P) + C_5{}^* \sin(5P) + C_6{}^* \sin(6P)$$

$$K_c = 1.05^* \sin(0.572) + 0.0675^* \sin(2^*0.572) + 0.0109^* \sin(3^*0.572) + 0.00348^* \sin(4^*0.572) - 0.00854^* \sin(5^*0.572) = 0.640$$

In Cell O56, the Kc is 0.64.

The ETo on that day is 9.75. Thus, the ETc is 0.64 * 9.75 = 6.25, which agrees with the value in Cell P56.

17. According to the Fourier series GDD equations, calculate the rooting depth for melons when GDD = 500 and 1,000. AWC is 10 %, MAD is 0.5, and ET_c is 7 mm/day at the first rooting depth and 9 mm/day at the second rooting depth. What is the required frequency of irrigation at the two rooting depths.

Required frequency of irrigation at GDD = 500

$$Z_i = Z_{ini} + (Z_{max} - Z_{ini})^*(GDD_{cum}/GDD_{root})$$

$$Z_{500} = 0.15 + (1.0 - 0.15)(500/1,500) = 0.43 \text{ m}$$

$$RAW = AWC^*Z^*MAD = 0.1^*0.43^*0.5 = 0.021 \text{ m} = 2.1 \text{ cm.}$$

If ET_c is 7 mm/day, then irrigation must take place every 3 days.

Required frequency of irrigation at GDD = 1,000

$$Z_i = Z_{ini} + (Z_{max} - Z_{ini})^*(GDD_{cum}/GDD_{root})$$

$$Z_{500} = 0.15 + (1.0 - 0.15)(1,000/1,500) = 0.72 \text{ m}$$

$$RAW = AWC^*Z^*MAD = 0.1^*0.72^*0.5 = 0.021 \text{ m} = 3.6 \text{ cm.}$$

If ET_c is 9 mm/day, then irrigation must take place every 4 days.

18. ***For the 2008 Fourier series Tucson weather data, and the Fourier series crop coefficient for alfalfa,***

input the cutting cycle with peak ET into the High frequency alfalfa Irr. Worksheet. Then determine the required depth of irrigation per day with a center pivot irrigation system in order to keep up with crop water needs. Check to make sure that the percent depletion does not exceed 50 %. Use TAW = 140 mm. Assume a leaching fraction of 10 % and irrigation efficiency of 90 % in order to calculate the required gross application rate (divide net application rate in Worksheet by efficiency and (1 – LF)).

Add the alfalfa parameters into a new fourier worksheet.

Set the root depths in column J = 1.8 m(Cell \$D\$2).

For the alfalfa GDD calc in column I, the equation for setting GDD back to zero after each cutting must be modified as follows.

$$= \mathrm{IF}(\mathrm{I6} + \mathrm{G7} > \$\mathrm{M}\$2, 0, \mathrm{I6} + \mathrm{G7})$$

After these modifications are made, the Fourier spreadsheet appears as follows.

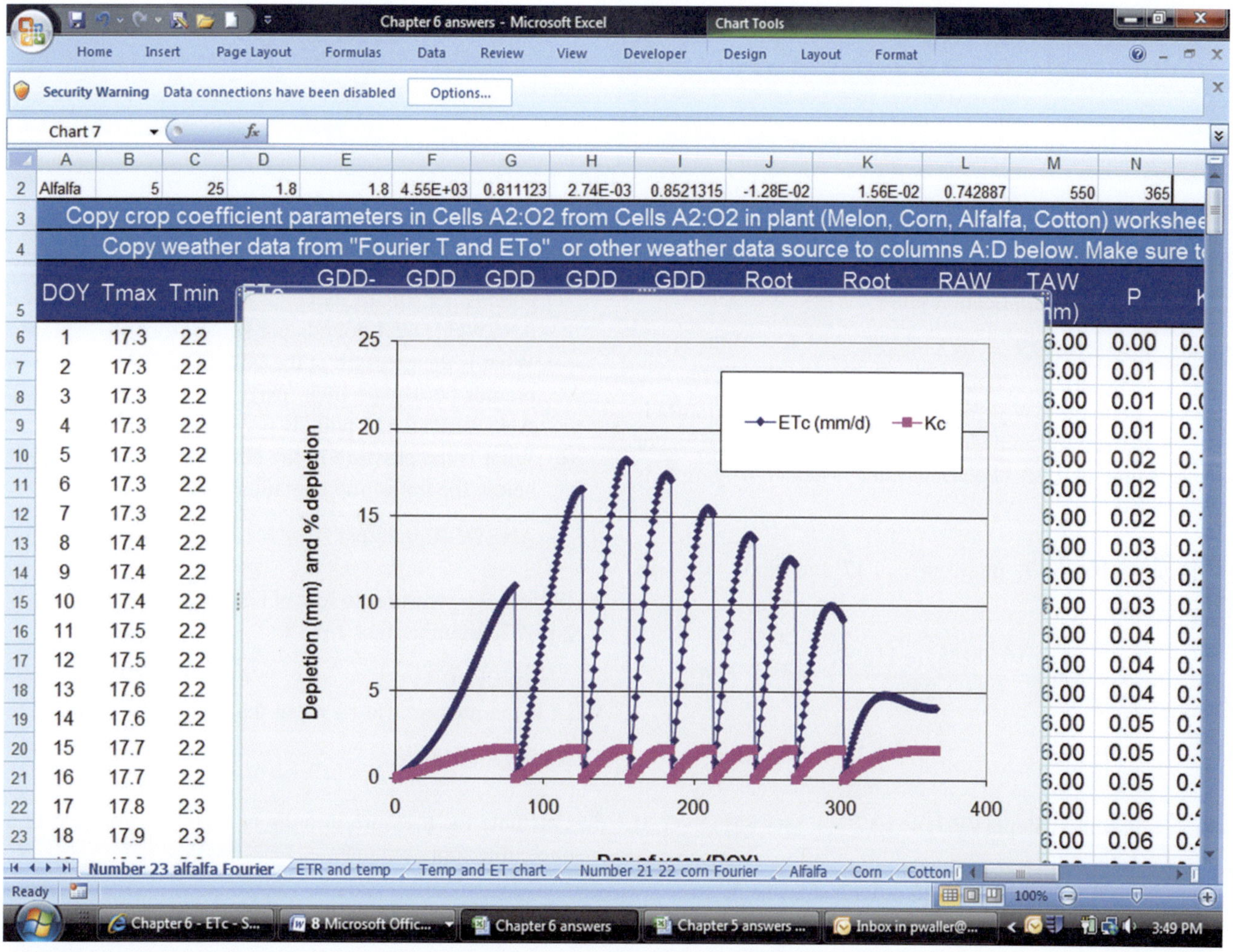

It appears that the peak water user is the third curve, so this data will be inserted into the high frequency alfalfa Worksheet. This period begins with DOY 122. The application rate is adjusted to 13.2 such that the final depletion percentage is approximately 50. There is not quite enough readily available water to carry the crop through the cycle. This is known because the curve drops below 0 % depletion. Perhaps a higher application rate is needed during the late cycle in order to carry the crop through. This worksheet is not equipped to evaluate this scenario.

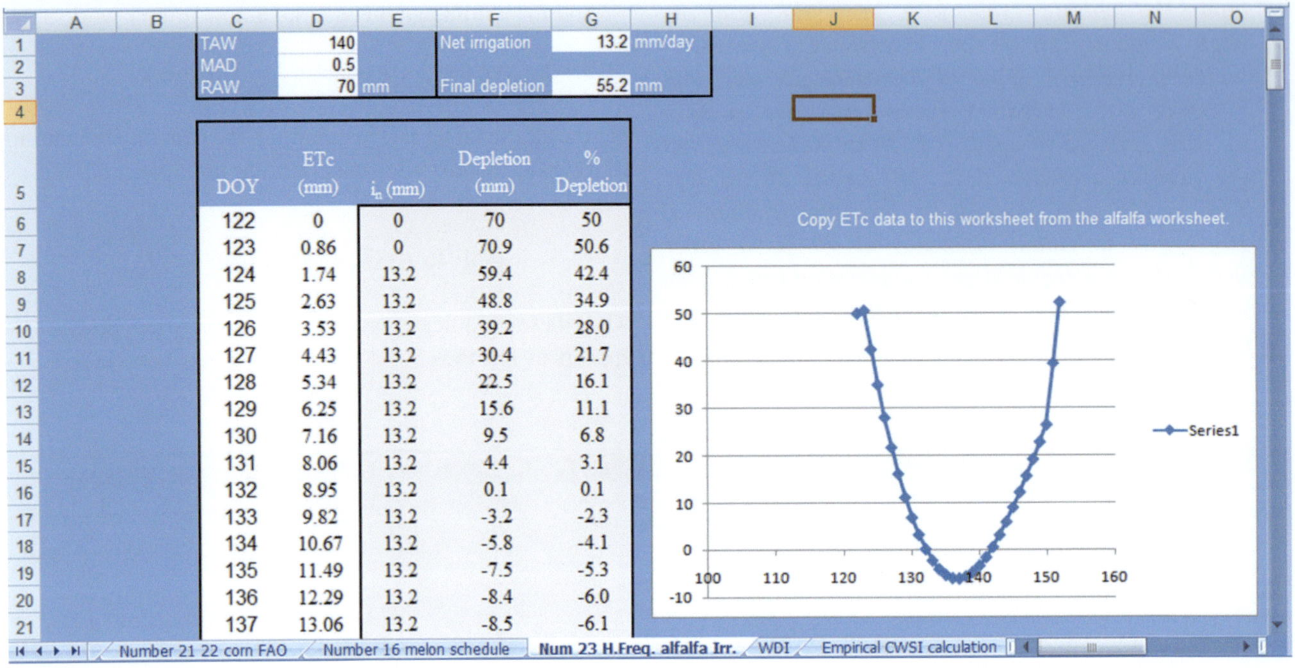

TAW 140; MAD 0.5; RAW 70 mm; Net irrigation 13.2 mm/day; Final depletion 55.2 mm

DOY	ETc (mm)	i_n (mm)	Depletion (mm)	% Depletion
122	0	0	70	50
123	0.86	0	70.9	50.6
124	1.74	13.2	59.4	42.4
125	2.63	13.2	48.8	34.9
126	3.53	13.2	39.2	28.0
127	4.43	13.2	30.4	21.7
128	5.34	13.2	22.5	16.1
129	6.25	13.2	15.6	11.1
130	7.16	13.2	9.5	6.8
131	8.06	13.2	4.4	3.1
132	8.95	13.2	0.1	0.1
133	9.82	13.2	-3.2	-2.3
134	10.67	13.2	-5.8	-4.1
135	11.49	13.2	-7.5	-5.3
136	12.29	13.2	-8.4	-6.0
137	13.06	13.2	-8.5	-6.1

Copy ETc data to this worksheet from the alfalfa worksheet.

If the required net application rate is 13.2 mm/day, then the required gross application rate is

$$13.2/0.9/0.9 = 16.2 \text{ mm/day}.$$

19. Calculate the CWSI for potatoes if the air temperature is 25 °C, canopy temperature is 24 °C, and actual vapor pressure is 1.32 kPa. Determine whether irrigation is needed.

From Table 6.17, intercept = 1.17 and slope = −1.83 for potatoes

$$e_{s-30} = \exp\left(\frac{16.78(25) - 116.9}{25 + 237.3}\right) = 3.17 \ kPa$$

$$\text{VPD} = \text{e}_\text{s} - \text{e}_\text{a} = 3.17 - 1.32 = 1.85 \ \text{kPa}$$

$$\text{dT}_\text{l} = \text{Intercept} + \text{Slope(VPD)} = 1.17 - 1.34(1.85) = -1.31$$

$$\text{T}_\text{a} + \text{intercept} = 25 + 1.17 = 26.17$$

Chapter 7: Solution

1. One cubic foot of water weighs 62.4 pounds. Assume you have a 1 ft × 1 ft × 1 ft container filled with water.
   (a) What is the pressure at the bottom of the container in pounds per square foot?
   ANSWER: 62.4 pounds per square foot
   (b) What is the pressure at the bottom of the container in pounds per square inch (psi)?
   ANSWER: 0.433 pounds per square inch (psi)
   (c) What is the pressure in the container at a depth 0.5 ft below the top of the container (psi)?

   ANSWER: 0.22 psi

2. What is the pressure (in feet of head) when the pressure is 2 psi? Remember that 1 psi = 2.31 ft of head.

   ANSWER: 0.22 psi

What is the pressure (in ft) when the pressure is 50 psi?

   ANSWER: 4.62 ft .

3. Calculate the pressure in units of feet at the bottom of a one cubic foot container.

   ANSWER: 1 ft of head

4. Calculate the pressure (psi and ft of head) at the bottom of the swimming pool (at sea level) that is 9 ft deep.

   ANSWER: 3.9 psi

5. Using the table below, write down the static pressure in psi and ft of head at each of the fittings found in Fig. 7.1:

	PSI	FT
A	8.66	20
B	8.66	20
C	32.5	75
D	32.5	75
E	39.4	91
F	39.4	91

6. What is the minimum acceptable inside pipe diameter for a Class 200 PVC pipe with a flow rate 90 gpm?

ANSWER: 2.7 inches is calculated, but in reality, you can't purchase that size pipe, so the real answer is 3 inches

7. Using Fig. 7.1, calculate the dynamic pressure (in ft of head) at each point of the fittings if the flow rate is 30 gpm through a 2″ Schedule 40 PVC pipe.

ANSWER:

Ft of Head	
A	20
B	20.14
C	75.75
D	75.93
E	92.11
F	105.3

8. Describe the relationships between the Z distribution, standard deviation, coefficient of variation, and mean, and describe how you can use those concepts to generate a normal distribution with mean equal to average sprinkler flow rate or application. Ignore pipe hydraulics, and use an equation to convert from Z values to the sprinkler distribution.

The Z distribution is the normal (0, 1) distribution, which means that it has mean zero and standard deviation 1. In order to generate this distribution in Excel, you use Z = NORMSINV(+RAND()), which generates a uniform 0,1 distribution and then converts it to the Z distribution. The Z distribution can be converted to a distribution of sprinkler application depths by $i = \bar{i} + Z^*CV^*\bar{i}$ where i is the observed application depth and CV (coefficient) is the observed standard deviation over the mean of application depths.

9. Generate 50 values that are normally distributed with mean 5 and standard deviation 1.

Use this equation = NORMINV(RAND(), mean, SD)
Copy the equation into 50 cells,
= NORMINV(RAND(), 5, 1)

10. Generate 50 values that are normally distributed with mean 50 and coefficient of variation 5 %.

Use this equation = NORMINV(RAND(), mean, mean*CV
Copy the equation into 50 cells,
= NORMINV(RAND(), 50, 50*0.05)

11. Changing no more than 5 pipe sizes, adjust the pipe sizes in the *Irrigation Simulation with VBA* worksheet (column D) such that the energy cost is equal to or less than pipe cost. State which pipe lengths were changed. Describe how the process of selecting pipe sizes might be automated in a computer program so that total annual energy cost + pipe cost was a minimum.

Sizes for lengths 20, 21, 9, 10, and 11 were increased by one size.

If I designed a computer program for the purpose of selecting pipe sizes, then the computer program would begin with setting pipe sizes based on maximum flow velocity, such as 1.5 m/sec. Then pipes at the boundaries between sizes would be sequentially increased or decreased by one pipe size until the sum of annual value of total pipe cost and energy cost were minimized.

12. Observe the spatial variation of yield for sprinkler CV (cell E1) 0.05 and 0.5 in the *Irrigation Simulation with VBA* worksheet and observe the effect on profit, rounding to the nearest $/ha. In the high variability case, you might get a flow rate that is negative. Put an IF statement in the spreadsheet that prevents negative flow rates. In each case, observe whether the major cause of application variability and yield variability is due to pressure variation in the pipeline or nonuniformity of application.

CV	Profit
0.05	$124/ha
0.5	$104/ha

In the first case, the major cause of variation in pressure variation. In the second case, the major cause of variation is nonuniformity of application.

13. Using the *Standard deviation Fig. 7.15* worksheet, increase the standard deviation to 200 LPH, and report the profit. If the standard deviation is 200, then what is the coefficient of variation at the first sprinkler (column C). How does using the standard deviation rather than the coefficient of variation change the distribution along the pipeline. In your estimation, which is more representative of variation along a pipeline and why.

$116/ha

The coefficient of variation is the standard deviation over the mean: 200/1000 * 100 % = 20 %. With the coefficient of variation, the standard deviation increases with sprinkler flow rate along the pipeline. The standard deviation, the deviation is not a function of sprinkler flow rate. In most cases, it would seem that the coefficient would be a better representation since higher overall application would have a correspondingly larger variation in application rates.

14. Using the *Standard deviation Fig. 7.15* worksheet, decrease the standard deviation to 20 LPH, and increase the pipe sizes so that energy cost decreases to $37/hr and pipe cost increases to $61. Report the profit. Does the overall profit change significantly from question 7–12, with low CV? Why or why not.

$122/ha

The profit is approximately the same because the savings in energy cost is offset by the increase in pipe cost.

15. Increase the number of sprinklers in the *Irrigation Simulation with VBA* worksheet to 40 by changing the value in cell A7 and copying row 36 downward. Keep end pressure the same (cell B3) and redo the pipe sizes so that the inlet pressure is no more than 20 m (no more than 20 % pressure variation). Try to vary pipe sizes such that flow rate decreases linearly along the pipeline. The next largest pipe size above 100 is 150. Report the number of sections with each pipe size.

$59/ha

Answers will vary:

150 mm	9 sections
100 mm	15 sections
75 mm	13 sections
50 mm	3 sections

16. Adjust the depth in cell E2 in the *Irrigation Simulation with VBA* worksheet, with CV = 0.05, to the nearest integer value such that profit is maximized. The value is between 65 and 70 cm. Also, make a graph of profit vs. depth applied at 5 cm intervals between 40 and 80 cm. Click the Depth Opt. button and make a graph for profit vs. depth applied (Columns S and T). Explain the cause of the curve.

69 cm The yield variability is primarily due to pressure loss.

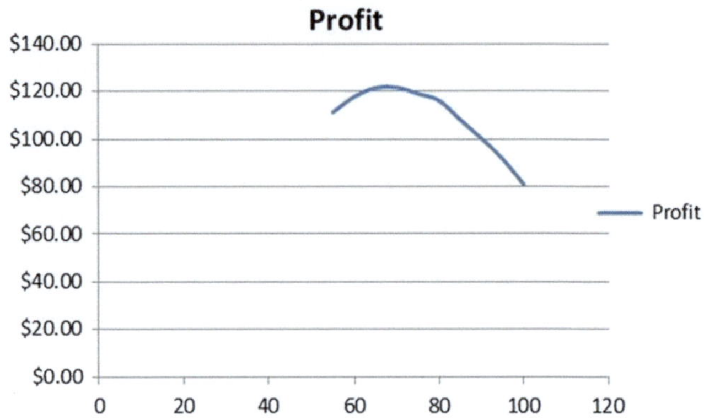

At low applied depth, the yield is reduced, and at high applied depth, the yield is reduced and expenses are high.

17. Using the *Irrigation_simulation with VBA* worksheet, observe the flow rate vs. distance plot with all pipe diameters equal to 75 mm. Explain the difference from the original graph with varying pipe size. Why wouldn't you want to use constant pipe size if friction loss is less?

With constant pipe size, there is almost no pressure loss at the end of the pipeline. Thus the curve is more of a parabola with constant pipe size. With constant pipe size, excess money is spent on large pipe at the end of the pipeline.

18. Insert actual inside pipe diameters for schedule 40 pipe in rather than the nominal pipe diameters column D rather than the nominal diameters in the *Irrigation_simulation with VBA* worksheet. Actual inside pipe diameters can be found in Table 8.1. Compare pressures in cell B36 and calculate total pressure loss in both cases, and calculate the percent difference in pressure loss by subtracting them from each other and dividing by the mean. State whether this is a significant difference and whether it is important to use actual pipe diameters in pressure loss calculations rather than nominal pipe diameters.

	Pressure	Pressure loss
Nominal pipe diameter pressure:	25.84 m	25.84–16.2 = 9.64
Actual pipe diameter pressure:	23.78 m	23.78–16.2 = 7.58

Percent different is 24 %

Based on this analysis, it is very important to use actual pipe diameters rather than nominal pipe diameters.

19. The roughness of pipes is characterized by the Hazen-Williams C value, which is included in the equation in column F in the spreadsheet. As pipes age, the roughness can increase, which decreases the C value. Using the actual pipe sizes from question 18, for the equations in column F, decrease the C value from 140 to 100 and report the total pressure loss. Compare to the pressure losses in question 18. What does this example say about the need to estimate long term pipe roughness in sprinkler systems.

The pressure loss increases to 15.58 m. It is approximately double the pressure loss with smooth pipe. Thus, it is very important to estimate the long-term pipe roughness. In many cases, the pipe will remain smooth, but there are cases in which the roughness will increase due to water chemistry or pipe quality, and this might cause major problems with uniformity.

20. Beginning with the spreadsheet with actual pipe sizes from question 18, change the Sprinkler K to 100 and x to 0.5 and input new pipe diameters such that pressure loss from one end of the lateral to the other is 20 %. Calculate the percent difference based on the average of the inlet and distal end pressure. Report the decrease in pipe cost and copy the pipe sizes into the answer.

The pipe cost for the unchanged pipeline is $49.45/ha/year.

$$0.2 = (\text{inlet} - 16.2)/(\text{inlet} + 16.2)*2$$
$$0.1(\text{inlet} + 16.2) = \text{inlet} - 16.2$$
$$0.1(\text{inlet} + 16.2) + 16.2 = \text{inlet}$$

Solve by iteration

Inlet pressure = 19.8 m

Total pipe cost is $42.75/ha/year

Answers will vary:

102.3 mm	3 sections
77.9 mm	13 sections
52.5 mm	9 sections
26.6 mm	5 sections

21. Write a VBA program to optimize energy cost and pipe sizing in the Irrigation Simulation worksheet.

Answers will vary

22. With the *Irrigation Simulation* worksheet, write a VBA program to calculate lateral end pressure if lateral input pressure is known. This will probably require an iterative procedure.

Answers will vary

23. Increase the slope to 3 % in the *Irrigation with slope* worksheet. Using the nominal pipe sizes, adjust the pipe sizes so that the difference between maximum and minimum pressure in the pipeline is no more than 2 m. Report on the number of each pipe length.

100 mm	1 sections
75 mm	16 sections
50 mm	11 sections
25 mm	2 sections

24. Make an algorithm in the Irrigation Simulation worksheet that calculates the percent difference between the maximum and minimum pressure in the lateral.

Answers will vary

25. Write a VBA program that evaluates and changes pipe sizes to column D in the *Irrigation Simulation* worksheet such that maximum pressure difference in the lateral is no more than 20 % and pipe cost is minimized.

Answers will vary

26. Using the *Leaching analysis* worksheet, run the "leach cost evaluation" for CV values of 5 %, 25 %, and 50 %, all at 80 cm applied depth. Copy the graphs of profit vs. leach price and evaluate. The current title of the graph is "Leach price analysis at 90 cm applied water." Make sure that you paste pictures of the figures into the Word document. Otherwise, they will be automatically updated when you rerun the simulation. You might need to rerun the simulation a few times if you get NAN for the 50 % CV. Determine whether the decrease in profit is primarily due to lost yield or increased leaching cost.

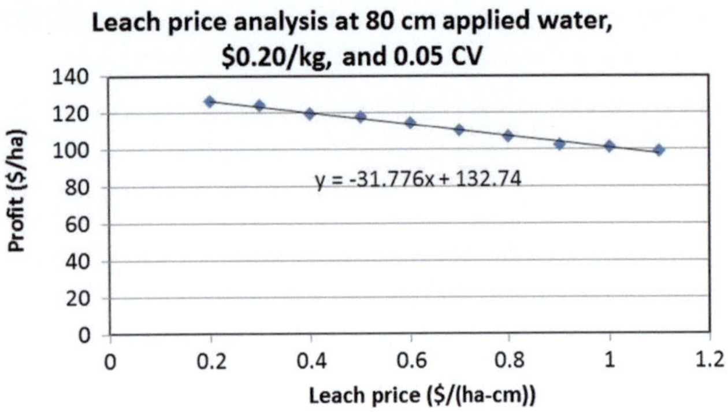

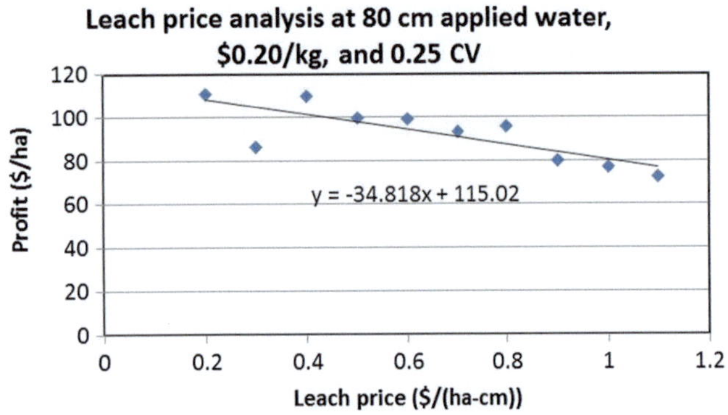

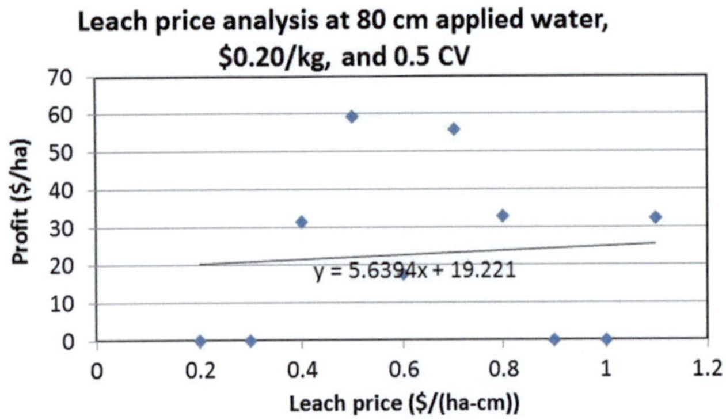

There is much more variability in profit with high CV values. The primary cause of decreased profit with high CV is decreased yield rather than leaching cost. There is little relationship between leaching cost and profit at high CV.

27. Using the Salinity analysis worksheet, copy the plots of saturated paste extract salinity vs. distance for CV values of 5 %, 25 %, and 50 %. Let irrigation water salinity equal 6 dS/m. State whether variability is primarily due to hydraulics or spatial variation of application in each case

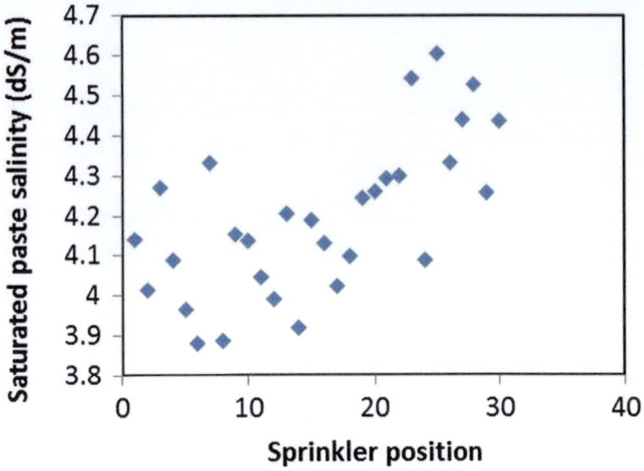

CV = 0.05

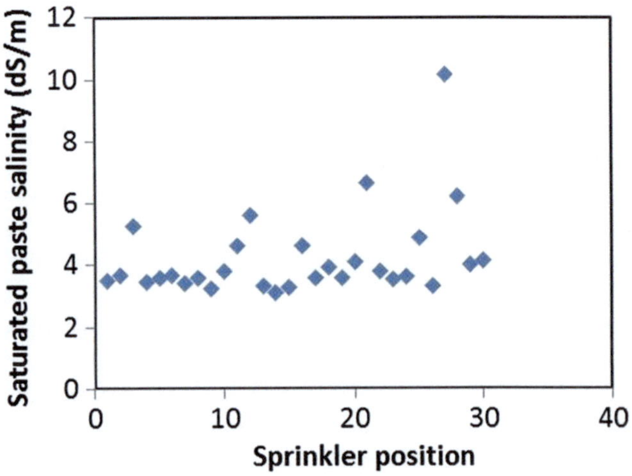

At CV = 0.25

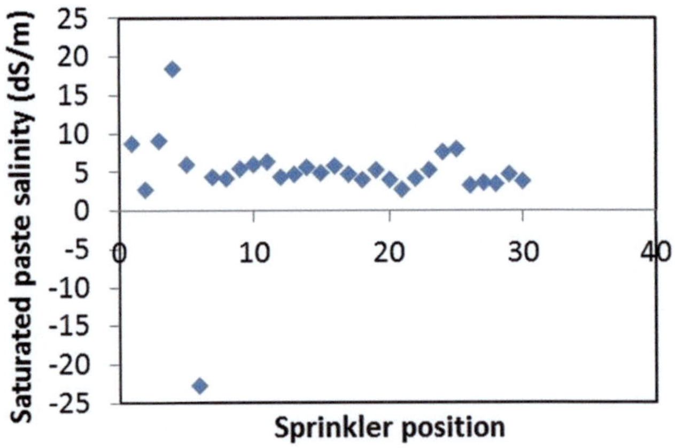

At CV = 0.5

For the low CV, the primary cause of variation is hydraulics. For CV = 0.25 and 0.5, the main cause of variation is spatial variation of application.

28. Write a VBA program that evaluates and changes pipe sizes in column D in the *Irrigation Simulation* worksheet, such that profit is maximized.

Answers will vary

Chapter 8: Solution

1. On the following drawing of a drip irrigation system network, label the type in the well, pump station, submains, mainlines, and laterals.

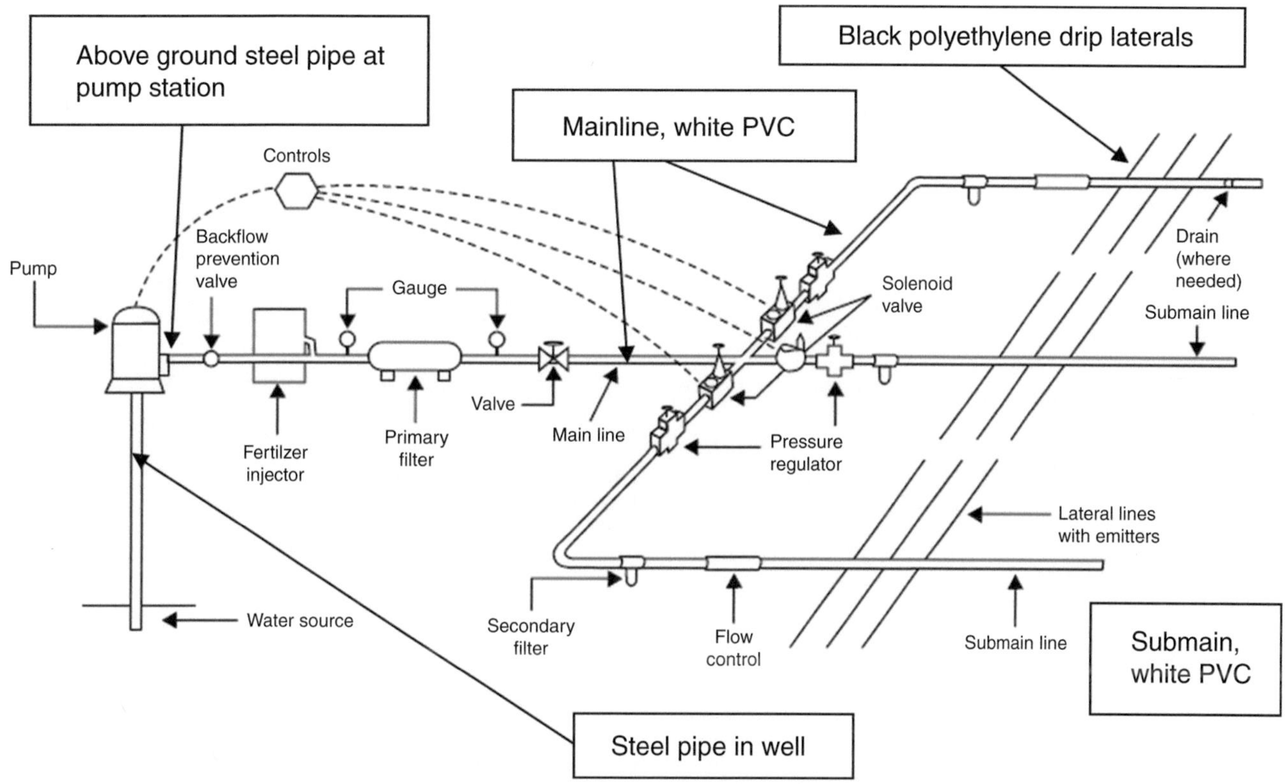

2. What is the hydraulic head and total energy of water in a pipe that is 5 m above the datum with pressure 350 kPa and water velocity 1.5 m/sec?

Velocity head $= 1.5^2/(2*9.81) = 0.114$ m
Pressure head $= 350,000/(9.81*1000) = 35.7$ m
Hydraulic head $= 35.7 + 5 = 42.7$ m
Total energy $= 42.7$ m $+ 0.114$ m $= 42.8$ m.

3. When does nonsteady state flow occur, and what are the possible hazards associated with nonsteady state flow?

Non steady state flow takes place when valves are opened or closed or pumps are turned on or off. When the system is turned and the pipe is filled the possible hazard is water hammer. When the system is turned off and the pipe is drained the possible hazard is creating a negative pressure in the pipe, which may lead to collapse of the pipe.

4. Maximum allowable flow velocity in PVC irrigation pipes is typically specified as 1.5 m/sec. What is the kinetic energy of water at this velocity? Express your answer in terms of m (length) and kPa (pressure). If the irrigation system operates at 350 kPa, then what percent of the energy is kinetic?

$KE = v^2/(2g) = 1.5^2/(2*9.8 \text{ m/sec}) = 0.11 \text{ m} = 1.1 \text{ kPa}.$
$KE/Pressure = 1.1/350*100\% = 0.31\%$

5. What two forces are included in the Reynolds number, and why is turbulent flow observed at higher Reynolds numbers?

Reynolds number is the ratio of momentum to viscous forces. Higher momentum (larger pipe diameter and higher velocity) and Reynolds number tends to propagate turbulent eddies while higher viscosity (lower Reynolds number) dampens out eddies and leads to laminar flow.

6. Calculate the friction loss in 1,000 m of 50 mm nominal diameter class 125 pipe. Calculate for flow velocities of 0.05, 1, and 3 m/sec with the Hazen-Williams and Darcy-Weisbach equations. Use C = 140 and 150 in the Hazen-Williams equation. Show your work.

Velocity = 1 m/sec
Analysis with Darcy-Weisbach.

ID = 56.6 mm.
$A = \pi D^2/4 = \pi 0.056^2/4 = 0.00246\ m^2$
$Q = V*A = 1*0.00246 = 0.00246\ m^3/sec$

$$Q = \left(\frac{0.0246\ m^3}{sec}\right)\left(\frac{3,600\ sec}{hr}\right)\left(\frac{1,000\ L}{m^3}\right) = 8,860\ L/hr$$

$$Re = \frac{vD}{\nu} = \frac{1*0.0566}{1*10^{-6}} = 56,600$$

The flow is turbulent so the Blasius equation is used to calculate f

$$f = \frac{0.316}{Re^{1/4}} = \frac{0.316}{56,600^{1/4}} = 0.0205$$

$$h_f = 6.377 fL\frac{Q^2}{D^5} = 6.377*0.0205*1000*\frac{8,860^2}{56.6^5} = 18.45\ m$$

Analysis with Hazen-Williams for v = 1 m/sec, C = 140, $h_f = 20.95$

$$h_f = kL\frac{\left(\frac{Q}{C}\right)^{1.85}}{D^{4.87}} = 1.22*10^{10}*1000*\frac{\left(\frac{2.516}{140}\right)^{1.85}}{56.6^{4.87}} = 20.95\ m$$

Analysis with Hazen-Williams for v = 1 m/sec, C = 150, $h_f = 18.43$

Thus, for v = 1 m/sec, which is a normal flow velocity in irrigation pipes, the Hazen-Williams equation agrees with the Darcy-Weisbach equation with C = 150.

Analysis with Darcy-Weisbach, for v = 3, $h_f = 126$ m

Analysis with Hazen Williams for v = 3, C = 140

$$h_f = kL\frac{\left(\frac{Q}{C}\right)^{1.85}}{D^{4.87}} = 1.22*10^{10}*1000*\frac{\left(\frac{7.548}{140}\right)^{1.85}}{56.6^{4.87}} = 160\ m$$

Analysis with Hazen-Williams for v = 3, C = 150

$$h_f = kL\frac{\left(\frac{Q}{C}\right)^{1.85}}{D^{4.87}} = 1.22*10^{10}*1000*\frac{\left(\frac{7.548}{150}\right)^{1.85}}{56.6^{4.87}} = 141\ m$$

In this range, which is greater than normal irrigation flow velocity, the Hazen-Williams equation overpredicts the friction loss when compared with the Darcy-Weisbach equation.

Analysis with Darcy-Weisbach, for v = 0.05 m/sec,

$$Re = \frac{vD}{\nu} = \frac{0.03*0.0566}{1*10^{-6}} = 1,700$$

$$f = 64/Re = 64/1,700 = 0.0377 \rightarrow h_f = 0.03\ m.$$

Analysis with Hazen-Williams for v = 0.03 m/sec, C = 150, $h_f = 0.028$

Analysis with Hazen-Williams for v = 0.03 m/sec, C = 140, $h_f = 0.032$

7. Redo Example 8.2, but the pipe discharges into a pond with a water surface elevation that is 100 meters below the upper pond water surface elevation. The pipe inlet projects into the upper pond. Use the Hazen-Williams equation.

This problem is first solved by hand and the Worksheet is at the end.

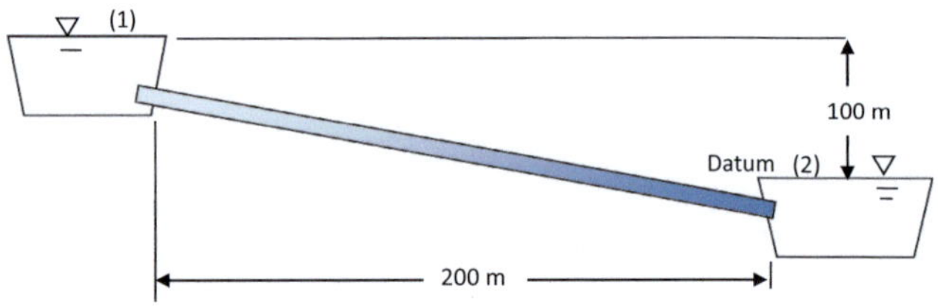

Set the control points at the water surfaces as shown in Fig. 8.3. The minor loss coefficient for a pipe projecting into the upper pond is 0.78 and for discharge to a reservoir is 1.0. The inside diameter of 2″ Class 125 pipe is 56.6 mm (Table 8.5).

$$\frac{v_2^2}{2g} + \frac{P_2}{\rho g} + z_2 = \frac{v_1^2}{2g} + \frac{P_1}{\rho g} + z_1 - h_f - h_m + H_p$$

Rearrange and solve for friction and minor losses.

$$h_f + H_m = \frac{v_1^2}{2g} + \frac{P_1}{\rho g} + z_1 - \left(\frac{v_2^2}{2g} + \frac{P_2}{\rho g} + z_2\right) + H_p$$

$$= \frac{0^2}{2g} + \frac{0}{\rho g} + 100 - \left(\frac{0^2}{2g} + \frac{0}{\rho g} + 0\right) +$$

$$0k_1 L\frac{\left(\frac{Q}{C}\right)^{1.85}}{D^{4.87}} + K_{entrance}\frac{v^2}{2g} + K_{exit}\frac{v^2}{2g} = 100$$

Calculate the Hazen-Williams friction loss in terms of velocity because the minor loss term also uses velocity. Use units of L/sec and convert to m3/sec by multiplying by 1,000. The entrance coefficient for a pipe projecting into the pond is 0.78, and the discharge coefficient is 1.0.

$$1.22*10^{10}*200\frac{\left(\frac{vA*1000}{C}\right)^{1.85}}{D^{4.87}}+(0.78+1)\frac{v^2}{2g}=100m$$

$$\frac{1.22*10^{10}*200\frac{\left(v\left(\frac{\pi*0.0566^2}{4}\right)*1000\right)^{1.85}}{150^{1.85}}}{56.6^{4.87}}+(1.78)\frac{v^2}{2*9.81}$$

$$=100m$$

The problem must be solved iteratively by guessing a velocity and then successively adjusting the velocity until the losses on the left side of the equation = 100 m.

A reasonable first guess is 1.5 m/sec since this is the maximum allowable velocity in many (closed end) pipe systems.

$$1.22*10^{10}*200\frac{\frac{\left(1.5\left(\frac{\pi*0.0566^2}{4}\right)*1000\right)^{1.85}}{150^{1.85}}}{56.6^{4.87}}+(1.78)\frac{1.5^2}{2*9.81}$$

$$=8.01\ m$$

In the Hazen-Williams equation, where pressure loss varies primarily as velocity to the 1.85 power, the solution converges most rapidly by taking the ratio of the desired loss to calculated loss to the 1/1.85 power. It converges in one iteration as follows.

$$v_2=v_1\left(\frac{actual\ \ losses}{calculated\ \ losses}\right)^{(1/1.85)}=1.5\left(\frac{100}{8.01}\right)^{(1/1.85)}$$

$$=5.87\ m/\sec$$

Substitute 1.286 for v in order to check the answer.

$$1.22*10^{10}*200\frac{\frac{\left(5.88\left(\frac{\pi*0.0566^2}{4}\right)*1000\right)^{1.85}}{150^{1.85}}}{56.6^{4.87}}+(1.78)\frac{(5.87)^2}{2*9.81}$$

$$=100\ m$$

Thus, 5.87 m/sec is the correct answer as calculated by the Hazen-Williams equation. The flow rate at an average velocity of 5.87 m/sec is

$$Q=1,000*3,600*5.87*\left(\frac{\pi*0.0566^2}{4}\right)=53260.30\ L/hr$$

$$\mathbf{Q=\ 14.77\ L/s}$$

The problem can also be solved in the Worksheet

	A	B	C
1	Hazen-Williams calculation of velocity		
2	Minor + vel. losses Km	1.78	
3	Pot + Press. Energy diff	100	m
4	Pipe parameters		
5	C	150	
6	Pipe length	200	m
7	Inside pipe diameter	0.0566	m
8	Cross-sectional area	0.0025	m2
9	Adjust cell C10 until cell C13 = cell C3		
10	Initial main velocity guess	1.5	m/sec
11	Initial energy diff	8.01	m
12	Adjusted velocity	**5.8705**	m/sec
13	Calculated energy diff	100.579	m
14	Flow rate	14.77	LPS
15	Velocity head	1.7565	m
16			
17			
18	Input data in white cells		

	E	F	
1	Darcy-Weisbach calculation of velocity (main pi		
2	Minor + vel. losses Km	1.78	
3	Pot + Press. Energy diff	100	m
4	Kinematic viscosity	1.00E-06	m2/se
5	Surface roughness K	0.00000	mm
6	Pipe length	200	m
7	Inside pipe diameter	0.0566	m
8	Cross-sectional area	0.0025	m2
9	Laminar flow?	FALSE	
10	Velocity from Hazen-W	5.871	m/sec
11	Reynolds number	332270	
12	f factor	0.0132	
13	Initial energy diff	84.8185	m
14	Adjusted velocity	**6.417**	m/sec
15	Adjusted f factor	0.0129	
16	Adjusted flow rate	16.15	LPS
17	Calculated energy diff	99.20	m
18	Velocity head	2.0987	m

Note that the Hazen-Williams answer is significantly lower than the Darcy-Weisbach answer at the high flow rate.

8. For the parameters in Example 8.3, find the flow rate with the Hazen-Williams equation: the inside diameter is 56.6 mm, the pipe length is 200 m, and the Hazen-Williams C value is 150. Recalculate flow rate with the Darcy-Weisbach equation. The inlet pipe projects into the reservoir.

The minor loss coefficient, K_m, for a square edged inlet is 0.5. Thus, total K = 1.5.

As a first guess, try velocity = 1.5 m/sec, since this is a typical PVC pipe design velocity.

$$1.22{*}10^{10}{*}200\frac{\left(\frac{1.5{*}0.002516{*}1000}{150}\right)^{1.85}}{56.6^{4.87}}+(1.5)\frac{1.5^2}{2{*}9.801}$$
$$=7.98\ m$$

In a problem where pressure loss varies as velocity squared or 1.85, the solution converges in one iteration as follows.

$$v_2=v_1\left(\frac{actual\ \ losses}{calculated\ \ losses}\right)^{1.85}=(1.5)\left(\frac{1}{7.98}\right)^{1.85}$$
$$=0.488\ \ m/\sec$$

$$1.22{*}10^{10}{*}200\frac{\left(\frac{0.488{*}0.002516{*}1000}{150}\right)^{1.85}}{56.6^{4.87}}+(1.5)\frac{0.488^2}{2{*}9.801}$$
$$=0.997\ m$$

Thus, the answer, according to the Hazen-Williams equation is v = 0.488 m/sec → 1.23 L/sec.

The final answer from the Hazen-Williams equation is used as the first guess for the Darcy-Weisbach equation. Results can be calculated in the Worksheet. The D-W solution is very close to the H-W solution.

	Hazen-Williams calculation of velocity		
2	Minor + vel. losses Km	1.5	
3	Pot + Press. Energy diff	1	m
4	Pipe parameters		
5	C	150	
6	Pipe length	200	m
7	Inside pipe diameter	0.0566	m
8	Cross-sectional area	0.0025	m2
9	Adjust cell C10 until cell C13 = cell C3		
10	Initial main velocity guess	1.5	m/sec
11	Initial energy diff	7.98	m
12	Adjusted velocity	**0.4881**	m/sec
13	Calculated energy diff	0.997	m
14	Flow rate	1.23	LPS
15	Velocity head	0.0121	m
16			
17			
18	Input data in white cells		
19			

Darcy-Weisbach calculation of velocity (main pi		
Minor + vel. losses Km	1.5	
Pot + Press. Energy diff	1	m
Kinematic viscosity	1.00E-06	m2/se
Surface roughness K	0.00000	mm
Pipe length	200	m
Inside pipe diameter	0.0566	m
Cross-sectional area	0.0025	m2
Laminar flow?	FALSE	
Velocity from Hazen-W	0.488	m/sec
Reynolds number	27628	
f factor	0.0245	
Initial energy diff	1.0700	m
Adjusted velocity	**0.471**	m/sec
Adjusted f factor	0.0247	
Adjusted flow rate	1.18	LPS
Calculated energy diff	1.00	m
Velocity head	0.0113	m

9. Calculate the pressure rating in metric units and convert to psi for 4 inch (100 mm) class 160 PVC (1120) pipe.

The dimension ratio for Class 160 pipe is 26. S = 13.8 MPa for PVC

$$PR=\frac{2S}{DR-1}=\frac{2{*}13.8}{26-1}=1.102\,\mathrm{MPa}$$
$$=1102\,\mathrm{kPa}/6.91\,\mathrm{kPa/PSI}=160\ \mathrm{PSI}$$

10. Ten inch (250 mm) Class 160 bell end pipe has expansion joints. Calculate the velocity and magnitude of the pressure wave if the operating velocity is 2.3 m/sec, and a valve suddenly closes. If the operating pressure is 50 psi (345 kPa), then what is the maximum surge pressure?

The answer is calculated in the surge worksheet. GPM is set to 1820 so that v = 2.3 m/sec in right column.

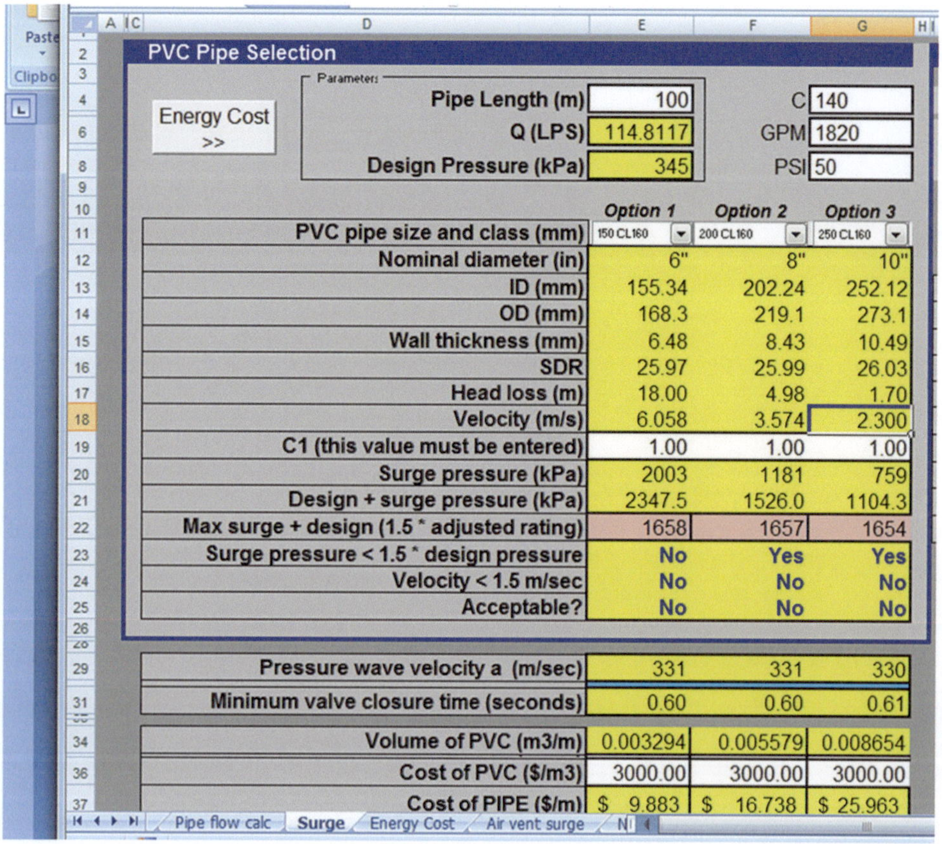

PVC Pipe Selection

Energy Cost >>

Parameters

Pipe Length (m)	100	C	140
Q (LPS)	114.8117	GPM	1820
Design Pressure (kPa)	345	PSI	50

	Option 1	Option 2	Option 3
PVC pipe size and class (mm)	150 CL160	200 CL160	250 CL160
Nominal diameter (in)	6"	8"	10"
ID (mm)	155.34	202.24	252.12
OD (mm)	168.3	219.1	273.1
Wall thickness (mm)	6.48	8.43	10.49
SDR	25.97	25.99	26.03
Head loss (m)	18.00	4.98	1.70
Velocity (m/s)	6.058	3.574	2.300
C1 (this value must be entered)	1.00	1.00	1.00
Surge pressure (kPa)	2003	1181	759
Design + surge pressure (kPa)	2347.5	1526.0	1104.3
Max surge + design (1.5 * adjusted rating)	1658	1657	1654
Surge pressure < 1.5 * design pressure	No	Yes	Yes
Velocity < 1.5 m/sec	No	No	No
Acceptable?	No	No	No
Pressure wave velocity a (m/sec)	331	331	330
Minimum valve closure time (seconds)	0.60	0.60	0.61
Volume of PVC (m3/m)	0.003294	0.005579	0.008654
Cost of PVC ($/m3)	3000.00	3000.00	3000.00
Cost of PIPE ($/m)	$ 9.883	$ 16.738	$ 25.963

Pipe flow calc | Surge | Energy Cost | Air vent surge

The maximum allowable surge pressure is 1654, and the design + surge pressure is 1104. Thus, the design is OK even though the velocity is greater than 1.5 m/sec.

The answer can also be calculated manually.

Calculate the velocity of the pressure wave.

$$a = \frac{\left[\frac{K}{\rho}\right]^{0.5}}{\left[1 + \left(\frac{K}{E}\right)\left(\frac{D}{t}\right)C_1\right]^{0.5}} = \frac{\left[\frac{2.2*10^9}{1,000}\right]^{0.5}}{\left[1 + \left(\frac{2.2*10^9}{2.76*10^9\ Pa}\right)\left(\frac{252.1}{10.49}\right)*1\right]^{0.5}}$$

$$= 330\ m/\sec$$

Calculate the magnitude of pressure wave.

$$\Delta H = \Delta v\frac{a}{g} = 2.3*\frac{330}{9.8} = 77\ m$$

Calculate the surge pressure

$P + \Delta H = 50$ psi (6.9 kPa/psi) + (77)(10.2 kPa/m) = 1130 kPa, which is the same as 1104 kPa calculated in the spreadsheet.

11. Flow rate is 10 GPM (37.9 L/min) and the design pressure is 50 psi (345 kPa). Select a pipe class and diameter that does not exceed the maximum allowable surge pressure. Use the surge equations in this case and not just the 1.5 m/s rule. Also perform an economic analysis for the best pipe diameter. Project parameters are $3,000/m^3 PVC, 20 year, 8 %, 1440 hr/year, $0.1/kW-hr, pump efficiency = 80 %.

The 18 mm (3/4 inch) pipe is acceptable.

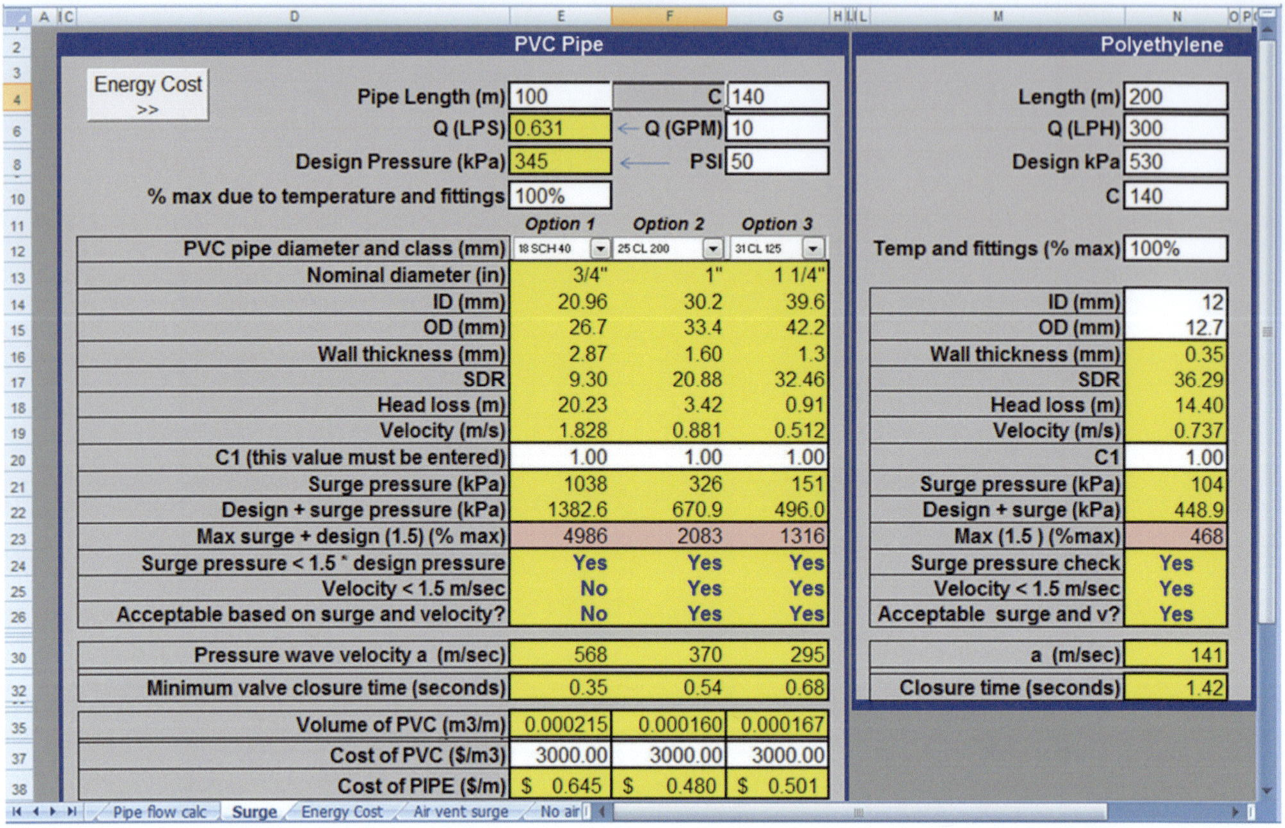

It is also interesting to evaluate the problem from an economic point of view. In order to do this, click on the energy cost button. For the assumed project parameters ($3,000/m^3 PVC, 20 year, 8 %, 1440 hr/year, $0.1/kW-hr, pump efficiency = 80 %), the total cost including energy and capital is much lower for the 1 ¼ pipe.

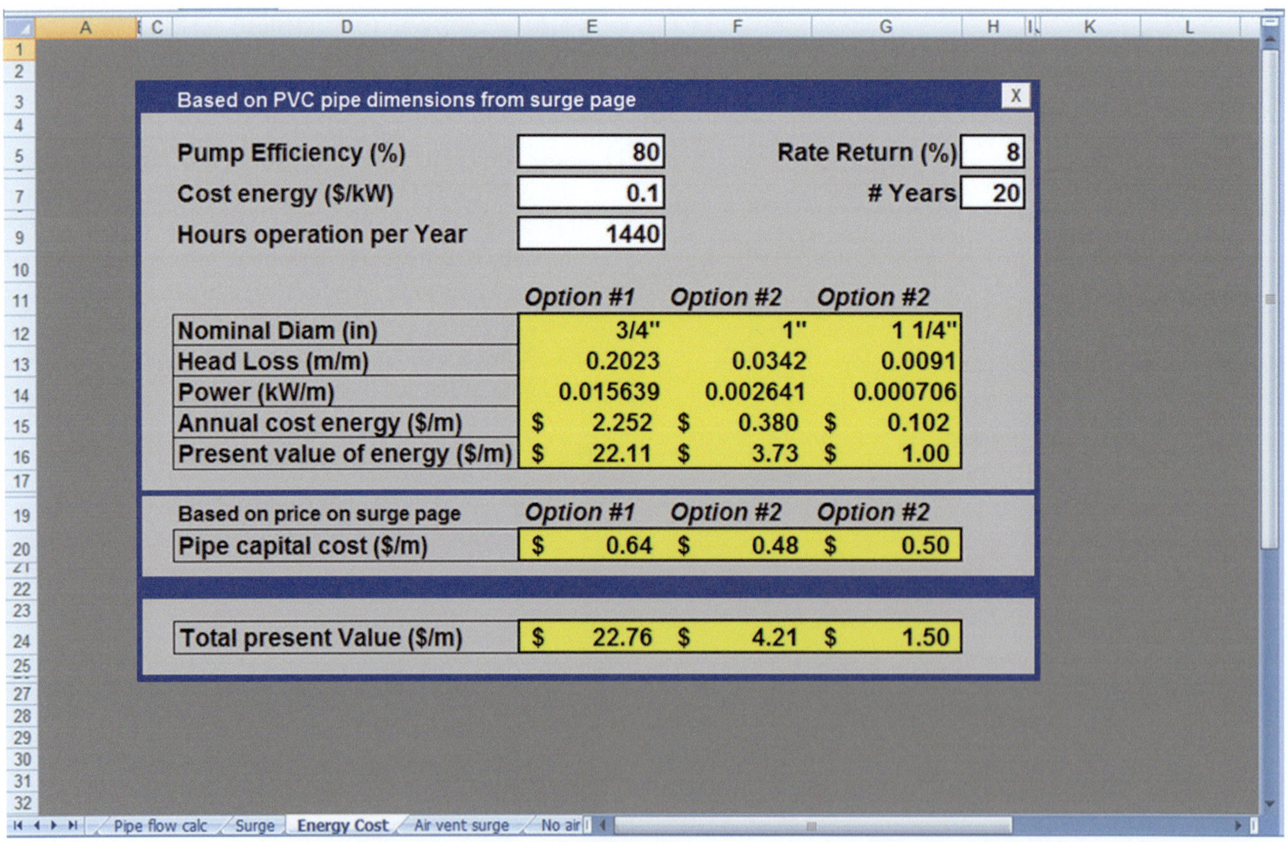

12. Calculate the maximum surge pressure in a long (100 m) drip irrigation tube (12 mm ID polyethylene). Solve the problem in two ways: assume an air vent and no air vent. Inlet pressure is 210 kPa. Conclude by discussing whether pressure relief valves and air vents are needed on long drip irrigation tubes and state the reason for your answer. Also consider sprinkler laterals (PVC) with and without sprinklers along the pipe.

The air vent surge is calculated as shown on the next page. The K value from cell K2 was used in Cell G6. Keep the inlet pressure the same as in the previous problem, but it should probably be lower for a drip system. Also note that we are neglected any pressure dissipation that may occur from having drip emitters on the lateral.

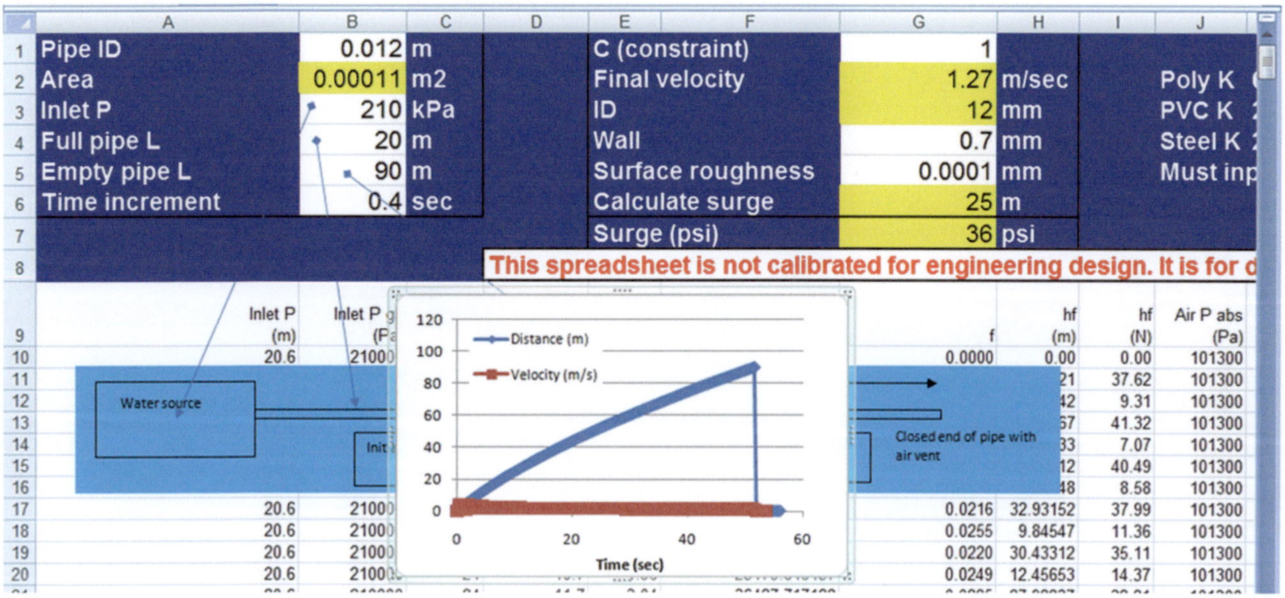

The no air vent surge is calculated as shown below. Note that you must change the increment in time in order to get the result shown in the next screen capture. Also note that the surge associated with drip tubing is not very large. This is why you don't need air vents on small drip tubes.

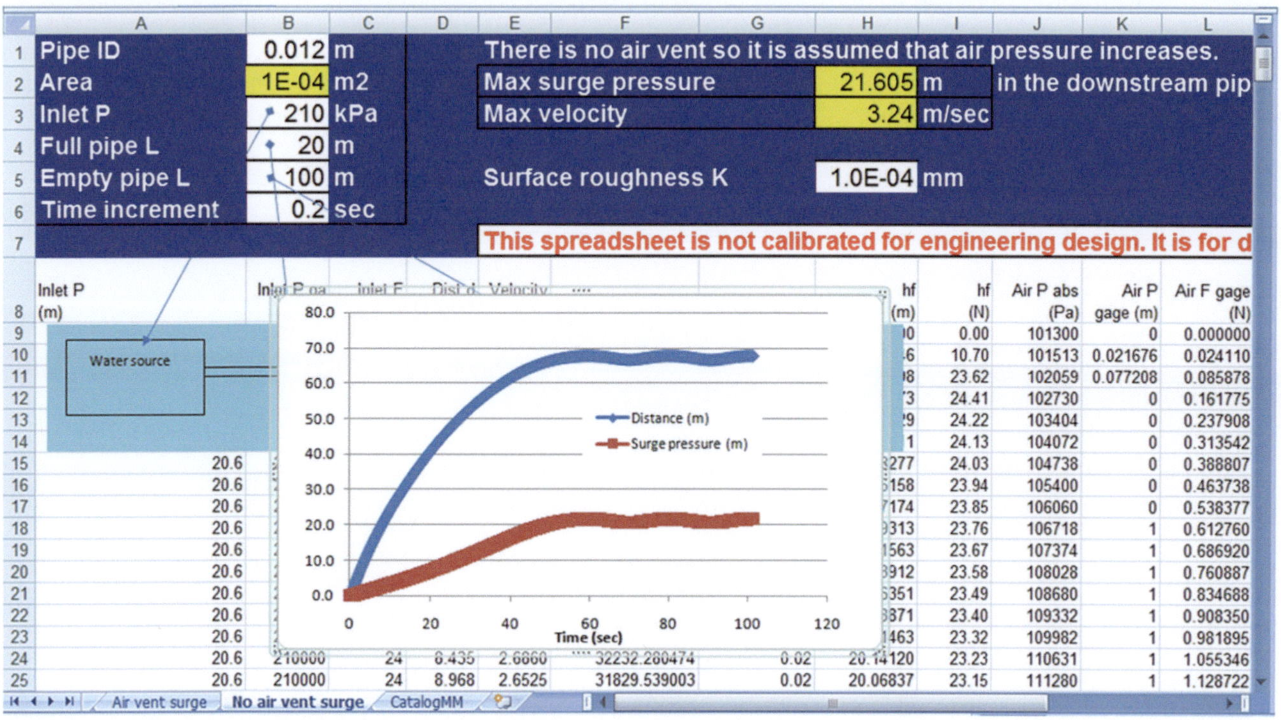

13. For the following parameters, find the best economic pipe size. The price of PVC is \$3,000/m^3. Project parameters are 20 year, 8 %, \$0.1/kW-hr, pump efficiency = 80 %.

2 pivots operating at one time (191 L/sec):	294 hr
1 pivot operating alone (96 L/sec):	2,076 hr

For the surge pressure check, the surge Worksheet was used. The following three pipe sizes and classes are acceptable, and of course larger sizes are also acceptable with respect to surge pressure.

12 inch class 200
14 inch schedule 40
16 inch schedule 40

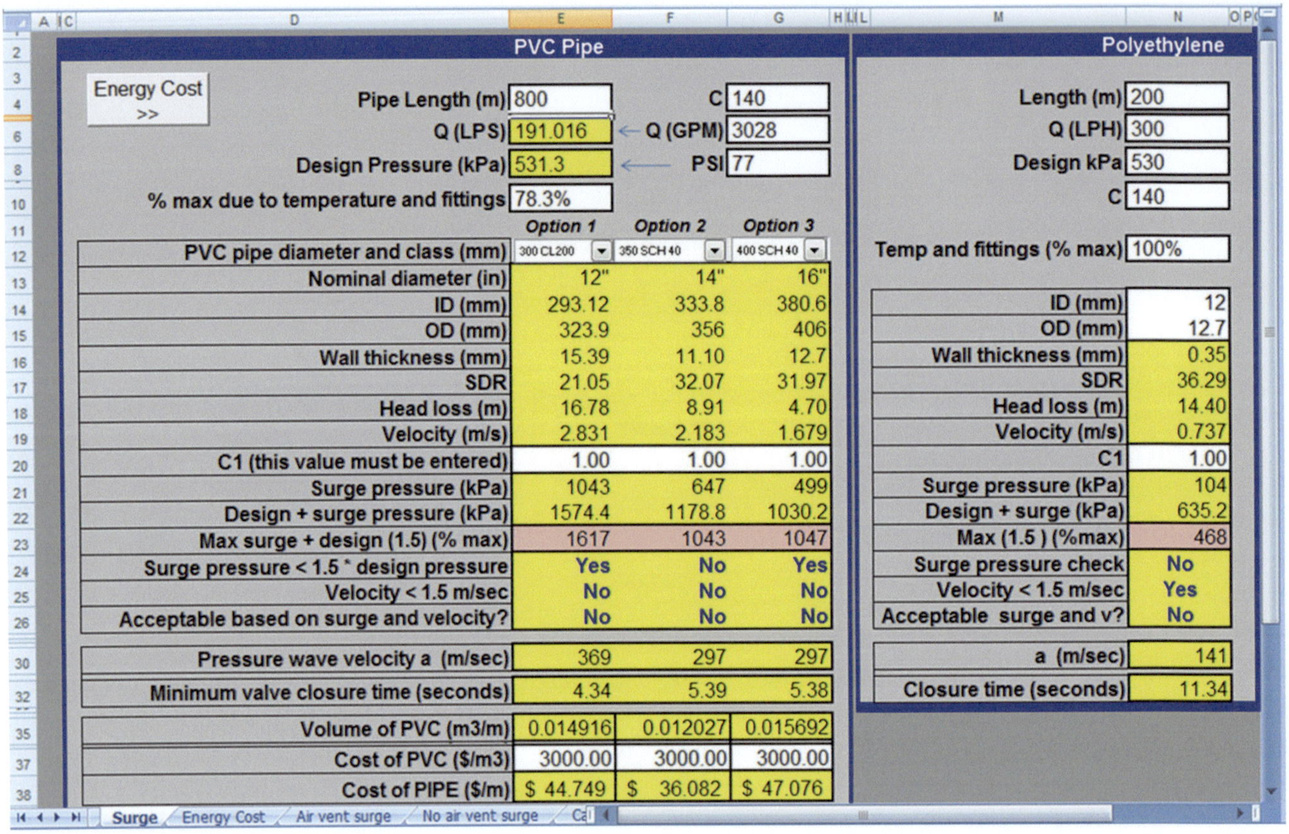

PVC Pipe

Energy Cost >>

Pipe Length (m)	800	C	140
Q (LPS)	191.016	Q (GPM)	3028
Design Pressure (kPa)	531.3	PSI	77
% max due to temperature and fittings	78.3%		

	Option 1	Option 2	Option 3
PVC pipe diameter and class (mm)	300 CL200	350 SCH 40	400 SCH 40
Nominal diameter (in)	12"	14"	16"
ID (mm)	293.12	333.8	380.6
OD (mm)	323.9	356	406
Wall thickness (mm)	15.39	11.10	12.7
SDR	21.05	32.07	31.97
Head loss (m)	16.78	8.91	4.70
Velocity (m/s)	2.831	2.183	1.679
C1 (this value must be entered)	1.00	1.00	1.00
Surge pressure (kPa)	1043	647	499
Design + surge pressure (kPa)	1574.4	1178.8	1030.2
Max surge + design (1.5) (% max)	1617	1043	1047
Surge pressure < 1.5 * design pressure	Yes	No	Yes
Velocity < 1.5 m/sec	No	No	No
Acceptable based on surge and velocity?	No	No	No
Pressure wave velocity a (m/sec)	369	297	297
Minimum valve closure time (seconds)	4.34	5.39	5.38
Volume of PVC (m3/m)	0.014916	0.012027	0.015692
Cost of PVC (\$/m3)	3000.00	3000.00	3000.00
Cost of PIPE (\$/m)	\$ 44.749	\$ 36.082	\$ 47.076

Polyethylene

Length (m)	200
Q (LPH)	300
Design kPa	530
C	140
Temp and fittings (% max)	100%
ID (mm)	12
OD (mm)	12.7
Wall thickness (mm)	0.35
SDR	36.29
Head loss (m)	14.40
Velocity (m/s)	0.737
C1	1.00
Surge pressure (kPa)	104
Design + surge (kPa)	635.2
Max (1.5) (%max)	468
Surge pressure check	No
Velocity < 1.5 m/sec	Yes
Acceptable surge and v?	No
a (m/sec)	141
Closure time (seconds)	11.34

Surge / Energy Cost / Air vent surge / No air vent surge

The next step is to use economics to select the best pipe diameter. Click the Energy Cost button on the Surge Worksheet. First evaluate at the high flow rate for 294 hours.

Based on PVC pipe dimensions from surge page

Pump Efficiency (%)	80	Rate Return (%)	8
Cost energy ($/kW)	0.1	# Years	20
Hours operation per Year	294		

	Option #1	Option #2	Option #2	Option #2
Nominal Diam (in)	12"	14"	16"	18"
Head Loss (m/m)	0.0210	0.0111	0.0059	0.0033
Power (kW/m)	0.491069	0.260781	0.137648	0.077365
Annual cost energy ($/m)	$ 14.437	$ 7.667	$ 4.047	$ 2.275
Present value of energy ($/m)	$ 141.75	$ 75.27	$ 39.73	$ 22.33

Based on price on surge page	Option #1	Option #2	Option #2	Option #2
Pipe capital cost ($/m)	$ 44.75	$ 36.08	$ 47.08	$ 59.66

Total present Value ($/m)	$ 186.50	$ 111.36	$ 86.81	$ 82.00

Evaluate the low flow rate

Based on PVC pipe dimensions from surge page

Pump Efficiency (%)	80	Rate Return (%)	8
Cost energy ($/kW)	0.1	# Years	20
Hours operation per Year	2076		

	Option #1	Option #2	Option #2	Option #2
Nominal Diam (in)	12"	14"	16"	18"
Head Loss (m/m)	0.0058	0.0031	0.0016	0.0009
Power (kW/m)	0.068015	0.036119	0.019065	0.010715
Annual cost energy ($/m)	$ 14.120	$ 7.498	$ 3.958	$ 2.225
Present value of energy ($/m)	$ 138.63	$ 73.62	$ 38.86	$ 21.84

Based on price on surge page	Option #1	Option #2	Option #2	Option #2
Pipe capital cost ($/m)	$ 44.75	$ 36.08	$ 47.08	$ 59.66

Total present Value ($/m)	$ 183.38	$ 109.70	$ 85.94	$ 81.51

The sum of present value costs are shown below. The pipe diameter with the lowest present value cost for the sum of energy and capital is the 20″ pipe.

Pipe Diam.	Capital	Low flow rate	High flow rate	Total
12 inch	44.75	138.63	141.75	Astronomical
14 inch	36.08	73.62	75.27	Astronomical
16 inch	47.08	38.86	39.73	$126/m
18 inch	59.86	21.84	22.33	$104/m
20 inch	70.15	12.84	13.12	$95/m***
21 inch	76.54	10.13	10.36	$96/m

Chapter 9: Solution

1. The revolutionary speed of electric pumps is slightly less than divisors of 3600. Typical pump rpm's are 875, 1750, and 3500. Why are most pumps manufactured with these revolutionary speeds?

Electrical current is 60 cycles per second, which is 3,600 cycles per minute. Pump motors have sets of windings that provide torque to the pump by varying the electric field. If there are two windings, then the pump rpm with no load is 3,600 rpm. If there are four windings, then the pump rpm is 1,800. If there are six windings, then the pump rpm is 900. The load on the motor from the pump slows the motor down slightly.

2. What would be a typical TDH for a centrifugal pump with flow rate 1,000 m^3/hr based on the typical specific speed for a centrifugal pump? Recalculate for pump flow rates of 100 m^3/hr and 10 m^3/hr. What type of pump would be appropriate for a very high flow rate and very low head?

$$1000\ \mathrm{m^3/hr} = 16,700\ \mathrm{L/min}$$

$$\begin{aligned} N_s &= .2108\mathrm{N}*\left(\mathrm{Q}^{.5}/\ \mathrm{H}^{.75}\right) = 500 \quad \text{for centrifugal pump.} \\ 500 &= .2108(1800)*\left(16,700^{.5}/\mathrm{H}^{.75}\right) \\ 1.32 &= \left(16,700^{.5}/\mathrm{H}^{.75}\right) \\ \mathrm{H} &= 451\ \mathrm{m} \end{aligned}$$

$$\begin{aligned} \text{At } \mathrm{Q} &= 100\ \mathrm{m^3/hr}, \mathrm{H} = 97\ \mathrm{m} \\ \text{At } \mathrm{Q} &= 10\ \mathrm{m^3/hr}, \mathrm{H} = 21\ \mathrm{m} \end{aligned}$$

An Archimedes screw pump would be appropriate for a very high flow rate and very low head.

3. Using the equations for the relationships between power, flow rate, and head, describe the shape of the head/capacity curve if efficiency was constant over a range of flow rates?

The shape would be concave up

4. Verify that the water horsepower generated by the 5.9375 impeller curve in Fig. 9.4 corresponds with the efficiency and brake horsepower curves. Calculate at the point of highest efficiency.

Water HP output is calculated with the following equation.

HP = (Q TDH)/(3960 * Eff) where Q is gpm and H is ft.

The highest efficiency is found at Q = 60 gpm. The head and efficiency at this flow rate are 28 ft and 69 %, respectively.

$$\mathrm{HP} = 60*28/(3960*0.69) = 0.61\ \mathrm{HP}$$

The corresponding point on the HP curve (darkest line) is also 0.61 HP

5. Describe the relationship between efficiency and flow rate in Fig. 9.4.

The efficiency is a maximum in the middle of the curve. It drops off slowly on either side of the high efficiency point but then drops off rapidly at the ends of the flow curve.

6. An irrigation system requires 600 gpm and 160 ft head. Select the best impeller for this application on the B4JPBH (Fig. 9.5) pump curve.

The 12 3/8″ impeller provides the required total dynamic head.

7. What is the maximum allowable flow rate of a B4JPBH pump (Fig. 9.5) with a 12 3/8″ impeller and a 40 HP motor? What is the maximum flow rate for the 50 HP motor with the same impeller?

The 40 HP curve exceeds the head-capacity curve above 1,000 gpm. The flow rate should never be allowed to exceed this value. The 12 3/8″ impeller curve never exceeds the 50 HP motor curve; however, the impeller curve ends at 1240 gpm so this is the maximum allowable flow rate. If necessary, the pump should be started with the throttling valve partially closed until the pipe system is pressurized in order to prevent high flow rates when discharging into an empty pipe. If the flow exceeds the maximum value, then amperage will become high and the pump will overheat.

8. An irrigation system requires TDH = 168 ft and Q = 600 gpm. Select an impeller diameter (trimmed if necessary) and select a motor HP with the B4JPBH pump.

The operating point is slightly above the 30 HP motor curve so a 40 HP motor must be selected. In order to select an impeller diameter, follow the slope of the efficiency curve down to the next smaller impeller. The point at which the slope crosses the 12 3/8″ curve is at Q = 580 gpm and TDH = 157 ft. Calculate the impeller diameter based on the difference in TDH since that is more sensitive to impeller diameter than flow rate at the operating point; however, it could also be calculated based on the difference in flow rate.

$$D_{\mathrm{Im}-2} = D_{\mathrm{Im}-1}\sqrt{\frac{H_2}{H_1}} = 12.375\sqrt{\frac{168}{157}} = 12.8''$$

9. The 16BZ pump (Fig. 9.7) with a 5¾″ impeller is used to run a sprinkler system. There is a 2 m pressure loss in pump fittings and filters. Find the operating point. Plot the two curves and verify that the calculated point is the correct point. The 5¾″ head-capacity curve and the irrigation system curve are:

$$\mathrm{TDH(m)} = -0.00170\ \mathrm{Q}^2 + 0.0743\ \mathrm{Q} + 43.76$$

$$\mathrm{Q_{system}}\left(\mathrm{m^3/h}\right) = 14.175\left(\mathrm{H_{system}}\right)^{0.531}$$

Adjust the system curve to account for pump fitting and filter losses ($\mathrm{H_{system}}$ = TDH − 2 m).

$$\mathrm{Q_{system}}\left(\mathrm{m^3/h}\right) = 14.175(\mathrm{TDH} - 2)^{0.531}$$

Substitute the head capacity equation for TDH in the system equation

$$\mathrm{Q} = 14.175\left(-0.00170\ \mathrm{Q}^2 + 0.0743\mathrm{Q} + 43.76 - 2\right)^{0.531}$$

Solve for Q by iteration, Q = 92.8 $\mathrm{m^3/hr}$, TDH = 36.1 m

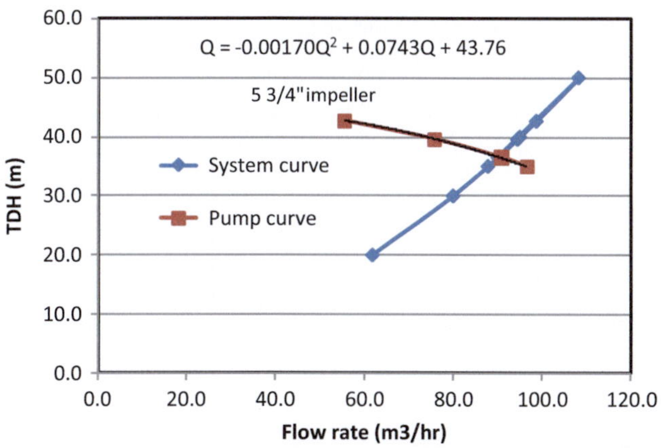

10. In Example 9.6, change the elevation of pivot 2 to 100 m elevation and pivots 3 and 4 to 120 m elevation. Select the pump operating pressure. Each pivot requires 100 L/sec. Determine the number of pumps, flow rate, and TDH of the pump station. Discuss options to reduce energy.

Four pumps must be installed. Each must have a flow rate of 100 L/sec, and TDH = 59 m. In order to reduce energy cost, booster pumps could be installed after pivot 2 in order to provide extra pressure for pivots 3 and 4. This would enable the four pumps at the pump station to have a TDH = 30 m.

Pressure requirement (All in meters)	Pivot 1	Pivot 2	Pivot 3	Pivot 4
Sprinkler pressure (+ elevation) required	15	15	15	15
Elevation difference (max pivot – pump elev.)	5	2	22	22
Pressure loss in pivot pipeline	2.8	5.4	15	15
Screen filter	2	2	2	2
Pump fittings losses	3	3	3	3
Safety factor	2	2	2	2
Total pressure requirement (TDH)	29.8	29.4	59	59

11. A variable speed pump controller is used to vary the flow rate of the 16BZ pump with the 5 3/4″ impeller. The revolutionary speed is lowered from 3500 to 3000 RPM. The system curve is $\mathrm{Q_{system}}$ ($\mathrm{m^3/h}$) = 14.175 $(\mathrm{H_{system}})^{0.531}$. There is 3.5 m head loss in the pump fittings and filters. Find the operating point TDH and flow rate.

Using the *variable speed pump* spreadsheet, the following curve was obtained.

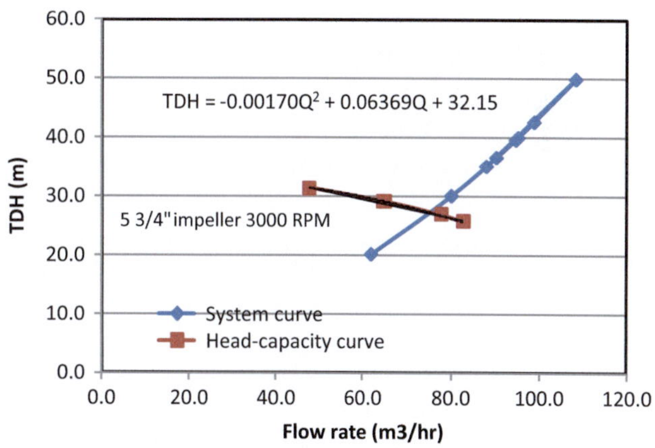

After iteration, the flow rate and TDH are 70 $\mathrm{m^3/hr}$ and 23 m, respectively.

12. Imagine that a new technology was developed that enabled farmers to produce biodiesel from crop residue. The biodiesel production unit has a capital equipment cost of $50,000; a labor, maintenance, and energy cost of $0.30/L, and produces 15,000 L of biodiesel per year. Calculate whether this would be a less expensive alternative than the electric pump system in Example 9.9. Use the *Fuel and pump costs* worksheet in Chapter 9 Excel program.

The annual costs of biodiesel production would be $14,729 L * $0.30/L = $4,400

Present value for 20 yr supply of fuel = PV(0.08, 20, 4,400) = $44,000

Assume that diesel pump capital and replacement costs are the same

Total PV cost of the **biodiesel system** is $44,000 + $50,000 + $21,000 + $876 + $5,106 = **$122,000**

The biodiesel system is not economically competitive with the electric system.

13. Redo Example 9.9 with a solar powered pump. Based on the cost of materials and the service life and replacement cost of solar components, the solar panel array provides electrical energy at a cost of $0.08/kW-hr for the 20 year project life. The solar pump can only be used during daylight; thus a larger pump is required and a reservoir must be constructed for storage. Increased capital cost of hydraulic components is $50,000 and replacement and maintenance costs remain the same as Example 9.9. Recalculate if carbon credits for the system are worth $1,000/yr.

Electric power costs.

The cost of energy is $0.08/kW-hr

BHP required by the pump is 36 HP. Convert to electrical power units:36 * 0.746 = 26.9 kW

Electric power required = (26.9/Motor efficiency) = 26.9/0.9 = 29.89 kW = 40 HP

The pump runs for 1,800 hours to the energy required is 29.89 * 1800 = 53,800 kW-hr

Annual cost of electricity is 53,800 kW-hr * $0.08/kW-hr = $4,297

Present value for 20 yr supply of energy = PV(0.08, 20, 4,297) = $42,188

Electric pump costs

Initial purchase and installation of equipment = $6,000 + $50,000 = $56,000

The present value of a centrifugal pump purchased in year 16 is $3,000 (1 + 0.08) $^{-16}$ = $876

Annual maintenance cost is $400 so present value of maintenance cost is $3927

Total present value cost of the solar powered system is $56,000 + $42,188 + $876 + $3,927 = **$102,991**

The cost of the system without carbon credits is more than the conventional electric supply system in Example 9.9. The present value of carbon credits is PV (0.08,20,1,000) = $9,818. This would bring the cost of the system down to **$93,173**, which would be less expensive than the electric system ($98,265).

14. A pump sucks water from a canal and discharges to a reservoir 100 m above the canal. Pump station valves and fittings are the same as in Example 9.11 except that the pipe diameters are 6″, 3″, 2.5″, and 4″ instead of 4″, 2″, 1 ½″, and 3″. Two other changes are that the eccentric angle is 50^{0} and the cone angle is 40°. Flow rate is 20 L/sec. All pump station pipe is 6 gage steel, and the mainline pipe is 4″ SCH 40 and is 500 m long. Assume an open discharge to the upper reservoir. Calculate the pump TDH required. Show calculations for the pressure loss in the eccentric reducer and the concentric cone. Calculate the percent of required TDH due to pump station losses, and the percent of total friction loss that is due to pump station losses. (Use worksheet).

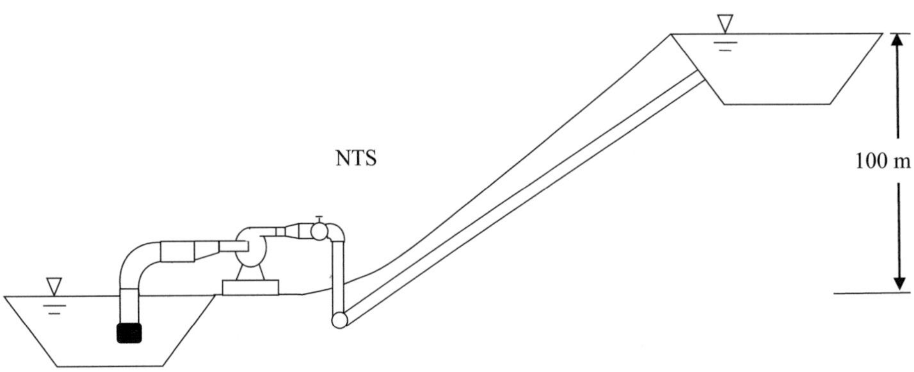

Head loss is calculated in the *Centrifugal pump calc* worksheet as follows. The required TDH is 129.56 m.

The outside diameter of 6 inch pipe is 168.3 mm. Six gage pipe has a wall thickness equal to 5.16 mm. Thus, the inside diameter of 6 inch 6 gage pipe is 168.3–10.3 = 158.0 mm. The outside diameter of 2 ½ inch 6 gage pipe is 73.0 mm. Thus the inside diameter is 73.0–10.3 = 62.7 mm.

Use 507 m length for mainline to account for PVC fittings.

Answer. Pump station losses are 1.74. As a percentage of TDH, they are (1.74/130) (100 %) = 1.3 %

As a percentage of friction losses (1.74/29.56) (100 %) = 5.9 %

	A	B	C	D	E	F
1	Component		Head requirement			Input data in
2	Required pump TDH			129.56	m	white cells
3	Total friction loss			29.56	m	
4	Mainline friction loss			27.81	m	
5	Total pump station friction			1.74	m	
6	Elevation gain			100	m	
7	Flow rate	20	L/sec			72 m3/hr
8	Steel pipe C	120		Head loss		
9	Suction pipe total			0.1415	m	
10	Suction pipe length	2	m	0.0174	m	Pipe friction
11	Suction pipe diameter	0.158	m			
12	Basket strainer K	1.3		0.0690	m	Fittings
13	Entrance coefficient K	0		0.0000	m	
14	Foot valve K	0.8		0.0424	m	
15	Sweep 90 K	0.24		0.0127	m	
16	Cross-sectional Area	0.0196	m3			
17	Flow velocity	1.02	m/sec			
18	Inlet pipe total			0.2256	m	
19	Inlet pipe length	0.2	m	0.0521	m	Pipe friction
20	Inlet pipe diameter	0.0786	m			
21	Eccentric angle (degrees)	60				
22	Eccentric K	0.20		0.1734	m	Fittings
23	Cross-sectional Area	0.004852	m2			
24	Flow velocity	4.12	m/sec			
25	Outlet pipe total			0.8022	m	
26	Outlet pipe length	0.04	m	0.0313	m	Pipe friction
27	Outlet pipe diameter	0.0627	m			
28	Cone angle	40				
29	Cone expansion K	0.36		0.7708	m	Fittings
30	Other fittings K	0.0		0.0000	m	
31	Cross-sectional Area	0.003088	m2			
32	Flow velocity	6.48	m/sec			
33	Discharge pipe total			0.5747	m	
34	Discharge pipe length	4	m	0.2666	m	Pipe friction
35	Discharge pipe diameter	0.104	m			
36	Discharge valve K	0.09		0.0254	m	Fittings
37	Discharge loss K (kinetic)	1		0.2826	m	
38	Angle bends K (sum)	0		0.0000	m	
39	Cross-sectional Area	0.008495	m2			
40	Flow velocity	2.35	m/sec			
41	Mainline pipe total			27.8149	m	
42	Mainline pipe length	507	m	27.5323	m	Pipe friction
43	Mainline pipe diameter	0.1023	m			
44	Mainline pipe C	140				
45	Mainline pipe total K	1		0.2826	m	Fittings
46	Cross-sectional Area	0.008219	m2			
47	Flow velocity	2.43	m/sec			

Steel pipe diameter

6 gage	OD	t	ID (mm)
6	168.3	5.16	158.0
4	114.3	5.16	104.0
2.5	73	5.16	62.7
2	60.3	5.16	50.0
1.5	48.3	5.16	38.0
3	88.9	5.16	78.6

PVC mainline diameter

Sch 40	OD	t	ID (mm)
4	114.3	6.02	102.3

15. Redo question 15, but discard the eccentric. Suction pipe is 75 mm (3 in.) pipe. Second, use a bushing on the discharge side (sudden expansion) rather than a cone expansion. Determine which change results in the greatest increase in head loss.

Use the Centrifugal pump calc worksheet to calculate the head loss on the suction side.

	A	B	C	D	E	F	G	H	I	J	K	L
1	Component		Head requirement			Input data in			Steel pipe diameter			
2	Required pump TDH			133.16	m	white cells			6 gage	OD	t	ID (mm)
3	Total friction loss			33.16	m				6	168.3	5.16	158.0
4	Mainline friction loss			27.81	m				4	114.3	5.16	104.0
5	Total pump station friction			5.35	m				2.5	73	5.16	62.7
6	Elevation gain			100	m				2	60.3	5.16	50.0
7	Flow rate	20	L/sec			72	m3/hr		1.5	48.3	5.16	38.0
8	Steel pipe C	120		Head loss					3	88.9	5.16	78.6
9	Suction pipe total			2.5485	m							
10	Suction pipe length	2	m	0.5213	m	Pipe friction			PVC mainline diameter			
11	Suction pipe diameter	0.0786	m						Sch 40	OD	t	ID (mm)
12	Basket strainer K	1.3		1.1262	m	Fittings			4	114.3	6.02	102.3
13	Entrance coefficient K	0		0.0000	m							
14	Foot valve K	0.8		0.6930	m							
15	Sweep 90 K	0.24		0.2079	m							
16	Cross-sectional Area	0.0049	m3									
17	Flow velocity	4.12	m/sec									
18	Inlet pipe total			0.0521	m							
19	Inlet pipe length	0.2	m	0.0521	m	Pipe friction						
20	Inlet pipe diameter	0.0786	m									
21	Eccentric angle (degrees)	0										
22	Eccentric K	0.00		0.0000	m	Fittings						
23	Cross-sectional Area	0.004852	m2									
24	Flow velocity	4.12	m/sec									
25	Outlet pipe total			2.1707	m							
26	Outlet pipe length	0.04	m	0.0313	m	Pipe friction						
27	Outlet pipe diameter	0.0627	m									
28	Cone angle	0										
29	Cone expansion K	0.00		0.0000	m	Fittings						
30	Other fittings K	1.0		2.1394	m							
31	Cross-sectional Area	0.003088	m2									
32	Flow velocity	6.48	m/sec									
33	Discharge pipe total			0.5747	m							
34	Discharge pipe length	4	m	0.2666	m	Pipe friction						
35	Discharge pipe diameter	0.104	m									
36	Discharge valve K	0.09		0.0254	m	Fittings						
37	Discharge loss K (kinetic)	1		0.2826	m							
38	Angle bends K (sum)	0		0.0000	m							
39	Cross-sectional Area	0.008495	m2									
40	Flow velocity	2.35	m/sec									
41	Mainline pipe total			27.8149	m							
42	Mainline pipe length	507	m	27.5323	m	Pipe friction						
43	Mainline pipe diameter	0.1023	m									
44	Mainline pipe C	140										
45	Mainline pipe total K	1		0.2826	m	Fittings						
46	Cross-sectional Area	0.008219	m2									
47	Flow velocity	2.43	m/sec									
48												

The head loss on the suction side increased from 0.1415 + 0.2256 = 0.37 m to 2.55 + 0.05 = 2.6 m Thus, the increase is approximately 2.2 m on the suction side. One the discharge side, the increase is 0.802 + 0.575 = 1.37 m to 2.17 + 0. 57 = 2.74 m. Thus, the increase is 1.37 m on the discharge side. There, the change in suction pipe geometry results in a greater change in losses than removing the cone. However, the cone removal is in the similar range of losses as changing the entire suction section. This shows that the expansion cone is very important.

16. Use the 16BZ pump with 5 3/4″ impeller (M) to deliver water to the upper reservoir for the system shown below. Select pipe diameters equal to 6″, 4″, 3″, and 4″ for the four pump station pipe sections. Use 4″ Schedule 40 PVC for the mainline, which is 493 m long. Draw a system curve (develop with *Centrifugal pump fittings* worksheet by inputting different flow rates and corresponding TDH requirement) and pump head-capacity curve based on Fig. 9.6. Find an exponential equation for the system curve and equation for the head-capacity curve, and calculate the point of intersection (operating point) for the system.

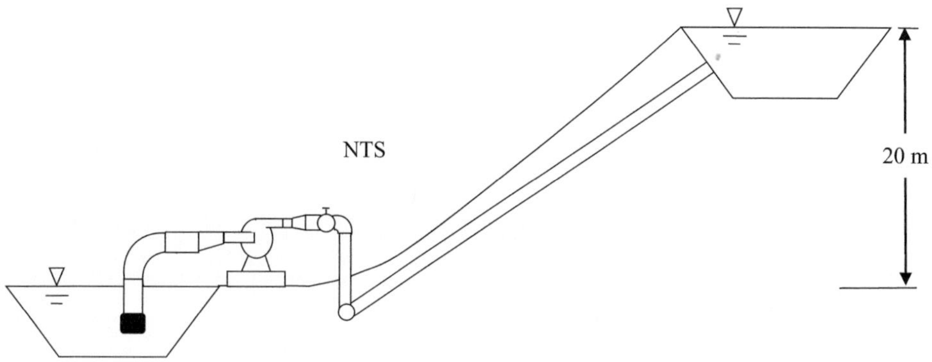

Based on Fig. 9.6 the pump discharge and inlet diameters are

Three flow rates were used to define the system curve with the *Centrifugal pump calc* worksheet.

L/sec	m^3/hr	m
10	36	30.6
15	54	42.8
20	72	59.2

Component			Head requirement		Input data in white cells
Required pump TDH			48.60	m	
Total friction loss			28.60	m	
Mainline friction loss			27.43	m	
Total pump station friction			1.17	m	
Elevation gain			20	m	
Flow rate	20	L/sec			
Steel pipe C	120		Head loss		
Suction pipe total			0.1681	m	
Suction pipe length	2	m	0.0174	m	Pipe friction
Suction pipe diameter	0.158	m			
Basket strainer K	1.3		0.0690	m	Fittings
Entrance coefficient K	0.5		0.0265	m	
Foot valve K	0.8		0.0424	m	
Sweep 90 K	0.24		0.0127	m	
Cross-sectional Area	0.0196	m3			
Flow velocity	1.02	m/sec			
Inlet pipe total			0.0428	m	
Inlet pipe length	0.2	m	0.0133	m	Pipe friction
Inlet pipe diameter	0.104	m			
Eccentric angle (degrees)	50				
Eccentric K	0.10		0.0295	m	Fittings
Cross-sectional Area	0.008495	m2			
Flow velocity	2.35	m/sec			
Outlet pipe total			0.1521	m	
Outlet pipe length	0.04	m	0.0104	m	Pipe friction
Outlet pipe diameter	0.0786	m			
Cone angle	40				
Cone expansion K	0.16		0.1417	m	Fittings
Other fittings K	0.0		0.0000	m	
Cross-sectional Area	0.004852	m2			
Flow velocity	4.12	m/sec			
Discharge pipe total			0.8037	m	
Discharge pipe length	1.5	m	0.1000	m	Pipe friction
Discharge pipe diameter	0.104	m			
Discharge valve K	0.09		0.0254	m	Fittings
Angle bend K (sum)	2.4		0.6783	m	
Cross-sectional Area	0.008495	m2			
Flow velocity	2.35	m/sec			
Mainline pipe total			27.4348		
Mainline pipe length	500	m	27.1521	m	
Mainline pipe diameter	0.1023	m			
Mainline pipe C	140				
Mainline pipe total K	1		0.2826	m	
Cross-sectional Area	0.008219	m2			

The system curve and pump curve were graphed and equations were found with Trendline in Excel.

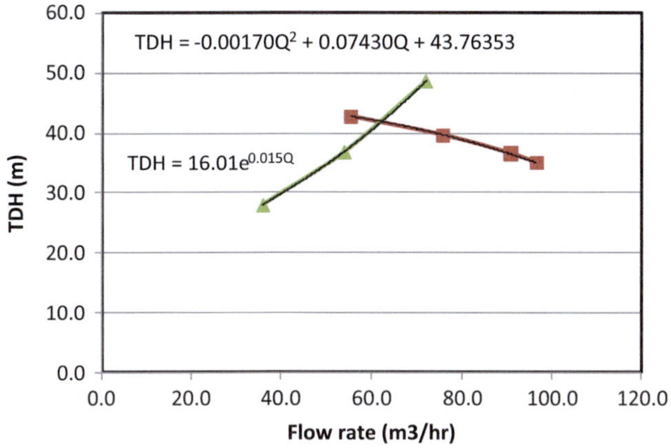

There are two equations and two unknowns, and substitution was used to solve for Q

$$TDH = -0.00170Q^2 + 0.07430Q + 43.76353$$
$$TDH = 16.01e^{0.015Q}$$
$$16.01e^{0.015Q} = -0.00170Q^2 + 0.07430Q + 43.76353$$
$$e^{0.015Q} = \left(\left(-0.00170Q^2 + 0.07430Q + 43.76353\right)/16.01\right)$$
$$0.015Q = \ln\left(\left(-0.00170Q^2 + 0.07430Q + 43.76353\right)/16.01\right)$$
$$Q = \left(\ln\left(\left(-0.00170Q^2 + 0.07430Q + 43.76353\right)/16.01\right)\right)/0.015$$

Iteration was used to find Q (note that the iterative solution does not converge if the equation is not converted from exponential to logarithmic form).

The operation point is 63.7 m^3/hr at 42.8 m TDH.

17. Venturi injectors are designed based on the principle that if water velocity increases, then pressure decreases as shown by the Bernoulli equation. A narrow throat increases the velocity at the suction point. Concentric cones are used to gradually increase flow rate to the throat and decrease flow rate from the throat. Based on what you know about concentric cones, draw a Venturi injector geometry that has minimum head loss.

The Venturi has a much smaller angle on the discharge side because expansion cone pressure loss is much higher than reducer cone (inlet cone) pressure loss. If the angle on the discharge side is minimized, then head loss on the discharge side is minimized.

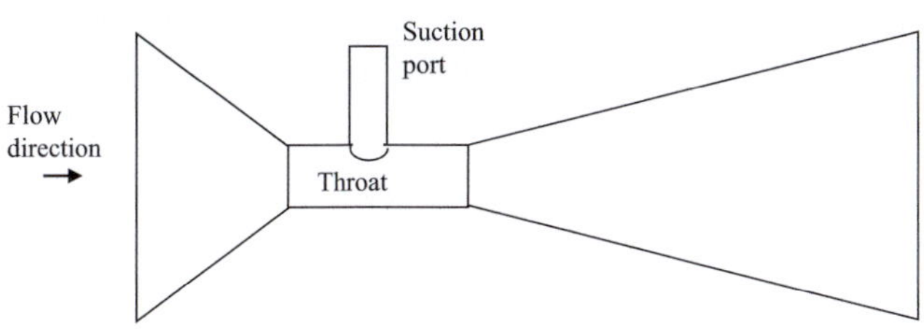

18. Some people recommend creating the pressure differential across a Venturi by restricting mainline flow. It is a much better idea to have a separate centrifugal pump provide the pressure differential, as shownin this example. Mainline flow rate is 200 L/sec, and Venturi flow rate is 0.90 L/sec. Venturi injection time is 1,000 hours per year. The required pressure differential across the Venturi is 283 kPa. The cost of energy is \$0.10/kW-hr. Calculate the energy cost per year for providing the required pressure differential across the Venturi by constricting the mainline flow with a valve. Calculate the energy cost of using a centrifugal pump in the bypass line to provide the pressure differential needed by the Venturi.

The power requirement for the case with a valve in the mainline that creates a pressure differential

Power calculations		
Pump flow rate	12000	LPM
Pump head	28.83	m
WHP	74.67	mhp
Efficiency	75.0 %	
BHP	99.56	mhp
Power (kW)	75.07	kW

The energy requirement is $0.75\ \text{kW}*1,000\ \text{hr} = 75,000\ \text{kW} - \text{hr}$.

The cost of energy is $75,000\ \text{kW-hr}*\$0.10/\text{kW-hr} = \$7,500/\text{yr}$

The power requirement for the centrifugal pump in the bypass line.

Power calculations		
Pump flow rate	54	LPM
Pump head	28.83	m
WHP	0.34	mhp
Efficiency	75.0 %	
BHP	0.45	mhp
Power (kW)	0.34	kW

The energy requirement is $0.34\ \text{kW}*1,000\ \text{hr} = 340\ \text{kW-hr}$.

The cost of energy is $340\ \text{kW-hr}*\$0.10/\text{kW-hr} = \$34/\text{yr}$

19. How could a Venturi be used within a pump to lift groundwater up to the surface in a well. (Hint: look up jet pumps).

A jet pump uses a Venturi as follows. A Venturi is placed below the water level in a well. The pump pushes water down one pipe to the Venturi. The Venturi then sucks in groundwater and the water is then forced back up another pipe to the ground surface.

The following are extra problems that weren't on the homework. They won't be on the exam, but you may be interested in them.

20. Calculate throat pressure (gage pressure and absolute pressure) and discharge pressure in a Venturi injector that that has a 30 mm internal diameter at both ends and that has a flow rate of 0.9 L/sec. The length of the entire Venturi is 15 cm and the length of the throat is 2 cm; however, assume that the equivalent length of the throat is 10 cm due to flow entering the throat through the suction tube. The upstream pressure is 300 kPa. The reducer cone angle (inlet side), θ, is 30°, and the expansion cone angle (discharge side) is 15°. Assume that Hazen-Williams C in the throat is 100. The inside diameter of the throat is 7 mm.

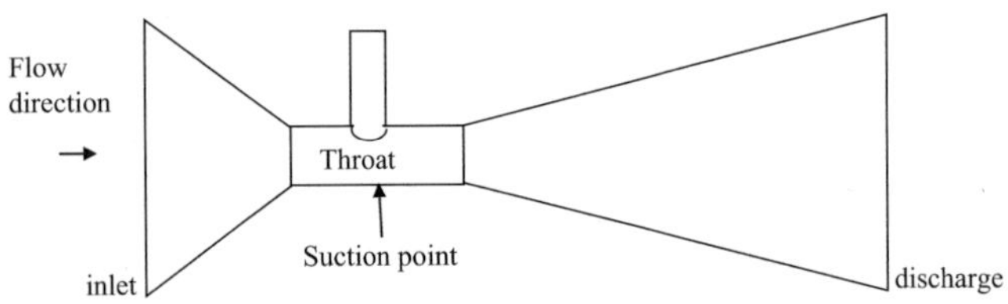

Friction loss in the throat is calculated as follows.

$$h_f = 1.22{*}10^{10}{*}0.1\ m\left(\frac{\left(\frac{0.9/1000}{100}\right)^{1.852}}{7^{4.87}}\right) = 15.2\ m$$

Calculate inlet cone head loss. The large and small diameters are 0.03 m and D_t, respectively

$$\theta < 45^0 \qquad K = 0.8\ \sin\frac{\theta}{2}\left(1-\frac{d_1^2}{d_2^2}\right)^2$$

$$= 0.8\ \sin\frac{30}{2}\left(1-\frac{0.007^2}{0.03^2}\right)^2 = 0.185$$

Flow velocity for expansion and reduction cone losses is based on flow within throat. Flow rate is

$$v_t = \frac{Q}{A} = \frac{0.9\ L/\sec\ /\ 1{,}000}{\pi\ 0.007^2/_4} = 23.4\ m/s$$

$$h_{mi} = K_i\frac{v_t{}^2}{2g} = 0.185\frac{23.4}{2{*}9.8} = 5.16\ m$$

Calculate discharge cone head loss

$$\theta < 45° \quad K = 2.6\ \sin\left(\frac{15(\pi/180)}{2}\right)\left(1-\frac{0.007^2}{0.03^2}\right)^2 = 0.30$$

$$h_{md} = K_i\frac{v_t{}^2}{2g} = 0.30\frac{23.4}{2{*}9.8} = 8.49\ m$$

Total pressure loss is the sum of losses

$$= 8.49 + 5.16 + 15.2 = 28.85$$

Discharge pressure is inlet pressure – pressure loss.

$$= 300 - 28.85{*}9.8 = 17\ \text{kPa}$$

Pressure in the throat is calculated based on the Bernoulli equation and the friction loss in the inlet cone and half of the throat.

$$\frac{v_1^2}{2g} + \frac{P_1}{\gamma} = \frac{v_S^2}{2g} + \frac{P_S}{\gamma} + h_{mi} + 0.5h_f$$

$$P_S = \gamma\left(\frac{P_1}{\gamma} + \frac{v_1^2}{2g} - \frac{v_S^2}{2g} - h_{mi} - 0.5h_f\right)$$

where h_f is throat pressure loss.

$$P_S = 9.8\left(\frac{300}{9.8} + \frac{1.27^2}{2{*}9.8} - \frac{23.39^2}{2{*}9.8} - 5.16 - 0.5{*}15.2\right)$$

$$= -10.0\ m$$

Thus, gage pressure is −98 kPa.

If gage pressure is 101.3 absolute pressure, then the absolute pressure in the throat is 3.3 kPa.

21. Redo question 20 but optimize the inlet and discharge angle in order to minimize pressure loss across the Venturi. Keep the same throat dimension and Venturi length. Derive an equation based on the geometry of the system that calculates discharge angle as a function of inlet angle.

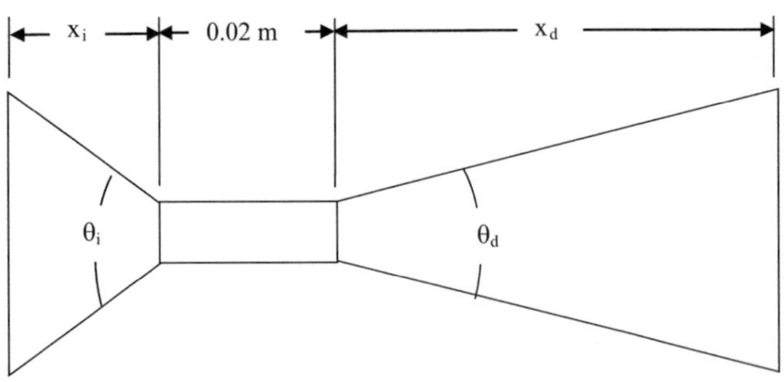

The length of the system is the sum of in two cones and the throat.

$$x_i + 0.02 + x_d = 0.3\ \text{m}$$

$$x_i = \frac{0.015 - \frac{D_t}{2}}{\tan\left(\frac{\theta_i}{2}\right)}$$

$$x_d = \frac{0.015 - \frac{D_t}{2}}{\tan\left(\frac{\theta_d}{2}\right)}$$

Where D_t is the throat diameter.

Solve for the discharge angle as a function of the inlet angle

$$x_i + 0.02 + x_d = 0.3 \text{ m}$$

$$\frac{0.015 - \frac{D_t}{2}}{\tan\left(\frac{\theta_i}{2}\right)} + 0.02 + \frac{0.015 - \frac{D_t}{2}}{\tan\left(\frac{\theta_d}{2}\right)} = 0.3$$

$$\frac{0.015 - \frac{D_t}{2}}{\tan\left(\frac{\theta_d}{2}\right)} = 0.3 - \frac{0.015 - \frac{D_t}{2}}{\tan\left(\frac{\theta_i}{2}\right)} - 0.02$$

$$\tan\left(\frac{\theta_d}{2}\right) = \frac{0.015 - \frac{D_t}{2}}{0.3 - \frac{0.015 - \frac{D_t}{2}}{\tan\left(\frac{\theta_i}{2}\right)} - 0.02}$$

$$\theta_d = 2\tan^{-1}\left(\frac{0.015 - \frac{D_t}{2}}{0.3 - \frac{0.015 - \frac{D_t}{2}}{\tan\left(\frac{\theta_i}{2}\right)} - 0.02}\right)$$

A spreadsheet was set up to calculate pressures as a function of inlet and discharge angle. Solver was used to maximum discharge pressure as a function of inlet cone angle.

Inlet diameter	0.03	m
Throat diameter	0.007	m
Throat length	0.02	m
Device length	0.15	m
Reductiion cone angle (full angle)	28.43	degrees
Discharge angle (full angle)	15.48	degrees
	Length (m)	
Reduction cone	0.0454	m
Throat	0.02	m
Expansion cone	0.0846	m
Total length	0.15	m
Flow rate (L/sec)	0.9	
Inlet flow velocity (m/sec)	1.27	m/sec
Throat velocity (m/sec)	23.39	m/sec
Inlet cone Km	0.176	
Pressure loss reduction cone	4.90	m
Throat equivalent length	0.1	m
Throat pressure loss	15.2	m
Expansion cone Km	0.313	
Expansion cone pressure loss	8.73	m
Total pressure loss	28.83	
Inlet pressure	300	kPa
Inlet pressure	30.59	m
Total friction loss in Venturi	28.83	m
Discharge pressure	1.76	m
Discharge pressure	17.24	kPa
Pressure difference across Venturi	283	kPa
Pressure at suction	−9.7	m
Absolute zero pressure	−101	kPa
Absolute zero pressure	−10.31	m
Difference between throat pressure and abs. zero	5.844	kPa

(continued)

Chapter 10: Solution

1. What are the five major types of aquifers?

Unconsolidated sand and gravel aquifers, semi-consolidated sand and gravel aquifers, volcanic aquifers, sandstone and carbonate rock aquifers, and sandstone aquifers.

2. Describe four different types of aquifers commonly used for irrigation and give an example of each.

There are three types of unconsolidated sand and gravel aquifers: basin fill, blanket sand and gravel, and glacial-fill. Basin fill aquifers are formed as sediment fills in basins between mountains. Basin fill aquifers are common in the Southwest United States. Blanket sand aquifers are formed by wind blown sand. The Ogallala aquifer is a blanket sand aquifer. Glacial-fill aquifers are deposited by glaciers and are common in the Northern Midwest. Semiconsolidated sand aquifers have sloping layers of sand and aquitards and are found on the East Coast.

3. Draw a confined and unconfined aquifer.

See Fig. 10.2.

4. Discuss groundwater recharge and discharge components for a basin

Groundwater recharge components include surface infiltration (rainfall or irrigation), mountain front recharge, underflow, interbasin flow, and surface infiltration.

5. Discuss the impact of overpumping on stream flow in arid climates.

Overpumping depletes the aquifer and lowers the water table. A stream that used to be supplied by flow from the aquifer loses water to the aquifer and stops flowing.

6. Discuss the natural condition and impact of man on the Ogallala aquifer.

Natural history:

The High Plains Aquifer (contains the Ogallala aquifer) in Nebraska, Colorado, Kansas, Oklahoma, Texas, South Dakota, Wyoming, and New Mexico (Fig. 10.11) is a blanket sand and gravel aquifer. These aquifers are primarily formed from wind blown sand, unconfined by impermeable layers above, and confined below by bedrock (Fig. 10.12).

Impact of man:

Some parts of the High Plains Aquifer in Nebraska have such high levels of nitrogen from nitrogen leaching in agricultural fields that farmers in the area do not even have to add nitrogen fertilizer to the pumped groundwater. Although the lack of need for fertilizer is nice for farming, the farmers cannot drink the water because it is not potable.

Because the rate of pumping has exceeded the recharge rate, primarily in the southern part of the Ogallala aquifer, water tables have declined by more than 30 m (100 ft) in some locations (Fig. 10.11). However, in regions of the aquifer where irrigation water is supplied by surface streams, the level of the aquifer has risen by 10–20 m. The water balance for the High Plains aquifer was 24 million cubic feet per day in and 24 million cubic feet per day out in the natural state (Fig. 10.12). Irrigation began during a drought in the 1930s and eventually the recharge rate (irrigation leaching and runoff) increased to 510 million cubic feet per day while the rate of extraction rose from zero to 810 million cubic feet per day. The lateral discharge dropped from 24 to 10 million cubic feet per day. The difference between recharge and extraction grew very large, 330 million cubic feet per day. As the aquifer became depleted in some areas such as Texas, the rate of withdrawal began to decrease in the 1970s.

7. Discuss the impact of irrigation on water table elevation and salinization.

Irrigation, especially inefficient irrigation, adds water to the aquifer and raises the water table. Eventually, the water table rises to the soil and subsurface drainage systems must be installed in order to leach salts from the soil. If subsurface drainage is not installed, then the soil becomes salinized and unfarmable.

8. Calculate the porosity, storativity, specific yield, and specific retention for an aquifer that has 40 % water below the water table, and yield of 1.2 m of water for every 5 m drop in water table elevation.

Porosity is equal to saturated water content = 40 %
Specific yield is 1.2/5 * 100 % = 24 %
Specific retention is equal to porosity – specific yield = 40 % – 24 % = 16 %
Field capacity is equal to specific retention = 16 %

9. A coarse sand aquifer has a water table slope of 1 m/100 m. Evaluate at the upper and lower limits of coarse sand hydraulic conductivity. What is the Darcy velocity of the water in the aquifer? The cross sectional area of the aquifer is 100 m × 1,000 m. What is the volume of water flow in 1 year? Convert water volume to acre-ft. How many acres of cotton could be irrigated with this volume per year? Also calculate for a silt aquifer with a hydraulic conductivity at the lower end of silt conductivities.

Low end of coarse sand hydraulic conductivity 10 m/day.

$$v = K\,dH/dz = 10*0.01 = 0.1\ \text{m/day}.$$
$$Q = vA = 0.1\ \text{m/day}\ (100\ \text{m x}\ 1,000\ \text{m}) = 10,000\ \text{m}^3/\text{day}$$
$$V = Q\,t = 10,000\ \text{m}^3/\text{day}*365 = 3.65\ \text{million m}^3$$

$$3.65\ \text{million m}^3*0.00081\ \text{m}^3/\text{ac} - \text{ft} = 3,000\ \text{ac} - \text{ft}.$$

This volume of water would irrigate approximately 1,000 acres of cotton, assuming that the gross depth of application would be 3 ft/year

Upper end of coarse sand hydraulic conductivity 1,000 m/day.

$$v = K\,dH/dz = 1,000*0.01 = 10\ \text{m/day}.$$
$$Q = vA = 10\ \text{m/day}\ (100\ \text{m x}\ 1,000\ \text{m}) = 1,000,000\ \text{m}^3/\text{day}$$
$$V = Q\,t = 1,000,000\ \text{m}^3/\text{day}*365 = 365\ \text{million m}^3$$

$$365\ \text{million m}^3*0.00081\ \text{m}^3/\text{ac} - \text{ft} = 300,000\ \text{ac} - \text{ft}.$$

This volume of water would irrigate approximately 100,000 acres of cotton, assuming that the gross depth of application would be 3 ft/year

Lower end of silt hydraulic conductivity 0.01 m/day.

$$v = K\,dH/dz = 0.01*0.01 = 0.0001\ \text{m/day}.$$
$$Q = vA = 0.0001\ \text{m/day}\ (100\ \text{m x}\ 1,000\ \text{m}) = 10\ \text{m}^3/\text{day}$$
$$V = Q\,t = 10\ \text{m}^3/\text{day}*365 = 3,650\ \text{m}^3$$

$$3,650\ \text{m}^3*0.00081\ \text{m}^3/\text{ac} - \text{ft} = 3\ \text{ac} - \text{ft}.$$

This volume of water would irrigate approximately one acre of cotton, assuming that the gross depth of application would be 3 ft/year.

10. Find the hydraulic gradient and the direction of flow with the East axis (x-axis) as zero degrees for the following three wells. Show your work (work it by hand), and check your work with the *Groundwater* program.

	East	North	Elevation
Well 1	50	600	104
Well 2	400	250	105
Well 3	200	50	108

Step a. The first step is to find the well with the intermediate water level. This is well 2, and the elevation is 105.

Steps b and c. Find the ratio of the high and low well elevations and the point at which the contour line crosses the line between the high and low wells.

$$\mathrm{Ratio_{H,L}} = ((z_H - z_M)/(z_H - z_L) = (108 - 105)/(108 - 104) = 0.75$$

$$x_{Int} = x_H - (x_H - x_L)^*\mathrm{Ratio_{H,L}} = 200 - (200 - 50)^*0.75 = 87.5\ \mathrm{m}$$

$$y_{Int} = y_H - (y_H - y_L)^*\mathrm{Ratio_{H,L}} = 50 - (50 - 600)^*0.75 = 462.5\ \mathrm{m}$$

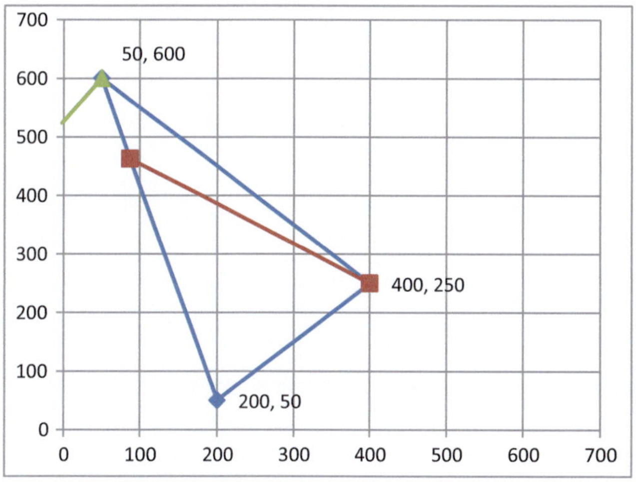

Determination of flow direction based on water table elevations in three wells.

The next step is to find the equation for the contour line.

$$m = (y_{Int} - y_M)/(x_{Int} - x_M) = (462.5 - 250)/(87.5 - 400) = -0.68$$
$$y = mx + b \quad b = y_{Int} - mx_{Int}$$
$$b = 462 - (-0.68)^*87.5 = 522$$
$$y = -0.68\ x + 522$$

The slope of the line that is perpendicular to this line, which is the direction of flow is

$$m_{Flow} = -(-1/-0.68) = 1.47$$

The y-intercept of the flow direction is found with the coordinates of the well with the lowest elevation.

$$b_{Flow} = y_L - m_{Flow}x_L = 600 - 50^*1.47 = -526.5$$

The intersection point of the two lines is found by solving the two equations simultaneously

$$y = (-\ m_{Flow}{}^*b/m + b_{Flow})/(1 - m_{Flow}/m)$$
$$y = (-1.47^*522/(-0.68) + 526.5)/(1 - 1.47/(-0.68)) = 523.4$$
$$x = (y - b)/m = (523.4 - 522)/(-0.68) = -2.1$$

The distance between the point and the line is found with the Pythagorean Theorem.

$$\mathrm{Distance} = \left((50 - (-2.1))^2 + (600 - 523.4)^2\right)^{0.5} = 92.6$$

The hydraulic gradient, dH/dL is the elevation difference between the contour line and the lowest well divided by the distance. The elevation of the contour line is 104 m.

$$dH/dL = (104 - 100.8)/92.6 = 0.0108\ \mathrm{m/m}.$$

11. If the aquifer in question 10 is a coarse sand aquifer with hydraulic conductivity equal to 100 m/day and porosity of 0.40, then calculate the Darcy velocity and the rate that a contaminant plume would travel through the aquifer.

The Darcy velocity is calculated with the Darcy equation.

$$v_{Darcy} = -K\frac{\Delta H}{L} = 100\ m/day^*0.0108\ m/m = 1.08\ m/day$$

The actual velocity is the Darcy velocity divided by the porosity.

$$v = \frac{v_{Darcy}}{\phi} = 1.08/0.4 = 2.7\ m/day$$

12. What is the reason that aquifer pollution is much more difficult to correct than surface water pollution?

The rate of exchange in aquifers is much slower than for surface waters so the pollutants are not flushed out of an aquifer.

13. What are the primary pollutants from agriculture that have contributed to aquifer pollution?

Nutrients and pesticides.

14. What often happens to shallow aquifers when irrigation is introduced to a region?

Excess water from irrigation leaches to the aquifer and the water table rises. Eventually, the water table rises to near to soil surface, which makes it impossible to leach salts out of the soil profile. The soil becomes salinized unless subsurface drainage systems are installed.

15. Find the transmissivity and conductivity of confined aquifer in which the flow rate to a well is 400 gpm, and observation wells at distances of 100- and 200-m from the pumping well have depths to the water table of 100-m and 98-m, respectively. The upper surface of the aquifer is 140 m below ground and the aquifer is 40 m thick. Check your calculations with the Confined aquifer Worksheet.

	Distance	Ground surface elevation	Depth to water table	Water table elevation above aquitard
Pumping well	0	180		
Observation well 1	100	180	100	80
Observation well 2	200	180	98	82
Lower boundary with aquitard	0	m		
Well flow rate	400	GPM		
Flow rate	2180	m3/day		
Transmissivity	120	m2/day		
Aqufer thickness	40	m		
Conductivity	3.006	m/day		

16. Find the conductivity and transmissivity in an unconfined aquifer in which the flow rate to a well is *400 gpm,* and observation wells at distances of 100- and 200-m from the pumping well have depths to the water table of 100-m and 98-m, respectively. The lower boundary of the aquifer (upper surface of aquitard) is 150 m below the ground.

	A	B	C	D	E	F
1		Distance	Ground surface elevation	Depth to water table	Water table elevation above aquitard	
2	Pumping well	0	150			
3	Observation well 1	100	150	100	50	
4	Observation well 2	200	150	98	52	
5			above datum			
6	Lower boundary of aquifer	0	m			
7	with respect to datum					
8				This section calculates K and T based on Q and h		
9	Well flow rate (U.S. units)	400	GPM	for an unconfined aquifer		
10	Well flow rate (metric units)	2180	m3/day			
11						
12	Conductivity	2.36	m/day			
13	Average aquifer thickness	51.00	m			
14						
15	Transmissivity	120.26	m2/day			

17. Calculate the transmissivity and storativity of a confined aquifer. The following drawdown data was collected from an observation well 100 m from the pumping well. The well flow rate was 2,000 m^3/day.

Time after initiation of pumping	Drawdown s (m)
1 min	0.44
2 min	0.55
4 min	0.66
8 min	0.77
0.01 day	0.86
0.02 day	0.97
0.04 day	1.08
0.08 day	1.19
0.16 day	1.30
0.32 day	1.41
0.64 day	1.52
1 day	1.59

Convert to t/r2.

t/r^2 (day/m^2)	Drawdown s (m)
6.9E-08	0.44
1.4E-07	0.55
2.8E-07	0.66
5.6E-07	0.77
0.000001	0.86
0.000002	0.97
0.000004	1.08
0.000008	1.19
0.000016	1.30
0.000032	1.41
0.000064	1.52
0.0001	1.59

Now use the scroll bar. You can extend the theoretical curve to include smaller values of t/r2 by lowering the initial value in cell A5: enter 0.00000001. Change the flow rate in cell B1 to 2,000 m^2/day. The scroll bars are adjusted until the storativity is equal to 0.00001, and the transmissivity is equal to 1,000 m^2/day. Results are shown below.

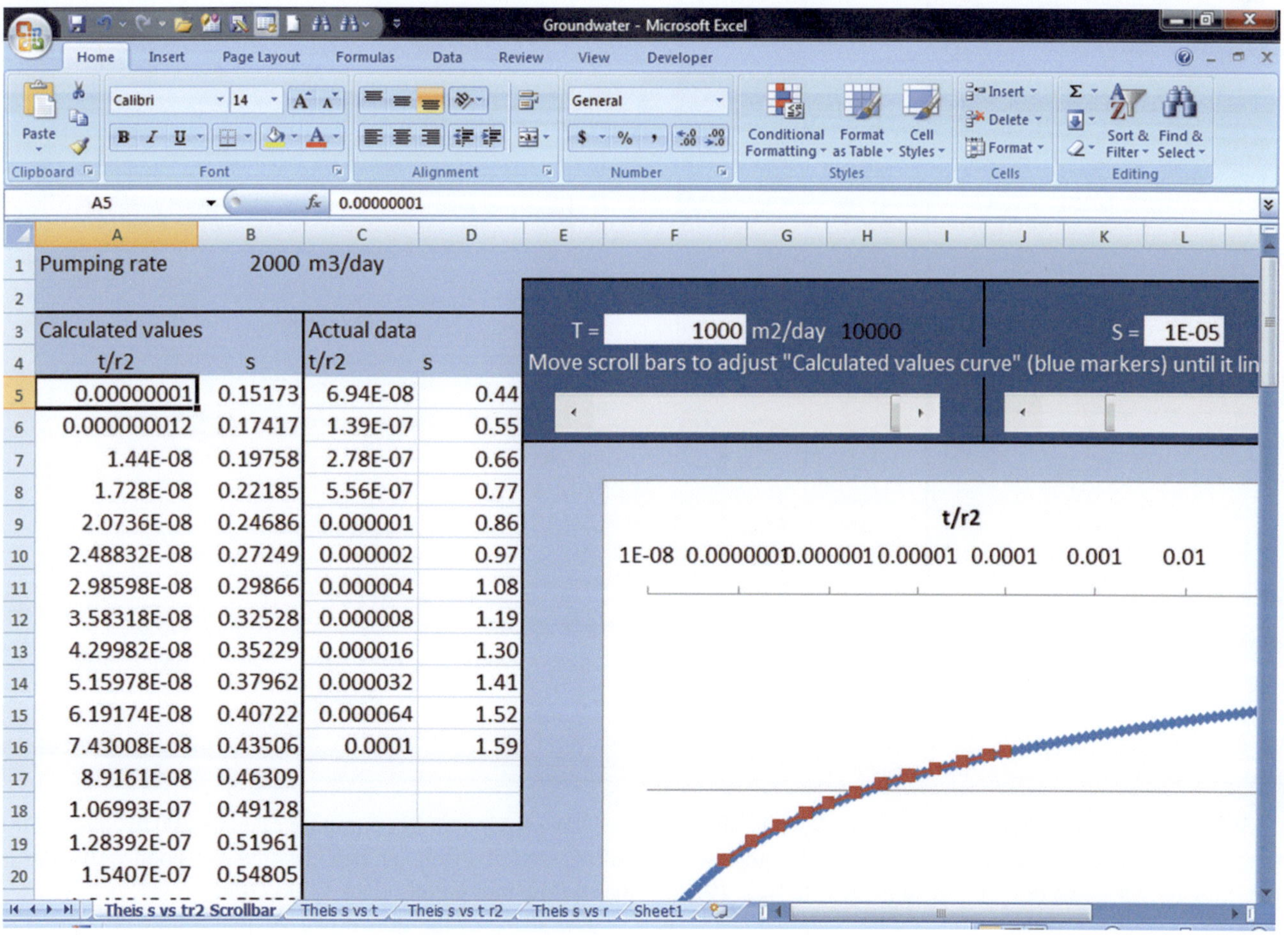

18. For the aquifer in question 17, calculate the drawdown in the well for a series of points between 1 hr and 1 week. The well diameter is 50 cm (use r = 0.25 m in the *Theis s* vs. *t* worksheet). The pump flow rate is 3,000 m^3/day.

The well radius is 0.25 m.

Time	Drawdown (m)
1 hr	4.49
1 day	5.25
7 days	5.72

Hand calculation

$$u = r^2S/(4Tt)) = 0.00001*0.25\wedge 2/(4*1,000*7) = 2.23*10^{-11}$$

$$W(u) = \left[-0.5772 - \ln\ u + u - \frac{u^2}{2*2!} + \frac{u^3}{3*3!} - \frac{u^4}{4*4!}\right] = 23.9$$

$$s = \frac{Q}{4\pi T}W\left(\frac{r^2S}{4Tt}\right) = \frac{3,000}{4\pi(1,000)}*23.9 = 5.72\ m$$

19. The well described in questions 17 and 18 has a maximum acceptable drawdown of 20 m inside the well. The pressure loss in the casing is flow rate (m^3/day)/4,600. Calculate the maximum allowable pump flow rate and the drawdown at 80 %, 60 %, 40 %, and 20 % of maximum. Plot the drawdown vs. flow rate curve. What is the shape of the curve? Convert the maximum flow rate to units of GPM and report whether this well would be considered a good well. Assume that the drawdown after 7 days is the steady state drawdown.

Input the aquifer parameters into the *Theis s* vs. *t* Worksheet and adjust Cell B4 until the drawdown is approximately 18 m after 7 days.

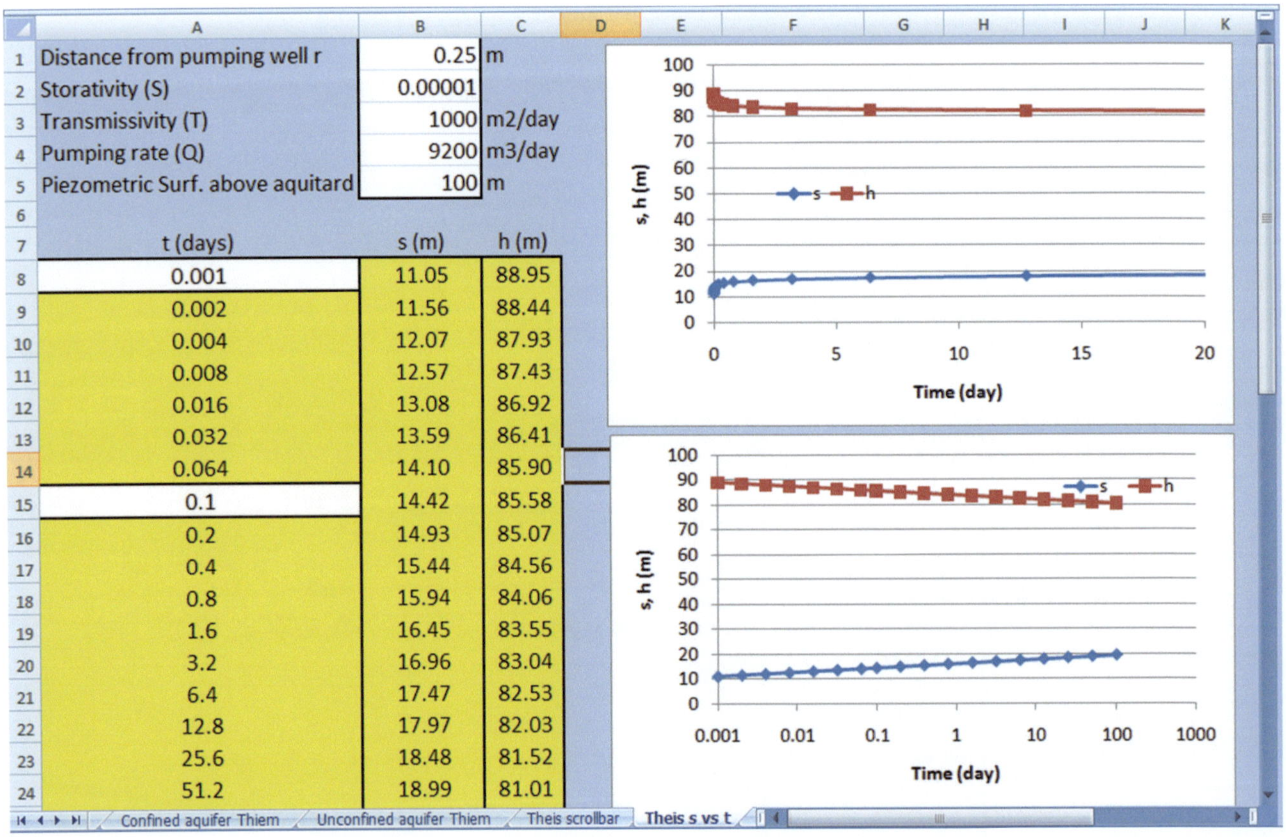

When Cell B4 is 9200 m^3/day, the drawdown is just less than 18. Two m head loss takes place in the well screen when the drawdown outside the casing is 18 m. (9200/4600) = 2 m. Thus, the total drawdown is 17.5 + 2 = 19.5. Close enough.

Thus, the maximum pumping rate is 9,200 m^3/day (1,700 GPM), and this well can be classified as a good well since it has a flow rate greater than 1,000 GPM.

The shape of the drawdown vs. pump flow rate curve is nearly linear. The upper line is outside the casing, and the lower line is inside the casing. The difference is the energy loss within the casing. (2 m at 9200 m3/day).

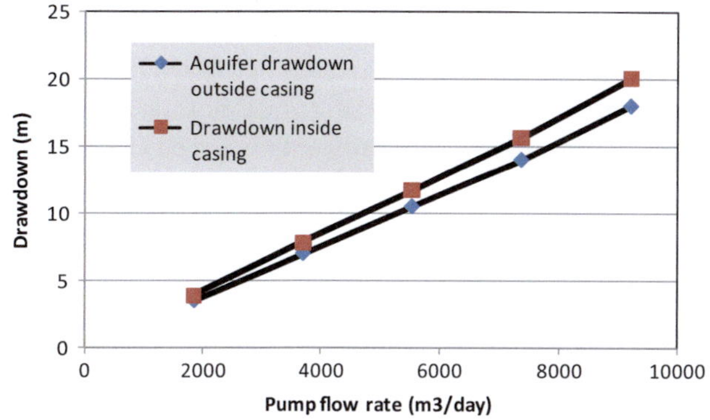

Figure. Drawdown vs. flow rate.

20. The static water table in the aquifer described in problems 17–19 is 40 m below the ground surface. The pump in the well has a pump curve as shown below. The pump has an open discharge 0.5 m above the ground surface, and the sum of minor losses (K) = 3.9 (including velocity head losses at the discharge). The pump hangs on a 12 inch pipe (Schedule 40) at an elevation 80 m below the ground surface, and there is a 2 m section of pipe above the ground surface (total 82 m pipe). The pipe has a Hazen Williams C = 100. Include the minor losses. Calculate the discharge flow rate.

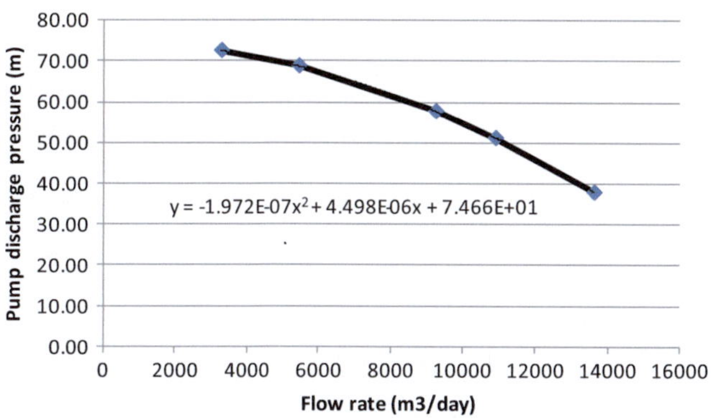

12 inch Schedule 40 pipe has 0.406 inch thickness, and outside diameter = 12.75 inches. Thus, the inside diameter is 12.75–0.812 = 11.94 in. (this figures are found in both Tables 8.2 and 10.3.) The metric ID is 303.2 mm.

The minor losses and pipe losses are subtracted from the total head produced by the pump.

$$H = -1.972*10^{-7}*Q + 4.498*10^{-6}*Q + 74.66 - \left(k_1 L \frac{\left(\frac{Q/3600*24}{C}\right)^{1.85}}{D^{4.87}} + K_m \frac{v^2}{2g} \right)$$

$$H = -1.972*10^{-7}*Q^2 + 4.498*10^{-6}*Q + 74.66 - \left(1.22*10^{10}*82 \frac{\left(\frac{Q/3600*24}{100}\right)^{1.85}}{303^{4.87}} + 3.5 \frac{\left(\frac{Q}{3600*24*\pi*(0.303/2)^2}\right)^2}{2*9.8} \right)$$

An equation was calculated with the Trendline function for drawdown vs. pumping rate.

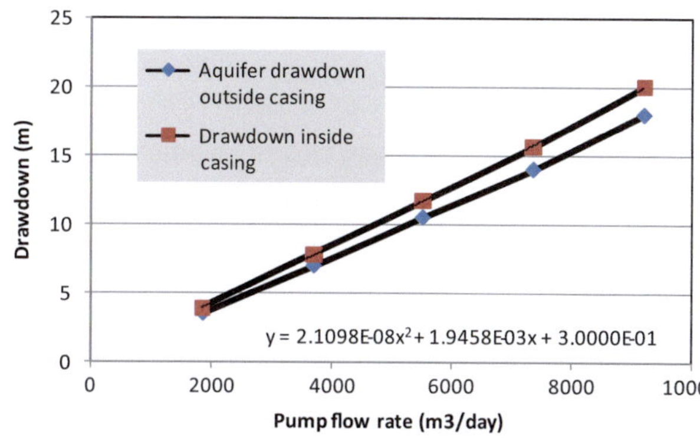

The distance of water in the well from the outlet at the surface (also equal to total head H as calculated in the pump and pipe equation above) is

$$H = 40\text{ m} + 0.5\text{ m} + 2.1098E - 8*Q^2 + 1.9458E - 3*Q + 0.3$$

Set the two equations equal to each other in order to solve for the flow rate.

$$40.8 + 2.1098E - 8*Q^2 + 1.9458E - 3*Q = -1.972*10^{-7}*Q^2 + 4.498*10^{-6}*Q + 74.7 - \left(1.22*10^{10}*62 \frac{\left(\frac{Q/3600*24}{100}\right)^{1.85}}{303^{4.87}} + 3.5 \frac{\left(\frac{Q}{3600*24*\pi*(0.303/2)^2}\right)^2}{2*9.8} \right)$$

Set the problem up to be solved by iteration in Excel (solve for Q).

$$Q = \left(-2.183 * 10^{-7} * Q^2 + 33.9 - 1.22*10^{10}* 82 \frac{\left(\frac{Q/3600*24}{100}\right)^{1.85}}{303^{4.87}} - 3.5 \frac{\left(\frac{Q}{3600*24*\pi*(0.303/2)^2}\right)^2}{2*9.8} \right) \Big/ 1.941*10^{-3}$$

$Q = 8010\ m^3/\text{day}$.

The pressure loss in the pipe and minor losses are calculated as follows.

$$h_f = 1.22*10^{10}*82 \frac{\left(\frac{8010/3600*24}{100}\right)^{1.85}}{303^{4.87}} - 3.5 \frac{\left(\frac{8010}{3600*24*\pi*(0.303/2)^2}\right)^2}{2*9.8}$$

$$= 0.295\ m$$

21. What are the two types of groundwater pollution? Which comes from field agriculture?

Point source and nonpoint source. Nonpoint source comes from field agriculture.

Chapter 11: Solution

1. ***Describe the different types of canals.***

- Main canals: Large canals constructed by the Federal Government that carry water between regions.
- Irrigation district main canals: the large canals that carry water to different parts of irrigation districts.
- Secondary (subsystem) canals: these canals deliver water from main canals to lateral (tertiary) canals.
- Irrigation district lateral canals: canals that deliver water to farm turnouts.
- On-farm irrigation ditches: irrigation ditches distribute water in fields. They deliver water to fields in a variety of ways such as open discharge to basins, alfalfa valves connected to spiles (pipes under the canal), or siphon tubes which are placed over the canal bank.
- Drainage channels: these channels remove drainage water from the project. Drainage channels are normally installed below grade in order to remove drainage water by gravity flow.

2. ***How is water diverted from a main canal to a lateral canal?***

The diversion of water from a main canal to a lateral canal has several design features. A diversion structure is placed just downstream from the turnout in order to increase upstream depth. This technique is also used in irrigation head ditches where water is blocked just downstream from the part of the field that is irrigated.

3. ***How is water diverted from a lateral canal to a field?***

Water is diverted from a lateral canal to a field through irrigation ditches. They can deliver water to fields in a variety of ways such as open discharge to basins, alfalfa valves connected to spiles (pipes under the canal), or siphon tubes which are placed over the canal bank. The canals are often blocked just downstream from where water is applied in order to increase canal elevation and flow rate to the field.

4. ***What is the reason for energy dissipation structures in canals and in canal outlets?***

Energy is dissipated in canal outlets to fields in order to prevent the erosion of soil. Energy is dissipated in structures within canals in order to have subcritical flow on a slope that would otherwise have supercritical flow (steep slope).

5. ***Calculate the conveyance efficiency and water duty for a canal that is 20 km long, has a wetted top width = 20 m, wetted perimeter = 26 m, and cross-sectional area = 100 m²? Average canal flow velocity is 1 m/sec. Reference ET is 10 mm/day. Average seepage rate is 5 mm/day. In addition to reporting the water duty and efficiency, convert the seepage rate to L/m²/day.***

$$Q = Av$$
$$Q = 100*1 = 100m^3/sec$$
$$Q = \frac{100m^3}{sec} * \frac{3600sec}{1hr} * \frac{24hr}{1\ day} = 8,640,000m^3/day$$

Assuming that the canal is a rectangular channel, then b = top width.

$$V_{Evap/day} = (20,000\ m)(20\ m)(0.010\ m/day) = 4,000\ m^3/day$$
$$V_{Seepage/day} = (20,000\ m)(26\ m)(0.005\ m/day) = 2,600\ m^3/day$$

The percent evaporation is $(4,000/8,640,000)(100\%) = 0.046\%$.

The percent evaporation is $(2,600/8,640,000)(100\%) = 0.030\%$.

$$\begin{aligned} Total\ water\ duty &= \%seepage\ losses \\ &\quad + \%evaporation\ losses \\ &= 0.046\% + 0.030\% = 0.076\%. \end{aligned}$$

Conveyance efficiency is $100\% - 0.076\% = 99.92\%$.

$$\begin{aligned} V_{Seepage/day} &= 2,600\ m^3/day \\ &= 2,600,000\ L/day/(20,000*26)\ m^2 \\ &= 5\ L/m^2/day. \end{aligned}$$

6. ***Calculate the conveyance efficiency to field 1 in Fig. 11.9 from the point of water diversion to the irrigation district. The conveyance efficiency of the irrigation district up to the farm turnout is 80 %. The main concrete canal on the farm has a conveyance efficiency of 80 %, and the earth-lined canal has a conveyance efficiency of 80 %.***

$$Efficiency = (0.8)(0.8)(0.8)(100\%) = 51.2\%$$

7. ***Redo Example 11.4, but the ditch is constructed in loam and sandy loam soils (Fig. 11.8) with a seepage rate of 1.5 m^3/m^2/day.***

The wetted perimeter is

$$P = b + 2{*}y{*}\left(1 + z^2\right)^{0.5} = 0.3 + 2{*}0.75{*}\left(1 + 2^2\right)^{0.5} = 3.65\ m$$

The seepage rate is 1,500 L/m^2/day. Then,

$$\begin{aligned} V_{Seepage/day} &= 3.65\ m{*}1,000\ m{*}1500\ L/m^2/day \\ &= 5,475,000\ L/day/km \\ V_{Seepage/yr} &= 5,475,000\ \ L/day{*}60\ days{*}0.001\ \ m^3/ \\ &\quad L{*}0.01\ \ ha - cm/m^3 = 3,285\ \ ha - cm/yr \\ \$_{Seepage/yr} &= 3,285\ ha - cm/yr{*}\$3.27/ha - cm \\ &= \$10,741.95/year. \end{aligned}$$

The present value of $10,741/yr for a 20 yr project at 8 % interest is approximately $110,000/km (multiply annual value by 10 for this period and interest rate). Therefore, it is profitable to line the irrigation ditch because the present value is bigger than the liner cost ($60,000/km).

8. ***Describe the method used to run canal water past a road, drainage ditch, or river valley.***

Pipes (culverts or siphons) are used in irrigation districts to go under or over obstructions, and to deliver water to farmers. Huge siphons are sometimes used on large canals to move water past river valleys.

9. ***A concrete lined trapezoidal channel (Fig. 11.14) has a slope of 0.3 % = 0.003 m/m. Flow rate in the channel is 300 L/sec, and the Manning's roughness coefficient, n, of the channel is 0.015. Calculate the depth of flow in the channel. The bottom width is 1 m and side slope z = 1.5.***

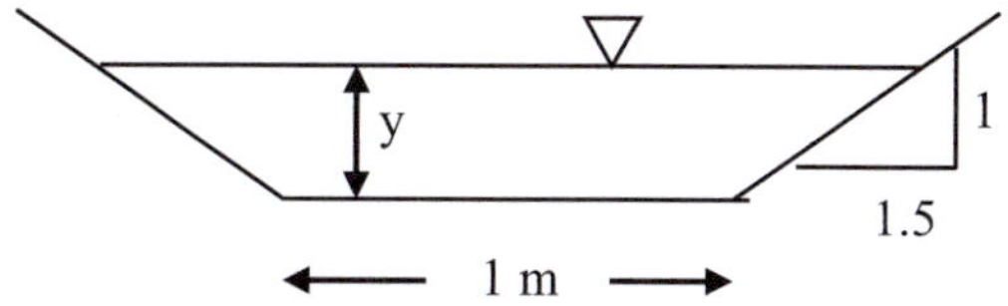

Solve for left side of Eq. 11.9. The following procedure is in the *Canal Depth* Worksheet.

$$\frac{Qn}{S_0^{0.5}} = \frac{0.3{*}0.015}{0.003^{0.5}} = 0.082$$

Use iteration to find y. First try y = 30 cm = 0.30 m.

$$AR^{2/3} = (b + zy)y\left(\frac{(b + zy)y}{b + 2y\sqrt{1 + z^2}}\right)^{2/3}$$

$$AR^{2/3} = (1 + 1.5{*}0.30){*}0.30\left(\frac{(1 + 1.5{*}0.30){*}0.30}{1 + 2(0.30)\sqrt{1 + 1.5^2}}\right)^{2/3}.$$

$$AR^{2/3} = 0.15$$

Adjust y as follows (this is a fast iteration procedure).

$$y_2 = y_1\left(\frac{\left(AR^{2/3}\right)_{actual}}{\left(AR^{2/3}\right)}\right)^{0.5}$$

$$y_2 = 0.3\left(\frac{0.082}{0.15}\right)^{0.5} = 0.21$$

Another iteration results in 0.21. Thus, final answer is 0.21 m = 21 cm water depth.

Add 25 % freeboard elevation.

$$0.21{*}1.25 = 0.266\ m$$

2	Q	0.30000	m3/sec	Trapezoidal or rectangular canal only			
3	z	1.5					
4	n	0.015					
5	b	1.000	m				
6	S	0.003	m/m				
7							
8	SF	0.082158					
9	yguess	0.3000	0.3000	0.2197	0.2133	0.2126	Final y value (depth in m)
10	AR23	0.0822	0.15318	0.0871632	0.0826873	0.0822162	
11							
12	Freeboard	25%					
13	Canal depth	26.6	cm				
14							
15	Area	0.280	m2				
16							
17	Velocity	1.070	m/sec				
18							
19	Top width	1.638	m				
20							
21	Q Froude	0.826					
22							
23	y Froude	0.826					

10. ***Calculate the Froude number for the previous problem, and determine whether the channel has supercritical or subcritical flow.***

You must use the Froude number equation for a trapezoidal canal.

$$Fr = \frac{v}{(A/_T g)^{0.5}}$$

$$v = \frac{Q}{A}$$

$$A = (b + zy)y = (1 + 1.5{*}0.21){*}0.21 = 0.28m^2$$

$$T = b + 2\ zy = 1 + 2{*}1.5{*}0.21 = 1.64\ m$$

$$v = \frac{0.3\ m^3/sec}{0.28\ m^2} = 1.07\ m/sec$$

$$Fr = \frac{1.07}{((0.28/_{1.64})(9.81))^{0.5}} = 0.826$$

Thus, the flow is subcritical.

11. ***If a canal were to travel down a relatively steep slope, what strategy could be used to prevent supercritical flow in the canal?***

In order to maintain subcritical flow over the majority of the canal, energy is dissipated in a series of structures such as steeply sloped chutes, weirs, concrete apron or blocks in order to dissipate energy.

12. ***For the canal dimensions described in problem 9, calculate the chute slope required to have a stable hydraulic jump with Froude number = 6.***

Using the spreadsheet, the chute slope required is: 0.216 m/m (21.6 %). In this problem, you cannot solve for the depth of the downstream hydraulic jump with the information provided in this chapter since the channel is trapezoidal. This was an error in the problem statement.

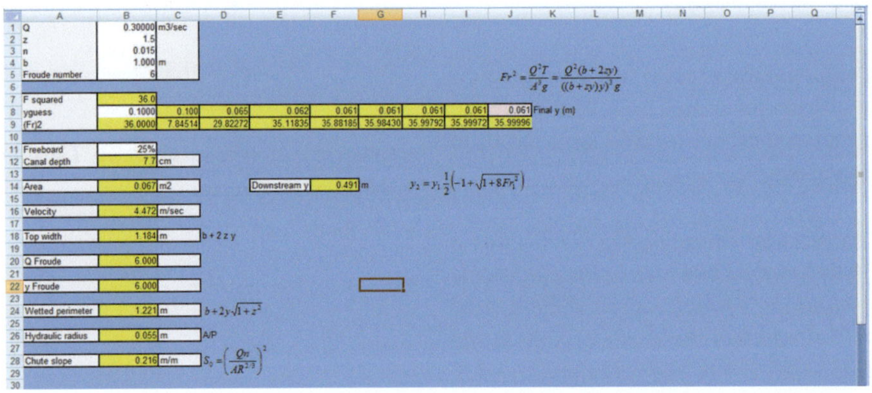

	A	B	C	D	E	F	G	H	I	J	K
1	Q	0.30000	m3/sec								
2	z	1.5									
3	n	0.015									
4	b	1.000	m								
5	Froude number	6								$Fr^2 = \frac{Q^2T}{A^3g} = \frac{Q^2(b+2zy)}{((b+zy)y)^3g}$	
6											
7	F squared	36.0									
8	yguess	0.1000	0.100	0.065	0.062	0.061	0.061	0.061	0.061	0.061	Final y (m)
9	(Fr)2	36.0000	7.84514	29.82272	35.11835	35.88185	35.98430	35.99792	35.99972	35.99996	
10											
11	Freeboard	25%									
12	Canal depth	7.7	cm								
13											
14	Area	0.067	m2		Downstream y	0.491	m	$y_2 = y_1\frac{1}{2}\left(-1+\sqrt{1+8Fr_1^2}\right)$			
15											
16	Velocity	4.472	m/sec								
17											
18	Top width	1.184	m	b + 2 z y							
19											
20	Q Froude	6.000									
21											
22	y Froude	6.000									
23											
24	Wetted perimeter	1.221	m	$b+2y\sqrt{1+z^2}$							
25											
26	Hydraulic radius	0.055	m	A/P							
27											
28	Chute slope	0.216	m/m	$S_0 = \left(\frac{Qn}{AR^{2/3}}\right)^2$							
29											
30											

13. ***Describe why a Froude number 6 is desirable before a hydraulic jump.***

An upstream Froude number between 4.5 and 9.0 maintains a stable hydraulic jump and high energy loss so this is the desired range of Froude number for upstream flow.

14. ***Calculate the flow over a standard contracted rectangular weir. The head over the weir is 0.13 m and the weir blade is 0.6 m across.***

$$Q = 1.74H^{3/2}(L_b - 0.2H)$$

$$= 1.74(.13m)^{3/2}(.6m - (0.2)(.13m)) = .047\frac{m^3}{\text{sec}}$$

Chapter 12: Solution

Question 12.1. Discuss the different types of sprinkler nozzles and systems used on center pivots. Discuss the strengths and weakness of the systems.

Top mounted impact sprinklers
Strengths: Large wetted diameter leads to lower runoff. Lower cost. Suitable for humid regions. Easy to detect if sprinkler is malfunctioning. No need for furrows or planting in circle.
Weaknesses: High evaporation leads to loss of efficiency. Wetting of foliage (may be disadvantage). Higher pressure required so energy cost is higher. Affected by wind. Droplet impact may cause crusting. May not be suitable for application to foliage with salty water.

LEPA
Strengths: Low energy requirement. Low evaporation loss. Don't wet foliage. Suitable for all climates (arid and humid). Suitable for salty water use, however, must be careful of salt buildup in beds if water applied to furrows.
Weaknesses: Higher cost of closely spaced drop tubes and nozzles. Need dikes and furrows. Harder to see emitter operation.

LESA
Strengths: Evaporation is relatively low so suitable for ARID climates. Suitable for broadcast crops that are not planted in rows.
Weaknesses: High cost of closely spaced drop tubes and nozzles. Low wetted diameter so may need multiple passes per day. May not be suitable for use with salty water if foliage is wetted.

MESA
Strengths: Larger wetted diameter than LESA, especially with rotors. Relatively low pressure requirement. Water distribution less affected by wind. More even application pattern. Relatively easy to see is sprinklers are not working. No need to plant in circle or construct furrows. Lower evaporation than impacts.
Weaknesses: Smaller wetted diameter than impact sprinklers. More evaporation than LESA or LEPA. May not be suitable for ARID climates. May cause soil crust formation by droplet impact. May not be suitable for use with salty waters.

Question 12.2. What is the difference between a linear move and a center pivot irrigation system?
Linear moves travel in a straight line and irrigate square fields. Linear moves irrigate from a canal or use a movable hose system.
Center pivots irrigate in a circle and draw water from a central pipe. Center pivots can only irrigate field corners with an arm at the end of the pivot or partially with a big gun that turns on in the corners.

Question 12.3. Calculate the percent evaporation from sprinkler droplets for the parameters in Example 12.1 part 1 except that relative humidity is 50 %. If the application depth is 25 mm to a mature corn crop from overhead impact sprinklers, then what is the total depth of evaporation + canopy interception loss?

$$e_s - e_a = 0.61\ \exp\left(\frac{17.27(40)}{40 + 237.3}\right)(1 - 0.5) = 3.68\ kPa$$

$$\begin{aligned} L_e &= \left[(1.98(3.57)^{-0.72} + 0.22(3.68)^{0.63} \right. \\ &\quad \left. + 3.6*10^{-4}(490)^{1.16} + 0.14(4.44)^{0.72}\right]^{4.2} = 26\% \end{aligned}$$

Estimate that canopy interception would be 10 % (2.5 mm). 26 % of 25 mm would be 6.5 mm. Thus, total loss would be 9 mm or 36 %.

$$e_s - e_a = 0.61\ \exp\left(\frac{17.27(25)}{25 + 237.3}\right)(1 - 0.5) = 1.58\ kPa$$

$$\begin{aligned} L_e &= \left[(1.98(3.57)^{-0.72} + 0.22(1.58)^{0.63} \right. \\ &\quad \left. + 3.6*10^{-4}(490)^{1.16} + 0.14(4.44)^{0.7}\right]^{4.2} = 12\% \end{aligned}$$

Estimate that canopy interception would be 10 % (2.5 mm). 12 % of 25 mm would be 3 mm. Thus, total loss would be 5.5 mm or 22 %.

Question 12.4. Calculate the percent evaporation from overhead sprinklers for the parameters in Example 12.1

part 2 except that relative humidity is 60 %. What is the depth of evaporation for an application depth of 25 mm to a mature corn crop from overhead impact sprinklers? Consider canopy interception and droplet evaporation?

$$e_s - e_a = 0.61\ \exp\left(\frac{17.27(40)}{40+237.3}\right)(1-0.6) = 2.94\ kPa$$

$$L_e = \Big[(1.98(3.57)^{-0.72} + 0.22(2.94)^{0.63} + 3.6{*}10^{-4}(490)^{1.16} + 0.14(4.44)^{0.7}\Big]^{4.2} = 23\%$$

Estimate that canopy interception would be 10 % (2.5 mm). 23 % of 25 mm would be 5.75 mm. Thus, total loss would be 8.25 mm or 33 %.

$$e_s - e_a = 0.61\ \exp\left(\frac{17.27(25)}{25+237.3}\right)(1-0.6) = 1.26\ kPa$$

$$L_e = \Big[(1.98(3.57)^{-0.72} + 0.22(1.26)^{0.63} + 3.6{*}10^{-4}(490)^{1.16} + 0.14(4.44)^{0.7}\Big]^{4.2} = 11\%$$

Estimate that canopy interception would be 10 % (2.5 mm). 11 % of 25 mm would be 2.75 mm. Thus, total loss would be 5.25 mm or 21 %.

Question 12.5. Calculate the flow rate of a center pivot that has a length of 350 m, and gross application depth 15 mm/day. The pivot operates for 21 hours/day.

$$A = \pi r_{max}^2 = \pi(350^2)(1\ \text{ha}/10{,}000\ \text{m}^2) = 38.5\ \text{ha}$$

$$Q_p = 0.116\frac{i_g A}{1 - L_{r-m}} = 0.116\frac{(16.2\ mm)\ (38.5\ ha)}{1-(3/24)} = 76.5\ L/s$$

	A	B	
1	Pivot length (radius)	350	m
2	Pivot area	38.5	ha
3	Hours per day operation	21	hr
4	Gross application rate	15	mm/day
5	Pivot flow rate	76.5	L/sec

Question 12.6 Calculate the maximum application rate for the parameters in question 12.5. The sprinkler wetted diameter is 4 m and the percent evaporation is 14 %. Then, show that the maximum application rate is the same if the pivot rotates three times per day (7 hour rotation). Show calculation and explain why the maximum application rate is the same in both cases.

Case 1: 1 rotation

$$T_a = T_r\left(\frac{D_w}{2\pi\ r}\right) = 21\ hr\left(\frac{4\ m}{2\pi(350\ m)}\right) = 2.29\ \text{min}$$

$$i_a = i_g(L_e) = 15\ \text{mm/day}\ (1-0.14) = 12.9\ \text{mm}$$

$$\frac{di}{dt}_{max} = \left(\frac{4}{\pi}\right)\left(\frac{i_a}{T_a}\right) = \left(\frac{4}{\pi}\right)\left(\frac{12.9}{2.28}\right) = 7.2\ mm/\text{min}$$

D	E	F	G	H	I
Sprinkler distance	350	m	Fraction evap	0.14	
Wetted diameter	4	m	ia	12.90	mm/day
Rotation rate	21	hr/day	net depth/pass	12.90	mm/pass
Ta	2.29	minutes	di/dt-max	7.17	mm/min

Case 2: 3 rotations

$$T_a = T_r\left(\frac{D_w}{2\pi\ r}\right) = 7\ hr\left(\frac{4\ m}{2\pi(350\ m)}\right) = 0.76\ \text{min}$$

$$\frac{di}{dt}_{\max} = \left(\frac{4}{\pi}\right)\left(\frac{i_a}{T_a}\right) = \left(\frac{4}{\pi}\right)\left(\frac{12.9/3}{0.76}\right) = 7.2\ mm/\text{min}$$

D	E	F	G	H	I
Sprinkler distance	350	m	Fraction evap	0.14	
Wetted diameter	4	m	ia	12.90	mm/day
Rotation rate	7	hr/day	net depth/pass	4.30	mm/pass
Ta	0.76	minutes	di/dt-max	7.17	mm/min

The maximum is the same because the sprinkler sizes and spacings are the same. One-third of the application depth is applied in one-third of the time.

Question 12.7. Using the parameters in question 12.6, calculate the depth of runoff and maximum application rate during each pass for an intake family 3 soil with 2 mm surface storage. Include the 0.6985 initial infiltration depth. Calculate for one revolution and three revolutions per day. The percent evaporation is 14 %. Use Chapter 12 Center pivot program and hand calculations.

Case 1: 1 revolution/day. There is 3.51 mm runoff.

	A	B		C	D	E
1	Pivot length (radius)	350	m		Sprinkler distance	350 r
2	Pivot area	38.5	ha	Hide graph	Wetted diameter	4 r
3	Hours per day operation	21	hr		Rotation rate	21 h
4	Gross application rate	15	mm/day	Show graph	Ta	2.29 r
5	Pivot flow rate	76.5	L/sec			
6						Input data is
7						
8	Surface storage	2	mm		Number steps	
9	Intake family for infiltration functions (write intake family number in cell B7)				40	
10	Intake family	3	Put intake family to left		fraction	
11	NRCS_a	0.3650	a and b are automatically		0.025	
12	NRCS_b	0.8160	updated		0.05	
13	Include the initial depth	TRUE	This is the 0.6985 mm infilt		0.075	
14	Runoff	3.51	mm		0.1	
15	Infiltration + storage	9.40	mm		0.125	
16	Sprinkler application	12.92	mm		0.15	

Case 2: 3 revolutions/day. There is zero runoff

	A	B		C	D	E	F
1	Pivot length (radius)	350	m		Sprinkler distance	350	m
2	Pivot area	38.5	ha	Hide graph	Wetted diameter	4	m
3	Hours per day operation	21	hr		Rotation rate	7	hr/day
4	Gross application rate	15	mm/day	Show graph	Ta	0.76	minutes
5	Pivot flow rate	76.5	L/sec				
6						Input data is entered	
7							
8	Surface storage	2	mm		Number steps		
9	Intake family for infiltration functions (write intake family number in cell B7)				40		
10	Intake family	3	Put intake family to left		fraction		
11	NRCS_a	0.3650	a and b are automatically		0.025		
12	NRCS_b	0.8160	updated		0.05		
13	Include the initial depth	TRUE	This is the 0.6985 mm infilt		0.075		
14	Runoff	0.00	mm		0.1		
15	Infiltration + storage	4.31	mm		0.125		
16	Sprinkler application	4.31	mm		0.15		

For one revolution per day, total area of columns to right is 6.21. Subtract 2 mm surface storage and 0.6985 from 6.21 mm. This equals 3.5 mm, the answer in Cell B14. There is no runoff for three revolutions per day.

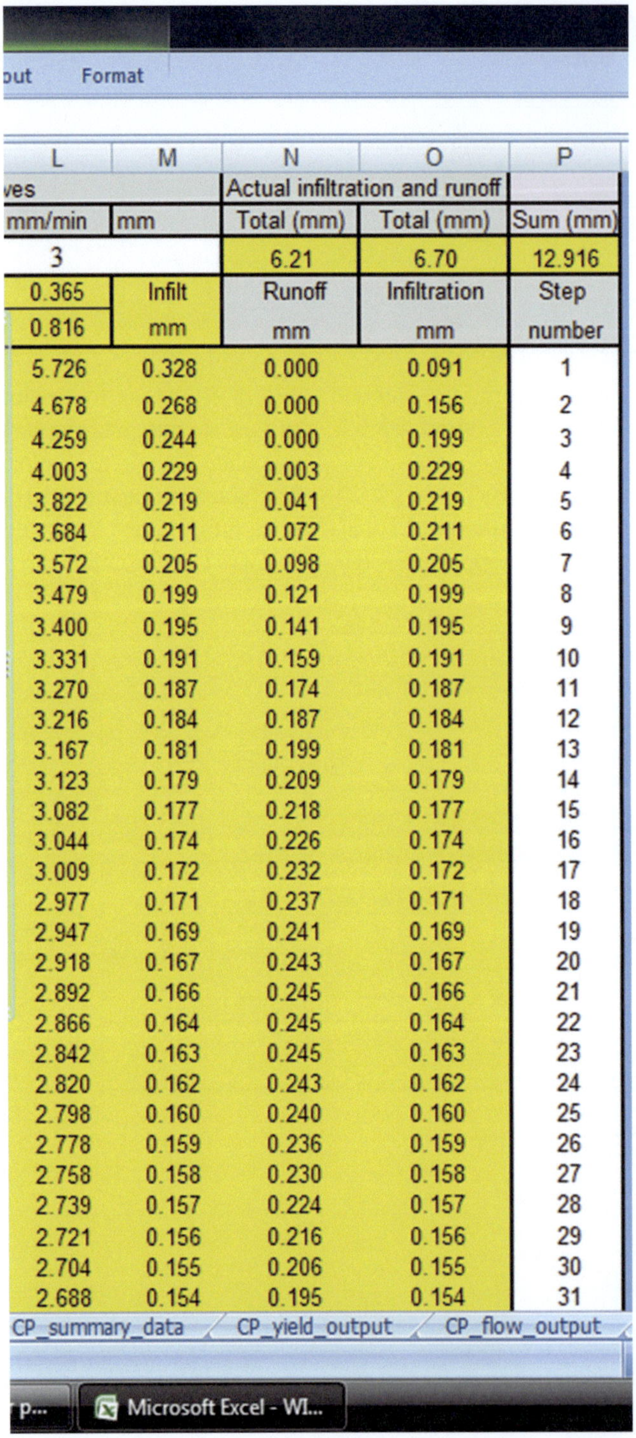

L	M	N	O	P
ves		Actual infiltration and runoff		
mm/min	mm	Total (mm)	Total (mm)	Sum (mm)
3		6.21	6.70	12.916
0.365	Infilt	Runoff	Infiltration	Step
0.816	mm	mm	mm	number
5.726	0.328	0.000	0.091	1
4.678	0.268	0.000	0.156	2
4.259	0.244	0.000	0.199	3
4.003	0.229	0.003	0.229	4
3.822	0.219	0.041	0.219	5
3.684	0.211	0.072	0.211	6
3.572	0.205	0.098	0.205	7
3.479	0.199	0.121	0.199	8
3.400	0.195	0.141	0.195	9
3.331	0.191	0.159	0.191	10
3.270	0.187	0.174	0.187	11
3.216	0.184	0.187	0.184	12
3.167	0.181	0.199	0.181	13
3.123	0.179	0.209	0.179	14
3.082	0.177	0.218	0.177	15
3.044	0.174	0.226	0.174	16
3.009	0.172	0.232	0.172	17
2.977	0.171	0.237	0.171	18
2.947	0.169	0.241	0.169	19
2.918	0.167	0.243	0.167	20
2.892	0.166	0.245	0.166	21
2.866	0.164	0.245	0.164	22
2.842	0.163	0.245	0.163	23
2.820	0.162	0.243	0.162	24
2.798	0.160	0.240	0.160	25
2.778	0.159	0.236	0.159	26
2.758	0.158	0.230	0.158	27
2.739	0.157	0.224	0.157	28
2.721	0.156	0.216	0.156	29
2.704	0.155	0.206	0.155	30
2.688	0.154	0.195	0.154	31

Question 12.9. Calculate the flow rate and runoff at the middle sprinkler (175 m from pivot point) for the pivot described in questions 12.6 through 12.9 for a single 21 hour rotation per day. Sprinkler spacing is 2 m at the middle of the center pivot. Fraction evaporation is 14 %. Sprinkler wetted diameter is 3.5 m. The soil intake family is 1. Surface storage is 4 mm.

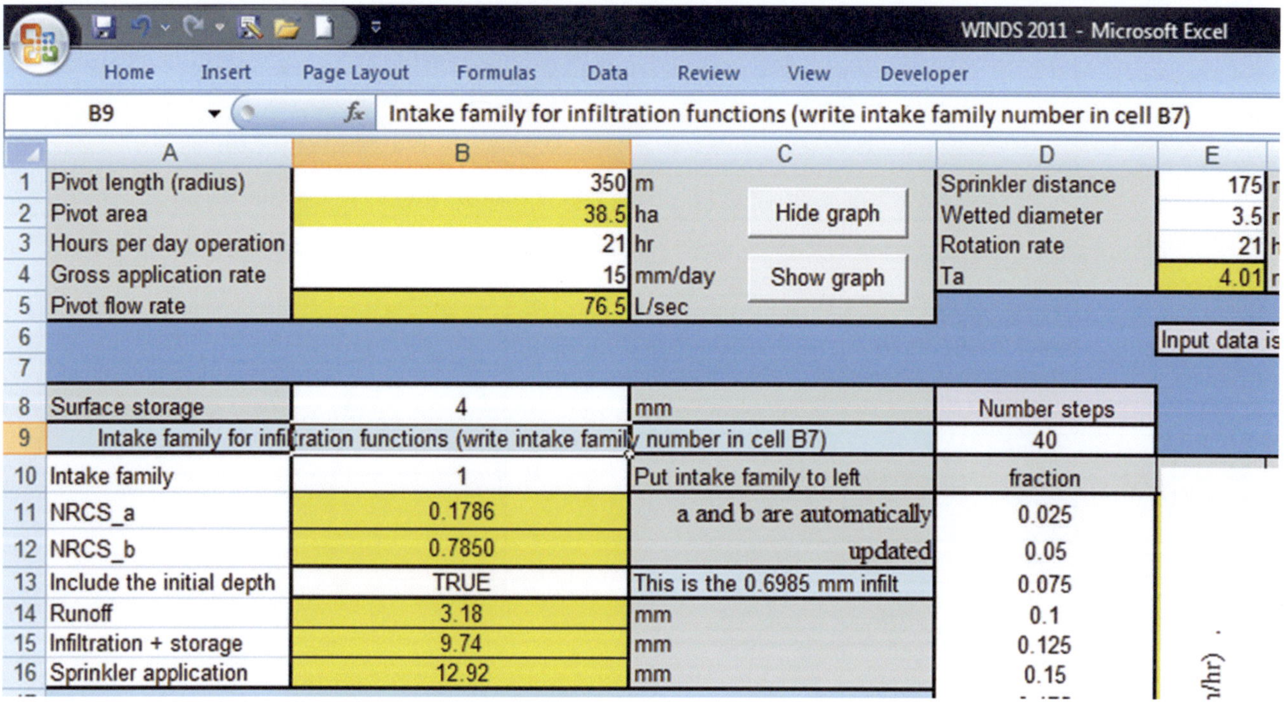

Question 12.10. Derive the Christensen's F factor in Eq. 12.12 by assuming that a center pivot has four sprinklers (1/4, 1/2, 3/4, and full length down the pivot).

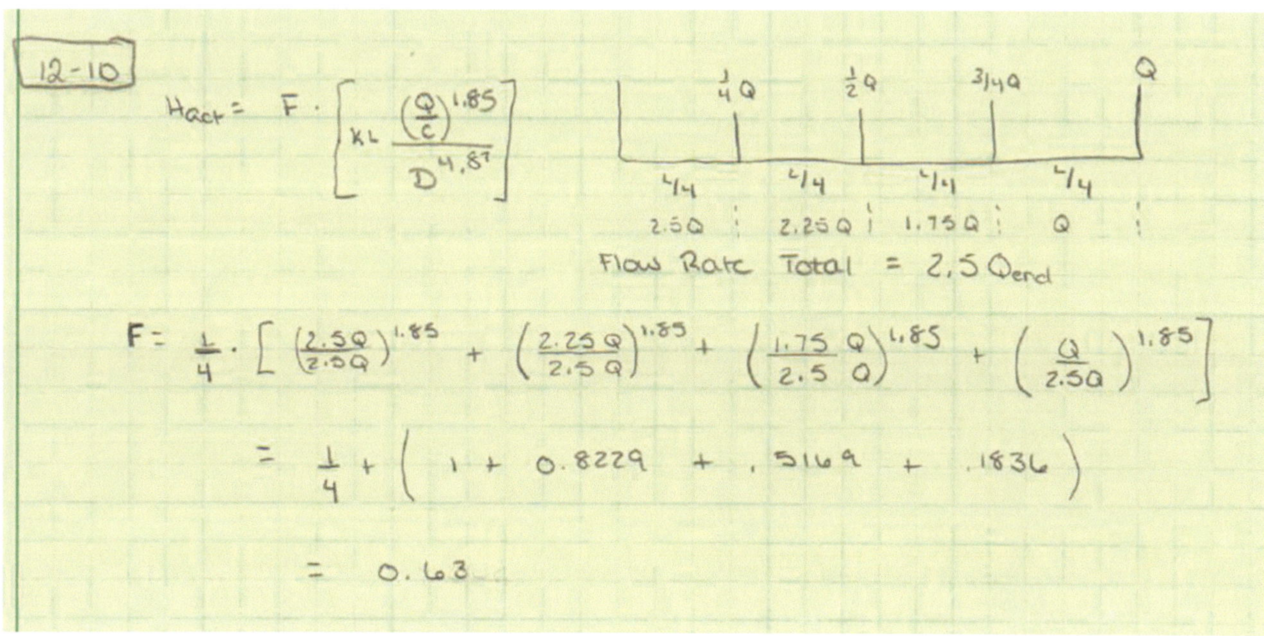

This is not the same as the pivot F factor; however, if 30 sprinklers were used in the calculation, then, the answer would be the same. The difference is due to truncation error.

Question 12.11. Low-pressure sprinkler nozzles at Paradise Cattle Company have a wetted diameter (Dw) of 2 m. The average rotation period of the center pivot is 8 hours. The pivot is 400 m radius and the flow rate is 95 L/sec. 5 % of water is lost to evaporation. Calculate the daily gross application rate, and plot the instantaneous application rate as a function of time at the 400 m.

Answer: max application rate is 102 cm/hr

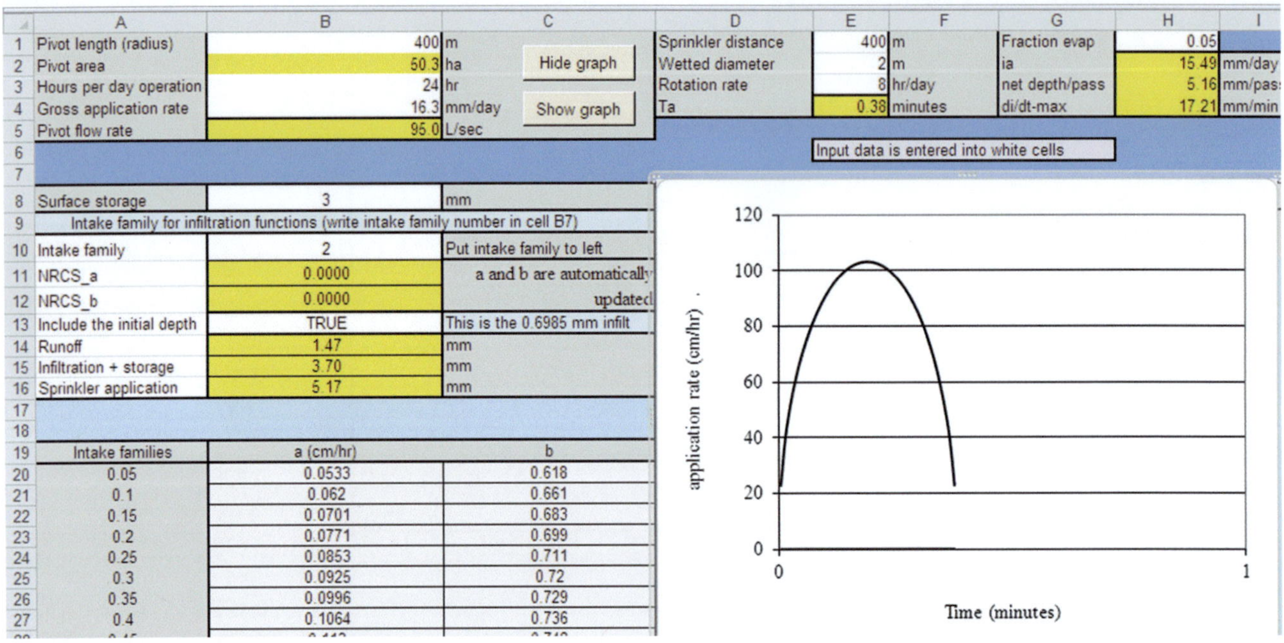

Question 12.12. Plot the instantaneous application rate vs. time for at 200 m for the same parameters as in question 12.11. Answer: max application rate is 51 cm/hr.

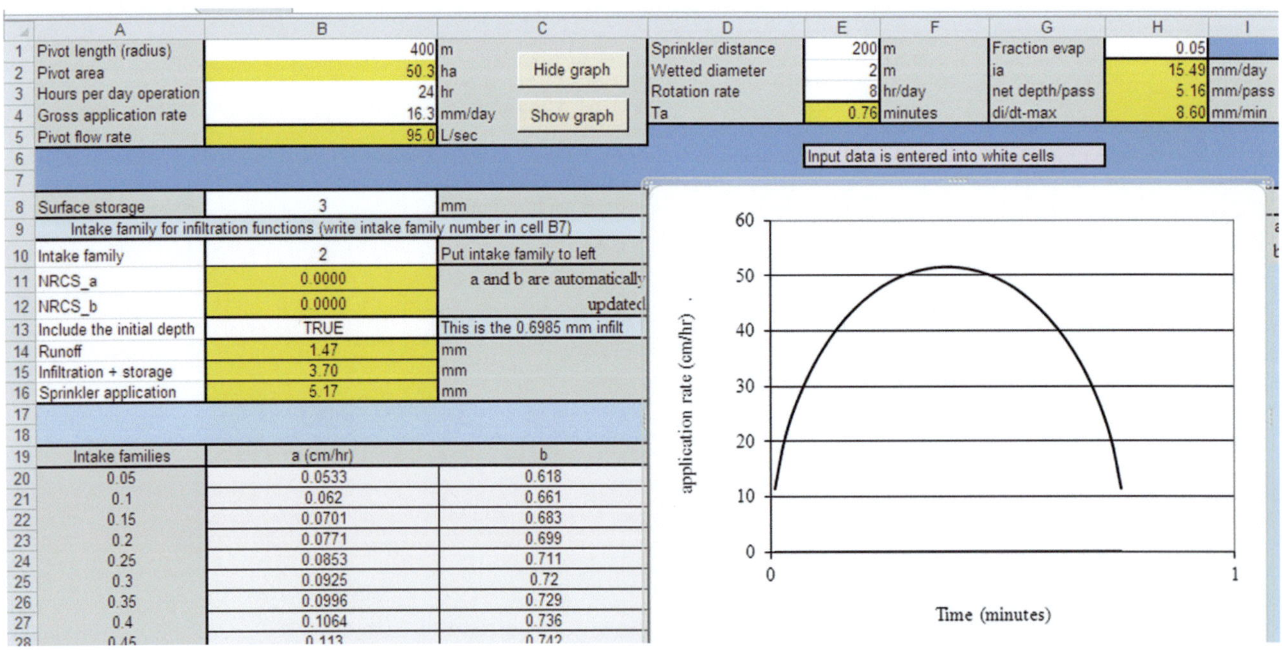

Question 12.13. Calculate the pressure loss in a center pivot that has a length of 350 m, and gross application depth applied during each pass is 5 mm. Time of rotation is 8 hours. Use 198 mm pipe. There is no down time.

With three rotations, the daily gross application depth is 15 mm in 24 hours. The pivot flow rate is 67 L/sec.

	A	B	
1	Pivot length (radius)	350	m
2	Pivot area	38.5	ha
3	Hours per day operation	24	hr
4	Gross application rate	15	mm/day
5	Pivot flow rate	67.0	L/sec

Calculate head loss with Hazen-Williams equation.

$$H_L = kL\left(\frac{\left(\frac{Q}{C}\right)^{1.852}}{D^{4.87}}\right) = 1.22{*}10^{10}{*}350{*}\left(\frac{\left(\frac{67}{130}\right)^{1.852}}{197^{4.87}}\right) = 7.8\ m$$

$H_{act} = H_{LP}F = (7.8)(0.54) = 4.20\text{m}$

Question 12.14. A sprinkler has a flow rate of 5 GPM at 20 PSI. What is the flow rate at 25 PSI?

$$Q_2 = Q_1\sqrt{\frac{H_2}{H_1}} = Q_1\sqrt{\frac{25}{20}} = 1/0.91\ \ (5) = 5.6\ GPM$$

Question 12.15. Use the *Chapter 12 Center pivot* model to find the optimal water application depth for CV values of 0.1 and 0.3.

For CV = 0.3, the maximum profit is found at 66 cm average depth applied.

Total	64	66	68	70	72	74
Yield (kg/ha)	1144.04	1156.59	1167.45	1176.07	1184.64	1191.67
Leached (cm)	12.90	14.19	15.56	17.00	18.51	20.08
ECe (dS/m)	0.00	0.00	0.00	0.00	0.00	0.00
Ks-salt	1.01	1.01	1.01	1.01	1.01	1.01
Yield ($/ha)	1052.52	1064.06	1074.06	1081.99	1089.87	1096.34
Water ($/ha)	220.38	227.26	234.15	241.04	247.93	254.81
Envir. ($/ha)	12.90	14.19	15.56	17.00	18.51	20.08
Energy ($/ha)	90.26	93.08	95.90	98.72	101.54	104.36
Profit ($/ha)	728.98	729.53	728.45	725.22	721.89	717.08
Pro.(no Env.)	741.88	743.72	744.00	742.22	740.40	737.16

For CV = 0.1, the maximum profit is found at 68 average depth applied.

Total	64	66	68	70	72	74
Yield (kg/ha)	1239.02	1252.47	1264.12	1274.04	1282.30	1288.97
Leached (cm)	10.89	12.10	13.44	14.91	16.51	18.25
ECe (dS/m)	0.00	0.00	0.00	0.00	0.00	0.00
Ks-salt	1.01	1.01	1.01	1.01	1.01	1.01
Yield ($/ha)	1139.90	1152.27	1162.99	1172.12	1179.72	1185.85
Water ($/ha)	229.19	236.35	243.52	250.68	257.84	265.00
Envir. ($/ha)	10.89	12.10	13.44	14.91	16.51	18.25
Energy ($/ha)	90.26	93.08	95.90	98.72	101.54	104.36
Profit ($/ha)	809.56	810.73	810.13	807.81	803.82	798.24
Pro. (no Env.)	820.44	822.83	823.57	822.71	820.33	816.49

Question 12.16. Is it worth adding pressure regulators for the elevations shown in Fig. 12.23? Regulators cost is $5.00 per. This is the same as question 12.15, but add the regulators.

First try 0.3 CV with regulators. For the case that considers environmental cost, the maximum profit with pressure regulators is found at 70 cm but it is only $727.29, which is less than the maximum profit with no pressure regulators, $729.53. Thus, pressure regulators are not worth it for 0.3 CV.

10	seasonal		
11	simulation	Pressure Regulator?	TRUE

Total	64	66	68	70	72	74
Yield (kg/ha)	1138.08	1152.02	1164.25	1174.83	1183.83	1191.29
Leached (cm)	12.79	14.06	15.41	16.84	18.32	19.85
ECe (dS/m)	0.00	0.00	0.00	0.00	0.00	0.00
Ks-salt	1.01	1.01	1.01	1.01	1.01	1.01
Yield ($/ha)	1047.03	1059.86	1071.11	1080.85	1089.12	1095.99
Water ($/ha)	218.61	225.44	232.27	239.11	245.94	252.77
Envir. ($/ha)	12.79	14.06	15.41	16.84	18.32	19.85
Energy ($/ha)	90.48	93.31	96.14	98.97	101.79	104.62
Profit ($/ha)	725.15	727.05	727.29	725.94	723.06	718.75
Pro.(no Env.)	737.94	741.11	742.70	742.77	741.39	738.60

0.1 CV with pressure regulators. The maximum profit with pressure regulators is 804.01 at 72 cm applied depth. This is less than the maximum profit of 810.73 without pressure regulators.

Total	64	66	68	70	72	74
Yield (kg/ha)	1196.33	1213.46	1228.90	1242.70	1254.93	1265.65
Leached (cm)	8.18	9.01	9.92	10.93	12.03	13.24
ECe (dS/m)	0.00	0.00	0.00	0.00	0.00	0.00
Ks-salt	1.01	1.01	1.01	1.01	1.01	1.01
Yield ($/ha)	1100.62	1116.39	1130.59	1143.28	1154.54	1164.40
Water ($/ha)	210.40	216.97	223.55	230.12	236.70	243.27
Envir. ($/ha)	8.18	9.01	9.92	10.93	12.03	13.24
Energy ($/ha)	90.48	93.31	96.14	98.97	101.79	104.62
Profit ($/ha)	791.56	797.10	800.98	803.27	804.01	803.26
Pro.(no Env.)	799.74	806.10	810.90	814.20	816.05	816.50

Question 12.17. A center pivot irrigation system requires 200 kPa sprinkler pressure. There are five pivots each with a flow rate of 110 L/sec. Pressure regulators are used. Sprinklers are 1 m above the land surface. There is a 4 m pressure loss in the pivot pipeline, and 10 m head loss in the pipe network between the pumps and the worst case pivot. The maximum elevation of the land surface is 20 m higher than the reservoir. Make all other necessary assumptions. How many pumps are required? At what pressure and flow rate should the pumps operate?

The total pump pressure required is the sum of:

- **Sprinkler pressure: 200 kPa = 20 m**
- **Sprinklers elevation: 1 m**
- **pressure loss in the pivot pipeline: 4 m**
- **head loss in the pipe network between the pumps and the pivot: 10 m**
- **maximum elevation of the land surface (higher than the reservoir): 20 m**

The total pressure required = 20 + 1 + 4 + 10 + 20 = 55

If an extra 2 m pressure is added to account for degradation of the pump over time and if the pivot requires a flow rate of 95 LPS then five pumps with 57 m pressure and 95 LPS flow rate should be used.

Chapter 13: Solution

1. Describe the major categories of turf sprinklers
   - Fixed spray sprinklers cover small areas (<5 m radius) and, as the name indicates, have a fixed spray that does not rotate. They generally have flow rates of 8, 4, and 2 LPM for full, half, and quarter circle sprinklers, respectively. This results in application rates of between 25 and 50 mm/hour.
   - Rotors have gear mechanisms that slowly rotate the sprinkler. These sprinklers cover a larger area (>5 m radius), and thus have a lower application rate (<25 mm/hr).
2. Which turf sprinklers have the highest application rate?
   - Fixed spray sprinklers
3. What is the maximum precipitation rate in mm/hr for a coarse sandy loam with a 6 % slope?
   - With cover: 51 mm/hr
   - Bare: 38 mm/hr
4. Discuss the reason for placing sprinklers in corners of turf areas.
   - Sprinkler application rate decreases with distance from the sprinkler; thus, if there would be no sprinkler in the corner, plants in the corner would receive a small depth of water in comparison to locations closer to sprinklers.
5. Describe head to head coverage.
   - Sprinklers should be designed for head-to-head coverage, which means that the wetted area from one sprinkler just touches the next sprinkler. The need for head to head coverage is based on the fact that sprinklers have a triangular pattern of application depth vs. distance from the sprinkler.
6. Why must sprinklers with different application rates be placed on different valves?
   - All sprinklers with the same application rate must run for the same length of time.
7. What are the advantages of swing joints?
   - Sprinklers should be installed on swing joints for two reasons:
     - First, people step on sprinklers and equipment runs over sprinklers, and this force can break the lateral pipe if the sprinklers are installed directly over the pipe. Swing joints allow the sprinkler to be pushed down without breaking the pipe.
     - The second advantage of swing joints is that they make positioning sprinklers easy during installation.
8. What is the application rate for a 1.8 m spray SRS spray head with half circle coverage? Assume 1.8 × 1.8 m

spacing. Comment on whether this is a low, average, or high application rate.

- 3.71 l/min, spacing = 1.8 m
- mm/hr = lpm * 120/(m * m) = 3.7 * 120/(1.8 * 1.8) = 137 mm/hr

This is an high application rate. For most soils, the sprinkler cannot be left on for an extended period without runoff.

9. What is the application rate for a full circle 5 GPM (19 LPM) sprinkler on 50 ft × 50 ft (15.2 m × 15.2 m) spacing? Calculate for U.S. (Imperial) and metric units.

Full-circle sprinkler

$$in/hr = \frac{GPM*96.3}{ft*ft}$$

$$mm/hr = \frac{LPM*60}{m*m}$$

$$\text{in/hr} = ((5\text{gpm}*96.3)/(50\text{ft}*50\text{ft})) = 0.1926 \text{ in/hr}$$

$$\text{mm/hr} = 19*60/(15.2*15.2) = 5 \text{ mm/hr}$$

10. What is the application rate for the sprinkler in question 9 with half circle coverage?

Half-circle sprinkler

$$in/hr = \frac{GPM*193}{ft*ft}$$

$$mm/hr = \frac{LPM*120}{m*m}$$

$$\text{in/hr} = ((5 \text{ gpm}*193)/(50\text{ft}*50\text{ft})) = 0.386 \text{ in/hr}$$

11. Lay out sprinklers in zones in a 45 ft × 45 ft (13.7 m × 13.7 m) turf area with the SRS spray heads with 4.6 m radius (Table 13.2). City water pressure is 40 PSI (276 kPa) and maximum flow rate for the system is 12 GPM (45.4 LPM). Divide the turf area into zones so that the maximum flow rate is not exceeded. You do not need to show pipes in the drawing. Reference ET rate is 12 mm/day and the crop coefficient for turf is 0.7. Scheduling coefficient is 1.3 and there is 7 % evaporation from sprinkler droplets. Determine the watering time per day.
    - Image see below

The 4.6 m diameter sprinklers have a maximum flow rate of 15.6, 7.8, and 3.9 LPM for the full, half, and quarter circle sprinklers, respectively. Use the higher flow rate in the table because the city water pressure is higher than the maximum listed pressure in the table (241 kPa).

Space the sprinklers on 4.6 m (15 ft) spacing, thus, the configuration in the image below. Calculate the application rate based on the full circle sprinkler. All other sprinklers have the same application rate.

$$\text{mm/hr} = 15.6*60/(4.6*4.6) = 44 \text{ mm/hr}(1.7 \text{ in/hr})$$

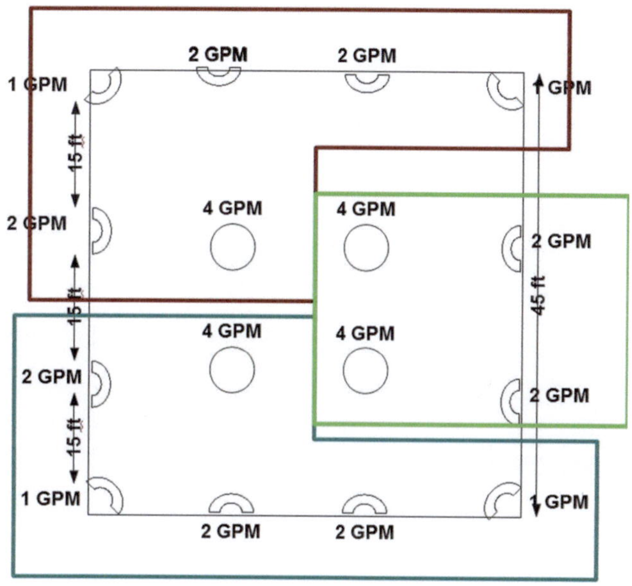

- ET_{rate} * crop coefficient = actual ET_{rate}
- 12 mm/d * 0.7 = 8.4 mm/d = $^{1}/_{3}$ in/d
- Watering time per day:
    - ET_{rate}/application rate = watering time per day
    - $^{1}/_{3}$ in/d/1.7 in/hr = 0.176 hrs = 10.58 min

12. A pipe system has 10 sprinklers that are 15 m apart and each sprinkler has a flow rate of 20 L/min. The field is level. Select appropriate pipe sizes for each section so that total pressure loss in the system is no more than 5 PSI. Show a hand calculation for two sections but you can use the Simple Lateral Worksheet for the entire lateral design. Repeat for a 2 % uphill slope and a 2 % downhill slope with the Simple Lateral Worksheet.

5 PSI pressure loss is the same as 3.5 m.
The 0 % slope solution

	E	F	G	H	I	J
›e design - starts at the end of the lateral (Row 15)					by Armando Barreto and Peter Waller	
C Value	150					
›rinklers	10					
) (+ = up)	0					
›ral (kPa)	441					
ıteral (m)	45.0					
	Select diameter	Inside diameter	Spr flow (LPM)	Distance (m)	Friction loss (m)	Head, next (m)
›nkler 10	25 SCH 40	26.64	20.00	15.00	0.255	45.25
Spr. 9	25 CL 200	30.2	20.00	15.00	0.500	45.75
Spr. 8	37 CL 125	45.32	20.00	15.00	0.147	45.9
Spr. 7	37 CL 125	45.32	20.00	15.00	0.250	46.2
Spr. 6	37 CL 125	45.32	20.00	15.00	0.378	46.5
Spr. 5	37 CL 125	45.32	20.00	15.00	0.529	47.1
Spr. 4	50 CL 125	56.58	20.00	15.00	0.239	47.3
Spr. 3	50 CL 125	56.58	20.00	15.00	0.306	47.6
Spr. 2	50 CL 125	56.58	20.00	15.00	0.381	48.0
Spr. 1	50 CL 200	54.56	20.00	15.00	0.552	48.5

The 2 % downhill solution

	E	F	G	H	I	J
›e design - starts at the end of the lateral (Row 15)					by Armando Barreto and Peter Waller	
C Value	150					
›rinklers	10					
(+ = up)	-2					
ral (kPa)	441					
teral (m)	45.0					
	Select diameter	Inside diameter	Spr flow (LPM)	Distance (m)	Friction loss (m)	Head, next (m)
›nkler 10	25 SCH 40	26.64	20.00	15.00	0.255	44.95
Spr. 9	25 CL 200	30.2	20.00	15.00	0.500	45.15
Spr. 8	25 CL 200	30.2	20.00	15.00	1.059	45.9
Spr. 7	31 CL 125	39.6	20.00	15.00	0.482	46.1
Spr. 6	31 CL 125	39.6	20.00	15.00	0.728	46.5
Spr. 5	31 CL 125	39.6	20.00	15.00	1.021	47.2
Spr. 4	37 CL 125	45.32	20.00	15.00	0.704	47.6
Spr. 3	37 CL 125	45.32	20.00	15.00	0.902	48.3
Spr. 2	50 CL 125	56.58	20.00	15.00	0.381	48.3
Spr. 1	50 CL 125	56.58	20.00	15.00	0.463	48.5
n/a	75 CL160					

The 2 % uphill solution

	E	F	G	H	I	J
e design - starts at the end of the lateral (Row 15)					by Armando Barreto and Peter Waller	
C Value	150					
inklers	10					
+ = up)	2					
al (kPa)	441					
eral (m)	45.0					
	Select diameter	Inside diameter	Spr flow (LPM)	Distance (m)	Friction loss (m)	Head, next (m
kler 10	37 CL 125	45.32	20.00	15.00	0.019	45.32
Spr. 9	37 CL 125	45.32	20.00	15.00	0.069	45.69
Spr. 8	50 CL 125	56.58	20.00	15.00	0.050	46.0
Spr. 7	50 CL 125	56.58	20.00	15.00	0.085	46.4
Spr. 6	50 CL 125	56.58	20.00	15.00	0.128	46.9
Spr. 5	75 CL125	83.42	20.00	15.00	0.027	47.2
Spr. 4	75 CL125	83.42	20.00	15.00	0.036	47.5
Spr. 3	75 CL125	83.42	20.00	15.00	0.046	47.9
Spr. 2	100 CL125	107.28	20.00	15.00	0.017	48.2
Spr. 1	100 CL125	107.28	20.00	15.00	0.021	48.5
n/a	75 CL 160					

Much larger pipe is needed on the uphill slope to conserve energy. It is cheaper to run the submain along the top of the hill and then to run sprinkler laterals downhill.

13. What is the scheduling coefficient for the U of A high pressure sprinkler in the *Chapter 13 Sprinkler Uniformity* program on the following grids? Discuss the reasons for the highest SC with the last spacing.

14 m × 14 m square. SC 1.51

13 m × 15 m triangle SC 1.50

16 × 16 m square SC 1.66

15 × 17.3 triangle SC 2.10

20 × 20 square SC 1.68

19 × 22 triangle SC 1.33

This is the best spacing of all. Most of the area is low. Thus, the SC is not high. Other spacings had a single or limited low spot, so the entire area was overwatered because of the low spot.

14. Use the *Economic summary* worksheet for the U of A high pressure sprinkler. Cost of each sprinkler head installation is \$30. Cost of trenching and pipe is \$5.00/m. Life of the project is 10 years. Interest rate is 9 %. Required water depth/year is 2 m. Evapotranspiration rate is 25 %. Cost of water is \$0.50/m^3. Determine the lowest present value for the square or triangular spacing. Which is lowest?

This is a strange result, but the optimal spacing is 8 × 9.2 with the triangular grid. The SC is the main criterion since the water cost is so high.

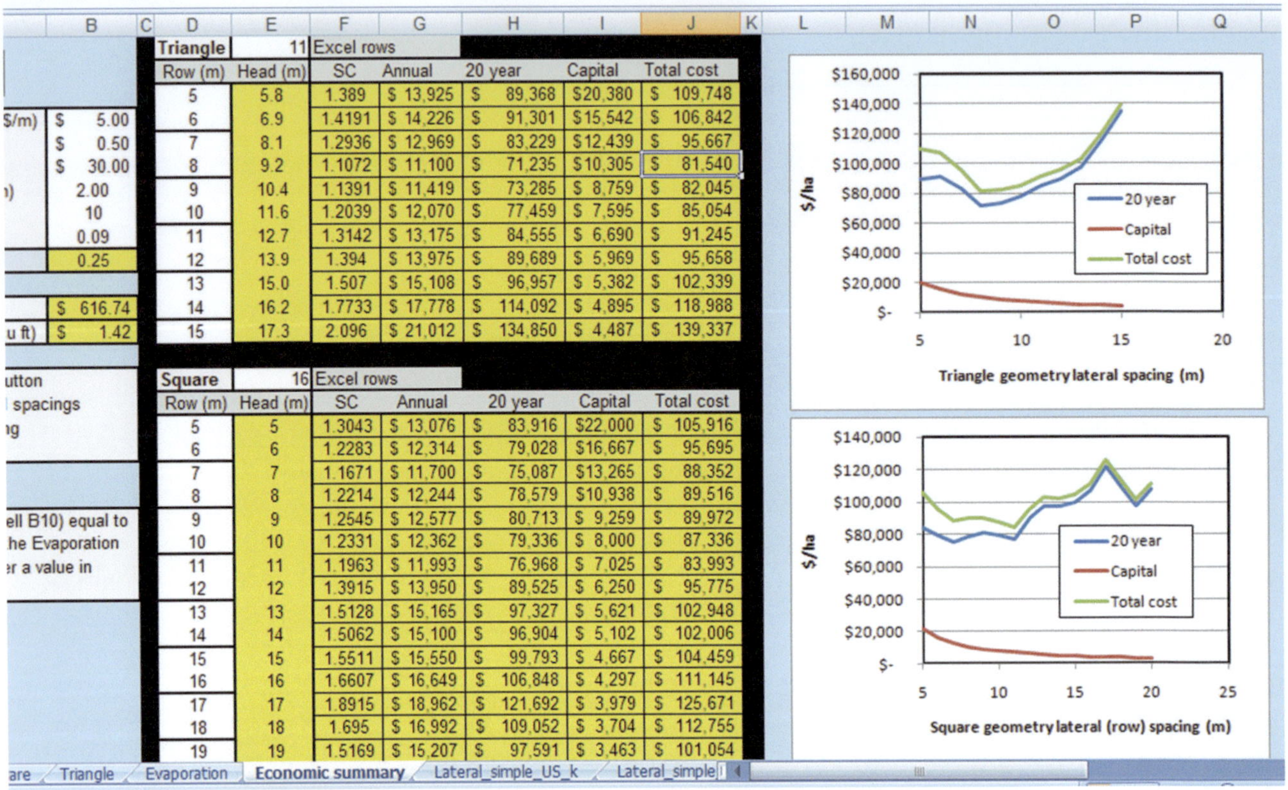

Triangle	11	Excel rows				
Row (m)	Head (m)	SC	Annual	20 year	Capital	Total cost
5	5.8	1.389	$ 13,925	$ 89,368	$20,380	$ 109,748
6	6.9	1.4191	$ 14,226	$ 91,301	$15,542	$ 106,842
7	8.1	1.2936	$ 12,969	$ 83,229	$12,439	$ 95,667
8	9.2	1.1072	$ 11,100	$ 71,235	$10,305	$ 81,540
9	10.4	1.1391	$ 11,419	$ 73,285	$ 8,759	$ 82,045
10	11.6	1.2039	$ 12,070	$ 77,459	$ 7,595	$ 85,054
11	12.7	1.3142	$ 13,175	$ 84,555	$ 6,690	$ 91,245
12	13.9	1.394	$ 13,975	$ 89,689	$ 5,969	$ 95,658
13	15.0	1.507	$ 15,108	$ 96,957	$ 5,382	$ 102,339
14	16.2	1.7733	$ 17,778	$ 114,092	$ 4,895	$ 118,988
15	17.3	2.096	$ 21,012	$ 134,850	$ 4,487	$ 139,337

Square	16	Excel rows				
Row (m)	Head (m)	SC	Annual	20 year	Capital	Total cost
5	5	1.3043	$ 13,076	$ 83,916	$22,000	$ 105,916
6	6	1.2283	$ 12,314	$ 79,028	$16,667	$ 95,695
7	7	1.1671	$ 11,700	$ 75,087	$13,265	$ 88,352
8	8	1.2214	$ 12,244	$ 78,579	$10,938	$ 89,516
9	9	1.2545	$ 12,577	$ 80,713	$ 9,259	$ 89,972
10	10	1.2331	$ 12,362	$ 79,336	$ 8,000	$ 87,336
11	11	1.1963	$ 11,993	$ 76,968	$ 7,025	$ 83,993
12	12	1.3915	$ 13,950	$ 89,525	$ 6,250	$ 95,775
13	13	1.5128	$ 15,165	$ 97,327	$ 5,621	$ 102,948
14	14	1.5062	$ 15,100	$ 96,904	$ 5,102	$ 102,006
15	15	1.5511	$ 15,550	$ 99,793	$ 4,667	$ 104,459
16	16	1.6607	$ 16,649	$ 106,848	$ 4,297	$ 111,145
17	17	1.8915	$ 18,962	$ 121,692	$ 3,979	$ 125,671
18	18	1.695	$ 16,992	$ 109,052	$ 3,704	$ 112,755
19	19	1.5169	$ 15,207	$ 97,591	$ 3,463	$ 101,054

15. Redo Example 13.3. In this case, the city can provide 70 gal/min (fewer zones are required). Set up a spreadsheet to calculate lateral losses. Leave sprinklers in the same positions.

It would not be impossible to divide the field into two zones because there is a total 142 GPM out of all the sprinklers. This is just slightly higher than 2 * 70 GPM. All of the central sprinklers should be run as one zone, and they should not be mixed with the outer sprinklers. The total flow of outer sprinklers is almost 100 GPM. Thus, the outer sprinklers must be split into two zones. The new sprinkler layout is shown in the following figure.

The hydraulics of each of the three zones is shown in the copied Worksheets after the drawing of the sprinklers and pipes.

Scheduling is not a problem since there was no problem with 5 zones in the example.

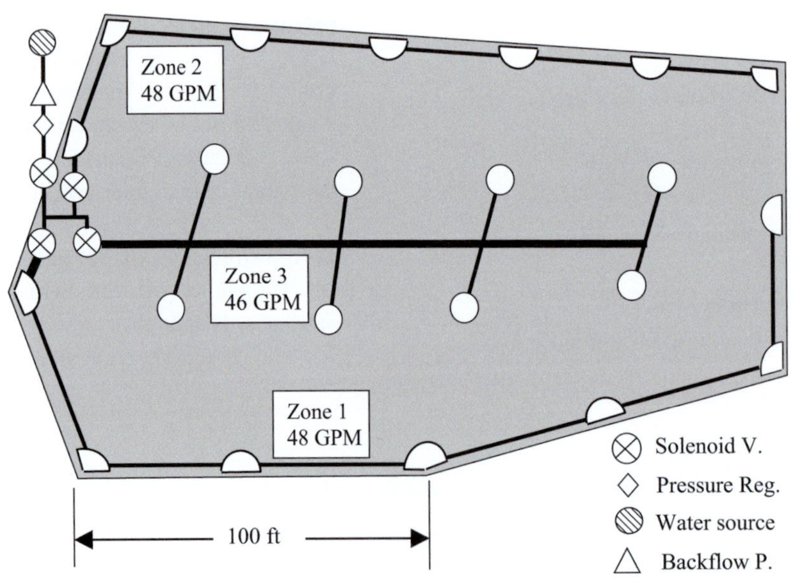

Zone 1 hydraulics.

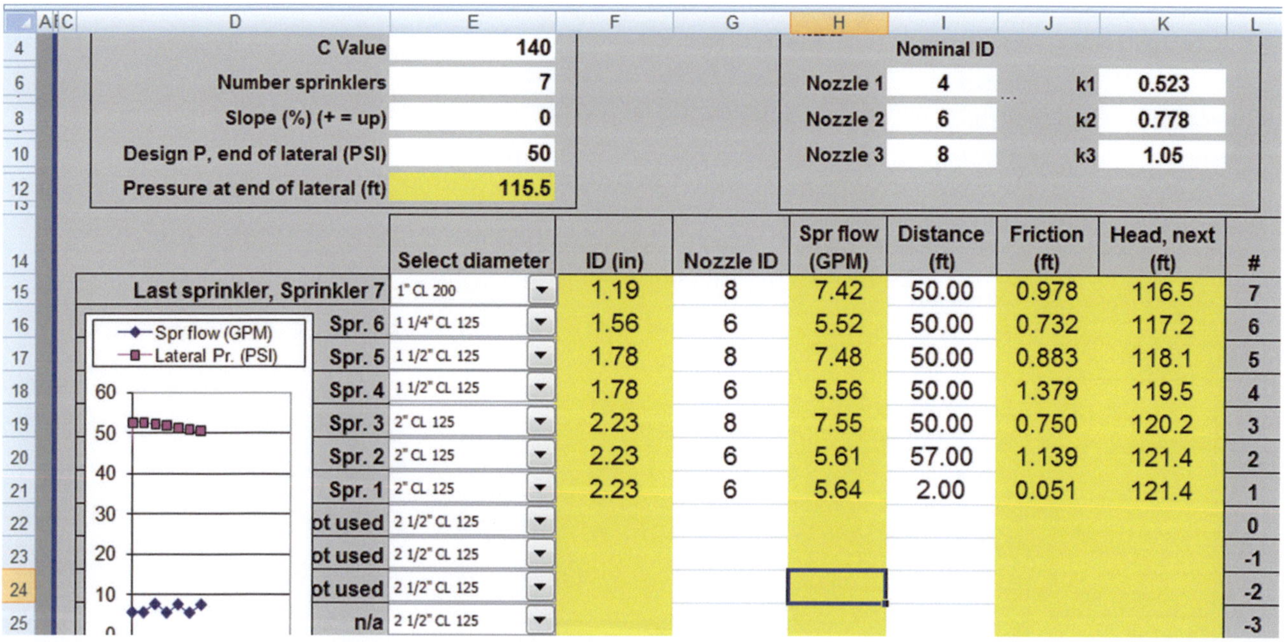

C Value	140
Number sprinklers	7
Slope (%) (+ = up)	0
Design P, end of lateral (PSI)	50
Pressure at end of lateral (ft)	115.5

Nominal ID			
Nozzle 1	4	k1	0.523
Nozzle 2	6	k2	0.778
Nozzle 3	8	k3	1.05

	Select diameter	ID (in)	Nozzle ID	Spr flow (GPM)	Distance (ft)	Friction (ft)	Head, next (ft)	#
Last sprinkler, Sprinkler 7	1" CL 200	1.19	8	7.42	50.00	0.978	116.5	7
Spr. 6	1 1/4" CL 125	1.56	6	5.52	50.00	0.732	117.2	6
Spr. 5	1 1/2" CL 125	1.78	8	7.48	50.00	0.883	118.1	5
Spr. 4	1 1/2" CL 125	1.78	6	5.56	50.00	1.379	119.5	4
Spr. 3	2" CL 125	2.23	8	7.55	50.00	0.750	120.2	3
Spr. 2	2" CL 125	2.23	6	5.61	57.00	1.139	121.4	2
Spr. 1	2" CL 125	2.23	6	5.64	2.00	0.051	121.4	1
ot used	2 1/2" CL 125							0
ot used	2 1/2" CL 125							-1
ot used	2 1/2" CL 125							-2
n/a	2 1/2" CL 125							-3

Zone 2 hydraulics.

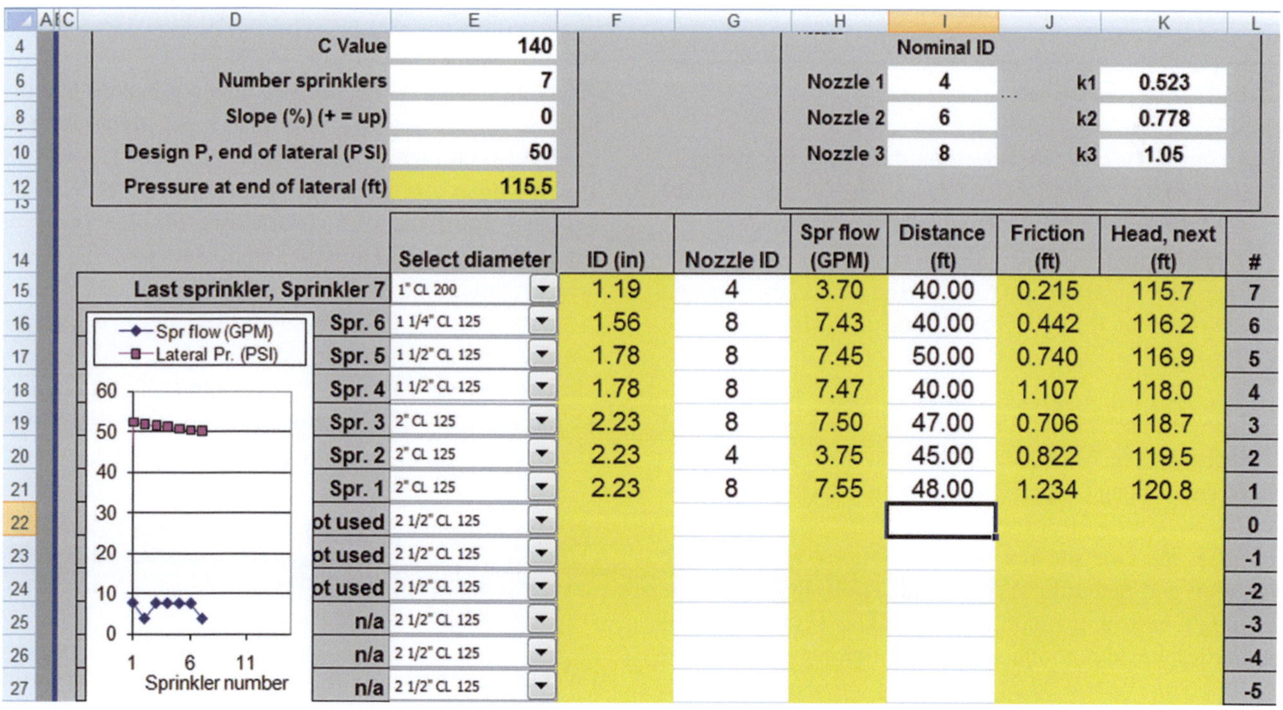

C Value	140
Number sprinklers	7
Slope (%) (+ = up)	0
Design P, end of lateral (PSI)	50
Pressure at end of lateral (ft)	115.5

Nominal ID			
Nozzle 1	4	k1	0.523
Nozzle 2	6	k2	0.778
Nozzle 3	8	k3	1.05

	Select diameter	ID (in)	Nozzle ID	Spr flow (GPM)	Distance (ft)	Friction (ft)	Head, next (ft)	#
Last sprinkler, Sprinkler 7	1" CL 200	1.19	4	3.70	40.00	0.215	115.7	7
Spr. 6	1 1/4" CL 125	1.56	8	7.43	40.00	0.442	116.2	6
Spr. 5	1 1/2" CL 125	1.78	8	7.45	50.00	0.740	116.9	5
Spr. 4	1 1/2" CL 125	1.78	8	7.47	40.00	1.107	118.0	4
Spr. 3	2" CL 125	2.23	8	7.50	47.00	0.706	118.7	3
Spr. 2	2" CL 125	2.23	4	3.75	45.00	0.822	119.5	2
Spr. 1	2" CL 125	2.23	8	7.55	48.00	1.234	120.8	1
ot used	2 1/2" CL 125							0
ot used	2 1/2" CL 125							-1
ot used	2 1/2" CL 125							-2
n/a	2 1/2" CL 125							-3
n/a	2 1/2" CL 125							-4
n/a	2 1/2" CL 125							-5

Zone 3 hydraulics.

C Value	140
Number sprinklers	8
Slope (%) (+ = up)	0
Design P, end of lateral (PSI)	50
Pressure at end of lateral (ft)	115.5

	Nominal ID		
Nozzle 1	4	k1	0.523
Nozzle 2	6	k2	0.778
Nozzle 3	8	k3	1.05

	Select diameter	ID (in)	Nozzle ID	Spr flow (GPM)	Distance (ft)	Friction (ft)	Head, next (ft)	#
Last sprinkler, Sprinkler 8	1 1/4" CL 125	1.56	6	5.50	0.20	0.001	115.5	8
Spr. 7	1 1/4" CL 125	1.56	6	5.50	50.00	0.541	116.0	7
Spr. 6	1 1/2" CL 125	1.78	6	5.51	0.20	0.002	116.0	6
Spr. 5	1 1/2" CL 125	1.78	6	5.51	50.00	1.015	117.1	5
Spr. 4	1 1/2" CL 125	1.78	6	5.54	0.20	0.006	117.1	4
Spr. 3	1 1/2" CL 125	1.78	6	5.54	50.00	2.159	119.2	3
Spr. 2	2" CL 125	2.23	8	7.54	0.20	0.004	119.2	2
Spr. 1	2" CL 125	2.23	6	5.59	50.00	1.360	120.6	1
ot used	2 1/2" CL 125							0
ot used	2 1/2" CL 125							-1
n/a	2 1/2" CL 125							-2
n/a	2 1/2" CL 125							-3
n/a	2 1/2" CL 125							-4

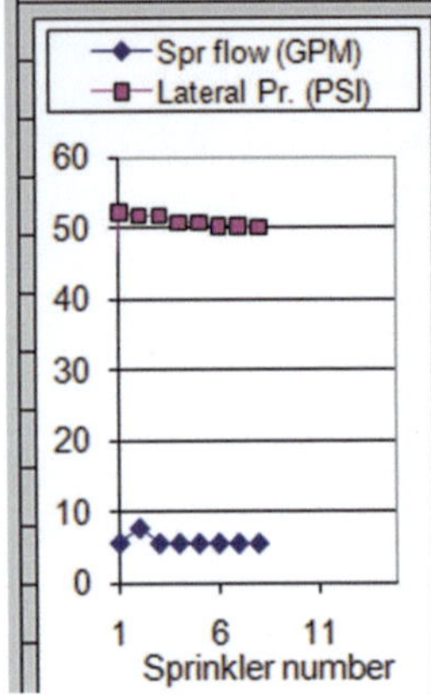

16. Answers will vary

Chapter 14: Solution

1. An orchard is on a hill. How should the mainline, submain, and laterals be positioned so that all the sprinklers on the property have nearly the same operating pressure?

The mainline should be run to the top of the hill, the submain should run along the top of the hill, and all the laterals should run downhill.

2. True or false. The two main causes of nonuniformity of application in wheel-line sprinkler systems are hydraulic variation along the pipeline and spatial variability of soils.

False, the two sources of nonuniformity are hydraulic variation and nonuniformity of application. In theory, you shouldn't have any runoff with sprinkler systems so soil properties would not influence distribution uniformity

3. True or false. Water hammer is a major problem in sprinkler laterals.

False, each of the sprinklers acts like an air vent and pressure relief valve.

4. The last two sprinklers on a PVC pipeline have a flow rate of 20 L/min. The end pressure is 350 kPa. The distance between sprinklers is 15 m. The slope is 3 % downhill. Select pipe sizes for the last two pipe sections so that the three sprinklers have no more than a 0.2 m variation in head between them. Use a Hazen-Williams C value of 150. You can use the Worksheet to find the pipe sizes, but also make the two calculations of pipe pressure losses and change in pressure from one sprinkler to the next by hand.

The pressure at the last sprinkler is 35.7 m (350 kPa)

The two pipe sizes that will work are 18 (3/4″, 23.66 mm) Class 200 and 25 (1″, 30.2 mm) Class 200

The friction loss in the last pipe section is

$$H_L = 1.22{*}10^{10}{*}15\ m\left(\left(\frac{20/60}{150}\right)^{1.852}/23.66^{4.87}\right)$$
$$= 0.454\ m$$

Calculate pressure at the next to last sprinkler with Eq. 14.1.

$$\begin{aligned}H_{n-1} &= H_n + H_L + (s_L)(S/100)\\ &= 35.7 + 0.454 + 15(-0.03) = 35.7\ \text{m}\end{aligned}$$

The friction loss in the next to last pipe section is

$$H_L = 1.22*10^{10}*15\ m\left(\left(\frac{40/60}{150}\right)^{1.852}/30.2^{4.87}\right) = 0.500\ m$$

$$\begin{aligned} H_{n-1} &= H_n + H_L + (s_L)(S/100) \\ &= 35.7 + 0.5 + 15(-0.03) = 35.75\ \text{m} \end{aligned}$$

5. A ¼ mile wheel-line has a downhill slope of 2 %. Use 3/16″ nozzles (ID = 4.8 mm). Calculate an equation for sprinkler flow rate vs. pressure. Then, determine whether pressure would have less variation with a 4 in (97.9 mm) or 5 in (123 mm) pipe.

$$\begin{aligned} Q &= 0.0666\ D^2H^{0.5}C_d = 0.0666(4.8^2)(H^{0.5})(0.97) \\ &= 1.49\ H^{0.5}. \end{aligned}$$

The pressure graph with 4 inch pipe and 2 % slope would be:

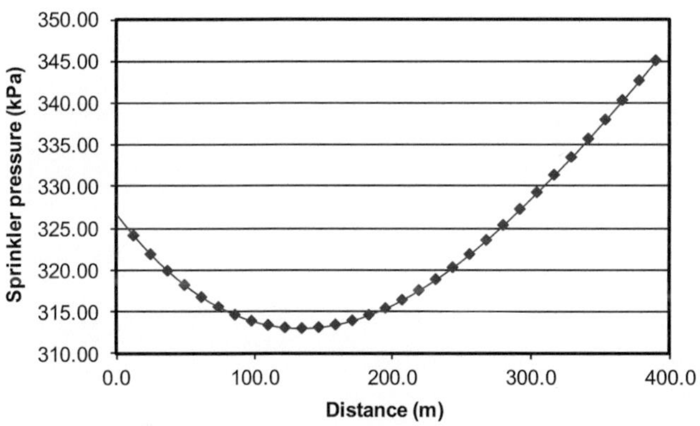

The pressure graph with the 5 inch pipeline and 2 % slope would be as follows. The pressure would be much more uniform with the 4 inch pipe.

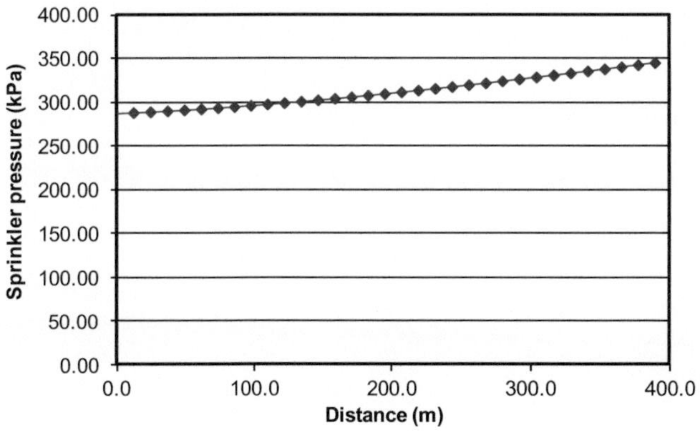

6. For the parameters in Example 14.2, calculate the seasonal depth of water application at the last sprinkler, 6.1 m row position, 0 m head position. Then calculate the depth of water application at the 2.4 m row position, 0 m head position for the first sprinkler. Are these the extremes of application depth? What is the percent difference between the maximum and minimum application rates? The average seasonal depth of water application to a field is 75 cm. Calculate the application depths at the maximum and minimum positions. Evaporation rate is 10 %.

From the worksheet, the relative depth of application at the first sprinkler is 1.04. The relative depth in at the 6.1 × 0 position is 1.45. Thus, the maximum relative application rate is found by multiplying the sprinkler relative rate by the maximum relative application position: 1.04 (1.45) = 1.51

The relative depth of application for the last sprinkler is 1.0. The relative depth at the 2.4 × 0 position in Table 14.4 is 1.0. Thus, the relative depth at this position is (1)(1) = 1.

These are the extremes in the system. Thus, there is approximately 50 % difference in the high and low application depths.

The next step is to find the average relative application depth. Ave_rel = 1.205. This can be found in Cell M2 in the Spatial_data_output or it could be found by taking the average of all the data points in Columns B:K.

The average gross application depth is 75 cm. Calculate the application depth at the maximum and minimum positions. Remember to include the (1 − Le).

$$\begin{aligned} AW_{max} &= (\max/\text{average})(1 - Le)(i) \\ &= 1.51/1.205(0.9)(75\ \text{cm}) = 84.6\ \text{cm} \\ AW_{min} &= 1/1.205(0.9)(75\ \text{cm}) = 56.0\ \text{cm} \end{aligned}$$

If you set up the Crop_data_and_summary sheet to only run 75 cm (max and min = 75), then you can find these values in Cells O330 and O7, respectively.

7. Redo Examples 14.1 and 14.2, but don't offset the wheel-line positions with respect to the hydrants. Compare the total profit, and the optimal depths with those found in Example 14.2. Recalculate the energy and water costs.

Everything is the same as Example 14.2 except that Cell A2 is changed to FALSE.

	A	B	C	D	E	F	G	H	I	J	K	L
1	Offset	K	1.018	Kpa LPM	Kh	1.22E+10	coefficent	Diameters	97.9	4-inch	Pipe length	390.10
2	FALSE	x	0.5		Slope	0	m/m	(mm) ID	123.0	5-inch	Hide flow graph	
3	Inlet Pr.	End Press	345	Kpa	CV	9%	sd/mean	Extra Press.	40	kPa	Show flow graph	
4	375.64	Spr. Spac	12.20	m	C	130	H-W	Pump eff.	70%		Hide pressure graph	
5		Hyd. Spac	18.3	m	Evap	10	%	Total flow	631.36	LPM	Show pressure graph	
6												
7	Sprinkler	Distance	Pressure (m)	Pressure (kPa)	Flow (LPM)	Diameter (mm)	Pipe flow (LPM)	Fric loss (m)	% diff flow	Rel	CV simulation	Z-dist
8	33	390.1	35.20	345.00	18.91	97.90	18.91	0.00043	0.00%	1	19.0968149	0.113425
9	32	377.9	35.20	345.00	18.91	97.90	37.82	0.00155	0.00%	1.0000061	20.8171915	1.149616

The results are shown on the following worksheet. Max profit is found at average gross applied depth equal to 100 cm.

$$\text{Water cost} = 100\ \text{cm}(\$0.02/\text{m}^3)((100\ \text{m}^3)/(\text{ha-cm})) = \$200.00/\text{ha}$$

$$E(kW-hr/ha) = \frac{0.0272\ (i_{mm})(h)}{Eff} = \frac{0.0272\ (1000mm)(42.3m)}{0.7} = 1,640\ kW-hr/ha$$

$$\text{Energy cost} = 1,640(\$0.10/\text{kW-hr}) = \$164.00/\text{ha}$$

	A–B	C	D	E	F	G	H	I	J	K
1	Eff. Precipitation	7.50	cm	CWPF (tons/cm)		x^1	Alfalfa	Wheel line		
2				Coefficients		0.083	NW USA			
3	Minimum Gross AW	75.00	cm		Yield value	100.0	$/ton			
4	Maximum Gross AW	120.00	cm		AWlimit	100.0	cm	Max benefit depth		
5	AW interval	5	cm		Energy cost	0.10	kW-hr	Ave net depth (cm)		90.00
6	Number columns	10			Leach cost	1.00	$/ha-cm	Ave gross depth		100.00
7	Number rows	10			Water cost	0.0200	$/m3			
8	Number sprinklers	33	Average gross depth applied (cm)							
9			75	80	85	90	95	100	105	110
10	Yield (kg/ha)		6.20	6.51	6.80	7.08	7.33	7.54	7.70	7.85
11	Leached depth(cm)		0.				4.63	6.63	9.20	11.88
12	Yield benefit ($/ha)		$ 619.				.47	$754.19	$770.22	$ 785.36
13	Water cost ($/ha)		$ 150.				.00	$200.00	$210.00	$ 220.00
14	Energy cost ($/ha)		$ 123.				.46	$164.69	$172.93	$ 181.16
15	Environmental cost ($/ha)		$ 0.				.63	$ 6.63	$ 9.20	$ 11.88
16	Total benefit ($/ha)		$ 345.				.38	$382.86	$378.09	$ 372.32
17	Total benefit w/o Env. ($/ha)		$ 346.				.01	$389.49	$387.29	$ 384.20

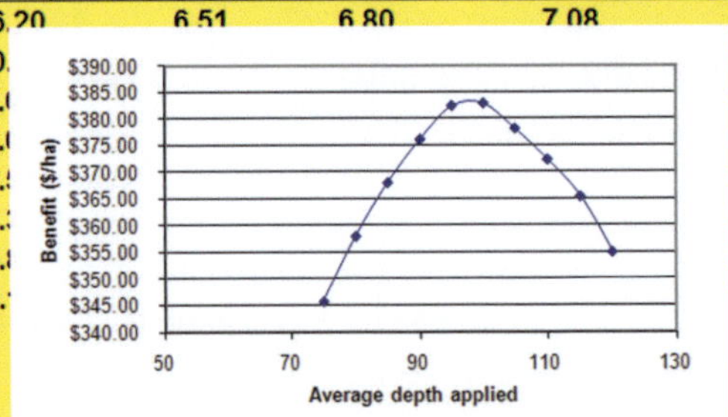

Main | Principal | Square | Triangle | Evaporation | Wheel-line | Crop_data_and_summary | Spatial_data_output | Lateral design | Submain design | Lateral des

Notice the much deeper drop in application rates between laterals with the wider lateral spacing. The profit is much lower, \$378 instead of \$440. The optimal depth is 105 cm instead of 100 cm.

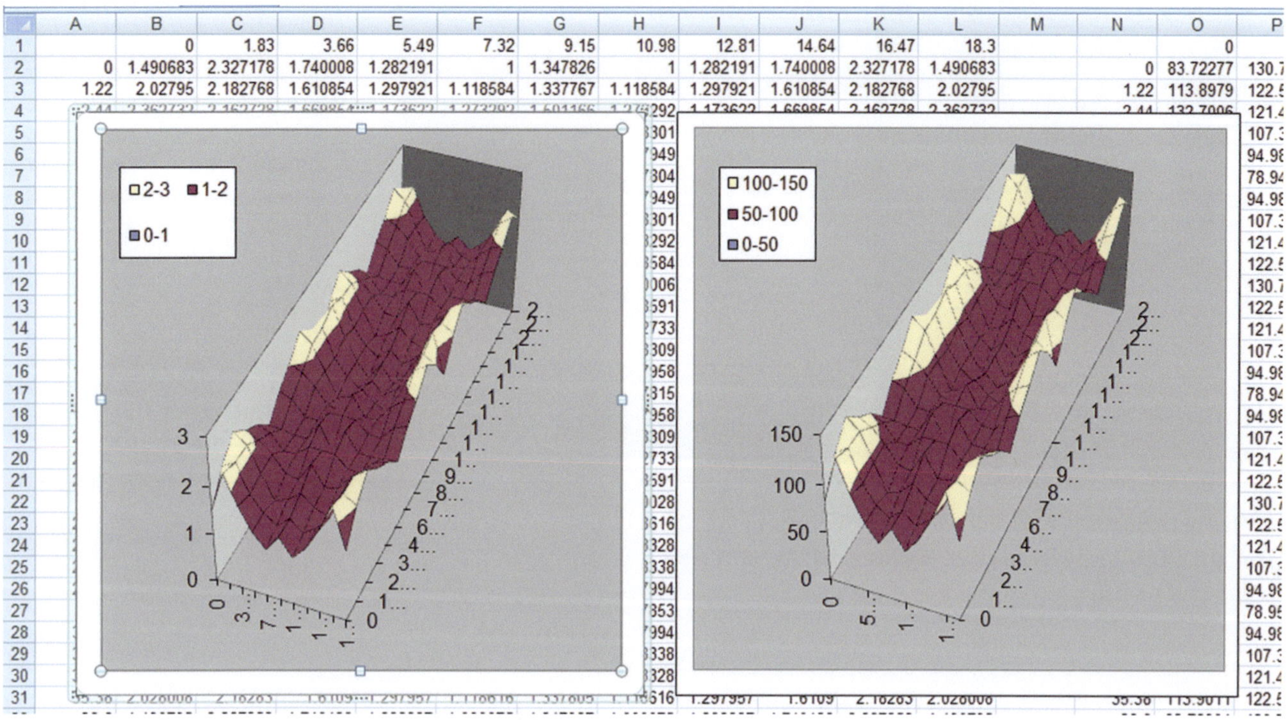

8. Redo Examples 14.1 and 14.2, but use handlines. The normal design for handlines is 40 ft along the mainline and 30 ft between nozzles. Select a nozzle and flow rate from catalogs at the following web sites. The length of the run is 1/8 mile long and the handlines use 3″ aluminum pipe. The slope is 0.005 m/m in the downhill direction. Use the same evapotranspiration, precipitation, power, and production functions as in Example 14.1. Don't offset the handlines. Select sprinklers based on catalog data below. Show the variability due to hydraulics and variation in wetting due to sprinkler patterns. Maximum application rate in 0.3 in/hr. Operate the handlines at 45 PSI pressure.

http://www.rainbird.com/ag/products/impacts/30H.htm
http://www.rainbird.com/documents/ag/chart_20JH.pdf.

Based on the 30 H catalog data, a 9/64″ nozzle operating at 45 PSI would have an application diameter of 42 ft (12.8 m) and flow rate of 3.80 GPM (14.4 LPM or 0.863 m^3/hr).

The maximum application rate can be found with Eq. 14.4.

$$\left(\frac{di}{dt}\right)_{\max} = \frac{0.863\ m^3/hr}{\frac{\pi}{3}\ (12.8)^2} = 5.03\ mm/hr$$

The maximum application rate is 0.3 in/hr (7.6 mm/hr). Because there is very little overlap, there is very little chance that there would be runoff since the maximum application rate would be slightly greater than 5.0 mm/hr.

Assume that the application is a perfect wedge with 13 m diameter.

The equation for the wedge would be as follows.

di/dt = (5.03) (13 − x)/(13) where × is the distance from the sprinkler. This results in an extra column in the Principal Worksheet. An extra line can be added to the list and the ComboBox range.

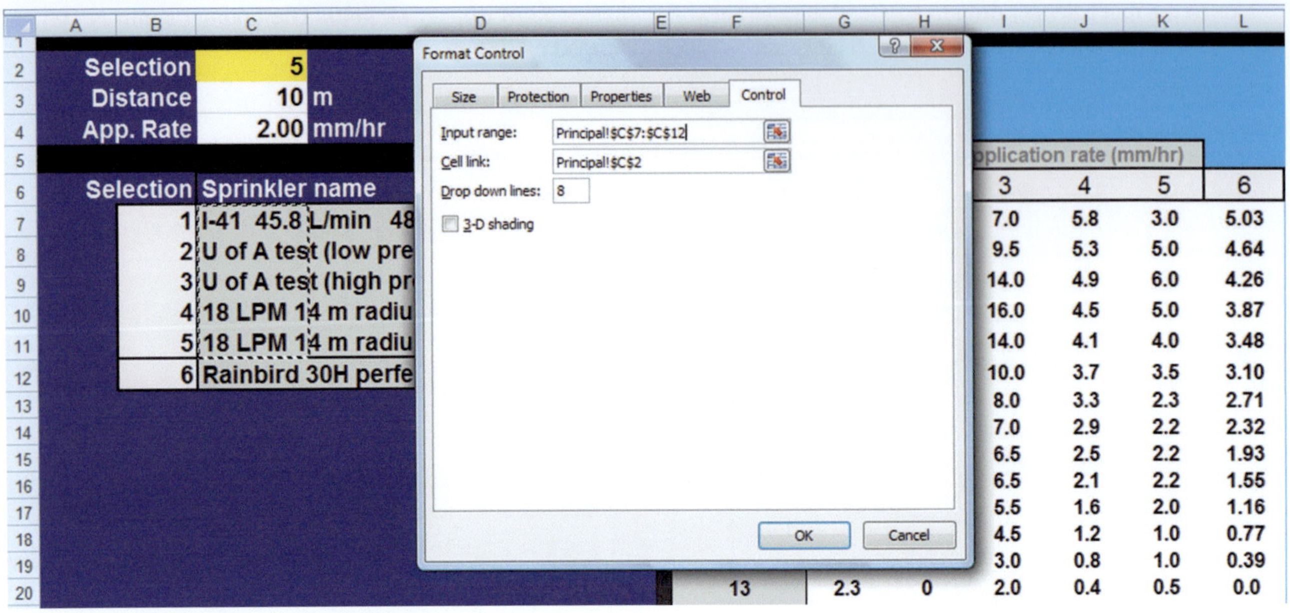

The new line has been added to the Worksheet for the Rainbird 30H sprinkler pattern represented by a perfect wedge (not realistic).

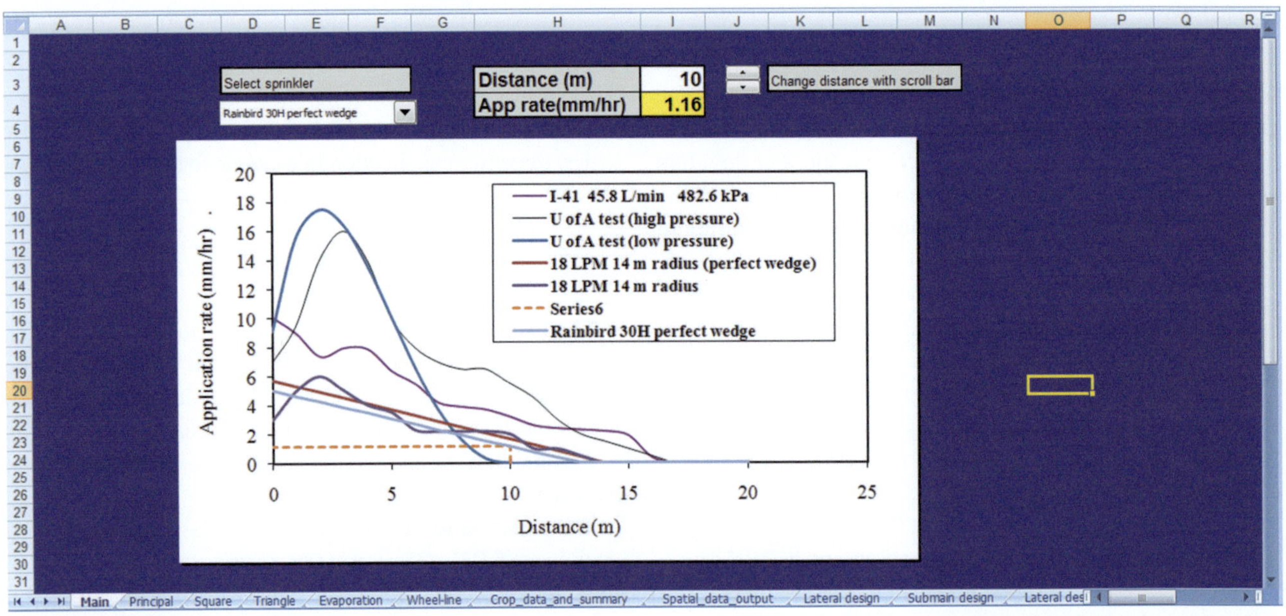

The next step is to calculate the k value.

This will be done with the flow data. 45 PSI/0.145 = 310 kPa.

$Q = kH^x$

$K = Q/(H^x) = 14.4\ \text{LPM}/(310^{0.5}) = 0.818$

$1/8\ \text{mile} = 201\ \text{m}.$

The inside diameter of 3 inch pipe is 74.0 mm. The C value is low because of the couplings and damage to pipes. Estimate as 90. The Wheel-line worksheet is set up as follows.

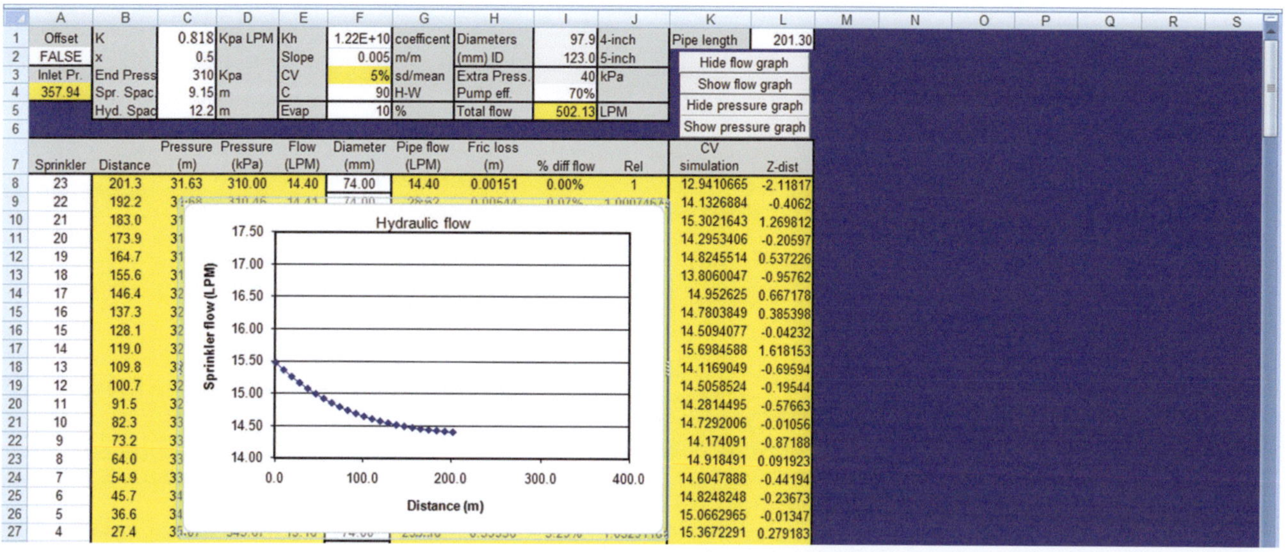

There are 23 sprinklers on the line so the Crop_data_and_summary Worksheet is set up as follows. The best average gross depth applied is 100 cm.

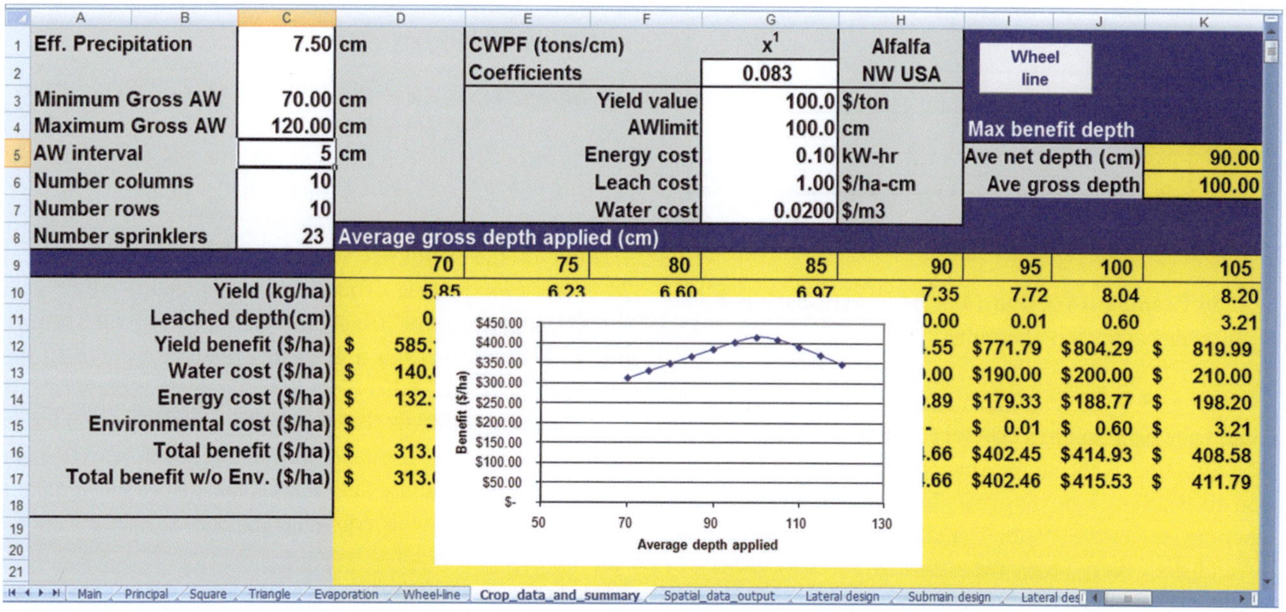

The variability due to sprinkler patterns is low since the wedge application pattern was specified. The CV is only 5 %.

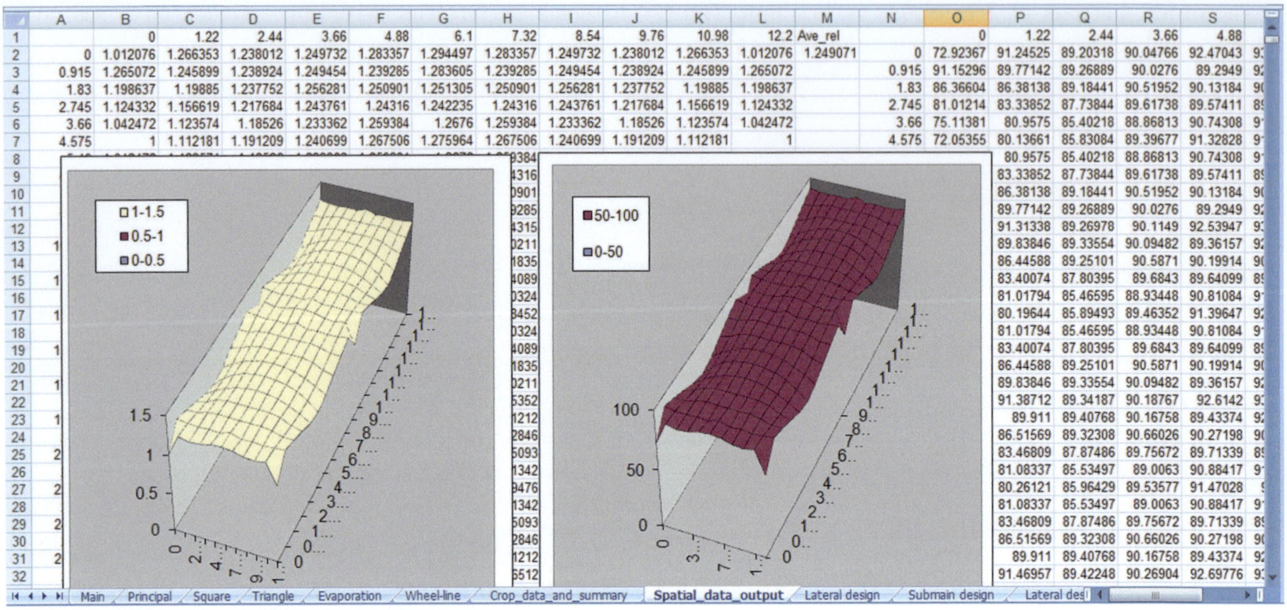

Even though the variability due to sprinkler application rates is low, it still looks to be more significant than the hydraulic variation.

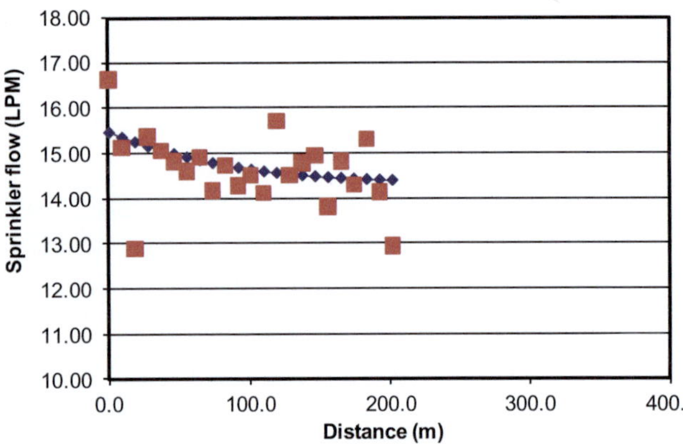

9. Redo Example 14.3 with 14 × 14 ft tree spacing. Leave all other parameters the same.

Select a 28 × 28 ft sprinkler spacing; all of the sprinklers would have head to head coverage. The farmer drives in the same direction as the slope. Thus, the installation will either need to be slotted or diagonal. In order to save pipe and trenching, select the slotted spacing with pipes every other row and slots every other tree.

The application rate for the 2.58 GPM (1/8″ nozzle) sprinkler would be too high (>0.3 in/hr)

$$di/dt = \frac{(2.58)(96.3)}{(28)(28)} = 0.32\ in/hr$$

The application rate with the 7/64″ nozzle would be

$$di/dt = \frac{(1.98)(96.3)}{(28)(28)} = 0.24\ in/hr$$

The application rate with the LF sprinkler would be

$$di/dt = \frac{(1.63)(96.3)}{(28)(28)} = 0.20\ in/hr$$

The application rates would be

$$(7/64'' \text{nozzle}) \qquad i = (0.24\ \text{in/hr})(24\ \text{hours})(1 - 0.05) = 5.5\ \text{inches in 24 hours.}$$

$$(\text{LF}) \qquad i = (0.20\ \text{in/hr})(24\ \text{hours})(1 - 0.05) = 4.6\ \text{inches in 24 hours.}$$

Both of the sprinklers would apply too much water in 24 hours (>4 inches). Thus, the farmer could either wait slightly longer between irrigation events or irrigate for a shorter period. If the farmer has a 12 hour irrigation cycle, then the 7/64″ nozzle is preferable since it would apply more water. If the irrigation period is 24 hours, then the LF sprinkler is preferable. Possibly, a smaller LF nozzle could be used with a reduced wetted radius. This is acceptable since the sprinkler significantly exceeds the requirement of head to head coverage.

10. Calculate a microsprinkler irrigation schedule for an orchard with 4 m × 3 m tree spacing and microsprinkler spacing. Each tree has a peak summer water use rate of 25 L/day. Microsprinklers have a 0.7 L/min flow rate and a 3 m diameter wetted area. Rooting depth is 1 m and allowable MAD is 0.35. The AWC is 12 % for a sandy loam soil, the expected loss to evaporation is

12 %, and the irrigation efficiency is 90 %. Specify an irrigation schedule for his system.

Calculate the schedule

$$S = 780 \quad Z \quad AWC \quad MAD \quad D_b{}^2$$
$$= 780(1\ m)(0.12)(0.35)(3^2) = 295\ L$$

$$(295\ \text{L})/(25\ \text{L/day}) = 12\ \text{days}$$

Calculate the irrigation run time

$$Q_{net} = Q_{gross}(1 - L_e)(Eff) = 0.7\ \text{L/min}(1 - 0.12)(0.9)$$
$$= 0.55\ \text{L/min}(295\ \text{L})/(0.55\ \text{L/min})$$
$$= 536\ \text{min} = 9\ \text{hours}$$

11. Repeat Example 14.5, except use 12 laterals by 12 sprinklers geometry, and the pump curve is $H = -5.18 * 10^{-5}\ Q^2 + 0.00828 * 10^{-3}\ Q + 900$. Find the operating point of the system.

The Lateral spreadsheet is set up first

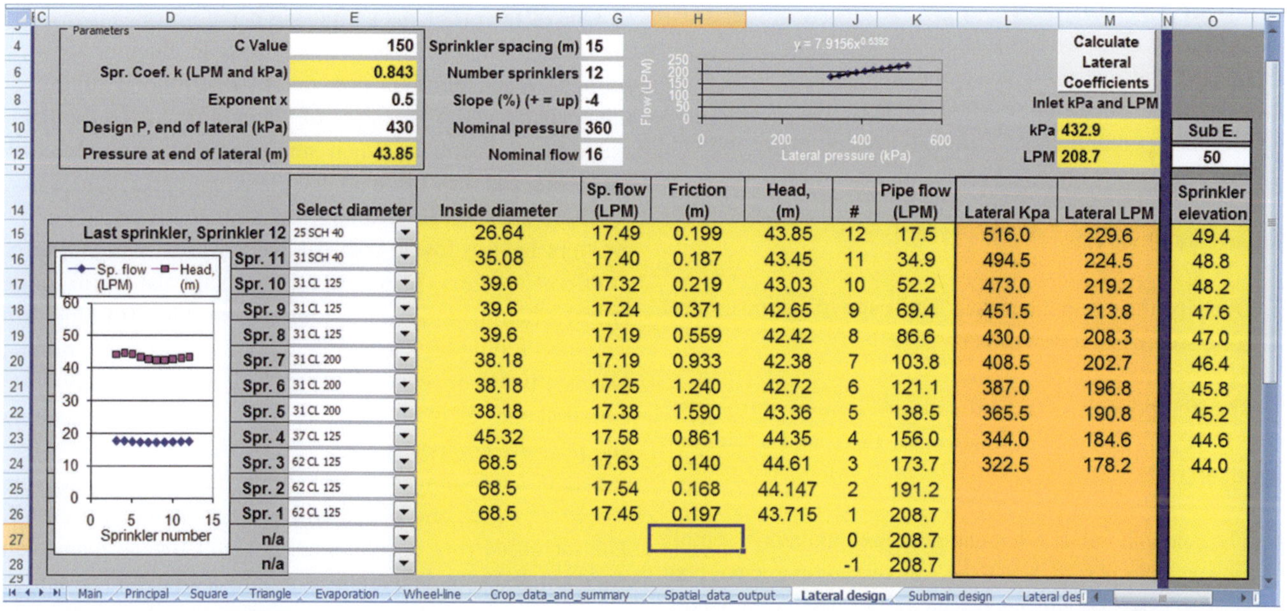

Parameters	
C Value	150
Spr. Coef. k (LPM and kPa)	0.843
Exponent x	0.5
Design P, end of lateral (kPa)	430
Pressure at end of lateral (m)	43.85
Sprinkler spacing (m)	15
Number sprinklers	12
Slope (%) (+ = up)	-4
Nominal pressure	360
Nominal flow	16
Calculate Lateral Coefficients — Inlet kPa and LPM: kPa	432.9
LPM	208.7
Sub E.	50

	Select diameter	Inside diameter	Sp. flow (LPM)	Friction (m)	Head, (m)	#	Pipe flow (LPM)	Lateral Kpa	Lateral LPM	Sprinkler elevation
Last sprinkler, Sprinkler 12	25 SCH 40	26.64	17.49	0.199	43.85	12	17.5	516.0	229.6	49.4
Spr. 11	31 SCH 40	35.08	17.40	0.187	43.45	11	34.9	494.5	224.5	48.8
Spr. 10	31 CL 125	39.6	17.32	0.219	43.03	10	52.2	473.0	219.2	48.2
Spr. 9	31 CL 125	39.6	17.24	0.371	42.65	9	69.4	451.5	213.8	47.6
Spr. 8	31 CL 125	39.6	17.19	0.559	42.42	8	86.6	430.0	208.3	47.0
Spr. 7	31 CL 200	38.18	17.19	0.933	42.38	7	103.8	408.5	202.7	46.4
Spr. 6	31 CL 200	38.18	17.25	1.240	42.72	6	121.1	387.0	196.8	45.8
Spr. 5	31 CL 200	38.18	17.38	1.590	43.36	5	138.5	365.5	190.8	45.2
Spr. 4	37 CL 125	45.32	17.58	0.861	44.35	4	156.0	344.0	184.6	44.6
Spr. 3	62 CL 125	68.5	17.63	0.140	44.61	3	173.7	322.5	178.2	44.0
Spr. 2	62 CL 125	68.5	17.54	0.168	44.147	2	191.2			
Spr. 1	62 CL 125	68.5	17.45	0.197	43.715	1	208.7			
n/a						0	208.7			
n/a						-1	208.7			

The next step is to enter the lateral exponential equation and inlet pressure at the last lateral into the submain worksheet.

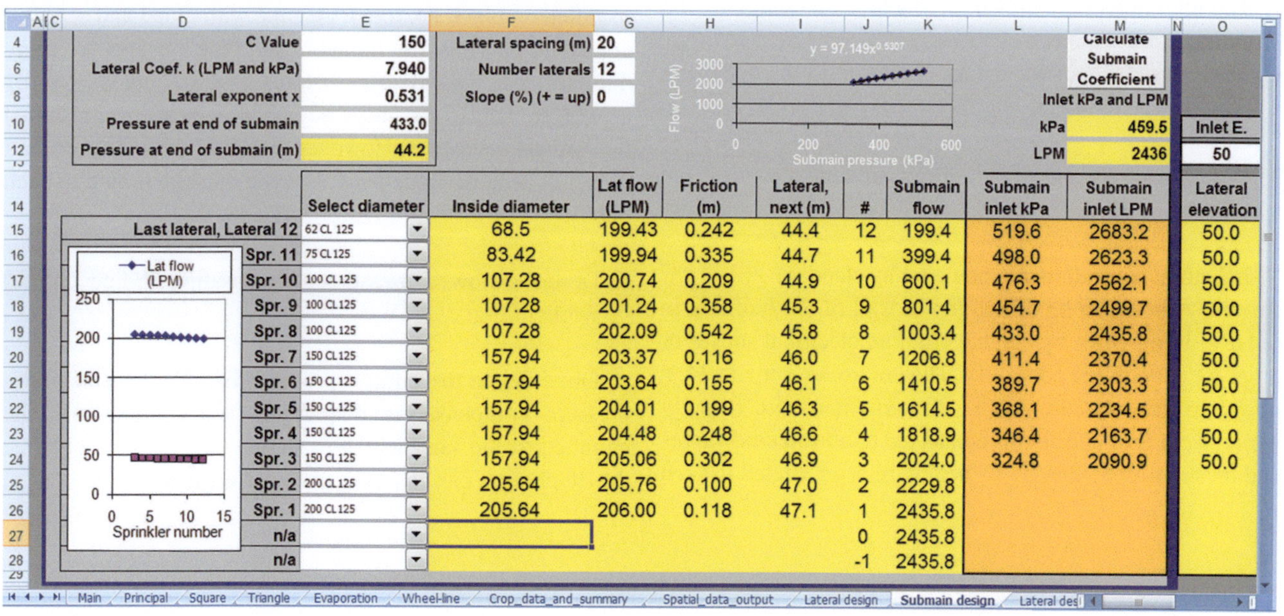

Parameters	
C Value	150
Lateral Coef. k (LPM and kPa)	7.940
Lateral exponent x	0.531
Pressure at end of submain	433.0
Pressure at end of submain (m)	44.2
Lateral spacing (m)	20
Number laterals	12
Slope (%) (+ = up)	0
Calculate Submain Coefficient — Inlet kPa and LPM: kPa	459.5
LPM	2436
Inlet E.	50

	Select diameter	Inside diameter	Lat flow (LPM)	Friction (m)	Lateral, next (m)	#	Submain flow	Submain inlet kPa	Submain inlet LPM	Lateral elevation
Last lateral, Lateral 12	62 CL 125	68.5	199.43	0.242	44.4	12	199.4	519.6	2683.2	50.0
Spr. 11	75 CL 125	83.42	199.94	0.335	44.7	11	399.4	498.0	2623.3	50.0
Spr. 10	100 CL 125	107.28	200.74	0.209	44.9	10	600.1	476.3	2562.1	50.0
Spr. 9	100 CL 125	107.28	201.24	0.358	45.3	9	801.4	454.7	2499.7	50.0
Spr. 8	100 CL 125	107.28	202.09	0.542	45.8	8	1003.4	433.0	2435.8	50.0
Spr. 7	150 CL 125	157.94	203.37	0.116	46.0	7	1206.8	411.4	2370.4	50.0
Spr. 6	150 CL 125	157.94	203.64	0.155	46.1	6	1410.5	389.7	2303.3	50.0
Spr. 5	150 CL 125	157.94	204.01	0.199	46.3	5	1614.5	368.1	2234.5	50.0
Spr. 4	150 CL 125	157.94	204.48	0.248	46.6	4	1818.9	346.4	2163.7	50.0
Spr. 3	150 CL 125	157.94	205.06	0.302	46.9	3	2024.0	324.8	2090.9	50.0
Spr. 2	200 CL 125	205.64	205.76	0.100	47.0	2	2229.8			
Spr. 1	200 CL 125	205.64	206.00	0.118	47.1	1	2435.8			
n/a						0	2435.8			
n/a						-1	2435.8			

The submain exponential equation is $97.14\ H^{0.531}$

The pump and submain equations are now solved for pump flow rate. (Use Solver in Excel)

$$0 = (97.14)\left(-5.18*10^{-5}Q^2 + 0.00828*10^{-3}Q + 900\right)^{0.531} - Q$$
$$Q = 2,670\ \text{LPM}.$$

Solve for the head.

$$H = (Q/97.14)^{1/0.531} = (2670/97.14)^{1/0.631} = 513\ \text{kPa}.$$

Chapter 15: Solution

1. Discuss the questions presented in Exercise 15.1. How would the scientific method change the results and people's perceptions? (Write at least 1 page, double-spaced).

Answers will vary.

2. Describe the components in a landscape irrigation control zone and the function of each part.

The components of a landscape irrigation control zone are a manual ball valve, a solenoid valve and a pressure regulator.

- The solenoid valve, filter, and pressure regulator assembly are typically placed below the soil surface in a valve box.
- A ball valve should be placed before the components, thus if any repair or modification is required, the system can be turned off.
- The solenoid valve is connected to the controller and turns the zone on and off.
- The filter prevents solids from reaching and plugging the emitters.
- The pressure regulator reduces pressure to the operating range of the emitters.

3. Describe how a controller is wired and how it controls the valves.

Irrigation controllers are powered by standard 110 V AC current. A hole can be drilled in the wall of the building in order to bring wires by conduit from an electrical socket or other location to the controller. The controller outputs 24 V power to the solenoid valves. One common wire (usually white) from the controller is connected to all solenoid valves and the controller. One 24 V "hot" wire is connected to each valve. Protection of electrical wiring is important. Metal conduit should be used to protect 110 V wire from the building to the controller. The 24 V valve wires should also be encased above ground. However, once the wires (i.e., #14 AWG) are in the ground they can buried without a conduit. The wires should be placed below the water pipes in the trench in order to protect them; thus, they are laid in the trench before the water pipes.

4. Compare the advantages and disadvantages of barbed fitting emitters, line source drip irrigation and bubbler irrigation.

One of the problems with using barbed fittings is that the connections are not as secure as PVC with glued or screwed fittings. When emitters pop out of distribution tubing or tubing is damaged, a stream of water sprays onto the landscape like a fountain. When single emitters have plugging problems, and they often do, the plant may receive no water.

There are several advantages of inline emitters: (1) inline emitters have a lower cost per wetted area than single or multiport emitters, (2) it is easy to replace the entire inline lateral when the emitters or tubing wear out, (3) inline drip irrigation tubing lasts longer than point source drip emitters, (4) even if one or two emitters go bad, there are many other emitters, and (5) many inline drip emitters are self-flushing and pressure compensating.

Another advantages of inline tubing is that it creates a line source of water, and plants will grow roots along the tube. The increased root development is better for the plant than the restricted root development around a single point source emitter.

Bubblers operate at a much higher flow rate (0.5-, 1-, or 2-GPM; 2-, 4-, or 8-LPM) than drip irrigation systems; thus, they have larger orifices and don't require filtration. The disadvantage is that more zones are needed. They are installed on black flexible PVC pipe with glued and screwed connections. Thus, the bubblers and pipe systems are more durable than typical drip systems and are not as readily broken by landscape tools, degraded in heat, or consumed by animals or insects. However, bubblers are more difficult to install because trenched PVC pipe must be routed to each bubbler.

5. Describe how a pressure compensating in-line emitter works.

One inlet to the emitter provides backpressure behind a pressure compensating diaphragm, and the other is the entrance to the turbulent flow path. As pressure increases, the diaphragm closes down and restricts flow.

6. Answer in-class Exercise 15.2. Using Table 15.2, determine the maximum distance between 4 LPH emitters in order to create a line source wetting pattern in a sandy soil? Would you select 30, 45, or 60 cm spacing?

The wetted diameter is 0.6 m; thus, ¾ of this distance would be 45 cm spacing.

7. Draw out a yard that you know about or can imagine and lay out the location of sprinklers and drip emitters. You can do this on a piece of graph paper where 1 inch = 10 ft or some other appropriate scale. You could also draw it out in a computer program. Locate positions of valves and pipes as well as define zones. For bubblers, let flow rate be 2 GPM, and for emitters let flow rate be 2 GPH. Use PVC or polyethylene tubing where appropriate. Make sure to group similar emitters in zones. Select pipe and calculate pipe friction losses.

Answers will vary

8. Evaluate the economic costs and benefits of a landscape irrigation system. Rate of return is 8 % and project life is 8 years.
   - Water costs for irrigation are \$500/yr
   - Capital cost for installation is \$1,500
   - Home selling price is \$200,000 and landscaping adds 17 % to the value of home. Home will be sold in 8 years and home price is expected to decrease in value by 10 % over the 8 year period.
   - Irrigation system maintenance is \$250/yr
   - Plant replacement costs is \$100/yr

Annual costs are sum of water, maintenance, and plants: $\$500 + \$250 + \$100 = \$850$.

The present value of \$850 payments for 8 years at an 8 % rate of return is \$4,885.

Total present value of costs is $\$4,885 + \$1,500 = \$6,385$.

Benefits

The value of the house in 8 years will be $\$200,000(1.1) = \$220,000$

Landscaping will add 17 % to the selling price $\$220,000(0.17) = \$37,400$.

Present value of benefits is $\$37,400(1 + 0.08)^{-8} = \$20,200$

The present value of benefits is 3 times greater than the present value of costs.

9. Which types of irrigation devices would be appropriate for four 5 m diameter, widely spaced, trees?

Either bubbler or inline emitters would cover a large wetted area and would provide sufficient water for the trees

10. Which types of irrigation devices would be appropriate for 10 × 50 ft planter with ground cover?

Inline emitters would cover a large ground surface area and could extend along the length of the planter.

Chapter 16: Solution

1. List the 5 steps in turf and landscape irrigation system design.
   - Owner interview.
   - Site survey.
   - Determine maximum flow rate and working pressure.
   - Plant water requirements and application rates.
   - Zoning and hydraulics.

2. What issues should be discussed with the property owner before designing the irrigation system?

The designer should explain to the owner the most suitable irrigation system alternatives and discuss installation, operation and maintenance costs, life expectancy, safety features (backflow and master valve), operation and design parameters (irrigation scheduling), distribution uniformity and, if a pump is required, energy consumption and flow metering. Some optional features like chemical injection, filtration and pressure regulations should be also discussed depending on owner's level of expertise and willingness to invest time or money into system management.

3. What aspects of the landscape should be included in the site plan?

The site plan should include a map of the area featuring water source, including static pressure and flow meter diameter, buildings and constructions, relevant landscape elements, like plants, turf areas, rocks, etc., existing irrigation elements, wind direction and speed.

4. What is the pressure loss in a 5/8″ municipal water meter at 12 GPM? Use the *Service line friction loss* worksheet. Is it acceptable?

Select 5/8" water meter in cell E15. Convert from GPM to LPM in cell N8 and enter flow rate in cell J15. Friction loss is found in cell L15 (3.58 m). Use cell N10 to convert value to **5.1 psi**. The water meter is acceptable.

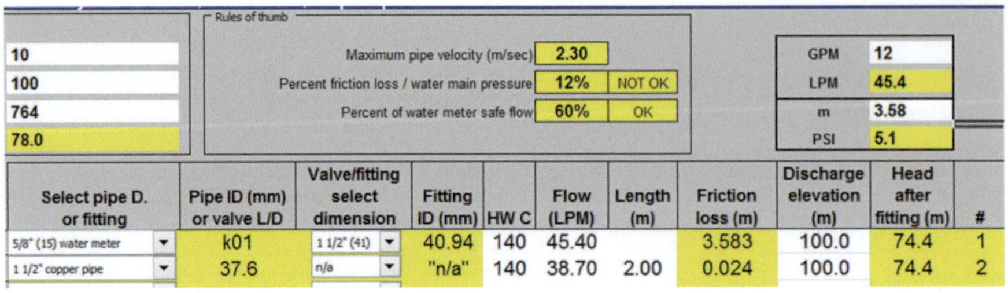

10	Maximum pipe velocity (m/sec)	2.30	
100	Percent friction loss / water main pressure	12%	NOT OK
764	Percent of water meter safe flow	60%	OK
78.0			

GPM	12
LPM	45.4
m	3.58
PSI	5.1

Select pipe D. or fitting	Pipe ID (mm) or valve L/D	Valve/fitting select dimension	Fitting ID (mm)	HW C	Flow (LPM)	Length (m)	Friction loss (m)	Discharge elevation (m)	Head after fitting (m)	#
5/8" (15) water meter	k01	1 1/2" (41)	40.94	140	45.40		3.583	100.0	74.4	1
1 1/2" copper pipe	37.6	n/a	"n/a"	140	38.70	2.00	0.024	100.0	74.4	2

5. Redo Example 16.1 but increase flow rate to 45.5 LPM. Use 1¼ inch copper pipe and fittings prior to the valves and water meter. Then use the 5/8 inch water meter and 5/8 inch valve and 5/8 inch copper in the last section. Use the *Service line friction loss* worksheet. Is the design acceptable?

The design is not acceptable because the friction loss is too high, 19 %.

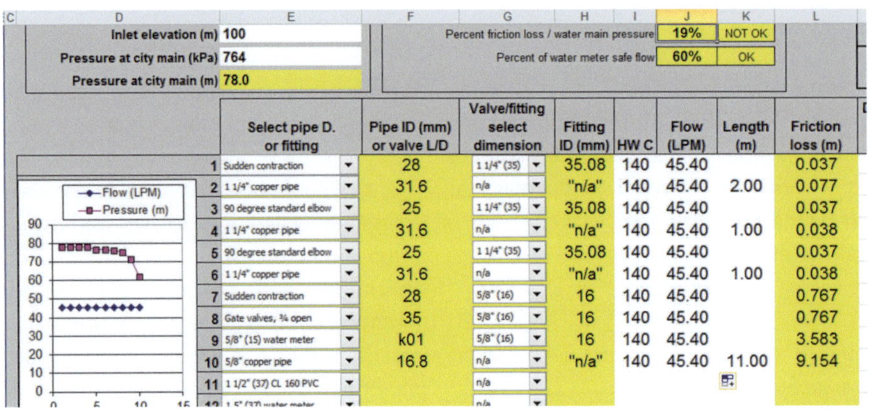

Inlet elevation (m)	100
Pressure at city main (kPa)	764
Pressure at city main (m)	78.0

Percent friction loss / water main pressure	19%	NOT OK
Percent of water meter safe flow	60%	OK

	Select pipe D. or fitting	Pipe ID (mm) or valve L/D	Valve/fitting select dimension	Fitting ID (mm)	HW C	Flow (LPM)	Length (m)	Friction loss (m)
1	Sudden contraction	28	1 1/4" (35)	35.08	140	45.40		0.037
2	1 1/4" copper pipe	31.6	n/a	"n/a"	140	45.40	2.00	0.077
3	90 degree standard elbow	25	1 1/4" (35)	35.08	140	45.40		0.037
4	1 1/4" copper pipe	31.6	n/a	"n/a"	140	45.40	1.00	0.038
5	90 degree standard elbow	25	1 1/4" (35)	35.08	140	45.40		0.037
6	1 1/4" copper pipe	31.6	n/a	"n/a"	140	45.40	1.00	0.038
7	Sudden contraction	28	5/8" (16)	16	140	45.40		0.767
8	Gate valves, ¾ open	35	5/8" (16)	16	140	45.40		0.767
9	5/8" (15) water meter	k01	5/8" (16)	16	140	45.40		3.583
10	5/8" copper pipe	16.8	n/a	"n/a"	140	45.40	11.00	9.154
11	1 1/2" (37) CL 160 PVC		n/a					

6. Define the species factor, microclimate factor, and density factor.
   - Species factor (K_s): This factor accounts for the characteristics of a particular plant species.
   - Microclimate factor (K_{mc}): This factor accounts for the specific site conditions (wind, surrounding heat-absorbing surfaces or reflective surfaces, etc.).
   - Plant density factor (K_d): This factor accounts for the collective leaf area of all plants in the landscape.

7. Look in the WUCOLS guide and determine the species factor for oleander in Blythe, California. Use the low desert classification.

The classification is moderate. In the user manual, the percentage of reference ET is 40–60 %.

8. Calculate the LPD requirement for a 2 m diameter oleander in Tucson, AZ Reference ET is 12 mm/day. The microclimate and density factors are 1.0. Irrigation efficiency is 85 %

Multiplying Eq. 16.2 by K_L, which in this case is 0.5 (0.5*1*1), and assuming an efficiency fraction of 0.85, one gets:

$$Requirement = \frac{12*(1*1*0.5)\left(2^2\pi\right)}{0.85*4} = 22\ LPD$$

9. Calculate the LPD requirement for a 5 m diameter orange tree in Blythe, CA. The landscape, microclimate, and density coefficients are 1.0. Reference ET is 14 mm/day. Irrigation efficiency is 85 %.

$$Requirement = \frac{14*(1*1*1)(2.5^2\pi)}{0.85} = 323LPD$$

10. Calculate S under an emitter that has a 1.0 meter diameter wetted area with a 1.0 m deep root zone. The MAD is 0.5. Field capacity is 20 % and permanent wilting point is 10 %.

Use Eq. 16.4.

$$S = \frac{\pi*1^2}{4}*1*0.1*0.5*1000 = 39L$$

11. The orange tree in problem 9 has 6 emitters with the water storage per emitter as calculated in problem 10. The reference ET rate is 10 mm/day. Calculate a watering schedule for the orange tree. Emitter flow rates are 4 LPH. Note: this problem demonstrates the problem with having few emitters on a large tree.

Six emitters will have a water storage = 234 L. Because the storage volume is less than the daily plant requirement (323 LPD), the system will need to run every day to prevent water stress.

$$Irrigation\ time = \frac{LPD}{LPH} = \frac{323LPD}{24LPH} = 13.5hr/day$$

12. Calculate the watering schedule for an oleander hedge with 2 m width. Reference ET rate is 10 mm/day. Wetted width is 0.6 m. Rooting depth is 2 m. Emitter flow rate is 2 LPH. Use a landscape coefficient of 0.5 and MAD = 0.5. Distance between emitters = 0.3 m. Root depth is 2 m. Soil is sandy loam with AWC = 0.12. Microclimate and density factors are 1.0. Irrigation efficiency is 80 %.

$$ET_L = 10\ mm/day*0.5*1*1 = 5\ mm/day$$

Calculate the LPD requirement per emitter

$$LPD = \frac{5\frac{mm}{day}*2*0.3}{0.80} = 3.75LPD$$

$$S = 1000*2*0.6*0.3*0.12*0.5 = 21.6L$$

$$Irrigation\ frequency = \frac{21.6L}{3.75LPD} = 5.8days$$

Thus, irrigation should take place every 5 days.

$$\begin{aligned} Irrigation\ run\ time &= (LPD*Irrigation\ interval)/ \\ &\quad Emitter\ flow\ rate \\ &= (3.75\ LPD*5Days)/2\ LPH \\ &= 9.4\ hr \end{aligned}$$

13. Use Landscape Irrigation program to calculate number of screw turns for bubblers irrigating 4 orange trees with 5, 6, 7, and 8 m diameter canopy. There is one bubbler per tree. The pressure is 280 kPa (adjusted in Device data page). Determine the irrigation schedule. Soil is loamy sand, density factor is 1.0, microclimate factor (exposure) is 1.2. Leaching fraction is 0.15. There is no overlap in wetting patterns. Application efficiency is 95 %.

Using the landscape irrigation worksheet, the following screw adjustments were calculated

Canopy diameter	Name	Type	Flow (LPM) Bub (LPH) Em	Line space (m)	Wetted Diam (m)	Tubing length (m)	Number of screw turns
8	HunterB	Bubbler kx	8.57		2.000		1.25
7	HunterB	Bubbler kx	6.56		2.000		0.71
6	HunterB	Bubbler kx	4.82		2.000		0.44
5	HunterB	Bubbler kx	3.34		2.000		0.32

Using the Step 4, Calculate schedule button, the following output is produced. Because the wetted area is small in comparison to the size of the tree, there are 7 irrigations per day of approximately 15 minutes per irrigation.

Statistics

Scheduling coefficient	1.00
Distribution uniformity	100.0%
Statistical uniformity	99.0%
Allowable % underirrigated	0%
Daily applied volume (L)	2382.00
Daily wasted volume (L)	0.00
	Step 4 Calculate schedule
Precipitation (mm/day)	0
Reference ET (mm/day)	12
MAD	50%
Days between irrigations	0
Irrigation run time (min)	14.66
Vol. App. per irrigation (L)	341.62
Number of irrigations/day	6.97

14. Calculate the number of bubblers per plant for 20 orange trees laid out in a 5 (NS) × 4 (EW) grid. Tree diameters are shown in meters. Then redo with spiral inline emitter configuration. Compare pipe sizes and number of valves. Application efficiency is 90 %. Leaching fraction is 0.15. There is no overlap in wetting patterns. Maximum number of bubblers per plant is five. Microclimate and density factors are 1.0. For the same soil and weather parameters as in Example 13, calculate the irrigation schedule.

5	4	3.4	6
2	4	6	3
3	4	6	4
1	3	4	4
4	4	4	6

Number of emitters per plant is shown in the following table.

Plant number	Plant name	Diam. or width (m)	Root depth (m)	Length (m)	Foliage area (m)	Water use (0.2-desert, 1 -high)	Plant type H, P, A	Number emitters/ plant
1	Orange	1.00	1		0.785	1	Plant	1
2	Orange	2.00	1		3.142	1	Plant	1
3	Orange	3.00	1		7.069	1	Plant	1
4	Orange	3.40	1		9.079	1	Plant	2
5	Orange	4.00	1		12.57	1	Plant	2
6	Orange	5.00	1		19.63	1	Plant	3
7	Orange	6.00	1		28.27	1	Plant	5

Slightly more than one irrigation is needed per day. The run time is 15.7 minutes.

Scheduling coefficient	1.22
Distribution uniformity	84.2%
Statistical uniformity	37.0%
Allowable % underirrigated	0%
Daily applied volume (L)	5723.00
Daily wasted volume (L)	1298.00
	Step 4 Calculate schedule
Precipitation (mm/day)	0
Reference ET (mm/day)	12
MAD	50%
Days between irrigations	0
Irrigation run time (min)	15.71
Vol. App. per irrigation (L)	6030.65
Number of irrigations/day	1.16

Chapter 17: Solution

1. How many 4 ft (1.22 m) diameter sand filters are needed for a 260 ha (640 ac) drip irrigated farm? Crop ET is 11 mm/day. Irrigation efficiency is 90 %.

Approximate the irrigation requirement as 11 mm/day/0.9 = 12 mm/day.

$$Q = 2{,}600{,}000\ \mathrm{m^2}{*}0.012\ \mathrm{m/day}{*}(1\ \mathrm{hr}/3{,}600\ \mathrm{sec}){*}(1\ \mathrm{day}/24\ \mathrm{hrs})$$
$$Q = 0.36\ \mathrm{m^3/s} = (0.36\ \mathrm{m^3/s})(1\mathrm{m^3}/1000\ \mathrm{L}) = 360\ \mathrm{LPS}$$

Maximum flow rate through sand filters should be between 10 and 18 LPS per square meter of tank cross-sectional area.

$$D_{\mathrm{filter}} = 1.22\ \mathrm{m}$$
$$A = \pi D^2/4 = 3.1416{*}(1.22\ \mathrm{m})^2/4 = 1.17\ \mathrm{m^2}$$

At 18 LPS, the number of filters is calculated.

$$\text{Number of filters} = (360\ \mathrm{LPS}{*}1.17\ \mathrm{m^2})/18\ \mathrm{LPS} = 23.3\ \text{filters}$$
$$\text{Number of filters} = 24$$

Because this is the upper range, and flow might vary, it is probably wise to increase the number of filters to 30.

2. Design particle size for a settling basin is 25 microns. Irrigation system flow rate is 1,000 GPM. What are the dimensions of the settling basin?

$$V_p = 3.43{*}10^{-5}D^2(SG - 1) = 3.43{*}10^{-5}{*}25^2{*}1.65 = 0.035\ \mathrm{m/min}$$

$$Area = 0.001{*}F{*}\left(\frac{Q}{V_p}\right) = 0.001{*}2.0{*}\left(\frac{1{,}000}{0.035}\right) = 57\ m^2$$

$$W = \sqrt{\frac{Area}{5}} = \sqrt{\frac{57}{5}} = 3.4\ m \quad L = 3.4{*}5 = 17\ \mathrm{m}$$

3. Calculate head loss and emission uniformity in a 120 m length of 18 mm ID tubing. x = 0.57 and k = 0.15. Inlet pressure is 200 kPa. Emitters are spaced at 0.2 m. The crop is carrots. Determine if the design is acceptable based on a criteria of 90 % emission uniformity. Use the analytic solution method and check your answer with the

Single feed lateral spreadsheet. The manufacturer's coefficient of variation is 0.07 or 7 %, and there are 2 emitters per plant. There is no slope.

200 kPa and 120 m length

Make an initial guess for lateral flow rate is based on the flow from the first emitter.

$$Q = 0.15^*(200\ \text{kPa})^{0.57} = 3.07\ \text{LPH}.$$

There are 5 emitters per m and the length of tubing is 120 m. Thus, there are 600 emitters and the total flow rate per lateral is (3.07 LPH) (120 m) (5 em/m) = 1,842 LPH. Calculate friction loss in fully flowing pipe

$$v = \frac{Q}{A} = \left(\frac{1,842\ Lph}{\pi(9/1000)^2\ m^2}\right)\left(\frac{m^3}{1,000\ L}\right)\left(\frac{hr}{3,600\ \ \sec}\right)$$
$$= 2\ m/\sec$$

$$\text{Re} = \frac{vD}{\nu} = \frac{2^*18/1,000}{1^*10^{-6}} = 36,000$$

$$f = \frac{0.316}{\text{Re}^{1/4}} = \frac{0.316}{36,000^{1/4}} = 0.023$$

$$h_f = 6.377 fL\left[\frac{Q^2}{D^5}\right] = 6.377^*0.030^*120\ m\left[\frac{1842^2}{18^5}\right]$$
$$= 31.6\ m$$

$$\text{h}_{\text{ac}} = \text{h}_{\text{f}}\text{F} = 31.6^*0.338 = 10.7\ \text{m}.$$

$$H_o = H_a + 0.74\ h_{ac} + \frac{S_e L}{2}$$
$$H_a = H_0 - 0.74\ h_{ac} = 20 - 0.74(10.7) = 12\ m$$

I am going to adjust upward to 15 m because the initial estimate of average pressure made the estimate of friction loss too high.

$$Q = 0.15^*(150\ \text{kPa})^{0.57} = 2.61\ \text{LPH}$$

Using the same calculation procedure as above, the average pressure is 14.1 m.

$$Q = 0.15^*(141\ \text{kPa})^{0.57} = 2.52\ \text{LPH}$$

Using the same calculation procedure as above, the average pressure is 14.43 m.

$$Q = 0.15^*(14.43/0.102\ \text{kPa})^{0.57} = 2.52\ \text{LPH}$$

Thus, the flow does not change and the iteration stops. Find the end pressure.

$$H_d = 14.43/0.102 + (0.26)(7.55) = 12.45$$

Inserting 12.45 as the distal end pressure in the Single Feed Lateral spreadsheet results in an inlet pressure of 20 m. See following screen copy. Make sure to press the Make Calcs button in order to find the terms for the emission uniformity.

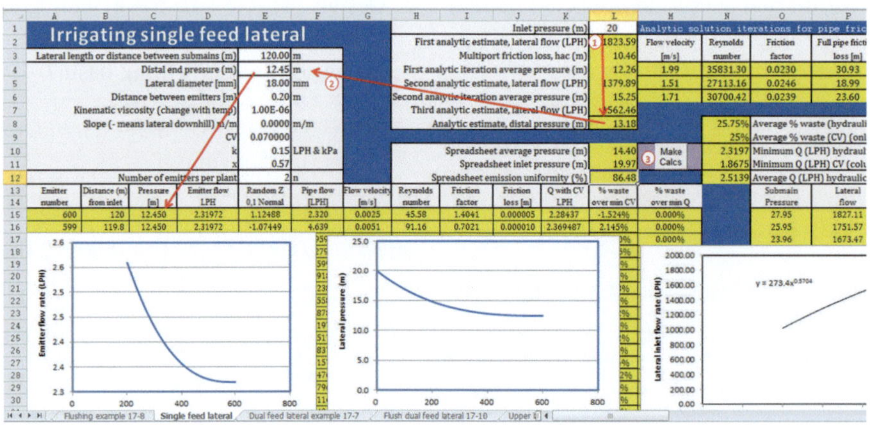

Spreadsheet

$$q_{min} = 0.15*(12.45/0.102)^{0.57} = 2.32 \text{ LPH}$$
$$q_{ave} = 0.15(14.4) = 0.15*(14.4/0.102)^{0.57} = 2.51 \text{ LPH}$$

$$U_e = 100\left[1 - \left(\frac{1.27}{\sqrt{2}}\right)0.07\right]\frac{2.32}{2.51} = 86.5\%$$

This design is not acceptable according to the criterion of 90 % emission uniformity. It is quite surprising that the emission uniformity is so high given the high pressure loss in the system.

4. Calculate head loss and emission uniformity in an 80 m length of 18 mm ID tubing. x = 0.57 and k = 0.15. Inlet pressure is 50 kPa. Emitters are spaced at 0.2 m. The crop is carrots. Determine if the design is acceptable based on a criteria of 90 % emission uniformity. Use the analytic solution method and check your answer with the *Single feed lateral* spreadsheet. The manufacturer's coefficient of variation is 0.07 or 7 %, and there are 2 emitters per plant. There is no slope.

50 kPa and 80 m length

Make an initial guess for lateral flow rate is based on the flow from the first emitter.

$$Q = 0.15*(50 \text{ kPa})^{0.57} = 1.38 \text{ LPH}.$$

There are 5 emitters per m and the length of tubing is 80 m. After a few iterations, the average pressure is 4.43 and flow is 1.29 (80 m) (5 em/m) = 515 LPH. Calculate friction loss in fully flowing pipe

$$v = \frac{Q}{A} = \left(\frac{515\ Lph}{\pi(9/1000)^2\ m^2}\right)\left(\frac{m^3}{1{,}000\ L}\right)\left(\frac{hr}{3{,}600\ \sec}\right)$$
$$= 0.56\ m/\sec$$

$$\text{Re} = \frac{vD}{\nu} = \frac{0.9*18/1{,}000}{1*10^{-6}} = 10{,}100$$

$$f = \frac{0.316}{\text{Re}^{1/4}} = \frac{0.316}{10{,}100^{1/4}} = 0.0315$$

$$h_f = 6.377fL\left[\frac{Q^2}{D^5}\right] = 6.377*0.0315*80\ m\left[\frac{515^2}{18^5}\right] = 2.25\ m$$

$$h_{ac} = h_f F = 2.25*0.338 = 0.76 \text{ m}.$$

$$H_o = H_a + 0.74\ h_{ac} + \frac{S_e L}{2}$$
$$H_a = H_0 - 0.74\ h_{ac} = 5 - 0.74(0.76) = 4.43\ m$$

New average flow calculation

$$Q = 0.15*(4.43/0.102 \text{ kPa})^{0.57} = 1.28 \text{ LPH}$$

Find the end pressure.

$$H_d = 4.43 - (0.26)(76) = 4.24$$

Inserting 4.24 as the distal end pressure in the Single Feed Lateral spreadsheet results in an inlet pressure of 20 m. See following screen copy.

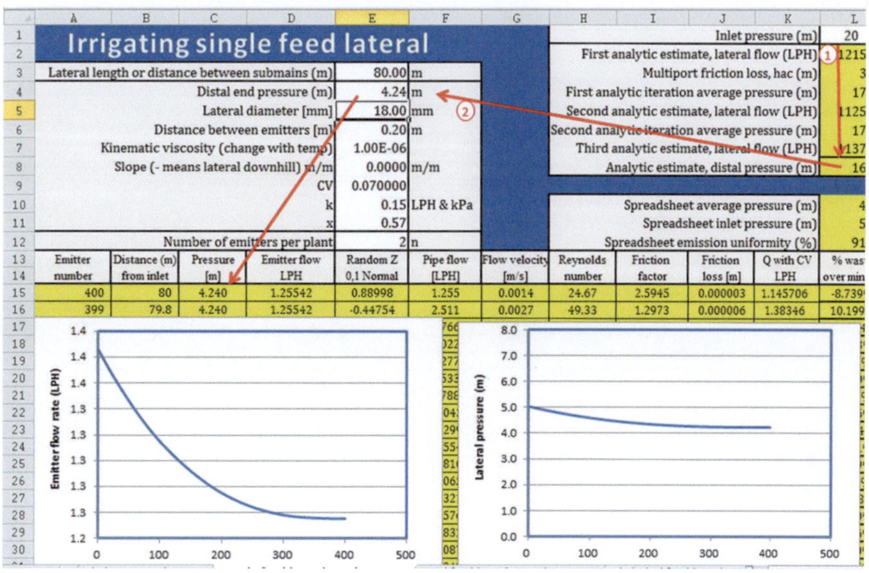

Spreadsheet

$$U_e = 100\left[1 - \left(\frac{1.27}{\sqrt{2}}\right)0.07\right]\frac{1.25}{1.29} = 91\%$$

This design is acceptable according to the criterion of 90 % emission uniformity. Just over half of the variability in this system is due to manufacturers coefficient of variation.

5. Calculate head loss and emission uniformity in a 120 m length of 18 mm ID tubing. x = 0.57 and k = 0.15. Inlet pressure is 200 kPa. Emitters are spaced at 0.2 m. The crop is carrots. Determine if the design is acceptable based on a criteria of 90 % emission uniformity. Just use the *Single feed lateral* spreadsheet for the calculation. The manufacturer's coefficient of variation is 0.07 or 7 %, and there are 2 emitters per plant. Slope is 5 % downhill.

50 kPa and 80 m length, 5 % downhill slope

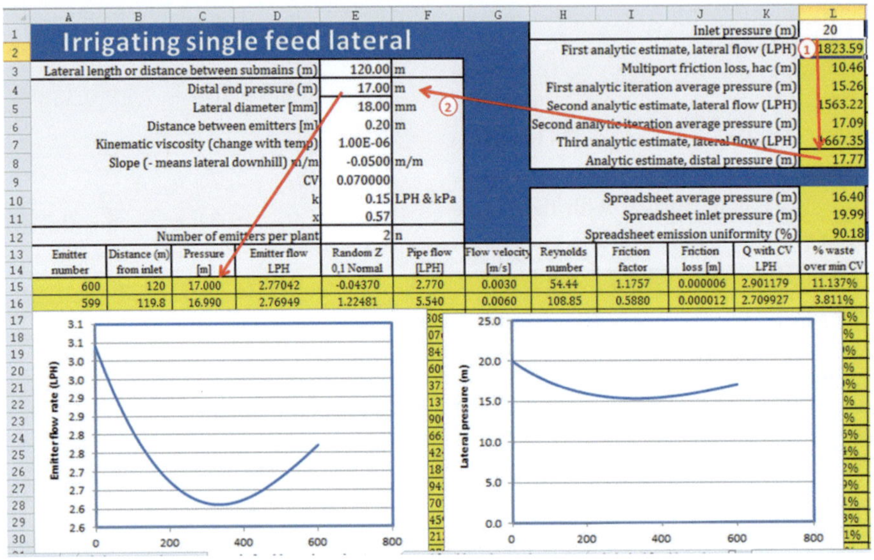

The emission uniformity is 90.2 % so the design is acceptable. The slope reduced the amount of pressure variation.

6. For the parameters in problem 3, calculate the inflow rate and pressure needed to flush the 120 m length tube.

Setting the distal end pressure at 2 m and flow at 200 LPH in order to obtain a velocity of 0.5 m. The following flow and pressure curves are obtained. The flow is 20 % less and the pressure is 20 % higher so this would probably be posible with the same pump.

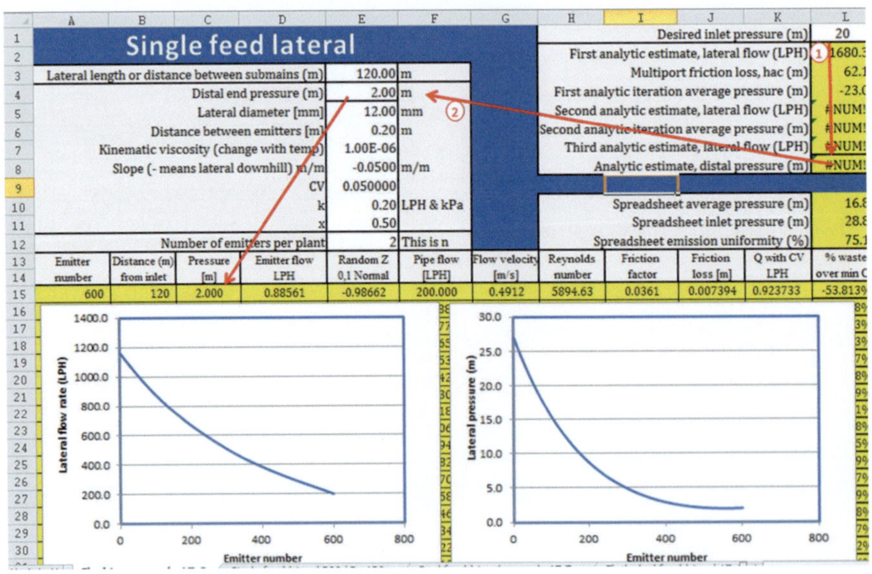

7. For the parameters in problem 5, calculate the inflow rate and pressure needed to flush the 120 m length tube.

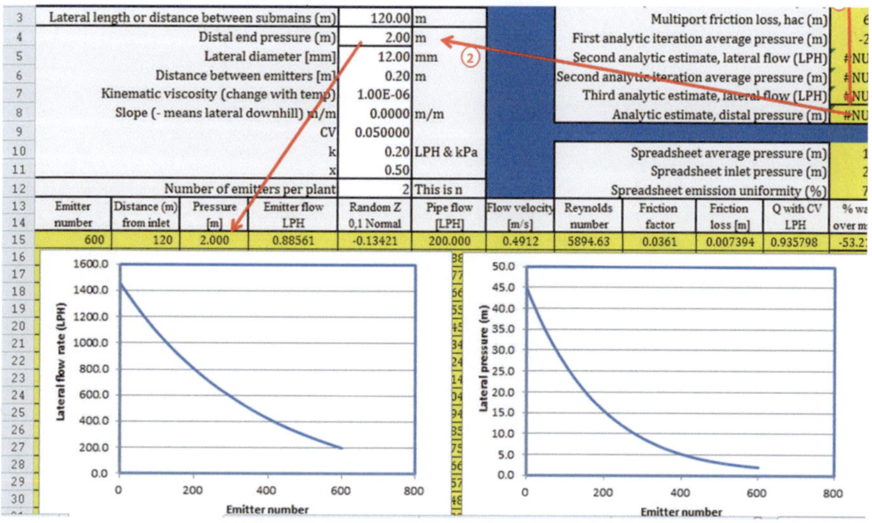

The pressure is very high in this case, and the flow rate to the lateral is not much less than the normal flow rate so it may be difficult to flush this lateral. A large booster pump would be needed.

8. Emitters are spaced at 1 m along the plant row and 1 m between plant rows, and emitter flow rate is 2 lph. For a daily plant water requirement of 12 mm/day, calculate the application rate and the application time if plants are watered on a daily basis. Assume 85 % efficiency.

$$(di/dt)_g = \left(\frac{Emitter\ flow\ rate}{area\ per\ emitter}\right) = \left(\frac{Q_e}{s_l{*}s_d}\right) = \left(\frac{2}{1{*}1}\right)$$

$$= 2\ mm/hr$$

The net application rate is the product of the gross application and the system efficiency:

$$(di/dt)_a = (di/dt)_g{*}\text{Eff.} = 2\ \text{mm/hr}{*}0.85 = 1.70\ \text{mm/hr}$$

Divide ET by the application rate to find required hours of application per day:

$$\text{hr/day} = 12\ \text{mm/day}/1.7\ \text{mm/hr} = 7.06\ \text{hr/day}$$

9. Two submains are 200 m apart and supply a dual feed lateral. Tubing diameter is 12 mm ID. x = 0.5 and k = 0.2. Inlet pressure is 100 kPa. Emitters are spaced at 0.5 m. There is no slope. Use the *upper lateral* worksheet to plot the hydraulic and flow variation. Make sure to press the Make calcs button.

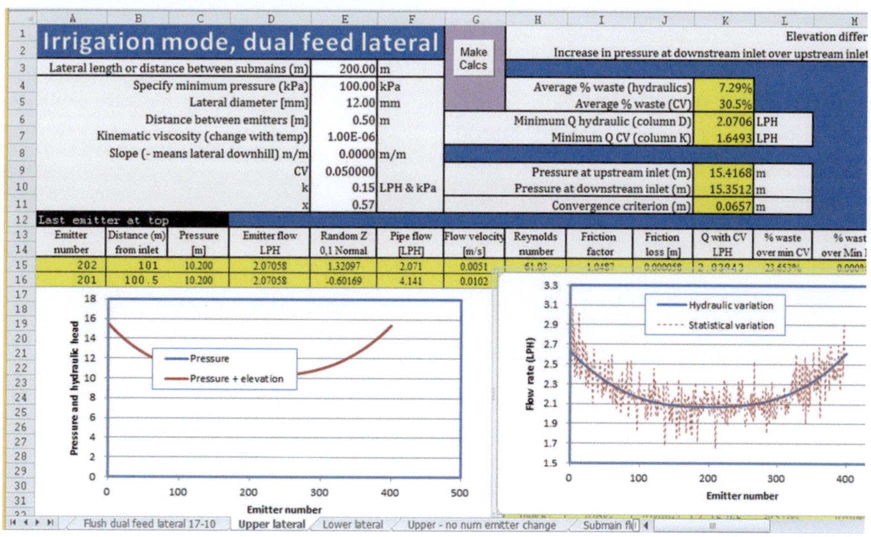

10. Two submains are 200 m apart and supply a dual feed lateral. Tubing diameter is 12 mm ID. x = 0.5 and k = 0.2. Inlet pressure is 100 kPa. Emitters are spaced at 0.5 m. There is 1 % slope. Use the *upper lateral* worksheet to plot the hydraulic and flow variation. There is no pressure difference between the uphill and downhill inlets. Explain the difference between the red and blue lines in the pressure graph. Which line is the hydraulic head?

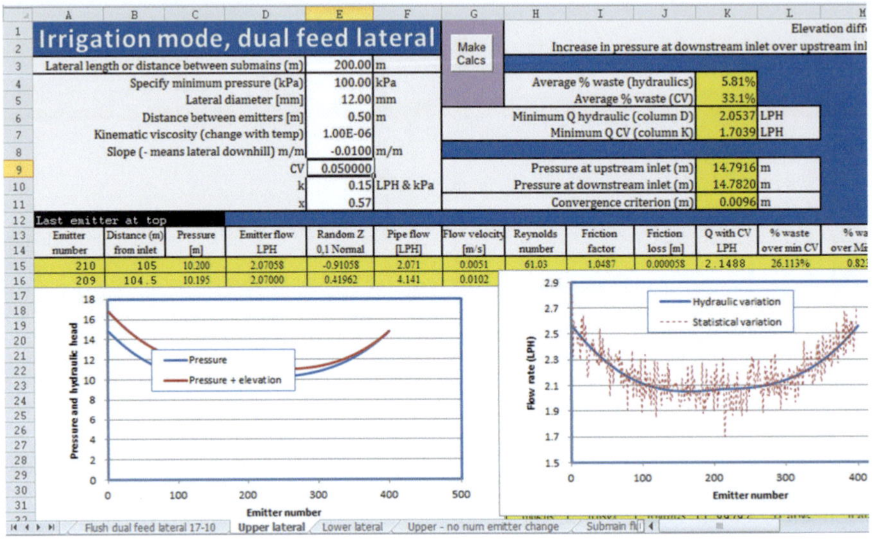

The blue line is the pressure line. The red line is the pressure + elevation. The upper end is higher so it has greater hydraulic head. The red line is the hydraulic head.

11. Design problem, answers will vary.

Chapter 18: Solution

1. A 90 m long submain supplies 12 mm ID laterals that are 90 m long. This is a single feed system. The laterals are spaced 1 m apart. Emitters are spaced every 0.3 m, k = 0.25, and x = 0.41. Slope of laterals is 1 % downhill. Manufacturer's coefficient of variation is 5 % and number of emitters per plant is 2. Verify that the lateral has at least 90 % emission uniformity. If not, then increase the pipe diameter. The submain is on level ground. Find the lateral flow rate vs. pressure equation (Fig. 18.3) and design the submain (select diameters). Because of flushing, minimum allowable size of the submain is 100 mm. The minimum acceptable pressure is 80 kPa. Find the required inlet pressure for the submain. Also, determine whether the emission uniformity is greater than 90 % for the entire zone.

The following data was entered into the *Downslope lateral* worksheet.

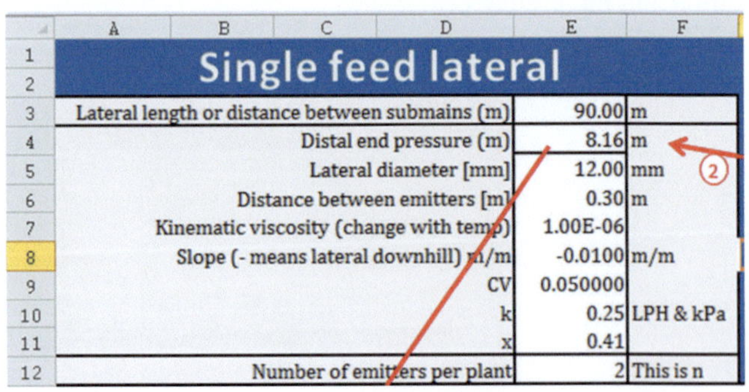

The *Make calcs* button was pressed, and the equation was developed as the program varied distal end pressure.

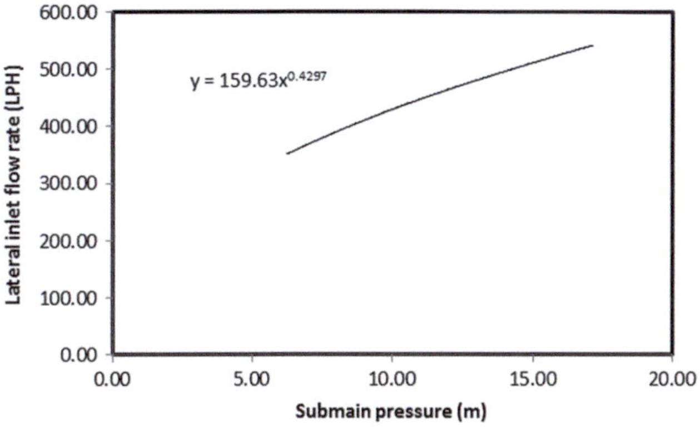

The required submain inlet pressure in order to maintain at least 8.16 m pressure at the distal end of the lateral is reported in cell L11, 12.43 m. Thus, this is the distal end pressure for the submain. The emission uniformity is 91.1 %.

$$U_e = 100\left[1 - \left(\frac{1.27}{\sqrt{2}}\right)0.05\right]\frac{1.50}{1.57} = 91.1\%$$

Spreadsheet average pressure (m)	9.07
Spreadsheet inlet pressure (m)	12.43
Spreadsheet emission uniformity (%)	91.10

The lateral flow pressure equation was entered into the *submain irrigation upper* worksheet, and 100 mm diameter pipe was entered in column E.

	A	B	C	D	E	F
1–2	Submain					
3				Length of submain (m)	100.00	m
4				Lateral inlet pressure at end of submain (m)	12.43	m
5						
6				Distance between laterals [m]	1.00	m
7				Kinematic viscosity (change with temp)	1.00E-06	
8				Slope (- means submain downhill) m/m	0.0000	m/m
9				CV		
10				k	159.63	LPH & kPa
11				x	0.43	

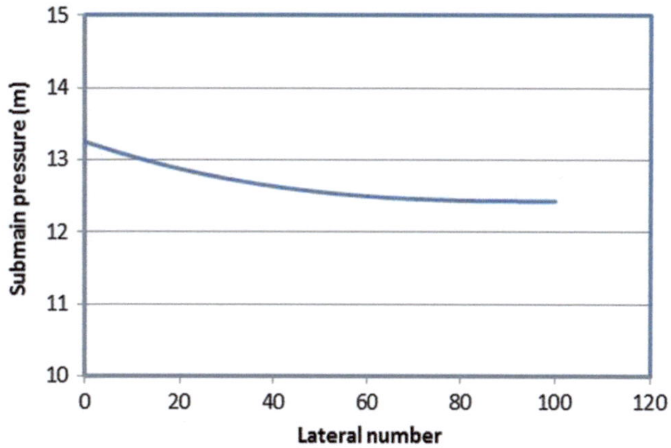

The required pressure at the submain inlet is 13.2 m. The average pressure in the submain is 12.64 m, which is 0.2 m higher than the distal end pressure. Thus the average pressure in the emission uniformity equation should be raised by 0.2 m. The average pressure in the last lateral was 9.07 m. Thus, the average pressure in the zone is approximately 9.27 m. Flow at this average pressure is calculated as follows.

$$\text{Qave} = 0.25(9.27/0.102)^\wedge 0.41 = 1.59$$

Substitute this flow into the emission uniformity equation. The emission uniformity of the entire zone is barely above 90 %.

$$U_e = 100\left[1 - \left(\frac{1.27}{\sqrt{2}}\right)0.05\right]\frac{1.50}{1.59} = 90.1\%$$

2. Repeat problem 1; however, there is no slope on the lateral. Evaluate emission uniformity on the individual lateral and in the zone. Is the emission uniformity above or below 90 %. Compare the exponent and the coefficient in the lateral flow pressure equation to that of problem 1. Explain the differences and similarities.

	A	B	C	D	E	F	G	H	I	J	K	L
1	Single feed lateral										Desired inlet pressure (m)	14.6
2											First analytic estimate, lateral flow (LPH) ①	574.01
3				Lateral length or distance between submains (m)	90.00	m					Multiport friction loss, hac (m)	7.12
4				Distal end pressure (m)	8.16	m ②					First analytic iteration average pressure (m)	9.33
5				Lateral diameter [mm]	12.00	mm					Second analytic estimate, lateral flow (LPH)	477.80
6				Distance between emitters [m]	0.30	m					Second analytic iteration average pressure (m)	10.78
7				Kinematic viscosity (change with temp)	1.00E-06						Third analytic estimate, lateral flow (LPH)	506.88
8				Slope (- means lateral downhill) m/m	0.0000	m/m					Analytic estimate, distal pressure (m)	9.29
9				CV	0.050000							
10				k	0.25	LPH & kPa					Spreadsheet average pressure (m)	9.55
11				x	0.41						Spreadsheet inlet pressure (m)	13.47
12				Number of emitters per plant	2	This is n					Spreadsheet emission uniformity (%)	89.81
13–14	Emitter number	Distance (m) from inlet	Pressure [m]	Emitter flow LPH	Random Z 0,1 Normal	Pipe flow [LPH]	Flow velocity [m/s]	Reynolds number	Friction factor	Friction loss [m]	Q with CV LPH	% waste over min CV

The emission uniformity on the single lateral is 89.91, slightly less than 90 %.

The lateral flow pressure coefficient (150) is approximately 7 % lower than 159, but the exponent has little change. The decreased coefficient reflects the fact that the average pressure in less with no slope so the flow is less. The exponent is reflective of the emitter exponent and turbulent pipe flow equation exponent (0.5), which are both in the range of the lateral exponent.

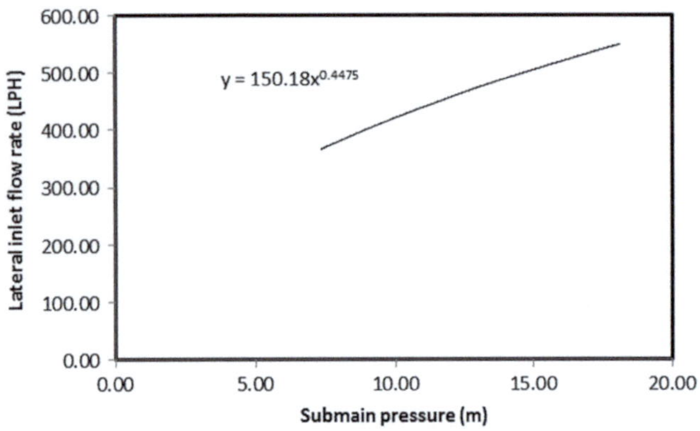

The required pressure at the lateral inlet is 13.47. The average pressure in the submain is 13.69. Thus, the average pressure in the zone should be increased by 0.22 m above the lateral average. The lateral average pressure is 9.55 so the average zone pressure should be 9.77

$$\text{Qave} = 0.25(9.77/0.102)^\wedge 0.41 = 1.62$$

Substitute this flow into the emission uniformity equation. The emission uniformity of the entire zone is 88.4 %, which is below the 90 % criterion.

$$U_e = 100\left[1 - \left(\frac{1.27}{\sqrt{2}}\right)0.05\right]\frac{1.50}{1.62} = 88.4\%$$

The inlet pressure to the submain is 14.3 m.

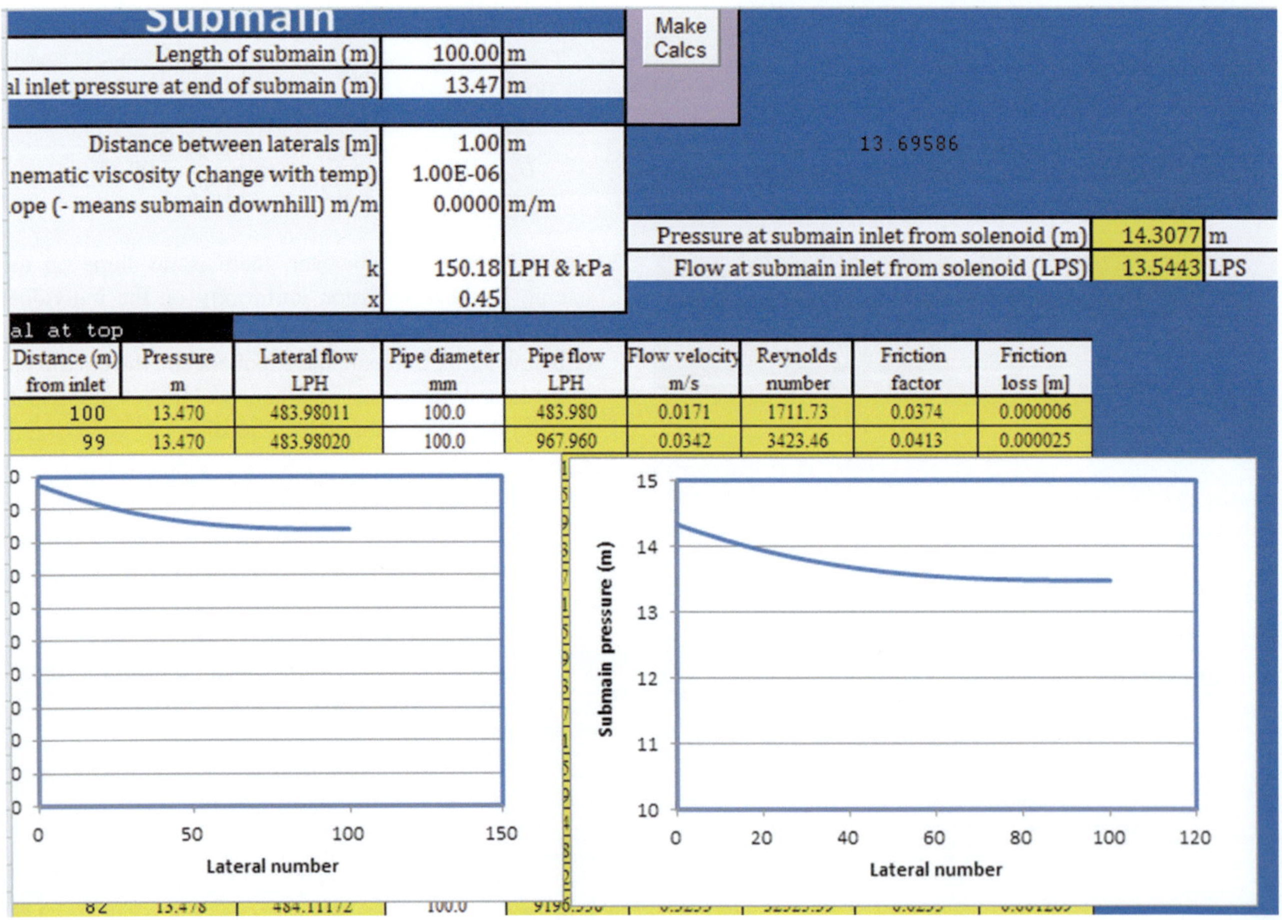

3. Repeat problem 2, but change to dual feed laterals with submains that are 180 m apart. For the purpose of hydraulic calculations, laterals are 90 m long to the midpoint. Emission uniformity should be greater than 90 % for any zone; thus, increase the pipe diameter to 14 mm, and determine whether this diameter results in an emission uniformity that is greater than 90 % for the entire zone. If the emission uniformity is more than 1 % greater than 90 %, then there is no need to make an additional calculation for the zone, because the zone will drop the uniformity by less than 1 %. There is no need to show all the graphs and equations. Just the results are sufficient.

Because of symmetry (no slope), we can use the single feed lateral worksheet that was used in the previous problems. Changing the diameter from 12 to 14 increases the emission uniformity to 92.6 %. Thus, the emission uniformity is high enough that the zone emission uniformity does not need to be checked. It is probably about 92 %.

4. Based on the parameters in problems 2 and 3, calculate the inlet pressure needed for the submains. Find the equation for submain inlet flow vs. pressure.

The required lateral inlet pressure is 10.64 (cell L11)

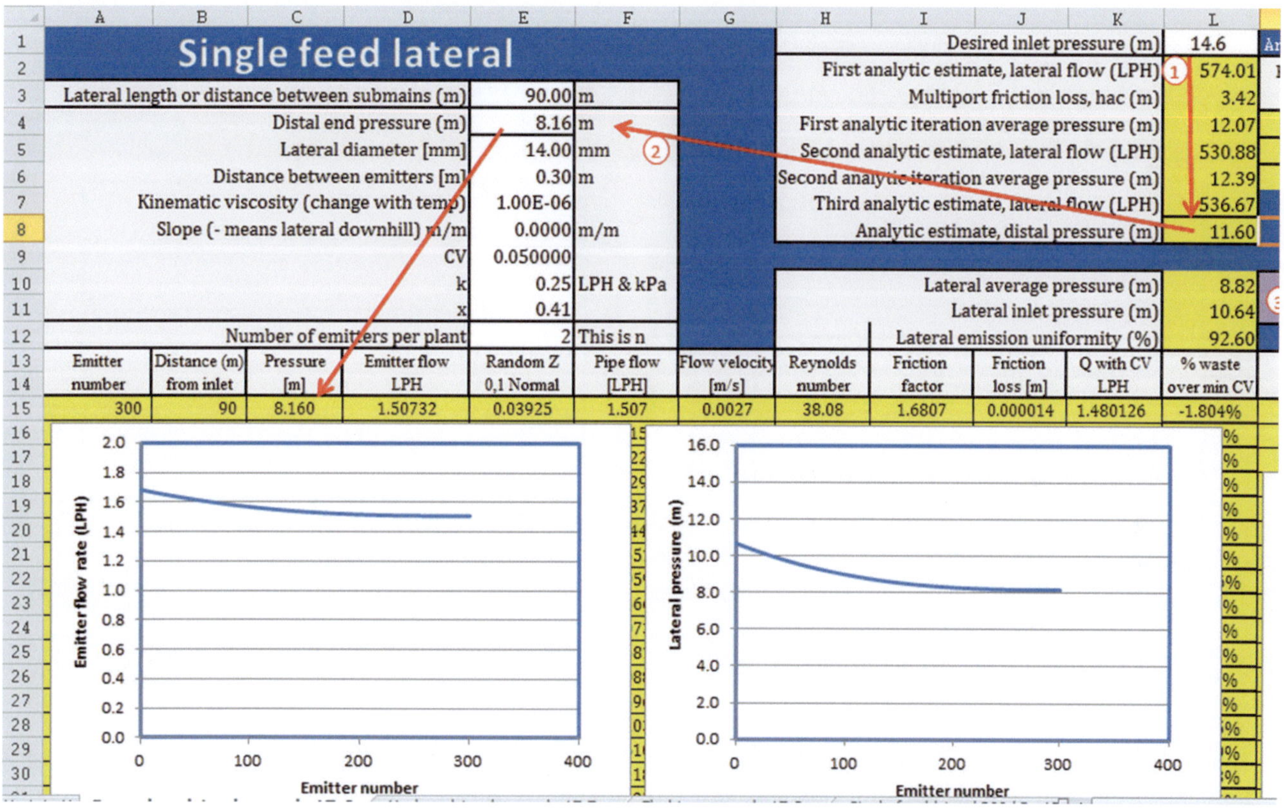

Single feed lateral

Lateral length or distance between submains (m)	90.00 m
Distal end pressure (m)	8.16 m
Lateral diameter [mm]	14.00 mm
Distance between emitters [m]	0.30 m
Kinematic viscosity (change with temp)	1.00E-06
Slope (- means lateral downhill) m/m	0.0000 m/m
CV	0.050000
k	0.25 LPH & kPa
x	0.41
Number of emitters per plant	2 This is n

Desired inlet pressure (m)	14.6
First analytic estimate, lateral flow (LPH)	574.01
Multiport friction loss, hac (m)	3.42
First analytic iteration average pressure (m)	12.07
Second analytic estimate, lateral flow (LPH)	530.88
Second analytic iteration average pressure (m)	12.39
Third analytic estimate, lateral flow (LPH)	536.67
Analytic estimate, distal pressure (m)	11.60
Lateral average pressure (m)	8.82
Lateral inlet pressure (m)	10.64
Lateral emission uniformity (%)	92.60

Emitter number	Distance (m) from inlet	Pressure [m]	Emitter flow LPH	Random Z 0,1 Normal	Pipe flow [LPH]	Flow velocity [m/s]	Reynolds number	Friction factor	Friction loss [m]	Q with CV LPH	% waste over min CV
300	90	8.160	1.50732	0.03925	1.507	0.0027	38.08	1.6807	0.000014	1.480126	-1.804%

The submain inlet pressure requirement is 11.33, and the flow rate is 12.18 LPS. The flow pressure equation at the submain inlet is $4.04H^{0.455}$.

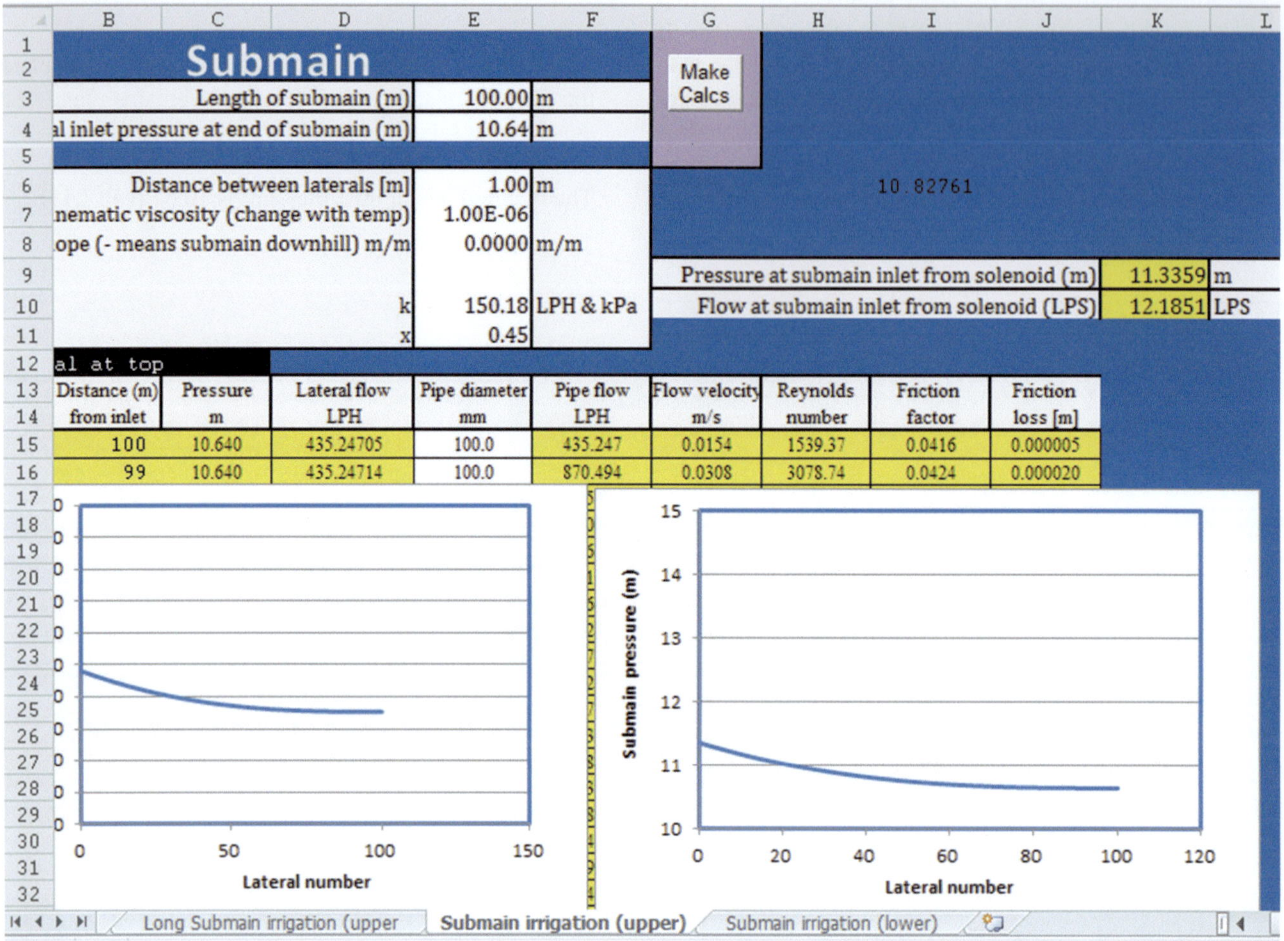

5. Based on the parameters in problems 2–4, evaluate flushing in the dual feed lateral with submains spaced on 180 m intervals. Use the *Flush dual feed lateral* worksheet. Find the inlet pressure required and the equation for lateral flow vs inlet pressure.

End pressure and flow should be set at 0.5 m and 280 LPH in order to generate sufficiency flow velocity for scouring (0.5 m/sec).

The equation for lateral flow rate vs. inlet pressure is $242H^{0.473}$ Inlet pressure to the lateral is 24.3 m.

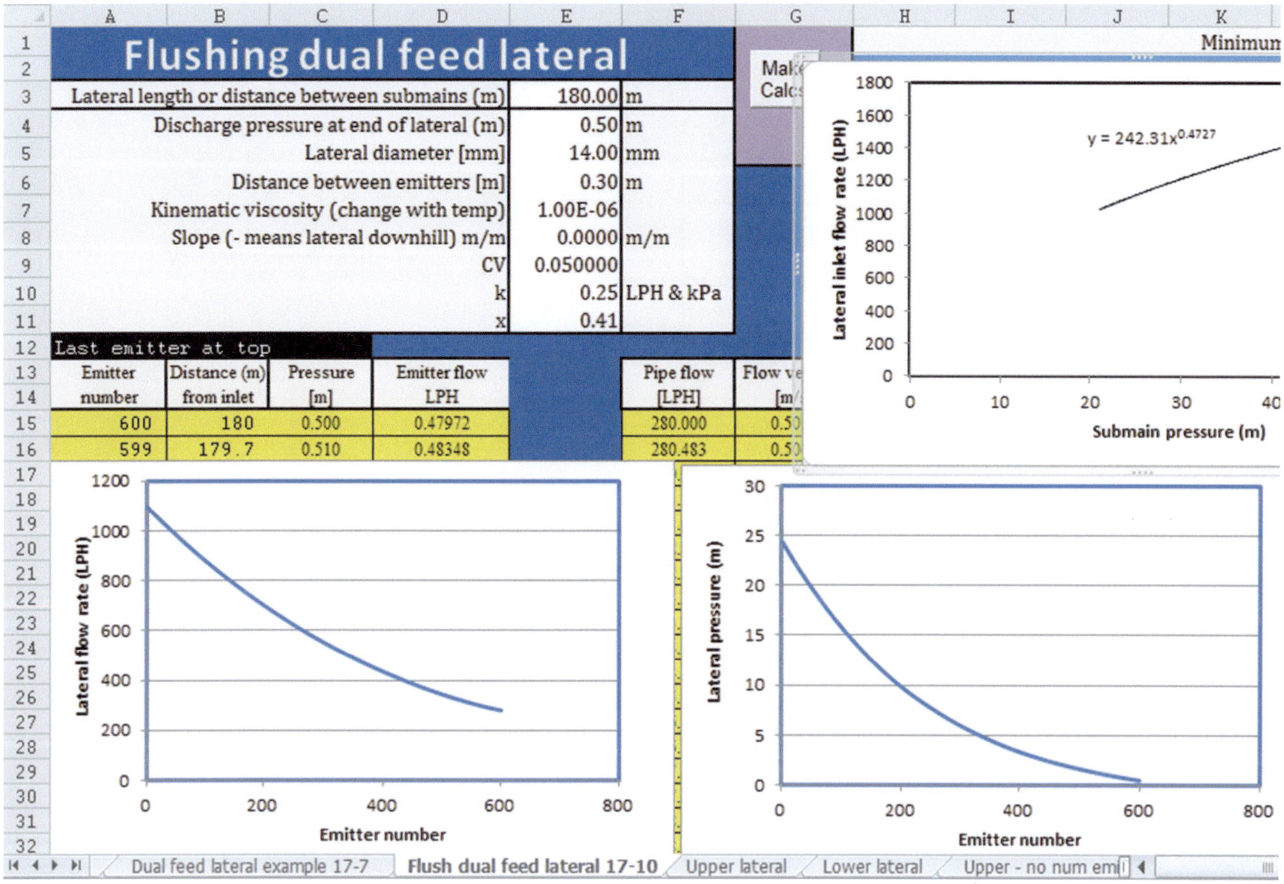

6. Based on the parameters in problems 2–5, evaluate flushing in the submain. There are 100 laterals spaced 1 m apart. Use the *Submain flushing* worksheet. Check that flow velocity is not excessive. Pipe flow velocity restrictions (normally < 1.5 m/sec) can be relaxed for flushing mode, as long as the owner slowly closes valves during the flushing process, keep the velocity below 2 m/sec in this problem. Find the required inlet flow velocity and pressure. Find the equation for inlet pressure vs. flow rate.

In order to keep flow velocity below 2 m/sec, pipe size is 150 mm (6 inch) on the first half of the submain, and 100 mm (4 inch) on the second half of the submain. Required inlet pressure is 25.2 m and flow is 30.36 LPS. The equation (CFS and m) is $6.6H^{0.47}$

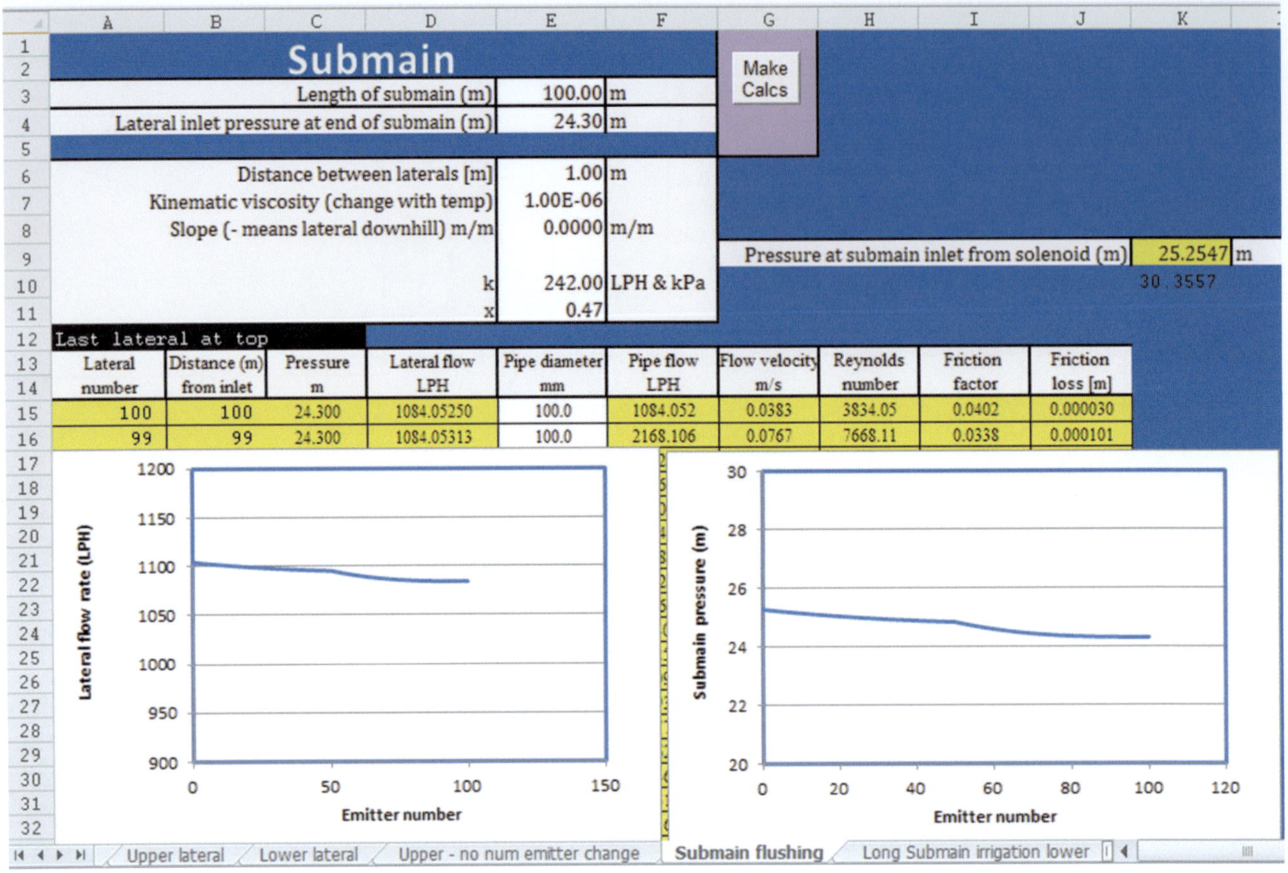

7. Using the information compiled in problems 1–6, design a pump, filter, and mainline system for a level field that has dimensions 400 m × 400 m. Allow 20 m for a central road so the length of laterals (distance between submains) is 180 m. Use the 100 m × 180 m zones that you have already designed. The road travels in the EW direction, and the pump is in the NW corner. Using the structure in Fig. 18.1, specify the required pipe sizes in mains 33–38 and 43–45. The irrigation schedule allows for 8 zones so each zone is run by itself. For example, pipes 1 and 5 are activated at the same time, etc… Mains 33–36 are used for flushing because flushing originates in submains 1, 2, 3, 4, 9, 10, 11, and 12. Use the *Chapter 18 Mainline* workbook to find flow velocities and head losses. Specify the required pump flow rate and pressure. You do not need to pick a particular pump (unless you want to). Zero year project life and 8 % ROR. The cost of energy is \$0.10/kW-hr, and the cost of water is \$3.27/ha-cm. Annual ETc is 1 m/y. Irrigation efficiency is 90 % and pumping efficiency is 80 %. Sand filter losses are 7 m and pump station losses are 3 m. Solenoid valve losses are 2 m. Do not worry about the flushing flow rate or pressure or the booster pump that would be required for flushing.

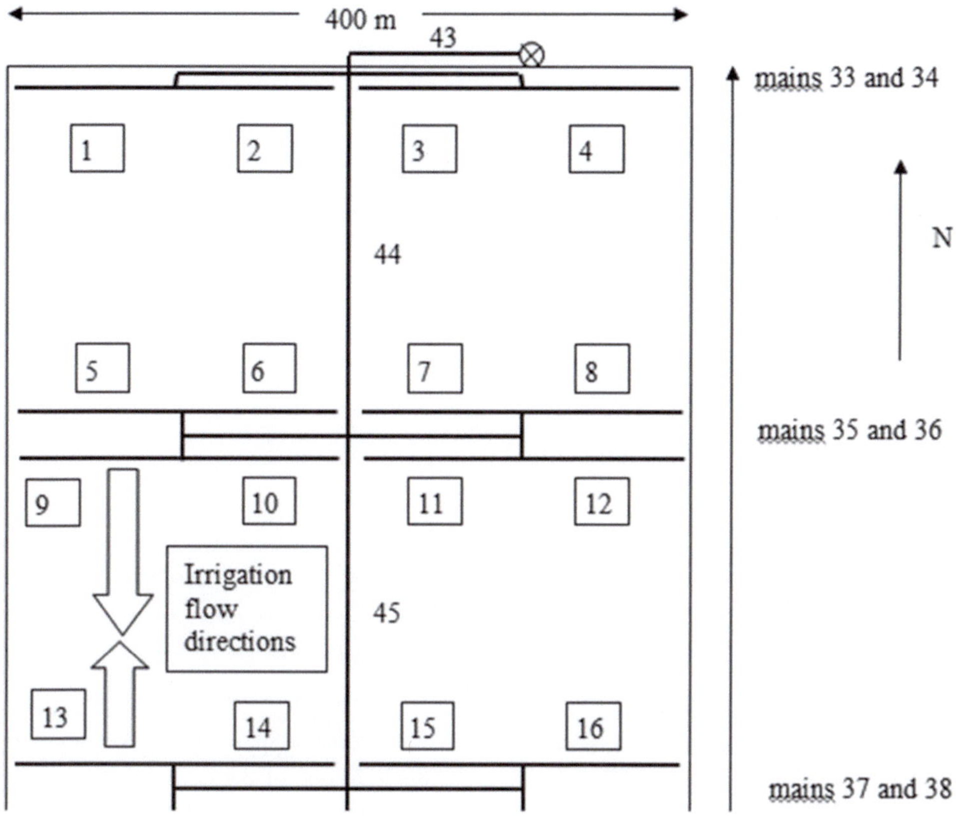

Mainlines are constrained to the 1.5 m/sec maximum velocity. Mains 43, 44, 33, 34, 35, and 36 must supply the flushing velocity, which is 30 LPS. Mains 45, 37, and 38 only supply the normal submain flow, which is 12.1 LPS.

Six inch (150 mm) flow velocity at 30 LPS is greater than 1.5 ft/sec. However, mains 37, 38 and 45 only carry 12.1 LPS, so they are suitable for 6 inch pipe. Designate the 300 m distance with mains 45 and 14 as 6 inch.

Flow velocity is 0.9 m/sec in 8 inch class 125. Because all pipes must carry the same volume, the entire mainline system is uses 8 inch class 125. The longest distance from pump to solenoid valve is 600 m. This distance and value are entered into the *Chapter 18 Mainline* workbook. During the flushing process, the head loss in 8 inch (200 mm) class 125 is 2.294 m, and the head loss in 10 inch (250 mm) class 125 is 0.785 m.

Select pipe D. or fitting	Pipe ID (mm) or valve L/D	Valve/fitting select dimension	Fitting ID (mm)	HW C	Flow (LPM)	Length (m)	Friction loss (m)	Discharge elevation (m)
8" (200) CL125 PVC	205.64	n/a	"n/a"	140	1800.00	600.00	2.294	100.0
10" (250) CL125 PVC	256.28	n/a	"n/a"	140	1800.00	600.00	0.785	100.0

During normal operation, the flow rate is 24.2 LPS, with each of two EW mains receiving 12.1 LPS. Because EW mains receive such a small flow in comparison to the flushing design flow, it is obvious, that all of the EW mains will use 8 inch pipe. In addition, main 45 will only carry a maximum of 12.1 LPS, so it is also an 8 in (200 mm) pipe. The only pipes that should possibly be 10 in (250 mm) are 44 and 43.

	Pipe	12.1 LPS		24.2 LPS		
Pipe size	cost	h_f (m)	ΔP (m)	h_f (m)	ΔP (m)	Cost for 300 m
6 (158 mm)	$5.87/m	.772	0.772			$1761
8 (206 mm)	$9.71/m	.213	.213	.771	.771	$2913
10 (256 mm)	$18.6/m			.264	.264	$5580

There is an increase of 0.5 m pressure in each comparison.

$$E = \frac{0.0272(1,110\ mm)(0.5\ m)}{0.8} = 18.9\ kW - hr/ha$$

$$(\$/ha) = \left(\frac{18.9\ kW - hr}{ha}\right)\left(\frac{\$0.10}{kW - hr}\right) = \$1.89/ha$$

The area of the field is 400*380/10000 = 15.2 ha

Total yearly cost for 15.2 ha = $1.89/ha * 15.2 = $28.7/yr.

Present value of energy saving with 206 mm (8 in. pipe (20 year, 8 %) is $282

The increased cost of pipe is over $1,000 to increase from 6 to 8 and from 8 to 10. Thus, larger pipe is not justified.

Use 6 in (200 mm) for sections 45, 37, and 38, and use 8 in (250 mm) for sections 43,44, and 33–36.

Calculate the total energy required at the pump for normal operation. Total pressure required at the pump is 24.9 m. Flow rate is 1452 LPM.

Number of pipe sections	2	Sand filter losses (m)	7	GPM	12
Inlet elevation (m)	100	Pump station losses (m)	3	LPM	45.4
Solenoid valve losses (m)	2.0	Head at mainline inlet	14.9	m	3.58
ired at end of last main (m)	11.3	Required pump pressure (m)	24.9	PSI	5.1

	Select pipe D. or fitting	Pipe ID (mm) or valve L/D	Valve/fitting select dimension	Fitting ID (mm)	HW C	Flow (LPM)	Length (m)	Friction loss (m)	Discharge elevation (m)	Head after fitting (m)
1	6" (150) CL125 PVC	157.94	n/a	"n/a"	140	726.00	300.00	0.772	100.0	14.1
2	8" (200) CL125 PVC	205.64	n/a	"n/a"	140	1452.00	300.00	0.771	100.0	14.9

8. Open the *Chapter 18 Economic analysis* workbook and *Cotton Drip lateral CV analysis* worksheet. Reduce the plant CV in cell E7 to 0.05. Select cotton as the crop in cell A2. In the range E1:E14, change the tubing diameter to 12, the plant CV to 0.05, the emitter coefficient to 0.2, and the emitter exponent to 0.5. Note, the Monte Carlo simulation program changes these values during the simulation. Plot the emitter flow rates in column D vs. emitter number in column A. Plot the 40 cm application depth in column K vs. emitter number in column A. Explain why some of the application depths are less than 40. You can highlight one of the cells in Column K and look at the equation in order to find the answer. Plot the yield vs. emitter number curve for the 50 cm depth in column AA and the 75 cm depth in column AE. Explain the shapes of the yield curves.

Some values are less than 40 because the random number generator in column B is used to randomize the depths, and half of these values are negative.

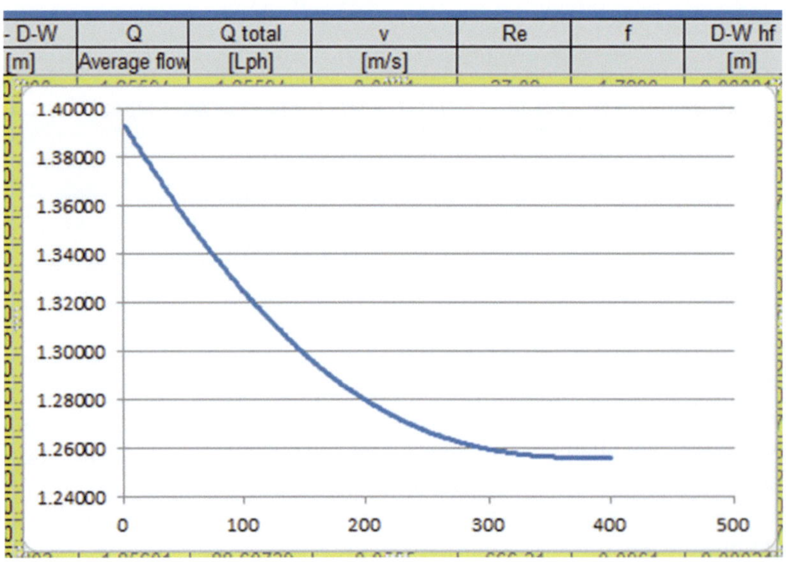

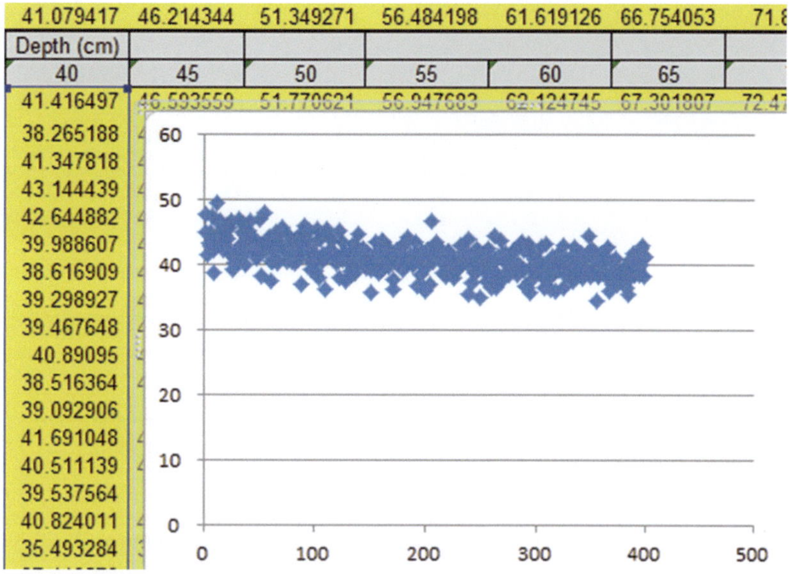

The upper yield curve is at 75 cm minimum. There is less variation than the 50 cm curve because it is near the peak of the yield curve and yield is not as sensitive to change in depth as in the 50 cm range. Yield decreases with distance in the 50 cm curve because yield is sensitive to decreased water application caused by hydraulics at the lower part of the curve.

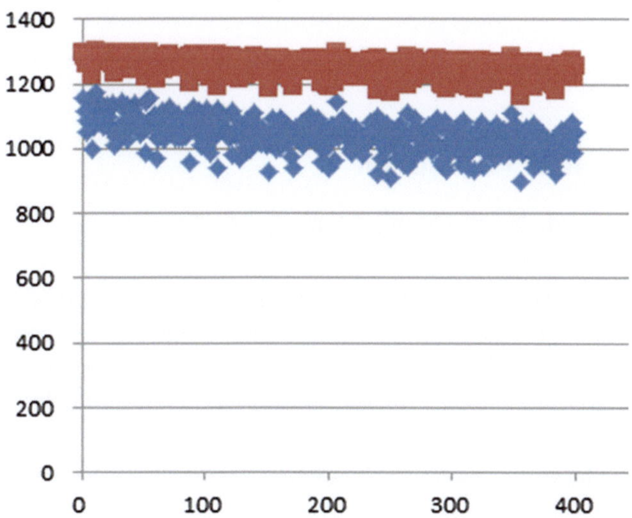

9. Open the *Chapter 18 Economic analysis* workbook and *Cotton Drip lateral CV analysis* worksheet. Move the graph away from table T5:X13. Clear cells T5:X13. Click the Monte Carlo Cotton button in cell P1. Watch what happens in cells H2:W13. Then click the Monte Carlo Cotton button in cell A1 and watch what happens in column E. and explain how the algorithm works. How many simulations are run at each tubing option and CV value (count the number of blinks in the formula bar for each condition)?

The VBA program changes the values in column E. Each time that a change is made, the random number generator changes all the values. The value in cell O9 is the maximum value from column N. This value is then placed in the corresponding CV and tubing category in table T1:X13.

The program runs each condition 10 times, with the random number generator changing the values for each run. This is Monte Carlo analysis.

10. Open the *Chapter 18 Economic analysis* workbook and *Cotton Drip lateral CV analysis* worksheet. Select cotton as the crop in cell A2. As shown below, change the replacement period for the 8 mil tape to 2 years (column AN), and run the Monte Carlo simulation by clicking the Monte Carlo button in cell B1. Note: the Monte Carlo simulation requires several minutes running time. Notice that the VBA program changes the parameters in the range E1:E14. Make Trendlines for each of the curves in the profit vs. CV graph in the range T1:X13. Compare with the equations in Fig. 18.23. If they are different, then explain why. Explain why options 1–2 have higher profit vs. CV than options 3–5. Explain why option 3 has higher profit than option 4.

	Type	Cost ($/m)	Repl. (yr)	diam (mm)	k	x	spacing	Min Press (kPa)	type
1	8 mil tape	0.09	2	16	0.1	0.5	0.25	50	Turbulent
2	8 mil tape,	0.07	2	16	0.2	0.5	0.5	50	Turbulent
3	15 mil tape	0.12	10	12	0.2	0.5	0.5	50	Turbulent
4	15 mil tape	0.2	10	16	2	0	0.5	100	Pressure comp
5	15 mil tape	0.24	10	16	1	0	0.25	100	Pressure comp

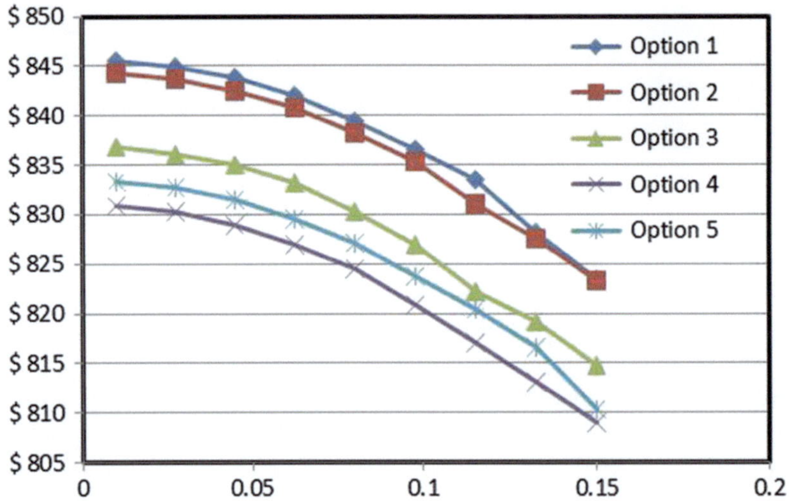

The equations are different; however, observation of the curves indicates that the difference in equations is just due to statistical differences in the Monte Carlo analysis and that the curves are in basically the same positions in both cases. The reason that there is no difference is that the replacement period does not change the relationship between CV and profit during each year.

Options 1–2 have higher profit than 3–5 because the tubing cost is lower. Option 3 has higher profit than option 4 because tubing cost is lower.

11. In the *Cotton financial calcs* worksheet, change the CV values for the every other year replacement scheme in columns B and D for options 1 and 2, as described in problem 10. Add installations costs every other in cells G15:H25. How does every other year replacement affect the annual benefit in rows 2:11 (also shown in the graph)? How does every other year replacement affect the overall profit of the system as shown in row 40? What is the only remaining option with positive

Replacing the tubing every other year improves the annual benefit from the crop; however, the cost of installation makes it prohibitive. The only option with a positive cash flow is option 3.

Chapter 19: Solution

ABE/CE 456/556 Chapter 19 homework solutions.

1. **Determine a fertilizer injection schedule for N,P, and K for bell peppers. Make same irrigation and soil assumptions as for cantaloupe example. The bell pepper nutrient curves and coefficients are shown below.**

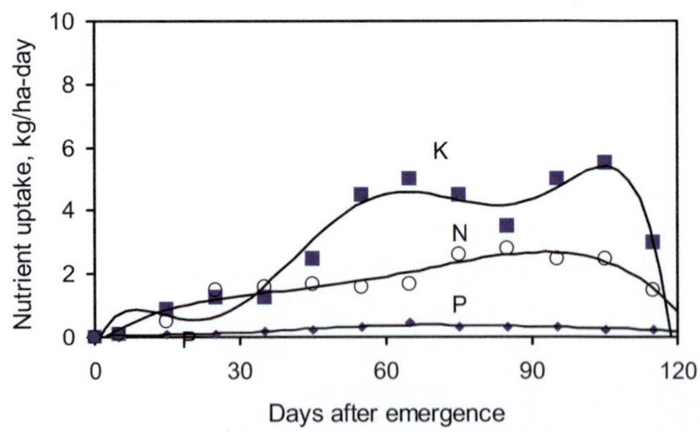

Nutrient uptake rate of N, P, and K as related to days after emergence for bell pepper. Data from Bar-Yosef (1999).

Calculated fertilizer requirements and fertilizer application times are shown below. Assume that 28 kg/ha N and 7 kg/ha P are available and that potassium is not limiting.

Table HW11-1. Nutrient application times for Bell Peppers.

	Nitrogen N (Kg/ha)		Phosphorus (kg/ha)		Poassium K (kg/ha)	
Stage	Required	Applied	Required	Applied	Required	Applied
1 (0–25 days)	17	17	2	0	4	0
2 (25–50 days)	36	35	5	0	55	0
3 (50–75 days)	50	40	9	23	166	0
4 (75–100 days)	65	50	9	0	150	0
5 (100–120 days)	39	36	5	0	14	
	206	178	30	23	389	0
		28		7		389

Multiply nutrient application rates in table HW11-1 by 15 to find nutrient application rates per block per stage. Assume that 20 kg/ha Ca is applied during the last 2 stages.

Table HW11-1. Nutrient application rates per block.

Stage	Nutrients -kg/block/stage				Water –			
	N	P	K	Ca	Irr. events per stage	Fertigation h/stage	Irrigation h/stage	Water applied per block per stage – (L)
1 (0–25 days)	255	0	0	0	6	12	24	$1.44 * 10^7$
2 (25–50 days)	525	0	0	0	12	24	48	$2.88 * 10^7$
3 (50–75 days)	600	345	0	0	16	32	64	$3.84 * 10^7$
4 (75–100 days)	750	0	0	300	16	32	64	$3.84 * 10^7$
5 (100–120 days)	540	0	0	300	16	32	64	$3.84 * 10^7$

Urea-ammonium nitrate (UAN32) is selected during the first 2 stages because only nitrogen is required. For the third stage, liquid formulations of UAN32 and ammonium polyphosphate (10-34-0) are selected and will be injected at two ports into the irrigation pipeline. For the last two stages, the grower decides to use CAN17 as the calcium source. CAN17 cannot be injected with urea-ammonium nitrate so the grower decides to use ammonium nitrate as the nitrogen source if CAN17 does not supply the nitrogen required during the fourth and fifth stages.

- **Stage 1**

UAN32 will supply 255 kg N per block over 12 h of fertigation (Table 27.7) during the first stage. The density of UAN32 is 1.33 kg/L and the formulation is 32-0-0 (Table 27.5).

$$\begin{aligned}&\textit{Volumetric N content of UAN32}\\&\quad = 1.33\ kg/L \times 0.32 = 0.425\ kg/L\ of\ N\\&\textit{UAN32 volume per block during stage 1}\\&\quad = 255\ kg/0.425\ kg/L = 600\ L\ UAN32\end{aligned}$$

Injection rate = 600 L/12 h of fertigation = 50 L/h. of UAN32

- **Stage 2**

UAN32 will supply 525 kg N per block over 24 h of fertigation during stage 2.

$$\begin{aligned}&\textit{UAN32 volume per block during stage 2}\\&\quad = 525\ kg/0.425\ kg/L = 1,235\ L\ UAN32\end{aligned}$$

Injection rate = 1,235 L/24 h of fertigation = 51.5 L/h. of UAN32

- **Stage 3**

Ammonium polyphosphate will supply 345 kg P per block over 32 h of fertigation during the third stage. The density of ammonium polyphosphate (APP) fertilizer is 1.37 and the formulation is 10-34-0 (Table 27.5). Thus, 34 % of the mass of the fertilizer is phosphorous as P_2O_5. P_2O_5 is multiplied by 0.44 (Table 27.4) to convert to elemental P (PO_4-P).

$$Volumetric\ P\ content\ of\ APP = 1.37\ kg/L \times 0.34 \times 0.44 = 0.205\ kg/L\ of\ P$$
$$APP\ volume\ per\ block\ during\ stage\ 3 = 345\ kg/0.205\ kg/L = 1,683\ L\ of\ APP$$

APP Injection rate = 1,683 L/(32 h of fertigation) = 53 L/h of APP

The nitrogen supplied by the ammonium polyphosphate during stage 3 is calculated as follows:

$$Volumetric\ N\ content\ of\ APP = 1.37\ kg/L \times 0.10 = 0.137\ kg/L\ of\ N$$
$$Mass\ N\ applied\ as\ APP = 1,683\ L \times 0.137\ kg/L = 231\ kg\ of\ N$$

If 231 kg N is applied as ammonium phosphate, then the remainder of the nitrogen requirement (600–231 = 370 kg/block) during stage 3 must be supplied by UAN32.

$$UAN32\ volume\ per\ block\ during\ stage\ 3 = 370\ kg/(0.425\ kg/L) = 870\ L\ of\ UAN32$$

Injection rate = 870 L/(32 h of fertigation) = 27 L/h of UAN32

- **Stage 4**

Calcium ammonium nitrate (CAN17) will supply 300 kg Ca per block over 32 h of fertigation during the fourth stage and fifth stages. The density of calcium fertilizer is 1.55 and the formulation is 17-0-0-8.8 Ca (Table 19.6). Thus, 8.8 % of the mass of the fertilizer is calcium.

$$Volumetric\ Ca\ content\ for\ CAN17 = 1.55\ kg/L \times 0.088 = 0.136\ kg/L\ of\ Ca$$
$$CAN\ 17\ volume\ per\ block\ during\ stage\ 4 = 300\ kg/(0.136\ kg/L) = 2,205\ L\ of\ CAN17$$

Injection rate of CAN17 = 2,205 L/(32 h of fertigation) = 69 L/h of CAN17

The amount of nitrogen supplied by CAN17 during stage 4 is calculated as follows.

$$Volumetric\ N\ Content\ of\ CAN17 = 1.55\ kg/L \times 0.17 = 0.264\ kg/L\ of\ N$$
$$Mass\ N\ applied\ as\ CAN17 = 2,205\ L\ CAN17 \times 0.264\ kg/L = 582\ kg\ of\ N$$

If 600 kg N is applied as CAN17, then the remainder of the nitrogen requirement (750–582 = 168 kg/block) during stage 4 must be supplied by ammonium nitrate.

Volumetric N content for Ammonium nitrate = 1.29 kg/L × 0.2 = 0.26 kg/L of N

Ammonium nitrate volume per block during stage 4 = 168 kg/(0.26 kg/L) = 646 L of Ammonium nitrate

Injection rate = 646 L/(32 h of fertigation) = 20 L/h of Ammonium nitrate

- **Stage 5**

$$Volumetric\ Ca\ content\ for\ CAN17 = 1.55\ kg/L \times 0.088 = 0.136\ kg/L\ of\ Ca$$
$$CAN\ 17\ volume\ per\ block\ during\ stage\ 5 = 300\ kg\ /\ (0.136\ kg/L) = 2,205\ L\ of\ CAN17$$

Injection rate of CAN17 = 2,205 L/(32 h of fertigation) = 69 L/h of CAN17

The amount of nitrogen supplied by CAN17 during stage 4 is calculated as follows.

$$Volumetric\ N\ Content\ of\ CAN17 = 1.55\ kg/L \times 0.17 = 0.264\ kg/L\ of\ N$$
$$Mass\ N\ applied\ as\ CAN17 = 2,205\ L\ CAN17 \times 0.264\ kg/L = 582\ kg\ of\ N$$

In this stage the plants only need 540 kg. Thus, no excess fertilizer is required.

2. **Plot carbonate, bicarbonate, and carbonic acid concentration as a function of pH for alkalinity as calcium carbonate equal to 120 mg/L.**

Molecular weight $CaC0_3$ = 100.1 g/mol

Then: 120 mg/l = 0.12 g/l; we have $\frac{0.12g/l}{100.1g/mol} = 0.0012mol/l$

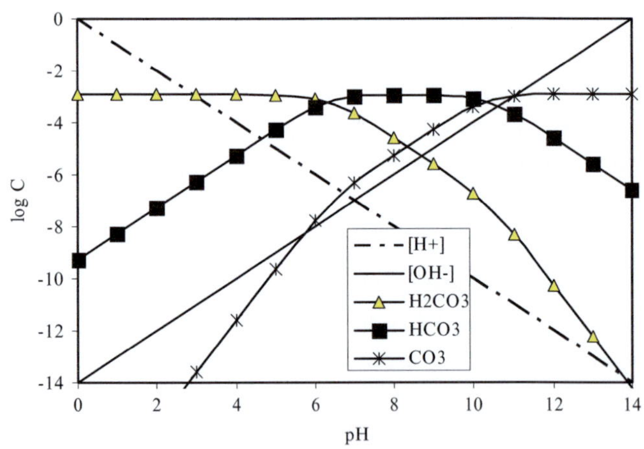

Figure HW11-2. Carbonate species vs. pH.

3. **Calculate the amount of 98 % sulfuric acid required to drop the pH from 7.9 to 6.8 for water with alkalinity as calcium carbonate equal to 120 mg/L.**

Water has $CaCO_3$ of approximately 120 mg/L. This is the same as 60 mg/L CO_3^{-2}. The molecular weight of carbonate (CO_3^{-2}) is 60 g/mole. The calculation of carbonate molarity is as follows.

$$\left(\frac{120\ mg\ CaCO_3}{L\ CAP\ water}\right)\left(\frac{0.6\ mg\ CO_3^{-2}}{1\ mg\ CaCO_3}\right)\left(\frac{g\ CO_3^{-2}}{1{,}000\ mg\ CO_3^{-2}}\right)\left(\frac{1\ mole\ CO_3^{-2}}{60\ g\ CO_3^{-2}}\right) = 0.0012\ mole/L$$

Hydroxyl ion concentration at $pH(7.9) = 10^{-(14-7.9)} = 2*10^{-6} mole/L$.

Calculate initial HCO_3^- molarity at pH (7.9).

$$[HCO_3^-] = \frac{[C][K_1][H^+]}{[H^+]([H^+]+K_1)+K_1K_2} = \frac{[0.0012][4.47*10^{-7}][1.26*10^{-8}]}{[1.26*10^{-8}]([1.26*10^{-8}]+4.47*10^{-7})+4.47*10^{-7}*4.68*10^{-11}} = 1.16*10^{-3} mol/L$$

Calculate the initial CO_3^{-2} molarity at pH (7.9)

$$[CO_3^{-2}] = \frac{[C][K_1][K_2]}{[H^+]([H^+]+K_1)+K_1K_2} == \frac{[0.0012][4.47*10^{-7}][4.68*10^{-11}]}{[1.26*10^{-8}]([1.26*10^{-8}]+4.47*10^{-7})+4.47*10^{-7}*4.68*10^{-11}} = 4.32*10^{-6}\ mol/L$$

Calculate the total alkalinity at pH (7.9)

$$\sum(HCO_3^- + CO_3^{2-} + OH^-) = 1.16*10^{-3} + 2*4.32*10^{-6} + 2*10^{-6} = 0.00117\ eq/L$$

Now calculate the alkalinity at the final pH (6.8).

Hydroxyl ion concentration at pH (6.8) = $10^{-(14-6.8)}$ = $1.58 * 10^{-7}$ mole/L.

Calculate final HCO_3^- molarity at pH (6.8).

$$[HCO_3^-] = \frac{[C][K_1][H^+]}{[H^+]([H^+]+K_1)+K_1K_2} = \frac{[0.0012][4.47*10^{-7}][1.58*10^{-7}]}{[1.58*10^{-7}]([3.16*10^{-7}]+4.47*10^{-7})+4.47*10^{-7}*4.68*10^{-11}} = 8.86*10^{-4}\ mol/L$$

Calculate final CO_3^{-2} molarity at pH (6.8)

$$[CO_3^{-2}] = \frac{[C][K_1][K_2]}{[H^+]([H^+]+K_1)+K_1K_2} == \frac{[0.001][4.47*10^{-7}][4.68*10^{-11}]}{[1.58*10^{-7}]([1.58*10^{-7}]+4.47*10^{-7})+4.47*10^{-7}*4.68*10^{-11}} = 2.62*10^{-7}\ mol/L$$

Calculate final total alkalinity at pH (6.5)

$$\sum\left(HCO_3^- + CO_3^{2-} + OH^-\right) = 8.86*10^{-4} + 2*2.62*10^{-7} + 1.58*10^{-7} = 0.000886 \text{ eq/L}$$
Change in alkalinity = initial alkalinity − final alkalinity
$= 0.00117$ eq/L − 0.000886 eq/L $= 2.84*10^{-4}$
$Q_aC_a = Q_wC_w \rightarrow 37{,}560$ meq/L : * acid flow rate = 0.284 meq/L*44 L/ sec
Acid flow rate = 0.284*44/37,560 = 0.0003327 L/ sec = 1.2 L/hr

4. **Calculate the LSI for the following water analysis at 25 °C and 35 °C. Measured pH is 7.96.**
   - alkalinity as calcium carbonate equal to 120 mg/L.
   - calcium as calcium carbonate equal to 65 mg/L
   - TDS equal to 1,000 ppm

   Calculate for 25 °C

$$A = (Log10[1000] - 1)/10 = 0.2$$
$$B = -13.12 \times Log10(25^oC + 273) + 34.55 = 2.088$$
$$C = Log10[65] - 0.4 = 1.41$$
$$D = Log10[120] = 2.1$$
$$pH_s = (9.3 + A + B) - (C + D)$$
$$pHs = (9.3 + 0.2 + 2.088) - (1.41 + 2.1) = 8.08$$

$LSI = pH - pH_s = 7.96 - 8.08 = -0.12$ (The LSI is close to zero, so precipitation may occur).

Calculate for 35 °C

$$A = (Log10[1000] - 1)/10 = 0.2$$
$$B = -13.12 \times Log10(35^oC + 273) + 34.55 = 1.9$$
$$C = Log10[65] - 0.4 = 1.41$$
$$D = Log10[120] = 2.1$$
$$pH_s = (9.3 + A + B) - (C + D)$$
$$pHs = (9.3 + 0.2 + 1.9) - (1.41 + 2.1) = 7.9$$

$LSI = pH - pH_s = 7.96 - 7.9 = 0.06$ (The LSI is very close to zero: add acid).

5. **Calculate the injection rate of 10 % chlorine bleach required in order to develop a concentration of 3 ppm elemental chlorine in an irrigation system with a 200 LPM flow rate.**

$$200\ LPM = 3.33\ LPS$$

Liquid (liquid flow rate, use 10 % chlorine)

$$\left(\frac{3.34\ L\ water}{\sec}\right)\left(\frac{3\ mg\ Cl_2}{L\ water}\right)\left(\frac{100\ mg\ bleach}{10.1\ mg\ Cl_2}\right)$$
$$\left(\frac{kg}{1*10^6\ mg}\right)\left(\frac{L}{1.15\ kg}\right)\left(\frac{3,600\ \sec}{hr}\right) = 0.31\ L/hr$$

Chapter 20: Solution

1. Describe the three phases of surface irrigation

Surface irrigation events have 3 phases: advance, storage, and recession. During advance, the wetting front (Fig. 20.1) moves down the channel. The second phase of surface irrigation is the storage phase (Fig. 20.2). After the advance reaches the end of the field, the water must remain ponded for a sufficient length of time for the end of the field to receive the required depth of water. The length of the storage phase depends on the required depth of infiltration, and the soil infiltration rate. It may last from several hours to 24 hours. If the storage phase is long, then a significant quantity of water may run off the end of the field. Also, significant leaching may occur at the upstream end. After irrigation water is turned off at the time of cutoff, recession begins: ponded water infiltrates or moves down the furrow and the upper end dries (Fig. 20.2). In furrow irrigation systems, the upper end of the furrow dries immediately after the time of cutoff, and then the dried section increases as the water infiltrates and moves off the end of the field. However, recession does not begin immediately for border and level basin systems because there is a much greater ratio of water on the field surface to wetted soil area than in furrows.

2. Answer the following questions true or false.
   (a) Uniformity is generally high if the storage phase is relatively small in comparison to the advance phase. false
   (b) The advantage of the two-point method is that the infiltration rate during the storage phase can be extrapolated from the infiltration rate calculated during the advance phase. false
   (c) The Kostiakov equation includes steady state infiltration. false
   (d) The two-point volume balance method can be expanded to find the coefficient b in the steady state term by using another point. false
   (e) The vertical infiltration rate as calculated from a double ring infiltrometer can be adjusted for furrow

irrigation by taking the width of the furrow divided by the distance between furrows. False

3. Using the Merriam and Clemmens approach, calculate k and a if the time to infiltrate 100 mm is 8 hours. Make the calculation by hand and in the *Infiltration* worksheet. Compare to the closest NRCS (SCS) curve number: the curve that is closest to 100 mm over 8 hours.

$$a = 0.675 - 0.2125\ \mathrm{LOG}_{10}(T_{100})$$
$$= 0.675 - 0.2125\ \mathrm{LOG}_{10}(8) = 0.483$$
$$k = 100/8^{0.483} = 36.6$$

The closest NRCS curve number is the 0.35 intake family.

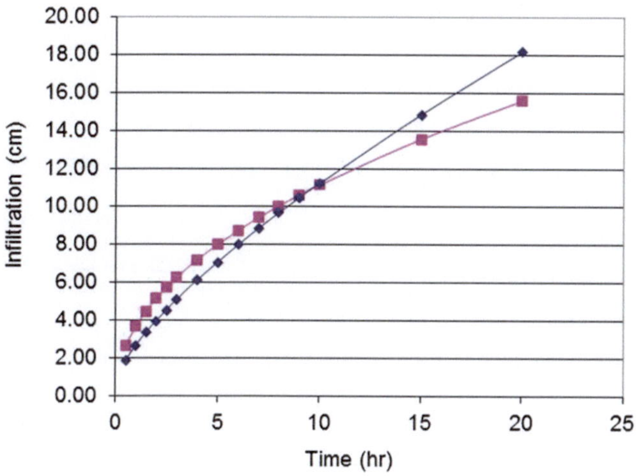

4. Flow rate = 2.0 L/s, z = 1.7, b = 0.2 m, Manning's n = 0.05, furrow slope = 0.001 m/m. Calculate the depth of flow by hand and with the *Furrow* worksheet. Using the Clemmens k and a values from problem 6, adjust the k value for the wetted perimeter and a furrow spacing of 1 m. Make calculations by hand and in furrow worksheet.

Calculate the section factor (Chap. 10)

$$\frac{Qn}{S_f^{1/2}} = \frac{(0.002\ m^3/\sec)\ (0.05)}{0.001^{1/2}} = Section\ factor$$
$$= 0.003162$$

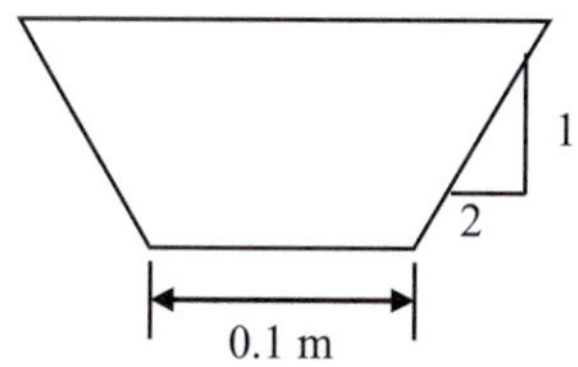

$AR^{2/3}$ is calculated with the following equation.

$$AR^{2/3} = A(A/P)^{2/3}$$
$$= (b + zy)y\left(((b + zy)y)/\left(b + 2y\left(1 + z^2\right)^{\wedge 0.5}\right)\right)^{2/3}$$

Make an initial guess, y = 0.04 m.

$$AR^{2/3} = (0.2 + (1.7)(0.04)(0.04)\left(\left(0.2 + (1.7)(0.04)\right)(0.04)/(0.2 + (2)(0.04)\left(1 + 1.7^2\right)^{\wedge 0.5}\right)\right)^{2/3}$$

Iterate by inputting the required section factor (0.003612) and the first guess for a section factor into the following equation. This equation is designed for rapid convergence for this iteration.

$$y = \text{initial } y\left(SF/AR^{2/3}\right)^{0.5} = 0.04(0.003162/0.00103)^{0.5} = 0.07\ \text{m}.$$

$$AR^{2/3} = (0.2 + (1.7)(0.07)(0.04)\left(\left(0.2 + (1.7)(0.07)\right)(0.07)/(0.2 + (2)(0.07)\left(1 + 1.7^2\right)^{0.5}\right)\right)^{2/3} = 0.0029$$

0.003162 and 0.0029 → y = 0.0730.

$$AR^{2/3} = (0.2 + (1.7)(0.073)(0.04)((0.2 + (1.7)(0.073))(0.073)/\left(0.2 + (2)(0.073)\left(1 + 1.7^2\right)^{\wedge 0.5}\right))^{2/3} = 0.00315$$

$$0.003162 = 0.00315 \rightarrow y = 0.0732.$$

Calculate wetted perimeter

$$P = b + 2\ y(1+z^2)^{0.5} = 0.1 + (2)(0.0732)(1+1.7^2)^{0.5} = 0.488\ \text{m}$$

Adjust the infiltration rate for the equivalent furrow infiltration width.

$$\frac{0.488\ \text{m} + 0.213\ \text{m}}{1.0\ m} 36.6 = 25.7$$

5. Using the information from problem 4, calculate s and h and plot the advance curve for a 500 m long furrow with 1 m spacing between furrows. Inflow rate is 2.0 L/s. Use a convergence criterion of less than 1 min difference for advance time between iterations. Make calculations by hand and with the furrow worksheet.

Calculate cross-sectional area at the furrow inlet, A_0.

$$\begin{aligned} A_0 &= (b + z\ y)y \\ &= (0.1\ \text{m} + (1.7)(0.0732\ \text{m}))(0.0732\ \text{m}) \\ &= 0.0237\ \text{m}^2 \end{aligned}$$

The design procedure starts with a guess for the advance time to ½ the field length and to the end of the field. For this example, we arbitrarily guess 100 and 250 minutes, respectively. For a flow rate of 1.0 L/s, calculate inflow volumes during these two periods.

$$V_{L/2} = Qt = 100\ \text{min}(60\ \text{sec/min})(0.001\ \text{m}^3/\text{sec}) = 6\ \text{m}^3$$

$$V_L = Qt = 250\ \text{min}(60\ \text{sec/min})(0.001\ \text{m}^3/\text{sec}) = 15\ \text{m}^3$$

$$h = \log(t_{L/2}/t_L)/\log(1/2) = \log(100/250)/\log(1/2) = 1.32$$

The next step is to calculate the subsurface shape factor.

$$\sigma_z = \frac{h + a(h-1) + 1}{(1+a)(1+h)} = \frac{1.32 + 0.483(1.32-1) + 1}{(1+0.483)(1+1.32)} = 0.72$$

The next step is to calculate subsurface storage. The infiltrated depths at the upper end of the field at 100 and 250 minutes are calculated.

$$d_{L/2} = kt^a = 25.7(100/60)^{0.46} = 32.9\ \text{mm infiltrated.}$$
$$d_L = kt^a = 25.7(250/60)^{0.46} = 51.2\ \text{mm infiltrated.}$$

Calculate subsurface storage at $t_{L/2}$ (time to reach L/2) and t_L

$$Vz_{L/2} = d_0\sigma_z Wx = 32.9/1{,}000\ \text{mm})(0.72)(1.0\ \text{m})(250\ \text{m}) = 5.9\ \text{m}^3$$
$$Vz_L = d_0\sigma_z Wx = (51.2/1{,}000\ \text{mm})(0.72)(1.0\ \text{m})(500\ \text{m}) = 18.4\ \text{m}^3$$

Calculate surface storage at $t_{L/2}$ and t_L.

$$V_{s\ L/2} = \sigma_y A_0 x = (0.75)(0.0237\ \text{m}^2)(250\ \text{m}) = 4.45\ \text{m}^3.$$
$$V_{s\ L} = \sigma_y A_0 x = (0.75)(0.0237\ \text{m}^2)(500\ \text{m}) = 8.9\ \text{m}^3.$$

Calculate total storage at $t_{L/2}$ and t_L.

$$V_{T\ L/2} = 5.9 + 4.45 = 10.36\ \text{m}^3$$
$$V_{T\ L} = 18.4 + 8.9 = 27.31\ \text{m}^3$$

The advance times are adjusted with Eq. 20.16

$$t_{m+1} = t_m\left(\frac{V_T}{Q\ t_m}\right)^{1.4} = 100\left(\frac{10.36}{12}\right)^{1.4} = 81\ \text{min}$$

$$t_{m+1} = t_m\left(\frac{V_T}{Q\ t_m}\right)^{1.4} = 250\left(\frac{27.31}{30}\right)^{1.4} = 219\ \text{min}$$

The procedure is then repeated for the next iteration with $t_{L/2} = 81$ min, and $t_L = 219$ min.

These times are close to the actual value. The next iteration leads to the final times 83 and 221

Iteration steps for two point volume balance method.

	t_1	t_2	h	σ_z	$V_{in-t\ L/2}$	$V_{in-t\ L}$	$V_{TL/2}$	V_{TL}
Initial guess	100	250	1.32	0.72	12	30	10.4	27.3
Iteration 1	81	219	1.43	0.73	9.77	26.3	9.90	26.5
Iteration 2	83	221	1.42	0.73	9.95	26.4	9.93	26.5

$$t = sx^h \quad s = t/x^h \quad s = 221/500^{1.42} = 0.0332$$

$$t = 0.03332\ x^{1.42}$$

The advance curve is shown below

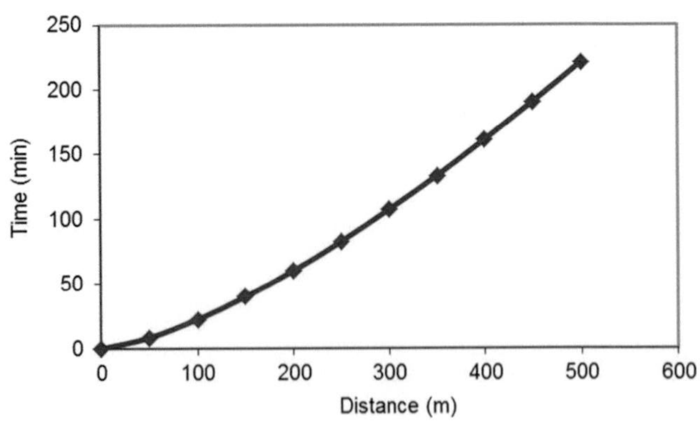

6. Find the time of cutoff for the parameters in problem 5. The depth required is 76 mm. You should calculate a total irrigation time of approximately 12 hr.

   Calculate the required intake opportunity time.

   $$d = kt^a$$
   $$t = (d/k)^{1/a} = (76/25.7)^{1/0.483} = 9.45\ \text{hr} = 567\text{min}$$

Calculate the recession time.

$$t_{rec} = V/Q = \left(0.0237\ \text{m}^2\right)(500\ \text{m})(0.8)/\left(0.12\ \text{m}^3/\text{min}\right) = 79\text{min} = 1.32\ \text{hr}$$

The advance time was already calculated as 220 min

$$t_{co} = t_{adv} + IOT_{req} - t_{rec} = (567\text{min} + 220\text{min} - 79\text{min})/60 = 11.8\ \text{hr}$$

7. Use the Furrow times worksheet to find the depth of infiltration every 50 m down the furrow for the parameters in problems 5–6. Calculate the depth of infiltration at 150 m by hand and compare to the worksheet. Calculate the DU LQ by hand and compare to the value in cell G33. By hand, calculate the DP%, RO %, and efficiency based on applied volume and infiltrated volume reported in the worksheet. In this problem, do not use cutback irrigation. Adequately irrigate the entire field (minimum required depth is applied to end of furrow).

Find the intake opportunity time at 150 m

$$\text{Advance time to 150 m} = sx_j^{\,h} = \left(0.0332{*}150^{1.42}\right) = 40\ \text{min}$$

$$\text{Time that water recedes from 150 m} = t_{co} + x_j/L\ t_{rec} = (10.8 + 150/500{*}1.32){*}60 = 733\ \text{min}$$

$$IOT = 733\ \text{min} - 40\ \text{min} = 693\ \text{min} = 11.55\ \text{hr}$$

$$d_{120} = kt^a = 25.7(11.52)^{0.483} = 83.7\ \text{mm}$$

Same calculation in one equation:

$$d_{120} = k\left(t_{co} + x_j/L\ t_{rec} - sx_j^{\,h}\right)^a = 25.7\left((10.8\ \text{h} + (150/500)(1.32\ \text{h}))(60) - \left(0.0332{*}150^{1.42}\right)/60\right)^{0.483} = 83.7\ \text{mm}$$

Calculate the DU LQ and compare to the value in the *Furrow times* worksheet

$$\text{Total volume} = 40.8\ \text{m}^3(\text{from cell G30})$$
$$\text{Field average} = (40.8/500\ \text{m}/1\ \text{m}){*}1000\ \text{mm/m} = 81.7\ \text{mm}$$
$$\text{Low quarter average} = (78.8 + 77.5 + 76{*}0.5)/2.5 = 77.7\ \text{mm}$$
$$\text{DU LQ} = 77.7/81.7(100\%) = 95\%$$

Calculate the DP% and RO%

$$\text{Applied volume} = 10.8\ \text{hr}(2\ \text{LPS})(1\ \text{L}/1000\ \text{m}^3)(3600\ \text{sec/hr}) = 85.07\ \text{m}^3$$
$$\text{RO\%} = (\text{Applied volume} - \text{infiltrated volume})/(\text{applied volume}) = \left(85.1\ \text{m}^3 - 40.8\ \text{m}^3)/85.1\ \text{m}^3\right)\ (100\%) = 52\%$$
$$\text{Required volume} = 500\ \text{m}(76\ \text{mm})/(1000\ \text{mm/m})(1\ \text{m furrow width}) = 38\ \text{m}^3$$
$$\text{DP\%} = (\text{Infiltrated volume} - \text{required volume})/\text{Applied volume} = \left((40.8\ \text{m}^3 - 38\ \text{m}^3)/85.1\ \text{m}^3\right)\ (100\ \%) = 3.3\%$$
$$\text{Efficiency} = \text{Water used}/(\text{water applied})(100\%) = 38\ \text{m}^3/\left(85\ \text{m}^3\right)(100\%) = 45\%$$

8. For the parameters in problems 5–7, observe how the parameters for cutback irrigation are calculated in the cutback worksheet. Can cutback take place as soon as water reaches the end of the furrow? What will be the cutback flow rate (half of initial flow rate). Find the following parameters and report the cell number in which they are found: the average infiltration at the end of the furrow when water reaches the end of the furrow, the volume infiltration rate when water reaches the end of the furrow (LPS), the average infiltration rate (mm/hr) 10 minutes after water reaches the end of the furrow, and the time when cutback can take place. Compare the total

depth applied to the depth applied without cutback, and the RO%, DP%, and efficiency

Cutback cannot take place as soon as water reaches the end of the furrow. The cutback flow rate will be 1 LPS.

- the average infiltration at the end of the furrow when water reaches the end of the furrow
  - cell F13, 10.36 mm/hr
- the total infiltration in the field when water reaches the end of the furrow (LPS)
  - cell F16, 1.44 LPS (exceeds the average infiltration 10 minutes after water reaches the end of the furrow)
  - cell G13, 9.35 mm/hr
- the time when cutback can take place
  - cell S1, 33 minutes after advance reaches the end of the field

The total depth applied is 54.5 m^3, which is far less than the 85 m^3 applied without cutback. The RO% is 25 % instead of 52 %. The DP% is higher (4.75 %), but this is just due to less total water applied. The efficiency is 70 % compared to 45 %.

9. Calculate an irrigation schedule for the parameters in previous problems for peak ETc 10 mm/day. Leaching fraction is 10 %.

The entire soil profile is filled during the irrigation; however, 10 % of the water is used for leaching. Thus, only 90 % is available for crop water uptake.

Infiltration is 76 mm. The depth left after leaching for crop uptake is 76 mm * 0.9 = 68.4 mm.

The crop could survive for approximately 7 days between irrigations so irrigate weekly.

10. Design a runoff-recovery system for the parameters in problems 3–7 and 9. Do not worry about economic analysis or comparison to cutback irrigation unless asked by instructor. Determine the required supply flow rate, Q_s, to the head ditch. The field width is 800 m wide. The delivery efficiency of the head ditch is 90 %. The collection system efficiency of the runoff recovery system, prior to water reentering the head ditch, is 87 % (3 % lost to evaporation and 10 % lost to seepage in the tailwater ditch and reservoir). Assume that the reservoir is full at the beginning and end of the irrigation cycle.

Supply source water is pumped from a well with a dynamic water table depth of 50 m, and the well pump efficiency is 80 %. Reuse water is pumped a distance of 1,300 m (800 + 500), and the difference in elevation between the head ditch inlet and the reuse reservoir water surface is 3 m.

Part 1 solution: Design of reuse system.

The furrow is designed to apply 76 mm, d_{req} at end of furrow, per irrigation. The interval between irrigation events is 7 days, and the field should be divided as shown in Fig. 20.18, except that all of the zones would be the same size. The number of zones is

$$N = (70\ \text{mm})/(10\ \text{mm/day}) = 7\ \text{sets}$$

The reuse delivery efficiency, Eff_{dr}, is the same as the pumped water delivery system efficiency since reuse water is added to the head ditch at the same point, 0.9. The reuse collection system efficiency, Eff_{rcs}, is 0.87. The required system flow rate, Q_s, with the reuse system is found by rearranging Eq. 20.33.

$$Q_s = \frac{FQ_f}{N\left(Eff_{ds}/_{100}\right)\left(1 + \left(\frac{RO\%}{100}\right)\left(Eff_{dr}/_{100}\right)\left(Eff_{rcs}/_{100}\right)\right)}$$

$$Q_s = \frac{800*2.0\ L/s}{7(0.9)(1 + (0.52)(0.9)(0.87))} = 180\ L/s$$

$$\begin{aligned} Q_r &= Q_s(Eff_{ds}/100\%)(Eff_{rcs}/100\%)(RO\%/100\%) \\ &= 180(0.9)(0.87)(0.52) = 74\ L/s \end{aligned}$$

Check results with Eq. 20.31

$$F = 7\left(\frac{180(0.9)}{2} + \frac{74(0.9)}{2}\right) = 800\ furrows$$

Number of furrows irrigated per day is 800 furrows/7 sets = 114 furrows per irrigation. Supply system flow rate, Q_s, without reuse is 114 furrows (1 L/s-furrow)/0.9 = 254 L/s.

Let the reservoir hold a 1-day water supply.

$$\begin{aligned} &(73.6/0.87\ \text{L/s})(24\ \text{hr})(3{,}600\ \text{sec/hr})/1{,}000\ \text{L/m}^3 \\ &= 7305\ \text{m}^3 \end{aligned}$$

If the reservoir is 2 m deep, then the average reservoir storage area is 2,000 m^2. Average dimensions of 200 m × 36 m would be adequate. If the side slopes are 2:1, and the bottom width is 32 m, then the top (water surface) width would be 40 m, for an average width of 36 m. Likewise, the bottom length should be 193 m and the top length (water surface) should be 207 m for an average length of 100 m. The top of the reservoir should be 0.3 m higher to account for filling of the bottom and an additional 0.3 m for freeboard. Thus, the width of the reservoir with a 2:1 side slope would increase by 1.2 m to 41.2 m. The length would increase by 0.3 m (2) + 0.3 m (5) = 2.1 m. Thus, the length of the reservoir is 209.1 m.

11. The required intake opportunity time is 17 hours, advance is 2 hours, and recession time is 1 hour. What is the t_{co}?

$$t_{co} = t_{adv} + IOT_{req} - t_{rec}$$
$$tco = 2 + 17 - 1$$
$$\mathbf{tco = 18\ hr}$$

12. Volume of deep percolation is 20 m^3, volume of runoff is 25 m^3, and volume used in the soil profile is 60 m^3. What is the irrigation efficiency, inflow volume, deep percolation percentage, and runoff percentage.

Irrigation Efficiency = **(60) / 105** = **57%**
Inflow Volume = **60 + 20 + 25** = **105m³**
Deep Percolation % = **20 / 60 * 100** = **33%**
Runoff % = **25 / 60 * 100** = **42%**

13. The average depth of infiltration in a field is 100 mm, and the average depth of infiltration over the last 25 % of the field is 90 mm. What is the DU?

Low quarter / Furrow Average
90 / 100 = **90%**

14. Why can't one just increase the flow rate to any velocity in order to get water across the field as quickly as possible?

High runoff percentage will lead to wasted water. Erosion may take place

15. Why does surge irrigation improve irrigation efficiency in some soils?

In medium to coarse textured soils, the surface tends to consolidate and seal up once the water is turned off. Then when water is reintroduced in the furrow, infiltration is very low, and the advance phase is very fast. With surge irrigation, the advance time can be nearly as fast as continuous flow, but use half as much water. Surge valves can also be effective because they give the farmer more control over the irrigation event.

16. Download WinSRFR onto your computer and copy a screen showing that it is open on your computer.

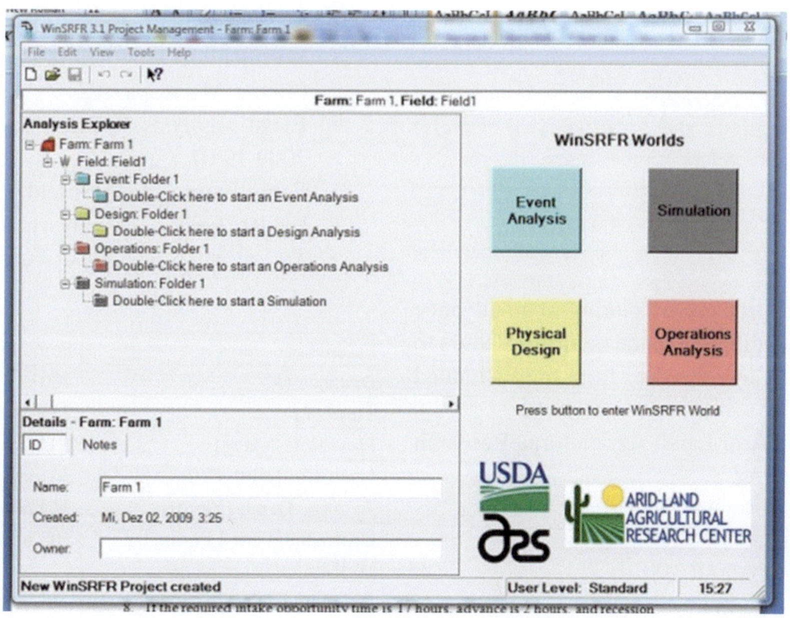

17. Put the data in Figs. 20.16, 20.17, 20.18, 20.19, 20.20, 20.21 and 20.22 into WinSRFR and get the same result shown in the chapter. Copy the data from the Estimated Function page and paste to this document. Include Two point advance per furrow data and Kostiakov a and b data.
    - WinSRFR 3.1 – Arid-Land Agricultural Research Center, Maricopa, AZ
    - Event Analysis Results – Mittwoch, 2. Dezember 2009 16:04
        - Farm: Farm 1, Field: Field1
        - Folder: Folder 1, Analysis: Analysis 1
    - *Two-Point Advance* per Furrow

Parameter	Point 1	Point 2
Distance (X)	200 m	400 m
Time (T)	1,35 hr	3,85 hr
Flow Rate (Qavg)	1 l/s	1 l/s
Upstream Depth (Y0)	52 mm	53 mm
Average Depth (Yavg)	42 mm	42 mm
Upstream Wetted Perimeter (WP0)	335 mm	336 mm
Average Wetted Perimeter (WPavg)	288 mm	290 mm
Upstream Flow Area (A0)	0,01075 m^2	0,01086 m^2
Average Flow Area (Aavg)	0,00798 m^2	0,00803 m^2
Surface Shape Factor (sy)		
User Entered	0,76	0,76
Calculated	0,743	0,739
Inflow Volume (Vin)	4,86 m^3	13,86 m^3
Surface Volume (Vy)	1,63 m^3	3,3 m^3
Vy/Vin (%)	33,6 %	23,8 %
Infiltrated Volume (Vz)	3,23 m^3	10,56 m^3
Power Advance Exponent	(r) = 0,66142	
Power Advance Constant (p) = 10,93252 m/min^r		
Estimates for Kostiakov k, a & b		
k: 18,791 mm/hr^a		
a: 0,46996		
b: 0 mm/hr		

18. Modify the previous problem by putting in an advance time to 2 hours to the midpoint in the field and 5 hours to the end of the field. Copy the data from the Estimated Function page.
    - WinSRFR 3.1 – Arid-Land Agricultural Research Center, Maricopa, AZ
    - Event Analysis Results – Mittwoch, 2. Dezember 2009 16:10
        - Farm: Farm 1, Field: Field1
        - Folder: Folder 1, Analysis: Analysis 1
    - *Two-Point Advance* per Furrow

Parameter	Point 1	Point 2
Distance (X)	200 m	400 m
Time (T)	2 hr	5 hr
Flow Rate (Qavg)	1 l/s	1 l/s
Upstream Depth (Y0)	52 mm	53 mm
Average Depth (Yavg)	45 mm	46 mm
Upstream Wetted Perimeter (WP0)	335 mm	336 mm
Average Wetted Perimeter (WPavg)	300 mm	304 mm
Upstream Flow Area (A0)	0,01075 m^2 0,01086 m^2	
Average Flow Area (Aavg)	0,00867 m^2 0,00885 m^2	
Surface Shape Factor (sy)		
User Entered	0,76	0,76
Calculated	0,806	0,815
Inflow Volume (Vin)	7,2 m^3	18 m^3
Surface Volume (Vy)	1,63 m^3	3,3 m^3
Vy/Vin (%)	22,7 %	18,3 %
Infiltrated Volume (Vz)	5,57 m^3	14,7 m^3
Power Advance Exponent	(r) = 0,75647	
Power Advance Constant	(p) = 5,34798 m/min^r	
Estimates for Kostiakov k, a & b		
k: 28,207 mm/hr^a		
a: 0,30328		
b: 0 mm/hr		

19. Modify the previous problems by changing the furrow bottom width to 200 mm and the slope to 0.003. Copy the data from the Estimated Function page.
    - WinSRFR 3.1 – Arid-Land Agricultural Research Center, Maricopa, AZ
    - Event Analysis Results – Mittwoch, 2. Dezember 2009 19:07
    - Farm: Farm 1, Field: Field1
    - Folder: Folder 1, Analysis: Analysis 1
    - Two-Point Advance per Furrow

Parameter	Point 1	Point 2
Distance (X)	200 m	400 m
Time (T)	2 hr	5 hr
Flow Rate (Qavg)	1 l/s	1 l/s
Upstream Depth (Y0)	36 mm	37 mm
Average Depth (Yavg)	31 mm	31 mm
Upstream Wetted Perimeter (WP0)	363 mm	363 mm
Average Wetted Perimeter (WPavg)	337 mm	339 mm

(continued)

Parameter	Point 1	Point 2
Upstream Flow Area (A0)	0,00992 m^2	0,00997 m^2
Average Flow Area (Aavg)	0,00808 m^2	0,00822 m^2
Surface Shape Factor (sy)		
User Entered	0,76	0,76
Calculated	0,815	0,825
Inflow Volume (Vin)	7,2 m^3	18 m^3
Surface Volume (Vy)	1,51 m^3	3,03 m^3
Vy/Vin (%)	20,9 %	16,8 %
Infiltrated Volume (Vz)	5,69 m^3	14,97 m^3
Power Advance Exponent	(r) = 0,75647	
Power Advance Constant	(p) = 5,34798 m/min^r	
Estimates for Kostiakov k, a & b		
k: 28,854 mm/hr^a		
a: 0,29884		
b: 0 mm/hr		

Note: make the following modifications, but don't run the program for problems 5–7 (it won't work)

20. Modify the furrow side slope to 1 and do a screen capture of the System geometry page, showing the modified furrow shape.

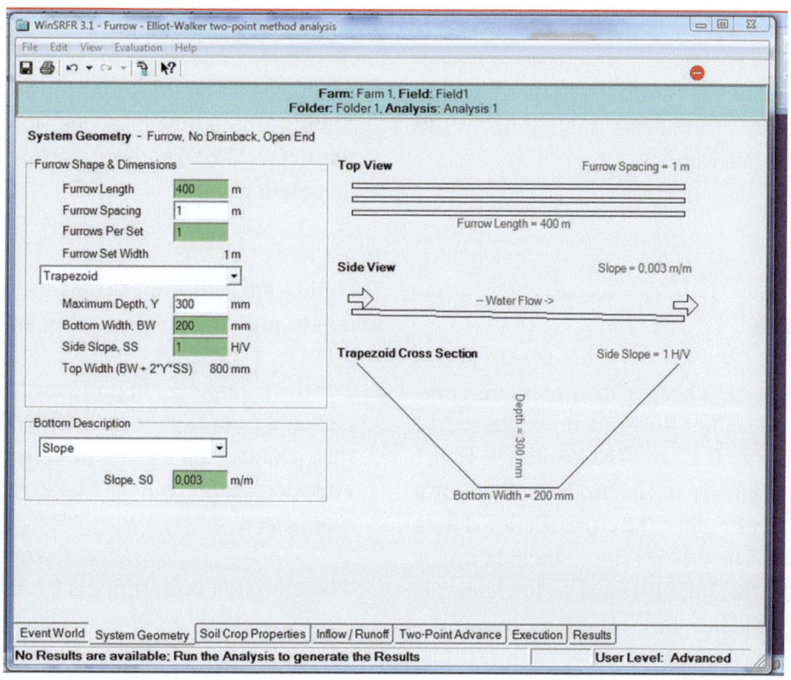

21. Change the furrow shape to the Power Law shape and use exponent M = 0.4. Do a screen capture of the System Geometry page, showing the modified furrow shape.

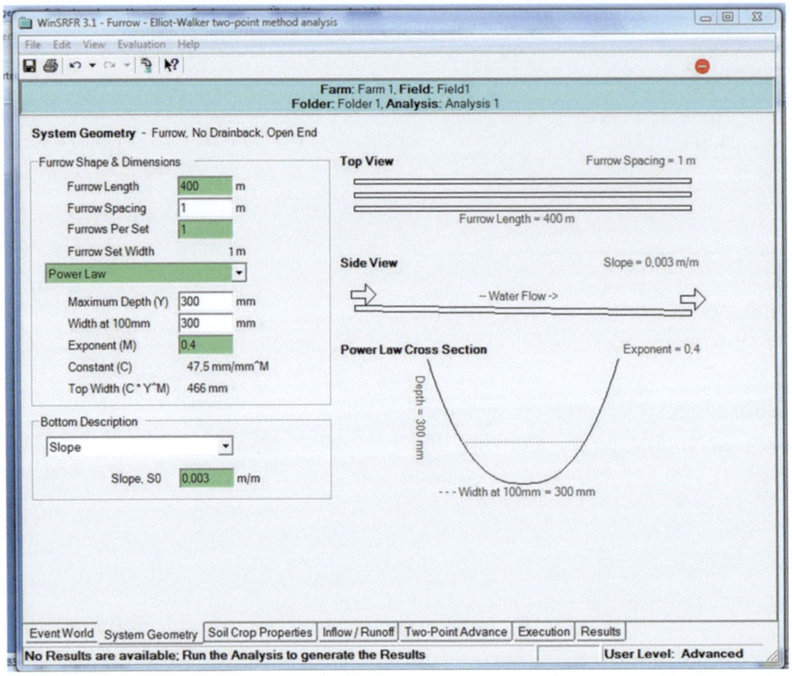

22. Change the Manning n on the Soil/Crop Properties page to that for alfalfa, mint, or broadcast small grain. Write down this value as the answer.
    – n = 0.15

Chapter 21: Solutions

1. Turbulent flow emitters are cheaper than pressure compensating emitters. The emitter flow rate flow rate is 2 L/hr, and emitter spacing is 0.17 m. The emitter k is 0.2 and × is 0.5. Tubing diameter is 12 mm. Manufacturer's CV is 3 %. Barbed fittings reduce the cross-sectional area of the tube to 95 mm^2. There is no slope. Include minor losses due to emitter barbs. Inlet pressure to the lateral is 120 kPa. Calculate whether the emission uniformity exceeds a 95 % threshold for hydroponic drip irrigation. Should the grower switch to pressure compensating emitters? Use the Drip lateral worksheet to make the calculation.

The emission uniformity is 93 %. Thus, the grower should switch to pressure compensating emitters.

2. For the parameters in problem 1, switch to pressure compensating emitters with 2 LPH flow rate. In order to do this, change the values in cells E10:E11. Calculate the emission uniformity. Make a screen copy of the worksheet.

   The emission uniformity is 97.3 %, which is acceptable.

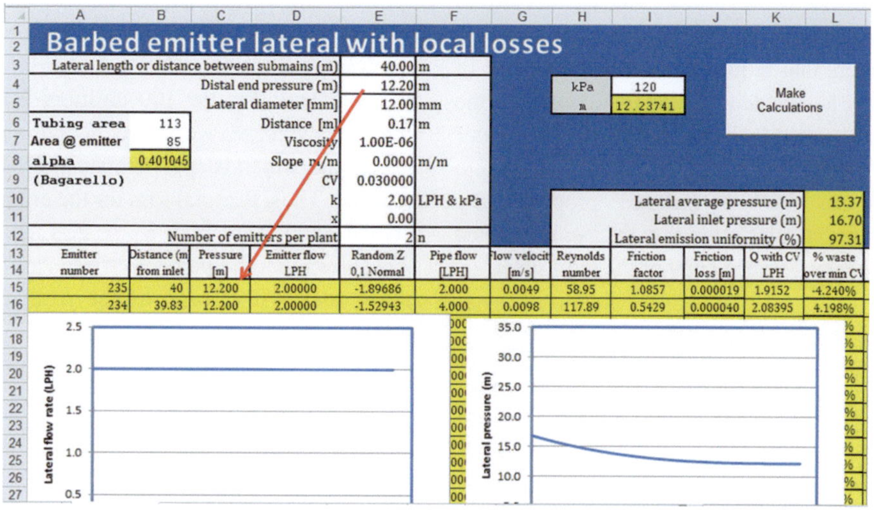

3. A 12 mm diameter irrigation lateral has pressure compensating inline emitters that reduce the internal diameter of the pipe to 10 mm. Flow velocity is 1.2 m/sec. Calculate the friction loss due to 1 emitter by hand. Use the *Inline emitter local losses* worksheet to calculate the friction loss due to 100 emitters, where flow is reduced proportionally vs. distance along the lateral. Compare to the loss calculated in Example 21.2. Make a screen copy of the worksheet.

$$\alpha = 0.116\left[\left(\tfrac{D}{d}\right)^{13.87} - 1\right] = 0.116\left[\left(\tfrac{12}{10}\right)^{13.87} - 1\right] = 1.33$$

$$\Delta H_s = \alpha \frac{V^2}{2g}$$

$$= 1.33\frac{1.5^2}{2g} = 0.15\, m$$

Friction loss due to 1 emitter at 1 m/sec flow rate is 0.15 m, which is 10 times greater than the head loss in Example 21.1.

The total head loss due to minor losses is 5.2 m, which is more than 10 times the loss calculated in Example 21.1.

The calculation for 100 emitters is made in the *Inline emitter local losses* worksheet.

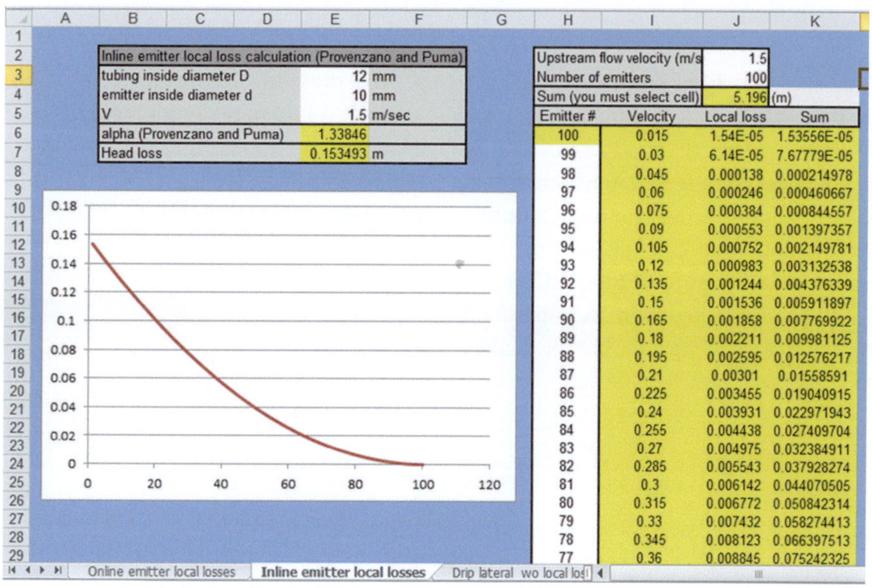

4. A 12 mm diameter irrigation lateral has pressure compensating online emitters that reduce the internal diameter of the pipe to 90 mm^2. Flow velocity is 0.5 m/sec. Calculate the friction loss due to 1 emitter by hand. Use the *Online emitter local losses* worksheet to calculate the friction loss due to 100 emitters, where flow is reduced proportionally vs. distance along the lateral. Compare to the loss calculated in Example 21.2. Make a screen copy of the worksheet.

$$\alpha = 1.68\left[\left(\frac{A_p}{A_g}\right) - 1\right]^{1.29} = 1.68\left[\left(\frac{113}{90}\right) - 1\right]^{1.29} = 0.29$$

$$\Delta H_s = \alpha \frac{V^2}{2g}$$

$$= 0.29 \frac{0.5^2}{2g} = 0.0037\ m$$

Friction loss due to 1 emitter at 1 m/sec flow rate is 0.0037, which is 37 % of the value in Example 21.2: 0.010 m.

The calculation for 100 emitters is made in the *Online emitter local losses* worksheet. Total loss due to local losses is 0.125 m, which is approximately 37 % of the losses in the example. Thus, the total loss for the entire lateral and the loss for one emitter are both 37 % less than for Example 21.2. This shows that the percent difference for a single emitter is approximately the same as the percent difference for the lateral.

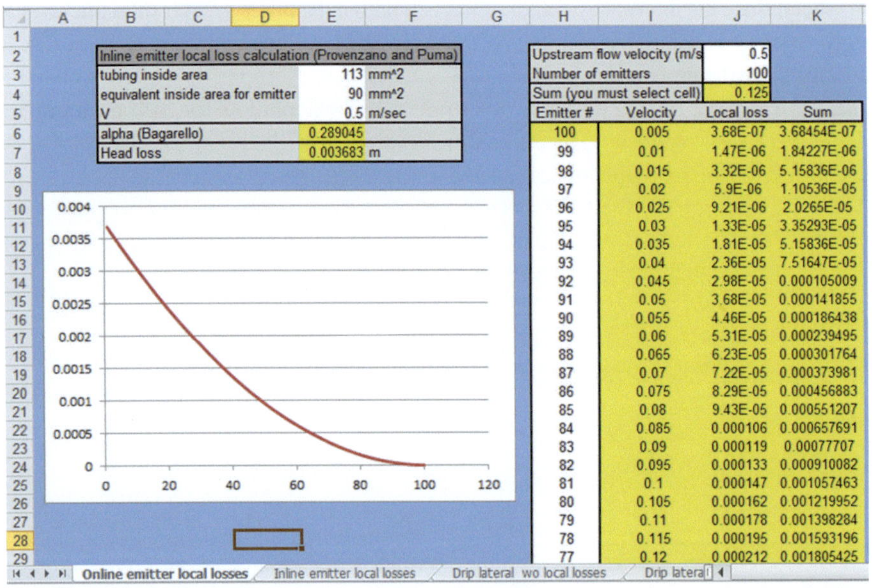

Inline emitter local loss calculation (Provenzano and Puma)		
tubing inside area	113	mm^2
equivalent inside area for emitter	90	mm^2
V	0.5	m/sec
alpha (Bagarello)	0.289045	
Head loss	0.003683	m

Upstream flow velocity (m/s	0.5
Number of emitters	100
Sum (you must select cell)	0.125

Emitter #	Velocity	Local loss	Sum
100	0.005	3.68E-07	3.68454E-07
99	0.01	1.47E-06	1.84227E-06
98	0.015	3.32E-06	5.15836E-06
97	0.02	5.9E-06	1.10536E-05
96	0.025	9.21E-06	2.0265E-05
95	0.03	1.33E-05	3.35293E-05
94	0.035	1.81E-05	5.15836E-05
93	0.04	2.36E-05	7.51647E-05
92	0.045	2.98E-05	0.000105009
91	0.05	3.68E-05	0.000141855
90	0.055	4.46E-05	0.000186438
89	0.06	5.31E-05	0.000239495
88	0.065	6.23E-05	0.000301764
87	0.07	7.22E-05	0.000373981
86	0.075	8.29E-05	0.000456883
85	0.08	9.43E-05	0.000551207
84	0.085	0.000106	0.000657691
83	0.09	0.000119	0.00077707
82	0.095	0.000133	0.000910082
81	0.1	0.000147	0.001057463
80	0.105	0.000162	0.001219952
79	0.11	0.000178	0.001398284
78	0.115	0.000195	0.001593196
77	0.12	0.000212	0.001805425

5. Emitters (4 LPH) are spaced 0.34 m in a 12 mm pipe that is 50 m long. Emitters are turbulent with k = 0.4 and x = 0.5. Slope is zero. Coefficient of variation is 5 %. Full tubing area without emitters is 113 mm^2 and tubing area at the emitters is 85 mm^2. Calculate emission uniformity and head loss. Include one calculation with local losses and one calculation without local losses. Compare to the friction loss due to local losses in Example 21.3. Make sure to press the *Make calculations* button. Make a screen copy of the worksheet with local losses.

The pressure loss in the pipe without considering local losses is 14.84 m and with local losses is 17.15 m. Thus, the head loss increases from 4.6 to 7 m with local losses. The head loss due to local losses is 2.4 m, which is half of that calculated in Example 21.3.

The emission uniformity drops from 90.4 % to 88.4 % with local losses.

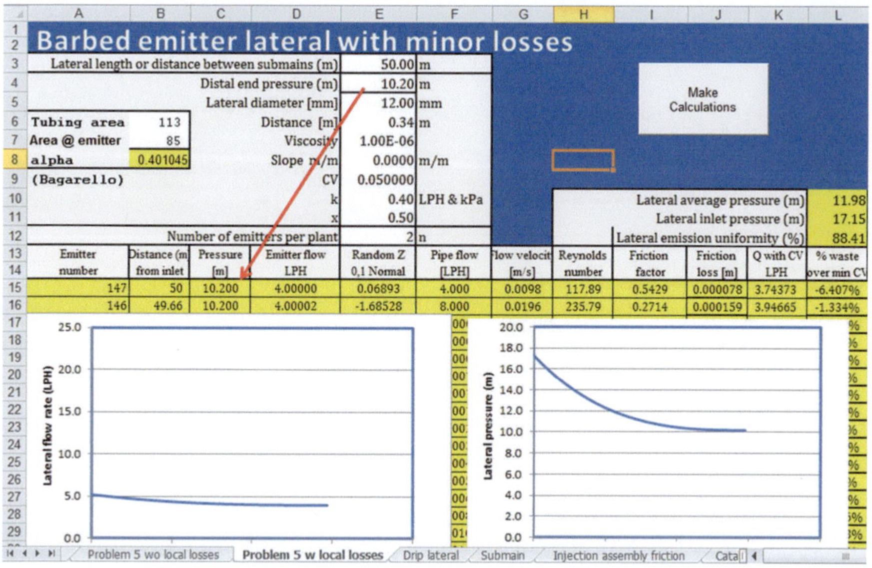

6. A submain supplies the lateral described in problem 5. The submain is 100 m long and laterals are spaced every 1.5 m. The inside diameter of the submain is 100 mm. Using the *Submain* worksheet, find the pressure loss in the submain and make a screen copy of the worksheet. Make sure to press the *Make calcs* button.

The required inlet pressure to the submain is 17.7 m. The pressure loss is 17.7 m − 17.2 m = 0.5 m.

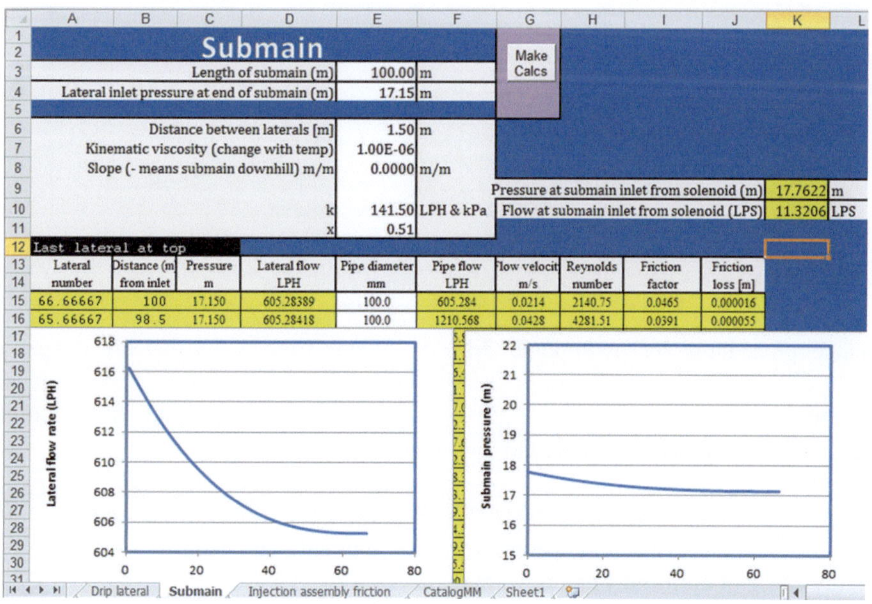

7. For the parameters in problem 6, calculate the required pump pressure and flow rate. Chemical injection is from diaphragm pumps, which cause no pressure loss in the irrigation pipeline. Ring filters are used to filter the water, which have 4 m maximum pressure drop. A large pressure regulator controls pressure to the system and has 4 m pressure drop across the regulator. Total fittings losses at the pump station are 5 m. Solenoid valves to each submain have a 3 m pressure drop.

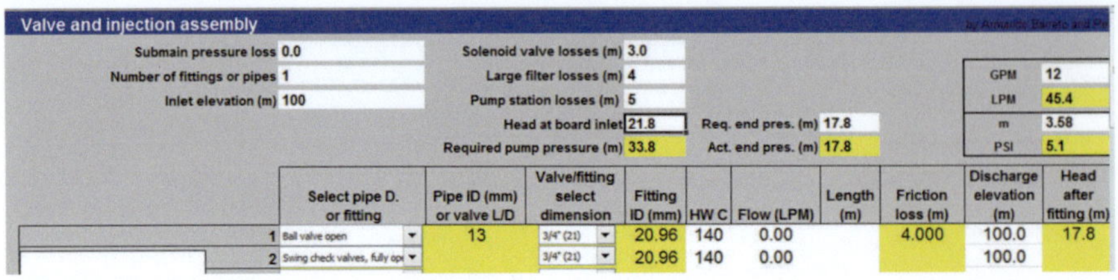

8. Determine whether the lateral shown in Fig. 21.15 will flush with the required 0.5 m/sec flow velocity at the end of the lateral.

The required inlet pressure is 5.51 so the lateral will flush. Emitters should be checked to make sure that the flushing mode at low pressure will not disrupt the process.

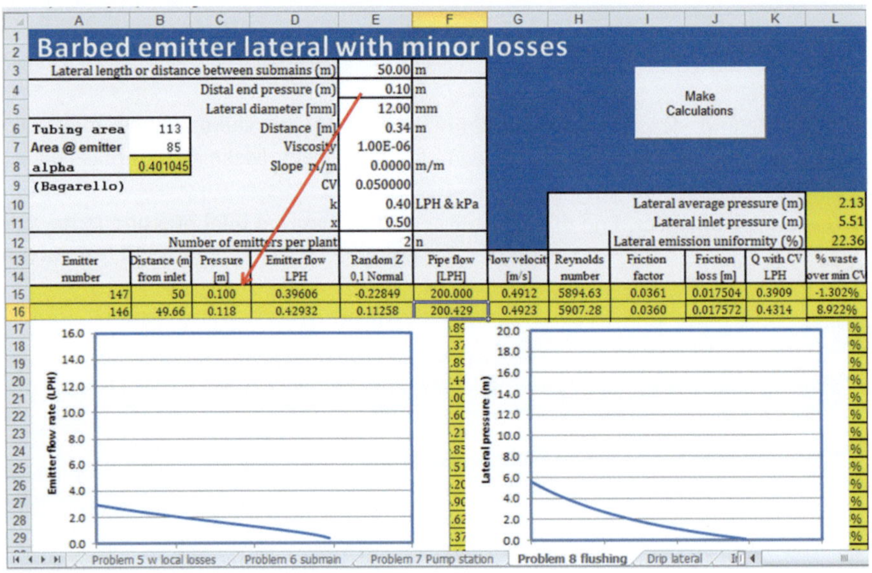

9. List the maintenance procedures for hydroponic drip irrigation.
   1. Injection of biocides
   2. Injection of acids to prevent precipitation of salts
   3. Flush drip laterals regularly
   4. Filtration
   5. Daily check of pump station pressure and operation
   6. Weekly or monthly check of uniformity and flow rates of drip emitters

10. Why is it desirable to have water from a deep well for irrigation of greenhouse crops?

Water from wells that are deeper than 60 m (200 ft) will probably not have appreciable amounts of bacteria or organic carbon, and disinfection is probably not necessary

11. Describe typical irrigation practices at dawn and dusk in a tomato greenhouse.

For tomatoes, there is normally no irrigation for the first 1½ to 2 hours after sunrise and the last 1½ to 2 hours before sunset in order to prevent fruit cracking.

12. Develop an irrigation schedule for the following solar radiation intensity. The units for energy are those given by the Arizona agricultural weather network. You will need to convert to J/cm^2.

Time (hours)	Solar radiation (MJ/m^2)
7	0.15
8	0.41
9	0.94
10	1.61
11	2.99
12	3.33
13	3.4
14	3.3
15	2.7
16	2.44
17	0.65
18	0.5
19	0.06

Hour	MJ/m^2	J/cm^2	Number of irrigations
7	0.15	15	
8	0.41	41	
9	0.94	94	1
10	1.61	161	2
11	2.99	299	4
12	3.33	333	4
13	3.4	340	4
14	3.3	330	4
15	2.7	270	4
16	2.44	244	4
17	0.65	65	1
18	0.5	50	
19	0.06	6	

13. Calculate daily depth applied for the irrigation schedule in question 8. Each water application volume is 100 ml. The area represented by each plant/emitter is $0.27\ m^2$.

Each irrigation supplies 100 ml to the plant. There are 27 irrigations. Thus, 2,700 ml is supplied to the plant.

$$\text{Depth} = \text{volume}/\text{area} = 0.0027/0.25 = 0.01\ \text{m}/\text{day} = 10\ \text{mm}/\text{day}$$

14. List the macronutrients and micronutrients

- Macronutrients: N, P, K, Ca, Mg, S
- Micronutrients: Fe, Mn, B, Zn, Cu, Mo, Cl.

15. Why were the Sachs and Knopf nutrient experiments unsuccessful? Why or why not?

They were not successful because they did not supply oxygen to the roots.

16. Who was the first plant nutritionist in the United States, and where did he work?

Dennis Hoagland from UC Berkeley.

17. Based on molarity, which three nutrients in the Hoagland solution (Table 21.2) are required in the greatest amount? Does this agree with Table 21.3? Does this agree with Table 21.4?

Nitrogen is included in the first three fertilizers so it is required in the greatest amount. Based on Table 21.3, nitrogen, potassium, and calcium have approximately the same concentration; however, nitrogen has a lighter molecular weight so a greater number of moles of nitrogen are required in comparison to potassium and calcium. In Table 21.4, potassium and calcium are required in much higher concentration than nitrogen during the early part of the season; however, the greater molecular weights indicate that the molar amounts are approximately the same.

18. T/F. Testing water for nutrients prior to determining a fertigation regime in the greenhouse is not necessary because greenhouse plants require so much fertilizer anyway?

False, you need to subtract the nutrients in the source water from the fertigation requirement.

19. Calculate the amount of manganese chelate in the B tank (Table 21.3) if the injection rate is 1/50th of the greenhouse irrigation flow rate.

$$5.8 * 50 = 290\ \text{mg/L}$$

20. T/F. Adding too much of one cation can decrease the uptake of other cations.

True

21. Calculate the concentrations of the fertilizers required to formulate the nutrient mix shown in the "Week 6–12" column in Table 21.4. Mix enough fertilizer to dissolve in a 500 L tank when the water flow rate is 200 X greater than the fertilizer injection rate. The source water contains 20 mg/L Ca. Perform calculations by hand and check answer with fertigation calculator.

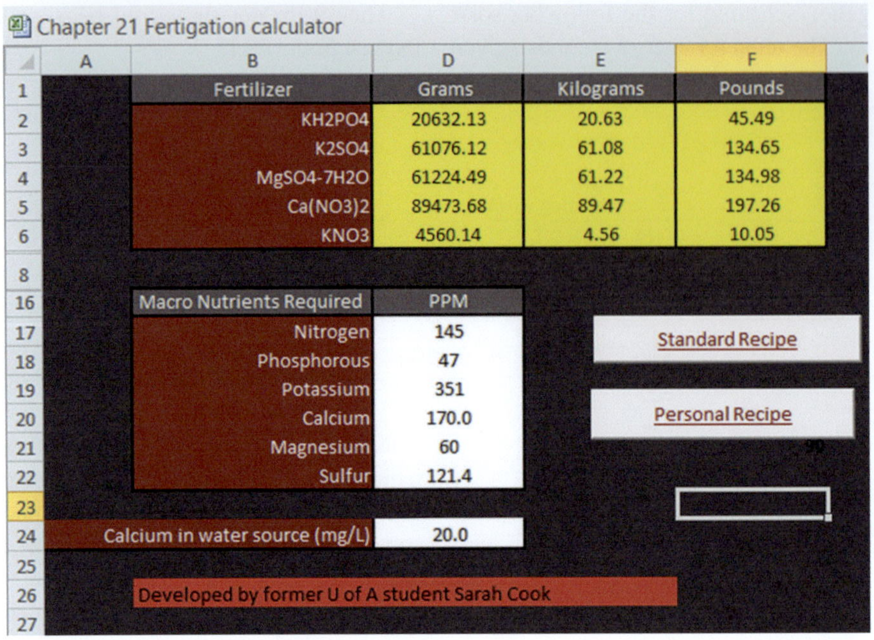

Chapter 21 Fertigation calculator

	Fertilizer	Grams	Kilograms	Pounds
2	KH2PO4	20632.13	20.63	45.49
3	K2SO4	61076.12	61.08	134.65
4	MgSO4-7H2O	61224.49	61.22	134.98
5	Ca(NO3)2	89473.68	89.47	197.26
6	KNO3	4560.14	4.56	10.05

Macro Nutrients Required	PPM
Nitrogen	145
Phosphorous	47
Potassium	351
Calcium	170.0
Magnesium	60
Sulfur	121.4

Standard Recipe

Personal Recipe

Calcium in water source (mg/L) 20.0

Developed by former U of A student Sarah Cook

Chapter 22: Solutions

1. What are the advantages and disadvantages of low-head gravity bubbler system?

Answer:

Advantage:

Works with minimum head as low as one meter
Low energy requirement i.e., no pumping required
Surface systems can easily be converted
No filtration required as the delivery tubes are larger in diameter to let small particles pass through them.

Disadvantages:

Initial cost is somewhat high mainly due to the pipe distribution system.
Airlock in the system can affect the efficiency
Limited to orchard crops, vines and tree crops
Requires basins around the trees in heavy soils
Rodents sometimes eat the delivery tube

2. What is the cause of airlocks in low-head gravity bubbler systems

Answer:

Airlocks are caused by air accumulation at the crest of undulating pipes.

3. What are the effects of airlocks?

Answer:

Airlocks absorb the flow energy and may partially or entirely block the flow of water

4. How do you avoid airlocks?

Answer:

By minimizing undulation of the flexible pipe components of the system such as the delivery hoses.
Keep the hydraulic gradient above the pipeline
Installation of air relief valves downstream of the crest (not applicable for bubbler systems)
Putting pipe stands at the crest
Proper installation of the system

5. What is the main cause of non-uniformity of flow along the lateral in bubbler systems?

Answer:

The delivery hose elevation not following the hydraulic gradient line. This happens mainly during installation and as time goes by the delivery hoses move.

6. The Hazen-Williams empirical equation for flow in pipes is given by

$$Q = 0.849CAR^{0.63}\left(\frac{h_f}{L}\right)^{0.54} \text{ metric system}$$

Where Q is flow rate, A is cross-sectional area, R is the hydraulic radius, L is length of pipe, and h_f is the friction head loss. The friction head loss h_f can be written from the above relationship as $h_f = K\left[\frac{Q}{C}\right]^{1.852}D^{-4.87}L$

For h_f in meter, D in meter, L in meter, and Q in cubic meters per second determine the value of K.

Solution:

$$Q = 0.84CAR^{0.63}\left[\frac{h}{L}\right]^{0.54}$$
$$A = \pi\frac{D^2}{4}$$
$$R = \frac{D}{4}$$

Then

$$Q = 0.849C\pi\frac{D^2}{4}\left[\frac{D}{4}\right]^{0.63}\left[\frac{h_f}{L}\right]^{0.54}$$
$$Q = 0.849\frac{\pi}{4}\left[\frac{1}{4}\right]^{0.63}CD^{2.63}\left[\frac{h_f}{L}\right]^{0.54}$$
$$Q = 0.2784CD^{2.63}\left[\frac{h_f}{L}\right]^{0.54}$$

$$\frac{h_f^{0.54}}{L^{0.54}} = \frac{Q}{0.2784CD^{2.63}}$$

Or

$$h_f = \frac{Q^{1/0.54}L}{(0.2784)^{1/0.54}C^{1/0.54}\left(D^{2.63}\right)^{1/0.54}}$$
$$h_f = \frac{Q^{1.852}L}{(0.2784)^{1.852}C^{1.852}D^{4.87}}$$
$$h_f = \frac{Q^{1.852}L}{0.9365C^{1.852}D^{4.87}}$$
$$h_f = 10.7\left[\frac{Q}{C}\right]^{1.852}LD^{-4.87}$$

Thus K = 10.7

Rearranging and solving for h_f

7. For microirrigation design we use the Hazen-Williams equation more commonly than the Darcy-Wiesbach equation even though the later is more accurate for different fluid and flow conditions and the Hazen-Wiliams equation is easier to use for water. (a) Develop a relationship between Darcy-Wiesbach resistance coefficient (f) in Equation – and the Hazen-Williams (C) in Equation – and (b) develop a relationship between C and Reynolds number. Assume water at 20 °C.

Solution

Part (a)

The Hazen-Williams and Darcy-Wiesbach equations can be written, respectively as:

$$h_f = k_1L\frac{\left(\frac{Q}{C}\right)^{1.852}}{D^{4.87}} \qquad (1)$$

$$h_f = k_2fL\frac{Q^2}{D^5} \qquad (2)$$

To develop the relationship between f and C equate Eq. 1 to Eq. 2 and solve for f or C

Then

$$k_1L\frac{\left(\frac{Q}{C}\right)^{1.852}}{D^{4.87}} = k_2fL\frac{Q^2}{D^5} \qquad (3)$$

Or

$$k_1\frac{\left(\frac{Q}{C}\right)^{1.852}}{D^{4.87}} = k_2f\frac{Q^2}{D^5} \qquad (4)$$

Then

$$\frac{k_1}{k_2}\frac{1}{C^{1.852}} = f\frac{Q^{(2-1.852)}}{D^{(5-4.87)}} = f\frac{Q^{0.148}}{D^{0.13}} \qquad (5)$$

Or

$$f = \frac{k_1}{k_2}\frac{D^{0.13}}{Q^{0.148}C^{1.852}} \qquad (6)$$

In equations 1 and 2 k_1 and k_2 for water at 20 °C, Q in Liters/second (L/s), D in millimeters (mm) and L in meters (m) are

$$k_1 = 1.22 \times 10^{10}$$

And

$$K_2 = 8.27 \times 10^7$$

Thus substituting these values Eq. 6 gives

$$f = 147.65\frac{D^{0.13}}{Q^{0.148}C^{1.852}} \qquad (7)$$

Or

Solving for C

$$C = \frac{D^{0.072}}{f^{0.54}Q^{0.08}} \qquad (8)$$

Part (b)
Using Eq. 8 and rearranging it

$$C = \frac{14.84}{f^{0.54}}\left(\frac{D}{Q}\right)^{0.08} D^{-0.008} \quad (9)$$

For water at 20 °C the Reynolds number can be computed as

$$R_n = 1.26 \times 10^6 \frac{Q}{D} \quad (10)$$

Or

$$\frac{D}{Q} = \frac{1.26 \times 10^6}{R_n} \quad (11)$$

Substituting Eq. 11 into Eq. 9 for D/Q and solving for C we have

$$C = \frac{14.84}{f^{0.54}}\left(\frac{1.26 \times 10^6}{R_n}\right) D^{-0.008} \quad (12)$$

Or

$$C = 45.64 \frac{D^{-0.008}}{f^{0.54} R_n^{0.08}} \quad (13)$$

The term $D^{-0.008}$ varies from 0.982 to 0.962 for D ranging from 10 to 125 mm. If we are to be conservative and take 0.962 for the term and use it in Equation 13 we will have a simple relationship between C and Reynolds number, R_n, as

$$C = 43.9 f^{-0.54} R_n^{-0.08} \quad (14)$$

8. The friction drop ratio a for a microirrigation lateral can be given by

$$R_x = \frac{h_{fx}}{h_{fL}}$$

where h_{fx} is the head loss due to friction from the head end to any point x along the lateral and h_{fL} is the head loss for the entire length of the lateral. It can also be shown that R_x can be calculated by

$R_x = 1 - \left(1 - \frac{x}{L}\right)^{m+1}$ where x is the distance from the head end to any point along the lateral, L is the length of the lateral, and m is the exponent for the velocity or flow varying in the friction loss equation Ex. m = 2 for Darcy-Wiesbach equation and 1.852 for Hazen-Williams equation. The pressure head, H_x at any given point x on the lateral neglecting the velocity head term can be computed as

$$H_x = H_o - h_{fx} - h_{Zx}$$

where H_x is the pressure head at a point located at a distance x from the pipeline inlet

H_0 is pressure head at the pipeline inlet and

H_{Zx} is difference in elevation between point x, and pipeline inlet

Using these relationships show that the average pressure head, H_a, can be calculated using $H_{av} = H_0 - \frac{m+1}{m+2} h_{fL} - \frac{H_{ZL}}{2}$ and

Solution:

The average pressure head in a lateral can be obtained by integrating $H_x = H_o - h_{fx} - h_{Zx}$ and dividing by the total length L.

then,

$$\frac{1}{L}\int_0^L H_x dx = \frac{1}{L}\left(\int_0^L H_0 dx - \int_0^L h_{fx} dx - \int_0^L H_{Zx} dx\right)$$

but h_{fx} can be substituted using the pressure drop ratio, R_x and the total head loss in the lateral h_{fL} as

$$H_{av} = H_0 - \frac{1}{L}\left\{h_{fL}\int\left[1 - \left(1 - \frac{x}{L}\right)\right]^{m+1} dx - \int S_0 x \, dx\right\}$$

$$H_{av} = H_0 - \frac{h_{fL}}{L}\left[\int_0^L dx + L\int\left(1 - \frac{x}{L}\right)^{m+1}\left(-\frac{1}{L}dx\right)\right] - \frac{S_0 L}{2}$$

$$H_{av} = H_0 - \frac{h_{fL}}{L}\left[L - \frac{L}{m+2}\left(1 - \frac{x}{L}\right)^{m+2}\right] - \frac{H_{ZL}}{2}$$

$$H_{av} = H_0 - \left[L - \frac{L}{m+2}(1)^{m+2}\right] - \frac{H_{ZL}}{2}$$

Note:

$$\left(1 - \frac{1}{m+2}\right) = \frac{m+1}{m+2}$$

Therefore the average pressure in a lateral can be expressed by

$$H_{av} = H_0 - \frac{m+1}{m+2} h_{fL} - \frac{H_{ZL}}{2}$$

9. Determine the size of a PVC manifold for low-head gravity bubbler system with each lateral to carry a flow of with 2 l/s flow to meet the crop demand and is to be laid in a field 100 m wide and 200 m long. The field is going to be used to grow orchard with tree spacing is 5 m by 5 m. The standpipe is to be at the middle of the manifold. The allowable head loss in the manifold is 0.2 meters.

Solution:
The length of the manifold is 50 m-5 m = 45 m
The Christiansen F factor for 5 outlet = 0.40

The energy gradient is $h_f/L_m = h_{fa}/FL_m = 0.2/(0.4 \times 45) = 0.011\ m/m$

The flow in the manifold is $5 \times 2\ L/s = 10\ L/s$

From figure – for a head loss gradient of 0.011/m/m and a flow rate of 10 L/s the size will be 102 mm or 4 inches.

10. Design a bubbler system to irrigate a citrus orchard with tree spacing of 6 m by 6 m. The field is level and has a dimension of 120 m by 96 m. The water source is a low-head pipeline located at the edge of the field. Assume the design head at the constant head device is 1.2 m and the maximum and minimum delivery hose elevations are 1 m and 0.3 m respectively. Also assume the laterals are laid midway between two rows of trees. To prevent air locks, assume the delivery hose flow rate of 0.047 L/s.

Solution:

Refer to field layout in the figure below.

Given:

Field length, L = 120 m

Field Width, W = 96.0 m

Field Slopes, $S_L = S_W = 0$ S_W is the slope along the width of the field

The design head, H_d, is given as 1.2 m.

Tree plant and row spacing, $S_p = S_r$ =6 m

Maximum delivery hose elevation, H_{max} = 1.0 m

Minimum delivery hose elevation, H_{min} = 0.3 m

Assuming the lateral is laid mid-way between two rows of trees and that two delivery hoses are installed per row spacing

Delivery hose spacing, $S_e = S_p$ = 6 m

Lateral spacing, $S_l = 2S_r$ = 12 m

Depth of lateral, d_l = 0.46 m

Length and number of pipes

Length of mainline = not needed

Length of manifold, L_m = (Width, W/2) – S_r = (96/2)-6 = 42 m

Length of lateral, $L_l = L - (S_p/2) = 120 - (6/2) = 117\ m$

Number of delivery hoses per lateral, $N_e = 2\ L/S_e = 240$ m/6 m =40 delivery hoses

Total number of trees, $N_t = N_e \times N_l = 40 \times 8 = 320\ trees$

Design flow rates

Flow per lateral, $q_l = q_{dh} \times N_e = 0.0473(40) = 1.893\ L/s$

Flow in Manifold, $q_m = q_l \times N_l = 1.893(4) = 7.572\ L/s$

Flow in mainline, $Q_s = q_{dh} \times N_t = 0.047(320) = 15.0\ L/s$

Sizing Pipelines

Manifold Diameter

Total allowable head loss $h_{fa} = H_d - H_{min} = 1.2 - 0.3\ 0.9\ m$

Assume 50 % of the total allowable head loss is within the manifold

Then

$$h_{fam} = 0.5(h_{fta}) = 0.5(0.9m) = 0.45\ m$$

The Christiansen reduction factor, F, is equal 0.41 for 4 outlets with the first outlet spaced one-half the spacing from the lateral inlet

The manifold head loss gradient is then

$H_f/L = h_{fam}/(FL_m) = 0.45/(0.41(42m)) = 0.026$ From Figure – with a head loss gradient of 0.026 m/m and flow of 7.6 L/s, gives a manifold diameter of 102 mm (4 inches) PVC pipe.

Size of laterals

Assuming minor losses to be zero the manifold friction losses between laterals are calculated using Equation – and are 0.047 m, 0.057 m, 0.028 m and 0.008 m respectively. The lateral inlet pressures are thus as follows

$$\begin{aligned} &Laterals\ 4\ and\ 5\ (center\ of\ field) = H_d - h_{fm} \\ &\quad = 1.20 - 0.047 - 1.15\ m \\ &Laterals\ 3\ and\ 6\ (next\ to\ center) = 1.15 - 0\ 057 = 1.10\ m \\ &Laterals\ 2\ and\ 7\ (next\ to\ the\ edges) = 1.10 - 0.028 = 1.07\ m \\ &Laterals\ 1\ and\ 8\ (edges\ of\ the\ field) = 1.07 - 0.008 = 1.06\ m \end{aligned}$$

The allowable head loss within the laterals and delivery hoses are calculated as the difference between the lateral inlet and the minimum delivery hose elevation, H_{min}

For lateral number 1

$$\begin{aligned} h_{fal} &= h_{fal} + h_{fadh} = (H_u - H_{do}) - \Delta Z = H_u - H_{min} \\ &= 1.06 - 0.3\ 0.76\ m \end{aligned}$$

For sizing the laterals and delivery hoses use 50 % of the allowable head loss within both laterals and delivery hoses. Thus,

$$h_{fal} = h_{fadh} 0.50(0.76\ m) = 0.38\ m$$

The Christiansen reduction coefficient, F, for 20 multiple outlets with one-half spacing for the first outlet is 0.36

The head loss gradient for laterals and delivery hoses are then,

$$\begin{aligned} h_f/L &= h_{fal}/FL_l = 0.38\ m/(0.36(117m)) \\ &= 0.009\ m/m\ laterals \\ h_f/L &= h_{fadh}/L_{dh} = 0.38m/4.46m = 0.085, delivery\ hoses \end{aligned}$$

Using the design flow rates for laterals of 1.89 L/s and 0.047 L/s for delivery hoses and using the design chart (see below) we get 63 mm for the lateral and 10 mm for the delivery hose

After the lateral and the delivery hose diameters are decided, the elevation of the delivery hoses can be calculated as presented in the table below.

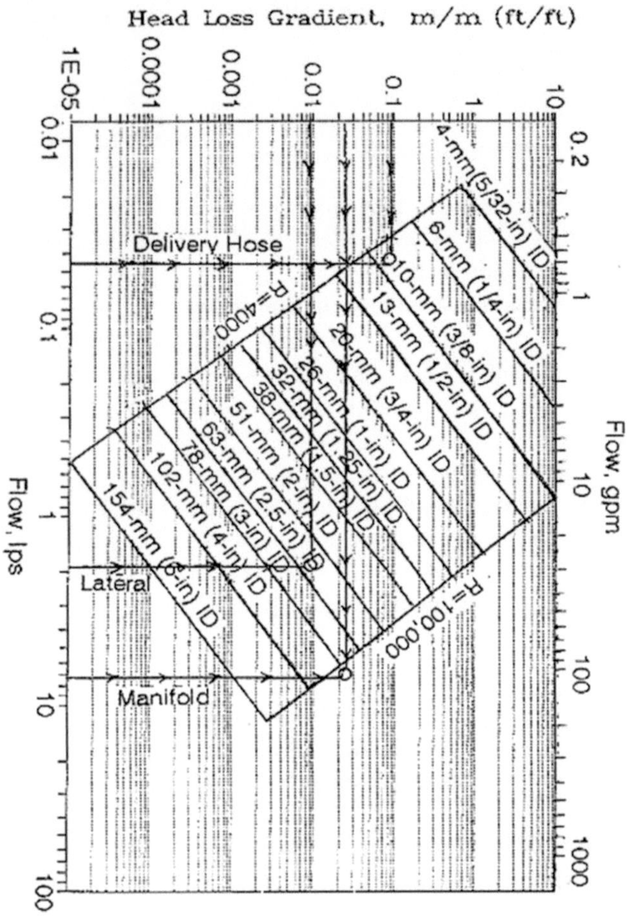

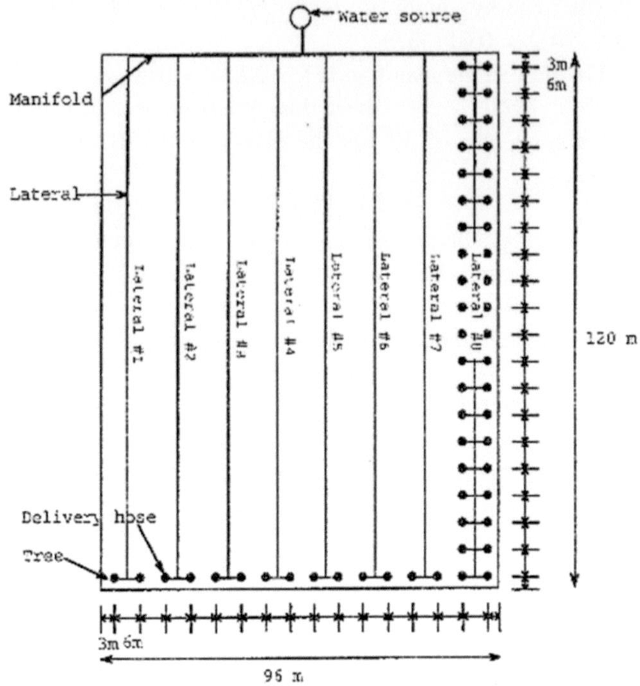

Figure 1. Field Layout for Level Ground Design

Table 2. Delivery Hose Elevation Tabulation (Level Field Design)

Lateral #1

Lateral diameter	=	63	mm
Lateral diameter length	=	117	m
Lateral pipe flow range	=	7.54-0.095	lps
Delivery hose diamater	=	10	mm
Delivery hose length	=	4.5	m

Out/ Sect. (1)	Flow (lps) (2)	Section Length (m) (3)	Total Length (m) (4)	Ground Elev. (m) (5)	Lateral Friction Losses (m) (6)	Lateral HGL (m) (7)	Hose Friction Losses (m) (8)	V/Head +Minor Losses (m) (9)	Total Losses (m) (10)	Hose HGL (m) (11)	Hose Elev. (m) (12)	Refer Datum (m) (13)
0			0.0	0.00		1.040				0.72		0.00
1	1.890	3.0	3.0	0.00	0.020	1.020	0.30	0.00	0.320	0.70	0.72	-0.50
2	1.796	6.0	9.0	0.00	0.037	0.982	0.30	0.00	0.337	0.67	0.68	-0.53
3	1.701	6.0	15.0	0.00	0.034	0.948	0.30	0.00	0.334	0.63	0.65	-0.57
4	1.607	6.0	21.0	0.00	0.031	0.917	0.30	0.00	0.330	0.60	0.62	-0.60
5	1.512	6.0	27.0	0.00	0.028	0.890	0.30	0.00	0.327	0.57	0.59	-0.63
6	1.418	6.0	33.0	0.00	0.025	0.865	0.30	0.00	0.324	0.55	0.57	-0.65
7	1.323	6.0	39.0	0.00	0.022	0.843	0.30	0.00	0.322	0.53	0.54	-0.67
8	1.229	6.0	45.0	0.00	0.019	0.824	0.30	0.00	0.319	0.51	0.52	-0.69
9	1.134	6.0	51.0	0.00	0.017	0.807	0.30	0.00	0.316	0.49	0.51	-0.71
10	1.040	6.0	57.0	0.00	0.014	0.792	0.30	0.00	0.314	0.48	0.49	-0.72
11	0.945	6.0	63.0	0.00	0.012	0.780	0.30	0.00	0.312	0.46	0.48	-0.74
12	0.851	6.0	69.0	0.00	0.010	0.770	0.30	0.00	0.310	0.45	0.47	-0.75
13	0.756	6.0	75.0	0.00	0.008	0.762	0.30	0.00	0.308	0.45	0.46	-0.75
14	0.662	6.0	81.0	0.00	0.007	0.755	0.30	0.00	0.306	0.44	0.46	-0.76
15	0.567	6.0	87.0	0.00	0.005	0.750	0.30	0.00	0.305	0.43	0.45	-0.77
16	0.473	6.0	93.0	0.00	0.004	0.747	0.30	0.00	0.303	0.43	0.45	-0.77
17	0.378	6.0	99.0	0.00	0.002	0.744	0.30	0.00	0.302	0.43	0.44	-0.77
18	0.284	6.0	105.0	0.00	0.001	0.743	0.30	0.00	0.301	0.43	0.44	-0.77
19	0.189	6.0	111.0	0.00	0.001	0.742	0.30	0.00	0.300	0.43	0.44	-0.77
20	0.095	6.0	117.0	0.00	0.000	0.742	0.30	0.00	0.300	0.43	0.44	-0.77

Total lateral friction loss = 0.298 m
Est. lateral friction loss = 0.330 m

Chapter 23: Solutions

1. What is the fecal-oral pathway of disease transmission?

Pathogens are spread through wastewater by the fecal-oral route. They multiply in the intestinal systems of humans and animals, and are then excreted. When pathogens from humans and livestock enter the water system, they are transmitted through water systems to drinking water or food unless wastewater treatment blocks their transmission.

The process of disease causing organisms entering into a water supply as human waste from which they are ingested by others through drinking is called fecal-oral transmission. Wastewater and water treatment systems are designed limit the possibility of fecal-oral transmission.

2. Pathogens come from four kingdoms of life: list and describe them.
   - Monera/bacteria (Prokaryote). All pathogens labeled as bacteria are part of the monera kingdom. The monera kingdom also includes organisms that are important components of wastewater treatment. They reduce or oxidize waste, decrease biological oxygen demand, convert ammonia to nitrate and ultimately nitrogen gas, and kill harmful pathogens in wastewater.
   - Protista (Eukaryote). All of the eukaryotes that don't fit in the animal, plant, or fungi kingdoms. This kingdom includes the protozoa, some of which consume bacteria in latter phases of wastewater treatment, and others that cause many waterborne diseases.
   - Fungi (Eukaryote). Responsible for primary decomposition of organic waste. Fungi grow roots into organic matter and suck out the nutrients. Some waterborne diseases are caused by fungi.
   - Animalia (Eukaryote) – helminth worms are carried in wastewater and cause disease.

3. List viruses that cause waterborne disease.
   - Poliovirus
   - Coxsackie virus
   - Echovirus
   - Enteroviruses
   - Hepatitis A
   - Hepatitis E
   - Caliciviruses
   - Rotaviruses
   - Reoviruses

4. List bacteria that cause waterborne disease.
   - Salmonella
   - Campylobacter
   - Shigella
   - E. coli O157:H7
   - Vibrio cholera

5. Describe the story of Typhoid Mary.

The most famous case (of Typhoid infection) was Typhoid Mary in New York: Mary Mallon was a cook for rich families. One of the families, which contracted typhoid fever, hired a private investigator to determine the cause of the typhoid fever in their family. The private investigator found that 22 people contracted typhoid fever at 7 different jobs where Mary worked from 1900 to 1906. Mary refused to believe that she was the source of typhoid fever. In fact, when officials asked to take urine and stool samples, Mary attacked them with a carving knife. Eventually, public health authorities confined her to Brother Island. She was released after 2 years with the understanding that she would give up cooking. However, an outbreak of typhoid fever (25 people) occurred 5 years later at a location where a Mrs. Brown was the cook. Mrs. Brown turned out to be Mary Mallon. New York authorities then confined her to Brother Island for the last 23 years of her life.

"North Brother Island is an island in the East River situated between the Bronx and Riker's Island. Its companion, South Brother Island, is a short distance away. The island was uninhabited until 1885, when Riverside Hospital moved there from the island now known as Roosevelt Island. Riverside Hospital was founded in the 1850s as the Smallpox Hospital to treat and isolate victims of that disease; its mission eventually expanded to other quarantinable diseases. Typhoid Mary was confined to the island for over two decades until she died there in 1938. The hospital closed shortly thereafter." ~ Wikipedia

6. List protozoan parasites that cause waterborne disease.

   Protozoans fed off of the Human Gut.

- Giardia lambia
- Cryptosporidium
- Cyclospora
- Microsporidia
- Toxoplasma gondii

7. Why are indicator organisms used and how do they distinguish between different sources of disease?

Rather than test for specific pathogens in wastewater, it is much cheaper to test for the presence of indicator organisms or surrogates that are produced in large quantities by humans or animals. It is assumed that if an indicator organism such as fecal coliform is absent, then other pathogens are absent. The ratio of different coliforms can indicate the source of water pollution (animal or human) because the ratio of fecal coliform to fecal streptococcus varies between animals and humans

8. List the factors that determine wastewater treatment plant discharge pathogen concentration.

The concentration of pathogens in the incoming wastewater stream is one factor that determines the effluent pathogen concentration. In general, developing countries have much higher pathogen counts in raw wastewater because there is a higher incidence of disease. Other factors that influence pathogen concentrations in wastewater are socio-economic status, per capita water use, and time of year.

9. Calculate the required contact time in order to remove 99.9 % of remaining *E-coli* at a

chlorine concentration of 3 mg/L.

Typically C T Values for E coli range between 0.001 and 0.01.

$$CT = 0.001$$

$$T = 0.001\ /\ 3\ \text{mg/L} = 0.00033\ \text{minutes}$$

10. What was the key factor that reduced typhoid fever in the United States.

The introduction of chlorine as a disinfectant in 1910 was the key factor that resulted in major declines in typhoid fever outbreaks.

Historically, waterborne diseases were common until the introduction of chlorine as a disinfectant for public water systems; the incidence of typhoid fever in the United States dropped from an average of 25 incidences per 100,000 people per year to approximately 400–500 cases per year (less than 0.2 incidences per 100,000 people). Thus, the per capita disease frequency dropped by 100 times due to the introduction of chlorine into public water supplies.

11. Why aren't ozone and chlorine used together?

Ozone oxidizes chlorine.

12. What environmental factors influence pathogen fate in the environment?

Environmental factors that decrease pathogen survival time are high temperature, low water content in soils, antagonistic microflora, and extreme pH (<3 or >9). Soil or organic matter that adsorbs organisms increases survival time but also decreases transport, which may decrease the hazard to the environment since organisms are not leached to transported to groundwater or surface water.

13. Give a brief summary (one paragraph) of how the body fights pathogens.

The human immune system is responsible for fighting off harmful pathogens. The key weapon for the human immune system is the ability to recognize self and non-self molecules. This allows the body to differentiate between good and bad cells (pathogens) in the body. One the bad cells are found, immune cells (primarily lymphoid and myeloid cells) create antibodies that attack the antigen (in this case considered the pathogen cells although an antigen can be bacteria, viruses or foreign cells). Once these antibodies attach to the antigen, the antigen is destroyed by the antibody complement system.

14. How do vaccines help the body fight disease?

The body builds up immunity to disease because some of the T cells and B cells used in the initial defense remain after the battle as memory cells. The next time that an individual encounters that same antigen, the immune system is primed to destroy it quickly. Long-term immunity can be stimulated not only by infection but also by vaccines made from infectious agents that have been inactivated or, more commonly, from minute portions of the microbe.

15. Why must acid be injected at the same time as chlorine during disinfection?

"Chlorine exists in two forms within water, hypochlorous acid (HOCl) and hypochlorite (OCl-)."

The hypochlorous acid and hypochlorite, OCl-, equilibrium in water is $HOCl \Leftrightarrow H^+ + OCl^-$.

Hypochlorous acid (free chlorine) is 80 times more effective at killing bacteria than hypochlorite because the hypochlorite has a charge and is repelled from the cell. Hypochlorous acid is the primary form of chlorine in acidic waters, pH <6.5. Thus, if the initial pH is high, acid must often be added in concert with chlorine in order to drop the pH to 6.5.

16. Redo the risk assessment problem. The expected number of salmonella bacteria in wastewater is 10^5 MPN/100 ml, and 200 children are expected to play for 16 hours in the landscaped area.

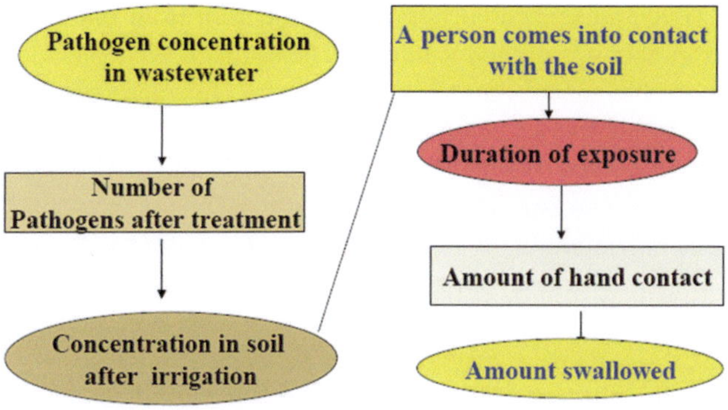

100,000 MPN/100 ml water * 15 ml water/100 ml soil = 15,000 MPN/100 ml of soil.

If children ingest 500 mg of soil every 8 hours, then they will ingest 1000 mg of soil in 16 hours.

Assume that the soil bulk density is 1.15 mg/ml.

1,000 mg soil ingested/1.15 mg/ml soil = 870 ml soil

870 ml soil * 15,000 MPN/100 ml soil = $1.31 \times 10^5 \approx$ 1,300,000 Salmonella bacteria

This number is in the range of 70–80 % infection. Thus, approximately 150 children will become sick.

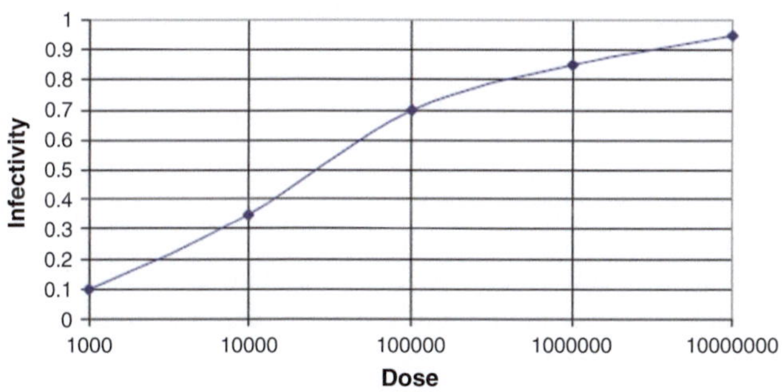

17. What are the 3 steps of Risk Analysis?

The process of Risk analysis involves three steps: Risk Assessment – determining the probability that an adverse event will occur and its magnitude, Risk Management – considers various regulatory options to minimize the risk, and Risk Communication – transfer of risk information to experts and non-experts.

18. What are the 4 steps of Risk Assessment?

Risk Assessment has four basic steps:

1. Hazard Identification – identifying the contaminant (i.e. Salmonella)
2. Dose–response Assessment – relationship between the number of organisms ingested and the probability of becoming infected (i.e. how many does it take to make you sick)
3. Exposure Assessment – Determining the concentration of a pathogen in the water and estimating amount of contact and possibly ingestion.
4. Risk Characterization – Estimating the potential impact (infection, disease) of a pathogen based on the severity of its effects.

19. What is BOD and why is high BOD detrimental when wastewater is discharged to streams?

Biochemical Oxygen Demand (BOD) is defined as the amount of oxygen required for the bacterial decomposition (oxidation) of organic matter under aerobic conditions at a standard temperature and incubation time.

If wastewater with a heavy organic load is added to surface waters, the dissolved oxygen concentration can decrease from the normal 5 to 7 mg/L to 2 mg/L or less, a point at which fish die.

20. Describe the difference between oxidation and synthesis and endogenous respiration.

Oxidation Synthesis

$$\begin{aligned} &COHNS + O_2 + \text{Nutrients} \rightarrow CO_2 + NH_3 \\ &\quad + C_5H_7NO_2 + \text{Other End Products} \end{aligned} \qquad (2--1)$$

Endogenous Respiration

$$C_5H_7NO_2 + 5O_2 \rightarrow 5CO_2 + 2H_2O + NH_3 + \text{Energy} \qquad (2--2)$$

Organic matter (CHONS) is oxidized and the resulting bacterial cells have the formula: C5H7NO2. Dead bacterial cells are then consumed by protozoa in a process called endogenous respiration. The result of endogenous respiration is stable, nontoxic, byproducts.

21. Calculate the BOD5 if the dilution factor is 20–1, the initial dissolved oxygen concentration is 6 mg/L, and the final dissolved oxygen concentration is 2 mg/L.

$$BOD = \left(DO_i - DO_f\right)\frac{V_b}{V_s} = \Delta DO(DF) \qquad (2--3)$$

where

DO_i = initial dissolved oxygen concentration, mg/L,
DO_f = final dissolved oxygen concentration, mg/L,
DF = dilution factor : volume of the bottle divided by volume of the sample,
V_b = volume of the bottle, ml,
V_s = volume of sample added to the bottle, ml,

(Pg. 21)

$$BOD_5 = (DO_i - DO_f)(DF) = (6 - 2)(20/1) = \mathbf{80 mg/L}$$

22. List the steps in the nitrogen cycle in soils.

Bacteria sequentially transform nitrogen from one form to another in soils and wetlands. The nitrogen cycle includes anaerobic (no oxygen) bacteria, aerobic (oxygen) bacteria and facultative (anaerobic or aerobic) bacteria: ammonification, followed by nitrification, and denitrification. The steps from conversion of organic N to NO3 – are called mineralization.

organic matter → amines	→	ammonium → nitrate	→	nitrate	→	nitrous oxide or nitrogen gas
	anaerobic		aerobic		anaerobic or aerobic	
	ammonification		nitrification		denitrification	

23. How is the nitrogen cycle carried to completion in wetlands with herbaceous aquatic plants?

Herbaceous aquatic plants that pump oxygen into the water facilitate nitrification and denitrification. Aerobic nitrification takes place near cattail and bulrush roots that add oxygen to the water and provide an aerobic zone for aerobic bacteria. Anaerobic denitrification takes place in water that is not close to plant roots and in the deep zones of wetlands that do not include plant material.

Chapter 24: Solutions

1. Explain the meaning of Eq. 24.1 in a sentence.

The rate of change of concentration is directly proportional to the concentration.

2. Integrate Eq. 4.1 and derive Eq. 4.2.

$$\frac{dC}{dt} = -kC$$

$$\frac{dC}{C} = -kdt$$

$$\int_{C_1}^{C_2} \frac{dC}{dt} = \int_0^t -kC$$

$$\ln(C_2) - \ln(C_1) = -kt \quad \ln\left(\frac{C_2}{C_1}\right) = -kt$$

$$\frac{C_2}{C_1} = e^{-kt} \quad C_2 = C_1 e^{-kt}$$

3. Explain the similarity between Eqs. 24.2 and 24.3, and explain the justification for using Eq. 24.2 for calculating effluent concentration from a wetland (Eq. 24.3).

Free water surface and subsurface flow wetlands can be modeled as plug flow systems. Thus, the outlet concentration can be modeled with the same equation as is used for a batch reactor with time equal to the length of time that water remains in the wetlands. It is like a batch reactor is moving through the wetlands.

4. Show how Eq. 24.5 is derived from Eq. 24.4.

$$T = V/Q$$

$$V = A^*d$$

$$T = A^*d/Q$$

5. Explain the meaning of Eq. 24.6.

The equation calculates the microbial activity with a baseline of 20 °C and microbial activity varies exponentially with temperature from the baseline.

6. Repeat the area based calculations of Example 24.1 but use 25 °C and 10 °C instead of 20 °C and 4 °C

Effluent is discharged from septic tanks into a subsurface flow system wetland. Assume that total nitrogen in septic tank effluent is 36 mg/L and that water temperature in a subsurface flow system wetland is 25 °C; calculate the total nitrogen, TN, in wetlands effluent if the wetlands hydraulic loading rate is 25,000 m^3/yr and the wetlands surface area is 1,071 m^2. Recalculate for 4 °C. Assume that the void fraction in the gravel is 0.33 and the depth of flow is 0.4 m. The maximum acceptable total nitrogen discharge concentration is 10 mg/L. Calculate a time-based coefficient based on the wetland volume and the area-based rate coefficient. Then, determine whether the wetland has an adequate size with a 5 day detention time at 10 °C.

Calculation of effluent concentration at 25 °C

The irreducible background concentration for total nitrogen is not temperature dependant and is a constant, 1.5 mg/L. The area-based rate constant, k_{20}, is 27 m/yr (Table 4.1).

$$k = k_{20}\ \theta_k^{(T-20)} = 27{*}1.05^{(25-20)} = 34\ m/yr$$

$$C_{out} = C^* + e^{\left(\frac{-kA}{Q}\right)}\left(C_{in} - C^*\right) = 1.5 + e^{\left(\frac{-34*1,070}{25,000}\right)}(36 - 1.5)$$
$$= 9.5\ mg/L\ TN$$

The wetland is adequate at this because the treatment goal of 10 mg/L of total nitrogen is met.

Calculation of effluent concentration at 10 °C

$$k = k_{20}\theta_k^{(T-20)} = 27{*}1.05^{(10-20)} = 16.6\ m/yr$$

$$C_{out} = C^* + e^{\left(\frac{-kA}{Q}\right)}\left(C_{in} - C^*\right) = 1.5 + e^{\left(\frac{-16.6*1,070}{25,000}\right)}(36 - 1.5)$$
$$= 18.4\ mg/L\ TN$$

The wetland is even more inadequate at 10 °C.

Calculate an equivalent time-based rate constant (e^{-kT}).

Volume of water in wetland

$$V_{cv} = d{*}A{*}q = 0.4\ m{*}1,071\ m^2{*}0.33 = 141\ m^3$$

Hydraulic detention time

$$T = V/Q = 141\ m^3/25,011\ m^3/yr(365\ day/yr) = 2\ days$$

Calculate time based k based on hydraulic detention time

$$\frac{-k_{area}A}{Q} = -k_{time}T = \frac{-k_{time}V_{cv}}{Q}$$

$$k_{time} = \frac{k_{area}A}{V_{cv}} = \frac{k_{area}}{d\theta} = \frac{(27\ m/yr)}{0.4{*}0.33}\left(\frac{1}{365\ day/yr}\right)$$
$$= 0.56\ d^{-1}$$

Calculate effluent TN with 5-day detention time and 25 °C ($A = 1,071 * 5/2 = 2,678\ m^2$)

$$k = k_{20}\theta_k^{(T-20)} = 0.56{*}1.05^{(25-20)} = 0.91/day$$

$$C_{out} = C^* + e^{(-kT)}\left(C_{in} - C^*\right) = 1.5 + e^{(-5*0.91)}(36 - 1.5)$$
$$= 1.9\ mg/L\ TN$$

Calculate effluent TN with 5-day detention time and 10 °C ($A = 2,678\ m^2$)

$$k = k_{20}\theta_k^{(T-20)} = 0.56{*}1.05^{(10-20)} = 0.34/day$$

$$C_{out} = C^* + e^{(-kT)}\left(C_{in} - C^*\right) = 1.5 + e^{(-5*0.34)}(36 - 1.5)$$
$$= 8\ mg/L\ TN$$

7. Treated wastewater has a BOD_5 of 100 mg/L, and flow rate is 200 L/min. What is the required area of a FWS wetland? Calculate based BOD load and hydraulic load. Design for an effluent BOD of 12 mg/L.

Calculation of effluent concentration at 20 °C

There is no change with temperature. k is 35 and C^* is 6. Convert flow rate to m^3/yr.

$$200\ L/min = 0.001\ m^3/L{*}60min/hr{*}24\ hr/day{*}$$
$$365\ days/year = 105,000\ m^3/yr.$$

The irreducible background concentration for total nitrogen is not temperature dependant and is a constant, 1.5 mg/L. The area-based rate constant, k_{20}, is 27 m/yr (Table 4.1).

$$C_{out} = C^* + e^{\left(\frac{-kA}{Q}\right)}\left(C_{in} - C^*\right) = 6 + e^{\left(\frac{-36*A}{105,000}\right)}(100 - 6)$$
$$= 12\ mg/L\quad TN$$

With respect to BOD, the wetlands would be adequate with a surface area of 8,000 m^2.

The optimal hydraulic loading rate is 200 m^3/ha/day.

At 105,000 m^3/yr, the hydraulic loading rate is

$$105,000\ m^3/yr/(0.8\ ha)/365 = 360\ m^3/day$$

This is within the recommended hydraulic loading rate of 150–500 m^3/day.

8. Why is there a minimum acceptable BOD_5 loading rate for wetlands?

   BOD (organic matter) provides carbon for denitrification.

9. Calculate the size of a wetland required to treat an animal waste effluent stream. The desired TN concentration on the discharge side of the wetland is 30 mg/L, the waste flow rate is 50 m^3/day, and the TN concentration in the dairy waste effluent is 200 mg/L.

The daily load of nitrogen is 200 mg/L * 1000 L/m^3 * 50 m^3/day * 10^{-6} kg/mg = 10 kg/day.

The Loading Rate is calculated as follows for an effluent concentration of 30 mg/L.

$$LR(kg/ha/day) = 0.68(C_{out}) - 7.88 = (0.68){*}30 - 7.88$$
$$= 12.5\ kg/ha/day.$$

$$A = TN/LR = 10/12.5 = 0.8\ ha.$$

10. A storage pond is expected to receive 10,000 m^3 of manure, clean water, and wastewater during a 6 month storage period (winter) (50 m^3/day * 200 days). The pond is only pumped once every 2 years and is expected to have settling of 600 m^3 of solids during the 2 year

interval. The normal depth of evaporation during the winter is 0.5 m, and the expected precipitation during the winter is 0.25 m. The depth of the 25-year 24-hour storm is 10 cm. The dimensions of the base of the pond are 20 m × 20 m. The side slope of the pond is 2.5: 1. (2.5 run × 1 rise). Calculate the depth of the pond.

$$A_{base} = \text{Area of base} = 20*20 = 400\ m^2$$
$$A_{top} = \text{Area of top} = (20 + 5*D)*(20 + 5*D)$$
$$V_{Pond} = \text{Volume pond} = (A_{base} + A_{top})/2*D$$
$$V_{Freeboard} = (A_{top} + ((20 + 5*(D - 1))*(20 + 5*(D - 1)))/2*0.3\ m$$
$$V_{PE} = \text{Volume precipitation} - \text{evaporation} = (\text{Area of top})*(0.25 - 0.5)$$
$$V_{WW} = 10{,}000\ m^3$$
$$V_{Solids} = 600\ m^3$$
$$V_{Storm} = 0.1\ m*A_{top}$$
$$V_{Pond} = V_{Solids} + V_{WW} + V_{PE} + V_{Freeboard} + V_{Storm}$$

A depth of 6.9 m results in the volume of the pond equal to the required volume.

Total slope	5	run over rise
D	6.9	m
Base W	20	m
Base L	20	m
Abase	400	m^2
Atop	2970.25	m^2
Vpond	11627.36	m^3
Vfreeboard	813.075	m^3
P – ET	−0.25	m
Vpe	0	m^3
Vww	10000	m^3
Vsolids	600	m^3
V required	11413.08	m^3

11. A dairy farm has 500 lactating dairy cows in Central Arizona (Maricopa County). Average mass/cow is 500 kg. Cows are kept in open lots, and 95 % of manure is dried. Five percent of manure from the milking parlor is washed into an anaerobic waste storage pond with 6 L water/day/cow. The maximum pond operating depth is 4 m. Side slopes are 2.5:1 (2 horizontal by 1 vertical). The 25 year – 24 hour storm is 7 cm. Annual precipitation is 15 cm, and mean annual evaporation is 200 cm. Soils on the sites for waste application are well drained sandy loams and have a leaching index of 6 (6 inches (15 cm) percolates below the root zone). The organic matter content is <2 %. The soils are flood irrigated. Soil slopes are close to dead level. Crop is cotton. Required annual nitrogen addition is 150 kg/ha nitrogen (N). Assume that no extra phosphorous or potassium are required, and that there is no environmental hazard associated with overapplication of phosphorous or potassium in these soils and in this hydrologic setting with no adjacent surface water bodies. Manure is applied by truck in spring just before planting when soil is warm and dry. Assume that the storage period is 365 days in order to account for the average evaporation rate during the year. Manure has been applied for several years every spring and is incorporated into the soil by plowing within 1 day of application. Calculate application rates for truck application of dried manure and for sprinkler application of liquid manure from the waste storage pond. Assume that 150 m^3/yr remain on the bottom of the pond each year.

Partition of liquid and solid manure and N load:
Step 1. Manure load and washwater volume/day

$$\text{Total manure mass/day is } 80\ \text{kg/day}/1{,}000\ \text{kg}*500\ \text{cows}*500\ \text{kg/cow} = 20{,}000\ \text{kg manure/day}$$
$$\text{Total manure volume/day is } 81\ \text{L/day}/1{,}000\ \text{kg}*500*500 = 20{,}000\ \text{L manure/day}$$

$$\text{Total volume of manure in the milking parlor is } 20{,}000\ \text{L manure}*0.05 = 1{,}000\ \text{L/day}$$

Assume that 3,000 L is used to wash the milking parlor at 6 L/day/cow.

Thus, a total of 4,000 L/day (1,500 m^3/year) is washed into the waste storage pond.

An Excel spreadsheet is set up to calculate the dimensions of the pond. The required dimensions are a base width of 6 m, base length of 10 m, and depth of 3.5 m. At this depth, the evaporation volume is 700 m^3/year. Because 1,500 L is added per year in washwater, and the volume of manure is 500 m^3/year, the dilution volume is 1,500–500 = 1,000 m^3/year.

Total slope	5	run over rise
D	3.5	m
Base W	6	m
Base L	10	m
Depth ET	200	cm
Depth precipitation	15	cm
Vww	1500	m^3
Vsolids	150	m^3
Abase	60	m^2
Atop	646.25	m^2
Vfreeboard	159	m^3
Vevaporation	706	m^3
Vprecipitation	96.9375	m^3
Vpond	1235.938	m^3
V required	1200	m^3

The dilution ratio is volume of water over volume of manure (1,000/500) = 2. Thus, there is greater than 50 % dilution for nitrogen reduction calculation.

Estimate the total nitrogen (N) in the excreted manure. Phosphorous and potassium are not calculated since the application rate is only calculated based on the nitrogen content.

Nutrients per storage period = Number of animals × mass (kg) × daily nutrient production (kg/day/1,000 kg) x storage period (days).

Yearly nutrient values for as excreted dairy cow manure (use Table 2.6 for kg/d/1,000 kg)

$$N = \frac{500{*}500{*}0.45{*}365}{1,000} = 41,050\ kg$$

Of this mass, 95 % is dried, 41,050–2,050 = 39,000 kg is dried, and 2,050 kg of N is added to the waste storage pond.

Step 3. Subtract nutrients lost during storage.

Dried manure retains 75 % of nitrogen. Thus, 9,750 kg remains.

Manure in a storage pond with greater than 50 % dilution retains 30 % nitrogen. Thus, 600 kg remains.

Step 4. Determine the plant available nutrients by mineralization in the soil.

The N mineralization rate after 3 years for waste stored in an open lot in a hot arid region is 53 %. Thus, 5,000 kg N is available as plant nutrients.

The N mineralization rate after 3 years for waste stored in a pond with greater than 50 % dilution is 49 %. Thus, 300 kg is available as plant nutrients

Step 5. Compute the plant nitrogen requirement.

$$\mathrm{N}_{\text{-plant}} = 150\ \mathrm{kg/ha}$$

Adjust the plant N requirement based on denitrification, leaching, and volatilization.

Add denitrification losses to the plant nitrogen requirement.

From Table 24.8, a well drained soil with an organic matter content of 0 % has an annual denitrification rate of 3–9 %. Because arid soils with low organic matter have very low denitrification rate, estimate the lowpoint of this range, 3 %. The values in Table 24.8 must be doubled for manure application so estimate the denitrification rate as 6 %. The plant nitrogen requirement is 150 kg/ha so the field nitrogen required is

$$\mathrm{N}_{\text{-required-denitrification}} = 150\ \mathrm{kg/ha}/0.94 = 160\ \mathrm{kg/ha}$$

Add leaching losses to the plant nitrogen requirement

The plant nitrogen requirement must be increased to replace anticipated leaching losses. As stated above, a leaching index of 6 (6 inches of annual percolation below the root zone), results in an annual nitrate loss of 10 percent.

$$\mathrm{N}_{\text{-required-denitrification and leaching}} = 160\ \mathrm{kg/ha}/0.9 = 177\ \mathrm{kg/ha}$$

Add application (volatilization) losses to plant nitrogen requirement.

Dry manure will be incorporated so there is no volatilization of ammonia.

The answer to question 1 is that the required application rate is

$$\mathrm{N}_{\text{-plant}} = 177\ \mathrm{kg/ha}$$

Step 6. Compute the area on which dry manure can be applied.

$$\text{Area} = (5,000\ \text{kg/year N})/(177\ \text{kg/ha N}) = 28\ \text{ha}$$

Step 6. Solution for application of storage pond water by irrigation:

For application of waste by sprinklers, the nitrogen requirement should be recalculated based on surface volatilization expected from sprinkler application (Table 24.7).

$$\mathrm{N}_{\text{-required-denitrification, leaching, and volatilization}} = 177\ \mathrm{kg/ha}/0.75 = 236\ \mathrm{kg/ha}$$

Area for full nitrogen application

$$= (300\ \text{kg/year N})/(236\ \text{kg/ha N}) = 1\ \text{ha} = 2\ \text{acres}$$

It isn't worth setting up a sprinkler system for one ha so just dilute the manure in surface irrigation canal and apply to a large field. If the field is 10 ha, then the application rate would be 30 kg/ha. The total volume to be applied is 1,500–700 = 700 m^3.

For a 10 ha area, this would be a depth of 700/100,000 = 0.007 m = 0.7 cm depth. If the total irrigation is 4 cm depth, then approximately 0.7/4 = 17 % would come from wastewater and 83 % would come from the normal surface irrigation source.

12. Calculate the dry manure salinity application rate (kg/ha) for problem 11.

20,000 kg manure per day
20 tons fresh manure per day
7,200 + 100 = 7,300 tons fresh manure per year
95 % of manure is dried 7300 * 0.95 = 6935 tons fresh manure per year
Applied to 28 ha
6935/28 = 247 tons fresh manure per ha
EC = 18.8 dS/m
18.8(640) = 11,500 mg/L = 1.15 % salt in dried manure
0.0155 (250) = 3.87 tons /ha = 3,870 kg/ha

13. Calculate the salinity application rate if 20 t/ha poultry manure (dry weight basis) is applied to a field.

There is 25 % solids in chicken manure.
23.7 ds/m(640 mg/L/ds/m) = 15168 mg/L = 1.517 % by weight salts in fresh manure
.01517/.25 kg dry manure/ kg wet manure = 0.06067 kg salts/kg dry manure
App rate = (20 ton manure/ha) (1000 kg/ton) (0.06067 kg salt/kg d. manure) = 1213 kg/ha
% salts = 2(3.9 + .91 + 9.6 + .72) = 30 % salts in dry manure = .30 kg salts/kg dry manure
.30(.25) = .075 kg salts/kg manure
App rate = (20 ton/ha) (1000 kg/ton) (.075 kg/kg) = 1,500 kg/ha

14. Calculate the required blend of groundwater/wastewater to provide 200 kg/ha nitrogen to the field. The municipal wastewater concentration after secondary treatment is 15 mg/L N with 75 % ammonia in an arid region. The crop requires 1.2 m depth of water. Groundwater has 6 mg/L nitrate.

Assume all ammonia in wastewater is converted to nitrate within 1 day.

$$1.2\ \text{m depth}*10{,}000\ \text{m}^2/\text{ha} = 12{,}000\ \text{m}^3/\text{ha}$$

$$200\ \text{kg/ha}/12000\ \text{m}^3/\text{ha} = 17\ \text{g/m}^3 = 17\ \text{mg/L}$$

$$C_T = C_{WW}F_{WW} + C_{GW}(1 - F_{WW})$$

$$17 = 15\ F_{WW} + 6(1 - F_{WW})$$

$$F_{WW} = 11/9 = 1.2.$$

Actually, supplemental nitrogen from fertilizer will be needed since there is not enough nitrogen in the wastewater. This is why the fraction is greater than 1.0.

Chapter 25: Solutions

1. An application of 100 kg/ha of NO_3-N is incorporated to a depth of 0.15 m. The 100 refers to only the N portion of nitrate. All of the fertilizer is dissolved on the day that it is incorporated. Calculate the change in the concentration of fertilizer in the upper 0.15 m per soil volume and in soil water. Water content is 0.3 L/L.

$$\Gamma_{fer} = K_{fer}\frac{A_{fer}}{10D_{fer}} = \frac{1.0}{10}\left(\frac{100\ kg/ha}{0.15\ m}\right) = 67\ mg/L$$

If water content is 30 %, then concentration of nitrate in water in the upper 0.15 m of soil is 67/0.3 = 222 mg/L.

2. Annual Tucson temperature data at 10 and 50 cm depth is available in the Chapter 25 WINDS salinity and nitrogen workbook. Develop an annual sin wave Eq. 25.16 based on this data, and plot temperatures at 5, 10, 50, and 100 cm.

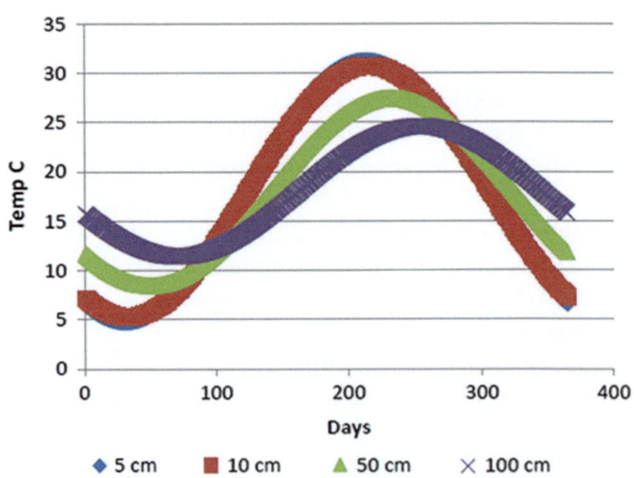

3. Hourly data is available in the *Chapter 25 WINDS salinity and nitrogen* workbook for 3 days in January and 2 days in June. Develop a diurnal sin wave Eq. 25.16 based on this data, and plot temperatures at the surface and at 5, 10, 50, and 100 cm for January and June. Compare the lag times calculated with the diffusivity to the lag times observed in the figures. Compare the equations that were derived based on data for the two different seasons.

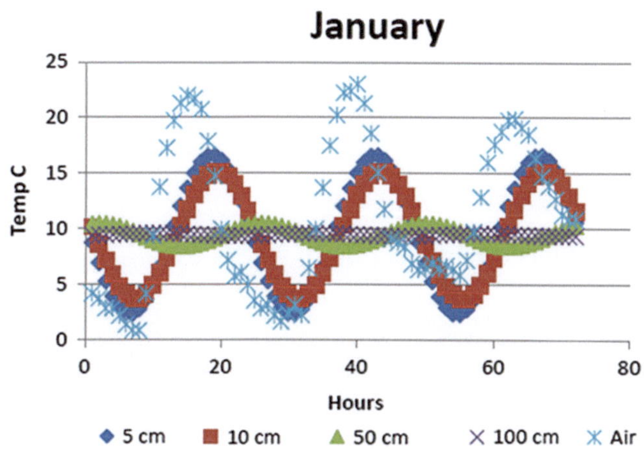

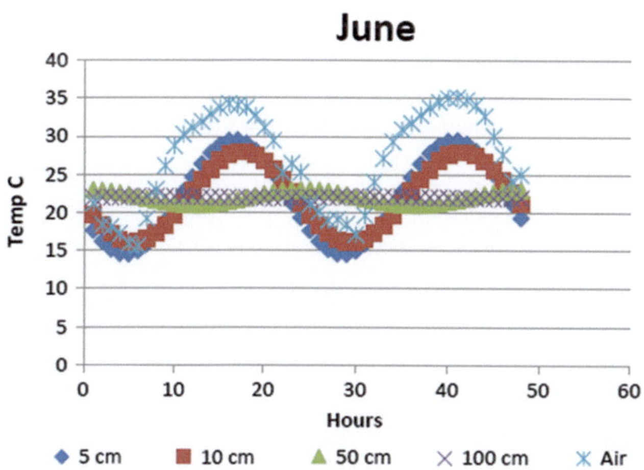

In January, the lag time appears to be about 4 hours based on the graphs. This is quite a bit longer than the predicted value of 0.21 hours. In June, however, the graph's lag time appears to be much shorter (about 1 hour). This is closer to the predicted lag time of 0.22 hours.

4. Find the mineralization over a 2 day period in a sandy loam soil with θ_{pwp} θ_{low}, θ_{high}, and θ_{sat} are equal to 0.1, 0.17, 0.26, and 0.42, respectively. Find the final concentration of nitrate in soil water if the initial concentrations are 15, 8, and 3 mg/L in layers 1, 2, and 3 respectively assuming no water content changes and no other sources or sinks of nitrate. Surface organic matter content is 2,000 μg/g, $K_{mnl} = 0.00005\ d^{-1}$, $\alpha = 0.021$. The average water contents and temperatures in the top three layers are

Water content	Cell elevations		Temperature
Layer 1	0.16	0–40 cm,	27 °C
Layer 2	0.28	40–80 cm,	22 °C
Layer 3	0.25	80–120 cm,	20 °C

Surface organic matter content is 2,000 μg/g, $K_{mnl} = 0.00005\ d^{-1}$, $\alpha = 0.021$.

Find the water content adjustment factors.

Water content is between θ_{pwp} and θ_{low} in the 0–40 cm depth increment:

$$f_{mnl\theta} = \left(\frac{\theta - \theta_{pwp}}{\theta_{low} - \theta_{pwp}}\right) = \left(\frac{0.16 - 0.10}{0.17 - 0.1}\right) = 0.86$$

Water content is between θ_{high} and θ_{sat} in the 40–80 cm increment

$$f_{mnl\theta} = 0.6 + 0.4\left(\frac{\theta_{sat} - \theta}{\theta_{sat} - \theta_{high}}\right)$$
$$= 0.6 + 0.4\left(\frac{0.42 - 0.28}{0.42 - 0.26}\right) = 0.95$$

Water content is between θ_{low} and θ_{high} 80–120 cm increment

$$f_{mnl\theta} = 1.0$$

Find the temperature adjustment factors

The temperature adjustment factor in the upper layer, with average temperature equal to 27 °C, is

$$f_{temp} = Q_{10}^{\left(\frac{T-t_b}{10}\right)} = 3^{\left(\frac{27-20}{10}\right)} = 2.16$$

The temperature adjustment factor in the 40–80 cm increment, with average temperature equal to 22 °C, is

$$f_{temp} = Q_{10}^{\left(\frac{T-t_b}{10}\right)} = 3^{\left(\frac{22-20}{10}\right)} = 1.25$$

The temperature adjustment factor in the 80–120 cm increment, at 20 °C, is 1.0

Find the average organic matter concentration in each cell.

Layer 1

$$O_n = O_{n_{max}} e^{-0.021z} = 2{,}000\ e^{-0.021\ *20} = 1{,}314\ \mu g/g$$

Layer 2

$$O_n = O_{n_{max}} e^{-0.021z} = 2{,}000\ e^{-0.021\ *60} = 567\ \mu g/g$$

Layer 3

$$O_n = O_{n_{max}} e^{-0.021z} = 2{,}000\ e^{-0.021\ *100} = 245\ \mu g/g$$

Find the mineralization rate in each of the cells and the change in concentration in cell water

The saturated water content is 0.42. Thus, the porosity is approximately 0.42. Soil bulk density is.

$$\rho = \rho_p(1 - \phi) = 2.65(1 - 0.42) = 1.54.$$

Mineralization rate and change in nitrate concentration in water in layer 1 are calculated below. The change in nitrate concentration in the soil solution is the change in nitrate concentration in the soil volume divided by the water content.

$$\Gamma_{mnl} = K_{mnl} f_{mnl\theta} f_{temp} \rho O_n = 0.00005{*}0.86{*}2.16{*}1.54{*}1,314 = 0.188 \text{ mg/L soil/d}$$
$$dN_1 = \Gamma_{mnl}/\theta = 0.188/0.16 = 1.18 \text{ mg/L water/day}$$

Total mineralization over 2 days in the upper layer is 2.35 mg/L water

Layer 2

$$\Gamma_{mnl} = K_{mnl} f_{mnl\theta} f_{temp} \rho O_n = 0.00005{*}0.95{*}1.25{*}1.54{*}567 = 0.052 \text{ mg/L soil/d}$$
$$dN_1 = \Gamma_{mnl}/\theta = 0.052/0.28 = 0.19 \text{ mg/L water/day}$$

Total mineralization over 2 days is 0.38 mg/L water.

Layer 3

$$\Gamma_{mnl} = K_{mnl} f_{mnl\theta} f_{temp} \rho O_n = 0.00005{*}1.0{*}1.0{*}1.54{*}245 = 0.019 \text{ mg/L soil/d}$$
$$dN_1 = \Gamma_{mnl}/\theta = 0.052/0.28 = 0.038 \text{ mg/L/day}$$

Total mineralization over 2 days is 0.076 mg/L water

Change in nitrate concentration in soil water.

	$\Gamma * 2$	$N_{initial}$ (mg/L_{water}/day)	Final concentration (mg/L water)
Layer 1	2.35	15	17.35
Layer 2	0.38	8	8.38
Layer 3	0.038	3	3.038

5. Calculate the denitrification rate in the sandy loam soil described in problem 14 at 60 cm depth. Assume that initial nitrate concentration in soil water is 8 mg/L, and that the denitrification rate constant is 0.002 d^{-1}. Let the depth adjustment factor equal 0.021. Calculate the change in concentration within the soil volume and change in concentration within the soil water. Consider both the denitrification and mineralization to calculate the final concentration in layer 2 after 2 days.

$$\theta_{den} = 0.6{*}\theta_{sat} = 0.6{*}0.42 = 0.25$$

Calculate the water content adjustment factor.

$$f_{den\theta} = \left(\frac{\theta - \theta_{den}}{\theta_{sat} - \theta_{den}}\right)^2 = \left(\frac{0.28 - 0.25}{0.42 - 0.25}\right)^2 = 0.031$$

Calculate the temperature adjustment factor:

$$f_{temp} = Q_{10}^{\left(\frac{T - t_b}{10}\right)} = 3^{\left(\frac{22 - 20}{10}\right)} = 1.25$$

Calculate depth adjustment factor, f_z.

$$f_z = e^{-0.021{*}60} = 0.28$$

Calculate net loss of nitrate due to denitrification in layer 2.

$$\Gamma_{den} = K_{den} f_{den\theta} f_{temp} f_z \theta N$$
$$= 0.002{*}0.031{*}1.25{*}0.28{*}0.28{*}8$$
$$= 4.9{*}10^{-6} \; mg/L_{soil}/day$$

Calculate change in concentration within soil water

$$dN = \text{mg}/L_{soil}/\text{day}/\theta = 4.9{*}10^{-6}/0.28$$
$$= 0.000017 \text{ mg/L/day}$$

The amount of denitrification is trivial compared to the amount of mineralization so the final concentration is still 8.38 mg/L water

6. Determine the seasonal removal of nitrate from the root zone (mg/L_{soil}) for a crop that has a yield of 8,000 kg/ha and has a nitrogen percentage of 1.6 %. Root zone depth is 1.5 m.

$$\Gamma_{upt} = \frac{Y \ \ fN}{10 \ \ dz} = \frac{8,000*0.016}{10*1.5}$$
$$= 8.53 \ mg/L_{soil}/season$$

7. N_{req} = 1.4 kg/ha, N_{min} = 2 mg/kg soil, and N_{max} = 1.8 kg/ha. The fraction of nitrogen taken up as nitrate is 1.0. The optimal level of nitrate in the soil is 30 mg/kg. Find Km, plot the uptake rate vs. soil nitrogen concentration, and calculate the uptake of nitrate at a soil nitrate concentration of 20 mg/kg.

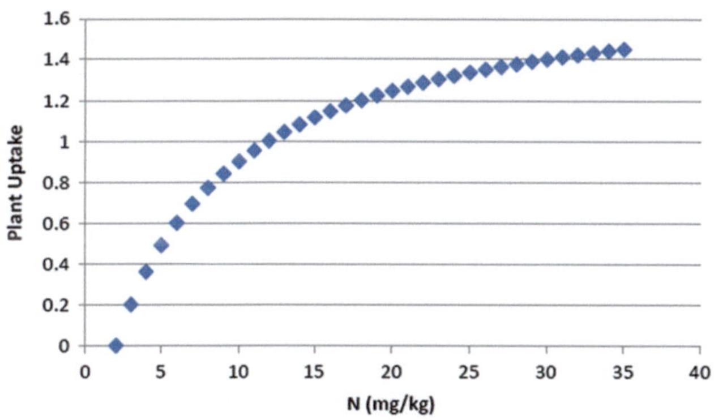

Plant uptake at 20 mg/kg = 1.25 kg/ha (from the figure)

Chapter 26: Solutions

1. **Repeat Example 26.1, but change the infiltration from the first storm to 3 cm, and the field capacity to 0.25. As before, infiltration from the second storm is 4 cm.**

$$d_{initial} = \theta \ \Delta z = 0.17*1.2 = 0.204 \ \ m$$

Calculate the maximum depth of water that the soil profile can hold

$$d_{cv-max} = \theta_{fc}\Delta z = 0.25*1.2 = 0.3 \ \ m$$

Find final depth of water in cell after the 3 cm storm.

$$d_{final} = d_{initial} + i = 0.204 + 0.03 = 0.234 \ \ m$$

d_{final} does not exceed $d_{cv\text{-}max}$, thus, there is no deep seepage and d_{final} = 0.234 m. Water content after the first storm is $\theta_{cv} = d_{cv}/dz = 0.234/1.2 = 0.195$

Now, repeat the calculation for the second storm. Initial depth of water in the soil (before the second storm) is

$$d_{initial} = 1.2 \ \ m*0.195 = 0.234 \ \ m$$

Depth of water added by second storm = 4 cm = 0.04 m.

$$d_{initial} + \Delta d_{cv} = 0.234 + 0.04 = 0.274 \ \ m$$

Thus, there is no deep seepage since the final water content is less than field capacity

Final water content after the second storm

$$\theta_{cv} = d_{cv}/dz = 0.274/1.2 = 0.228$$

2. **Redo question 1, but divide the soil into three layers of 0.4 m depth.**

For all cells, the initial depth of water is

$$d_{cv} = \theta \ \Delta z = 0.17*0.4 = 0.068 \ \ m$$

The maximum water depth held by each cell is

$$d_{cv-max} = \theta_{fc} \ \ \Delta z = 0.25*0.4 = 0.1 \ \ m$$

Infiltrated depth during the first storm is 3 cm (0.03 m).

For the first cell

$$d_{final-1} = d_{initial-1} + i_1 = 0.068 + 0.03 = 0.098 \ \ m$$

$$\theta_1 = d_{final-1}/dz = 0.098/0.4 = 0.245$$

Because the final depth does not exceed the maximum depth, no water is leached below the first cell. Water content in cells 2 and 3 remains the same.

The second storm infiltration is 4 cm (0.04 m). The first cell can hold 0.002 m, so 0.038 m drains to the second cell, and water content in the first cell is 0.25. The second cell can hold 0.032 m so the final water depth in cell 2 is 0.1 m, and water content is 0.25. Finally, the third cell receives 0.06 m so the final water depth in cell 3 is 0.074 m.

The final water content in cell 3 is

$$\theta_3 = d_{final-3}/dz = 0.074/0.4 = 0.185$$

3. **Redo Example 26.3, but change the upper layer FC to 0.26, and the lower layer FC to 0.24. Change the percent of ET removed from the upper layer to 70 % and the percent of ET removed from the lower layer to 30 %.**

Upper layer : $\theta_{FC} = 0.26, \theta_{PWP} = 0.1,$ cell thickness, dz $= 60$ cm, TAW $= (0.26 - 0.1)*60 = 9.6$ cm

Lower layer : $\theta_{FC} = 0.24, \theta_{PWP} = 0.11,$ cell thickness, dz, $= 40$ cm, TAW $= (0.24 - 0.11)*40 = 5.2$ cm

Water content before second storm.

Upper layer, $D_r = 4$ days$*0.7$ cm/day $= 2.8$ cm $\Delta\theta = -2.8/60 = -0.0467$

$\theta_{final} = 0.26 - 0.0467 = 0.213$

Lower layer, $D_r = 4$ days$*0.3$ cm/day $= 1.2$ cm $\Delta\theta = -1.2/40 = -0.03$ $\theta_{final} = 0.24 - 0.03 = 0.21$

Water content after second storm.

Upper layer, storm adds 3 cm, which exceeds the 2.8cm depletion : $\theta_{final} = 0.26$

Lower layer, 0.2 cm leaches to the lower layer. $\Delta\theta = 0.2/40 = 0.005$: $\theta_{final} = 0.21 + 0.005 = 0.215$

What is the percent depletion before and after the second storm in each layer?

$$D_{r\ \text{upperl ayer–before storm}} = 2.8\ \text{cm}\ \ \%\text{depletion} = 2.8\ \text{cm}/9.6\ \text{cm}*100 = 29\%$$
$$D_{r\ \text{lower layer–before storm}} = 1.2\ \text{cm}\ \ \%\text{depletion} = 1.2\ \text{cm}/5.2\ \text{cm}*100 = 23\%$$
$$D_{r\ \text{upper layer–after storm}} = (0.2 - 0.2)60 = 0\ \text{cm}\ \ \%\text{depletion} = 0\ \text{cm}/9.6\ \text{cm}*100 = 0\%$$
$$D_{r\ \text{lower layer–after storm}} = (0.24 - 0.215)40 = 1\ \text{cm}\ \ \%\text{depletion} = 1\ \text{cm}/5.2\ \text{cm}*100 = 19\%$$

---

4. **Redo Example 26.4 with the WINDS model and by hand, but lower the leaching fraction to 0.05. Make calculations for the irrigation on the third day for the upper two layers by hand. Next, use the WINDS model to calculate EC for 100 days. There are only two field sections in the *WINDS Chapter 26* workbook. The sections are organized with respect to their irrigation zones in the *spatial data* worksheet. Add another G01 section in column C and write "3" in the same row in column. In cell K7, specify that the number of cells is 3 and click the *Make new sections* button. This process adds the C_3 worksheet to the end of the workbook. The next step is to populate the date in the *Crop_data* worksheet for section 3. You can do this in the A*ctive Data* worksheet or just copy the cells from section 2 (column C) to section 3 (column D) in the *Crop_data* worksheet. If you use the *Active_data* worksheet, then the copy the information from section 2, "Copy data from crop data," and then copy rows 3–450 to section 3 (specified in cells G13:G16) and click the "Copy data to crop data" button. After calculating the required application depth for 0.05 leaching fraction, add the calculated fraction of baseline irrigation to the section 3 column in the G01 worksheet. Go to the Main worksheet. In cell G2, specify that three sections will be evaluated. After clicking Run, select position 3 in the Get Data combo box (upper right side of the worksheet). Find the "Water content" graph and the "Irrigation, rain depth, and leaching" graph with the Selection form. If rainfall appears in the graph, remove the rainfall from the *Active year weather* page for the first 100 days. Find the soil water salinity graph in the *Salinity* worksheet. Compare to the salinity levels in Example 26.4. Copy and paste the worksheets or graphs into this document. Use the graphs to assess the processes.**

The depth of irrigation is 7 cm/0.95 = 7.371 cm (0.074 m) /irrigation for the 0.05 LF. Available water capacity is (0.2–0.184)(0.5) = so Eq. 26.38 is not valid.

Layer 4, irrigation day salinity calculation for 0.05 LF

$$EC_{\text{final}} = \frac{d_{in}EC_{in} + EC_{initial}\Delta z\theta_{initial}}{d_{in} - ET_{layer} + \Delta z\theta i_{initial}^{\ initial}}$$

$$EC_{\text{final}} = \frac{(0.0737\ \text{m})(1) + 2.17\ (0.49)\ (0.184)}{0.0737m - 10\ mm{*}0.4/1000 + 0.49\ \text{m}(0.184)} = 1.68\ dS/m$$

Layer 3, irrigation day salinity calculation for 0.05 LF

The initial moisture content in the upper layer is 0.184. Thus, the storage depth during the irrigation is calculated as (0.2–0.184) (0.5) +0.004 = 0.012 m. This means that 0.0737–0.012 = 0.0617 m passes through layer 4 and into layer 3.

$$EC_{\text{final}} = \frac{(0.0617\ \text{m})(1.68\ \text{dS/m})\ + 2.13(0.5)(0.188)}{0.0617\ m - 10\ mm(0.3)/(1000\ mm/m) + 0.5(0.188)} = 1.99\ dS/m.$$

Solution with WINDS model for 100 days

The fraction of baseline irrigation is 1/0.95 = 1.053. Thus, the seasonal depth applied is 1,053 mm for leaching fraction equal to 0.05.

$$LF = \frac{i - ET}{i} = \frac{1,053 - 1,000}{1,053} = 0.05$$

Change the irrigation depths in the G01 worksheet to 1.053, and run the model from the Main page.

	A	B	C	D	E	F
1	DOY	Ref (mm)	Sec_1	Sec_2	Sec_3	Sec_4
2	1					
3	2					
4	3	70	1	1.053	1.053	
5	4					
6	5					
7	6					
8	7					
9	8					
10	9					
11	10	70	1	1.053	1.053	
12	11					
13	12					
14	13					
15	14					
16	15					
17	16					
18	17	70	1	1.176	1.053	
19	18					
20	19					
21	20					
22	21					
23	22					
24	23					
25	24	70	1	1.176	1.053	
26	25					

Get Data from section 3 in the Main worksheet. Then click the *View water content data* button. Select the water content graph and the irrigation and leaching graph from the *Water graphs* form.

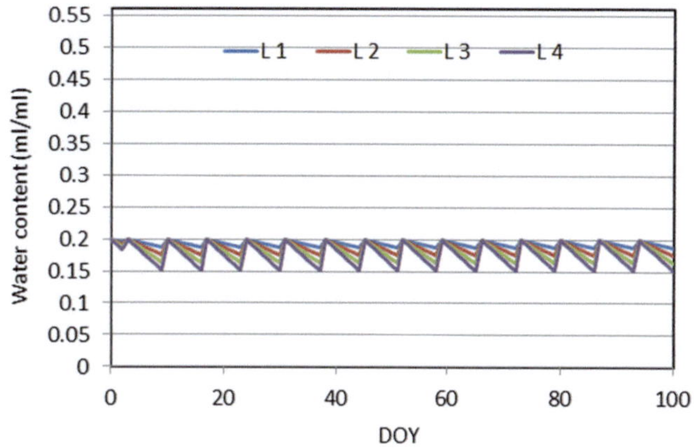

Water content vs. DOY graph

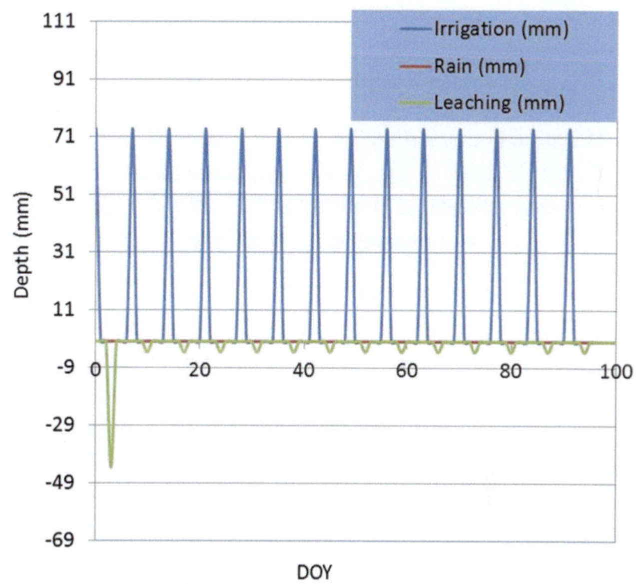

Irrigation, rain depth, and deep leaching graph from Water Content worksheet.

Go back to the Main worksheet and click the *View salinity data* button.

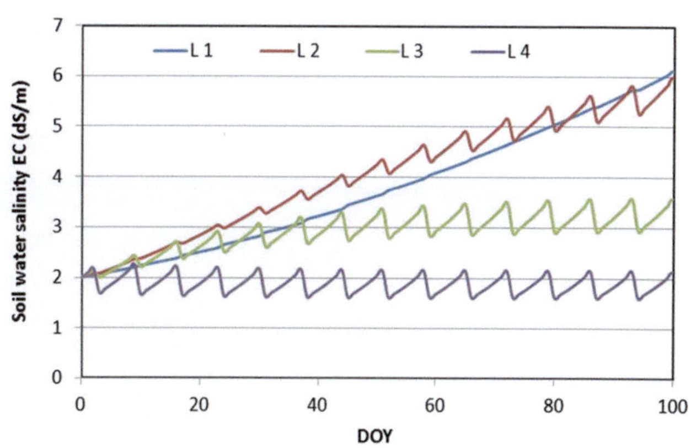

Soil water salinity graph.

Assessment:

The irrigation and leaching graph shows that leaching was excessive on day 3 so irrigation was not needed. The soil water content graph shows that irrigations returned the soil to field capacity after each irrigation. A small excess was applied, which resulted in the leaching events seen in the irrigation and leaching graph. The salinities are in between the no leaching and 15 % leaching cases in Example 26.4. Salinity built up in the lower part of the root zone, but the upper part of the root zone was leached. There is no jagged line for salinity in the lower layer because water content change was minimal between irrigation events; thus, the jagged change between irrigation events for the upper layers was due to water content change and resultant concentration of the existing salinity in less water.

5. **Calculate leaching fraction for irrigation water salinity 2 dS/m and required EC$_e$ 1.5 dS/m?**

$$LF = \frac{EC_{iw}}{5*EC_e - EC_{iw}} = \frac{2}{5*1.5 - 2} = 0.36$$

6. **What are the ratios EC$_e$/EC$_{ave}$, and EC$_e$/ EC$_{dw}$ in Example 26.4? EC$_{dw}$ is the leachate salinity. Discuss the importance of understanding these ratios with respect to crop management decisions?**

The ratio of $EC_e / EC_{ave} = 0.5$

The ratio of $EC_e / EC_{dw} = 1.6/6.7 = 0.24$

If one assumed that the drainage water salinity was the same as the ECe defined for crop management, then using the drainage water salinity as ECe would result in a much higher leaching fraction than necessary for the health of the crop. Similarly, using the model for salinity management would be incorrect if the ECave was used as the ECe.

7. **Redo Example 26.5, but change the fertilizer application to 40 kg/ha on the first day application and change nitrate concentration in the irrigation water to 20 mg/L. Make a new hand calculation of the changes due to fertilizer application and irrigation during the first three days in the upper cell. Run the WINDS simulation for 100 days with the higher irrigation water nitrate concentration and higher fertilization rate on day 3. The irrigation rate will be the same as problem 4. You can change the nitrogen data in the Active_data worksheet and copy it to section 3 in the *Crop_data* worksheet. Make sure that cell G5 in *Main* worksheet is marked True. Run the simulation from the *Main* worksheet. Select 3 in the *Get_data* combobox in the Main worksheet. Click the *View Nitrogen* data button. Copy the following graphs into your homework document: *Nitrate* (mg/kg) in layers, *Irrigation and drainage nitrate* (you might need to update both x and y axes from the selection form or from the axes), *Reactions*, and *Cumulative leaching, nitrate and reactions*. Assess the processes by looking at the graphs.**

Calculate the change in nitrate concentration due to fertilizer application (day 2, 1 day after application) on day 2

$$m_{fer} = \frac{App\ rate(kg/ha)}{10} = \frac{40\ kg/ha}{10} = 4\ mg/L_{soil}\ m$$

Calculate new nitrate concentration on day two with 40 kg/ha fertilizer with Eq. 26.39

$$N_{final} = \frac{d_{in}N_{in} - d_{out}N_{initial} + m_{min} - m_{den} + m_{fer} - m_{upt} + \Box \times \Box_{initial}\ \Delta z\ \theta_{initial}}{\Delta z\ \theta_{final}}$$

$$N_{final} = \frac{(0)N_{in} - (0)N_{initial} + 0.0455 - 0 + 4 - 0 + 296\ (0.49)\ (0.192)}{0.49\ (0.184)} = 354\ mg/L$$

Next, calculate the concentration in the upper cell after application of irrigation water on the third day. In this case, the complete mixing equation must be used because water passes through the layer.

$$N_{final} = \frac{d_{in}N_{in} + m_{min} - m_{den} + m_{fer} - m_{upt} + {}_{initial}\Delta z\ \theta_{initial}}{d_{in} - ET_{layer} + \Delta z\ \theta i_{initial}^{initial}}$$

$$N_{final} = \frac{(0.0737\ m)(20\ mg/L) + 0.0455 - 0 + 0 - 0 + 354(0.49)\ (0.184)}{0.0737\ m - 10\ mm*0.4/1000 + 0.49\ m\ (0.184)} = 209\ mg/L$$

There is no change in cell 3.

Data for section 3 is copied back into the *Active data* worksheet. Fertilization and irrigation parameters are changed in the lower left part of the *Nitrogen* form, which is accessed in the *Active Data* worksheet. Data is then copied back to the *Crop_data* worksheet.

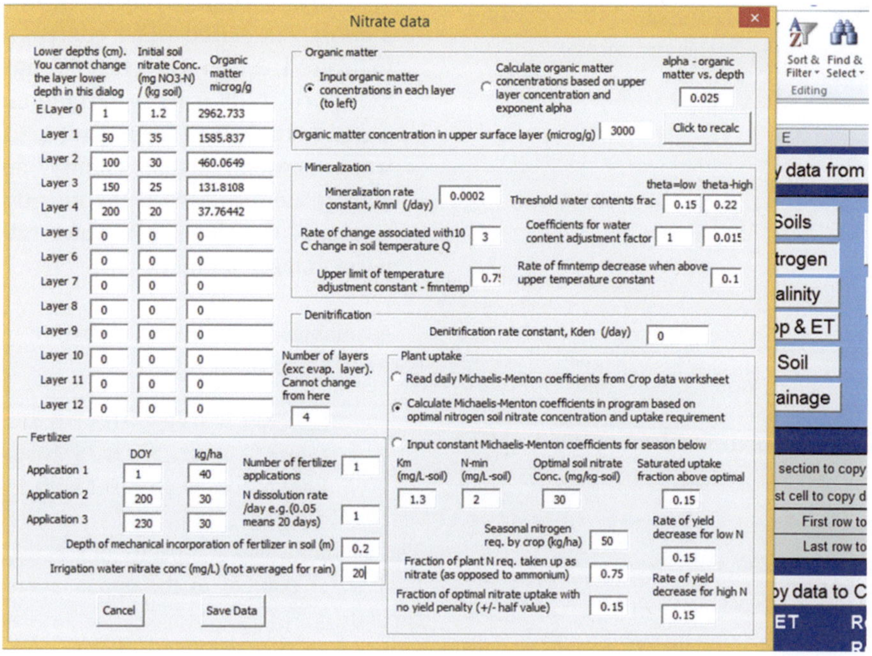

Nitrogen form

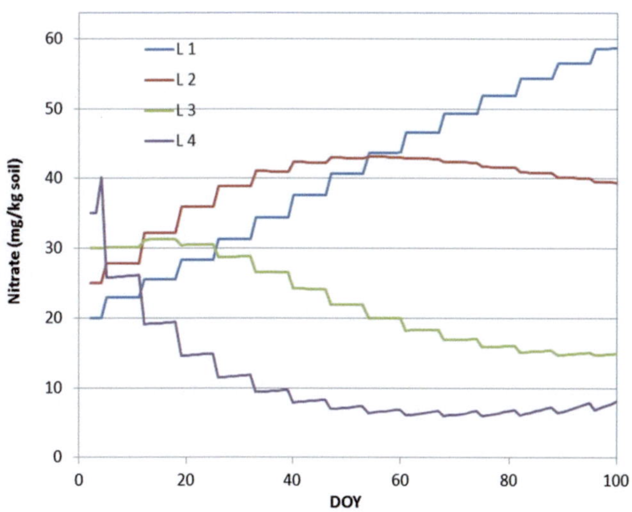

Nitrate in layers graph

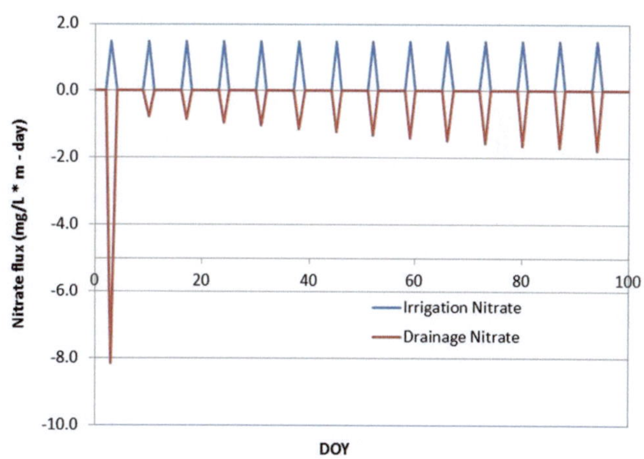

Irrigation and drainage nitrate graph

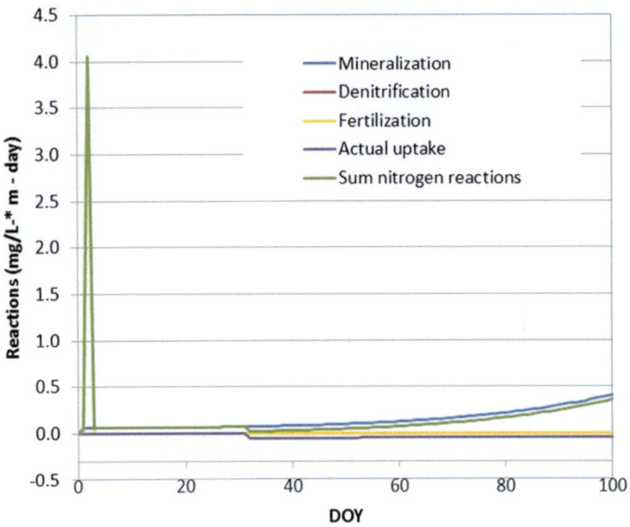

Mineralization, denitrification, fertilization, and uptake graph

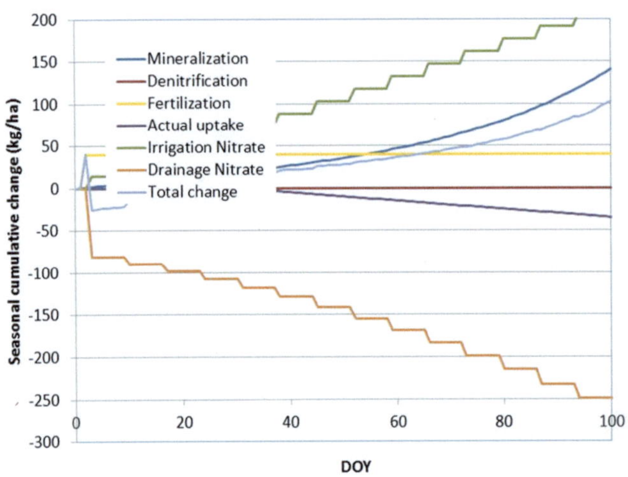

Cumulative mineralization, denitrification, fertilization, and uptake graph

Assessment:

There was heavy leaching during the first irrigation event; however, unavoidable leaching took place after every irrigation event since nitrate was concentrated in the lower part of the soil profile. The primary source of nitrogen was irrigation water. Nitrate concentration was low in the upper part of the soil profile, even below the optimal 30 mg/kg concentration. However, it reached the 50–60 mg/kg range in the lower part of the soil profile. Mineralization was a significant part of the total nitrogen mass balance during the season; however, denitrification was insignificant.

8. **A soil has three 0.4 m layers, numbered 1, 2, and 3 from the bottom, with field capacity in all layers equal to 0.25. The initial water salinity in layers 1, 2, and 3 is, 23-, 7-, and 5-dS/m, respectively. ET is 10 mm/day with 20 %, 30 %, and 50 % of ET in layers 1, 2, and 3 respectively. Irrigation water salinity is 2 dS/m. The initial water content on the previous day in layers 1, 2, and 3 is 0.18, 0.15, and 0.10, respectively. Soil porosity is 0.4. An irrigation event adds 11 cm water to the soil in the morning. Compare to the final water content, actual salinity, and saturated paste extract salinity before the morning irrigation event. Compare the changes in water salinity and saturated paste extract salinity.**

Available water capacity before irrigation (to select between equations).

$$\text{Layer } 3 = (0.25 - 0.1)*0.4 = 0.06 \text{ m}$$
$$\text{Layer } 2 = (0.25 - 0.15)*0.4 = 0.04 \text{ m}$$
$$\text{Layer } 3 = (0.25 - 0.18)*0.4 = 0.028 \text{ m}$$

The available water capacity is less than the storm depth so final water content in the upper layers after storm is field capacity

$$\text{Layer } 3, \theta = 0.25$$
$$\text{Layer } 2, \theta = 0.25$$

10 mm is added to layer 1 by irrigation, and 2 mm is lost by ET $\theta = 0.18 + 0.01/0.4–0.002/0.4 = 0.2$

Salinity calculations. Must use equation that accounts for irrigation water passing through the layer for the upper 2 layers. No water passes through layer 1. The final salinity calculation must include the ET water that is lost from the soil during the day.

$$EC_{\text{final}} = \frac{d_{\text{in}}\text{EC}_{\text{in}} + EC_{initial}\Delta z\theta_{initial}}{d_{in} - ET_{layer} + \Delta z\theta i_{initial}{}^{initial}}$$

$$EC_{\text{final}-3} = \frac{(0.11\text{m})(2) + 5(0.4)\ (0.1)}{0.11\ m - 10\ mm*0.5/1000 + 0.4\ (0.1)} = 2.90\ dS/m$$

$$EC_{\text{final}-2} = \frac{(0.05\ \text{m})(2.90) + 7(0.4)\ (0.15)}{0.05\ m - 10\ mm*0.3/1000 + 0.4\ (0.15)} = 5.28\ dS/m$$

$$EC_{final} = \frac{i\ EC_{iw} - d_{seepage}EC_{initial} + EC_{initial}\Delta z\ \theta_{initial}}{\Delta z\ \theta_{final}}$$

$$EC_{final-1} = \frac{0.01*5.28 - 0 + 23\ *0.4*0.2}{0.4*0.2} = 23.7\ dS/m$$

$$EC_e = EC\frac{\theta}{\theta_{Sat}}$$

$$EC_{e3-initial} = 5*(0.1/0.4) = 1.25\ dS/m$$
$$EC_{e2-initial} = 7*(0.15/0.4) = 2.62\ dS/m$$
$$EC_{e1-initial} = 23*(0.18/0.4) = 10.35\ dS/m$$

$$EC_{e3-final} = 2.90*(0.25/0.4) = 1.81\ dS/m$$
$$EC_{e2-final} = 5.28*(0.25/0.4) = 3.30\ dS/m$$
$$EC_{e1-final} = 23.7*(0.2/0.4) = 11.8\ dS/m$$

The water salinity changed most in the upper layer. The saturated paste extract salinity increased in all layers, most dramatically in the upper layer. The reason for the increase in saturated paste extract salinity is that salt was added to all layers. The reason for the decrease in water salinity in the upper layer is that water in the layer was diluted by irrigation water with lower salinity.

9. **During a 1 day period, the upper layer of soil, 0.4 m depth, has a mineralization rate of 0.1 mg/L * m, a denitrification rate of 0.05 mg/L*m, and plant uptake of 1 *kg/ha*. One cm (average for the field) depth of water is added to the layer by drip irrigation and the irrigation water has a nitrate concentration of 20 mg/L. Transpiration removes 1.4 cm from the layer. No water leaches to the next layer. The initial water content is 0.18, and the initial nitrate concentration in the soil water is 15 mg/L. Calculate the final water content and nitrate concentration in the water. Calculate the kg/ha nitrate in the layer at the end of the day.**

Final water content is $0.18 + (0.01 - 0.014)/0.4 = 0.179$

Plant uptake $- 1\ \text{kg/ha} = 0.1\ \text{mg/L*m}$

$$N_{final} = \frac{d_{in}N_{in} - d_{out}N_{initial} + m_{min} - m_{den} + m_{fer} - m_{upt} + {}_{initial}\Delta z\ \theta_{initial}}{\Delta z\ \theta_{final}}$$

$$N_{final} = \frac{0.01*20 - 0 + 0.1 - 0.05\ + 0 - 0.1\ +\ 15\ *0.4*0.18}{0.4*\ 0.179} = 17.2\ mg/L$$

Final $\text{kg/ha} = (17.2\ \text{mg/L})\ (0.4\ \text{m})*10\ \text{kg/ha/(mg/L*m)} = 172\ \text{kg/ha}$

Chapter 27: Solutions

1. How is energy lost as water flows through pipelines, soils, and channels?

In soils, heat is generated and energy is lost as water molecules slide past pore walls and slide past each other. As water moves through pipelines and channels, turbulent eddies in the flow cause water molecules to move past each other and release heat.

2. Convert 20 m water hydraulic head to units of kPa, atmospheres, bars, J/kg, ft head, and PSI. Approximate values are acceptable. Try to memorize the approximate relationships.

$$20\ m = 2\ atm = 2\ bar = 200\ kPa = 200\ J/kg$$
$$= 20/0.3048 = 66\ ft\ /2.31 = 28\ PSI$$

3. Water flows through a 1 m long column at a rate of 2 m/d and the pressure differential from one end of the column to the other is 1 m. Calculate the hydraulic conductivity of the media in the column. What would the flow rate be if the pressure differential was 100 kPa?

The energy difference, ΔH, is 1.0

$$L = 1.0\quad Darcy\ velocity = 2\ m/d\quad k = vL/DH = 2\ m/d$$

Part 2.

$$100\ kPa = 10\ m \quad v = k\Delta H/L = 2\ m/d(10/1) = 20\ m/day$$

4. Calculate the kinetic energy of the water in problem 4 for the case of 1 m pressure differential. Is calculation of kinetic energy necessary in soil water calculations?

$$E = v^{\wedge 2}/2g = (2m/day/60/60/24)^2/(2*9.8)$$
$$= 2.73e - 11\ Joules$$

Calculation of kinetic energy is not necessary in soils calculations.

5. Draw the energy diagram for Example 27.2 (part 1)

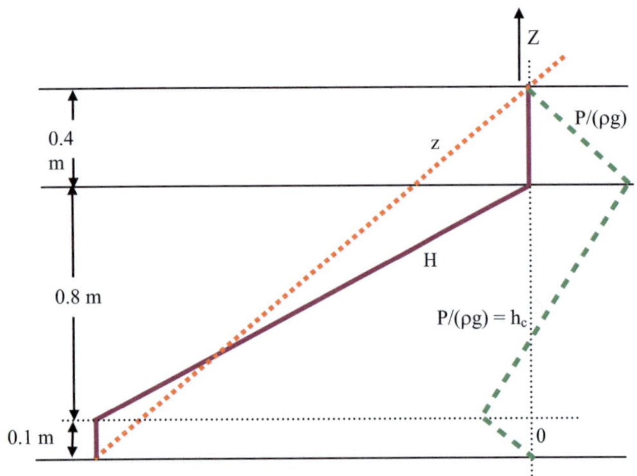

6. For Example 27.2 (part 1), place the lower water surface at the same elevation as the upper boundary of the sand (0.4 m below the elevation of the upper tank water surface). Draw the energy diagram.

$$H_f = 0.4\ m$$

Calculate the velocity of flow with Darcy's law.

$$v = k\frac{H_f}{L} = 5\ cm/hr\frac{0.4\ m}{0.8\ m} = 2.5\ cm/hr$$

Calculate the flow rate Q through the sand filter.

$$Q = vA = v\left(pd^2/4\right)$$
$$= (2.5\ cm/hr)(1\ m/100\ cm)p\left(0.7^2/4\right)\left(1000L/m^3\right)$$
$$= 9.6\ L/hr.$$

If the porosity of sand is 0.38, then calculate the actual velocity of water through the filter.

$$v_{actual} = v_{Darcy}/porosity = 2.5\ cm/hr/0.38 = 6.6\ cm/hr.$$

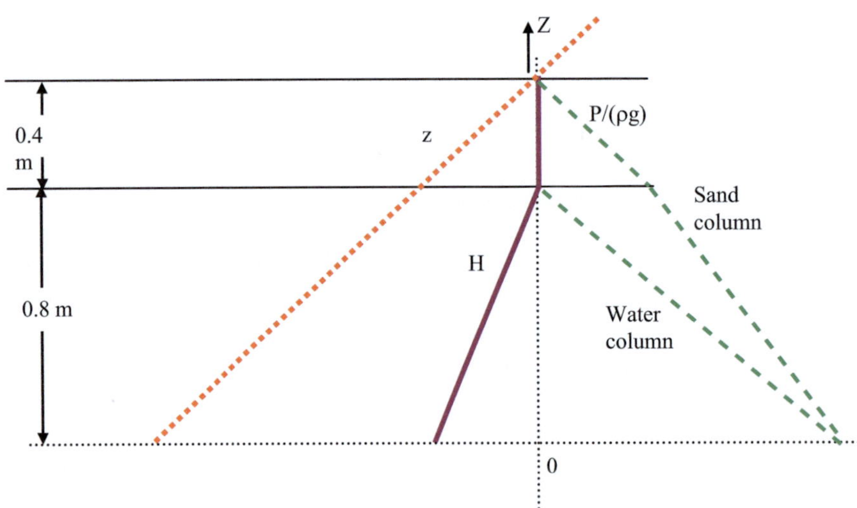

7. For Example 27.5, change the conductivities in cells 1, 2, and 3 to 5-, 4-, and 3-cm/hr, and change the heights of the cells to 0.3, 0.4, and 0.5-m, respectively. Draw the energy diagram. Calculate the flow velocity (Darcy velocity) in the column and energy loss in each layer?

$$K_e = \frac{1.2}{0.3/_5 + 0.4/_4 + 0.5/_3} = 3.67\ cm/hr \mathrm{v} = K_e\frac{H_f}{L} = 3.67\ cm/hr\frac{1.5\ m}{1.2\ m} = 4.59\ cm/hr$$

$$H_{f1} = \frac{\mathrm{v}L}{K} = \frac{4.59*0.3}{5} = 0.28\ m \quad H_{f3} = \frac{\mathrm{v}L}{K} = \frac{4.59*0.4}{4} = 0.46\ m$$

$$H_{f2} = \frac{\mathrm{v}L}{K} = \frac{4.59*0.5}{3} = 0.77\ m$$

Start with known energy at position A and work downwards through the column.

Position	Total energy (H)	Elevation energy (z)	Matric potential (h or P)
A	**1.6 m**	**1.2 m**	**1.6–1.2 = 0.4 m**
B	**1.6–0.28 = 1.32 m**	**0.9 m**	**1.32–0.9 = 0.42 m**
C	**1.32–0.46 = 0.86 m**	**0.5 m**	**0.86–0.5 = 0.36 m**
D	**0.86–0.77 = 0.1 m**	**0 m**	**0.1–0.0 = 0.1 m**

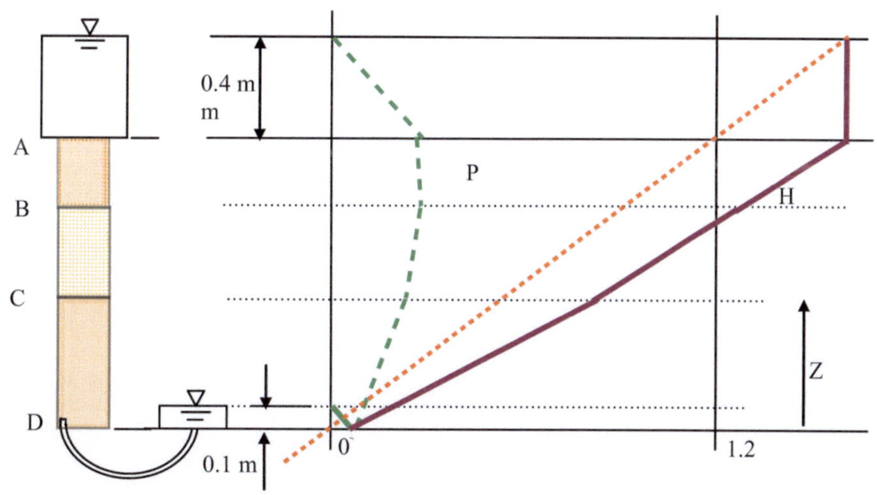

8. For Example 27.6, change the conductivities in cells 1, 2, and 3 to 5-, 4-, and 3-cm/hr, and change the heights of the cells to 0.3, 0.4, and 0.5-m, respectively. Draw the energy diagrams at the midelevation for each of the three cells in the direction parallel to flow (rotate the axes).

$$K_e = \frac{5*0.3 + 4*0.4 + 3*0.5}{1.2} = 3.83\ cm/hr \quad \mathrm{v} = K_e\frac{H_f}{L} = 3.83\ cm/hr\frac{1.5\ m}{1.2\ m} = 4.8\ cm/hr$$

$$\mathrm{v}_1 = K_1\frac{H_f}{L} = 5\ cm/hr\frac{1.5\ m}{1.2\ m} = 6.25\ cm/hr$$

$$\mathrm{v}_2 = 4\ cm/hr\frac{1.5\ m}{1.2\ m} = 5\ cm/hr \quad \mathrm{v}_3 = 3\ cm/hr\frac{1.5\ m}{1.2\ m} = 3.75\ cm/hr$$

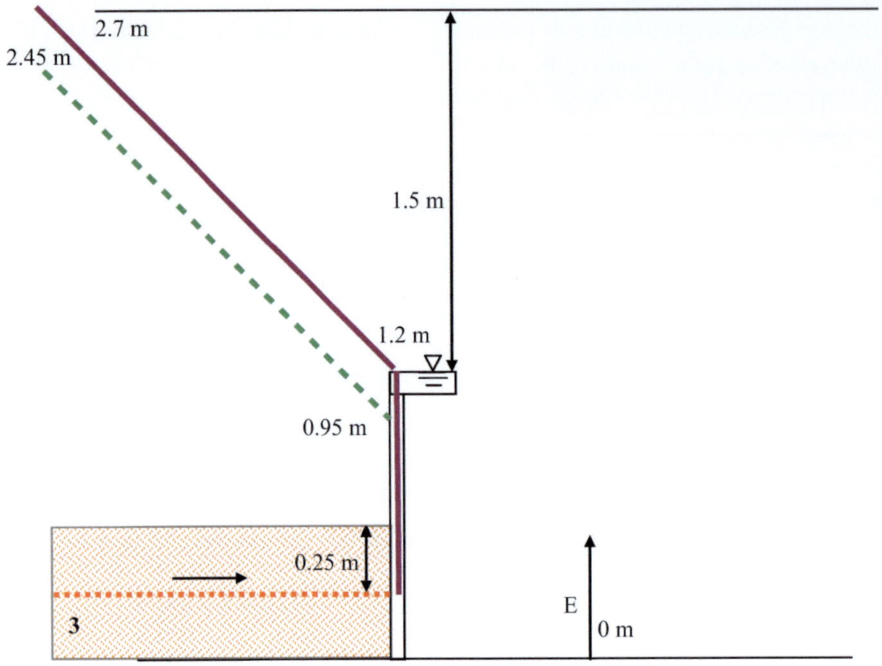

9. A column has two soil layers. Layers 1 and 2 have saturated hydraulic conductivities of 1 and 2 cm/hr, respectively. The datum is at the bottom of the column. Calculate the energy, elevation, and pressure potential at points A, B, and C. Draw the energy potential lines. You do not need to draw any lines below the datum. Calculate the flow velocity (Darcy velocity) in the column and energy loss in each layer?

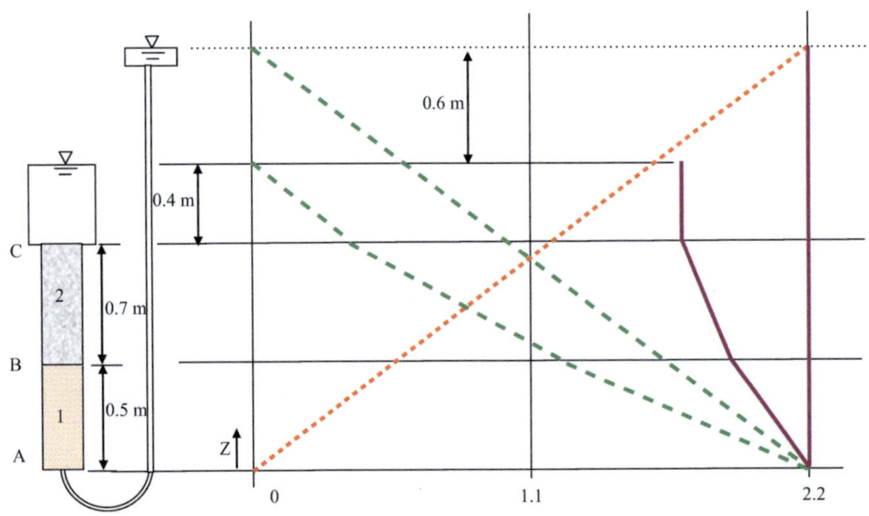

$$K_e = \frac{1.2}{0.5/1 + 0.7/2} = 1.41 \ cm/hr$$

$$v = K_e \frac{H_f}{L} = 1.41 \ cm/hr \frac{0.6 \ m}{1.2 \ m} = 0.705 \ cm/hr$$

$$H_{f1} = v_1 \frac{L}{K_1} = 0.705 \ cm/hr \frac{0.5 \ m}{1 \ cm/hr} = 0.352 \ m$$

$$H_{f2} = v_1 \frac{L}{K_1} = 0.705 \ cm/hr \frac{0.7 \ m}{2 \ cm/hr} = 0.247 \ m$$

Position	Total energy (H)	Elevation energy (z)	Matric potential (h or P)
A	**2.2 m**	**0 m**	**2.2 m**
B	**2.2–0.35 = 1.85 m**	**0.5 m**	**1.85–0.5 = 1.35 m**
C	**1.85–0.25 = 1.6 m**	**1.2 m**	**1.6–1.2 = 0.4 m**

10. Draw the energy potential lines for this soil column, and calculate the total and matric potential energy at points A, B, and C. The conductivity of layer 1 is 2 cm/day, and the conductivity of layer 2 is 0.5 cm/day.

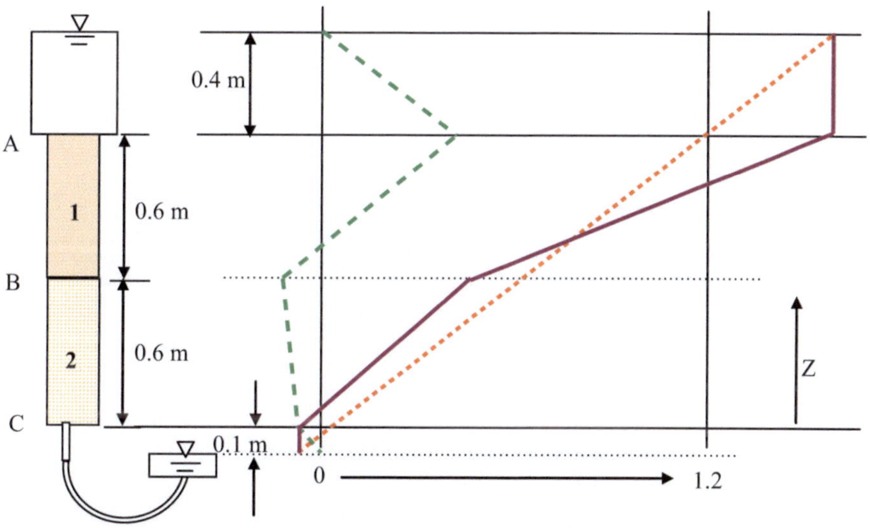

$$K_e = \frac{1.2}{0.6/1 + 0.6/2} = 1.33 \ cm/hr$$

$$v = K_e \frac{H_f}{L} = 1.33 \ cm/hr \frac{1.7 \ m}{1.2 \ m} = 1.88 \ cm/hr$$

$$H_{f1} = v_1 \frac{L}{K_1} = 1.88 \ cm/hr \frac{0.6 \ m}{1 \ cm/hr} = 1.13 \ m$$

$$H_{f2} = v_1 \frac{L}{K_1} = 1.88 \ cm/hr \frac{0.6 \ m}{2 \ cm/hr} = 0.57 \ m$$

Position	Total energy (H)	Elevation energy (z)	Matric potential (h or P)
A	**1.6 m**	**1.2 m**	**0.4 m**
B	**1.6–1.13 = 0.47 m**	**0.6 m**	**0.47–0.6 = –0.13 m**
C	**0.47–0.57 = –0.1 m**	**0 m**	**–0.1–0 = –0.1 m**

11. Using Eq. 27.25, calculate the change in mass in cell 2, which has an energy of 4 J, For this example, E (meters) = 2 * mass (kg). Cell 1 E = 2 J and Cell 3 E = 3 J. The conductivity between cells is 6 m/sec, and the cross-sectional area of cells, A, is 0.1 m^2. The length of cells is 0.4 m. The length of time steps is 1 second. Calculate final mass and energy.

$m_{initial}$ = mass = E/2 = 2 J.

$$m_{2=final} = m_{2=initial} - AK\frac{E_2 - E_1}{gL}\Delta t + AK\frac{E_3 - E_2}{gL}\Delta t$$

$$m_{2=final} = 2 - (0.1)(6)\frac{4-2}{9.8*0.4}(1) + (0.1)(6)\frac{5-4}{(9.8)(0.4)}(1)$$

$$= 2 - (0.1)(6)\frac{-1}{(9.8)(0.4)}$$

$$m_{2=final} = 2 + 0.15 = 2.15$$

Final energy = 4.3 J

12. Redo Example 27.7 except let the initial masses be equal to 1-, 4-, 5-, 6-, 8-, and 10-kg. Print out the energy graph and discuss results.

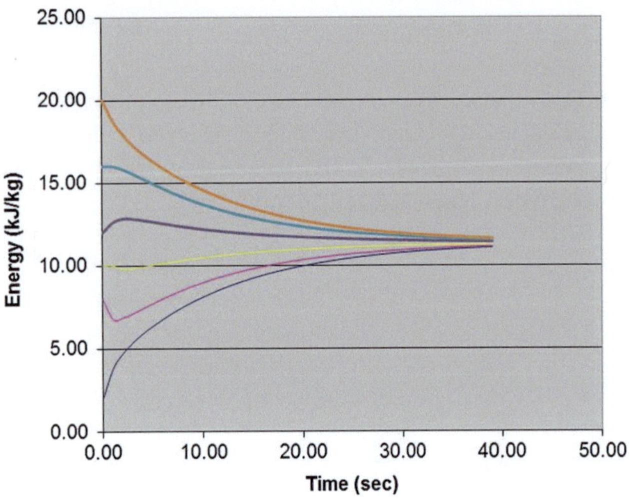

Because the energy gradient is not even, flow is initially in the direction of cells that have a relatively low mass.

13. Repeat problem 12, but change the time step to 10 seconds. Print out the energy graph and discuss results.

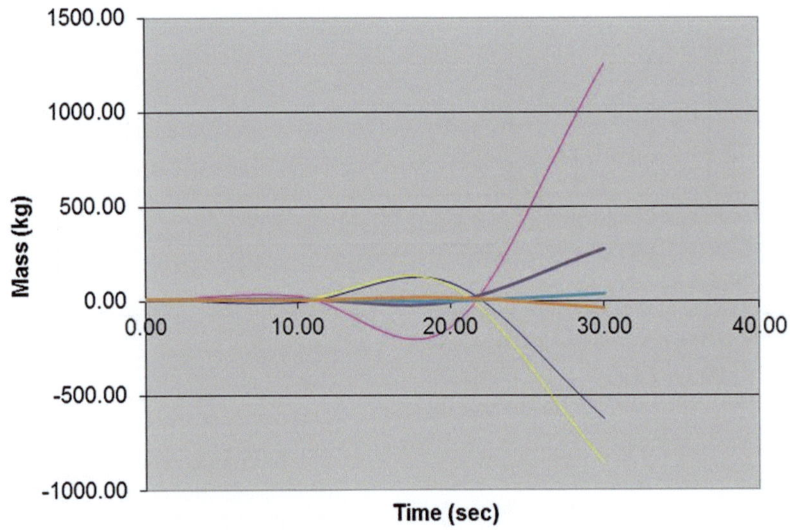

The solution is unstable because the time steps are too large relative to the space steps.

14. Change the *Calculations in worksheet* so that 1 kg mass is continually added each second to the left of the control volume (cell 1). Make the time step 0.2 seconds. Then make a second modification so that 0.4 kg is removed from the right side whenever the mass on the left side is greater than 8. Simulate for 60 seconds and show in a graph. Show the equations that you used. Make sure that the A,K, and L values are the same as those in the Mass and Energy worksheet: ($A = 0.1$, $k = 6$, and $L = 0.4$).

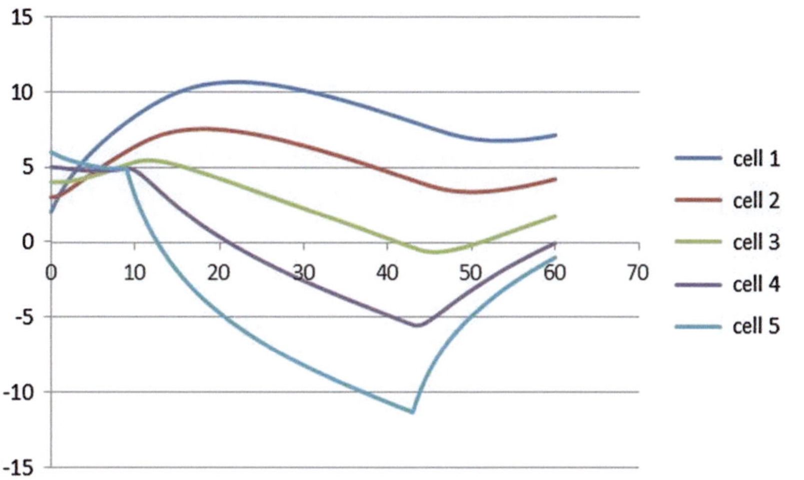

Left side equation Column C

$$= (K9 - J9)*K*A/L/9.806*T + C9 + Added\_mass$$

Right side equation in Column G

$$= -(N9 - M9)*K*A/L/9.806*T + G9 + IF(C10 > 8, -0.4, 0)$$

15. Repeat Problem 14 in *Functions in Worksheet*. Show the equations/functions as well as the graph. You will need to add the Right_mass() function as well as make the modification on removed mass. Also, let the criteria be that the left mass must be greater than 10.

Left side equation Column C

$$= Left\_mass(K9, J9, K, A, L, T, C9) + 0.2$$

Right side equation in Column G

$$= Right\_mass(M10, N10, K, A, L, T, G10) - if(C10 > 8, 0.4, 0)$$

16. Repeat Problem 14 but make the changes in the VBA Mass_Energy subroutine so that the results are shown in the Mass and Energy worksheet. Show the parts of the code that you changed. The graph should have the same pattern as the previous graphs.

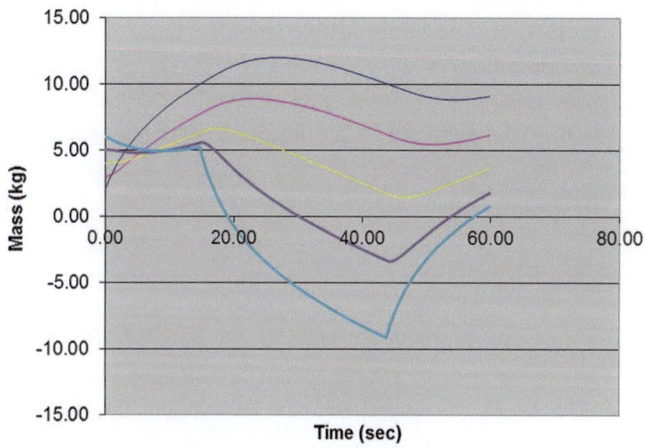


```
Code:
  Final_time = 60
    Num_Cells = Range("D1")
    T = 0.2
    Num_Times = Final_time / T

    For i = 1 To Num_Times
      For j = 1 To Num_Cells
        If j = 1 Then
          Mass(i, j) = Left_mass(Energy(i -
          1, j + 1), Energy(i - 1, j), K, A, L, T,
          Mass(i - 1, j)) + 0.2
        ElseIf j < Num_Cells Then
```

```
        Mass(i, j) = Middle_mass(Energy(i - 1,
        j - 1), Energy(i - 1, j), Energy(i - 1,
        j + 1), K, A, L, T, Mass(i - 1, j))
      Else
        Mass(i, j) = Right_mass(Energy(i - 1, j
        - 1), Energy(i - 1, j), K, A, L, T, Mass
        (i - 1, j))
        If Mass(i, 1) > 10 Then
          Mass(i, j) = Mass(i, j) - 0.4
        End If
      End If
      Energy(i, j) = 2 * Mass(i, j)
    Next j
  Next i
```

17. A SSF wetland is run at 0.3 m depth, 10 m wide, and 20 m long. The hydraulic conductivity of the gravel in the wetland is 10^{-3} m/sec (360 cm/hr), and the porosity is 0.38. The impermeable liner has a 0.5 % slope in the direction of flow. What are the Darcy velocity, flow rate, pore volume, and hydraulic detention time within the wetland? Calculate the percent reduction in flow from that of the wetland described in Example 27.6.

The difference in impermeable liner elevation between the beginning and end of the wetland is 0.005 * 20 m = 0.1 m. Calculate Darcy velocity,

$$\mathrm{v} = K\frac{\Delta H}{L} = 360\frac{0.1}{20} = 1.8\ cm/hr$$

$$\mathrm{Q} = \mathrm{vA} = (1.8\mathrm{cm/hr})(0.01\ \mathrm{m/cm})(10\ \mathrm{m})(0.3\ \mathrm{m})(1000\ \mathrm{L/m^3}) = 54\ L/hr$$

$$\mathrm{V_{cv}} = \mathrm{A\,L\,f} = 10{*}0.3{*}20{*}0.38 = 23\ m^3$$

$$HDT = \frac{V}{Q} = \frac{23\ m^3}{54\ L/h}\left(\frac{1{,}000\ L}{m^3}\right)\left(\frac{day}{24\ hr}\right) = 18\ days$$

$$\textit{Percent reduction in flow capacity} = (1 - (168 - 54)/168){*}100 = 68\%$$

18. Evaluate your answer in problem 17 with the SSF workbook. Determine whether 54 L/hr (0.054 m^3/hr) results in steady 0.3 m depth flow. Inlet flow rate is specified in cell B5. Set all cells at an initial depth of 0.3 m. Set discharge depth (cell L4) equal to 0.3 m. Inlet bot. difference (cell G3) is 0.1 m.

Parameters are set as shown below

	A	B	C	D	E	F	G	H	I	J	K	L
1	Wetland width	10	Num_Cells	5			Cell 1	Cell 2	Cell 3	Cell 4	Cell 5	Cell 6
2	K (m/hr)	3.6	Time_step	0.1		Initial surface	0.39	0.37	0.35	0.33	0.31	
3	Wetland length	20	Time_unit	hours		Inlet bot. diff (m)	0.1					
4	Length unit	m	Final time	100		Porosity	0.38		Discharge depth (m)			0.3
5	Inlet flow rate	0.054	m3/hr	Water depth in cell (m)						Water surface elevation (r		
6		hours	Cell 1	Cell 2	Cell 3	Cell 4	Cell 5			Cell 1	Cell 2	Cell 3
7	Click to run SSF Macro	0.00	0.3000	0.3000	0.3000	0.3000	0.3000		0.00	0.3900	0.3700	0.3500
8		0.10	0.3000	0.3000	0.3000	0.3000	0.3000		0.10	0.3900	0.3700	0.3500
9		0.20	0.3000	0.3000	0.3000	0.3000	0.3000		0.20	0.3900	0.3700	0.3500

All cells remain at 0.3 m depth so the SSF model is verified by the problem 16.

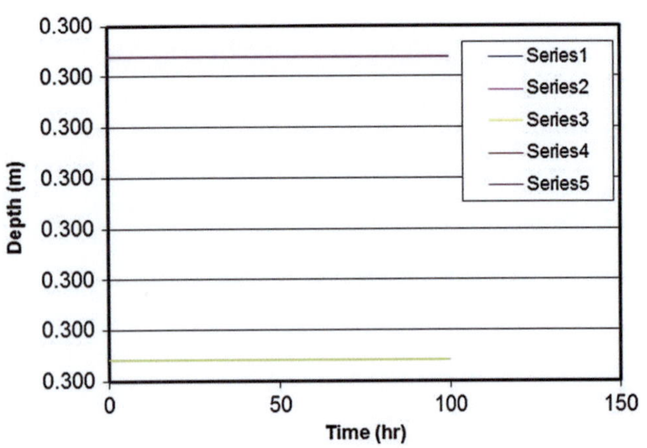

19. With initial conditions described in problem 17, investigate the effect of increasing the discharge depth to 0.5 m and increasing the flow rate to 90 L/hr with the *SSF* model. Run for 200 hours. Verify with the equations in problem 16 that this flow depth results in steady state depth = 0.5 m.

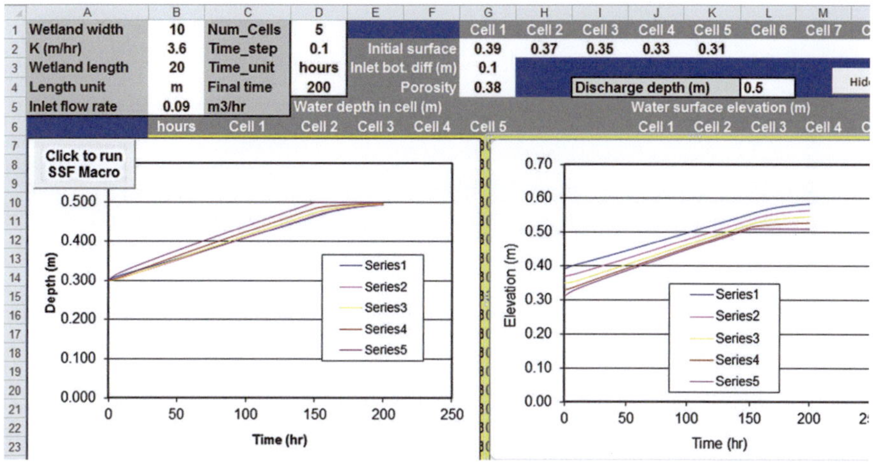

20. With the initial conditions described in problem 17, investigate the effect of keeping the discharge depth at 0.3 m and increasing the flow rate to 90 L/hr with the *SSF* model. Run for 200 hours. Show the graph and describe the final conditions after 200 hours. Look at the hydraulic gradient and discuss its effect on flow rate. Discuss the ability of SSF wetlands to handle variations in flow rate.

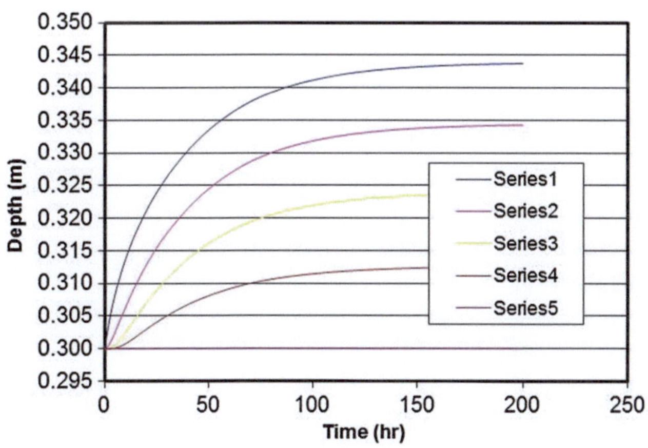

The upper end of the raceway is only 0.4 m deeper than at the 0.54 flow rate. The lower end remains at 0.3 m depth. This increased depth at the upper end results in a greater hydraulic gradient across the raceway, which drives the increased flow. An increase of almost 100 % flow rate results in very little change in depth in the raceway, which shows that the raceway is able to handle large differences in flow rate due to increasing the hydraulic gradient.

21. Plot the water characteristic curves for sandy loam and clay soils with the van Genuchten equation for the terms in Table 27.2. Make three graphs for each soil. 0–50 cm matric potential with linear scale, 0–15 bar matric potential with linear scale, and 0–15 bar matric potential with logarithmic scale on the x-axis. For logarithmic scale, the matric potentials must be expressed as positive values. Describe the differences in the curves. Make sure that the water content scale on all graphs begins at zero in order to make the comparison.

The clay soil has very little decrease in water content over the first 50 cm; however, the sandy loam soil drops significantly over the first 50 cm. The clay soil retains significantly more moisture, even at a matric potential of 15 bar. The logarithmic curves have the same shape for both soils.

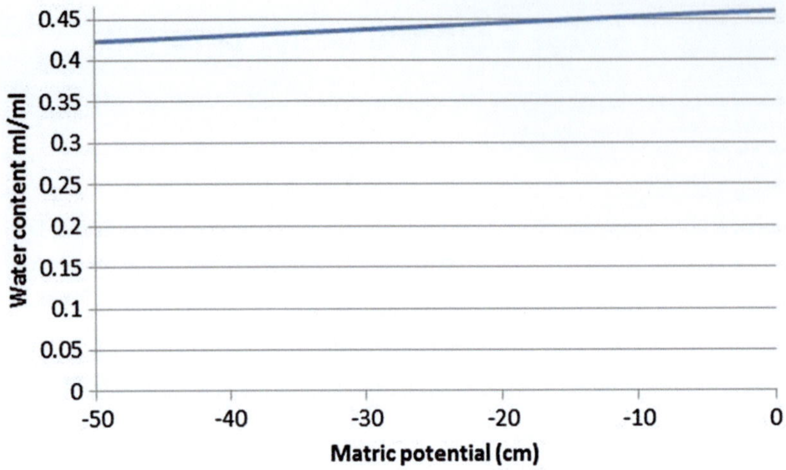

Clay soil (0 to −50 cm)

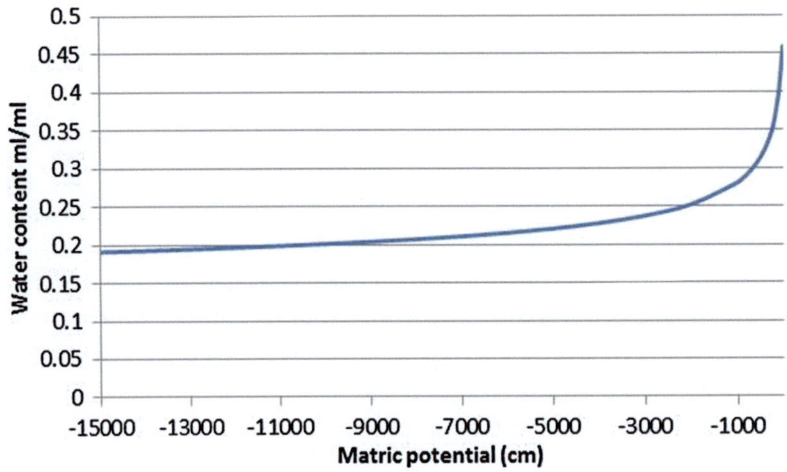

Clay soil (0 to −15000 cm)

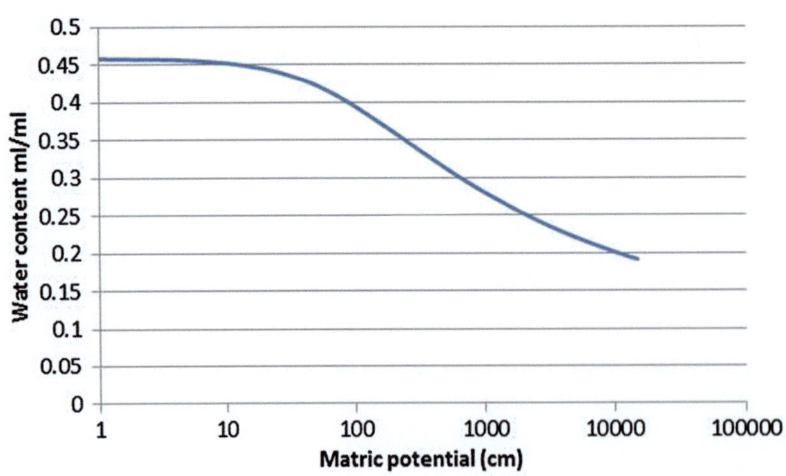

Clay soil (logarithmic axis)

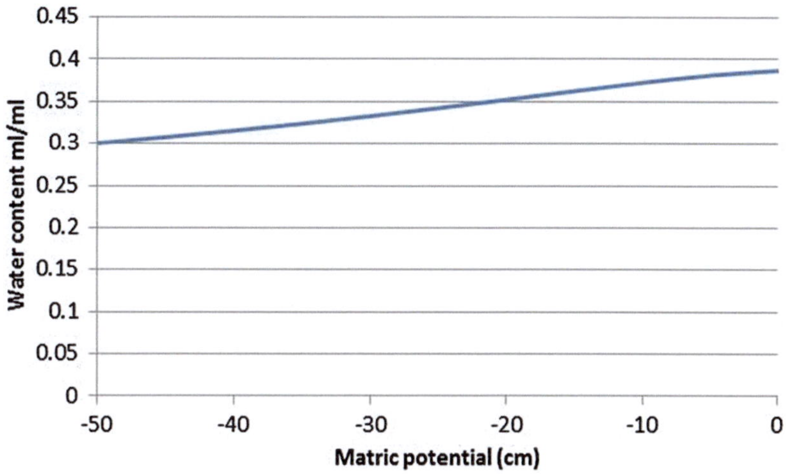

Sandy loam soil (0 to −50 cm)

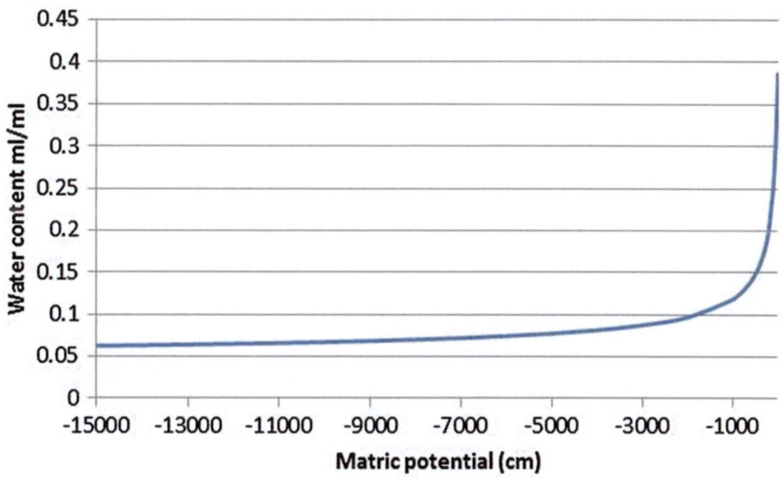

Sandy loam soil (0 to −15000 cm)

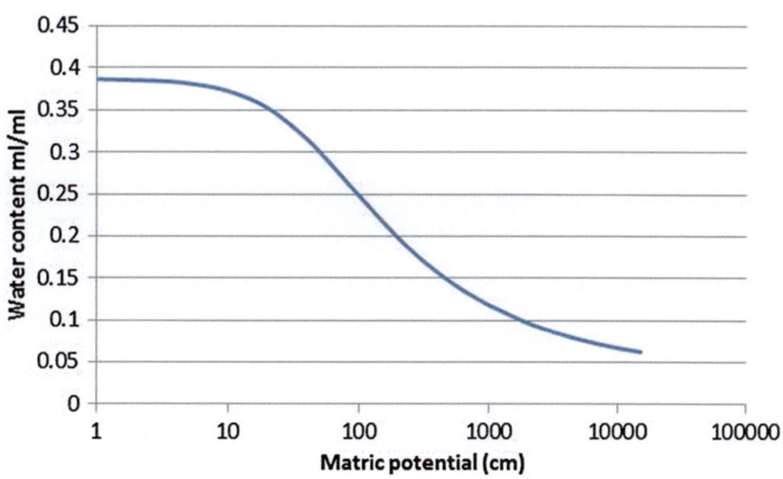

Sandy loam soil (logarithmic axis)

22. Find λ and hb for the Goldsboro Sandy Loam water content vs. matric potential in the table below. Show initial log-log plot, adjusted log-log plot, and show final graph with a comparison of the experimental water characteristic curve and the Brooks-Corey water characteristic curve. (Note: the graph does not align quite as well as the example in the book).

Ψm (m)	−Ψm (m)	θ
0.00	0.00	0.365
− 0.10	0.10	0.355
− 0.30	0.30	0.320
− 0.50	0.50	0.275
− 0.60	0.60	0.260
− 0.80	0.80	0.235
− 2.00	2.00	0.188
− 2.50	2.50	0.182
− 3.00	3.00	0.178
− 3.50	3.50	0.170

Ψm (m)	−Ψm (m)	θ	θ_e	S	S_e	B-C Eq.
0.00	0.00	0.365	1.000	1.000	1.000	1.000
− 0.10	0.10	0.355	0.953	0.973	0.953	1.000
− 0.30	0.30	0.320	0.791	0.877	0.791	1.000
− 0.50	0.50	0.275	0.581	0.753	0.581	0.635
− 0.60	0.60	0.260	0.512	0.712	0.512	0.571
− 0.80	0.80	0.235	0.395	0.644	0.395	0.482
− 2.00	2.00	0.188	0.177	0.515	0.177	0.282
− 2.50	2.50	0.182	0.149	0.499	0.149	0.247
− 3.00	3.00	0.178	0.130	0.488	0.130	0.222
− 3.50	3.50	0.170	0.093	0.466	0.093	0.203

Numbers in yellow were used for calculation of pore size distribution index.

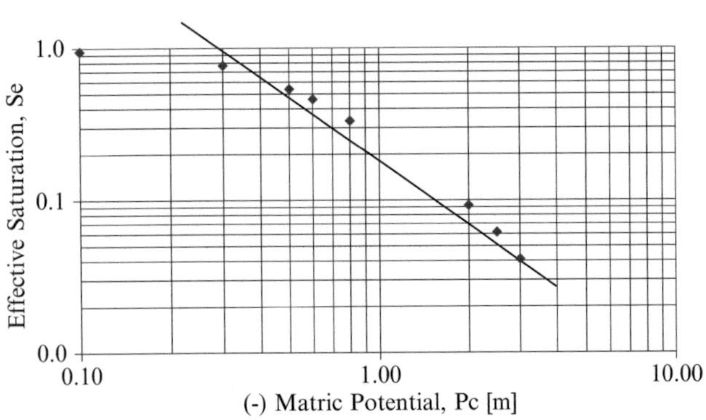

Initial log-log plot with $\theta_r = 0.17$

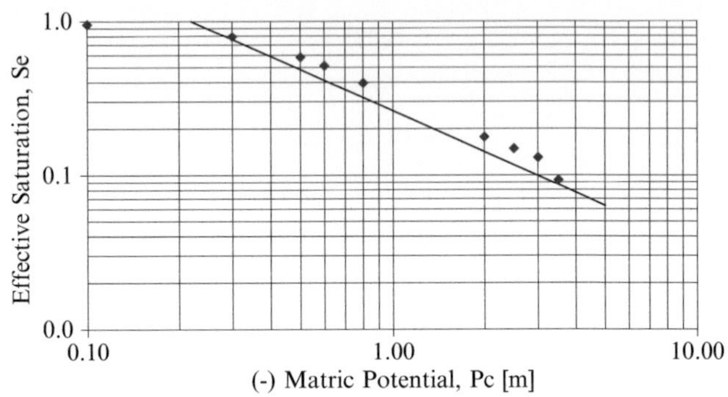

Adjusted log − log plot with $\theta_r = 0.15$

The Goldsboro soil does not fit the Brooks-Corey model very well. The Wagram loamy sand Brooks-Corey curve had a very close match with the experimental curve; however, the Goldsboro soil is not so close.

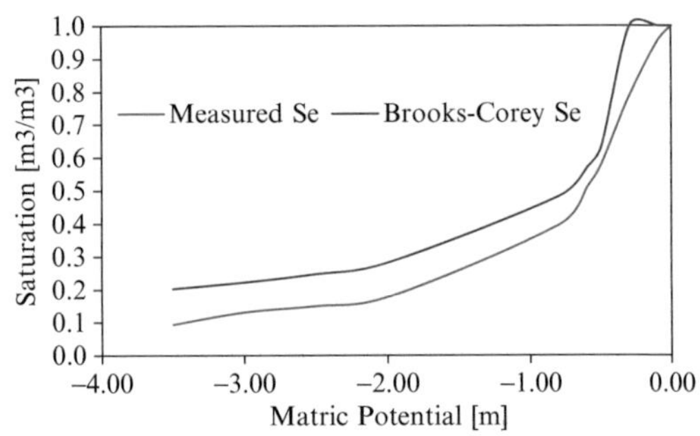

The values in yellow were selected for calculation of the pore size distribution index.

$$\lambda = \frac{\log(0.953) - \log(0.130)}{\log(0.10) - \log(3.00)} = 0.585$$

The bubbling pressure is the intersection of the slope of the pore size distribution index on log-log paper and the $S_e = 1.0$ line, −23 cm.

h_b =	−0.230
λ =	0.585
θs =	0.365
θr =	0.150
S_r	0.411

23. Replace the simplistic relationship between mass and energy in the Mass and Energy worksheet VBA code with the Van Genuchten equation for water content (mass) and matric potential (energy) relationship. Also, include the Van Genuchten relationship between matric potential (energy) and conductivity instead of the constant conductivity that is included in the problem. In order to simplify the problem, you can assume that the cells are horizontal in order to avoid the calculation of elevation energy. The solution is basically the VBA code used in the next chapters, but without the vertical change in elevation energy.

Chapter 28: Solution

1. **Derive Eq. 28.3 from Eq. 27.20**

$$\Delta V_2 = -AK\frac{H_2 - H_1}{L}\Delta t + AK\frac{H_3 - H_2}{L}\Delta t$$
$$\frac{\Delta V_2}{A} = \frac{-AK\frac{H_2 - H_1}{L}\Delta t + AK\frac{H_3 - H_2}{L}\Delta t}{A}$$
$$\Delta d = \left(-K\frac{H_2 - H_1}{L} + K\frac{H_3 - H_2}{L}\right)\Delta t$$
$$\Delta d = \Delta\theta\Delta x$$
$$\Delta\theta\ \Delta x = \left(-K\frac{H_2 - H_1}{L} + K\frac{H_3 - H_2}{L}\right)\Delta t$$

2. **Define H in terms of matric potential and elevation, and then rearrange Eq. 28.3 so that it has the same format as Eq. 28.2.**

$$\Delta\theta_j\Delta z = \left(-K_{e(j\&j-1)}\frac{H_j - H_{j-1}}{\Delta z} + K_{e(j\&j+1)}\frac{H_{j+1} - H_j}{\Delta z}\right)\Delta t$$
$$\frac{\Delta\theta_j}{\Delta t} = \frac{1}{\Delta z}\left(-K_{e(j\&j-1)}\frac{h_j - h_{j-1} + z_j - z_{j-1}}{\Delta z} + K_{e(j\&j+1)}\frac{h_{j+1} - h_j + z_{j+1} - z_j}{\Delta z}\right)$$
$$\frac{\Delta\theta_j}{\Delta t} = \frac{1}{\Delta z}\left(-K_{e(j\&j-1)}\left(\frac{h_j - h_{j-1}}{\Delta z} + \frac{z_j - z_{j-1}}{\Delta z}\right) + K_{e(j\&j+1)}\left(\frac{h_{j+1} - h_j}{\Delta z} + \frac{z_{j+1} - z_j}{\Delta z}\right)\right)$$
$$\frac{\Delta\theta_j}{\Delta t} = \frac{1}{\Delta z}\left(-K_{e(j\&j-1)}\left(\frac{h_j - h_{j-1}}{\Delta z} + 1\right) + K_{e(j\&j+1)}\left(\frac{h_{j+1} - h_j}{\Delta z} + 1\right)\right)$$

3. **Include osmotic potential as a function of soil water EC in Eq. 8.5. Leave only the change in water content (Δ θ) on the left side of the equation. Use Eq. 5.2 to calculate osmotic potential based on soil water EC.**

The osmotic potential, $\psi_s = -3.6 * \mathrm{EC}$ in units of meters. Because the other terms are in units of meters, the equation can be substituted directly into the equation. Use the equation derived in question 1.

$$\Delta\theta_j = \frac{\Delta t}{\Delta z}\left(\begin{array}{l} -K_{e(j\&j-1)}\left(\frac{h_j - h_{j-1}}{\Delta z} + \frac{-3.6*(EC_j - EC_{j-1})}{\Delta z} + 1\right) \\ +K_{e(j\&j+1)}\left(\frac{h_{j+1} - h_j}{\Delta z} + \frac{-3.6*(EC_{j+1} - EC_j)}{\Delta z} + 1\right)\end{array}\right)$$

The EC in this equation is soil water EC and not ECe.

4. **Data on three cells (3, 4, 5) is listed below. Write the equation derived in question 3 in terms of cell 4.**

Calculate the change in water content in cell 4 with and without the influence of salinity. Use a 1 day time step.

z_3 = 1.0 m	h_3 = -2 m	$EC_{w\text{-}3}$ = 4 dS/m
z_4 = 1.4 m	h_4 = -3 m	$EC_{w\text{-}4}$ = 3 dS/m
z_5 = 1.8 m	h_5 = -5 m	$EC_{w\text{-}5}$ = 6 dS/m

The effective hydraulic conductivity between cells 3 and 4 is 0.01 m/day

The effective hydraulic conductivity between cells 4 and 5 is 0.001 m/day

Equation written for cell 4.

$$\Delta\theta_4 = \frac{\Delta t}{\Delta z}\left(\begin{array}{l} -K_{e(3,4)}\left(\dfrac{h_4 - h_3}{\Delta z} + \dfrac{-3.6*(EC_4 - EC_3)}{\Delta z} + 1\right) \\ +K_{e(4,5)}\left(\dfrac{h_5 - h_4}{\Delta z} + \dfrac{-3.6*(EC_5 - EC_4)}{\Delta z} + 1\right) \end{array}\right)$$

Calculation with salinity

$$\Delta\theta_4 = \frac{1}{0.4}\left(\begin{array}{l} -0.01\left(\dfrac{-3-(-2)}{0.4} + \dfrac{-3.6*(3-4)}{0.4} + 1\right) \\ +0.001\left(\dfrac{-4-(-5)}{0.4} + \dfrac{-3.6*(6-3)}{0.4} + 1\right) \end{array}\right)$$

$$\Delta\theta_4 = \frac{1}{0.4}\left(\begin{array}{l} -0.01\left(\dfrac{-1}{0.4} + \dfrac{3.6}{0.4} + 1\right) \\ +0.001\left(\dfrac{1}{0.4} + \dfrac{-10.8}{0.4} + 1\right) \end{array}\right)$$

$$= \frac{1}{0.4}\left(-0.01(3) + 0.001\left(\frac{-9.4}{0.4}\right)\right) = -0.133$$

Calculation without salinity

$$\Delta\theta_4 = \frac{1}{0.4}\left(\begin{array}{l} -0.01\left(\dfrac{-3-(-2)}{0.4} + 1\right) \\ +0.001\left(\dfrac{-4-(-5)}{0.4} + 1\right) \end{array}\right)$$

$$\Delta\theta_4 = \frac{1}{0.4}\left(\begin{array}{l} -0.01\left(\dfrac{-1}{0.4} + 1\right) \\ +0.001\left(\dfrac{1}{0.4} + 1\right) \end{array}\right)$$

$$= \frac{1}{0.4}(-0.01(-1.5) + 0.001(3.5)) = 0.046$$

5. **Calculate the water content, effective water content, effective saturation and hydraulic conductivity of Wagram loamy sand at −1 bar matric potential.**

1 bar = 1,000 cm.

$$K_{rw} = \left(\frac{-30}{-1{,}000}\right)^{2+3(1.27)} = 1.42*10^{-9}$$

$$S_e = \theta_e = \left(\frac{h_b}{h_c}\right)^{\lambda} = \left(\frac{-30}{-1{,}000}\right)^{1.27} = 0.012$$

$$\theta = (\theta_s - \theta_r)\left(\frac{h_b}{h_c}\right)^{\lambda} + \theta_r$$

$$= (0.31 - 0.044)\left(\frac{-30}{-1{,}000}\right)^{1.27} + 0.044 = 0.047$$

6. **Calculate the Darcy velocity between two points in unsaturated Wagram loamy sand with matric potential values of −1,000 cm (point 1) and −2,000 cm (point 2) and elevations of 10 cm (point 1) and 20 cm (point 2), respectively. Assume that the saturated hydraulic conductivity of Wagram loamy sand is 0.6 cm/hr. The distance between the two points is 50 cm. Use the Brooks Corey model for hydraulic conductivity. Use the geometric mean to calculate effective hydraulic conductivity.**

$$K_{rw} = \left(\frac{-30}{-1{,}000}\right)^{2+3(1.27)} = 1.42*10^{-9}$$

$$K_{rw} = \left(\frac{-30}{-2{,}000}\right)^{2+3(1.27)} = 2.53*10^{-11}$$

$$K(h_{1000}) = K_S K_{rw\ 1000} = 0.6\ \text{cm/hr}*1.4*10^{-9} = 8.5*10^{-10}\text{cm/hr}$$

$$K(h_{2000}) = K_S K_{rw\ 2000} = 0.6\ \text{cm/hr}*2.5*10^{-11} = 1.5*10^{-11}\text{cm/hr}$$

$$K_e = \frac{L_1 + L_2}{\dfrac{L_1}{K_1} + \dfrac{L_2}{K_2}} = \frac{25 + 25}{\dfrac{25}{8.5*10^{-10}} + \dfrac{25}{1.5*10^{-11}}} = 3.0*10^{-11}\ cm/hr$$

$$h_f = \Delta H = h_2 + z_2 - h_1 - z_1$$
$$= -2{,}000 + 20 - (-1000) - 10 = -990\ \text{cm}$$

The fact that the energy gradient is negative means that water flows from point 1–2.

$$v = -K_e\frac{\Delta H}{L} = -3*10^{-11}\frac{-990}{50} = 5.9*10^{-10}\ cm/hr$$

7. **Three cells numbered 1 through 3 from bottom to top have 0.5 m depth. Use the van Genuchten equations to calculate water contents after 1 day and 2 days. Initial water content in all cells is 37 %. n = 1.31, θ_r = 0.095, θ_S = 0.41, α = 0.019, L = 0.5, and K_0 = 6.24 cm/day. There is no infiltration and no seepage of water below the control volume. Calculate effective conductivity between two cells with the geometric mean. Calculate water contents after the first day and after the second day. It is impossible for any cell to have greater than the saturated water content.**

The middle cell water content will remain the same after the first day since the input from above equals the drainage below. The upper cell water content is calculated as follows.

$$\theta_{n\text{-}final} = \theta_{n-initial} + \frac{\Delta t}{\Delta z^2}\left(-K_{e(n\,\&\,n-1)}(H_n - H_{n-1})\right) + \frac{i}{\Delta z}$$
$$\theta_{3\text{-}final} = \theta_{3-initial} + \frac{\Delta t}{\Delta z^2}\left(-K_{e(3\,\&\,2)}(h_3 - h_2 + \Delta z)\right) + \frac{i}{\Delta z}$$
$$\theta_{3\text{-}final} = 0.37 + \frac{1}{0.5^2}(-0.0018(0.5)) + 0 = 0.3663$$

The lower cell water content is calculated as follows.

$$\theta_{1\text{-}final} = \theta_{1-initial} + \frac{\Delta t}{\Delta z^2}\left(K_{e(1\&2)}(h_2 - h_1 + \Delta z)\right)$$
$$\theta_{1\text{-}final} = 0.37 + \frac{1}{0.5^2}(0.0018(0.5)) = 0.3737$$

Day 2
Calculate initial effective water content in each cell

$$\theta_{e-1} = \frac{\theta - \theta_r}{\theta_s - \theta_r} = \frac{0.3737 - 0.095}{0.41 - 0.095} = 0.885$$
$$\theta_{e-2} = \frac{\theta - \theta_r}{\theta_s - \theta_r} = \frac{0.37 - 0.095}{0.41 - 0.095} = 0.873$$
$$\theta_{e-3} = \frac{\theta - \theta_r}{\theta_s - \theta_r} = \frac{0.3663 - 0.095}{0.41 - 0.095} = 0.861$$

Calculate matric potential in each cell

$$h_{c-1} = -\frac{\left(\theta_e{}^{-1/m} - 1\right)^{1/n}}{\alpha} = -\frac{\left(0.885^{-1/0.236} - 1\right)^{1/1,31}}{0.019}$$
$$= -39.1\ cm h_{c-2}$$
$$= -\frac{\left(\theta_e{}^{-1/m} - 1\right)^{1/n}}{\alpha}$$
$$= -\frac{\left(0.873^{-1/0.236} - 1\right)^{1/1,31}}{0.019}$$
$$= -43.3\ cm h_{c-3}$$
$$= -\frac{\left(\theta_e{}^{-1/m} - 1\right)^{1/n}}{\alpha}$$
$$= -\frac{\left(0.861^{-1/0.236} - 1\right)^{1/1,31}}{0.019}$$
$$= -47.7\ cm$$

Calculate hydraulic conductivity in each cell

$$K(\theta)_1 = K_0\theta_e{}^{1/2}\left[1 - \left(1 - \theta_e{}^{1/m}\right)^m\right]^2 = 6.24{*}0.885^{0.5}\left[1 - \left(1 - 0.885^{1/0.236}\right)^{0.236}\right]^2 = 0.218\ cm/d$$
$$K(\theta)_2 = K_0\theta_e{}^{1/2}\left[1 - \left(1 - \theta_e{}^{1/m}\right)^m\right]^2 = 6.24{*}0.873^{0.5}\left[1 - \left(1 - 0.873^{1/0.236}\right)^{0.236}\right]^2 = 0.18\ cm/d$$
$$K(\theta)_3 = K_0\theta_e{}^{1/2}\left[1 - \left(1 - \theta_e{}^{1/m}\right)^m\right]^2 = 6.24{*}0.77^{0.5}\left[1 - \left(1 - 0.77^{1/0.236}\right)^{0.236}\right]^2 = 0.157\ cm/d$$

Calculate effective hydraulic conductivities with the geometric mean

$$K_{e-2-3} = \frac{2}{\frac{1}{K_j} + \frac{1}{K_{j+1}}} = \frac{2}{\frac{1}{0.18} + \frac{1}{0.157}} = 0.17\ cm/day = 0.00170\ m/day$$
$$K_{e-1-2} = \frac{2}{\frac{1}{K_j} + \frac{1}{K_{j+1}}} = \frac{2}{\frac{1}{0.18} + \frac{1}{0.219}} = 0.20\ cm/day = 0.0020\ m/day$$

Calculate final water contents in each cell

$$\theta_{3-final} = \theta_{3-initial} + \frac{\Delta t}{\Delta z^2}\left(-K_{e(3\,\&\,2)}(h_3 - h_2 + \Delta z)\right) + \frac{i}{\Delta z}$$

$$\theta_{3-final} = 0.3663 + \frac{1}{0.5^2}\left(-0.001696(-0.477 - (-0.433) + 0.5)\right) + 0 = 0.3632$$

$$\theta_{1-final} = 0.3737 + \frac{1}{0.5^2}(0.002004*(-0.433 - (-0.391) + 0.5)) = 0.3774$$

$$\theta_{2-final} = \theta_{2-initial} + \frac{\Delta t}{\Delta z^2}\left(-K_{e(2\&\,1)}(h_2 - h_1 + \Delta z) + K_{e(2\,\&\,3)}(h_3 - h_2 + \Delta z)\right)$$

$$\theta_{2-final} = 0.37 + \frac{1}{0.5^2}(-0.0020*(-0.433 - (-0.391) + 0.5) + 0.0017(-0.477 - (-0.443) + 0.5)) = 0.3694$$

---

8. **Repeat problem 7 but allow water to drain below the lower cell. There is no water table. Use Eq. 28.27 for cell 1.**

Day 1

Other calculations remain the same as in question 7 except for the final water content in cell 1.

$$\theta_{final} = \theta_{initial} + \frac{\Delta t}{\Delta z^2}\left(K_{e(1\,\&\,2)}(h_2 - h_1 + \Delta z) - K_1 \Delta z\right)$$

$$\theta_{1-final} = 0.37 + \frac{1}{0.5^2}(0.0018(0.5) - 0.0018*0.5) = 0.37$$

Day 2

$$\theta_{2-final} = \theta_{2-initial} + \frac{\Delta t}{\Delta z^2}\left(-K_{e(2\&\,1)}(h_2 - h_1 + \Delta z) + K_{e(2\,\&\,3)}(h_3 - h_2 + \Delta z)\right)$$

$$\theta_{2-final} = 0.37 + \frac{1}{0.5^2}(-0.0018*(-0.433 - (-0.433) + 0.5) + 0.0017(-0.477 - (-0.443) + 0.5)) = 0.3694$$

The water content in cell 1 will remain the same (0.37) at the end of the second time step since the water content in cell 2 is still 0.37 at the end of the first time step; however, the water content in cell 2 will change.

$$\theta_{2-final} = \theta_{2-initial} + \frac{\Delta t}{\Delta z^2}\left(-K_{e(2\&\,1)}(h_2 - h_1 + \Delta z) + K_{e(2\,\&\,3)}(h_3 - h_2 + \Delta z)\right)$$

$$\theta_{2-final} = \theta_{2-initial} + \frac{1}{0.5^2}(-0.0020*(-0.433 - (-0.391) + 0.5) + 0.0017(-0.477 - (-0.443) + 0.5)) = 0.3694$$

---

9. Repeat problem 7 but restrict the downward movement of water to 0.1 cm/day from layer 1 (as with subsurface drainage and a water table).

Day 1

Other calculations remain the same as in question 7 except for the final water content in cell 1.

$$\theta_{final} = \theta_{initial} + \frac{\Delta t}{\Delta z^2}\left(K_{e(1\,\&\,2)}(h_2 - h_1 + \Delta z) - \Delta z * d_{seepage}\right)$$

$$\theta_{1-final} = 0.37 + \frac{1}{0.5^2}(0.0018(0.5) - 0.5*0.001) = 0.3716$$

Day 2

Calculate the new effective water content, matric potential and hydraulic conductivity for cell 1.

$$\theta_{e-1} = \frac{\theta - \theta_r}{\theta_s - \theta_r} = \frac{0.3716 - 0.095}{0.41 - 0.095} = 0.878$$

$$h_{c-1} = -\frac{\left(\theta_e{}^{-1/m} - 1\right)^{1/n}}{\alpha}$$

$$= -\frac{\left(0.878^{-1/0.236} - 1\right)^{1/1,31}}{0.019} = -41.5\ cmK(\theta)_1$$

$$= K_0\theta_e{}^{1/2}\left[1 - \left(1 - \theta_e{}^{1/m}\right)^m\right]^2$$

$$= 6.24{*}0.878^{0.5}\left[1 - \left(1 - 0.878^{1/0.236}\right)^{0.236}\right]^2$$

$$= 0.199\ cm/d$$

Calculate effective hydraulic conductivities with the geometric mean

---

$$K_{e-1-2} = \frac{2}{\frac{1}{K_j} + \frac{1}{K_{j+1}}} = \frac{2}{\frac{1}{0.1848} + \frac{1}{0.199}} = 0.1915\ cm/day = 0.001915\ m/day$$

$$\theta_{2-final} = \theta_{2-initial} + \frac{\Delta t}{\Delta z^2}\left(-K_{e(2\&1)}(h_2 - h_1 + \Delta z) + K_{e(2\&3)}(h_3 - h_2 + \Delta z)\right)$$

$$\theta_{2-final} = 0.37$$

$$+\frac{1}{0.5^2}(-0.001915{*}(-0.433 - (-0.415) + 0.5) + 0.0017(-0.477 - (-0.443) + 0.5)) = 0.3694$$

The water content in cell 1 will remain the same (0.37) at the end of the second time step since the water content in cell 2 is still 0.37 at the end of the first time step; however, the water content in cell 2 will change.

$$\theta_{2-final} = \theta_{2-initial} + \frac{\Delta t}{\Delta z^2}\left(-K_{e(2\&1)}(h_2 - h_1 + \Delta z) + K_{e(2\&3)}(h_3 - h_2 + \Delta z)\right)$$

$$\theta_{2-final} = \theta_{2-initial}$$

$$+\frac{1}{0.5^2}(-0.001915{*}(-0.433 - (-0.391) + 0.5) + 0.0017(-0.477 - (-0.443) + 0.5)) = 0.3694$$

$$\theta_{1-final} = 0.3716 + \frac{1}{0.5^2}(0.001915(0.5) - 0.001{*}0.5) = 0.3734$$

---

10. **Thirty cm is removed from the soil as the water table is lowered from 2.5 to 0.5 m. Calculate the incremental specific yield between 0.5 and 2.5 m.**

$$SY = \frac{\Delta d_d}{\Delta DTWT} = \frac{0.3}{2} = 0.15 = 15\%$$

11. **The water table drops from 1.0 to 1.5 m below the ground surface, and the specific yield of the soil is 10 %. Calculate the depth of water removed from the soil profile.**

$$\Delta z_{WT} = -\Delta DTWT = -(1.5\ \text{m} - 1.0\ \text{m}) = -0.5\ \text{m}$$

$$d_d = -SY{*}\Delta z_{WT} = -0.1{*}(-0.5) = 0.05\ m = 5\ cm$$

12. What do the right and left terms on the right side of Eq. 28.17 represent? What is the hatched area in Fig. 28.8? What does θ(z) represent in the Eq. 28.17? Why is the Brooks-Corey calculation of θ(z), which is based on matric potential substituted into the equation derivation? Explain why the upper limit of integration is $z_t - z_{WT}$ and the lower limit is h_{ba}.

The left term represents the depth of water in the soil when the soil is saturated. The right term represents the depth of water at each elevation based on the matric potential at that elevation. The hatched area in Fig. 8.7 is the difference between the two. The equation solves for water content at each elevation which is θ_z. The Brooks-Corey equation for water content as a function of matric potential is substituted into the equation because matric potential is directly proportional to elevation. The upper limit of integration represents the matric potential at the soil surface and the lower limit of integration represents the matric potential at the bubbling pressure elevation.

13. **Calculate the depth of water drained and specific yield for a water table that drops from the soil surface to 0.8 m above the datum in Wagram loamy sand. Place the datum 1.5 m below the soil surface.**

The bubbling pressure elevation is 0.8 m + 0.3 m = 1.1 m.

$$d_d = (\theta_s - \theta_r)\left(z_t - z_{hb} - \frac{h_{ba}^{\lambda}}{-\lambda + 1}\left((z_t - z_{hb} + h_{ba})^{-\lambda+1} - (h_{ba})^{-\lambda+1}\right)\right)$$

$$d_d = (0.305 - 0.044)\left(1.5 - 1.1 - \frac{0.3^{1.27}}{-1.27 + 1}\left((1.5 - 0.8)^{-1,27+1} - (0.3)^{-1,27+1}\right)\right) = 0.045\ m$$

$$SY = \frac{\Delta d_d}{\Delta DTWT} = \frac{0.045}{0.7\ m} = 0.064 = 6.4\%$$

14. If the problem 13 water table dropped from 0.7 m below the surface to 0.8 m below the surface, then how much water would you expect would be drained from the soil during this 0.1 m drop in water table elevation? Calculate the depth drained in two ways: use the specific yield that you calculated in problem 13 and also calculate the actual depth drained at 0.8 m depth with Eq. 28.18. Then use the two drained depths to calculate the incremental specific yield between 0.7 m and 0.8 m depth below the soil surface.

At 6.4 % specific yield, the depth drained would be 0.1 m * 0.064 m/m = 0.0064 m = 0.64 cm

The actual drained depth at 0.8 m depth below the surface is calculated as follows

$$d_d = (\theta_s - \theta_r)\left(z_t - z_{hb} - \frac{h_{ba}^{\lambda}}{-\lambda + 1}\left((z_t - z_{WT})^{-\lambda+1} - (h_{ba})^{-\lambda+1}\right)\right)$$

$$d_d = (0.305 - 0.044)\left(1.5 - 1.0 - \frac{0.3^{1.27}}{-1.27 + 1}\left((1.5 - 0.7)^{-1,27+1} - (0.3)^{-1,27+1}\right)\right) = 0.063\ m$$

Thus, the actual drained depth between 0.7 and 0.8 m is

$$0.063 - 0.045 = 0.018 \text{ m} = 1.8 \text{ cm}.$$

This drained depth is much greater than that calculated based on the specific yield in question 13. The low specific yield in the upper soil region is caused by the fact that there isn't very much water lost during the initial drop in water table elevation because the bubbling pressure is still near the soil surface.

Thus, the incremental specific yield between 0.7 m and 0.8 m is 0.018/0.1 = 0.18 = 18 %.

15. **Use Eq. 28.19 to find the incremental specific yield at 0.75 m depth below the soil surface for Wagram loamy sand and compare to the incremental specific yield that was calculated in problem 14.**

$$SY = \frac{d(d_d)}{d(z_{WT})} = -(\theta_s - \theta_r)\left[-1 + \left(\frac{h_{ba}}{DTWT}\right)^{\lambda}\right]$$

$$= -(0.305 - 0.044)\left[-1 + \left(\frac{0.3}{0.75}\right)^{1.27}\right] = 18\%$$

The incremental specific yields calculated with the two methods are exactly the same.

16. **Find the depth drained if the water table drops from the soil surface to 50 cm depth, from the soil surface**

to 100 cm depth, and from the soil surface to 150 cm depth. Use the Brooks-Corey Goldsboro sandy loam parameters that you calculated question 27–23. Set the datum at 3 m. Find the specific yield associated with the water table drop from the surface to each depth

The volume drained with the water table at 50 cm below the soil surface is calculated as follows.

$$d_d = (0.365 - 0.15)\left(3 - 2.73 - \frac{0.23^{1.27}}{-0.59+1}\left((3-2.5)^{-0.59+1} - (0.23)^{-0.59+1}\right)\right) = 0.013\ m$$

The volume drained with the water table at 100 cm below the soil surface is calculated as follows.

$$d_d = (0.365 - 0.15)\left(3 - 2.23 - \frac{0.23^{1.27}}{-0.59+1}\left((3-2)^{-0.59+1} - (0.23)^{-0.59+1}\right)\right) = 0.0658\ m$$

The volume drained with the water table at 150 cm below the soil surface is calculated as follows.

$$d_d = (0.365 - 0.15)\left(3 - 1.73 - \frac{0.23^{1.27}}{-0.59+1}\left((3-1.5)^{-0.59+1} - (0.23)^{-0.59+1}\right)\right) = 0.133\ m$$

$$\begin{aligned} SY - 50 &= 1.3/50*100\% = 2.6\ \% \\ SY - 100 &= 6.5/100*100\% = 6.5\ \% \\ SY - 150 &= 13.2/150*100\% = 8.8\ \% \end{aligned}$$

17. **Derive Eq. 28.31 from Eq. 28.20**

$$d_{total} = \theta_s(z_{hb}) + \int_{h_b a}^{Zt-ZWT} \theta(z)dz$$

$$\theta(z) = \theta_r + (\theta_s - \theta_r)\left(\frac{h_b\, c}{h_c}\right)^{\lambda} = \theta_r + (\theta_s - \theta_r)\left(\frac{h_{ba}}{z}\right)^{\lambda}$$

$$d_{total} = \theta_s(z_{hb}) + \left(\int_{h_{ba}}^{Zt-ZWT} \left(\theta_r + (\theta_s - \theta_r)\left(\frac{h_{ba}}{z}\right)^{\lambda}\right)dz\right)$$

$$d_d = \theta_s(z_{hb}) + (z_t - z_{hb})\theta_r + \frac{(\theta_s - \theta_r)h_{ba}^{\lambda}}{-\lambda+1}\left((z_t - z_{WT})^{-\lambda+1} - (h_{ba})^{-\lambda+1}\right)$$

18. **Use the integrator at http://integrals.wolfram.com/index.jsp to integrate the left hand side of Eq. 28.24.**

The answer should be the right side of Eq. 8.37.

19. **Find the hypergeometric function page at the Wolfram website listed in question 18 and find the transformation shown between Eqs. 28.24 and 28.25. List the line number of the transform on the Wolfram page to prove that you found the transform. Find the series solution to the hypergeometric function on the same page and list the line number. Explain how you would implement the series solution in a computer code or spreadsheet.**

Line 35. Transformation

Line 7. Series solution. Each term in the series includes an extra term that is multiplied by the previous term and then added to the series. The series solution is implemented in the hg function as follows within the *van Genuchten* Excel/VBA program.

```
Function hg(a, b, c, z) As Single
  Dim i As Integer
  Dim value As Single
  value = 1
  hg = value
  For i = 1 To 40
    value = value * (a + i - 1) * (b + i - 1) / (c + i - 1) /
    i * z
    hg = hg + value
  Next i
```

20. **A soil has the following parameters: $n = 1.5$, $\alpha = 0.06$, $z_t = 2$ m, $z_{WT} = 1$ m, $\theta_s = 0.45$, and $\theta_r = 0.08$. Calculate the a, b, c, and w terms for the transformed Gauss hypergeomtric function in Eqs. 28.25. A function is included in the Van Genuchten Excel/VBA program called hg that calculates the hypergeometric series solution. Calculate the series solution with the function hg by calling it from a worksheet with the following: "=hg(a,b,c,w/(w-1))". Make sure to write z in units of cm in the calculation of w since α has units of 1/cm. Finally, calculate the total depth of water in the soil profile from the datum to the soil surface.**

$$a = 1,$$
$$b = 1 - 1/n = 1 - 1/1.5 = 0.333$$
$$c = 1 + 1/n = 1 + 1/1.5 = 1.667$$
$$w = -((a(z_t - z_{WT}))^n) = -\left((0.06*(200 - 100))^{1.5}\right)$$
$$= -\left(6^{1.5}\right) = -14.7$$
$$hg(a, b, c, w/(w - 1)) = 1.49$$

$$d_{total} = \theta_s(z_{WT}) + \theta_r(z_t - z_{WT}) + (\theta_s - \theta_r)(z_t - z_{WT})(1 - w)^{1/n-1}{}_2F_1\left(1, 1 - \frac{1}{n}; 1 + \frac{1}{n}, \frac{w}{w-1}\right)$$
$$d_{total} = 0.45(1) + 0.08(2 - 1) + (0.45 - 0.08)(2 - 1)(1 - (-14.7))^{-0.33}*1.49 = 0.75\ m$$

21. **Redo problem 20, but place the water table at 0.1 m above the datum. Calculate the specific yield if the water table drops from 1 m to 0.1 m above the datum.**

$$w = -((a(z_t - z_{WT}))^n) = -\left((0.06*(200 - 10))^{1.5}\right) = -38.5$$
$$hg(a, b, c, w/(w - 1)) = 1.60$$

$$d_{total} = 0.45(1) + 0.08(2 - 1) + (0.45 - 0.08)(2 - 1)(1 - (-38.5))^{-0.33}*1.60 = 0.53\ m$$
$$SY = (0.75 - 0.53)/0.9 = 25\%$$

22. **Find the average water content in two layers. The upper cell has lower and upper boundaries that are 1 and 1.2 m above the datum. The lower cell has lower and upper boundaries that are 0.8 and 1.0 m above the datum. Both cells are in hydraulic equilibrium with the water table. The matric potential, h_c, at the center of the upper cell is – 0.5 m. Use the Brooks-Corey parameters for Wagram loamy sand.**

The water table elevation is 1.0 m below the center of the cell. The center of the cell is 1.1 m above the datum so the water table is located 0.6 m above the datum.

The bubbling pressure is 0.3 m so the bubbling pressure elevation is 0.9 m above the datum.

The cell is above the bubbling pressure elevation so Eq. 27.27 can be used to estimate the average water content in the cell.

$$\theta = (\theta_s - \theta_r)\left(\frac{h_b}{h_c}\right)^{\lambda} + \theta_r$$
$$= (0.305 - 0.044)\left(\frac{0.3}{0.5}\right)^{1.27} + 0.044 = 0.18$$

The cell below contains the bubbling pressure elevation so calculate as follows.

$$d_{cell} = \theta_s(z_{hb} - z_L) + \theta_r(z_u - z_{hb}) + \frac{(\theta_s - \theta_r)h_{ba}^{\lambda}}{-\lambda + 1}\left((z_u - z_{WT})^{-\lambda+1} - (h_{ba})^{-\lambda+1}\right)$$

$$d_{cell} = 0.305(0.9 - 0.8) + 0.044(1.0 - 0.9) + \frac{(0.261)0.3^{1.27}}{-1.27 + 1}\left((1 - 0.6)^{-0.27} - (0.3)^{-0.27}\right)$$

$$d_{cell} = 0.0566\ m$$

$$\theta = d_{cell}/\Delta z = 0.0566/0.2 = 0.28$$

23. **Calculate the average water content in a cell that has upper and lower limits that are 1.4 m and 1.2 m above the datum. Use the parameters listed in question 20.**

$$a = 1,$$
$$b = 1 - 1/n = 1 - 1/1.5 = 0.333$$
$$c = 1 + 1/n = 1 + 1/1.5 = 1.667$$
$$w_L = -((a(z_L - z_{WT}))^n) = -\left((0.06(1.2 - 1.0)*100)^{1.5}\right) = -1.31$$
$$w_u = -((a(z_u - z_{WT}))^n) = -\left((0.06(1.4 - 1.0)*100)^{1.5}\right) = -3.72$$

$$hg(a, b, c, wl/(wl - 1)) = 1.661$$
$$hg(a, b, c, wu/(wu - 1)) = 1.299$$

$$d_{cell} = \theta_r(z_u - z_L) + (\theta_s - \theta_r)(z_u - z_{WT})(1 - w_u)^{1/n-1}{}_2F_1\left(1, 1 - \frac{1}{n}; 1 + \frac{1}{n}, \frac{w_u}{w_u - 1}\right)$$
$$-(\theta_s - \theta_r)(z_L - z_{WT})(1 - w_L)^{1/n-1}{}_2F_1\left(1, 1 - \frac{1}{n}; 1 + \frac{1}{n}, \frac{w_L}{w_L - 1}\right)$$
$$d_{cell} = 0.08(1.4 - 1.2) + (0.45 - 0.08)(1.4 - 1)(1 - (-3.72))^{1/1.5-1}*1.299$$
$$-(0.45 - 0.08)(1.2 - 1)(1 - (-1.31))^{1/1.5-1}*1.661 = 0.0654\ m$$

$$\theta = d_{cell}/\Delta z = 0.0654/0.2 = 0.327$$

---

24. **A crop requires 0.006 m/day evapotranspiration. Calculate the maximum distance between the water table and the bottom of the root zone for sub irrigation without water stress. Use Wagram loamy sand parameters. Compare the discretized solution (Eq. 28.32) to Anat's Eq. (28.31). The saturated hydraulic conductivity of Wagram loamy sand is 0nnnn144 s**

Based on the discretized solution, if the water table is set at 0.52 m depth below the root zone, then the matric potential is zero at the water table.

The Anat equation calculation is between 0.54 m and 0.55 m below the root zone for an upward flux rate of 0.006 m/day.

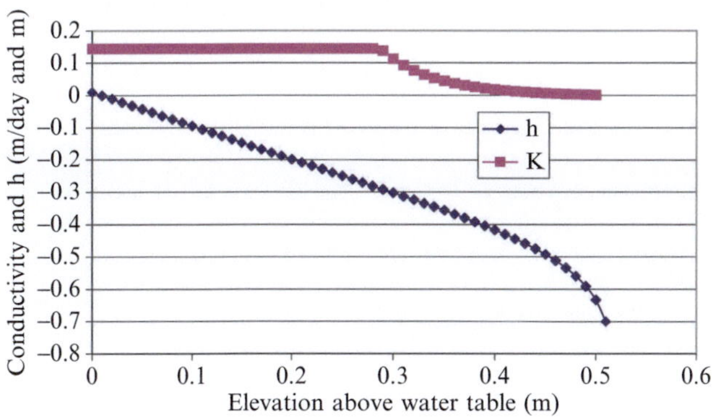

25. Repeat Example 28.3; however, use the parameters for sandy loam in the *WINDS* model. Make sure to change the initial water content in the Active Data page to field capacity (0.21) before copying the data to Crop_data. Make sure that the final DOY is set to 200 days (cell E3) in order to run the simulation for the entire period. Show the water content, matric potential, and matric potential + elevation graphs. The Matric potential graph is in the *Matric potential* worksheet. Discuss results.

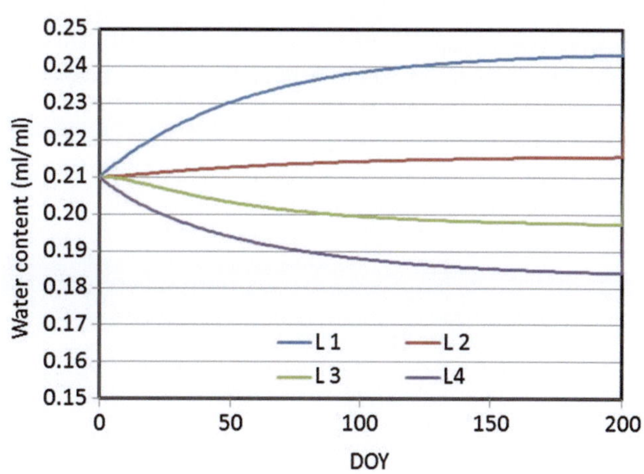

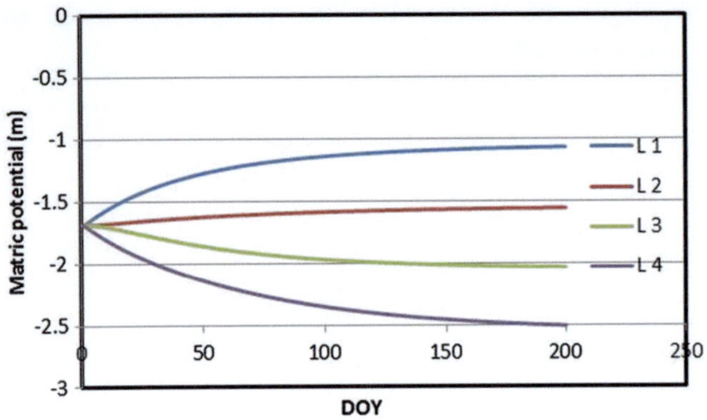

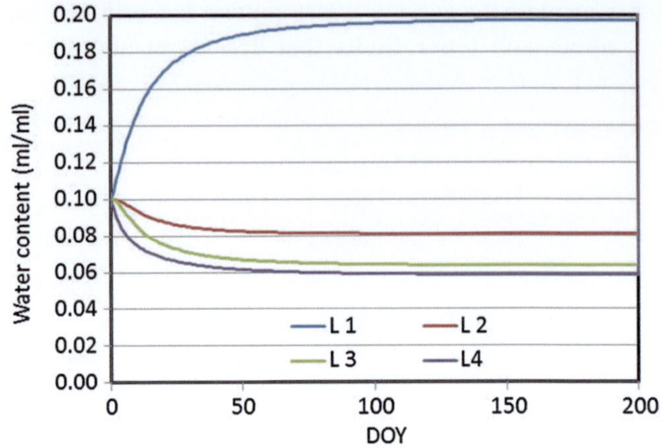

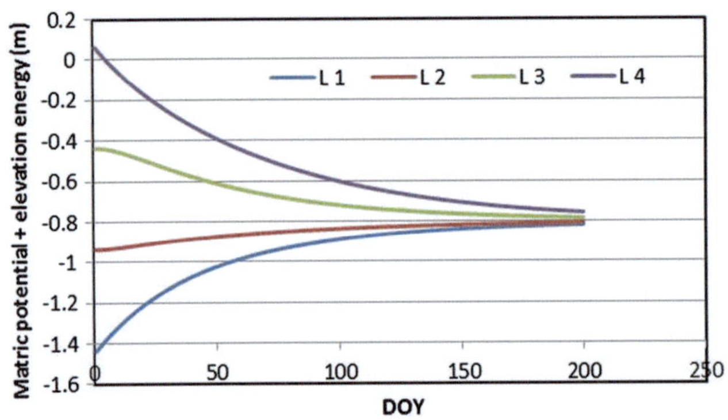

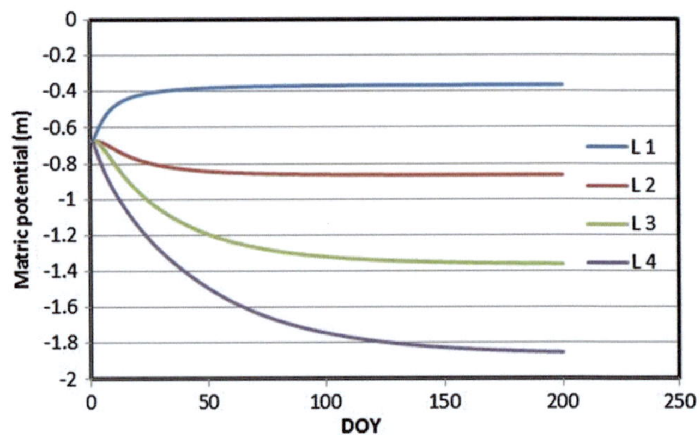

As with Example 28.3, redistribution takes a long time at field capacity. The reason for this is that the matric potential is approximately −2 m. The hydraulic conductivity at the low matric potential is very low. The redistribution ceases when the matric potential + elevation is the same in all cells.

26. Repeat Example 28.3; however, use the parameters for sand in the *WINDS* model. Make sure to change the initial water content in the Active Data page to field capacity (0.1) before copying the data to Crop_data. Show the water content, matric potential, and matric potential + elevation graphs. The Matric potential graph is in the *Matric potential* worksheet. Discuss results. Is there reason to believe that the field capacity estimate may be too high?

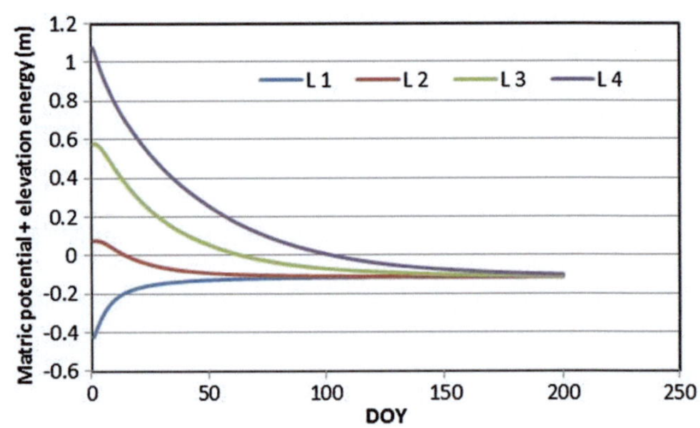

Although more water drained to the lower cells in this example, the field capacity estimate may be reasonable. The drainage did not occur immediately, and field capacity is the amount of water held by the soil 2 days after irrigation. Redistribution ceased after equilibrium.

27. Repeat Example 28.10; however, use the parameters for sand in the *WINDS* model. Show the water table graph and the water content graph. Evaluate drainage rate multipliers 0.01 and 0.02. The multiplier can be changed in the *Drainage* form, which is accessed from the *Active Data* worksheet.

The first two graphs are for the 0.01 multiplier. The change in water table elevation vs. time is approximately the same as Example 28.10. However, the soil drains to a much lower water content once the water table moves downward.

The drainage rate is much faster for the 0.02 multiplier (following pages).

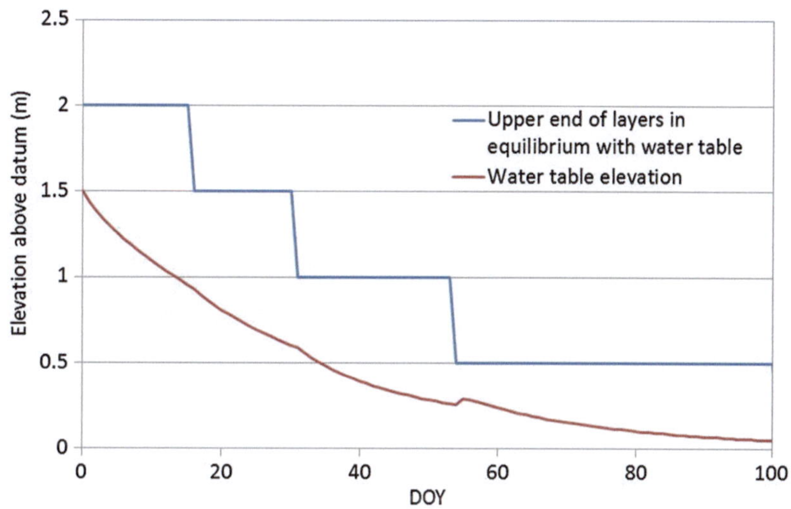

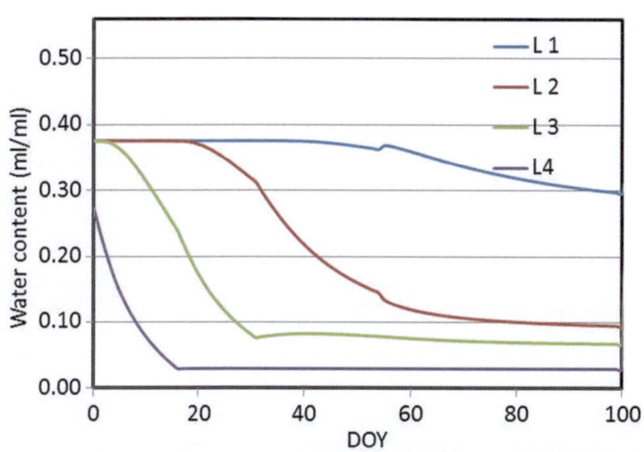

The following graphs are for the 0.02 multiplier

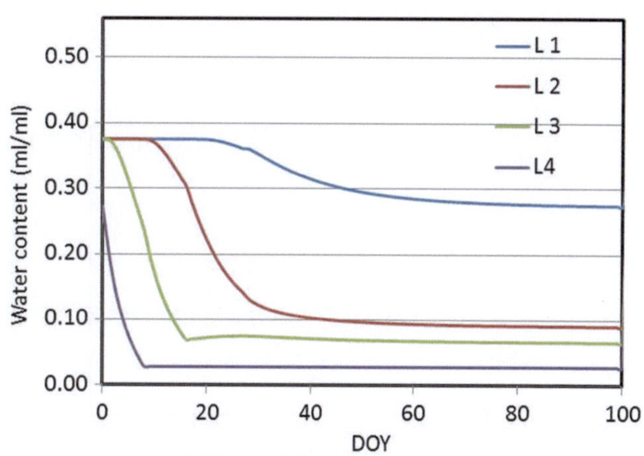

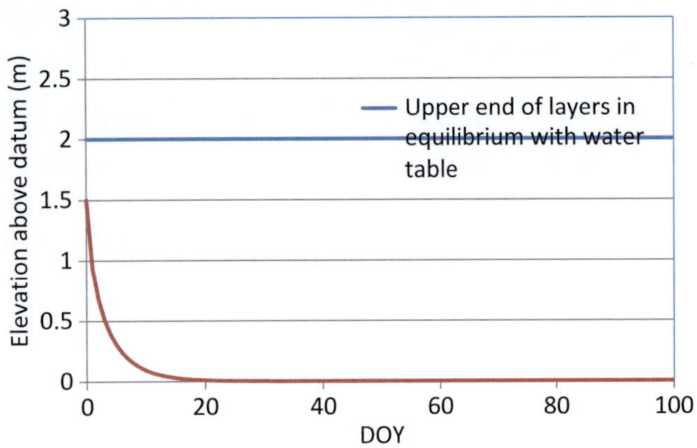

28. Repeat Example 28.10; however, use the parameters for clay in the *WINDS* model. Show the water table graph and the water content graph. Evaluate drainage rate multiplier 0.02. The multiplier can be changed in the *Drainage* form, which is accessed from the *Active Data* worksheet. Explain the results.

The clay soil did not release water as the water drained. Thus, the water table was lowered quickly, but the soils remained saturated.

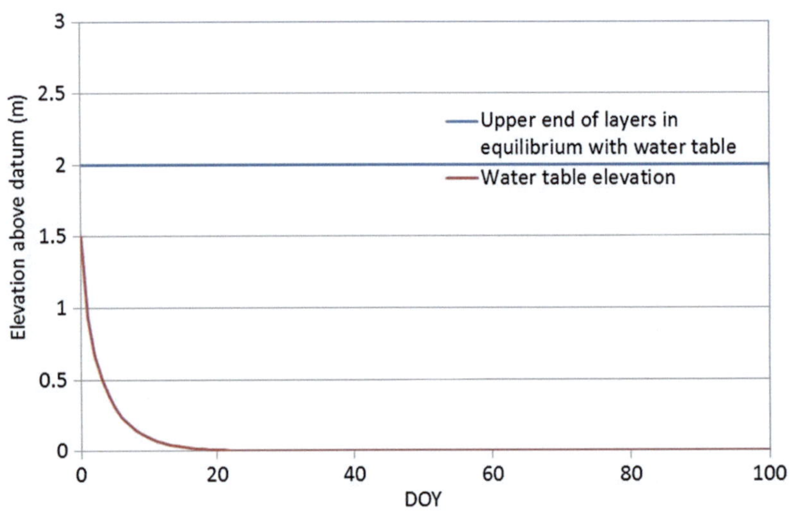

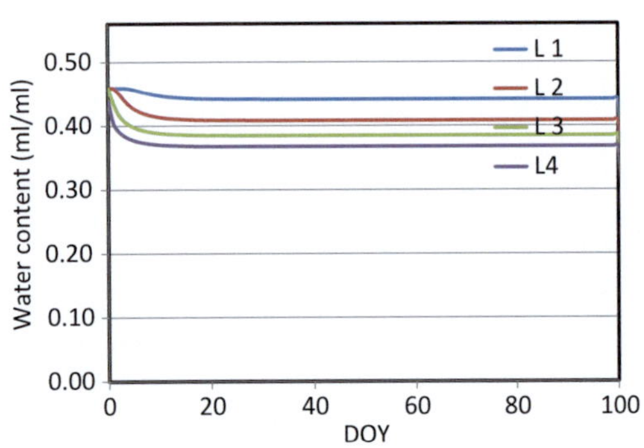

29. Repeat Example 28.12; however, use the parameters for sandy loam in the *WINDS* model. Increase the drainage rate multiplier to 0.03. Set the initial elevation of the water table at 0.2 m. Show the water content graph and the water table elevation graph.

The high drainage rate keeps the soil dry. If not for the high rate, the model would bomb due to the water table being at the soil surface. Just being honest.

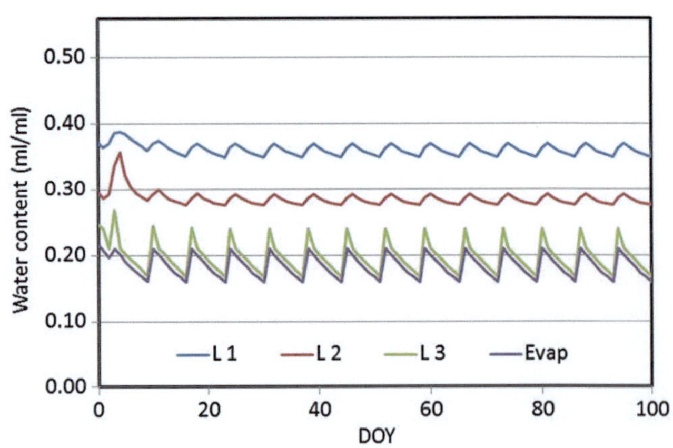

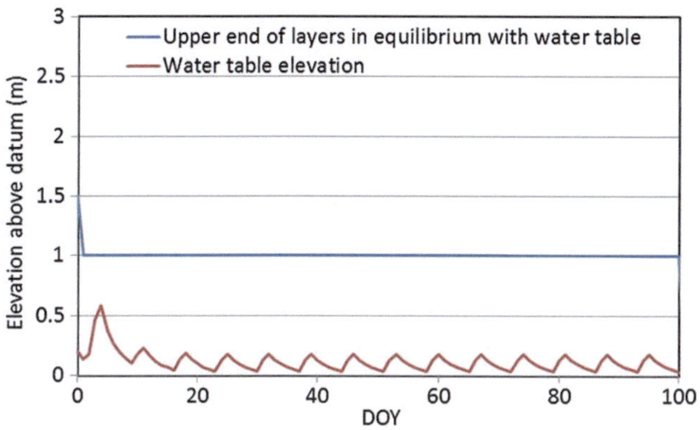

Chapter 29: Solutions

1. Based on Fig. 29.8, estimate the midseason crop coefficient (around DOY 200). Also estimate the maximum soil evaporation coefficient early in the season and during the midseason.

The crop coefficient is 1.0 on DOY 200 since the ET is the same as reference ET.

2. Why would the dual crop coefficient provide a better comparison between two irrigation methods such as infrequent surface irrigation of a crop and frequent minisprinkler irrigation?

The dual crop coefficient accounts for surface wetting. Frequent wetting leads to greater evaporation.

3. Based on Fig. 29.8, when would the use of a dual crop coefficient be superior to a single crop coefficient, during early season or midseason? How could you adjust a single crop coefficient to reflect the frequency of rainfall and irrigation events during the early season?

The dual crop coefficient is superior during the early season. FAO56 accounts for frequency of irrigation in the surface crop coefficient.

4. Calculate crop evapotranspiration. Reference evapotranspiration is 10 mm/day. The basal crop coefficient is 0.8 and the soil evaporation coefficient is 0.1. The water stress factor 0.9.

$ET_0 = 10\ \text{mm/day}$
$ET = 10*0.8*0.9 + 0.1*10 = 7.2 + 1 = 8.2\ \text{mm/day}$

5. Use the following data to calculate the percent of maximum potential yield due to water stress. There is no salinity stress. The sum of Ky values during the season is 150.

DOY	Actual transpiration (mm/day)	Potential transpiration (mm/day)	Crop sensitivity to water stress	
185	8	8	1.5	(1–8/8) = 0
186	7	8	1.5	(1–7/8) 1.5 = 0.1875
187	6	8	1.5	(1–6/8) 1.5 = 0.375
188	5	9	1.5	(1–5/9) 1.5 = 0.67
189	4	9	1.5	(1–4/9) 1.5 = 0.83
190	3	10	1.5	(1–3/10) 1.5 = 1.05
191	7	9	1.5	(1–7/9) 1.5 = 0.33
192	8	8	1.5	(1–8/8) 1.5 = 0

Total potential transpiration during this period is 69 mm. Total reduction = 3.44

$$(1 - 3.44/150) = .977 \rightarrow 97.7\%\ \text{of maximum yield}$$

6. Calculate the depths of water extracted and the final water content for the four layers in the table. MAD is 0.4, ET_0 is 6 mm, K_{cb} is 0.6, and K_w is 0.05. Field capacity is 0.19 and permanent wilting point is 0.10. Assume that there is no evaporation due to full canopy cover.

Layer number	Lower boundary (m)	Initial water content	Fraction transpiration
4	0.1	0.13	0.05
3	0.4	0.17	0.6
2	0.8	0.18	0.3
1	1.2	0.22	0.05

Calculate the threshold water content

$$\theta_t = \theta_{FC} - (\theta_{FC} - \theta_{PWP})\,p = 0.19 - (0.19 - 0.10)*0.4 = 0.154$$

Layer	Nominal fraction	Adjustment	First estimate	Final fraction
4	0.05	(0.13–0.10)/ (0.154–0.10) =0.55	0	0
3	0.6	(0.17–0.10)/ (0.154–0.10) > 1	0.6	0.6/ 0.95 = 0.632
2	0.3	(0.18–0.10)/ (0.154–0.10) > 1	0.3	0.3/ 0.95 = 0.316
1	0.05	(0.22–0.10)/ (0.22–0.10) > 1	0.05	0.05/ 0.95 = 0.052
			Sum 0.95	Sum 1.0

The average percent depletion is less than p so there is no reduction in ET

$$\text{Transpiration} = ET_0 K_{s-water} K_{cb} = 6\ \text{mm/day}*1*0.6 = 3.6\ \text{mm}$$

Layer	Fraction	Transpiration (mm)	Final water content (ml/ml)
4	0	3.6(0) = 0 mm	0.13–(0/1000)/0.1 = 0.13
3	0.632	3.6(0.632) = 2.28 mm	0.17–(2.28/1000)/0.3 = 0.162
2	0.316	3.6(0.316) = 1.14 mm	0.18–(1.14/1000)/0.4 = 0.177
1	0.052	3.6(0.052) = 0.18 mm	0.22–(0.18/1000)/0.4 = 0.2196

7. Run the *WINDS Chapter 29 surface irrigation model.* Use the data from Example 29.3, but redo the irrigation schedule (IRR_01 worksheet) and depths such that no more than 14 mm water is lost to leaching after each irrigation (use the irrigation leaching graph in the Water Content worksheet or column CX) and no stress occurs (MAD is 0.5, percent depletions are listed in column DF and the *percent depletion* graph) before DOY 240. Let the fraction wetted (*ET_fractions worksheet*) be 1.0 for the entire growing season. Print out the *ETc*, *irrigation and leaching*, and *water content* graphs from the *Water_content* worksheet. Discuss how you might automate this adjustment process in the computer program. Discuss whether the required number of irrigations is practical. Consider whether the stress is overestimated in the early season. Also consider the effect of frequent irrigations on ETc (see ETc graph).

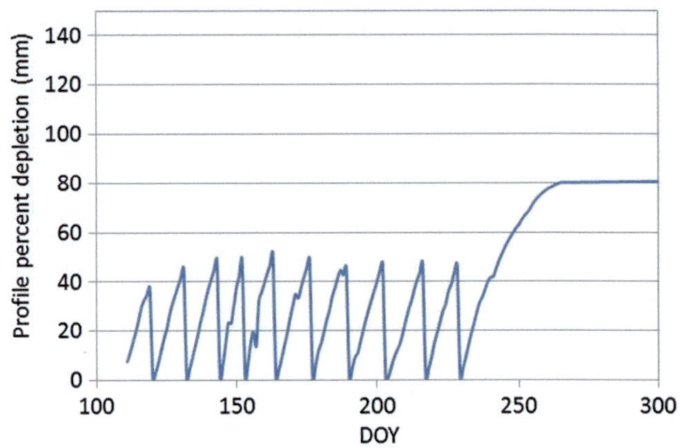

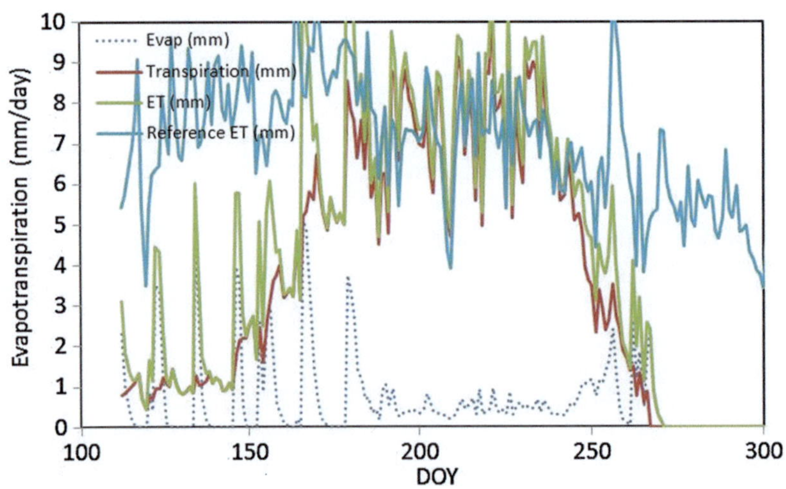

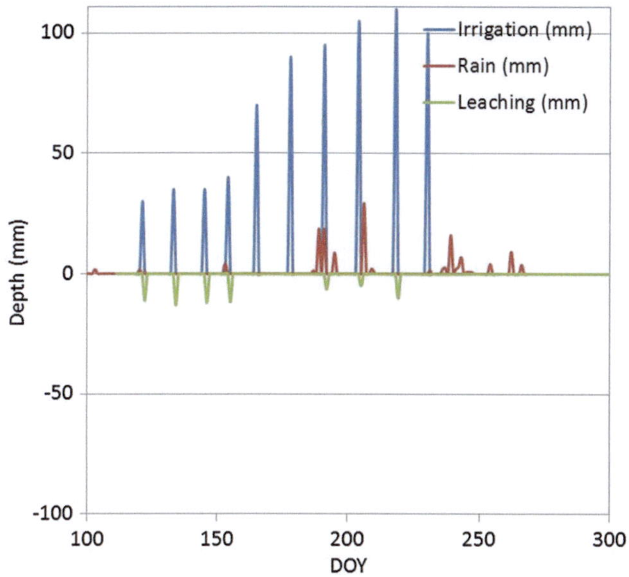

The program could be adjusted by requiring irrigation every time that depletion reaches 50 %. Depths could be adjusted based on the calculated leaching depth.

Ten irrigation events would require extra labor. It may not be economically practical. Also efficiency might be reduced with frequent shallow irrigation events.

It is possible that roots would just grow down to water and the plant would not experience any stress early in the season with the original irrigation schedule.

Frequent irrigations in the early season does lead to increased evaporation loss.

8. Run the *WINDS Chapter 29 surface model*. Change the soil type to clay loam. Print out the percent depletion graph from the *Water_content* worksheet. Compare to problem 7 and discuss the effect of soil type on irrigation scheduling and deep leaching. How could the deep leaching have been prevented with the same large irrigation depths?

The irrigation schedule had no stress, but the problem is that the soil water was not depleted before the large irrigation events so a similar amount of wasted water is observed. The irrigation events should have been spaced further apart to prevent deep leaching.

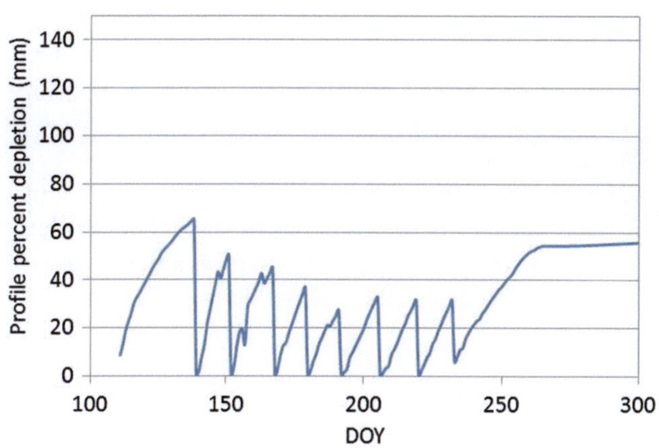

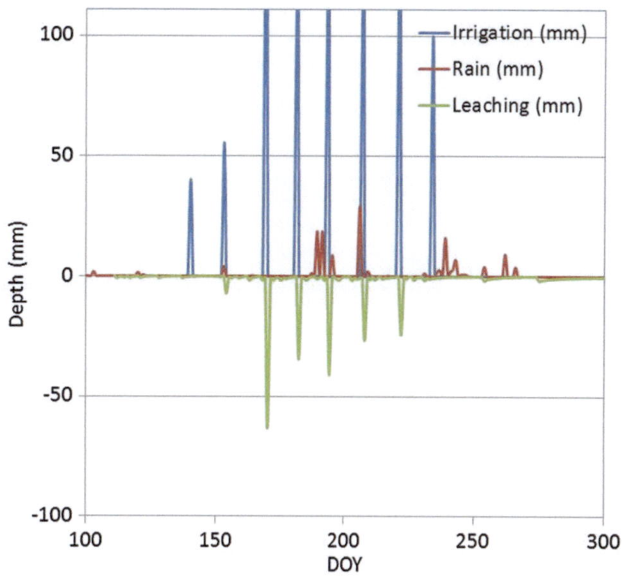

9. Initial water content in the soil profile before flood irrigation is 0.15, Ks = 0.2 cm/hr, and depth of ponded water is 10 cm. Plot the infiltration rate vs. time curve for ponded infiltration. Calculate infiltration rate every 0.1 hours with a spreadsheet. Calculate the infiltration rate and depth for 3 hours. Also plot the cumulative infiltration.

10. An irrigation event is 12 hours long. The depth of ponded water is 5 cm. Bubbling pressure = 0.3 m and saturated water content = 0.305. The antecedent water content is 15 %. Saturated hydraulic conductivity = 0.2 cm/hr. Calculate by hand for the first two time steps, Use time steps = 0.1 hr. Run the *1 layer ponded* worksheet to find the depth infiltrated after 12 hours. You may need to adjust the columns in order to make it work correctly. One of the challenges of Green-Ampt infiltration is to find an assumed depth of infiltration before the first time step, which results in a smooth overall curve. Adjust the initial infiltrated depth until the curve is smooth. Investigate the sensitivity of the estimate of total infiltration over the entire irrigation event to the initial assumed depth of infiltration. Print out the infiltration rate in cumulative infiltration curves. Based on the graph, what term in the Green-Ampt equation does the final rate of infiltration approach.

$$v_{Darcy} = K_s \frac{h_f}{L_f} = K_s \frac{H_0 + S_{av} + L_f}{L_f} = \frac{(H_0 + S_{av})K_s}{L_f} + K_s$$

$\theta_S - \theta_I = M = 0.305 - 0.15 = 0.155$

$S_{av} = 0.76 * h_b = 0.76 * 0.30\ \text{m} = 0.23\ \text{m} = 23\ \text{cm}$

$H_0 = 5\ \text{cm}$

Initial depth of infiltration (before the first calculation) equals Ks, 0.2, resulting in a smooth curve. With M = 0.15, to the initial depth of the wetting front is 1.33.

$$\frac{di}{dt}=\frac{(H_0+S_{av})K_s}{L_f}+K_s$$

$$=\frac{(5+23)0.2\ cm/hr}{1.33\ cm}+0.2\ cm/hr=0.83\ cm/hr$$

Depth of infiltration

$$\mathrm{i}=0.83\ \mathrm{cm/hr}{*}0.1\ \mathrm{hr}=0.083\ \mathrm{cm}$$

Total depth of infiltration after first time step

$$\mathrm{i}=0.2\ \mathrm{cm}+0.083\ \mathrm{cm}=0.283\ \mathrm{cm}$$

Depth of wetting front

$$\mathrm{L}=\mathrm{i}/\mathrm{M}=0.283/0.15=1.89\ \mathrm{cm}$$

Next time step. Calculation of infiltration rate in the second time step.

$$\frac{di}{dt}=\frac{(H_0+S_{av})K_s}{L_f}+K_s$$

$$=\frac{(5+23)0.2\ cm/hr}{1.89\ cm}+0.2\ cm/hr=0.65\ cm/hr$$

Depth of infiltration

$$\mathrm{i}=0.65\ \mathrm{cm/hr}{*}0.1\ \mathrm{hr}=0.065\ \mathrm{cm}$$

Total depth of infiltration after first time step

$$\mathrm{i}=0.283\ \mathrm{cm}+0.065\ \mathrm{cm}=0.35\ \mathrm{cm}$$

Depth of wetting front

$$\mathrm{L}=\mathrm{i}/\mathrm{M}=0.35/0.15=2.31\ \mathrm{cm}$$

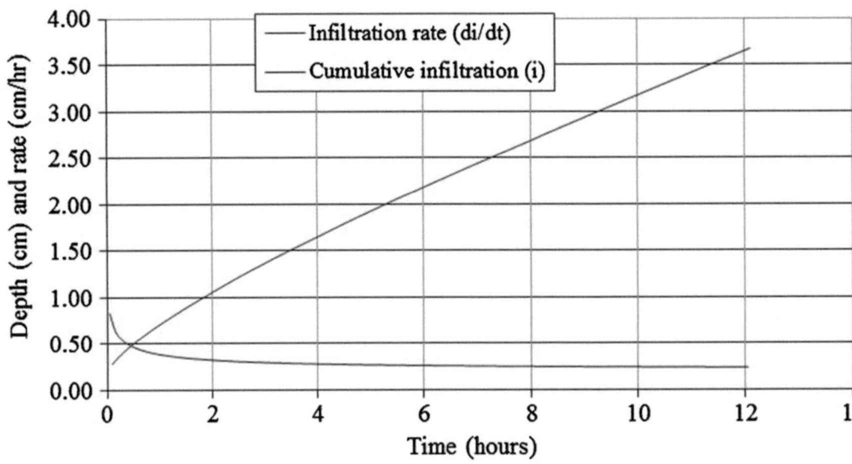

Investigation of sensitivity

Initial infiltration (cm)	Total infiltration (cm)
0.2	3.67
0.5	3.79
1	4.09

There is about a 10 % change in total infiltration with an increase of 5 times in the initial depth estimate.

The final rate of infiltration approaches the saturated hydraulic conductivity.

11. For the parameters in question 10, change the depth of ponding to 10 cm. Evaluate whether increasing the depth of ponding plays a significant role in infiltration in the case of typical depths used in surface irrigation events.

The infiltration increased from 3.67 to 3.80. Thus, depth of ponding has very little effect in the range of typical surface irrigation depths.

12. For the parameters in question 10, change the initial water content to 10 % and 20 %, and 25 %. Evaluate whether antecedent water content plays a significant role in infiltrated depth and depth of the wetting front.

Changing antecedent water content has a significant effect on infiltration and even more on depth to the wetting front.

Antecedent water content	M	Depth infiltrated (cm)	Depth to wetting front (cm)
10 %	0.205	4.25	20
15 %	0.155	3.72	24
20 %	0.105	3.23	31
25 %	0.055	2.82	52

13. A soil has two layers, both 20 cm depth. The upper layer (0–20 cm) has a drainable porosity equal to 0.2. The lower layer (20–40 cm) has a drainable porosity equal to 0.10. The suction at the wetting front in both layers is 20 cm, and the depth of ponding is 10 cm. The saturated hydraulic conductivity in both layers is 0.6 cm/hr. Plot the infiltration rateand cumulative infiltration for 12 hours of ponding. Compare to a uniform drainable porosity in the entire soil profile of 0.2. Is there a significant different in infiltration rate once the wetting front reaches the 2nd layer due to the changed drainable porosity?

Use the *Two layers* worksheet but set the storm intensity to a high number so there was no limitation on water.

There is little change due to the changed drainable porosity in the lower layer. Infiltration with M = 0.1 in lower layers is 13.67 cm. Infiltration with M = 0.2 is 14.85 cm

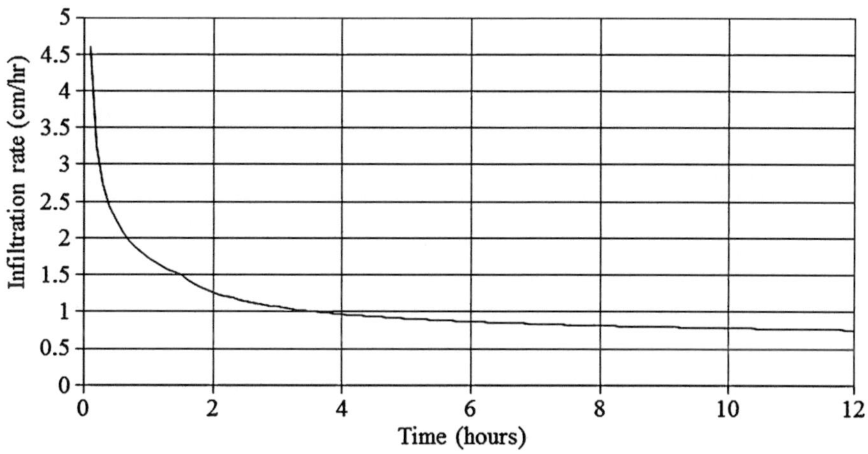

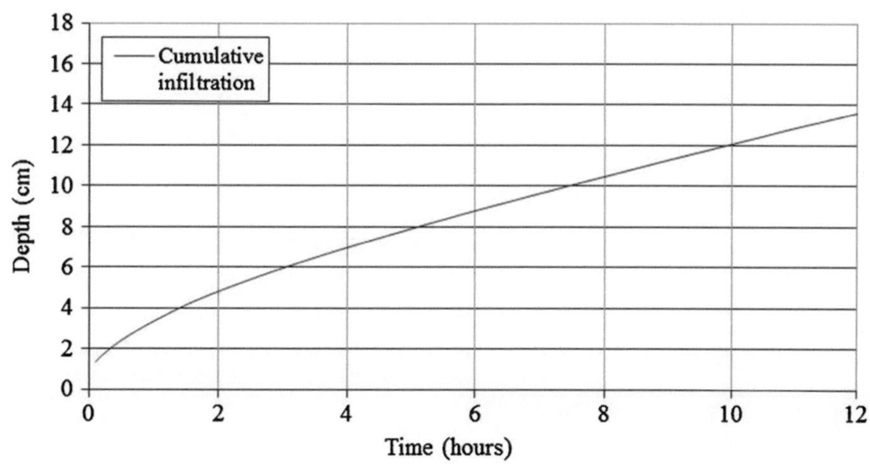

14. Redo problem 13, but let the conductivity of the lower layer equal 0.3 cm/hr and M = 0.1. Calculate effective hydraulic conductivity as the geometric mean and show a sample calculation of the weighted conductivity. Plot the infiltration rate and the depth of the wetting front vs. time. Evaluate the effect of decreased conductivity in the lower layer on cumulative infiltration and how this might influence irrigation practices.

This if statement calculates the weighted conductivity.
UBtwo is depth of top of lower layer
L13 is depth of the wetting front
Ks is saturated hydraulic conductivity of upper layer
Kstwo is saturated hydraulic conductivity of lower layer

$$= \text{IF}(\text{L13} < \text{UBtwo}, \text{Ks}, \text{L13}/(\text{UBtwo}/\text{Ks} + (\text{L13} - \text{UBtwo})/\text{Kstwo}))$$

Example, depth is the wetting front is at 39.2 cm with Ks = 0.6 and Kstwo = 0.3.

$$\text{Keff} = 39.2/(20/0.6 + (39.2 - 20)/0.3) = 0.40$$

The effect of decreased conductivity on infiltration is significant, dropping total infiltration to 10.59 cm. One possible method of dealing with a low conductivity lower layer might be to have more frequent and shallower irrigation events.

15. Redo Example 29.5 (*WINDS Chapter 29 tropical*), but use a sandy loam soil. Print the ET/transpiration graph and the water content graph. Compare the crop water stress during the season with and without water harvesting. Make sure to change the initial water content to field capacity.

With water harvesting

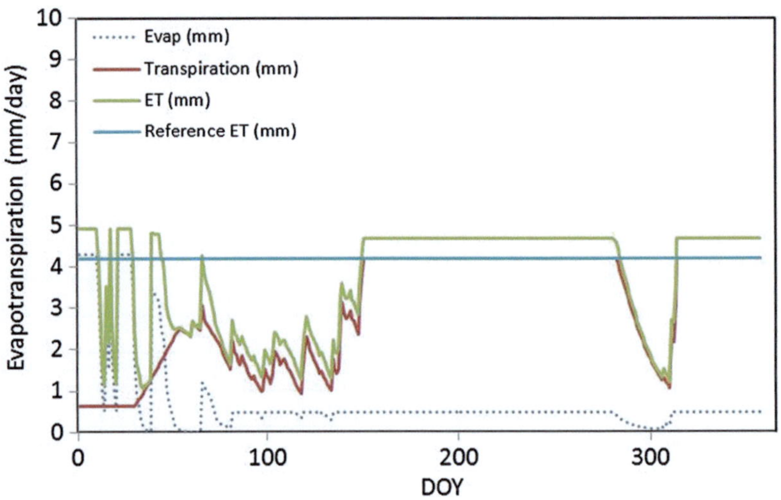

With water harvesting

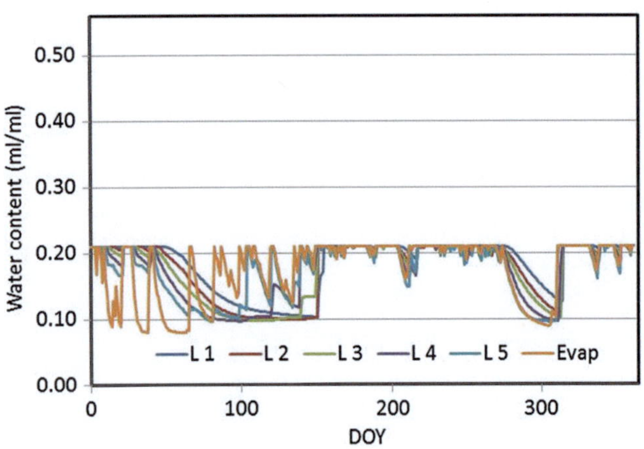

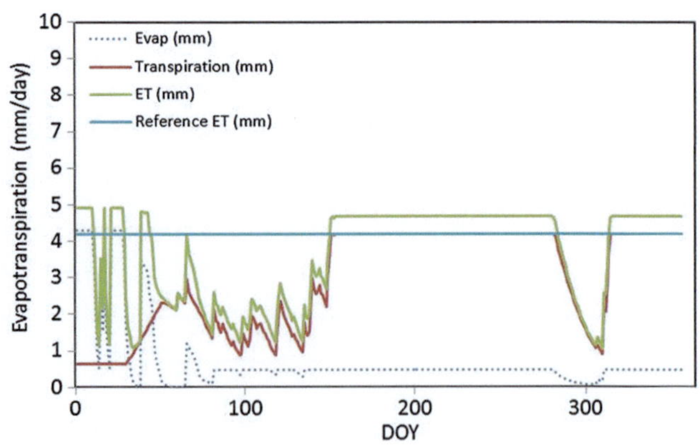

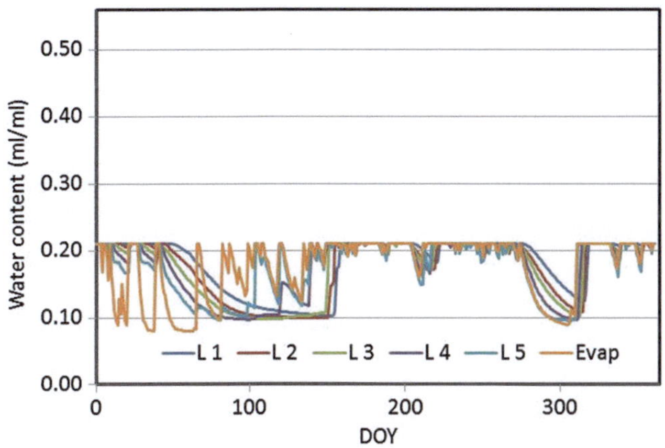

There is almost no difference in water stress with and without water harvesting in the high infiltration rate soil.

Chapter 30: Solution

1. List the positive and negative environmental aspects of subsurface drainage.

   - The impact of water table management practices is primarily on receiving surface water.
   - Subsurface drainage systems reduce erosion. Overland flow is reduced while the total surface and subsurface drainage increases at the edge of the field.
   - Subsurface drainage results in significant reduction of sediment and pollutants (transport to streams), primarily phosphorous and potassium not in solution.
   - Subsurface drainage sometimes results in increased nitrate nitrogen concentrations delivered to receiving water; however, controlling subsurface drainage can reduce the nitrate nitrogen delivered.
   - Water table management provides the capability to enhance best management practices that deal with field surface nutrient management and erosion control.
   - Subsurface drainage intercepts downward percolating water, which allows for monitoring water quality and providing treatment if needed.
   - Water table management practices can contribute to reducing nonpoint pollution alone and with other practices.

2. What measures are recommended by the NRCS recommends to reduce the impact of drained water from farmland on the environment.
   - Implement wetland restoration areas, denitrifying ponds, or managed riparian zones where drainage water could be "treated" to remove excess nitrate-N before discharge into drainage ditches or streams.
   - Design new subsurface drainage systems or retrofit existing drainage systems to manage soil water and water table levels through controlled drainage or subirrigation, lowering concentrations of nitrate-N in shallow ground water. The cost of retrofitting existing systems for subirrigation must be compared to the benefit of increased yields.
   - Use alternative cropping systems that contain perennial crops to reduce nitrate-N losses. Obtaining a market and a satisfactory economic return presents some barriers.
   - Fine tune fertilizer N management. Research shows that applying the correct rate of N at the optimum time substantially affects the reduction of nitrate-N losses.
   - Improved management of animal manure would contribute to lowering nitrate-N losses in livestock-producing areas. Knowing the nutrient content and application rate of the manure, spreading it uniformly, and incorporating it in a timely manner would all lead to better management and confidence in manure N as a nutrient source.

3. List the features that should be noted in a drainage reconnaissance survey.
   - Location and extent of any wetlands.
   - The areas in which crops show damage, as pointed out by the farmer, indicated by the aerial photograph, or noted in personal observations.
   - Personal observations of unique landscape features, ecologically significant areas, land use patterns, operation (land management) aspects, and site visibility.
   - Topography and size of the watershed area
   - Size, extent, and ownership of the area being considered for drainage.
   - Location of the drainage outlet and its condition.
   - Location, condition, and approximate size of existing waterways.
   - Presence of cultural resources.
   - Potential impacts outside the area being evaluated.
   - General character of soil throughout the area needing drainage, including land capability, land use, crops and yields, and salinity or sodicity.
   - High-water marks or damaging floods and dates of floods.
   - Utilities, such as pipelines, roads, culverts, bridges, and irrigation facilities and their possible effect on the drainage system (see NEM part 503).
   - Sources of excess water from upslope land or stream channel overflow and possible disposal areas and control methods.
   - Condition of areas contributing outside water and possible treatment needed in these areas to reduce runoff or erosion.

- Condition of any existing drainage system and reasons for failure or inadequacy. Old subsurface drainage systems that have failed because of broken or collapsed sections may well be the cause of a wet area.
- Estimate of surveys needed.
- Type and availability of construction equipment.
- Feasibility.

4. Based on the geometry of Fig. 30.1 and the derivation of Eq. 30.1, are drains designed for saturated flow over the top of the drain?

No.

5. Derive a drainage coefficient equation for flow with the water table directly over the drain. Compare the ratio of flow rates based on your equation and Eq. 30.4. Discuss why drains are not designed based on the flow rate that takes place when the water table is directly over the drain.

The cross sectional area is 2 π D, and the energy gradient is 1.0

$$q_{WT} = KA\frac{\Delta H}{\Delta x} = 2\pi KD$$
$$q_d = KD\pi\frac{m_o}{L/2} = \frac{2\pi K m_o D}{L} \quad (11)$$

The ratio is m/L, which can be in the range of 50 or 100. Thus, drains would need to carry flows that are 50–100 times greater if they were designed for the water table directly over the drain.

6. Derive Eq. 30.6 from Manning's equation assuming that the pipe is full and half full and determine whether the equation assumes that the drain is flowing half full of full.

$$Q = \frac{AR^{2/3}S_0^{0.5}}{n} = \frac{\pi D^2/_4\left(\pi D^2/_{4\pi D}\right)^{2/3}S_0^{0.5}}{n} = \frac{\pi D^{2.67}/_{4^{1.67}}S_0^{0.5}}{n}$$
$$Q = AD_c$$
$$D^{2.67} = \frac{4^{1,67}nQ}{\pi S^{0.5}} = \frac{4nQ}{\pi S^{0.5}} = 3.23^{0.375}(nAD_c)^{0.375}S^{-0.1875}$$
$$= 1.55(nAD_c)^{0.375}S^{-0.1875}$$

Note: A is used both for field area and Pipe inside area. D is used for inside diameter and D_c refers to the drainage coefficient

Unit conversion

Manning's equation uses units of m/sec. Convert to mm/day

$$1.55(nAD_c)^{0.375}S^{-0.1875}$$
$$= 1.55\left[\left(\frac{m^3}{\text{sec}}\right)\left(\frac{ha}{10,000\ m^2}\right)\left(\frac{1,000\ mm}{m}\right)\left(\frac{24*60*60\ sed}{day}\right)\right]^{0.375}$$
$$ID = 1.55*29.9*(nAD_c)^{0.375} = 46.5*(nAD_c)^{0.375}S^{-0.1875}$$

The ID is slightly overpredicted by Eq. 11.6 if the pipe is flowing full as in this derivation. Thus, the pipe is not completely full.

Calculate the diameter required for a half full pipe. A half full pipe would have the following equation (area is half but wetted perimeter is unchanged). The level of pipe filling can be found by .

Section	Area, A	Wetted perimeter, P	Hydraulic radius, R	Top width, T
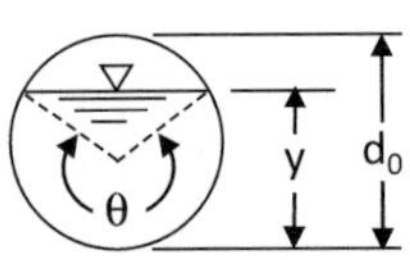 Culvert	$\frac{1}{8}(\theta - \sin\theta)d_0^2$	$\frac{1}{2}(\theta)d_0$	$\frac{1}{4}\left(1 - \frac{\sin\theta}{\theta}\right)d_0$	$\left(\sin\frac{\theta}{2}\right)d_0$

$$Q = \frac{AR^{2/3}S_0^{0.5}}{n} = \frac{\pi D^2/_8\left(\pi D^2/_{4\pi D}\right)^{2/3}S_0^{0.5}}{n} = \frac{\pi D^{2.67}/_{2*4^{1.67}}S_0^{0.5}}{n}$$
$$2.0(nAD_c)^{0.375}S^{-0.1875} = 2.0\left[\left(\frac{m^3}{\text{sec}}\right)\left(\frac{ha}{10,000\ m^2}\right)\left(\frac{1,000\ mm}{m}\right)\left(\frac{24*60*60\ sed}{day}\right)\right]^{0.375} \backslash$$
$$ID = 2.0*29.9*(nAD_c)^{0.375} = 59.8*(nAD_c)^{0.375}S^{-0.1875}$$

The coefficient is 51.7 which is approximately halfway between the full and half pipe coefficient. Thus, Eq. 11.6 assumes that the pipe is somewhere in the range of 3/4 full.

7. A 5 ha area has a drainage coefficient of 19.1 mm/day. The drain slope is 0.3 %. Calculate the required drain size.

$$ID = 51.7(D_c*A*n)^{0.375}s^{-0.1875}$$
$$= 51.7(19.1*5*0.015)^{0.375}0.003^{-0.1875} = 176 \text{ mm}$$

Recalculate based on half full drain and n = 0.017 with Eq. 11.6.

$$ID = 51.7(D_c*A*n)^{0.375}s^{-0.1875}$$
$$= 51.7(19.1*5*0.015)^{0.375}0.003^{-0.1875} = 213 \text{ mm}$$

The ID calculated based on the half full drain and slightly higher n is still slightly lower than that calculated with the figure, but it is closer than the calculation with the unmodified Eq. 11.6.

8. Calculate the required diameter of a subsurface drain based for the following parameters.

Let Manning's n	0.017
Drain elevation above impermeable layer	3 m
Drain slope	s = 0.2 m/100 m
Hydraulic conductivity	K = 2 m/day
Drain spacing	L = 60 m.
Length of drain pipe	L_d = 500 m
Maximum WT height above drains:	m_0 = 2 m - 0.48 m = 1 m.
Average depth of flow to drain	D = d + m_0/2 = 3 + 1/2 = 3.5 m.

Calculate drainage coefficient.

$$D_c = \frac{2,000\pi K m_o D}{L^2} = \frac{2,000\pi*2*2*3.5}{60^2} = 24.4 \ mm/day$$

Calculate pipe diameter

$$ID = 51.7(D_c*A*n)^{0.375}s^{-0.1875}$$
$$= 51.7(24.4*500*60/10,000*0.017)^{0.375}*0.002^{-0.1875}$$
$$= 180 \ mm$$

The next larger drain diameter is 205 mm (8 in).

9. Can a geotextile filter be used in a loam soil? What mesh is recommended ?

Either a sand and gravel or a geotextile filter is acceptable.

A loam soil has an even distribution of sand, silt, and clay with a smaller percentage of clay than the others. The Number 200 sieve only catches sand particles (Fig. 11.10). Thus, most of the particles in the soil would pass the Number 200 sieve. If this is the case, the filter mesh size should be the same as a number 50 sieve, 0.297 mm.

Chapter 31: Solution

1. Give two reasons for installing subsurface drainage in a field and discuss its importance.

Leach salts from the soil, and lower the water table. Drainage removes excess water from the soil profile. Evapotranspiration removes water and leaves much of the salts in the water in the soil, salinizing the soil. If the water table is perched, then it is impossible to leach these salts from the soil. Water logging of soils can restrict the diffusion of oxygen from the atmosphere to the roots. Lack of oxygen will decrease plant yield

2. Describe the two different types of flow to subsurface drains

Just after a storm or irrigation, water flows directly into the drains from above. After the water table intersects the drain, water flows to the drain from the side and below the drain as water flows in response to the hydraulic gradient of the water table.

3. List the water sources and sinks for drained soils.

Water Sources:

Rainfall
Irrigation
Groundwater

Water Sinks:

Evaporation
Transpiration
Drainage
Deep Seepage or leaching

4. Describe the Dupuit Forchheimer assumption and draw flow lines to a subsurface drain according to the Dupuit-Forchheimer assumption.

The Dupuit-Forchheimer assumption is that all flow is horizontal.

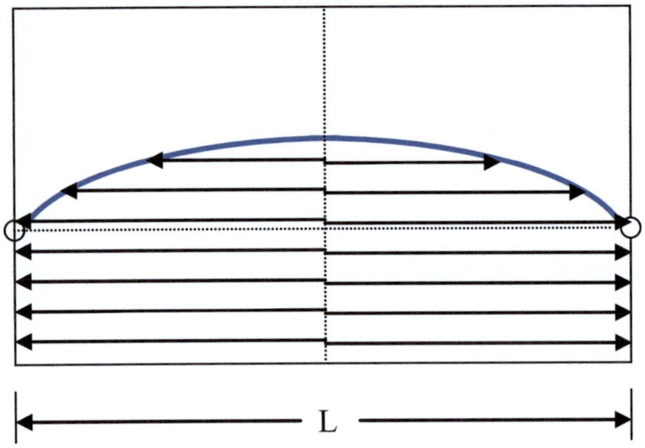

5. Discuss the following drainage systems: open parallel ditches, conventional drainage, water table control, and subirrigation.

Open ditch drainage include open ditches that drain to rivers. They can be very deep and surround the outer parts of fields or they can be shallow and installed within the field.

Conventional subsurface drainage systems are corrugated plastic perforated pipes installed between 1 and 2 m below the soil surface.

Water table control systems require additional structures in the collection system, but they can prevent excessive drainage in order to preserve water in the soil and supply the plant during dry periods.

Subirrigation systems provide water to the crop by adding water from an outside source. These systems can result in the highest crop yield because they remove water when it is not needed but add water when it is needed.

6. Calculate the yield reduction for corn with the following water table data. Assume that CS is 0.12 from DOY 135–143 and is 0.10 from DOY 144–160. YRDMAX = 102 and DSLOPE = 0.75. When does the storm occur?

DOY	DTWT	DOY	DTWT
135	5	143	5
136	15	144	15
137	24	145	24
138	32	146	32
139	39	147	39
140	44	148	44
141	46	149	46
142	0	150	47

DOY	DTWT	CS	SEW-30	CS SEW
135	5	0.12	25	3
136	15	0.12	15	1.8
137	24	0.12	6	0.72
138	32	0.12	0	0
139	39	0.12	0	0
140	44	0.12	0	0
141	46	0.12	0	0
142	0	0.12	30	3.6
143	5	0.12	25	3
144	15	0.1	15	1.5
145	24	0.1	6	0.6
146	32	0.1	0	0
147	39	0.1	0	0
148	44	0.1	0	0
149	46	0.1	0	0
150	47	0.1	0	0

The stress day index is 14.2

The percent of maximum yield is YRDMAX – DSLOPE * SDI = 102 – 0.75 * 14.2 = 91.3 %

The storm takes place on DOY 142.

7. Why does the water table elevation in Figure 31.5 increase rapidly after a rainstorm and then decrease more and more slowly over time? Discuss the influence of energy gradients.

The water table increases rapidly during a storm (can rise a meter within a matter of minutes). If the soil was recently drained, then very little water is required to fill the pores and cause the water table to rise. If the soil is dry prior to the storm, then the water table will not rise as quickly and possibly not at all. During a storm, infiltration is a vertical process; thus, infiltration is driven by a relatively large energy gradient. However, the energy gradient in the horizontal direction of an elliptical water table is low. If the water table at the midpoint between drains is 1 m higher than the drain and the drains are 100 m apart, then the gradient moving water to subsurface drains may be 1 to 50 (1/50 m/m). As the water table declines, then the gradient also decreases, and water table moves to the drains even slower.

8. Rainfall is 0.02 m / day, and drain spacing is 40 m. Calculate flow rate q_5, at a distance of 10 m from the drain. Find the flow rate into the drain per unit length of drain tubing.

$$q_x = \left(\frac{L}{2} - x\right)R = \left(\frac{40}{2} - 10\right)*0.02\ m/day = 0.2\ m^2/day$$

The total flow rate per unit length, q_p, into the drain is

$$q_d = LR = 40*0.02 = 0.8\ m^2/day$$

9. A drain is placed in the field in the shape of a circle with radius 20 m. Derive equations for the flow rate to a unit length of the drain (m^2/day) at any distance from the drain. Let R represent the radius of the drain circle, r = distance from the center of the circle, D = distance from the drain, and P represent the precipitation rate. Only derive an equation for the flow on the inside of the circle.

First derive an equation for the inside of the drain.

The circumference at any distance from the center of the circle

C = 2 π r

The area enclosed by the distance r is

A = π r^2

The volume rate of precipitation is PA

The flow per unit area at distance r is

AP/C = π r^2 P / (2 π r) = rP/2

At the drain, the unit flow to the drain is RP/2

10. Calculate the required spacing between drains. The farmer wants to maintain the water table at least 0.7 m below the soil surface. Yearly rainfall is 1 m/yr. The impermeable layer is 2.5 m below the soil surface. Hydraulic conductivity of the soil is 1 m/day. The conventional practice in the region is to install drains at 1.1 m depth below the soil surface. Drains are standard 4 in (10 cm) diameter drains. Effective drain radius for the 4 in drain is 0.51 cm

Average rainfall (1.0 m/year)/(365 days/year) = 0.00274 m/day.

The drains are 1.1 m below the soil surface. Thus, the water table elevation above the drain, m, is

m = 1.1 m − 0.7 m = 0.4 m.

The elevation of the drain, d, above the impermeable layer is

d = 2.5 m − 1.1 m = 1.4 m.

Make an initial guess of L = 40 m and solve for d_e with equation 10–12 because d/L is much less than 0.3. Spreadsheet calculated values are shown below

L	c	de	new L
40	3.496	1.015	37.68
37.68	3.493	0.998	37.41
37.41	3.493	0.996	37.38

11. Set up a finite difference solution in a spreadsheet in order to solve for water table elevation vs. distance from the drain for the conditions in question 10. Solve for Δz with the finite difference solution at each position by rearranging Darcy's law and solving for Δz, and sequentially calculate the increasing drain elevation beginning at the drain and working toward the midpoint between drains. For the location close to the drain (horizontal distance from drain is less than distance from drain to impermeable layer), set up the flow to the drain as a quarter circle (or slightly more as water table increases with distance from the drain), and calculate head loss as a function of distance from the drain. Where it intersects the drain, let the water table be at the midpoint elevation of the drain. Use Dupuit-Forchheimer solution far from the drain. You should have less than 10 % difference between your solution and the 0.4 m elevation m at the midpoint between drains calculated in question 10.

For the distance from the drain to 1.1 m from the drain the following spreadsheet solution is obtained for the circular geometry.

	A	B	C	D	E	F
1	Distance to midpoint	18.69	m			
2	Precipitation	0.00274	m/day			
3	Hydraulic conductivity	1	m/day			
4	Circle zone dx	0.05	m/day			
5	Distance to impermeable layer	1.1	m			
6						
7	m	q	Water table elevation	cross section	Cross-sectional	dz = 1/K v dx
8	Distance (m)	Flow (m2/day)	above drain midpoint (n	(radians)	length (m)	Head loss (m)
9	0.05	0.0510736	0	1.570796327	0.117809725	0.021676309
10	0.1	0.0509366	0.021676309	1.784257068	0.178425707	0.014273896
11	0.15	0.0507996	0.035950205	1.806027398	0.27090411	0.009375937
12	0.2	0.0506626	0.045326143	1.793662362	0.358732472	0.007061335
13	0.25	0.0505256	0.052387477	1.777357402	0.44433935	0.005685474
14	0.3	0.0503886	0.058072952	1.762007889	0.528602367	0.00476621
15	0.35	0.0502516	0.062839162	1.748444113	0.61195544	0.004105822
16	0.4	0.0501146	0.066944984	1.736621915	0.694648766	0.00360719
17	0.45	0.0499776	0.070552174	1.726312941	0.776840823	0.003216721
18	0.5	0.0498406	0.073768894	1.717277378	0.858638689	0.002902303
19	0.55	0.0497036	0.076671198	1.70930589	0.940118239	0.002643476
20	0.6	0.0495666	0.079314674	1.702225432	1.021335259	0.002426559
21	0.65	0.0494296	0.081741232	1.695895367	1.102331989	0.002242047
22	0.7	0.0492926	0.083983279	1.690201705	1.183141194	0.002083124
23	0.75	0.0491556	0.086066403	1.685051747	1.26378881	0.001944771
24	0.8	0.0490186	0.088011175	1.680369655	1.344295724	0.001823207
25	0.85	0.0488816	0.089834382	1.676092947	1.424679005	0.00171553
26	0.9	0.0487446	0.091549912	1.672169761	1.504952785	0.001619473
27	0.95	0.0486076	0.093169385	1.668556733	1.585128897	0.001533238
28	1	0.0484706	0.094702623	1.665217348	1.665217348	0.001455384
29	1.05	0.0483336	0.096158007	1.662120645	1.745226678	0.001384737
30	1.1	0.0481966	0.097542744	1.659240212	1.825164234	0.001320336

For the 1.1 m distance to the midpoint between drains, the following solution is found for the linear geometry. The elevation m is 0.428, which is within 10 % of the elevation m = 0.4.

linear dx	1	m		
	q	Water table elevation	Cross-sec	dz = 1/K v
Distance (m)	Flow (m2/day)	above drain (m)	length (m)	Head loss (
1.1	0.0481966	0.097542744	1.197543	0.040246
2.1	0.0454566	0.13778899	1.237789	0.036724
3.1	0.0427166	0.17451302	1.274513	0.033516
4.1	0.0399766	0.208029037	1.308029	0.030562
5.1	0.0372366	0.238591508	1.338592	0.027818
6.1	0.0344966	0.266409255	1.366409	0.025246
7.1	0.0317566	0.291655425	1.391655	0.022819
8.1	0.0290166	0.314474723	1.414475	0.020514
9.1	0.0262766	0.33498877	1.434989	0.018311
10.1	0.0235366	0.353300132	1.4533	0.016195
11.1	0.0207966	0.369495411	1.469495	0.014152
12.1	0.0180566	0.383647615	1.483648	0.01217
13.1	0.0153166	0.395818026	1.495818	0.01024
14.1	0.0125766	0.40605764	1.506058	0.008351
15.1	0.0098366	0.414408316	1.514408	0.006495
16.1	0.0070966	0.420903658	1.520904	0.004666
17.1	0.0043566	0.4255697	1.52557	0.002856
18.1	0.0016166	0.42842542	1.528425	0.001058

12. Redo problem 11, but include a sloping geometry ($\theta = 10^0$) for the impermeable layer as shown below. Let the elevation of the drain be 1.1 m above the impermeable layer at the position of the drain. Assume Dupuit-Forchheimer flow in the region far from the drain. Compare the midpoint water table elevation to that calculated in question 11.

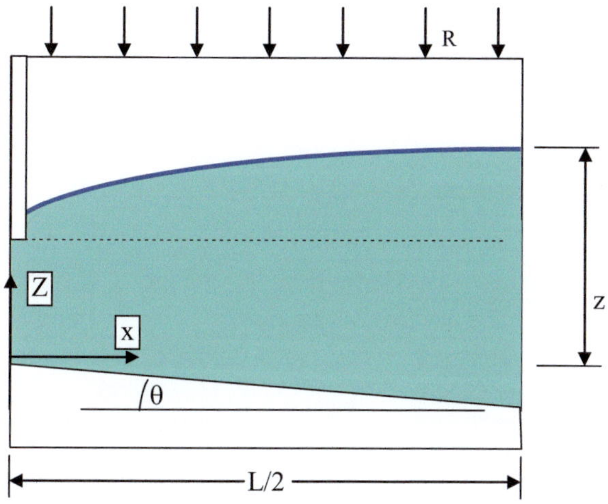

The following equation was placed in column K in order to account for the increasing cross-sectional area due to the sloping impermeable layer, where Imp refers to the depth of the impermeable layer below the drain, directly under the drain.

=J10 + Imp + TAN(2*PI()/360*10)*H10

linear dx	1	m		
theta	10	degrees		

Distance (m)	q Flow (m2/day)	Water table elevation above drain (m)	Cross-sectional length (m)	dz = 1/K v dx Head loss (m)
1.1	0.0481966	0.097542744	1.391502423	0.034636375
2.1	0.0454566	0.132179119	1.602465779	0.028366659
3.1	0.0427166	0.160545778	1.807159418	0.023637428
4.1	0.0399766	0.184183205	2.007123826	0.019917356
5.1	0.0372366	0.204100562	2.203368163	0.016899854
6.1	0.0344966	0.221000415	2.396594998	0.014394005
7.1	0.0317566	0.23539442	2.587315983	0.012273955
8.1	0.0290166	0.247668375	2.775916919	0.010452979
9.1	0.0262766	0.258121354	2.962696878	0.008869149
10.1	0.0235366	0.266990503	3.147893008	0.007476938
11.1	0.0207966	0.274467441	3.331696926	0.006242044
12.1	0.0180566	0.280709485	3.514265952	0.005138086
13.1	0.0153166	0.285847571	3.695731018	0.004144403
14.1	0.0125766	0.289991974	3.876202402	0.003244567
15.1	0.0098366	0.293236542	4.05577395	0.002425332
16.1	0.0070966	0.295661874	4.234526263	0.00167589
17.1	0.0043566	0.297337764	4.412529134	0.000987325
18.1	0.0016166	0.298325089	4.58984344	0.000352212

The midpoint water table elevation is 0.3, which is much lower than the midpoint water table elevation in question than down. Compare to the midpoint water table elevation calculated in question 12.

13. Redo question 12, but let $\theta = -4^0$ which means that the impermeable layer slopes upward from the drain rather

The midpoint water table elevation is 0.6, which is double the elevation question 12.

linear dx		1	m		
theta		-4	degrees		
		q	Water table elevation	Cross-sectional	dz = 1/K v dx
Distance (m)		Flow (m2/day)	above drain (m)	length (m)	Head loss (m)
	1.1	0.0481966	0.097542744	1.120623251	0.043008745
	2.1	0.0454566	0.140551489	1.093705184	0.041562023
	3.1	0.0427166	0.182113512	1.065340395	0.040096668
	4.1	0.0399766	0.22221018	1.035510251	0.038605702
	5.1	0.0372366	0.260815882	1.004189141	0.037081261
	6.1	0.0344966	0.297897143	0.97134359	0.035514313
	7.1	0.0317566	0.333411456	0.936931091	0.033894275
	8.1	0.0290166	0.367305731	0.900898554	0.03220851
	9.1	0.0262766	0.399514241	0.863180252	0.030441614
	10.1	0.0235366	0.429955855	0.823695054	0.02857441
	11.1	0.0207966	0.458530264	0.782342652	0.02658247
	12.1	0.0180566	0.485112734	0.73899831	0.024433885
	13.1	0.0153166	0.50954662	0.693505383	0.022085769
	14.1	0.0125766	0.531632389	0.645664341	0.019478542
	15.1	0.0098366	0.551110931	0.595216071	0.016526099
	16.1	0.0070966	0.567637031	0.541815358	0.01309782
	17.1	0.0043566	0.580734851	0.484986366	0.008982933
	18.1	0.0016166	0.589717784	0.424042487	0.003812354

14. Redo question 10 but let hydraulic conductivity of the soil within the top 1.1 m equal 0.3 m/day and below the drain equal 1 m/day.

Find the effective conductivity. Let D_1 be equal to half of the elevation, m, above the drain, 0.3 m, in order to reflect the fact that part of the upper region is submerged and is a function of distance from the drain.

$$K_e = \frac{K_1 D_1 + K_2 D_2}{D_1 + D_2} = \frac{0.3*0.2 + 1*1.4}{0.2 + 1.4} = 0.91 \ m/day$$

Spreadsheet iterations are shown below

L	c	de	new L
40	3.496	1.015	35.44
35.44	3.490	0.981	34.86
34.86	3.489	0.976	34.78

15. With information from question 10, calculate the change in water table elevation over two weeks with the modified Hooghoudt equation. The initial elevation is 0.5 m, and the specific yield, SY_{mid}, is 15 %.

The water table falls at the rate shown in the following figure.

	A	B	C	D	E	F	G	H	I	J	K
1	L	37.38	m								
2	de	0.996	m								
3	initial m	0.5	m								
4	Sm	0.15									
5	Drain depth	1.1									
6											
7											
8		day	m	qp	ddrainage	dm	final m	DTWT			
9		0	0.5	0.133	0.0036	-0.024	0.48	0.6			
10		1	0.48	0.126	0.0034	-0.022	0.45	0.62			
11		2									
12		3									
13		4									
14		5									
15		6									
16		7									
17		8									
18		9									
19		10									
20		11									
21		12									
22		13									
23		14									
24											
25											
26											
27											
28											
29											

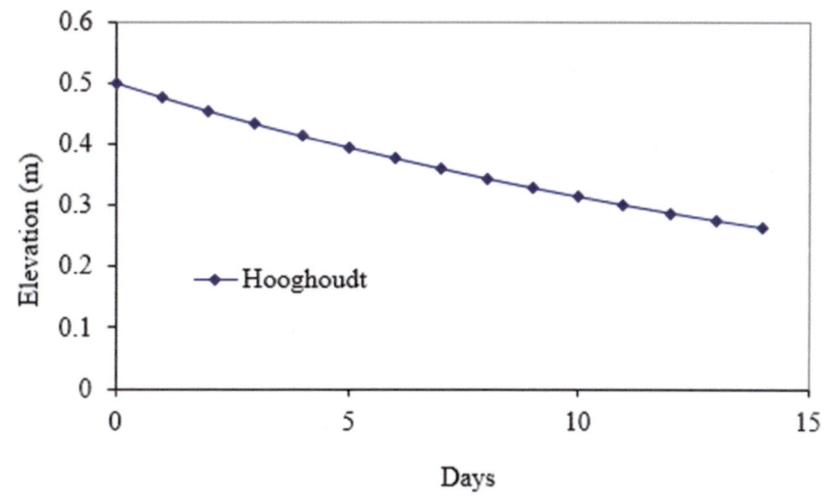

16. With parameters from question 10 and question 15 (d_e = 1.0), calculate the change in water table elevation over time with the USBR equation. The initial elevation is 0.5 m, and the specific yield, SY_{mid}, is 15 %. Remember to calculate D with the effective depth. Plot results and compare to the Hooghoudt transient graph from question 15. There is very little agreement because the KDt/SL^2 value is at the limit of the USBR curve from which it is derived.

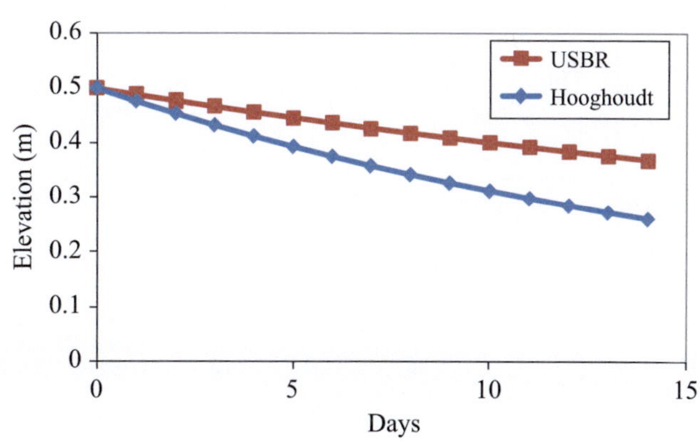

17. Redo question 16 but let the equivalent depth, de = 3 for both the Hooghoudt and USBR solutions. The solutions are close because the USBR equation is in the central part of the USBR curve.

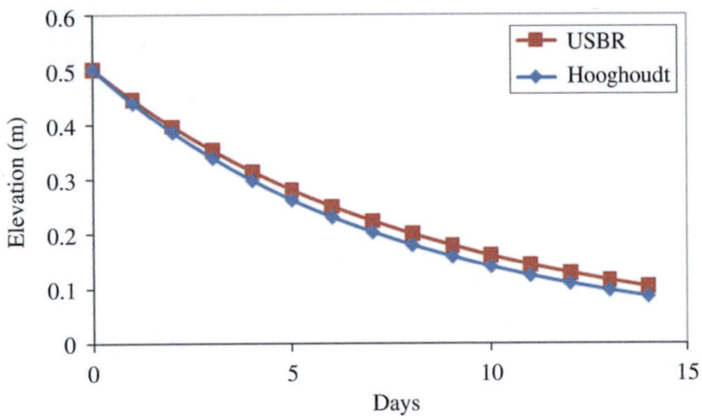

18. A field has an infiltration rate of 38 mm/hr. An irrigation with a gross application depth (volume applied divided by field area) of 10 cm is applied, and the fraction of runoff is 0.2. Calculate the change in water table elevation for a drainage system with $d_e = 3.0$ m and K = 0.5 m/day. Plot the water table elevation (m) vs. time for four weeks of irrigation events that take place once each week. Remember to include the specific yield in the calculation of change in water table elevation and also to include drainage on the days that irrigation water is added to the water table.

The fraction of deep percolation from Table 31.3 is 24 %

$d_{dp\text{-}i} = i \ *(1\text{-}f_{RO})*(f_{dp}) = 0.1 \ * \ (1 - 0.2) \ * \ 0.24 = 0.019$ m = 1.9 cm

The change in water table elevation is

$\Delta m = d_{dp} / SY = 0.019 / 0.15 = 0.13$ m = 13 cm.

The simulation of four irrigation events is shown below .

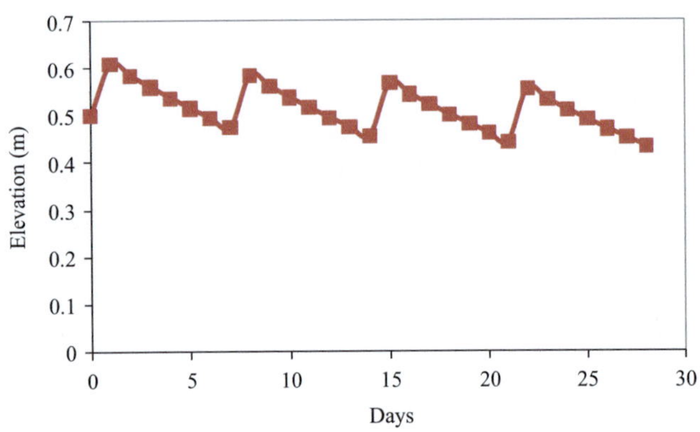

19. Redo question 18, but add 20 cm gross application depth during each irrigation event. Plot the water table elevation vs. time (m).

The water table elevation vs. time is shown below. The water table rises from irrigation to irrigation, and eventually nears the soil surface. This demonstrates that overirrigation results in waterlogging of the soil whereas the correct application depth in question 18 results in no long-term rise of the water table. In many regions with inefficient irrigation systems, the primary cause of water table rise is inefficient irrigation.

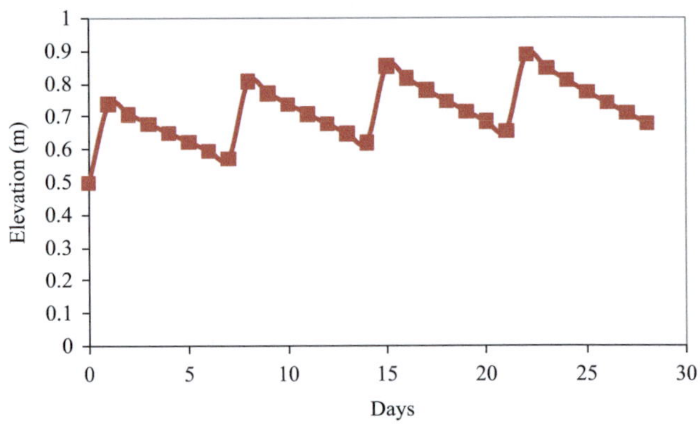

20. Redo Example 31.9, but call drainage companies for drainage alternatives and prices in your region.

Answers will vary

21. Redo Example 31.10, but evaluate a field soil and drainage scenario in your area, specified by the instructor.

Answers will vary

Index

P. Waller, M. Yitayew, *Irrigation and Drainage Engineering*, DOI 10.1007/978-3-319-05699-9

MIX
Papier aus verantwortungsvollen Quellen
Paper from responsible sources
FSC® C105338

If you have any concerns about our products,
you can contact us on
ProductSafety@springernature.com

In case Publisher is established outside the EU,
the EU authorized representative is:
Springer Nature Customer Service Center GmbH
Europaplatz 3, 69115 Heidelberg, Germany

Printed by Libri Plureos GmbH
in Hamburg, Germany